2nd International Conference on Mathematics, Science, Education and Technology (ICOMSET 2017)

IOP Conference Series: Materials Science and Engineering
Volume 335

Padang, Indonesia
5 - 6 October 2017

Editors:

Dr. Ramli
Dr. Minda Azhar
Dr. Ramadhan Sumarmin
Dr. Festiyed
Dr. Yulkifli
Alizar
Ananda Putra
Dr. Dwi Hilda Putri

ISBN: 978-1-5108-6189-3
ISSN: 1757-8981

Printed from e-media with permission by:

Curran Associates, Inc.
57 Morehouse Lane
Red Hook, NY 12571

Some format issues inherent in the e-media version may also appear in this print version.

Printed by Curran Associates, Inc. (2018)

Institute of Physics
Dirac House, Temple Back
Bristol BS1 6BE UK

Phone: 44 1 17 929 7481
Fax: 44 1 17 920 0979

techtracking@iop.org

Additional copies of this publication are available from:

Curran Associates, Inc.
57 Morehouse Lane
Red Hook, NY 12571 USA
Phone: 845-758-0400
Fax: 845-758-2633
Email: curran@proceedings.com
Web: www.proceedings.com

TABLE OF CONTENTS

PHYSICS

BIOLOGY

CHEMISTRY

MATHEMATICS

TECHNOLOGY

EDUCATION

IOP Conf. Series: Materials Science and Engineering **335** (2018) 011001 doi:10.1088/1757-899X/335/1/011001

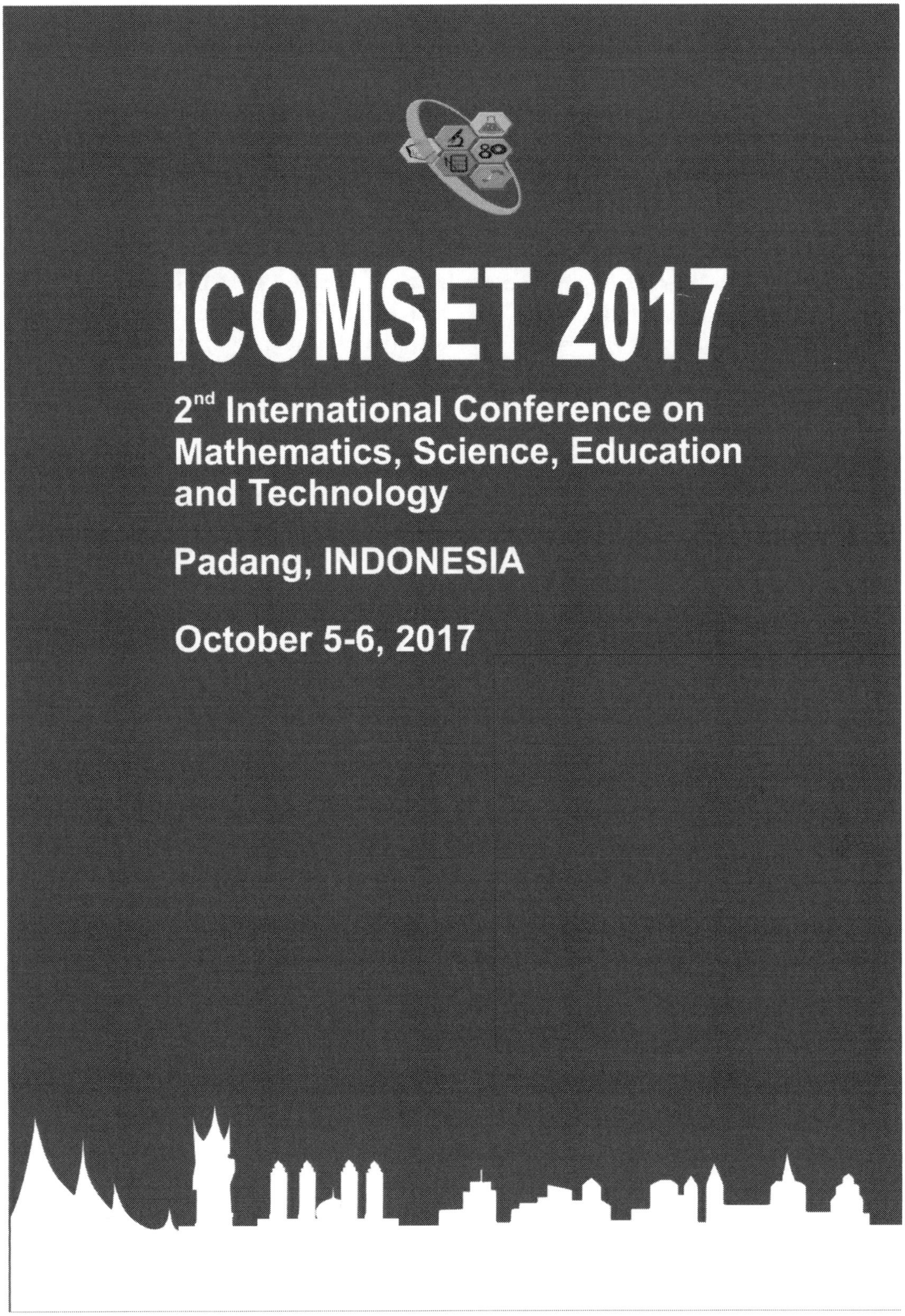

IOP Conf. Series: Materials Science and Engineering **335** (2018) 011001 doi:10.1088/1757-899X/335/1/011001

PREFACE

On behalf of the Steering Committee, I would like to thank you for your participation in the 2nd International Conference on Mathematics, Science, Education and Engineering (ICOMSET2017) which has been held at Grand Inna Muara Hotel and Convention Center in Padang, West Sumatera, Indonesia from October 5 (Tuesday) through 6 (Friday), 2017.

This 2nd ICOMSET is organized by the Faculty of Mathematics and Natural Science, Universitas Negeri Padang. The main objective of this conference is to provide an international platform for researchers, Academicians as well as industrial professionals from all over the world to present Their research results in Mathematics, Science, Education, Technology, and other related fields. This conference also provides opportunities for the delegates to exchange new ideas and application experiences, to establish research relations and to find partners for future collaboration.

I would like to express my sincere appreciation to all the participants, financial sponsors, exhibitors, supporting organizations and all the committee members who has made ICOMSET 2017 successful. With these strong support, we are sure ICOMSET will be beneficial to all the participants, and you enjoy Padang.

We are looking forward to meeting you in the next ICOMSET.

IOP Conf. Series: Materials Science and Engineering **335** (2018) 011002 doi:10.1088/1757-899X/335/1/011002

EDITORS

Dr. Ramli
Kampus FMIPA UNP, Universitas Negeri Padang,
Jl. Prof. Dr. Hamka, Air Tawar, Padang, West Sumatera, Indonesia
ramli@fmipa.unp.ac.id

Dr. Minda Azhar
Kampus FMIPA UNP, Universitas Negeri Padang,
Jl. Prof. Dr. Hamka, Air Tawar, Padang, West Sumatera, Indonesia
minda_azhar@yahoo.com

Dr. Ramadhan Sumarmin
Kampus FMIPA UNP, Universitas Negeri Padang,
Jl. Prof. Dr. Hamka, Air Tawar, Padang, West Sumatera, Indonesia
ramadhan_unp@yahoo.com

Prof. Dr. Festiyed
Kampus FMIPA UNP, Universitas Negeri Padang,
Jl. Prof. Dr. Hamka, Air Tawar, Padang, West Sumatera, Indonesia
Festiyed@fmipa.unp.ac.id

Dr. Yulkifli
Kampus FMIPA UNP, Universitas Negeri Padang,
Jl. Prof. Dr. Hamka, Air Tawar, Padang, West Sumatera, Indonesia
yulkifliamir@gmail.com

Alizar, Ph.D
Kampus FMIPA UNP, Universitas Negeri Padang,
Jl. Prof. Dr. Hamka, Air Tawar, Padang, West Sumatera, Indonesia
alizarulianas@yahoo.com

Ananda Putra, Ph.D
Kampus FMIPA UNP, Universitas Negeri Padang,
Jl. Prof. Dr. Hamka, Air Tawar, Padang, West Sumatera, Indonesia
anandap@fmipa.unp.ac.id

Dr. Dwi Hilda Putri
Kampus FMIPA UNP, Universitas Negeri Padang,
Jl. Prof. Dr. Hamka, Air Tawar, Padang, West Sumatera, Indonesia
dwi_hildaputri@yahoo.com

IOP Conf. Series: Materials Science and Engineering **335** (2018) 011003 doi:10.1088/1757-899X/335/1/011003

ORGANIZING COMMITTEE

Steering Committees
- Ganefri
- Lufri
- Yulkifli
- Hendra Syarifuddin
- Hardeli

General Chair
- Yohandri

Co-Chairs
- Rahadian Zainul

Secretary
- Dezi Handayani
- Umar Kalmar Nizar

Technical Program
- Dwi Hilda Puteri
- Fitri Arsih

Secretariat
- Alizar
- Zulhamidi
- Yusuf,
- Dani Mustika
- Doni Fisko
- Dian Nurtasari
- Yulianti
- Ceng Sri Yunita

Publication
- Ramli
- Minda Azhar

Treasurer
- Herda Susanti

Finance
- Indang Dewata
- Ali Asmar
- Syamsi Aini
- Budi Oktavia

Accommodation
- Meira Parma Dewi
- Iryani
- Usman
- Nofri Hardisal

Documentation
- Toni Supriadi
- Edi Kurnia

IOP Conf. Series: Materials Science and Engineering **335** (2018) 011004 doi:10.1088/1757-899X/335/1/011004

Peer review statement

All papers published in this volume of *IOP Conference Series: Materials Science and Engineering* have been peer reviewed through processes administered by the proceedings Editors. Reviews were conducted by expert referees to the professional and scientific standards expected of a proceedings journal published by IOP Publishing.

Magnetic Susceptibility and Heavy Metals in Guano from South Sulawesi Caves

H Rifai[1]*, R Putra[1], M R Fadila[1], E Erni[2], C M Wurster[3]

[1] Physics Department, Faculty of Mathematics and Natural Sciences, Universitas Negeri Padang, Indonesia
[2] Physics Department, Faculty of Mathematics and Natural Sciences, Universitas Sriwijaya, Indonesia
[3] School of Earth and Environmental Science and Centre for Tropical Environmental Science and Sustainability, James Cook University, Australia

*rifai.hamdi@gmail.com

Abstract: Measurement of some magnetic properties have been performed on vertical profile from South Sulawesi caves (Mampu and Bubau) by using low cost, rapid, sensitive and non destructive magnetic method. The aim is to attempt to use magnetic characters as a fingerprint for anthropogenic pollution in the caves. Guano samples were collected every 5 cm at a certain section of Mampu and Bubau cave, South Sulawesi, starting from surface through 300 cm in depth of mampu Cave and 30 cm of Bubau Cave. The magnetic parameters such as magnetic susceptibility and percentage frequency dependence susceptibility were measured using the Bartington MS2-MS2B instruments and supported by X-Ray Fluoroscence (XRF) to know their element composition. The results show that the samples had variations in magnetic susceptibility from 3.5 to 242.6 x 10^{-8} m^3/kg for Mampu Cave and from 8.6 to 106.5 x 10^{-8} m^3/kg for Bubau Cave and also magnetic domain. Then, the XRF results show that the caves contain several heavy metals. Magnetic and heavy metal analyses showing that the magnetic minerals in caves are lithogenic (Fe-bearing minerals) in origin and anthropogenic (Zn content) in the caves.

1. Introduction

Guano is one of the deposits on the cave floor from the rest of the process of digestion (feces) of bats or birds containing mineral component [1]. Deposits of guano mostly contain carbon, nitrogen, phosphate or sulphate minerals [2]. Not infrequently chitin substance ($C_{18}H_{26}N_2O_{10}$) could be found in bat guano in large quantities [3]. Chitin derived from insect fragment, the structural component from bat feces. The sediment of guano will undergo deposition a few meters after thousands of years [1].

During the process of precipitation in the cave for thousands of years, guano could record environmental changes that occur for pasttime [3]. Environmental changes stored in the cave can be seen from variation of magnetic mineral concentration of guano. Despite the presence of magnetic minerals in sediments is very small, but has a great influence and can serve as proxy for environmental change [4]. The magnetic minerals transport process in the cave occur in wet and or dry state through water, wind and animal cave [5].

The magnetic mineral properties can be determined by using magnetism method [6]. Magnetic susceptibility is described as the ratio of total magnetization induced to the intensity of the magnetic field generating magnetization [7,8]. Specific mass magnetic susceptibility (χ) is a function of concentration, grain size and mineralogical mineral magnetic found in the sediment [9]. Measurement of magnetic properties using magnetic susceptibility can identify magnetic mineral concentrations due to rapid, cost-effective, efficient, non-destructive, and affordable measurements [7,4]. Magnetism methods have been utilized into a wide range of environments to provide a fast, low cost and sensitive characterization of sediments [10]. Environmental alterations have been well documented from lake sediments [11,17], marine sediments [12,16] and archaeological [13], based on magnetic mineral properties.

Magnetic susceptibility may be affected by elemental composition especially Fe. The Fe content in the guano sample is proportional to the magnetic susceptibility value increase [7]. In this study, we measured the guano from Mampu and Bubau caves using magnetic susceptibility test and X-Ray Fluorescence (XRF) test. Then the magnetic parameters will be compared with the chemical method. The measurement result is expected to be a proxy for environmental change.

2. Methods

Guano sampling was carried out from two sites in South Sulawesi. The both sites are located in Mampu cave (4° 32' S, 120° 22'E) and Bubau cave (3° 52' S, 119° 89'E). The samples were taken vertically with 5 cm intervals for Mampu cave depth until 300 cm and Bubau cave depth until 30 cm.

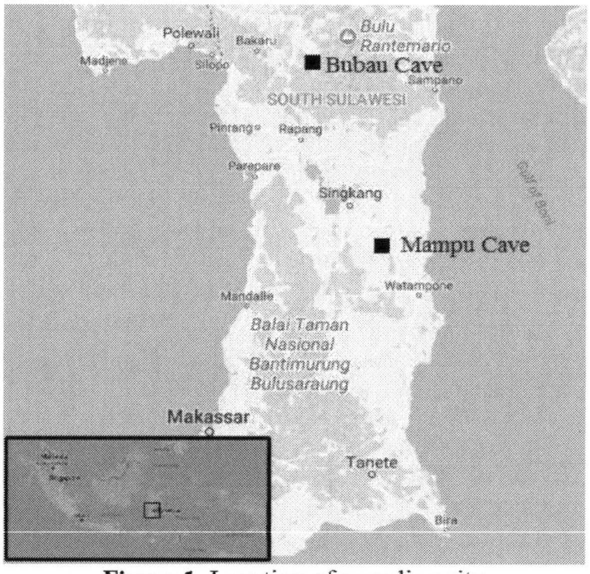

Figure 1. Location of sampling site

The samples were packed into polyethylene bags then shipped to Geophysics Laboratory in State University of Padang. In the laboratory, guano samples were opened from polyethylene bags and placed inside a tightly secured cylindrical plastic holder.

Firstly, all samples were treated for mass-specific magnetic susceptibility measurements by Bartington MS2B instrument. Measurements of magnetic susceptibility were performed at two different frequencies, 0.47 kHz for low frequency magnetic susceptibility (χ_{LF}) and 4.7 kHz for high frequency magnetic susceptibility (χ_{HF}) with the 1.0 sensitivity setting. Each sample was measured three times to obtain average value with an air reading before and after each series for drift correction.

The mass spesific frequency dependent susceptibility χ_{FD} could be obtained from percentage ratio $\chi_{LF} - \chi_{HF}$. This parameter can detect superparamagnetic grains.

To be compared with magnetic susceptibility data, selected guano samples were examined with chemical method using X-Ray Fluorescence (XRF). The analysis of XRF was performed in order to obtain element composition of mineral magnetic containing in guano. XRF measurements were carried out by using PANalitical type Epsilon 3 at Chemical Laboratory in State University of Padang. The XRF samples were guano from representative samples from upper section of the cores from Mampu cave and Bubau cave.

3. Results

3.1 Magnetic Susceptibility

Figure 2 shows relation of magnetic susceptibility profile in low frequency (χ_{LF}) and frequency-dependent susceptibility (χ_{FDS}) from Mampu and Bubau cave. The χ_{LF} profile of Mampu cave could be divisible by two zones. High susceptibility zones were represented in upper from surface until 10 cm. It shows range value between 242.6 x 10^{-8} m^3/kg and 124 x 10^{-8} m^3/kg. Higher magnetic susceptibility value could be founded in surface. Low susceptibility zones could be seen from depth of 20 cm until 300 cm with range between 37.9 x 10^{-8} m^3/kg and 3.5 x 10^{-8} m^3/kg. Lower magnetic susceptibility value could be founded in 300 cm. Most of magnetic susceptibility values below 100 x 10^{-8} m^3/kg. After 100 cm depth, values of χ_{LF} has a relative steady with low value and there s no significant change. While values of χ_{LF} in Bubau cave is different than Mampu cave. Magnetic susceptibility in surface tends to be low value of 20.1 x 10^{-8} m^3/kg. In contrast to 20 and 25 cm depth, χ_{LF} value tend to increase strongly with high value of 98.6 x 10^{-8} m^3/kg. Then decreased significantly at a depth of 30 cm with lowest value of 7.4 x 10^{-8} m^3/kg.

Guano samples from Mampu cave has varying χ_{FDS} values with lowest value 1.65 % to highest value 35.53 %. Frequency dependent susceptibility value (χ_{FDS}) on Figure 2 shows a more complicated trend across section. There are many maximum peaks of χ_{FDS} and fluctuative each sequence on Mampu cave. The Maximum χ_{FDS} values (> 10%) depicted at 5cm, from 30 cm to 40 cm, 55 cm, 65 cm, 110 cm, 140 cm, 175 cm to 185 cm, 220 cm, 240 cm, 270 cm, 280 cm, and 300 cm.The are two peaks value (>30%) observed at 215 cm and 250 cm depth. This case could be considered as erroneous measurements [7]. The χ_{FDS} profile of Bubau cave shows variation values from 0 % to 8.25 %. It seems different with χ_{FDS} values with Mampu cave. There is no significant change of χ_{FDS} values in the Bubau cave. Generally χ_{FDS} values increase from 10 cm to 25 cm then decrease strongly until value of 0 %. On χ_{FDS} value of 0 %, it also could be considered as error in measurement.

Figure 3 and 4 explain relation of χ_{FDS} % versus χ_{LF} showing guano samples in each Mampu cave and Bubau cave. The χ_{FDS} values which are smallest or closer to zero (< 2%), usually dominated by multidomain (MD) grains. Values range between 2 - 10% indicates mixture of stable single domain (SSD) and superparamagnetic (SP). Values which are greater than 10 % indicates superparamagnetic (SP) [7].

Dissemination of grains size each section in Mampu cave evenly. Generally, most of grains size in Mampu cave and Bubau cave on the low magnetic susceptibility values. Grains size of sample from Mampu cave are largely dominated by mixture of SSD and SP magnetic grains, while grains size of sample from Bubau cave consist of MD and mixture of SSD and SP magnetic grains.

IOP Conf. Series: Materials Science and Engineering **335** (2018) 012001 doi:10.1088/1757-899X/335/1/012001

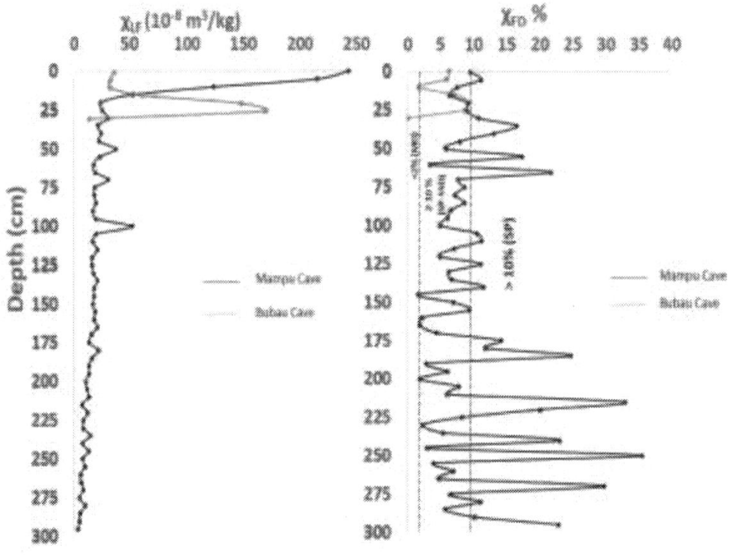

Figure 2. The magnetic susceptibility value in low frequency (χ_{LF}) and frequency-dependent susceptibility (χ_{FDS}) versus depth from Mampu and Bubau cave.

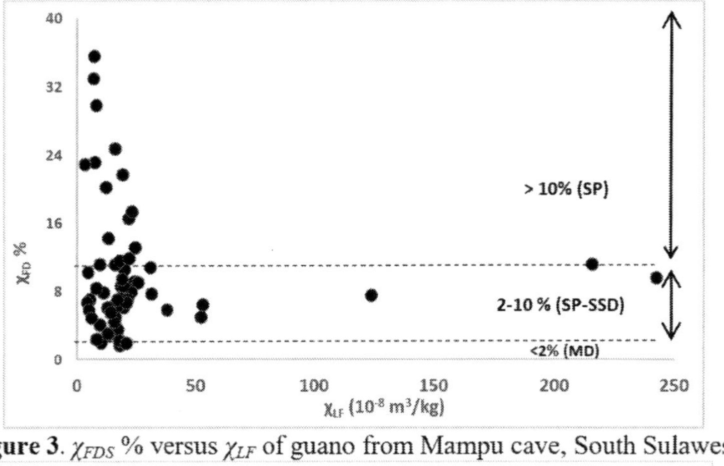

Figure 3. χ_{FDS} % versus χ_{LF} of guano from Mampu cave, South Sulawesi.

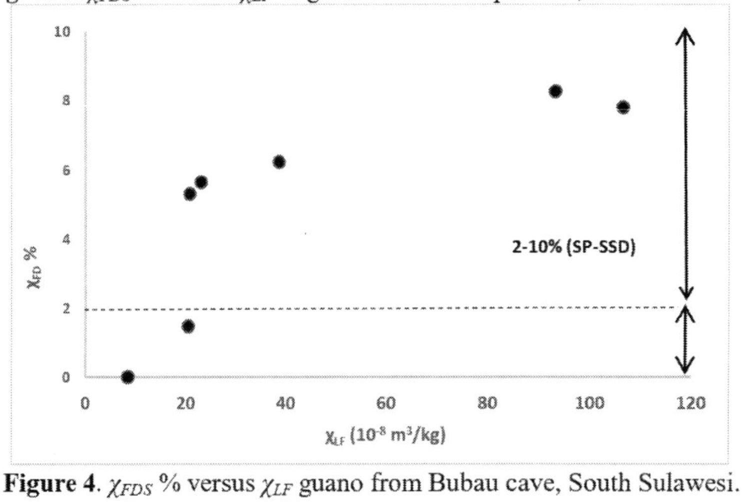

Figure 4. χ_{FDS} % versus χ_{LF} guano from Bubau cave, South Sulawesi.

3.2 Geochemistry

Table 1 shows XRF results that represent variation of element composition of surface sample from both caves. Data from XRF shows presence of element such as (Fe and Ni) which are part of ferromagnetic elements, paramagnetic (K, Mg, Mn, Ti, Al and Ca) and diamagnetic (Pb, Cu and Zn). Ca is a high concentration element with percentage 54.822% in Mampu cave and 33.294% in Bubau cave. While Pb is a low concentration element found in the Mampu cave with percentage 0.004% and 0.002 % Bubau cave. Concentration S is not present in the Mampu cave than Bubau cave with percentage 3.372%. Concentration of Fe in the Mampu cave is greater (7.45%) than Bubau cave (3.379 %).

Table 1. XRF results showing element composition.

Composition Element	Mampu Cave %	Bubau Cave %
Mg	2.415	6.665
Al	6.256	3.243
Si	8.552	14.153
P	15.067	16.428
S	-	3.372
Cl	0.118	0.713
K	0.399	16.936
Ca	54.822	33.294
Ti	0.707	0.472
Cr	0.041	0.011
Mn	0.387	0.189
Fe	7.45	3.379
Ni	0.021	0.013
Cu	0.405	0.134
Zn	0.712	0.284
Pb	0.004	0.002

4. Discussion

As comparation to magnetic susceptibility value of above 250 x ^{8} m^3/kg 10 in surface obtained from cave sand in Western Cape, South Africa [14]. High magnetic susceptibility in Mampu cave (above 100 x 10^{-8} m^3/kg) in upper section could be indicated high concentration of ferrimagnetic in the guano deposits. By looking at the condition in the Mampu cave, the cause may come from anthropogenically altered material occurred through human activities. It also confirmed by any archaeogical remains in the cave. The enhancement of magnetic susceptibility value could relate human occupation that took place in the Maronia cave, Greece where tt 15 cm and 100 cm depth have peaks of magnetic susceptibility value of 53 x 10^{-8} m^3/kg and 52.2 x 10^{-8} m^3/kg. These layers also could be indicated by records of the anthropogenic influence. Moreover, magnetic susceptibility could be good proxy for intensity of human activity and degree of magnetic susceptibilty enhancement appears to be a good indicator of occupation longevity and intensity. The low magnetic susceptibility values in Mampu cave occur naturally without influenced by human activity [14].

The magnetic susceptibility pattern in Bubau cave look different and not similar than Mampu cave. The Low magnetic susceptibility value (χ_{LF}) founded in surface. Bubau cave seems has not any archaeological remains. It is difficult to determine magnetic susceptibility pattern because the samples taken only to depth of 30 cm and not the same as the depth of Mampu cave. But it looks that Bubau cave is natural and not found evidence which is affected by anthropogenic influence. In other words, guano samples of Bubau cave formed by climate alteration.

Data from geochemical analysis by using XRF, shows relatively high concentration of calcium in sediment. Calcium is one of element of gypsum ($CaSO_4 .2H_2O$) found in the cave sediment [13]. Mampu cave and Bubau cave are mostly composed of several high concentration elements including paramagnetic. In addition, there is good correlation between magnetic susceptibility value with concentration or abundance of magnetic mineral particularly Fe. Concentration of Fe in the Mampu cave is greater than Bubau cave. This is what causes susceptibility value of Mampu cave higher than Bubau cave. Heavy metal shows as a carrier of magnetic properties into the cave. Generally, it could be conclude that when magnetic susceptibility and metal concentrations are both high particularly Fe, mostly the source indicated from anthropogenic influence [18].

5. Conclusion

Magnetic susceptibility value on guano samples from mampu cave is higher than Bubau cave. This is influenced by concentration of Fe that causes the increase of magnetic susceptibility value. Variation of susceptibility from Mampu cave tend to decreasing after 10 cm depth. Grains size from both caves predominantly consist of mixture SSD and SP magnetic grains.

Acknowledgments

The authors would like to thank to James Cook University for providing samples Geophysics and Chemical Laboratory State University of Padang for facilities of samples measurements.

References

[1] Bird M I, Boobyer E M, Bryant C, Lewis H A, Paz V and Stepehens W E 2007 A Long Record of Environmental Change from Bat Guano Deposits in Makangit Cave, Palawan, Philipines *J. Earth and Environmental Science Transactions of the Royal Society of Edinburgh*. **98** 59-69.

[2] Wurster C M, Munksgaard N, Zwart C and Bird MI 2015 The Biogeochemstry of Insectivorous Cave Guano : A Case Study from Insular Southeast Asia *J. Biogeochemistry*. **124** 163-17.

[3] Wurster C M, McFarlane D A, Wassenar L I, Hobson K A, Athfield N B and Bird M I 2008 Stable Crbon and Hydrogen from Bat Guano in the Grand Canyon, USA, Reveal Younger Dryas and 8.2 ka Event *J. Geology*. **36** 683-686.

[4] Huliselan E K and Bijaksana S 2007 Identifiksi Mineral Magnetik pada Lindi (*Leachate*) *J. Geofisika*. **2** 8-13.

[5] Rifai H 2010 Konsistensi Sifat Magnetik Guano dari Dua Gua Kelelawar di Kabupaten 50 Kota Sumatera Barat *Prosiding dan Rapat Tahunan BKS-PTN Wilayah Barat ke 21*. 1-9.

[6] Rifai HV, Rusli N G D, Mufit F 2014 Kaitan Komposisi Unsur Dasar Penyusun Mineral Magnetik dengan Nilai Suseptibilitas Magnetik Guano dari Gua Bau-Bau Kalimantan Timur *J. Pillar of Physics*. **4** 49-56.

[7] Dearing J A 1999 Environmental Magnetic Susceptibility: Using the Bartington MS2 System *British Library Cataloguing in Publication Data*.

[8] Kanu M O, Meludu O C and Oniku S A 2013 Measurement of Magnetic Susceptibilityof Soils in Jalingo, N-E Nigeria: A Case Study of the Jalingo Mechanic Village *J. World Applied Sciences*. **24** 178-187.

[9] Sroubek P, Diehl J F, Kadlec J and Valoch K 2001 A Late Pleistocene Palaeoclimate Record Based on Mineral Magnetic Properties of the Entrance Facies Sediments of Kulna Cave, Czech Republic *J. Geophysics Int*. **147** 247-262.

[10] Evans M E and Heller F 2003 Environmental Magnetisme: Principles and Applications of Enviromagnetics *Academic Press*. USA.

[11] Paasche O, Lovlie R, Dahl SO, Bakke J and Nesje A 2004 Bacterial Magnetite in Lake Sediments: Lake Glacial to Holocene Climate and Sedimentary Changes in Northern Norway *J.*

IOP Conf. Series: Materials Science and Engineering **335** (2018) 012001 doi:10.1088/1757-899X/335/1/012001

Earth and Planetary Science Letters. **223** 319-333.

[12] Yang X, Grapes R, Zhou H and Yang J 2008 Magnetic Properties of Sediments from The Pearl River Delta, South China: Paleoenvironmental Implication *J. Science in China Series D: Earth Sciences*. **51** 56-66.

[13] Marwick Ben 2005 Element Concentration and Magnetic Susceptibility of Anthrosol: Indicators of Prehistoric Human Occupation in the Inland Pilbara, Western Australia *J. Archaeological Science*. **32** 1357-1368.

[14] Herries Andy I R and Fisher E C 2010 Multidimensional GIS Modeling of Magnetic Mineralogy as A Proxy for Fire Use and Spatial Patterning: Evidence from the Middle Stone Age Bearing Sea Cave of Pinnacle Point 13 B (Western Cape, South Africa) *J. Human Evolution*. **59** 306-320.

[15] Aidona E, Pechlivanidou S and Pennos Ch 2013 Environmental Magnetism: Application to Cave Sediments *Bulletin of the Geological Society of Greece* **47** 892-900.

[16] Aidona E and Liritzis I 2012 Magnetic Suceptibility and Radioactivity Changes of Aegean and Ionian Sea Sediments during Last Glacial/Interglacial: Climatic and Chronological Markers *J. Coastal Research*. **28** 342-353.

[17] Tamuntuan G, Bijaksana S, Gaffar E, Russel J, Safiuddin L O and Huliselan EK 2010 The Magnetic Properties of Indonesian Lake Sediment: A Case Study of a Tectonic Lake in South Sulawesi and Maar Lakes in East Java *J. ITB Science*.**42** 31-48.

[18] Canbay M 2010 Investigation of the Relation Between Heavy Metal Contamination of Soil and its Magnetic Susceptibility *J. Physical Sciences Int.* **5** 393-400.

IOP Conf. Series: Materials Science and Engineering **335** (2018) 012002 doi:10.1088/1757-899X/335/1/012002

Development Radar Absorber Material using Rice Husk Carbon for Anechoic Chamber Application

Z Zulpadrianto[1*], Y Yohandri[1], A Putra[2]

[1]Graduate Program of Physics, Faculty of Mathematics and Natural Science, Universitas Negeri Padang, Jl. Prof. Hamka, Sumatera Barat, Indonesia 25131

[2]Chemistry Department, Faculty of Mathematics and Natural Science, Universitas Negeri Padang, Jl. Prof. Hamka, Padang, Sumatera Barat, Indonesia 25131

zpadrianto@gmail.com

Abstract. The developments of radar technology in Indonesia are very strategic due to the vast territory and had a high-level cloud cover more than 55% of the time. The objective of this research is to develop radar technology facility in Indonesia using local natural resources. The target of this research is to present a low cost and satisfy quality of anechoic chambers. Anechoic chamber is a space designed to avoid reflection of EM waves from outside or from within the room. The reflection coefficient of the EM wave is influenced by the medium imposed by the EM wave. In laboratory experimental research has been done the development of material radar absorber using rice husk. The rice husk is activated using HCl and KOH by stirring using a magnetic stirrer for 1 Hours. The results of rice husk activation were measured using a Vector Network Analyzer by varying the thickness of the ingredients and the concentration of the activation agent. The VNA measurement is obtained reflection coefficient of -12dB and. -6.22dB for 1M HCL and KOH at thickness 10mm, respectively.

1. Introduction

Indonesia is in a region spanning the disaster. Based on data and information National Disaster Management Agency of Indonesia from early 2017 to 11 April 2017, there are 884 cases of disaster. Among them are floods, landslides, tidal waves/abrasion, earthquakes, forest and land fires, transportation accidents, and tornadoes. To minimize the impact of the disaster, it is necessary to see the rotation of the disaster that occurred. Accurate observations and data sets, such as mapping from affected areas, are urgent to be met. Reliable technology in mapping the region is one of the remote sensing [1].

Remote sensing is science and art in obtaining information about an object, area, symptoms through analysis of data obtained by means without direct contact with objects, areas, and observed phenomena [2]. Radar is a sensor of remote sensing. Radar is one application of the electromagnetic spectrum (microwave) with a frequency range of 0-300 GHz or wavelength 1m to 1mm. Radar (radio detection and ranging) is an electromagnetic wave system for detecting and determining the location of objects with long distances that work by emitting electromagnetic waves and processing the reflected signal to know the character of the object. Several technologies for radar have been developed in previous studies [3]-[7]. Almost in all the development of remote sensing technology requires a very expensive facility one of them is anechoic chambers. Anechoic chambers are spaces designed to absorb electromagnetic wave reflections to reduce reflection and external noise in measuring the characteristics of an antenna or radar component. Anechoic chamber provides precise measurement and controlled electromagnetic environment, and prevents EM interference from inside and outside the room [8].

In general, the anechoic chamber is built using an electromagnetic wave absorber. However, the availability of current electromagnetic wave absorber materials still has some disadvantages such as expensive, high density and low absorptive [9]. Therefore it is necessary to develop electromagnetic wave absorbers with low cost, low density and high level of absorbency based on local resources. Through physical and chemical processing, carbon materials have the potential to be used as EM absorbents. Carbon material is a good electromagnetic energy absorber.

Carbon as radar absorber material can be generated from local resources such as rice husks. This material content of silica (SiO2) in rice husk ash is 94 - 96% [10]-[12]. Rice is the main product of agriculture in agrarian countries, including Indonesia. Rice husk is an abundant by-product of rice milling, and has been used only as a fuel for burning stone. Unsafe rice husk handling will lead to environmental pollution. 20% of the weight of rice is rice husk, and varies from 13 to 29% of the chaff composition is the husk ash which is always produced each time the husk is burned [12]

The technology of electromagnetic wave absorbing materials is an important topic for military purposes and is also a potential product in the field of business. In particular, the development of material radar absorbers (RAM), electromagnetic wave absorbers in the radar frequency region, at the working frequency of GHz, has been actively studied for a long time. The basic design theories of RAM, such as Salisbury screen theory, Jaumann's absorber theory, etc. were published from the early 1950s onwards [13]. To reduce the radar cross section of the target RAM can be used to cover its surface. Decreased radar cross section (RCS) can be achieved when the RAM is sure to cover all material that allows cross section [13].

In an effort to minimize the reflection from the surface of the medium by considering the physical equations that represent the reflection process. There are three conditions that affect minimum reflectivity, namely: reflection coefficient, impedance, and wave attenuation. For the reflection coefficients can be solved by the following equation:

$$r = \frac{\eta_M - \eta_0}{\eta_M + \eta_0} = \frac{Z_M - Z_0}{Z_M + Z_0}$$

Here, r is the reflection coefficient, η_0 = wave scattering in air, η_M = wave scattering on the medium. The equation can also be converted into an intrinsic impedance ($Z = 1 / \eta$).

The reflection coefficient will be zero if $\eta_M = \eta_0$. In other words, the material on the screen has the same impedance as the waveform medium [14]. The intrinsic impedance in the freeway can be calculated by the following equation:

$$Z_0 = \frac{E}{H} = \sqrt{\frac{\mu_0}{\varepsilon_0}} \approx 377\ ohms$$

Where: E and H are electric and magnetic fields. μ_0 and ε_0 are permeability and permittivity respectively. Material with impedance of 377 ohm will not reflect microwaves if wave is applied to the material.

The mechanism of electromagnetic wave absorption in the material is generally influenced by two factors: thickness and material type. The thickness factor occurs in all the materials and the thicker the absorbing material is also greater. While the absorption of electromagnetic radiation on the magnetic material in addition to the thickness factor also occurs another interaction of electromagnetic waves from the outside will rotate the magnetic dipole so that the material impedance occurs. The interaction can also occur when the frequency of the electromagnetic wave corresponds to the frequency generated so that the magnetic material will absorb electromagnetic waves only at specific frequencies.

IOP Conf. Series: Materials Science and Engineering **335** (2018) 012002 doi:10.1088/1757-899X/335/1/012002

2. Method

2.1. Synthesis of activated carbon based on rice husk

The starting material used in the research is rice husk. The rice husk cleared from other materials soil and gravel. Carbonization of the rice husk is carried out by an imperfect Pyrolysis process of a carbonaceous material. The carbonization result mashed using a blender and sieved using 100 meshes. The Activation process is conducted using HCl and KOH with variation of concentration 1 M to 5 M. Amount of 20 grams of materials dissolved in activator (KOH and HCL) based on the variation of concentration with constant stirring for 1 hour. The activation result is precipitated for 48 hours. It is then neutralized by adding the KOH solution wisely to the solution with HCl until pH 6-7 reached, whereas in the KOH solution it is neutralized by adding HCL solution wisely until pH 6-7 reached. After neutralization, the material is washed with aquades and then filtered using filter paper until the impurity water is separated from the activated material. The ingredients were then dried in an oven at 110 °C for 2 hours and the activated carbon obtained.

2.2. Characterization

Vector Network Analysis (VNA) is performed to measure the absorptive performance of the activated carbon. The Measurement was carried out with variations in thickness of 4 mm, 6 mm, 8 mm, and 10 mm.

3. Result and discussion

The Absorption of radar in materials of concentration HCl and KOH of 1M, 2M, 3M, 4M and 5M were identified using a Vector Network Analyzer in the 4-8 GHz frequency range. The effect of concentrate activation on the absorption of wave generated at each thickness variation can be seen from the maximum reflection coefficient (RL) in dB. The result of characterization using VNA with HCL activator is shown in Figure 1-3.

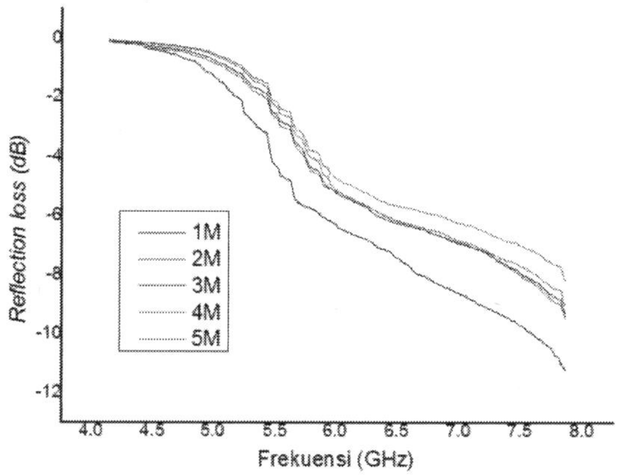

Figure 1. Reflection coefficient plotted as a function of frequency for variation of HCL activator concentration at thickness 6mm.

The RL value and the minimum reflection coefficient of rice husks were -11.2 db, 0.275. These results were obtained from VNA measurements with activation agents using HCL with variation of concentrations of 1-5M and at thickness 6mm. RL and minimum reflection coefficient are obtained when using HCL 1M activation agent.

IOP Conf. Series: Materials Science and Engineering **335** (2018) 012002 doi:10.1088/1757-899X/335/1/012002

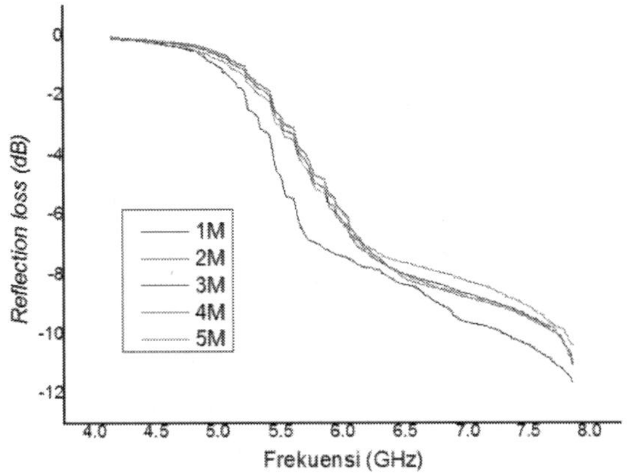

Figure 2. Reflection coefficient plotted as a function of frequency for variation of HCL activator concentration at thickness 8mm.

Figure 2 shows the activation of rice husk using HCL substance with variation of concentration 1-5M at 8mm thickness. The minimum reflection coefficient is obtained at 1M HCL concentration, which the reflection coefficient of -11.60 db.

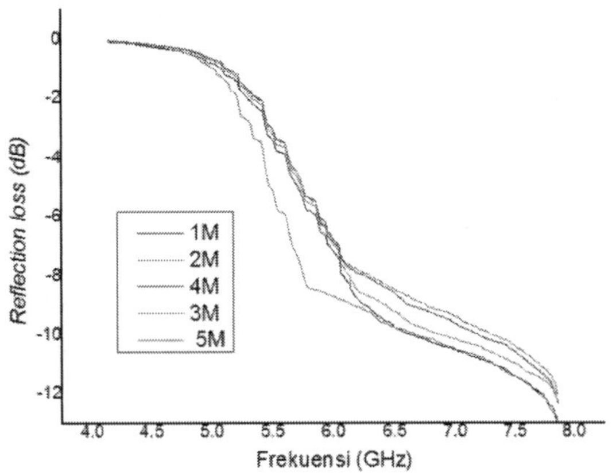

Figure 3. Reflection coefficient plotted as a function of frequency for variation of HCL activator concentration at thickness 10 mm.

The largest absorption of radar wave occurs at the concentration of HCL 1 M with a thickness of 10 mm that can be seen on Figure 3. The measured results of reflection loss and reflection coefficient for 1M HCL activation agent are obtained -12db and 0,251, respectively. The data explained that the material with a thickness of 10 mm has the largest reflection loss value. The thickness of 10 mm of the material has a large volume so that more particles that interact with radar waves that cause absorption increased.

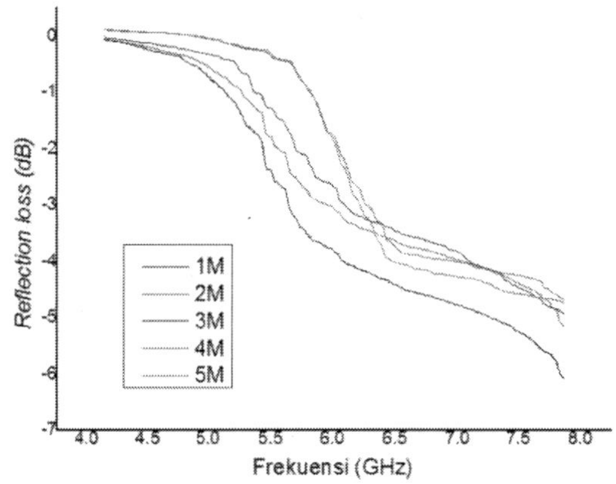

Figure 4. Reflection coefficient plotted as a function of frequency for variation of KOH activator concentration at thickness 6mm.

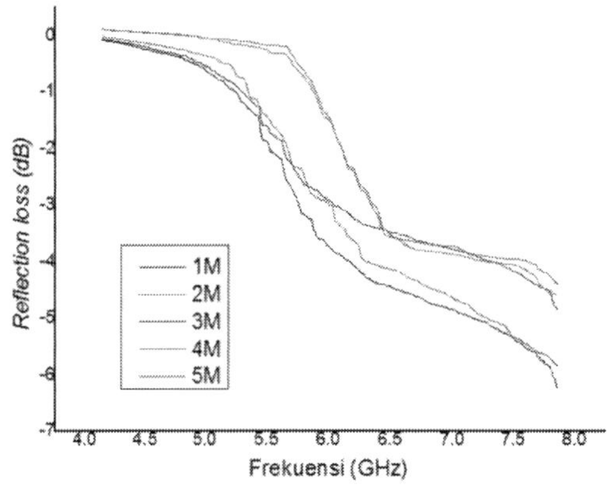

Figure 5. Reflection coefficient plotted as a function of frequency for variation of KOH activator concentration at thickness 8mm.

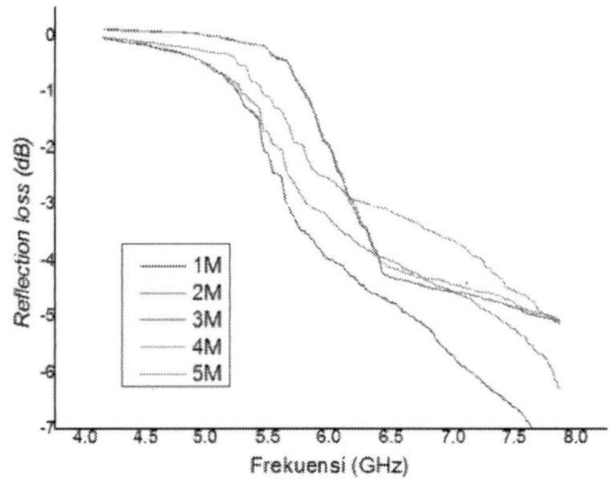

Figure 6. Reflection coefficient plotted as a function of frequency for variation of KOH activator concentration at thickness 10mm.

The largest absorption of radar wave occurs at the concentration of KOH 1 M with a thickness of 10 mm. Result of VNA measurement is obtained reflection loss and reflection coefficient -6.22 db and 0.488 respectively for KOH 1M activation agent. The data explained that the material with a thickness of 10 mm has the largest reflection loss value. The thickness of 10 mm of the material has a large volume so that more particles that interact with radar waves that cause absorption increased.

4. Conclusion

Test VNA characterization showed a maximum absorber value of -12.90 dB at a thickness of 10 mm with a HCl activator of 1 M. On the other hand, for KOH activator, the maximum absorber is obtained for 1M concentration at 10 mm thickness. Based on this works, the HCL activator and 10 mm thickness of the material is satisfy for the radar absorber material.

References

[1] Xie, Yichun, Zongyao Sha, and Mei Yu. "Remote sensing imagery in vegetation mapping." Journal of Plant Ecology, 2008.
[2] Lillesand.T.M, dan Kiefer R.W. 1979. Remote Sensing and Image Interpretation. New York: John Willey and Sons.
[3] M. Hussein, Yohandri, J.T. Sri Sumantyo and A. Yahia, 2013, A low sidelobe level of circularly polarized microstrip array antenna for CP-SAR sensor, Journal of Electromagnetic Waves and Applications, Vol. 27, No. 15, pp. 1931-1941. DOI: 10.1080/09205071.2013.828577
[4] M.Z. Baharuddin, Y. Izumi, J.T. Sri Sumantyo, and Yohandri, "Side-Lobe Reduced, Circularly Polarized Patch Array Antenna for Synthetic Aperture Radar Imaging," IEICE Trans. Electron., vol. E99.C, no. 10, pp. 1174–1181, 2016.
[5] Yohandri, Asrizal, and J. T. Sri Sumantyo, "Design of tilted beam circularly polarized antenna for CP-SAR sensor onboard UAV," in 2016 International Symposium on Antennas and Propagation (ISAP), 2016, pp. 658–659.
[6] Yohandri, J. T. Sri Sumantyo and H. Kuze, "Circularly polarized array antennas for synthetic aperture radar," in Progress in Electromagnetics Research Symposium, 2011, pp. 1244–1247.
[7] Yohandri, B. Setiadi; Y.K. Chan and V.C. Koo, "A ground-based synthetic aperture radar system using portable VNA for landslide monitoring," in 2017 International Symposium on Antennas and Propagation (ISAP), 2017, pp. 1–2.
[8] Aswoyo, Budi. 2010. "Pengukuran Koefisien Refleksi Dinding "Anechoic Chamber" Pada Frekuensi Gelombang Mikro Band-X Menggunakan Teknik Gelombang Berdiri." Surabaya: Published by EEP. ISISSN: 2088-0596.
[9] Y. H. Zou, H. B. Liu, L. Yang, and Z. Z. Chen, J. Magn. Magn. Mater. 302, 343, (2006).
[10] Houston, D.F. 1972. "Rice Chemistry and technology." American Association of Cereal chemist Inc. Minnesota.
[11] Prasad C.S., Maiti K,N., Venugopal R. 2000. "Effect of rice husk ash in whiteware compotitions." Ceramic International, 27.
[12] Putro, Andhi Laksono. 2007. "Abu Sekam Padi Sebagai Sumber Silika Pada Sintesis Zeolit ZSM-5 Tanpa Menggunakan Templat Organik." Akta Kimindo Vol. 3 No. 1 Oktober 2007 : 33 - 36 .
[13] Oh, Jung-Hoon, Kyung-Sub Oh, Chun-Gon Kim, dan Chang-Sun Hong. 2004. "Design of radar absorbing structures using glass/epoxy composite containing carbon black in X-band frequency ranges." Composites: Part B 35 (2004) 49–56.
[14] Saville, Paul. 2005. Review of Radar Absorbing Materials. Canada: Defence R&D Canada – Atlantic.

IOP Conf. Series: Materials Science and Engineering **335** (2018) 012003 doi:10.1088/1757-899X/335/1/012003

Crystal Structure Analysis of Electromagnetic Wave Absorber Material $BaFe_{12-x}Ti_{x/2}Zn_{x/2}O_{19}$ Based

M Delina[1*], N Nenni[1], W A Adi[2]

[1]Department of Physics, State University of Jakarta, East Jakarta, Indonesia
[2]Center for Science and Technology of Advanced Materials, National Nuclear Energy Agency (BATAN), Puspiptek Serpong, South Tangerang, Indonesia

*mutia_delina@unj.ac.id

Abstract. The optimization of $BaFe_{12-x}Ti_{x/2}Zn_{x/2}O_{19}$ (x=2.2; 2.4; 2.6; 2.8) single phase composition have been performed. The materials were synthesized by solid state reaction method through mechanical milling technique. The materials were made from the mixture of oxide materials, which are $BaCO_3$, Fe_2O_3, TiO_2 and ZnO. The mixture was milled for five hours using a High Energy Milling (HEM), was dried at 100^0C in the Oven and then was sintered at 1000^0Cfor five hours in the Furnace. The phase identification of $BaFe_{12-x}Ti_{x/2}Zn_{x/2}O_{19}$ (x=2.2; 2.4; 2.6; 2.8) were carried out by using a Match Program while the crystal structure analysis were investigated by using a General Structure Analysis System (GSAS) program. The refinement results of x-ray diffraction pattern showed that the sample of $x \leq 2.4$ have a $BaFe_{12}O_{19}$ single phase while the sample of x> 2.4 have two phases, which are $BaFe_{12}O_{19}$ and $ZnFe_2O_4$ phases. The surface morphology of sample and the element of sample were identified through an analysis of Scanning Electron Microscope (SEM) and Energy Dispersive Spectroscopy (EDS) data.

1. Introduction

Barium Hexaferrite ($BaFe_{12}O_{19}$) has received great attention because of its applications in permanent magnet, high density magnetic recording media, and microwave devices [1,2]. This material ($BaFe_{12}O_{19}$) has a high saturation magnetization which is 72 emu/gr, high coercivity field which is 6700 Oe, high anisotropy field, high Curie temperature which is 450^0C, and high magnetic permeability [3-5]. Some characteristics of the material are suitable for application of electromagnetic wave absorber material. However, the high coercivity field cause its absorption ability is reduced because there is no resonance occurs between the magnetic moment with magnetic field and electric field within electromagnetic wave. So, to increase its absorption ability, the coercivity field of Barium Hexaferrite must be reduced [6]. One of the techniques to change its magnetic properties is modification of Barium Hexaferrite through ionic substitution.

Fe^{+3} ions in Barium Hexaferrite are substituted by one trivalent ion or by combination of one divalent ion and one tetravalent ion [7] that have the similar size. The efficiency of substitution is depend on the differenceof electronegativity and the ionic radius between the dopant and host element [8] (in this case, the host element is Fe^{+3} ions). The similar ionic radius and the electronegativity make dopant to replace half of Fe^{+3} ions position surely. In this research, Ti^{+4} and Zn^{+2}ions are used to replace Fe^{+3} ions position partially within Barium Hexaferrite. The ionic radius of Ti^{+4} andZn^{+2} ions are 0.605 Å and 0.68Å that have similar ionic radius with Fe^{+3} (0.645 Å), while their electronegativity are 1.54, 1.7 and 1.83 [7-10].

In previous research, Nenni et al have analyzed about effect of substitution Ti-Zn ions on magnetic and electromagnetic wave absorption properties of$BaFe_{12-x}Ti_{x/2}Zn_{x/2}O_{19}$ material system [10]. The research reported that$BaFe_{12-x}Ti_{x/2}Zn_{x/2}O_{19}$(x=2) material has single phase of $BaFe_{12}O_{19}$. Therefore, there will be optimization of $BaFe_{12-x}Ti_{x/2}Zn_{x/2}O_{19}$ composition in this research. The crystal structure analysis of $BaFe_{12-x}Ti_{x/2}Zn_{x/2}O_{19}$ material (with various composition x=2.2;2.4;2.6 and 2.8) using solid state reaction method will be carried out. The addition of Ti^{+4} and Zn^{+2}ions are predicted to replace Fe^{+3}ions position partially without change the crystal structure of Barium Hexaferrite. The effect of substitution Ti^{+4} and Zn^{+2} ions to crystal structure and lattice parameter of samples are acknowledged by X-Ray Diffractometer (XRD). The surface morphology and the element of samples are observed by Scanning Electron Microscope (SEM) and Energy Dispersive Spectroscopy (EDS).

2. Material and Method

$BaFe_{12-x}Ti_{x/2}Zn_{x/2}O_{19}$ (with various composition x=2.2;2.4;2.6 and 2.8)material were synthesized by solid state reaction method through mechanical milling technique. The materials were made from the mixture of $BaCO_3$, Fe_2O_3, TiO_2 and ZnO. Mass of each material was determined based on stoichiometric calculation then was weighed and was placed in the stainless steel vial. The mixture was added with five balls mill and ethanol. The ratio of the sample and five balls mill was 1:2. The mixture was milled for five hours using High Energy Milling (HEM) and then was dried at 100^0Cin the Oven. The dried sample was mashed and was sintered at 1000^0Cin the Furnace.

The X-Ray diffraction pattern of modified samples were measured by X-Ray Diffractometer (XRD) using radiation of$CuK\alpha$ (λ = 1.5406 Å). The formedphase of the sampleswere analyzed using Match and General Structure Analysis (GSAS)programqualitatively and quantitatively. After that, the surface morphology and the componentof the sampleswereobserved by Scanning Electron Microscope (SEM) and Energy Dispersive Spectroscopy (EDS).

3. Results and Discussion

The x-ray diffraction pattern of the samples were matched with x-ray diffraction pattern of standard $BaFe_{12}O_{19}$ phase on Match program. Figure 1 shows the results of x-ray diffraction pattern of $BaFe_{12-x}Ti_{x/2}Zn_{x/2}O_{19}$ samples (with various composition x=2.2; 2.4; 2.6 and 2.8). The results of phase identification show that the x-ray diffraction pattern of samples x=2.2 and x=2.4 have the similar pattern with x-ray diffraction pattern of standard $BaFe_{12}O_{19}$ phase. Therefore, both of them have single phase of $BaFe_{12}O_{19}$. The result of $BaFe_{12}O_{19}$ phase identification is referred fromCrystallography Open Database (COD) with the 96-900-8138 card number. However, there are another diffraction peak at the angle of 30^0 and between 60^0 up to 70^0 in the samples x=2.6 and x=2.8, which are $ZnFe_2O_4$ phase. It is caused by unbalanced reaction in the sample so that the samples x=2.6 and x=2.8 have two phases, which are $BaFe_{12}O_{19}$ and $ZnFe_2O_4$ phases.

For the sample x=2.6, identification of $BaFe_{12}O_{19}$and $ZnFe_2O_4$ phases are indicated from COD: 96-901-4002 and COD: 96-900-510. On the contrary, identification of $BaFe_{12}O_{19}$ and $ZnFe_2O_4$ phases for the sample x=2.8 are indicated from COD: 96-900-8138 and COD: 96-900-6896. But, Match program has not given the information of structure parameters and the amount of mass fraction of $BaFe_{12}O_{19}$and $ZnFe_2O_4$ phases so that quantitative analysis is required by using GSAS program. Figure 2 shows the refinement results of x-ray diffraction pattern of the sample (with various composition x = 2.2; 2.4; 2.6 and 2.8) using GSAS program.

Figure 2(a) and 2(b) shows that the x-ray diffraction pattern of samples x=2.2 and x=2.4 have similar pattern with the x-ray diffraction pattern of calculation. The whole diffraction peaks of samples x=2.2 and x=2.4 has been fitted and the difference between the diffraction peaks of the sample and calculation are relatively slight. It shows that both of the sample have single phase which is $BaFe_{12}O_{19}$

phase. Figure 2(c) and 2(d) inform the x-ray diffraction pattern of samples x=2.6 and x=2.8. Blue and red dots show the diffraction peaks of $ZnFe_2O_4$ phase, so that both of those samples have two phases.

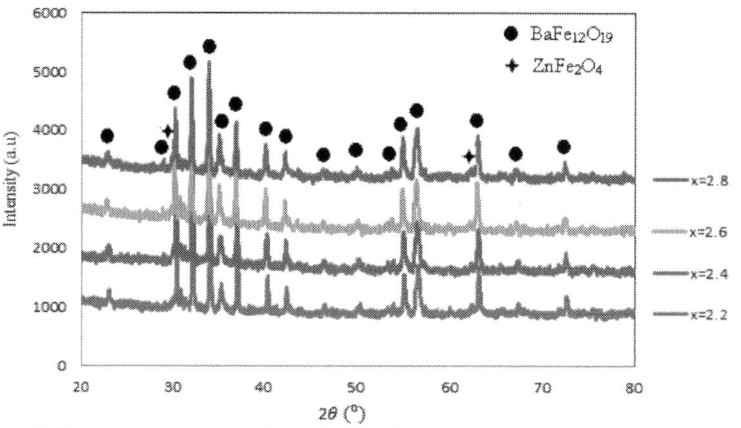

Figure 1.X-ray diffraction pattern of $BaFe_{12-x}Ti_{x/2}Zn_{x/2}O_{19}$sample with various composition (x=2.2; 2.4; 2.6 and 2.8).

Based on these results, Fe^{+3} ions have been partially substituted by Ti^{+4} and Zn^{+2} ions. However, this substitution has the optimum composition, which is x=2.4. It means that not all of Ti^{+4} and Zn^{+2} ions can replace part of Fe^{+3} ions in Barium Hexaferrite. The substitution of Ti^{+4} and Zn^{+2} ions changes the structure parameter of $BaFe_{12-x}Ti_{x/2}Zn_{x/2}O_{19}$. The information about the change of structure parameter can be looked at Table 1.

Barium Hexaferrite has hexagonal structure with space group $P6_3/mmc$, lattice parameter a = b = 5.89Å, and c = 23.17 Å, $\alpha = \beta = 90^0$, $\gamma = 120^0$[11]. Based on table 1, there are changes of lattice parameter on each phase and sample. It is indicated thatFe^{+3} ions have been partially substituted byTi^{+4} and Zn^{+2} ions. The alteration of lattice parameter on the samples x=2.2 and x=2.4 still maintain hexagonal crystal structure of Barium Hexaferrite. There are unbalance reaction on the samples x=2.6 and x=2.8 so that Fe_2O_3 and ZnO ions were bound then formed the $ZnFe_2O_4$ phase. The qualities of fitting sample are determined by R factor and χ^2 where the value of χ^2 is less than 1.3 [12]. Based on Table 1, the value of χ^2 from each sample is corresponding to 1.3 thus we can say that the samplesare fitted well.

Mass fraction of each phase of $BaFe_{12-x}Ti_{x/2}Zn_{x/2}O_{19}$(x=2.2; 2.4; 2.6 and 2.8) is shown in Figure 3. Based on Figure 3, samples x=2.2 and x=2.4 have 100% $BaFe_{12}O_{19}$ phase. However, 3.87% $ZnFe_2O_4$ phase emerge on the sample x = 2.6 and increased up to 7.9% on the sample x=2.8. It shows that x=2.4 is the optimum composition of Ti^{+4} and Zn^{+2} ions substitution.

The results of refinement are also supported by observation of the particle's surface morphology using SEM, as shown in Figure 4. The surface morphology of particles x=2.2 and x=2.4 are relatively uniform and evenly distributed in the entire of sample surface. The particle sizes are less than 5 μm, respectively, as shown in Figure 4(a) and 4(b). The results are equal with phase identification result by Match and GSAS that the samples x=2.2 and x=2.4 have single phase. While Figure 4(c) and 4(d) show the aggregation in the particles of the samples x=2.6 and x=2.8 because of other phases. The particle size of both samples is also less than 5μm.

IOP Conf. Series: Materials Science and Engineering **335** (2018) 012003 doi:10.1088/1757-899X/335/1/012003

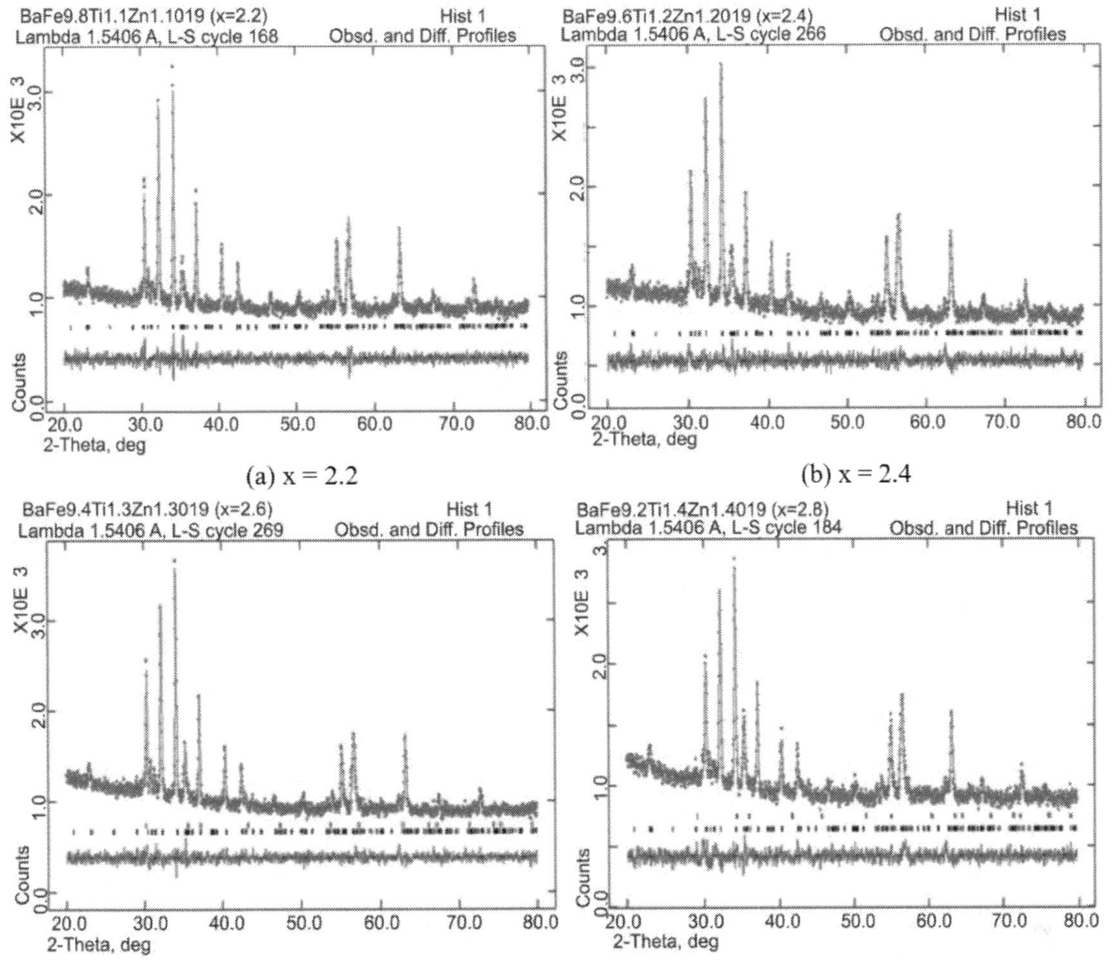

(a) x = 2.2 (b) x = 2.4

(c) x=2.6 (d) x=2.8

Figure 2. The refinement results of x-ray diffraction pattern of $BaFe_{12-x}Ti_{x/2}Zn_{x/2}O_{19}$. a) x=2.2, b) x=2.4, c)=2.6, d)=2.8. The red, green and purple curves express the x-ray diffraction pattern of sample, x-ray diffraction pattern of calculation and the difference of them. The black, red and blue dots show the diffraction peaks of $BaFe_{12}O_{19}$ and $ZnFe_2O_4$ phases.

Table 1.Structure Parameter, RFactor, and Goodness of Fit (χ^2). R_{wp} (%) and χ^2 are the quality of fitting factor.

Sample (*x*)	**Phase**	**Lattice parameter (Å)** *a*	*b*	*c*	**Crystal structure**	**V (Å^3)**	**ρ (g/cm^3)**	**R_{wp} (%)**	χ^2
2.2	$BaFe_{12}O_{19}$	5.891307	5.891307	23.201977	Hexagonal	697.395	11.775	3.54	1.26
2.4	$BaFe_{12}O_{19}$	5.891109	5.891109	23.19949	Hexagonal	697.274	11.338	3.47	1.28
2.6	$BaFe_{12}O_{19}$	5.874818	5.874818	23.135712	Hexagonal	691.516	13.529	3.54	1.34
	$ZnFe_2O_4$	8.364467	8.364467	8.364467	Cubic	585.214	5.427		
2.8	$BaFe_{12}O_{19}$	5.892707	5.892707	23.206532	Hexagonal	697.864	11.141	3.65	1.38
	$ZnFe_2O_4$	8.663953	8.663953	8.663953	Cubic	650.352	4.924		

IOP Conf. Series: Materials Science and Engineering **335** (2018) 012003 doi:10.1088/1757-899X/335/1/012003

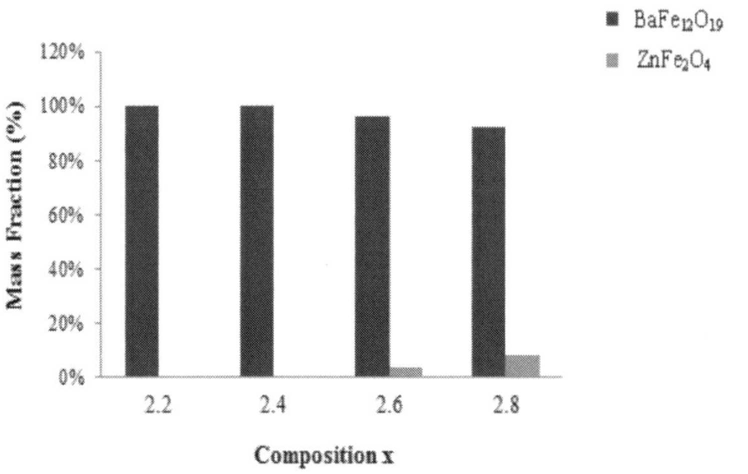

Figure 3.Mass fraction as a function of composition.The blue and green graphare the accretion of mass fraction of $BaFe_{12}O_{19}$ and $ZnFe_2O_4$ phases.

The elements of $BaFe_{12-x}Ti_{x/2}Zn_{x/2}O_{19}$ sample are known by the measurement using EDS as shown in Figure 5 and Table 2. The carbon elements in each sample are originated from carbon tip during the preparation so that they are not part of the elements of the four samples. The sample with composition of x=2.2 and x=2.4 have single phase and the similar crystal structure although their composition are different.

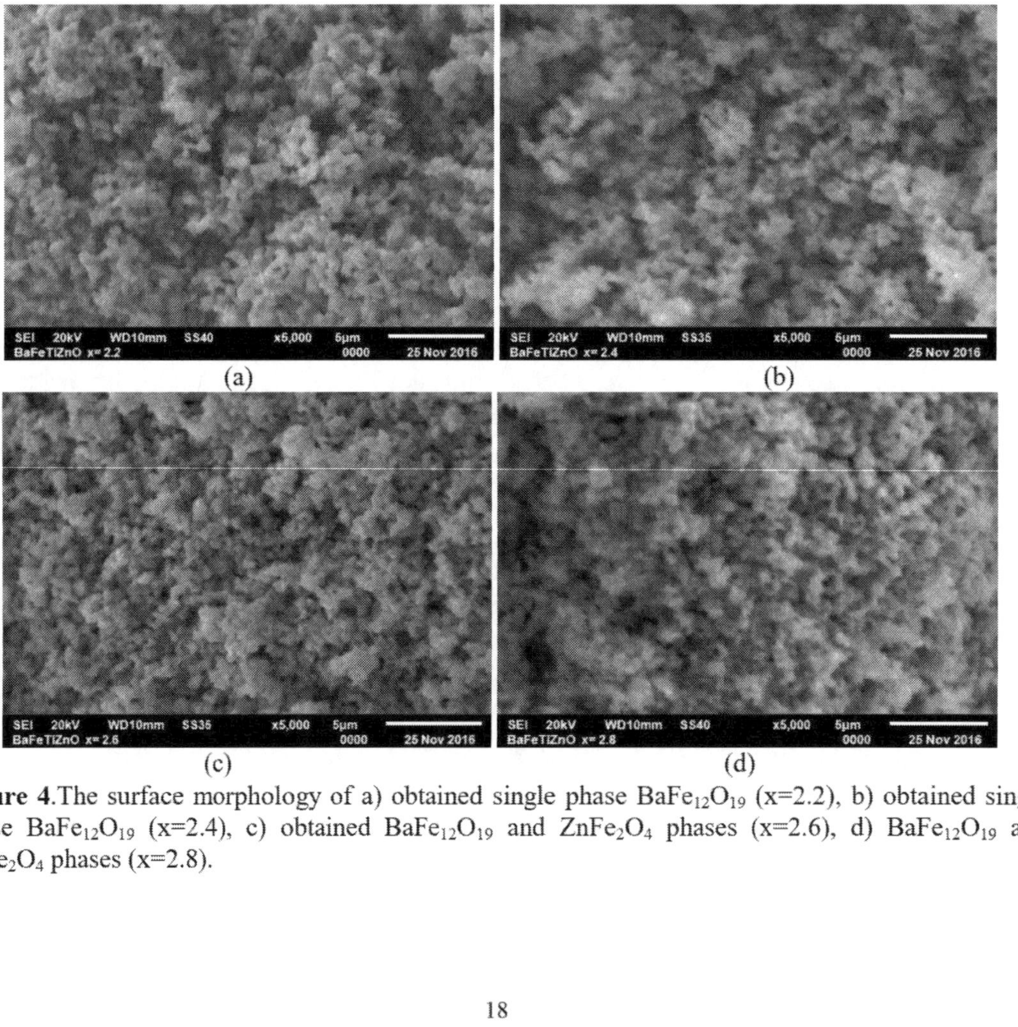

Figure 4.The surface morphology of a) obtained single phase $BaFe_{12}O_{19}$ (x=2.2), b) obtained single phase $BaFe_{12}O_{19}$ (x=2.4), c) obtained $BaFe_{12}O_{19}$ and $ZnFe_2O_4$ phases (x=2.6), d) $BaFe_{12}O_{19}$ and $ZnFe_2O_4$ phases (x=2.8).

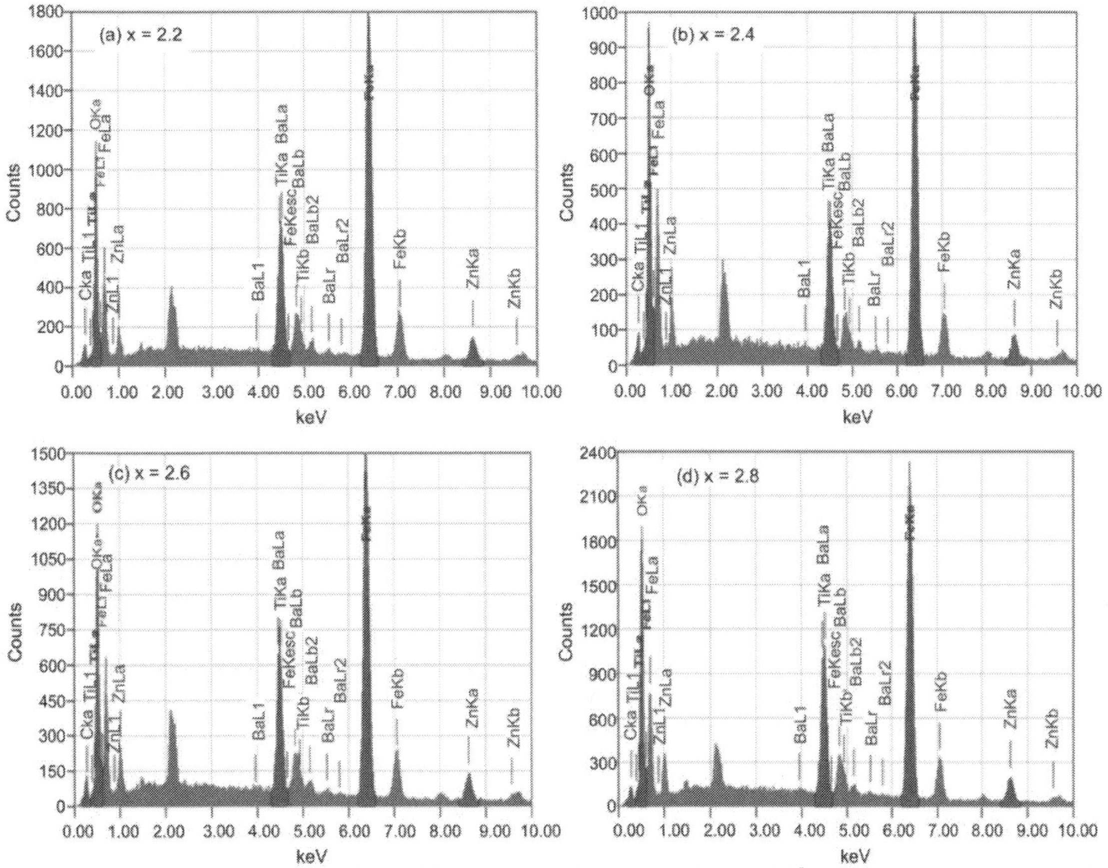

Figure 5. EDS Spectrum of $BaFe_{12-x}Ti_{x/2}Zn_{x/2}O_{19}$ after sintered at 1000^0C for 5 hours a) x=2.2, b) x=2.4, c) x=2.6, d) x=2.8

Table 2.Element distribution on the $BaFe_{12-x}Ti_{x/2}Zn_{x/2}O_{19}$(x = 2.2;2.4;2.6;2.8) was measured by EDS. Each of four samples consist of Barium (Ba), Iron (Fe), Oxygen (O), Titanium (Ti), Zink (Zn) and Carbon (C) unsur which are in 4.464 keV, 6.398 keV, 0.525 keV, 4.508 keV, 8.630 keV, 0.277 keV energy, respectively.

		$BaFe_{12-x}Ti_{x/2}Zn_{x/2}O_{19}$							
Unsure	keV	x=2.2		x=2.4		x=2.6		x=2.8	
		wt.%	at.%	wt.%	at.%	wt.%	at.%	wt.%	at.%
Barium (Ba)	4.464	16.72	4.71	15.28	3.66	16.48	4.30	15.67	4.12
Iron (Fe)	6.398	51.18	35.43	43.36	25.53	46.56	29.91	46.18	29.84
Oxygen (O)	0.525	17.47	42.23	26.76	54.98	21.20	47.55	22.71	51.23
Titanium (Ti)	4.508	3.89	3.14	3.82	2.62	4.25	3.19	5.27	3.97
Zink (Zn)	8.630	7.63	0.60	7.29	3.67	7.92	4.35	8.05	4.45
Carbon (C)	0.277	3.10	9.99	3.48	9.54	3.58	10.70	2.13	6.39

4. Conclusion

The synthesis of $BaFe_{12-x}Ti_{x/2}Zn_{x/2}O_{19}$with various composition (x = 2.2; 2.4; 2.6; 2.8) have been successfully performed. The refinement results of the x-ray diffraction pattern show that the samples x=2.2 and x=2.4 have single phase while the samples x> 2.4 have two phases. Thus the substitution of Ti^{+4} and Zn^{+2} ions into Fe^{+3} ions are only of 2.4 at.% without altering the hexagonal crystal structure of Barium Hexaferrite. The surface morphology of samples x=2.2 and x = 2.4 are uniform and spread evenly.

Acknowledgments

The authors would like thank to National Nuclear Energy Agency (BATAN) Indonesia for the equipment for XRD, SEM, and EDS measurements.

References

[1] Mosleh Z, Kameli P, Poorbaferani A, Ranjbar M, Salamati H. 2016. Structural, Magnetic and Microwave Absorption Properties of Ce-Doped Barium Hexaferrite. Journal of Magnetism and Magnetic Materials397: 101-107.

[2] Bsoul I, Mahmood SH. 2010. Magnetic and structural properties of BaFe12−xGaxO19 nanoparticles.Journal of Alloys and Compounds489: 110-114.

[3] Singh P, Babbar VK, Razdan A, Goel TC, Srivastava SL. 2004. Magnetic, Dielectric and Microwave Absorption Studies of Ba-CoTi Hexaferrite-Epoxy Composites. Indian Journal of Pure and Applied Physics, Vol. 42, pp 221-228.

[4] Soman V, Mehare MD. 2016. Structural and Electrical Properties of Co-Sn Substituted $BaFe_{12}O_{19}$. International Journal of Modern Engineering Research (IJMER), Vol. 6.ISSN: 2249-6645.

[5] Radwan M, Rashad MM, Hessien MM. 2007. Synthesis and Characterization of Barium Hexaferrite Nanoparticles. Journal of Materials Processing Technology181: 106-109.

[6] Quiroz P, Halbedel B, Bustamante A, González JC. 2011. Effect of Titanium Ion Substitution in the Barium Hexaferit Studied by Mössbauer Spectroscopy and X-ray Diffraction. Hyperfine Interactions: Springer Science and Business Media B.V.

[7] Li J, Zhang H, Harris VG, Liao Y, Liu Y. 2015. Ni-Ti Equiatomic Co-Substitution of Hexagonal M-Type Ba(NiTi)xFe12-2xO19Ferrites. Journal of Alloys and Compounds649: 782-787.

[8] Xie GC, Fang L, Peng LP, Liu GB, Ruan HB, Wu F, Kong CY. 2012. Effect of In-doping on the Optical Constants of ZnO Thin Films. Physics Procedia32: 651-657.

[9] Ahmed, M.A., Rady, K.E.-S., El-Shokrofy, K.M., Arais, A.A. and Shams, M.S. 2014. "The Influence of Zn2+ Ions Substitution on the Microstucture and Transport Properties of Mn-Zn Nanoferrites". Materials Sciences and Applications, 5, 932-942.http://dx.doi.org/10.4236/msa.2014.513095.

[10] Nenni, Delina M, Adi WA, Taryana Y. 2017. Effect of Substitution Ti-Zn Ionson Magnetic and Electromagnetic Wave Absorption PropertiesofBaFe12-xTix/2Znx/2O19Material System. SPEKTRA: Journal of Physics and Its Applications, Vol 2, No.2.

[11] Maswadeh Y, MahmoodSH, Awadallah A, Aloqaily AN. 2015.Synthesis and Structural Characterization of non-Stoichiometric Barium Hexaferrite Materials with Fe:Ba ratio of 11.5–16.16. IOP Conf. Series: Materials Science and Engineering, 92012019.doi:10.1088/1757-899X/92/1/012019.

[12] Gunanto YE, Jobiliong E, Adi WA. 2016. Microwave Absorbing Properties of Ba0.6Sr0.4Fe12-zMnzO19 (z = 0–3) Materials in X Band Frequencies. J. Math. Fund. Science, Vol 48(1): 55-65.

IOP Conf. Series: Materials Science and Engineering **335** (2018) 012004 doi:10.1088/1757-899X/335/1/012004

Effect Of Milling Time on Particle Size of Forsterite (Mg_2SiO_4) from South *Solok* District

S Sarimai[1], R Ratnawulan[1*], R Ramli[1], A Fauzi[1]

[1]Department of Physics, Faculty of Mathematics and Sciences, Universitas Negeri Padang, Indonesia

*ratnawulan@fmipa.unp.ac.id

Abstract. West Sumatra has considerable serpentine mineral resources, including the *Jorong Sungai Padi Nagari Lubuak Gadang Sangir* Subdistrict, South *Solok* District. Exploitation of minerals of serpentine is still processed in raw or semi-finished material so that it has a low selling value. Serpentine minerals contain forsterite minerals that have higher economic value if in the form of nanoparticles. The manufacture of forsterite nanoparticles has been done using synthetic materials, while synthetic materials are expensive and require a long process to make them. The treatment of temperature variations of calcination to serpentine minerals, obtained results found forsterite phase that dominates at a temperature of 800 °C. Serpentine minerals can be used as alternative ingredients for the nanoparticle makers of forsterite that are easy to find in the deep, and do not require expensive to make them. The purpose of this study was to investigate the effect of milling time on the microstructure and grain size of the serpentine forsterite mineral nanoparticles in the form of crystal structure, crystal size, and particle size. The results of the study showed grain size of 5, 10, 20, and 40 hours milling time are 579, 478, 451, and 385 nm respectively. Based on the research that has been done can be drawn conclusion Time milling effect on the grain size of forsterite mineral serpentine from South Solok District, the longer milling time the size of forsterite grains smaller. Optimum milling time to produce nano forsterite is 40 hours with a grain size of 385 nm.

1. Introduction

Serpentine has an important role in the steel industry as a material supporting the establishment of the steel industry. West Sumatra has considerable serpentine mineral resources, one of which is *Jorong Sungai Padi Nagari Lubuak Gadang, Sangir* Subdistrict, South Solok District. The exploitation of serpentine minerals is still processed in raw material or semi-finished material so it has a low selling value.

Serpentine is a metamorphic rock formed from serpentine minerals due to basalt changes in the high pressure seabed at low temperatures. Serpentine minerals belong to the silicate minerals class Phyllosilicates.

Serpentine has the appearance of various physical properties. The physical properties of serpentine minerals are green to dark green, exhibiting fractures and commonly encountered with silica minerals, calcites that fill rock cavities [1].

Serpentine contains an economically valuable oxide compound with varying levels in each region. Serpentine is an oxide compound, in which there are forsterite mineral types (Mg_2SiO_4), hematite

(Fe_2O_3), clinoenstatite ($MgSiO_3$), and Quartz (SiO_4) [2]. Serpentine can be used as a raw material for making Forsterite which has high economic value if it is made into nanoparticles.

Forsterite is a member of the olivine mineral group and is the richest of magnesium minerals. Geologically, the forsterite is present in and forsterite frost rocks also present in dolomite marble which is the result of a limestone metamorphosis with high magnesium content.

Nanoparticles can occur naturally or through the process of making by humans. Preparation of nanoparticles is significant for the manufacture of particles of a size less than 100 nm and as well as changing their properties or functions [3]. People generally want to understand more deeply why nanoparticles can have different properties or functions than similar materials in bulk.

The properties that change in nanoparticles are usually related to the following phenomena. First, quantum phenomena as a result of the limitations of electrons and other charge carriers in particles. This phenomenon impacts on some material properties such as transmitted color change, transparency, mechanical strength, electrical conductivity, and magnetization. Second, the change in the ratio of the number of atoms that occupy the surface against the total number of atoms. This phenomenon impacts on changes in boiling point, freezing point, and chemical reactivity. These changes are expected to be superior to nanoparticles compared to similar particles in bulk state. The researchers also believe that we can control those changes in the desired direction.

The making of nanoparticles can be carried out in solid, liquid, or gas phases. The manufacturing process can take place physically or chemically. The physics-making process does not involve chemical reactions. What happens is only the breaking of large materials into nanometer-sized materials, or the incorporation of very small materials, such as clusters, into nanometer-sized particles without changing the properties of materials. The process of making chemically involves the chemical reaction of a number of precursors so that another nanometer material is produced. An example is the formation of salt nanoparticles by reacting corresponding acids and bases [3].

In general, the manufacture of nanoparticles will fall into two major groups. The first way is to break up large particles into nanometer-sized particles.

This approach is called a top-down approach. The second approach is to start from the atoms or molecules or clusters that are formulated to form the desired nanometer-sized particle. Top down is the manufacture of nanostructures by minimizing large materials, while the bottom up is a way of assembling atoms or molecules and combining them through chemical reactions to form nanostructures. An example of a top down method is grinding with a milling device, while bottom-up technology is using sol gel technique, and chemical precipitation [4].

The preparation of forsterite nanoparticles with variation of milling time on the mixing of talc and MgO influenced the formation of forsterite nanoparticles formed by mechanical activation method for 20 hours, 40 hours and 60 hours using ball milling with annealing temperature of 1200 °C obtained by nanocrystalline crystal size of 40 nm. While 5 hours of milling using ball milling with annealing temperature 1200 °C obtained crystal size forsterite 60 nm [1]. One of the milling tools is High Energy Milling (HEM) which is a ball roller tool used to perform small-scale mechanical integration process in the laboratory.

Milling is mechanically a simple and effective method for growing solid crystals (the size of the crystalline grains to be smaller) without going through the vapor phase or chemical reactions, as is usually required in other nanoparticle-making processes [6]. This smoothing machine is able to turn a hard and breakable sample into a powdered analytical sample. The manufacture of forsterite nanoparticles has been using talc and magnesium oxide material with ball mill method. The milling time variations used were 5, 10, 20, 40, and 60 hours, to obtain forsterite nanoparticles with grain size.

The manufacture of forsterite nanoparticles has been done by several researchers, but the material used is a synthetic material. Synthetic materials are expensive and require a long process to make them. Treatment of temperature variations of calcination to serpentine minerals, obtained results found forsterite phase that dominates at a temperature of 800 °C [2].

Serpentine minerals can be used as alternative materials for nanoparticle forsterite makers that are easily available in nature, and do not cost a fortune to make them. This research is conducted by milling time variation on serpentine mineral using High Energy Milling (HEM) method. High energy milling is a unique technique by using the collision energy between the crushing balls and the chamber wall that is rotated and driven in a certain way. The advantage of high energy milling is that it can make relatively large amounts of nanoparticles in a relatively short time [4]. In addition, the milling technique is one of the techniques for growing solid crystals without going through an evaporate phase or chemical reaction treatment as commonly required in the synthesis process in general [2]. This article discusses the process of Mechanical Milling, powder mixed particles will undergo a process of cold welding and repeated destruction, when the ball collide with a number of powders will be trapped between the two balls.

The impact load given by the ball will make the powder deformed and eventually destroyed. The surface of the newly formed mixed powder particles allows the process of cold welding back between the fellow particles to form a new particle size larger than its original size. The particles will then again experience a collision and eventually re-destroyed, and so on until it reaches the nano's size. Mechanical Milling process of material properties also affect the final result. Here are some structures that affect the variation of milling time.

Phases are materials that have structures and compositions different from others. The phase difference that occurs can not be separated from the energy effects of the atoms for the diffusion process. Figure 3 can be seen the mechanism of phase formation. The XRD peak width is a function of particle size, then the size of the crystal (D) is expressed in the following Scherrer equation [3]:

$$D = K\frac{\lambda}{B\cos(\theta_B)} \qquad (1)$$

D is the size (diameter) of crystalline, λ is the wavelength at 1.54 Å, θ_B is the Bragg angle, B is FWHM of the selected peak, and K is the material constant whose value is less than one. Commonly used values for K $\approx$0.9. Scanning Electron Microscope (SEM) was conducted to determine the sample morphology in various fields. The principle is the wave nature of electron diffraction at very small angles. The electrons can be dispersed by the charged samples, for the non-conducting samples coated with carbon, gold or gold alloys, which serve to drain excessive electrons in the sample to ground. The pattern formed illustrates the structure of the sample.

2. Research Methods

This article is a type of experimental research using high energy milling method to make nanoparticles, then investigated the effect of milling time on grain size of forsterite nanoparticles from serpentine mineral found in *Jorong Sungai Padi Nagari Lubuak Gadang Sangir*, South *Solok* District. The elements contained forsterite are analyzed using x-ray diffraction, while the grain size using scanning electron microscope. This research will perform several steps, namely: making nanoparticle, characterization, and data analysis.

3. Results

Table 1 Forterite test results that have been milled with variations of time 5, 10, 20, and 40 hours using XRD detected elements detected silicon elements, Pericles compounds, and forsterite. The result is obtained by using High score Plus software with ICDD database. The x-ray Diffraction test can be used to determine the crystalline size of the test sample. Table 2 can be seen the crystalline size of the serpentine forsterite mineral samples with various time variations of milling on forsterite compounds can be determined using high score plus software.

The result of the measurement is obtained by crystalline size from forsterite phase, for the time of milling for 5 hours up to 10 hours crystalline size of forsterite is increasing, this is because the samples have agglomeration in milling process. The forsterite particles are broken or broken by the balls so that small size then occurs grouping until the time of milling 10 hours this occurs due to the powder that has been solved and small sized compaction which resulted in the powder together and resulted in crystalline from forsterite become large, but at the time milling for 20 hours up to 40 hours of crystalline size from the forsterite shrank back this because the forsterite sample with a milling time variation of 5 hours to milling time for 40 hours of powder has a collision force with balls that have strong energy so that the powder becomes eroded again and become small. In addition to the collision style the powder also experiences other styles such as friction and compression. So at the time of milling for 40 hours from forsterite has reached the point of fracture and the size is smaller again. Grain size can be known by surface morphology obtained from SEM tool on forsterite with variation of milling time 5, 10, 20, and 40 hours. SEM is used to see grain size and surface morphology after milling process.

Table 1 .Fosterite structure with variation of milling time

Milling time	Mineral	Crystal structure							
		a (Å)	a (Å)	a (Å)	α	β	γ	Space group	Crystal system
5 hours	Fosterite	10.219	5.99	4.758	90	90	90	Pnma	Orthorombic
	Periclase	4.42	4.42	4.42	90	90	90	Fm-3m	Cubic
	Silicon	6.636	6.636	6.636	90	90	90	Ia-3	Cubic
10 hours	Fosterite	10.219	5.99	4.758	90	90	90	Pnma	Orthorombic
	Periclase	4.42	4.42	4.42	90	90	90	Fm-3m	Cubic
	Silicon	6.636	6.636	6.636	90	90	90	Ia-3	Cubic
20 hours	Fosterite	10.219	5.99	4.758	90	90	90	Pnma	Orthorombic
	Periclase	4.42	4.42	4.42	90	90	90	Fm-3m	Cubic
									-
40 hours	Fosterite	10.219	5.99	4.758	90	90	90	Pnma	Orthorombic

Table 2. The crystalline size of the forsterite phase

Milling time	Crystalline size (nm)
5 hours	22,43
10 hours	28,72
20 hours	28,32
40 hours	17,76

The result of characterization using SEM for several time milling can be seen in Figure 1. Figure 1 shows that grain sizes of 5, 10, 20, and 40 h of milling time are 579 nm, 478 nm, 451 nm and 385 nm, respectively.

4. Discussion

The milling time affects the size of the crystal grains in the forsterite, this is because during the milling process, the powder particles will undergo repeated welding and repeating process when the ball collides with each other, the powder will be trapped between the two spheres and will result in deformed powder and then disintegrate. The surface of the newly formed powder particles allows the process of cold welding back between fellow particles to form new particles of larger size than their original size. Then the particle will again experience a collision and eventually re-destroyed, and so on until it reaches the nano size.

The morphology of forsterite for all samples is generally the same i.e uneven rounded shape. The results of the research were obtained for foresterite samples whose grain size was smaller than the previous study, which obtained a nanocrystalline grain size of 500 nm [4]. The morphology of the forsterite surface is indicated by the SEM of the forterite grain size, the smoother as the milling time increases. Forsterite forms for all samples are generally equally spherical, this is because forsterite comes from anthropogenic processes. Naturally, anthropogenic processes occur at high temperatures

IOP Conf. Series: Materials Science and Engineering **335** (2018) 012004 doi:10.1088/1757-899X/335/1/012004

in the process and production of Mg-related materials, in these high-temperature processes typically producing rounded minerals.

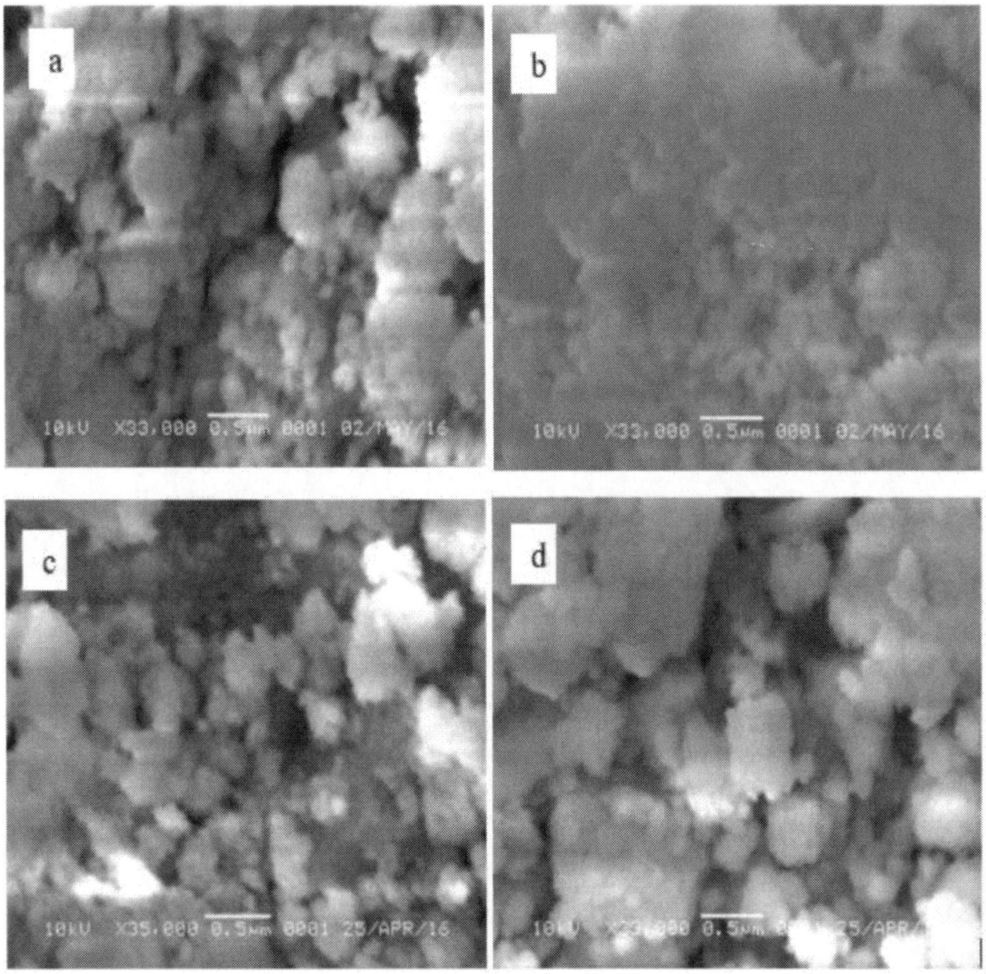

Figure 1. Comparison of SEM forsterite results of various time variations with magnification 33000x a) 5 hours, b) 10 hours, c) 20 hours, and d) 40 hours

5. Conclusion

The grain size of 5, 10, 20, and 40 hour milling time is 579, 478, 451, and 385 nm, respectively. Based on the research that has been done can be drawn conclusion Time milling effect on the grain size of forsterite mineral serpentine from South Solok District, the longer milling time the size of forsterite grains smaller. Optimum milling time to produce nano forsterite is 40 hours with a grain size of 385 nm.

Reference

[1] Tavangarian, F dan R. Emadi. 2010. Synthesis Of Pure Nanocrystalline Magnesium Silicate Powder. *Ceramics-Silikaty*. 54(2). pp 122-127.
[2] Febrini, Vivi.2014. "Pengaruh Temperatur Sintering Terhadap Sifat Fisis Mineral Serpentin Yang Terdapat Di Jorong Sungai Padi Nagari Lubuak Gadang Kecamatan Sangir Kabupaten Solok Selatan".UNP: Padang.
[3] Abdullah, M. 2008. *Sintesis Nanomaterial*. Bandung. ITB Bandung

[4] Kumar, CSSR., Hormes, J., and Leuschner, C. 2005. *Nanofabrication Towards Biomedical Applications*. Wiley-VCH, Weinheim. Germany
[5] Tavangarian. F., & R. Emadi. 2010. Sintesis of nanocrystalline forsterite (Mg2SiO4) powder by combined mechanical action and thermal treatment. *Materials Research Bulletin* 45(4) pp. 388-391
[6] Nurul T.R. 2007. "HEM type E3D".Alat penghancur pembuat fungsional nanometer dan gerakan elips 3 dimensi, patent no p00200700207.

IOP Conf. Series: Materials Science and Engineering 335 (2018) 012005 doi:10.1088/1757-899X/335/1/012005

Early Warning System of Flood Disaster Based on Ultrasonic Sensors and Wireless Technology

W Indrasari[1*], B H Iswanto[1], M Andayani[1]

[1]Physics Department, Faculty of Mathematics and Science, Universitas Negeri Jakarta, Jakarta 13220, Indonesia

widyafisikaunj@gmail.com

Abstract. A flood disaster provides considerable losses to the people who live around the river. To mitigate losses of material due to flood disaster required an early warning system of flood disaster. For that reason, it necessary to design a system that provide alert to the people prior the flood disaster. And this paper describes development of a device for early detection system of flood disasters. This device consists of two ultrasonic sensors as a water level detector, and a water flow sensor as a water flow velocity sensor. The wireless technology and GSM is used as an information medium. The system is designed based on water level conditions in the Katulampa Dam, Bogor. Characterization of water level detector showed that the device effectively works in a range of water level of 14-250 cm, with a maximum relative error of 4.3%. Meanwhile the wireless works properly as far as 75 m, and the SMS transmission time is 8.20 second.

1. Introduction

Indonesia is an archipelago country that has high rainfall in the world. Many areas in Indonesia are prone to flood disaster. Flood disasters could occur anytime and give adverse impacts to society, especially people who lived around the river. For that reason, it necessary to design a system that provide alert to the people prior the flood disaster. The purpose of the warning system is to mitigate number of victims and material losses suffered by the society.

Nowadays, there are many early warning systems of flood have been developed. Komaludin et al [1] used ultrasonic sensor-based microcontrollers to detect maximum and minimum water levels. Sumarno et al [2] developed a flood early warning system based on microcontroller Atmega16 and buzzer, and used short message service as a flood alert indicator.Other method was also developed to improve the effectiveness of information delivery of alert flood indicator due to geographical and distance factors, such a telemetry method using wireless technology [3], and information broadcasting through social media [4]. An early warning system requires a location as a water level reference, it can be either a water gate or a dam, because each water gate or dam has a standby status with different water levels.

In this paper, we describe the development of a prototype flood early warning system using wireless technology and short message service (SMS) as an information broadcasting medium.Three sensors are used to increase a measurement accuracy, there are two ultrasonic sensors which used to measure water level, and a water flow sensor to measure the flow rate of water. We used a Katulampa Dam as a water level reference.

IOP Conf. Series: Materials Science and Engineering **335** (2018) 012005 doi:10.1088/1757-899X/335/1/012005

2. Standby Status of Water Gate

Flood alert status in Jakarta was obtained by information analyzing of the Water-surface level observation stations (TMA) of the rivers that lead to Jakarta (PUSDALOPS BPBD DKI Jakarta, 2016).Each water gate has different warning criteria depend on the depth of the river.

There are four types of warning status: (1) Standby-IVstatus, there has not been a significant increase in the flow of water, (2) Standby-IIIstatus, a condition when the rain causes the water debit increasing at the water gates but the condition is still not critical or dangerous. However, if alert standby III has been officially stated, the public should be alert and prepare everything from the various possibilities of flood disaster, (3) Standby-IIstatus, the condition when the rain causes the flow of water began to expand, (4) Standby-Istatus, the condition when the water-flow remain at standby II condition morethan six hours, or even higher.At Water Monitoring Station of Katulampa-Dam has been defined the water level each status condition: Standby IV-status <79 cm, Standby III-status 80 - 149 cm, Standby II -status 150-199 cm, and Standby I-status ≥ 200cm.

3. Experimental

The developed prototype was composed of two ultrasonic sensors type of HC-SR04as a water level detector sensor, a water-flow sensor type of yf-s201 to measure flow rate, a microcontroller as control system, xbee wireless module as transmitter and receiver, a GSM sim900A module as a short message sender, and buzzer as an indicator of water level.

3.1. Characterization of Ultrasonic Sensor

This characterization was conducted to determine the range of the sensors, and to identify resolution of the ultrasonic sensor.The characterization scheme is shown in Figure 1a. From this measurement, the distance range of both sensors is obtained 14-250 cm with a maximum relative error of 4.33%, meanwhile the minimum distance detected by the sensor is 0.5 cm with a maximum relative error of 1.2%. Furthermore, the optimization of the prototype scale to the water level conditions at each standby status was carried out. From the optimization obtained the best ratio result of prototype is 1: 5. This ratio would be used to adjust the distance of the sensor and water surface.

3.2. Hardware Design

Using the ratio of 1:5, water level on standby-I status would be ≥ 40 cm.Thus, a water reservoir of 30 cm x 40 cm x 50 cm was created. Both ultrasonic sensors are mounted on a 60 cm long-stick, 10 cm above the water-maximum surface. The distance of the sensors to the water surface for each Standby-status is shown in Table 1.

Table 1. Water level on the Prototype

Status	Sensor to water surface distance (cm)
Normal	≥48 cm
Standby-IV	44 cm – 47 cm
Standby-III	43 cm – 30 cm
Standby-II	29 cm – 20 cm
Standby-I	≤ 19 cm

3.3. Designing Software

There are two systems involved in the prototype: monitoring system as a transmitter or data sender system, and a remote warning system as a receiver system. The remote communication uses a pair of module, xbee series 2. For that reason, there are two programs used on this prototype, namely the transmitter program on the monitoring system, and the receiver program on the remote warning system. The flow chart of transmitter and receiver program shown in Figure 1b.

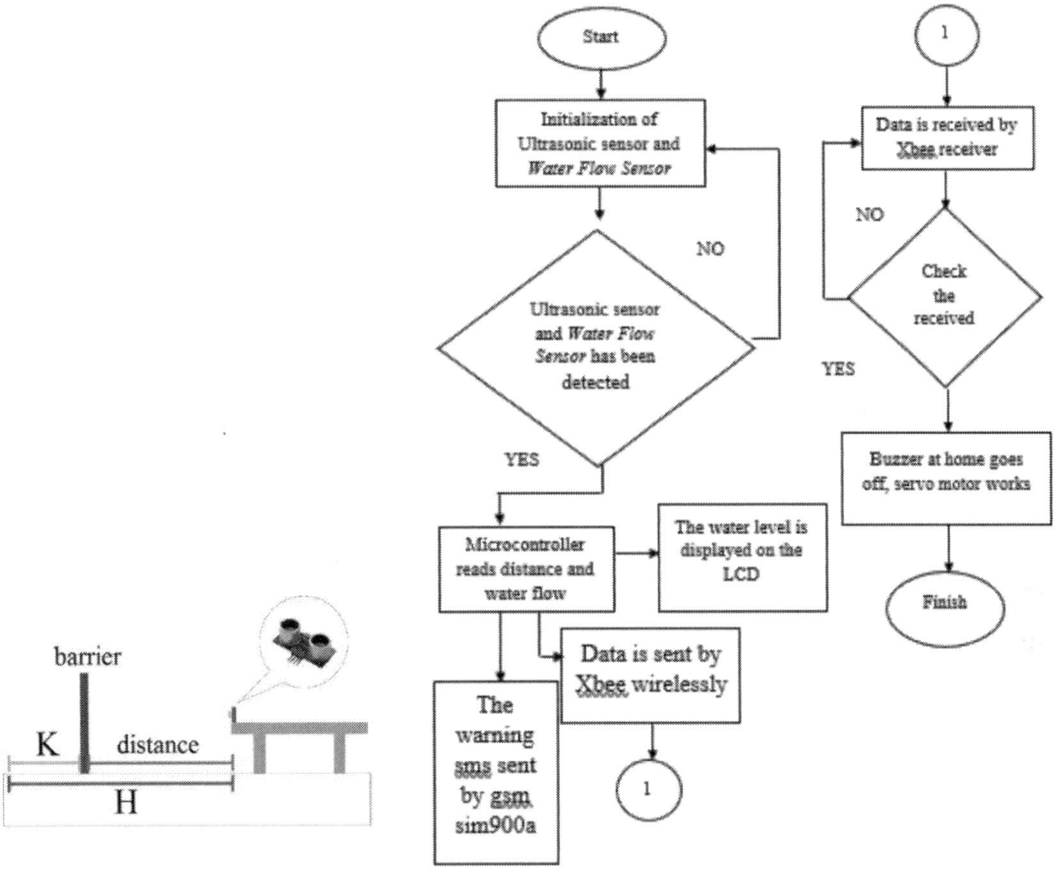

(a) (b)

Figure 1. (a) The characterization of ultrasonic sensor. (b) The flow chart of transmitter and receiver program

4. Result and Discussion

4.1. Measurement of Ultrasonic Sensor

The measurement of ultrasonic sensor system to the water surface of the reservoir is carried out in three different conditions: condition-1 uses a water source. Condition-2 uses two water sources, and condition-3 uses a water source and a sinkhole (moving surface water).Water surface at condition-1 and condition-2 tend to be stable, meanwhile the water surface at the condition-3 is moving and unstable, thus it more resembles the actual condition. The results of the measurements of sensors 1 and 2 at all three test conditions are shown in Figure 2. The actual distance is measured using a ruler as a standard gauge.

IOP Conf. Series: Materials Science and Engineering **335** (2018) 012005 doi:10.1088/1757-899X/335/1/012005

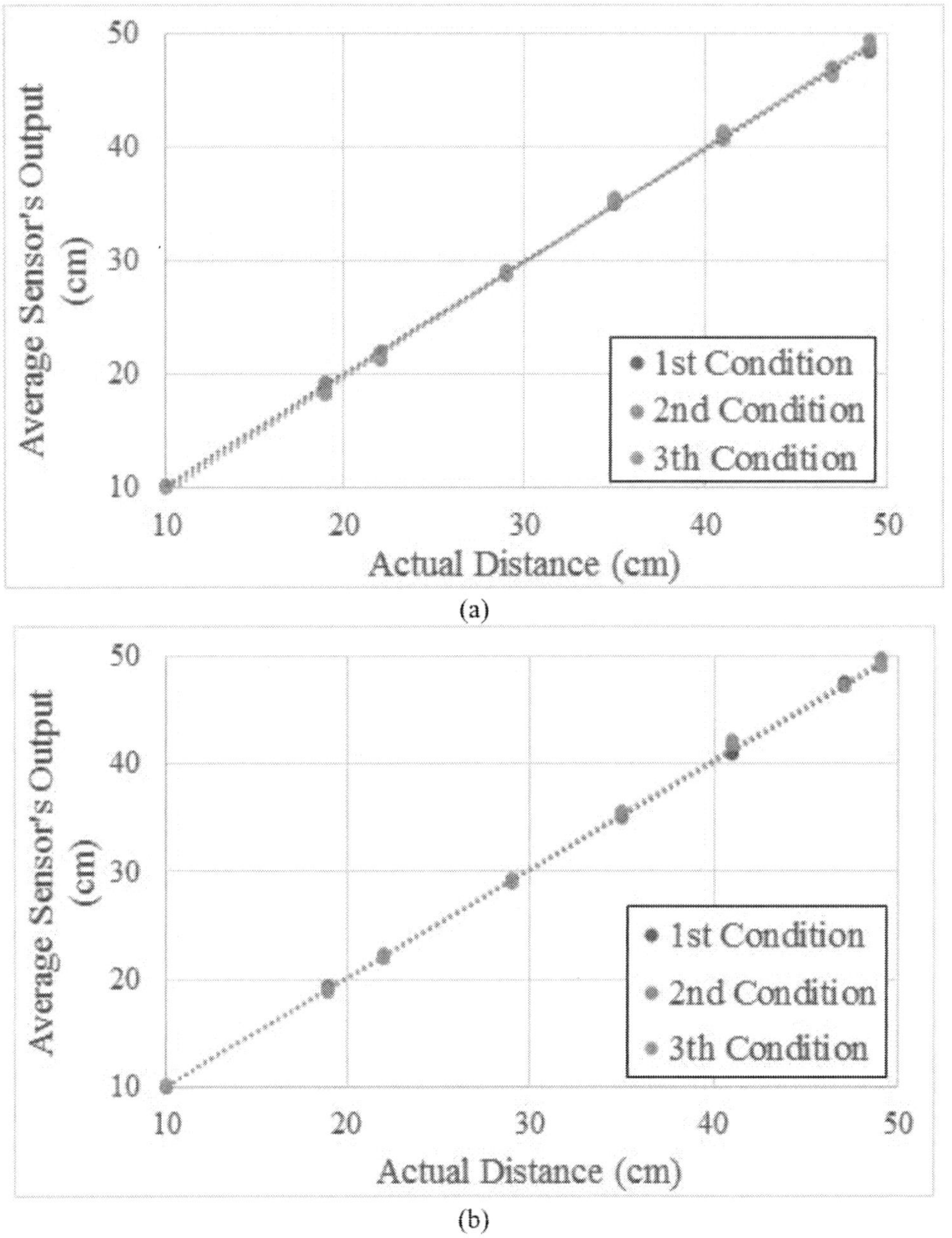

Figure 2. Measurement of ultrasonic sensors at condition 1, 2 and 3. (a) Ultrasonic Sensor 1 (b) Ultrasonic Sensor 2

Figure 2 shown that both sensors can work properly to detect the changes of water level in the reservoir. But the measured deviations on condition-3 is greater than the measurement on other conditions. Because basically the sensor does a better reading when the object surface is flat. Ultrasonic sensor will give the best reading at hard and smooth surface. This surface reflects larger signal and produce a strong echo signal than the rough or soft surface. This condition would reduce either the operating distance of ultrasonic sensor and the accuracy of a measurement [5].

The other factor which affects to sensor measurement is the distance of the sensor to the object. The distance enhancement of the sensor to the object, will prolong the response time of sensor to send the reading result, and it makes the echo signal received by the sensor is getting weaker. Otherwise, the

IOP Conf. Series: Materials Science and Engineering **335** (2018) 012005 doi:10.1088/1757-899X/335/1/012005

closest distance between sensor and object would produce the stronger echo signal. The relative error of the distance measurement at Standby-I status is described in Table 2. It can be seen that in the Condition-3, the moving water surface, the measurement error of sensors is quite small. The overall, using two ultrasonic sensors would improve effectiveness of the sensor reading, as water level readings are more precise, especially when the water surface conditions are unstable.

Table 2. The relative error of sensors reading

Relative error at Standby-I Status	Condition-1	Condition-2	Condition-3
Sensor 1	3.1 %	2.3 %	2.3 %
Sensor 2	2.4 %	2.7 %	1.0 %

4.2. Measurement of Water-flow Sensor

In this prototype water flow sensor is used to provide information of water flow stability. The result of sensor measurement is shown in Figure 3. In Figure 3a show the current water flow decreases by increasing water level. In the other hand, Figure 3b shows that the current water flow is fixed. It means the sensor output tends to be stable in the 36.3 ml - 37.8 ml range. The decrease or increase of the water flow through the sensor depends on the flow rate of the water source. Water flow sensor YF-S201 works based on the Hall Effect phenomenon [6]. The water flow through the sensor will rotate the impeller in the sensor. The rotation of the impeller will create a magnetic field on the coil in the water flow sensor. The magnetic field will be converted by Hall Effect into pulse signal. The pulses are then converted to water flow rates.

The decreasing or increasing of water flow rate reads by the sensor has no effect on the water level change. Since the rate of water flow has no indication to the change of warning status, it just gives an additional information of the water flow rate. If the flow rate of water suddenly rises drastically, and continuously to be of high value, it can be a warning of flooding.

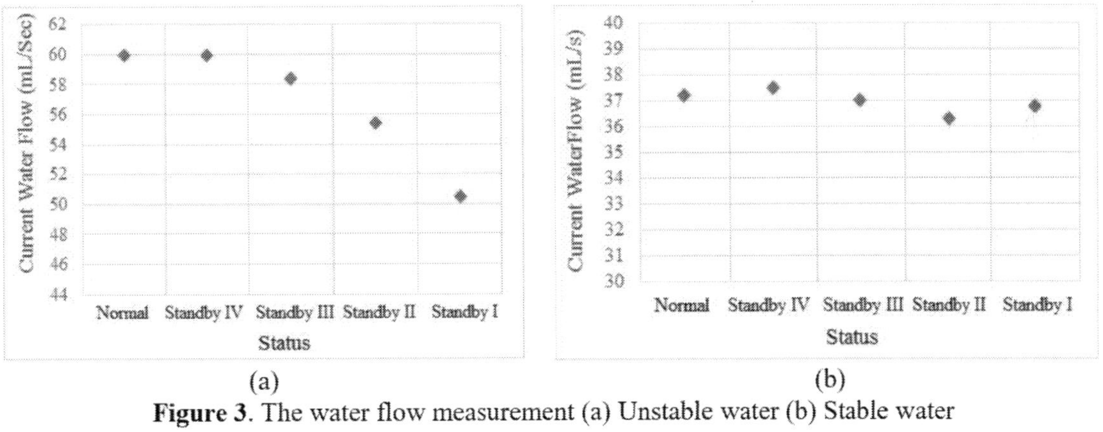

Figure 3. The water flow measurement (a) Unstable water (b) Stable water

4.3. The XBee Module Testing in Receiving Data

The test was done on the two conditions,indoor without barrier and outdoor with a wall barrier. The test results show the wireless capability. The xbee module works well under barrier conditions up to a maximum distance of. As for conditions with a barrier, xbee is only capable of receiving data as far as 24.3 m.

4.4. The GSM SIM900A Testing

The tests were performed on 5 different cell phone numbers with different operators, the message reception speed was calculated from the time the system issued the "A" character on the monitor serial which meant Standby-I.

Table 3. The SMS Speed Delivery Test

Phone Number	Status	Time of SMS delivery (1st delivery)	Time of SMS delivery (2nd delivery)
+6285710619759	Sent	8.2 s	33.3 s
+6281314546611	Sent	14.1 s	9.5 s
+6287785579476	Sent	22.1 s	24.6 s
+6289604192268	Sent	8.5 s	10.2 s
+6285881000626	Sent	17.3 s	16.9 s

The fastest time of receiving SMS based on the test is 8.2 s and the longest time is 33.3 s. The duration of sending SMS was not always constant. It changes any time depending on the speed of the GSM communication line at the time of sending SMS. The phone number receiving the standby message will get an SMS containing the Standby-I status as well as a water level height report when the SMS is sent from a water level monitoring system (Figure 4).

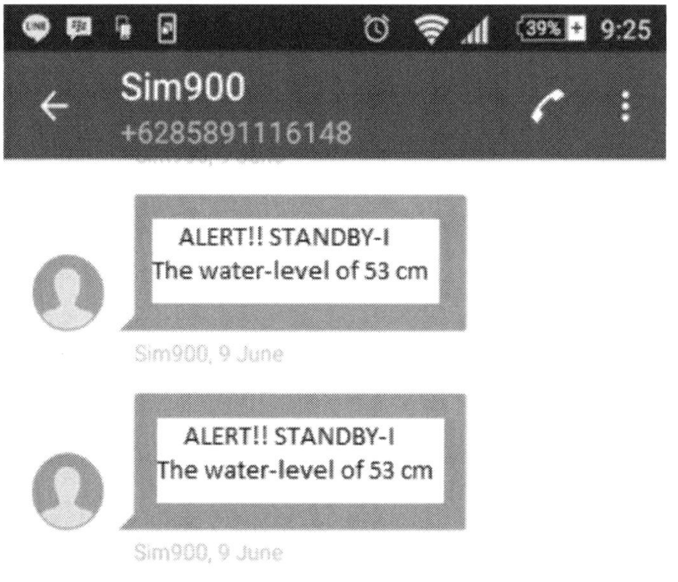

Figure 4. The contents of the received warning SMS

4.5. The Overall System Test Result

The test was performed to determine the overall system performance in accordance with the design software created. Every distance change specified according to the standby status was seen to be appropriately run in accordance with the design of the software.Table 4 shows that the tests performed have been in accordance with the design of the software.

This prototype has high capability work on unstable water level conditions with the Ultrasonic sensor-1 accuracy of 97.6%, and the Ultrasonic sensor-2 accuracy of 97.2%. The using of multiple sensors can improve the accuracy of this prototype, especially at moving water-surface conditions. The incorporation of wireless technology, using xbee and GSM modules, increase the efficiency of warning-message delivery.

IOP Conf. Series: Materials Science and Engineering **335** (2018) 012005 doi:10.1088/1757-899X/335/1/012005

Table 4. The Overall System Test Result

Character sent/received	SMS status	LED	LCD	Buzzer	Automatic gate
A	Sent	Red	Standby-1	On	Closed
B	-	Yellow	Standby-II	On	Open
C	-	Green	Standby-III	Off	Open
D	-	Off	Standby-IV	Off	Open
E/N	-	**Off**	**Normal**	**Off**	**Open**

5.Summary

Development of early warning system of flood disaster that utilizes the wireless technology, using xbee and GSM modules, as a long-distance warning has been carried out successfully. The ultrasonic sensor can work effectively in measure the moving water surface conditions. Meanwhile, the measurement of water flow sensor can be used as additional information, changes in water flow velocity in early warning system of flood disaster.

Acknowledgment

The authors would like to thank to Universitas Negeri Jakarta that has funded this research.

References

[1] Komaludin, Garliaji., (2014), Prototipe Pendeteksi Ketinggian Permukaan Air Maksimum dan Minimum Menggunakan Sensor Ultrasonik Berbasis Mikrokontroler, Jurnal LPKIA, Vol.1 No.1

[2] Sumarno, Irawan, B., Brianorman, Y., (2013), Sistem Peringatan Dini Bencana Banjir Berbasis Mikrokontroler Atmega 16 Dengan Buzzer Dan Short Message Service (SMS), Jurnal Sistem Komputer, FMIPA Universitas Tanjungpura

[3] Susanto, H., Pramana, R., Mujahidin, M., (2013), Perancangan Sistem Telemetri Wireless Untuk Mengukur Suhu Dan Kelembaban Berbasis Arduino Uno R3 Atmega328p Dan Xbee Pro, Jurnal Teknik Elektro, Fakultas Teknik, Universitas Maritim Raja Ali Haji

[4] Attabibi, M L., Husni, M., Ciptaningtyas, H T., (2013), Peringatan Dini Mengenai Tinggi Air Sungai Melalui Media Jejaring Sosial Menggunakan Mikrokontroler, Jurnal TEKNIK POMITS Vol. 2, No. 1, ISSN: 2337-3539

[5] M, Suleiman, etc., (2015), Ultrasonik Fluid Level Measuring Device, International Journal of Research in Science Vol 1

[6] Nugraheni, N, et al., (2014) Efek Hall, Laboratorium Fisika Material, Departemen Fisika Fakultas Sains dan Teknologi Universitas Airlangga Surabaya

IOP Conf. Series: Materials Science and Engineering **335** (2018) 012006 doi:10.1088/1757-899X/335/1/012006

A PIPO Boost Converter with Low Ripple and Medium Current Application

S Bandri*, A Sofian, F Ismail

Department of Electrical Engineering, Padang Institute of Technology
JL. Gajah Mada Kandis Nanggalo Padang, West Sumatra, Indonesia

*sepannurbandria@yahoo.com

Abstract. This paper presents a Parallel Input Parallel Output (PIPO) boost converter is proposed to gain power ability of converter, and reduce current inductors. The proposed technique will distribute current for n-parallel inductor and switching component. Four parallel boost converters implement on input voltage 20.5Vdc to generate output voltage 28.8Vdc. The PIPO boost converter applied phase shift pulse width modulation which will compare with conventional PIPO boost converters by using a similar pulse for every switching component. The current ripple reduction shows an advantage PIPO boost converter then conventional boost converter. Varies loads and duty cycle will be simulated and analyzed to verify the performance of PIPO boost converter. Finally, the unbalance of current inductor is able to be verified on four area of duty cycle in less than 0.6.

1. Introduction

A photovoltaic (PV) power generation is introduced to deploy green energy system on our environment. This system requires battery bank for energy storage and also direct connected toward load. Both methods have been used an energy conversion to deliver electric power. Dc-dc converter regulated voltage about 20,5V from PV panel to 28,8V dc voltage to charge the battery bank. Conventional dc – dc boost converter is usually appliedd for low input voltage to produce high output voltage. Limited voltage and current gain depend on value of inductance and capacitance and also ability of selecting the switching component.

Boost converter has to consider to current and voltage ripple on input and output side, then also voltage stress on switching component. Voltage and current ripple can be eliminated by calculating inductance and capacitance components to maintain value properly that depends on switching frequency. When voltage multiplier cell method and topology are applied on boost converter therefore voltage stress on semiconductor is able be minimized. Interleaved boost converter is also simulated by to eliminate the output ripple on high power application. By coupling inductor in one core also produces high power dc-dc converter.

Input-parallel output-series dc-dc boost converter has been implemented by. The proposed converter achieved wide range input with high gain output by adopting interleaved dc-dc converter for fuel cell application. Synchronous boost converter with input parallel and output parallel (IPOP) provided a current ripple reduction and high efficiency. 400V input and 800V output which were implemented on PIPO bidirectional converter for charging battery bank.

IOP Conf. Series: Materials Science and Engineering **335** (2018) 012006 doi:10.1088/1757-899X/335/1/012006

In this paper, synchronous PIPO boost converter is developed to distribute current that divided by a number of parallel modules. It shows a feasible solution to apply low current MOSFET into medium power of converter for charging battery. It has approximately 20,5V input voltage from parallel connection solar panels which will convert to maximum 28.8V/1kW output converter

2. Proposed Topology and Operational Principle

Figure.1 shows a proposed boost converter that applied for high power with low current ripple. Four-parallel basic-boost converter is expected to reduce current ripple, and reduce voltage stress on switching component. Figure.1 shows a proposed boost converter that applied for high power with low current ripple. Four-parallel basic-boost converter is expected to reduce current ripple, and reduce voltage stress on switching component.

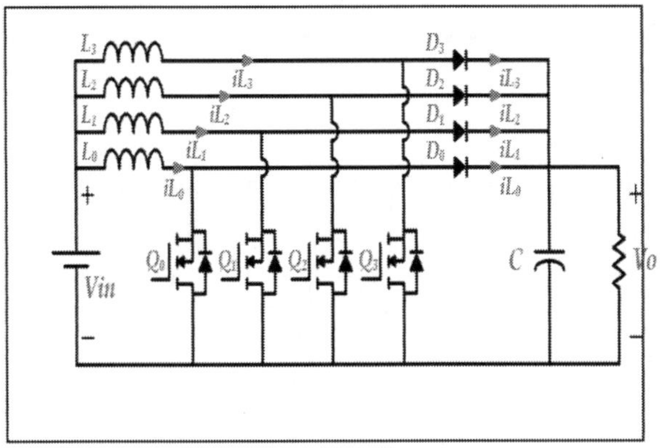

Figure 1. N-Parallel input parallel output boost converter

It refers to basic boost converter applied single inductor, diode, and switching component.Thus, output voltage will describe as,

$$\frac{V_o}{V_i} = \frac{1}{1-D} \tag{1}$$

But n-parallel input and parallel output boost converter are supported for high current input, while the output voltage is similar with output voltage of basic boost converter. A number of parallel boost converters which will distribute current input into n-parallel inductor and switching component. Hence, the current and voltage stress on switching component can be reduced. Relation between input current, output current and duty cycle are explained as,

$$I_{in} = \frac{1}{1-D} I_o \tag{2}$$

$$I_{L1} = I_{L2} = I_{L3} = I_{L4} = \frac{1}{n(1-D)} I_o \tag{3}$$

The switching components of PIPO boost converter operates at sequential time with different starting phase angle. The angle depends on n-parallel of switching components. When four-parallel boost converters are applied on proposed battery charging, therefore the starting angle for every MOSFET has a 90° difference. It was displayed on Figure 2 and Figure 3.

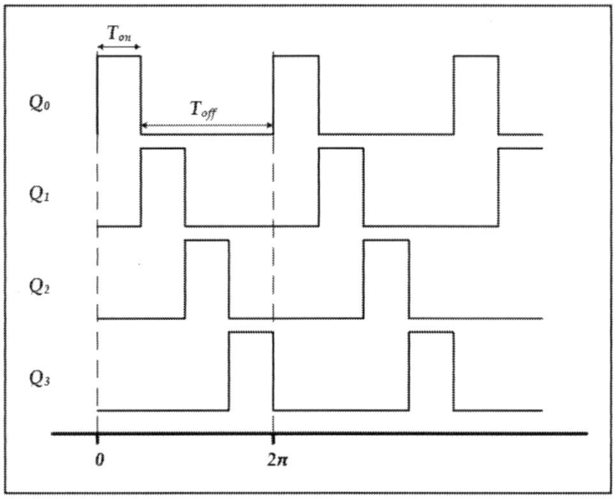

Figure 2. Proposes switching method for PIPO converter

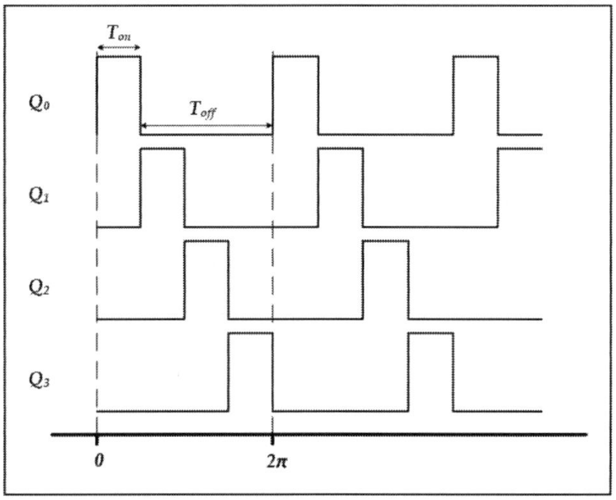

Figure 3. Proposes switching method for PIPO converter

This converter is high current application with output voltage suitable for charging 24 volt battery. To determine the components value of boost converter, parameter calculation and simulation are required. The 20.5 volt input voltage will be converted into 28.8 volt output voltage with power 1kW. Simulation is important to verify all parameters before developing the hardware. In this case, a PSIM software application has been used to analyze PIPO Boost Converter. Figure 4 shows circuit simulation used to PSIM. Similar PWM is applied toward four components switching on PIPO converter. Four-parallel inductor will distribute current and reduce temperature of inductor. Hence, low current MOSFET is able to be implemented for high current converter.

IOP Conf. Series: Materials Science and Engineering **335** (2018) 012006 doi:10.1088/1757-899X/335/1/012006

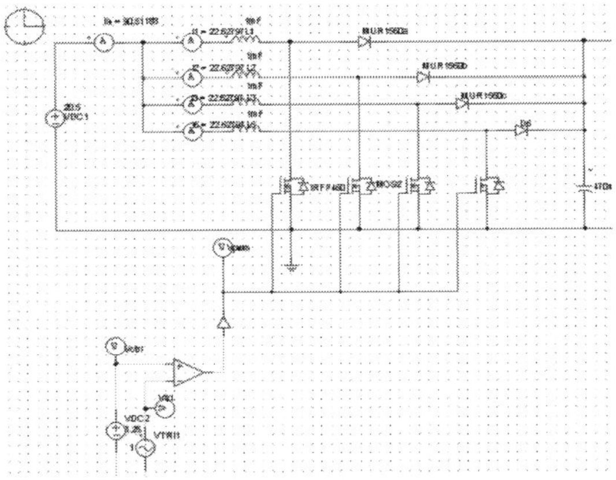

Figure 4. PIPO converter with Similar PWM

Another way, separating drives MOSFET with phase-shift PWM can be implemented to reduce current ripple on output side. Figure 5 displays a circuit PIPO converter with separated switching drive for implementing phase shift PWM.

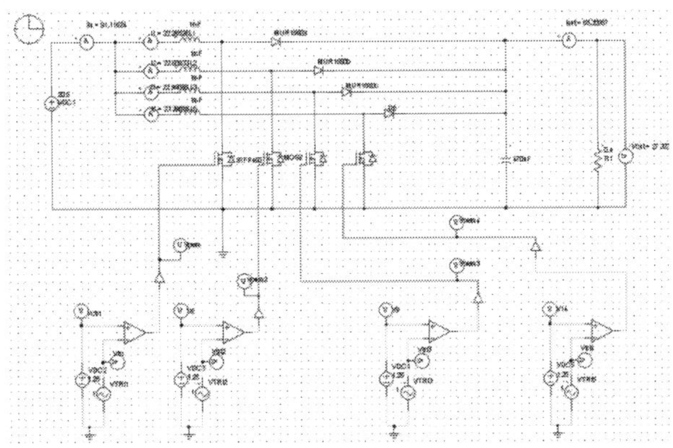

Figure 5. PIPO Converter with phase shift PWM

It refers to [7],the PIPO boost converter will be investigated about current balancing on n-parallel inductor, and also output current ripple. As a better performance in order to get 1kW output with 28.8volt, PIPO boost converter has to have the following specifications.

Table 1. Characteristic of Components

Parameters	Value
Input Voltage	20.5V
Minimum Output Voltage	25.5V
Maximum Output Voltage	28.8V
L1,L2,L3,L4	1Mh
Capacitor	470μf
Resistive Load	0.8 ohm
Diode	MUR1560
MOSFET	IRFP460
Frequency Switching	20kHz

3. Simulation Results

It refers to Figure 4 and specification, PIPO boost converter with single PWM has been simulated. The simulation results displayed on Figure 6. It has been driven by 0.2 duty cycle with similar PWM. The ripple current on inductor L1, L2, L3, and L4 are similar with average 0.8 ampere. While ripple of input current is 3.2 ampere and ripple of output current is 3.5 ampere. The result displayed high current ripple on PIPO boost converter.

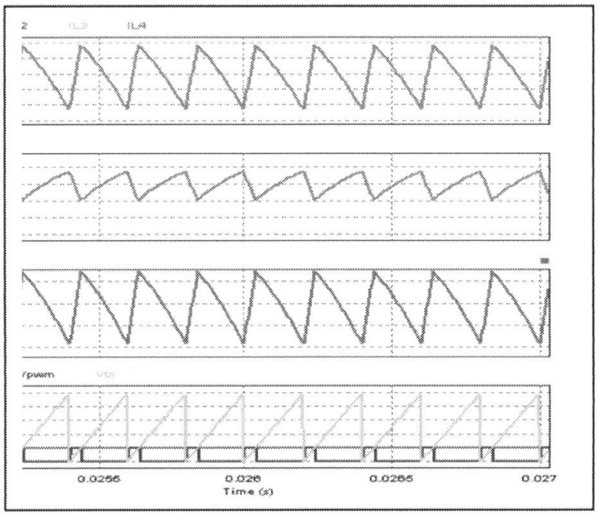

Figure 6. Conventional single PWM PIPO Boost converter on 0.2 duty cycle

Figure 7 explains about simulation of PIPO converter result that applied phase shift PWM. The circuit simulation is displayed by Figure 5 have a four MOSFET as switching component. This MOSFETs will be driven by PWM that explained by Figure 2. The simulation shows a low current input and current output when PIPO converter was applied by 0.2 duty cycle. The average ripple of input current is 0.04 ampere and average ripple of output current is 0.2 ampere.

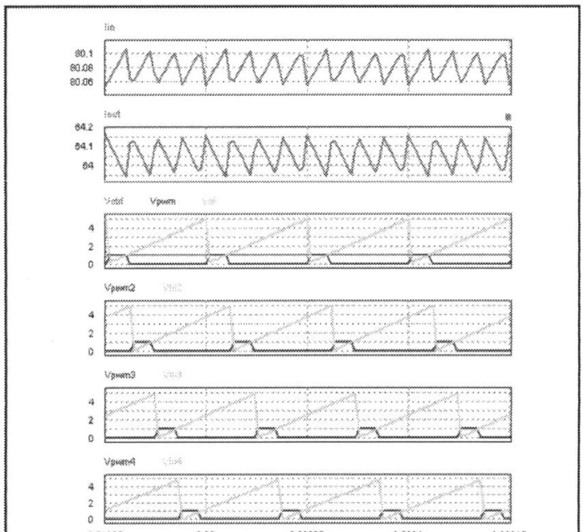

Figure 7. Phase shift PWM PIPO Boost converter on 0.2 dutycycle

IOP Conf. Series: Materials Science and Engineering **335** (2018) 012006 doi:10.1088/1757-899X/335/1/012006

From the results, it can be concluded that proposed PIPO boost converter is more suitable to be applied with phase shift PWM. The better performance with low ripple current input and output was presented on 0.25 duty cycle. It displayed on Figure 8.

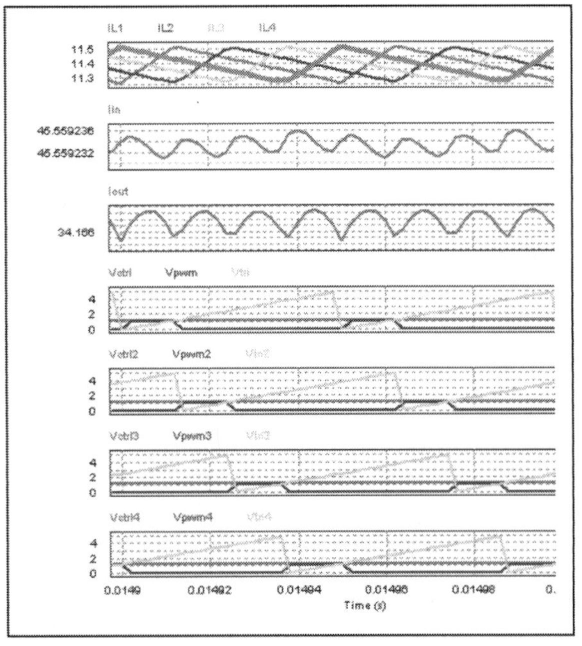

Figure 8. Balance inductor current on 0.25 dutycycle

Another experiment was observed toward PIPO boost converter by implementing variation duty cycle for similar value of load. There were unbalancing inductors current with ripple distortion that occurred during intersection two or more pulses for MOSFET. This cycle repeated during intersection time between pulses. It shows on Figure 9.

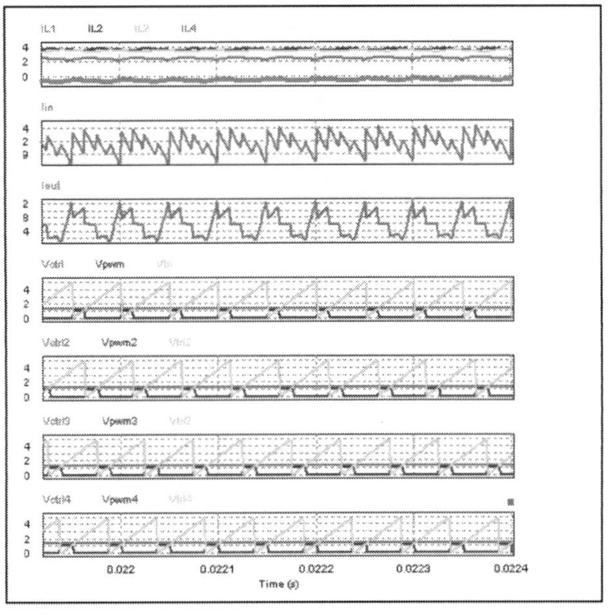

Figure 9. Unbalance inductor current on 0.27 dutycycle

Table 2. Parameter observation by varies duty cycle

No	D	IL1	IL2	IL3	IL4	I_{in}	I_{out}	V_{out}
1	0.218	20.83	20.91	20.99	21.06	83.80	65.60	24.24
2	0.220	28.90	27.13	25.76	2.12	83.90	66.17	26.47
3	0.248	18.70	24.5	24.05	23.75	91.00	68.18	27.27
4	0.250	22.65	22.74	22.8	22.90	91.11	86.33	27.33
5	0.258	22.65	22.73	22.82	22.90	91.11	68.33	27.33
6	0.260	16.40	18.55	43.11	13.90	91.99	68.40	27.36
7	0.288	23.13	23.92	24.56	29.60	101.22	72.00	28.80
8	0.290	25.27	25.43	25.48	25.55	101.64	72.30	28.90
9	0.468	45.00	45.26	45.42	45.30	181.0	96.50	38.60
10	0.470	60.00	52.66	49.36	19.25	181.2	96.87	38.75
11	0.498	45.45	56.30	52.95	51.87	204.6	102.18	40.87
12	0.500	51.00	51.25	51.50	51.25	205.0	102.5	41.00
13	0.508	51.00	51.25	51.50	51.25	205.0	102.5	41.00
14	0.510	47.95	80.80	35.9	42.90	207.63	103.4	41.36
15	0.538	57.96	59.38	66.61	56.70	240.6	111.08	44.43
16	0.540	60.25	60.54	60.80	60.55	242.14	101.67	44.67

The unbalancing of current inductor occurred during transition pulse with duty cycle less than 0.6. There were four unbalance areas verified when phase shift PWM implemented for PIPO boost converter (marked by yellow background). Table 1 and Figure 10 explain about unbalancing inductor current. During PIPO boost converter running on unbalancing of inductor current, power loss occurred on inductor side with over current flow as the result, it increased temperature of inductor.

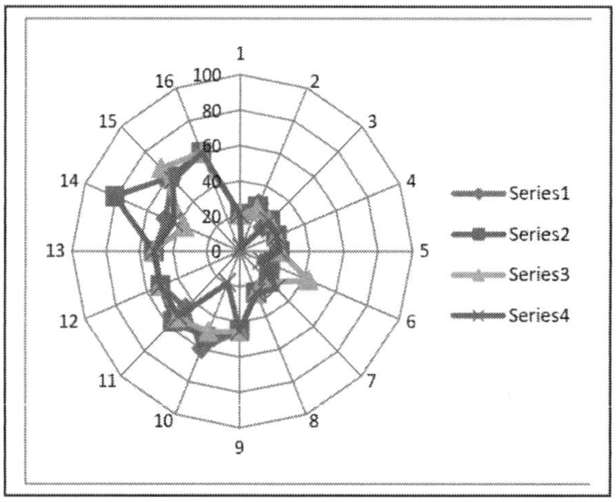

Figure 10. Unbalance inductor current on 0.27 dutycycle

4. Conclusion

PIPO boost converter with a low current ripple was presented in this paper that the Phase shift PWM method used to drive PIPO boost converter was success to reduce ripple current become 1mA as compared with single pulse PWM. The second observation was also concluded that four parallel input and parallel output boost converter would be implemented, as the result, four areas duty cycle would be noticed.

Acknowledgment
I extend a thank you to the team at the laboratory of electrical engineering of the 'Institut Teknologi Padang' who have helped this research run smoothly. I would also like to thank for the '*Kopertis Wilayah* X' of Indonesia that has funded this research.

References

[1] Tohid Nouri,et.all., 2014," Generalized Transformerless Ultra Step-Up DC-DC Converter with Reduced Voltage Stress on Semiconductor", *IET Power Electronics*, vol.7, issue 11, pp.2791-2805.
[2] Po-Wa Lee,et.all., 2000," Steady-State Analysis of An Interleaved Boost Converter with Coupled Inductor", *IEEE Transaction On Industrial Electronics*, vol 47.No.4, pp 787-795, 2000.
[3] Wang, P. et al., 2017. Input-parallel Output-series DC-DC Boost Converter with a Wide Input Voltage Range , for Fuel Cell Vehicles. , 9545(c).
[4] Vazquez, A. et al., 2017. Different Modular Techniques Applied in a Synchronous Boost Converter with SiC MOSFETs to Obtain High Efficiency at Light Load and Low Current Ripple. *IEEE Transactions on Industrial Electronics*, 0046(c), pp.1–1.
[5] Cadirci, I. et al., 2002. Unity power factor boost converter with phase shifted parallel IGBT operation for medium power applications. Electric Power Applications, *IEE Proceedings* -, 149(x), pp.237–244..
[6] Corcau, J.I. et al., 2015. Multisim implementation and analysis of a DC to DC boost converter with two parallel stages. 2015 *International Conference on Electrical Drives and Power Electronics*, EDPE 2015 - Proceedings, pp.88–93.
[7] Sen, G. et al., 2013. Comparison of current balancing configurations for Primary Parallel Isolated Boost Converter. *International Aegean Conference on Electrical Machines and Power Electronics, ACEMP 2011 and Electromotion 2011 Joint Conference*, (September), pp.449–454.

IOP Conf. Series: Materials Science and Engineering **335** (2018) 012007 doi:10.1088/1757-899X/335/1/012007

Adsorption and Pore of Physical-Chemical Activated Coconut Shell Charcoal Carbon

E Budi[1*], U Umiatin[1], H Nasbey[1,2], R A Bintoro[1], Fi Wulandari[1], E Erlina[1].

[1]Physics and Physics Education Program, Faculty of Mathematics and Science, Universitas Negeri Jakarta, Jakarta 13220, Indonesia
[2]School of Science and Technology, Gunma University, Japan.

*esmarbudi@unj.ac.id

Abstract. The adsorption of activated carbon of coconut shell charcoal on heavy metals (Cu and Fe) of the wastewater and its relation with the carbon pore structure was investigated. The coconut shell was pyrolized in kiln at temperature about 75 - 150 ^{0}C for about 6 hours to produce charcoal and then shieved into milimeter sized granule particles. Chemical activation was done by immersing the charcoal into chemical solution of KOH, NaOH, HCl and H_3PO_4, with various concentration. The activation was followed by physical activation using horizontal furnace at 400^{0}C for 1 hours in argon gas environment with flow rate of 200 kg/m^3. The surface morphology of activated carbon were characterized by using Scanning Electron Microscopy (SEM). Wastewater was made by dissolving $CuSO_4.5H_2O$ and $FeSO_4.7H_2O$ into aquades. The metal adsorption was analized by using Atomic Absorption Spectroscopy (AAS). The result shows that in general, the increase of chemical concentration cause the increase of pore number of activated carbon due to an excessive chemical attack and lead the increase of adsorption. However it tend to decrease as further increasing in chemical activator concentration due to carbon collapsing. In general, the adsorption of Cu and Fe metal from wastewater by activated carbon increased as the activator concentration was increased.

1. Introduction

Water contamination from industrial wastewater due to toxic heavy metals content is one of the severe environmental problems [1]. Toxic heavy metals such as Fe and Cu are the main water pollutant due to undegradeble biologically into environment. However, agricultural waste by product can be used as an absorbent [2]. Coconut shell charcoal, as an agricultural waste, with elements content of Carbon (45.03%), Hydrogen (6.94%), Oxygen (47.47%) and ash (0.56%) as an agriculture waste product has been used as effective and inexpvensive adsorbent of heavey metal wastewater. Furthermore, the production of activated coconut shell charcoal carbon consume less electrical energy, human toxicity and global warming thus the process is friendly to environment [3]. Naturally, the chemical composition of coconut shell consists of lignin (29.35%), cellulose (24.20%), hemicellulose & others (38.56%), moisture (7.7-3%) and ash (0.68%) [4]. The shell carbon content increases up to 76.32 % as the shell was pyrolized into charcoal [5]. The carbon structure of cocout shell charcoal has high porosity and large surface area and this properties are important in adsorption process.

The coconut shell charcoal can be activated by physical and or chemical activation. The chemical reagent The chemical activation was performed by impregnating the charcoal into activating regeant such as KOH, NaOH, HCl and H_3PO_4 [6]. The reagents were reactive in opening the pore wider. The chemical activation has advantages than physical activation due to its higher yield lower activation temperature. The process can reduce drastically empty space and lower weight loss thus increase packing density and proportion of mesopores [7]. The physical activation can tailor the pore size distribution more accuratly but lower packing density. However the process is more friendly to enviroment since it use no any toxic chemical solution. The granular activated carbon can be used as excellent alternative, efective and inexpensive absorbent to remove high amounts of heavy metal such as Fe and Cu from waste water [8, 9].

The objective of this paper is to investigate the adsorption of activated coconut shell charcoal carbon on heavy metal of Fe and Cu of waste water. The pore and surface morphology of activated carbon were also discussed.

2. Experiment

The coconut shell as raw material was collected from natural waste of local markets in Jakarta. The shell was cleaned from coir and then dried under sun light for about three days. The coconut shell charcoal was produced by pyrolysing 100 kg shell in the kiln at temperature about 75-150 ^{0}C for about 6 hours. After the process, the coconut shell was converted into coconut shell charcoal. The charcoal then was cleaned by using aquades and dried by using furnace at temperature of 120 ^{0}C for about 12 hours. After that, the charcoal was sieved by using mortar to form milimeter size granule and cleaned again by using aquades before cleaned by using ultrasonic cleaner in 96% alcohol at room temperature for 30 minutes and dried by using hot plate for 60 minutes.

The chemical activation was performed by immersing the charcoal for 24 hours into KOH solution with various concentration of 30, 40, 50 and 60% ; NaOH solution with various concentration of 1%, 2%, 4%, 7% and 11%; HCl and H_3PO_4 solutions with both various concentration of 2%, 4% and 6%. After that the charcoal then washed by using aquades and dried by using hot plate for 3 hours. All the charcoal samples was then physically activated by using horizontal furnace at fixed temperature of 400 ^{0}C in argon gas environment with its pressure of 200 kg/m^3 for an hour.

The surface morphology and pore structure of activated carbon was characterized by using Scanning Electron Microscopy (SEM) analysis (Jeol, JED-2300). Batch (static) and dynamic experiments were conducted at room temperture to study the adsorption performance. Two wastewater containing 5 mg/lit $CuSO_4.5H_2O$ and $FeSO_4.7H_2O$ were used for the adsoprtion performance of 3 gram sample of activated carbon in 25 ml of the wastewater. In batch experiment The adsorption of activated carbon on Fe and Cu wastewater was analyzed by using Atomic Adsorption Spectroscopy (AAS).

3. Results and discussion

3.1. Pore and surface morphology of activated carbon

The pore and surface morphology of activated carbon at various KOH and NaOH activator concentration are showed in Figure 2 and Figure 3, respectively. While, the pore and surface morphology of activated carbon H_3PO_4 activator is showed in Figure 4. It shows that in general, the pores are distributed on the carbon surface and its distribution is influenced by chemical agent concentration.

It shows that the chemical agent attact the charcoal carbon surface and develop suplemental pores and internal surface. The process leads to the increase of the pore number and size [5]. The agent also can react with hydrocarbon and develop some pores in microcrystalline carbon of charcoal [10]. However, the pore number then decrease at further increasing chemical agent concentration due to the incipient carbon collapsing after excessive chemical agent attact [11]. The pore size increase as the chemical activator concentration is increased lead the decrease of micropore number and increase of the mesopore and macropore numbers. At the same time, the distribution of pore size become narrower and wider [12].

During activation process the residual organic molecules such as hydrocarbon was dehydrated by the chemical activator and was prevented to be deposited on the carbon surface. The hydrocarbon was removed during the physical activation at high temperature and inert gas condition [11,13]. The chemical activator also inhibit the formation of volatile matter, tar and other liquid that might deposit in the pore [14].

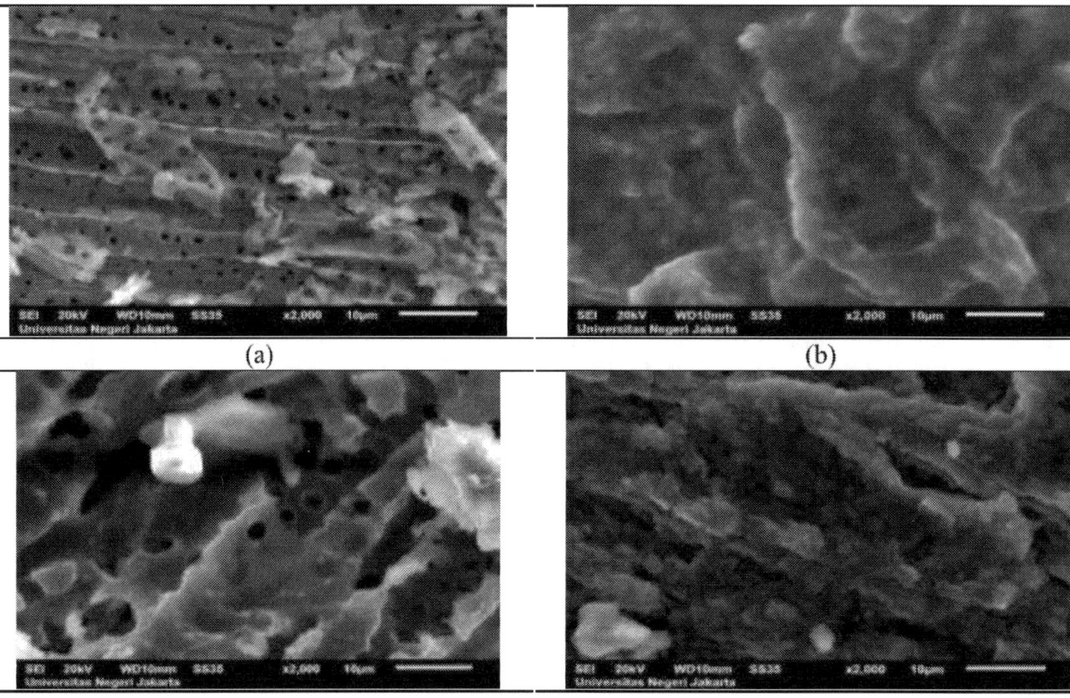

(a) (b) (c) (d)

Figure 1. Pore and surface morphology of activated carbon at various KOH activator concentration of (a) 30%, (b) 40%, (c) 50% and (d) 60%.

(a) (b) (c)

Figure 2. Pore and surface morphology of activated carbon at NaOH activator concentration of (c) 4%, (d) 7% and (e) 11%.

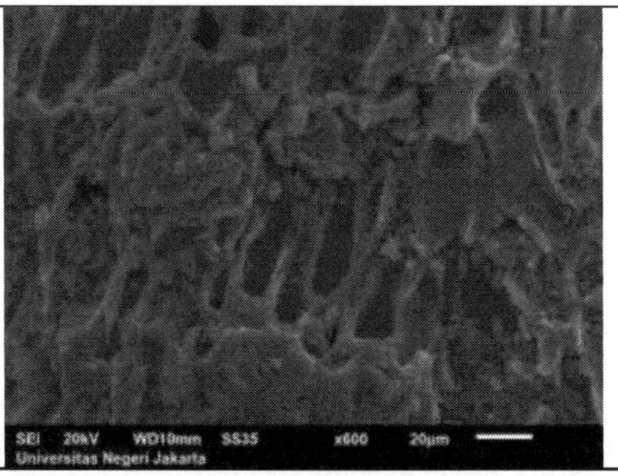

Figure 3. Pore and surface morphology of activated carbon at H_3PO_4 activator concentration of 4%.

3.2. Adsorption of activated carbon

The experimental results on the heavy metal adsorption of activated carbon are shown in following tables. Generally, the adsorption increase as the chemical activator concentration is increased.

Table 1. AAS analysis for adsorption of Cu and Fe metals by samples activated at various KOH concentration. Initial Cu and Fe concentration is 2.4512 ppm and 2.3613 ppm, respectively.

KOH concentration (%)	Cu concentration (ppm)	Fe concentration (ppm
30	0.4703	0.6809
40	0.4615	0.5831
50	0.4027	0.704
60	0.6626	0.5839

Table 2. AAS analysis for adsorption of Cu and Fe metals by samples activated at various NaOH concentration. Initial Cu and Fe concentration is 2.4512 ppm and 2.3613 ppm, respectively.

NaOH concentration (%)	Cu concentration (ppm)	Fe concentration (ppm)
1	0.4687	1.8032
2	0.5212	2.124
4	0.5316	1.6942
7	0.5399	1.6405
11	0.4677	1.8770

Table 3. AAS analysis for adsorption of Cu metals by samples activated at various HCl concentration. Initial Cu concentration is 2.4512 ppm.

HCl concentration (M)	Cu concentration (ppm)
2	0.2862
4	0.0786
6	0.0276

Table 4. AAS analysis for adsorption of Cu metals by samples activated at various H_3PO_4 concentration. Initial Cu concentration is 2.4512 ppm.

H_3PO_4 concentration (M)	Cu concentration (ppm
2	1.0488
4	1.386
6	1.6127

The adsorption process depend on the cation charge density and diameter of hydrated cation where the large diameter of ion has minimum adsorption [8]. Its known that the cation charge density of Fe^{+2} and Cu^{+2} is the same but both ions have different atomic radius which radius of Cu^{+2} is larger than that of Fe. However, since the pore size of activated carbon in this investigation is about 50 – 125 nm that bigger than metal ion radius size thus both Fe and Cu metal ions should be easy to be penetrated into the pore. In this experimental results, however, the efficiency of Cu heavy metal adsorption is higher than Fe.

The driving force of metal adsorption mechanism onto the activated carbon depend on the solubility and affinity of adsorbate [15]. It may be due to the bonding type between metal ion and activated carbon. In general the adsorption capacity of Cu is higher than Fe, thus it might be that the Fe ion is easier to be deposited causes the adorption process is not completely.

4. Conclusion

The physical-chemical activation on coconut shell charcoal can evolute and modify the pore structure and eventually improve its adsorption capacity on the heavy metal such as Fe and Cu. In general, the adsorption of activated carbon on Cu heavy metal is higher than that of Fe heav metal.

Ackowledgement

The authors gratefully acknowledge research facilities support from Laboratorium of Material Physics, Physics and Physics Education Programs, Faculty of Mathematics and Science, Universitas Negeri Jakarta. This research was financialy supported by DIKTI-Ministry of Education with contract No. 21/UN 39.9/PL/Hibah Bersaing/III/2014.

References

[1] Ibrahim, W. M. Hassan, A. F., Azab, Y. A. 2016 *Egyptian J. Basic and Appl. Sci.* 3 pp 241–249
[2] Hegazi, H.A. 2013 *HBRC Journal* 9 pp 276–282
[3] Arena, N., Lee, J., Clift, R. 2016 *J. Cleaner Prod.* 125 pp 68-77
[4] Liyanage, C.D. and Pieris, M. 2015 *Procedia chem.* 16 pp 222-228
[5] Mozammel, H.M., Masahiro, O., Bhattacharya, S.C. 2002 *Biomass and Bioenergy* 22 pp.397–

IOP Conf. Series: Materials Science and Engineering **335** (2018) 012007 doi:10.1088/1757-899X/335/1/012007

400
[6] Budi, E. et al. 2016 *AIP Con. Proc.* 1712 050003
[7] Prauchner , M.J., Reinoso, F.R. 2012 *Microporous and Mesoporous Mat.* 152, pp 163–171
[8] Goher, M.E., Hassan, A.M., Abdel-Moniem, I.A., Fahmy, A.H., Abdo, M.H., El-sayed, S.M. 2015 *Egyptian J. Aquatic Res.* 41, pp 155–164
[9] Borhan, A., Abdullah, N.A., Rashidi, N.A., Taha, M.F. 2016 *Procedia Eng.* 148, pp 694–701
[10] Wu, L., Zhang, X., Liu, D., Peng, H., Long, T. 2014 IERI Proc. 9 33–42
[11] Fierro, V., Fernandez, V.T., Celzard, A. 2007 *Microporous and Mesoporous Mat.* 101, pp 419–431
[12] Babuponnusami, A., Velmurugan, S., Sundararaman, T.R. 2011 *Int. J. Res. Chem. Environment*, pp 135-141
[13] Khah, A.M., Ansari, R. 2009 *Int. J. Chem. Tech. Res.* 1 pp.859-864
[14] Anisuzzaman, S.M., Joseph, C.G., Taufiq-Yap, Y.H., Krishnaiah, D., Tay, V.V. 2015 *J. King Saud Univ. Sci.*
[15] Banerjee, S., Mukherjee, S., LaminKa-ot, A., Joshi, S.R., Mandal, T., Halder, G. 2016 J. *Adv. Res.* 7 pp.597–610.

IOP Conf. Series: Materials Science and Engineering **335** (2018) 012008 doi:10.1088/1757-899X/335/1/012008

Characterization of Silica Sand Due to The Influence of Calcination Temperature

R Ratnawulan[1*], A Fauzi[1], A E S Hayati[1]

Physics Department, Universitas Negeri Padang, Indonesia

*ratnawulan@fmipa.unp.ac.id

Abstract. The characterization of natural silica from *Tanah Datar* region has been done due to variation of calcination temperature. Characterization is done using X-Ray Fluorescence (XRF) and X-Ray Diffraction (XRD). The results of the XRF test show that the content of silica content is 88.957%. Based on XRD test it is found that temperature increase causes structural changes in natural silica. At a temperature of 26 ° C the silica has a quartz phase and in the temperature range of 800 ° C to 1000 ° C there is a new phase, cristobalite. As for the temperature of 1100 oC appears a tridymite phase. Furthermore, the higher the given temperature causes of the crystal size of the crystal be small. In addition to phase and crystal size, there is also a change of lattice parameters.

1. Introduction

Silica (SiO_2) has wide application in various technology applications. Utilization of silica is most widely used for raw materials manufacture glass and glass industry, cement mixture, detergent manufacture, welding coating, cleaning and electrolyte. The last few years the utilization of silica and calcium is made in nano size into bioactive materials for repair of bone tissue. Other applications in the industry are for the production of pigments, pharmaceuticals, ceramics and catalysts, waterglass manufacture, as welding coatings, cleaning, electrolysis, hard cement manufacture, heat-resistant cement, building cement and acid-resistant cement [1,2,3].

Each country has a different silica concentration. The difference is in accordance with the state of the natural structure of each country. Quartz sand known as white sand is one of the weathered rocks that contain the main mineral of silica. Thus, the natural state and weathering processes that occur is one cause why the concentration and silica structure is different.

Materials that are heated to a certain temperature will affect the structure of the material. When a material is heated at a fixed rate of warming there will be a physical change such as phase change [4]. Silica has various crystalline forms such as α-quartz, β- quartz, α-cristobalite, β-cristobalit and tridymite [5]. Other forms are stishovite and fluorite, but rarely appear. With the many structures that have made silica widely used in the industrial world.

Temperature plays an important role in the structural change of silica. At room temperature (± 28 ° C) the silica will be amorphous crystals. At 200-300°C temperature, the cristobalite phase is seen. At a temperature of 163°C, the tridymite phase is observed. Silica glass has an amorphous crystalline structure at a temperature of 500-600°C [3]. At temperatures of 800-900° C, it is quartz and when the temperature is increased to 1000 ° C it will change to tridymite and critobalite. This structure difference is due to the presence of a silica-forming material that depends on the occupied region.

Temperatures in addition to causing a change in phase also cause changes that occur in crystal size and lattice parameters.

This article discusses the structural changes of natural silica due to variation of calcination temperature. This information is important to know the emergence of new phases in the valuable silica economically at a certain temperature.

2. Methods

2.1. Sample Preparation:

The sample used is silica sand material obtained from *Tanah Datar* District, Indonesia. The resulting silica sand is separated from the impurities of the garbage, before being smoothed the quartz sand first, the quartz sand smoothing is done using a bowl mill. After smoothing the sieving is done with an automatic sieve. The sample is heated using the furnace. The crushed quartz sand is heated using 800^{0}C, 900^{0}C, 1000^{0}C, 1100^{0}C temperatures with a duration of 2 hours.

2.2. Characterization

The refined silica sand is characterized using XRF with type Epsilon 3 to determine the content and the element content of the quartz sand. Then the heated sample was characterized using XRD Type CubiX3 Cement (CuKα 1,5606 Å, 40 mA, 45 kV) to determine the characteristics of the silica oxide in terms of the phase and crystal structure that emerged. The process of identifying phase and structure is obtained from search match using high score plus program. The crystal size was calculated using the Scherrer formula, $D = 0.9\ \lambda\ /\ \beta\ \cos\theta$, where λ was the wavelength of X-ray radiation, β was the full width at half maximum (FWHM) of the peaks at the diffracting angle θ [6].

3. Results and Discussion

3.1 Silica Sand Concentration

The result of the measurement obtained from X-Ray Fluoresence (XRF) on silica sand from *Tanah Datar* District, Indonesia is in the form of peak intensity of element with potential energy. The XRF test results show that the silica concentration is 88.957%. The silica content of this area is different from other regions due to different natural conditions and weathering processes and the amount of impurities in the soil and rock layers of the area [7].

3.2 Effect of temperature on silica structure

3.2.1 Phase

Based on XRD data analysis for each sample it can be seen that the variation of the calcination temperature influences the natural silica phase that is formed. The silica phase formed for each temperature is shown in Figure 1.

Figure 1 illustrates the effect of temperature on changes in the natural silica phase. At temperature 26 ° C silica is quartz. At a temperature of 800 ° C there appears a new phase of cristobalite beta at an angle of 50.40 with a hkl [331] and an angle of 64.10 with a hkl plane [034]. At a temperature of 900 ° C there appears to be a change in the silica phase in which the crystal quartz is missing or does not appear. Temperature 1000°C visible cristobalite beta phase appears at an angle of 43.13 with the hkl plane [222]. When the temperature of 1100 °C occurs the addition of a new phase of tridymite alpha with hkl [311] and re-emerges the quartz crystal field [022] in the diffraction pattern. To see see the phase formed can be seen in Table 1.

IOP Conf. Series: Materials Science and Engineering **335** (2018) 012008 doi:10.1088/1757-899X/335/1/012008

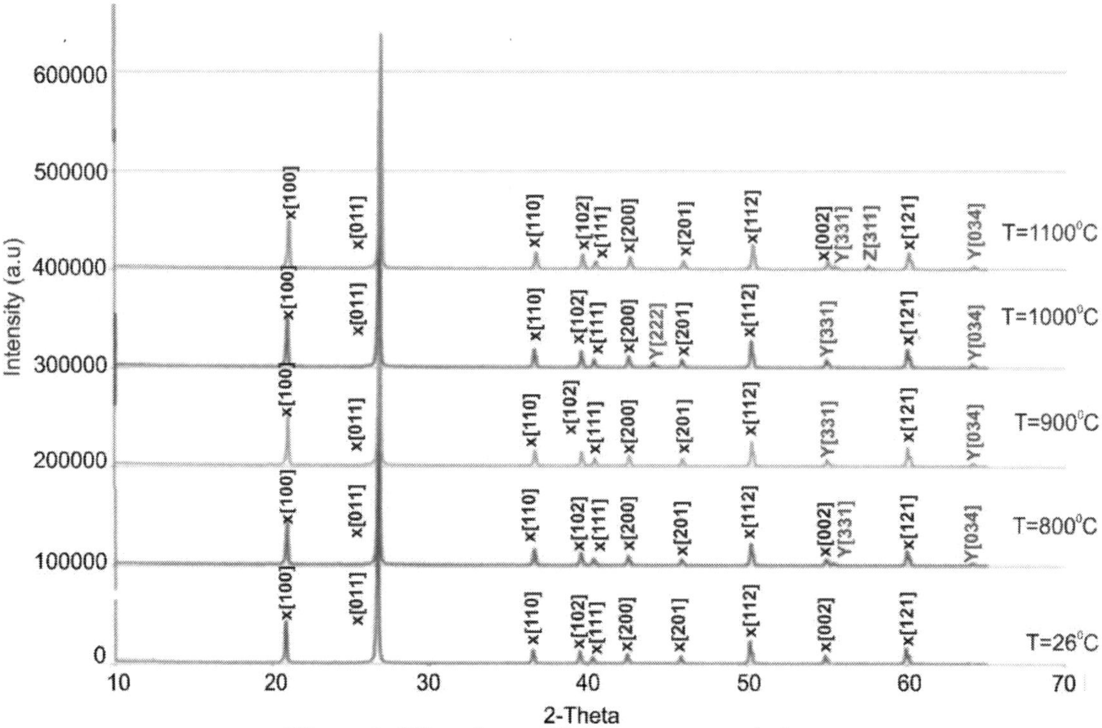

Figure 1. Silica phase with temperature variation

Table 1. The silica phase formed on each sample

Temperature	Quartz phase	Cristobalite phase	Tridymite phase
26 °C	√	-	-
800 °C	√	√	-
900 °C	√	√	-
1000 °C	√	√	-
1100 °C	√	√	√

In Table 1, the quartz phase appears at each temperature. For the cristobalite phase begins at Temperature 800 ° C, the tridymite phase appears only at 1100 ° C. In accordance with the silica phase diagram, the beta tridymite will appear at elevated temperatures. Temperature can affect the growth of silica crystals, such as the emergence of new phases. The effect of temperature on phase change on silica can be seen from figure 2. At room temperature of 26^{0}C phase that appears only quartz phase. This phase arises because at room temperature the silica is in an amorphous state [5]. At normal temperatures quartz will be in quartz trigonal structure (α-quartz) and then turn into hexagonal β-quartz at temperature 573 °C. In this case, the emergence of new phase began to be seen at 800^{0}C temperature in the presence of cristobalite phase at the angle of 55.4^0 and 64.1^0 as well as the temperature 900^{0}C. At a temperature of 1000^{0}C arise trydimite. Based on the phase diagram of silica, at temperature 1000^{0}C will appear trydimite. In certain circumstances tridimite does not appear at low temperatures because tridymite is not an original form of β-quartz. So that tridymite can be replaced by β-quartz [1]. Silica under atmospheric pressure has three crystalline forms: quartz, tridymite and cristobalite. Tridymite appears at temperatures of 870^{0}C to 1470^{0}C [5].

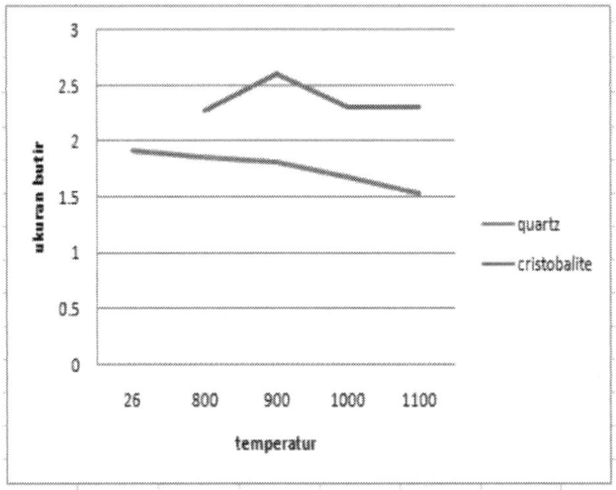

Figure 2. Relation temperature and crystal size

Figure 2 shows only two silica minerals quartz and cristobalite, whereas tridymit is not clearly influenced by temperature because tridymite only appears at 1100^{0}C only. Quartz crystal size can be seen that its value decreases with increasing temperature. For cristobalite the size of the crystal is raised at 900^{0}C and then decreases as the temperature is raised.

Seen in Figure 2 the size of the quartz crystals decreases as the temperature increases, the higher the given temperature the greater the FWHM value the smaller the crystal size becomes. The higher the intensity formed the smaller the crystal size of the crystal. This is reinforced by the X-ray diffraction pattern where as the temperature of the diffraction peak increases, it is higher and sharper. The cause of temperature affect the crystal size is due to rising temperatures so the diffusion of atoms is also increasing, thus increasing the formation of the same new phase nuclei. Between these nuclei there is a crystal boundary that separates it and is separated by pores. Temperature is continuously erected, causing the reduction / removal of the pores tesebut. When the pores are so small they can no longer lock the crystal boundaries against the effect of the crystal growth forces [4]. Increasing heating temperatures result in smaller crystal size.

3.2.2 Lattice parameters

In the XRD measurement data that has been matched with JCPDF data, the composite phase of quartz sand is obtained. It also obtained a crystal structure which includes lattice parameters (α, β, γ, a, b, and c), group space, crystal system contained in silica minerals from quartz sand. Changes in crystal structure due to temperature variation can be seen in Table 2 for quartz silica minerals.

Based on Table 2 it is seen that the lattice parameters of silica with temperature variations. Where at temperature 26°C the lattice parameter of quartz a = b = 4,9124 Å, c = 5,4039 Å with hexagonal crystal form, for lattice parameter of cristobalit a = b = c = 7,4170 Å with cubic crystal form. At 800 ° C the lattice parameters of quartz a = b = 4,9019 Å, c = 5,3988 Å, with hexagonal crystal form, for lattice parameters of cristobalite a = b = c = 7,2679 Å with cubic crystal form. At 900°C the lattice parameters of quartz a = b = 4,8915 Å, c = 5,3885Å, with hexagonal crystal form, for lattice parameters of cristobalite a = b = c = 7,2679 Å with cubic crystal form. At temperature 1000°C, the lattice parameters of quartz a = b = 4,900 Å, c = 5,400Å. At the temperature of 1100°C the lattice parameters of quartz a = b = 4,8715 Å, c = 5.3885Å, lattice parameters of critobalit a = b = 5,1190 Å, c = 7,1683 Å with tetragonal crystal form whereas for tridymite the lattice parameter is a = 18,4940 Å b = 4,9910 Å and c = 25,832 Å with its monoclinic crystal form. Increased temperatures cause the phase change of a material, it also causes the structure of the material also to change. The structural changes that occur are hexagonal structured quartz phases with lattice parameters a = b = 4.9124 Å c = 5.4039 Å which dominantly shrinks as temperature increases. Likewise with cristobalite lattice parameters that begin a = b = c = 7.4170 Å tends to decrease with increasing temperature.

Table 2. The lattice parameters of silica to temperature

T (°C)	Phase	Lattice paramater							
		a (Å)	a (Å)	a (Å)	α	β	γ	Space group	Crystal system
26	Quartz	4.9124	4.9124	5.4039	90	90	120	P3221	Hexagonal
	Cristobalite	-	-	-	-	-	-	-	-
	Tridymite	-	-	-	-	-	-	-	-
800	Quartz	4.9124	4.9019	5.3988	90	90	120	P3121	Hexagonal
	Cristobalite	7.2679	7.2679	7.2679	90	90	90	P213	Cubic
	Tridymite	-	-	-	-	-	-	-	-
900	Quartz	4.8915	4.8915	5.3885	90	90	120	P3221	Hexagonal
	Cristobalite	7.2679	7.2679	7.2679	90	90	90	P213	Cubic
	Tridymite	-	-	-	-	-	-	-	-
1000	Quartz	4.9000	4.9000	5.4000	90	90	120	P3221	Hexagonal
	Cristobalite	7.2679	7.2679	7.2679	90	90	90	P213	Cubic
	Tridymite	-	-	-	-	-	-	-	-
1100	Quartz	4.8715	4.8715	5.3885	90	90	120	P3221	Hexagonal
	Cristobalite	5.1190	5.1190	7.1683	90	90	90	P41212	Tetragonal
	Tridymite	18.494	4.9910	25.832	90	117	90	Cc	Monoclinic

4. Summary

In Summary, the concentration of silica from *Tanah Datar*, Indonesia is 88.957%. Increased calcination temperature affects the structure of natural and emerging phases of silica. At a temperature of 26 ° C the silica has a quartz phase and in the temperature range of 800°C to 1000°C there is a new phase, cristobalite. As for the temperature of 1100°C appears tridymite phase. In addition, the increase in calcination temperature affects the size of the crystalline silica. The higher the given temperature the smaller the crystal size of the crystal. In addition, there is also a change of lattice parameters of quartz phase hexagonal structures with lattice parameters a = b = 4.9124 Å c = 5.4039 Å which dominantly becomes smaller as temperature increases. Likewise with cristobalite lattice parameters that begin a = b = c = 7.4170 Å tends to decrease with increasing temperature.

Acknowledgment

The authors thank to RISTEK DIKTI for financial support through *Hibah MP3EI* 2016-2017 for this work.

References

[1] Zhongkui H Liu A Li C and Xuesi C 2009 Preparation of bioactive glass ceramic nanoparticles by combination of sol–gel and coprecipitation method. *Journal of Non-Crystalline Solids* 355 pp. 368–372
[2] Nozawa K H Gailhano L.Raison P Panizza H and Usiki 2005 *Langmuir* 21 pp.1516-1523.
[3] Chang H and Sun S Q 2014 Silicon nanoparticles: Preparation, properties, and applications *Chin. Phys. B* 23(8) p.088102
[4] Wongduan M Supon A Tawee T Prasak T and Sukon P 2001 Effect of Calcination Temperature on Phase Transformation and Particle size of Barium Titanate Fine Powders Synthesized by the Catecholate Process *Science Asia* 27 pp.239-243
[5] Silvi B and D'Arco P 2002 *Remodeling of material and silicate material.*USA Cluwer academic phublisher.
[6] Riazian M and Bahari A 2012 *Int. J.Nano Dimens* 3(2) pp.127-139.
[7] Teguh P 2009 Survei Potensi Pasir Kuarsa Di Daerah Ketapang Propinsi Kalimantan Barat. *Jurnal Sains dan Teknologi Indonesia* 11 pp. 126-132.

IOP Conf. Series: Materials Science and Engineering **335** (2018) 012009 doi:10.1088/1757-899X/335/1/012009

An Investigation of Seismicity for the West Sumatra Region Indonesia

S Syafriani

Departement of Physics, Faculty of Mathematic and Natural Sciences, Universitas Negeri Padang, Indonesia

syafri@fmipa.unp.ac.id

Abstract. The purpose of this research was to investigate the seismicity of the West Sumatra region in the coordinates area of 94° E – 104° E and 2° N - 4° S. Guttenberg-Richer magnitude-frequency relation and seismic risk have been computed. Historical data of earthquakes used from year of 1970 to 2017 with magnitude higher than 4. The study area was divided into 8 sub-regions based on seismotectonic characteristics, plate tectonic and geological models. The determination of seismotectonic characteristics was based on the level of seismic activity in a region (a value) and rock stress condition (b value). High a value was associated with high seismic activity, whereas high b values were associated with low stress rock conditions, and vice versa. Based on the calculation results, a and b values were obtained in the interval of 5.5-11.3 and 0.7-2. The highest b value was obtained in the sub region 5 (Nias islands), while the lowest b value was obtained in sub region 7 (the Mentawai islands). The sub region 7, Mentawai Islands was indicated as the seismic risk potential areas.

1. Introduction

The Western Region of Indonesia is part of the Eurasian plate and also an active tectonic area since it is at the confluence of several large and small plates [1]. The collision of two large plates, the Indo-Australian and the Philippine plate controlled by the Eurasian plate led to a change in the strong lateral variations that reflect dynamic processes in the past to the present. The collision between the Indo-Autralian and Eurasian plates varies annually, for example near Sumatra about 60 mm / year and 78 mm / year in the eastern Sunda arc [2]. In addition, the Indonesian archipelago is an unique physiography, trenches, arc-trench, gaps, gravity anomalies, volcanic arcs and a series of young mountains with characteristics of earthquakes depth spreads along the subduction zone, high seismicity as well as sequences active volcanoes.

West Sumatra became part of the dynamics of the plate structure area. This is characterized by the high level of seismicity in this region and the presence of a series of active volcanoes in the land area. Besides, this region is also traversed by active fault that formed along Sumatera Island, which is also the source of the cause of earthquake. Major earthquakes have been recorded several times in West Sumatera, in 1926, 1943 and 2007, in the form of shallow earthquakes in land area, while the 2009 earthquake was sourced in the ocean, which has the potential to generate tsunamis [3].

The high intensity of the earthquake that occurred in this region of West Sumatra, requires an understanding of the tectonic conditions of this region. Based on the earthquake that happened in the

IOP Conf. Series: Materials Science and Engineering **335** (2018) 012009 doi:10.1088/1757-899X/335/1/012009

past, it can be figured how is the condition of local tectonic stress and seismic activity of this region. Also the amount of energy that still contained in the rock can be determined.

The local stress condition of the rock in an area can be determined by calculating some values of seismotectonic parameters. Characteristics of seismotectonic parameters can be identified by looking at the spatial distribution of b-values and a-values in the region based on the frequency-magnitude distribution relationship (FMD). The value a represents the seismic activity level of a region, whereas the b-value reflects the tectonic conditions associated with the region's stress conditions. Based on spatial distribution of low-a value means low seismic activity and indicates the presence of energy accumulation (asperity) [4], otherwise for a high-a-value. Whereas the spatial high-b values are associated with low rock stress conditions [5], [6]. High b-values have high heterogeneity medium conditions, but low b-values correlate with high stress rock conditions and have low heterogeneity medium [7]. A region with high stress, stress has a b-value of 0.4-0.9, while under low stress conditions the b-value is about> 1.2 [8]. This seismotectonic parameter value information is used to investigate seismicity at region of West Sumatra.

Based on the information obtained from these seismotectonic parameters can be understood how the current condition of seismotectonic parameters of West Sumatra region. The current seismotectonic condition of West Sumatra is seen by comparing it with other seismotectonic parameters of the observation period. Another observation period used as a comparison here is the observation period 1980 - 2016 [9].

2. Data and Methods

Investigation of seismicity in West Sumatera is done by looking at the relationship between the seismotectonic parameters of earthquake events. This seimotectonic parameter is obtained from viewing the relationship of frequency-magnitude-distribution relationship that satisfies the equation as follows [9].

$$\text{Log } N(M) = a - b\,M \qquad (1)$$

where N (M) = number of cumulative frequencies of earthquake with magnitude M; a and b constants of seismotectonic parameters, showing the area and length of observation time and M = earthquake magnitude determined on the Richter scale. Statistically, b value indicates the slope of the log N ~ M of the straight-line equation that formed from equation 1. The value of b also indicates the heterogeneity of the medium where the high b value is associated with a low rock stress condition, and vice versa. Meanwhile, the high of a value is relates to seismic activity in a region.

The data used in this study was taken from the USGS/NEIC earthquake catalogues. The study area lies between latitude 2°LU and 4°LS and longitude 94°BT and 104°BT with depth of earthquakes less than 350 km and magnitude bigger than 4.0 SR. The data is also removed from foreshock and aftershock in the decluttering process. In addition, the cumulative number curveof earthquake is also plotted to determine the trend of earthquake events in a matter of years. The value of Mc is determined by plotting the earthquake frequency distribution curve. Meanwhile, spatial distribution analysis of seismotectonic parameters is digested by using gridding technique with a space of grid 0.1° x 0.1° with the expected number of earthquakes for each grid is N = 50.

3. Results and Discussion

For this study, all earthquakes in the periods of 1970-2017 have been used. The amount of data that has been through the process of the cluster into 4728 events. The seismicity of the West Sumatera region are shown Figure 1. The density of seismicity of West Sumatera is most visible around the island of Nias (Zone 5), south of Siberut Island, around the Mentawai Islands (Zone 7), along the active fault zone of Sumatra and also in the subduction path. This high seismic density was caused by tectonic activity beneath Sumatra island. These tectonic activities include subduction zones located in the western part of Sumatra Island and stretching along its coastline, also the active fault activities

IOP Conf. Series: Materials Science and Engineering **335** (2018) 012009 doi:10.1088/1757-899X/335/1/012009

such as the Mentawai fault, and Sumatran Fault. Most of the earthquakes caused by the subduction zone and the active fault of Sumatra.

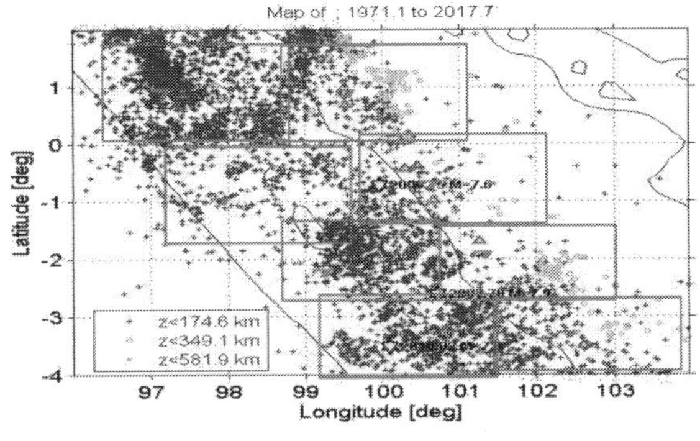

Figure 1. Seismicity of the West Sumatra Region

The tendency of earthquake occurrence in West Sumatera is shown by the cumulative number curve in Figure 2. This cumulative number shows the occurrence of earthquake with time for data that has been done declustering process. The trend of earthquake events increases from year to year, as can be seen in figure 2. In period between 1971 and 2005 there was little change in seismic activity, whereas after 2005 seismic activity changed significantly. Figure 3 shows the earthquake frequency distribution curve in West Sumatera region. The value of Magnitude Completeness (Mc) in the West Sumatra region according to the catalog used is 4.5. This indicates that the earthquake catalog used to record well the earthquake with the smallest magnitude of 4.5 in the region of Sumatra

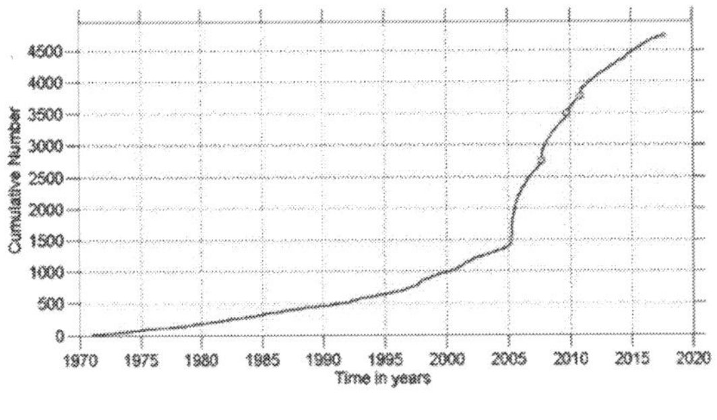

Figure 2. Cumulative number of occurrences of earthquake with time in the period from 1971 to 2017..

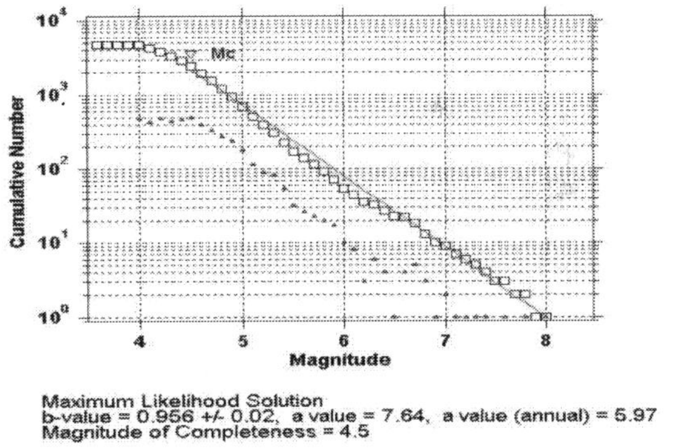

Figure 3. Cummulative number and Magnitude distribution

The spatial distribution of b-values for the observation period in this study, 1971-2017 can be seen in Figure 4. The low-b values was represented by blue. The low b-values are in the south of Nias Island, subduction lines, and south of Siberut Island. These three areas are seismic zones (regime stress) are quite dynamic and has the potential to cause significant and destructive earthquake. A region with a low-b value is an active seismic area that is still locked called a locked zone. Locked zone can someday bring significant and destructive earthquakes. Thus, areas with low-b values as mentioned above need to be aware of significant earthquakes. While areas with a high b-value as seen in the north of Nias Island are referred to as creeping zones, are active fault zones that do not experience energy accumulation and do not experience slippage. This is in accordance with previous research that low b-values are associated with rock stress conditions in the area high, and vice versa.

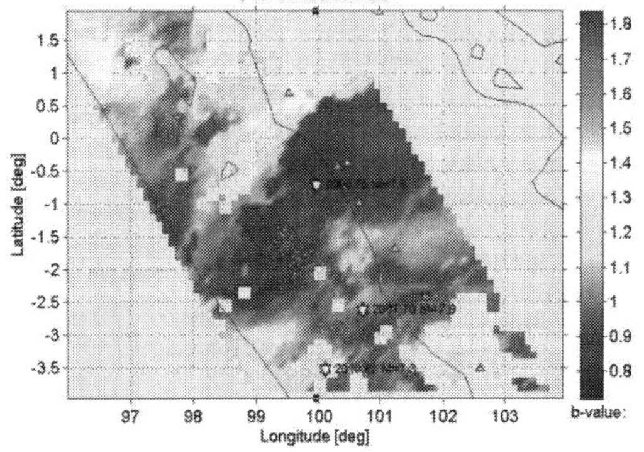

Figure 4. Spatial Distribution Map of b-values in West Sumatra in the observation period from 1971 to 2017.

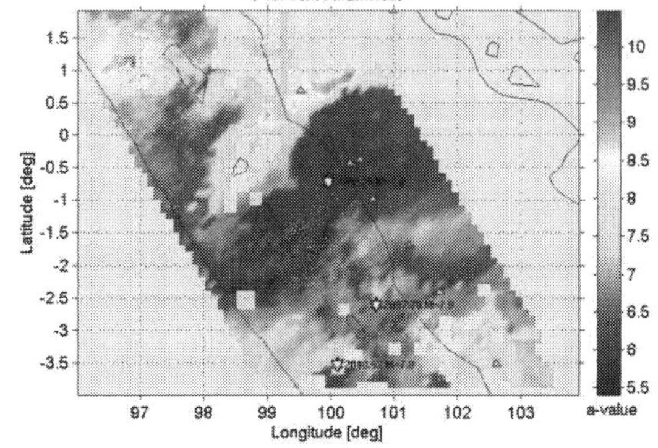

Figure 5. Spatial Distribution Map of a-values in West Sumatra in the observation period from 1971 to 2017.

The spatial distribution of a-values for the observation period in this study 1971-2017 can be seen in Figure 5. The low a-values represented by blue that clearly seen in the area around the subduction path, south of Nias Island, and south of Siberut Island. Areas with low a-value are indicated to have relatively low seismicity. This low level of seismicity indicates the accumulation of energy (asperities) in the area. Low seismicity or low seismic activith is associated with a low a-value spatial distribution. Areas with low seismic activity that potentially significant earthquakes will be happend. Low seismic activity means the accumulation of energy in the region was doing. In west Sumatra region, the area that has high quake level are north of Nias Island. This condition is in accordance with the spatial distribution of earthquake density of West Sumatra as shown in Figure 6 and the distribution of Magnitude of Completeness (Mc) in Figure 7.

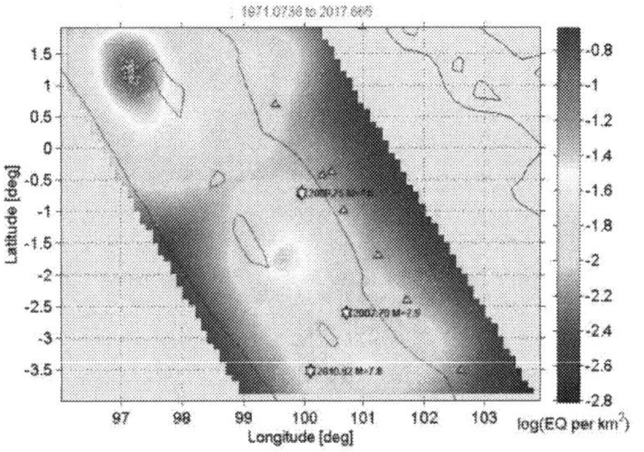

Figure 6. Earthquake density in West Sumatra Region in period time 1971 to 2017

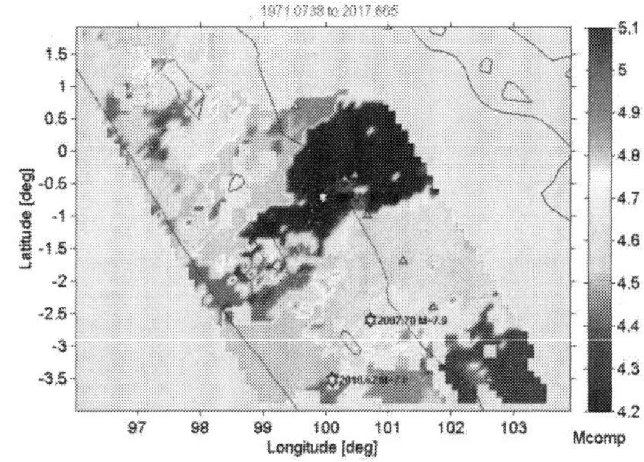

Figure 7. Magnitude Distribution of Completeness (Mc) in the observation period from 1971 to 2017.

The distribution of magnitude completeness (Mc) values in West Sumatera shows that the area around the Nias Island is dominated by an earthquake with magnitude between 4.5 - 4.8 SR. The area around Batu island is dominated by a magnitude 4.8 - 4.9 SR. as well as the area around Siberut Island, the earthquake that often happens with magnitude 4.2 - 4.6 SR. Variations of Mc values were analyzed

IOP Conf. Series: Materials Science and Engineering **335** (2018) 012009 doi:10.1088/1757-899X/335/1/012009

using the position of the recording station used. In addition, other factors that affect the variation of Mc values are real time data from the earthquake recording station.

4. Conclusion

In this paper, the spatial distribution of a-values and b-values in West Sumatra and surrounding areas, proposed. It is found that significant areas of potential earthquake are Siberut Island and Southern part of Nias Island. Although the b-value in this area during this period is still higher than in the previous period. This area still needs to be wary of significant and destructive earthquakes as the overall b-value of this region is still low that mean the potentially significant and damaging eathquake in the future. The distribution of density of earthquake in this area also figured and found that the highest seismic density was Nias Island. This is because in this area often occur small earthquake.

Acknowledgement

This work was supported DIPA Universitas Negeri Padang Tahun Anggaran 2017 no. 1649/UN35.2/PG/2017. We also thank the operators of the BMKG and USGS-NIED data for making their earhquake catalog.

References

[1] Natawidjaja, D.H., and W. Triyoso. (2007) The Sumatran fault zone: from source to hazard, 1 (No.1), 21-47.

[2] Putra, R.R,. Kiyono, J. Shaking Characteristic of Padang City, Indonesia . Graduate School of Engineering, Kyoto University, Japan . Ono Y. Department of Urban Social System and Civil Engineering, Tottori University, Japan . 15 WCEE. LISBOA 2012

[3] Wiemer, S., (2001), A software package to analyze seismicity: ZMAP. Seismological Research Letters, 72(2):373–382.

[4] Katsumata, K. 2011. A Long Term Seismic Quiescence Started 23 Years Before The 2011 of The Pacific Coast Of Tohoku Earthquake (M=9.0). Earth Planet Space. 63.709-712.

[5] Sunardi, Bambang dkk. 2013. Analisis Periodesitas dan Perubahan Tingkat Kegempaan Wilayah Jawa Barat Berbasis Statistik. Conference Paper. BMKG Pusat.

[6] Rohadi, dkk. (2007). Variasi Spasial Seismisitas di Zona Subduksi Jawa. Jurnal Meteorologi dan Geofisika. Vol. 8 No.1 Juli 2007 : 42 – 47.

[7] Scholz, C.H. 1968. The frequency-magnitude relation of microfracturing in rock and its relation to earthquakes, Bull. Seis. Soc. Amer., 58, 399- 415.

[8] Wyss, M., S. Wiemer, and S. Tsuboi. 1973. Mapping b-value anomalies in the subducting slabs beneath Japan, using the JUNEC data, Bull. Seism. Soc. Am., submitted.

[9] Gibowicz, S.J. 1973, Variation of the frequency-magnitude relation during earthquake sequence in New Zealand, Bull. Seimol. Soc. Am, 63, 517 – 528.

IOP Conf. Series: Materials Science and Engineering **335** (2018) 012010 doi:10.1088/1757-899X/335/1/012010

Optical Properties of Fe_3O_4 Thin Films Prepared from the Iron Sand by Spin Coating Method

N Yulfriska[1], D Rianto[1], F Murti[1], Y Darvina[1], R Ramli[1*]

[1]Physics Department, Faculty of Mathematics and Sciences, Universitas Negeri Padang, Indonesia

*ramli@fmipa.unp.ac.id

Abstract. Research on magnetic oxide is growing very rapidly. This magnetic oxide can be found in nature that is in iron sand. One of the beaches in Sumatera Barat containing iron sand is Tiram Beach, Padang Pariaman District, Sumatera Barat. The content of iron sand is generally in the form of magnetic minerals such as magnetite, hematite, and maghemit. Magnetite has superior properties that can be developed into thin films. The purpose of this research is to investigate the optical properties of transmittance, absorbance, reflectance and energy gap from Fe_3O_4 thin films. This type of research is an experimental research. The iron sand obtained from nature is first purified using a permanent magnet, then made in nanoparticle size using HEM-E3D with milling time for 30 hours. After that, the process of making thin film with sol-gel spin coating method. In this research, variation of rotation speed from spin coating is 1000 rpm, 2000 rpm and 3000 rpm. Based on XRD results indicated that the iron sand of Tiram beach contains magnetite minerals and the SEM results show that the thickness of the thin films formed is 25μm, 24μm and 11μm. The characterization tool used for characterizing optical properties is the UV-VIS Spectrophotometer. So it can be concluded that the greater the speed of rotation the thickness of the thin layer will be smaller, resulting in the transmittance and reflectance will be greater, while the absorbance will be smaller. Energy gap obtained from this research is 3,75eV, 3,75eV and 3,74eV. So the average energy gap obtained is 3,75eV.

1. Introduction

In the field of science and technology, one of the most rapidly growing researches today is the study of magnetic oxides. Magnetic oxide can be found in nature that is in iron sand. Magnetic minerals are the main compositions contained in iron sand such as magnetite (Fe_3O_4), hematite (α- Fe_2O_3), and maghemit (γ- Fe_2O_3) [1].

One of the beaches in West Sumatra containing iron sand is *Tiram* Beach in *Padang Pariaman* District of West Sumatera. Where based on the results of XRD characterization shows that iron sand Tiram beach mineral oxide containing iron in the form of magnetite in addition to other oxide minerals. Fe_3O_4 in the form of nano has superior properties so has a great opportunity in the field of industry and electronics for the manufacture of thin films, as semiconductor devices. The simplest and often used method of preparation for growing a thin films is the sol-gel method because it has the advantage of high homogeneity of the material [2].

To grow this thin film there are several methods such as DC magnetron sputtering [3], Screen printing [4], and pyrolysis spray [5]. But this method has the disadvantage of requiring sophisticated instruments, requiring high operating costs, and the small formation area of film produced [6].

Coating method such as spin coating is the easiest and simplest growth method. Because this method has advantages such as can produce a thin films that has a pretty good quality and also the cost of making a relatively cheap.

A thin films is a material of nanometer size formed from the arrangement and incorporation of the atomic properties. Growing technique is by way of growth of particles or atoms on the surface of the substrate with the magnitude of thickness in the order of micrometer. This thin layer is made of metal, inorganic, organic, or organic metal mixture and has the properties of semiconductors, insulators and conductors. The thin films has optical properties that are heavily dependent on the energy gap of the properties of the thin film material to be deposited.

The optical properties of the material are the response form of the material to the electromagnetic especially visible light. The UV-VIS spectrophotometer is used to look at the optical properties of the material, when light is passed on the material then some of this light will be reflected, some will be absorbed, transmitted and scattered, which is mathematically. The incoming light can be written in the equation:

$$I_o = R_e + T + A_b + S = 100\% \quad (1)$$

With I_o is the intensity of light coming, R_e is reflectance, T is transmittance, A_b is absorbance and S is scattered. The energy gap is the amount of energy required to move from the valence band to the conduction band. The energy gap can be searched using UV-Vis diffuse reflectance spectrophotometry based on the intensity measurements that are reflected by the sample. To know the energy gap can be used Kubelka-Munk method. Energy gap is obtained from the graph of the intermediate relationship:

$$h\nu \text{ (eV) vs } (F(R'\infty)h\nu)^{1/2} \quad (2)$$

Eg is an energy gap (eV), h is planck constant ($6{,}626\times10^{-34}$ Js), c is the speed of light (3×10^{8}m/s) and λ is the wavelength (nm). The amount of $h\nu$ at the moment $(F(R'\infty)h\nu)^{1/2} = 0$, Obtained from the linear regression equation is the amount of energy gap [7].

Fe_3O_4 is a material with semiconductor material properties, reported by Ghandoor, H.E, et al, 2012 [8]. Semiconductor materials are materials having electrical conductivity between conductors and insulators with large energy gaps < 6 eV [9].

2. Experimental

This research is a type of experimental research, with the flow diagram shown in Figure 1. The sample used is iron sand samples taken from *Tiram's* Beach in *Padang Pariaman* District, West Sumatra. In this research, the tools used are HEM-E3D, magnetic stirrer, spin coating, furnance, permanent magnet, digital scales, spatula, measuring cylinder, erlenmeyer tube, beaker, stirrer, drop pipette, glass substrate and UV-VIS spectrophotometer. While the material is iron sand, aquabidest, oxalic acid, nitric acid and ethylene glycol.

The shape of the iron sand of *Tiram's* Beach is shown in Figure 2. Iron sand is then purified by pulling iron sand using permanent magnet 30 times to separate iron sand (ferrite) with other material mixed in iron sand (residue). After that iron sand washed using aquabidest. It is then dried and retracted using a permanent magnet 20 times. Then the sample is milling using HEM-E3D for 30 hours. Which shows that at milling time 30 hours visible loss of hematite phase so that only one phase left is Fe_3O_4.

After completion of the process of purification and manufacture of nanoparticles Fe_3O_4. Furthermore, the process of making a thin layer with spin coating is prepared using sol-gel method with precursor ($Fe(NO_3)_3.9H_2O$).

IOP Conf. Series: Materials Science and Engineering **335** (2018) 012010 doi:10.1088/1757-899X/335/1/012010

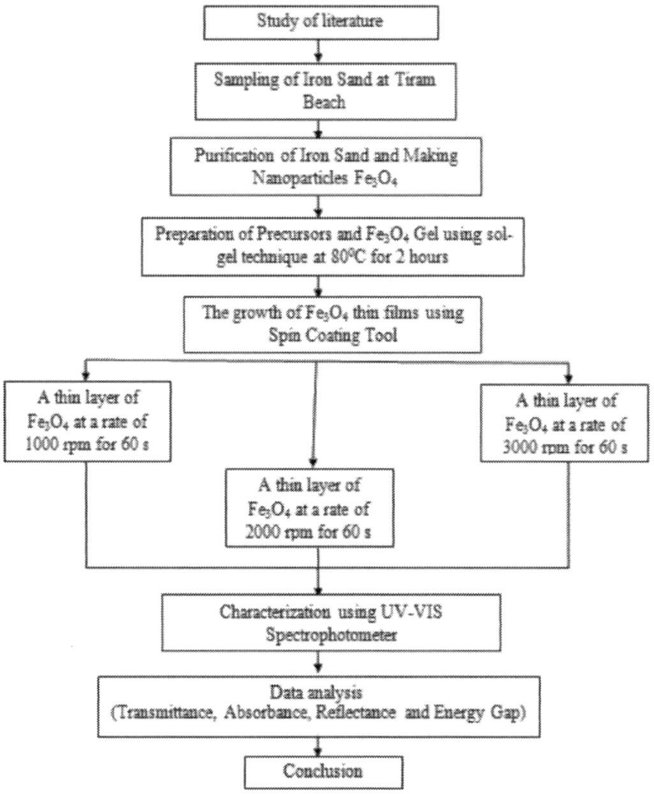

Figure 1. Flow Diagram of Research

Figure 2. Iron sand from *Tiram's* Beach

Precursor ($Fe(NO_3)_3.9H_2O$), prepared at a temperature of 110^0C by reacting 17,4 gr Fe_3O_4 and 4,5 gr oxalic acid ($C_2H_2O_4$) with 42 mL of nitric acid (HNO_3). Furthermore, the process of sol-gel synthesis of magnetite as in [10]. by dissolving 14 gr precursor produced with 55 gr ethylene glycol. The solution is then heated at a temperature of 80^0C for 2 hours while stirring using a magnetic sitrer at a constant speed.

Then a thin films of Fe_3O_4 is grown using a spin coating on a glass substrate that has been using ethanol. The magnetite gel solution is dripped on the glass substrate and then rotated with a spin coating velocity variation. Variations of spin coating speed used are 1000 rpm, 2000 rpm and 3000 rpm with 60 seconds duration. The length of time the coatings were used according to previous research conducted by Eken, 2009 [10]. Furthermore, each thin films formed was dried at 110^0C for 15 minutes, after which it was annealed at 300^0C, then characterized by UV-VIS Spectrophotometer.

3. Result and discussion

Based on the result of characterization of Fe_3O_4 thin film using UV-VIS spectrophotometer, optical properties such as transmittance, adsorbance and reflectance value of the reflectance value can be obtained energy gap. The SEM results show that the thickness of the thin film for the rotation speed of 1000 rpm, 2000 rpm and 3000 rpm respectively is 25μm, 24μm and 11μm.

The results of the transmittance measurements of the Fe_3O_4 thin film at a rotation speed of 1000 rpm, 2000 rpm and 3000 rpm are respectively shown by Figure 3.

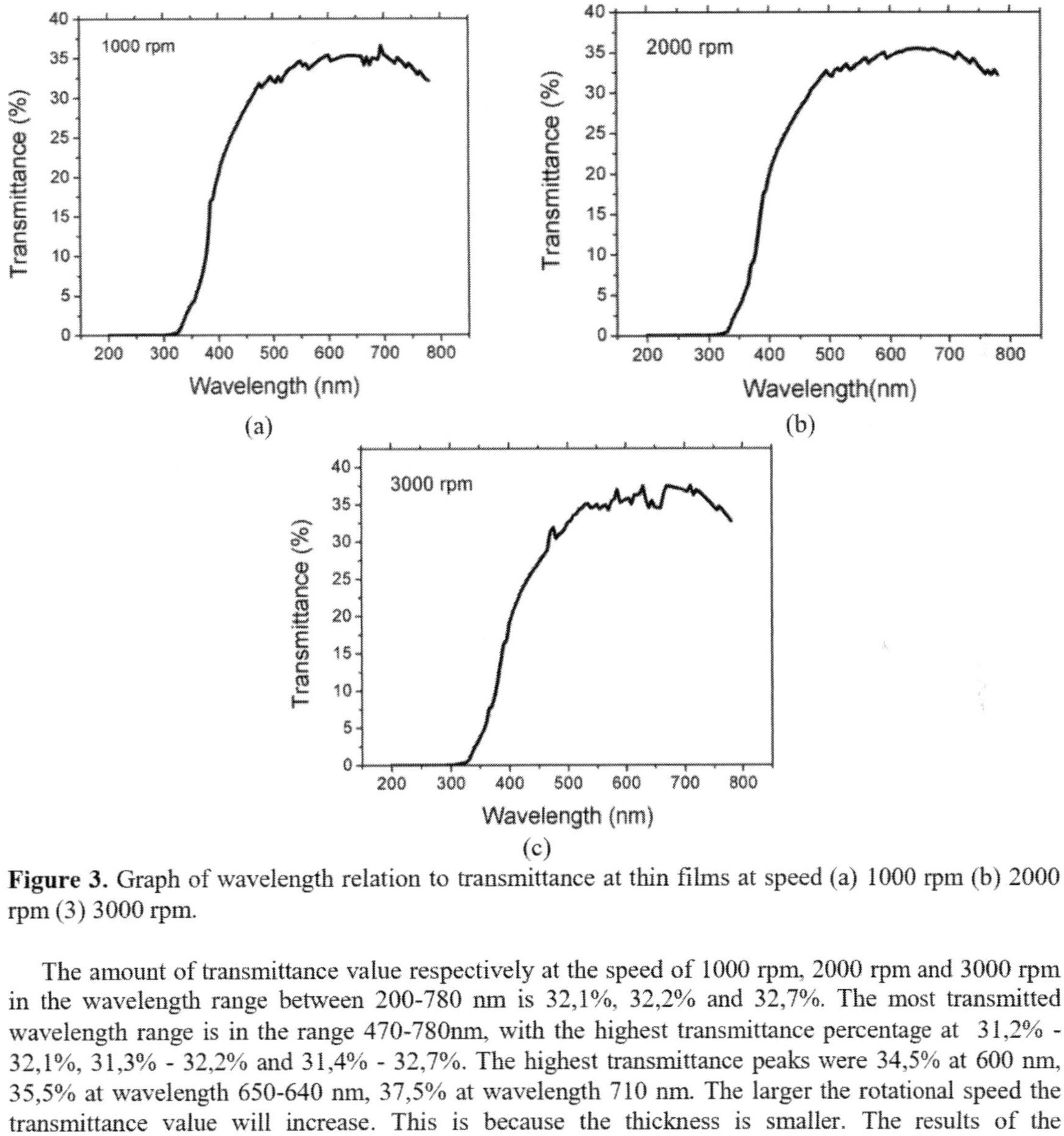

Figure 3. Graph of wavelength relation to transmittance at thin films at speed (a) 1000 rpm (b) 2000 rpm (3) 3000 rpm.

The amount of transmittance value respectively at the speed of 1000 rpm, 2000 rpm and 3000 rpm in the wavelength range between 200-780 nm is 32,1%, 32,2% and 32,7%. The most transmitted wavelength range is in the range 470-780nm, with the highest transmittance percentage at 31,2% - 32,1%, 31,3% - 32,2% and 31,4% - 32,7%. The highest transmittance peaks were 34,5% at 600 nm, 35,5% at wavelength 650-640 nm, 37,5% at wavelength 710 nm. The larger the rotational speed the transmittance value will increase. This is because the thickness is smaller. The results of the absorbance measurements of the Fe_3O_4 thin films by Figure 4.

The amount of absorbance value successively in the wavelength range between 200-780 nm that is equal to 96,721% - 31,247 % , 95,428% - 33,785 % and 93,54% - 33,551%. The most abundant wavelength range is in the range of 200-280 nm, with the highest absorbance percentage of 96,721% - 100%, 95,428% - 95,207% and 93,54% - 93,109%. It can be seen that the absorbance will decrease with the reduced thickness of the thin film. This is in accordance with Beer's Law which states that the

IOP Conf. Series: Materials Science and Engineering **335** (2018) 012010 doi:10.1088/1757-899X/335/1/012010

magnitude of the absorbance value is proportional to the thickness of the material / medium penetrated by light. The result of the reflectance measurement of the Fe_3O_4 thin film is shown in Figure. 5.

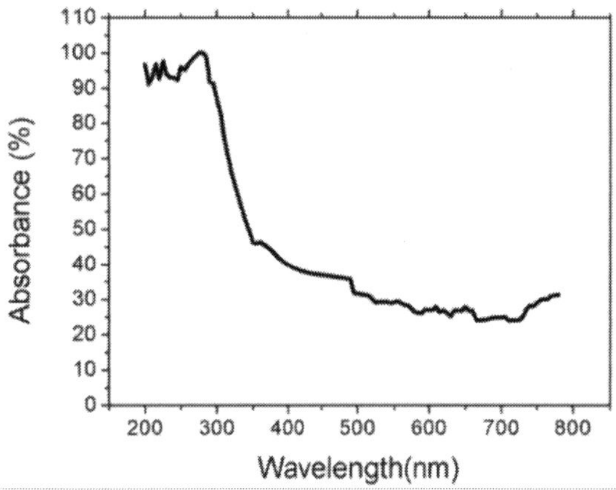

(a)

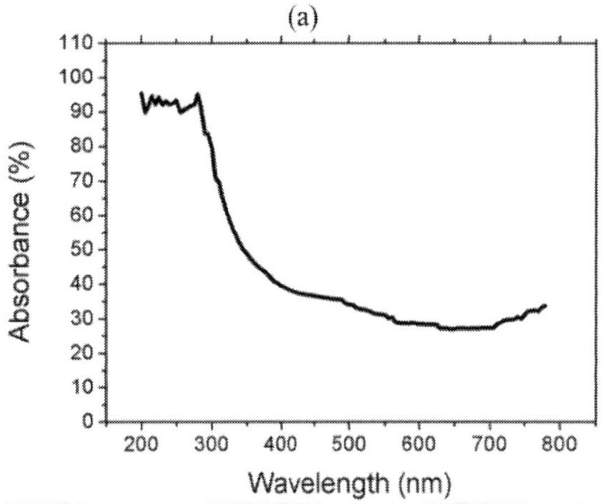

(b)

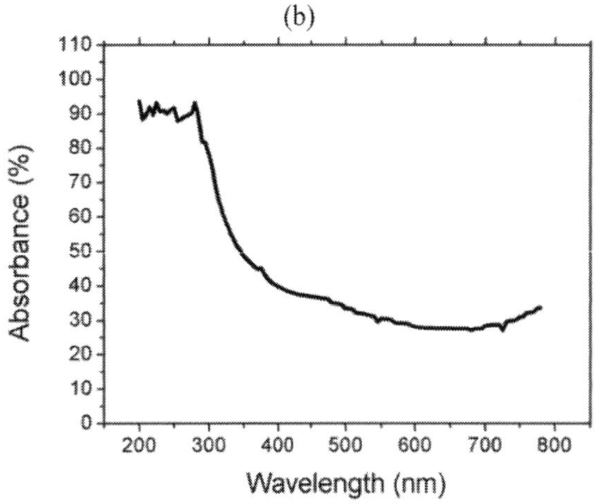

(c)

Figure 4. Graph of wavelength relation to absorbance in thin film at speed (a) 1000 rpm (b) 2000 rpm (c) 3000 rpm

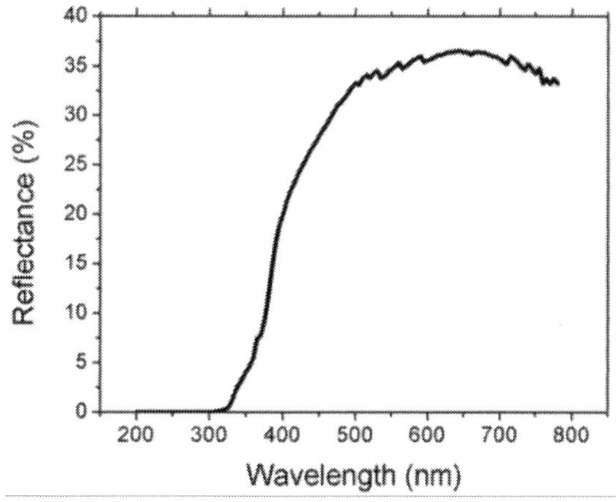

(a)

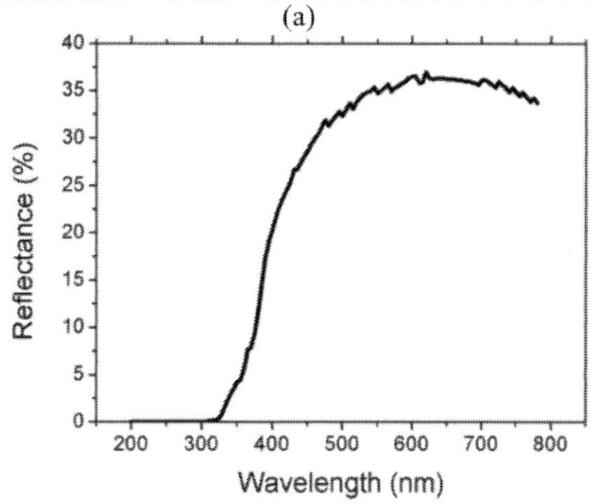

(b)

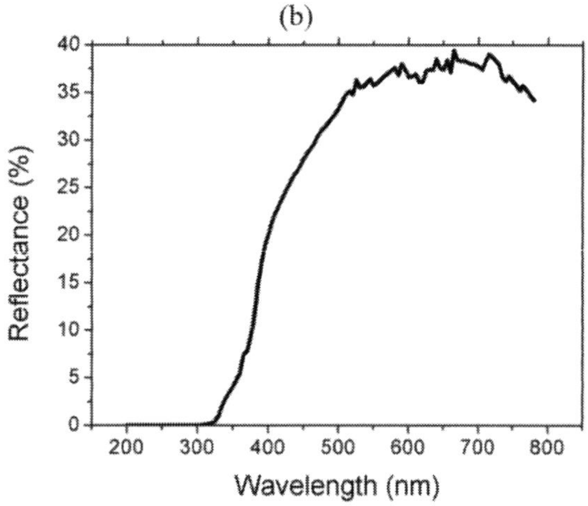

(c)

Figure 5. Graph of the wavelength relation to the reflectance at the thin film at a rate of (a) 1000 rpm (b) 2000 rpm (c) 3000 rpm.

IOP Conf. Series: Materials Science and Engineering **335** (2018) 012010 doi:10.1088/1757-899X/335/1/012010

The magnitude of the reflectance value in the wavelength range between 200-780 nm is 33.2%, 33,7% and 34,2%. The most reflected wavelength range is in the range 490-780 nm, with the highest percentage of transmission is 32.3% - 33,2%, 32,3% - 33,7% and 32,3% - 34,2%. The highest peak reflectance was 36,5% at 645 - 640 nm, 37% at 620 nm, 39,4% at 665 nm wavelength. Based on the above results indicate that if the spin coating rotation is greater then the reflectance value will also be greater. Because the rotational speed resulted in the smaller film thickness. So the value of reflectance will be greater.

The decrease of thickness of a thin films, then the absorbance is also greater, while the reflectance or the forwarded will be smaller and vice versa, the thinner the thickness the greater the light is transmitted and reflected, while the light absorbed will be smaller. The magnitude of the energy gap of the Fe_3O_4 thin film is shown in Figure 6.

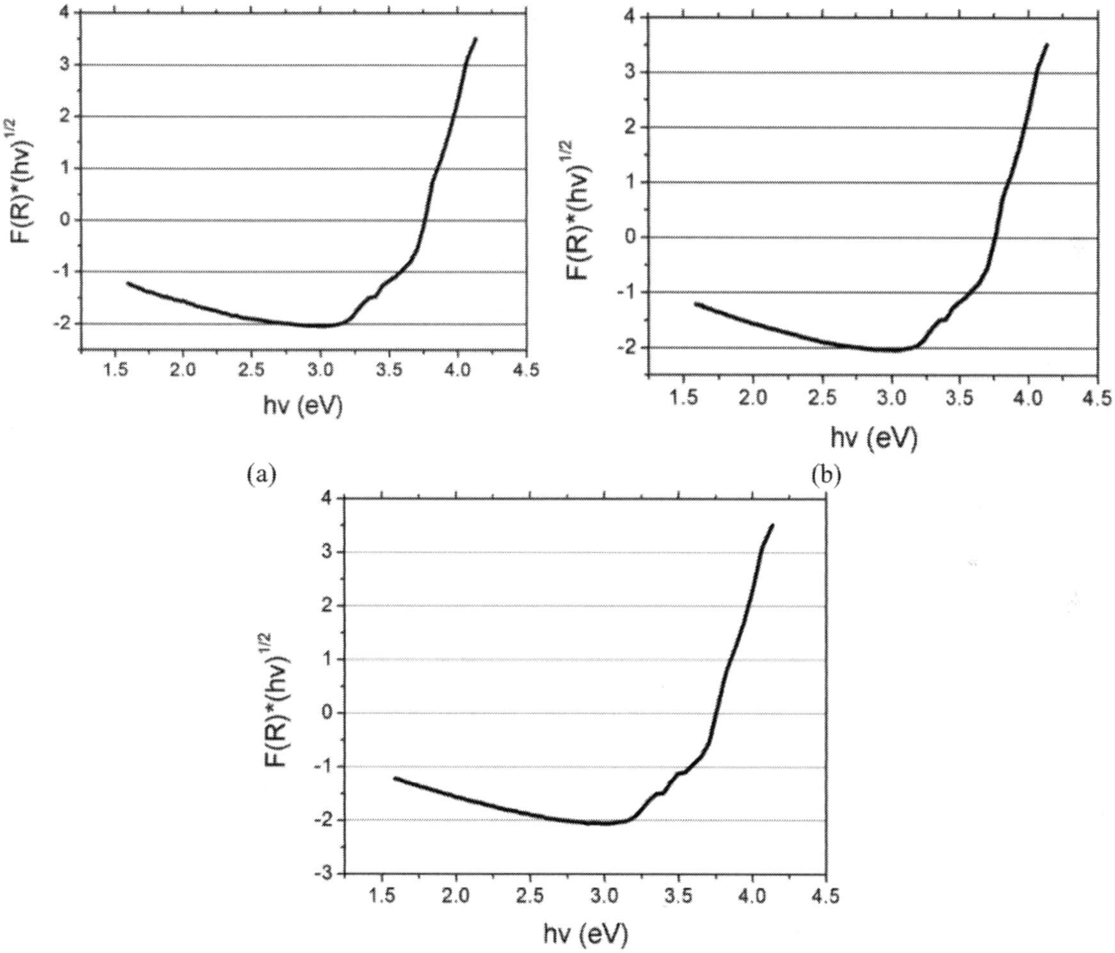

(c)

Figure 6. Graph of energy gap at a thin films with speed (a) 1000 rpm (b) 2000 rpm (c) 3000 rpm

The purpose of analyzing the energy gap of a material serves to determine a material that is conductor, semiconductor or insulator. The smaller the energy gap value of a material the energy band gap (between the conduction band and the valence band) becomes smaller, so that the electrons in the valence band will be easier to jump (excitation) to the higher band of conduction band [11]. The magnitude of the Fe_3O_4 thin films energy gap value at 1000 rpm, 2000 rpm and 3000 rpm is 3,75 eV,

3,75 eV and 3,74 eV, respectively. So the average value of the energy gap in the Fe_3O_4 thin films is 3,75 eV.

4. Conclusion

Based on the research, the optical properties of the Fe_3O_4 thin films are transmittance, absorbance, reflectance and energy gap of Fe_3O_4. This shows that the rotational velocity affects the transmittance, absorbance and reflectance. The greater the rotational speed the transmittance and the reflectance will also be greater, while the absorbance will decrease. Because the thickness of the thin films will decrease with increasing rotational speed. While the amount of energy gap obtained is 3,75 eV, 3,75 eV and 3,74 eV. So the average value of energy gap on Fe_3O_4 thin films is 3,75 eV.

Acknowledgement

The author would like to thank for Universitas Negeri Padang that has funded this research through PNBP Universitas Negeri Padang, under contract number 1650/UN35.2/PG/2017. The authors are also grateful for laboratory analyst of physics and chemical department of Universitas Negeri Padang.

References

[1] Yulianto, A and Bijaksana, S. 2002. Karakterisasi Magnetik dari Pasir Besi Cilacap. *Jurnal Fisika Indonesia*, Vol A5, No 0527.

[2] Bhushan, B. 2007. *Handbook of Nanotechnology* . Ohio State University. Columbus USA. p. 316

[3] Ramli., Hartono, A., Sanjaya, E., Aminudin, A., Khairurrijal., Haryanto, F., Imawan, C., and Djamal, M.. 2016. Novel Ternary CoFe2O4/CuO/CoFe2O4 as a Giant Magnetoresistance Sensor. *Journal Mathematics and Fundamental Sciences*., 48 (3), pp. 230-240

[4] Ito, S., Chen, P., Comte, P., Nazeeruddin, M.K., Liska, P., Péchy, P., and Grätze, M.. Fabrication of Screen-Printing Pastes From TiO2 Powders for Dyesensitised Solar Celss. *Progress in Photovoltaics* 15 (7), pp. 603-612.

[5] Isac, L., Duta, A., Kriza, A., Manola, S., and Nanu, M. 2007. Copper Sulfides obtained by spray pyrolysis possible absorbers in solid-state solar cells. *Journal Thin Solid Films* 515(15), pp. 5755-5758.

[6] Fitriani, P., Merissa, S., Iskandar, F., Mukti, R.R., Khairurijal and Abdullah, M. 2014. Sintesis Nanokomposit Fe3O4 / Zeolit dengan Metode Pemanasan Microwave dan Pengujian Sifat Katalitiknya pada Penurunan Viskositas Minyak Berat. *Journal of Research and Development on Nanotechnology in Indonesia*.1. pp. 23-26

[7] Shinde.V.R , Gujar, T.P., and Lokhande, C.D. 2007. Studies on growth of ZnO thin films by a novel chemical method. *Solar energy Material and Solar Cell*. 91(12), pp.1055-1061.

[8] Ghandoor, H.E., Zidan, H.M,. Khalil, M M.H, and Ismail, M. I. M. 2012. Synthesis and Some Physical Properties of Magnetite (Fe3O4) Nanoparticles. *Int. J. Electrochem. Sci*., 7, pp.5734 - 5745.

[9] Oktaviani, Y & Astuti. 2014. Sintesis Lapisan Tipis Semikonduktor dengan Bahan Dasar Tembaga (Cu) Menggunakan Chemical Bath Deposition. *Jurnal Fisika Unand* 3(1), pp. 53-58.

[10] Eken, A. E & Macit Ozenbas. 2009. Characterization of Nanostructured Magnetite Thin Films Produced by Sol-gel Processing. *Journal of Sol-Gel Science and Technology* 50(3), pp 321–327.

[11] Liman, J., Harsono, B, Rohman, T.T., Trimukti, U., Khalid, M., Roharti, E., and Irzaman. 2015. Uji Sifat Optik Film Tipis Ba0,55Sr0,45TiO3 di Atas Substrat Corning Glass 7059. *Jurnal Fisika Indonesia* 19, pp 45-48.

IOP Conf. Series: Materials Science and Engineering **335** (2018) 012011 doi:10.1088/1757-899X/335/1/012011

Analysis of Seismotektonic Patterns in Sumatra Region Based on the Focal Mechanism of Earthquake Period 1976-2016

F P Indah [1], **S Syafriani** [1*], **Z S Andiyansyah** [2]

[1]Physics Department, Faculty of Mathematics and Science, Universitas Negeri Padang, Indonesia
[2]Badan Meteorologi Klimatologi dan Geofisika, Pusat GempaBumi Regional VI Padang Panjang, Sumatera Barat, Indonesia

Fitripermataindah33@gmail.com

Abstract. Sumatra is in an active subduction zone between the indo-australian plate and the eurasian plate and is located at a fault along the sumatra fault so that sumatra is vulnerable to earthquakes. One of the ways to find out the cause of earthquake can be done by identifying the type of earthquake-causing faults based on earthquake of focal mechanism. The data used to identify the type of fault cause of earthquake is the earth tensor moment data which is sourced from global cmt period 1976-2016. The data used in this research using magnitude $m \geq 6$ sr. This research uses gmt software (generic mapping tolls) to describe the form of fault. From the research result, it is found that the characteristics of fault field that formed in every region in sumatera island based on data processing and data of earthquake history of 1976-2016 period that the type of fault in sumatera fault is strike slip, fault type in mentawai fault is reverse fault (rising faults) and dip-slip, while the fault type in the subduction zone is dip-slip.

1. Introduction

Indonesia is at the center of the collision of large tectonic plates, the collision of the Indo-Australian tectonic plates in the south, the Eurasian Plate in the North and the Pacific Plate in the Northeast to the southwest [11]. The collision between the Eurasian Plate and the Indo-Australian Plate affects western Indonesia (off the coast of Sumatra, Java and Nusatenggara). In addition, the territory of Indonesia is also skipped magmatic arc.

The island of Sumatra is located in an active seismic subduction region as a result of its location which lies on the boundary of the plate between the Indo-Australian plate to the north that dips down the Eurasian plate. This situation also resulted in the island of Sumatra has many volcanoes and hills known as hill ranks. In addition, this situation also resulted in the island of Sumatra has a fault or fracture that stretches along the mainland of Sumatra from the bay semangko in Lampung to Banda Aceh in Aceh Province.

The movement of the Indo-Australian plate initially had a speed of 86 mm / year and then decreased drastically to 40 mm / year as a result of the collision process. This decrease continues to occur up to 30 mm / year at the beginning of a new tectonic configuration process. Furthermore, the rate of return has increased significantly up to 76 mm / year [6].

Based on the explanation of the above tectonic causes the formation of three earthquake-prone zones namely subduction zone, Mentawai fault, and Sumatera faults. The subduction zone is located in the northwestern part of Sumatra Island, the Mentawai fault is located around small islands in the northwestern part of Sumatra island (Sunda strait - Nias island), and the Sumatera fault lies on the mainland of the western island of Sumatra from Semangko Bay (Sunda Strait) to the northern part of the island of Sumatra (Aceh). Natawidjaja, (2007) says that the potential "megathrust" earthquakes of subduction zones vary along the path depending on the fault segmentation and the size of the locked zone, as well as the history of the earthquake recurrence. These three parameters determine where there is a large accumulation of strain energy (tectonics) that can produce large earthquakes later on. The great earthquake of Aceh-Andaman (2004, Mw 9.15) [14] and also the Nias-Simelue earthquake (2005, Mw 8.7) [3] were previously characterized by seismic zone gaps.

Next row of two large earthquakes in the northern part of this subduction megathrust zone of Sumatra, the next megathrust zone is the Mentawai fault [10]. The last major quake on this Mentawai fault was the year 1797 (Mw8.4) and 1833 (Mw 8.9) [8]. These two large earthquakes resulted in a large tsunami occurring on the coasts of western Sumatra and Bengkulu. Paleoseismic and paleogeodetic research using microatol coral method reveals the fact that there was a great earthquake in the 1300s and 1600s similar to that of 1833 [7]. This shows that Mentawai megathrust has an earthquake cycle of about 200 years [8].

On Sumatra land, the Sumatera fault extends along the Bukit Barisan Mountains, from Semangko Bay in the Sunda Strait to the northern part of Aceh [13]. Already about 20 major and devastating earthquakes have occurred along the Sumatra Fault in the last 100 years [9]. In other words, a major earthquake in Sumatera fault occurs on average in every five years. Thus, unlike in the Sumatera fault that has the potential to release a large earthquake with magnitude of> 8 but only about 2-3 times in the span of 100 years, the earthquake in Sumatra fault magnitude <7.7 but often and the source closer to the population [10].

The Sumatra fault consists of large segments. The fault segments are the Sundanese segment ($6,75^0$ S ~ $5,9^0$ *S*),semangko segment ($5,9^0$ S ~ $5,25^0$ *S*), Kumering segment ($5,3^0$ S ~ $4,35^0$ *S*), Manna segment ($4,35^0$ S ~ $3,8^0$ *S*), Musi segment ($3,65^0$ S ~ $3,25^0$ *S*), Ketaun segment ($3,35^0$ S ~ $2,75^0$ *S*), Dikit segment ($2,75^0$ S ~ $2,3^0$ *S*), Siulak segment ($2,25^0$ S ~ $1,7^0$ *S*), Suliti segment ($1,75^0$ S ~ $1,0^0$ *S*), Sumani segment ($1,0^0$ S ~ $0,5^0$ *S*), Sianok segment ($0,7^0$ S ~ $0,1^0$ *N*), Barumun segment ($0,3^0$ N ~ $1,2^0$ *N*), Angkola segment ($0,3^0$ N ~ $1,8^0$ *N*), Toru segment ($1,2^0$ N ~ $2,0^0$ *N*), Renun segment ($2,0^0$ N ~ $3,55^0$ *N*), Tripa segment ($3,2^0$ N ~ $4,4^0$ *N*), Aceh segment ($4,4^0$ N ~ $5,4^0$ *N*), Seulimeum segment ($5,0^0$ N ~ $5,9^0$ *N*) [9]. The fault segments in Sumatera Island have made the tectonic structure of Sumatera Island to become active, so it is necessary to analyze the activity of fault section to know the type of faults in the area.

The activity of the earthquake fault field can be identified through the focal mechanism of the earthquake. Focal Mechanism Solutions (FMS) or focus mechanism is a fault analysis that is formed due to tectonic activity in a region based on the moment tensor value of the earthquake.

Moment tensor values used in analyzing focus mechanisms can be obtained from Global Centroid Moment Tensor (CMT). The value of tensor moment is then used as input or input on Generic Mapping Tools (GMT) software to get a beach ball which represents the movement of the fault field formed due to tectonic activity or earthquake in an area

2. Data and methods

The research data used is earthquake data sourced from Global CMT. The data used is the Global CMT in the Sumatra earthquake period 1976-2016. The study area is Sumatra with coordinates of 6 ° LU-11 ° LS and 95°BT-109°BT. Earthquake magnitude used in this research is $M \geq 5$ with the depth of the earthquake 0-300 km.

Data processing technique using Focal Mecanishm to produce fault forms of earthquake in Sumatra. For the painting of earthquake focus mechanism can use GMT software. Fault type of earthquake can be analyzed based on earthquake focus mechanism generated. The focal mechanism will display the causative type of earthquake based on the parameters of the fault is strike, dip, and rake. Thus, from these parameters will be obtained type fault earthquake causes in earthquake-prone areas.

Aki and Richard (1980) [1] determine the type of fault based on strike, dip, and rake parameters as follows:

a) Strike slip fault, if $\delta = 90^0$ and $\lambda = 0^0$ (left shear) or $\lambda = 180^0$ (right shear).
b) Normal fault, if $\delta \neq 0^0$ and $\delta \neq 90^0$ and $-180^0 \leq \lambda \leq 0^0$.
c) Thrust fault, if $\delta \neq 0^0$ and $\delta \neq 90^0$ and $0^0 \leq \lambda \leq 180^0$.

Based on the above analysis can be identified fault type of earthquake based on the value of fault parameters.

3. Results And Discussion

3.1 Research Results

The data used in this study comes from the Global CMT in the form of data from the moment tensor earthquake value that occurred in the region of Sumatra period 1976-2016. Data derived from Global CMT is used to describe seismicity maps of earthquake distribution and focal mechanism to describe the type of fault cause of earthquake. The data format for seismicity map depiction consists of latitude, longitude, depth, and magnitude. Earthquakes with small magnitudes to large magnitudes often occur due to the tectonic order of the island of Sumatra. Figure 1 shows the seismicity map of earthquake distribution with magnitude M <5 SR and M ≥ 5 SR Period 1976-2016.

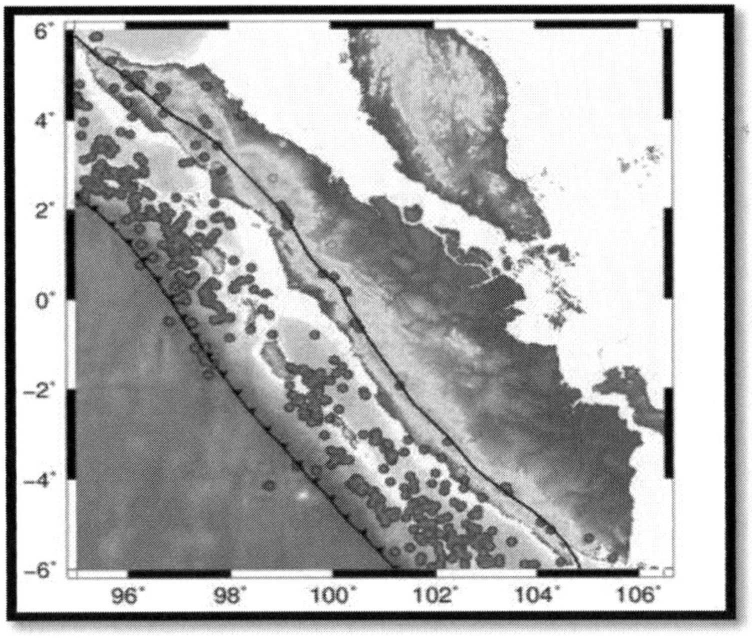

Figure 1. Map of Seismicity of Sumatra Region Period 1976-2016

Figure 1 shows the seismicity map of earthquake distribution period 1976-2016 with magnitude M <5 SR and M ≥ 5 SR. The focal mechanism depiction in each earthquake active zone can be seen in Figure 2.

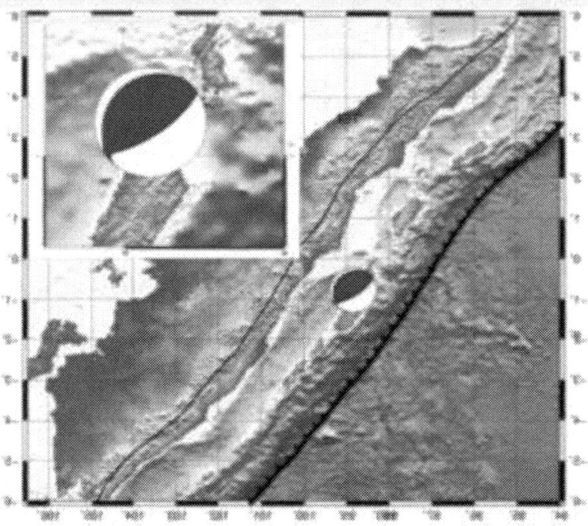

Figure 2. Focal Mechanism of Earthquake 25 April 1987 with Magnitude 6.4 SR

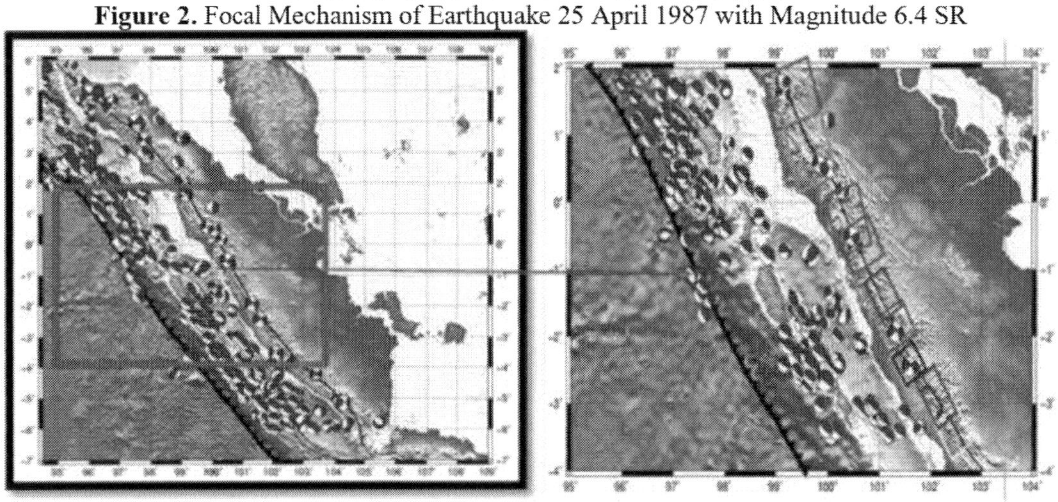

(a) (b)

Figure 3. Map of Focal Mechanism Distribution of Earthquake Magnitude ≥ 5 SR Period 1976-2016 (a) Sumatera Region; (b). West Sumatra Region

Figures 2 represent the shape of an earthquake focal mechanism on the Mentawai fault and the Sumatra fault. Based on the data of moment tensor values obtained by the form of focal mechanism earthquake distribution based on earthquake data with magnitude ≥ 5 SR period 1976-2016 as shown in Figure 3. Figure 3 shows the shape of the earthquake focal mechanism in the subduction zone, the Mentawai fault, and the Sumatra fault. If we zoom the map of focal mechanism distribution of earthquake in West Sumatra Region in figure 3(b).

3.2Discussion

In this study, data from Global CMT is processed using Generic Mapping Tools (GMT) software to obtain earthquake focal mechanism based on earthquake data of 1976-2016 period for three earthquake-causing zones in Sumatera region (subduction zone, Sumatera fault, and Mentawai fault).

Based on Figure 1 it can be seen that earthquakes have occurred in Sumatra with magnitude <5 SR and magnitude ≥ 5 SR. This shows that the tectonic order of Sumatra is very active which is characterized by many earthquakes occurring in Sumatra. In Figure 1 it can be seen that the most prevalent earthquakes are shallow earthquakes with depths of 0-70 km distributed in subduction zones and Mentawai faults. While a small part of the earthquake in this period occurred at a depth of 71-300 km. Based on the depth and distribution of the earthquake it can be seen that the tectonic order of Sumatra is very active. The area of Sumatra in the marine region (Mentawai fault and subduction zone) is more active than inland Sumatra (fault / Sumatera faults).

An earthquake on Mentawai fault may also be caused by subduction activities where the Indo-Australian plate infiltrated under the Eurasian plate. In some earthquake events, the cause of earthquake in Mentawai fault is caused by subduction zone. One study that explains it is research on the Mentawai earthquake report on October 25, 2010 conducted by the Meteorology Climatology and Geophysics Agency, 2010 [2] which states that "Mentawai Earthquake on October 25, 2010, 7.2 magnitude, 3.61 LS - 99.93 BT, 78 km Southwest Pagai, Mentawai-western Sumatra meets the criteria as an earthquake that causes TSUNAMI (Magnitude ≥ 7.0 SR, position at sea, depth <33 km). This earthquake is a tectonic earthquake caused by the collision of the Indo-Australian Ocean plate to the Eurasian continental plate with a source mechanism of thrust fault. " Therefore, it is necessary to identify the type of fault causing the earthquake.

In this study for the description of the type of fault causing earthquake using focal mechanism based on the data of moment tensor value. The result of the seismotectonic pattern formed in the form of a focal sphere used to display an earthquake focus mechanism that explains the fault parameters of the earthquake. Moment tensor can be used to measure the strength of the earthquake by using the seismic moment parameter (M_0) shown in equation (14). Determination of the type of fault can be done through the analysis of fault parameters that cause earthquake strike (φ), dip (δ), and rake (λ). From the results of the analysis of fault parameters known fault type causes earthquake.

Figure 2 shows an analysis of the type of faults that cause the earthquake on April 01, 1998 with magnitude 6.9 SR based on strike (φ), dip (δ), and rake (λ) obtained from Global CMT. An earthquake on April 01, 1998 has a strike value of 320, dip = 21, and slip = 105 on the nodal 1 and strike = 124, dip = 69, and slip = 84 fields of nodal 2 obtained from Global CMT. After analysis based on strike parameter (φ), dip (δ), and rake (λ) it is known that the type of earthquake fault of April 01, 1998 is reverse fault where Aki and Richard, 1980 [1] stated that the reverse fault, if $\delta \neq 0^0$ and $\delta \neq 90^0$ and $0^0 \leq \lambda \leq 180^0$. The seismic moment of earthquake on April 01, 1998 is 3.32 obtained using equation (14).

The depiction of the focal mechanism of the entire earthquake with the magnitude of M ≥ 5 SR in the period 1976-2016 is shown in Figure 4. Figure 4 shows the shape of the earthquake focal mechanism in the subduction zone, Mentawai fault, and Sumatra fault. The shape of the earth's focal mechanism has the same overall shape in each of the active zones.

The shape of focal mechanism in Sumatera fault based on the data of moment tensor earthquake of the period 1976-2016 shown in Figure 4 shows that the type of faults that cause earthquake in the fault of Sumatra is generally strike slip. This is in accordance with previous studies conducted by Sieh and Natawidjaja, (2000) [13] stating that "The Summit Fault is a stratified curvature strike slip consisting of 20 major segments along the backbone of Sumatra". The shape of this pattern is one of them caused by the influence of the geological condition of the island of Sumatra. The Indo-Australian Plate infiltrated in an oblique direction into the Eurasian plate. The oblique tilting caused the formation of SFZ is a dextral strike slip fault zone that extends from the northern tip to the southern tip of Sumatra Island.

The form of focal mechanism in Mentawai fault based on earthquake data of 1976-2016 period shown in Figure 4 is dominated by reverse fault. Mukti et al. (2012)[5] conducted a study of seismic cross section under the Mentawai Fault and stated that there are anticillin (rising rock structures) located on Mentawai fault. This is due to the encouragement of the Indo-Australian plate toward the Eurasian plate causing anticillin. The result of anticillin on Mentawai fault resulted in the formation of reverse fault pattern on Mentawai fault. In addition, in this study also obtained the form of strike slip fault on Mentawai fault. Samuel and Harburry (1996) [12] say that for the Mentawai fault that extends to Nias island has a fault pattern of stike slip.

The focal mechanism of the subduction zone based on earthquake data of the period 1976-2016 is a dip-slip consisting of normal fault and reverse fault. Madrinovella et al. (2011) [4] states that geologically, the pattern of the fault plane formed in the subduction zone is due to the subduction of the Ocean. The Indo-Australian Plate moves north-northeast and pushes the Eurasian plate at a rate of 5 cm per year. Madrinovella et al (2011) [4] also mentioned that "the convergent boundaries on two rigid plates between the Eurasian plate and the Indo-Australian plate led to a dip-slip fault pattern occurring in the subduction zone of Sumatra."

4. Conclusions

Based on the research results obtained conclusion as follows:

1. Sumatran tectonic order is very active which is characterized by the frequent earthquake that is earthquake with magnitude <5 SR and earthquake with magnitude ≥ 5 SR in period 1976-2016.
2. Seismotectonic patterns formed on each region of Sumatra Island based on the data processing and earthquake history data from 1976-2016 is the type of fault in the fault of Sumatra is strike slip, the type of fault on the Mentawai fault is reverse fault and strike slip, and the type of fault in the subduction zone is dip-slip.
3. Characteristics of the fault area formed in each region on the island of Sumatra one of the earthquake on April 1st, 1998 has strike value = 320, dip = 21, and slip = 105 on nodal 1 and strike = 124, dip = 69, and slip = 84 on nodal 2 obtained from Global CMT has a fault type reverse fault.

Acknowledgments

This research is part of a research grant of the Superior Research Scheme of Higher Education Fiscal Year 2017 with a research team headed by Syafriani, Ph. D based on the Letter of Assignment of Research No. 1649 / UN35.2 / PG / 2017 dated May 31, 2017. The authors also thank to the Meteorology Climatology and Geophysics Agency (BMKG) Geophysics Station of Klas I Padang Panjang and Global CMT for the data used in this study. Additionally thanks to GMT Software for maps or pictures.

References

[1] Aki, K. and P. G. Richard. 1980. *Quantitative Seismology and Method*. San Fransisco: W. H. Freeman And Company.

[2] BMKG, 2010. *Mentawai Earthquake Report*, October 25, 2010. Jakarta.

[3] Briggs, R., dkk. 2006. Deformation and slip along the Sunda megathrust in the Great 2005 Nias-Simeulue earthquake.: Science, v. 311, p. 1897-1901.Earthquake Event in 2013. *Journal of Science and Arts POMITS* Vol. 3, No. 1, (2014) 2337-3520.

[4] Mardrinovella, I., Widiyantoro, S., Irwan, M., 2011. Relocation of Padang Earthquake Hyposenter 30 September 2009 Using Double Difference Method. *JTM* Vol.XVIII No. 1.

[5] Mukti, M. M., et al. 2012. Structural evolution of backthrusting in the Mentawai Fault Zone, offshore Sumatran forearc. *Geochem. Geophys. Geosyst.*, 13, Q12006.

[6] Natawidjaja, D.H. 1994. Quantitative Geological Assessments of Liwa Earthquake 1994. *Proceeding of Annual Convention of Indonesian Association of Geophysicists* (HAGI) 1994.

[7] Natawidjaja, D.H. 2003. *Neotectonics of the Sumatran fault and paleogeodesy of the Sumatran subduction zone*. Ph.D thesis, California Institute of Technology (Caltech).

[8] Natawidjaja, D.H., et al. 2006. Source Parameters of the Great Sumatran Megathrust Earthquakes of 1797 and 1833 Inferred from Coral Microatolls. *J. Geophys. Res.*, v. 111.

[9] Natawidjaja, D. H., and W. Triyoso. 2007. The Sumatran Fault Zone: from Source to Hazard, 1 (No. 1),2147.

[10] Natawidjaja, D. H., et al. 2007. Interseismic Deformation Above the Sunda Megathrust

IOP Conf. Series: Materials Science and Engineering **335** (2018) 012011 doi:10.1088/1757-899X/335/1/012011

Recorded in Coral Microatolls of the Mentawai Islands, West Sumatra. *J. Geophys. Res.*
[11] Plummer, C et al. 2003. *Physical Geology*. New York: McGraw Hill Companies.
[12] Samuel, M. A. and Harbury, N. A., 1996. *The Mentawai Fault Zone and Deformation of the Sumatran Forearc in the Nias Area*. Geological Society, London, Special Publications. v.106, p337-351.
[13] Sieh, K., and Natawidjaja, D.H., 2000. Neotectonics of the Sumatra fault, Indonesia. *Journal of Geophysical Research*, 105, B12, pp. 28, 295- 28,326.
[14] Subarya, C., et al. 2006. Plate-boundary deformation associated with the great Sumatra-Andaman earthquake: *Nature*, v. 440, p. 46-51

Analysis of Crystal Structure of Fe_3O_4 Thin Films Based on Iron Sand Growth by Spin Coating Method

D Rianto[1], N Yulfriska[1], F Murti[1], H Hidayati[1], R Ramli[1*]

[1]Department of Physics, Faculty of Mathematics and Sciences, Universitas Negeri Padang, Indonesia

*ramli@fmipa.unp.ac.id

Abstract. Recently, iron sand used as one of base materials in the steel industry. However, the content of iron sand can be used as starting materials in sensor technology in the form of thin films. In this paper, we report the analysis of crystal structure of magnetite thin film based on iron sand from Tiram's Beach. The magnetic content of sand separated by a permanent magnet, then it was milled at 30 hours milling time. In order to increase the purity of magnetite, it washed after milling using aquades under magnetic separation by a magnet permanent. The thin film has been prepared using iron (III) nitrate by sol–gel technique. The precursor is resulted by dissolving magnetite in oxalic acid and nitric acid. Then, solution of iron (III) nitrate dissolved in ethylene glycol was applied on glass substrates by spin coating. The X-Ray Diffraction is operated thin film characterization. The structure of magnetite has been studied based on X-Ray Peaks that correspond to magnetite content of thin films.

1. Introduction

Iron is the second most plentiful metal in the ground and present in sediment form. The sediment of iron contained in sedimentary rocks in form of sand, known as iron sand. Distribution of iron sand in Indonesia have encountered in various islands began the West Coast of Sumatra, Kalimantan, Sulawesi, region of Nusatenggara, Maluku Islands and South Beach of Java [1].

The iron sand is commonly called as *black sands* due to its own colour. The colour is affected by the iron oxides content like magnetite (Fe_3O_4), maghemite (γ- Fe_2O_3) and hematite (α-Fe_2O_3). Magnetite (Fe_3O_4) has become a priority in research activities because it has more much advantages compared with the other mineral compounds. The structure of Fe_3O_4 formed in cubic inverse spinel. According to ICDD with the code 01-089-0688known that Fe_3O_4has a *space group* F d-3m with No. 227 and the same lattice parameter a = b = c of 8.396 Å and the angle $\alpha = \beta = \gamma = 90^0$ [2]. Detail of the structure of $Fe_3O_4$is shown in Figure 1.

Recently, Research on magnetic oxide has progressed rapidly. Nanostructure of oxide magnetic is important in development of advance and new functional materials [3]. Magnetic characteristic of magnetite is superior in respond external magnetic field compared to the other magnetic compounds contained in the iron sands [4]. Magnetite is a potential candidate as materials for ceramics, energy storage, magnetic data storage, ferofluids, and magnetic sensor.

Thin film is a layer made from organic, inorganic, metal, or a compound of an organic metal material which can have physical properties as conductors, semiconductors, superconductors, and insulator. The film is made using atomic or particle deposition technique. The thickness of film can be

IOP Conf. Series: Materials Science and Engineering **335** (2018) 012012 doi:10.1088/1757-899X/335/1/012012

obtained in micrometer even nanometer scale. The common characteristic of thin film is different with powder or bulk materials, it caused by the differences of preparation process, composition, morphology and structure.Various materialshave potential as semiconductors, photo-catalyst, magnetic storage and the dielectrichave beenstudied.The magnetitefilmcan be used as a data storage memory, magnetic sensors for mitigation of earthquake disaster and for DNA detection [5].

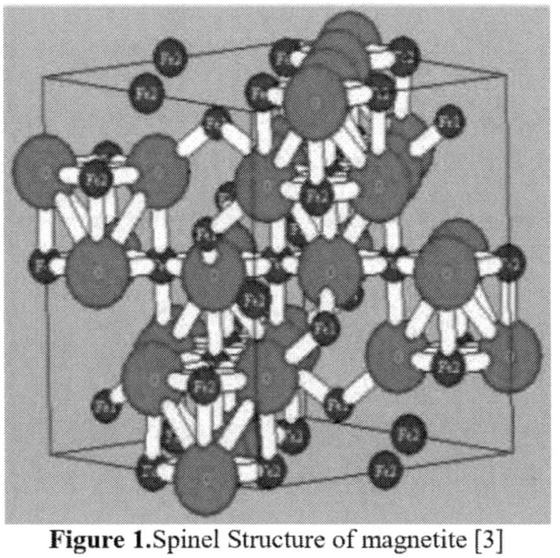

Figure 1. Spinel Structure of magnetite [3]

The following physical thin film growth method : *sputtering* [2], DC *magnetron sputtering* [6], spray *pirolisis* [7] *slip casting* [8], electrophoresis [9], and *screen printing* [10] have disadvantage like lack of deposition areas, requiring sophisticated instruments, high operating costs, and complicated cleaning systems [11]. Chemical method such as chemical vapour deposition has inadequate for film production on a large scale because it requires complicated additional tools[12].

The coating method is the simplest and easiest method in thin filmdeposition. Differentproperties of thin film obtained in difference type of coating methods. Spin coating is one of the coating methods in whichdeposition done over the substrateat a certain rotary speed.Solution or gel that dropped above the substrate flow toward the edge cause of centripetal force that resulted by rotation. Then the solvent evaporated in which condensation reaction take place. These processes will result solid thin film above the substrate. Spin coating disc sketched in Figure 2. The coated gel is produced using sol-gel techniqueinvolving chemical process such as hydrolysis and condensation reactions.In addition, the sol-gel technique is easy to produce homogenous materials and the composition of materials can be controlled as desired [13]. Spin coating process conducted in four steps consisting of deposition, spin-up and spin-off and evaporation. These steps will determine the thickness of thin film [14]. Spincoating process can be explained using Figure 3.

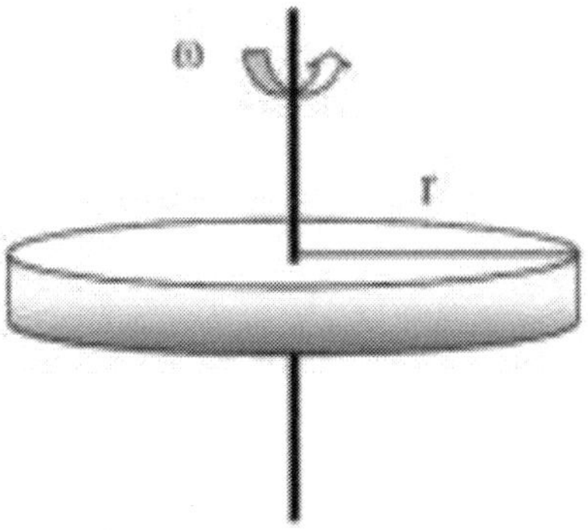

Figure2. Spinning disc of Spin Coating

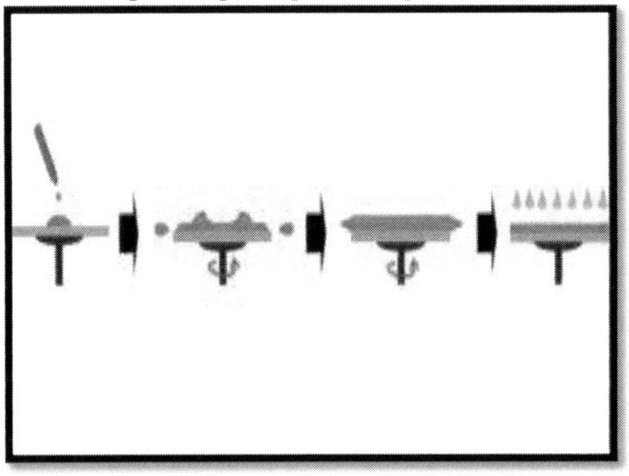

Figure 3. Stages scheme of Spin Coating

The first step of coating is dripping or flowing of the gel on the substrate.At the deposition ste, the substrate has not been rotated.Then at the next step the substrate begins to rotate.Since centrifugal force resulted from circular motion the liquid become dispersed out of the disc centre toward the edge.At this step the substrate accelerates and the next two steps the rotation rate begins to be constant, meaning no angular acceleration on the substrate.At the*spin-off*step an amount of excess liquid or gel will go to the edge and eventually detached.The thinner film deposited the less droplets are wasted.This affected by the addition of flow resistance and viscosity when the layer gets thinner.The last step is evaporation as the significant process of depletion film [15].

One of important things in film deposition is thickness caused it will affect the quality of crystal, different thickness will produce different crystal quality.Thin film thickness determined by coated angular velocity used.The velocity is one of the most important factors in the coating process.The velocity will affect the angle of the centrifugal force on the liquid. For more specific, the angular velocity will determine the crystal structure [16]. In this paper will discuss the influence of the coated angular velocity on crystal structureof Fe_3O_4thin films.

2. Experimental

The iron sand is taken from Tiram's Beach, Padang Pariaman West Sumatra, which then separate using magnetic stick for 30 times, this process aims to sort out the iron content from impurities. Then it washed using aquades, dried and retracted using a permanent magnet for 20 times. The iron sand 6 gr is milled using HEM-E3D for 30 hours. Then it washed using distilled water to purge or eliminate impurities. Drying introduced as final purification process.

In this study, Fe_3O_4thin film is prepare using sol-gel techniques with precursor iron (III) nitrate ($Fe(NO_3)_3.9H_2O$). The precursorresulted by dissolving 17,4 gr magnetite in oxalic acid 4,5 gr ($C_2H_2O_4$) and nitric acid (HNO_3) 42 mL, at 110^{0}C. Then, solution of ($Fe(NO_3)_3.9H_2O$) dissolved in ethylene glycol at 80 ^{0} C for 2 h under constant stirring. The coating solution was dropped onto glass substrate, which was rotated at 1000 rpm, 1500 rpm, 2000 rpm, 2500 rpm and 3000 rpm. The coating process conducted in 60 s. After deposition by spin coating, the film was dried at 110^{0}C for 15 minutes and followed annealing process at 300^{0}C, for 3.5 hours. Magnetite thin films then characterized using X-Ray Diffraction (XRD) to analyze the crystal structure of magnetite. The results then compared with standard ICDD for magnetite.

3. Result and discussion

The peak intensity of XRD graph at plane (311) is more dominant than the other planes of Fe_3O_4crystals (see Fig. 4). According to ICDD standards (International Center of Diffraction Data), the orientation intensity of the crystal plane (311) is higher than the other orientations of magnetite crystal. The XRD data also give information that magnetite phase lattice parameters ranged from 0.8366 nm to 0.8367 nm, slightly lower than standard value of 0.84045 nm. The graph also show the presence of a peak that correspond to nitric acid (HNO_3) at 2θ = 27.90^{0}at almost of XRD Graphs except at 200 rpm. Nitric acid (HNO_3) used as reagent in synthesis of precursors ($Fe(NO_3)_3.9H_2O$). An amount of nitric acid does not react completely with magnetite powder so it still leaves in the solution. Although thin films have been given heat processes such as drying and annealing, but not enough to evaporate all of the nitric acid.

The XRD results show the magnetite films grown in polycrystalline crystal indicated bythree until six magnetite peaks at each X-Ray pattern with variation of angular velocity. Yet the crystal of thin films at 2000 rpm having three magnetite peaksless than the others. This mean that order of crystal orientation at 2000 rpm is more uniform than the others. Thefilm deposited at 2000 rpm also has higher purity rate compared to the others. This statement is supported by XRD data showing no peak HNO_3 at 2000 rpm. So it can be conclude that magnetite thin film that deposited at 2000 rpm has a better quality of crystal compared to the other thin films.

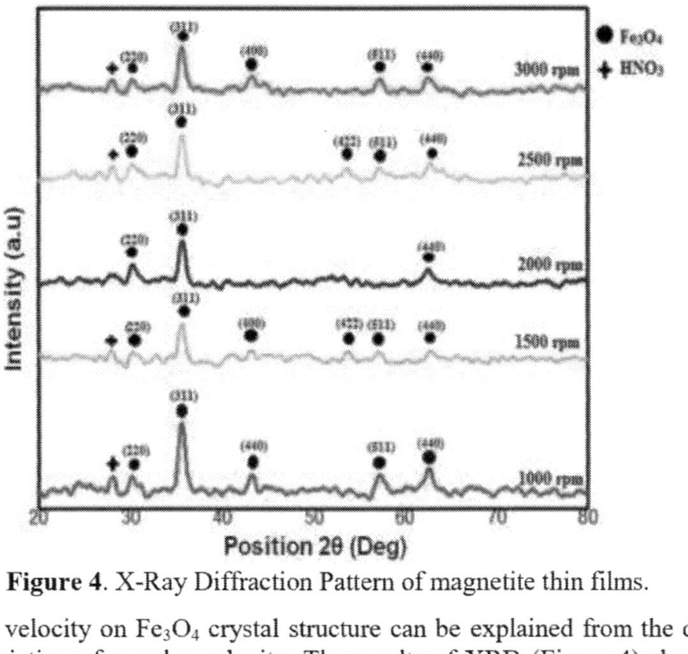

Figure 4. X-Ray Diffraction Pattern of magnetite thin films.

Effect of coating velocity on Fe_3O_4 crystal structure can be explained from the diffraction pattern produced at each variation of angular velocity. The results of XRD (Figure 4) show that number of orientation intensity of the Fe_3O_4 crystal planes decreases as increasing the angular velocity, 1000 rpm up to 1500 rpm and 2500 rpm up to 3000 rpm. Similar results were also found in the PZT thin film study [17], that found decreasing the orientation intensity of the crystal planes of PZT thin film. Thedecrease in the intensity of the crystal orientation of the thin film is due to the higher angular velocity rate resulting high magnitude of centrifugal force. So amount of gel scattered out the substrate. This means that the number atoms in film-deposition decreased, so the probability of crystalline formation decreases that affects the intensity of XRD results[9].

XRD graphs for certain angular velocity indicate an increase of intensity value. The increasing intensity, 1500 rpm up to 2500 rpm, caused bymass of geldropped is not the same (due to difficulty in microgram scale) so that the number of Fe_3O_4 atoms dropped is different for each coating process. Increasing of Fe_3O_4gel massmake more gel attached to the substrate in the coating process. As a result, the number of the atoms making up the crystalline plane of the Fe_3O_4 thin films increases, so the probability of Fe_3O_4 formation also increases and directly affects the peak intensity. Thereare factors that cause the result of thin films characterization is not obey to theory : homogeneity of thin films, flatness of thin films that is uneven, and the difference in the massof dropped gel[18].

Notonly thevelocity effect the intensity of the crystal orientation, but also it caused change of FWHM at each diffraction peak. FWHM is a function of the particle size. So difference of FWHM gives different crystallite size. Diameter of crystallite (D) expressed in the following Scherrer equation [19].

$$D = K\frac{\lambda}{B\cos\theta_B} \tag{1}$$

The microstrain (e) satisfies the equation [18],

$$e = \frac{b}{4\tan\theta_B} \tag{2}$$

λ (lamda) is the wavelength 1.54 Å, θ_B is the Bragg angle, B is FWHM of the selected peak, and K is the material constant that value is less than one, Commonly K ~ 0.9.The FWHM value determines the size of the crystals and the value of the microstrain of the magnetite crystals formed. These changes can be seen in the graph of the relation of the angular velocity to the crystallite size and the microstrain in Figure 5.

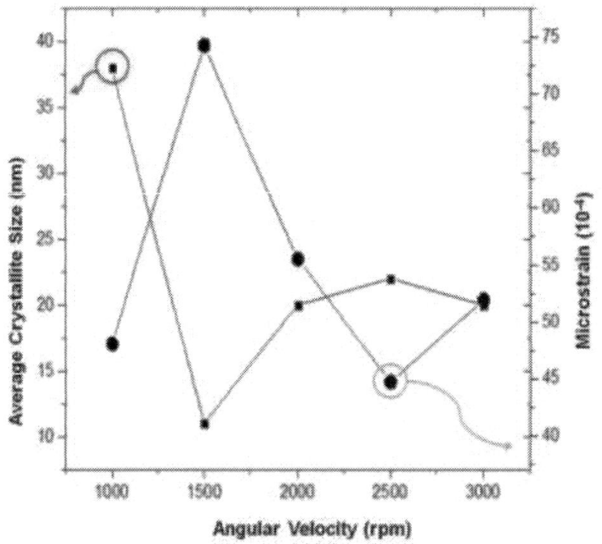

Figure 5. Effect angularvelocity of spin coating oncrystallitesize and microstrain of magnetite thin film.

The average magnetite crystallite size decreases at 1000 rpm up to 1500 rpm and at 2500 rpm and 3000 rpm. Decreased crystal size correlates with decreasing film thickness. Study ofCobalt Ferrite [3] also found that the crystalline sizebecomesa decreaseby depletion of film thickness from 48 to 40 nm.

The change in FWHM values also affects the value of the microstrains. The microstrain of the magnetite thin films shows an increase in value at 1000 rpm up to 1500 rpm and at 2500 rpm and 3000 rpm. This is due to the diffraction peaks marked by the increase of the FWHM value. On the other hand the angular velocity of 1500 rpm to 2500 rpm indicates a decrease in the value of microstrains caused by the FWHM value is getting smaller.

4. Conclusion

The results of this study indicate that at angular velocity of 1000 rpm to 1500 rpm and at 2500 rpm to 3000 rpm there is a decrease in the XRD peak intensity value of crystal field. On the other hand, the microstrains have increased in that range. The XRD graph also shows a thin film grown at 2000 rpm has better crystal quality compared to other thin layers.

Acknowledgement

This work was funded by the BOPTN Universitas Negeri Padang, under contract number 1650/UN35.2/PG/2017. The authors are also grateful for laboratory analyst of the physics department of Universitas Negeri Padang, laboratory analyst of mechanical Engineering Universitas Kristen Indonesia.

References

[1] Mufit, F., F, Amir, dan S, Bijaksana. 2006. Kajian tentang Sifat Magnetik Pasir Besi dari Pantai Sunur Pariaman Sumatera Barat. *Jurnal Geofisika*. Vol 1, pp 2-5

[2] Fuad. A, et al. 2010. *Sintesa dan Karakterisasi Sifat Struktur Nano Partikel* $Fe_{3-x}Mn_xo_4$ *dengan Metode Korespirasi*. Semarang : Universitas Negeri Malang

[3] Ramli, Jonuarti, R., Hartono, A. 2017. Analisis Struktur Nano dari Lapisan Tipis *cobalt Ferrite* Yang Dipreparasi dengan Metode *Sputtering*. Eksakta Vol. 18 No. 1

[4] Jonal, La. 2014. *Pengaruh Frekuensi Gelombang Ultrasonik Terhadap Struktur Kristal Dan Sifat Magnet Nano Partikel Magnetit* (Fe_3O_4). Universitas Islam Negeri Sunan Kalijaga : Yogyakarta.

[5] Siyambalapitiya, Chamila S., Growth And Physical Properties of Magnetite Thin Films , (Graduate School Theses and Dissertations. Paper 2705, 2006).

[6] Ramli, et al. 2016. Novel Ternary $CoFe_2O_4$/CuO/$CoFe_2O_4$ as a Giant Magnetoresistance Sensor. Journal Math. Fund. Sci., Vol. 48, No. 3, 230-240

[7] Isac, L. et al. 2007. Copper Sulfides obtained by spray pyrolysis possible absorbers in solid-state solar cells. Journal *Thin Solid Films, Elsevier*. 5755-5758

[8] Nuryadi, R. 2011. Efek adsorpsi Dye ke dalam Lapisan TiO_2 dengan metode Elektroforesisi : DSSC Berbasisi Lapisan TiO_2 terbuat dengan Metode *Slip Casting* dan Metode Elektroforesis. *Jurnal Rekayasa Kimia dan Lingkungan,* Vol. 8, No.1, Hal. 35-40.

[9] Nuryadi, R. 2011. Efek adsorpsi Dye ke dalam Lapisan TiO_2 dengan metode Elektroforesisi : DSSC Berbasisi Lapisan TiO_2 terbuat dengan Metode *Slip Casting* dan Metode Elektroforesis. *Jurnal Rekayasa Kimia dan Lingkungan,* Vol. 8, No.1, Hal. 35-40.

[10] Ito, S. et al. Fabrication of Screen-Printing Pastes From TiO_2 Powders for Dyesensitised Solar Celss. *Journal Progress in Photovoltaics : Research and Applications.* Vol. 15, Issue 7, 603-612.

[11] Shinde.V.R et. al. (2007). "Studies on growth of ZnO thin films by a novel chemical method". *Solar energy Material & Solar Cell*. 91, 1055-1061

[12] Cho, S.I., C. H. Chung, and S. H. Moon, 2002, Surface decomposition mechanism of Ti(OC3H7)4 on a platinum surface, Thin Solid Films, 409, 98.

[13] Bhushan, B. 2007. Handbook ofNanotechnology . Ohio State University. Columbus USA.

page. 316

[14] Luurtsema, G.A. 1997. Spin Coating for Rectangular Substrates (Thesis). University of California. Barkeley. hal. 4-7.

[15] Rustami, S. 2008. Sistem Kontrol Kecepatan Putar *Spin Coating* Berbasis Mikrokontroler ATmega8535. Bogor : Institut Pertanian Bogor

[16] Purwanto, Romli. & Prajitno G. (2013). *Variasi kecepatan dan waktu pemutaran spin coating dalam pelapisan TiO untuk pembuatan dan karakterisasi prototipe DSSC dengan ekstraksi kulit manggis (garcinia mangostana) sebagai dye sensitizer*, ITS Surabaya

[17] Naat, Jhonson Nune, et al. 2014. Pengaruh Kecepatan Putar Deposisi Terhadap Struktur Kristal,Ketebalan Dan Morfologi Lapisan Tipis Timbal Zirkonat Titanat (Pzt) Dengan Metode Spin Coater. *Chimica et Natura Acta Vol.2 No.2, Agustus 2014:115-119*

[18] Sulastri, S. 2010. *Pengukuran Sebaran Lapisan Tipis Hasil Spin Coating dengan Metode Interferometrik*. Surakarta : Universitas Sebelas Maret.

[19] Suryanarayana. C., Norton. M., (1998): *X-Ray Diffraction: A Practical Approach*, Plenum Press, New York.

IOP Conf. Series: Materials Science and Engineering **335** (2018) 012013 doi:10.1088/1757-899X/335/1/012013

In Vivo Tes of Dicofol on Cocoon Production and Viability of Earthworm *Pontoscolex corethrurus* Fr. Mull

R Sumarmin[1]*, N K Huda[2], E Yuniarti[1]
[1]Zoology Laboratory of Biology Department, Universitas Negeri Padang, West Sumatera, Indonesia
[2]Embryology Laboratory of Biology Department of Andalas University, West Sumatera Indonesia

* ramadhan_unp@yahoo.com

Abstract. The uncontrol using of pesticides, harmful to the environment, health, and it would have impact to non-target animal as earthworm. This study describes the effect of the Dicofol to cocoon production and viability of earthworm Pontoscolex corethrurus Fr. Mull., has been done in-July - Augustus 2016 at the zoology laboratory of Biology Department of Universitas Negeri Padang. The experiment used the Completely Randomized Design (4 treatments 6 replications). The treatments are with 0 g / l (P1), 0.002 g / L (P2), 0.004 g / L (P3), and 0.006 g / L (P4) and 0.008 g / L of Dicofol that diluted to water. The Data of production and viability of earthworm cocoons Pontoscolex corethrurus Fr. Mull collected during 30 days in alternate day. Data analyzed by ANOVA and Duncan New Multiple Range Test at $p < 0.05$. The results Showed that the average number of cocoons production at P1 30 cocoons (the highest), 16 cocoons P2, P3 7 cocoons, and the P4 and P5 0 cocoons (the Lowest). The average percentage of cocoons viability were highest in P1, and P2 (100%); P3 (10%) and the cancel at P4 and P5 (0%). It can conclude that the pesticide Dicofol decreased the production and viability of the earthworm cocoons Pontoscolex corethrurus Fr. Mull.

1. Introduction

The use of Pesticides has been able to increase agricultural production in the world, to reach approximately US $ 25 billion per year [1]. Pesticides are often used by farmers exceed the recommended dose for farmers think with more and more pesticides are used then it will get better results.

Dicofol pesticides is one of the insecticides that are often used by farmers in Indonesia. Dicofol insecticides include but are not specific class of organophosphate certain kill insects, but can also cause deadly poisoning or other organisms or non-target animals. Various forms of organophosphorus compounds are often used to control insects and use more frequently and more than other groups of insecticides. The result is the market is dominated marketing of pesticide compounds [2].

This insecticide achieve the goals coming in through the mouth, skin or respiratory. Insecticides Dicofol is one pesticide to poison insects, including organophosphates group containing carbon and phosphorus, and can be neurotoxic in humans. Dicofol very effectively used to combat and eradicate plant pests such as aphids, rice weevils, flies, leafhoppers and so on [3].

Nonetheless, the unwitting use of insecticides Dicofol continuous negative impact on the environment both humans, animals and soil biota. The impact of these insecticides can be death, obstacle metabolic activity, inhibition of behavior, reproduction and hatchability soil biota. One of them is the earthworm is the soil biota that are often found on agricultural land and have a beneficial role in the soil ecosystem. Earthworms and microbs on agricultural land have function to degradation substancies in the process of decomposition and mineralization of organic matter. The decomposition process of organic matter causes changes in soil structure so that it can improve soil aeration and water holding ability of the soil [3,4].

Some research has been conducted regarding the effect of an insecticide on earthworms as practiced by [5] about the use of Insecticide Endosulfan Effect on Production and Viability of Cocoon Earthworm *Pontoscolex corethrurus* Fr. Mull and [6] about the use of Insecticide Carbofuran Effect on Production and Viability of Cocoon Earthworm *Pontoscolex corethrurus* Fr. Mull. In both of these studies found a decline in production and quality earthworm cocoons.

Based on this background, this reseach aim to investigate the effect of insecticides on production and viability Dicofol earthworm cocoons *Pontoscolex corethrurus* Fr. Mull.

2. Methods

This study was conducted in July-October 2016 in the Laboratory of Zoology Department of Biology, Universitas Negeri Padang. The study design used Completely Randomized Design (CRD), with 5 treatments each treatment consisted of 6 replicates. The concentration of treatment based on dose conversion from laboratory scale to field scale are:

P1. 0 g / l (control) (Solvent / water)
P2. 0.002 g / l Dicofol
P3. 0.004 g / l Dicofol
P4. 0.006 g / l Dicofol
P5. 0.008 g / l Dicofol

a. Dicofol solution

Concentration is fluid density Dicofol insecticidal active ingredient. Dicofol insecticide concentration used was 60%. To make Dicofol insecticide concentration of 0.002 g / l is to enter by using a micro pipette Dicofol as 2,5μ then enter it into the measuring cup and add the solvent until the volume becomes 10 ml, this will continue to make further treatment.

b. Observations earthworm Cocoon Production

Production To detect earthworm cocoons *Pontoscolex corethrurus* Fr. Prepared Mull 25 Plastic jars the size of 5 liters, then filled with Growing Media Standard 1 kg of the next volume mencapurkan Dikofol solution according to treatment and stir until smooth and each jar was added polybags and labeled. Each jar Plastic included 5 earthworms and on the ground placed rice bran. Then plastic jars covered using gauze. 1 time every 5 days each plastic jars sorted and counted the number of earthworm cocoons produced.

c. Observations earthworm cocoon Viability

Cocoon produced by the worms of each treatment was calculated, separated, and placed on filter paper moistened with water that has been deposited for 3 days in a petri dish container. Then once placed on a hard strain and covered with filter paper that has been moistened. Every day spraying by using distilled water 2 times a day, as well as to see the cocoon hatch. Total cocoon hatch is calculated every day until the 15th day of incubation.

d. Data Analysis

Data are number of cocoons produced and cocoon hatch in each treatment group were tested by ANOVA, followed by DNMRT test *(Duncan'sNew Multiple Range* Test).

3. Results and Discussion

Based on research conducted at the Laboratory of Zoology Department of Biology, Universitas Negeri Padang, found that Dicofol have an influence on production and viability earthworm cocoons *Pontoscolex corethrurus* Fr. Mull, as shown in Table 1.

Table 1. The average yield earthworm cocoon production *Pontoscolex corethrurus* Fr.Mull

No.	Treatment	Total average cocoon
1.	P1	30^{a}
2.	P2	16^{b}
3.	P3	7^{c}
4.	P4	0^{d}
5 .	P5	0^{d}

Description: the numbers in the same column followed by the same Superscript letters, not significantly different (P <0.05) in DNMRT test

The results showed that the production and viability of earthworms *Pontoscolex corethrurus* Fr. Mull., Conducted over 30 days of observation with various concentrations showed a highly significant difference between treatments. The higher concentration of insecticide Dicofol given further reduces the number of cocoons produced.

The highest number of cocoons found in P1 treatment (control) is 30 cocoon. While the number of cocoon lowest is 0 at P4 and P5 treatment because there are treatments applied to turn off all the test animals. Total cocoon at significantly different P1 with P2, and P3. Total cocoon on P3 is the lowest of all experimental units were still alive. It also shows that there are significant differences among treatment and decrease the number of cocoon in line with the increase in dose.

It most likely cause a decrease in the number of cocoon production between treatments is active insecticidal compounds of Dicofol can enter the body worms either through the mouth or skin and such compounds to also on the reproductive system. In addition the nature of systemic Dicofol also stored and circulated in the body of the worm.

The decline in fecundity due to disruption of the process of oogenesis mainly on the process for entrainment Dicofol to vitelogenesis cause poisoning when opened or can not be found will be dissected cocoon prepared for ovulation, while in treatment P1 (control), P2, and P3 will still find many cocoons or eggs.

According to [5] fecundity on earthworm cocoons *Pontoscolex corethrurus* Fr. Mull., A decline due to the influence insecticides are toxic to the body tissues of animals and systemic works as a contact poison, stomach poison, and the poison can breath in through digestion and respiration. Cocoon production is a subject that can adjust to some kinds of conditions, especially in response to food. Total cocoon issued a the connecting link between one generation to the next generation.

This result contrary to the result of research [7], that the earthworms *Pontoscolex corethrurus* Fr. Mull., which produces the highest number compared to other worm cocoons of 98 cocoons within one year under laboratory conditions. In his research explains that *P. corethrurus, Dichogaster affinis, Metaphire houlleti and Octochaetona beatrix* has the ability to self-fertilization or parthenogenetic species can produce a cocoon without mating. Almost all exotic species of earthworms in the humid tropics that reported at least facultative parthenogenic, while local species only produce cocoons in case of marriage.

The existence or hormone production in the earthworm is strongly influenced by the feed given so that will affect to the amount of cocoons produced. According to the researh result [5] that the earthworm population depends on the strongly type and concentration of insecticide using. The differences between the number of the cocoon of earthworms hatched caused by changes in temperature and limited sources of food reserves in the cocoon and the ability detoxication or excrete toxicant associated with the concentration of insecticide.

Table 2. The average yield Viability earthworm cocoons *Pontoscolex corethrurus* Fr.Mull.

No.	Treatment	Mean percentage (%) hatching cocoon
1.	P1	100[a]
2.	P2	100[a]
3.	P3	10[b]

Description: The numbers in the same column followed by the same Superscript letters, not significantly different ($P < 0.05$) in DNMRT test

Based on Table 2 are known percentage of the earthworm *Pontoscolex corethrurus* Fr.Mull. cocoon hatchability, in the treatment of P1, P2, and P3, there are differences in average percentage of hatching cocoon. In the treatment of P1 and P2, have an average percentage of the cocoon hatchability as same as namely 100%. While on treatment P3 cocoon hatching percentage as much as 10%.

The percentage hatchability earthworm cocoons *Pontoscolex corethrurus* Fr. Mull., which was seen among treatment P1 and P2 is 100%. While the lowest percentage of cocoon hatchability seen in treatment P3 is 10%. It is suspected in Dicofol P3 has been able to be embryotoxic, causing death. It could be Dicofol which also neurotoxic properties has led to the failure of the embryo develops for failing the formation of the central nervous system. The percentage difference in hatchability for Dicofol pesticides are systemic and embryos are also toxic.

Suspected poisoning has occurred against the ovum in the process of oogenesis. If indeed this is the case then there is no development of the embryo from the beginning [5]. While on treatment P1 and P2 do not affect the viability cocoon hatchability or suspected in the concentration Dikofol though present in the worm body tissue but still on his tolerance threshold.

The decreased amount due to insecticide Dicofol cocoon has the power to kill broad spectrum against insects and earthworms. These compounds can enter the body through physical contact, digestive system, and respiration [5]. Soil moisture and temperature also affect the hatching success earthworm cocoons and removal of the field and laboratory conditions [7].

References

[1] Ceballos A.I., and Gonzales J.G. 2009. Influence of Adult *Pontoscolex corethrurus* on Development of Cocoons and Hatchlings. *Dynamic Soil, Dynamic Plant*. 3(2): 119-121.

[2] Bhatacharjee, S., Dey, A., and Chaudhuri, P.S. 2014. Growth and Reproduction of *Pontoscolex corethrurus* (Muller) in the Mineral Soils of Different Age Groups of Rubber (*Havea brasiliensis*) plantations Under Laboratory Conditions. *Annals of Biological Research*. 5(7): 1-9.

[3] Ning, D. 2001. In Ex situ bioremediation Diazinon Using microbial *Indigenous* isolates B3. Journal of Chemistry: Institut Pertanian Bogor.

[4] Nath, S., and Chaudhuri, P.S. 2014. Growth and Reproduction of *Pontoscolex corethrurus* (Muller) with Different Experimental Diets. *Tropical Ecology*. 55(3): 305-312.

[5] Narayan S.P., Sathrumithra S., Christopher G., Thomas A.P., and Julka J.M. 2016. Current distribution of the invasive earthworm *Pontoscolex corethrurus* Fr.*Mull*. after a century of its first report from Kerala state India. *Opus. Zool. Budapest*. 47(1): 101-107

[6] Nofyan,. E. 2012. Effect of Insecticide Carbofuran against Cocoon Production and Viability of Earthworm *Pontoscolex corethrurus* Fr.Mull. *Science Research Journal: Sriwijaya University, South Sumatra*, Indonesia. 09: 044-047

[7] Darmawan, A., Atmowidi, T., Manalu, W., and Suryosubroto, B. 2015. *Pontoscolex corethrurus* Fr.*Mull*. (Oligocheta) Glossoscolicidae in Forest Transformation System in Bungku Village, Jambi, Indonesia. *Biodiversity Journal*. 6(2): 505-512.

IOP Conf. Series: Materials Science and Engineering **335** (2018) 012014 doi:10.1088/1757-899X/335/1/012014

Tristania Sumatrana Effect On Female Mus Musculus Fertility

S Syamsurizal

Biology Department, Faculty of Mathematics and Science, Universitas Negeri Padang, Indonesia

syam_unp@fmipa.unp.ac.id

Abstract. The use of traditional medicinal plants are generally based on empirical experience, therefore it is necessary scientific approach in order to bring traditional medicine into medical practice and the formal health services. *Tristania sumatrana* Miq. is one of the traditional medicinal plants are often used as a contraceptive for women in West Sumatra *Tristania sumatrana* Miq. extract can prolong the estrous cycle in mice to eleven days. This study aimed to influence *Tristania sumatrana* Miq. extract treatment on fertility of female mice. Experiments conducted a randomized block design 5x2. Five dose groups: control with no treatment, placebo, treatment with doses of 600, 900 and 1200 mg/kgbw and the old two treatment groups: 10 and 20 days. Fertility parameters studied were ovarian weight follicle Graaf number, the corpus luteum and fetal life. The research proves of *Tristania sumatrana* Miq. extract treatment causes a decrease very significantly the ovarian weight (treatment of 900 mg/kgbw for 10 days), follicles Graaf number (900 mg/kgbw for 20 days), the corpus luteum (600 mg/kgbw for 10 days) and live fetuses (900 mg/kgbw for 10 days). *Tristania sumatrana* Miq. extract treatment can lead to decreased fertility of female mice.

1. Introduction

The use of herbs as medicine have been done a long time as a remedy for family planning (1). This situation is supported by geographic and demographic factors. The Indonesia archipelago in tropical regions rich in various species of flora. Of the forty thousand species of flora in the world, thirty thousand of them grow in Indonesia. Indonesia population mostly live in rural areas, so that equitable health services difficult. This situation causes not all residents can enjoy modern health services, including family planning services. The Government supports the traditional medical treatment can be responsible for equitable distribution of health services. WHO also recommends ways to use traditional medicine as complementary to equitable health services to the public (2).

The use of traditional medicine is generally based on empirical experience alone. It required a scientific approach in order to bring traditional medicine into the medical practice and the formal health services. These conditions make traditional medicines have priority to be developed to support autonomy in drug procurement indispensable for equitable distribution of health services. Step-by-step development of traditional medicine include: testing drug efficacy; research and development of traditional medicine as well as the cultivation of sources of natural medicinal substances.

Already the collected 87 species of Indonesian plants have anti-fertility effect with an active ingredient of steroids such as diosgenin, hekogenin and stigmasterol (3). About 50 species have been scrutinized anti fertility effects in experimental animals and human females. Results of the study showed that the plant's proficiency level can act as anti-gonadotropic, anti-implantation or anti ovulation (4).

The survey in the Valley Bawan village, West Sumatra obtained information that *Tristania sumatrana* Miq. is often used as a contraceptive for women. How to use it is to make lotion bark steeping dried, then drank the water once in two days after the menstrual period for 10-14 days. After drinking water steeping the bark lotion usually they do not menstruate, or if menstruation is not much and not for long.

Tristania sumatrana Miq. extract can prolong the estrous cycle of white mice *(Mus musculus)* to eleven days (5). While the mice were normal estrous cycle is 4-5 days. The phenomenon is an early indication of the presence of a disorder of the reproductive system in animals after administration of wood extract lotion. A plant substance that can show its activity as an anti-fertility in female animals, generally associated with the ability to intervene in the reproductive system that includes the organs of the hypothalamus, anterior pituitary and ovaries (6) chemical analysis bark *Tristania Sumatrana* Miq. steroids obtained from the type of β-sitosterol and stigmasterol (7) which are anti-fertility (8). More detailed biological activity has not been fully studied, such as its effect on ovarian weight, number of follicles, primary, secondary, tertiary, Graaf, atresia, corpus luteum, the child (fetus), and resorption.

Based on the background formulated research problem: Is *Tristania sumatranaMiq.* extract effect on the fertility of female mice which include ovarian weight, number Graaf follicle, corpus luteum, and live fetuses. Hypothesis: The *Tristania sumatrana* Miq. extract can decrease the fertility of female Swiss Webster mice. This study aimed to extract the effect of *Tristania sumatrana Miq.* on fertility of female mice. This research is expected to be useful in providing information about the effects of extract of *Tristania sumatrana Miq.*on fertility of female mice which include ovarian weight, number Graaf follicle, corpus luteum, and live fetuses as well as provide information to explore the possibility of the use of *Tristania sumatrana* Miq. as a raw material for contraceptives.

2. Method

The study design is a randomized block design 5 x 2 factorial pattern with two independent variables. The first independent variable is the dose to 5 treatments. The second independent variable is the duration of the 2 treatments. For more details, kind of treatment given to the female experimental animals can be seen in Table 1.

Table 1 Kind of treatment given to the female experimental animals

Factor		B				
	Stage	b 1	b 2	b 3	b 4	b 5
A	a 1	a 1 b 1	a 1 b 2	a 1 b 3	a 1 b 4	a 1 b 5
	a 2	a 2 b 1	a 2 b 2	a 2 b 3	a 2 b 4	a 2 b 5

Legend:

A	:	Long treatment	a1b1	:	Treatment of 10 days, 600 mg/kg bw
a1	:	Treatment of 10 days	a1b2	:	Treatment of 10 days, 900 mg/kg bw
a2	:	Treatment of 20 days	a1b3	:	Treatment of 10 days, 1200 mg/kg bw
B	:	Dose	a1b4	:	Positive control of 10 days, CMC 1%
b1	:	Treatment of dose 600 mg/kg bw	a1b5	:	Negative control of 10 days
b2	:	Treatment of dose 900 mg/kg bw	a2b1	:	Treatment of 20 days, 600 mg/kg bw
b3	:	Treatment of dose 1200 mg/kg bw	a2b2	:	Treatment of 20 days, 900 mg/kg bw
b4	:	Positive control of CMC 1%	a2b3	:	Treatment of 20 days, 1200 mg/kg bw
b5	:	Negative control	a2b4	:	Positive control of 20 days, CMC 1%
			a2b5	:	Negative control of 20 days

Number of replications of each treatment four times, are determined using a formula Federer. Animal experiments used female mice were healthy, had never been pregnant, have regular estrous cycles 4-5 days. 20-30 gr body weight and 2-3 months old. Animals females were selected based on vaginal smears every day at the same time for 3 estrous cycle. Mice with regular estrous cycles 4-5 days designated as experimental animals. Having obtained 40 head of experimental animals with normal estrous cycles, then grouped randomly into 10 cages. Each cage contained 4 mice.

Wood extract dose *Tristania Sumatrana Miq.* given in experimental animals refers to Musdja who has researched the effect of lotion timber on the estrous cycle in mice taking doses of 300 and 600 mg/kgbw. Based on that, to see its effect on oogenesis, the initial dose of the initial dose used in this study is there a 600 mg/kgbw (b1), for a dose of 300 mg / kb bb yet to show tangible results. The next treatment groups, the dose was increased to 900 mg/kgbw (b2) and 1200 mg/kgbw (b3). The volume of extract for once-feeding is 0,20 ml.

Tristania sumatrana Miq. extract in experimental animals conducted over 10 days in the first five groups, five other groups is extended to 20 days. Increasing the dose and administration time extension aims to further assure the effect of treatment of experimental animals and to determine the dose and duration of administration effective to suppress fertility.

Tristania Sumatrana Miq. obtained in Agam, West Sumatra manufacture bark Tristania sumatrana Miq. extract by taking 100 grams of powdered bark of *Tristania sumatrana Miq.* then dissolved in 1 L of 95% ethanol for 24 hours (maceration). Extracts were formed was concentrated by *rotary vacuum evaporator* (rovapor). Concentrated extract obtained is evaporated with a water bath, stirring until dry extract (39), then extract from the bark of *Tristania sumatrana* Miq. ready to be made according to the dosage solution.

Doses of 600 mg/kgbw was prepared by dissolving bark *Tristania sumatrana* Miq. extract as much as 15.024 mg in 0.2 ml of CMC 1%. The calculations are as follows: the average weight of female mice strain Swiss Webster 2,5- age 3 months after the weighing is 25.04 g. Many extracts are used for a dose of 600 mg/kgbw determined by calculation 25.04 / 1000 x 60 = 15.024 mg. The same was done way in determining many extracts used in doses of 900 and 1200 mg / kg bw. Doses of 900 mg/kgbw was prepared by dissolving bark *Tristania sumatrana* Miq. extract A total of 22, 536 mg in 0.2 mL of 1% CMC. A dose of 1200 mg/kgbw was prepared by dissolving bark *Tristania sumatrana* Miq. extract. A total of 30.048 mg in 0.2 ml 1% CMC.

Chemicals used include: fixative Bouin's fixation organ ovary, ether to kill animals, paraffin for infiltration and printing (blocking) stocks histology of the ovary, staining for staining smears of the vagina, benzyl benzoate to remove any residual alcohol in the ovaries, serial alcohol 50%, 70%, 80%, 95%, and 100% for the elimination of water (dehydration) from the ovary, a solution of hematoxylin-eosin to stain preparations histology of the ovary, distilled water, aquades, albumin to attach the specimen to the glass slide and xylol to clear stocks histology.

Tools experiments. Enclosures and accessories; one set of extraction: *rotary vacuum evaporator,* water bath, the type Sartorius 4202 balance to the nearest 0.1 mg, goblets, aluminum foil, a pipette, a tool grinder *Milley Mill.* A set of surgical tools; one set of tooling and observations preparations: microtome, the bottles are 30 ml for spot fixing, the dye bath*(staining jar),* a light microscope, the scales of mice that *Triple Bean OHAUSS Balance Scale Corp. Union NJ USA* no. 2610 to the nearest 0,1g.

Treatment of experimental animals. Treatment done every morning at 08.00 am. Giving begins at the stage of diestrus to the experimental animals were treated 10 days and at the estrous stage to the experimental animals were treated 20 days. Animal experiments are grouped as in Table I. Completed treatment (female mice), the experimental animals mated with female mice by placing first on stage with a 1 male-fertile pro estrus in one cage a maximum of 10 days. During this period the observed presence or absence of the vagina as a sign stopper has occurred copulation. When copulation has occurred then the female mice were separated from the male and maintained for 18 days from the occurrence of copulation. After that, the mice were sacrificed pregnant or not pregnant.

Anti-fertility drug action in living beings females can be divided into a number of purposes, including preventing the maturation of the egg, preventing ovulation, preventing fertilization. The parameters studied from the *Tristania sumatrana* Miq. extract fertility test animals are 1) Weight ovary; both ovaries are left and right ovaries weighed. 2) The number of follicles Graaf; follicle Graaf studied through histological preparations of ovaries. 3) The number of corpus luteum; the amount of the corpus luteum can be obtained in two ways. First, the corpus luteum of the ovary directly observed macroscopically. The corpus luteum will appear as bright red bumps. Second, the observation of a microscopic amount of the corpus luteum of the ovary and through histological preparations necessary to ensure the observations made by microscopic. 4) The number of live fetuses; female mice that were pregnant or not pregnant anesthetized with ether on the 18th day after the marriage. After the death of experimental animals dissected through the ventral abdominal area to open the uterus and took the fetus, then calculated the number of live fetuses.

Data obtained from observations of normality was tested using the method of Shapiro and Wilk. Homogeneity Barlett test data. If both tests are met, then the data were normally distributed population and its variations homogeneous, followed by a two-way ANOVA test. If the normality and homogeneity is not met then the transformation $\sqrt{(x + ½)}$ in the non-parametric statistics were analyzed with Friedman. If based on the Friedman test there is significant then forwarded the data analysis with the Advanced Friedman test.

3. Results

3.1. Ovary weight

Weighing ovaries results of mice left and right on surgery day 18 of pregnancy are presented in Table 2. The results of normality and homogeneity test proved that ovary weight data were normally distributed with a homogeneous variation. Based on ANOVA test is known there are differences in ovarian weight were highly significant (P <0.01) for all doses of treatment. BNT obtained through further test weight differences ovarian highly significant (P <0.01) between the group treated extract *Tristania sumatrana* Mig. at doses of 600, 900 and 1200 mg/kg/bw for 10 and 20 days. In addition, there are differences in ovarian weight were not significant (P> 0.05) for the duration of the treatment, which is between the treatment groups for 10 days compared with 20 days treatment at all doses. Similarly, the dose level with the old administration (T1T2) using extracts *Tristania sumatrana* Miq. do not interact significantly (P> 0.05) in ovarian weight. Test multiple ovarian weight BNT can be seen in Table 3.

Table 2. Mice ovarian weight (mg) after treatment with *Tristania sumatrana* Mig. Extract at doses of 600, 900, and 1200 mg/kg bw for 10 and 20 days.

Group		600/10	900/10	1200/10	PC/10	NC/10	600/20	900/20	1200/20	PC/20	NC/20
Repeat	1	23.05	18.35	23.70	33.70	36.80	25.04	26.15	25.47	30.75	31.68
	2	33.84	26.10	28.43	30.42	42.30	21.91	26.79	24.92	32.70	32.30
	3	20.00	17.50	27.30	33.25	45.90	16.20	18.80	16.20	33.50	33.30
	4	26.73	20.60	18.00	31.97	31.90	19.70	19.00	24.60	33.50	31.44
Avg		25.91	20.64	24.36	32.34	39.23	20.71	22.69	22.80	32.61	32.18
SD		5.16	3.35	4.06	1.27	5.33	3.22	3.79	3.82	1.12	0.72

Description: 600/100, treatment group treated with *Tristania sumatrana Mig*. ekstact with dose 600 mg/kg bw for 10 days. Likewise 900/10, 1200/10, PC/10 (positive control), and NC (negative control). 600/20, treatment group treated with *Tristania sumatrana Mig*. ekstact with dose 600 mg/kg bw for 20 days. So did 900/20, 1200/20, PC/20 and NC/20.

Table 3. Test multiple ovarian weight BNT

Doses	N T	600/10	900/10	1200/10	PC/10	NC/10	600/20	900/20	1200/20	PC/20	NC/20
600/10	25.91	-									
900/10	20.64	5.27	-								
1200/10	24.63	1.55	3.72	-							
PC/10	27.89	1.99	7.52*	3.53	-						
NC/10	32.94	7.04*	12.30**	8.58*	5.05	-					
600/20	20.71	5.19	0.07	3.65	7.18*	12.23**	-				
900/20	22.69	3.22	2.05	1.67	5.21	10.26**	1.97	-			
1200/20	22.80	3.11	2.16	1.56	5.10	10.15**	2.08	0.11	-		
PC/20	29.48	3.57	8.84*	5.12	1.58	3.47	8.76*	6.79*	6.68*	-	
NC/20	29.56	3.65	8.92**	5.20	1.66	3.39	8.84*	6.87*	6.76*	0.08	-

* = $P < 0.05$; ** = $p < 0,01$ P_{table} 5% = 6.57; 1% = 8.88

3.2. The number of Graaf follicles

A number of Graaf follicles in the ovaries of mice left and right are presented in Table 4. Results of normality test data on the number of Graaf follicles turns distribution is not normal, then the data is transformed. The test results of normality and homogeneity tests showed that the amount of data that has been transformed Graaf follicles normally distributed, with variance homogeneity. Two-way ANOVA test results showed that: There are differences in the number of Graaf follicles highly significant (P <0.01) in the combination of Tristania sumatrana Miq. extractnamely between doses of 600, 900 and 1200 mg/kgbw with a duration of 10 and 20 days. Based on BNT multiple test proved no difference Graff follicles were highly significant (P <0.01) between the group treated with the extract of *Tristania sumatrana* Miq. at doses of 600, 900 and 1200 mg/kgbw for 10 and 20 days compared to controls. Similarly, the dose level with the old treatment (T1T2) using extracts *Tristania sumatrana Mig.* do not interact significantly (P>0.05) in reducing the number of follicles Graaf. Test multiple BNT number of follicles Graaf for the combination treatment can be seen in Table 5.

Table 4. A number of Graaf follicles in the ovaries of mice left and right

Group	600/10	900/10	1200/10	PC/10	NC/10	600/20	900/20	1200/20	PC/20	NC/20
Repeat										
1	12	16	7	16	33	17	8	7	35	8
2	15	14	14	28	23	11	12	10	22	30
3	16	15	23	22	23	17	3	8	23	23
4	14	14	4	17	10	7	11	9	20	20
Avg	14.25	14.75	12.00	20.75	22.25	13.00	9.00	8.50	25.00	20.25
SD	1.48	0.93	7.31	4.76	8.17	4.24	2.74	1.12	5.87	7.95

Table 5. BNT multiple test of Graaf follicles for combination of *Tristania sumatrana* Mig. Extract with doses of 600, 900 and 1200 mg/kg/bw of 10 and 20 days.

Doses	NT	600/10	900/10	1200/10	PC/10	NC/10	600/20	900/20	1200/20	PC/20	NC/20
600/10	3.77	-									
900/10	3.84	0.07	-								
1200/10	3.30	0.47	0.54	-							
PC/10	4.53	0.76	0.69	1.23*	-						
NC/10	4.62	0.85	0.79	1.33	0.10	-					
600/20	3.55	0.22	0.29	0.26	0.97	1.07*	-				
900/20	2.96	0.81	0.88	0.33	1.56**	1.66**	0.59	-			
1200/20	2.91	0.86	0.93	0.39	1.62**	1.72**	0.64	0.05	-		
PC/20	4.97	1.20*	1.13	1.67**	0.44	0.34	1.42**	2.01**	2.06**	-	
NC/20	4.39	0.62	0.55	1.10*	0.13	0.23	0.84	1.43**	1.48**	0.58	-

* = P < 0,05 ** = P < 0,01 P table 5% = 1,04 ; 1% = 1,41

3.3. Total corpus Luteum

Calculation result of ovarian corpus luteum in the number of left and right mouse are presented in Table 6. The results of normality and homogeneity test data on the number of corpus luteum normal distribution with variance homogeneity. The results of a two-way ANOVA test showed: There are differences in the number of corpus luteum were highly significant (P <0.01) in the combination of *Tristania sumatrana* Miq. extract namely between doses of 600, 900 and 1200 mg/kg/bw with a duration of 10 and 20 days. Based on BNT multiple test proved no difference in the number of corpus luteum were highly significant (P <0.01) between the group treated with the extract of *Tristania sumatrana Mig.* at doses of 600, 900 and 1200 mg/kgbw for 10 and 20 days compared to controls. Test multiple BNT number of corpus luteum can be seen in Table 7.

Table 6. Number of corpus luteum after treatment with *Tristania sumatrana* Mig. Extract at doses of 600, 900 and 1200 mg/kg/bw of 10 and 20 days.

Group		600/10	900/10	1200/10	PC/10	NC/10	600/20	900/20	1200/20	PC/20	NC/20
Repeat	1	10	12	6	11	24	7	11	4	11	17
	2	10	9	11	18	14	8	10	8	12	11
	3	8	10	8	12	11	11	9	7	13	12
	4	9	2	8	14	13	11	10	4	9	12
Avg		9.25	8.25	8.25	13.75	15.50	9.25	10.00	5.75	11.25	13.00
SD		0.83	3.77	1.79	2.68	5.03	1.79	0.71	1.79	1.48	2.35

IOP Conf. Series: Materials Science and Engineering **335** (2018) 012014 doi:10.1088/1757-899X/335/1/012014

Table 7. BNT multiple test corpus luteum number for combination of Tristania sumatrana mig. Extract with doses of 600, 900 and 1200 mg/kg/bw of 10 and 20 days

Doses	NT	600/10	900/10	1200/10	PC/10	NC/10	600/20	900/20	1200/20	PC/20	NC/20
600/10	9.25	-									
900/10	8.25	1.00	-								
1200/10	8.25	1.00	0.00	-							
PC/10	13.75	4.50*	5.50*	5.50*	-						
NC/10	15.50	6.25**	7.25**	7.25**	1.75	-					
600/20	9.25	0.00	1.00	1.00	4.50*	6.25**	-				
900/20	10.00	0.75	1.75	1.75	3.75	5.50*	0.75	-			
1200/20	5.75	3.50	2.50	2.50	8.00**	9.75**	3.50	4.25*	-		
PC/20	11.25	2.00	3.00	3.00	2.50	4.24	2.00	1.25	5.50*	-	
NC/20	13.00	3.75	4.75*	4.75*	0.75	2.50	3.75	3.00	7.25**	1.75	-

* = P < 0,05 ; P = < 0,01 P table 5% = 4,25 ; 1% = 5,74

3.4. Life Fetus amount

Calculation result of the number of live fetuses in the uterus of mice left and right are presented in Table 8. The results indicate that the homogeneity test data on the number of live fetuses that have been transformed not homogeneous. Furthermore, the data is processed using Friedman's non-parametric statistical test. Friedman test results showed differences in the number of live fetuses were highly significant (P <0.01) in the combination of *Tristania sumatrana* Miq. extract between doses of 600, 900 and 1200 mg/kgbw with a duration of 10 and 20 days. Friedman test results the number of live fetuses can be seen in Table 9.

Table 8. The number of live fetuses after treatment with Tristania sumatrana Mig. Extract at doses of 600, 900 and 1200 mg/kgbw of 10 and 20 days

Group		600/10	900/10	1200/10	PC/10	NC/10	600/20	900/20	1200/20	PC/20	NC/20
Repeat	1	8	12	6	11	13	4	10	0	10	12
	2	10	9	11	11	12	9	10	10	12	11
	3	7	8	8	11	12	10	9	1	12	11
	4	10	2	4	13	12	8	9	0	8	12
Avg		8.75	7.75	7.25	11.50	12.25	7.75	9.50	2.75	10.50	11.55
SD		1.30	3.63	2.59	0.87	0.43	2.28	0.50	4.21	1.66	0.50

IOP Conf. Series: Materials Science and Engineering **335** (2018) 012014 doi:10.1088/1757-899X/335/1/012014

Table 9. Friedman test number of live fetus for combination of Tristania sumatrana Mig. Extract with doses 600, 900, and 129

Dosis	NT	600/10	900/10	1200/10	PC/10	NC/10	600/20	900/20	1200/20	PC/20	NC/20
600/10	17.0	-									
900/10	14.5	1.5	-								
1200/10	16.5	0.5	2.0	-							
PC/10	31.5	14.5	17.0*	15.0	-						
NC/10	37.5	20.5*	23.0**	21.0*	6.0	-					
600/20	14.0	3.0	0.5	2.5	17.5*	23.5**	-				
900/20	20.5	3.5	6.0	4.0	11.0	17.0*	6.5	-			
1200/20	7.0	10.0	7.5	9.5	24.5**	30.5**	7.0	13.5	-		
PC/20	29.0	12.0	14.5	12.5	2.5	8.5	15.0	8.5	22.0**	-	
NC/20	31.5	14.5	17.0*	15.0	0.0	6.0	17.5*	11.0	24.5**	2.5	-

* = P < 0,05 ; ** = P < 0,01 R table 5% = 16,78 ; 1% = 22,01

4. Discussion

The results of the two-way ANOVA test showed differences in ovarian weight, number of follicles, corpus luteum and Graff very significant in the combination of extract *Tristania sumatrana* Miq. at doses of 600, 900 and 1200 mg/kgbw with a duration of 10 and 20 days. The LSD shows these extracts award at doses of 600, 900 and 1200 mg/kgbw for 10 and 20 days led to a significant reduction of ovarian weight, number Graaf follicle and corpus luteum. Weight loss can be due to the reduced number of ovarian follicles, primary, secondary, tertiary and Graaf. Weight will decrease if there is ovarian follicle development barriers (9, 10) .

The Tristania sumatrana Miq. extract turned out to cause a decrease in primary follicles, secondary and tertiary (11). *Tristania sumatrana Miq*. known to contain steroids from the class of β-sitosterol and stigmasterol (1, 8, 12). Steroids are one of the factors that may hinder the development of the follicle through suppressing FSH levels (13). Plants that contain estrogenic steroid generally so that it can affect the menstrual cycle and follicular development (11, 14, 15).

The decline in the number of follicles can also be caused by steroid β-sitosterol and stigmasterol of *Tristania sumatrana Miq*. These steroids can disrupt the balance shaft hypothalamus, pituitary and ovaries. Steroid β-sitosterol and stigmasterol thought to cause increased levels of estrogen in the blood, thus inhibiting the secretion of FSH and LH. FSH regulates the development and the number of follicles. Decreased levels of FSH may lead to follicular development is inhibited from primary follicles, secondary, tertiary and Graaf (6). β-sitosterol and stigmasterol in Lichenes can also suppress fertility albino mice (13).

Another possibility is the active substance contained in *Tristania sumatrana Miq*. directly inhibit follicle development in the ovary. This was shown by the declining number of primary and secondary follicles (9, 16).

Decrease in the number of follicles in the treatment of 600 mg/kgbw for 10 days, has reached an optimal effect against ovarian weight, number Graaf follicle and corpus luteum. Increasing the dose to 900 and 1200 mg/kgbw and the addition of the treatment time to 20 days did not significantly decrease the research parameters. Drug response in organisms such as sinusoidal curve, dose escalation will lead to increased response of the organism. Optimal doses above the curve going up exponentially but tends to be linear (17) (10).

The optimal dose may also occur due to the active compound contained in *Tristania sumatrana Miq*. Exceed the number of specific receptors (17). Although the concentration of β-sitosterol and stigmasterol in the body is very high but it can not form a complex with the receptor to be brought to the target cells. Thus increased levels of steroid β-sitosterol and stigmasterol in blood plasma through increasing the dose or duration of administration not necessarily add to its effectiveness.

The interaction between the old dose administration (T1T2) is not meaningful for all parameters. This means that between increasing doses (600 to 900 and 1200 mg / kg bw) and duration of administration (10 and 20 days) Tristania sumatrana Miq. extract not interdependent in affecting ovarian weight, number Graaf follicle and corpus luteum (18). Based on advanced Friedman test showed that the majority of the extract *Tristania sumatrana*Miq.at doses of 600, 900 and 1200 mg/kgbw for 10 and 20 days can reduce significantly the number of live fetuses. Tristania sumatrana Miq. extract can extend into the eleventh day of the estrous cycle (5, 19). Extension of the estrous cycle is an indication of their reproductive system disorders as a result of disruption of the balance gonadotropin (17, 20).

Based on the analysis that has been done against suspected follicle β-sitosterol and stigmasterol derived from extracts of *Tristania sumatrana Miq*. reduce the number of follicles Graaf. The decline in the number of follicles causes a decrease in ovulatory proven to be a highly significant decrease in the number of corpus luteum. Decreased ovulation will minimize the amount of fertilization, thus reducing the number of fetuses.

Increasing doses of the Tristania sumatrana Miq. extract from 600 to 900 and 1200 mg/kgbw and duration of administration of the addition of 10 to 20 days to show the difference in the number of fetuses that are not significant compared to the treatment at a dose of 600 mg/kgbw for 10 days. As has been earlier that the treatment at a dose of 600 mg/kgbw for 10 days is optimum, as a result of an increase in the dose and duration of administration is no longer statistically significant effect on the reduction in the number of fetuses.

The interaction between the old dose administration (T1T2) is not meaningful for the parameter number of fetuses. This means that between increasing doses (from 600 to 900 and 1200 mg/kg bw) and duration of administration (10 and 20 days) *Tristania sumatrana* Miq. extract not inter dependent in influencing the number of fetuses.

5. Conclusion

Based on the results of this study concluded that the Tristania sumatrana Miq. extract can decrease the fertility of female mice *Mus musculus*L. Swiss Webster strain, which include: a decrease in ovarian weight; a decrease in the number of follicles Graaf; a decrease in the number of corpus luteum and decreased number of live fetuses.

References

[1] Mahima Sharma DA, Kiran Bhagour, R.S. Gupta. Natural Aphrodisiac and Fertility Enhancement Measures in Males. Current Medicine Research and Practice. 2017:1-9.

[2] WHO. Guidlines for the Assesment of Herbal Medicine, Program on Traditonal Medicine. Geneva. 1991.

[3] Tati Herlina US, Ukun M.S. Soedjanaatmadja. Biologically active natural products from Indonesian Erythrina plants Proceeding of The International Seminar on Chemistry 2008;1(1):204-7.

[4] Rashaduz Zaman MP, Md. Sekendar Ali and Mohammed Abu Sayeed. Evaluation of Antifertility Effect of Methanolic Bulb Extract of Allium cepa on Swiss Albino Male and Teratogenic Effect on Female Mice Advances in Biological Research 2015;9(2):128-32.

[5] Yarneli Gani NM. Pengaruh Sari Kayu Kasai terhadap Daur Estrus Mencit. Universitas Andalas. 1988:44-8.

[6] Syahrum MH. Sistem Reproduksi Betina. Fakultas Kedokteran Universitas Indonesia. 1994:25-37.

[7] Widjaja H. Isolasi Steroid dari Kulit Batang Kayu Kasai. Universitas Andalas. 1982:34.

[8] Suryawanshi JAS. Neem - Natural Contraceptive for Male and Female. International Journal of Biomolecules and Biomedicine. 2011;1(2):1-6.

[9] Syamsurizal. Pemberian Ekstrak Tristania sumatrana Miq. untuk Menurunkan Folikel Primer, Sekunder dan Tersier Mencit Swiss Webster. BioScience. 2017;1(2):1-7.
[10] Zambarana. Ovarium and plasental weight of various number of fetus in the rat. Bio Reprod. 1991;4(1):212-22.
[11] Gupta M. A review of pharmacological properties, pharmacognosy and therapeutic actions of Putranjiva roxburghii Wall. (Putranjiva). International Journal of Herbal Medicine. 2016;4(6):104-8.
[12] Marlinza JSaR. Effect of Pasak Bumi (Eurycoma longifolia Jack) Root In Precopulation Stage to the Fertility of Female Mouse (Mus musculus L.). Biospecies. 2009;2(2):24-30.
[13] Kumud Tanwar JMaJBSK. Phytochemical Investigation and Anti-Fertility Activity of Lichen Parmelia Perlata International Journal of Recent Research and Review. 2015;8(4).
[14] Kumar J. Evaluation of Flavonoid Contents. International Research Journal. 2009;2(6):27-31.
[15] Vanithakumari TMaG. Comparative Progesterone Study of The Effects of Beta-Sitosterol, Estradiol and on Selected Biochemical Parameters of the Uterus of Ovariectomised Rats. Journal of Ethnopharmacology. 1992;36(1):51-5.
[16] Sukanya Jaroenporn SM, Kingkaew Wattanasirmkit, Hataitip Trisomboon, Gen Watanabe, Kazuyoshi Taya, and Wichai Cherdshewasart. Effects of Pueraria mirifica, an Herb Containing Phytoestrogens,on Reproductive Organs and Fertility of Adult Male Mice. Endocrine. 2006;30(1):93-101.
[17] Meffin. Fundamental of clinical pharmacology, how drug act. Medical Progres. 1990;51(1):64-71.
[18] K SSAS. A Study on Phytochemical Evaluation of Traditional Rice Variety of Tamil Nadu - 'Maappillai Samba' by Gc-Ms. International Journal of Pharma and Bio Sciences. 2015;6(3):606-11.
[19] Sarvani Manthri CSK, Manjula Talluri. Pharmacognostic,Phytochemical and Pharmacological Review of Dendrophthoe Falcata. Journal of Phytology. 2011;3(3):18-25.
[20] Gahlot DSaK. Pharmacognostical Standardization and Phytochemical Screening of Bauhinia Vahlii Leaves. World Journal of Pharmacy and Pharmaceutical Sciences. 2017;6(8):2635-43.

IOP Conf. Series: Materials Science and Engineering **335** (2018) 012015 doi:10.1088/1757-899X/335/1/012015

Morphometry and Lens of Eyes *Bilih* Fish (*mystacoleucus padangensis*, Bleeker) from Lake Toba, North Sumatra and Lake Singkarak, West Sumatra

A Razak[1,2*]

[1] Biology Department, Faculty of Mathematics and Sciences, Universitas Negeri Padang, Padang 25131, Indonesia

[2] Environmental Department, Graduate Program, Universitas Negeri Padang, Padang 25131, Indonesia

*Ar710322@gmail.com

Abstract. This research has been carried out 2015. Bilih fish today need conservation and attention for sustainability. Habitat this fish is treated by human activities in Lake Singkarak, West Sumatera and Lake Toba in North Sumatera. The objectives of the research are describes morphometry of the body and relation with lens of eyes. The methods of the reasearch for measure all parts of surface body fish according www. fishbase.org. For measure and chemical composition of lens of eyes Bilih Fish (M. padangensis) are according Razak (2005). T he result of the research are indicated the size of morphology body Bilih Fish from Lake Toba and from Lake Singkarak is diffrent. Furthermore, diameter of lens is trend linier follow the growth of the body Bilih Fish from Lake Singkarak and Lake Toba. The chemical composition of lens of eyes Bilih Fish from Lake Singkarak contains Sulfur until 73.77% per 100 ppm, another substances like Calcium, Silicone, Magnesium, Phosporus 4.09%-4.83% per 100 ppm. The chemical composition of lens of eyes Bilih Fish from Lake Toba contains Sulfur only 50.08% per 100 ppm, another substances like Kalium, Calcium, Silicone, Magnesium, Phosporus 1.09%-10.43% per 100 ppm. Kalium substance only found in lens of eyes Bilih Fish from Lake Toba. As conclusion, morphometry body Bilih Fish from Lake Toba is bigger better than Bilih Fish from Lake Singkarak and chemical composition lens of eyes Bilih Fish from Lake Toba is influenced by environmental waters factors.

1. Introduction

Bilih fish (Mystacoleucus padangensis) is an important economic freshwater fish. Bilih fish is an endanger species only found in Lake Singkarak [1, 2, 3, 4]. Bilih fish at Lake Singkarak are overfishing and over exploitation, this is danger condition for Bilih fish life. Introduced bilih fish to Lake Toba is an alternative solution for conservation.

Since 2001, Bilih fish it was introduced in Lake Toba. Since then, Bilih fish population growing in Lake Toba until 2015. The introduce of bilih fish to Lake Toba, North Sumatra is done after some of the fisheries researchers consider the results of the study of Bilih fish in their natural habitat. Since there have been efforts to introduce bilih fish to Lake Toba in 2001, this fish can live and breed there. The habitat is suitable and the fish is much larger than the existing bilih fish in its own region. This fish is taken from the Lake Singkarak where on the basis of the results of studies of fisheries researchers conducted the distribution of fish bilih to Lake Toba, North Sumatra. as a water candidate for the introduction of bilih fish. On January 3, 2003, there were 2,840 fishes with a total length of between 4.1 and 5.7 cm and weighing between 0.9 and 1.5 grams were scattered into Lake Toba [5].

Today, one problem at Lake Toba is overfisihing. The research about morphometry and lens of eyes bilih fish at different habitat in Lake Toba and Lake Singkarak are important for sustainibility and conservation bilih fish in the future. This matter related with objectives of the research are describes morphometry of the body and relation with lens of eyes bilih fish.

2. Materials and Methods

2.1. Collection Samples

We collected 100 fish samples from Lake Toba (2° 36' 15.3" (2.6042°) N ; 98° 49' 24.2" (98.8234°) E and Lake Singkarak (1000 31020.10 " E and 000 38025.20 S).

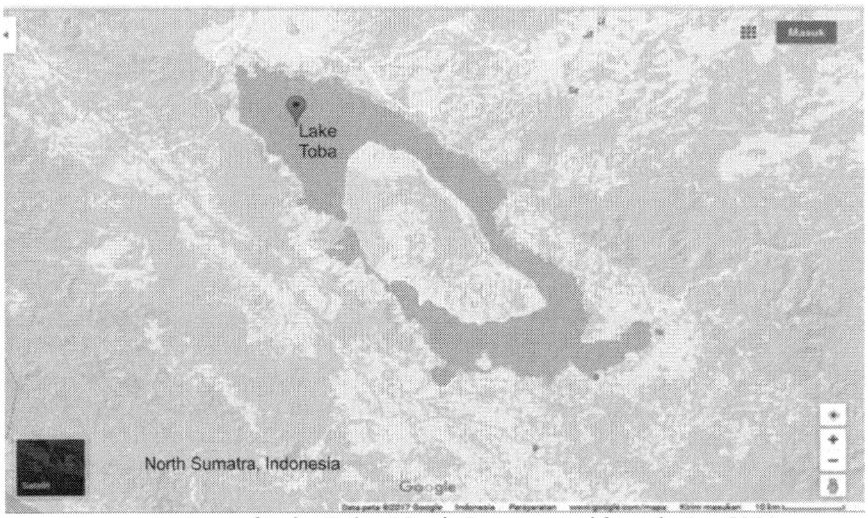

Figure 1. Maps of Lake Toba, North Sumatera with scale 1 : 5000 m (Source : www.google map.com, 2017)

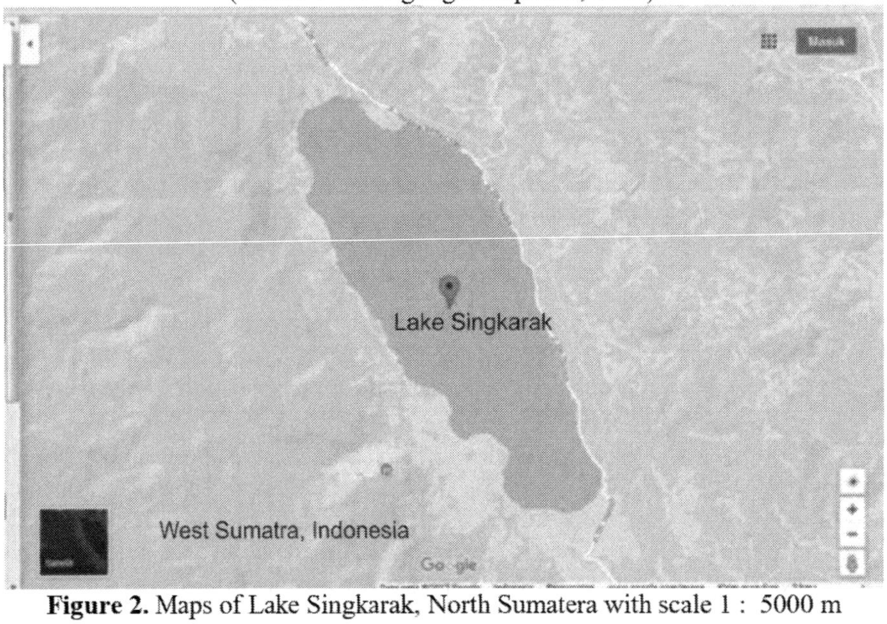

Figure 2. Maps of Lake Singkarak, North Sumatera with scale 1 : 5000 m (Source : www.google map.com, 2017)

2.2. Methods for Measurements morphological character

We measure character morphology with digital clippers We measured 12 morphological character using digital calipers to an accuracy 0.01 mm following Zahorska [6]. We also measure relation size of body, diameter of lens Bilih fish and interaction both to environmental condition of waters in Lake Toba and Lake Singkarak.

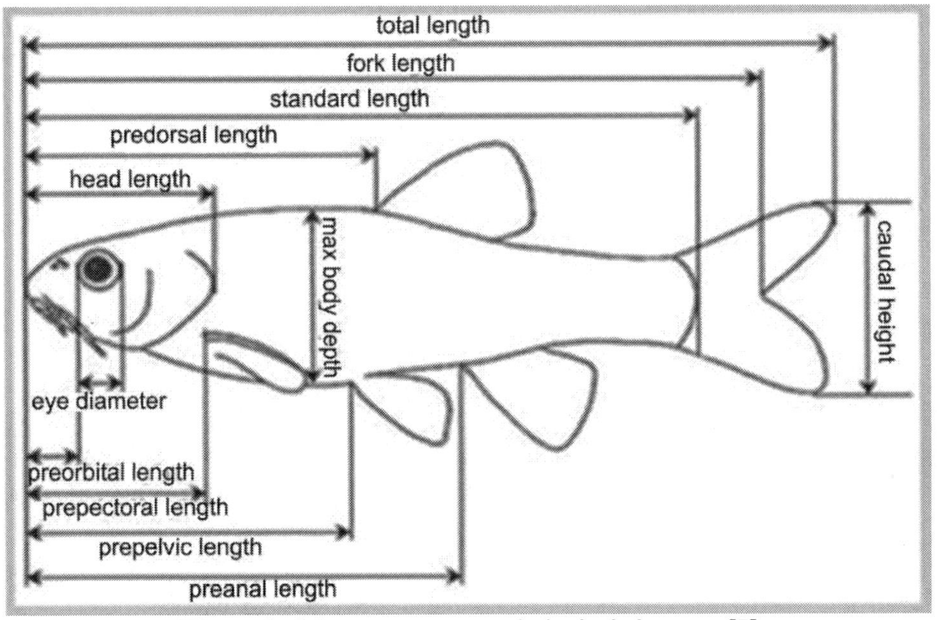

Figure 3. Measurements morphological character [7]

3. Results and Discussions

3.1. Mophometry of Body

The data of morphometric measurement of Bilih fish from Lake Singkarak and Lake Toba shows the difference. From the sample observed body size of Bilih fish from Lake Toba is bigger than Bilih fish from Lake Singkarak. Bilih fish Lake Toba long of body 2.80 cm and Bilih fish from Lake Singkarak only 2.40 cm. In general, the size of TL (Total Length, total body length ranges from 9.20 to 11.50) Bilih fish from Lake Toba. Bilih fish from Lake Singkarak 9.20-11.00 cm as seen in Table 1. For another character we know from Table 1. Table 1 is indicate all character morphometri Bilih fish from Lake Toba bigger than Bilih fish from Lake Singkarak. This condition showed Bilih fish was introduced to Lake Toba success use natural resources from water Lake Toba [5]. In other side condition from Lake Singkarak have damage from many human activities including pollution. This condition big problem in Lake Singkarak and cause reduce growth of bilih fish in this lake.

Table 1. Average measure Morphometri Character of Body *Bilih* Fish Toba and *Bilih* Fish Singkarak Following (www.fishbase.org, 2014)

Location	TL	FL	SL	PreL	HL	eL	PorL	PecL	PeLL	PeaL	MB	CH
Bilih Singkarak	9.90	8.40	7.40	3.50	1.90	0.40	0.50	2.60	5.00	5.00	2.80	2..20
Bilih Toba	10.30	8.90	10.30	4.40	270	0.60	1.60	4.20	5.80	5.80	2.40	2.00

Data in Table 2 , that is fact, Bilih fish from Lake Toba bigger than Bilih fish from Lake Singkarak. Bilih fish that introduce to Lake Toba can adaptation and success life in new habitats. The same condition was found [8] when she and the team reserach bilih fish from 3 location : Lake Toba, Lake *Dibawah* and *Anai* river. The size of Bilih fish from Lake Singkarak is smallest than bilih fish from Lake Toba, Lake *Di Bawah* and *Anai* River. This condition is influenced by condition waters in Lake Singkarak. Overfishing and many human activities triggered some problems in this lake. Another factors, According Kartamihardja and Sarnita [5] in Lake Toba many natural foods and wide areal spawning Bilih fish. This is condition life and trigger growth performance *Bilih* fish in Lake Toba.

Table 2. Distribution Size of Bilih Fish in Two Location (L Toba and L.Singkarak)

Location	Length			Weight	
	Min (mm)	Max (mm)	Average ± SD	Max (mm)	Min (mm)
Lake Toba	92	123	111±4.90	16.60	7.40
Lake Singkarak	92	110	101±10.02	14.00	7.20

The data in Table 2 shows that the growth of fish body of Bilih fish is allometric positive. This condition is according researchs Efendie [9]. Allometric positive weight growth is more rapid growth compared to long growth. Furthermore, the length and weight of Bilih fish in Lake Toba is greater than Bilih fish located in Lake Singkarak. Bilih fish population in Lake Toba get more optimum some environmental factors. The factors are main menu foods of Bilih Fish, habitat and wider areal in Lake Toba more condusive than Lake Singkarak.

Many things can affect the growth and survival of the larvae, factors that can affect the feed and the environment. Environmental factors are very influential on the growth of fish. According to Boyd and Tucker [10], water quality is the main thing to consider in cultivation because water is a living environment of fish. Furthermore, water quality will affect the growth and survival of fish. Poor water quality can reduce the survival and growth. Therefore, good water management efforts are needed to accelerate growth and improve the quality of life of fish. One of the environmental factors that can affect the growth and survival of the fish is the flow. The movement of water causes an even distribution of oxygen, otherwise the current is capable of supplying natural feed.

3.2. Chemicals Composition of Lens of Eyes Bilih Fish

The diameter and chemicals composition of Lens Eyes Bilih Fish From Lake Toba and Lake Singkarak see in Table.3. Data in Table 3. shows that the range diameter of lens Bilih fish from Lake Toba 0.0030 to 0.0120 mm and diameter of Bilih fish from Lake Singkarak 0.0046-0.0079 mm. Furthermore, chemicals composition of lens eyes bilih fish from Lake Toba and Lake Singkarak have 8 element and higher prosentage are Sulfur, but smallest prosentage element is Kalsium (Ca) in lens of eyes Bilih fish Toba.

For lens of eyes Bilih Fish Singkarak smallest procentage s element Kalium (K). This condition is influenced by growth performance individual of Bilih fish and related environmental condition waters in Lake Toba and Lake Singkarak. Conditions of waters in Lake Toba better than the environmental conditions in the Lake Singkarak which pressure over exploitation by man around this lake. Some environmental factors associated with the growth of Bilih fish include lens diameter and its chemical content are water temperature, air temperature, pH, and hardness [8].

Table 3. Diameter and Chemicals composition of Lens Eyes Bilih Fish From Lake Toba and Lake Singkarak

Location	Diameter Lens			Chemicals Composition of Lensa (% per 100) ppm)		
	Min (mm)	Max (mm)	Average ± SD	number of elements	Max	Min
Lake Toba	0.0030	0.0120	0.0006±0.0021	8	S: 50.8	K : 1.07
Lake Singkarak	0.0046	0.0079	0.0011±0.0009	6	S: 74.8	Ca : 1.81

3.3. Interaction Water Quality to Growt Body and Lens of Eyes Bilih Fish

Any some environmental factors diffrence gowth of Bilih fissh habitat in Lake Toba and Lake Singkarak There are main natural foods, hardness, turbidity and water velocity. The growth of fish is affected by external factors in the form of food and environment [11]. Lake Singkarak has value the highest hardness (67.00 mg / L), followed by Lake Toba (25.63 mg / L) and the lowest of its hardness is Below Lake (7.73 mg / L). The hardness parameter is negatively correlated with a view of growth. Meaning that, Lake Singkarak contains high levels of hardness and have display growth of Bilih fish is isometric . For Bilih fish in the Lake Below and Lake Toba with relatively lower hardness has growth view allometic positive. This result leavesback with the existing theory, that of water the more sumptuous the better for health of the fish as it can provide calcium and other ions present in waters [8]. Hardnesses containing calcium, magnesium and carbonate ions and salinity containing the dominant elements of sodium and chloride ions result in the exchange of ions in the gill epithelium as the dividing membrane between blood and water. The exchange of Na + / H + and Cl- / HCO3- increases the excretion of ammonia so that its concentration in water increases. In addition, calcium and chloride ions cause competitiveness to compete in absorption by gill epithelium because gill epithelium tends to bind calcium and chloride ions rather than nitrites (Wedemeyer, 1996). This causes the concentration of nitrite in water to be high. The more lime contained in water that has a higher hardness, less good for fish growth, so that growth decreases sharply [12]. Relation turbidity and growth fish, increased turbidity will cause a sharp reduction of sunlight either in lakes, rivers or in any waters. Turbidity is associated with primary productivity and decreased zooplankton as fish food. This is influences growth of fish. In Lake Toba, the waters have many sulfur compound, it is important for high indexrefractive index of lens. If the lenses of Bilih fish in Lake Toba have lenses with high index refraction then the ability is very useful to get more food and this is related to the growth rate of Bilih fish in Toba lake is greater in comparison with the lake singkarak. This is in the opinion according [13] recent research has shown that the high abundance of methionine (a protein containing Sulfur) is consistent with the need for refractive index crystallin on the lens. Methione plays a role in high index refraction and reduces osmotic pressure.

4. Conclusion

Morphometry body Bilih Fish from Lake Toba is bigger better than Bilih Fish from Lake Singkarak and chemical composition lens of eyes Bilih Fish from Lake Toba is influenced by environmental waters factors. Environmental waters at Lake Toba is better than at Lake Singkarak.

References

[1]. Kottelat, M., A. G. Whitten, S.N. Kartika Sari dan S. Wirjoatmodjo. 1993. Freshwater Fishes of Western Indonesia dan Sulawesi. Perplus Eds.(HK) Ltd and EMDI. Indonesia

[2]. Syandri, H. 1996. Some Biological Aspects of Bilih Fish (M. Padangensis) in Open Water Lake Singkarak Bogor Agriculture Institut.Indonesia.
[3]. Syandri, H., Usman, 1997. The effects of Artificial Plankton on Survival Rate of Larvae Bilih Fish (M. Padangensis). Padang, Research Institute and Community Services. Bung Hatta University, 26 pp. (in Indonesia).
[4]. Syandri, H. J Resultunaidi., Azrita, 2011. Manajemen of Result Bilih Fish (M. Padangensis) Based on Local Wisdom in Lake Singkarak. Journal of Indonesia Fisheries Policy. 3 : 11-18
[5]. Kartamihardja, E.S & A.S. Sarnita. 2008. Populasi Ikan Bilih di Danau Toba. Pusat Riset Perikanan Tangkap, Badan Riset Kelautan dan Perikanan, Departemen Kelautan dan Perikanan. Jakarta
[6]. Zahorska E, Kovac V, Falka I, Beyer K, Katina S, Copp GH et al Morphological variability of the Asiatic cyprinid, topmouth gudgeon Pseudorasbora parva,in its introduced European range. Journal of Fish Biology. 2009; 74:167-185.
[7]. Bleeker., (1852), *Mystacoleucus padangensis*, Available at: www. FishBase: 2014. Accessed: March, 2017.
[8]. Nofrita, D., Syandri, H., & Tjong, D. H. (2015). Morphological differentiation between Bilih Fish (Cyprinidae: Mystacoleucus).
[9]. Effendi, 2002. Metode Biologi Perikanan. Penerbit Djambatan. Jakarta.
[10].Boyd, C. E. 1982. Water Quality Management for Pond Fish Culture. Elsevier Scientific Publishing Co., New York
[11].Huet, M. 1971. Textbook of Fish Culture and Cultivation of Fish. Fishing New Book, Ltd., London
[12].K. Nirmala, R. Wulandari dan D. Djokosetiyanto, 2013. Effect of Hardness in 3 ppt of Water Salinity on growth and survival rates of Barb Fish (Barbus conhonius Hamilton-Buchanan) Jurnal Akuakultur Indonesia, 4 (1): 17–24 (2005)
[13].B. Mahler, Y. Chen, J. Ford, C. Thiel, Graeme Wistow,‡ and Zhengrong Wu, 2013. Structure and Dynamics of the Fish Eye Lens Protein, γM7-Crystallin. Biochemistry. HHS Publish Access.

IOP Conf. Series: Materials Science and Engineering **335** (2018) 012016 doi:10.1088/1757-899X/335/1/012016

The Addition of Several Mineral Sources on Growing Media of Fluorescent Pseudomonad for the Biosynthesis of Hydrogen Cyanide

L Advinda*, M Fifendy, A Anhar

Dept. of Biology, Faculty of Mathematics and Natural Science, State University of Padang, Indonesia

*linda_advinda@yahoo.com

Abstract. All Fluorescent pseudomonad is a group of rhyzobacteria which these days often utilized on plant disease control. The growing media is an absolute requirement which needs to be considered for the growth and cultivation of bacteria. The mineral source contained in growing media of bacteria may affect the production of hydrogen cyanide compound. The objectives of the research were to obtain the best source of minerals for biosynthesis of cyanide acid compounds by fluorescent pseudomonad isolates PfPj1, PfPb1, PfPj2, Kd7, Cas, Cas3, and LAHp2. This research is a qualitative experimental research including observation of hydrogen cyanide compound produced after the growing media of fluorescent pseudomonad bacteria added with several mineral sources. The treatments were given: A = ZnSO4.7H2O 0.5 mM addition, B = CoCl2.6H2O 0.5 mM addition, and C = Fe2SO4.7H2O 0.5 mM addition. From the result of the research, it was concluded that the addition of ZnSO4.7H2O mineral resources on the growing media of fluorescent pseudomonad isolate Cas and Cas3 produced the best hydrogen cyanide. Whereas addition of CoCl2.6H2O mineral source on the growing media showed poor hydrogen cyanide production for all fluorescent pseudomonad isolates

1. Introduction

Fluorescent pseudomonad is a biocontrol agent that can be isolated from the surface area of plant roots. This group of bacteria is non-pathogenic, which can suppress pathogenic pathogens through rhizosphere colonization, producing siderophores, antimicrobial hydrogen cyanide(HCN), IAA, phosphate solvent compounds, and induced systemic resistance. Paul and Sarma [1] reported fluorescent pseudomonad can suppress root rot disease in black pepper plants caused by Phytophthora capsici. According to Advinda [2], fluorescent pseudomonad isolates PfPj1 can inhibit the growth of Ralstonia solanacearum bacteria, and also able to increase the growth of banana plants. Furthermore, Advinda [3] reported fluorescent pseudomonad isolates PfPj1, PfPj2, PfPj3, PfPb1, PfPb2, PfPb3 and PfPm1 can suppress the development of Blood Disease Bacteria (BDB) in banana seedlings. According to Doktycz, et al., [4] Pseudomonas fluorescens GM30 strains can be used for Populus delloides plant defense against pathogens, as well as increasing their growth.

The growth and propagation of biocontrol agents require growing media. The chemical composition contained in the growing media may affect the growth of the biocontrol agent. Antimicrobial/antibiotic production by biocontrol agents is influenced by chemical compositions or mineral salts that constitute a growing media. The production of antibiotics from Streptomyces sp.

IOP Conf. Series: Materials Science and Engineering **335** (2018) 012016 doi:10.1088/1757-899X/335/1/012016

MS-266 Dm4 is affected by various concentrations of mineral salts provided such as: K2HPO4, MgSO4.7H2O, and KCl. The maximum antibiotic concentration was produced at K2HPO4 concentrations of 0.2 g/100 mL, MgSO4.7H2O 0.15 g/100 mL, and KCl 0.05 g/100 mL [5]. Addition of Zn2+, Cu2+, and NH4Mo2+ on growing media of P. fluorescens CHAO can produce 2,4-diacetyl phloroglucinol (DAPG) antimicrobial compounds [6]. This study explored the biosynthesis of hydrogen cyanide by fluorescent pseudomonad isolates PfPj1, PfPb1, PfPj2, Kd7, Cas, Cas3, and LAHp2 that is grown on growing media with the addition of different mineral sources.

2. Material and methods

This research is a qualitative experiment research including observation of hydrogen cyanide compounds produced after the growing medium from pseudomonad fluorescent were added with several mineral sources. Fluorescent pseudomonad used are isolates PfPj1, PfPb1, PfPj2, Kd7, Cas, Cas3, and LAHp2. Mineral salts added in King's B growing medium are ZnSO4.7H2O 0.5 mM, CoCl2.6H2O 0.5 mM, and Fe2SO4.7H2O 0.5 mM.

2.1. Rejuvenation and propagation of fluorescent pseudomonad

Fluorescent pseudomonad isolates PfPj1, PfPb1, PfPj2, Kd7, Cas, Cas3, and LAHp2 were rejuvenated on solid King's B medium, and incubated for 2x24 hours. The inoculum propagation was carried out by taking a pure culture ose in petri, then bred in 25 mL King's B liquid medium, and is shake for 24 hours.

2.2. Preparation of growing medium of fluorescent pseudomonad

The making of King's B + ZnSO4.7H2O 0.5 mM media was done by weighing pepton protease by 20 g, 1.5 g of K2HPO4, 1.5 g of MgSO4.7H2O, 10 mL of glycerol, 10 mL of ZnSO4.7H2O 0,5 mM, and 18 g of gelatin. Then the mixture of the ingredients is inserted into beaker glass, and distilled water added to a volume of 1,000 mL. Sterilization of the media was carried out using an autoclave at 121 ° C and 15 psi of pressure for 15 min. The preparation of 2 other media (addition of CoCl2.6H2O 0.5 mM, and Fe2SO4.7H2O) was carried out in the same manner..

2.3. The making of hydrogen cyanide production trial media

The production media of hydrogen cyanide consist of 4.2 g of glycine and 10 g of TSA which is dissolved into aquadest to 1 Liter volume. Then sterilized in autoclave at 121° C and pressure of 15 psi for 15 minutes. The indicator of hydrogen cyanide compound production is a solution consisting of 2 g of picric acid and 8 g of sodium carbonate dissolved into aquadest to 200 mL volume [7].

2.4. The cultivation of isolates PfPj1, PfPb1, PfPj2, Kd7, Cas, Cas3, and LAHp2

The fluorescent pseudomonad isolates PfPj1, PfPb1, PfPj2, Kd7, Cas, Cas3, and LAHp2 were grown on each prepared media by taking 1 mL of pseudomonad fluorescent suspension (1 Mc Farland's scale). Furthermore, the suspension is inserted into a sterile petri dish, and poured the media according to the treatment. Bacteria cultures were incubated at room temperature for 2x24 hours. The ability of each isolate to grow on each treatment medium given was observed.

2.5. The propagation of isolates PfPj1, PfPb1, PfPj2, Kd7, Cas, Cas3, and LAHp2

Propagation of isolates was carried out for the trial of hydrogen cyanide production. Isolates that have been grown in each mineral addition to the media treatment, are propagated using the same media according to the treatment but with no addition of gelatin (liquid media). Inoculum propagation was performed by taking an ose culture in petri, then bred in 25 mL of liquid medium, and was shake for 24 hours.

2.6. The production of hydrogen cyanide

Isolates that have grown were implanted in glycine media in a petri dish with a cast method. Pieces of filter paper that have been spilled 1 mL hydrogen cyanide detector solution were embedded on the lid of petri dish. Bacteria cultures were incubated at room temperature for 2x24 hours. A change of color of the filter paper from yellow to light brown, brown or reddish-brown was recorded as weak (+), moderate (++) or strong (+++) reaction respectively.

3. Results

3.1. The living ability of the fluorescent pseudomonad in a growing media with addition of several minerals

The best living ability in King's B + ZnSO4.7H2O growing media is found in 4 fluorescent pseudomonad isolates, i.e. isolates PfPj1, Kd7, Cas, and Cas3. The four isolates had the highest bacterial count compared to the other three isolates. The PfPj1 isolate had the best living ability in the growing media of King's B + CoCl2.6H2O, but the isolates of PfPb1 were not able to grow. In the growing media of King's B + Fe2SO4.7H2O, Cas3 isolates are not able to grow. The other six pseudomonas isolates were able to grow on King's B + Fe2SO4.7H2O (Table 1). The growing media ZnSO4.7H2O, CoCl2.6H2O, and Fe2SO4.7H2O used can be seen in Figure 1.

Table 1. The number of bacteria that grow on several different minerals added media (CFU / mL).

No	Isolate	Media type		
		KB+Zn	KB+Co	KB+Fe
1	PfPj1	30.10^{11}	202.10^{10}	$101.\ 10^{10}$
2	PfPb1	35.10^{10}	-	$160.\ 10^{10}$
3	PfPj2	202.10^{10}	110.10^{9}	$176.\ 10^{10}$
4	Kd_7	30.10^{11}	3.10^{9}	$122.\ 10^{10}$
5	Cas	30.10^{11}	50.10^{9}	$176.\ 10^{10}$
6	Cas3	30.10^{11}	3.10^{9}	-
7	LAHp2	225.10^{10}	6.10^{9}	$16.\ 10^{10}$

(a) (b) (c)

Figure 1. (a) ZnSO4.7H2O, (b) CoCl2.6H2O, and (c) Fe2SO4.7H2O media

3.2. The production of hydrogen cyanide

Study analysis of hydrogen cyanide production showed the addition of ZnSO4.7H2O mineral source is best in producing hydrogen cyanide by fluorescent pseudomonad isolate Cas and Cas3. In the CoCl2.6H2O-added growth media, all isolates showed poor hydrogen cyanide production from none to weak. Isolate PfPb1, PfPj2, and LAHp2 did not produce hydrogen cyanide on growth media with addition of CoCl2.6H2O and Fe2SO4.7H2O (Table 2.). The production of hydrogen cyanide by fluorescent pseudomonad will be seen by the differences in color seen on the filter paper. The color of the yellow filter paper indicates that no hydrogen cyanide is produced, the light brown is produced by hydrogen cyanide but the reaction is weak, the brown color generated by moderate reaction hydrogen cyanide, while the strong reaction exhibits a reddish brown color (Figure 2).

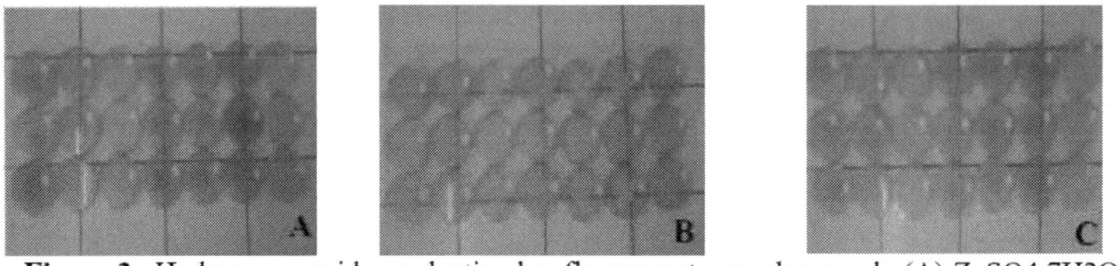

Figure 2. Hydrogen cyanide production by fluorescent pseudomonad. (A) ZnSO4.7H2O growing media, (B) CoCl2.6H2O growing media, and (C) Fe2SO4.7H2O growing media

Table 2. The production of hydrogen cyanide from pseudomonad fluorescent.

No	Isolate	Media type		
		KB+Zn	KB+Co	KB+Fe
1	PfPj1	++	+	++
2	PfPb1	+	-	-
3	PfPj2	++	+	++
4	Kd_7	+++	+	++
5	Cas	+++	+	++
6	Cas3	+++	+	++
7	LAHp2	+	-	-

- None
+ Weak
++ Moderate
+++ Strong

4. Discussion and Conclusion

The role of hydrogen cyanide in suppressing various plant diseases has been reported by several researchers. Hydrogen cyanide and ammonia synthesis were observed in all the bacterial isolates. Kumar et al., [8] reported that some isolated pseudomonad produce hydrogen cyanide as the primary mechanism of biocontrol. Prasad et al., [9] states that hydrogen cyanide is not only produced by biocontrol agents of Pseudomonas fluorescens, but also by Bacillus subtilis, Trichoderma viride, and T. harzianum. Screening of six isolates of P. fluorescens found only one isolate producing strong hydrogen cyanide, while the other five were moderate. According to Ramette et al., [10] in addition to increasing plant growth, Pseudomonas can release hydrogen cyanide to the rhizosphere of tobacco, thereby suppressing Thieviopsis basicola, causal agent of black root rot of tobacco.

The addition of Zn and Fe to the growing media, can enhance the ability of bacteria as a biocontrol agent. Addition of ZnSO4 to soil as a growing media of fluorescent pseudomonad isolates UTPF5 and MPFM1 could significantly increase the growth of bean crops, while for UTPF76 isolates there was a decrease in growth. Zn is an essential co-factor for changing the tryptophan amino acid into auxin (growth stimulant hormone) [12].

From the research, it was concluded that the addition of ZnSO4.7H2O mineral resources to the growing media of fluorescent pseudomonad isolate Cas and Cas3 produced the best hydrogen cyanide. While the addition of CoCl2.6H2O mineral source on growing media showed poor hydrogen cyanide production for all pseudomonad fluorescent isolates.

5. References

[1]. Paul D and Sarma, Y R. 2006. Plant Growth Promoting Rhizobacteria (PGPR)-Mediated Root Proliferation in Black Pepper (Piper ningrum L.) as Evidenced Trough GS Root ® Software. Archives of Phytopathology and Plant Protection. Month 2006; 39(0):1-4.

[2]. Advinda L. 2004. Growth Response of Immunized Banana Plants with Fluorescent Pseudomonads Against Ralstonia solanacearum. Faculty of Mathematics and Natural Science. State University of Padang
[3]. Advinda L. 2009. Physiological Response of Banana Plant Introduced with Pseudomonad Fluorescent Formula Against Blood Disease Bacteria (BDB). Dissertation. Postgraduate Program. Andalas University. Padang.
[4]. Doktycz M J, Pelletier D A, Schadt C W, Tuskan G A and Weston D. 2014. Plant growth promoting rhizobacterium, Patent US20140206539.
[5]. Ababutain I M, Aziz Z K A and AL-Meshhen N A. 2013. Optimization of environmental and nutritional conditions to improve growth and antibiotic productions by Streptomyces sp. Isolated from Saudi Arabia Soil. International Research Journal of Microbiology (IRJM). Vol. 4(8) pp. 179-187, September, 2013.
[6]. Duffy B K, Keel C and Défago G. 2004. Potential Role of Pathogen Signaling in Multitrophic Plant-Microbe Interactions Involved in Disease Protection. Applied And Environmental Microbiology, March. 2004, Vol. 70, No. 3, p. 1836–1842.
[7]. Reetha A K, Pavani S L and Mohan S. 2014. Hydrogen Cyanide Production Ability by bacterial antagonist and their Antibiotics Inhibition Potential on Macrophomina phaseolina (Tassi.) Goid. 2014. Int.J.Curr.Microbiol.App.Sci (2014) 3(5): 172-178.
[8]. Kumar Aj, Kumar Am, Devi S, Patil S, Payal C and Negi S. 2012. Isolation, screening and characterization of bacteria from Rhizospheric soils for different plant growth promotion (PGP) activities: an in vitro study. Recent Research in Science and Technology. 2012, 4(1): 01-05.
[9]. Prasad M R, Sagar B V, Devi G U, Triveni S, Koteswar Rao S R and Chari K D. 2017. Isolation and Screening of Bacterial and Fungal Isolates for Plant Growth Promoting Properties from Tomato (Lycopersicon esculentum Mill.). Int.J.Curr.Microbiol.App.Sci (2017) 6(8): 753-761
[10]. Ramette A, Moenne-Loccoz Y and Defago G. 2006. Genetic diversity and biocontrol protection of fluorescens pseudomonas producing phloroglucinols and hydrogen cyanide from swiss soils naturally suppressive or conducive to Thieviopsis basicola mediated black rot of tobacco. FEMS Microbiol Ecol 55 (2006) 369–381.
[11]. Salamiah and Wahdah R. The utilization of Plant Growth Promoting Rhizobacteria (PGPR) in controlling local rice tungro diseases in South Kalimantan. 2015. PROS SEM NAS MASY BIODIV INDON Volume 1, no. 6 September 2015 ISSN: 2407-8050 Page: 1448-1456.
[12]. Omidvari M, Sharifi R A, Ahmadzadeh M and Dahaji P A. 2010. Role of Fluorescent Pseudomonads Siderophore to Increase Bean Growth Factors. Journal of Agricultural Science Vol. 2, No. 3; September 2010.

IOP Conf. Series: Materials Science and Engineering **335** (2018) 012017 doi:10.1088/1757-899X/335/1/012017

Growth and Tomato Nutrition Content with Bandotan (*Ageratum Conyzoides* L) Bokashi Applied

A Anhar, R Junialdi, A Zein, L Advinda and I Leilani
Biology Department, Faculty of Mathematics and Sciences, Universitas Negeri Padang, Indonesia

*azwiranhar@fmipa.unp.ac.id

Abstract. Bandotan (*A.conyzoides*) is one of the most common weeds in dry land. Weeds can reduce the results of a variety of crops. On the contrary, the use of this weed as the organic material would be able to increase the nutrient content of the soil. The problem is, the decomposition of this weed is naturally longer than the process in the form of Bokashi. This study aims to determine the effect of Bandotan applied in the form of Bokashi on growth and nutrient content of tomato plants. This study was an experimental study and completely randomized design was used with 5 treatments and 3 replications. The treatment were rate of bandotan bokashi those 100, 120, 140, 160 g / polybag and 0,6g NPK / polybag as a control. The research was conducted in the Screen House of Biology Department, Faculty of Mathematic and Sciences, Universitas Negeri Padang. Tomato growth observed was high, wet weight, biomass and weight of the fruit. While the nutritional quality of tomatoes was vitamin C and A. Data were analyzed using ANOVA and a further test DNMRT at 5% level. The results showed that bokhasi bandotan 120g / polybag give best effect to the weight of tomatoes. However, bandotan bokashi do not give effect to the high, wet weight, biomass, vitamin C and vitamin A of tomato. Bokashi bandotan can be utilized as a substitute for synthetic fertilizer NPK for tomato plants.

1. Introduction

No doubt that in its habitat, the plant will interact with a variety of abiotic and biotic factors. The form of interaction can be beneficial or harmful. Adverse interactions occur as a result of competition in utilizing the resources available in the environment. In addition, some plants also secrete a compound residues which can inhibit the growth of other plants.

Ageratum conyzoides is one of the plants that are potentially harmful plants. *A.conyzoides* including 15 large plant pests in Southeast Asia [1]. *A.conyzoides* is one of the most important weeds found on agricultural land in Indonesia [2]. The amount of the main crop yield loss due to weed competition varies according to the type of plant. *A.Conyzoides* significantly affect the growth and yield curly chilli [3] Loss of grain yield of rice plants because of infestations of *A. conyzoides* ranged from 25-47% and 13-38% in the straw [4].

Harmful of *A. conyzoides* not only due to competition interspesies, but a chemical compound that is produced and released into the environment is also unfavorable plant main crops. *A. conyzoides* contains chemical compounds known as alelopati [5]. Residues generated by toxic weeds can be derived from the root exudation or excretion, volatilasi leaf through the stomata, or leaching from fresh leaves by rain or dew, late of the litter that has been decomposed, and the transformation of soil microorganisms [6]. Extract of *A. conyzoides* can suppress weed growth and yield of tomato plants. Giving fluids *A. conyzoides* leaf juice at a concentration of 100 g / l is already able to inhibit the germination of mustard. Even at a concentration of 500 g / l was able to inhibit the germination of 100% [7]. Inhibition also occurs in plant height, leaf number and dry weight of mustard [8].

Although *A.Conyzoides* can reduce main crops as a result of competition or because they contain residues, but herbaceous plants also potentially be used as a source of plant nutrients. *A.conyzoides* containing organic carbon 42.11%, total N 3.78% and P 0.21% with the C / N ratio of 11.15. The nutrient content of the essential nutrients that are needed by the plant growth and development. *A.conyzoides* can be utilized in the form of organic matter or compost [9]. Application *A.conyzoides* embedded into the ground can significantly increase the number of leaves and dry weight of upland rice plants at the age of 9 weeks after planting [10]. Mulching bandotan (*A conyzoides* L.) 300 g / polybag significantly increased plant height, number of leguminous crops, and no significant effect on weed dry weight and plant dry weight green beans [11]. Application *A.conyzoides* on three types of plants in the form of compost higher influence on plant height, basal area, productivity and biomass allocation shoots than in the form of mulch. It is thought to relate to the contribution of nutrients provided in the form of compost is higher than the administration in the form of mulch [12].

Various studies indicate that application of organic material effect on plant growth after administration of over 8 weeks. This was caused, organic materials need sufficient time for decomposition. The release of nutrients in the decomposition of roots *A. conyzoides* at the initial stage more slowly than the next stage [13]. The length of time depends on the type of decomposition of organic materials used. Lignin and cellulose negatively correlated with the initial decomposition rate coefficient significantly. In contrast, the N concentration was positively correlated with the rate of decomposition coefficients. Microbial populations and diversity of fungi vary during the period of decomposition in the field and mainly depends on the nutrient content of the waste. The diversity of fungal species on the decomposition of roots A. *conyzoides* lower than in the leaves. Thus, mulching leaves would be more profitable than with plant roots

Based on the above it is known that *A.conyzoides* on the one hand able to inhibit the growth of other plants. Instead, the herbaceous plants are also able to increase the nutrient content of the soil, which in turn can be beneficial to the growth of staple crops. Giving *A.conyzoides* in a form that has been decomposed turned out better influence on plants, especially after a few weeks after planting. In order to accelerate the effect on plant growth it is necessary to attempt organic matter in the form of Bokashi. Bokashi is an organic fertilizer produced by the fermentation of organic materials. It contains both decomposed and undecomposed organic matter, microbial biomass, and the intermediate and ultimate substances produced by microbes during fermentation [14].

Application of bokhasi (*Tithonia diversifolia*) can increase the number of leaves, number of tillers per hill, plant height, number of plant roots, fresh weight and dry weight of onion crop [15]. Application bokhasi Gliricidia (*Gliriciadia sepium* (jacq.) Kunth) can increase the growth of cocoa seedlings especially number of leaves, seedling height, stem diameter, wet weight and dry weight [16]. Meanwhile application of *lamtoro* bokhasi can increase the weight of the fruit of tomatoes [17]. This study aimed to determine the effect of bandotan Bokashi on growth and tomato nutrition content

2. Research Methods

The research was conducted in screen house and plant physiology of biology department, Universitas Negeri Padang. The experiment arranged in completely randomized design with five treatment and three replications. The treatments were 0,6 g NPK, 100 g, 120 g, 140 and 160 g bandotan bokashi per polybag. Planting media used is garden soil. The dry garden soil was crushed and sieved using a 5 mm sieve. Put 6 kg of soil into polybags with diameter 35x40 cm [16]. The seeds of the savior varieties F1 soaked for 1x24 hours. The seeds are then grown in polybags in the nursery polybags and covered with thin soil. After reaching 30 cm high, the seedlings are ready for transplanting [18]. Bokashi bandotan is made by mixing 500 g of bran, 2 kg husk, 4 kg *bandotan*, 3 liters of sugar water with 1 percent concentration and 200 ml EM4. All material is stirred until homogeneous. Make sure the material has a water content of 30-40%. Put the mixture into the sack, cover and incubate for seven days. Maintain material temperatures between 30-50 ° C. After seven days, bokashi bandotan is ready for treatment [19]. Seed tomato transplanting to polybags by making a hole in the ground as deep as 10-15cm [20]. One week after planting the plant is given an stake to help the plant grow upright and reduce the physical damage of plants, wind blow and facilitate the plant mentanaince [21]. Bokhasi given one week before planting in accordance with the dose of treatment. Bokashi is given 5 cm around the stem. [16]. Watering is done twice a day in the morning and afternoon. Weeding done from planting until before harvest by removing weeds that grow around the plant . Observations were made on plant height, plant wet weight, fruit weight, biomass, vitamin and vitamin A [22]). The data obtained were analyzed by Analysis of variance and further test with DNMRT at 5% level [23].

3. Results and discussion

3.1. High Tomato Plant

The results showed that the bokashi dose did not show a linear correlation with plant height. Application of bokashi with a dose of 100 g showed the same effect with the control of NPK fertilizer. Increasing the dose of bokashi to 120 g actually resulted in reduced plant height. Even if the dose is increased to 140, the plant height is significantly reduced compared to the 100 gram bokashi dose (Table 1). Thus it can be seen that the application of bokashi can replace the role of NPK synthetic fertilizer. Increased plant height is associated with organic compounds application. Organic compounds derived from bandotan are decomposed by microorganisms contained in EM4.) Microorganisms on EM4 can ferment organic materials into anorganic, sugars, alcohols, amino acids and various biogenic substances that can be absorbed directly by the roots of plants to acelerates the growth and development of plant [25]. This is related to the needs of plants on macro nutrients in the form of N, P and K. Nutrition needs of these nutrients can be met from the source of organic materials used as a bokashi materials. A.conyzoides contained total N, 3.78% and P total of 0.21% [9].

Table 1. Effect of *bandotan* bokhasi on tomato plant height (8 weeks)

Treatment	High plant average (cm)
A. 0,6 g NPK	94,56 c
B. 100g	90,90 bc
C. 120g	85,67 ab
D. 140 g	80,87 a
E. 160 g	79,80 a

The numbers followed by the same letters show no significant difference at the 5% level by DNMRT

Nitrogen is an essential element required for successful plant growth [21]. NPK fertilizer contains high nitrogen, which is 15% in the form of NH3 [26]. Nitrogen-deficient plants had significant lower leaf nitrogen and chlorophyll a contents. They also showed a significant reduction in their photosynthetic capacity [28]. Plant height pepper was were influenced by N fertilizer [29]. Therefore, it can increase the growth of plant height. Whereas *bandotan* bokhasi contains only 6.3% nitrogen [30], so the high growth of tomatoes tends to be low. However, the height of the plants on the controls gives no significant effect compared with 100 g bokhasi applications.

Plant growth is related to the activity of apical meristematic cells. When cell division in the meristematic region takes place, the tip moves upward, leaving the cells formed by the division. These cells then increase in size so that it becomes a part of the elongation area that slowly occurs differentiation and maturation. As a result of this activity, growth becomes faster and plant height may increase during the growing season [31]. The application of organic matter increases the height of tomato plants at the age of 4 and 8 weeks [32].

3.2. Wet Weight Tomato Plant.

The wet weight of the tomato plant ranges from 71.3 on the treatment of 160 g bokashi to 85.8 on the 100g bokashi treatment. Nevertheless, the application of bokashi did not have a significant effect on the wet weight of plant crops. The average wet weight of tomato plants can be seen in Table 2.

Table 2. Effect of bandotan bokhasi on wet weight of tomato plants

Treatment	Wet weight plant (g)
B. 100 g bokhasi	85,8
C .120g bokhasi	83,6
D .140g bokhasi	73,3
A .0,6 g NPK	71,5
E .160g bokhasi	71,3

The wet weight indicates the moisture present in the plant (28). The wet weight of the plant is a reflection of the water content and nutrient composition of the plant tissue. Absorption of nutrients and water by plants is crucial to the success of plants to grow (32). Although not showing significant differences, the dosage of bokhasi is directly proportional to the wet weight of the plant. The higher the dose given the greater the wet weight of the plant. This is thought to be related to the soil's ability to absorb water. One of the advantages of organic fertilizer is its ability to absorb water higher than inorganic fertilizer. The availability of sufficient water results in increased nutrient solubility, so that the availability for the plant is also increasing. This is what causes the growth of plants to be better. Soil that contains enough water, then the roots of plants will grow well, so that it will affect the nutrients, which in turn increase the wet weight of plant (33). Another thing that causes no significant effect of doses on wet weight of tomato plants is the need of nutrients is sufficient. Thus, increasing doses will not have a significant effect on wet weight. The wet weight of the plant is affected by genetic and environmental interactions. Environmental factors that affect are the elements of nutrients, water and light.

3.3. Weight Fruit

Bandotan bokhasi application affects the weight of tomatoes. An increase in dosage from 100 g to 140 is directly proportional to the weight of tomatoes. An increase in bokhasi dose exceeding 140 did not have a significant effect on the weight of fruit (Table 3)

Table 3. Effect of bandotan bokhasi on the weight of fruit

Treatment	Fruit weight average (g)
D. 140g	2,2 b
C. 120g	2,19 b
E. 160g	2,06 ab
B. 100g	1,99 a
A.0,6g NPK	1,98 a

The numbers followed by the same letters show no significant difference a the 5% level by DNMRT

Based on the data in Table 3, it appears that bokhasi applications exceeding 100 g can replace the role of NPK fertilizer. Conversely, an increase in doses exceeding 140 actually leads to a decrease in the weight of the fruit and gives the same yield as NPK. This happens because the decomposition of organic material not only supplies macro nutrients, but also micro nutrient elements are needed in metabolism. In addition, bandotan bokhasi has advantages in improving soil structure, increasing humus and stimulating the life of soil microorganisms (34). Microorganisms that live in the soil increase the growth and production of plants, because of their role in fixing N2 from the air and converting ammonium to nitrate (35). The availability of organic compounds, good soil structure, high humus and life of soil microorganisms contained in bandotan bokhasi this is causing the increase of fruit weight.

3.4. Tomato Plant Biomass

The highest biomass of plants was obtained at 100g and the lowest dose of 120g bokhasi. Bokashi application proved to replace the role of NPK. Nevertheless, a dose increase of up to 160 g of bokashi has not been able to increase the biomass of plants (Table 4). Biomass in the form of dry weight of plants reflects the accumulation of organic compounds, water and carbon dioxide. Total biomass olive was strongly reduced under nitrogen deficiency [28]. The nutrients at the root will be used for synthesis of organic compounds and which contribute to the increase of dry weight of plants [33]. Bokhasi applications are able to change the soil chemical properties for the better, especially the increased availability of phosphorus. The higher the phosphorus uptake by plant roots the higher the plant biomass [36]. Phosphorus is also important for cell division and enlargement, thus plant growth is reduced when the supply of P is too low [37].

Table 4. The influence of bandotan bokhasi on tomato plant biomass

Treament	Biomass (g)
B. 100g	11,7
E. 160 g	10,4
A. 0,6g NPK	9,43
D. 140g	9,36
C. 120g	9,33

3.5. Vitamin C

The highest application of botanical fertilizer was obtained at a dose of 160g and the lowest at 100g treatment. Nevertheless, the results show that bokashi or NPK applications do not affect the vitamin C content (Table 5)

Table 5. Effect of bandotan bokhasi on vitamin C tomato content

Treatment	vitamin C content (%)
E . 160g	0,84
C . 120g	0,82
D . 140g	0,71
A . 0,6g NPK	0,69
B . 100g	0,56

Vitamin C is a water-soluble white crystal that has many benefits for humans [38]. Bandotan bokhasi application has no significant effect on tomato vitamin C levels. Vitamin C is influenced by genetic and environment including nutrients and pH of planting medium. Nutrients that play a role in the formation of vitamin C is manganese (Mn). Manganese required by the plant for the formation of proteins and vitamins especially vitamin C. The availability of manganese for the plant depends on soil pH. Availability of Mn is low at neutral to alkaline pH [39]. There was no significant difference in vitamin C content among treated plants pepper with different nitrogen levels at mature stage [27]. Vitamin C content in tomato was not confirmed by the application of vermicompost and chemical fertilizers. (40) Potassium rates supplied to the tomato plant did not affect the vitamin C concentration in the fruits [41]. This is why the bokhasi application does not affect the levels of vitamin C tomatoes.

3.6. Vitamin A

The content of vitamin A tomatoes ranged between 0.60 at a dose of 140 g bokhasi to 1.11 at a treatment of 160 g. Although bokhasi doses do not significantly affect vitamin A content, bokashi applications may replace the role of NPK fertilizers (Table 6).

Table 6. Effect of bokhasi bandotan on vitamin A tomato content

Treatment	vitamin A content (ppm)
E. 160 g	1,11
B . 100g	1,08
A . 0,6g NPK	0,78
C . 120g	0,73
D. 140g	0,60

No significant effect of bokashi dose on vitamin A content is thought to be related to nutrients needed for vitamin formation. Elements that play a role in the formation of vitamin A is copper (Cu). Copper plays an important role in carbohydrate catabolism an enzyme activator in the process of storing food reserves and is one of the elements in the formation of vitamin A which indirectly play a role in the formation of chlorophyll [21]. An increase in dose of up to 160 g is not sufficient to improve the vitamin A synthesis. It can also be observed in NPK fertilizer applications that do not contain Cu elements. Thus, the role of bokashi is only limited as a substitute for NPK synthetic fertilizers and has not been able to increase the content of vitamin A.

4. Conclusion

Bokashi *bandotan* can be used as a substitute for NPK chemical fertilizer in tomato plants. Bokashi application gives the same or higher result compared with NPK fertilizer. The application of bokashi can increase the height and weight of tomato fruit. In contrast, bokashi application has not been able to increase the content of vitamin C and A in tomato plants.

References

[1] Waterhouse , D.F. 1993. The Major Arthropod Pests and Weeds of Agriculture in Southeast Asia: Distribution, Importance and Origin . Monograph 21. Australian Centre for International Agricultural Res. Canberra.
[2] Moenandir, J. 1990. Pengantar Ilmu dan Pengendalian Gulma. Buku I. Rajawali Press. Jakarta
[3] Syam, Z., S.Yenni dan Khainur. Pengaruh Kerapatan Gulma Siamih (Ageratum Conyzoides) terhadap Tanaman Cabe Keriting .Prosiding Semirata FMIPA Unila, Lampung.
[4] Munandar. 2004. Pupuk Organik. Jakarta: Penebar Swadaya.
[5] Aini, B. 2008. Pengaruh Ekstrak Alang-alang (Imperata cylindrica), Bandotan (Ageratum conyzoides) dan Teki (Cyperus rotundus) Terhadap Perke-cambahan Beberapa Varietas Kedelai (Glycine max L). Skripsi. Universitas Islam Negeri Malang. Malang
[6] Kilkoda, A.K.,2015. Respon Allelopati Gulma Ageratum Conyzoides dan Borreria Alata Terhadap Pertumbuhan dan Hasil Tiga Varietas Kedelai (Glycine Max). Jurnal Agro Vol. II, No. 1, Juli 2015
[7] Fitria, Y., D. Guntoro dan J.G.Kartika. 2011. Pengaruh Alelopati Gulma Cyperus rotundus, Ageratum conyzoides dan Digitaria adscendens. Ptosiding Seminar Nasional PERHORTI. Lembang 23-24 November 2011
[8] Hafsah,S., M.Abduh Ulim, dan Cut Mutia Nofayanti. 2012. Efek Alelopati Ageratum Conyzoides Terhadap Pertumbuhan Sawi. J. Floratek 8: 18 – 24
[9] Pratikno, H., E.Arisoesilaningsih dan E. Handayanto. 2004. Pemanfaatan Biomasa Tumbuhan Liar di Lahan Berkapur DAS Brantas untuk meningkatkan Ketersediaan P tanah. Fakultas Pertanian, Universitas Brawijaya.
[10] Suprijono, E. 2004. Potensi Berbagai Tanaman Pengganggu Sebagai Bahan Organik Tnaha Terhadap Pertumbuhan dan Hasil Padi Gogo . Laporan Penelitian. Fakultas Pertanian, Universitas Bengkulu

[11] Herlin, S. F. 2010. Uji Pemberian Mulsa Bandotan (Ageratum conyzoides L.) Terhadap Pertumbuhan Gulma dan Hasil Tanaman Kacang Hijau (Phaseolus radiates L.). Skripsi. Padang: UNAND.

[12] Roy,S. K.Arunachalam, B.M. Duta, and A. Arunachalam. 2010. Effect of organic amendment of soil on growth and productivity of three common crops viz.Zea mays, Phaseolus vulgaris and Abelmoschus esculentus. Appl. Soil Ecol.45:78-84

[13] Majumder, M. A K. Shukla and A. Arunachalam . 2008. Nutrient release and fungal succession during decomposition of weed residues in a shifting cultivation system . Communications in Biometry and Crop Science Vol. 3, No. 1, 2008, pp. 45–59.

[14] Yamada, K, Kato, S, Fujita, M, Xu, HL, Katase, K, & Umemura, H 2003, Investigation on the Properties of EM Bokashi and Development of Its Application Technology, EMRO,Japan, pp 1-12

[15] Sari, I.G. 2009. Pengaruh pemberian bokhasi(Tithonia diversifolia) terhadap pertumbuhan vegetatif bawang merah (Allium ascalonicum) pada tanah ultisol. Skripsi. Padang: UNP.

[16] Saputra, A. 2009. Pengaruh Pemberian Bokhasi Gliricidia sepium Terhadap Perumbuhan Kakao (Theobroma cacao) pada tanah ultisol. Skripsi. Padang: UNP.

[17] Deswita, W. P. 2011. Pengaruh Pupuk Bokhasi Lamtoro Terhadap Pertumbuhan dan Mutu Tomat (Lycopersicum esculentum Mill.). Skripsi. Padang: UNP.

[18] Pracaya. 2007. Bertanam Tomat. Yogyakarta: Kanisius

[19] Aswandi, A, dan O'eng. 2004. Semai. Jurnal. Diakses 22 Mei 2012.

[20] Benton-Jones J. (2008). Tomato Plant Culture in the Field, Greenhouse, and Home Garden. Boca Raton, FL: CRC Press.

[21] Annisava, A.R.,L. Anjela and B. Solfan.2014. Respon Tanaman Sawi (Brassica juncea L.) Terhadap Pemberian Beberapa Dosis Bokashi Sampah Pasar Dengan Dua Kali Penanaman Secara Vertikultur. Jurnal Agroteknologi, Vol. 5 No. 1, Agustus 2014 : 17 -24

[22] AOAC. 1970. "Official Methods of Analysis," Ed. Horwitz. W., p. 769. Assoc. Of Official Anal Chem., Washington, D.C.

[23] Hanafiah, K.I. 2008. Rancangan Percobaan teori dan aplikasi. Edisi ketiga. Jakarta: Rajawali Press

[24. Soesilo. 1998. Teknologi Effective Microorganisms. Jakarta: Pusat Penyuluhan Kehutanan

[25] Pratikno, H., E.Arisoesilaningsih dan E. Handayanto. 2004. Pemanfaatan Biomasa Tumbuhan Liar di Lahan Berkapur DAS Brantas untuk meningkatkan Ketersediaan P tanah. Fakultas Pertanian, Universitas Brawijaya.

[26] Liu, C.W., Y. Sung., B.C. Chen and HJ.Y. Lai. 2014. Effect of Nitrogen Fertilizer on the Growth and Nitrogen content of Lettuce (Lectuca sativa L). Int J Environ Res Public Health. Vol.11 (4): 4427-4440

[27] Pirngadi, K. dan Abdulrachman S. 2005. Pengaruh Pupuk Majemuk NPK (15-15-15) Terhadap Pertumbuhan dan Hasil Padi Sawah. Jurnal Agrivigor Vol 4 No 3. Diakses 20 Desember 2012.

[28] Boussadia, O., K. Steppe., H. Zgallai., S. Ben El Hadj., M. Braham., R. Lemeur and M.C. Van Labeke. 2010. Effects of nitrogen deficiency on leaf photosynthesis, carbohydrate status and biomass production in two olive cultivars 'Meski' and 'Koroneiki'. Scientia Horticulturae 123 (2010) 336–342

[29] Aminifard, M.H., H. Aroiee , H. Nemati , M. Azizi & M. Khayyat . 2012. Effect Of Nitrogen Fertilizer On Vegetative And Reproductive Growth Of Pepper Plants Under Field Conditions. Journal of Plant Nutrition. Volume 35, 2012 - Issue 2

[30] Suwahyono, U. 2011. Petunjuk Praktis Penggunaan Pupuk Organik Secara Efektif dan Efisien. Jakarta: Penebar Swadaya.

[31] Clark, D., K.R.Stern.,D. Vodopiich and R. Moore. 1995. Botany. William C Brown Pub.

[32] Maryanto dan Abdul Rahmi. 2015. Pengaruh Jenis dan Dosis Pupuk Organik Terhadap Pertumbuhan dan Hasil Tanaman Tomat (Lycopersicum esculentum Mill) Varietas Permata. Jurnal AGRIFOR Volume XIV Nomor 1.

[33] Taiz, L., E. Zeiger., I.M. Moller and A. Murphy. 2015. Plant Physiology and Development. Six edition. Sinaur Associate.

[34] Mihrani. 2008. Evaluasi Penyuluhan Penggunaan Bokhasi Kotoran Sapi Terhadap Pertumbuhan dan Produksi Rumput Gajah. Gowa: STTP. Jurnal agrisistem Vol 4 No 1.

[35] Islami, T. dan Wani. H. U. 1995. Hubungan Tanah, Air dan Tanaman. Semarang: IKIP semarang Press.

[36] Sutedjo, M.M., Kartasapoetra, S. dan Sastroatmodjo. 1991. Mikrobiologi Tanah. Jakarta: Rineka Cipta

[37] Zekri, M and T. A. Obreza. 2015. Plant Nutrients for Citrus Trees. http://edis.ifas.ufl.edu/pdffiles/SS/SS41900.pdf

[38] Almatsier, S. 2009. Prinsip-prinsip Ilmu Gizi. Jakarta: Gramedia Pustaka Utama

[39] Clark, R.B., Baligar, V.C. 2000. Acidic and alkaline soil constraints on plant mineral nutrition. In: R.E. Wilkinson (ed.) Plant-Environment Interactions. Marcel Dekker Inc, New York, pp: 133-177.

[40] Mukta, S. M. M.Rahman and M. G .Mortuza.2015. Yield and Nutrient Content of Tomato as Influenced by the Application of Vermicompost and ChemicalFertilizer
J. Environ. Sci. & Natural Resources,8(2): 115-122, 2015

[41] Fontes, P. C.,R A. Sampaio and F. L. Finger. 2000. Fruit Size, Mineral Composition and Quality Of Trickle-Irrigated Tomatoes as Affected By Potassium Rates. Pesq. agropec. bras., Brasília, V.35, No.1, p.21-25, Jan. 2000.

IOP Conf. Series: Materials Science and Engineering **335** (2018) 012018 doi:10.1088/1757-899X/335/1/012018

Ethnobotany in Traditional Ceremony at *Kanagarian Sontang Cubadak Padang Gelugur* Subdistrict, *Pasaman* District

Des M[1*], R Rizki[2], H Hidayati[2]

[1]Department of Biology, Faculty of Mathematics and Sciences, Universitas Negeri Padang, Indonesia
[2]Department of Biology Education, STKIP PGRI West Sumatera, Indonesia

*des.unp@gmail.com

Abstract. *Kanagarian Sontang Cubadak* has two distinct customary cultures namely *Minang* and *Mandailing* customs. Each costum has a sequence of processions performed on the customs of marriage, birth, coronation of community elders, the events of new society, salvation and death. In the implementation of the community using plants in the ceremonial procession of the ceremony. To know the types, meanings and ways of using plants in this traditional ceremony, conduct ethnobotany research on traditional ceremonies of marriage, birth, and death. Research conducted in May - June 2016 in *Kanagarian Sontang Cubadak Padang Gelugur* Subdistrict, *Pasaman* District, West Sumatra, Indonesia. Sampling using descriptive survey method, by interviewing village elders and the community. The sample identification was done in the botanical laboratory of Universitas Negeri Padang and STKIP PGRI West Sumatra. The results obtained 30 species from 19 family. In the *Minang* tribe ceremony found 16 species of 10 family, while in the *Mandailing* tribe 26 species of 18 family. Utilization of plants that eventually found in species *Oryza sativa L.* with a benefit value of 0.56. The meaning of plants in traditional ceremonies, relationships kinship, related social relations, related prayers and related aesthetics. The use of plants in traditional ceremonies is used in a raw state and some are processed before use.

1. Introduction

Indonesia has a high level of cultural diversity, and is contained in the motto of *Bhineka Tunggal Ika*. The slogan depicts the existence of 336 cultural tribes and 250 local languages [1]. One element of regional culture that is universal is an element related to traditional ceremonies in an area. Each region has various events or rituals in their respective cultures.

Traditional ceremony is a ceremony which carried out a related hereditary, in particular rules on the basis of customs. Traditional culture that is allegedly many have the wisdom of the environment has experienced a terrible erosion, so most of the younger generation now have no idea and do not care anymore with the ancestral heritage. Therefore it is necessary to dig the cultural and cultural practices to strengthen the community base in maintaining their culture [2].

People usually use some kind of plant as material in carrying out certain ceremony. Utilization of this plant is called Ethnobotany. Ethnobotany is a science that studies the traditional use of plants by primitive tribes or human relationships with plants and the environment [3]. Ethnobotany derives the word *"ethno"* studies about society and *"botany"* studies of plants. Ethnobotany is defined as the study

IOP Conf. Series: Materials Science and Engineering **335** (2018) 012018 doi:10.1088/1757-899X/335/1/012018

of the direct linkage between humans and plants. Thus ethnobotany recognizes the reciprocal and dynamic nature of human and plant relations [4].

Ethnobotany is a science that studies plants traditionally by primitive tribes. The term Etnobotany was first coined by an American botanist named Jhon W. Harshberger in 1895 [5]. Ethnobotany as a science that studies about nature or traditional relationships between humans and plants around them. Ethnobotany includes all the material, cultural, medical and religious relationships of people with plants in an ecosystem. Includes crops for food, medicine, timber, animal feed, fiber, bamboo and other materials in need, and beliefs related to the type of individual with the local ecosystem [6].

The *Minangkabau* people perform a variety of *Minang* ceremonial such as birth ceremonies, deaths, weddings, religious ceremonies consisting of circumcision ceremony (*Sunat Rasul*), *akikah*, *Khatam Qur'an*, and commemorate the big days of Islam, repelling ceremonies, angry ceremony is a ceremony of tradition that is held during down the rice field to sow seeds, during outbreaks of disease, when experiencing drought and so on. With the aim of asking God to be saved and kept away from all sorts of disasters [7]. *Minangkabau* wedding ceremony is not known dowry, which there is money pickup from the side of the female family to the family of men. *Minangkabau* traditional wedding procession consists of *jopuik manjopuik, betang maninang, batuka tando, akad nikah, baralek gadang, jalang manjalang* and so on. Each traditional ceremony in *Minangkabau* usually uses customs and speeches. There are five customary ceremonies that commonly use an addition, namely inaugurating the appointment of *penghulu*, marriage ceremony, death ceremony, house building ceremony and birth ceremony [8].

Mandailing tribe originated from North Sumatera area that is *Mandailing* area. *Mandailing* tribe include *Batak Toba* culture. Almost the whole community is religious mandailing are *Islam*, as well as custom ceremony is patterned of *Islam*. The ceremony is started from *manulak sere, mangalehen mangan, makobar, mangalab boru, mangolat boru, manyantani, and indahan pasaiboru*. The traditional ceremony of *mangupa* or *mangupa dohot tondi badan* is done to restore and strengthen the spirit and the body. While the ceremonies of death on the mandailinng custom is called *siluluton*, the procession of death and the release of the corpse is more religiously dominated than the customs.

The results of research conducted by Maulidiah, in 2015 [9] the marriage ceremony of *Horja Haroan Boru* society *Batak Angkola* tribe in *Padangsidempuan* obtained 28 species of plants in 14 family used. The research results Suswita [10] with title Etnobotani Studies and Forms of Plant Conservation Efficiency used in *Kenduri* Traditional Ceremony in Several Districts In *Kerinci* District, *Jambi*, found 37 species of plants in 22 family.

Kenagarian Sontang Cubadak is one of the *Kanagarian* are located in *Padang Gelugur* Subdistrict, *Pasaman* District, West Sumatra, Indonesia. Until now, people still use plants in the procession of traditional ceremonies. Based on the observation, in *Kenagarian Sontang Cubadak* there are two customary cultures namely, *Minang* and *Mandailing*. Some people use the original custom and some also use a mixture between the two customs.

In order to obtain ethnobotany data in traditional ceremony in *Kenagari Sontang Cubadak*, *Padang Gelugur* subdistrict, *Pasaman* District has been conducted research to know the species of plants are used in traditional ceremonial procession, the meaning of each species and how the use of plants in the procession of the ceremony.

2. Research Method

This research was conducted in May 2016, on 4 *Jorong* existing in *Kanagaraian Sontang Cubadak Padang Gelugur* Subdistrict, *Pasaman* District. The tool used is interview guide, field book, ruler, pencil, digital camera, hanging label, herbarium label, duct tape, plastic bag size 50 kg, plastic bag size 50 kg, machetes, scissors, thick cardboard, newsprint, , tracing paper, monting paper, sewing needles, and materials for making plant specimens used in ceremonies and alcohol 96%.

The research method used is survey method, interview and collection of plants in the field. Further observation and recording of data in the field. The data were collected using interview technique using closed questionnaire and open questionnaire.

Determination of respondents is done by using purposive sampling method. The criteria of respondents interviewed were *ninik mamak*, *hatobangon*, community, and people who took part in the ceremony.

From the collected data, the collecting of plant species, for the making of herbarium and done shooting. Identification is done in the Botany Laboratory of Universitas Negeri Padang and STKIP West Sumatra. The data analysis used is qualitative and quantitative. Qualitative data include type types of plants are used, how to use herbs and meaning of the plant species are used, while the data is quantitative, include the use/benefits of herbs used. To determine the use value/value of benefits a species used method of informant consensus. This analysis uses the formula Cunningham [11] which has been modified, as follows:

$$Uvis = \frac{\sum Uis}{Nis} \qquad (1)$$

where :

UVis : value for a species in traditional ceremonies

Uis : the use of a plant species in a series of events in a traditional ceremonial procession

Nis : Total series of events in the procession of traditional ceremonies

3. Results and Discussion

Based on the research that has been done, there are found 30 species in 19 family of plants used in traditional ceremony in *Kanagarian Sontang Cubadak Padang Gelugur*, *Pasaman* District as shown in Table 1.

Table 1. Species of plants used in traditional ceremonies in *Kenagarian Sontang Cubadak Padang Gelugur* subdistrict of *Pasaman* District

No	Species	Local Name	Familia	Habitus	Status	Md	Mn
1	*Justicia gendarussa* Burm. f.	*Sipilin*	1.Achantaceae	Shrubs	Cultivation	√	_
2	*Ruellia tuberosa* L.	*Bunga terompet*		Shrubs	Cultivation	√	_
3	*Celosia argentea* Linn	*Bunga pasir*	2. Amaranthaceae	Terna	Cultivation	√	
4	*Areca catechu* L.	*Pining/pinang*	3. Arecaceae	Tree	Cultivation	√	√
5	*Arenga pinnata* (Wurmb) Merr	*Bargot/anau*		Tree	Illegal	√	√
6	*Cocos nucifera* L.	*Arambir/karambia*		Tree	Cultivation	√	√
7	*Kalanchoe laciniata* (L.) DC.	*Dingin-dingin/sidingin*	4. Crassulaceae	Terna	Cultivation	√	√
8	*Saccharum officinarum* L	*Tabu merah*	5. Graminae	grass	Cultivation	_	√
9	*Oryza sativa* L.	*Dahanon/bareh*		grass	Cultivation	√	√
10	*Oryza sativa* L./f. Glutinosa A uct	*Sipulut/puluik*		grass	Cultivation	√	√
11	*Dendrocalamus asper* (Schult.) Backer	*Bulu*		grass	Illegal	√	_
12	*Themeda gigantea* (Cav.) Hack. ex Dutie	*Sanggar/pimpiang*		grass	Cultivation	√	√
13	*Ocimum sanctum* L.	*Sanggul*	6. Lamiaceae	Terna	Cultivation	√	_
14	*Cordyline fruticosa* (L.) A. Chev	*Hatunggal*	7. Liliaceae	Shrubs	Illegal	√	_
15	*Lawsonia inermis* L	*Inai*	8. Lythraceae	Shrubs	Cultivation	_	√
16	*Hibiscus rosa-sinensis* L.	*Bunga raya*	9. Malvaceae	Shrubs	Cultivation	√	_
17	*Artocarpus elasticus* Reinw. ex Blume	*Torop*	10. Moraceae	Tree	Illegal	√	_
18	*Ficus benjamina*var. variegata.	*Haruaya*		Tree	Illegal	√	_
19	*Musa balbisiana* Colla	*Pisang sitambatu/pisang sambantu*	11. Musaceae	Terna	Cultivation	√	√
20	*Musa* sp.	*Pisang ratus*		Terna	Cultivation	_	√

Table 1. Species of plants used in traditional ceremonies in *Kenagarian Sontang Cubadak Padang Gelugur* subdistrict of *Pasaman* District (continue)

No	**Species**	**Local Name**	**Familia**	**Habitus**	**Status**	**Md**	**Mn**
21	*Bougainvillea spectabilis* Willd.	*Bungo kertas/bungo karateh*	12. Nygtaginaceae	Shrubs	Cultivation	√	√
22	*Pandanus tektorius* Parkinson ex Du Roi	*Pandan duri*	13. Pandanaceae	Shrubs	Illegal	√	_
23	*Piper betle* L.	*Burangir/siriah*	14. Piperaceae	Liana	Cultivation	√	√
24	*Rosa centifolia* L.	*Bunga ros*	15. Rosaceae	Shrubs	Cultivation	√	
25	*Uncaria gambir* (Hunter) Roxb	*Gambir/gambia*	16. Rubiaceae	Shrubs	Illegal	√	√
26	*Ixora cocsinia* L.	*Asoka*		Tree	Cultivation	√	-
27	*Citrus hystrix* DC.	*Utte mungkur*	17. Rutaceae	Tree	Cultivation	√	
28	*Citrus aurantifolia* (Cristm. & penz.) Swingle	*Limau kapeh*		Tree	Cultivation	_	√
29	*Nicotiana tabacum* L.	*Timbako/timbakau*	18. Solanaceae	Shrubs	Cultivation	√	√
30	*Curcuma longa* L.	*Unik*	19.Zingiberaceae	Terna	Cultivation	√	_

Explanation : - Md : Mandailing Customs
- Mn: Minang Customs

In Table 1 looks familia Graminae most widely used consisting of 5 species. In *Mandailing* Customs there are 18 family with 26 species, while in *Minang* 10 family with 16 species. Plants are more widely used in traditional ceremonies obtained from the cultivation of the community, which is 23 species, while the wild plants 7 species. All species used in traditional ceremonies in *Kanagarian Sontang Cubadak* have meaning for the ceremony. To see the meaning of plants used in traditional ceremonies and the number of plants used per procession of traditional ceremonies can be seen in Table 2.

In Table 2 shows the meaning of plant species used in traditional ceremonies of *Mandailing* and *Minang* teribe. Utilization of the most plants in the procession of this ceremony is *Oryza sativa* (*dahanon/bareh*) and *Piper betle* (*burangir/siriah*). In *Mandailing* traditional wedding ceremony, *Oryza sativa* is used 4 times and at birth ceremony used once, while at *Minang* wedding ceremony is not used. *Oryza sativa* is only used in the ceremony as much as 2 times and the death of 1 time. *Piper betle* in *Mandailing* traditional wedding ceremony used 4 times and 1 birth. While at the traditional wedding ceremony of *Minang*, *Piper betle* only used as much as 2 times and at the ceremony of water used as much as 1 times.

Every use of plants has meaning in the procession stage of traditional ceremony and there is also how to use it. How to use the plants and what parts are used in the stages of wedding ceremonies procession can be seen in Table 3.

Table 2.The meaning contained in the use of plants in traditional ceremonies in *Kanagarian Sontang Cubadak Padang Gelugur, Pasaman* District

Species	*Mandailing*				*Minang*						TPS
	P	TP	KL	TP	P	TP	KL	TP	KM	TP	
1	2	3	4	5	6	7	8	9	10	11	12
Oriza sativa L.	Provide safety and health	4	Sense of sharing	2		-	A sense of wanting to share happiness	2	Give alms	1	9
Piper betle L.	Fraternity and strong kinship	4	Love	1	Humans have strong blood and kinship	2	Humans have strong blood and kinship	1		-	8

IOP Conf. Series: Materials Science and Engineering **335** (2018) 012018 doi:10.1088/1757-899X/335/1/012018

Table 2.The meaning contained in the use of plants in traditional ceremonies in *Kanagarian Sontang Cubadak Padang Gelugur, Pasaman* District (continue)

1	2	3	4	5	6	7	8	9	10	11	12
Areca catechu L.	Community representatives	3	Community representatives	1	Completeness of human life in society and civilized	2	Completeness of human life in society and civilized	1		-	7
Nicotiana tabacum L.	Completeness of custom	3	Completeness of custom	1	Completeness of custom	2	Completeness of custom	1		-	7
Uncaria gambir (Hunter) Roxb	Completeness of custom	3	Completeness of custom	1	Completeness of custom	2	Completeness of custom	1		-	7
Cocos nucifera L.	Spent his youth	3	Supplementary ingredients of sagun-sagun	1		-	Added ingredients of rice porridge	1	Added ingredients of rice chips and rendang pulut	1	6
Musa balbisiana Colla	Away from distress	4		-		-	Provides comfort	2		-	6
Arenga pinnata (Wurmb) Merr	Sweeteners food	2	Sweeteners food	1		-	Sweeteners food	1	Sweeteners food	1	5
Bougainvillea spectabilis Willd	Sign of the girl	2		-		-	Ojong-ojong decoration	1		-	3
Kalanchoe laciniata (L.) DC.	Gives coolness	2		-		-	Gives coolness	1		-	3
Oriza sativa L./f. Glutinosa A uct	Agreement that has been rounded	2		-		-		-	Gratitude	1	3
Themeda gigantea (Cav.) Hack. ex Dutie	For the household to last	2		-		-	Ojong-ojong decoration	1		-	3
Curcuma longa L.	Rice dye	2		-		-		-		-	2
Ficus benjamina var. variegata.	In order to have many offspring	2		-		-		-		-	2
Artocarpus elasticus Reinw. ex Blume	In order for the bonds of brother hood between the two sides to be close	1		-		-		-		-	1
Celosia argentea Linn	As an ornament	1		-		-		-		-	1

IOP Conf. Series: Materials Science and Engineering **335** (2018) 012018 doi:10.1088/1757-899X/335/1/012018

Table 2.The meaning contained in the use of plants in traditional ceremonies in *Kanagarian Sontang Cubadak Padang Gelugur*, Pasaman District (continue)

1	2	3	4	5	6	7	8	9	10	11	12
Citrus aurantifolia (Cristm. & penz.) Swingle		-		-	Fragrance of henna	1		-		-	1
Citrus hystrix DC.	Fragrance of pangir	1		-				-		-	1
Cordyline fruticosa (L) A. Chev	All customary rules must be enforced	1		-		-		-		-	1
Dendrocalamus asper (Schult)Backer	Told the party	1		-		-		-		-	1
Hibiscus rosa-sinensis L.	In order for life to be great	1		-		-		-		-	1
Ixora cocsinia L.	Stems, leaves, flowers		tied at the end of rice wrap (tappurobu sprouts) as jagar-jagar.								22
Justicia gendarussa Burm. f.	Materials for pangir	1		-		-		-		-	1
Lawsonia inermis L.		-		-	Expressing family affection	1		-		-	1
Musa sp.		-		-		-		-	Indicates a hundred day	1	1
Ocimum sanctum L.	Materials for pangir	1		-				-		-	1
Pandanus tectorius Parkinson ex Du Roi	Custom mat materials	1		-		-		-		-	1
Rosa centifolia L.	As a jagar-jagar decoration	1		-		-		-		-	1
Ruellia tuberosa L.	As an ornament	1		-		-		-		-	1
Saccharum officinarum L.								-	As souvenirs are brought back home	1	1

Traditional ceremonies: P (Wedding), KL (Birth), KM (Dead),
Use of plants : TP (Total use of plant species in traditional ceremonies), TPS (Total use of plant species in traditional ceremonies)

Table 3. How to use and parts of plants used in the procession of wedding ceremony in *Kanagarian Sontang Cubadak Padang Gelugur, Pasaman* District

Species	Part used	*Mandailing* Customs (A)						*Minangkabau* Customs (B)		
		Procession sequence						Procession sequence		
		1A	2A	3A	4A	5A	6A	1B	2B	3B
1	2	3	4	5	6	7	8	9	10	11
Oriza sativa L.	Seed	_	Cooked for rice wrap (tappurobu sprouts), and made rice flour sagun-sagun	Made yellow rice, rice flour for it pohul-pohul	Made yellow rice	_	Rice banquet	_	_	_
Piper betle L.	Leaf	Arranged on plate as burangir na hombang (opening event)	Arranged on plate as (opening event) which will be submitted to the hatobangon and hung on jagar-jagar	_	Arranged on a plate to give to the hatobangon	Tied in the middle of custom mats as decoration	_	Arranged on plate to show to ninik mamak as opening event	Arranged on the plate as the opening act	_
Areca catechu L.	Seed	Arranged on plate as burangir na hombang (opening event)	Arranged on the plate as the opening event will be submitted to Hatobangon	_	Arranged on a plate to be handed to hatobangon	_	_	Arranged on plate to be shown to ninik mamak as opening event	Arranged on plate as opening event	_
Nicotiana tabacum L.	Leaf	Arranged on plate as burangir na hombang (opening event)	Arranged on a plate as (opening event) which will be submitted to the hatobangon	_	Arranged on a plate to give to the hatobangon	_	_	Arranged on plate to be shown to ninik mamak as opening event	Arranged on plate as opening event	_
Uncaria gambir (Hunter) Roxb	Leaf	Arranged on plate as burangir na hombang (opening event)	Arranged on a plate as (opening event) which will be submitted to the hatobangon	_	Arranged on a plate to give to the hatobangon	_	_	Arranged on plate to be shown to ninik mamak as opening event	Arranged on plate as opening event	_
Cocos nucifera L.	Flesh of fruit, leaves	_	In the grater then cooked to core nuts and sagun-sagun	Shredded then cooked it pohul-pohul	_	Young leaves are decorated into bamboo-like gates	_	_	_	_
Musa balbisiana Colla	Leaves, stems	_	Dumped bone leaves and wrapped into rice from the base to the tip of the leaf	The banana tree skin is placed along the door of the house	_	The bark of the stem is taken as wide as the palm of the hand for a watering device to the bridal head (pamispis)	Parrot cover	_	_	_

Table 3. How to use and parts of plants used in the procession of wedding ceremony in *Kanagarian Sontang Cubadak Padang Gelugur, Pasaman* District (continue)

1	2	3	4	5	6	7	8	9	10	11
Arenga pinnata (Wurmb) Merr	Exudate	–	Cooked into the core and sugar for the sagun-sagun	Sugar cooked makes itak pohul-pohul	–	–	–	–	–	–
Bougainvillea spectabilis Willd.	Stems, leaves, flowers	–	tied at the end of rice wrap (tappurobu sprouts) as jagar-jagar.	–	–	Tied in all parts of the bamboo gate as decoration and adornment of custom mats	–	–	–	–
Kalanchoe laciniata (L.) DC.	Stems, leaves	–	–	Put on the banana tree skin to be trampled before entering the men's house.	–	partially sliced for pangir and partly tied together for pamispis	–	–	–	–
Oriza sativa L./f. Glutinosa A uct	Seed	–	Cooked into a core pulp	Made in powdery flour makes itak pohul-pohul	–	–	–	–	–	–
Themeda gigantea (Cav.) Hack. ex Dutie	Stems, leaves, seeds, flowers	–	tied at the end of rice wrap (tappurobu sprouts) as jagar-jagar.	–	–	Tied to be distributed to female invited guests as a sign of a woman	–	–	–	–
Curcuma longa L.	Tubers	–	–	In the puree then smeared into rice (yellow rice)	Refined and smeared as a yellow dye	–	–	–	–	–
Ficus benjamina var. variegata.	Stems, leaves	–	tied at the end of rice wrap (tappurobu sprouts) as jagar-jagar.	–	–	Tied to be distributed to female invited guests as a sign of a woman	–	–	–	–
Artocarpus elasticus Reinw. ex Blume	Leaf	–	tied at the end of rice wrap (tappurobu sprouts) as jagar-jagar	–	–	–	–	–	–	–
Celosia argentea Linn	Stems, leaves, flowers	–	–	–	–	Tied in the middle of a custom mat as an ornament	–	–	–	–
Citrus aurantifolia (Cristm. & penz.) Swingle	Fruit	–	–	–	–	–	–	–	–	Squeezed and extracted

IOP Conf. Series: Materials Science and Engineering **335** (2018) 012018 doi:10.1088/1757-899X/335/1/012018

Table 3. How to use and parts of plants used in the procession of wedding ceremony in *Kanagarian Sontang Cubadak Padang Gelugur, Pasaman* District (continue)

1	2	3	4	5	6	7	8	9	10	11
Citrus hystrix DC.	Fruit	_	_	_	_	Cut into 3 parts and put together with other pangir ingredients	_	_	_	_
Cordyline fruticosa (L) A. Chev	Stems, leaves	_	_	_	_	partially sliced for pangir and partly tied together for pamispis	_	_	_	_
Dendrocalamus asper (Schult.) Backer	Stem	_	_	_	_	Made like a gate	_	_	_	_
Hibiscus rosa-sinensis L.	Stems, leaves, flowers	_	tied at the end of rice wrapped as jagar-jagar.	_	_	_	_	_	_	_
Ixora cocsinia L.	Stems, leaves, flowers	_	tied at the end of rice wrap (tappurobu sprouts) as jagar-jagar.	_	_	_	_	_	_	_
Justicia gendarussa Burm. f.	Stems, leaves	_	_	_	_	Partially sliced for pangir and partly tied together for pamispis	_	_	_	_
Lawsonia inermis L.	Leaf	_	_	_	_	_	_	_	_	Finely ground to color the bridal nails
Ocimum sanctum L.	Stems, leaves	_	_	_	_	partially sliced for pangir and partly tied together for pamispis				
Pandanus tectorius Parkinson ex Du Roi	Leaf	_	_	_	_	Spread across at the front gate as a custom mat for banners of congratulations and horas to the bride	_	_	_	_
Rosa centifolia L.	Stems, leaves, flowers	_	tied at the end of rice wrap (tappurobu sprouts) as jagar-jagar.	_	_	_	_	_	_	_

Table 3. How to use and parts of plants used in the procession of wedding ceremony in *Kanagarian Sontang Cubadak Padang Gelugur, Pasaman* District (continue)

1	2	3	4	5	6	7	8	9	10	11
Ruellia tuberosa L.	Stems, leaves, flowers	_	_	_	_	Tied in the middle of a custom mat as an ornament	_	_	_	_
Pandanus tectorius Parkinson ex Du Roi	Leaf	_	_	_	_	Spread across at the front gate as a custom mat for banners of congratulations and horas to the bride	_	_	_	_
Rosa centifolia L.	Stems, leaves, flowers	_	tied at the end of rice wrap (tappurobu sprouts) as jagar-jagar.	_	_	_	_	_	_	_

Explanation : Custom used : *Mandailing* (A), *Minang* (B)
Custom wedding procession *mandailing* : 1A (*ManjugubkonHatobangon*), 2A (*Maebat*), 3A (*Manyantan Boru*), 4A (*Margondang*), 5A (*Patuaekkon*), 6A (*Parjamuan*)
Traditional wedding procession *Minang* : 1B (*Batimbang Tando*), 2B (*Duduak Niniak Mamak*), 3B (*Malam Ba Inai*)

In Table 3 shows how the use of plants in a series of ceremonial procession Mandailings wedding and Minang wedding and parts of plants used. In Mandailing custom there are 6 processions performed, namely *manjugubkon hatobangon, maebat, manyantan boru, margobdang, patuaekkon*, and *parjamuan* with 26 species. While in Minang custom consists of 3 procession that is *batimbang tando, duduak niniak mamak,* and *malam ba ianai* with 6 species. The way to use plants not only exist at the wedding ceremony, but also in the birth ceremony. How to use plants in the birth ceremony, and see parts of the plants used in *Kanagarian Sontang Cubadak* can be seen in Table 4.

Table 4. The way to use and parts of plants used in the procession of birth ceremonies in *Kanagarian sontang cubadak Padang Gelugur*, Pasaman District

Spesies	**Parts used**	**Mandailing (A) Procession sequence**		**Minang (B) Procession sequence**			
		1A	**2A**	**1B**	**2B**	**3B**	**4B**
1	2	3	4	5	6	7	8
Oriza sativa L.	Seed	Made rice flour for sagun-sagun	Indahan manjamu	Cooked into rice	Cooked to a porridge for distribution to neighbors	_	_
Piper betle L.	Leaf	Burangir custom and given to the children, mothers, and fathers who held the event	_	_	Arranged on plate as opening event	_	_
Areca catechu L.	Seed	Burangir custom	_	_	Split in half and suspended on the plate as the opening act	_	_

Table 4. The way to use and parts of plants used in the procession of birth ceremonies in *Kanagarian sontang cubadak Padang Gelugur*, Pasaman District (continue)

1	2	3	4	5	6	7	8
Nicotiana tabacum L.	Leaf	Burangir custom	_	_	Arranged on plate as opening event	_	_
Uncaria gambir (Hunter) Roxb	Leaf	Burangir custom and given to the children, mothers, and fathers who held the event	_	_	Arranged on plate as opening event	_	_
Cocos nucifera L.	Flesh of fruit	Shredded for sagun-sagun	_	_	Added slurry ingredients for distribution to neighbors		
Musa balbisiana Colla	Leaves, stems	_	_	Wrapped into ojong-ojong rice	_	_	Leather bark rod when bathed
Arenga pinnata (Wurmb) Merr	Exudate	Sugar for the sagun-sagun	_	_	sweetener porridge to be distributed to neighbors	_	_
Bougainvillea spectabilis Willd.	Stems, leaves, flowers	_	_	Tied to the edge of ojong-ojong rice	_	_	_
Kalanchoe laciniata (L.) DC.	Stems, leaves	_	_	_	_	_	Inserted into baby shower water
Themeda gigantea (Cav.) Hack. ex Dutie	Stems, leaves, seeds, flowers	_	_	Tied to the edge of ojong-ojong rice	_	_	_

Explanation : Custom used: Mandailing (A), Minang (B)
The procession of indigenous birth mandailing: 1A (Parjamuan Hatobangon), 2A (Manjamu Tondi Dohot Badan),
The procession of Minang customary birth: 1B (7 Bulanan), 2B (Duduak Niniak Mamak), 3B (Mambubur), 4B (Turun Mandi)

In Table 4 can be seen how the use of plants at the birth ceremony and bangian used. There are 11 species of plants used in ceremonial skills. Seven species are used in Mandailing traditional birth ceremonies and 10 species are used in traditional Minang birth ceremonies. In addition to the procedures for the use of plants in the ceremony of marriage and birth, there is the procedure of the use of plants in the ceremony of death. Mortuary rites in *Kanagarian Sontang Cubadak* can be seen in Table 5.

In Table 5 can be seen how the use of plants in custom ceremonies of death and parts of plants used. The ceremony of death is only done on Minang customs that is at the ceremony of 100 days. There are 6 species that are used with parts used from stems, seeds, and fruit.

Every plant used in traditional ceremonies has a value in the traditional ceremony. There are the most important plants that always exist in traditional ceremonies and there is also a plant that only sebahagai additional to add aesthetic value in traditional ceremonies. The benefit of value of the plants used in traditional ceremonies in *Kanagarian Sontang Cubadak* can be seen in Table 6.

Table 5. How to use plants in the procession of death ceremony in *Kenagarian Sontang Cubadak Padang Gelugur,* Pasaman District.

No	Species	Parts used	Mandailing	Minang
			-	*Prosesi 100 hari* (ceremony of 100 days)
1	*Saccharum officinarum* L.	Stem	-	Cut along 10 cm, wrapped with banana and chips
2	*Oriza sativa* L.	Seed	-	Rice roasted, crushed, banana with leather for chips
3	*Oriza sativa* L./f. Glutinosa A uct	Seed	-	In roaster for rendang pulut
4	*Arenga pinnata* (Wurmb) Merr	Exudate	-	Red sugar as a sweetener
5	*Cocos nucifera* L.	Flesh of fruit	-	Rasped and made coconut milk
6	*Musa* sp.	Fruit	-	Banana 1 seed wrapped with sugar cane and chips

Table 6. Value of the benefit plants used in traditional ceremonies at *Kanagarian Sontang Cubadak Padang Gelugur*, Pasaman District

No	Species	Use Value
1	*Oryza sativa* L.	0,56
2	*Piper betle* L.	0,5
3	*Areca catechu* L.	0,44
4	*Nicotiana tabacum* L.	0,44
5	*Uncaria gambir* (Hunter) Roxb	0,44
6	*Cocos nucifera* L.	0,38
7	*Musa balbisiana* Colla	0,38
8	*Arenga pinnata* (Wurmb) Merr	0,31
9	*Bougainvillea spectabilis* Willd.	0,19
10	*Kalanchoe laciniata* (L.) DC.	0,19
11	*Oryza sativa* L./f. Glutinosa A uct	0,19
12	*Themeda gigantea* (Cav.) Hack. ex Dutie	0,19
13	*Curcuma longa* L.	0,13
14	*Ficus benjamina* var variegata.	0,13
15	*Artocarpus elasticus* Reinw. ex Blume	0,06
16	*Celosia argentea* Linn	0,06
17	*Citrus aurantifolia* (Cristm. & penz.) Swingle	0,06
18	*Citrus hystrix* DC.	0,06
19	*Cordyline fruticosa* (L) A. Chev	0,06
20	*Dendrocalamus asper* (Schult.) Backer	0,06
21	*Hibiscus rosa-sinensis* L.	0,06
22	*Ixora cocsinia* L.	0,06
23	*Justicia gendarussa* Burm. f.	0,06
24	*Lawsonia inermis* L	0,06
25	*Musa* sp.	0,06
26	*Ocimum sanctum* L.	0,06
27	*Pandanus tektorius* Parkinson ex Du Roi	0,06
28	*Rosa centifolia* L.	0,06
29	*Ruellia tuberosa* L.	0,06
30	*Saccharum officinarum* L.	0,06

In Table 6 shows that the value of plant benefits used in the procession of traditional ceremony in *Kanagarian Sontang Cubadak* there are 7 groups with a range of values 0.06-0.56. Species that have the highest value (Use Value) is Oryza sativa, with a value of 0.56. This indicates that the species is used in nine custom ceremonial processions of 24 total entire ritual ceremonies in *Kanagarian Sontang Cubadak*. Then followed by Piper betle 0.5, and *Areca catechu, Nicotiana tabacum, and Uncaria gambir* each with a UV value of 0.44.

4. Conclusion

In the ethnobotany of traditional ceremonies in *Kanagarian Sontang Cubadak*, it can be concluded :

1. *Kanagaraian Sontang Cubadak* society utilize 30 species of plants with 19 family that still survive to this day. The highest plant utilization in *Oryza sativa* L. species with a benefit value of 0.56.
2. The meaning of plants used in traditional ceremonies is the meaning related to familial relationships, to social relations, to prayer and to aesthetics.
3. To use plants in traditional ceremonies depends on the customs and processions performed. The use of this plant is used in the raw state and some are processed or cooked first.

5. References

[1]. Supriatna, J. 2008. Preserving Nature Indonesia. Yayasan Obar Indonesia: Jakarta.
[2]. Rahyuni, E. Yniati dan R. Pitopang. 2013. "Ethnobotany Study of Tribe Tribe Ritual Plant of Kasimbar Village of Parigi Moutong Regency". Journal of natural science, 2(2):46-54.
[3]. Suryadarma, IGP. 2008. *Etnobotani*. Diktat Kuliah. Biology Education Studies Program FMIPA. University of Yogyakarta.
[4]. Alexiades, M. N. 1996. *Selected Guidelines For Ethnobotanical Research A Field Manual.* Botanical Garden : The New York
[5]. Austin, D.F. 2006. Florida Ethnobotany. CRC Press. Boca Raton London New York Wasington D.C.
[6]. Ahmed, J. U. 2010. *Development Vision Of North-East India*. Concept Publishing Company (P) TLD, New Delhi-110059: India.
[7]. Achroni, D. 2008. *Ceremony of Indigenous Archipelago*. CV The Sound of Median Sejahtera: Surakarta.
[8]. Djamaris, E. 2001. *Introduction to the Minangkabau People's Literature*. Yayasan Obor Indonesia: Jakarta.
[9]. Maulidiah, R. 2015. *Ethnobotany Traditional Ceremony of Haorja Haroan Boru Marriage Community of Angkola Batak Tribe in Padangsidimpuan*. Essay. Department of Biology. Fmipa. University of Andalas Padang.
[10]. Suswita, D. 2013. *Studi Ethnobotany And Forms Of Plant Preservation Efforts Used In Kendo Sko Rituals In Some Of Kerinci District, Jambi.* Thesis. Postgraduate Program of FMIPA. University of Andalas Padang.
[11]. Chunningham, A.B. 2001. *Applied Ethnobotany (People, Wild Plant Use And Conservation)*. Earthscan. London.

IOP Conf. Series: Materials Science and Engineering **335** (2018) 012019 doi:10.1088/1757-899X/335/1/012019

Pollen Morphology of *Caesalpinia pulcherrima* (L.) Swartz in Highland and Lowland West Sumatra

R Fitri[1*], Des M.[1]

[1]Biology Department, Faculty of Mathematics and Natural Sciences, Universitas Negeri Padang, Indonesia

*rahmadhanifitri@fmipa.unp.ac.id

Abstract. Determine the morphology structure of pollen on some variation colour of corolla *Caesalpinia pulcherrima* L. (Swartz) in highland and lowland West Sumatra has been conducted. The result reveals that topography and variation colour of corolla *C. pulcherrima* L. (Swartz) affects the shape of pollen. Pollen *of C. pulcherrima* L. (Swartz) has single grains or monad, isopolar polarity, radial symmetry, and size categories large. The length of polar axis (P) 58.16 to 74.11 μm, the length of the equatorial diameter (E) 59.86 to 75.97 μm, so that pollen can be classified into sub-spheroidal sub-oblate, spheriodal sub-spheroidal oblate, and sub-spheroidal prolate. Ornamentation of *C. pulcherrima* (L.) Swartz was reticulate. The pollen has aperture 3, the type pore and located in equatorial. From these data can be concluded that pollen from varying colour of corolla *C. pulcherrima* (L.) Swartz has same in terms of unit, polarity, symmetry, size, and type aperture, but it different in terms of shape.

1. Introduction

Caesalpinia pulcherrima (L.) Swartz is a species of sub-family Caesalpinoideae. The Caesalpinoideae consists of 150 genera, 2700 species and one of them is *C. pulcherrima* (L.) Swartz. Distribution of Caesalpinoideae includes tropical, subtropical and temperate regions [1]. *C. pulcherrima* (L) Swartz has variation of corolla. Coloured corolla of *C. pulcherrima* (L.) Swartz is red or yellow. Many *C. pulcherrima* (L) Swartz cultivated ornamental plants and sometimes as wild plants [2]. Based on observations of previous researchers found that corolla of *C. pulcherrima* (L.) Swartz has 3 colours (Figure 1). Colour corolla *C. pulcherrima* (L.) Swartz is orange, yellow, and pink [3].

Pollen morphology can be used as the taxon identification at the level of family, genera, species, and even varieties [4, 5]. The morphological differences of pollen found in Bauhinia L. and Phanera Lour. They are members of the subfamily of Caesalpinioideae. Due to the difference in the pollen structure, the two genera separated [6]. The outer pollen structure, including shape and size of the pollen can be used to distinguish genera and species levels of the Rhizophoraceae mangrove plant [7].

Differences in morphological structure of plants can be influenced by several factors such as genetic factors and environmental factors. Environmental factors that can affect changes in plant morphology include climate, temperature, soil type, soil conditions, altitude, and humidity. If environmental factors are stronger than genetic factors, then plants in different places with different environmental conditions will have varying morphology [8]. Examples of plant-morphological variations that are influenced by the environment are *sambiloto* (*Andrographis paniculata*). In this plant shows that there are variations of leaf size, stem height, number of branches, and also number of leaves. this occurs because of differences in location grow (there are differences in height of the growing location or topography). Based on this, it can be predicted that pollen morphology can also be influenced by environmental factors such as topography [9]. The present study is based on variation pollen morphology *C. pulcherrima* (L.) Swartz in highland and lowland West Sumatra.

Figure 1. Colour Variation of Corolla of *C. pulcherrima* (L.) Swartz.

2. Materials and Methods

Pollen samples were obtained from highland (Bukittinggi City) and lowland (Padang City) West Sumatra. The pollen grains were prepared for light (LM) and Scanning Electron Microscope (SEM) by the standard acetolysis methods [10]. Observed the pollen unit, polarity, symmetry, size, shape, type of aperture, and surface shape of pollen. The size of the pollen was measured by polar axis and equatorial axis. The terminology used is in literature and book identification like *Pollen Terminology* (Hesse *et al.*, 2009) [11] and *Pollen Morfology and Plants Taxonomy* (Erdmant, 1972) [12].

3. Results and Discussion

Observation of pollen morphology result study that pollen unit of *C. pulcherrima* (L.) Swartz were only monad (Figure 2). Others research found that pollen unit of *C. pulcherrima* (L.) Swartz also has unit pollen monad [3] [13] [14]. Pollen grains isopolar, radially symmetry, and size categories large (51–100μm) because diameters pollen is 58,16–74,11μm. Hesse et al. reveal that *C.pulcherrima* (L.) Swartz has pollen with isopolar polarity because they have exact same distal and proximal part [11].

Comparison of length axis polar (P) and diameter equatorial (E) in highland and lowland West Sumatra present in Table 1. Pollen of *C. pulcherrima* (L.) Swartz in the highland found have three shapes are *sub-spheroidal sub-oblate, sub-spheroidal oblate spheriodal,* and *sub-spheroidal prolate.* While in the lowland found have two shapes are *sub-spheroidal oblate spheriodal,* and *sub-spheroidal prolate*. The length axis polar (P) pollen in the highland ranges from 58,16–74,11μm and length diameter equatorial (E) ranges from 60,80–75,97μm. The length axis polar (P) pollen in the lowland ranges from 59,17–70,30μm and length diameter equatorial (E) ranges from 59,86–70,23μm.

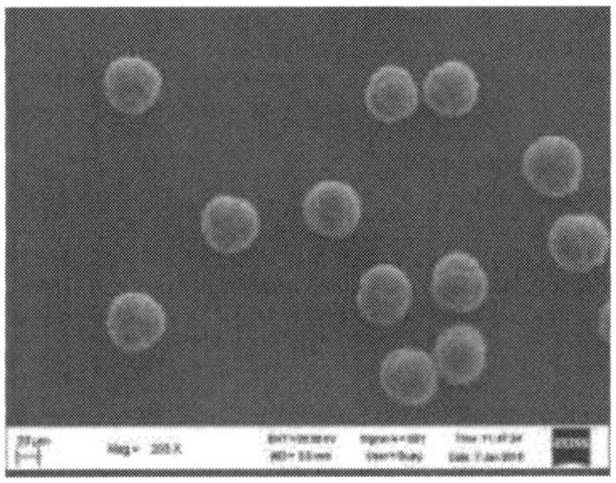

Figure 2. Pollen of *C. pulcherrima* (L.) Swartz have Unit Monad.

Table 1. Comparison of Pollen Shapes on Variation *C. pulcherrima* (L.) Swartz

No.	Colour Corolla	Length Axis P and Diameter E (μm)		Index P/E.100 (μm)	Shapes	Aperture		
		P	E			Total	Type	Position
A. *Caesalpinia pulcherrima* (L.) Swartz in Highland West Sumatra (Bukittinggi City)								
1.	Orange	58,64– 70,28	61,28– 65,97	95,69 – 106,53	Sub-spheroidal Oblate Spheriodal – Sub-spheroidal Prolate	3	Pore	Equatorial
2.	Yellow	58,16– 74,11	67,65– 75,97	85,97 – 97,55	Sub-spheroidal Sub-oblate – Sub-spheroidal Oblate Spheriodal	3	Pore	Equatorial
3.	Pink	62,93– 67,75	60,80– 66,53	101,83 – 103,50	Sub-spheroidal Prolate	3	Pore	Equatorial
B. *Caesalpinia pulcherrima* (L.) Swartz in Lowland West Sumatra (Padang City)								
1.	Orange	59,17– 67,08	61,58– 69,57	96,09 – 96,42	Sub-spheroidal Oblate Spheriodal	3	Pore	Equatorial
2.	Yellow	63,46– 70,30	64,65– 70,23	98,16 – 100,10	Sub-spheroidal Oblate Spheriodal – Sub-spheroidal Prolate	3	Pore	Equatorial
3.	Pink	60,47– 67,36	59,86– 66,03	101,02 – 102,01	Sub-spheroidal Prolate	3	Pore	Equatorial

In general, the pollen form (shape) all variations of *C. pulcherrima* (L.) Swartz includes sub-spheriodal (P/E × 100 = 75 – < 133μm) with 5 variations of sub-oblate, spheriodal oblate, spheriodal prolate, and prolate. Previous research results also support this research. Previous research has revealed that Spheriodal Oblate and Spheriodal Prolate included in the Subspheriodal types[13][15][16]. The form of pollen *C. pulcherrima* (L.) Swartz looks polar is circular.

The surface shape or ornamentation of *C. pulcherrima* (L.) Swartz is reticulate (Figure 3). The pollen has 3 apertures with pores type and located in the equatorial section (Figure 4). Previous research has revealed that ornamentation of its plant is reticulate[3][15][16]. However, other research results reveal ornamentation of *C. pulcherrima* (L.) Swartz is psilate perforate [13]. This aperture has three intertice each with double pores (tricolporate). The pollen morphology *C. pulcherrima* (L.) Swartz for all variations can be seen in Table 2.

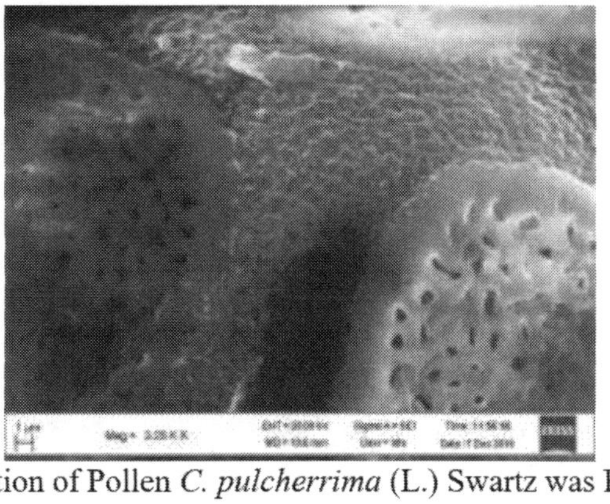

Figure 3. Ornamentation of Pollen *C. pulcherrima* (L.) Swartz was Reticulate with SEM.

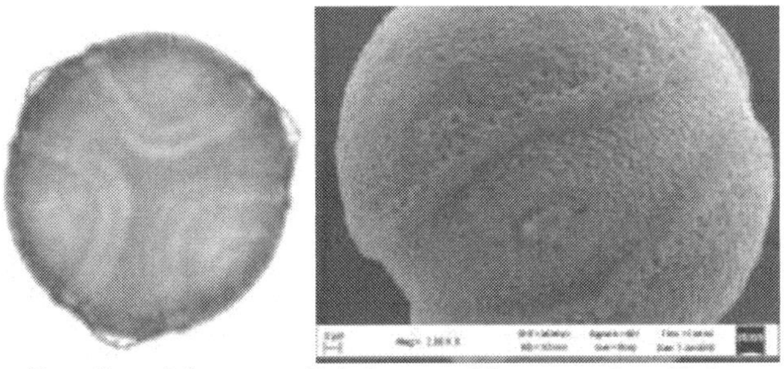

Figure 4. Aperture of pollen *C. pulcherrima* (L.) Swartz Observered on Polar and Equatorial with LM and SEM.

According to Table 1, it is knows that the pollen form (shape) of *C. pulcherrima* (L.) Swartz with the colour corolla variation of the pink flower from the highland and lowland is the same. While the colour corolla variation of orange and yellow flowers was different. The colour corolla variation of yellow flower from the highland is smaller than the lowlands, and the colour corolla variation of orange flower the highland is greater than the lowlands.

Table 2. Pollen Morphology of *C. pulcherrima* (L.) Swartz in Highland and Lowland West Sumatra (Analyze with SEM)

No.	**Position**	**Coloured Corolla**		
		Orange	**Yellow**	**Pink**
A.	***Caesalpinia pulcherrima*** **(L.) Swartz in Highland West Sumatra (Bukittinggi City)**			
1.	Equatorial	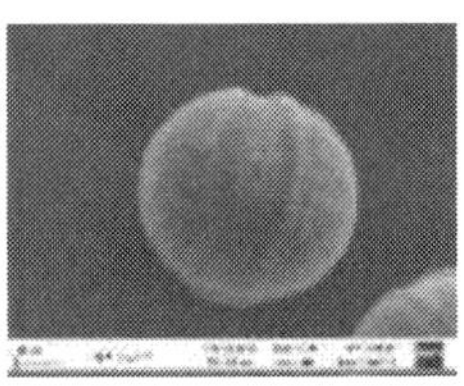	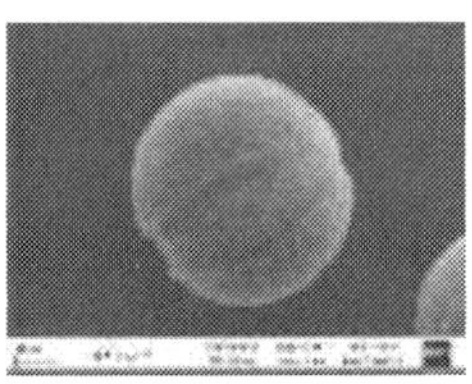	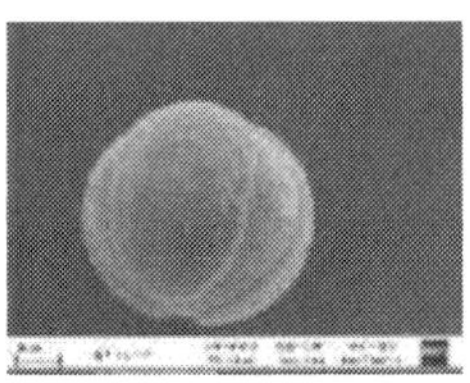
2.	Polar	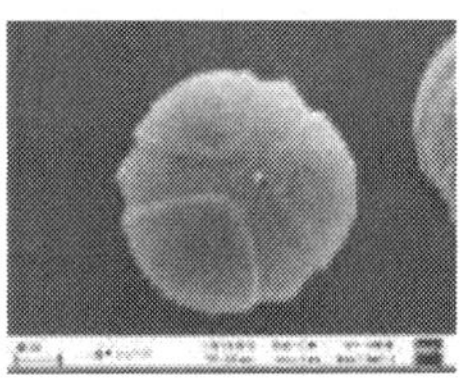	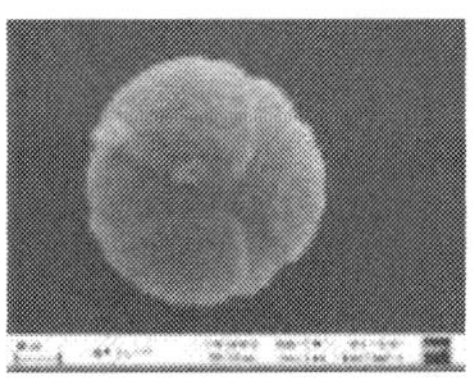	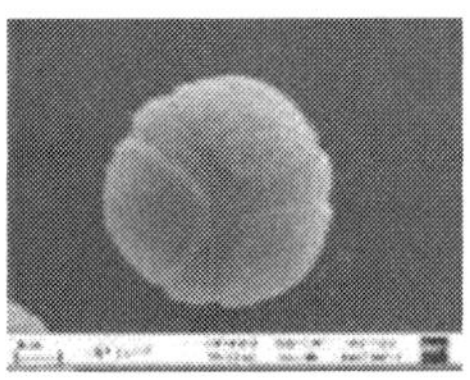
B.	***Caesalpinia pulcherrima*** **(L.) Swartz in Lowland West Sumatra (Padang City)**			
1.	Equatorial	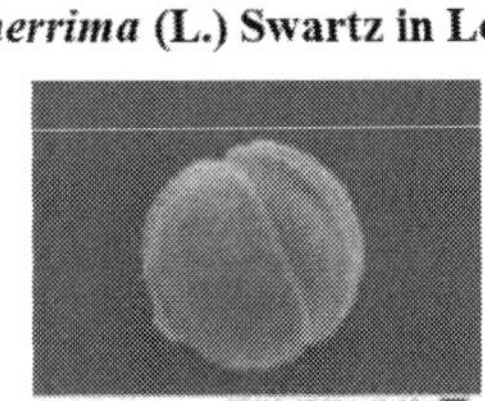	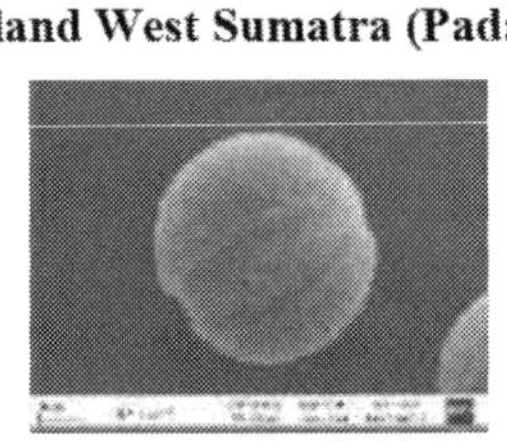	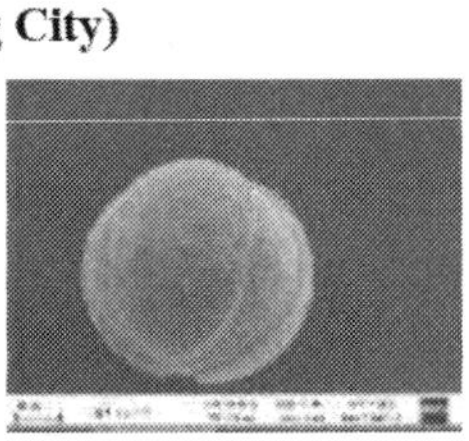
2.	Polar	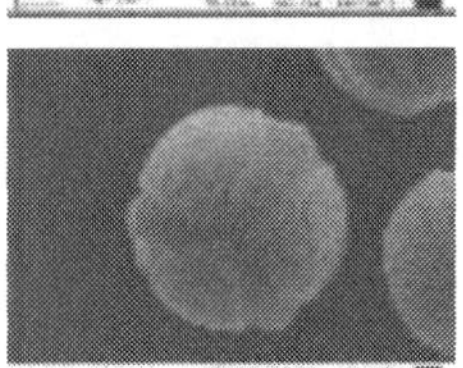	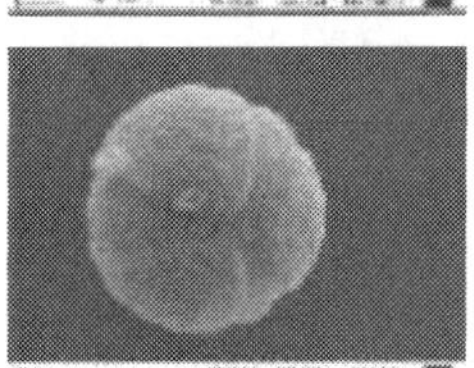	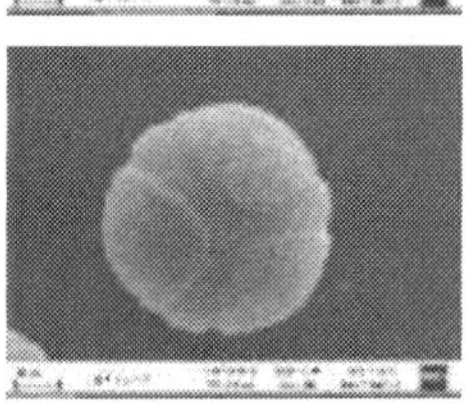

4. Conclusion

In the result reveals that the topography and variation colour of corolla *C. pulcherrima* L. (Swartz) affects the shape of pollen. Pollen *of C. pulcherrima* L. (Swartz) has single grains or monad, isopolar polarity, radial symmetry, and size categories large. The length of polar axis (P) 58.16 to 74.11 μm, the length of the equatorial diameter (E) 59.86 to 75.97 μm, so that pollen can be classified into sub-spheroidal sub-oblate, spheriodal sub-spheroidal oblate, and sub-spheroidal prolate. Ornamentation of *C. pulcherrima* (L.) Swartz was reticulate. The pollen has 3 aperture, the type pore and located in equatorial. From these data can be concluded that pollen from variations colour of corolla *C. pulcherrima* (L.) Swartz has same in terms of unit, polarity, symmetry, size, and type aperture, but it different in terms of shape.

References

[1] Singh, G. 2010. *Plant Systematics An Integrated Approach.* New Hamsshire: Science Publishers.

[2] Steenis, C. G. G. J. Van, *et al.* 2006. *Flora untuk Sekolah di Indonesia.* Jakarta: Pradnya Paramita.

[3] Fitri, R., Dezi H., and Des M. 2016. "Morfologi Serbuk Sari pada Beberapa Variasi Warna Mahkota Bunga *Caesalpinia pulcherrima* (L.) Swartz". *Eksakta*, Vol 2 Tahun XVII: 38-44.

[4] Bagu, F. S. 2003. "Taxonomy of Delphinium L. (Ranunculaceae) In Java Based on Pollen Morphology". *Eugenia*, Vol. 9 (1): 1-8.

[5] Lacy, J. G., P. B. Kaufman. 2006. *Botany Illustrated: Introduction to Plant, Major Groups, Flowering Plant Families.* New York: Springer.

[6] Santos, F. De A. R. Dos., D. De M. Novaes, L. P. De Queiroz. 2012. "Pollen of *Bauhinia* L. And *Phanera* Lour. (leguminosae-Caesalpinioideae) From The Brazilian Caatinga. *American Journal of Plant Science*, Vol 2: 909-920.

[7] Irawan, B., S. Muadz, A. Rosadi. 2013. "Karakterisasi dan Kekerabatan Tumbuhan Mangrove Rhizophoraceae Berdasarkan Morfologi, Anatomi, dan Struktur Luar Serbuk Sari". *Prosiding Seminar Nasional Sains dan Teknologi Nuklir*, PTNBR-BATAN, Bandung 4 Juli 2013.

[8] Widiyanti, Suranto, Suiyarto. 2008. "Studi Variasi Morfologi Biji, Serbuk Sari, dan Pola Pita Isozim Padi (*Oryza sativa*) Varietas Rojolele". *Bioteknologi* Vol.5, No.1: 18-25.

[9] Pujiasmanto, B., J. Moenandir, Syamsulbahri, Kuswanto. 2007. Kajian Agroekologi dan Morfologi Sambiloto (*Andrographis paniculata*) pada Berbagai Habitat. *Biodiversitas* Vol. 8 No. 4: 326-329.

[10] Aprianty, N. M. D., E. Kriswiyanti. 2008. "Studi Variasi Ukuran Serbuk Sari Kembang Sepatu (*Hibiscus rosa-sinensis* L.) dengan Warna Bunga Berbeda". *Jurnal Biologi*, Vol. 12, No. 1: 14-18.

[11] Hesse, M., et al. 2009. *Pollen Terminology: An Illustrated Handbook.* New York: Springer-Verlag.

[12] Erdmant, G. 1972. *Pollen Morphology and Plant Taxonomy. Angiosperm: An Introduction to Palynology I.* New York: Hafner Publishing Company.

[13] Kumaladita, L. 2014. "Hubungan Kekerabatan Jenis-jenis Tumbuhan Anggota Sub Famili Caesalpinoideae di Daerah Istimewa Yogyakarta Berdasarkan Berdasarkan Kajian Morfologi Serbuk Sari Sebagai Sumber Belajar Biologi Siswa SMA Kelas X". *JUPEMASI-PBIO*, Vo.1, No.1 Tahun 2014: 93-97.

[14] Colinvaux, P., Paulo E.D.O., Jorge E.M.P. 2005. *Amazon Pollen Manual and Atlas.* Netherlands: Hardwood Academic Publishers.

[15] Jumah, A. 1996. "Studies On The Morphology of Pollen Grains of The Leguminosae – The Caesalpinoideae". *Ghana Journal of Science*, Vol.31-36: 23-28.

[16] Perveen, A., and M. Qaiser. 1998. "Pollen Flora of Pakistan – X. Leguminosae (Subfamily: Caesalpinoideae)"*Tr. J. of Bonaty*, Vo. 22: 145-150.

IOP Conf. Series: Materials Science and Engineering **335** (2018) 012020 doi:10.1088/1757-899X/335/1/012020

Leaf Trichomes Morphology of Hyptis suaveolens (L.) Poit. (LAMIACEAE)

M Chatri[1*], A Baktiar[2], M Mansyurdin[3], P Periadnadi[3].

[1]Department of Biology, Faculty of Mathematics and Natural Sciences, Universitas Negeri Padang, Indonesia
[2]Department of Pharmacognosy, Faculty of Pharmacy, Andalas University, Padang Indonesia
[3]Department of Biology, Faculty of Mathematics and Natural Sciences, Andalas University, Padang, Indonesia

*moralitachatri@gmail.com

Abstract. *Hyptis suaveolens* L. Poit is one of the plants from family Lamiaceae and is an aromatic plant. The aroma contained in plants is usually secreted by certain structures in plants, such as glandular trichomes. At this plant has been carried out observations about the type and distribution of trichomes by using light microscopy and SEM (*Scanning Electron Microscopy*). The results showed that the leaves of this plant are non-glandular trichomes types and glandular, either on the surface abaxial and adaxial and on the veins. Non-glandular trichomes consist of the monoselular and multicellular trichomes. While the glandular trichomes consist of peltate type, capitate type I and type II.

1. Introduction

Hyptis suaveolens L. Poit. is a herbaceous, branched, aromatic, often woody plant at the base, height 0.4-2 m. Long-haired stems, oval leaves with a base of rounded heart shapes, irregularly serrated [1, 2], leaf surfaces have hair glands. Hair glandsular or glandular trichomes of aromatic plants common produce essential oil [3]. Glandular trichomes are one of the secretory structures found in plants [4] that can secrete secondary metabolite [5] especially in Labiatae, Verbenaceae and Geraniaceae [6] families. The scented leaves are caused by essential oils. Which are secreted by glandular trichomes. Many species of the Labiatae family have glandular trichomes [7] spread over vegetative and generative organs [8] and more prevalent in leaves and flowers [9]. The morphology, distribution and number of glandular trichomes may be as distinguishing characteristics at the subfamily level of Labiatae [10]. According to Cutler [11], when a plant has a trichome, it is usually in a form that can characterize the plant species.

Trichomes are additional cells or derivatives of the epidermis, [4, 12]. In some plants, the trichomes are persistent or may persist in their organs as long as the organ is alive or it may dry out and fall off at any one time [12]. Trichomes usually arise from asymmetric cleavage in a single protodermis of stem cells. Although in fact the cell division of the stem cell is symmetrical and more than one beginner cell is involved. In adult cells the possibility without protopasma, but the cells remain alive [13]. The cell wall of trichomes generally consists of cellulose and is covered with cuticles [14]. In general, trichomes can be divided into two groups, ie non glandular or glandular trichomes that have secretions [11, 15, 16].

The main function of trichomes in plants is as a defense [17]. Non-glandular trichomes in plants may, among other things, affect the physiological and ecological conditions of plants, whereas glandular trichomes may provide plant resistance to insects, herbivorous and pathogenic attacks based on the chemical compounds they contain [18]. The results of glandular trichome secretions may also be useful for humans, such as for the manufacture of pesticides, drugs, cooking spices and perfumes [19].

2. Methods

Plant material

H.suaveolens leaves are obtained from natural populations in the area of Air Tawar, Padang, Indonesia. Leaves taken are young, mature and old leaves to be observed anatomically by using light microscopes and electron microscopes.

Light microscope

Young, adult and old *H.suaveolens* leaves are taken in the field, then taken to the laboratory and then fed into the FAA solution (formalin, acetic acid and alcohol). The leaves are made of transverse incision and stained with safranin, then placed on a glass object and covered with a cover glass, then observed trichome form and trichome types contained on the leaf by light microscope model Primo Star iLED Miroscopes from Carl Zeiss MicroImaging GmbH P. OB.4041.37030, Gottingen, Germany.

Scanning Electron Microscopy (SEM).

Approximately 0.5 cm fresh leaves from *H.suaveolens* are placed on the top of the sample holder of the electron microscope, then immediately observed by Scanning Electron Microscopy Phenom Prox model. The observed leaf surface is the adaxial (base and leaf end) and abaxial (base and leaf end).

3. Results And Discussions

The glandular trichomes of *H.suaveolens* leaf was investigated by light microscope and Scanning Electron Microscopy (*SEM*) and was summarized in Figure 1 and Figure 2.

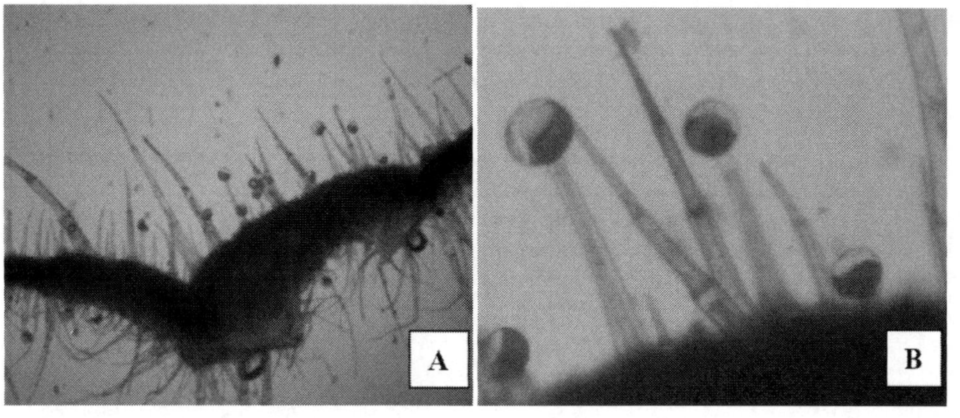

IOP Conf. Series: Materials Science and Engineering **335** (2018) 012020 doi:10.1088/1757-899X/335/1/012020

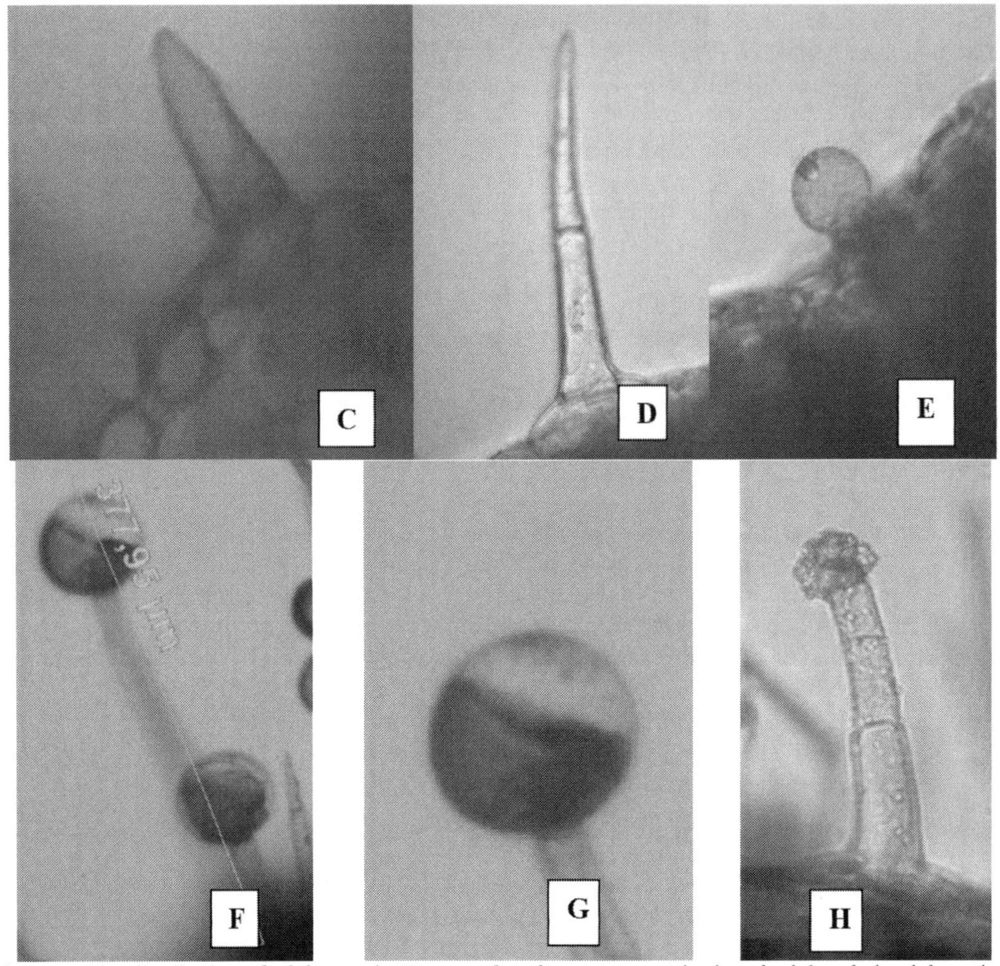

Figure 1. The distribution of trichomes in H.*suaveolens* leaves present in the adaxial and abaxial portions observed by light microscopy (A and B), unicellular non-glandular trichomes (C), multicellular non-glandular tricomes (D), glandular tricks of peltate type (E), Glandular trichomes capitate type I and capitate II (F), head cells consisting of secretory cells with subcutaneous cavities in glandular trichomes (G) and glandular trichomes that have secreted secretions (H).

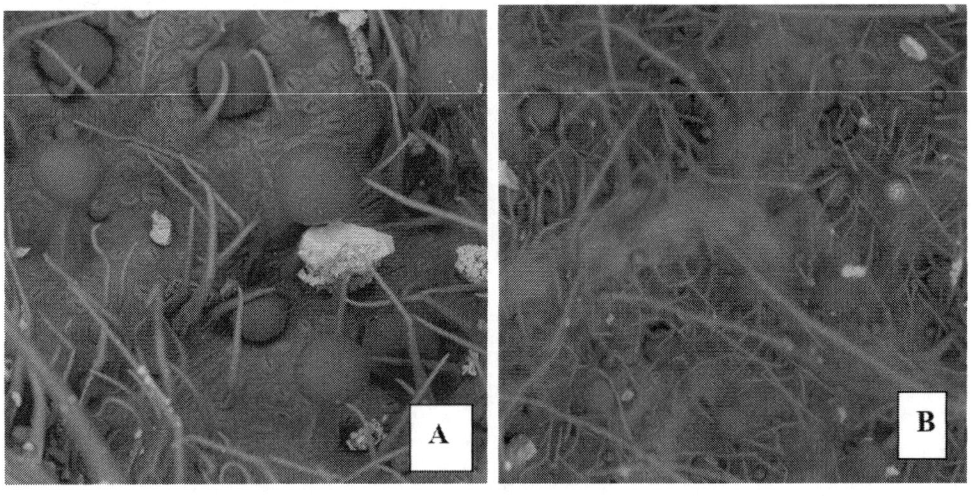

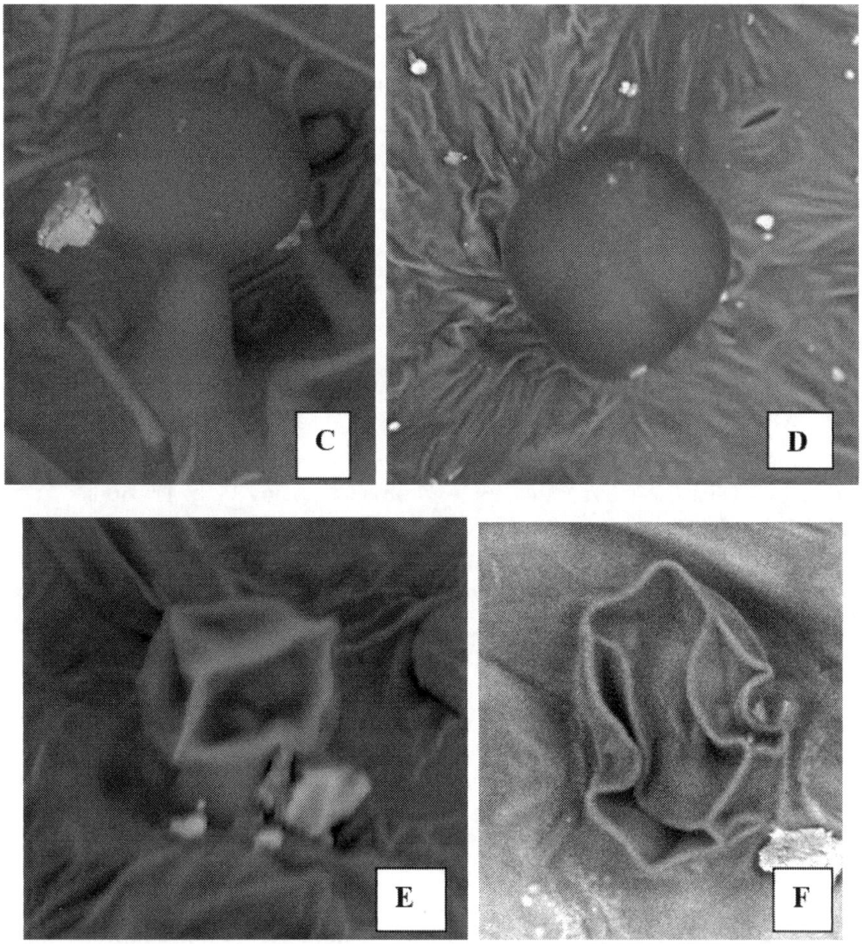

Figure 2. Trichomes observation of *H.suaveolens* leaf with SEM, visible type of non-glandular and glandular tricomes (A), non-glandular and glandular trichomes are also present in leaf bone (B), glandular capitate type I (C) trichomas, cell Head in a glandular trichome consisting of four cells (D), a glandular trichome that releases its secretory (E-F).

Trichoma capitate type II has a stalk with a size longer than the capitate type I. The capitate-type mice are commonly found in the Lamiaceae family [10], but with varying head length and head size. Commonly composed of one or two stalk cells with one or two round cells forming a secret cell. Cell secretion in glandular trichomes is assumed to be a volatile oil-producing cell in the familia Lamiaceae Cell secretion is related to the stem and cell base and is also associated with the secretion-subcutaneous oil subcutaneous cavity (Fig.

The glandular trichomes observed by SEM, visible non-glandular and glandular trichomes (Fig. 2A). Werker, et al. [18], explains that almost all plant body surfaces that emit the scent of the Lamiaceae family are covered by non-glandular trichomes and glandular trichomes or secretory trichomes. In Figure 2B, the spread of non-glandular trichomes and glandular trichomes is also present in the leaf bone. Glandular trichomial head cells are round with smooth surfaces. In the head cell there is a secretion cell consisting of one to several cells. The secretion results will be collected in the subcutaneous cavity, if this cavity is broken, the secretion results will come out and the glandular trichom cell head will contract (Fig. 2E-F). Malecci and Servettaz [20] explains, in adult trichomes, the results of secretions accumulate in the large subcutaneous space and form above the secretion cells. The release of this secretion results in a certain aroma. The leaves are scented because of the glandular trichomes that secrete essential oils.

4. Conclusion

From the results of the study, it can be concluded that in *H. suaveolens* L. leaf there are five types of trichomes, ie unicellular non-glandular and monocellular, glandular peltate, type I capitate and type II capitate. Both non-glandular trichomes and glandular trichomes are dispersed in abaxial and adaxial portion of the leaf and are present in the leaf vein.

References

[1]. Henderson.1959. Malayan Wild Flowers (Dicotyledons). Tien Wah Press ltd., Singapore.
[2]. Backer, C.A and R.C.B Van de Brink. 1965. Flora of Java. Vol. II. N.V.P Noordhoff-Graningen. Netherlands.
[3]. Simpson .G.2006.Plant Systematics.Elsevier Academic Press. USA.
[4]. Fahn, A. 1990. *Plant Anatomy*.4st Ed. New York. Pergamon Press
[5]. Schilmiller, A.L., R.L Last and E Pichersky. 2008. Harnessing Plant Trichome Biochemistry for the Production of Useful Compounds. The Plant Journal 54: 702-711.
[6]. Handa, S.S. 2008. An Overview of Extraction Techniques for Medicinal and Aromatic Plants. In Extraction Technologies for Medicinal and Aromatic Plants. International Centre for Science and High Technology: 21-54.
[7]. Singh G. 2010. Plant Systematics. An Intregated Approach. Third Edition. Science Publisher. USA.
[8]. Werker, E. 1993. Fuction of Essential Oil-Secreting Glandular Hairs in Aromatic Plants of Lamiaceae-a Review. Flavour and Fragrance Journal. Vol.8 (5): 249-255.
[9]. Baran P., C Ozdemir and K Aktas. 2010. Structural Investigation of the Glandular Trichomes of *Salvia argentea*. Biologia 65/1 : 33-38.
[10]. Ascensao L., N Marques and M.S Pais. 1995. Glandular Trichomes on Vegetatif and Reproductive Organs of *Leonotis leonurus* (Lamiaceae). Annals of Botany 75: 619-626.
[11]. Cutler D.F. 1978. *Applied Plant Anatomy*. Longman Inc. New York.
[12]. Varghese T.M. 1987. An Introduction to The Anatomy of Angiospermae. Allied Publisher Limited. New Delhi.
[13]. Dickison W.C. 2000. *Integrative Plant Anatomy*. Academic Press. USA
[14]. Mishra S.R. 2009. *Understanding Plant Anatomy*. Discovery Publishing House PVT. Ltd. New Delhi. India.
[15]. Mauseth J.D. 1988. *Plant Anatomy*. Pulishing Company Inc.Menlo Park, California.
[16]. Werker E. 2005. Trichome Diversity and Development. In Plant Trichomes (Hallahan D.L and J.C Gray, eds.). Advances in Botanical Research. Vol 31. New York. Academic Press.: 1-35.
[17]. Levin, D.A. 1973. *The Role of Trichomes in Plant Defense*. The University of Chicago Press.
[18]. Werker, E., U. Ravid and E. Putievsky. 1985. Struktur of Glandular Hairs and Identification of Main Component of Their Secreted Material in Some Species of Labiatae. Israel J Bot. 34:31-35.
[19]. Duke, S.O. 1994. Commentary Glandular Trichomes; A Focal Point of chemical and Structural Interaction. International Journal of Plant Science 155: 617-620.
[20]. Malecci, L.B and O. Servettaz. 1991. Morphology and Distribution of Trichomes in Italian species of *Teucrium* sect. *Chamaedrys* (Labiatae) - a Taxonomical Evaluation. Pl.Syst.Evol.174 : 83 – 91.

Screening of Thermophilic Bacteria Produce Xylanase from *Sapan Sungai Aro* Hot Spring South Solok

I Irdawati[1*], S Syamsuardi[2],AAgustien[2],Y Rilda[2]
[1]Departmentof Biology,Universitas Negeri Padang, Indonesia
[2]Departmentof Biology,Andalas University, Indonesia

*irdawati.amor40@gmail.com

Abstract - xylanase is one of the enzymes with great prospects as hemicellulose hydrolyzing enzyme. Global annual market demand for this enzyme reach US $ 200 million. This enzyme catalyzes the xylan (hemicellulose) reactions breaking into xilooligosakarida and xylose. Xylanase can be applied to various industrial sectors such as bread, sugar xylose, biofuels, especially in bleaching paper (bleaching) pulp. Xylanase Isable to replace conventional chemical bleaching using chlorine that is not friendly for the environment. Currently xylanase production is extracted from the thermophilic bacteria for enzyme stability at high temperatures that are suitable for industrial applications. Thermophilic bacteria can be isolated from a hot spring, one of the which is a source of Sapan Sungai Aro Hot Spring, located in the district South Solok. The aim of this study was to select and identification of thermophilic bacteria can produce xylanase.This roomates is a descriptive study, which was Carried out in the Laboratory of Microbiology, Mathematic and Science Faculty of Padang State University, and Laboratory of Bacteriology, BasoVeterinary Research Center. The research procedure consisted of the preparation and sterilization of materials and tools, medium manufacturing, regeneration, selection and identification. Selection is performed by using a semiquantitative screening plate that contains xylan substrate. Identification is based on microscopic and biochemical characteristics until the genus level.Selection results Showed 12 out of 16 isolates had xilanolitik activity, with the highest activity is SSA2 with xilanolitik index of 0.74. The top five index producehigestxilanolitik isolates that are SSA2, SSA3 and SSA4 identified as Bacillus sp. 1., and SSAS6 and SSA7 is Bacillus sp. 2.

1. Introduction

Enzymes are bio-catalyst. Enzymes improve chemical reactions that occur in the cells of organisms without changing the structure itself[1].Chemical reactions can walk 108 to 1020 times faster than normal in the presence of enzymes. Are specifically bound enzyme with its substrate in catalyzing the reaction and produce products[2].

In line with the advancement of applied biotechnology enzymes to catalyze chemical reactions outside the cells or their environment. Enzymes are used in various industrial sectors such as textiles, food, detergents, paper and cosmetics, as well as biofuels.Needs of the enzyme in the world market in 2012 was US $ 4.5 billion and nearly US $ 4.8 billion in 2013, it is estimated that the market needs in 2018 amounted to US $ 7.1 billion[3].

Xylanase enzyme is one of the enzymes that were instrumental in the industrial world. Market demand for the xylanase is also quite high, at US $ 200 million[4].This enzyme catalyzes the cracking reactions xylan (hemicellulose) into xilooligosakarida and xylose. Xylanase can be used in pulpbleaching,hemicellulose conversion into a source of raw material forproduction, biofueland the production of sugar xylose. Very important xylanase used in theprocess. pulp bleachingXylanasereplacing bleaching chemicallyconventionalusing chlorine unsustainable[5].In addition, in the production of bread xylanase also used to be a softer dough structure[6].

Production of xylanase enzyme for industrial use is extracted from various species, such as bacteria, fungi, yeast and plants. When this has been known to almost 4000 types of enzymes, 200 of which came from the microorganisms and used commercially[7].Microorganisms produce enzymes that can be utilized humans in the number and types vary widely. They are easy to be cultured and potential and in accordance to the industry[8].Microorganism one right choice for use in industrial processes, because it is thermostable. For example, brown stain removal process or bleaching of paper material(pulpbleaching),for optimal results of this process should use high temperatures and alkaline conditions[9].

Or termozim thermostable enzymes suitable for use in high-temperature reactions. Some of the main advantages of using high-temperature reaction is to reduce the risk of microbial contamination, lower viscosity, increases the rate of movement of the molecules, and increases the solubility of the substrate[10].Thermostable enzymes capable of increasing the speed of the reaction thus saving time, labor and operational costs, facilitate the separation of volatile compounds, and more stable at a longer storage period[11].

Thermophilic bacterium grows optimally at a temperature range of 60-108 ^{0}C. The bacteria can be found in various places in nature, either sea or land which is heated by geothermal energy, such as hot springs, volcanic sediments from the mainland, hydrothermal vent systems, and ventilation hydrothermal seafloor in[12].Indonesia is one area that is quite a lot of hot springs, so it has the potential to generate sources of microorganisms that can produce thermostable xylanase enzyme.

The microorganisms capable of producing thermostable xylanase enzyme is a valuable biological assets for the benefit of industrial biotechnology. Thermophilic bacteria are known to produce thermostable xylanase is Bacillus subtilis[7] and Pseudomonas[13],while the fungus known to produce this enzyme is Streptomyces[14], Aspergillusand Trichoderma.From 5 hot springs in the Northern Himalayas, Himachal Pradesh. 101 isolates were obtained and tested enzymatic activity. Through analysis of the 16S ribosomal RNA gene sequences identified genus Bacillus and Paenibacillus has the potential to produce xylanase[15].

Indonesia is a region that is pretty much had the hot water sources. One of these hot springs are hot springs *Mudiak Sapan* South Solok. Telahdilaporkanada19 thermophilic bacteria, selected 13 of them showed xylanase activity. The hot springs have temperatures up to 93 ^{0}C with a pH of 8. The obtained 18 isolates the overall shape and relatively Gram-positive bacilli and 1 isolate shaped Gram-negative bacilli[16].

In addition to hot springs *Mudiak Sapan*, in South Solok district also has other hot springs. The hot springs are a source of hot water *Sapan Aro* River located in the South Solok District. The hot springs have a temperature of 75^{0}C and pH 8. It was reported 16 isolates (7 comes from water samples and 9 of thesamples). sludgeFrom these hot springs obtained 16 isolates belonging to the shaped bacilli and gram-positive bacteria and produce endospores, and also showed amylolytic activity[17].Synthesis influenced by the presence of an enzyme substrate. So a bacterium that produces amylase in starch medium containing potentially produce xylanase on medium containing xylan [18].

Microscopic characteristics are very helpful in the identification phase. But the bacteria with the same shape as likely to have metabolic and physiological activity are different. Biochemical characteristics are also required to specify a type of bacteria. In fact, the most closely related bacteria that can be separated into a particular species with biochemical tests, one of which is determined by their ability to ferment the carbohydrate group selection[19].

Currently it has developed a wide range of biochemical tests, such as tests TSI(TripleSugarIron),gas production, the production of H2S, catalase, oxidase, motility, IMViC(Indole,,Methyl-red, Voges Proskauer,citrate),urease, oxidative-fermenteatif (OF), and fermentation of various kinds of carbohydrates (glucose, sucrose, lactose, mannitol, sorbitol, arabinose, and others), the hydrolysis of casein, gelatin and many other tests. The various tests can be used to determine the genus or even species of a bacterium[20].

2. Methods

Sterilization of Apparatus

All the tools that will be used first sterilized. Sterilization and medium performed by autoclaving at 121°C and a pressure of 15 psi for 15 minutes.

Medium Formation

Medium Nutrient Agar (NA)

Medium used that medium to regenerate NA thermophilic bacteria. A total of 10 g of NA was dissolved in 500 mL of distilled water and heated until homogeneous, then sterilized in an autoclave at 121°C at a pressure of 15 psi for 15 minutes.

Medium Production Beechwood Xylan

Bactopepton as much as 0.5%, yeast extract 0.1%, K2HPO40.1%, MgSO4.7H2O 0.02%, and 0.1% beechwood xylan, dissolved in 1 L of distilled water. Then heated us ing an electric stove and stir until homogenous. After boiling, medium poured into the erlenmeyer and sealed with gauze and aluminum foil. Sterilization in an autoclave at a temperature of 121°C and a pressure of 15 psi for 15 minutes.

1) The Spread Plate

The Medium used for isolating the thermophilic bacterium is sterilized NA, poured in a sterile petri. After 0.1 mL of hot water samples was solid, then it was distributed with hockey sticks on the surface of agar, it was incubated for 24 hours at a temperature of 60 ° C, each of hot water samples was made in duplicate. While for the isolation of "sludge" sediment process is done by weighing samples of sediment as much as 1 g. Then it was put into a test tube containing 9 mL of distilled water. Then it was divortek until the sediments homogeneous, then took 0.1 mL suspension using a micropipette and inserted into a petri plate which filled with NA medium.

2) The Streak Plate

For the purification of bacteria, each bacterial colonies which were growing in culture were previously taken with one inoculating loop an it was scratched intopatri plate to obtain pure cultures of bacteria.

a. Xylanase Activity

Xylanase activity test carried out by inoculating all pure isolates obtained from the previous stage. Isolates thermophilic bacteria grown in Nutrient Agar medium 24 hours. Then the isolates which have grown were tested in a suspension by taking a 1-2 inoculating loop of isolate bacterial culture into a sterile test tube which already contain physiological saline solution 0.85%. The mixture was homogenized with vortex mixture turbidity, and then it was compared with turbidity Mac Farland scale of 1 equivalent to 3.108CFU / mL. A total 0,1mL bacterial suspension was pipetted on Nutrient Agar medium + 0.5% Beechwood Xylan. The cultures were incubated for 72 hours at a temperature of 60 ° C [10]. Bacterial isolates were grown drops by 1% congo red solution and allowed to stand for 30 minutes, then it was dropped by 2M NaCl to see the clear zone. After that, it was measured to see the diameter of clear zone. The Isolates which produce xylanase indicated by a clear zone around bacterial colonies. Clear zone diameter of bacteria which have formed, then it is measured by using a caliper [11].

b. Catalase Test

Two drops of hydrogen peroxide (H2O2) 3 % are dropped on a glass slide, then placed one inoculating loop of thermophilic bacteria and flatted. The positive catalase test is characterized by the existence of air bubbles. The existence of air bubbles indicates the amount of oxygen is produced.

Observations
Observations are carried out by observing the characteristics of bacteria and enzyme activities of xylanase that is produced by termophilic bacteria in hot springs *Mudiak Sapan*, South Solok District.

1) The Spread Plate
The Medium used for isolating the thermophilic bacterium is sterilized NA, poured in a sterile petri. After 0.1 mL of hot water samples was solid, then it was distributed with hockey sticks on the surface of agar, it was incubated for 24 hours at a temperature of 60 ° C, each of hot water samples was made in duplicate. While for the isolation of "sludge" sediment process is done by weighing samples of sediment as much as 1 g. Then it was put into a test tube containing 9 mL of distilled water. Then it was divortek until the sediments homogeneous, then took 0.1 mL suspension using a micropipette and inserted into a petri plate which filled with NA medium.

2) The Streak Plate
For the purification of bacteria, each bacterial colonies which were growing in culture were previously taken with one inoculating loop and it was scratched intopatri plate to obtain pure cultures of bacteria.

a. Xylanase Activity
Xylanase activity test carried out by inoculating all pure isolates obtained from the previous stage. Isolates thermophilic bacteria grown in Nutrient Agar medium 24 hours. Then the isolates which have grown were tested in a suspension by taking a 1-2 inoculating loop of isolate bacterial culture into a sterile test tube which already contain physiological saline solution 0.85%. The mixture was homogenized with vortex mixture turbidity, and then it was compared with the turbidity Mac Farland scale of 1 equivalent to 3.108CFU / mL. A total 0,1mL bacterial suspension was pipetted on paper discs at Nutrient Agar medium + 0.5% Beechwood Xylan. The cultures were incubated for 72 hours at a temperature of 60 ° C [10]. Bacterial isolates were grown drops by 1% congo red solution and allowed to stand for 30 minutes, then it was dropped by 2M NaCl to see the clear zone. After that, it was measured to see the diameter of clear zone. The Isolates which produce xylanase indicated by a clear zone around bacterial colonies. Clear zone diameter of bacteria which have formed, then it is measured by using a caliper.

b. Catalase Test
Two drops of hydrogen peroxide (H2O2) 3 % are drop on a glass slide, then placed one inoculating loop of thermophilic bacteria and flatted. Positive catalase test is characterized by the existence of air bubbles. The existence of air bubbles indicates the amount of oxygen is producedBiochemical Characterization of thermoxylanolityc Bacteria.

To determine the type of bacteria producing thermophilic xylanase enzyme necessary to do some biochemical tests. Biochemical characterization performed on 5 isolate s showed the highest xylanase activity. Biochemicalcharacterization was conducted on the test: TSI (Triple Sugar Iron), gas production, the production ofH2S, oxidase,motility, IMViC (Indole, Methyl-red, Voges Proskauer, Citrate), urease, oxidative-fermenteative (OF) and fermentation of various kinds carbohydrates (glucose, sucrose, lactose and mannitol), and the hydrolysis of gelatin.

Triple Sugar Iron Test (TSI)
TSI testing using Slant Jelly Medium containing 1% lactose and sucrose, and 0.1% glucose. Then Phenol red was added to the medium that is used as a pH indicator which will detect the formation of acid from the fermentation of carbohydrates. Inoculation of bacteria on the surface of the medium in a

zigzag, then incubated for 18-24 hours. Some of the reactions that might occur in the medium are as follows.

(1) Base yellow (K) and red inclined surface (M), leading to fermentation of glucose (Phenol red indic ator turns yellow due to the presence of acid in the base). The sloping surface remains red (alkaline) due to limitations of glucose in the medium.

(2) Base yellow (K) and an inclined surface of red (M), refers to the fermentation of lactose and Sucrose (concentration of the glucose is high) which make acid formation occurred throughout the medium.

(3) Establishment of gas, occurs when the order wassplit by the presence of air.

(4) The formation of H2S, visible through the black color in order.

(5) The base and red inclined surface (M|M), indicating none of fermentable sugars, and no gas or H2S production

Oxidase Test

Testing is done by adding reagents oxidase tetramethyl-p-phenylenediamine dihydrochloride 1% on filter paper.Then the bacterial cultures were taken with ose and smeared on the filter paper. The paper will be blue, black or purple in areas be oiled by culture if positive test results.

Test of IMViC (Indole, Methyl-red, Voges-Proskauer,and Citrate) Test of Urea Indole Motile (MIU)

Colonies that will be identified, retrieved using ose. Colonies were inoculated on the test medium motile indole urea in the form of semi-solid manner in order to be stabbed. Media incubated at room temperature for 18-24 hours. The bacteria movement indicated by the spread of bacteria colonies around the puncture. Urea positive reaction indicated media color change to pink. Indole positive reaction is indicated by the addition of Kovac reagent will then produces a red ring on the surface and show negative if it produces an orange ring on the surface.

Test of Methyl Red (MR)

Colonies that will be identified retrieved using ose. Colonies were inoculated on media MR-VP. The media were incubated at 37°C for 18-24 hours. Then add three to four drops of methyl red indicator. The red color indicates a positive reaction.

Test of Voges-Proskauer (VP)

Colonies that will be identified retrieved using ose. Colonies were inoculated on media MR-VP. The media were incubated at 37°C for 18-24 hours. Then added reagent Barrit. The suspension was shaken for 20-30 seconds. Positive VP reaction, in case of acid formation marked change of medium color to pink after addition of Barrit reagents.

Test of Citrate

Colonies that will be identified retrieved using ose. Colonies implanted scratch zigzag on the seeding media simmons citratein form of slant jelly. Then the media were incubated at 37°C for 18-24 hours. Blue colonies showed positive results, while green colonies showed a negative reaction.

Urease Activity Test

Urease activity by inoculating bacteria on the medium Christensen's Urea Agar and the use of a pH indicator phenol-red. Incubation process in the tube for 24-48 hoursat 50°C. Positive test results will increase the pH which causes the indicator changes from red/orange to pink or dark purplish red. While negative if it does not change color.

OF Test

Test OF preceded by inoculation of cultures in twotubes of OF, which containing media Hugh and Leifson. One tube was added oil for making the anaerobe condition. Both of them were incubated at

50°C for 24-48 hours. The Positive fermentative result if there is a color change from green to yellow medium on both the tube and the positive oxidative if only a tube that is not covered in oil that changes color to yellow. Negative result if there is no change in the medium tube. The color could turn into a bluish due to the amine metabolism peptone by bacteria.

Carbohydrates Fermentation Test
Colonies that will be identified retrieved using ose. The colony was inoculated into tubes which each contain glucose, lactose, mannose, maltose, and sucrose. Into each of the sugar solution is added phenol red indicator (phenol red) and inserting to Durham tubes in upside down position. The microbial suspension was incubated 50°C for 18-24 hours. When the color of the medium turn yellow,it means the only farming acid from the fermentation of carbohydrates. When there are air bubbles in the Durham tube, means of fermentation gas are formed.

Test Hydrolysis Gelatin
This test is performed by inoculating cultured in a nutrient medium gelatin. Then the medium was incubated for 24 hours at 50°C. Before the test medium is incubated again in the refrigerator at temperature of 40°C, medium will,medium will remain liquid after cooling. If not hydrolyzed gelatin medium will remain solid.

3. Results and Discussion

3.1 Screening of thermophylic bacteria

Xylanase activity journal in this study was conducted using Congored,selected 12 of the 16 isolates showed xylanolytic activity. This activity is characterized by the formation of clear zone around bacterial colonies. The measurement results xylanolytic index (IX) can be seen in Table 1.

This clear zone arising from the hydrolysis of xylan by xylanase activity extracellular thermophylic bacteria which grow on the medium xylanolytik. Xylan is contained in the medium decomposed by xylanase into xylose and xilooligosakarida. Therefore,dyes Congo red that bind strongly to xylan hydrolysis is not found in areas[29].

Clear zone formed in this study ranged from 7.09 mm to 10.45 mm with xylanolytic index in the range of 0.18 to 0.74. Isolates SSA 2 clear zone most (8.77 mm) to 0.74 xylanolytic index than other isolates as shown in appendix 1 of the clear zone formed includes a small if compared with Winanda study[16] with the isolates from Sapan Mudiak hot springs, South Solok regency, which has xylanolytic index range between 0.54 to 1.39. So also with the results obtained by Susilowati[13] from the hot springs Sonai, Southeast Sulawesi, with xylanolytic index range between 0.4 to 1.8. But xylanolytic Index thermophilic bacteria from hot springs *Sapan Aro* River, South Solok District is still greater than the results obtained Habibie[30] of mud Lapindo Sidoarjo, with xylanolytic index range from 0.3 to 0.66.

Of the 12 isolates, xylanase enzyme hydrolytic activity in xylan hydrolyze show different values. Firstly it relates to the variation of xylanase genes of each of these bacteria. Differences in the amino acid sequence making up a protein (enzyme xylanase) cause differences activity. For example Thoyib et al[31]reported the activity of isolates identified as Pseudomonas sp. xilanolitik have higher activity than the isolates were identified as *Bacillus sp*.

Second, the difference is certainly related activities with factors that affect the action of the enzyme. Some of the factors that affect the activity of enzymes include enzyme concentration, substrate concentration, temperature, pH, and inhibitors. The enzyme concentration is directly proportional to the speed of the reaction, but the concentration of the enzyme is the basic factor that is dependent on other factors. The higher the concentration of substrate which will increase reaction speed to maximum speed is achieved.

Table 1. Xylanolytic Index xylanase-producing thermophilic bacteria from hot springs *Sapan Aro* River, South Solok District, West Sumatra Indonesia

No.	ISOLATE	Xilanolytic index
1	**SSA2**	0.74
2	**SSA4**	0.61
3	**SSA3**	0.58
4	**SSAS6**	0.46
5	**SSA7**	0.40
6	**SSA5**	0.37
7	**SSA15**	0.37
8	**SSA1**	0.35
9	**SSA8**	0.31
10	**SSA16**	0.31
11	**SSA13**	0.23
12	**SSA14**	0.18
13	**SSA9**	0
14	**SSA10**	0
15	**SSA11**	0
16	**SSA12**	0

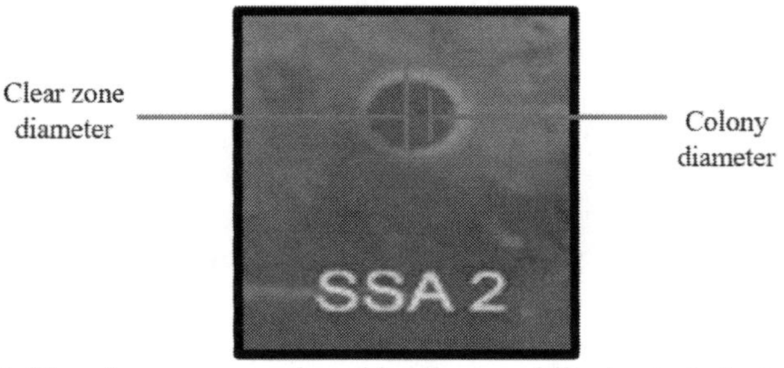

Figure 1. The clear zone produced by thermophilic bacteria isolates SSA 2

Temperature and pH are the two factors that work specifically, has a certain optimum value. For example, experimental results Susilowati et.al [13]states that the optimum temperature and pH for xylanase activity *Pseudomonas sp.* is 50 °C and pH 9.0. While[30] reported that thermophylic bacteria isolates that she gets having xylanase activity optimum at a temperature of 50 °C and pH 7.0. The presence of an enzyme inhibitor capable of inhibiting the action of the enzyme. Likewise xylanase enzyme, a compound in wheat(Triticumaestivum)called TLXI able to inhibit xylanase activity. TLXI is a non-competitive inhibitor[32].

B. Identification of Thermofilic bacteria

We has been found five isolates have xilanolitik index tertinggiyaitu SSA2, SSA3, SSA4, SSAS6, and SSA7 been selected for identification. Identification of the five isolates based on microscopic and biochemical characters can be seen in Table 2.

Characteristics of the five isolates of the same microscopic, that belonged to the Gram-positive, and have shaped bacillus endospores. It is also confirmed from the results of previous studies that have been done by Yenti [17].These results are slightly different from those obtained Habibie[30],of 14 isolates obtained from Lumpur Lapindo Sidoarjo 11 shaped bacilli and the other 3 are cocci, but all isolates obtained also included a group of Gram-positive and has endospores. Endospores is a special structure that is created when the environmental conditions are not favorable, as a strategy to survive when extreme environmental changes occur such as changes in extreme heat, toxic chemicals, and water shortages. This structure is commonly encountered in Gram-positive bacteria, such as the genus Bacillus and Clostridium.Therefore, these bacteria are often found in the environment termofil[19].

Table 2. Characterization of xylanase-producing bacterium ternofilik hot springs Sapan Aro River, South Solok

No.	Test	Isolates	
		SSA2, SSA3 and SSA4	**SSAS6 and SSA7**
1	Aerobic / Anaerobic	A	A
2	GRAM; shape	(+); basil	(+);bacillus
3	spores	(+)	(+); basil
4	TSIA	m | m	m | m
5	Gas	(-)	(-)
6	H2S	(-)	(-)
7	catalase	(+)	(+)
8	Oxidase	(-)	(-)
9	Motility	(+)	(-)
10	Indole	(-)	(-)
11	Urease	(+/-)	(-)
12	Citrate	(+/-)	(-)
13	Lactose	(-)	(-)
14	Glucose	(-)	(-)
15	Sucrose	(+/-)	(-)
16	Mannitol	(+/-)	(-)
15	MR	(-)	(-)
16	VP	(+)	(-)
17	OF	(-)	(-)
18	Gelatin	(+)	(-)
Genus		*Bacillus sp. 1*	*Bacillus sp. 2*

Notes: A: Aerobic, m | m: red - red (negative test result), (-) negative test results, (+): positive test results, (+/-): results of the test can be positive or negative.

Based on the characteristics in Table 3, the identification referring to Bergey's Manual of Bacteriology Systemathics. The identification results show that all isolates are similarities with the genus Bacillus.This is in accordance with the characteristics of this genus of gram-positive bacilli, forming endospore, and aerobic[33].3 isolates namely SSA2, SSA3 and SSA4 have similarities in terms of its ability to decompose H2O2, motile, asetonin production, and gelatin hydrolysis expressed as Bacillus sp.1.While SSAS6 similar to SSA7, only showed a positive test on catalase test is expressed as Bacillus sp. 2. Thoyib [31],reported that it has gained 5 alkalofilik bacterial genus that has xilanolitik activity of calcareous soils Krakitan limestone hill region of Klaten. It consists of 5 genus Pseudomonas, Bacillus, Flavobacterium, Micrococus, Paracoccus, and Alcaligens.Sharma et al.,[22]reported that the genus Bacillus and Paenibacillus has the potential to produce xylanase of 5 hot springs in the Northern Himalayas, Himachal Pradesh.

Bacillus sp. 1 and Bacillus sp.2 have catalase or peroxidase enzymes to break down hydrogen peroxide (H2O2)to form oxygen and water. This capability is needed bacteria to protect themselves from product O2, which is toxic, because it is a strong oxidizing agent that can be destructive to the metabolism in cells. Bacteria that have this enzyme generally is that aerobic or facultative anaerobes[34]. Bacillus sp. 1 shows the positive results of the motility test, this leads to ownership structures which help the movement, such as flagella. This genus generally have flegel peritriks, the flagellum located around the edge of the cell as in B. subtilis, B. cirlculans, B. brevis and B. polimyxa[35].

Gelatin hydrolysis capability was also demonstrated by Bacillus sp.1.This capability is based on the presence of the enzyme gelatinase. This enzyme is able to hydrolyze gelatin condensed into liquid (liquid). Sometimes these traits also indicate the pathogenicity of a particular bacterium. Gelatinase production relates to the ability of bacteria damages the collagen and into the host body [20].

The genus Bacillus have a high abundance in ecosystems hot springs[22] [30][15].Abundance is related danganproperties Bacillus which make it capable of living in this ecosystem. One important property is the ability to form spores. When the environmental support to the necessities of life begin to disappear, or when changes in extreme environmental conditions, such as temperature changes, bacteria will soon form resistant structures called endospores. When the environment back ends meet bergerminaasi forming endospores will soon return vegetative cells[35].Willey et al.[36]added that adaptation of thermophilic bacteria which makes it able to withstand the high temperature environment are as follows: a. Have a good protein stability against heat, especially enzymes. This protein has a hydrophobic amino acid organization that is very good in the interior and more hydrogen bonding, as well as bonding nonkovalennya. The number of amino acid proline more so that the polypeptide chain is more rigid and stable to heat. This protein is also stabilized and assisted in pelipatannya by a chaperon protein; b. A kind of histone proteins also help in stabilizing the thermophilic bacteria DNA to heat; c. Thermophilic bacterial membrane lipids tend to saturate, more branches and larger molecular weight, it is able to increase its melting point.

4. Conclusion

From the results of research that has been done can be concluded as follows.
1. It has been selected 12 of 16 isolates having xylanolytic activity, with the highest activity being SSA2 with xylanolitic index 0.74.
2. The top five isolates producing the highest indexxilanolitik were SSA2, SSA3, and SSA4 identified as Bacillus sp. 1., and SSAS6 and SSA7 are Bacillus sp. 2..

References

[1] Palmer, T., and PL Bonner. 2007. *Enzymes: Biochemistry, Biotechnology and Clinical Chemistry, 2nd Ed.* Cambridge : Woodhead Publishing.
[2] Madigan, MT, JM Martino, DA Stahl, and DP Clark. 2012. *Biology of Microorganisms, 13rd Ed.* San Francisco : Pearson Education Inc.
[3] Dewan, SS 2014. Report Overview: Global Markets for Enzyme in Industrial Application. *BBCresearch*: June 2014.

[4] Sarethy, IP, Y. Saxena, A. Kapoor, M. Sharma, SK Sharma, V. Gupta dan S. Gupta. 2011. Alkaliphilic Bacteria: application in Industrial Biotechnology. *J. Ind. Microbiol Biotechnology*. 38, p. 769
[5] Szendefy, J. 2005. Solid State Fermentation and Utilization of Xylanase in Pulp Bleaching. *Thesis*. Budapest University of Technology and Economic.
[6] Whitehurst, RJ, and BA Law (*Eds*). 2002. *Enzymes in Food Production*. Sheffield: Sheffield Academic Press.
[7] Li S., X. Yang, S. Yang, M. Zhu and X. Wang. 2012. Technology Prospecting on Enzymes: Application, Marketing and Engineering. *Computational and Structural Biotechnology Journal*. 2(3), p. e201209017
[8] Palmer, T. 1991. *Understanding Enzymes 3rd Ed.* Chichester : Ellis Horwood.
[9] Singh, R., V. Kapoor and V. Kumar. 2012. Utilization of Agro-Industrial Wastes for The Simultaneus Production of Amylase and Xylanase by Thermophilic Actinomycetes. *Brazillian Journal of Microbiology*, 43(4), pp.1545-1552.
[10] Bruins, ME, A. E Janssen, and RM Boom, 2001. Thermozymes and Their Application: A Review of Recent Literature and Patents. *Appl Biochem Bioetechnol*. 90 (2),pp 155-186.
[11] Trismilah dan DR Waltam. 2009. Produksi Xilanase Menggunakan Limbah Pertanian dan Perkebunan. *J. Tek. Ling*. 10 (2),pp 137-144.
[12] Kumar, R. and R. Nusinov. 2001. Review: How do Thermophilic Protein Deal with Heat. *Cell Molecular LS*. Vol. 58, pp: 1216-1233.
[13] Susilowati, PM, S. Raharjo, D. Kurniati, R. Rahim, Sumarlin, dan Ardiansyah. 2012. Produksi Xilanase dari Isolat Sumber Air Panas Sonai, Sulawesi Tenggara menggunakan Limbah Pertanian. *Jurnal Natur Indonesia*. 14(3),pp 199-204.
[14] Singh, V., VC Pandey, DC Pathak and S. Agrawal. 2012. Purification and Characterization of Lacella sacchari starain B42 Xylanase and its Potential for Pulp Biobleaching. *African Journal of Microbiology Research*. 6(7),pp 1397-1410.
[15] Sharma, N., G. Vias and S. Phatania. 2013. Culturable Diversity of Thermophilic Microorganisms Found in Hot Springs of Northern Himalayas and to Explore Their Potential for Production of Industriallly Important Enzymes. *Scholar Academic Journal of Bioscience (SAJB)*. 1 (5) pp 165-178.
[18] Burn, RG, and RP Dick. *Enzymes un the Environment, Activity, Ecology, and Application*. Basel: Marcel Dekker, Inc.
[19] Tortora, GJ, BR Funke and CL Case. 2013. *Microbiology, An Introduction, 11st Ed.* San Franscisco: Pearson Education Inc.
[20] Prescott, H. 2002a. *Laboratory Exercise in Microbiology, 5th Ed.* New York : McGraw Hill.
[21] Pelczar, MJ, dan ECS Chan. 2005. *Dasar-dasar Mikrobiologi.* Terjemahan oleh Hadioetomo. Jakarta: UI Press.
[22] Sianturi, DC 2008. Isolasi Bakteri dan Uji Aktivitas Amilase Termofil Kasar dari Sumber Air Panas Penen Sibirubiru Sumatera Utara. *Tesis*. USU Medan.
[23] Shields, P., and C. Lambert. 2010.Oxidase Test Protocol. *American Sociaty For Microbilology, MicrobeLibrary*. http://www.microbelibrary.org/library/laboratory+test/3229-oxidase-test-protocol. Acessed 24 june 2015.
[24] Brink, Bennita. 2010. Urease Test Protocol. *American Sociaty For Microbilology, MicrobeLibrary*.http://www.microbelibrary.org/library/laboratory+test/3223-urease-test-protocol. Acessed 24 june 2015.
[25] MacWillams, MP 2009a. Indole Test Protocol. *American Sociaty For Microbilology, MicrobeLibrary*.http://www.microbelibrary.org/component/resource/laboratory-test/3202-indole-test-protocol . Acessed 24 June 2015.
[26] McDevitt, Sylvia. 2013. Methyl Red and Voges-Proskauer Test Protocols. *American Sociaty For Microbilology, MicrobeLibrary*.http://www.microbelibrary.org/component/resource/laboratory-test/3204-methyl-red-and-voges-proskauer-test-protocols. Acessed 24 June 2015.
[27] MacWillams, MP 2009. Citrate Test Protocol. *American Sociaty For Microbilology*, MicrobeLibrary.http://www.microbelibrary.org/component/resource/laboratory-test/3203-citrate-test-protocol. Acessed 24 june 2015.
[28] Hanson, Anne. 2008. Oxidative Fermentatif Test Protocol. *American Sociaty For Microbilology, MicrobeLibrary*. http://www.microbelibrary.org/component/resource/laboratory-test/3151-oxidative-fermentative-test-protocol. Acessed 24 june 2015.

[29] Teather, RM, and PJ Wood. 1982. Use of Congo Red Polysaccharide Interaction in Enumeration and Characterization of Cellulolytic Bacteria from the Bovine Rumen. *Applied and Enviromental Microbiology*. 43(4), pp777-780.

[30] Habibie, FM, DP Sigres, dan LN Islami. 2014. *Isolasi dan Karakterisasi Mikroba Termofilik Penghasil Xilanase dari Lumpur Panas Lapindo sebagai Alternatif Pengganti Klorin pada Industri Kertas*. Jurusan Teknologi Hasil Pertanian, Fakultas Teknologi Pertanian Universitas Brawijaya.

[31] Thoyib, H., dan R. Setyaningsih, Suranto. 2007. Seleksi dan Identifikasi Bakteri Alkalifilik Penghasil Xilanase dari Tanah Bukit Krakitan, Bayat, Klaten. *Bioteknologi*: 2 (1), pp. 6-12.

[32] Fierens, E., S. Rombolist, K. Gebreuer, H. Goeseart, K. Brijs, J. Beugrand, G. Volkaert, SV Campenhout, P. Proost, CM Courtin, and JA Delcour. 2007. TLXI, a Novel Type of Xylanase Inhibitor from wheat (*Triticum aestivum*) Belonging To The Thaumatin Family. *Biochem J. Vol* 403, pp 583-591.

[33] Prescott, H. 2002b. *Microbiology, 5th Ed*. New York : McGraw Hill.

[34] Goldman, E. and LH Green. 2009. *Practical Handbook of Microbiology 2nd Ed*. New York: CRC Press.

[35] Todar, K. 2014. *Online Textbook of Bacteriology*. http://textbookofbacteriology.net/Bacillus_2.html. Acessed 3 Juli 2015.

[36] Willey, J., LM Sherwood, and CJ Woolverton. 2009. *Prescott's Principles of Microbiology*. New York : McGraw Hill

IOP Conf. Series: Materials Science and Engineering **335** (2018) 012022 doi:10.1088/1757-899X/335/1/012022

Empowering Farmer's Wife Makes Catfish Feed (Pellet) in *Nagari Limau Gadang Pesisir Selatan* District, Indonesia

A Armen

Department of Biology Universitas Negeri Padang, Indonesia

armenimik@ymail.com

Abstract. The life of *Nagari Limau Gadang, Pesisir Selatan* District, West Sumatra, Indonesia is poor. They conduct economic activities in Kerinci Seblat National Park (TNKS), so that TNKS become damaged. Catfish farming can serve as the main effort to increase family income. Economically catfish farming is very profitable. Catfish farming is not difficult and can be done by farmers who do not usually cultivate freshwater fish. The right family member is empowered to improve the family economy through the cultivation of catfish is the farmer's wife. Farmers' wives in *Nagari Limau Gadang* tend not to improve the family economically, they play the role of taking care of children, doing housework and sometimes delivering food to the fields and to the fields. How to research, selected 20 poor farmer's wife, then trained to make ponds and pellets. The pool made are the main pond, spawning pond, nursery pool and pond enlargement. Pellet made from raw trash fish, quail feces, fine bran and cassava. The pellet formulation is made of 4 kinds. Each formulation produces a quality pellet. Based on laboratory test produced pellet contains very good nutrition for enlargement catfish. The nutritional range of protein is 21,79%-34,60%, fat is 1,95%-2,32% and carbohydrate is 26,55%-39,65%. The goal to be achieved in this research is for the wife of farmers can contribute to improve the family economy. Specific targets to be achieved in this study is the wife of skilled farmers cultivate catfish.

1. Introduction

Conducting economic activities (earn a living) in TNKS high risk because the distance to pick up wood to the TNKS area reached 14 km and the terrain is very dangerous. Based on interviews of researcher with 20 residents, who carry on daily activities as transporters of processed wood, all of them say bored to the forest, but circumstances force. Conducting economic activities in the village is not sufficient to meet the needs of life.

According to the researchers, Nagari Limau Gadang residents can survive economic activities outside TNKS if the government or the care of TNKS is able to create a choice economic business, such as poultry farming, gardening outside TNKS and freshwater fish farming. All these businesses are very promising to increase the economy of the population.

The right family member is empowered to improve the family economy through the cultivation of catfish is the farmer's wife. Farmers' wives in *Nagari Limau Gadang* tend not to improve the family economy, they play the role of taking care of children, doing housework and sometimes delivering food to the fields and to the fields. The activities of the farmer's wife are only fulfilling the wife's obligation to the husband in married life. Farmers' wives are economically unproductive in improving the family economy. Generally their activities only stay at home and only help cultivate the land at any given time.

The role of the wife can be utilized to improve the family economy. Economic activity that can be done by a farmer's wife is catfish farming. Catfish farming is not complicated and can be done by the farmer's wife while taking care of the household, generally the farmers have a decent land to be used as a catfish pond. Catfish farming is quite profitable. Fish farmers are profitable enough. The benefits of cultivation can still be improved if farmers can make their own feed. The profit earned will reach 40-50%. If using factory feed gain is only 20-25%.

Specific objectives to be achieved in this research is for the farmer's wife can contribute to improve the family economy.The urgency of this study, if the family economy increases then TNKS avoid the pressure of the population. Innovative findings found in the study are appropriate formulations and qualified self-feeding skills.

2. Literature Review

This study focuses on determining the right formula to make a quality catfish (pellet) feed. This fish pellet is expected to be made by farmer's wife. Wife of farmer who has skilled to make feed will be able to cultivate catfish. Cultivation of catfish by the farmer's wife will increase the farm household's economic. The cultivation of catfish by the farmer's wife will improve the farm household's economic. Increasing the family economic will reduce farmer visits to TNKS [1].

Women (farmers' wives) do not contribute directly to the family's income to support households. The wife of the farmer always helps the husband to do the job as a farmer. Economically, farmers 'wives do not ease the burden of husbands' earnings.

The activities of Nagari Limau Gadang residents to earn a living in TNKS, such as farming, hunting and gathering of natural resources will damage the ecosystem and cause disaster. As evidence of ecosystem damage TNKS Nagari Limau Gadang, there have been flash floods in 1987, 1997, and 2005. According to [2] damage ecosystem of TNKS can be overcome by way of economic activity. One of the economic activities that can be done by the community is freshwater fish farming, because land and water resources are very supportive. Yusran, [2], said that fish farming will improve people's economy so they will not enter the forest area.

Poverty can literally be said to be a state of not having enough. In various views, there are three types of poverty that are often expressed are structural poverty, relative poverty and absolute poverty. Structural poverty is understood as poverty arising as a result of government policies and corporate behavior that make the poor, have little or no access to the productive economic. Relative poverty is a poverty that arises not only in terms of income alone but also living conditions in the social environment, while the absolute poverty according to [3] is poverty measured from the level of ability to finance a minimum living in accordance with the dignity of human life.

Environmental damage is caused by many factors, especially human activities that are not friendly with the environment itself. Humans should be responsible for preserving the environment, but they are destroying the environment. They tend to take the natural wealth at will, causing damage and pollution. Once the natural wealth is used, they do not care about the needs of future generations who also have the right to enjoy it. Needs often encourage people to take TNKS natural resources on a large scale regardless of the impact. One of the main factors causing the destruction of TNKS environment is poverty [3].

Many experts argue that poverty is one of the main causes of environmental destruction in TNKS. The environmental damage caused by the poor tends to be influenced by their mindset because they are squeezed by poverty, their minds are focused only on the food they can get to survive today. It is this narrow mindset that drives them to damage the environment and deprives the natural resources without allowing time for nature to renew its resources [4].

Associated with fish farming, fish farmers have problems in getting cheap feed, because fish feed prices tend to increase. According to [5] if the breeder uses artificial feed from factory, its value can reach 70% from all cost component. One effort that can be done for successful freshwater fish cultivation is to make their own feed. Most fish feed raw materials are available in Nagari Limau

Gadang.

Based on research that researchers have done about the cultivation of tilapia as a model of economic activity options overcome the dependence of the population on TNKS. The results showed that the activity of tilapia farming can decrease the level of population visit to TNKS [1].

Researchers have also conducted training to make pellets for tilapia against youth and women in the village of Nagari Limau Gadang. The results showed 95% of the participants trained to be skilled at making the feed [6].

Researchers have conducted a study on the ratio of the effect of homemade feed to artificial feed to the weight of tilapia. The results showed no difference means the influence between homemade pellets with factory-made pellets [7]. Listening to the findings of relevant research, which once proposed to do, the catfish cultivation research is very possible to improve the family economy.

Research activities conducted are to guide, train, test and compare to get the best quality pellet. The best quality pellet produced is used as feed for catfish cultivation. To understand the implementation of the research it is necessary to understand about catfish and feed ingredients in need. Catfish began to be known in Indonesia around 1986, the catfish live wild in swamps, freshwater rivers, but now the cultivation has been carried out intensively because it turns out this fish has a high nutrient content that consists of 17-37% protein, 4 , 8% fat, 1.2% minerals, 1.2% vitamins and 75.1% water, rapid growth, the fans are increasing due to the delicious and delicious taste of meat [8].

One of the most crucial elements of the growth and mortality of the nourished fish is the element of adequate feeding availability, in addition to the existing natural feed, so to increase or accelerate the growth of fish should be given nutritious feed. Provision of nutritious feed can increase the production of fish maintenance up to three times compared with that is not given nutritious feed.

The nutrient feed given to fish contains at least protein, carbohydrates and fat. All three of these substances will be converted into energy that is necessary to perform its activities, but the protein contained in a feed more determine the growth of fish than carbohydrates and fats [9].

The function of nutritious food for fish is to maintain the body, replace the damaged cells, after which the remaining excess feed is used for growth. In line with Dewi's opinion, [10] says that the protein present in the feed is needed by the fish as a source of energy, to replace damaged cells and to grow.

Feed has a very important role in the growth of fish. To obtain optimal growth required amount and quality of feed available in sufficient condition. The purpose of feeding is to produce as much meat as possible in the shortest time possible. Giving of nutrient feed can increase the production of fish that is kept up to 3 times, compared with that is not given nutritious food, then added by [8] that with the feeding of nutritious, fish weight gain can be 25-35% every month from the beginning.

Raw materials used to make catfish pellets are raw materials are widely available in *Nagari Limau Gadang* and in the surrounding *Nagari*. This raw material in collect into catfish feed. The raw materials needed to make catfish pellets are (1) Trash fish. Trash fish is a fish that is not used. Trash fish can be used as a source of protein to make fish feed (pellet). Trash fish contains proteins that can be standard as a source of protein for fish and poultry feed (Rinoto, 2014). (2) Quail quail, Quail's feces. High protein and carbohydrate content contained quail feces, then quail feces can be used as a source of protein and carbohydrates to make fish feed. (3) Cassava is the raw material prioritized in this research, because it is easy to obtain has never been used as a feed raw material. In addition to raw materials, cassava also serves as a glue [7]. Cassava used is a smoothing of cassava that has been boiled.

It is stated that the nutritional cassava is: protein, 1.2%, fat 0.3%, carbohydrate 34.7%, water 63% [11]. (4) The selected bile has a fine texture (grain), not moist, no musty odor, and has a fresh color. Fine bran contains nutrients: 11.35% protein, 12.15% fat, 28.62% carbohydrate, 10.5% ash, crude fiber 24.46% and water 10.15% [10].

3. Research Methods

The 20 poor peasants' wives are grouped into four groups and designated as catfish farmers. They are nurtured and directly involved in research. Research activities in Year I: 1. Cultivating groups of cultivators (poor farmer's wife) to create a holding pond. The parent's pool is made permanent with 2x3 meter size as much as two pieces. One for the male parent and one for the mother of the female; 2. Creating a permanent spawning pond with a size of 2x3 meters as much as one fruit; 3. Creating a pond nursery (not permanent) size 5x10 meters as much as two pieces; 4. Make a permanent magnification pond with the size of 5x6 meters as many as four pieces; 5. Make fish feed with raw materials, trash fish, quail stool, fine bran and cassava. All raw materials except cassava are dried and mashed. The cassava is boiled, then all the ingredients are put together into dough and printed using the printing press. Printed dough, dried in the sun; 6. Evaluation of guidance and guidance to group of farmers. If it doesn't succeed, the coaching will continue until it is complete. In Year I was also done to formulate the nutritional content of pellets and laboratory test. To test the quality of pellet to the growing power of catfish will be done in Year II.

In Year II conducted research activities: 1. Fostering a group of farmers formulates the nutritional content of fish feed; 2. Creating catfish feed based on formulation; 3. Practice making formulated feed; 4. Test the effect of homemade feed on weight gain of catfish. In this activity look for the right formulation to get protein, carbohydrate and fat levels. By considering the nutrient content of raw materials. To obtain a protein content of about 22.92% (minimum protein requirement of catfish by 20%) in one kilogram pellet, 30% trash fish required, 45% quail feces, 15% fine bran, 10% cassava, with the provision of trash fish protein 40%, 20% quail stool, 12% fine bran and 1.2% cassava.

Variations of nutrient content formulas are made as many as 4 kinds:

(a) 30% trash fish: 45% quail feces: 15% fine bran: 10% cassava
(b) 35% trash fish: 40% quail feces: 15% finely bran: 10% cassava
(c) 40% trash fish: 35% quail feces: 15% fine bran: 10% cassava
(d) 45% trash fish: 30% quail feces: 15% fine bran: 10% cassava

Of the 4 variations, one of the best variations will be compared to the factory-made feed. The main requirement of artificial pellets should contain high nutrition, easy to process and does not contain toxic, easy to obtain, affordable price. Ordinary pellets made with bran or rice bran, mas snail, salted fish BS Super, papaya leaves, vitamins, concentrates, yeast tempe. How to make pellet:

a. take the trash fish, quail stools, fine bran, cassava mixed into one and stir evenly, then added with tempe yeast as much as 3 tbsp or 125 gram. Stir gently and evenly;
b. dough closed tightly, let stand all night for the fermentation process can take place well;
c. fermented dough that has been settled one night mixed with salted fish and papaya leaves that have been ground first, then stir evenly;
d. prepare the pellet printer, and enter the dough. The resulting pellets are cut into a maximum of 1.5 cm, then dried;
e. the pellets are then dried in the sun by using a wide container to dry them all, drying under direct sunlight;
f. the dried pellets are placed in a plastic bag to be waterproof, then store in a room with good air circulation.

The third' year, conducted research activities to spread the seeds of catfish to pond enlargement. In the third year activity comparative influence between the best formulation of homemade feed with factory-made feed. Each group of farmers is required to compare the effect of homemade pellets with factory-made pellets. Each pond contains as many as 2000 seeds of catfish. Then we analyzed the difference of the effect of homemade pellet with the factory-made pellet by using T-Test formula, if the result show that the effect of homemade pellet with factory-made pellet does not mean or better, so homemade pellet is suitable for catfish farming.

4. Results and Discussion

Research in the first year is to prepare the facilities and infrastructure for research year II and year III. The result of research in the first year is 1 unit of spawning pond, 2 units of nursery pond, 2 units of parent pond and 2 units of enlargement pond. During the year I also produced a variety of pellet formulas, the level of caught and the skills of farmers' wives to make ponds and pellets.

Percentage of farmer's wife who understands and does not understand in making a pond and pellet (feed) before being built, can be seen in Table 1.

Table 1. Percentage of farmer's wife who understands and does not understand in making a pond and pellet (feed) before being scouted.

Activities	Sum of farmers wife	Not Understand	Percentage of un-understand	Understood	Percentage of understood
Parent Pond	20 people	18	90%	2	10%
Spawning Pond	20 people	18	90%	2	10%
Pond Nursery	20 people	19	95%	1	5%
Enlargement Pond	20 people	18	90%	2	10%
Making Pellet	20 people	20	100%	0	0%

Based on the low percentage of the wife of farmers in making a pond of parents, spawning ponds, nursery pools, ponds enlarging because farmers' wives have never / regularly cultivate freshwater fish (catfish). Their work only helps husband to cultivate fields in TNKS and cultivate rice fields. Percentage of farmer's wife who understands and does not understand in making pond and pellet (feed) after scouted, can be seen in Table 2.

Table 2. Percentage of farmer's wife who understands and does not understand in making pond and pellet (feed) after being scouted.

No.	Activities	Sum of famers wife	Not Understand	Percentage of ununderstand	Understanded	Percentage of understanded
1.	Parent Pond	20 people	4	20%	16	80%
2.	Spawning Pond	20 people	4	20%	16	80%
3.	Pond Nursery	20 people	3	15%	17	85%
4.	Enlargement Pond	20 people	4	20%	16	80%
5.	Making Pellet	20 people	3	15%	17	85%

Taking into account the percentage of farmers' wives in making ponds, spawning ponds, nursery pools, ponds of magnification show a very high understanding. The high percentage is strongly influenced by the guidance, attention and seriousness of the farmer's wife facing the guidance and training activities. Interest factor and interest in cultivating catfish to increase family income is very decisive. Good family income will be able to prosper the family.

The completed pond units: a. spawning pond as much as 1 fruit; b. pond main as much as 1 fruit; c. 2 nursery ponds; d. pond enlargement of 2 pieces; e. pellet for laboratory test; f. The best produced product is the formula number.

To produce catfish feed required raw materials containing nutrients. The necessary nutrients are proteins, fats, carbohydrates and minerals. Raw materials made into pellets are trash fish, quail stools, fine bran and cassava. The result of pellet formulation made from raw material is 4 variation. For more clearly the percentage of raw materials needed to make a quality pellet can be seen in Table 3.

Table 3. Percentage of raw materials for quality pellets

No.	Percentage of raw materials for quality pellets			
	Trash fish	Quail feces	Finely bran	Cassava
1.	30%	45%	15%	10%
2.	35%	40%	15%	10%
3.	40%	35%	15%	10%
4.	45%	30%	15%	10%

To make a quality pellet more preferred protein and carbohydrate nutrients. Trash fish contains 43% -50% protein, quail faces contains 18% -20% protein and 11% -15% carbohydrate, cassava contains 1.2% protein and 34.7% carbohydrate and fine bran contains 11.35 protein % and carbohydrates 28.62%.

Pellets made from 30% trash fish, 45% quail feces, 15% fine bran and 10% cassava obtained pellet containing 22.82% protein and 17% carbohydrate. This pellet is suitable for enlarging catfish.

Based on pellet laboratory test containing very good nutrition for catfish enlargement is pellet made from raw material trash 40%, quail 35% feces, 15% fine bran and 10% cassava.

5. Conclusions

5.1. Conclusion

1. The wife of farmers interested in cultivating catfish and can be fostered to make the main pond, spawning pool, pond nursery, pond enlargement and catfish feed.
2. The wife of the farmer does not have the skills or knowledge to make the pond and feed the catfish.
3. Making pond pond, spawning pond, pond nursery and pond enlargement can be done by farmer's wife with guidance.
4. The wife of capable and skilled farmers to make a quality catfish feed.
5. Raw fish raw materials, quail feces, fine bran and cassava can be made into a high quality catfish feed.

5.2. Suggestions

1. To make a quality fish feed must use high quality raw materials.
2. Raw materials used as fish feed dough should be made smooth using a tool that is able to smooth the raw material.
3. Mixing raw materials into dough, must be done evenly.
4. Form feed, should be accommodated with plywood size 20 cm x 80 cm so that pellets coming out of the print tool do not break.
5. Pellets that have been formed, dried to a minimum water content.
6. Pellet that has been formed, packed in plastic and stored.

References

[1]. Armen.2015. Budidaya ikan nila pilihan untuk mengatasi ketergantungan penduduk terhadap sumber daya hayati taman nasional kerinci seblat di nagari limau gadang lumpo, *Jurnal Sainstek Vol. VII No. 1:* pp 42-50.

[2]. Yusran.2011. *Konservasi Sumber Daya Pesisir dalam Meningkatkan Ekonomi Sumber Daya Alam.*[online]: http:jujungbandung.com

[3]. Ferryal, Marantha Bernard. 2010. *Kemiskinan dan Kerusakan Lingkungan.* http: maranathabernardferryal.blogspot.com/2010/04/kemiskinan-dan-kerusakan-lingkungan.

[4]. Altin, Darus. 2007. *Meningkatkan Perekonomian Masyarakat Melalui Budidaya Ikan.* [online]. http: fppb.ubb.ac.id/?Page=artikel_ubb&&id=184

[5]. Sinaga.2013. *Budidaya Ikan Air Tawar Dapat Meningkatkan Ekonomi Rakyat.* [online]. http:hariansemarang.net.
[6]. Armen. 2012. *Pembuatan Pakan Ikan Nila di nagari Limau Gadang Lumpo Kecamatan Iv Jurai Kabupaten Pesisir Selatan*.FMIPA : UNP
[7]. Armen.2010. Perbandingan Pengaruh Pemberian Pakan Buatan Pabrik dengan Pakan Buatan Sendiri Terhadap Pertumbuhan Ikan Nila. *Jurnal Eksakta*
[8]. Soetomo H.A Moch .2009. *Teknik Budidaya Ikan Lele Dumbo*.Bandung :Sinar Baru
[9]. Dewi, Ratna. 2008. *Pengaruh Presentase Pemberian Pakan Buatan Terhadap Pertumbuhan dan Kelangsungan Hidup Ikan Mas Koki*.Padang : Fakultas Perikanan Universitas Bung Hatta.
[10]. Mudjiman, Ahmad. 2000. *Makanan Ikan*. Jakarta : Penebar Swadaya.
[11]. Rinoto. 2014. "Membuat Pakan Lele dari Limbah Puyuh". [online]: http: catatanharianrinoto.blogspot.com. Acessed: 27 April 2014.

Potency of Amylase-producing Bacteria and Optimization Amylase Activities

G Indriati* ; R R P Megahati, E Rosba

Department of Biology Education, STKIP PGRI Sumbar, Padang, West Sumatera, Indonesia

*gustina_indriati@yahoo.com

Abstract. Enzymes are capable to act as biocatalyst for a wide variety of chemical reactions. Amylase have potential biotechnological applications in a wide range of industrial processes and account for nearly 30% of the world's enzyme market. Amylase are extracellular enzymes that catalyze the hydrolysis of internal α-1,4-glycosidic linkages in starch to dextrin, and other small carbohydrate molecules constituted of glucose units. Although enzymes are produced from animal and plant sources, the microbial sources are generally the most suitable for commercial applications. Bacteria from hot springs is widely used as a source of various enzymes, such as amylase. But the amount of amylase-producing bacteria is still very limited. Therefore it is necessary to search sources of amylase-producing bacteria new, such as from hot springs Pariangan. The purpose of this study was to isolation of amylase-producing bacteria from Pariangan hot spring, West Sumatera and amylase activity optimization. The results were obtained 12 isolates of thermophilic bacteria and 5 isolates of amyalse-producing bacteria with the largest amylolytic index of 3.38 mm. The highest amylase activity was obtained at 50°C and pH 7.5.

1. Introduction

Thermophilic bacteria can be isolated from various places such as crater areas, volcanoes and hot springs. In the field of food industry, amylolytic enzymes play a role in the manufacture of glucose syrup, bread making, and baby food. In the field of non-food industry, amylolytic enzymes play a role in the paper industry, leather tanning, pharmaceuticals, textiles and as detergent additives. This industrial process requires an amylolytic enzyme that is resistant to high temperatures of about 70-80° C [4]. Amylase has been used in the field of biotechnology, in industrial processes the use of amylase already reaching 30% of the world's enzyme market [10]. Amylase is an extracellular enzyme that hydrolyzes the internal bonds of α-1,4-glycosidic starch into dextrin and other carbohydrate molecules of the glucose unit [5]. Amylases are generally produced by animals and plants, but amylases of thermophilic bacteria are more attractive to be used commercially.

Pariangan hot springs is one hot spring located in *Nagari Pariangan Tanah Datar* District of West Sumatera province, Indonesia. Temperature of *Pariangan* hot spring between 47-50°C and pH ranged between 8.4-9.0. This caused *Pariangan* hot springs has a high level of bacterial diversity. In addition, bacteria from high pH environments will generally produce stable enzymes that are high in pH, such as amylase. Stable amylase at high pH is very attractive for use in various industries, especially the detergent industry. This study aims to isolate amylase-producing bacteria from *Pariangan* hot spring, West Sumatera and optimization of amylase activities.

2. Methodology

1. Isolation of Thermophile Bacteria

The hot water samples inside the bottle are shaken to homogeneous. Hot water samples are poured into a liquid NA medium and leave to freeze. Subsequently incubated at 50°C for 24 to 48 hours. Characteristics of the colony morphology observed include shape, edges, and elevation [3].

2. Purification of Thermophile Bacteria

Growing bacterial colonies were inoculated into NA medium that had been densely packed with quadrant methods. Subsequently incubated for 24 to 48 hours at a temperature of 50°C. Thus seen single growing colonies. Any pure isolate that can be grown is assumed to produce amylase on the starch-containing medium and each sample is repeated twice (duplo).

3. Selective Media of Thermophile Bacteria

Bacteria that have been successfully purified, then grown on a selective medium or medium for starch. Each Petri dish was placed in a different bacterial colony with one source of the same isolate then incubated for 24 hours at 50°C.

4. The Activity of Thermophilic Bacteria

Bacteria that successfully grow on selective medium then droping with iodine solution and waited a few minutes until really fused between iodine solution with starch so clear zone clear.

5. Isolation of Amylase

The amylase-producing isolates were grown into 25 ml basal medium (3 g/l K_2HPO_4, 3 g/l KH_2PO_4, 3 g/l $MgSO_4$, 5 g/ NaCl, and 10 g/l starch) with pH 7.5 and dishaker at 150 rpm at 50°C for 24 hours. When the bacterial growth is subsequently removed 5 ml bacterial cultures into 100 ml basal medium and centrifuged at 150 rpm for 24 hours at 50°C. Bacterial culture formed centrifuged at 5000 rpm for 10 minutes. The supernatant containing the thermostable amylase extract was taken by micropipette and inserted into the microcentrifius tube for the activity test.

6. Amylase Activity Test

The amylase activity test was performed by incubating 0.5 ml of starch 1% for 5 min at 50°C and then adding 0.5 ml of amylase then incubated again for 1 hour at 50°C. To stop the hydrolysis process heating the boiling water (100°C) for 20 minutes. Then added 1 ml of Samogy Nelson solution then divortex and reheated to boiling water for 20 minutes. Further cooled in ice water and added 1 ml of arsenomoliblad solution then shaken then sufficient volume to 10 ml by adding 7 ml aquadest. Measure absorbance at wavelength 540 nm [8].

7. Determination of Ph Optimum Amylase

Amylase activities was tested with pH variation with 1% starch in buffer pottasium phosphate (pH 6.5-8.5).

8. Determination of Temperature Optimum Amylase

Optimum temperature for amylase activities was determined by incubating amylase at different temperature (45-60°C) using substrate with 1.5% concentration and amylase activities were tested.

3. Results and Discussion

1. Isolation of Thermophile Bacteria

The morphological characterization of thermophilic bacteria showed that the bacteria originating from Pariangan hot spring were cream colored, smooth shallows, and flat elevation (Table 1).

Table 1. Morphology characterization of termophile bacteria.

Location	Isolates code	Colour	Shape	Shallows	Elevation
S1 (Temperature 50°C, pH 8)	S1.1	Yellowness	Round	Smooth	Arise
	S1.2	Crem	Round	Smooth	Flat
	S1.4	Crem	Round	Smooth	Arise
S2 (Temperature 48°C, pH 9)	S2.1	Yellowness	Round	Smooth	Arise
	S2.2	Crem	Round	Smooth	Arise
	S2.3	Crem	Round	Smooth	Flat
	S2.4	Yellow	Round	Smooth	Flat
S3 (Temperature 47°C, pH 8)	S3.1	Yellowness	Round	Smooth	Flat
	S3.2	Crem	Round	Unorderd	Flat
	S3.3	Crem	Round	Smooth	Arise
	S3.4	Yellowness	Round	Smooth	Arise

The results of amylase activity tests obtained 4 isolates of amylase-producing bacteria from 11 isolates (Figure 1). The amylolytic enzyme activity test was conducted qualitatively by considering the measurement of the amylolytic index (Table 2). S1.2 isolate have amylolytic index is 3.83 mm, S2.3 isolate have amylolytic index is 1.91. S3.1 isolate have amylolytic index is 3,38, S3.2 isolate have amylolytic index is 2,42. The largest of amilolytic index was 3.83 mm that produced of S1.2 isolate and the lowest activity was 1.91 mm that produced of S2.3 isolate.

Table 2. Amylolytic Index of bacteria isolates

Location	Isolates code	Clear zone diameter (mm)	Colony diameter (mm)	Amylolytic indeks (mm)
S1 (Temperature 50°C, pH 8)	S1.1	-	-	-
	S1.2	1,61	0,42	3,83
	S1.4	-	-	-
S2 (Temperature 48°C, pH 9)	S2.1	-	-	-
	S2.2	-	-	-
	S2.3	1,17	0,61	1,91
	S2.4	-	-	-
S3 (Temperature 47°C, pH 8)	S3.1	1,59	0,47	3,38
	S3.2	1,02	0,42	2,42
	S3.3	-	-	-
	S3.4	-	-	-

11 isolates of thermophilic bacteria originating from Pariangan hot spring 7 isolates did not produce amylase activity. According to [11] iodine solutions do not provide colour with carbohydrate polymers of less than five monosaccharide groups, eg glucose. The media surrounding bacteria colonies that do not produce amylase will be blue when iodine solution drops. This shows that the starch in the medium is not degraded to simple sugars which means that bacteria do not produce amylase.

According to [2] the small diameter of the clear zone formed from each isolate is different. This is caused the ability to hydrolyze the starch of each isolate is different. The difference between clear zones is also due to differences in the amylolytic gene that each thermophile bacterium possesses. According to [9] the magnitude of the resulting clear zone depends on the amount of glucose monomer produced from the process of starch hydrolysis. The larger the amount of glucose monomers produced the larger the clear zone formed around the colony.

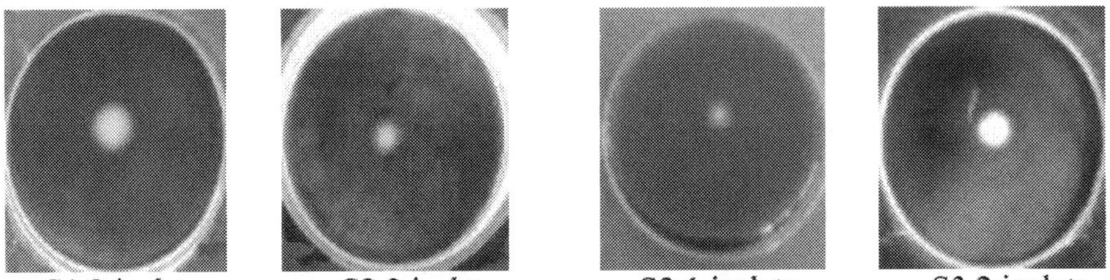

S1.3 isolate S2.3 isolate S3.1 isolate S3.2 isolate

Figure 1. Thermophilic Bacteria Isolates

2. Amylase Activity Test from Thermophilic Bacteria

From the research that has been done, obtained the results of different amylase activity of four different bacterial isolates. The highest amylase activity was obtained by S2.3 isolate that is 66,68 U/ml, followed by S1.3 isolate that is 64,54 U/ml, S3.2 isolate that is 43,76 U/ml, and S3.1 isolate that is 37.04 U/ml (Figure 2). The highest activity isolate was selected for further research that is S2.3 isolate.

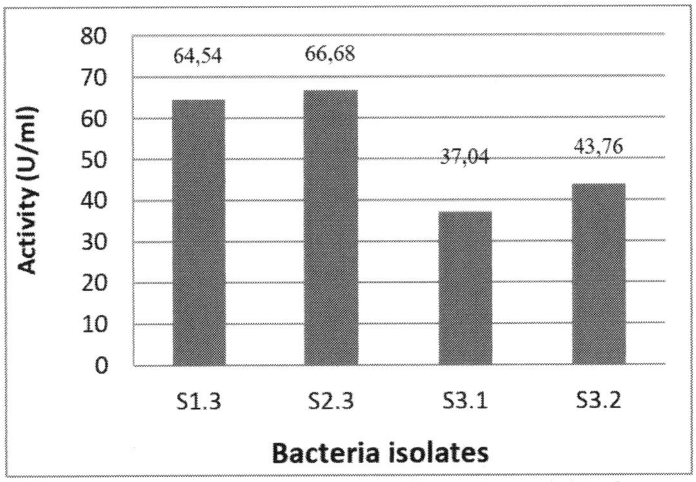

Figure 2. Amylase Activity Test of Four Bacterial Isolates

The presence of enzyme activity of different values each isolate caused by enzymes produced each type of different microorganisms will produce different enzymes of the amount and sequence of amino acids that form the enzyme. According to [1] that the specific activity of different enzymes from *Bacillus* sp. probably due to the amount of enzyme and amino acid enzyme protein produced by each isolate of *Bacillus* sp. different from each other.

a. pH Optimum

The highest amylase activity was obtained at pH 7.5 which was 81.9281,92 U/ml (Figure 3). According to Kristjonson (1999) the optimum pH of amylase-producing thermophilic bacteria lies between pH 6.0-9.0. The catalytic activity of enzymes within cells may be partially regulated by changes in the pH of the environment medium [7]

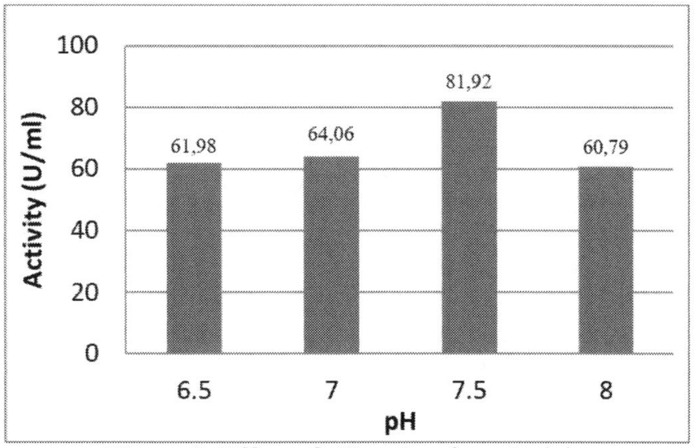

Figure 3. Effect of pH to Amylase Activity

b. Temperature optimum

The result of temperature optimization on amylase activity was obtained that the highest amylase activity was obtained at temperature 50°C (Figure 4).

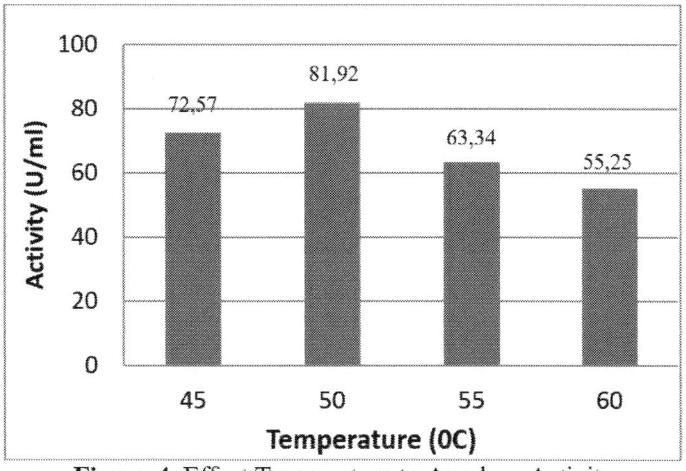

Figure 4. Effect Temperature to Amylase Activity

The highest activity was obtained at 50°C incubation temperature, because it was the temperature suitable for growth of thermophilic bacteria to produce amylase. According to [6] the optimum temperature of amylase-producing thermophilic bacteria lies between 40-70°C. The optimum temperature of amylase is usually always the same as the ambient temperature

References

[1]. Agustien, A. 2010. Protease Bakteri Termofilik. UNPAD Press: Bandung.

[2]. Ginting, J. 2009. Isolasi Bakteri Dan Uji Aktivitas Enzim Amilase Kasar Termofilik Dari Sumber Air Panas Semangat Gunung Kabupaten Karo. Sumatera Utara *Tesis*

[3]. Hadioetomo, S.R. 1993. Mikrobiologi Dasar Dalam Praktek. Gramedia Pustaka Utama: Jakarta

[4]. Hag, I. ; Ali, S. ; Javed, M.M ; Hameed, U. ; Saleem, A. ; Adnan, F. and Qadeer, M.A. 2010. Production of Alpha Amylase From A Randomly Induced Mutant Strain *Bacillus amyloliquefaciens* And Its Aplication As A Desizer In Textile Industry. Park. J. Bot **42(1):** 473-484.

[5]. Jugran, J. ; Rawat, N. and Joshi G.K. 2015. Amylase production by *Geobacillus* sp GJA1 isolated from a hot spring in uttarakhand *ENVIS. Bulletin Himalayan Ecology* **23**: 21-26.

[6]. Kristjonson, J.K. 1999. Thermophilic Bacteria. IRC Press: New York.

[7]. Lehninger, A.L. ; Nelson, D.L and Cox, M.M. 2005. Principles of Biochemistry. I. WH Freeman & Co: New York US.

[8]. Nelson, N. 1944. A photometric adaptation of the Samogyi method for the determination of glucose. *Journal of Biological Chemistry* **153**(2):375-380

[9]. Novitasari, Y.E and Herdyastuti, N. 2014. Screening Bakteri Termofilik Penghasil Enzim Amilase dari Sumber Air Panas Singgahan Tuban Jawa Timur. Universitas Negeri Surabaya. Jawa Timur.

[10]. Van der Maarel M. ; Van der Veen B. ; Uitdehaag, H. ; Leemhuis, H. and Dijkhuizen, L. 2002 Properties and application of starch converting enzymes of the α- amylase family J. Biotechnol **94**: 137-155.

[11]. Winarno, F.G. 1989. *Enzim Pangan*. Gramedia: Jakarta.

IOP Conf. Series: Materials Science and Engineering **335** (2018) 012024 doi:10.1088/1757-899X/335/1/012024

Synthesis of Magnesium Ferrites for the Adsorption of Congo Red from Aqueous Solution Using Batch Studies

E Erdawati[1*], D Darsef[1]

[1]Department of Chemistry, Universitas Negeri Jakarta, Jalan Rawamangun, Jakarta, Indonesia

*erda_wati_0912@yahoo.com

Abstract. A sol gel method with citric acid as an anionic surfactant was used to fabricate nano magnesium ferrites ($MgFe_2O_4$) under different calcination temperatures for 2h, respectively. The microstructure and surface morphology of magnesium ferrite powder were characterized by FTIR, XRD, SEM, and BET. The results of this study are useful for adsorption Congo red. The results showed that increasing solution pH and extending contact time are favorable for improving adsorption efficiency. with initial Congo red concentration of 50 mg/L and 100 mg/L. Adsorption data fits well with the Langmuir isotherm models with a maximum adsorption capacity (qm) and a Langmuir adsorption equilibrium constant (K) of 65.1 mg/g and 0.090 L/mg, respectively. The adsorption kinetic agrees well with pseudo second order model with the pseudo second rate constants (K2) of 0.0468 and0.00189 g/mg/min for solutions with initial congo red of 50 and 100 mg/L, respectively

1. Introduction.

The fabrication of nano ferrites has attracted much greater interest because of their large surface-to-volume ratio yielding remarkable material properties. A number of methods have been developed to fabricate ferrites including electrospinning [1], sol gel [2], coprecipitation [3], high energy ball milling [4], solid state reaction [5], and oxidative thermal decomposition [6]. Among the various methods, the sol–gel method yields more promising results in the fabrication of ferrites particularly at a fairly low temperature [7]. The sol gel method for preparing ferrites has many advantages with respect to good stoichiometric control and particle size distribution at a lower temperature [3].

Previous publications reported that ferrites were used for magnetic separation technology, as a rapid and effective technology for adsorption (Rashad, 2006). Due to large specific surface area, excellent physicochemical characteristics and easy separation under external magnetic fields, spinel ferrite magnetic nanoparticles have been regarded as a promising adsorbent and widely used for dye removal from wastewaters.

Zui Ding, et al [8] synthesized $CoFe_2O_4$ ferrite nanoparticles for removal Congo red from aqueous solution by an ethanol-assisted hydrothermal method, where the ethanol is mixed with water as the solution The maximum adsorption capacity for CR of $CoFe_2O_4$ nanoparticle is 190.5 mg g^{-1} when the volume ratio of ethanol to water is 4/3. In this work, magnesium ferrite was directly prepared by a sol gel method , and they were employed to absorb Congo red in water. The adsorption efficiency has been investigated in details.

2. Methods

2.1 Syntesis of magnesium ferrites

The fabrication of magnesium ferrites by a sol gel method using a solution of citric acid as an anionic surfactant included several steps: (i) a solution of ferric nitrate, magnesium nitrate , and citric acid was mixed; (ii) the mixture was constantly stirred and heated up to 100°C for 24h to form the gel; (iii) the brown, fluffy gel was calcinated at different temperatures (300°C, 600°C and 800°C) for respective 2h.

2.2. Adsorption experiments

Into the CR solution is inserted a magnesium ferrite. Agitation rate was held constant at 120 rpm.,filter , and concentration in supernatan was analyzed spectrophotometrically (Shimadzu UV 160) at 540 nm. The amount of CR adsorbed onto the compounds (%) was calculated as:

$$\frac{C_0 - C_t}{C_0}\ x\ 100\ \% \qquad (1)$$

where C_o is the initial CR concentration (mg/L), C_t is the concentration of CR (mg/L) at equilibrium time, t.. All the experiments were carried out in duplicates.

3.Results and Discussion

3.1 The FTIR and XRD Spectra

The absorption peaks of the FTIR spectra at 3393 cm^{-1}, 1624 cm^{-1} and 1384 cm^{-1} of the fabricated magnesium ferrites disappeared as the calcinations increased from 300°C to 800°C indicating the removal of the OH-, CO- and NO groups from the material of interest, which is shown in Figure 1. The absorption peak of the FTIR spectra (400 cm^{-1} to 800 cm^{-1}) was more obviously shown as the calcination temperatures were getting increased (from 300°C to 800°C) that indicated the formation of tetrahedral – octahedral structures of the metals oxide after the removal of all volatile matters. Pradeep *et al.* (2008) investigated the vibration of the tetrahedral and octahedral structures of metal complexes at 561 cm^{-1} and 437 cm^{-1} of the FTIR spectra of magnesium ferrite that was attributed to the shorter bond length of tetrahedral complex producing more stable structure.

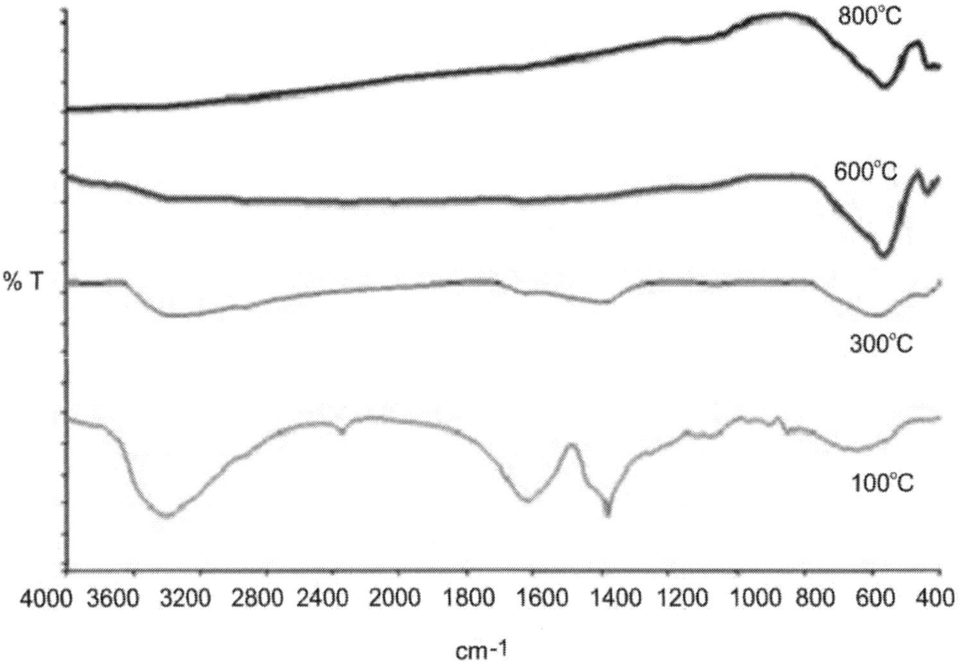

Figure 1. The FTIR spectra of magnesium ferrites at 100°C, 300°C, 600°C and 800°C.

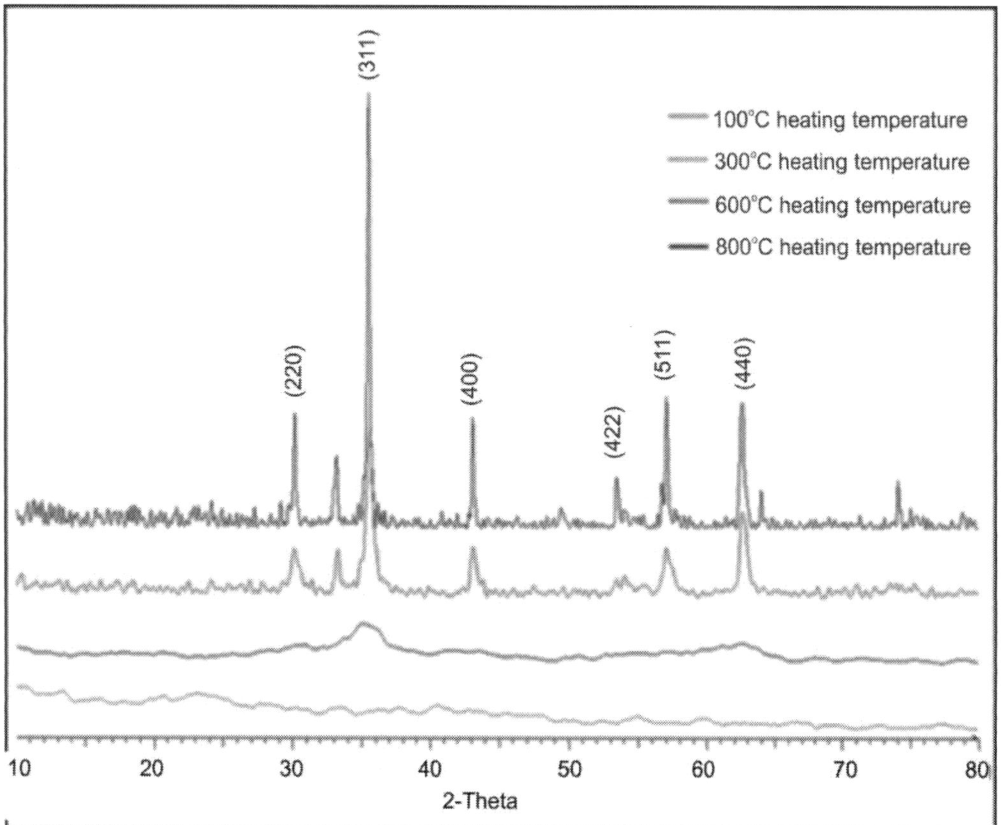

Figure 2. The XRD spectra of magnesium ferrites at 100^0 C, 300^0 C, 600^0 C and 800^0 C

Figure 2 show the difractograms for both $MgFe_2O_4$ at different temperatures (100, 300, 600 and 800°C) where Mg ferrite still existed as amorphous materials at 100°C. As the temperature increased to 300°C the material structure of Mg ferrite changed to semi-crystal with the appearance of some broaden peaks at certain 2θ (Bragg's angle), and when the calcinations continued to 600°C and 800°C Mg ferrites existed in crystalline form with the existence of sharp peaks in characteristic patterns of Mg-ferrite respectively. The higher thermal effect gave larger movement to atoms to do self rearrangement forming their crystalline structure.

3.2 Adsorption CR by magnesium ferrite nanoparticle

3.2.1 Effect of pH.

The pH of the aqueous solution is an important parameter in controlling the adsorption process of Congo red. When pH is increased from 7 to 12, the colour removal increases slightly from 80% to 85% and then decreases slightly up to pH 13. There is significant decrease of 19% in colour removal, in the pH range of 13-14. The effect of solution pH on the uptake of CR is presented in Figure 3. As shown in the figure, the capacity of magnesium ferrite decreased from pH 12 to pH 14 and reached 22.90 mg/g maximum capacity adsorption at pH 12. condition for percentage uptake of CR using magnesium ferrite . The solution of pH 12 was then selected for other optimization experiment.

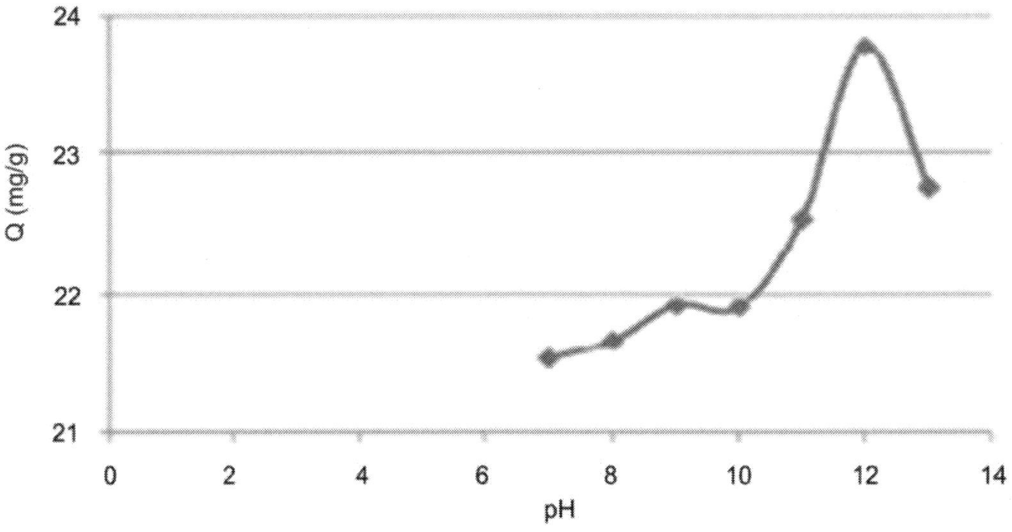

Figure 3. The effect of pH on the removal of Congo red solution using magnesium ferrite

3.2.2 Effect of Contact Time

The adsorption experiments to evaluate the contact time effect on percentage uptake of CR were brought within 5 minutes to 60 minutes. The experiments were conducted under the standard conditions; temperature of 298K, pH 12.0 and the concentration CR solution 100 ppm. As shown in Figure 4, percentage uptake of CR increased rapidly within 10 minutes and remained constant after 20 minutes indicating an equilibrium state. At the contact time of 20 minutes, the capacity adsoprtion of magnesium ferrite for CR 100 ppm is 21,8854 mg/g, respectively. A contact time at 20 minutes was then selected for other optimization experiment.

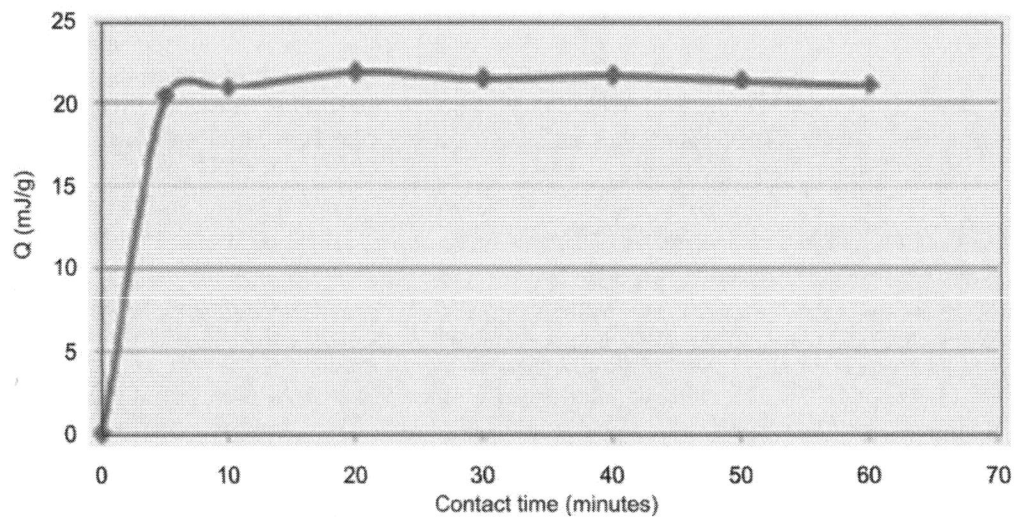

Figure 4: The effect of contact time on the removal of CR solution using magnesium ferrite

3.2.3 Effect of Initial Concentration

The adsorption study was completed with different concentration of CR at a fixed pH of 12, temperature of 298K, a contact time of 20 minutes using the same amount of magnesium ferrite 0,1 gram. The concentrations of CR were selected from 100 ppm, 200 ppm, 400 ppm, 600 ppm, 800 ppm and 1000 ppm. Figure 5 depicts the effect of different concentrations on percentage uptake of CR. It was found that dosage magnesium ferrite of 0.1 g, gained the capacity adsorption of 87,727 mg/g. for 700 ppm respective concentrations of CR. From the Figure 5, It can be concluded that the capacity adsorption of magnesium ferrite decreased with increment concentrations of CR.

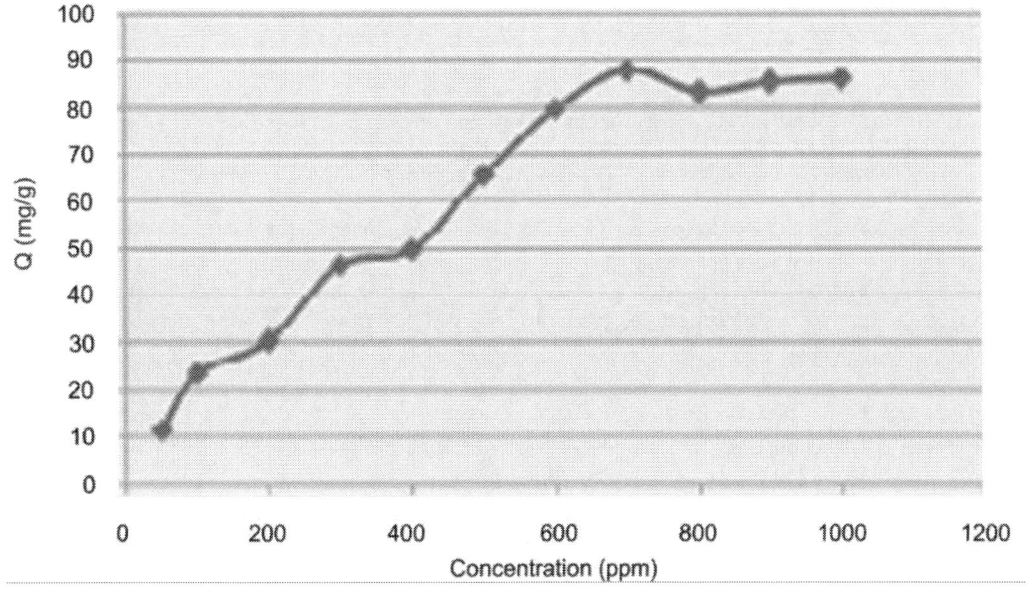

Figure 5. The effect of concentration magnesium ferrite on the removal of CR

3.2. 4 Adsorption Isotherm

Adsorption isotherm study is important to study in order to understand the interactions between adsorbent and adsorbate. The Langmuir model is based on assumption that uptake occurs on a homogenous surface by monolayer sorption without interaction between adsorbed molecules. It is expressed as:

$$\frac{C_e}{q_e} = \frac{1}{Q_0 * b} + \frac{C_e}{Q_0} \quad (2)$$

where, q_e is the amount of Congo red adsorbed per unit mass of adsorbent (mg/g); *Ce* is the equilibrium CR concentration (mg/L); and Q_0 and b are the Langmuir constants representing the monolayer adsorption capacity (mg/g) and the energy of adsorption (L/mg), respectively. The plot of C_e/q_e versus C_e gives a straight line with a slope of $1/Q_0$ and an intercept of $1/Q_0b$.The isotherm study percentage uptake of CR was generated in batch experiments using different initial concentrations range from 200 ppm to 1000 ppm at constant temperature of 298 K, a pH of 12.0 and 0.1 g dosage of $MgFe_2O_4$. The isotherm is presented in Figure 6 and from the Figure 6 it can be represented that equilibrium data fitted the Langmuir isotherm with high correlation coefficient (R^2) value of 0.9374. The calculated Langmuir isotherm constant for the maximum capacity adsorption silica gel z, Q_0 (mg/g) was 94.33 mg/g.

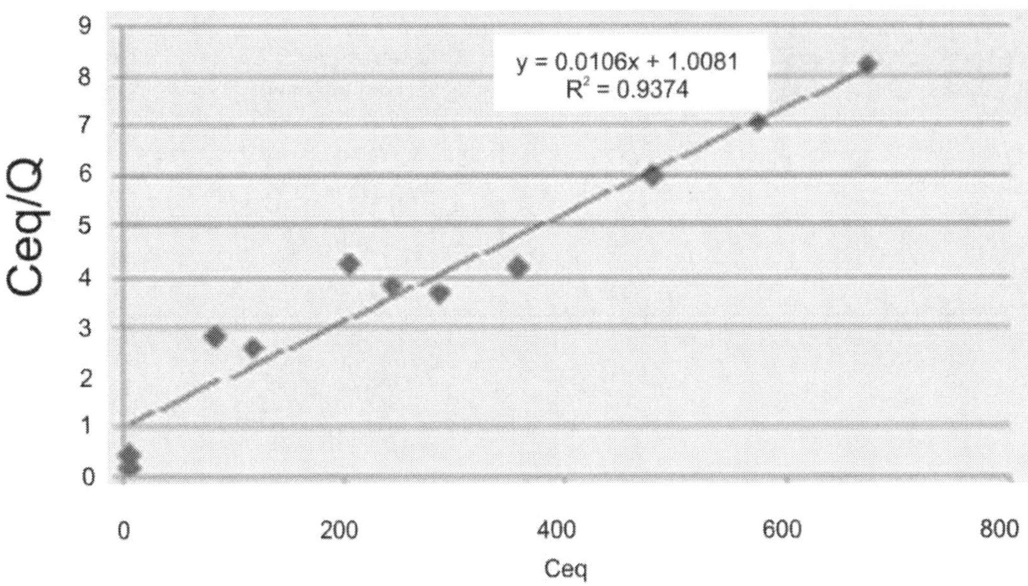

Figure 6. Langmuir Isotherm plot for the removal of CR solution using magnesium ferrite

3.3 Kinetic Study

Pseudo-first order and pseudo-second order kinetic model were introduced to the experimental data in order to investigate the adsorption kinetics of CR onto magnesium ferrite adsorbent. Pseudo-first order was introduced by Lagergren[31] and the equation can be written as:

$$\log (q_e - q_t) = \frac{\log q_e - K_1 t}{2.303} \quad (3)$$

Pseudo-second order kinetics can be expressed as:

$$\frac{t}{q_t} = \frac{1}{K_2 q_e^2} + \frac{1}{q_e} t \quad (4)$$

K_1 is pseudo-first order rate constant (1/min), K_2 (g/CR min^{-1}) is pseudo-second order rate constant. q_e (mg/g) is amount of solute absorbed on to at equilibrium, q_t (mg/g is the amount of solute adsorbed at any time, t. Both K_1 and k_2 values can be calculated from the slopes of the plots log (q_e-q_t) against time and plot t/q_t versus t, respectively.

Information about pseudo-first and pseudo-second order models are summarized in Table 1 which presents the comparison of adsorption kinetics data for both model with different initial CR concentrations. The correlation values for pseudo first order kinetic model obtained were relatively small whereas the experimental q_e values did not fulfil the calculated values gained from the linear plots illustrated in Table 1. Therefore, adsorption CR onto magnesium ferrite adsorbent did not follow pseudo-first order model.

Table 1. Comparison of the pseudo-first and second-order adsorption rate constants, calculated and experimental q_e values for different CR concentrations.

C_o CR	q_e, exp (mg/g)	First-order kinetic model			Second-order kinetic model		
		q_e, cal (mg/g)	K_1	R^2	q_e, cal (mg/g)	K_2	R^2
40 ppm	53.33	6.15	0.004	0.442	50.5	0.017	0.999
70 ppm	68.63	25	0.011	0.694	66.22	0.001	0.994
100 ppm	47.73	16.1	0.016	0.896	50.25	0.001	0.989

The correlation coefficient values for pseudo second order showed higher results as compared to the values of pseudo first order. The experimental and calculated q_e are very close, indicating the applicability of pseudo second order in order to describe the reaction of CR adsorbed onto magnesium ferrite adsorbent.

The kinetic study at 40 ppm concentration is presented in Figure 7. The equilibrium data fitted the second order kinetic model with high correlation coefficient (R^2) value of 0.999. The calculated q_e for the amounts of solute sorbed at equilibrium of CR, q_e was 50.5 mg/g.

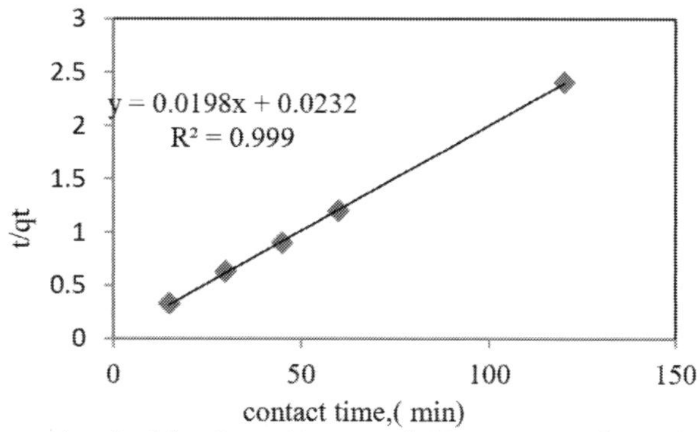

Figure 7. Pseudo-second order kinetic at 40 ppm of CR concentration using magnesium ferrite

3.4. Desorption studies

Desorption of the loaded Congo red were carried out at pH12.0. The elution of the Congo red from magnesium ferrite surface may be explained. Adsorption/desorption cycles, the elution effi- ciency was found to be 98.1 ± 1% for Congo red, respectively. Regeneration of the adsorbent was easily done using EDTA at pH 12.0 reaching efficiency of about 98%.

5.Conclusions

The present study shows that the crystal structure and morphology of the fabricated ferrites were influenced by the calcination temperatures. The FTIR spectra of magnesium ferrite showed the remarkable peak (400 cm^{-1} – 600 cm^{-1}) both at the 600°C and 800°C, because the tetrahedral and octahedral structures were obviously formed after all volatile matters removed at both the temperatures of interest. The XRD spectra of magnesium ferrite showed the formation of pure magnesium ferrites at 600^{0}C and 800^{0}C in the crystalline structure compared to the semi-amorphous structure of the related ferrite at 300^{0}C. The SEM micrograph of magnesium ferrite showed a flake-form structure which they were calcinated at 800°C. This study provides a meaningful correlation of the effects of calcination temperature on the crystalline structure of the ferrites of interest.

Also the present study shows that magnesium ferrite is suitable adsorbents for the removal of Congo red from aqueous solution. The removal of Congo red was shown to be dependent on pH, contact time, as well as initial concentration. The experimental data well fitted to Langmuir equation confirming the monolayer coverage of Congo red solution onto magnesium ferrite adsorbents. The reaction of Congo red adsorbed on magnesium ferrite was found closely to pseudo-second order model.

Acknowledgement

The authors would like thank to State university of Jakarta for providing research facilities and funding the project as well as to Ministry of Higher Education, Indonesia

References

[1]. Maensiri.S, M. Sangmanee, A.Wiengmoon,Magnesium ferrite ($MgFe_2O_4$) nanostructure fabricated by electrospinning. 2008. *Nanoscale Research Letters*, **4:** 221-228.

[2]. Liu.C.P., M.W Li, C.Zhong, J.R. Huang, Y.L. Tian, L. Tong, W.B., Mi.2007. Comparative study of magnesium ferrite nanocrystallites prepared by sol-gel and co-precipitation methods. *Journal of Material Science*, **42**: 6133-6138.

[3]. Siemons, M., Weirich, Th., Mayer, J., Simon, U. 2004. Preparation of nanosized perovskite-type oxide via polyol method. *Journal of Inorganic and General Chemistry*, **630**: 2083-2089.

[4]. Pradham.S.K., S. Bidb, M. Gateshki, V. Petkov.2005. Microstructure characterization and cation distribution of nanocrystalline magnesium ferrite prepared by ball milling. *Material Chemistry and Physics,* 93:224-230.

[5]. Mishra.S.,T.K.Kundu, K.C.Barick, D.Bahadur, D.Chakravorty, 2005. Prepare nanocrystalline $MnFe_2O_4$ by doping with Ti^{4+} ions using solid state reaction route. Journal of Magnetism and Magnetic Materials, **307** : 222-226.

[6]. Lick. L.D., D.B. Soria.2008. The synthesis of $MnFeO_3$ from the oxidative thermal decomposition of $Mn[Fe(CN)_5NO].2H_2O$. *Journal of the Argentine Chemical Society*, **97**:102-108.

[7]. Pradeep.A., G.Chandrasekaran.2005. FTIR study of Ni, Cu and Zn substituted nano-particles of MgFe2O4. *Material Letters*, **60** : 371-374.

[8]. Pradeep.A., P. Priyadharsini, G.Chandrasekaran, 2008.Sol gel route of synthesis of nanoparticles of MgFe2O4 and XRD, FTIR and VSM study. *Journal of Magnetism and Magnetic Materials*, **320** : 2774-2779.

[9]. Rashad.M.M.2006. Synthesis and magnetic properties of manganese ferrite from low grade manganese ore. *Materials Science and Engineering B*, **127**:123-129.

[10]. Zui Ding, Wei Wang, Yajun Zhang, Feng Li, J. Ping Liu. 2015. Synthesis, characterization and adsorption capability for Congo red of $CoFe_2O_4$ ferrite nanoparticles. *Journal of Alloys and Compounds,* **640** : 362-370.

Identification and Purification of Nyalo River Silica Sand as Raw Material for the Synthesis of Sodium Silicate

S Aini[1*], U K Nizar[2], A Amelia NST[3], J Efendi[4].

Chemistry Department, Faculty of Mathematics and Science, State University of Padang. Indonesia.

*syamsiaini@fmipa.unp.ac.id

Abstract. This research is on identification and purification of silica sand from Nyalo River. It will be used as a raw material for synthesis of sodium silicate. Silica sand was separated from clay by washing it with water, and then the existing alumina and iron oxide were removed by soaking the silica sand with 1 M HNO_3 solution. Qualitative and quantitative analysis of the silica sand with X-ray diffraction and X-ray fluorescence revealed that, silica sand existed in quartz form and contained a small amount of impurity oxide such as Al_2O_3, K_2O, MgO, CaO, Fe_2O_3 with percentage below the minimum threshold. The percentages of silica were 80.59% before purification. After three purificationsteps the silica percentage become 98.38%. It exceedsthe minimum threshold of silica percentage for industry.So, the silica sand from Nyalo River has high potency as a raw material for sodium silicate synthesizing.

1. Introduction

The Silica sand is one of Indonesia's natural wealth, the most abundant is in West Sumatra. The silica sand of West Sumatra is 82% of Indonesia's silica sand reserves, but has not been optimally utilized to meet the needs of the Sodium silicate industry [1]. Sodium silicate is a necessary reagent as a raw material and filler of various industries such as; industrial of silica gel, paper, soap, detergent, zeolite, mesoporous silica (for HPLC, metal nanoparticles host, etc)[2]. Sodium silicate for the Indonesia industry is still imported, while the industry demand on sodium silicate is increase continually.BPS Semarang reported that the increasing of sodium silicate import from 2012-2014 is 39.093,30 to 45.664,41 ton[3].

It is predicted for the coming year, the need for sodium silicate will continue to increase. The industrial demand on sodium silicate must be accompanied by the production of sodium silicate from local materials.While, the types and chemical composition of silica sand in Indonesia is quite varied, depending on the process and the influence of impurities mineral formed during sedimentation method. Therefore, the first step of the research is on the identification and purification of silica sand as raw material which can be used as raw material for synthesis of sodium silicate.

2. Experimental Section

In this study, the identification and purification of silica sand is carried out in four step of activity as follows:

2.1. *Sample collection and physical treatment*

The silica sand of 1 mm to 6 mm in size with high silica content collected from the Nyalo River in the southern coastal district, 80.5 km from Padang. The clear silica sand-like glass samples are physically separated from black sand, and washed with water to release clay, dried in an oven at 110 ° C for 3 hours, mashed and sieved with a 100 mesh sieve (150 μm). The third step, thedry silica sand separatedusingmagnet, placed in plastic bottles, put in desiccators, characterizeusing X-Ray fluorescence, X-ray diffraction, and treated for chemical purification.

2.2 *Chemical treatment*

Fifty grams of fineand clay-free silica sand and minimum quantity magnetic material were soaked in 100 ml of 1M HNO_3solution for 20 hours,then washed with water until neutral, dried in an oven at 110 °C for 3 hours, put in desiccators, and characterizedusing X-Ray fluorescence.

2.3 *Characterization*

X-Ray Fluorescence (XRF) type DWXRF minipal 4 PW403045B was used for characterization of chemical composition of silica sand. Meanwhile, the crystallinity of silica sand was characterizedusing X-ray diffractometer.

3. Results and discussion

The ease of retrieval silica sand, safe and cheap purification processes, high silica content, and low impurities are favorable offerings for the exploitation of silica sand for an industrial raw material [4].The silica sand deposits of the Nyalo River is located along the river to the mouth of the river, in accordance with the silica sand formation process (Fig. 1a). Based on location, silica sand can be taken through two attractive transport routes; first, road transport,through the coastal Bungus, the area of Pisang and Pinang River, until the Nyalo River. While, the second, sea transport through the Mandeh island tourism area. The silica sand deposits of the Nyalo River are the result of weathering of rocks, carried by water and undergoing sedimentation along the river, at the mouth of the river, and along the beach. Silica sand deposits of Nyalo River are quite large. The hills around the river, estuaries and beaches looked white (Fig.1). Silica sand in this area reflects light at night. So, this area is named Nyalo river (Nyalo = glow)(local opinion).

(a)

(b)

Figure 1. (a) Location of Sungai Nyalo (taken from Google map)
(b) Deposits of Silica sand at Nyalo River

Currently,the silica sand of West Sumatera has not been used for sodium silicate industrial. Therefore, the researchers have many opportunities to identify and purify it. Identification and purification product of silica sand is a reference point for all processing tests in the later steps of silica sand assessment procedure, for examplethe silica sands a raw material for the synthesis of sodium silicate[5].Sodium silicate can be used as raw material for paper, detergent, zeolite, and silica mesoporous industry.

There are 4 common steps for purification of silica sand; (a) mechanical process, (b) physical process, (c) chemical treatment, and (d) thermal treatment[4]. The first step has been done by separating the transparent sand from the black sand. The second step silica sand was washed to remove clay.The third step, clay-free silica sand was smoothed and sieved, to obtain larger surface of silica sand, and then it was easy to clean at the chemical treatment step, and facilitated magnetic separation. The impurity of silica sand as alumina and iron oxide had been removed by soaking the silica sand sample using 1M HNO_3 solution.

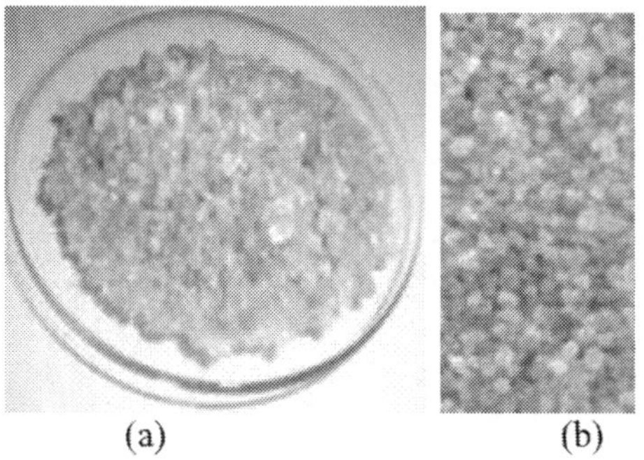

(a) (b)

Figure 2. (a) The clay-free silica sand after wash with water.
(b) Commercial Silica sand (Sam & Bros. Commercial brokers, Dubai)

Fig. 2 shows the Nyalo River silica sand after washing with water has the same morphology and color as commercial silica sand of Sam & Bros. Dubai production. The brown appearance of silica sand showed the presence of Fe_2O_3 as impurity compounds. Therefore, the silica sand needs to be mashed first to extend the touch area with the solvent, and so on dissolved the impurity compound with 1 M HNO3 solution. The silica sand before and after purified with HNO_3 can be seen in Fig. 3.The silica sand after magnetic separation and after wash with HNO_3 has a brighter color than silica sand before purification.

IOP Conf. Series: Materials Science and Engineering **335** (2018) 012025 doi:10.1088/1757-899X/335/1/012025

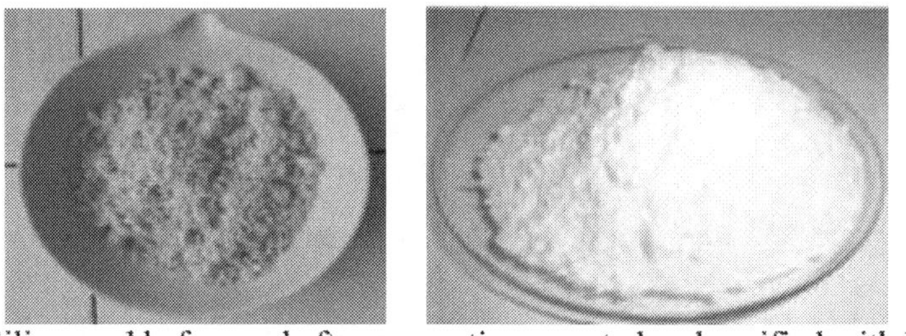

Figure 3. Silica sand before and after magnetic separated and purified with HNO_3 1M

Percentage of impurity oxide and SiO_2 of silica sand before and after purification can be seen in Table 1. Table 1 shows the three steps of purification of silica sand are capable to reduce impurity oxides such as Al_2O_3, Fe_2O_3, MgO, K_2O and Na_2O (<2%) in silica sand, with percentage below the minimum threshold, and obtained silica sand with SiO_2 percentage was 98.38%.This percentage exceeded the minimum threshold of silica percentage for sodium silicate industry[6,7].

Table 1. Chemical Composition of silica sand before and after purification

Oxide	Silica oxide (%)		
	Silica sand before purification	Silica sand after being purified with water and separated by magnet	Silica sand after purified with 0.1 M HNO_3
SiO_2	80,59	96.78	98,38
Al_2O_3	13,08	2.54	0,27
Fe_2O_3	5,92	0.06	0,06
CaO	0,12	0,09	0,01
MgO	0,20	0.10	0,09
K_2O	0,05	0.09	0,14

The type and crystallinity of silica sand wascharacterized using X-ray diffractometer.The x-ray diffraction pattern of silica sand (Fig.4) shows that, the silica sand of the Nyalo river is quartz in the presence of the main peak of quartz at 2θ; 20.857, 26.64, 36.542, 39.75, 50.141, 59.953, 68.147. These peaks are the same as diffraction pattern of standard quartz (Reference code: 01-085-0797). Meanwhile, 2θ peaks of alumina and iron oxide are invisible. The results of characterization using X-ray diffractometer are in line with x-ray fluorescence, supported by the colour of silica sand after washing. Hence, iron oxide and alumina as impurities of silica sand are small and unobserved.

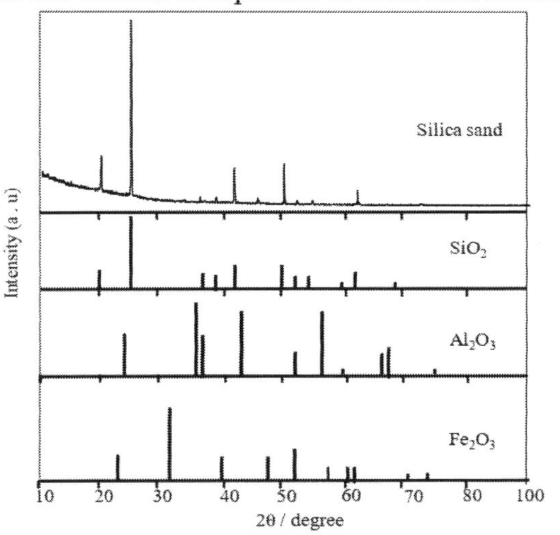

Figure 4. X-ray diffraction patterns for Fe_2O_3, Fe_3O_4, SiO_2 and silica sand of Nyalo River

Most sources of sand used by the sodium silicate industry are quartz. The existence of alumina at silica sand, and sodium silicate for zeolite and mesoporous silica industry can be allowed up to 1.5% , because, Al_2O_3can substitute some SiO_2 in three-dimensional networks[6,7].

4. Conclusion

The silica sand of the Nyalo River could be purified by only three steps. Silica sand was separated from clay by washing it with water,and then the existing alumina and iron oxide were removed by magnetic separations, soaked with 1 M HNO_3 solution. The silica sand of the Nyalo River are quartz.Silica sand collection can be done via road or sea transport.Physical purification results show that, the silica sand of Nyalo River has the same morphology as commercial silica. Characterization and purification study has proved that silica sand has the highest SiO_2percentage of 98.87% above the minimum threshold of industrial silica sand. The silica sand after purification has Al_2O_3, K_2O, CaO, MgO, Fe_2O_3 impurity oxides below the minimum threshold. The Nyalo river silica sand is very promising and can be utilized for a variety of industrial purposes after simple purification processes, especially as raw material for the synthesis of sodium silicate.

Acknowledgments

The author thankMinister of Research and Technology of Indonesia, and LP2M UNP who have provided research grants, Penelitian Unggulan Perguruan Tinggi (PUPT) 2017,for Synthesis and characterization of Fe_2O_3 immobilized in mesoporous silica based on green chemistry as a heterogeneous catalyst for Azo-dye decomposition study.

References

[1] Suhala, S. Dan Arifin, M. (1997).*Bahan Galian Industri Pusat Penelitian dan Pengembangan Teknologi Mineral,* Bandung.

[2] Lagaly, G., Tufar, W., Minihan, A., & Lovell, A. (2000). Silicates *Ullmann's Encyclopedia of Industrial Chemistry*: Wiley-VCH Verlag GmbH & Co. KGaA.

[3] Badan Pusat Statistik. 2014. *Statistic Indonesia*. www.bps.go.id

[4] Shaffer, N.R. (2006). The Time of Sands : quartz-rich sand deposits as a Renewable Resource. *Electronic Green Journal*. UCLA Library. 1(24). University of Idaho.

[5] Li, J.-s., Li, X.-x., Shen, Q., Zhang, Z.-z., & Du, F.-h. (2010). Further Purification of Industrial Quartz by Much Milder Conditions and a Harmless Method. *Environmental Science & Technology, 44*(19), 7673-7677.

[6] Flörke, O. W., Graetsch, H. A., Brunk, F., Benda, L., Paschen, S., Bergna, H. E., Schiffmann, D. (2000). Silica *Ullmann's Encyclopedia of Industrial Chemistry*: Wiley-VCH Verlag GmbH & Co. KGaA.

[7] Platias, S. Vatalis, K.I. 7 Charalampides, G. (2014). Suitability of Quartz Sands for Different Industrial Applications. *ScienceDirect. Procedia Economic and Finance*. 14. 491-498.

IOP Conf. Series: Materials Science and Engineering **335** (2018) 012026 doi:10.1088/1757-899X/335/1/012026

Partial Gene Cloning and Enzyme Structure Modeling of Exolevanase Fragment from *Bacillus subtilis*

M Azhar[1*], D Natalia[2], S Syukur[3], N Andriani[1], J Jamsari[4]

[1]Biochemistry Laboratory, Faculty of Mathematics and Natural Sciences, Universitas Negeri Padang, Jl. Prof. Hamka, Air Tawar, Padang 25131, Indonesia.
[2]Biochemistry Research Division, Faculty of Mathematics and Natural Sciences, Institut Teknologi Bandung, Jl. Ganesha10, Bandung 40132, Indonesia
[3]Biochemistry Laboratory, Faculty of Mathematics and Natural Sciences, Universitas Andalas, Padang 25163, Indonesia
[4]Biotechnology Laboratory, Faculty of Agriculture, Universitas Andalas, Padang 25163, Indonesia

*minda@fmipa.unp.ac.id

Abstract. Inulin hydrolysis thermophilic and thermotolerant bacteria are potential sources of inulin hydrolysis enzymes. Partial gene that encodes inulin hydrolysis enzymes had been isolated from *Bacillus subtilis* using polymerase chain reaction (PCR) method with the DPE.slFandDPE.eR degenerative primers. The partial gene was cloned into pGEM-T Easy vector with *E. coli* as host cells and analyzed using BLASTx, CrustalW2, and Phyre2 programs. Size of thepartial gene had been found539 bp that encoded 179aminoacid residues of protein fragment. The sequences of protein fragment was more similar to exolevanase than exoinulinase. The protein fragment had conserved motif FSGS, and specific hits GH32 β-fructosidase. It had three residues of active site and five residues of substrate binding. The active site on the protein fragment were D (1-WLNDP-5), D (125-FRDPK-129) and E (177-WEC-179). Substrate binding on the protein fragment were ND (1-*WLNDP*-5), Q (18- FYQY-21), FS (60-FSGS-63) RD (125-FRDPK-129) and E (177-WEC-179).

1. Introduction

Inulin hydrolysis thermophilic bacteria are potential sources of inulin hydrolysis enzymes. Inulin hydrolysis thermophilic bacteria has been screened and characterized as *Bacillus*genus[1]. Most of the thermophilic bacteria belong to the *Bacillus*genus [2].Nucleotide sequences of 16S rRNA partial gene of inulin hydrolysis thermophilic bacteria that expressed thermostable endoinulinase had been reported in 2008 by Gao [3]. The bacteria was identified as *Bacillus smithii* T7.

Thermophilic and thermotolerant bacteria can produce inulinase or levanase thermostable enzymes. Inulinase and levanase are belong to glycoside hydrolase family 32. The enzymes are very important in industrial aplications. The enzymes aremost appropriate choice as catalyst of inulin hydrolysis reaction toproduce fructose and Fructo-Oligosaccharides (FOS). The enzyme can also be used for production of various metabolites such as citric acid, lactic acid, ethanol, biofuels, butanediol [4].Another advantage, the use of the bacteria is the enzyme that expressed more easily purified due to heat treament. In addition, the enzymatic reaction at high temperature enhances inulin solubility, lowering the viscosity, reduces the risk of contaminants and accelerates reaction [5].Furthermore, inulin solubility was greater in hot water [6].

Inulinase and levanase gene has been isolated from bacteria, fungi and yeast. Inulinase gene that had Open Reading Frame (ORF) 1542 bphas been isolated from genomic DNA *Pichia guilliermondii* yeastusing PCR technique [7].The gene encodes 514 amino acid residues of the inulinase. Levanase gene has been isolated from genomic library *Gluconacetobacter diazotrophicus* SRT4[8].Tertiary structure of exoinulinase and endoinulinase has been found in PDB (Protein Data Bank, http://www.rcsb.org/pdb/ home/home.do), but tertiary structure of exolevanase was not available.

Inulin hydrolysis partial gene has been isolated from genomic DNA *Bacillus licheniformis* by using PCR technique with DPE.slF and DPE.eR degenerative primers[9]. This paper described isolation of inulin hydrolysis partial gene from genomic DNA *Bacillus subtilis* UBCT-030 by using the primer. The partial gene was compared with protein sequences in GenBank database and determined structure modeling of enzyme fragment.

2. Materials and Methods

2.1. Bacteria

Bacillus subtilis UBCT-030 that has been identified in phenotype and genotype was used as source of genomic DNA [10].The bacteria was cultured in the medium that contained inulin as the sole carbon source[11].

2.2. Isolation of Genomic DNA Bacteria

Isolation and purification of genomic DNA bacteria was carried out according to the procedure in the Wizard Genomic DNA Purification Kit.

2.3. Isolation of partial gene

The partial gene that encodes inulin hydrolysis enzyme was isolated from the genomic DNA *Bacillus subtilis* UBCT-030 by PCR method using DPE.slFandDPE.eR primers. The composition of the master mix as follows: 2.5 µL10xDreamTaqbuffer, 0.5 µL10 mMdNTPmix, $MgCl_2$ 3mM, 0.5 µL forward primer DPE.slF(TGG MTR AAY GAY CCM AAC GGA C)20 µM, 0.5 µL reverse primer DPE.eR (GGR CAT TCC CAY WCV CCR TC) 20 µM, 0.5 µL samples(75ng/µL), 0.125 µL DreamTaqDNA polymerase. ddH2O was added to volume of 25 µL. The PCR process was done at initial denaturation 94°C for 2 minutes, denaturation 94°C for 45 seconds, annealing 56°C for 30 minutes, elongation 72°C for 45 seconds, final elongation 72°C for 5minutes. PCR cycles were done29times. Amplicons were electrophoresed using 0.8% agarosegel in TAE buffer and DNA band were visualized by UV transillumination. DNA marker was used 1kb DNA ladder (Fermentas).

2.4. Partial gene cloning and sequencing

Partial gene that encodes inulin hydrolysis enzyme was ligatedtopGEM-T Easyvector[12]. Mole ratio ofpartial geneto thevectorwas 3:1. Mixture of ligationreaction were 5 µL of ligationbuffer 2 xrapidT4DNA ligase, 1 µL of pGEM-T Easy50ng, 1 µL of T4DNA ligase (3 Weissunits/µL), 1 µL of insert 20ng/µL, andddH$_2$O was added to final volume of 10 µL. There action mixture was incubated at room temperature at4°C for overnight. DNA recombinants (5 µL) were transformed in *E. coli*TOP10F'[13]. Recombinant DNA that carried the insert DNA was isolated from white colony transformant according to the procedure in the High-Speed PlasmidMiniKit(Geneaid). Recombinant DNA (1000ng)were sequenced by using dideoxy-Sanger method with T7 universal primer in Macrogen Korea.

2.5. Bioinformatics

Similarity with another protein in the GenBank database was determined by using the BLASTx program (http://www.ncbi.nlm.nih.gov). Determination of enzyme structure modeling was used Phyre2 program at http://www.sbg.bio.an.uk/phyre2 [14].

2.6. Accession number of nucleotide sequence

Nucleotide sequence of 16S rRNA gene from *Bacillus subtilis* UBCT-030 has assesion number KU358720. The sequences of partial genes from *Bacillus subtilis* UBCT-030 has been deposited in GenBank with accession number KU377724.

3. Results and Discussion

3.1.Amplification of the partial gene

Amplification of the partial gene encoding inulin hydrolysis enzyme from *Bacillus subtilis* UBCT-030 was performed at the optimum condition of PCR using DPE.slF and DPE.eR primers[9]. The PCR process was carried out at annealing temperature 56°C, 3 mM $MgCl_2$. The each cycle of PCR was set up denaturation94°C for 45seconds, annealing 56°C for 30 minutes, elongation 72°C for 45seconds. Finalelongation of PCR was 72°C for 5 minutes. Amplicon was purified and cloned in pGEM-T Easy vector. In the condition was produced amplicons about 550 bp (Figure 1).

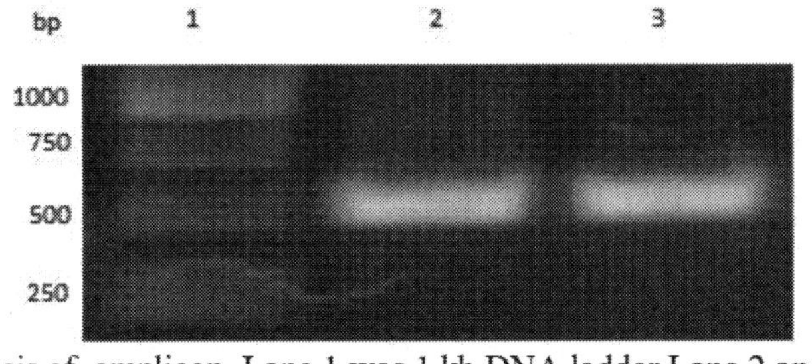

Figure 1. Electrophoresis of amplicon. Lane 1 was 1 kb DNA ladder.Lane 2 and 3 were amplicons of partial gene

3.2.Cloning of partial gene in pGEM-T Easy

The partial gene (about 550 bp) was ligated to the pGEM-T Easy vector to produce recombinant DNA. Recombinant DNA was transformed in *E. coli* TOP10F' cells by using heat shock method. Transforman cells were grown in LB medium containing ampicillin, IPTG and X-gal. There were white and blue transformant colonies on the surface of the media after 16 hours. White transformant colonies contained pGEM-T Easy with insert DNA, while blue colonies contained pGEM-T Easy without insert DNA (Figure2).

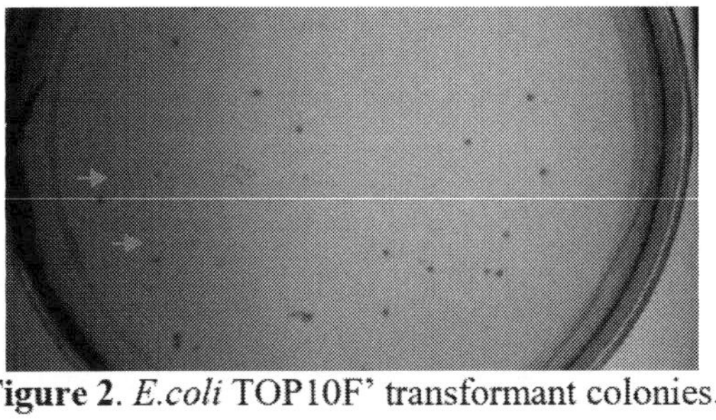

Figure 2. *E.coli* TOP10F' transformant colonies.

White colonies were transformants containingpGEM-T Easy with insert DNA (red arrows). Blue colonies were transformants containingpGEM-T Easywithout insert DNA.

The difference of transformants color were caused composition of bacteria media. Bacteria media was LB medium that contained IPTG (isopropyl-thiogalactoside), X-gal (5-bromo-4-chloro-3-indolyl-β-D galactopyranoside), and ampicillin. X-gal is β-galactosidase substrate, while IPTG is an inducer of lacZ gene. LacZ gene express β-galactosidase that catalyze breaking of bond on the X-gal produces galactose and 5-bromo-4-chloro-3-hydroxyindole which further oxidized to 5,5'-dibromo-4.4'-dichloro-indigo. The compound gives blue color in the transformant cell.

Expression of β-galactosidase will be disrupted if the pGEM-T Easy carry insert DNA. The insert DNA was ligated to the lacZ gene (on the cutting side of the EcoRV restriction enzyme) in the pGEM-T vector. Thymidine were added at the both ends of the vector [12]. In the pGEM-T Easy vector also contain gene that expressed ampicillin degrading enzymes. Therefore, the transformants that contained pGEM-T Easy vector can live in LB medium that contained ampicillin. Thus, white transformant colonies contained insert DNA.

White transformant colonies that contained desired insert DNA were screened by using colonyPCR method. The method is very fast and easy to screen transformant colonies that contained specific DNA sequences using PCR without DNA purification [15]. Products of colonyPCR that aligned with the insert DNA fragment was transformant colonies that contained desired insert DNA (Figure 3). Recombinant DNA from the colonies were isolated and purified. The DNA was used as a template for sequencing process of the insert DNA.

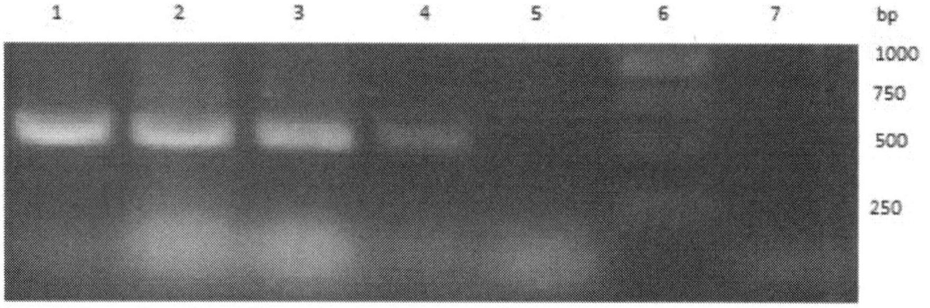

Figure 3. Electrophoresis of amplicon from white transformant colonies using colonyPCR. Lane 1 to 5 were amplicon colonyPCR using DPE.slFand DPE.eR primers. Lane 6 was the 1 kb DNA ladder marker. Lane 7 was primers (negative control).

3.3.The nucleotide sequences of partial gene and deduced amino acid sequences

The partial gene of *Bacillus subtilis* UBCT-030 was found 539 bp using the DPE.slF and DPE.eR primers (Figure 4). Deduction 539 bp of the partial gene produce protein fragment along the 179 amino acid residues (Figure 4b). In the protein fragment was found conserved domain A, B, B1, C, D and E. The domains were found in the group of Glycoside Hydrolase family (GH)32 enzyme [16]. Thus, partial gene is belong to GH32 enzymes. Domain A and E were position of DPE.slF, and DPE.eR primers [9]. The nucleotide sequences of the 539 bp was compared to the protein sequence in the GenBank database using BLASTx program. The protein fragment had specific hits GH32 β-fructosidase with three residues of the active and five residues of substrate binding (Figure 5).

Protein fragment along the 179 amino acid residues had high similarity with levanase from *Bacillus subtilis* (GenBank accession number WP041053668.1), and *Bacillus licheniformis* (GenBank accession number AGU89675.1) 98% and 97% respectively (Table 1). The protein fragment also had some similarities with the other GH32 enzymes, namely glycoside hydrolase from *Bacillus* sp. UNC41MFS5 (GenBank accession number WP026568421.1), SacC2 *Bacillusbataviensis* (GenBank accession number WP007084638.1), exoinulinase from *Aspergillus awamori* (PDB accession number 1Y9M|A), β-fructosidase from *Thermotoga maritima* (PDB accession number 1W2T|A), endoinulinase from *Aspergillus ficuum* (PDB accession number 3SC7|X), and levan fructotransferase from *Arthrobacter ureafaciens* (PDB accession number 4FFH|A) 65%, 64%, 49%, 41%, 39% and 34% respectively. Thus, the partial gene was isolated from genomic DNA *Bacillus subtilis* UBCT-030 was group of family GH32 enzymes. The protein fragment had closest resemblance to levanase than exoinulinase.

IOP Conf. Series: Materials Science and Engineering **335** (2018) 012026 doi:10.1088/1757-899X/335/1/012026

(a)

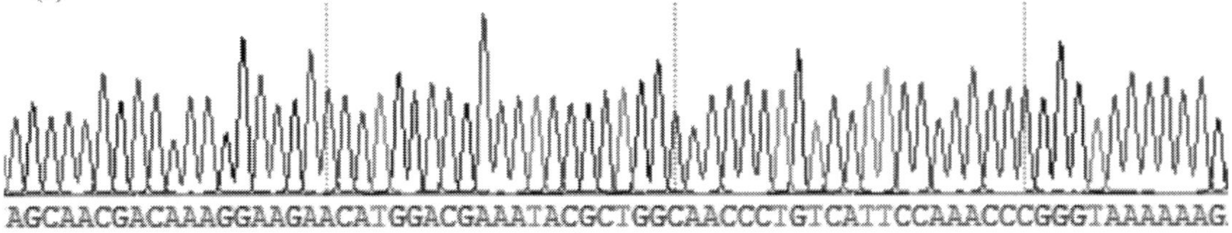

b)

```
TGGCTGAATG ATCCAAACGG ACTGGTATAT TATGCCGGGG AATATCACTT GTTCTATCAA   60
TACCATCCAT ACGGGCTTCA GTGGGGGCCC ATGCATTGGG GGCATGCCGT CAGCAAAGAT  120
TTGGTTACAT GGGAACACCT TCCTGTTGCG CTTTATCCGG ATGAAAAAGG CACGATCTTT  180
TCTGGAAGCG CAGTTGTAGA TAAAAATAAC ACAAGCGGTT TTCAAACAGG CAAAGAGAAG  240
CCGCTTGTGG CCATTTATAC ACAGGATCGG GAAGGCCATC AAGTGCAAAG TATTGCCTAT  300
AGCAACGACA AAGGAAGAAC ATGGACGAAA TACGCTGGCA ACCCTGTCAT TCCAAACCCG  360
GGTAAAAAAG ATTTTCGCGA TCCAAAAGTC TTTTGGTATG AGAAAGAAAA AAAGTGGGTG  420
ATGGTGCTTG CGGCCGGTGA CCGAATCCTC ATTTATACAT CAAAAAATCT GAAGCAGTGG  480
ACGTATGCAA GTGAATTTGG ACAGGATCAA GGAAGCCACG ACGGAGAATG GGAATGCCC    539
```

c)

```
WLNDPNGLVY YAGEYHLFYQ YHPYGLQWGP MHWGHAVSKD LVTWEHLPVA LYPDEKGTIF   60
SGSAVVDKNN TSGFQTGKEK PLVAIYTQDR EGHQVQSIAY SNDKGRTWTK YAGNPVIPNP  120
GKKDFRDPKV FWYEKEKKWV MVLAAGDRIL IYTSKNLKQW TYASEFGQDQ GSHDGEWEC   179
```

Figure 4.The nucleotide sequences of partial gene (a, b) and the amino acid residue of protein fragment (c). Electropherogram (a) is blue nucleotide sequences in (b). Red nucleotide sequences in (b) were DPE.slF and DPE.eR primers. Blue amino acid residues were conserved domain A, D and E respectively (c).

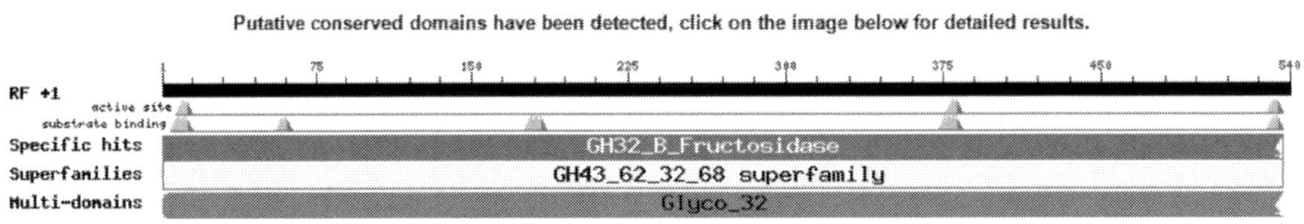

Figure 5. Group of the partial gene of *Bacillus subtilis* UBCT-030

Type action exo- of inulinase on conserved domain C in fungi have conserved motif Phe-Ser-Gly-Ser (FSGS), while the type action endo- had conserved motif FTGT[17,18]. In *Aspergillus niger* CBS13.88 was found exoinulinase that have conserved motif FSGS, while endoinulinase FTGT [16]. Conserved motif FSGS also was found in the protein fragment of *Bacillus subtilis* UBCT-030. Thus, the protein fragment of *Bacillus subtilis* UBCT-030 was a levanase that had type action exo-.

Table 1. Similarity of partial gene 539 bp with inulin hydrolysis enzymes from bacteria and yeast

GenBank Accession number	Description	Similarity (%)
WP041053668.1	levanase *Bacillus subtilis*	98
AGU89675.1	Lev1*Bacillus licheniformis*	97
WP026568421.1	glycoside hydrolase *Bacillus sp6*. UNC41MFS5	65
WP007084638.1	SacC2 *Bacillus bataviensis*	64
1Y9M|A	Exo-Inulinase from *Aspergillus awamori*	49
1W2T|A	Beta-fructosidase from *Thermotoga maritima*	41
3SC7|X	Endo-inulinase from *Aspergillus Ficuum*	39
4FFH|A	Levan fructotransferase from *Arthrobacter ureafaciens*	34

3.4. Structure modeling of exolevanase fragment

Tertiary structure of exolevanase was not available in the Protein Data Bank (PDB) database. Therefore, template for structure modeling of protein fragment from *Bacillus subtilis* UBCT-030 was used tertiary structure of exoinulinase from *Aspergillus awamori* in the PDB database at http://www.rcsb.org/pdb/home/ home.do. Identical the protein fragment with exoinulinase from *Aspergillus awamori* were 52%. Structure modeling of the protein fragment was built from 179 amino acid residues. Alignment of protein fragment from *Bacillus subtilis* UBCT-030 with exoinulinase from *Aspergillus awamori* was showed in Figure 6

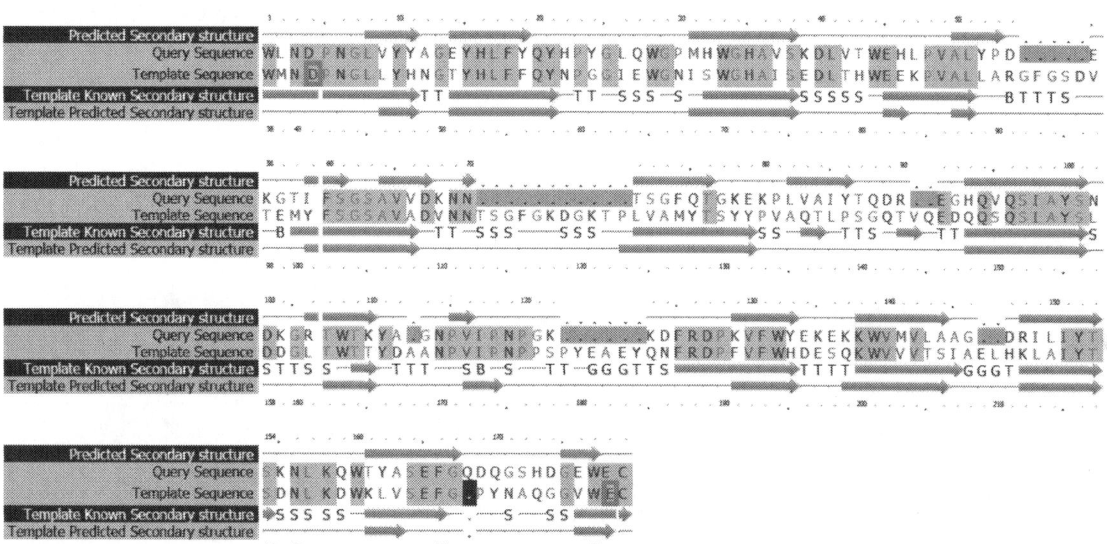

Figure 6. Alignment of exoinulinase with protein fragment of *Bacillus subtilis* UBCT-030 Catalytic sites were red box

Overall folding of the tertiary structure of the protein fragment was as folding of family GH32 enzyme. The family GH32 enzyme is composed two domains, namely β-propeller domain at the N terminal domain (Domain 1) and β-sandwich domain at the C terminal domain (Domain 2). Domain β-propeller consists of five blades. They are blade 1 (blue), blade 2 (light blue), blade 3 (light green), blade 4 (green), blade 5 (yellow) (Figure 8a). Domain β-propeller on exoinulinase *Aspergillus awamori* consists of 353 amino acid residues, from Phe20 to Gln372[19]. Protein fragment was

IOP Conf. Series: Materials Science and Engineering **335** (2018) 012026 doi:10.1088/1757-899X/335/1/012026

domain β-propeller was formed from amino acid residues 179. The protein fragment consist of three blades are blade 1 (yellow), blade 2 (yellow) and blade 3 (yellow) (Figure 7b).

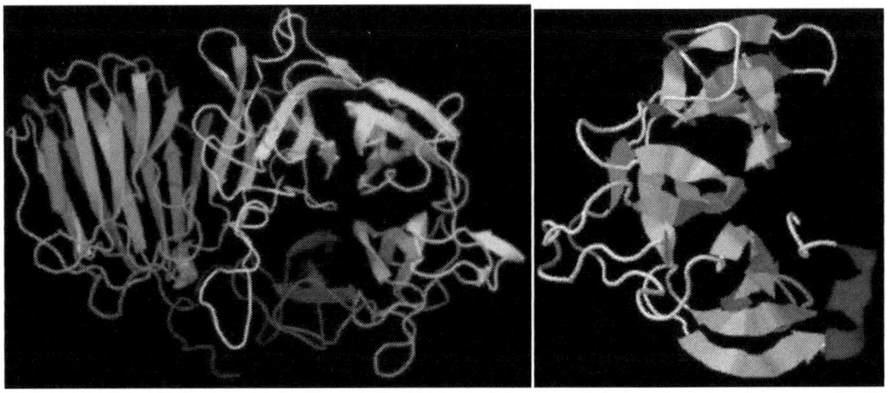

a. Exoinulinase of *Aspergilllusawamori* (PDB accession number c1y9gA) b. Exolevanase fragment of *Bacillus subtilis* UBCT-030

Figure 7. Structure modeling of exoinulinase and exolevanase fragment

The catalytic residues on the protein fragment were Asp4 (1-WMNDPNG-7), and Glu178 (177-WEC-179). The amino acid residues located on blade 1, and the blade 3 resvectively in the β-propeller domain of the protein fragment. The position of catalytic residues is equivalent position in the family GH32 enzyme. Catalytic residues of exoinulinase from *Aspergillus awamori* is Asp41 on blade 1 and Glu241 on blade 4. Position of Asp 41 was in motif 38-WMNDPNG-44, while position of Glu241 was in the motif 241-ECPGL-245 [19]. Active site on the protein fragment of *Bacillus subtilis* UBCT-030 were D (1-WLNDP-5), D (125-FRDPK-129) and E (177-WEC-179). Substrate binding were ND (1-WLNDP-5), Q (18- FYQY-21), FS (60-FSGS-63) RD (125-FRDPK-129) and E (177-WEC-179). Position of the residues in the blade was in Figure 8.

a. Catalytic site (red) b. Active site (red) c. Substrate binding (red)

Figure 8. Enzyme structure modeling of exolevanase fragment from *Bacillus subtilis* UBCT-030

4. Conclusions

Thepartial genehad been found539 bp from *Bacillus subtilis* UBCT-030 that encodes 179aminoacid residues of exolevanase fragment. The protein fragment had *specific hits* GH32 β-fructosidase. It had two residues catalytic site, three residues active site and five residues substrate binding

Acknowlegment

This research was funded by Hibah Doktor, from Directorate General for Higher Education, Ministry of National Education, Indonesia under contract number of 486/SP2H/PP/DP2M/VI/2010.

References

[1] Allais JJ, Hoyos-Lopez G, Kammoun S, Baratti JC. 1987. Isolation and characterization of the thermophilic bacterial strain with inulinase activity. *Appl. Enviromental Microbiol.***53**:942–945.

[2] Souza AN, Martins MLL. 2001. Isolation, properties, and kinetics of growth of a thermophilic Bacillus. *Brazillian J. Mirobiology***32:** 271–275.

[3] Gao W, Bao Y, Liu Y, Zhang X. 2008. Characterization of thermostable endo-inulinase from a new strain *Bacillus Smithii* T7. *Appl. Biochem Biotechnol***157**:498–506.

[4] Singh RH, Chauhan K, Kennedy JE. 2016. A panorama of bacterial inulinase: production, purification, characterization and industrial applications. *International Journal of Biological Macromolecules.* **96:** 312-322.

[5] Vieille C, Zeikus GJ. 2001. Hyperthermophilic enzymes: sources, uses, and molecular mechanisms for thermostability. *Microbiol. Mol. Biol. Rev.***65:** 1–43.

[6] Phelps CF. 1965.The physical properties of inulin solution. *Biochem.J***95:** 41–47.

[7] Zhang T, Gong F, Chi Z, Liu G, Chi Z, Sheng J, Li J and Wang X. 2009. Cloning and characterization of the inulinase gene from a marine yeast *Pichia guilliermondii* and its expression in *Pichia pastoris*. *Antonic van Leeuwenhoek***95:** 13–22.

[8] Menendez C, Hernandez L, Selman G, Mendoza MF, Hevia P, Sotolongo M, A. J. 2002.Molecular cloning and expression in Escherichia coli of an exo-levanase gene from the endophytic bacterium Gluconaceto-bacter diazotrophicus SRT4. *Curr. Microbiol.***45:** 5–12.

[9] Azhar M, Natalia D, Syukur S, Vovien, Jamsari. 2015. The gene fragment that encodes inulin hydrolysis enzymes from genomic *Bacillus licheniformis* by PCR technique using new primers. *Int. J. Biol. Chem.***9:** 59–69.

[10] Azhar M. 2013. Characterization of extracellular enzymes on inulin substrate and molecular levanase gene from thermophilic *Bacillus licheniformis* UBCT-007, local isolate of West Sumatra. (Universitas Andalas. Indonesia).

[11] Castro GR, Baigori MD, Sineriz F. 1995. A plate technique for screening of inulin degrading microorganisms. *J. Microbiol. Methods***22:** 51–56.

[12] PROMEGA. 2010. pGEM(R)-T and pGEM(R)-T Easy Vector Systems Technical Manual TM042 - pgem-t and pgem-t easy vector systems protocol.pdf.

[13] Sambrook J, Russell DW. 2001. *Molecular Cloning a Laboratory Manual*. pp.1.105-1.111. 3th Ed. Cold Spring Harbor Laboratory Press, Cold Spring Harbor, New York.

[14] Kelley LA, Mezulis S, Yates CM, Wass MN, Sternberg MJE. 2015. The Phyre2 Web Portal for Protein Modeling, Prediction, and Analysis. *Nature Protocols.* **10:** 845–858.

[15] Woodmen, M.E. 2008.*Current Protocols in Microbiology* A.3D.pp.1-6 (John Wiley & Sons, Inc.)

[16] Goosen C. 2007. Identification and characterization of glycoside hydrolase family 32 enzymes from *Aspergillus niger*. (Rijksuniversitet Groningen).

[17] Ohta K, Akimoto H, Moriyama S. 2004. Fungal inulinases: enzymology, molecular biology and biotechnology. *J.Appl.Glyosci***51:** 247–254.

[18] Adawiyah SR, Shuhaimi M, Yazid AMM, Manaf AA, Rosli N, S. 2011. Molecular cloning and sequence analysis of an inulinase gene from an *Aspergillus* sp. *World J Microbiol Biotechnol***27:** 2173–2185.

[19] Nagem RAP, Rojas AL, Golubev AM, Korneeva OS. 2004. Crystal structure of exo-inulinase from *Aspergillus awamori*: the enzyme fold and structural determinant of substrate recognition. *J.Mol.Biol***344:** 471–480.

IOP Conf. Series: Materials Science and Engineering **335** (2018) 012027 doi:10.1088/1757-899X/335/1/012027

Water Quality Assessment and Determining the Carrying Capacity of Pollution Load Batang Kuranji River

I Dewata[1,2*], and Z Adri[2]
[1]Chemistry Department, Faculty of Mathematics and Science, Universitas Negeri Padang, Indonesia
[2]Environmental Science, Postgraduate Program, Universitas Negeri Padang, Indonesia

i_dewata@yahoo.com

Abstract. This study aims to determine the water quality and carrying capacity of pollution load Batang Kuranji River in the headwaters, middle, and downstream. This research is descriptive quantitative parameters of pH, BOD, COD, TSS, and DOES Depictions of river water quality refer to RegulationNo.82/2001, while determination of carrying capacity of pollution load river refers to the Kep Men LHNo.10/2003.The result is Kuranji Batang River water quality upstream region included in either category who meet the quality standard first class ofPP82/2001. TSS concentrations at head waters of 21 mg/L, BOD1,6 mg/L, COD7,99mg/L and DO 7,845 mg/L. While the carrying capacity of pollution load river in upstream region included in both categories namely BOD of 4,4 kg/sec, COD 273,60 kg/sec, TSS906,00kg/sec, and DO parameters of 49.20 kg/sec. Middle region (point 2, 3, and 4) water quality Batang Kuranji River has exceeded the quality standard of 82/2001 for class II and class III. Meanwhile, carrying capacity of pollution load river in area included in ugly category. The calculation is done with application Qual2Kw show that carrying capacity of pollution load river of BOD -857.3 kg/sec, COD -777.40 kg/sec, TSS +9511.5 kg/sec, and DO +69.30 kg/sec.

1. Introduction

Padang City is one of the cities in West Sumatra that has several rivers, including Batang Arau River, Batang Kandis River, Batang Kuranji River, Air Dingin River and other rivers. One of the best quality river is Batang Kuranji River. Administratively Batang Kuranji River is located in 4 (four) sub-districts of Padang City, namely Pauh, Kuranji, Nanggalo, and districts of North Padang. Batang Kuranji River Basin has an area of 202.70 km^2 and the length of the river is 17 km. Upstream Batang Kuranji River is located in District Pauh and downstream in District North Padang. (BPS, 1997).

In general, the function of Batang Kuranji River is widely used by people around the watershed to meet the needs of life such as drinking water or PDAM water intake, bathing, washing, tourism, and sand and stone retrieval activities. Besides being widely used by the community, Batang Kuranji River is also used as a container of liquid waste or waste water coming from the activity. Activities located along the Batang Kuranji river flow that have the potential to produce liquid waste and flow into the river are tofu factories, bakeries, workshops, catering, restaurants, traditional markets, agriculture, human settlements, and others. The increase of population in the Batang Kuranji river basin also resulted in activities carried out by communities along the river flow also increased. Increasing quantity / amount of liquid waste in the absence of management can lead to degradation of river water quality. This is because the liquid waste output will disembogue and flow into the river. Thus the river that is currently still can be used or utilized by the community will not be utilized again by the community along the river in the future (Bapedalda, 2007).

Based on the description, to maintain the water quality of Batang Kuranji River in order not to decrease in the future, it is necessary to study the quality of river water in the existing condition, so it can determine the capacity of Batang Kuranji River. Capacity of the river is useful to know the potential or ability of the river in receiving the incoming liquid waste that causes the river to become uncontaminated. In addition, because water is one of the resources needed, so to anticipate future developments that the potential decrease of river water quality is very large then it is necessary determination of river capacity.

2. Study Method

This type of research is descriptive quantitative, while the determination of river capacity refers to KepMen LH. 110/2003 on Guidance on Capacity Determination of Water Pollution Load at Water Source through Qual2Kw method. Water quality research and determination of Batang Kuranji River capacity is carried out in Batang Kuranji river basin. The research begins in April 2012 until February 2013.

2.1. Data Collection

Secondary data required is water quality from related institutions and community activities along streams that produce both point source and non-point source liquid wastes. Primary data in the form of GPS(Global Positioning System) river, river's width, the details of incoming flow and outflow of Batang Kuranji River, Batang Kuranji River's water quality, etc. Physical parameters measured are temperature, solid, turbidity, and color. Parameters of water chemistry such as pH value, BOD (Biochemical Oxygen Demand), COD (Chemical Oxygen Demand), DO (Disolved Oxigen).

2.2. Data Analysis

The determination of river pollution load capacity is by Qual2Kw method by United States Environmental Protection Agency (US EPA). The data required are river flow, river quality, river length and width, air temperature, wind speed, cloud cover, water sampling coordinates, wastewater quality monitoring in accordance with river water parameters and discharge of water source or liquid waste into the river (KLH, 2009).

2.2.1. Quality of River Water. Data analysis on water quality of Batang Kuranji River refers to PP. No. 8/2001 on Water Quality Control and Water Pollution Management.

2.2.2. Load Pollution Capacity. The capacity of the river is determined based on KepMen LH. 110/2003 by using the application Qual2Kw. Simply, the above equations can be illustrated in equations (1), (2), and (3).

$$W_i = W_{i-1} + W_{in,i} - W_{ab,i} \quad (1)$$

$$W_i = \sum_{j=1}^{psi} Q_{ps,i,j}\, c_{psi,j} + \sum_{j=1}^{npsi} Q_{nps,i,j}\, c_{npsi,j} \quad (2)$$

$$Q = V\,x\,A \quad (3)$$

<u>Explanation :</u>

W_i = River mass flow rate from point i to point i+1, (m^3/dt)
W_{i-1} = River mass flow rate from point i-1, (m^3/dt)
$W_{in,i}$ = The total mass flow rate entering streams from point and nonpoint source, (m^3/dt)
$W_{ab,i}$ = The total mass flow rate out of the river from point and nonpoint source, (m^3/dt)
Q = River debit (m^3/dtk)

V = River flow rate (m/dtk)
A = River cross section area (m^2) (Pelletier G J.,et all,. 2005

3. Results and Discussion

3.1. Results

Padang city is located on the west coast of Sumatra Island at the position of 00° 44'00"- 10° 10' 35" SL and 100° 05'05"- 100° 05' 05" WE, with a height of 0 - 500 meters on the mdpl and the area around 694, 96 km^2. In terms of topography of the four sub-districts traversed by the River Batang Kuranji is at an altitude of 0-100 meters from sea level with a length of 17 km.

Land utilization in the four sub-districts traversed by the Batang Kuranji stem rivers, namely Pauh sub-district, still exist state forests and community forests which are conservation areas as protection for all watersheds. In addition, there are also rice fields, gardens and fields in the upstream of the river. Meanwhile, compared to the downstream area of the river (North Padang Subdistrict) there is no state forest that is used as a protection for the ecosystem of organism.

3.2. Indentify Pollutant Sources

Identification results of the Batang Kuranji River basin are found that pollutant sources are generally derived from residents' wastes, industrial activities such as tofu factories, restaurants, catering, workshops, car and motor washings, chicken farms, C excavations, traditional markets and agricultural activities.

3.3. River Water Quality

3.3.1 Batang Kuranji River Water Quality

The result of Batang Kuranji River water quality inspection from upstream to downstream based on primary data and secondary data presented in Tables 3 and 4 shows that there has been a change of water quality of Batang Kuranji River from upstream to downstream for TSS parameter, BOD, COD, DO, and pH. In the upstream area, the water quality of Batang Kuranji River is still considered good, that is still fulfill the quality standard of class 1 (one) PP 82/2001. Water in this class may be designated as raw water for drinking water or other designations in accordance with the same water quality requirements.

Table 1. The result of Batang Kuranji Water Quality Of Padang in sampling 1,2, and 3 Analysis : laboratory of Chemistry, FMIPA UNP Padang.

No	Para meter	sat	Sampling point 1			2			3		
			DP	DS	Rata^2	DP		mean	DP	DS	mean
1.	TSS	mg/L	12	30	21	6	90	48	13	180	96,5
2.	pH	-	6,9	7	6,9	6,8	7	6,9	6,9	6,4	6,65
3.	BOD	mg/L	1,2	2	1,6	2,3	6	4,15	1,9	27	14,4
4.	COD	mg/L	9,2	6,8	7,9	15,7	17	16,3	11,8	68,3	40,0
5.	DO	mg/L	6,7	9	7,84	7,08	7	7,04	7,08	5,5	6,29

Table 2. The result of Batang Kuranji Water Quality Of Padang in sampling point 4, 5, 2, dan6.

No	Para Meter	unit	Sampling ponit 4			5			6		
			DP	DS	mean	DP	DS	mean	DP	DS	mean
1.	TSS	mg/L	13	190	101,5	18	311	164,5	18	1450	734
2.	PH	-	6,3	6,6	6,45	6,9	6,8	6,85	5,8	6,8	6,3
3.	BOD	mg/L	1,73	35	18,36	1,12	40		2,37	64	33,18
4.	COD	mg/L	8,11	93,4	50,75	5,07	106	55,45	13,2	197	105,4
5.	DO	mg/L	6,13	6,1	6,115	4,37	6,5	5,435	5,95	5,2	5,575

Analysis: laboratory of chemistry Universitas Negeri Padang, DP is Primery Data and DS is Secondary Data

1 : Main stream of Kuranji River (0° 53' 56" S; 100° 27' 38" E)
2 : Limau Manis and Batu Busuak river position (0° 55' 19" S; 100° 26'21" E)
3 : Kalumbuk (0° 55' 14" S; 100° 23' 27" E)
4 : Intake PDAM (0° 54' 20" S; 100° 22' 26" E)
5 : Kuranji river and Batang Guo river intersection (0° 53' 58" S; 100° 21' 22" E
6 : Down stream of Kuranji River (0° 54' 20" S; 100° 20' 52" E)

Table 1 shows the quality of river water at each sampling site by comparison of water classes. BOD, COD, and TSS parameters tend to change their quality from upstream to downstream areas. Specifically, for TSS Parameters experienced significant changes in downstream area at the point of sampling location 5 that is TSS concentration 164,5 mg/L and point 6 TSS concentration 734 mg/L accumulated sediment due to river edge erosion in upstream and middle river area and sand or stone harvesting by communities will enrich suspended solids (TSS) in the area.

3.3.2 Batang Kuranji River Capacity

The calculation of Batang Kuranji River Capacity is carried out based on the monitoring result which has been done and the assumption of the determination of water class in each region or segment in Batang Kuranji River (Figure 3). Qual2Kw application review results known that the pollutionload capacity of Batang Kuranji River (BOD, COD, TSS and DO) in upstream areas is categorized as good, whereas pollution load capacity in downstream areas is categorized as poor. Next Figure shows the concentration of BOD monitoring results exceeding the threshold. The green curve with triangle points is the targeted BOD concentration when each water segment is defined according to PP 82 of 2001. While the red color curve (dashed line) is the result of monitoring the quality of Batang Kuranji river. Through Qual2kw application approach is calculated BOD pollution load into river from upstream to downstream is 2,169,30 kg/second. Pollution load that enter at existing condition in upstream area (segment 1) for BOD parameter is 68 kg/s, while based on PP 82/2001 or targeted is 72.40 kg/sec. This means that upstream of Batang Kuranji River is still able to accommodate pollution loads of 4.4 kg/sec on BOD parameters.

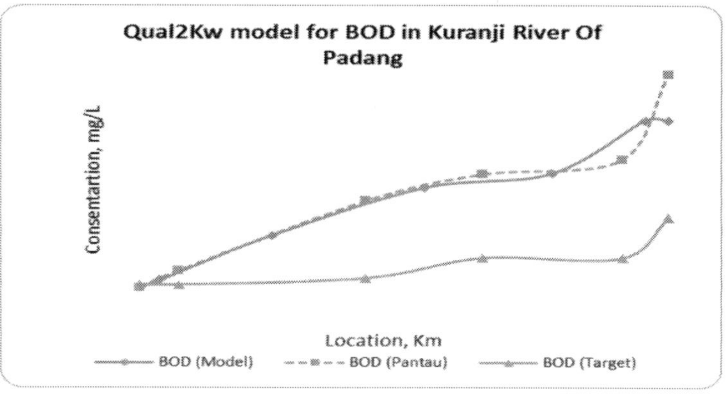

Figure 1. Qual2Kw modeling BOD kconcentration in Batang Kuranji River, Padang City.

Whereas respectively on the 2nd, 3rd, 4th and 5th segments the pollution load coming into the river has exceeded the targeted threshold in accordance with PP 82/2001. The 2, 3 and 4 segments are included in the middle of the Batang Kuranji river, while on the 5th section it includes the downstream

IOP Conf. Series: Materials Science and Engineering **335** (2018) 012027 doi:10.1088/1757-899X/335/1/012027

area of the river. The pollution load on area 2 has exceeded the 299.5 kg/s capacity threshold from the pollution load target of 54.6 kg/sec. In the third and fourth sections, it has exceeded 557.80 kg/sec of the total targeted pollution load of 218.40 kg/sec. Whereas in down stream, the pollution load that has entered has exceeded the target which is charged at 296,40 kg/sec. Description of Batang Kuranji River pollution load pollution curve for BOD parameter is presented in Figure1.

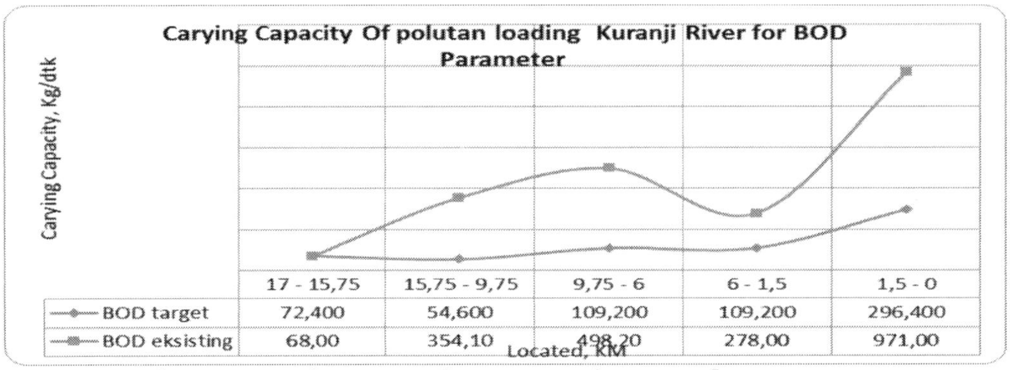

1 2 3 4 5

Figure 2. Capacity of pollution load of Batang Kuranji river in Padang, COD parameter.

Based on the above description and Figure 2, the pollution load capaity in the upstream area (segment 1) for the BOD parameter in good category. Meanwhile, for the middle area (segments 2, 3, and 4) are in bad condition, while in the downstream area (segment 5) the pollution load capacity with BOD parameter is included in bad category.

The pollution load capacity for COD parameter also shows the same with BOD as shown in next Figure. The green color curve (COD target) is the target of pollution load of Batang Kuranji River when the river class has been determined in accordance with PP 82 of 2001, meanwhile the blue curve (COD Model) is the output of river quality modeling from upstream to downstream. While the red colors curve (COD monitor) is the result of monitoring the quality of Batang Kuranji river in Figure 3.

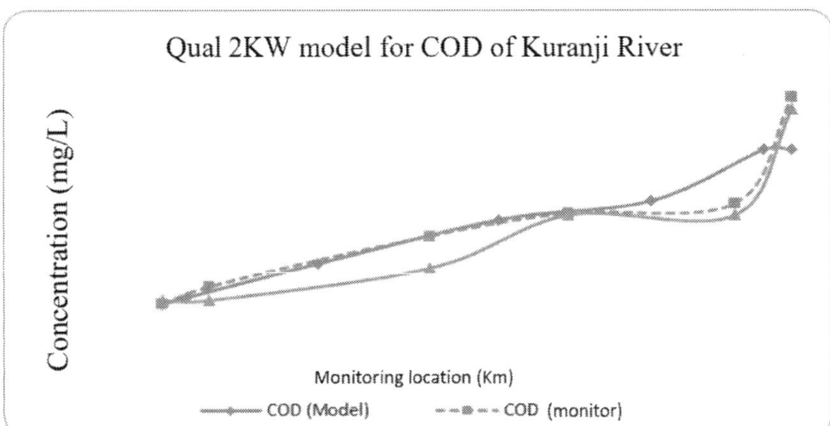

Figure 3. Qual2Kw modeling COD concentration in Batang Kuranji river, Padang city COD parameter.

In the upstream area (segment 1) the pollution load capacity of COD can still accommodate pollution loads of 273.60 kg/sec from the pollution load target of 482.00 kg/sec. The second segment area of the incoming pollution load has exceeded the targeted threshold for the class II class water of 455.00 kg/sec while the monitoring result on the second segment shows the pollution load is 868.00 kg/sec. In areas 3 and 4 designated as class III water, it has also exceeded the threshold of 364.40 kg/sec from the targeted 1,820 kg/sec. While downstream, the COD parameter has been overloaded by 413 kg/sec of the target set at 2,470 kg/sec. In this area is defined as the segment that has water with the standard grade 4. Based on the description, the pollution load capacity for upstream to COD parameters is good, middle area (segment 2, 3, and 4) load capacity of pollution load included in bad category, while the downstream area of its capacity is quite bad .

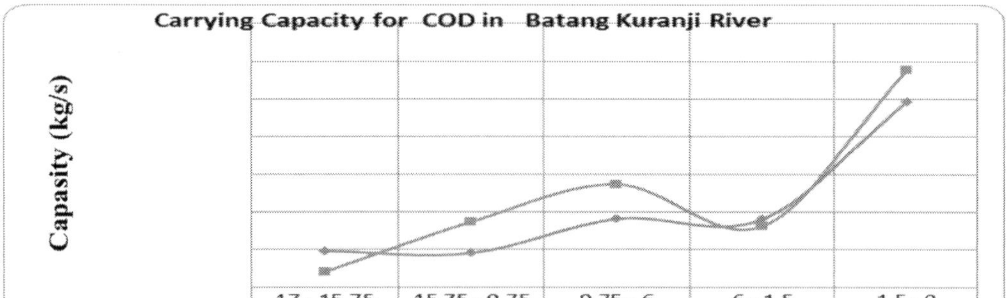

	17 - 15,75	15,75 - 9,75	9,75 - 6	6 - 1,5	1,5 - 0
COD target	482,00	455,00	910,00	910,00	2.470,00
COD eksisting	208,40	868,00	1.367,90	816,50	2.883,00

1 2 3 4 5

Figure 4. Capacity of pollution load of Batang Kuranji river, Padang city COD parameter.

The result of Qual2Kw application modeling for TSS parameter is shown in Figure below. The green color curve (TSS Target) is the concentration of TSS when the river class is determined in accordance with PP 82 of 2001. In the upstream area set class I, in the second segment is set class II, area 3 And 4 is designated water with class III, meanwhile for downstream area (segment 5) determined class IV. The red curve (monitored TSS) is the monitoring results of river quality from upstream to downstream, while the blue color curve (TSS model) is the modeling output issued by the Qual2Kw application. (Figure 4). Upstream river is still able to accommodate a TSS pollution load of 906.00 kg/sec from the targeted of 1,510 kg/sec. The upstream area (segment 1) is defined as water having a class I class. While in the middle (section 2) the pollution load is 2,000 kg/sec in excess of the target of 910 kg/sec. This area is designated as water with class II standard. In the 3 and 4 segments based on Qual2Kw modeling it can still accommodate pollution loads of 10601.50 kg/sec.

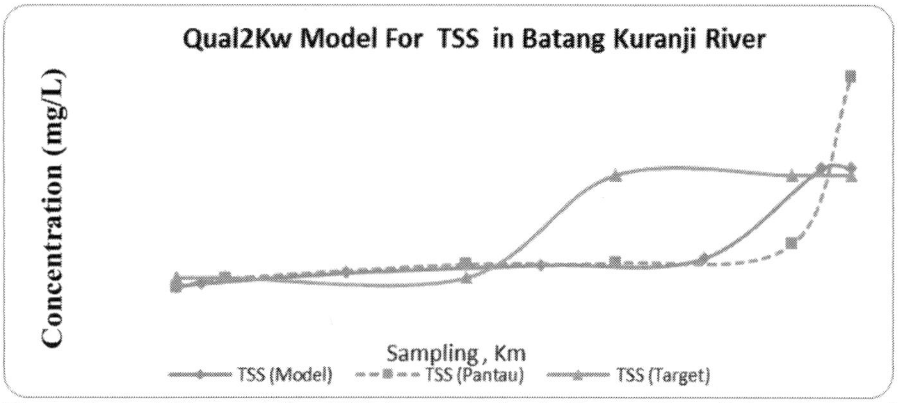

Figure 5. Qual2Kw modeling TTS Concentration in Batang Kuranji river Padang city.

This area is designated as standard class III water with the target of pollution load allowed to enter 14,560 kg/sec. Further, downstream is designated as class IV, the pollution load allowed for TSS parameter is 12,480 kg/sec. The monitoring results for the 5th segment show 25,659.00 kg/sec which exceeds 13,179.00 kg/sec of the charged target. For further details, the pollution load capacity of the Batang Kuranji River from upstream to downstream is presented in Figure 5.

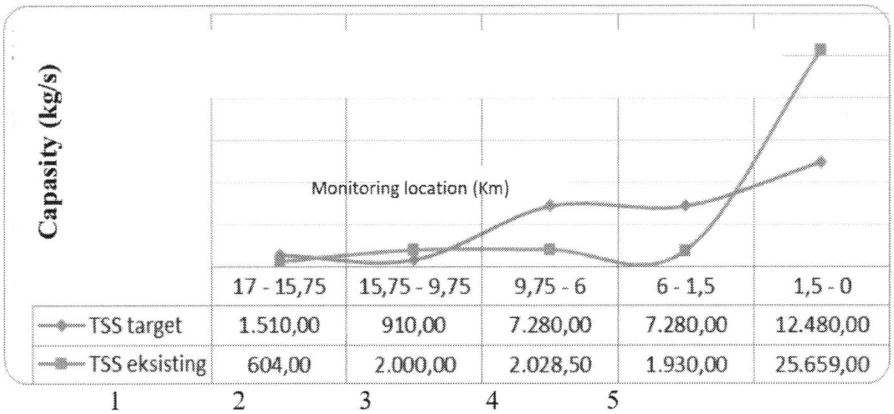

Figure 6. Capacity of pollution load of Batang Kuranji river in Padang City for TTS Parameter

Referring to the description, the pollution load capacity for the upstream area of the TSS parameter is included in good category. In the middle area (segments 2, 3, and 4) the pollution load capacity is included in good category because it can still accommodate pollution loads of 9,511.5 kg/sec. While in downstream area capacity of pollution load of *Batang Kuranji* River classified as poor. This can be seen from the total pollution load that has been entered has exceeded the threshold determined by PP 82/2001.

Furthermore, the parameters of DO pollution load coming from upstream to downstream areas still meet the quality standard threshold set by PP 82/2001. The results of modeling with Qual2Kw applications for the DO parameters are shown in Figure 6.

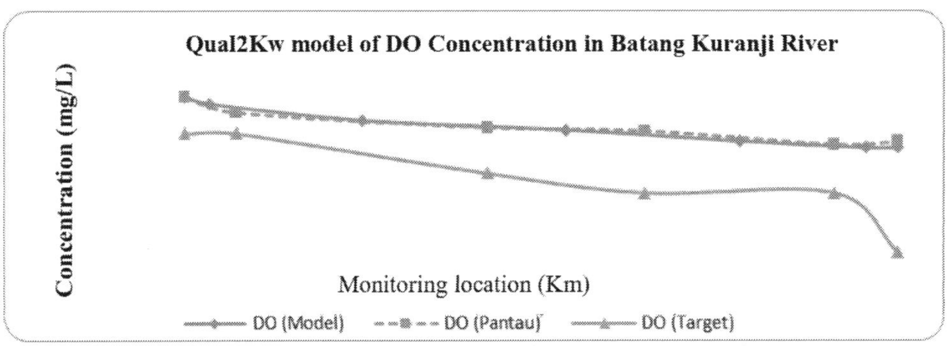

Figure 7. Qual2Kw modeling DO concentration in *Batang Kuranji* river

In upstream areas, the DO concentrations are still able to accommodate a pollution load of 49.20 kg/sec from the total target charged at 157.20 kg/sec. The upstream of the river is designated as water with a class I standard intended for raw water for drinking water. In the middle area (segment 2) the contamination load is 48.50 kg/sec, while the pollution load target for segment 2 is 72.80 kg/sec or still able to accommodate pollution load of 24.30 kg/sec. Segment 2 is defined as an area with a class II quality standard. Area of 3 and 4 segments which is intended for class III water quality, total pollution loads are 64.2 kg/sec. While the target of pollution load in this area is 109,20 kg/sec, so this area still able to accommodate the contamination load of 45 kg/sec. In the downstream area, the total burden of contamination of 90 kg/sec is designated for class IV water, which is not required to target pollution loads.

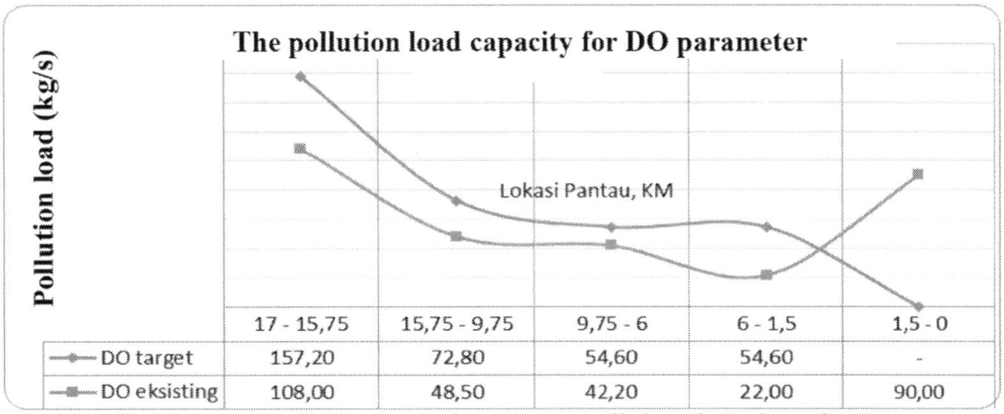

Figure 8. Capacity of pollution load of Batang Kuranji river in Padang city DO parameter

Based on the above description, the pollution load capacity of Batang Kuranji River for the DO parameters in the upstream, middle and downstream areas are classified as good (Figure 8)

4. Discussion

Water quality and pollution of Batang Kuranji River Padang hence obtained some findings as follows:

a. There has been a decrease in water quality for BOD, COD, TSS, DO and pH parameters in Batang Kuranji River from upstream to downstream areas caused by the increasing number of settlements in the downstream area. The water quality of Batang Kuranji River in upstream areas is still included in the category of raw water class I (one) which is required as drinking water water and / or other allotment which requires the same water quality as the use. Upper river areas are areas where there are only rice fields, plantations, protected forests as the center of conservation areas, and a small proportion of human settlements.

b. The middle of the transition area also has decreased the quality of river water. The middle area is designated for water with classes II and III in accordance with Government Regulation 82 of 2001. Field monitoring shows that this area is found in many agricultural, livestock, residential, commercial and excavation activities.

c. In the downstream areas, the water quality condition of Batang Kuranji River belongs to the category of class IV water (four) which is designated as water for irrigation, cultivation and or other allotment which requires the same water quality as the use. In this area, there are many settlements and activities that produce liquid waste. With such population data and monitoring results, it is found that downstream areas have decreased the quality of river water exceeding the threshold of class IV PP 82 year 2001.

The decrease of pollution load capacity of Batang Kuranji River River, it is necessary to manage the wastes that enter the river so that water quality will also increase.

a. The pollution load capacity of Batang Kuranji River in the upstream area is included in either category. This is evidenced by the calculation of pollution load using the application of Qual2Kw. For BOD parameters in upstream areas still able to accommodate pollution loads of 4.4 kg/sec, COD parameters 273,60 kg/sec, TSS 906 kg/sec, and for parameter of DO equal to 49,20 kg/sec. While in the downstream area of pollution load capacity based on Qual2kw application modeling included in the category of ugly. These conditions can be seen from the ability of the river to receive pollution loads.

b. In general, central area of Batang Kuranji River is included in bad category. This is indicated by the capacity for BOD and COD parameters to exceed the specified threshold (Table 13). As for TSS parameters on segment 2 exceeds the specified threshold. The calculation result shows that TSS capacity in area 2 is -1.090 kg/sec, this is because in this region class II is defined while the

IOP Conf. Series: Materials Science and Engineering **335** (2018) 012027 doi:10.1088/1757-899X/335/1/012027

activity in this area is C excavation which can lead to the increase of TSS. In areas 3 and 4 the water is designated with class III so that the capacity of the river is increased while the existing activities along this area based on the monitoring results do not significantly increase the TSS content. The activities that exist in this area are agriculture, traditional market, and residential area.

c. The BOD parameters in the downstream area have exceeded the capacity of 674,60 kg / s, COD 413,00 kg / s, TSS 13,179,00 kg / s, and for the DO parameter the incoming pollution load is 90 kg / s. In the downstream area, the total pollution load for the required DO parameters is not required. This is because the threshold on this area is included in the class 4 water quality standard, which DO requirement is 0 (zero). The decrease of pollution load capacity in the downstream area is triggered by the excavation activity C that is the taking of sand, stone, and gravel. So, with this activity there is an increase in TSS concentration in the river.

5. Conclusion

Conclusion of water quality and determination of Batang Kuranji river capacity research are:
On the Batang Kuranji River Upstream: Water quality is in good category and fulfill first class water quality standard 1 (one) according to PP 82 year 2001. TSS concentration in upstream area 21 mg/L, BOD 1,6 mg/L, COD 7,99 mg/L and for parameter DO 7.845 mg/L. The pollution load capacity is in good condition and able to accommodate incoming pollution load for BOD parameter of 4.4 kg/sec, COD parameter 273,60 Kg/sec, TSS 906,00 kg/sec, and DO parameter 49,20 Kg/sec. On the Batang Kuranji River Central: Point 2, TSS concentrations 48 mg/L, BOD 4.15 mg/L, COD 16.35 mg/L and DO 7.04 mg/L which has exceeded the threshold. Point 3, TSS concentrations of 96.5 mg/L, BOD 14.45 mg/L, COD 40.05 mg/L, and DO 6.29 mg/L have exceeded the class III threshold. Point 4, TSS concentration 101,5 mg/L, BOD 18,365 mg/L, COD 50,755 mg/L, and DO 6,115 mg/L have exceeded class III quality standard. The pollution load capacity of Batang Kuranji River in the ugly category, ie, segments 2, 3, and 4 are BOD -857.3 kg/sec, COD -777,40 kg/sec, for TSS +9.511.5 kg/sec, while DO + 69.30 kg/sec. On the Batang Kuranji River Downstream, Water quality of Batang Kuranji River for downstream area included in bad category, is TSS concentration in downstream area that is 734 mg / L, BOD 33,185 mg / L, COD, 105,39 mg / L, and DO concentration that is 5,575 mg / L. The river capacity downstream (section 5) for BOD parameters has exceeded -674.60 kg/sec, COD -413.00 kg/sec, TSS -13.179,00 and DO at 90 kg/sec.

References

[1] Biro Pusat Statistik, 2007 " *Padang Dalam Angka* ", BPS Kota Padang
[2] Bapedalda Kota Padang. 2007. *Penetapan Daya Tampung Beban Pencemaran Batang Arau Padang*
[3] Keputusan Menteri Lingkungan Hidup Nomor 110 Tahun 2003 tentang Pedoman Penetapan Daya Tampung Beban Pencemaran Air Pada Sumber Air
[4] Pelletier G J., Chapra S C., Tao H. 2005. *Qual2Kw – A Modeling Framework For Simulating River and Stream Water Quality (version 2.07)*
[5] Peraturan Pemerintah Nomor 82 Tahun 2001 tentang Pengelolaan dan Kualitas Air dan Pengendalian Pencemaran Air
[6] Undang-Undang Republik Indonesia Nomor 32 tahun 2009 tentang Perlindungan dan Pengelolaan Lingkungan Hidup

IOP Conf. Series: Materials Science and Engineering **335** (2018) 012028 doi:10.1088/1757-899X/335/1/012028

Synthesis and Characterization of C-Cinnamal Calix [4] Resorsinarena from Cinnamon Oil Waste West Sumatra

S B Etika*, E Nasra, I Rilaztika

Chemistry Department Faculty of Mathematics and Natural Science, Universitas Negeri Padang. Jln. Prof.Dr.Hamka Air Tawar Padang, Indonesia

*sribentietika67@gmail.com

Abstract. Synthesis and characterization of compound C-Cinnamal Calix [4] Resorsinarena (CCCR) of cinnamon oil waste have been done. This study was aimed to synthesis and characterize C- Cinnamal Calix [4] Resorsinarena from cinnamaldehyde violated cinnamon oil waste. C-Cinnamal Calix [4] Resorsinarena was synthesized by electrophilic substitution reaction of cinnamaldehyde isolated by the acid and resorcinol at 77oC temperature for 2 hour. The data analysis spectrum UV-VIS and FT-IR showed that the compound isolated cinnamaldehyde same as pure cinnamaldehyde compound. The characterization of C-Cinnamal Calix [4] Resorsinarena in the form of reddish-colored solids with melting point 3580C by using UV-VIS showed the presence of double bond, FT-IR showed the absorption at the wave number 3323,94 cm-1 indicating the –OH group, the wave number 1610,94 cm-1 showed the vibration C=C, the strong region absorption of 1500,86 cm-1 indicating the presence of an aromatic ring, the at 1442,88 cm-1 wave number indicating the presence of CH3.

1. Introduction

Environmental issues related to water pollution have been a hot topic in recent decades. One of the water pollution is cause by bad industrial disposal. The waste will cause the damage of the environment (Harera, 2015). One of the environmental parameters associated with water pollution is the heavy metal.

In the last decade, determination of low levels of heavy metals in sea-water has received particular attention as a way to assess the early impact of human activities on the marine environment. The determination of trace elements in sea-water is difficult. Electrothermal Atomic Absorption Spectrometry (ETAAS) and Inductively Coupled Plasma Atomic Emission Spectrometry (ICP-AES) have been widely used for the dertmination of trace elements in different samples because the available data are highly precise and accurate; however, with se-water, the co-volatilization of matrix salts produce severe interference owing to intense background absorption signals. Also, ICP-AES methods sometimes suffer from problems with signal suppression and clogging of the sample introduction system when the sample contains dissolved solids at concentrations greater than 0.2 % m/v. further, ICP-AES methods are less sensitive than ETAAS. (Rudner, P.C et al, 1998)

The use of preconcentration step is one of the more effective ways of increasing sensitivity. One of these preconcentration methods used to measure the heavy metals in very small concentration is solid phase extraction. Solid Phase Extraction (SPE) is one of the concentrating techniques through the sorption (adsorption-desorption) process which is still being developed.

The previous research used solid phase extraction methods with resin and'"ligating bracelets" as the supporting polymer material that can be used for copper ion analysis. But, the supporting polymer materials used in that research is expensive and limited.

A research based on natural products as a supporting polymer material for retention of heavy metals. One of can be an alternative to overcome the problem the natural products that can be used is cinnamon oil. Cinnamomum vermicelli (Cinnamomumverum) is derived from plants growing in different countries in Asia, such as Indonesia, West Sumatra. There are many benefits of cinnamon, such as immune system.

Cinnamon comes from Indonesia, which in trade is better known as cassiavera. Indonesian cinnamon is well known having a good smells. The main product of cinnamon is the outer skin of cinnamon used as a spice for food seasoning. In addition, the other products of cinnamon are powder, cinnamon essential oil and cinnamon oleo resin which is widely used in food beverage, pharmaceutical and cosmetics industry (Guenther,1972). The chemical comcomposition of with cassia, microscopically is simillar. Sclerenchyma cells in the cassia are present in the outer shell, The inner sclerenchyma cells are thicker and the thickening is detected in the cassia skin. The outer skin cells of cork are smaller than the inner skin. The amount of starch particles on the skin of the two types of cinnamon is equal, but slightly larger in cassia.

The main components of cinnamon are cinnamaldehyde, eugenol and some aldehydes in relatively small percentage quantity. Essential oils can be obtained by steam distillation and water for 4 -5 hours. First, the material in destillationprocest is chopped making oil evaporation faster. The wet and dry materials has no effect to the oil component. The highest essential oil concentration is in the stem, but there is also in the root leaves, and flowers in small percentege. The chemical composition in the volatile oil of cassia Indonesia are cinnamaldehyde, cinnamalacetate, salicylaldehyde, cinnamic acid, salsilic acid, ortho-methoxy, benzaldehyde (methyl salisaldehyde), methyl-orto–curamaldehyde (o-methoxy Cinnamaldehyde) and phenylpropyl acetate. The conectation of Indonesian cassia cinnamaldehyde is 80%-90%. This cinnnamaldehyde is a yellow liquid, volatile, slightly soluble in water, soluble in alcohol and ether, insoluble in petroleum ether and has a molecular formula C8H7CHO with Mr. 132,15 (Guenther, 1987).

Cinnamaldehyde was isolated from cinnamon essential oil in 1834 by Dumas and peligot and syntesis in the laboratory by the Italian chemist Luigi Chiozza in 1854. The natural product was trans-cinnamaldehyde. The cinnamaldehyde molecule consists of a benzene ring attached to an unsaturated aldehyde. Many synthesis methods are used in the laboratory, but the most efficient method is the distillation of cinnamon bark oil vapor. Cinnamaldehyde is a cinnamic derivative with is a major component of cinnamon oil. The chemical structure of cinnamaldehida consists of a benzene substituted by a conjugated carbonyl system. Cinnamaldehyde can be isolated from cinnamon using sodium bisulfite solvent. Some of the benefits of the cinnamaldehyde is flavoring in chewing gum, ice cream, candy, and beverage drinks. Usage levels range from 9 to 4900 ppm. It is also used in some natural, sweet, or fruit fragrance cinnamaldehyde can be used as food seasoning, it can be marketed as cinnamon powder.

The cinnamaldehyde reacted with resorcinol will give C-Cinnamal Calix [4] Resorsinarena (CCCR). CCCR has twelve benzene residues, eight hydroxyl groups, and four alkenyl groups (double bonds).

C-Cinnamal Calix [4] Resosinarena (CCCR) is a cyclic macromolecule compound contained aromatic group. The aromatic groups are connected to each other through a methylene bridge forming a cavity with an active group therein, this compound to be used as a host molecule for another guest, either anion, cation, or neutral compound. Based on the reaction mechanism, the CCCR synthesis reaction is an electrophilic substitution reaction (Handayani, 2014). The reaction between cinnamaldehyde and resorcinol produce C-Cinnamal Calix [4] Resorsinarena. CCCR has twelve benzene residues, eight hydroxyl groups, and four alkenyl groups (double bonds). The presence of electron pair in the hydroxyl group, as well as the presence of electrons in aromatic residues and double bonds is predicted to have special activity on heavy metal cations.

IOP Conf. Series: Materials Science and Engineering **335** (2018) 012028 doi:10.1088/1757-899X/335/1/012028

The syntesis of CCCR from cinamon oil can be used to retain heavy metal ions by solid phase extraction methods. The purpose of this study was to synthesize CCCR from cinnamaldehyde in cinnamon oil and characterize CCCR with FTIR, UV-VIS, RMI and Melting Point.

2. Experimental section

2.1. Materials and Equipments

The equipmens used in this research are reflux set, buchner funnel, FTIR (Jasco 460 plus Spectroscopy), rotary evaporator (HeidolphLaborota 4000), Desiccator,Ultravioletspectroscopic Agilent 8453 UV-Vis. The materials are NaOH (merck), HNO3 (merck), resorcinol (merck), cinnamaldehyde, HCl 5%, ethanol (merck), diethyl ether (merck), NaHSO3 (merck), Ether (merck), Na2SO4 (merck), Sweetwood Oil (PT STS).

2.2. Research Procedures

2.2.1. Isolation of cinnamaldehyde from cinnamon oil. Isolation of cinnamaldehyde from cinnamon oil has done by adding saturated sodium bisulfite until there is no more precipitation. The precipitate does not increase. The precipitate was filtered and washed with ethanol 10 ml and diethyl ether 10 ml. Then, the precipitate was added HCl 5% 250 ml and refluxed at 60 ° C for 30 minutes. The mixture was twice extracted with ether and the extraction coating was combined and dried with anhydrous Na2SO4. After the filtration, the solution was evaporated by using a rotary evaporator and produced cinnamaldehyde

2.2.2. Characterization of Cinnamaldehyde. Further examination, the double bond of the isolated cinnnamaldehyde was confirmed by UV-VIS Spectroscopy, the functional group was confirm by using FTIR, Isolated cinnamaldehyde the amont of hydrgen atoms used by RMI, was confirmed the boiling point used melting Point.

2.2.3. Synthesis of CCCR. The isolated cinnamaldehyde was mixed with a resistor with a ratio (1: 1) in ethanol 90% 50 ml and added of HCl 5 %. The solution was stirred and refluxed at 77 ° C for 2 hours. The precipitate was filtered and washed by using 10 ml ethanol and 10 ml of water and dried to obtain CCCR solids of 3.245 g.

2.2.4. CCCR Characterization. The double bonds in CCCR synthesis was confirmed by using UV-VIS spectroscopy, the functional groups was confirm by using FTIR, the number of hydrogen was confirm by using RMI and the melting point was conform Melting point.

3. Results and Discussion

3.1. The isolation of cinnamaldehyde from Cinnamon Oil

The cinnamaldehyde in the concentrated high cinnamon oil (51.6% - 56.0%) gives the easing isolation process take place quite easily. The isolation result was 8,234 g with 27,72% from 30 g of cinnamon oil rendemen.

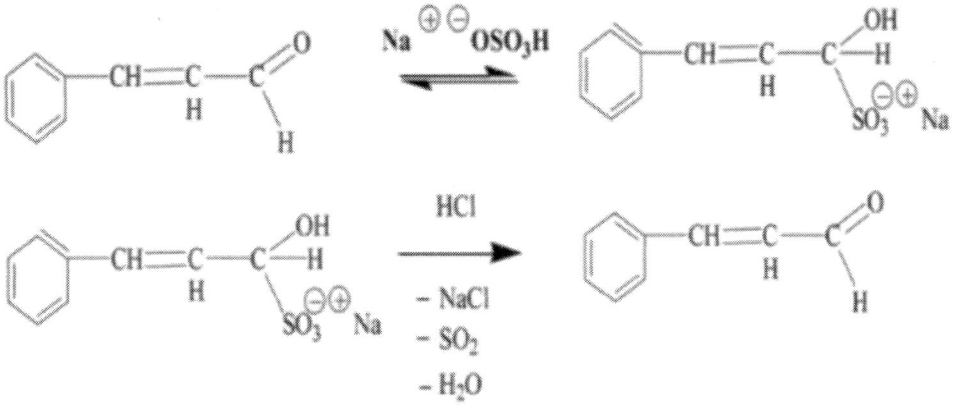

Figure 1. The reaction of cinnamaldehyde with sodium bisulfite (Ngadiwiyana, et al, 2004).

In Figure 1, the cinnamaldehyde can be removed from cinnamon oil by addition of saturated sodium bisulfite. Bisulfite the product is a salts and the reaction is reversible so reaction, the aldehyde can be produced by the addition of acids (Anam, Khairul, 2004).

3.2. The Characterization of Isolated Cinnamaldehyde

The isolation of cinnamaldehyde from30 g of cinnamon oilgives 8.234 g, 27.72 %. ^{1}H-NMR spectrum (400 MHz, acetone) δ 5.6 (1H, dd), 6.5-6.7 (1H, m), 7.0 (1H, d), 8.7 (1H, CHO, d).

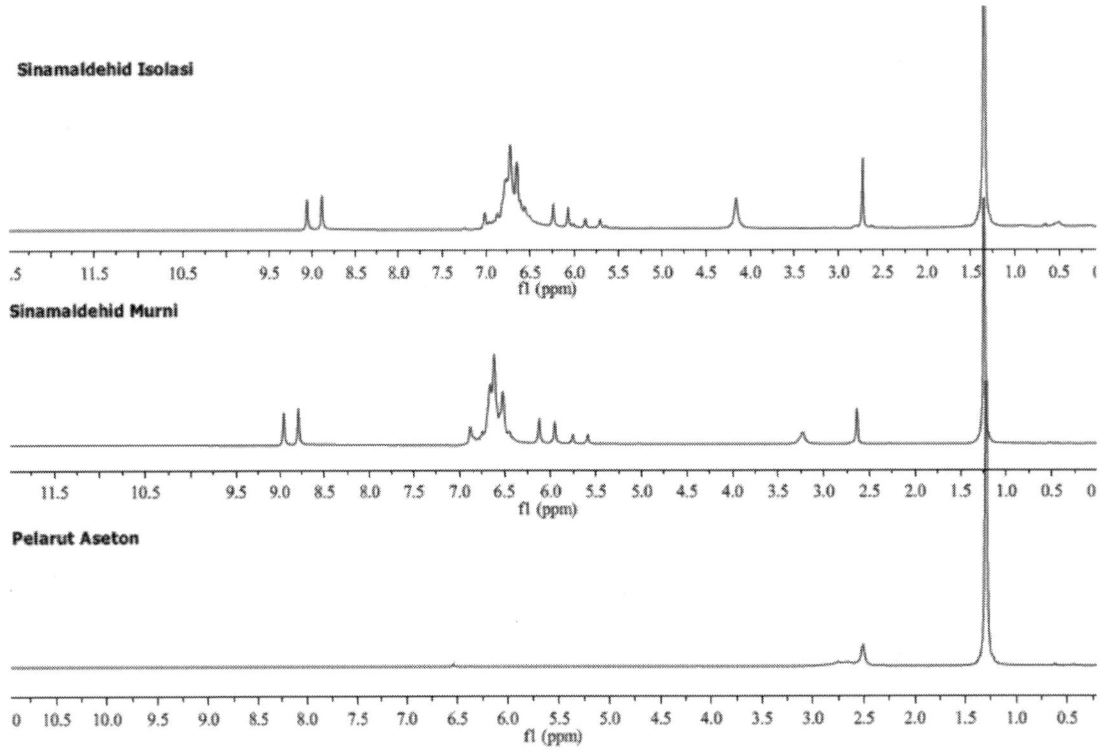

Figure 2. The ^{1}H NMR of Cinnamaldehye from isolation, pure, and the NMR solvent

The cinnamaldehyde was characterized by FTIR. From the IR spectra in Figure 3 there is sharp peak at 1668,72 cm^{-1} wavelength indicating the presence of carbonyl uptake (C=O). These structurally carbonyl clusters are conjugated with double bonds. The wave number 2929.48 cm^{-1}referes to (CH_2). The wave number 3044,89 cm^{-1} refers to (C-H) and amplify the isolation result. The existence of an aromatic ring is indicated by the wave number 1450.60 cm-1. The presence of a bond (C=C) is indicate by data on the fingerprint region in absorption of wave number 689.09 cm-1 (Sastrohamidjojo, 1992). This result is the same as the pure cinnamadelhyde FTIR spectrum.

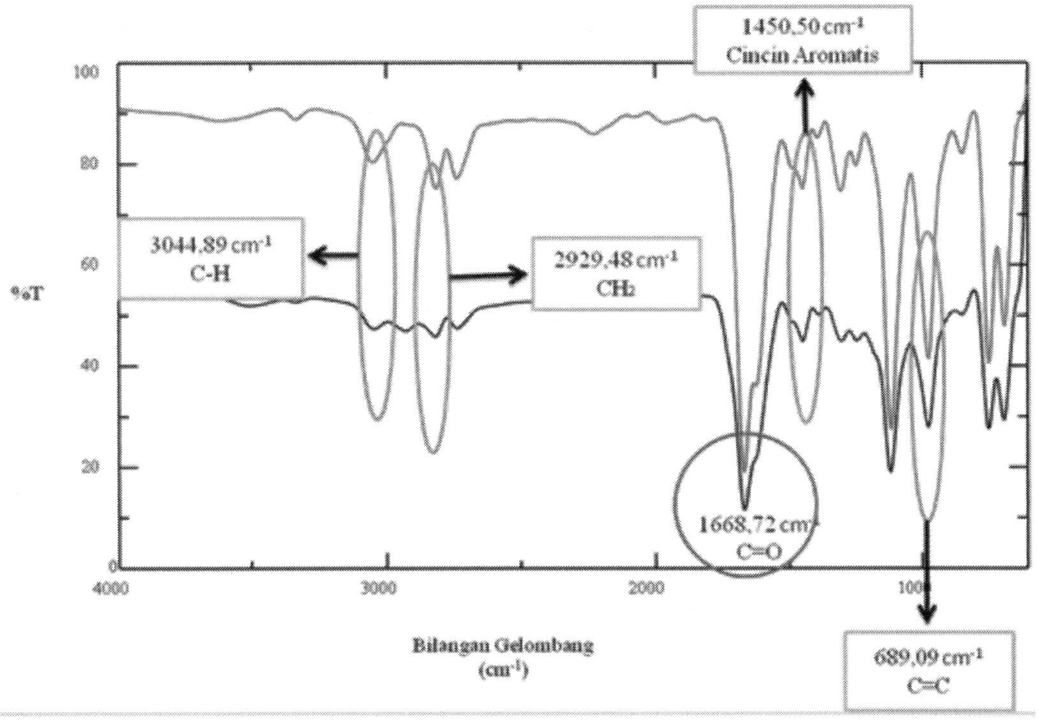

Figure 3. The IR Spectrum Cinnamaldehyde

In the other side, UV-VIS spectroctroscopy measurements of cinnamaldehyde were absorbed at 320 nm wavelengths with an absorbance of 1,72 indicating the presence of the conjugated double bond of the cinnamaldehyde compound shown in Fig 4. The results are similar to the UV-VIS cinnamaldehyde spectrum constructed from cinnamaldehyde pure.

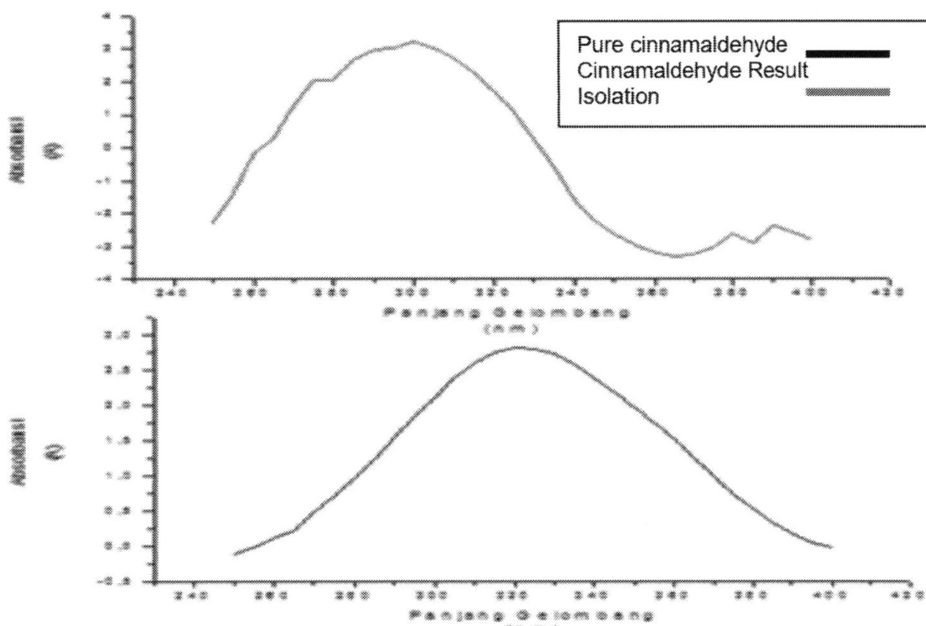

Figure 4. The UV-Vis Spectrum of cinnamaldehyde

3.3. The Synthesis of CCCR

Thesynthesis of CCCR is done by reacting cinnamaldehyde with resorcinol in the acidic atmosphere, (Figure 5). The synthesis of CCCR is prepared at temperature 77°C with in 2 hours.

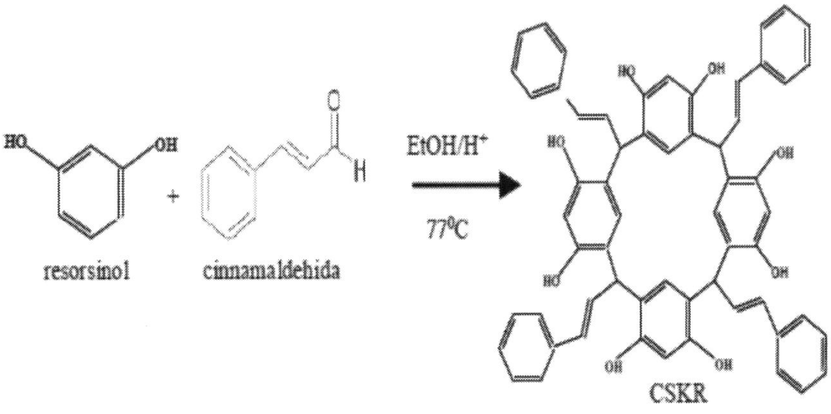

Figure 5. Synthesis of CCCR (Sardjono, 2008)

The synthesis product is directly removed from the reaction mixture as a reddish solid as much as 3.245 g. The CCCR synthesis results were characterized using FTIR, UV-VIS spectroscopy and melting Point.

3.4. The Characterization of CCCR

The synthesis of CCCR gives 3.245 g as a reddish solid. ^{1}H NMR spectrum (400 MHz, acetone and ethyl acetate) δ 0.7 (1H, d), 1.3 (1H, d), 2.5 (1H, s), 2.7 (1H,s), 6.6 (1H, OH).

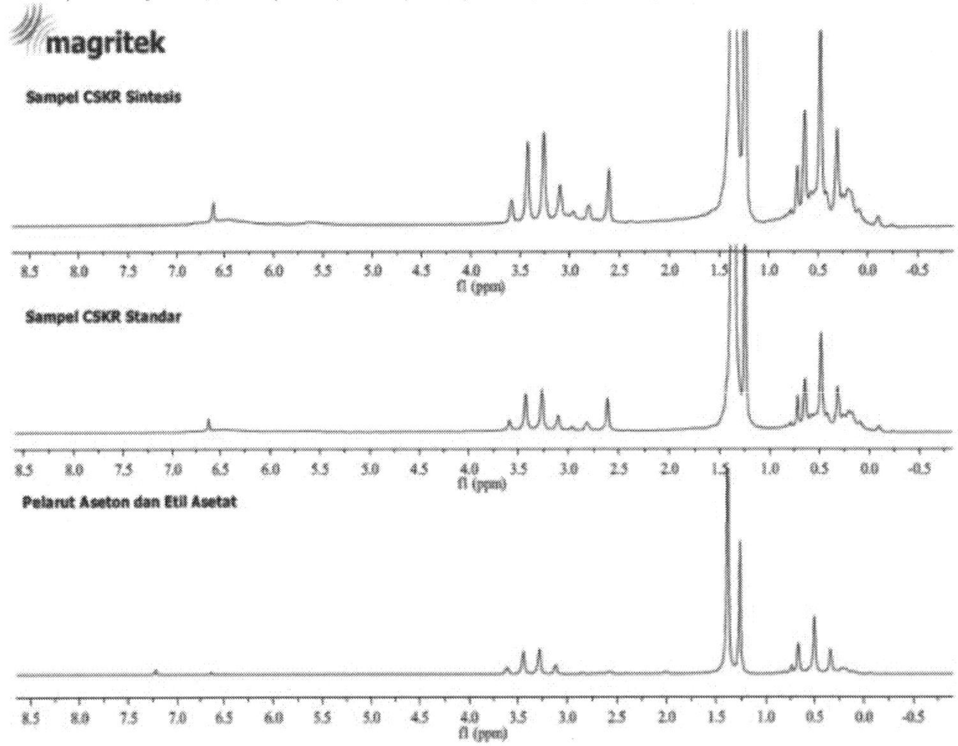

Figure 6. The ^{1}H NMR of synthesized CCCR, pure, and the NMR solvent

IOP Conf. Series: Materials Science and Engineering **335** (2018) 012028 doi:10.1088/1757-899X/335/1/012028

Thecharacterization of CCCR with FTIR shows a wide and strong peak at 3323.94 cm^{-1} wavelength indicating O-H span of vibration. The wave number 1610.94 cm^{-1} shows the vibration of (C = C). A strong absorption area of 1500.86 cm^{-1} denotes the existence of an aromatic ring. Furthermore, the IR spectrum also features absorption bands derived from vibrations (CH_2) located on the fingerprint 695.52 cm^{-1} - 839.60 cm^{-1} (Sastrohamidjojo, 1992), as shown in Figure 5. Characterization of CCCR of pure of cinnamadehyde shows the same results as CCCR from the synthesis,

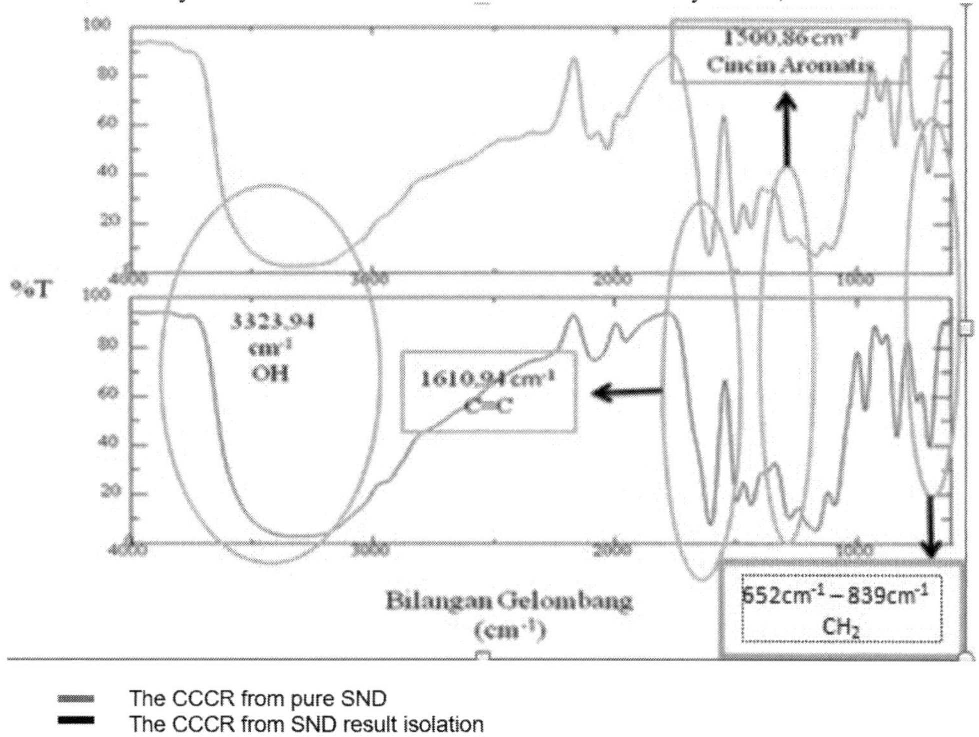

Figure 7. The Infrared Spectrum of CCCR

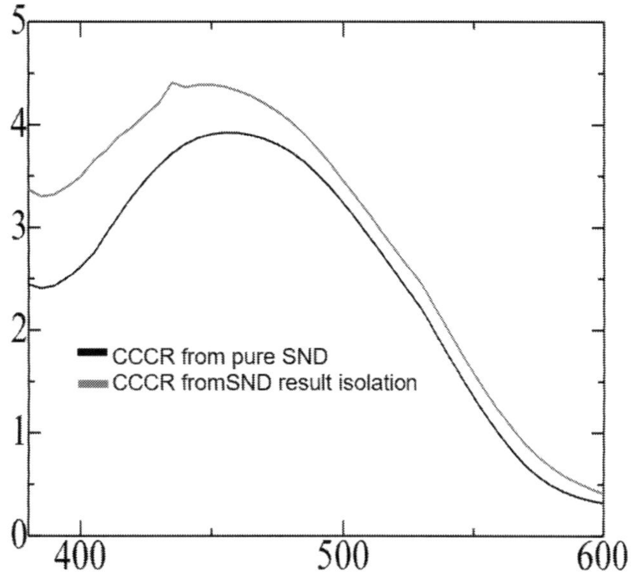

Figure.8. The Spectrum UV-VIS of CCCR

To support the result data of from cinnamaldehyde isolation, the melting point measurements were compared with the synthesis result from pure cinnamaldehyde. The synthesized CCCR melting point is 358 ° C and the melting point of CCCR of pure cinnamaldehyde is 359°C. In the other side, UV-VIS spectroscopy measurements of CCCR were absorbed at 440 nm wavelengths with an absorbance of 4.365 indicating the presence of the conjugated double bond of the CCCR compound shown in Figure 8. The results are similar to the UV-VIS CCCR spectrum constructed from cinnamaldehyde pure.

4. Conclusion

Based on the results of the synthesis and Characterization of C-CinnamalCalix [4]Resorsinarena from Cinnamon Oil, the conclusion are: 1. The isolation cinnamaldehyde produced as much (27,72%) 8,324 g from 30 g and 2. Synthesized CCCR produced as much as 3.245 g in the form of reddish solids

References

[1] Anam, Khairul. 2004. Pemanjangan Sistem konyugasiCinnamaldehid dan Uji Aktifitas Sebagai Bahan Aktif Takbir Surya. JKSA. Vol. VII. No. 1

[2] Guenther, Ernest.1987. Minyak Atsiri, Jilid 1. Jakarta: Universitas Indonesia.

[3] Handayani, Desi Suci. 2014. Sintesis Dan Karakterisasi C-HeksilKaliks [4] Resorsinarena Dari Kondensasi Resorsinol – Heptanal. Alchemy jurnal penelitian kimia. Vol. 10 No 1. Hal 49-53

[4] Harera, Lingga Rizal, et al. 2015. Sintesis Cu(II)-Imprinted Polymers Untuk Ekstraksi Fasa Padat Dan Prakonsentrasi Ion Tembaga (II) dengan Ligan Pengkhelat 4-(2-Pyridylazo) Recorcinol. Al Kimiya. Vol. 2 No. 1, 30-39.

[5] http://www.sciencelab.com/msds.php?mdsId=9923482

[6] Is Fatimah. 2000. Penggunaan Na-Zeolit alam Teraktivasi sebagai Penukaran Ion Cr+3 dalam Laritan. LOGIKA. Vol. 4 No 5.

[7] Kumar, G Mahesh, dkk. 2014. Inductively Coupled Plasma Atomic Emission Spectroscopy: An Overview. Internasional Journal of Pharmaceutical Research &Analysis. Vol. 4. No 8. Hal 470 - 477

[8] Meeviro. 2010. Logam berat dan lingkungan. Diakses 17 januari 2012.

[9] Ngadiwiyana, et al. 2004. Pemanjangan sistem terkonjugasi CinnamalCinnamaldehid dan uji aktivitas sebagai bahan aktif tabir surya. KSA. Vol. VII, No. 1, 24-29

[10] Sardjono, Ratnaningsih E et al. 2008. Sintesis Kaliks [4] Resorsin Arena DariMinyak Kayu Manis Dan Penggunaannya Untuk Ekstraksi FasaPadatLogam Berat Hg(II) dan Pb(II). Jurnal Pengajaran MIPA Vol.12 No.2.

[11] Sastrohamidjojo, H. 1992. Spektroskopi. Yogjakarta: Liberti

[12] Sholikh, RiskaAzizawati. 2013. Sintesis Dan Karakterisasi C-4-Hidroksi-3-Metoksifenil Kaliks [4] ResorsinarenaSulfonat. Alchemy Jurnal Penelitian Kimia. Vol 9 no 2. Hal 13-21.

[13] Rudner, P. C; Pavon, J. M. C; Rojas, F. S and de Torres, A. G. 1998. Use of Flow Injection Cold Vapour Generation and Preconcentration on Silica Functionalized with methylthiosalicylate for The Determination of Mercury in Biological Samples and Sea-water by Inductively Coupled Plasma Atomic Emission Spectrometry. J. Anal. At. Spectrom, 13, 1167-1171.

Solar Cell Polymer Based Active Ingredients PPV and PCBM

H Hardeli*, H Sanjaya, R Resikarnila, R Nitami H

Chemistry Department, Universitas Negeri Padang, Indonesia

*hardeli1@yahoo.com

Abstract. A polymer solar cell is a solar cell based on a polymer bulk heterojunction structure using the method of thin film, which can convert solar energy into electrical energy. Absorption of light is carried by active material layer PPV: PCBM. This study aims to make solar cells tandem and know the value of converting solar energy into electrical energy and increase the value of efficiency generated through morphological control, ie annealing temperature and the ratio of active layer mixture. The active layer is positioned above the PEDOT:PSS layer on ITO glass substrate. The characterization results show the surface morphology of the PPV:PCBM active layer is quite evenly at annealing temperature of 165 ° C. The result of conversion of electrical energy with a UV light source in annealing samples with temperature 165 ° C is 0.03 mA and voltage of 4.085 V with an efficiency of 2.61% and mixed ratio variation was obtained in comparison of P3HT: PCBM is 1: 3

1. Introduction

Solar cell is very suitable to be developed in Indonesia considering the location is very strategic that is located on the equator, where the intensity of sunlight is quite large. Indonesia is classified as a rich source of solar energy with a total intensity of irradiation on average 4.5 kWh per square meter per day, compared with Japan whose total intensity of irradiation is only 150-180 Wh per square meter per day. In addition, as it lies on the equator, the sun in Indonesia shines about 2,000 hours per year (Suherdiana, 2008). The utilization of solar energy in Indonesia is only limited to natural benefits, so very much from the energy entering the surface of the earth re-reflected into outer space (Mairizwan, Hendro, 2015).

Solar cells work using solar energy by converting solar radiation directly into electricity. Today's most widely used solar cells are silicon which is the result of the rapid development of inorganic semiconductor technology. Although solar cells are now dominated by silicon materials, the high cost of production makes them more expensive than fossil energy sources. For that, cheap solar cells with high cell performance and organic solar cells or polymer solar cells can be a solution. These cells are easily made from organic material, inexpensive, lightweight, flexible and colorful (Soetrisno, 2009).

The P-Phenylene Vinylene (PPV) polymer is a conjugated conductive organic polymer having a single double bond in the form of a repeating unit of combined benzene rings and transacetylene bonds. This type of bond allows the electrons to be delocalized along the polymer chain, thereby providing the semiconductor properties of the polymer. PPV has a conductivity of 10-9-10-13 S / cm (Bradley 1987).

PPV is often used as a material model to obtain a basic understanding of photophysic conjugated polymers. PPV polymers readily dissolve in ordinary organic solvents and can be made in the form of films by spin-coating techniques (Bahtiar, A. 2009). The structure of PPV can be seen in figure 1.

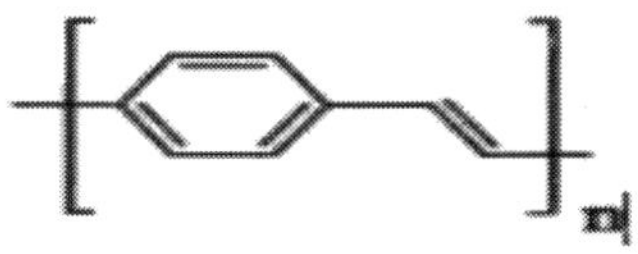

Figure 1. PPV Structure (Baskoroadi, G. 2011)

PPV has a ray absorption at a wavelength of 450-550 nm. The PPV material with wavelength absorption is likely to be solar cell material, since sunlight has the highest energy on the spectrum with a wavelength of 500 nm (Nurosyid, 2014). PPV polymeric materials are called electron donor materials (Aernouts 2006) and are also called p-type semiconductors. Orbital energy at -5.3 eV for HOMO (Highest Occupied Molecular Orbital) and LUMO (Lowest Unoccupied Molecular Orbital) at -3.2 eV (Oku et al., 2008).

The characteristics of photovoltaic effects of PPV solar cells with the addition of polymer [6.6] - phenyl C61 butyric acid methyl ester (PCBM) were also observed to obtain better organic solar cells. The PCBM polymer material was chosen because it has the same solvent as PPV and is acceptor (Aernouts, 2006) and also includes n-type semiconductors. PCBM has HOMO orbitals - 5.1 eV and LUMO - 3.7 eV (Zahlou et al., 2014). The structure of PCBM can be seen from Figure 2.

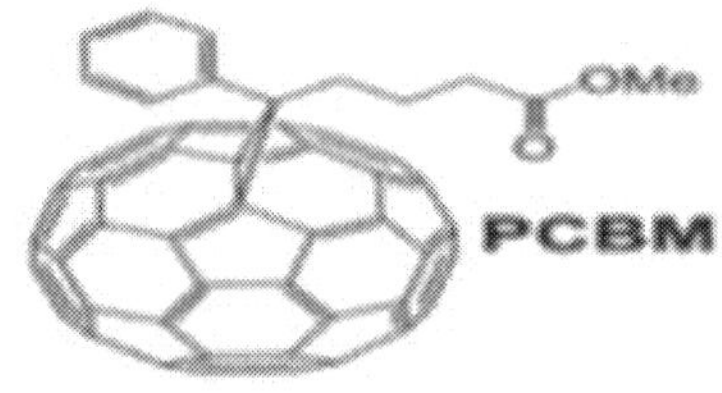

Figure 2. PCBM Structure (Bahtiar, A., et al. 2011).

Studies conducted on polymer solar cells typically use only one layer of polymer coating. therefore in this research will be done variation of PPV polymer coating amount: PCBM as effort to control morphology, that is 1,2, 3 and 4 coating. Thus, the expected polymer solar cells are capable of producing high efficiency.

2. Research Methodology.

A. Materials and Tools

Equipment used is XRD (X'Pert Powder PAN Alytical Pw 3060/40), FTIR (Perkinelmer type: FTIR Frontier Spectrometer), UV-VIS Agilent 8453, Digital Multimeter, 250 mL glass, measuring cylinder, dropper dropper, 5 and 10 mL, stirring rod, magnetic stirer, spatula, petri dish, lumpang alu, tape, separating funnel, furnace and digital multimeter

The materials used are TiO2 Degusa P-25, methanol pa, acetic acid pa, ethyl acetate pa, aquades, n-hexane pa, purple sweet potato, paper whatman, KI (potassium iodide), I2 (iodine), acetonitrile pa, cetyl trimethyl ammonium bromide, pencil powder, ITO (Indium Tin Oxide) glass, and polyethylene glycol (PEG), PPV (poly [2-methoxy-5- (2-ethyl-hexyloxy) -1,4-phenylene) PCBM ([6,6] -phenyl-c61-butyric acid methyl ester), PEDOT (poly (styrenesulfonate), PSS (poly (3,4-ethylenedioxythiophene).

B. Experiments

1. ITO lithography

The ITO substrate prior to use should be in a clean and sterile state of the impurity particles then the substrate was washed glass substrate soaked in a glass of water containing soapy water, then wash thoroughly. The glass is immersed in 50 mL of methanol and disonikasi with ultrasonic cleaner for ± 120 minutes. Then the glass is cleaned with aqudes and reconstituted in an ultrasonic cleaner for ± 120 minutes. Cleaned glass is stored in a bottle containing aquades for fat and dust free. Spray (spray) aquades onto the glass surface so no clumps of water are attached. Glass is dried in a vacuum oven at ± 110 ° C for ± 60 minutes when it is to be used

2. Determining the Annealing Temperature of Making Polymer Thin Films

The ITO mask is worked on the part that will not be removed using adhesive tape while the part to be removed is immersed into 50% HCl for ± 15min. 3 minutes.±3 minutes and water for ±Then the sample was dipped into 3 NaHCO3 for After that the samples were dried by removing the mask from the sample and ultrasonic cleaning using water, IPA (isopropyl alcohol), and aceton for 10 minutes.± Above the ITO layer is then superimposed PEDOT paste: PSS using screen printing techniques. The coating is then heated in a vacuum oven at 120°C for 60 minutes. PEDOT: PSS is used as a hole transporter and exciton blocker, and prevents ITO diffusion into the active coating polymer. then above the PEDOT layer: PSS is regenerated the active layer of P3HT polymer: PCBM. A total of 200mg of P3HT was dissolved into 10ml chlorobenzene and stirred until homogeneous solution, and also 200mg. PCBM dissolved into 10ml chlorobenzene is stirred until homogeneous solution. Using a pipette plastic, the two solutions are mixed into one bottle and stirred until homogeneous. Coating process using spin coating technique at 600rpm rotation speed for 30 seconds with active area of 2.6 cm2. Furthermore, the polymer layer is left overnight in the nitrogen environment. After that the annealing process of the active layer at a temperature of 120 0C and 150 0C using a vacuum oven, then deposition metal contact Al. Al contact coating is done by thermal evaporation technique. Then the capsulation is done by covering the top surface of the cell with glass using a sealant as its adhesive medium and then heating at a temperature of 120 °C for 10 minutes. To determine the I-V characteristic of polymer solar cells, measurements were made by irradiating the cells using a Xenon light source on a 60 mW / cm2 light irradiation and a temperature of 25 °C.

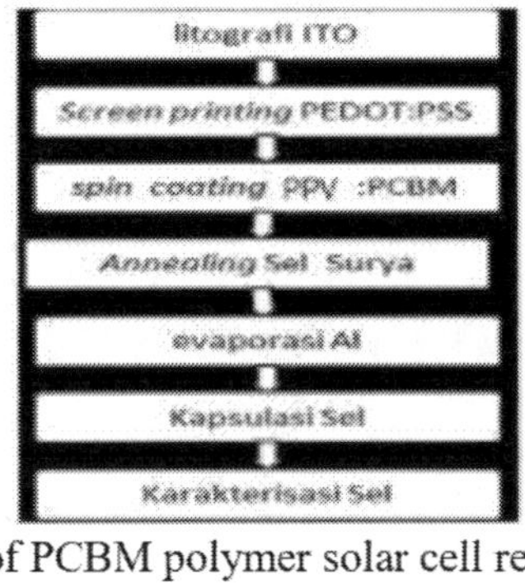

Figure 3. Flow chart of PCBM polymer solar cell research procedure: PPV

3. Conducting Morphological Control of Active Layers Through Annealing Treatment At Different Temperature Range

At sharing annealing temperatures SEM characterization was performed to obtain an active layer morphology image. Samples that are annealed at different temperatures will provide different surface morphologies. At this stage will be seen the surface of the annealing at what temperature is most homogeneous and evenly distributed

4. Preparation of mixture PPV: PCBM

Preparation of mixtue PPV: PCBM (3: 1)

A total of 30 mg of PPV was dissolved into 3 ml of chlorobenzene and stirred until homogeneous solution, and also 10 mg PCBM dissolved into 1 ml of chlorobenzene stirred until homogeneous solution. Both solutions are mixed and stirred until homogeneous (Pratiwi, Z. R. 2013).

5. Preparation of PEDOT: PSS

PEDOT: PSS weighed as much as 0.05 g, then dissolved in 1.5 mL of aquades, then stirred until a paste was formed

6. Preparation of Carbon Elektoda Counters
As the carbon source used graphite from pencil. The graphite is superimposed onto the ITO substrate on its conductive part and then heated at 450 ° C for 10 minutes to allow the graphite to form a good contact of carbon particles with the ITO substrate (Damayanti, Retno 2014)
7. Characterization of Thin Film on Glass Surface
a. SEM / EDAX
Characterization is done by using SEM / EDAX to see the surface morphology and elements present in thin films. The XRD characterization was performed to see the crystal structure of thin films on the glass surface as well as to determine the size of the crystals obtained. The size of the crystals in nanometer size will surely greatly help the production of the resulting solar cell. UV-Vis will be used to obtain the optical characteristics of the active layer of polymer solar cells
b. XRD
Characterization with XRD was performed to determine the crystal structure of the TiO2 catalyst powder by knowing the sample peaks and comparing with the standard peaks of the Hanawalt index. The characterization was done in Physics Laboratory of FMIPA UNP, using Philips PW 1710 instrument equipped with PW channel control voltage 40 kV, current 30 mA and diffraction angle velocity 2Φ = 200-1000 with observation speed 2,40 / minute.
8. Solar Flow Testing of Solar Cells
In solar cells that have been assembled testing, namely direct testing of measured voltage and current from solar cells using a digital multimeter. Light source used is direct sunlight. From the value of the current and the voltage obtained can be calculated the efficiency value of the solar cell made.

3. Result and Discussion

1. Solar Cells Based on Polymer Blends PPV: PCBM
Solar cells based on polymer blends can be produced by growing the active layer on a glassed ITO substrate glass. Components of solar cells made of ITO substrate lithography, paste PEDOT:PSS, active layer PPV: PCBM and Carbon electrode counter.

The ITO substrate resistance used in this study is 100 Ω and PEDOT: PSS layer that has been made in the form of dark blue paste. The active layer created by mixing PPV with PCBM has a brownish red color. This solar cell uses a carbon electrode made by superimposing a pencil powder (graphite) on the ITO glass substrate surface.

Solar cell with PPV active layer: PCBM can produce electric current because PPV and PCBM are polymer or organic semiconductive materials. In PPV and PCBM there is a double bond that is conjugated, so that the more bonds π, the greater the current generated. PPV and PCBM are capable of absorbing light from UV lamps and generating electrons during UV light on the surface of solar cells, then electrons will flow through a pair of electrodes ie ITO as the cathode to the carbon anode so as to produce an electric current that will be read on a direct test with a multimeter.

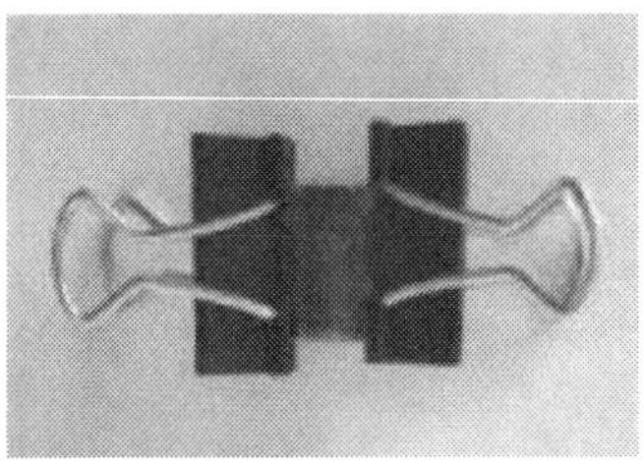

Figure 4. The hybrid polymer solar cell device is in reverse structure

The solar cell device is encapsulated by clamping the structure of the solar cell with the binder clips on the off set on the left and right sides. This polymer solar cell is tested for its ability to convert UV light that serves to irradiate solar cells into electrical energy. UV rays are used with intensity 24 W and then characterized the active layer morphology of PPV polymer mixture: PCBM.

A. Characterization of PPV Active Layer: PCBM with SEM

Samples characterized by SEM were polymer active layer PPV:PCBM which has been deposited on glass substrate ITO. The result of SEM characterization in the form of morphological picture of sample surface can be seen in Figure 5 below. Characterization is done at magnification up to 40,000 times.

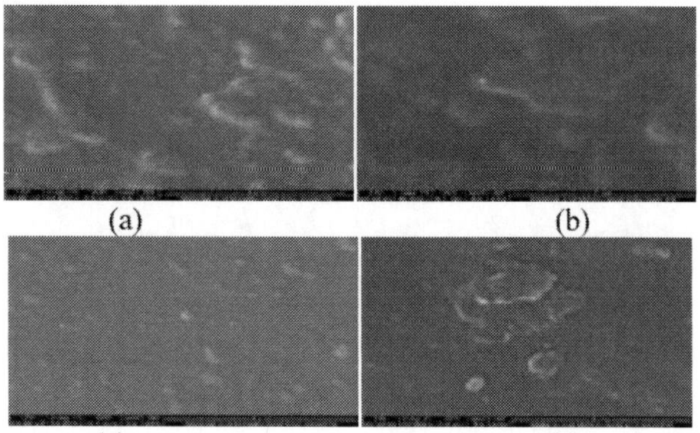

Figure 5. SEM results of PPV active layer: PCBM coating (a) 2 coatings, (b) 3 coatings, (c) 4 coatings and (d) 5 coatings.

The SEM results show the homogeneity of the surface of the active layer PCBM:PPV (3: 1) deposited above the thin layer of ITO, the most flat layer of the coating active layer 4 times, which only slightly holes appear.

The SEM results in the active coating of 2 and 3 coatings were obtained not too flat. In coating 2 and 3 coatings can be seen there is a stacking section. This is due to the incomplete growth of the coating during the spin coating process so that the polymer solution has not formed a flat layer. Another factor that causes incomplete layer is the rotational speed of the spin coating. In the coated 5-layer coating sample, there were more stacking parts compared to the active coating of 2, 3 and 4 coatings. This is because the thickness of the coating to get a layer that is not too flat.

B. Properties of Solar Cells Polymers

Polymer solar cells that have been successfully assembled then carried out direct testing, namely the ability to convert energy with the help of light sources used are UV lights. The irradiation performed with a large intensity of 24W / m2. Polymer solar cells successfully convert into electrical energy which is indicated by the value of voltage and electric current.

Treatment of annealing to active layer PCBM: PPV is carried out at 120 ° C, 135 ° C, 150 ° C, 165 ° C and 180 ° C above the transition glass temperature, where the glass transition temperature for PPV is about 110 ° C. When the process of annealing active polymer molecules will be free to move in the film, so PCBM will relax and re-structure the process (Pratiwi, Z.R. 2013). The resulting voltage of the solar cell on the annealing treatment can be seen in Table 1.

Table 1 Results of voltage measurement on PCBM: PPV (1: 1)

Annealing Temperature	Voltage (V)	Current Power (mA)
120 °C	1,077	0,01
135 °C	1,081	0,01
150 °C	4,009	0,02
165 °C	4,032	0,02
180 °C	1,086	0,01

Table 1 shows that the optimum annealing temperature for polymer blends PCBM:PPV is 165 0C with the highest voltage of 4,032 V and a current current of 0.02 mA. The resulting voltage decreases

IOP Conf. Series: Materials Science and Engineering **335** (2018) 012029 doi:10.1088/1757-899X/335/1/012029

at temperatures above 165 ° C, this is due to the effect of photon energy absorption on the active layer. Yunzhang lu, et al in Pratiwi, R.Z (2013) states that close to 200 ° C the active layer will undergo a significant color change which states the polymer reaches its melting point.

The strength of the resulting current is still relatively small. Cocima, et al (2012) states that the current strength is influenced by the photon light source. Small photon sources will produce a small current because only a small amount of photon energy is absorbed by the active layer. The small amount of current generated is also caused by a hole or stack on the active layer of PCBM: PPV that can inhibit the absorption of photons.

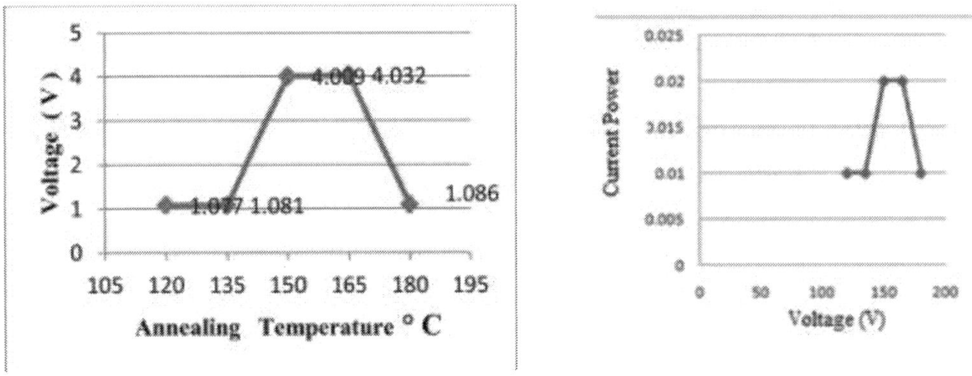

Figure 6. Annealing temperature comparison curve to (a) voltage (V) and (b) current strength

The annealing treatment is carried out to a temperature of 180 °C, Yunzhang Lu, et al in Pratiwi R.Z (2013) states that at temperatures close to 200°C there is a color change in the active layer which states that the active ingredient polymer reaches its melting point. Based on the measurement result, the current and voltage values obtained the highest efficiency at annealing temperature of 165 °C in Table 2

Table 2. Polymer solar cell parameters on PCBM: PPV (1: 1)

Temperature	Efficiency
120	0.23
135	0.23
150	1.71
165	1.72
180	0.25

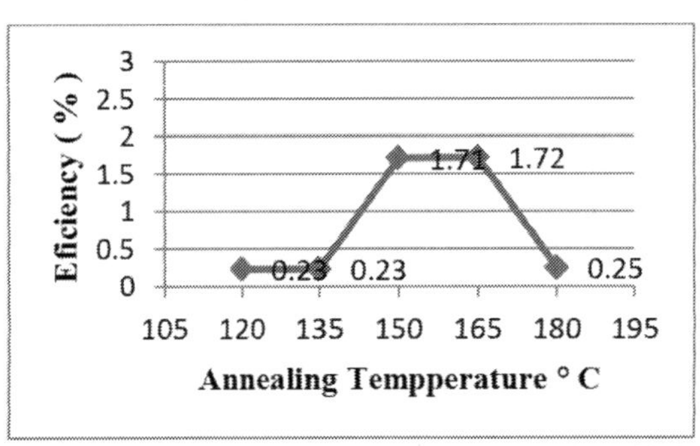

Figure 7. Annealing temperature comparison curve to the efficiency of solar cells

The highest efficiency value of solar cells was 1.72% using semiconductor polymer PCBM:PPV as the active layer, and shown in Figure 7 where it was obtained at 165°C with the highest efficiency. The efficiency produced is relatively small, but higher than previous research conducted by Pratiwi, RZ (2013) using polymer P3HT:PCBM obtained 0.003% efficiency and also research conducted by Chotimah, Triyana, K and Kartini, I obtained maximum efficiency of 0.81%.

This study also conducted an increase in the number of particles of PCBM in active ingredients with a ratio of 2: 1 and 3: 1 to PPV. Annealing was done at a temperature of 165 ° C, because on PCBM: PPV (1: 1) obtained the highest efficiency there was a temperature of 165 ° C. Increasing the number of PCBMs can increase the voltage and current strength of the solar cells, as in Table 3.

Table 3. Current and voltage measurement results at 165 ° C

[PCBM:PPV]	Voltage (V)	Current Power (mA)	Eficiency (%)
1:1	4,032	0,02	1.72
2:1	4,045	0,03	1,73
3:1	4,085	0,03	2,61

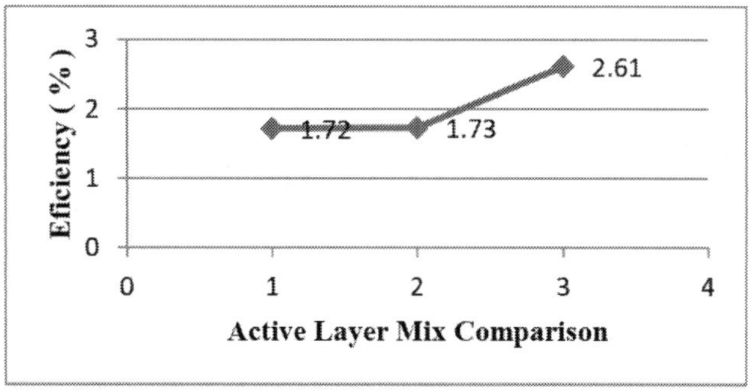

Figure 8. The ratio of the active ingredient mixture to the efficiency of the solar cell

Based on Table 3, it is seen that increasing the number of PCBM polymers in the active layer also improves the performance produced by the cells. PCBM in the active layer serves as an electron donor when UV irradiation is performed. Increasing the number of PCBMs in the active layer is capable of producing the conversion of light to better electrical energy, as more and more photon energy can be absorbed so that more electron-holes are generated (Nurussaniya, et al., 2013).

Increasing the number of donor particles in the active layer will increase the amount of photon energy absorbed by the active layer. At annealing temperature 165 ° C resulted in higher efficiency at 3: 1 ratio that is 2,61%. This is because more of the photons are absorbed on the surface of the active layer. Widiastuti, Nanik, et al (2015) stated that increasing the number of particles in the active layer will increase the absorption of photon energy.

The efficiency improvement in the 3: 1 ratio is very small, as revealed by Mokarromah (2016) that the absorption of the photon is proportional to the rate at which electron-hole is formed. Electron excitation will lead to the semiconductor coding band. The electrons in the free conduction band move, but if the electron is too much it will inhibit the electron flow rate. This causes the electrons to flow through the outer circuit to the weaker electrode counter.

The results of this study are still relatively low when compared with the current solar cell efficiency, but better than the previous research conducted by Bakhtiar, Ayi et al (2011) which produced 0.5 x 10-3%. Chotomah, et al (2012) yielded 0.81% by varying the intensity of xenon lights.

The low efficiency of solar cells is influenced by some of them by the presence of solvents and water vapor trapped in the active layer in the spin coating process. When annealed solvent and water vapor will evaporate, thus creating a hole and creating an uneven morphology of the active layer. The presence of oxygen also affects the performance of the solar cells, since oxygen can bind to the ultimate carbon from the polymer double chains, then the bonds form a reaction that will interfere with the P3HT conjugation (Pratiwi, R.Z. 2013). The small energy source of the photon used also affects solar cells, since the photon energy source is directly proportional to the performance of the solar cells (Chotimah, et al., 2013).

Conclusion

Based on the results of research and discussion, the following conclusions can be drawn:

1. The morphology of the more uniform and homogeneous active layer is produced at an annealing temperature of 165 ° C.
2. Highest current and voltage produced by solar cells using PCBM: PPV as active ingredient in a row is 0.02 mA and 4,032 V at annealing temperature of 165 ° C.
3. Efficiency of solar cells to the effect of annealing temperature obtained maximum at a temperature of 165 ° C that is 1.72%.
4. Efficiency of polymer solar cells increases with increasing amount of PCBM in coated active material, ie at 3: 1 at 2.61%.

Refferences

[1] Bakhtiar, A., Aprilia.,A., dan Fitrilawati., 2011, Sel-Surya Polimer: *State of Art dan* Progres Penelitiannya, Jurnal Material dan Energi Indonesia, Vol. 01, No. 01 (2011) 7 – 14

[2] Baskoroadi, Gesang. 2011. Sel Surya Titanium Dioksida Tersensitas *Polyphenylene Vinylene* dengan Elektrolit Polimer. Skripsi.FMIPA Institut Pertanian Bogor.

[3] Chotimah., Triyana, K.,dan Kartini, I., 2012, "Efek Intensitas Cahaya terhadap Efisiensi Konversi Daya Sel Surya Organik *Bulk Heterojunction* Berbasis Poly(3 hexylthiophene) dan Phenyl C61 butyric Acid Methylester, *Prosiding Pertemuan Ilmiah XXVI HFI Jateng & DIY*, Purworejo, 14 April 2012.

[4] DeBoisblanc, Jenna. 2010. Synthesis and Characterization of P3HT:PCBM Organic Solar Cells. Thesis.University of Catenburry.

[5] Fibilawati, F., Susilawati., T., Wulandari., P., dan Siregar., R.E., 2005, Pengaruh suhu konversi termal pada kualitas film tipis PPV, Jurnal Fisika Dan Aplikasinya, Vol. 01. No. 02

[6] Hardeli., Suwardani., Riky., Fernando, Maulidis, dan Ridwan., S., 2013, Dye Sensitized Solar Cells (DSSC) Berbasis Nanopori TiO_2 Menggunakan Antosianin dari Berbagai Sumber Alami. Proseding Seminar FMIPA Universitas Lampung.

[7] Hardeli, *et al.* 2016.*Synthesis And Electrical Properties Of Zno-ITO And Al-ITO Thin Film By Spin Coating Technique Through Sol Gel Process*. ***Journal of Chemical and Pharmaceutical Research.*** Vol. 8, No. 8 : 915-921.

[8] Krebs, C. F., 2008, Polymer Photovoltaik A Practical Approach. United States of America: SPIE.

[9] Maddu, Akhiruddin & Zuhri, Mahfuddin. 2007.Penggunaan ekstrak antosianin kol merah sebagai Fotosensitizer pada sel surya TiO_2 Nanokristal Tersensitisasi Dye.Jurnal Makara, Teknologi, Vol. 11, No. 2 : 78-84

[10] Nurussaniah, Cari dkk., 2013. Studi Pengaruh Penggunaan Poly(3-Hexylthiophene) P3HT Dan Grafit Terhadap Kinerja Sel Surya. Jurnal Fisika. Vol.3, No.1

[11] Pratiwi, ZR., Rose., E.S., Rusdiana., D., dan Shobih., 2013, Pengaruh Suhu Annealing Lapisan Aktif Polimer P3HT:PCBM Terhadap Unjuk Kerja Sel Surya Polimer Yang Ditumbuhkan Di atas Substrat Gelas, Fibusi(JoV) Vol.1 No.3, Desember 2013

[12] Widiastuti, Nanik, dkk., 2015. Karakteristik Sifat Optik Lapisan Tipis Material Pcbm Hibrid Klorofil Spirulina Sp. Dengan Metode Spin Coating.Jurnal Fisika Indonesia.Vol. 7. No. 1

[13] Yijie, Xia. 2011. *Highly Conductive Poly(3,4Ethylenedioxythiophene): Poly(Styrenesulfonate) (Pedot:PSS) Films And Their Application In Polymer Photovoltaic Devices*. Thesis. Department Of Materials Science and Engineering: National University Of Singapore.

[14] Yulia, Kusuma D.,A., Priatmoko., S dan Wahyuni., S., 2012. Elektroda Solar Cell Berbasis Komposit TiO_2/SiO_2 Sebagai Energi Alternatif Terbarukan.*Indonesia Journal Of Chemical Science*. Vol. 1, No. 2.

Degree of Acetylization Chitosan Gonggong Snail Shells

H Horiza[1], I Iskandar[2], N Aldo[3*]

1 Department Of Environmental Health, Ministry Of Health Polytechnic Tanjungpinang, e-mail: hevi220987@gmail.com
2 Department of Environmental Health Department, Ministry Of Health Polytechnic Tanjungpinang, e mail: iwan_kkp_tpi@yahoo.com
3 Department of Nursing Department, Ministry of Health Polytechnic Tanjungpinang,

*novian_aldotpi@yahoo.com

Abstract. Chitosan is a polysaccharide obtained from the deacetylation of chitin, which is generally derived from crustacean animal waste and animal skins other sea. One marine animals that have compounds that can be processed chitin chitosan is derived from the snail Gonggong marine waters of Riau Islands province. The purpose of this study was to determine the degree of chitosan from the shells of snails asetilisasi Gonggong. This research is an experimental research laboratory. The results of this study indicate that the degree of chitosan shell snail deasetilisasi Gonggong is 70.27%.

1. Introduction

Indonesian nation consists of 17 502 islands and a coastline of 81,000 km with a total area of fishing in the sea about 5.8 million km2,which consists of the archipelagic waters and territorial area of 3.1 million km2 as well as the waters of the Indonesian Exclusive Economic Zone (ZEEI) area of 2.7 million km2.These facts indicate that the prospects for development of fisheries and marine Indonesia is considered very bright and become one of the strategic economic activities. Live fish resources in the territorial waters of Indonesia has a level of biodiversity (bio-diversity) the highest.

Fishery potential in Riau Islands Province, including in the area of the South China Sea fishery management and Natuna through the last boundary line Indonesian Exclusive Economic Zone (ZEEI). In 2006, the fishery production amounted to 220,570.61 tons. Some 217,094.91 tons (99.5%) came from capture fisheries at sea. Followed by marine aquaculture production amounted to 3279.05 tons (0.4%), production of 174.66 tons of freshwater aquaculture and brackish water aquaculture production (farms) amounting to 21.99 tons (0.1%). Marine waters in Riau Islands have the potential of marine life is abundant and diverse (Ministry of Maritime Affairs and Fisheries, 2011). One of the potential of the Riau archipelago are found around the coast of Bintan and Batam is Snail Gonggong. Snail Gonggong in Tanjungpinang production in 2013 reached 147.6 tons, where the average searcher gonggong got less than 6 kg of snails barking day [1].

Gonggong snail is a typical food from Tanjungpinang. Gonggong snails are found in seafood restaurants in Tanjungpinang. Waste generated from Snail shell Gonggong reached 75% of the amount Gonggong snails are eaten. From 1 kg of waste obtained Snail shell Snail Gonggong ± 750 grams.
Waste shells of snails Gonggong only used for crafts such as key chains and ornaments on souvenirs, but it is not uncommon shell snail Gonggong dumped and left to rot which eventually cause negative

impacts on the environment such as causing odor that spoil the aesthetics of the environment and be garbage piling of the coast or under the house eating seafood. One alternative effort snail shell Gonggong waste utilization in order to have the value and usability of Gonggong Snail shell waste into products of high economic value is the process into chitin and chitosan [2].
Chitosan is poly (2-amino-2-deoxy-β- (1-4) -D-glucopyranose) with molecular formula (C6H11NO4)n which can be obtained from the deacetylation of chitin (Figure 2.2). Chitosan is also found naturally in some organisms [3].

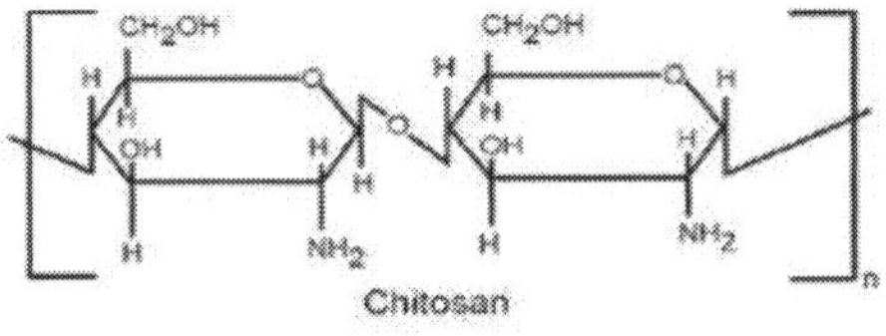

Figure 1. The structure of the polymer chitosan

Chitosan deacetylation process can be done by chemical or enzymatic. Chemical process using a base, for example NaOH, and can produce chitosan with a high degree of deacetylation, which reached 85 to 93% [3]. So that the physical and chemical properties of chitosan are not uniform. In addition, the chemical process can also cause environmental pollution, difficult to control and involves many side reactions that can decrease yield .Enzymatic processes can cover the shortage of the chemical process. Basically, the enzymatic deacetylation is selective and does not damage the structure of the chitosan chains, resulting in chitosan with more uniform characteristics in order to extend the field of application [3].

Chitosan has unique properties that can be used in a variety of ways and has a variety of uses, such as adhesives, additives for paper and textiles, drinking water purification, as well as to accelerate wound healing, and improve the binding properties of color. Chitosan is a powerful adhesive to transition metal ions [4].

Chitosan can be isolated from snail shell waste Gonggong through three stages namely deproteinization, demineralization and deasetilisasi. Deacetylation process aims to break the covalent bond between chitin acetamide group so it turns amine (-NH2).The magnitude of the removal of acetyl group in chitin acetamide group known as the degree deacetylation (DD). According to Khan [5] suggest that chitin with a DD of 75% or more commonly known as chitosan. Deacetylation degree is one of the most important chemical characteristics as DD chitosan affects performance in many applications [5].

2. Methods

This study was an experimental study carried out in laboratories Environmental Health Department of the Ministry of Health Polytechnic Tanjungpinang from March to June 2017. The chitosan FTIR radiation measurements carried out in the Laboratory of Chemistry, State University of Padang.
The tools used are Oven, filter paper, pH meter, universal indicators, flask, glass beaker, stir bar, glass funnel, measuring cup, pipette volume, analytical balance, hot plate. Materials used are NaOH, HCl, distilled water, Whatman filter paper, Paper Label, Gongong snail shell. Stages of this research is the sample preparation Gonggong snail shells, chitin isolation of snail shells Gonggong deasetilisasi and FTIR measurements.

1.Sample preparation

Waste Gonggong Snail shells cleaned of residual dirt and food debris attached. Snail shells Gonggong repeatedly washed until completely clean. Snail shells Gonggong dried at a temperature of 50°C oven for 6 hours and blended to a powder.

2.Isolation Chitin

Chitin isolation process is done by modifying the method Hang in Fahmi [6] and Susilowati [7]. Snail shell chitin isolation from Gonggong done through the stages of deproteinization and demineralization.

3.Deacetylation

Ten grams of chitinadded 200 mL of 50% NaOH. The reaction is conducted at a temperature of 120°C for 4 hours. The solution is separated from the residue and the residue was neutralized by washing using aquades do the repeatedly. The residue was dried in an oven for 4 hours at a temperature of 60°C. After drying blended until smooth with a size of 150 mesh. The residue is further characterized using FTIR.

4.FTIR measurements

Chitosan is stored in a desiccator for one day to remove the water content. Furthermore, the manufacture of pellets from powders containing samples of ± 1 mg and 10-100 mg KBr. The mixture is subjected to grinding until a homogeneous powder and pressed with a hydraulic pump. The results obtained are in the form of pellets, the pellets will be analyzed by FTIR spectrophotometer base-line method.

3. Results and Discussion

Waste bark snail shells cleaned and dried at 50 ° C using the oven for 6 hours and blended to a powder. A total of 500 g of powdered shells of snails Gonggong put in a glass beaker and added with 1.5 M HCl in the ratio 1: 5 (w / v) for 4 hours at a temperature of 65°C, stirring constantly. The continue cooled briefly and then filtered. Then washed with distilled water until neutral and tested withAgNO3 to detect residual Cl ions. Furthermore, the solids dried in an oven with a temperature of 80°C for 24 hours to obtain powdered shells of snails Gonggong without minerals.

Snail Shells Gonggong powder demineralized put in a glass beaker and added with NaOH 3.5% with a ratio of 1:10 (w / v) for 4 hours at a temperature of 65-70°C. Then cooled briefly and then filtered. Then washed with distilled water until neutral. Furthermore, the solids dried in an oven with a temperature of 80°C for 24 hours in order to obtain chitin.

Chitin obtained from entered deproteinization process into a glass beaker and added with NaOH 50% with a ratio of 1:20 (w / v) for 4 hours at a temperature of 120°C, stirring constantly. Subsequently cooled briefly and then filtered. Then washed with distilled water until neutral and tested withAgNO3 to detect residual Cl ions. Furthermore, the solids dried in an oven with a temperature of 60°C for 4 hours to obtain chitosan.

Figure 2. Chitosan shells of snails Gonggong dried.

Chitosan Deacetylation degree Snail Shells Gonggong degree of deacetylation of chitin is determined by the method of thebase-line.This method is based on comparison of the value of spectrum absorbance of infrared absorption bands at wave number 1655 cm -1 and 3450 cm-1. Absorbance (A) is expressed as the equation (1), while the value of DD is expressed as the equation (2).

$$A = \log P0/P \quad (1)$$

where:
P0 =% transmittance at base-line (maximum absorption)
P =% transmittance at the maximum

$$\text{absorption}\%DD = 100 - [(A1665 / A3450) \times 115] \quad (2)$$

Value of A1655 and A3450 is a value corresponding to an absorption band 1655 cm-1 and 3450 cm-1.Absorption band 1655 cm-1 is the absorption band of N-acetyl carbonyl group while the absorption band 3450 cm-1 is an absorption band group NH2 [5,8].

The results of calculations using the equations 2 FTIR spectrum acquired 70.27% DD. The process of removal of acetyl groups (-CH3COO-) of the acetamide group (-NHCOCH3) forming the amine group (NH 2) is also called deacetylation. If the acetyl group separated from the chitin is called deacetylation of chitin. The mechanism of deacetylation of chitin occurs in an alkaline solution. In alkaline solution an ester carbonyl carbon can be attacked by a nucleophile without prior protonation [9]. Process acetyl group at acetamide deacetylation of chitin can be explained as follows: carbon carbonyl group is attacked by a nucleophile OH-, resulting in an addition reaction to form intermediates. These intermediates eliminination subsequent reactions so that an acetyl group on the loose form of chitin acetamide acetate.

The process of release of acetyl groups from chitin acetamide groups associated with OH- ion concentration in the solution. OH- concentration will be greater in strong alkaline solution. The stronger a greater alkaline OH- concentration in the solution.
DD chitin isolation results obtained are 70.27% these results suggest that isolation already a chitosan because, according Terbojevich [10] if DD 40-100% called chitosan.

Conclusion
Acetylation study was to determine the degree of chitosan from the shells of snails Gonggong can be concluded that chitosan DD Gonggong Snail Shells 70.27%.

References
[1] Dinas Kelautan Perikanan Pertanian Kehutanan dan Energi Kota Tanjungpinang. *Data Perkembangan Perikanan Kota Tanjungpinang Tahun 2013*. Tanjungpinang. 2013.
[2] Rahmadani D.S, Soripada T.A, Silaban R. *Pemanfaatan kitosan dari limbah cangkang bekicot sebagai adsorban logam tembaga*. 2011
[3] Sugita Purwantiningsih, Sjahriza Ahmad, Wukirsari Tuti, Wahyono Dwi. *Kitosan sumber biomaterial masa depan*. PT Penerbit IPB Press: 2009.
[4] Meriatna. Penggunaan membran khitosan untuk menurunkan kadar logam Krom (Cr) dan Nikel (Ni) dalam limbah cair industri pelapisan logam. [Thesis]. Medan. Universitas Sumatera Utara.2008. Indonesia
[5] Khan, T.A., Peh, K.K. dan Ching, H.S. (2002). Reporting degree of deacetylation of chitosan: the influence of analytical method.
[6] Fahmi, R. (1997). Isolasi dan Transformasi Kitin menjadi Kitosan. *Jurnal Kimia Andalas*. Vo.3. No.1. 61-68.
[7] Susilawati A dan Reskiati. 2014. Pemanfaatan Limbah Udang sebagai Pengawet Produk Olahan Perikanan. *Jurnal Balik Diwa*. Vol 5, No 2.
[8] Mello, KGPC, Bernusso, L.C., Pitombo, R.N.M, dan Polakiewicz, B. (2006). Synthesis and physicochemical characterization of chemically chitosan by succcinic anhydride. *Brazilian archives of biology and technology*. Vol 49.no4:665668.
[9] Fessenden, R.J. and Fessenden, J.S. (1984). *Kimia Organik* Edisi kedua. Jakarta: Erlangga.
[10] Terbojevich, M. dan Muzzarelli, RAA. (2000). *Chitosan*. University of Ancona.

IOP Conf. Series: Materials Science and Engineering **335** (2018) 012031 doi:10.1088/1757-899X/335/1/012031

Active sites prediction and binding analysis E1-E2 protein human papillomavirus with biphenylsulfonacetic acid

I Iryani*, F Amelia, I Iswendi

Chemistry Department, Faculty of Mathematics and Science, Universitas Negeri Padang, Indonesia

*in.iryani@yahoo.co.id

Abstract. Cervix cancer triggered by *Human papillomavirus* infection is the second cause to woman death in worldwide. The binding site of E1-E2 protein of HPV 16 is not known from a 3-D structure yet, so in this study we address this issue to study the structure of E1-E2 protein from *Human papillomavirus* type 16 and to find its potential binding sites using *biphenylsulfonacetic acid* as inhibitor. Swiss model was used for 3D structure prediction and PDB: 2V9P (E1 protein) and 2NNU (E2 protein) having 52.32% and 100% identity respectively was selected as a template. The 3D model structure developed of E1 and E2 in the core and allowed regions were 99.2% and 99.5%. The ligand binding sites were predicted using online server meta pocket 2.0 and MOE 2009.10 was used for docking. E1-and E2 protein of HPV-16 has three potential binding site that can interact with the inhibitors. The Docking *biphenylsulfonacetic acid* using these binding sites shows that ligand interact with the protein through hydrogen bonds on Lys 403, Arg 410, His 551 in the first pocket, on Tyr 32, Leu 99 in the second pocket, and Lys 558m Lys 517 in the third pocket.

1. Introduction

Human papillomavirus (HPV) is a double-stranded DNA virus that can induce both benign and malignant tumors on the skin and mucosa epithelium. Nowadays, there are more than 150 types of HPV have been identified and 30 of them can cause genital warts and cancer risk. Various studies have shown that HPV initiate and induce cervical cancer [1].

Based on its ability to induce cervical cancer, HPV is divided into two types of high-risk types (high risk) and low risk (low risk). High risk HPV types are HPV-16, -18, -31, -33, -45 and -56 and low risk types are HPV-4, -11, -13, -27, -34, -44, -65 [2, 3].

Cervical cancer is the second leading cause of death in women with cancer worldwide. Every year, there is an increase of 450,000 cases with 50% of them can lead to death [3]. The higher number of the cervical cancer infection not only in Indonesia which has 11.78% cervical cancer patients but also in United States which has the higher number of women infection of this virus (80%) for woman in the 50 ages [4, 5]. Because of the high impact of HPV infections are harmless, and then the choice of therapy/ treatment is needed.

Planning drug use as a target viral replication requires knowledge of the genomic HPV. HPV gene Early (E), Late (L) and Long Control Region (LCR). E1 and E2 genes function as viral DNA replication, gene E4 and E5 genome amplification as well as E6 and E7 are encoprotein. E1 protein has three domains, namely the C-terminal ATPase / helicase domain which serve to release the ends of DNA, DNA-Binding Origin domain (ODB) to recognize viral origin, and the N-terminal domain. E2 protein consists of two domains, N-terminal transactivation domain (TAD) and C-terminal domains. The TAD is involved in transcriptional regulation and are directly related to E1 and C-terminal domains of DNA-binding/ dimerization domain (DBD) [1, 6].

The determination of binding site region is an important factor in an attempt to find a cure cervical cancer. To address this issue, we indicated the amino acid residues of E1-E2 protein HPV which have potential binding to ligand using *biphenylsulfonacetic acid* as inhibitor [7].

2. Experimental

The sequences of E1-E2 of human papillomavirus-16 were obtained from the PaVE: the papillomavirus knowledge source online web (http://pave.niaid.nih.gov/).

Online Swissmodel (http://swissmodel.expasy.org/) was performed to construct quaternary structure by homology modeling [8]. Selected model validated criteria as assessed by PROCHECK [9]. The binding site prediction of the protein model will be analyzed using metapocket 2.0 online [10, 11] in http://projects.biotec.tu-dresden.de/metapocket. Molecular docking has been done using the MOE 2008.10 software to validate the ligand binding.

3. Results and Discussion

3.1. Structure Analysis

The sequences of E1 and E2 protein HPV-16 has modelled using 2V9P and 2NNU templates respectively (Figure 1). E1 predicted protein has 52.32% similarity to 2V9P while the E2 predicted protein was 100% to 2NNU.

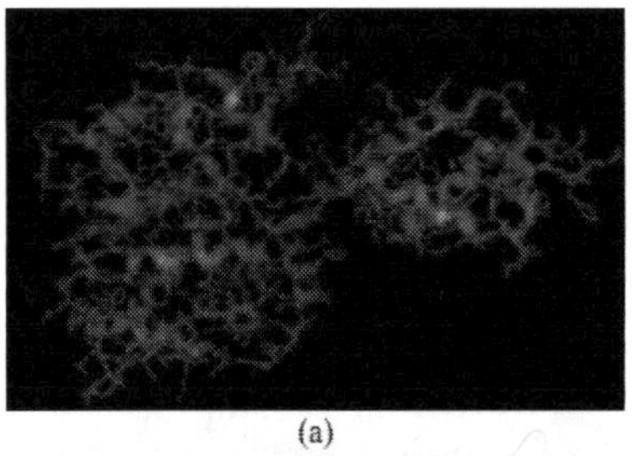

(a)

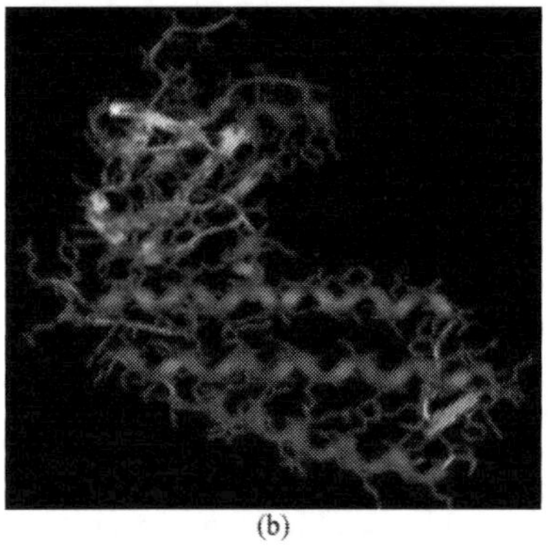

(b)

Figure 1. The predicted homology structure of (a) E1 (b) E2 protein HPV-16.

The predicted analysis of model of E1 and E2 using PROCHECK shows that 88.2 % and 95.7% respectively residues are in the most favorable region and 11.0% and 3.8% in additional allowed region. The percentage of the residues in disallow regions are low, 0.8% and 0.5% in the Ramachandran's plot (Figure 2). A good quality model would be expected to have over 90% the core and allowed regions[9]. Since 88.2% + 11.0% = 99.2% for E1 and 95.7% + 3.8% = 99.5% for E2, it can be conclude that this model has a good quality of three dimensional structure.

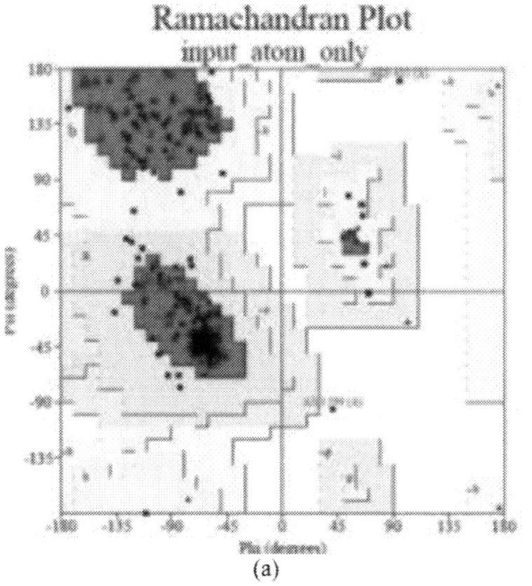

(a)

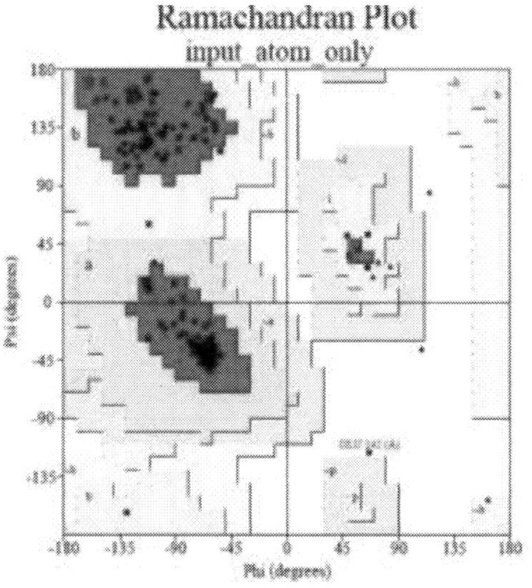

(b)

Figure 2. Ramachandran plot of the psi/phi distribution of the NS4A model as obtained by PROCHECK (a) E1 (b) E2 protein HPV-16.

3.2. Binding site protection

Metapocket 2.0 predict the binding site using four methods: LIGSITE (cs), PASS, Q-SiteFinder, and SURFNET to improve the prediction success rate. It calculate both the geometry and energy interaction between the proteins and ligand [10]. The pockets in E1 and E2 protein of HPV-16 were three sites.

Table 1. Protein Residues in Binding Site Analysis using Meta Pocket 2.0.

	Protein E1	Protein E2
Pocket 1	Polar V549 V555 P512 A514 M547 G496 A516 P553 L554 W425 V370 Y428 M420 A415 Y412 A406 Y365 A364 V402 F508 W509 I499 V498 A376 A399	Polar E74 T78 T81 C140 E141 E142 E80 C109 N84T107 H161
	Non Polar V549 V555 P512 A514 M547 G496 A516 P553 L554 W425 V370 Y428 M420 A415 Y412 A406 Y365 A364 V402 F508 W509 I499 V498 A376 A399	Non Polar L77 A143 Y102 G108 Y138 I139 P106 Y159
Pocket 2	Polar R447 Q441 E603 K483 N597 N599 E591 T565 R462 T616 R615	Polar S98 K68 Q71 C40 D22 T93 Q95 H18 E100 E39 Q12 S65 Q57
	Non Polar M444 Y448 Y602 F445 L604 I442 L446 F486 L485 G482 F592 V601 G598 P600 P593 F594 F456 I473 L474 L475 F586 F588 L460 F463 L490 L563 G487 M520 I564 A459 V584 V443 F453 F612 L457 W439 V451	Non Polar Y19 Y32 W33 M36 A72 L99 I15 L94 V97 Y43 P60 V64 A41 L16 V58
Pocket 3	Polar R447 Q441 E603 K483 N597 N599 E591 T565 R462 T616 R615	Polar N181 K182 C195 T197 S198 K172 S201
	Non Polar M444 Y448 Y602 F45 L604 I442 L446 F486 L485 G482 F592 V601 G598 P600 P593 F594 F456 I473 L474 L475 F586 F588 L460 F463 L490 L563 G487 M520 I564 A459 V584 V443 F453 F612 L457 W439 V451 L464 L620 L622 I467 M422 G466 P561P560 L557 L513 A516 A514 M420 I518 A415	Non Polar Y154 V183 Y155 W184 P196 V199 F200

In an effort to figure out the enzyme-substrate interaction and to determine the key residues responsible for interaction, the model of E1-E2 was docked with biphenylsulfonacetic acid using MOE 2008.10 software. The calculated free energy of binding of protein E1-E2 with biphenylsulfonacetic acid (Figure 3) in pocket 1, 2, and 3 were -15.43137, -14.08253, and -18.11686 kcal/mol respectively. The more negative and lower value of free energy means the more preferable the interaction between protein and ligan [12].

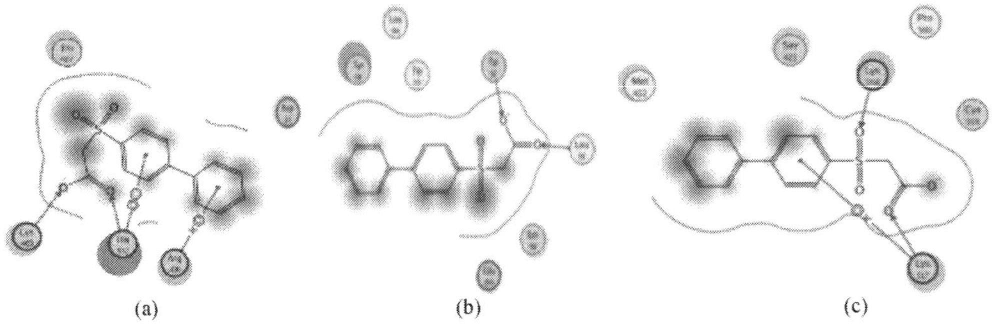

(a) (b) (c)

Figure 3. Interaction of E1-E2 Protein HPV-16 with Biphenylsulfonacetic Acid in (a) pocket 1, (b) pocket 2, and (c) pocket 3. Interaction residues with substrate are labeled using arrow.

The Docking biphenylsulfonacetic acid using these binding sites shows that ligand interact with the protein through hydrogen bonds on Lys 403, Arg 410, His 551 in the first pocket, on Tyr 32, Leu 99 in the second pocket, and Lys 558m Lys 517 in the third pocket.

4. Conclusion

The E1-E2 protein HPV-16 has been suggested to be a potential drug target for cervix cancer therapy. The lack of the structural information about this enzyme made us design of the three-dimensional model. The 3D model developed structure of E1 and E2 in the core and allowed regions were 99.2% and 99.5%. They were docked in three different potential binding sites to refine the energy minimization and interaction residues between biphenylsulfonacetic and three pockets. They were Lys 403, Arg 410, His 551 in the first pocket, on Tyr 32, Leu 99 in the second pocket, and Lys 558m Lys 517 in the third pocket.

5. Acknowledgements

This research was founded by the Ministry of Research and Technology and Higher Education of The Republic of Indonesia through Decentralized Competitive Research Fund Scheme (2015) contract no. 205/UN35.2/PG/2015

References

[1] Archambault, C.M.D.A.a.J., Small Molecule Inhibitors of Human Papillomavirus Protein - Protein Interactions. The Open Virology Journal, 2011. **5**: p. 80-95.
[2] Woodman, C.B.J., S.I. Collins, and L.S. Young, The natural history of cervical HPV infection: unresolved issues. Nat Rev Cancer, 2007. **7**(1): p. 11-22.
[3] Kanodia, S.F., Laura M. and W. Martin Kast Mechanisms Used by Human Papillomaviruses to Escape the Host Immune Response. Current Cancer Drug Targets, 2007. **7**(1): p. 79-89.
[4] http://www.depkes.go.id/. 28 Mei 2013.
[5] Braaten, K.P. and M.R. Laufer, Human Papillomavirus (HPV), HPV-Related Disease, and the HPV Vaccine. Reviews in Obstetrics and Gynecology, 2008. **1**(1): p. 2-10.
[6] Kalvatchev, Z.R., F, Human papilloma virus: realities and perspectives. Biotechnol & Biotechnol 2007. **21**(2): p. 137-143.
[7] White, P.W., et al., Biphenylsulfonacetic Acid Inhibitors of the Human Papillomavirus Type 6 E1 Helicase Inhibit ATP Hydrolysis by an Allosteric Mechanism Involving Tyrosine 486. Antimicrobial Agents and Chemotherapy 2005. **49**(12): p. 4834–4842.
[8] Biasini, M., et al., SWISS-MODEL: modelling protein tertiary and quaternary structure using evolutionary information. Nucleic Acids Research, 2014. **42**(Web Server issue): p. W252-W258.
[9] Laskowski R A, M.M.W., Moss D, Thornton J M, PROCHECK: a program to check the stereochemical quality of protein structures. J. Appl. Cryst., 1993. **26**: p. 283-291.
[10] Huang, B., metaPocket: a meta approach to improve protein ligand binding site prediction. Omics 2009. **13**(4): p. 325-330.
[11] Zengming Zhang, Y.L., Biaoyang Lin, Michael Schroeder and Bingding Huang, Identification of cavities on protein surface using multiple computational approaches for drug binding site prediction. Bioinformatics 2011. **27**(15): p. 2083-2088.
[12] Kitchen, D.B., et al., Docking and scoring in virtual screening for drug discovery: methods and applications. Nat Rev Drug Discov, 2004. **3**(11): p. 935-949.

The Study of Blocking Agent on Lengkeng (*Euphoria Logan* Lour) Fruit Shell and Seed for Adsorption of Pb (II) from Aqueous Solution

D Kurniawati[1*], R Zein[2], Z Chaidir[3], H Aziz[4]

[1)] Chemistry Dept. FMIPA Universitas Negeri Padang, Indonesia
[2) 3)4)] Chemistry Dept. FMIPA Universitas Andalas Padang, Indonesia

*desy.chem@gmail.com

Abstract. The study focuses on the roles played by mayor functional groups(carboxyl) in the lengkeng shell for sorption of Pb (II). The biosorbent was characterized by FTIR and elemental analyses. The parameters such as pH, initial concentration, particle sizes, adsorbent dose and flow rate were also studied. The results showed that the optimum condition was at pH = 3, concentration 400 mg/l, 250 µm particle sizes, adsorbent dose 0,5 g and 2 ml/min flow rate with adsorption capacity 4,8933 mg/g(shell) and 5,2720 mg/g(seed). It is show that ion exchange play as a more important role in the sorption of Pb (II) on lengkeng shell and seed. Blocking of COOH groups by chemical esterification resulted in Pb important reduction in metal binding.The result showed that adsorption capacity of lengkeng shell uncreases until 63.67 % and lengkeng seed uncreases 98.70%.

1. Introduction

Heavy metal ions, such as zinc, cadmium, nikel copper, lead ions at all are commonly found in Industrial effluents [1]. From the health and economical points of view. It is of great interest to be able to effectively remove these heavy metal ion from the waste waters before their discharging and if possible, separate them for recovery and reuse. Conventional methods used to remove toxic heavy metals from aquatic environment include reverse osmosis, electro-dialysis, ultra-filtration, ion exchange, and chemical preparation [2,3]. The method based on metal binding capacities of various functional groups present in biological materials. In this study, the biosorption of lead by lengkeng (*Euphoria logan* lour) shell and seed from single and binary mixtures were investigated.

The aim of the present work is to study blocking agent of fungsional group from lengkeng shell and seed as adsorbent for removal of Pb(II) cation from aqueous solution. The cation exchange properties of there residues may be attributed to the present of carboxylic fungsional group. To reach this aim, several parameters controlling the adsoption capacity of Pb(II) cation on to the surface of lengkeng shell and seed were study here.

2. Material and Methods

2.1. Treatment of lengkeng shell and seeds

In present work, biosorption experiments were conducted by using stock standard solution (1000 mg/L) of lead ion in 1 % HNO_3. Shell and seed of lengkeng (*Euphoria logan* lour) were used as biomaterial in this study. *Euphoria logan* lour fruit were colleted fro the market of Padang city. The sheel and seed of *Euphoria logan* lour were washed with deionized water, air-dried for 3 days and ground using crusher. The working powder was activated by a 20 gr biomass in excess of 80 ml HNO_3 0,01 M for 2 h, followed by washing thoroughly with deionized water and then air-dried [4]

2.2. Blocking Biosobent treatment procedures
The carboxyl group of the biosorbent was blocked with methanol blocking agent. Blocking was carried out by shaking, at room temperature, 8 g of the raw biosorbent in 100, 150 and 200 mL of 99.9% pure methanol and concentrated hydrochloric acid (HCl), given a final acidic concentration of 0.1 M HCl for 6 h at 200 rpm. Then the biosorbent was thoroughly washed with deionized water and then air-dried. The biosorbent was dried and ready to used.

3. Results and discussions
3.1. Effect of solution of pH on metal ion uptake
To study the effect of pH on sorption, the pH of the Pb(II) solution is adjusted to values in pH range of 2-6. Figure 1 shows the maximum pH at 3.

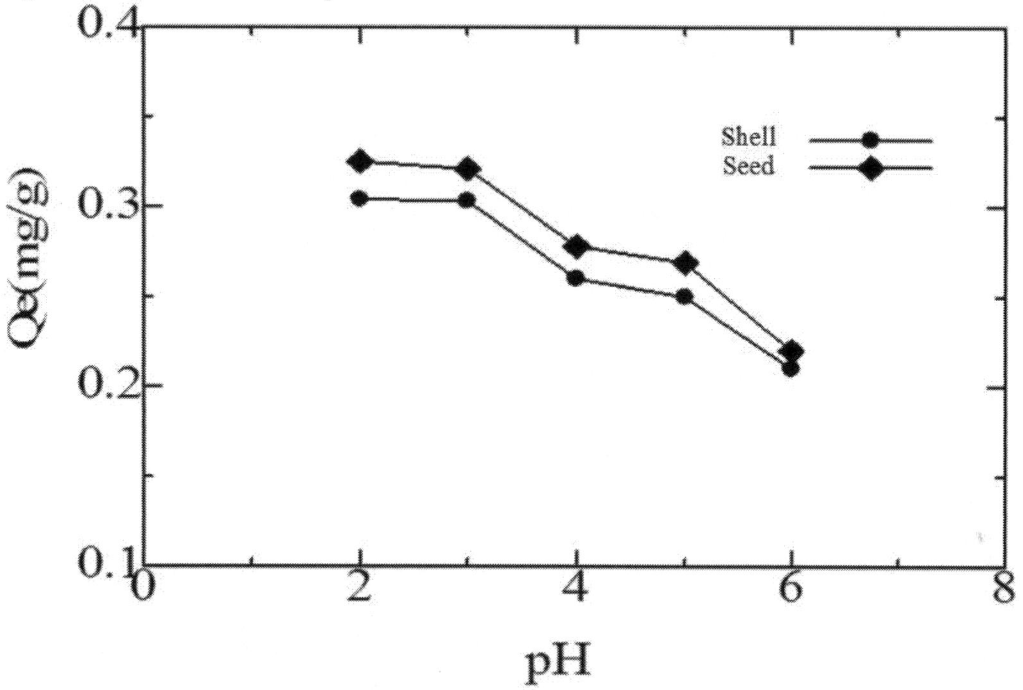

Figure 1. Effect of pH on Pb (II) biosorption by *Euphoria logan* lour seed and shell, particle size 150 μm, mass of biosorbent 0,5 gr, concentration 40 mg/L with column method.

It is well documend that solution pH is an important parameter that affects the biosoption of heavy metal ion [1, 5, 6]. At lower pH, the higher concentration of the hydrogen ions effectively leads to fewer ligands being available for the binding of metal ions [1]. In the present study, biosoption of Pb could not be carried out beyond pH 6 due to precipitation of $Pb(OH)_2$ at higher pH. The effect of solution pH on the uptake of lead ions from aqueous solution showed an decreased in adsobtion capacity removal from 0.3252 mg/g to 0.2275 mg/g of lead ion solution for lengkeng shell and seeds.

3.2. Effect of initial concentration on metal ion uptake
The effect of initial concentration on the sorption of lead ion is presented in Figure 2. Different concentrations of the adsorbate from 20 to 600 mg/L were use study the effect of initial lead concentration on the mechanisme of adsorption by lengkeng shell and seed. The sorbate removal increased with increase in the initial concentration until finally the constant at 400 mg/g on the lengkeng shell and seeds.

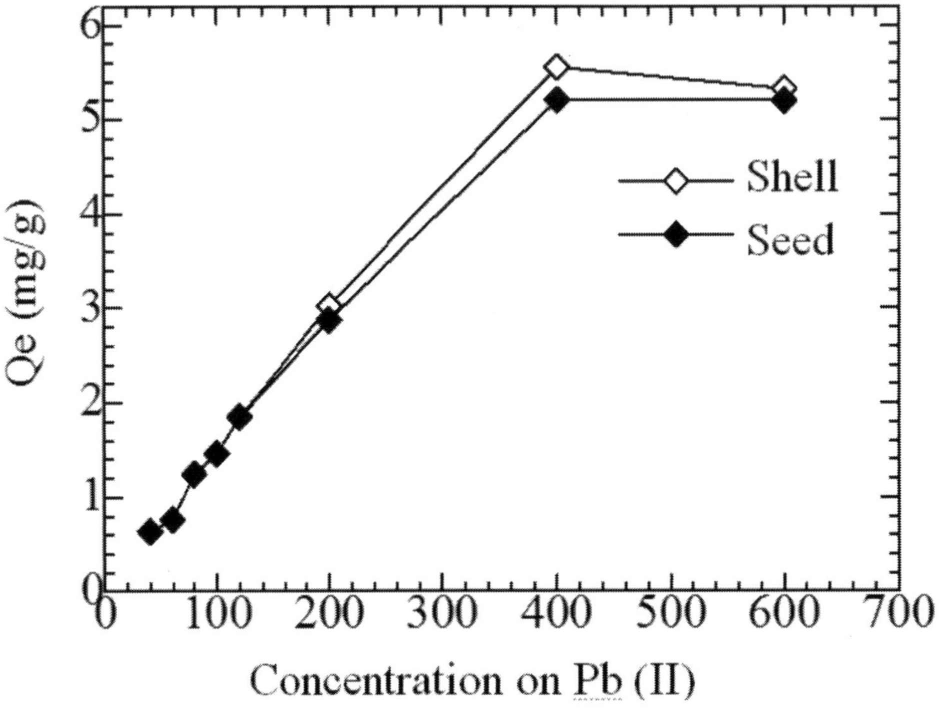

Figure 2. Effect of concentration on Pb (II) biosorption by *Euphoria logan* lour seed and sheel, particle size 150 μm, mass of biosorbent 0,5 gr, optimum pH 3 with column method.

The distribution of the solute Pb(II) metal ion between the liguid phase and solid phase can be described by several mathematical relationships such as the standard Langmuir-Freundlich isoterm model [7,8]. From the experimental data for binding metal ions lengkeng shell and seed system with the maximum sorption capacities is 4.8933 mg/g for shell and 5.270 mg/g for seed.

3.3. Effect of particle size of lengkeng shells and seeds

The effect of particle size on the sorption capacities has been carried out by using various sizes 106. 150. 250 and 425 μm. The result is shown in Figure 3. The result indicated that the amount adsorbed by each individual particle size showed that 250 μm particle size give optimum condition for effective adsorption on shell and seed adsorbent with sorption capacity 4.9050 mg/g for shell, 4.7710 mg/g for seed, respectively. The increase in size for lengkeng seed and shell corresponds to a higher adsorption capacity this may be caused due to the fibrous properties of lengkeng shell and seed wich may have given rise to better adsorption surface area [9, 10]. It is also influenced by the method of column because of style of ILCOVIC when metal ions is flowed into the column. The smaller the particle size cause the flow of metal ion is not about all the surface of the adsorbent. So the particle size of 250 μm the absorption of metal ion by sheel and seeds of lengkeng fruit are optimum at this study. Therefore for the next experiment 250 μm of particle size has been selected in blocking agent process.

IOP Conf. Series: Materials Science and Engineering **335** (2018) 012032 doi:10.1088/1757-899X/335/1/012032

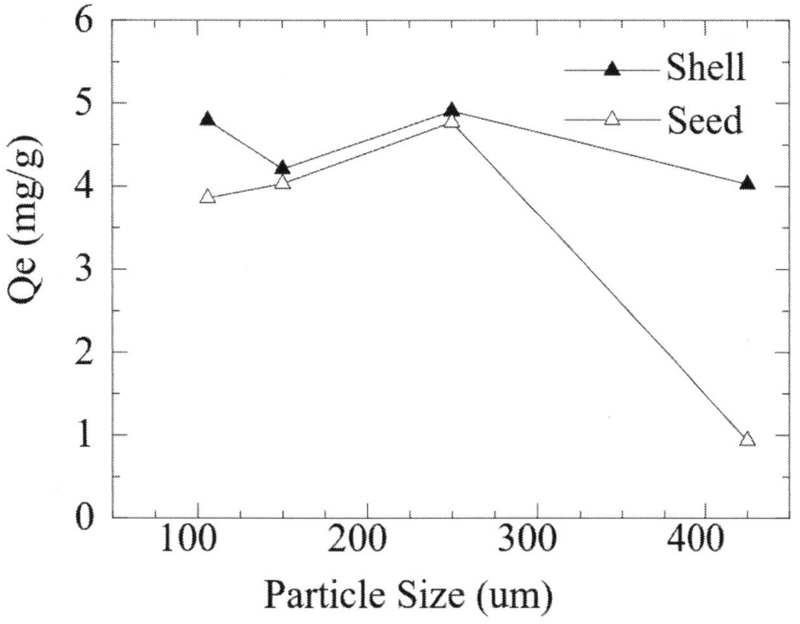

Figure 3. Effect of particle size on Pb (II) biosorption by *Euphoria logan* lour seed and shell, mass of biosorbent 0,5 gr, concentration 400 mg/L, pH 3 with column method.

3.4. Effect of adsorbent dose of lengkeng shells and seeds

The effect of dosage clearly plays a significant role on the ability of lengkeng shell and seed powder to remove metal ion from solution. A dosage increase of 0.1 g- 0.5 g give capacity increase from 13.03 % - 68.81 % for shell and 7.98 %-60.37 % for seed, but capacity adsorbent decreased from 14.2 mg/g- 5.58 mg/g for shell and 8.4 mg/g – 5.54 mg/g for seeds in Figure 4.

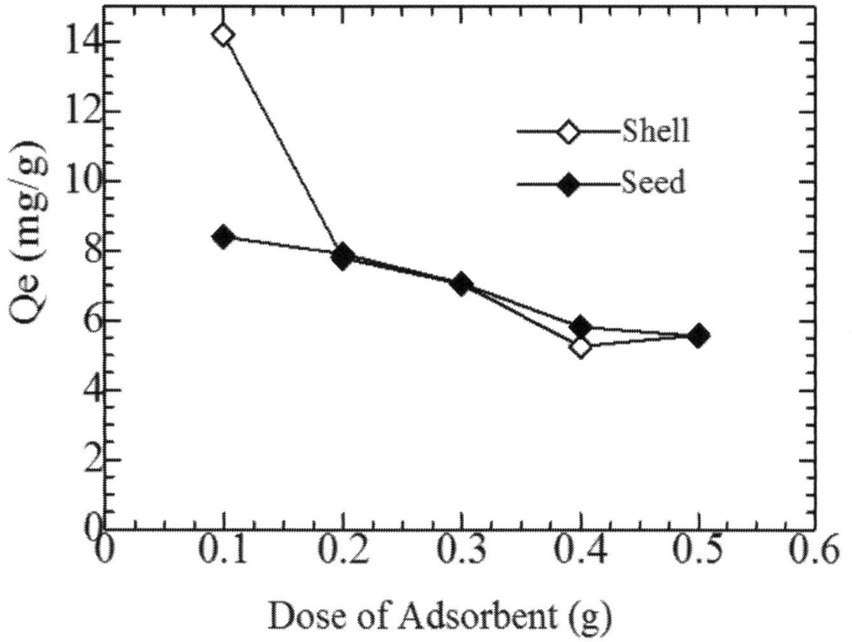

Figure 4. Effect of dose adsorbent on Pb (II) biosorption by *Euphoria logan* lour seed and shell, particle size 150 μm, mass of biosorbent 0,5 gr, concentration 400 mg/L, pH 3 with column method.

3.5. Effect of flow rate on metal ion uptake

The effect of flow rate on adsorption the metal ion on to the lengkeng shell and seed in Figure 5. The results indicated that absorption capacity is optimum at flow rate of 2 ml/min. There is no change in the significant at the flow rate of 1-6 ml/min.

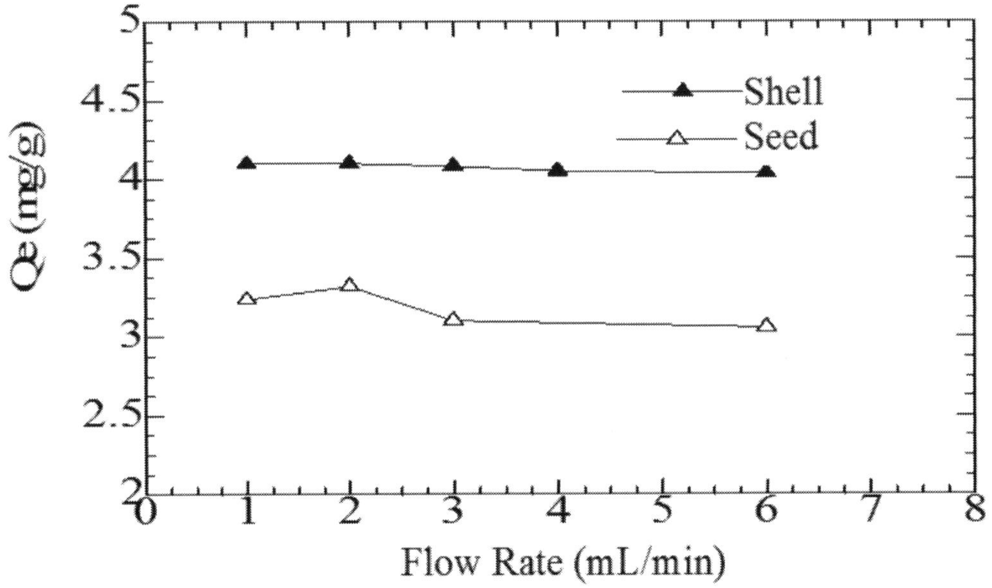

Figure 5. Effect of flow rate on Pb (II) biosorption by *Euphoria logan* lour seed and shell, mass of biosorbent 0,5 gr,concentration 400 mg/L, pH 3 with column method.

3.6. Effect of blocking agent of lengkeng shell and seed

The results obtained show that the blocking agent the functional group in the carboxyl group cause a decrease in the absorption capacity metal ion Pb(II) from 4.905 to 1.792 mg/g for shell and 4.771 to 0.062 mg/g for seeds.The optimization volume for blocking is taken at a volume of 150 mL for Pb^{2+} on the lengkeng shell (Figure 6) and seeds (Figure 7).

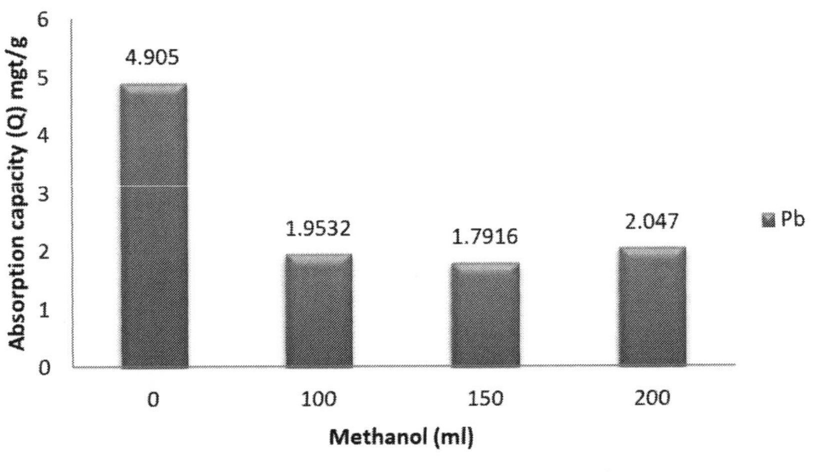

Figure 6. Effect of the volume blocking of carboxyl groups to biosorption capacity of Pb (II) ion by lengkeng shell.

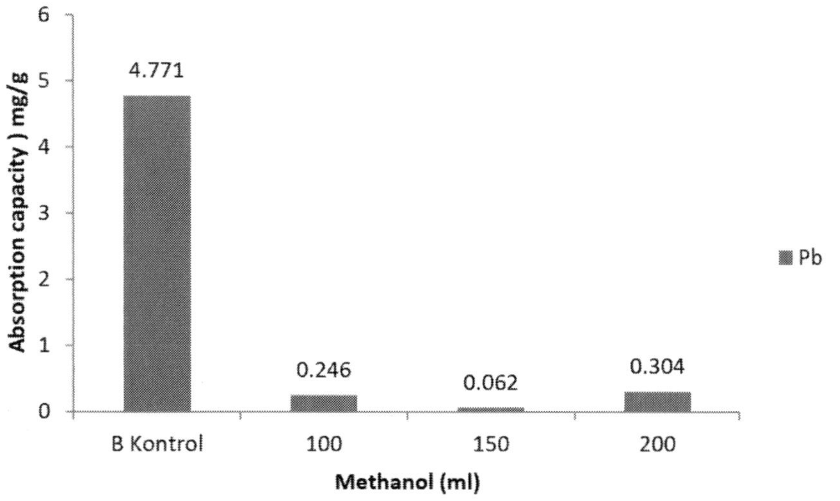

Figure 7. Effect of the volume blocking of carboxyl groups to biosorption capacity of Pb (II) ion by lengkeng seed.

This reasead that the carboxylic group from biosrben the shell and seeds of lengkeng has an important role in binding metal ions Pb(II) [11, 12]

4. Conclusion

The results that lengkeng shell and seed are an effective biosorbent to remove lead (II) from aqueous solution. The maximum biosorption capacity was found to be 4.8933 mg/g for lengkeng shell and 5.2720 mg/g for lengkeng seed in column experiments. Biosorption proses was affected by pH 3, concentration 400 mg/l, 250 µm particle size, 0,5 gr biosorbent and 3 ml/min flow rate. While in shell and seeds of lengkeng fruit which carboxyl functional groups are blocked with methanol showed that the metal ion absorption capacity decreased from 4,9050 mg/g to 1.7916 mg/g (63.67%) and 4.7110 mg/g to 0.0620 mg/g (98.7 %). Based on the research, the counclusion drawn are as follows carboxyl group in lengkeng shell and seed very important in the proses of biosorption. The study revealed that lengkeng shell and seed biosorbent could be used as an adsorbent for removal of other heavy metals on large scale.

References

[1] R.Zein, R, Suhaili, *et all.* 2010. *Removal of Pb(II), Cd(II) and Co(II) From Aqueous Solution using Garcinia mangostana L.fruit shell.* Journal of Hazardous Materials 181 (2010)52-56

[2] Adelaja O. A., Amoo I.A, dkk. 2011. *Biosorption of Lead (II) ions from aqueous solution using Moringa oleifera pods.* Archives of Applied Science Research.

[3] Anna Witek-Krowiak, Roman G. *et all.* 2011. *Biosorption of heavy metals from aqueous solutions onto peanut shell as a low-cost biosorbent.* Desalination 265 .126–134.

[4] Kurniawati.D., Lestari,I, Harmiwati, *et all.* 2015. *Biosorption of Pb(II) from aqueous using column method by lengkeng (Euphoria longan* Lour *) seed and shell.* Journal of Chemical and Pharmaceutical Research. Vol.7(12): 872-877.

[5] T.K. Naiya,P.Chowdhury,A.K. Bhattacharya,S.K.Das,.2009. *Saw dust and neem bark as low-cost natural biosorbent for adsorptive remove of Zn (II) and Cd(II) ions from aqueous solution.* J.Chem.Eng 148;2009; 68-79

[6] G. Blazquez,F.Hermainz, M.Calero,M.A.Martin-Lara, et all. 2009. *The effect of pH on biosorption of Cr(III) and Cr(VI) with olive stone*, J.Chem.Eng.148(2/3);473-479.

[7] S.Gupta,D.Kumar,JP.Gaur.2009. *Kinetic and isotherm modeling of lead (II) sorption onto some waste plant material.*J.Chem.Eng 148(2/3);2009;226-223

IOP Conf. Series: Materials Science and Engineering **335** (2018) 012032 doi:10.1088/1757-899X/335/1/012032

[8] E. Malkoc,Y.Nuhoglu.2007. *Potential of tea factory waste for chromium (VI) removal from aqueous solutions;thermodynamic and kinetic studies*,Sep.Purif.Tech-no.54.291-298

[9] Keshminder Singh., T.V,N.Padmesh.2013. *Biosorption of Dyes Onto Mango Seed and Pineapple Peel.* EURECA.45-46

[10] Sunday O.EZE, Jude C. IGWE and Darlington DIPO.2013. *Effect of particle on adsorption of heavy metals using chemically modified and unmodified fluted pumpkin and broad-leafed pumpkin pods.* Int.J.Biol.Chem.Sci 7(2);852-860, April

[11] Marin,A.B.P., J.F.Ortuno,M.I.Aguilar,V.F.Meseguer,J.Saez,M.Llorens.2010. *Use of chemical modification to determine the binding of Cd(II),Zn(II) and Cr(II) ions by orange waste.* Biochemical Engineering Journal.53;2-6

[12] Ch. Suresh, D. Harikisore, dkk. 2014. *Development of Wood Apple Shell (Feronia acidissima) Powder Biosorbent and Its Application for the Removal of Cd(II) from Aqueous Solution.* Hindawi Publishing Corporation The Scientific World JournalVolume 2014, Article ID 154809.

Monosodium Glutamate Analysis in Meatballs Soup

D Marlina, A Amran, A Ulianas*

Department of chemistry, Faculty of Mathematics and Science, Universitas Negeri Padang, Sumatera Barat, Indonesia

alizar_chem@yahoo.co.id

Abstract. The analysis of monosodium glutamate (MSG) in meatball soup using Cu^{2+} ion as a MSG complex by UV-Vis spectrophotometry has carried out. Reaction of MSG with Cu^{2+} ions have formed complex compounds $[Cu(C_5H_8NO_4)^2]^{2+}$ characterized by the color change of Cu^{2+} ion solution from light blue to dark blue. Maximum of complex absorbance $[Cu(C_5H_8NO_4)^2]^{2+}$ is at 621 nm wavelength. The results showed that, the greatest condition of complex $[Cu(C_5H_8NO_4)^2]^{2+}$ was at pH 10, concentration of Cu^{2+} 0.01 M, complex time is a 30 minute and stable for 170 minutes. Linear response and detection limit of MSG analysis with Cu^{2+} ions are 0.0005-0.025 M (R^2 = 0.994) and (LOD) 0.0003 M. repeatability and recovery method is quite good (% RSD = 0.89% and %recovery = 93%). The analysis of MSG content in meatball soup with MSG complex method was 0.00372 M in sample A and 0.00370 M in sample B.

1. Introduction

Flavouring is a food additive that can add flavour and aroma to the food. The need for food that tastes good and delicious causes the use of flavouring agents to increase over time such as monosodium glutamate (MSG). MSG was first isolated in the form of crystals from seaweed (Laminaria japonica) and identified as glutamic acid [1]. Excessive use of MSG can interfere with health such as dizziness and accelerate the growth of cancer cells [2]. The maximum dose of MSG allowed in humans is 60 mg / kilogram of body weight [3]. The results of a survey by the Indonesian Consumers Foundation (YLKI) found that the traders of meatball noodles, dumpling noodles and boiled noodles in Jakarta using MSG as much as 1840-3400 mg/bowl. Food Additive Organization (FAO) and World Health Organization (WHO) classified MSG as food additive with acceptable daily intake of 60 mg/kg body weight/day [3].

Methods that have used for MSG analysis are with HPLC, fluorescence, gas chromatography, and paper chromatography [4]. In this research has done MSG analysis by UV-Vis spectrophotometry because this method is simple, fast, economical and have good sensitivity and selectivity. MSG solution is a colourless solution, which has carboxyl and amino groups that cannot absorb visible light radiation, so it cannot measure absorbance in visible light areas. Condition of the compounds that can measured absorbance by UV-Vis spectrophotometer is a compound that can give coloured uptake. Therefore, we need a reagent that can give a coloured uptake spectrum with MSG. In this study used Cu^{2+} ion solution as MSG complex. Glutamate reacting with Cu^{2+} ions in aqueous solution will form a blue coloured complex compared to the colour of the Cu^{2+} ion solution before forming the complex, such as the following reaction.

$$Cu^{2+} + 2C_5H_7NO_4 \rightarrow [Cu(C_5H_7NO_4)_2]^{2+}$$

IOP Conf. Series: Materials Science and Engineering **335** (2018) 012033 doi:10.1088/1757-899X/335/1/012033

2. Experiment

2.1. Chemical

Copper (II) Nitrate [$Cu(NO_3)_2$], Sodium hydroxide (NaOH) and hydrogen phosphate (Na_2HPO_4) are produced by Emsure. L-Monosodium glutamate (MSG) and Sodium dihydrogen phospate (NaH_2PO_4) were obtained from Sigma Aldrich. All the chemical solutions were prepared in aquades.

2.2. Complexing MSG with Cu^{2+} Ion and Maximum Wavelength Determination

Solution of Cu^{2+} 0.01 M 1 ml mixed with 2 ml of MSG solution 0.01 M and then stirred until homogeneous. Furthermore, absorbance of complex $[Cu(C_5H_7NO_4)_2]^{2+}$ was tested with a UV-Vis spectrophotometer at a wavelength of 400-700 nm. Maximum absorbance wavelength $[Cu(C_5H_7NO_4)_2]^{2+}$ is the highest absorbance.

2.3. Complex MSG with Cu2 + Ion and Maximum Wavelength Determination

Solution of Cu^{2+} 0.01 M 1 ml mixed with 2 ml of MSG solution 0.01 M and then stirred until homogeneous. Furthermore, absorbance of complex $[Cu(C_5H_7NO_4)_2]^{2+}$ was tested with a UV-Vis spectrophotometer at a wavelength of 400-700 nm. Maximum absorbance wavelength $[Cu(C_5H_7NO_4)_2]^{2+}$ is the highest absorbance.

2.4. Complex optimization

The effect of pH on complex formation $[Cu(C_5H_7NO_4)_2]^{2+}$ was carried out by introducing 2 ml of MSG solution (0.01 M) into the test tube separately at varying pH (6-12), then into the test tubes added 1 mL of Cu^{2+} (0.01 M) ion solution and homogenized. Furthermore, absorbance of each solution measured at maximum wavelength with UV-Vis Spectrophotometer. The highest absorbance is the optimum pH of a more complete complex reaction between MSG and Cu^{2+} ions. While, influence of concentration of Cu^{2+} Ion Solution on complex formation $[Cu(C_5H_7NO_4)_2]^{2+}$ is done by mixing 2 ml of Cu^{2+} ion solution with varying concentration (0.0001; 0.001; 0.002; 0.004; 0006; 0.008; 0.01; 0.02; 0.06; 0.08 M) into the test tube separately, then into the Cu^{2+} ion solutions add 2 ml of MSG 0.05 M solution at optimum pH. Absorbance Each of these mixtures measured at maximum wavelength with a UV-Vis Spectrophotometer. Time and stability of complex done by mixing 2 ml of MSG solution (0.01 M) at optimum pH with 1 mL of Cu^{2+} 0.01 M (optimum concentration) solution. Complex absorbance measured at a maximum wavelength with time periods 0-180 minutes.

2.5. Analysis of MSG in meatball soup

2.5.1. Preparation of Calibration Curve and Limit of detection. The calibration curve prepared by mixing 2 mL of MSG solution (0.0005-0.025M) with 1 ml of Cu^{2+} 0.01 M ion solution separately at optimum pH and homogenized. Each of these mixtures measured its absorbance at a wavelength of 621 nm. The calibration curve prepared with a graph between MSG concentration versus absorbance value, then the linear regression equation determined. While the limit of detection (LOD) calculated based on calibration curve [5].

2.5.2. Reproducibility and Recovery. The Reproducibility test carried out by mixing 2 ml of MSG solution (0.01 M) with 1 mL of Cu^{2+} (0.01 M deep) ion solution in the reaction tube at optimum pH. Measure absorption at maximum wavelength with UV-Vis Spectrophotometer (n=7). Then calculate the percentage of relative standard deviation (% RSD) [6]. While the precision test performed by taking 5 mL of meatball extract and added with 5 mL of phosphate buffer pH 10. Then added 0.0253 gr MSG (MSG level 0.015 M) into solution, so on 2 mL the solution mixed with 1 mL Cu^{2+} ion solution (0.01 M and allowed to stand for 30 minutes. The absorption measured at a wavelength of 621 nm with a UV-Vis Spectrophotometer. Percent recovery (% R) calculated by comparing the concentrations obtained with real concentrations. The% recovery value (% R) can be calculated by the following equation [7], while, Cf = the total concentration of the sample obtained from the measurement, Ca = the actual sample concentration and C = concentration of the analyte added.

$$\% \mathrm{R} = \frac{\mathrm{Cf} - \mathrm{Ca}}{\mathrm{C}} X\ 100\ \%$$

MSG analysis in the meatball soup done by filtering 50 mL of meatball sample. The meatball soup filtrate was centrifuges for 10 min, then 5 mL of meatball soup extract added 5 mL of phosphate buffer pH 10. Then 2 mL mixed with 1 mL of Cu^{2+} 0.01 M ion solution and allowed to stand for 30 min. The solution measured uptake at a wavelength of 621 nm with a UV-Vis Spectrophotometer. The MSG concentration in the meatball soup calculated by the linear regression equation obtained from the calibration curve with the formula y = ax + b with, while y = UV-Vis absorbance of complex, a = slope of the calibration curve x = MSG concentration in sample and b = intercept.

3. Result and Discussion

3.1. Complex MSG with Cu^{2+} Ion and Maximum Wavelength Determination

Compound complexes $[Cu(C_5H_8NO_4)_2]^{2+}$ produced from MSG and Cu^{2+} ions form a planar quadrilateral geometry. One common phenomenon that arises when formation of complex compounds is the colour change. Glutamate is a non-essential natural amino acid that reacts with Cu^{2+} ions to form a deep blue complex [8]. Interaction of glutamate with Cu^{2+} ions will form a covalent bond of coordination between metal and ligand. Figure 2 shows the absorbance and wavelength of MSG Cu^{2+}ions before and after forming complexes. Presence of differences in absorbance before and after the reaction shows that there has been a complex reaction between MSG and Cu^{2+} ions, this is accordance with previously reported [8]. The results of this study found that the maximum absorbance complex $[Cu(C_5H_8NO_4)_2]^{2+}$ is at 621 nm wavelength.

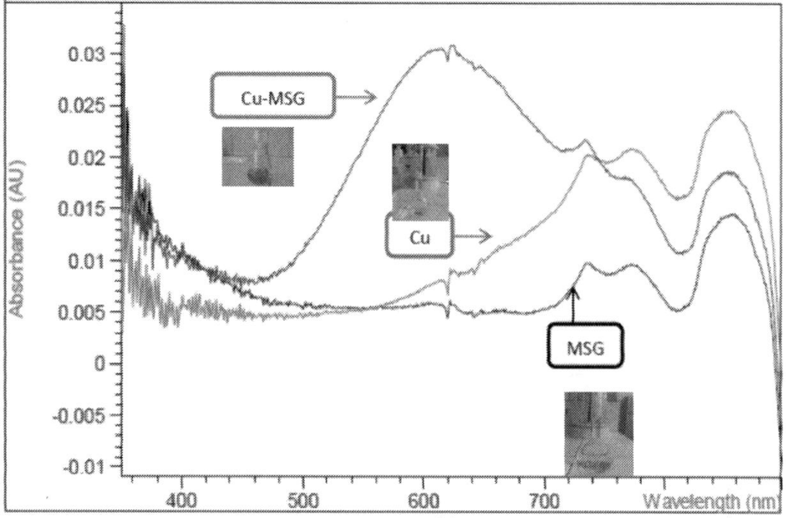

Figure 1. Absorbance and wavelength of MSG, Cu^{2+} ion and $[Cu(C_5H_8NO_4)_2]^{2+}$complex

3.2. MSG Complex Optimization

The degree of acidity (pH) is very influential on formation of complex $[Cu(C_5H_8NO_4)_2]^{2+}$, the greater the number of complexes formed the more intense color [8]. Figure 3 shows the effect of pH on complex formation $[Cu(C_5H_8NO_4)_2]^{2+}$. The results of the study found that formation of the complex $[Cu(C_5H_8NO_4)_2]^{2+}$ was more if the pH of the MSG solution increased from pH 6 to pH 10 and decreased after pH 10. In an acidic condition MSG solution contained more H + ions which could

IOP Conf. Series: Materials Science and Engineering **335** (2018) 012033 doi:10.1088/1757-899X/335/1/012033

interfere Cu^{2+} activity to react with MSG so that the little $[Cu(C_5H_8NO_4)_2]^{2+}$ complexes formed. While the MSG solution is in an alkaline condition, interruption of H + ions decreases to activity of Cu^{2+} ions and formation of $[Cu(C_5H_8NO_4)_2]^{2+}$ complexes increase, but if excessive MSG solution conditions at pH>10, OH- ion can interact with Cu^{2+} ions and decrease the amount of complex $[Cu(C_5H_8NO_4)_2]^{2+}$formed. So on, means that at pH 10 condition the complex $[Cu(C_5H_8NO_4)_2]^{2+}$ formed is the most widely compared to the other pH.

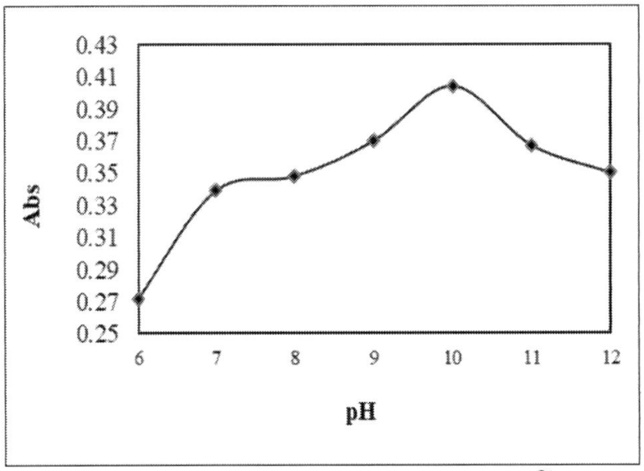

Figure 2. pH effect to complex of $[Cu(C_5H_8NO_4)_2]^{2+}$ formation

Influence of Cu2+ ion concentration to formation of complex compound [Cu(C5H8NO4)2]2+ is as shown in Figure 4. Higher complex absorbance with increasing Cu2+ ion concentration, indicates that addition of Cu2+ ion concentration can increase the amount of complex [Cu(C5H8NO4)2]2+ formed. While, the number of complexes [Cu(C5H8NO4)2]2+ does not increase if Cu2+ is more than 0.01 M ions, because it estimated that the all Cu2+ ion was complexed with MSG.

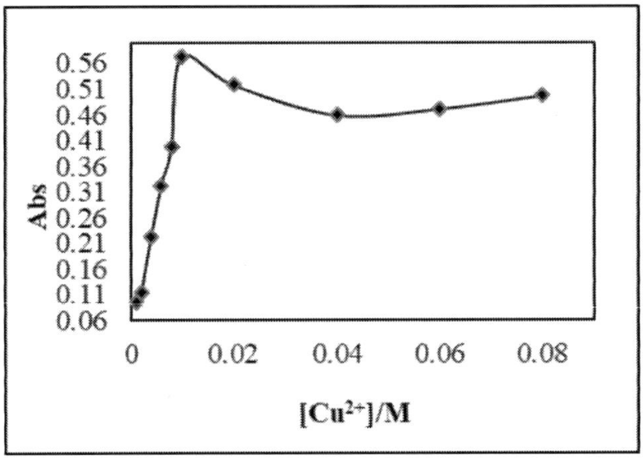

Figure 3. Concentration of Cu^{2+} ion effect to complex formed of $[Cu(C_5H_8NO_4)_2]^{2+}$

The time effects of complex formation $[Cu(C_5H_8NO_4)_2]^{2+}$ are as shown in Figure 5. The complex absorption correlates with the reaction time in which the number of complexes $[Cu(C_5H_8NO_4)_2]^{2+}$ formed increases until the reaction time reaches 30 minutes and remained until the reaction time of 170 minutes, subsequently decreased after reaction time of 170 minutes. This means that the time of complex forming of $[Cu(C_5H_8NO_4)_2]^{2+}$ is perfectly within 30 minutes, stability of the complex is up to

170 minutes. This study is accordance with what reported before that time greatly affects the formation and complex stability [9].

3.3. Analysis of MSG in meatball soup

3.3.1. Calibration Curve and Limit of detection. Figure 6 is an MSG calibration curve showing the relationship between MSG concentration and the complex $[Cu(C_5H_7NO_4)_2]^{2+}$ absorbance value. There is excellent linear stimulation at concentration of MSG 0.0005-0.025 M (R^2 = 0.9941) with the equation y = 11.259x - 0.0012. The value generated by the calibration curve is good if the value of the correlation coefficient (R^2) approaches 1.0. This means that the increase in MSG absorbance value is linear and significant with increasing concentration [8]. While the detection limits value (LOD) is 0.0003M.

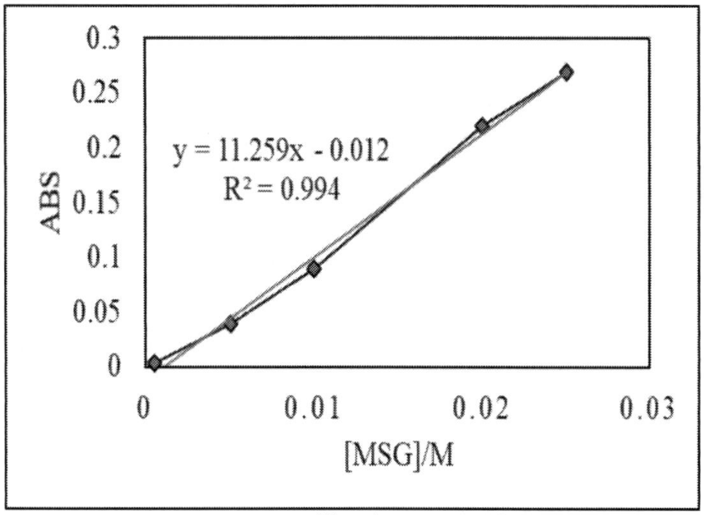

Figure 4. Calibration curve of MSG

3.3.2. Reproducibility and Recovery. From the research results obtained %RSD is 0.89% (n = 7) with standard deviation (SD) 0.005. The results show that the reproducibility value is eligible and the method used is valid, so this method can be used for MSG analysis. Further, the accuracy expressed as the retrieval test with the percent unit (%recovery) of the analyte added [7] . The results obtained a recovery value of 82% for sample A and 93% for sample B. According to Rahayu et al, (2010) A method has good accuracy if the acquisition price is 80-120%. Based on the% recovery value, this method can give the right results for MSG analysis. Next, the analysis of MSG content in meatball soup is 0.00372 M in sample A and 0.00370 M in sample B.

4. Conclusion

MSG analysis in meatball soup can be done by spectrophotometric method and using Cu^{2+} ion as complexing. The color change of Cu^{2+} ion solution from light blue to dark blue after added MSG showed a positive response for MSG analysis. The results of this study indicate that from some samples of meatball saup shows MSG levels that do not exceed the permitted threshold.

IOP Conf. Series: Materials Science and Engineering **335** (2018) 012033 doi:10.1088/1757-899X/335/1/012033

Acknowledgement
We would like to thank the Universitas Negeri Padang for a top-down research grand via research funds 1794/UN35.2/PG/2017.

References

[1] Linderman, B., Yoko, O. and Yuzo, N. 2002. The discovery of Umami, *Chemical Senses,* 27: 43–844.

[2] Mustafa, S., Yasir, S. and Samina H. 2015. Determination of Monosodium Glutamate Content in Selected Traditional Meat Dishes. *International Journal of Scientific & Engineering Research,* 6: 569-572.

[3] Walker, R. & Lupien, J.R. 2000. The safety evaluation of monosodium glutamate. *Journal of Nutrition,* 130: 1049-1052.

[4] Krisnha, V.N., Karthika, D., Surya, D.M., Vishalini, M. and Pradeepa, Y.J. 2010. Analysis of Monosodium l-Glutamate in Food Products by High Performance Thin Layer Chromatography. *J Young Pharm,* 2: 297-300.

[5] Skoog, D.A., Holler, F.J. & Nieman, T.A. 1998. *Principles of Instrumental Analysis*. 5th Edition. Australia: Thomson Learning, Inc.

[6] Ulianas, A., Heng, L.Y. and Ahmad, M. 2011. A Biosensor for Urea from Succinimide-Modified AcrylicMicrospheres Based on Reflectance Transduction. *Sensors,* 11: 8323-8338.

[7] Ling, T.L., Ahmad, M., Heng, L.Y. 2013. UV-vis spectrophotometric and artificial neural network for estimation of ammonia in aqueous environment using cobalt (II) ions. *Analytical Methods*. 5: 6709-6714.

[8] Prasetyo, E. 2012. Simple Method of Copper Analisis Using Monosodium Glutamate and its Aplication in Ore Analysis. *Mineralogia*. 43: 1-2.

[9] Ulianas, A., Heng, L.Yook., Hanifah, S.A., and Ling, T.L. 2012. An Electrochemical DNA Microbiosensor Based on Succinimide-Modified Acrylic Microspheres. *Sensors*, 12: 5445-5460.

Characterization of PCC Cement by Addition of Napa Soil from Subdistrict Sarilamak 50 Kota District as Alternative Additional Material for Semen Padang

M Mawardi*, D Deyundha, R Zainul, R Zalmi P

Chemistry Department, Faculty of Mathematics and Science, Universitas Negeri Padang, Jln. Prof.Dr.Hamka Air Tawar Padang 25131, Indonesia.

*mawardianwar@yahoo.com

Abstract. The study has been conducted to determine characteristics of the portland composite cement by the addition of napa soil from Sarilamak subdistrict, 50 Kota District as an alternative additional material at PT. Semen Padang. Napa soil is a natural material highly containing silica and alumina minerals so that it can be one of material in producing cement. This study aims to determine the effect of napa soil on the quality of portland composite cement. Napa soil used in the variation compositions 0%, 4%, 8%, 12% and 16%, for control of cement used 8 % of pozzolan and 0 % of napa soil. Determination of cement quality by testing cement characteristics include blaine test, sieving, lost of ignition or LOI, insoluble residue, normal consistency, setting time and compressive strength. Cement was characterized using XRF. Fineness of cement decreases with the addition of napa soil. Lost of Ignition of cement decreased, while the insoluble residue increased with the addition of napa soil. Normal consistency of cement increasing, so does initial setting time and final setting time of cement. While the resultant compressive strength decreases with the addition of napa soil ón 28 days, 342, 325, 307, 306, and 300 kg / cm2.

1. Introduction

Proportional amount will result the ability in binding fine and coarse aggregate to be a material called concrete the quality of good concrete depends on the quality of the cement. In processing of producing cement, it using addictive and additional material to get good quality and to minimize the defrayment. The additional material used in the producing of cement is gypsum that has function as retarder in hardening of cement, and pozzolan. Pozzolan can be divided into two, they are natural pozzolan and synthetic pozzolan Natural pozzolan is a pozzolan that comes from nature, meanwhile synthetic pozzolan is synthesized by combustion of clay, husk and coal forms fly ash [1]. Pozzolan contains silica and alumina that when it is mixed by water and free CaO will form cement compound [2].

Napa soil is natural material that is used by West Sumatra people as medicine of stomachache and diarrhea [3]. Napa soil highly contains SiO2 and Al2O3, the presentation is SiO2 63.20 % and Al2O3 16.55% [4]. Thus, napa soil can be used as alternative material as source of silica and it is hoped that it can be used as potential inorganic material as adsorbent, catalyst and additional material in cement industry [5].

In West Sumatera, there are several areas that lies napa soil, they are district of *50 Kota*, district of Tanah Datar, district of Pesisir Selatan, and district of Solok. Seeing the abundance of napa soil in west Sumatra so the research about the effect of the utilization of napa soil as alternative additional material in producing cement is very interesting to be done. Thus, in this research the writer will discuss about the effect of the utilization napa soil from *Sarilamak* subdistrict, *50 Kota* District as alternative additional material in producing cement.

2. Experimental

2.1. Materials and Instruments

The instrument used in this research are crusher brands ME-100 JAW CRUSHER 5"X8", mini mill, mixer machine Toni Technik ToniMI, mixer knife, vicat tools Zwick Roell Toni SET, furnace, analytical scale Sartorius BL 210 S, alpine air jet 200LS-N, cube mold, gloves, pressing equipment, oven brand Carbolite CWF 1300, crusher, ring tablet, blaining tool brand Toni Technik Toni Trol, glass equipment. Characterization of cement using X-ray fluorecence (xrf).

The materials used in this research are napa soil from subdistrictsarilamak, district 50 kota, clinker, gypsum, lime stone, ottawa sand, herzog pill, NaOH brands Merck, HCl brands KGaA, MM indicator, NH4NO3.

2.2. Metodology

2.2.1 Preparation of cement sample. The chunk of napa soil is crushed in crusher, then it is dried for a day. Sample is made by mixing clinker, gypsum, lime stone and pozzolan for control cement, and clinker, gypsum, lime stone and napa soil with the composition in Table 1 after being grinded using mini mill. The cement of napa soil is ready to be tested.

Table 1. The Composition of Cement (%)

Sample	Napa Soil	Clinker	Gypsum	Limestone	Pozzolan
Cement Control	-	76	4	12	8
Cement 1	4	76	4	16	-
Cement 2	8	76	4	12	-
Cement 3	12	76	4	8	-
Cement 4	16	76	4	4	-

Based on Table 1, The amount of composition of napa soil is varied. It is done because napa soil is the testing variable. Meanwhile the lime stone is the component that it does not influence enough to the quality of cement. The amount of sample must be 5000 g, so the composition of lime stone is possibly reduced to make the amount of sample in producing cement keeps 5000 g.

2.2.2 Characterization of cement. Characterization of cement done based on SNI 2049: 2015.

Blaine Analysis

2,8948 g sample is entered to blaining tool brand Toni Technik, then the fineness of the cement can be known.

Sieving Analysis

20 gram of sample is placed on 45μ sieve, the machine is switched on and set the timer to 3 minutes after three minutes, the rest of the sample found on the sieve is weighted, then the percentation can be calculated.

% rest = (mass of rest / mass before sieving) x 100 %

Lost of Ignition (LOI) Analysis

Analysis of LOI is aimed to determine the containing of water and CO2 that lost when the cement ignited. 1.000 gram of cement is ignited in temperature of 1000°C for 15 minutes, then it is placed in desiccator, then the LOI can be calculated.

%LOI= ((dish+sample)before being ignited-after being ignited)/(mass of sample)

Insoluble Residue Analysis
1 g of sample is placed in 250 ml beaker glass and added 25 ml of HCl and aquadest until 100 ml. Then the solution is boiled and filtered, then it is rinsed by aquadest. The filtering paper that containing the precipitation is placed in the beaker glass which containing 100 ml NaOH 1 %. The filter paper is broken then boiled. Then it is added by two drops indicator of red methyl and added by drops of HCl until the color of solution become red. That solution is filtered by filtering paper and rinsed by hot water and NH4NO3 solution to make sure that the filtering paper is perfectly rinsed. Filtering paper and its content is placed in crusible that has been measured its weight then it is placed in furnace in 1000 °C for 30 minutes, then calculate the insoluble part of cement.

Insoluble Part = (mass of precipitate)/(mass of sample) x 100%

Normal Consistency Analysis
650 gram of sample is placed in the mixing dish and it is added by 156 ml of water. Cement is added into the water then mix it for 30 seconds then stop mixing for 15 seconds. The paste that sticks on the bowl is collected, mix it again for 1 minutes, stop mixing and collect the paste that sticks on the dish. Make the shape of the paste to be the ball shape and push the ball through the vicat ring.

Setting Time Analysis
650 gram cement is placed based on its normal consistency. Cement is added by water to form a paste. The paste of cement is prepared to normal consitency analysis. After being printed, the analysis is done by penetrating test using vicat pin with 1 mm diameter in every 10 minutes. The first setting is marked when the pin penetrates more than 25 mm. The final setting is marked when the vicat pin can not penetrate the sample.

Compressive Strength Analysis
740 gram of sample, 2035 gram of standard sand and 410 ml of water. Cement is placed in the dish and it is added by water. Sample is mixed using mixer with low velocity (140±5) radiants per minutes for 30 seonds. Then the sabd is slowly added with low velocity. The mixing is stoped then the velocity is increased until (285±10) radiants per minutes for 30 seconds. The mixer is stopped and the mixer ;eft for 1.5 minutes. For the first 15 seconds, take the mixture that stick in the surface of the dish, then the dish is covered by its cover to prevent the water evaporation of the paste. The mixer is switched on for 60 seconds with medium velocity, then it is stopped for 90 seconds, then it is run for 15 seconds. The mixture is poured down to each cube, then it is slowly pressed to ± 4 mm for 32 times for one sample. Then the cube mold is fully fulfilled and pressed for 32 times. The excess of the mixture on the surface of the mold is cutted using the knife and the surface of the sample is softened using the spoon. After being molded, the sample is placed in the humid room for 20-24 hours and the surface of the sample is conditioned to have contact with the humid air, but it prevented to have contact with water. After one day, the sample is take out from the mold, the sample is soaked in the water containing calcium oxide and it is placed in the anti-rust humid room until the definite time for analyzing the sample. The compressive strength analysis is done for the sample in the age of 3 days, 7 days and 28 days. The sample is cleaned by the dry cloth. The compressing of sample is done by using compressive tools that has been calibrated.

XRF Analysis
XRF analysis is done to know the chemical composition of sample, they are napa soil cement and napa soil. 30-gram sample is added by 2 pellets of herzog pill, then is grinded at it is ready for XRF analysisto get the information about chemical composition of the sample.

3. Results and Discussion

3.1. The Effect of Utilizing of Napa Soil to The Chemical Composition of Cement
The Effect of Utilization of Napa Soil to The Chemical Composition of Cement Presented in Table 2.

Table 2. The Result of XRF Analysis to The Cement with Variation of Napa Soil Composition.

Composition of Napa Soil (%)	SiO_2 (%)	Al_2O_3 (%)	Fe_2O_3 (%)	CaO (%)	MgO (%)	SO_3 (%)	Other (%)
0	22.00	6.10	3.29	59.36	0.91	1.79	6.55
4	20.79	5.25	3.03	61.09	0.90	1.84	7.10
8	23.41	5.55	3.11	58.92	0.89	1.79	6.33
12	25.57	5.59	3.03	57.35	0.86	1.71	5.89
16	28.36	5.94	3.11	54.71	0.85	1.67	5.36

In the Table 2. It can be seen that it happens the changing of mineral composition of cement by addition af napa soil. The more % of napa soil addition, there will be an increasing of content of SiO_2, Al_2O_3, and Fe_2O_3while the content of CaO, MgO, and SO_3decrease. It happens because napa soil is a natural material that highly contains alumina and silica. The highest content of $SiO_2$is found in napa soil that contain 16 % of napa soil, it has 28,36 % of SiO_2. CaO, SiO_2, and $Al_2O_3$are the most chemical composition contained in the cement. The content of $SO_3$is limited in determining the cement quality. Because, if its content is more than the qualified standard , it will make the setting time of cement become longer. According to SNI 7064:2014 [6] SO_3, the maximum content of $SO_3$is 4 %.

3.2. The Effect of Utilizing of Napa Soil to Blaine Analysis.

Blaine analysis is conducted to determine the fineness of cement particle. The effect of utilizing of napa soil to blaine analysis can be seen in the Table 3.

Table 3. Result of Blaine Anlaysis of Cement

Variation of Napa Soil Composition (%)	The Fineness of Cement Particle (cm^2/g)	The Fineness of Cement Particle (m^2/kg)
0	4012	401,2
4	4027	402,7
8	3977	397,7
12	3934	393,4
16	3974	397,4

In Table 3, it can be seen that the fineness of cement particle produced by the addition of napa soil still qualifies the standard of SNI 7064:2014, it is minimally 280 m^2/kg. The curve of corelation between napa siol composition and the fineness of cement particle can be seen in Figure 1.

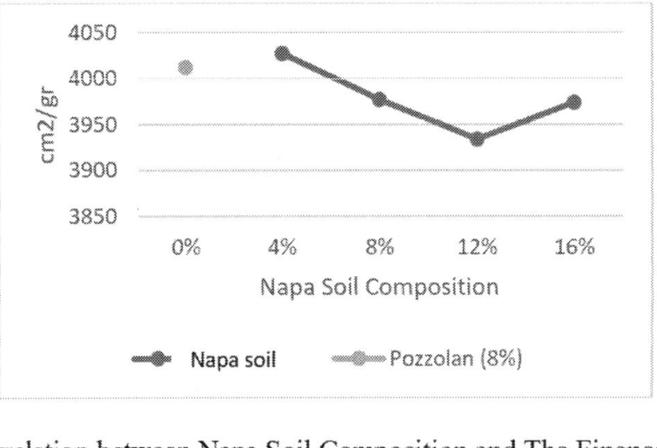

Figure 1. Curve of Correlation between Napa Soil Composition and The Fineness of Cement Particle.

Based on the curve above, it can be seen that the more quantity of napa soil in the cement the fineness of the cement will be smaller, it means that the cement particle produced is getting coarse.However, in the addition 16 % of napa soil for the cement, it increases the fineness of the cement particle. It happens because of unconstant grinding time of cement.

3.3. The Effect of Utilizing of Napa Soil to Sieving Analysis.

The effect of utilizing of napa soil to sieving analysis can be seen in the Table 4.

Table 4. The Result of Sieving Analysis

Variation of Napa Soil Composition (%)	Rest on The Sieve
0	13,71
4	15,16
8	15,67
12	15,86
16	15,98

From the data above, it can be seen that the rest on the sieve (%) for control cement is 13.71 %, it is much smaller than the cement that contain napa soil. It means that the particle size of control cement is smaller than napa soil cemnt. The curve of corelation between napa soil composition and sieve analysisi can be seen in Figure 2.

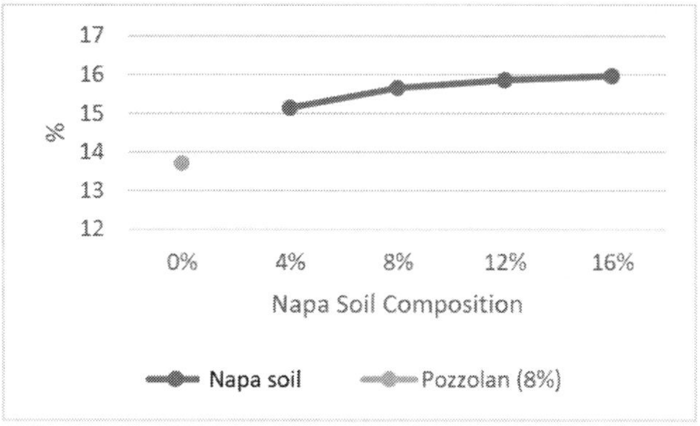

Figure 2. Curve of Corelation Between Napa Soil Composition and % 45 μ sieving Analysis

Based on Figure 2, it can be seen that the more quantity of napa soil composition in cement, the more percentage of sieving analysis. It means that the percentage of cement that can pass the sieve is getting smaller, because the particel sixe is getting biger, it makes the surface area smaller, it makes the compressive strength of the cement is getting lower.

3.4. The Effect of Utilizing of Napa Soil to Lost of Ignition Analysis

The effect of utilizing of napa soil to lost of ignition analysis can be seen in the Table 4.

Table 5. Test result of ignition lost

Variation of Napa Soil Composition (%)	Lost of Ignition (%)
0	6,48
4	7,65
8	6,32
12	4,85
16	3,37

IOP Conf. Series: Materials Science and Engineering **335** (2018) 012034 doi:10.1088/1757-899X/335/1/012034

From the data above, it can be seen that 0 % addition of napa soil result 6,48 % of lost of ignition, this value is almost the same with the lost of ignition in 8 % of napa soil addition, it is about 6, 32 %. The curve of corelation between napa soil composition and lost of ignition analysis can be seen in Figure 3.

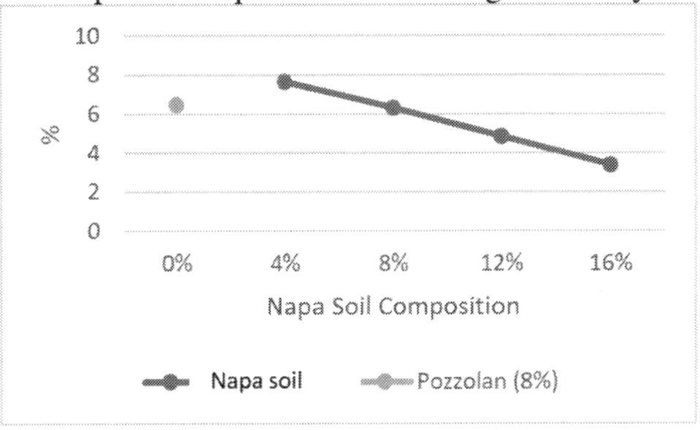

Figure 3. Curve of Corelation Between Napa Soil Composition and Lost of Ignition (%)

Based on the figure above, it can be seen that the more quantity of napa soil composition decreases the percentage of lost of ignition of cement. It is caused by the most component of napa soil is SiO_2, and the value of lost of ignition is influebced by the content of CaO ($CaCO_3$) in the limestone. In this study the percentage of limestone decreases as the increasing of percentage of napa soil.

3.5. The Effect of Utilizing of Napa Soil to Insoluble Part Analysis

The effect of utilizing of napa soil to the value of insoluble pasrt analysis can be seen in Table 5.

Table 6. The Result of Insoluble Part Analysis of Cement

Variation of Napa Soil Composition (%)	Insoluble Part (%)
0	7.72
4	4.94
8	8.11
12	11.76
16	15.19

From the data in the table 5, it can be seen that the value of insoluble part analysis of 0 % of addition of napa soil (pozzolan 8 %) is 7,72 %, it almost has the same value with 8 % of addition of napa soil, it is 8,11 %. The curve of corelation between napa soil composition and the value of insoluble part analysis can be seen in Figure 4.

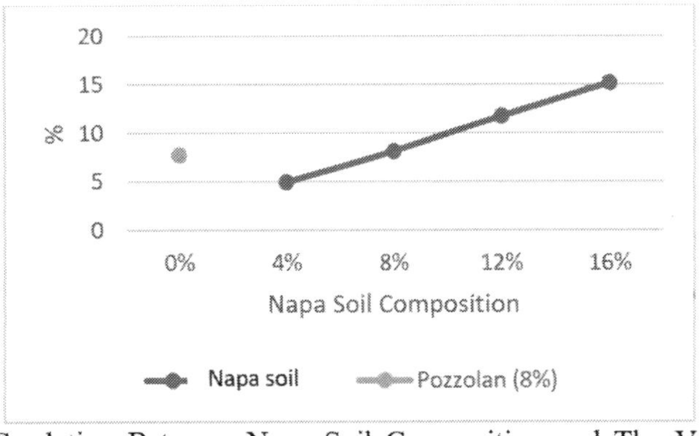

Figure 4. Curve of Corelation Between Napa Soil Composition and The Value of Insoluble Part Analysis

Based on the figure above, it can be seen that the increasing of napa soil composition in cement increases the value of insoluble part of cement. It is caused by the the impurities content such as organic material in limestone and napa soil.

3.6. The Effect of Utilizing of Napa Soil to Normal Consistency Analysis

Normal consistency analysis is done to determine the quantity of water needed to form cement paste. The effect of utilizing of napa soil to normal cosistency analysis canbe seen in Table 6.

Table 7. The Result of Normal Consistency Analysis of Cement

Variation of Napa Soil Composition (%)	Normal Consistency (%)
0	25,23
4	24,62
8	24,92
12	25,54
16	25,85

Based on the data in table 6, based on Table 6, 0 % of napa soil addition (pozzolan 8 %) results the lowest value of normal consistency, it is 25,23 %, so it needs less water in producing the cement paste. The curve of corelation between napa soil composition and the value of normal consistency can be seen in Figure 5.

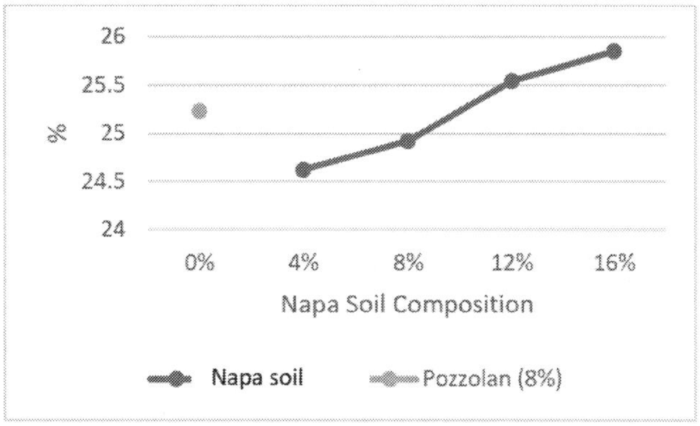

Figure 5. Curve of Ccorelation Between Napa Soil Composition and Value of Normal Consistency

Based on Figure 5 it can be seen that the increasing of napa soil composition in cement increases the value of normal consistency of cement. It means that the more quantity of water needed in producing the cement paste. The value of normal consistency is influenced by the particle size of cement that make the water distribution happen difficultly in cement because the low surface area of cement, thus it needs more water.

3.7. The Effect of Utilizing of Napa Soil to Setting Time Analysis

The setting time of cement is the time needed for cement hardening proces.

Table 8. The Result of Cement Setting Time Analysis

Variation of Napa Soil Composition %	Setting Time	
	Outset (minutes)	Final (minutes)
0	142	231
4	120	209
8	122	211
12	126	215
16	128	217

Based on Table 8 it can be seen that 0 % of napa soil addition results 142 as the highest value of outset setting time and 231 minutes as final setting time. Outset setting time is started when the mixture of cement made that is indicated by penetrating of vicate for less than 25 mm.Final setting time is started when the mixture of cement made and form full strength [7], it is indicated when the vicate can not penetrate the surface of harden cement. Curve of corelation between napa soil composition and cement setting time can be seen in Figure 6.

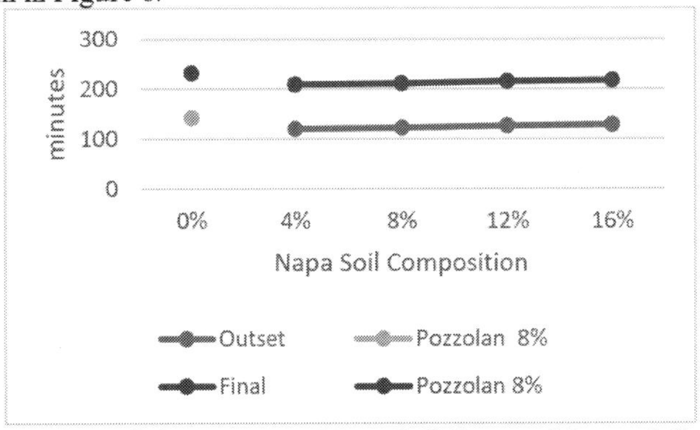

Figure 6. The Curve of Corelation Between Napa Soil Composition and Cement Setting Time

Based on Figure 6, it can be seen that the increasing of napa soil composition in cement increases the cement setting time. The cement setting time is influenced by gypsum utilizing that inhibits the hardening proces of cement. The composition of gypsum is same in each samples so that the influence can not be significantly seen.

The particle size of cement also influences the cement setting time. The bigger particle size of the cement the longer setting time of the cement. The setting time of the cement that contains napa soil still fulfill the standard of the cement SNI 7064:2014, the maximal outset setting time is 45 minutes and 375 minutes for the final setting time.

3.8. The Effect of Utilizing of Napa Soil to Compressive Strength Analysis

Compressive strength analysis is done in the day of 3, 7, and 28. The effect of utilizing of napa soil to the compressive strength of the cement can be seen in Table 9.

Table 9. The Result of Compressive Strength Analysis

Napa Soil Composition (%)	Compressive Strength (kg/cm^2)		
	Day 3	Day 7	Day 28
0	195	262	342
4	197	256	325
8	186	248	307
12	182	246	306
16	181	236	300

From the data above, it can be seen that the compressive strength of the cement containig 4, 8, 12, and 16 % of napa soil tend to decrease than the compressive strength of control cement, curve of corelation between napa soil composition and compressive strength is in Figure 7.

IOP Conf. Series: Materials Science and Engineering **335** (2018) 012034 doi:10.1088/1757-899X/335/1/012034

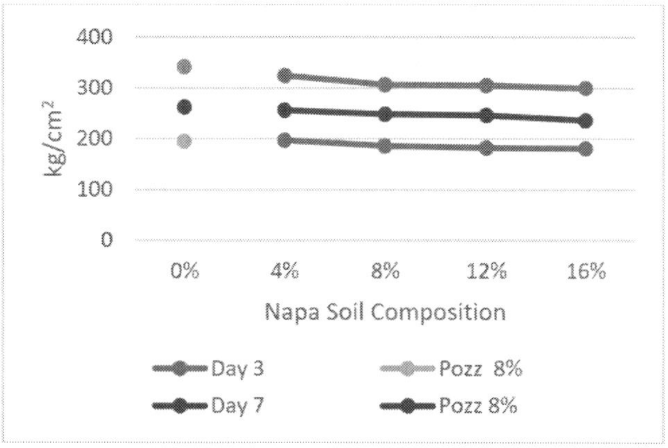

Figure 7. The Curve o Corelation Between Napa Soil Composition and Compressive Strength of Cement

Based on the data in Figure 7, it can be seen that the increasing of quantity of napa soil in the cement decreases the compressive strength of the cement. It is caysed by the content of SiO_2 in cement increases as the increasing of napa soil composition. The compressive strength of napa soil cement is still qualified for the cement standard SNI 7064:2014, the minimal compressive strength of day 3, 7 and 28 are 130, 200, and 280 kg/cm^2.

4. Conclusion

Based on the result of study that has been conducted, there are several conclusions. The addition of napa soil causes the high level of SiO2, Al2O3, Fe2O3 in cement, while the others composition in cement such as CaO, MgO, and SO3 decrease. The utilization of napa soil as alternative material as addition material in producing cement increases the normal consistency, the insoluble part and the setting time, while the rest on sieve, the loss of ignition and the compressive strength decreases.

References

[1] Al-Chaar, Ghassan K., MouinAlkadi, David A. Yaksic, and Lisa A. Kallemeyn. 2011. The Use of Natural Pozzolan in Concrete as an Additive or Substitute for Cement. *Engineer Research and Development Center.*

[2] Pati, S.L., J N Kale, dan S Suman. 2012. Fly Ash Concrete: A Technical Analysis For Compressive Strength. *International Journal of Advanced Engineering Research and Studies*. 2(1): 128-129.

[3] Mawardi, Hary Sanjaya, dan RahadianZainul. 2015. Characterization of napa soil and adsorption of Pb (II) from aqueous solutions using on column method.*Journal of Chemical and Pharmaceutical Research.*7(12):905-912

[4] Mawardi. 2012. Karakteristik Uji Blaine Konsistensi Normal dan Waktu Pengikatan Semen yang Menggunakan Tanah Napa sebagai Bahan Aditif. Jurnal Periodic Vol 1 No 1.

[5] Ningsih, Syukrya, Mawardi, dan Hary Sanjaya. 2013. Determination of Maximum Adsorption Capacity of the Napa Soil on Chromium Ions (III). Periodic.Vol 2 No 2.

[6] SNI 7064:2014. 2014. *Semen Portland Komposit*. Badan Standarisasi Nasional.

[7] Irawati, Nelvi, Nilda Tri Putri, dan Alexie Herryandie Ba. 2015. Strategi Perencanaan Jumlah Material Tambahan Dalam Memproduksi Semen Dengan Pendekatan Taguchi Untuk Meminimalkan Biaya Produksi (Studi Kasus Pt Semen Padang).*Jurnal Optimasi Sistem Industri.*Vol.14, No. 1.

IOP Conf. Series: Materials Science and Engineering **335** (2018) 012035 doi:10.1088/1757-899X/335/1/012035

Synthesis and Characterization of ZNO/MN Nanocomposite by using Sol-Gel Method

S K W Ningsih*, B Bahrizal, E Nasra, U K Nizar, R Farisya

Chemistry Department, Faculty of Mathematics and Science, Universitas Negeri Padang, Jl. Prof. Hamka, Air Tawar, Padang 25131, Indonesia

*sherly14@fmipa.unp.ac.id

Abstract. Zink oxide doped Mn nanocomposites were synthesized by simple sol-gel method at low temperature by using combination of aquadest with methanol as the solvent and ethylene glycol as the additive. Zink acetate dehydrate and manganese chloride tetrahydrate were used as the precursors. Composition dopants were 1,3,5,and 7%. The crystals were formed by drying at110°C for1 hour, after which they were heated at ± 500°C for 2 hours. The as-prepared ZnO/Mn nanocomposites were characterized by X-ray diffraction (XRD) and UV Diffuse Reflectance Spectrometer (UVDRS) . The XRD patterns of the ZnO nanocrystals showed that they are mostly hexagonal wurtzite with specific peaks at 2θ = 31, 34, 36, 47, 56, 63, 66 dan 69. The sizes of the ZnO doped Mn particles produced with 1%, 3%, 5% and 7% were18-95; 17-87; 18-96 19-98 nm, respectively. UVDRS analysis showed that the band gap of the ZnO were 2,60; 2,90; 2,99 dan 3,01 eV for 1%, 3%, 5% and 7% Mn respectively.

1. Introduction

Nanoparticles are particles measuring between 1-100 nanometers[1]. These particles are part of nanotechnology that is very popular and growing rapidly since early 2000[2]. Nano-sized materials have been applied in various fields as catalysts, coatings, semiconductors, pharmaceutical and electronic products[3].

One of the most widely synthesized materials into nanoparticles is Zinc Oxide (ZnO). ZnO exhibits exciting optical, acoustic and electrical properties that have a number of potential applications in the fields of electronics, optoelectronics and sensors. In addition, ZnO has several advantages including a stable chemical structure, non-toxic, and can be used as an additive into various materials, and availability in nature is very abundant so that the price is cheap[4].

Zinc Oxide (ZnO) is an important semiconductor material because it has a wide band gap, so it is widely used for applications such as transpsrent conductors, solar cell windows, gas sensors, photovoltaic devices[5]and photocatalysts[6] . The semiconductor material is a good photocatalyst for the degradation of pollutants as well as toxic compounds namely Zinc Oxide (ZnO), Titania (TiO_2), Tungtsen Oxide (WO_3), Zinc Stannate (Zn_2SnO_4)[7] . ZnO has a relatively large size and a small surface area.

The process of forming a nanomaterial will affect the properties of the material itself, in addition additive addition will also modify its properties. Additives are additives used to control yield morphology. With the addition of these additives, materials obtained that have microstructural homogenisity and small particle size are generated so that material reactivity increases[8].

Doping is an effective method for changing physical properties (e.g. optical, magnetic and electrical properties) in materials and will extend their application to the material from its basic properties[9]. Doping with 3d metals such as Mn, Ni, Fe, Co, Cr and the like will increase the surface area and reduce the particle size of ZnO nanoparticles. In this study, ZnO nanoparticles doped with Mn2+ (Manganese). Various studies have shown that Mn doped semiconductors have affected the physical, chemical and structural properties of pure ZnO nanoparticles. For example, the optical properties of pure ZnO nanoparticles especially on band gap tunings can be greatly increased at the nanoscale with doped Mn.

There are several parameters that will surely affect the particle size, shape and optical properties of ZnO doped Mn nanoparticles such as the effect of dopant ion concentration of manganese, pH and surfactant[10]. In this study only focused on the effect of dopant Mn concentration with percent weight per volume (w / w) of 1%, 3%, 5% and 7%.

This manganese metal is usually not used in a pure state but as a mixture. Manganese is metallic with a melting point of about 1244 ° C and a boiling point of 1962 ° C. In pure state, manganese metal is hard, breakable, and silver-white. Manganese is easily oxidized by air, reacts slowly with water, and forms various compounds with varying degrees of oxidation from +2 to + 7. In this study, other materials of relatively cheap price such as MnCl2.4H2O are found.

Attempts to synthesize nano ZnO have been widely practiced. Some of the most commonly used methods are Co-Precpitation[11], solvothermal[12], wet-chemichal[13]. These methods have advantages and disadvantages of each. One other method that has been developed is the sol-gel method. Sol gel technique is more commonly used in nanoparticle synthesis because it has several advantages as follows: based on the product produced by sol-gel process obtained better homogeneity. High purity and fast crystallinity formation process[14]. Based on the energy used, the sol-gel technique is quite economical[15]as it can take place at low temperatures. Because the reaction takes place at low temperatures, the separation phase and the process of rapid crystal formation hence in terms of operational costs on the sol-gel process is quite economical. In terms of environmental process sol-gel including environmentally friendly[16]because the waste generated is quite low[17].

Zinc oxide (ZnO) is one of the pupular of semiconductor that have a wide band gap of 3.37 eV with large exciton binding energy of 60 meV. ZnO have extensive application doe to its electrical and optical properties. ZnO can be applied in many applications, including gas sensors, generators, field emission transistors, ultraviolet photodetectors, in biomedical systems, biosensors, electric materials, light emitting diode, solar cellsand piezoelectric tranducer.

There are several ways to produce ZnO nanoparticles such as thermal decomposition, carbothermal reduction process, solid sate method, hydrothermal process, sonochemical methods, chemical vapor deposition (CVD), metal organic chemical vapor deposition (MO-CVD), polymerization method, precipitation process, and sol-gel method. Sol-gel process is one of the simplest and lowest cost (inexpensive).

Here, the synthesized ZnO doped Mn nanoparticles prepared by sol-gel method. Zinc acetate dihydrate was used as precursor, combination of aquadest and methanol was used as solvent, and ethylene glicol as the additives. The various compositions were used 1, 3,5 and 7%. Sol-gel method has a number of advantages over other methods such as inexpensive equipment, will produce small particle size and uniform distribution particle with highest homogenity[8].

2. Material

All the reagents were analytical reagent grade and were used without further purification. The precursors used in this research were zinc acetate hydrate{$Zn(CH_3COO)_2.2H_2O$} (Merck) and manganese cholride tetrahydrate{$MnCl_2.4H_2O$} (Merck). Combination of aquadest and methanol was used as the solvent and ethylene glicol was used the additive.

3. Synthesis of ZnO nanoparticles

A total of 2.744 g$Zn(CH_3COO)_2.2H_2O$ was dissolvedwith 50 mL of aquades and methanol mixture with a ratio of 1: 4 in a 50 mL beaker. Closed with aluminum foil paper and stirred using a magnetic stirrer for 40 minutes. In the solution was added manganese chloridatetrahidrathidrat [$MnCl_2.4H_2O$] as dopan with 1% concentration and distirer for 40 min. Mixed the positive is 1.4 mL of ethylene glicol and stirred for 90 minutes. The solution is allowed for one night. The second solution is made by adding 3%, 5% and 7% dopants in the same way. The soles obtained after stirring are then dried in the oven at 110 ° C for 1 hour.The results obtained are included in the furnace at a temperature of 500°C for 2 hours. The samples obtained are then crushed using mortar and pestle.Finally, the dried powder was ground in agate mortar. The synthesized ZnO nanoparticles were analyzed by using X-ray Diffraction (XRD) with a diffractometer by using monochromatic CuKα with λ = 1.5406 and band gap study was carried out by Ultra Violet Diffuse Refflactance Spectrometer (UVDRS).

4. Results and discussion

4.1. Sol ZnO doped Mn with various composition preparations

Table 1. Observation data of sol ZnO doped Mn with various compositions

Composition of ZnO *doped* Mn	Observations
1%	Zinc acetate dihydrate was easily dissolved in the solvent mixture produces a clear solution, after addition of manganese chloride tetrahydrate 1% yields clear color of solution, continued with ethylene glycol is added the color of the solution becomes clear.
3%	Zinc acetate dihydrate was easily dissolved in the solvent mixture produces a clear solution, after addition of manganese chloride tetrahydrate 3% yields clear color of solution, continued with ethylene glycol is added the color of the solution becomes clear.
5%	Zinc acetate dihydrate was easily dissolved in the solvent mixture produces a clear solution, after addition of manganese chloride tetrahydrate 5% yields clear color of solution, continued with ethylene glycol is added the color of the solution becomes clear.
7%	Zinc acetate dihydrate was easily dissolved in the solvent mixture produces a clear solution, after addition of manganese chloride tetrahydrate 7% yields clear color of solution, continued with ethylene glycol is added the color of the solution becomes clear.

Characterization of ZnO nanoparticles, XRD patterns

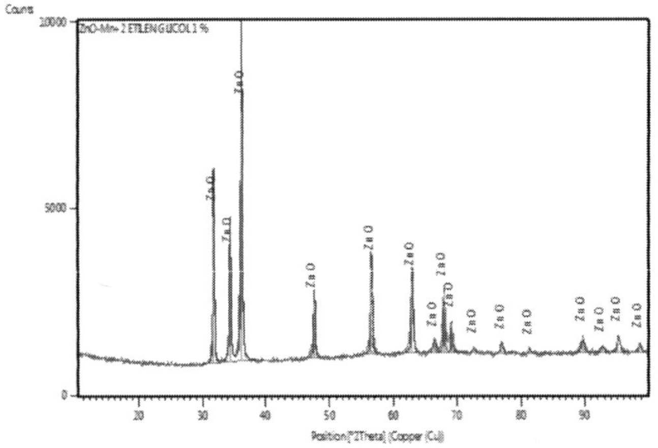

Figure 1. XRD pattern of ZnO doped Mn 1% nanoparticle synthesized by using ethylene glycol as additive

The X-ray diffraction pattern of the ZnO synthesized by using ethylene glycol as additive was shown in Fig.1. This data clearly shows distinct peaks at 2θ = 31.72; 34.38; 36.19; 47.48; 56.67;

62.84; 67.87 and 69.07. The peaks have been identified as peaks of hexagonal ZnO (*wurtzite*) crystallites with various diffracting planes [100], [002], [101], [102], [110], [103], [112] and [201], respectively. ZnO nanoparticle posses a high crystallinity since all the peaks was very sharp. All of the reflections in this pattern can be readily indexed to a hexagonal phase of ZnO which is in good agreement with the literature result(ICCD No. 01-080-0075). The average crystalline size of the synthesized ZnO doped 1% Mn nanoparticle prepared by using ethylene glycol as additive was calculated by using Scherrer to be about 32.15- 95.37 nm (Table 2).

Table 2. XRD data of ZnO doped Mn 1% nanoparticle prepared by using ethylene glycol as additive

Pos.[°2Th.]	*Height cts]*	*FWHMLeft [°2Th.]*	*d-spacing [Å]*	*Rel. Int. [%]*	Crystalite size (nm)
31.7229	3929.51	0.2047	2.82072	58.66	39.90
34.3811	2787.33	0.2558	2.60848	41.61	32.15
36.1984	6698.23	0.2303	2.48158	100.00	32.31
47.4852	1465.44	0.2184	1.91476	21.88	37.27
56.6755	2382.91	0.0936	1.62685	27.77	95.37
62.8493	1911.73	0.2808	1.47743	28.54	32.79
67.8753	1633.39	0.1872	1.37975	24.39	50.59
69.0721	810.16	0.1872	1.35873	12.10	50.95

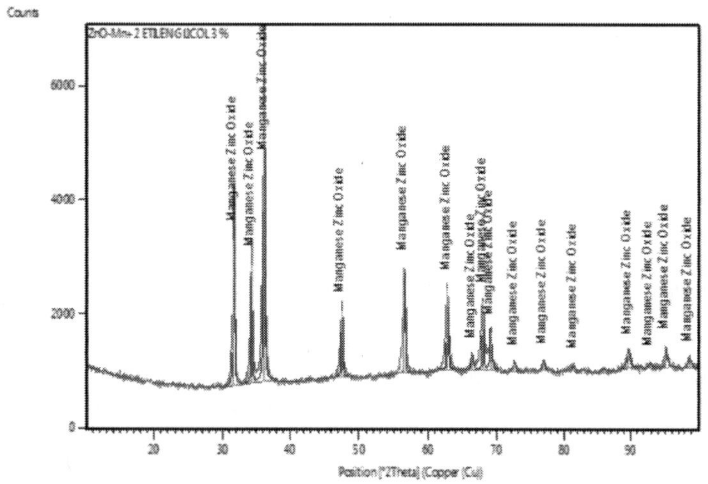

Figure 2. XRD pattern of ZnO doped Mn 3% nanoparticle synthesized by using ethylene glycol as additive

The X-ray diffraction pattern of the ZnO doped 3% Mn synthesized by using ethylene glycol as additive was shown in Fig.2. This data clearly shows distinct peaks at $2\theta = 31.71$; 34.37; 36.18; 47.48; 56.54; 62.79; 67.88 and 69.03. The peaks have been identified as peaks of hexagonal ZnO (*wurtzite*) crystallites with various diffracting planes [100], [002], [101], [102], [110], [103], [112] and [201], respectively. ZnO nanoparticle posses a high crystallinity since all the peaks was very sharp. All of the reflections in this pattern can be readily indexed to a hexagonal phase of ZnO which is in good agreement with the literature result(ICCD No. 01-079-0207). The average crystalline size of the synthesized ZnO doped 3% Mn nanoparticle prepared by using ethylene glycol as additive was calculated by using Scherrer to be about 35.47-87.20 nm (Table 3).

Table 3. XRD data of ZnO doped Mn 3% nanoparticle prepared by using ethylene glycol as additive

Pos.[°2Th.]	*Height [cts]*	*FWHMLeft [°2Th.]*	*d-spacing [Å]*	*Rel. Int. [%]*	Crystalite size (nm)
31.7126	2570.15	0.2303	2.82161	53.36	35.47
34.3740	2081.06	0.1535	2.60900	43.20	53.58
36.1846	4816.87	0.2047	2.48250	100.00	40.38
47.4844	1163.64	0.1791	1.91479	24.16	47.93
56.5411	1840.70	0.1023	1.62770	38.21	87.20
62.7995	1439.20	0.1535	1.47971	29.88	59.97
67.8886	1214.39	0.1535	1.38065	25.21	61.70
69.0344	651.42	0.2558	1.36051	13.52	37.28

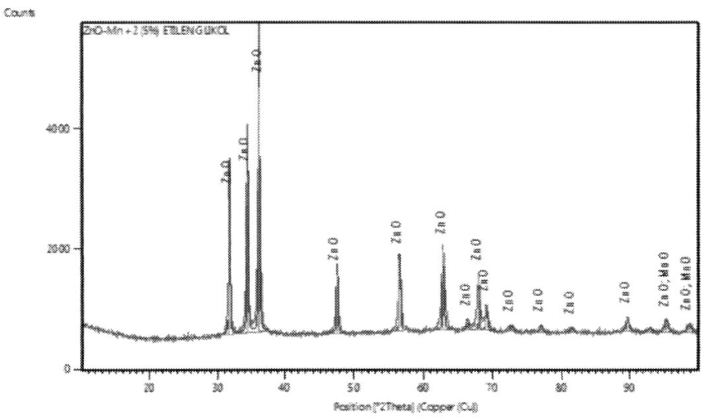

Figure 3. XRD pattern of ZnO doped Mn 5% nanoparticle synthesized by using ethylene glycol as additive

The X-ray diffraction pattern of the ZnO doped 5% Mn synthesized by using ethylene glycol as additive was shown in Fig.3. This data clearly shows distinct peaks at $2\theta = 31.72$; 34.40; 36.17; 47.48; 56.53; 62.77; 67.90 and 69.01. The peaks have been identified as peaks of hexagonal ZnO (*wurtzite*) crystallites with various diffracting planes [100], [002], [101], [102], [110], [103], [112] and [201], respectively. ZnO nanoparticle posses a high crystallinity since all the peaks was very sharp. All of the reflections in this pattern can be readily indexed to a hexagonal phase of ZnO which is in good agreement with the literature result(ICCD No. 01-080-0074). The average crystalline size of the synthesized ZnO doped 5% Mn nanoparticle prepared by using ethylene glycol as additive was calculated by using Scherrer to be about 40.18-101.20 nm (Table 4).

Table 4. XRD data of ZnO doped Mn 5% nanoparticle prepared by using ethylene glycol as additive

Pos.[°2Th.]	*Height [cts]*	*FWHMLeft [°2Th.]*	*d-spacing [Å]*	*Rel. Int. [%]*	Crystalite size (nm)
31.7263	2275.97	0.1791	2.82043	55.80	45.61
34.4053	2626.48	0.2047	2.60670	64.40	40.18
36.1766	4078.59	0.1791	2.48303	100.000	46.15
47.4828	972.56	0.2047	1.91485	23.85	41.93
56.5341	1214.47	0.1535	1.62789	29.78	58.12
62.7771	1350.67	0.1791	1.48018	33.12	51.39
67.9081	922.63	0.0936	1.37916	22.62	101.20
69.0110	404.11	0.2047	1.36091	9.91	46.58

IOP Conf. Series: Materials Science and Engineering **335** (2018) 012035 doi:10.1088/1757-899X/335/1/012035

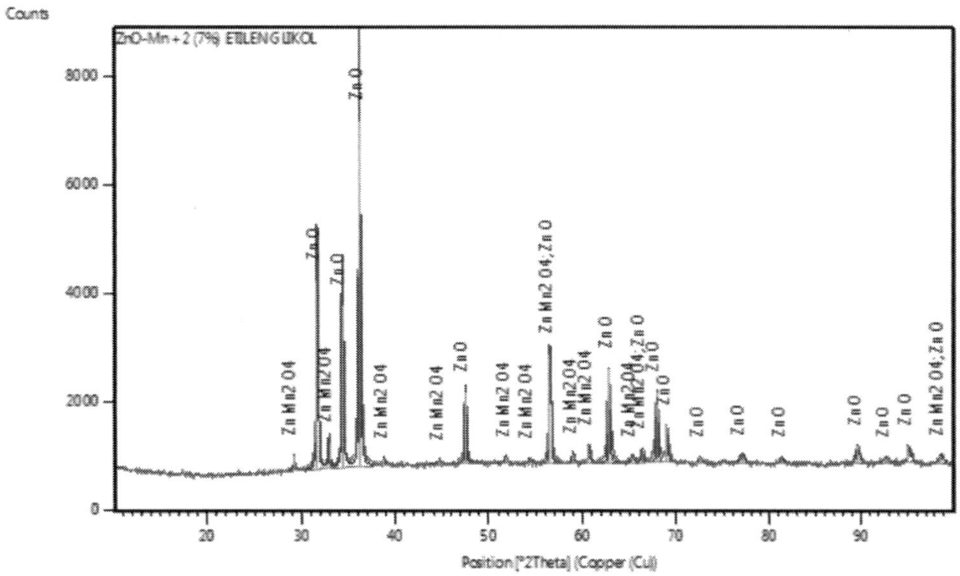

Figure 4. XRD pattern of ZnO doped Mn 7% nanoparticle synthesized by using ethylene glycol as additive

The X-ray diffraction pattern of the ZnO doped 7% Mn synthesized by using ethylene glycol as additive was shown in Fig.4. This data clearly shows distinct peaks at $2\theta = 31.70$; 34.37; 36.18; 47.45; 56.51; 62.78 and 67.87. The peaks have been identified as peaks of hexagonal ZnO (*wurtzite*) crystallites with various diffracting planes [100], [002], [101], [102], [110], [103], [112] and [201], respectively. ZnO nanoparticle posses a high crystallinity since all the peaks was very sharp. All of the reflections in this pattern can be readily indexed to a hexagonal phase of ZnO which is in good agreement with the literature result(ICCD No. 01-076-0704). The average crystalline size of the synthesized ZnO doped 7% Mn nanoparticle prepared by using ethylene glycol as additive was calculated by using Scherrer to be about 40.18-101.85 nm (Table 5).

Table 5. XRD data of ZnO doped Mn 7% nanoparticle prepared by using ethylene glycol as additive

Pos.[°2Th.]	*Height [cts]*	*FWHMLeft [°2Th.]*	*d-spacing [Å]*	*Rel. Int. [%]*	Crystalite size (nm)
31.7017	3504.03	0.1791	2.82256	54.78	45.60
34.3701	3013.57	0.2047	2.60929	47.11	40.18
36.1802	6396.43	0.1535	2.48279	100.00	46.15
47.4540	1312.93	0.1535	1.91595	20.53	55.91
56.5164	2018.06	0.1791	1.62835	31.55	49.80
62.7865	1750.14	0.1279	1.47998	27.36	71.97
67.8713	1302.54	0.1535	1.38096	20.36	61.69
68.9944	677.51	0.0936	1.36007	10.59	101.85

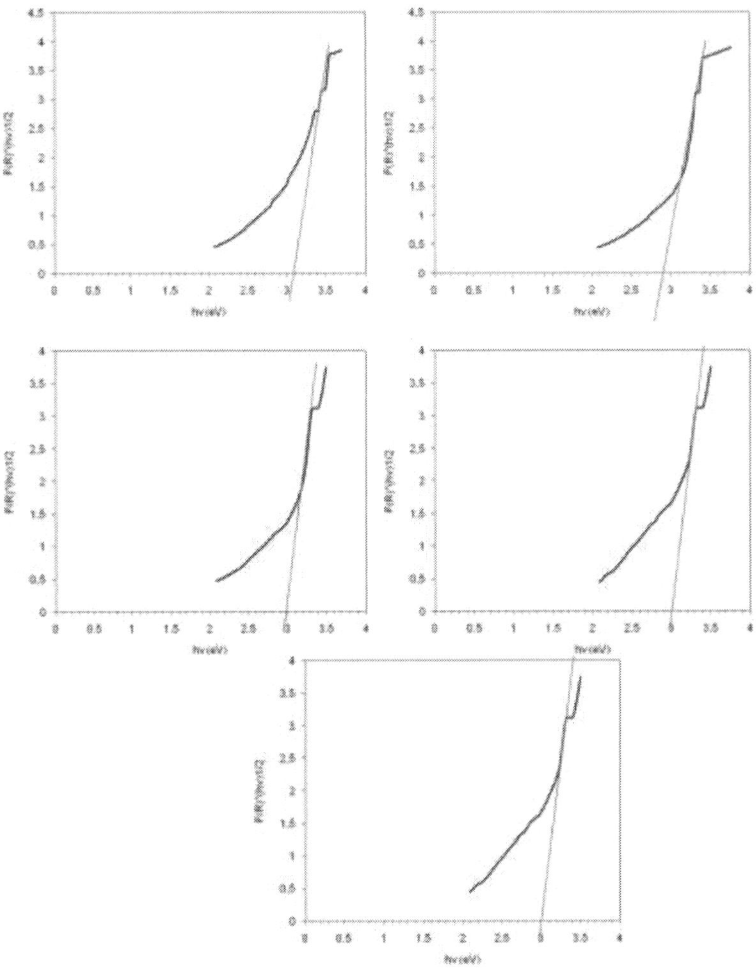

Figure 5. Band gap data of ZnO doped Mn 1, 3, 5, 7 % dopants

Figure 5 depicts the band gap data of ZnO doped Mn. The effect of various compositions were studied in order to obtain the band gap values of ZnO. The synthesized ZnO doped Mn nanoparticles prepared by using 1, 3, 5 and 7% dopant composition were 2.60, 2,90, 2,99 and 3,01 respectively. The composition dopant of 1% Mn has been the smallest band gap energy data.

5. Conclusion

ZnO doped Mn nanocomposites were successfully prepared by sol-gel method with various composition of dopants. The additives play a significant role on the crystalline size and morphology of the ZnO nanoparticles. XRD data for ZnO doped 1, 3, 5, 7% Mn prepared by using ethylene glycol shows the hexagonal (*wurtzite*) structure of ZnO with crystalline sizes in the range of 32.15- 95.37; 35.47-87.20; 40.18-101.20 and 40.18-101.85 nm, respectively. The band gap data of ZnO doped 1, 3, 5, 7% were 2.60, 2,90, 2,99 and 3,01 respectively.

Acknowledgements

The authors would like to thankMinistry ofResearch and Higher Education, Indonesia for providing financial support through of decentralization research, Penelitian Produk Terapan, 2017.

References

[1] Shahmiri, M., Nor, A.I., Norhazlin, Z., Nilofar, A., Bakhtyar, B., Zaharim, A & Sopian , K. 2013. *Effect of pH on the Synthesis of CuO Nanosheets by Quick Precipitation Method.* Vol. 9, Hal. 137-146, 2013 ISSN 2224-3496. Malaysia: Universiti Putra Malaysia.

[2] Suwanda, R. & Syamsul, M.M. 2013. *Pengembangan Inovasi Teknologi Nanopartikel Berbasis Pat Untuk Menciptakan Produk Yang Berdaya Saing. Jurnal Teknik Industri ISSN: 1411-6340 Volume 3 Nomor 2, Juli 2013.* Departemen Teknologi Industri Pertanian, Fateta IPB Bogor.

[3] Saez. V & Mason T.J. 2009. *Sonoelectrochemical Synthesis of Nanoparticles,* Molecules 14 2009 pp 4284-4299.

[4] Z. H. Astuti. 2003. *Kebergantungan Ukuran Nanopartikel TerhadapWarna yang Dipancarkan pada Proses Deeksitasi, Skripsi DepartemenFisika ITB.* Bandung.

[5] Bizarro, M. 2010. *High Photocatalytic Activity Of Zno And Zno:Al Nanostructured Film Deposited by Spray Pyrolysis. Applied Catalysis B: Environmental,* pp. 198-203.

[6] Bizarro, M., Sanches-Arches A., Garduno-Wilches I., Alonso J.C., Ortiz, A. 2011. *Synthesis and Characterization of ZnO and ZnO:Al by Spray Pyrolysis with High Photocatalytic Properties, Catalysis Today,* pp. 129-134.

[7] Baruah, S., Pal, S.K., Dutta, J. 2012. *Nanostructure Zinc Oxide for water treatment, Nanoscience & Nanotechnology-Asia.* Vol.2 No.2, pp. 90-102.

[8] Ningsih, S. K. W. 2016. *Sintesis Anorganik.* Padang: UNP Press.

[9] Li, Y., Meng, J. 2014. *Al-doping effects on structure and optical properties of ZnO nanostructures. MaterialsLetters,* pp. 260-262.

[10] Ling, Tan Tong. 2012. *Synthesis and Characterization of Mn-doped ZnO Nanoparticles.* Program of Resource Chemistry. Faculty of Resource Science and Technology,University Malaysia Sarawak.

[11] Noonuruk, R., Iw, Mekprasart, T., Supparattanasamai, W. Techitdheera, 1221, T. Kanyapan, and W. 2014. *Pecharapa College of Nanotechnology.* King Mongkut's Institute of techonology ladkrabang, Bangkok, 10520 Thailand.

[12] Viswanatha, R., Y. Arthoba, Nayak, T.G. Venkatesha, C.C Vidyasagar. 2013. *Characterization and Optical Properties of Sn-ZnO Nanoparticle.* Universal Research Publication. ISSN : 2278-1374.

[13] Ullah, R., Dutta, J. 2008. *Photocatalytic Degradation of Organic Dyes with Manganese-doped ZnO nanoparticles.* Journal of Hazardous Materials, 156 (1-3), 194-200.

[14] Ningsih, S. K. W., Miftahul K., and Silvi V. 2015. Synthesis *and Characterization of ZnO Nanoparticles by Sol-Gel Method with Various Additives.* The International Conference on Mathematics, Sciences, Education and Technology.

[15] Ningsih, S. K. W. and Miftahul K. 2017. *Synthesis and Characterization of NiO Nanocrystals by using Sol-Gel Method with Various Precursors.*Makara Journal of Science, 19-24.

[16] Sundari, R., TI Hua, M. Aziz and UK Nizar. 2014. *The Characterization Study Of Ferrites (Magnesium and Manganese) Using Sol-Gel Method.* The Malaysian Journal of Analytical Sciences, 18 (3), 485-490.

[17] Sookman, c., paisan, k, & waraporn, t. 2005. "*The effect of calcined temperature on the property of nickel oxide catalyst synthesized by sol-gel method".international symposium eco-energy and material science and engineering symposium.* Chiangmai, thailand.

IOP Conf. Series: Materials Science and Engineering **335** (2018) 012036 doi:10.1088/1757-899X/335/1/012036

The Effect of Titanium Tetrahedral Coordination of Silica-Titania Catalyst on the Physical Properties of Biodiesel

U K Nizar[1*], J Hidayatul[1], R Sundari[2], B Bahrizal[1], A Amran[1], A Putra[1], L Latisma DJ[1], I Dewata[1]
[1]Department of Chemistry, Universitas Negeri Padang, Jl. Prof. Hamka, Air Tawar, Padang, West Sumatera, Indonesia
[2]Departement of Mechanical Engineering, University of Mercu Buana, Jakarta

*umarjati77@gmail.com

Abstract. This study investigates the correlation of the number of titanium tetrahedral coordination and biodiesel production. The solid-state method has been used to synthesis of silica-titania catalyst for biodiesel production, which the precursors, i.e. silica and titania commercials were heated in the temperature range of 450 - 550°C. The characterization of the prepared silica-titania has been studied by FTIR and DR UV-Vis in order to identify and calculate the presence of titanium tetrahedral coordination in silica-titania catalyst. A very small peak at around 950 cm^{-1} indicated the presence of titanium tetrahedral coordination through Si–O–Ti bonds. Deconvolution of DR UV-Vis spectra showed the coordination of titanium in silica-titania is more octahedral. However, the number of titanium tetrahedral coordination of the prepared silica-titania is found higher than that of TiO_2 commercial. The increasing of titanium tetrahedral fraction in silica-titania affects the physical properties of biodiesel in terms of boiling point, viscosity and density, which is produced by the reaction of methanol and palm oil.

1. Introduction

The biodiesel production has developed on going with the reduction of natural resources from fossil fuel globally, high increase of world crude oil price and increasing environmental problem due to fossil fuel. Biodiesel is one of the most promising alternative fuels because it has several benefits over problems generated by fossil fuel. The added value of biodiesel is that its properties close to properties of diesel fuel; an engine using biodiesel does not need to be modified and lubricating property of biodiesel is better than that of diesel and therefore, it can overcome exhausted engine and maintenance cost of engine (1–3).

Palm oil is more promising source of biodiesel than other sources because its abundance in Indonesia is the most available in the world. The life time and production ability of palm trees are longer (until 25 years), and its resistance towards climate change as well. A report stated that the utilization of palm oil for biodiesel production not causing a reduction of palm oil available for consuming purpose (4,5).

The biodiesel production from transesterification of fatty acid and alcohol generally using a catalyst, either homogeneous or heterogeneous catalyst. Generally, the homogeneous catalyst is an acid catalyst (H2SO4) or base catalyst (NaOH and KOH). The utilization of homogeneous catalyst shows several drawbacks particularly in the separation process of catalyst and product. The product is mixed with the catalyst used, thus, the separation process is not simple and requires high cost (not economics). When the separation is not good the biodiesel yield is still mixed with the catalyst and causes corrosion of engine (6). Several obstacles found in homogeneous catalyst can be overcome by the utilization of a heterogeneous catalyst. The biodiesel product has different phase from that of the catalyst and therefore, the separation process is more simple and economics. In addition, the used catalyst can be reused for next reactions (1).

Silica-titania (SiO2-TiO2) is one of binary oxides, which is important in the heterogeneous catalyst group. The utilization of silica-titania as a catalyst in the synthesis of biodiesel has not been widely reported. Based on the literature study, the silica-titania reported as catalyst in biodiesel fabrication was the substituted silica-titania with sulfate group. In general, the reaction product shows the higher catalytic activity of silica-titania and titania-sulfate with a conversion achievement of 90%. However, the percentage increasing of biodiesel yielded is not proportional with the percentage of sulfate group in silica-titania(7).

Several important matters such as the stability of silica-titania with sulfate group during calcinations, the correlation of sulfate composition and percentage of conversion, the existence of tetrahedral framework in silica-titania and stability of sulfate group during reaction of biodiesel fabrication have not been elucidated in previous work. The tetrahedral framework of titanium is known very active and determines the catalytic activity of titanium in silica-titania. The formation of tetrahedral framework through Si–O–Ti bond causes acidic properties on silica-titania surface increased due to different in geometrical form and coordination of Si and Ti (8). The acidic properties are much required for a catalyst in the fabrication of biodiesel.

This study reported the effects of titanium tetrahedral framework in silica-titania on the physical properties of biodiesel yielded. The synthesis of silica-titania is based on the solid state method because it is more eligible for solid precursors and low cost fabrication, compared to another method i.e. sol-gel method(9–11). Characterizations of samples applied FTIR and DR UV-Vis. The fraction of titanium tetrahedral in the synthesized silica-titania was calculated applying deconvolution of UV spectra and examined in biodiesel fabrication from the reaction of palm oil and methanol. The physical properties of biodiesel product including boiling point, viscosity, and density were determined.

2. Experimental

2.1. Chemicals

Titanium oxide (Acros), silica (Sigma Aldrich), and toluene (Merck) were purchased and used in this study. Palm oil (Bimoli) and methanol (Merck) were used for biodiesel synthesis.

2.2. Synthesis of silica – titania

Synthesis of silica-titania was done by mixing respective 10 mmol silica and 10 mmol titania (1 : 1) in 10 ml toluene. The mixture was sonicated using ultrasonic equipment for 1h to disperse silica and titania. Then the mixture was dried in a fume hood at room temperature to evaporate toluene. The solid sample formed was then calcinated at various temperatures (450°C, 500°C and 550°C) for 8h. The products from calcinations were characterized by FTIR and DR UV-Vis, and were applied for synthesis of biodiesel as well.

2.3. Characterization of silica-titania catalyst

The FTIR characterization was used to identify type of chemical bonding in the mixture of silica and titania. The samples were analyzed in wave number range of 4000 – 400 nm^{-1}. The DR UV-Vis characterization was used to determine tetrahedral and octahedral coordination in the sample and also to determine the fraction of titanium tetrahedral based on spectral deconvolution. The samples were analyzed in wavelength range of 200 – 400 nm.

2.4. Application of silica-titania catalyst for synthesis of biodiesel l(1,6,11-12)
In general, the synthesis of biodiesel was done through transesterification reaction of commercial palm oil and methanol with addition of silica-titania catalyst. The mole ratio of methanol and palm oil was made as 6 : 1, and the amount of catalyst was taken as 10% from the weight of palm oil. The mixture was homogenized by a magnetic stirrer and heated for 3h at 65°C. Then the mixture was cooled and the separation process was done between the product with catalyst and excess methanol. The first separation was done for catalyst by centrifuging. The next separation was done for excess methanol using rotary evaporator at temperature above boiling point of methanol. The biodiesel product was characterized with FTIR and the physical properties were examined for boiling point, viscosity, and density.

3. Result and Discussion

3.1 FTIR spectra of silica-titania prepared by solid state method
FTIR spectra of TiO_2, SiO_2 and series of SiO_2-TiO_2 synthesized based of various calcinations at 450°C, 500°C, and 550°C are shown in Figure.1. As seen in Figure.1, the spectrum of TiO_2 shows band absorptions at 734 cm^{-1} and 2352 cm^{-1}. The band absorption at 2352 cm^{-1} indicates the absorption of Ti-O, while the absorption at 734 cm^{-1} is attributed to absorption of Ti-O-Ti. The spectrum of SiO_2 is attributed by the main band absorptions at wave numbers of 1058 cm^{-1} and 803 cm^{-1}, respectively. The band absorption at wave number of 1058 cm^{-1} is attributed to asymmetric stretching vibration of Si-O-Si, and that of 803 cm^{-1} is related to symmetric stretching vibration of Si-O-Si (7)(14).

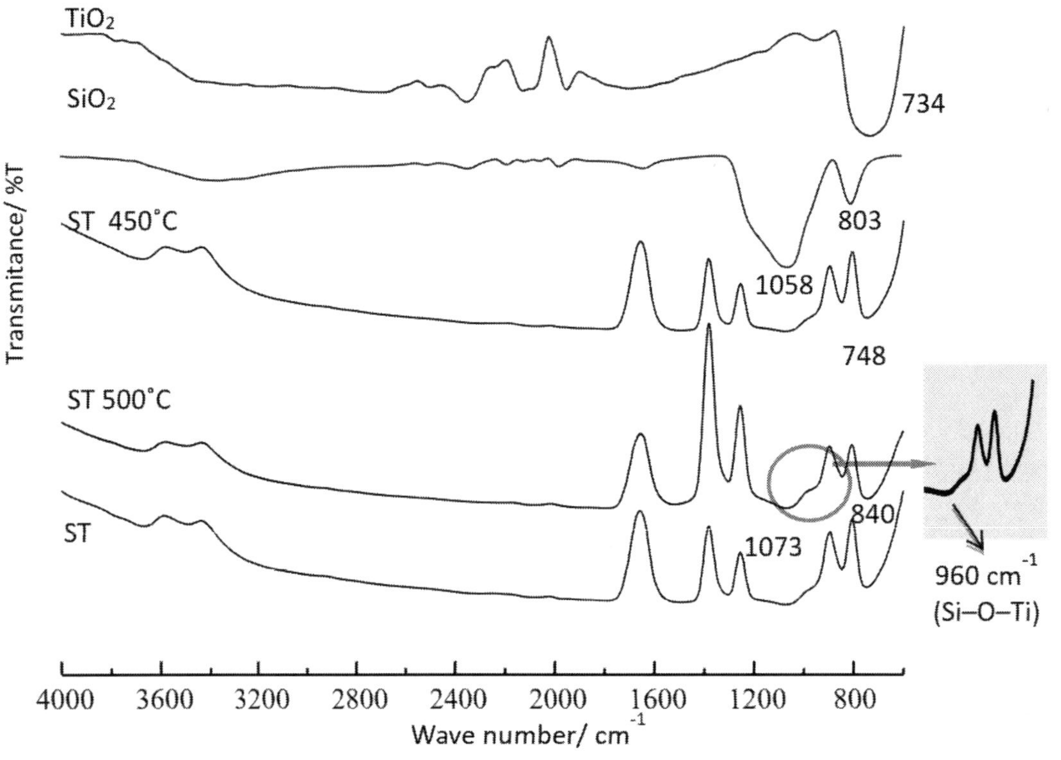

Figure 1. The FTIR spectra of TiO_2, SiO_2 and series silica-titania.

The FTIR spectra of SiO_2-TiO_2 series are presented by main band absorptions at respective wave numbers of 3668 cm^{-1}, 3429 cm^{-1}, 1073 cm^{-1}, 960 cm^{-1}, 840 cm^{-1} and 748 cm^{-1}. The broad band absorptions at wave number of 3668 cm^{-1} dan 3429 cm^{-1} show the –OH stretching vibration of sample surface. The –OH band absorption is due to the existence of water vapor generated during FTIR characterization process (10). Other band absorptions observed are the weak broad band at wave number of 1073 cm^{-1} and weak absorption band at 840 cm^{-1}. That band absorptions are related to asymmetric and symmetric stretching vibrations of Si-O-Si. In addition, a weak broad band absorption at wave number of 748 cm^{-1} is attributed to vibration of Ti-O-Ti (7). A very small band absorption at wave number of 960 cm^{-1} is due to vibration of Si–O–Ti bonding. This bonding is an indication of chemical interaction between SiO_2 and TiO_2. More bonding of Si–O–Ti formed indicated more titanium tetrahedral fractions generated in SiO_2 and TiO_2 (15). The formation of Si–O–Ti bonding is verified and confirmed by deconvolution band absorption of DR UV-Vis spectra (15–17).

3.2. Deconvolution of DR UV-Vis spectra

The deconvolution spectra of DR UV-Vis has been done to examine the effect of calcinations on titanium tetrahedral fraction and to evaluate the titanium tetrahedral fraction due to formation of Si–O–Ti bonding. According to literatures, the DR UV-Vis spectra in the range of 200 - <280 nm is attributed to tetrahedral region and that above than 280 nm is related to octahedral region as well as shown in Figure. 2 (10).

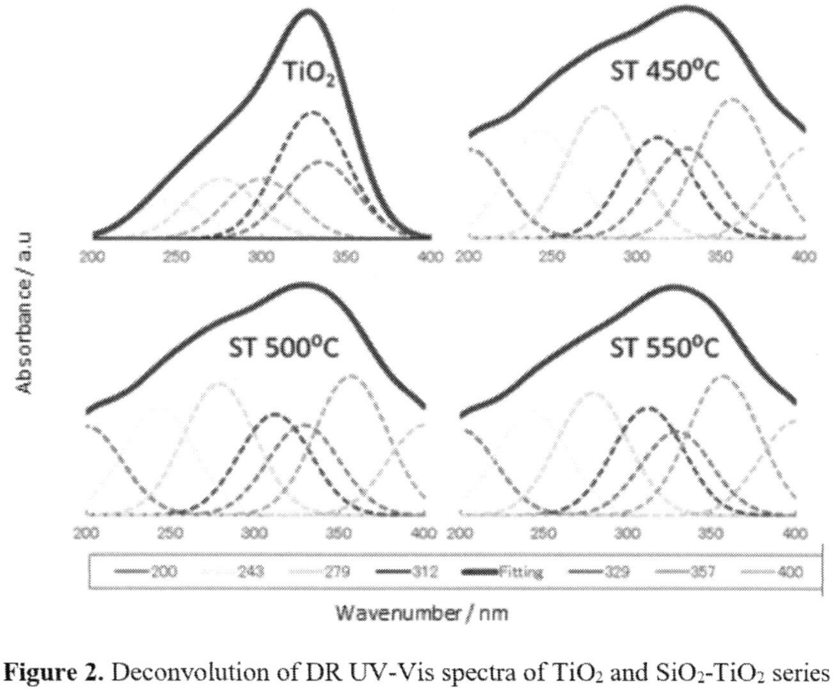

Figure 2. Deconvolution of DR UV-Vis spectra of TiO_2 and SiO_2-TiO_2 series

On the basis of deconvolution calculation the band absorptions at 200, 242 and 279 nm is corresponding to titanium tetrahedral fraction and that at 312, 329, 358 and 400 nm is related to titanium octahedral fraction. In general, silica-titania catalyst prepared by TiO_2 and SiO_2 solids is likely to form more octahedral fraction rather than tetrahedral fraction. However, the result of this investigation shows an increase of titanium tetrahedral fraction compared to TiO_2 commercial. The increasing of titanium tetrahedral fraction is shown in Table 1 (11) .

Table 1. Percentage of titanium tetrahedral based on deconvolution of DR UV-Vis spectra

Sample	Calcinations	Symbol	Tetrahedral
TiO2	-	TiO2	27,10 %
SiO2-TiO2	450 °C	ST 450°C	43,03 %
SiO2-TiO2	500°C	ST 500°C	42,11 %
SiO2-TiO2	550°C	ST 550°C	41,87 %

On the basis of deconvolution DR UV-Vis spectra the titanium tetrahedral fraction increased from 27,10% to 43,03%. This phenomenon verified the formation of Si–O–Ti bonding due to interaction between SiO_2 and TiO_2, which was verified by FTIR spectra through the existence of band absorption at 960 cm^{-1}. Thus, it can be concluded that both results of FTIR and deconvolution of DR UV-Vis spectra are in good agreement.

3.3. Biodiesel synthesis

Generally, transesterification reaction produces biodiesel or methyl ester in liquid phase with glycerol as by-product (3). However, transesterification is not occurred but saponification with soap as main product. In order to confirm the formation of biodiesel at this preliminary investigation, several physical properties of the product are compared with that of its crude oil (Table 2). The characterization included density, viscosity and boiling point.

Table 2. Physical properties of crude oil and biodiesel product.

Sampel	**Physical properties**		
	Density (g/ml)	**Viscosity (ml/s)**	**Boiling point (^{0}C)**
Palm Oil (Bimoli)	0,9055	0,0909	315
Biodiesel (ST 400°C)	0,8978	0,098	290
Biodiesel (ST 450°C)	0,8972	0,098	288
Biodiesel (ST 500°C)	0,8981	0,0961	290

Table 3 shows the results of density, viscosity, and boiling point. Based on the investigation, the biodiesel products have lower density and boiling point compared to that of palm oil. The result is in accordance with a literature that the target of this investigation is to reduce the physical properties of oil product through transesterification of palm oil. The synthesis process is to meet the purpose that the methyl ester produced by transesterification of palm oil can be advantaged as fuel in diesel engine without any change or modification of that engine. Figure. 3 shows the effect of titanium tetrahedral fraction on biodiesel properties

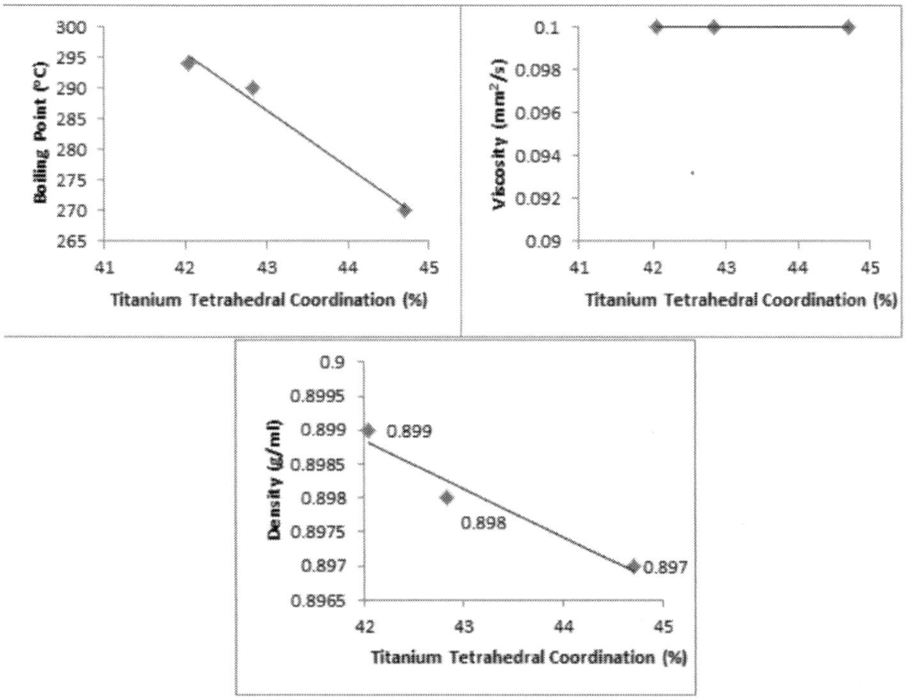

Figure 3. The correlation between titanium tetrahedral fraction in catalyst SiO_2-TiO_2 with physical properties of biodiesel.

Figure. 3 shows the increased titanium tetrahedral fraction in silica-titania catalyst causing decreased in both density and boiling point of biodiesel product but increased in its viscosity. This investigation is attributed to higher the quantity of titanium tetrahedral fraction causing better catalytic activity of silica-titania catalyst. The obtained results is in accordance with previous studies described that catalytic properties of silica-titania is influenced by its local structure, i.e. the tetrahedral as active local structure (10).

In addition, the FTIR apectra is also used to examine the change of functional groups from crude oil to methyl ester as shown in Figure.4 with respect to biodiesel product (450°C). The substantial absorption peaks of palm oil (bimoli) are appeared at wave number of 2922.15 cm^{-1}, 2859.99 cm^{-1}, 1742.80 cm^{-1}, 1453.60 cm^{-1} and 1160.60 cm^{-1} and the remarkable absorption peaks of biodiesel product (450°C) are attributed at wave number of 2922.02 cm^{-1}, 2859.89 cm^{-1}, 1742.45 cm^{-1}, 1453.49 cm^{-1} and 1160.54 cm^{-1}. The strong band absorption at wave number 1800 – 1700 cm^{-1} is attributed to C=O, while the band absorption at wave number 1170 – 1200 cm^{-1} is related to C-O-C vibration of methyl ester and weak absorption at 1600 – 1450 cm^{-1} is corresponding with vibration of C-C group. In addition, strong band absorption at 3000 – 2800 cm^{-1} is related to C-H vibration of methyl group (18).

IOP Conf. Series: Materials Science and Engineering **335** (2018) 012036 doi:10.1088/1757-899X/335/1/012036

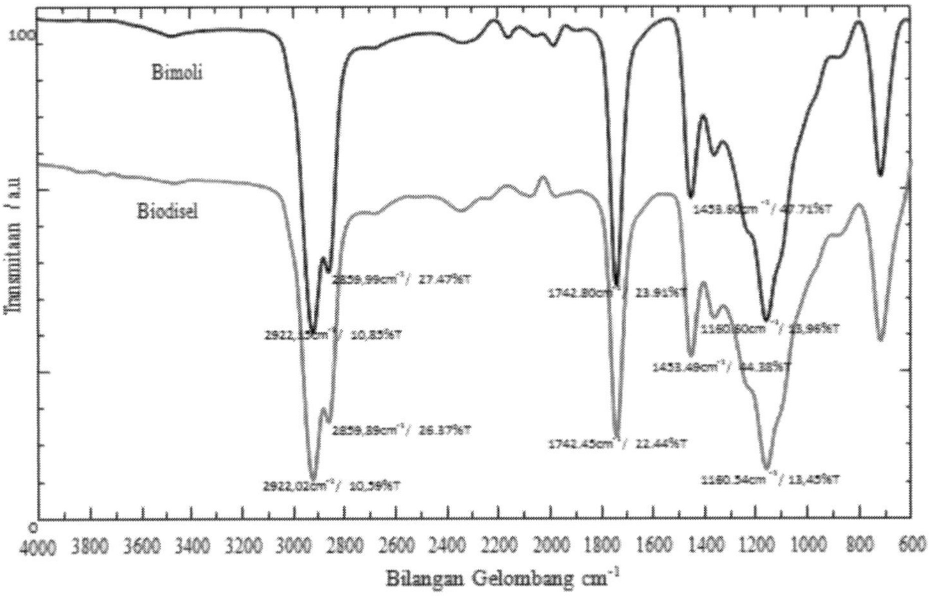

Figure 4. FTIR spectra of crude oil and biodiesel

4. Conclusion

The synthesis of silica-titania from solid precursors using solid-state method can increasing the number of titania tetrahedral framework in silica-titania, compared to the commercial titania. The highest number of titanium tetrahedral obtained by calcining the mixture of titania and silica precursors at 450°C for 8hr. The properties of biodiesel formed during transterification between palm oil and methanol affected by silica-titania. It was proved by the determination of boiling point, viscosity and density.

Acknowledgement

We gratefully acknowledge the funding from Ministry of Research, Technology and Higher Education of the Republic of Indonesia and ***Universitas Negeri Padang*** for financial support under PNPB 2017 research grant.

References

[1] Chouhan APS, Sarma AK. Modern heterogeneous catalysts for biodiesel production: A comprehensive review. Renew Sustain Energy Rev. 2011;15(9):4378–99.
[2] Lokman IM, Rashid U, Taufiq-Yap YH. Meso- and macroporous sulfonated starch solid acid catalyst for esterification of palm fatty acid distillate. Arab J Chem [Internet]. 2016;9(2):179–89.
[3] Ilmi M, Kloekhorst A, Winkelman JGM, Euverink GJW, Hidayat C, Heeres HJ. Process intensification of catalytic liquid-liquid solid processes: Continuous biodiesel production using an immobilized lipase in a centrifugal contactor separator. Chem Eng J. 2017;321:76–85.
[4] Axelsson L, Franzén M, Ostwald M, Berndes G, Lakshmi G, Ravindranath NH. Perspective: Jatropha cultivation in southern India: Assessing farmers' experiences. Biofuels, Bioprod Biorefining. 2012;6(3):246–56.
[5] Lam MK, Tan KT, Lee KT, Mohamed AR. Malaysian palm oil: Surviving the food versus fuel dispute for a sustainable future. Renew Sustain Energy Rev. 2009 Aug;13(6–7):1456–64.

[6] Talebian-Kiakalaieh A, Amin NAS, Mazaheri H. A review on novel processes of biodiesel production from waste cooking oil. Appl Energy. 2013;104:683–710.
[7] Shao GN, Sheikh R, Hilonga A, Lee JE, Park YH, Kim HT. Biodiesel production by sulfated mesoporous titania-silica catalysts synthesized by the sol-gel process from less expensive precursors. Chem Eng J. 2013;215–216:600–7.
[8] Chmielarz L, Gil B, Kuśtrowski P, Piwowarska Z, Dudek B, Michalik M. Montmorillonite-based porous clay heterostructures (PCHs) intercalated with silica–titania pillars—synthesis and characterization. J Solid State Chem. 2009;182(5):1094–104.
[9] Ningsih SKW. Synthesis and characterization of NiO nanopowder by sol-gel process. AIP Conf Proc. 2015;1677.
[10] Nizar UK, Efendi J, Yuliati L, Gustiono D, Nur H. A new way to control the coordination of titanium (IV) in the sol-gel synthesis of broom fibers-like mesoporous alkyl silica-titania catalyst through addition of water. Chem Eng J. 2013;222:23–31.
[11] Sundari R, Hua TI, Aziz M, Nizar UK. the Characterization Study of Ferrites (Magnesium and Manganese) Using Sol Gel Method. 2014;18(3):485–90.
[12] Najem S, Khurshid A. Biodiesel production by using heterogeneous catalyst Biodiesel production by using heterogeneous catalyst. 2014;(March).
[13] Jansri S. Preparation of Vegetable Oil as Biodiesel Feedstock Via Re-Esterification: A Suitable Catalyst [Internet]. Vol. 79, Energy Procedia. Elsevier B.V.; 2015. 143-148 p.
[14] Ruzimuradov ON. Formation of bimodal porous silica-titania monoliths by sol-gel route. IOP Conf Ser Mater Sci Eng. 2011;18:32004.
[15] Ren J, Li Z, Liu S, Xing Y, Xie K. Silica–Titania mixed Oxides: Si–O–Ti Connectivity, Coordination of Titanium, and Surface Acidic Properties. Catal Letters. 2008;124(3–4):185–94.
[16] Kholdeeva O a, Trukhan NN. Mesoporous titanium silicates as catalysts for the liquid-phase selective oxidation of organic compounds. Russ Chem Rev. 2007;75:411–32.
[17] Nur H. Modification of titanium surface species of titania by attachment of silica nanoparticles. Mater Sci Eng B Solid-State Mater Adv Technol. 2006;133(1–3):49–54.
[18] Tonucci L, Coccia F, Bressan M, D'Alessandro N. Mild photocatalysed and catalysed green oxidation of lignin: A useful pathway to low-molecular-weight derivatives. Waste and Biomass Valorization. 2012;3(2):165–74.

Optimation and Determination of Fe-Oxinate Complex by Using High Performance Liquid Chromatography

B Oktavia[1*], E Nasra[1], R C Sary[1],
[1]Chemistry Department, Faculty of Mathematics and Science, State University of Padang, Indonesia

*budhioktavia8@gmail.com

Abstract. The need for iron will improve the industrial processes that require iron as its raw material. Control of industrial iron waste is very important to do. One method of iron analysis is to conduct indirect analysis of iron (III) ions by complexing with 8-Hydroxyquinoline or oxine. In this research, qualitative and quantitative tests of iron (III) ions in the form of complex with oxine. The analysis was performed using HPLC at a wavelength of 470 nm with an ODS C18 column. Three methods of analysis were performed: 1) Fe-oxinate complexes were prepared in an ethanol solvent so no need for separation anymore, (2) Fe-oxinate complexes were made in chloroform so that a solvent extraction was required before the complex was injected into the column while the third complex was formed in the column, wherein the eluent contains the oxide and the metal ions are then injected. The resulting chromatogram shows that the 3rd way provides a better chromatogram for iron analysis.

1. Introduction

Iron metal in ordinary waters is found in the form of Fe (II) and Fe (III) cations. Fe (II) cations are less stable than Fe (III) cations because they will rapidly oxidize into Fe (III) cations (Arpadjan, 2012). Analysis of iron content and iron ion speciation is very important in determining the quality of a water. Octavia et al. (2008) have performed Fe (II) and Fe (III) simultaneous analyzes through conical formation with salicylic acid and fenantrolin using micro column on ion chromatography. The iron content of iron that exceeds the concentration of 0.31 mg / L in the river can cause rust on clothing, porcelain, water pipe and cause unpleasant taste in drinking water coming from the river (Achmad, 2004).

The use of a complex in quantitative metal analysis using HPLC may increase the sensitivity of separation and selectivity of metal ion separation (Pimrote, 2012). The metal can be converted into complex compounds by extraction of neutral chelate formation in an organic solvent containing a complex (Day 2002).

One of the complexing reactions is the formation of chelate complex compounds with organic reagents, ie 8-hydroxyquinoline which is often also called by the name of trivial oxine. Oxine is a widely used substance in the fields of analytical, inorganic and organic chemistry since the 1960s (Morrison, 1965 and Stary, 1963). Further Takeuchi, et al (2001) has also used oxine as a complex in the determination of magnesium and aluminum by using fluorescent HPLC.

Here is a description of iron and oxine. Iron has atomic number 26, density 7,874 g / cm3 (20°C), electron configuration [Ar] 3d6 4s2, melting point 1536°C, boiling point 2750 °C, atomic radius 126 pm (coordination number 12), and abundance in earth crust 62000 ppm (Sugiyarto, 2003). Pure iron is quite reactive. Fast iron is oxidized by moist air to form iron (III) oxide hydrate (rust). The metal is easily soluble in mineral acids. With non-oxidizing acids without air, Fe (II) is obtained. In the presence of air or when used HNO3, some iron becomes Fe (III). Extremely strong oxidizing media such as concentrated HNO3 or dichromate-containing acids make passive iron (Cotton, 1989).

Oxine has a molecular weight of 145.15 g / mol with the molecular formula C9H7ON, is a white crystalline compound with a melting point of 74-76 ° C. It is almost insoluble in water and ether, but is highly soluble in alcohol, chloroform, benzene and mineral acid solutions. Oxy is an amphoter in solution, because it has a nitrogen group and a phenolic hydroxy group. The chemical reaction between the oxide and the metal forms the inner complex compound, in which the metal ion forms a coordination bond with the Nitrogen (N) atom or oxygen atom (O) of the carboxyl group to form a stable fiveth circumference.

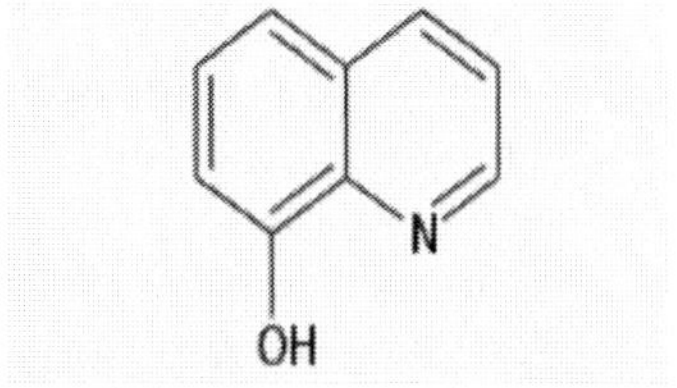

Figure1. Oxine Structure (Day, 2002)

The Fe (III) ion is extracted quantitatively with the oxine in a pH atmosphere between 1.9 to 12.5 producing Fe-Oxinate, wherein the complex dissolves in chloroform by forming a yellow color and gives absorption at a wavelength of 470 nm by a spectrophotometer (Stary, 1963). The reaction between Fe (III) with the oxin takes place with the reaction equation as follows:

$$Fe^{3+} + 3(HOx)_{org} \leftrightarrow (FeOx_3)_{org} + 3\ H^+$$

The oxine is very soluble in alcohol and alcohol can also dissolve easily in water. With this fact there can be several methods in the formation of Fe-oxinate complexes, ie if the oxides and metal ions can form complex compounds in alcohol, no solvent extraction step is required to separate the organic phase. Then the complex can also be formed directly in the chromatographic column and the latter complex is formed by using oxine in chloroform and then followed by solvent extraction before entering into the separation column. All Fe-oxinate analyzes were performed using HPLC and ODS C-18 columns.

In this research, we have determined the Fe (III) ion by using oxine as complexing by doing 3 complex forming method and analysis. A comparison of these three methods can be used to determine Fe (III) content indirectly using HPLC.

2. Experimental

2.1. Apparatus

The experimental apparatus was assembled from a HPLC Agilent 1120 with an injection-loop volume of 20 µl, UV-Vis detector, a ODS C-18column (150×4.6 mm I.D.). Spektronik Genesys 20 D, pH meter merk Hanna Instrument HI2211,analytical balance, separating funnel, glassware (measuring flask, test tube, beaker, dropper, stirring rod) and reagent bottle.

2.2. Chemicals

Analytical reagent grade chemicals were purchased from Merck, unless otherwise noted. $Fe(NO_3)_3.9H_2O$, Oxine, Chloroform, Methanol, Ethanol, Asetonitril, HNO_3, NaOH dan aquades.

IOP Conf. Series: Materials Science and Engineering **335** (2018) 012037 doi:10.1088/1757-899X/335/1/012037

2.3. Research Procedure

2.3.1. Determination of Wavelength of Fe-Oxinate Complex. A total of 1 mL of Fe (III) solution of 10 ppm was added to the reaction tube and 3.5 mL of 0.1 M oxine was dissolved in ethanol. The directly formed complex is measured using Spectronic at a wavelength of 400 - 600 nm.

2.3.2. Determination of Stability Time of Metal Oxinate Complex. The 10 ppm Fe (III) ion solution is inserted into the separating funnel. Then 0.1 M oxine is added to solution. Extract for 20 minutes for Fe (III) to water phase and separate organic phase. Taken its organic phase and measured its complex uptake with a variation of measuring time duration of 0-150 minutes with a span measurement of 10 minute absorbance value using Spectronic.

2.3.3. Determination of Fe-Oxinate complex analysis using HPLC. There are 3 methods for the analysis of Fe oxidation complex as follows :

- The Fe-Oxinate complex is formed by using oxine dissolved in ethanol. The complex formed does not need to be separated and then injected into HPLC for further analysis. The Fe-oxinate complex is injected 20 μl into the ODS C18 column at maximum wavelength using the mobile phase Ethanol: Water = 50: 50 with an eluent flow rate of 0.8 mL / min.
- Formation of Fe-oxinate complex using oxine dissolved in chloroform. Fe oxinate complex that is formed needs to be separated by solvent extraction. Then as much as 20 μL sample is injected into the injector. The analysis was performed using the ODS C 18 column at maximum wavelength. The mobile phase Ethanol: Water = 50: 50 with an eluent flow rate of 0.5 mL / min.

Fe (III) ion analysis is done by forming Fe Oxinate complex in column. Eluen used is 0.1 M oxine in ethanol. The Fe (III) ions injected into the column will then form the Fe oxinate complex when encountered with the oxin in the mobile phase.The analysis was performed using the ODS C 18 column at maximum wavelength. The mobile phase Ethanol: Water = 50: 50 with an eluent flow rate of 0.5 mL / min.

3. Results and Discussion

3.1. Determination of Wavelength of Fe-Oxinate Complex

The determination of the maximum wavelength of Fe-oxinate complex has been determined by the following conditions; as much as 1 ml of Fe (III) solution of 10 ppm inserted into the reaction tube added 3.5 mL of 0.1 M oxine which has been dissolved in ethanol. The complex is formed yellow. The formed complexes are not separated by solvent extraction because there is no phase difference between the water solvent and the ethanol solvent. With such method it is possible to save the analysis time without treating solvent extraction such as separation method with other complex formation. So that the formed complex can be directly measured using Spectronic UV-Vis at wavelength 400 - 600 nm.

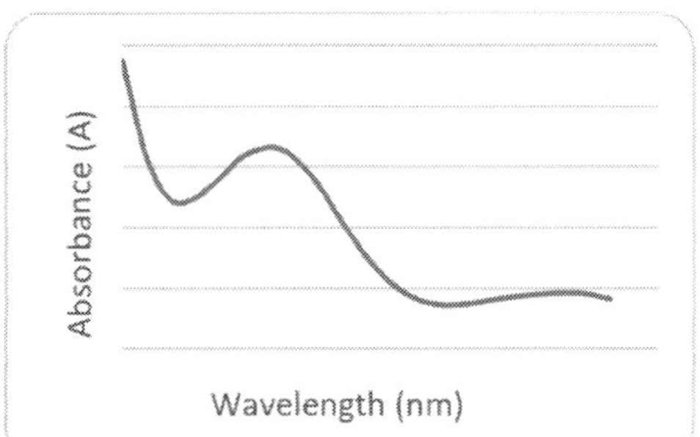

Figure 2. Wavelength curve of Fe-oxinate complex.

Based on the data of the maximum wavelength determination in figure 2 above, it was found that the Fe-oxinate complex formed in ethanol gave maximum wavelength in the 465-470 nm range. The wavelength 470 nm was chosen as wavelength in measurement by HPLC.

3.2. Determination of Wavelength of Fe-Oxinate Complex
At intervals of 110 to 150 minutes there is a stable uptake of the Fe (III) -Oxinate complexes formed. So that at 110 min is the optimum time of the stability of Fe (III) -Oxinate complex which is good for analysis.

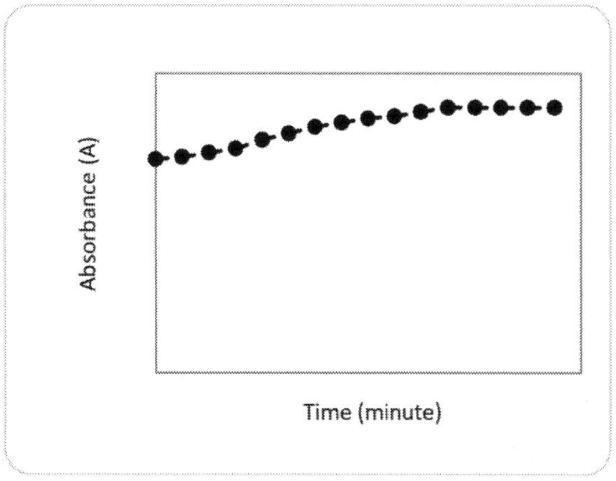

Figure 3. The curve of the relationship between complex stability time and absorbance.

3.3. Determination of of Fe-Oxinate complex analysis using HPLC

3.3.1. The first method. The Fe-Oxinate complex is formed by using oxine dissolved in ethanol. The complex is formed in ethanol, so there is no need for extraction. The formed complex is then injected 20 μl into the ODS C18 column at 470 nm wavelength using the mobile phase Ethanol: Water = 50: 50 with an eluent flow rate of 0.8 mL / min. The retention time of the Fe-Oxinate complex arises at minute 2.8

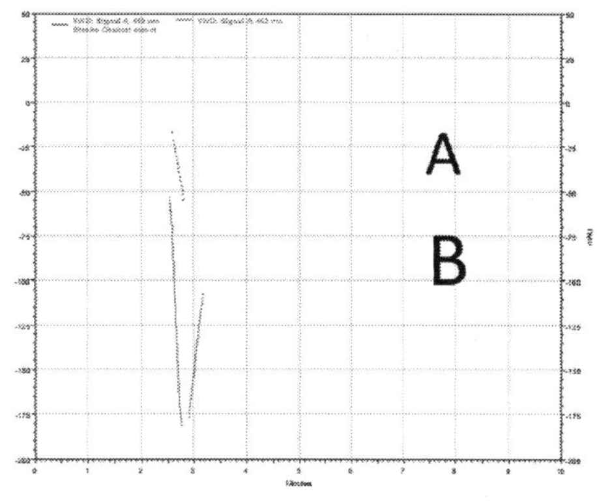

Figure 4. Chromatogram of Fe (III) metal ion analysis in the form of oxinate complex; A) Oxine blends in ethanol and B) Fe-oxinate Complex.

3.3.2. The second method. Formation of Fe-oxinate complex using oxine dissolved in chloroform. The complex is formed in chloroform, so there is need for extraction. The formed complex is then injected 20 μl into the ODS C18 column at 470 nm wavelength using the mobile phase Ethanol: Water = 55:

IOP Conf. Series: Materials Science and Engineering **335** (2018) 012037 doi:10.1088/1757-899X/335/1/012037

45 with an eluent flow rate of 0.5 mL / min. The retention time of the Fe-Oxinate complex arises at minute 7.1.

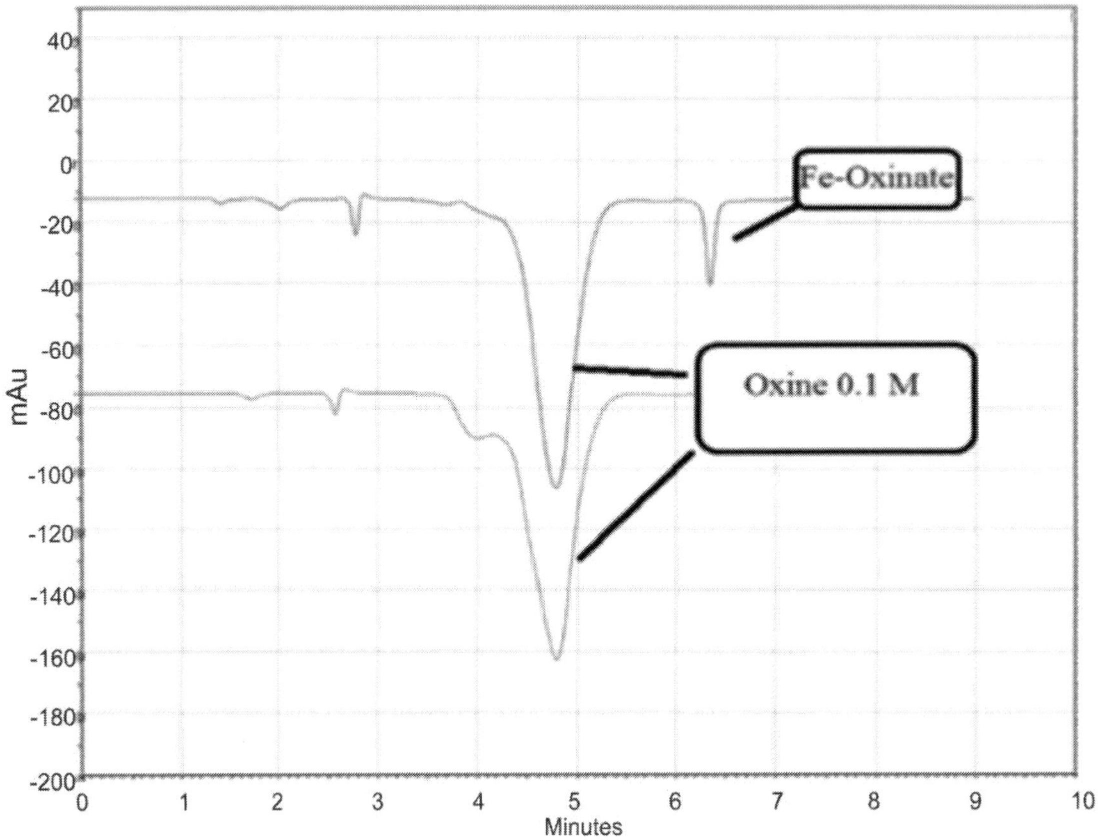

Figure 5. Chromatogram of Fe (III) metal ion analysis in the form of oxinate complex, by using chloroforms as solvent.

3.3.3. The last method. Fe (III) ion analysis is done by forming Fe Oxinate complex in column. The complex is formed in column HPLC, so there is no need for extraction. The ion Fe^{3+} is then injected 20 μl into the ODS C18 column at 470 nm wavelength using the mobile phase Ethanol: Water = 55: 45 with an eluent 0.1 M oxine in ethanol, flow rate of 0.5 mL / min. The retention time of the Fe-Oxinate complex arises at minute 2.8.

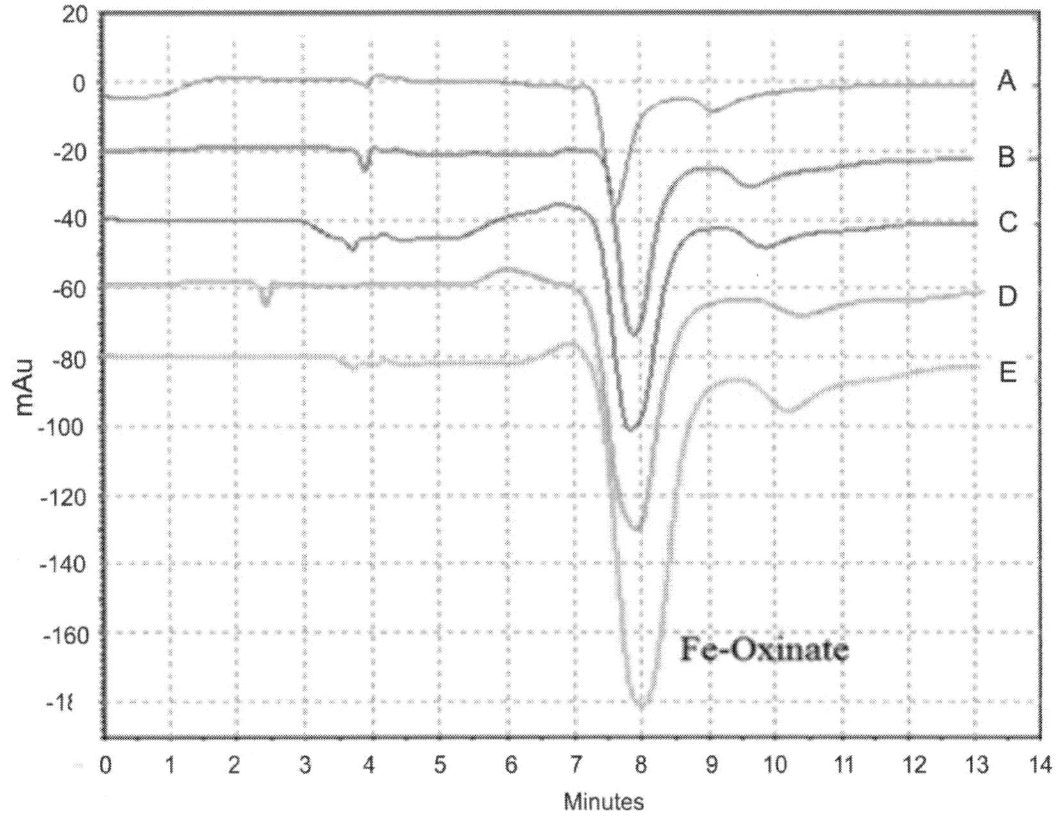

Figure 6. Chromatogram of Fe (III) metal ion analysis in the form of oxinate complex, by using oxine as mobile phase.

4. Conclusion

Based on the results obtained can be concluded that: The Fe-oxinate complex may form in an organic phase at a maximum wavelength of 470 nm. The optimum time of the stability of Fe (III) -Oxinate complex is 110 minutes. There are 3 methods of determining the concentration of Fe-oxinate complex. The complex formation in the chromatographic column provides the best chromatogram in the analysis.

References

[1] Achmad, Rukaesih. 2004. *Kimia Lingkungan*. Yogyakarta : ANDI.
[2] Arpadjan, S. 2012. Field Sampling, speciation, and determination of dissolved iron (II) and iron (III) in waters. Bulgarian chemical Communications, Vol 44, 299-306.
[3] Cotton, Albert. F. 1989. *Kimia Anorganik Dasar*. Jakarta : Universitas Indonesia Press.
[4] Day dan Underwood. 2002. *Analisa Kimia Kuantitatif.* Jakarta : Erlangga.
[5] Morrison, H.G., Freizer, H., 1965, Solvent Extraction in Analytical Chemistry, John Wiley and Sons, New York, 162-165.
[6] Oktavia, B., Lim, L.W., Takeuchi, T., 2008, Simultaneous determination of Fe(III) and Fe(II) via Complexation with Salicylic Acid and 1,10 Phenanthroline in Microcolumn Ion Chromatography, Analytical Sciences, Vol 24, 1487-1492.
[7] Pimrote, W. 2012. Simultaneous Determination of Heavy Metals In Drinking Waters And Wines By Methods : Ion Pair Reversed Phase High Performance Liquid Chromatography And

Capillary Zone Electrophoresis. *International Food Research Journal,* Vol.19, No.3, Halaman : 1079-1088.

[8] Stary, J., 1963, Systematic study of the solvent extraction of metal oxinates, Anal. Chim Acta, Vol 28, 132-149.

[9] Sugiyarto, Kristian H. 2003. *Dasar-dasar Kimia Anorganik Logam.* Yogyakarta : UNY.

[10] Takeuchi, T., Inoue, S., Yamamoto, M., Tsuji, M., Miwa, T., 2001, Fluometric determination of magnesium and aluminium via complexation with oxine in high performance liquid chromatography, Journal of Chrom., A., Vol 910, 373-376.

The Utilization of VCO (Virgin Coconut Oil) in Manufacturing of Solid Soap with Red Betel Leaf Extract Addition (Paper Crotum Ruiz &Pav)

M Sari*

Department of Chemistry Education, IAIN Batusangkar, Jl.Sudirman No.137.Batusangkar, Indonesia

*mayasari_1985@yahoo.com

Abstract. Soap is a compound of sodium or potassium with fatty acids from vegetable oils or solid animal fats, soft or liquid, and foamy. Considering the potential of VCO as the raw material for making soap and supported by the benefits of red betel leaves, then this research is done by making solid bath soap from VCO which is supplemented with Red betel leaf extract. The purpose of this research is to make solid soap from VCO with an extract of red betel leaf based on SNINo.06-3532199. Analyzing VCO oil, which is used for the manufacture of soap, consists of analysis of saponification figures, Iodine number and peroxide number. Has made solid soap from VCO oil with Red betel leaf extract. From the five quality standards established under SNI 06-3532-1994 only two quality standards that can be done that is water quality and the amount of acidity. The percent of water quality obtained is 10% meanwhile the amount of acidity obtained is 9,32%. According to the data, it can be concluded that the solid soap made was not fulfill SNI 06-3532-1994.
Keyword: soap, VCO, red betel leaf

1. Introduction

Soap is a compound of sodium or potassium with fatty acids from vegetable oils or solid animal fats, soft or liquid, and foamy[1]. The soap is produced by the process saponification, which is the hydrolysis of fat to fatty acid and glycerol under alkaline conditions [2]. Sodium Hydroxide (NaOH) and Potassium Hydroxide (KOH) are commonly used to make alkaline condition. The standard of soap quality has been established by the government in SNI no. 06-35321994. This provision is made to ensure the quality of the soap to be used by Community.

One of the main ingredients of soap manufacturing is oil. The oils that can be used such asolive oil, palm oil, fish oil, coconut oil[3]. It already produced a solid soap from VCO (Virgin Coconut Oil). The advantage of coconut oil is a high content of saturated fatty acids, especially lauric acid about 44-52%, so that coconut oil is resistant to oxidation that can cause rancid smell. Coconut oil also contains 13-19% fatty acids, acids palmitate 8-11%, 6-10% capric acid, 5-9% caprylic acid, 5-8% oleic acid, stearic acid 1- 3%, and 2% linoleic acid[3]

Indonesian medicinal plants were recently widely used is red betel *(Pipercrocatum)*. Phytochemical screening of red betel leaves shows the presence of the ingredients essential oil. Soap with extract red betel leaf can be used to reduce whitish complaints caused by bacteria *Candida sp* and *Staphilococcusepidermis*[4]. Antibacterial and antimicrobial effects of soap red betel extract is estimated originate from the content of flavonoids, alkaloids, saponins, tannins and essential oils. Flavonoids act as antibacterial by forming compounds complex to extracellular proteins that interfere with the integrity of bacterial cell membranes. [5]

Focusing on the potential of VCO as the raw material of soap manufacture and supported by The benefits of red betel leaf, then this research is done by making a solid soap bath Of VCO added with red betel leaf extract. This research is expected to obtained a solid soap with oil content of red betel leaf oil based on SNINo.06-35321994.

2. Method

2.1 Tools

An analytical balance, goblets, water bath, stirring glass, oven, separating funnel, erlenmeyer flask, Condensator, biuret, micro biuret, test tube, soap container, blender, Bottle weigh, the basic ingredients of coconut oil coconut meat are sold in Local market, red betel leaf.

2.2 Ingredients

NaOH (pa), glacial acetic acid, KI (pa), KOH (pa), PP solution, HCl, $Na_2S_2O_3$, H_2O_2

2.3 Squeezed Producing

10 coconuts prepared. The shredded coconut meat, mixed with hot water at 70 ° C with a ratio of 1: 1 (1 gram of coconut meat grated: 1 ml water) in the basin. After the coconut meat is squeezed with a cloth and filtered with a plastic filter, the obtained coconut milk is then allowed to stand for 2 hours in order to separate the cream and skim. Creams that contain lots of fat mixed with extracts red betel leaf for the manufacture of VCO contains flavonoid[5]

2.4 Preparation of red betel leaf extract

Selected fresh red betel leaves, which have been sorted based on color, fresh leaves, stems are also cut and then washed, weighed using an analytical balance sheet which weighing 100 grams. Red betel leaf that has been weighed was cut into pieces. Conducted pretreatment by immersion using ethanol 80% (ethanol 96% which has been diluted) of 500 mL for 10 minutes. Red betel leaves yield crushed with a *juicer*. The result was irradiated using *microwave* for 1 minute, with a moderate temperature of 40 O C and then filtered with filter. After being irradiated and filtered red betel leaf is evaporated using *a rotary evaporator* with the temperature of 40 C[6]

2.5 VCO flavanoid producing

red betel leaf extract 300 ml supplemented with 700 ml of coconut cream to a total mixture volume of 1000 ml (concentration of 30% red betel leaf extract in plastic container stirring.) The mixture is stirred and idle for 18 hours. Furthermore, separation between water on the layer Bottom with oil and blondo on the top layer, by opening on the container part. To separate the oil from the blondo is done centrifugation at a speed of 3000 rpm, for 15 minutes. The result blondo settles on the bottom of the tube while the oil is on the top layer. The oil obtained in the form of VCO containing flavanoid is used for the manufacture of solid soap bath.[5]

2.6 Acid Score Determining

A total of 0.1 g of oil was added to 200 ml beaker, then added 95% ethanol as much as 25 ml. The mixture is heated on temperature of 65°C while stirring to form solution. The solution was titrated with 0.1 N KOH with a 1% fenolphthalein indicator until it appears pink. Once it is done calculation of the amount of mg KOH used to neutralize free fatty acids in 1 gram oil sample[5]

2.7 Saponification Determining

2.5 grams of VCO which contain flavanoidred betel leaf in erlenmeyer 250 ml, added 25 NaOH0.5M. The form was made (the process is the same from the beginning, but without using oil samples). Neither experimental materials norformmade duplo (2x replicates). Then reflux up perfect sampling (approximately 30 minutes). The saponification reaction is complete if the drip results reflux in a test tube containing clear colored water. After cooling, the reflux results added 2 drops of phenoltalein indicator and was subjected to HCl until the color was correct lost[5]

2.8 Soap making

20 grams of flavanoid reacted VCO with NaOH. The amount and concentration reacted, determined based on the saponification number of VCO containing flavonoid of red betel leaf. The products of soap are made in 3 variations of NaOH concentration, ie 25%, 30%, 35%, calculated based on the number of saponification and the amount of water required for making each of the concentrations of NaOH. Each treats NaOH concentration. Every treatment of NaOH concentration made 3 replications. After the VCO is reacted with NaOH, The mixture is stirred slowly until thickened and homogeneous. The soap produced in a clean container that has been prepared and allowed to stand for 4 weeks. Furthermore, the resulting bath soap is tested for its quality based on SNI no. 06-3532-1994[5]

2.9 Peroxide Score determination

1 g sample of the oil was added 30 ml mixture solvent, consisting of glacial acetic acid and chloroform with a ratio of 3: 2 (v / v). Then heated under the boiling point for 2 minutes. Added 0.5 ml KI saturated and reheated for 2 minutes. Then added distilled water and titrated with Na 2 S 2 O 3 of 0.1 N or 0.01 N by using a 0.5% starch indicator. The same thing is done toward the blank. The results dictated in meq O 2/1000 g sample

3. Result and Discussion

The process of soap making is done by two processes, namely, *cold saponification process* and *Semiboiled saponification*. From both methods are done to make the soap like the picture above obtained conclusions about the characteristics of the soap made:

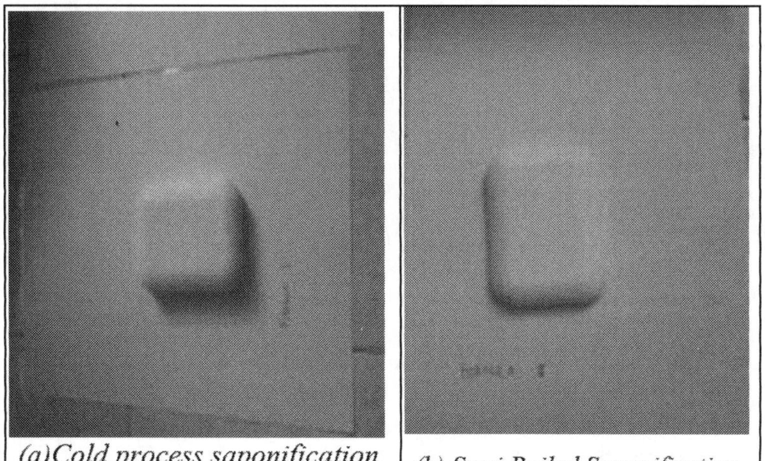

(a)Cold process saponification *(b) Semi Boiled Saponification*

Figure 1. The soap made by (a)*Cold Process Saponification* (b) *Semi Boiled Saponification.*

The comparison of physical characteristic of soaps can be shown in Table 1.

Table 1. The comparison of physical characteristics of soaps toward both methods used

Physical Parameters	Method	
	Cold Process Saponification	Semiboiled Saponification
Color	White	White
Texture	Rude	Smooth
Smell	Rancid	No smell

Measurements of saponification rates on VCO are performed before the soap is made by using the oil. The saponification rate obtained from the calculation were11.12 mg / g. Further testing of soap that has been made, the following is the test data shown in Table 2.

Table 2. The Result of Soap Quality Test

No	Test	Test result	SNI 06-3532-1994
1	Water content	10%	Max 15%
2	Degree of Acid	9.33%	70%

The method used in the process of soap making in this study is Batch Method. According to Wasitaadmadja on batch process, the tool used is a large container that serves as a boiling place for raw materials[7]. This boiling place is also called as a kettle, so the batch process on soap making is called as the boiler process. This round kettle is equipped with a heating coil. Raw materials are fed from top tool along with caustic soda (NaOH) and water for the process of making soap. Heating done for several hours, so that obtained pure soap that can be processed into various forms of soap. In batch process there are 2 (two) process developed, that is *Cold-Process saponification* and *Semiboiled saponification.* In this process the soap is produced containing impurities (impurities) of fatty residues[8]. Fat is simply melted in a vessel equipped with a stirrer andan amount of NaOH solution as the additional with rapid stirring. After the emulsion process and coagulation, the product is poured on the printing tool. The saponification process is enhanced by way of cooling and hardener. This process is a batch system saponification is the simplest way, because it does not require much equipment.

This method is adapted for making soap added with glycerin as well red betel leaf extract. Glycerin serves as a softener and betel leaf extract used to improve the quality of soap as an antimicrobial. Study that had done can be seen that the soap products are made with semi-boiled saponification process obtained better results if compared with cold process saponification. As shown in figure 1 and 2, the soap texture of the semi boiled method more refined when compared with cold process. The rancid smell coming from the saponification reaction is noticeably sharp on the soap by the cold process method. Probably caused by the amount of impurities or impurities contained in the soap, thus affecting the texture and aroma produced.

The addition of red betel leaf extract on the process of making soap is done with boiling technique. This technique is done by boiling red betel leaves to water thick red stew. Stew technique is chosen because it is simpler inside workmanship other than the reason unavailability of supporting laboratory instruments for doing maceration techniques that are considered to extract better. From research expected to be perfect. But in the study did not proceed with Photochemical testing to ensure the content of red betel leaf extract. Class of glycoside compounds, steroids, triterpenoids, flavonoids, tannins and antrakinonin red betel leaf extract by maceration technique[9]. Therefore, it is suspected that in red betel leaf extract produced in this study also contains the compounds. The addition of red betel leaf is done to improve soap quality because of the content of triterpenoid compounds, flavonoids and tannins shows that red betel plants have activity as an antimicrobial able to fight some gram positive and negative bacteria.

The antibacterial and antimicrobial effects of red betel extract soap are underestimated to originate from flavonoid content, alkaloids, saponins, tannins and essential oils. Flavonoids act as Antibacterial by forming complex compounds against extracellular proteins, which is disturbing the integrity of bacterial cell membranes. Alkaloids also have the ability as antibacterial, by interfering with the peptidoglycan component of the bacterial cell. While tannin works as antibacterial with its effect as astringent so can induce complex formation between tannin with microbial substrate. Saponinis one of the content in red betel extract that has antibacterial and antifungal activity. Research by Soetan in Farida Zubirshows that saponins have pharmacological activity against *Staphylococcus aureus* and *Candidaalbicans*[1].In the clinical assessment of Farida Zubir, red betel leaf soap was significant reducing the total score (skin redness, odor, mucus, edema and squamous). As well as subjective judgment by the patient shows that red betel soap extract can be reduced the whiteness score on usage for 1 week. Thus, it can be concluded that red betel leaf is recommended to be used as an antibacterial effect. However, clinical testing of red betel leaf soap effect was not performed in this study.

In this research, the manufacture of VCO is not done, considering the time and limited cost. The oil used to make soap in this study is VCO commercials oil that can be purchased at Pharmacies. VCO or Virgin Coconut Oil is coconut oil containing high fatty saturated acid content, especially lauric acid About 44-52%, so that coconut oil is resistant to odor-causing oxidation rancid. Coconut oil also contains 13-19% fatty myristic acid, palmitic acid 8-11%, 6-10% capric acid, 5-9% caprylicacid, 5-8% oleic acid, 1-3% stearic acid, and 2% linoleic acid[8].

In the research, there were 11.12 mg / g saplings, which means 11.12 Mg KOH to powder one gram of fat or oil. This number does not match allegations which supposedly because saponification numbers also explain the large amount of fatty acids, which are bounded as triglycerides or fatty acids in the oil. The shorter of the acid chain Fat, then the greater of number of saponings as well as vice versa. In fact, VCO has a carbon chains ≥ 20 in the form of tristerin so it should have a small sampling number. Here's a picture of the tristerin compound.[2]

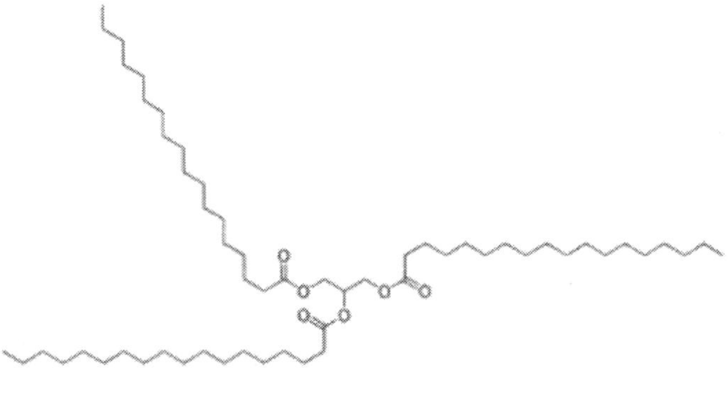

Figure 2. Structure of Tristerin Compound

The magnitude of this saponization value occurs possibly because of the small level purity of the VCO used. This cannot be ascertained because of the VCO used in the study are VCO purchased.
Table 2 is the result of testing of soaps made by the semiboiled saponification method. From the test results can be seen that the water content in soap is 10% while the SNI standard is a maximum of 15%. Water content will affect the hardness of soap. It can be concluded that the soap made in this study has moisture content, which is good because if more than 15%, the resulting soap will be so hard inefficient in use and the soap becomes easily soluble in water.

In table 2 also can be seen the amount of acidity of produced soap products. Amount the acidity of soap obtained was 9.33%. The amount of fatty acids is the total amount of all fatty acids in soap that have or have not yet reacted with alkali. This value is still far from expectation because of the quality standard for the amount of acidity of the perfect soap is 70% this means there are 70% fatty acids and ingredients added As a filler in soap making less than 30%. The goal is to improve the efficiency of cleaning process of dirt in the form of oil or fat when using the soap. Commonly added fillers are honey, perfume, glycerol, water glass, milk protein and so forth. The purpose of adding a filler to provide a compact and solid form, moisturizes, adds a nutrient required by the skin[10]. The amount of fatty acids in the soap produced is only 9.32% and the rest consists of other fillers. In fact, a good soap should have more fatty acids so that saponification reactions can be maximized.

Thus, it can be concluded that the soap made not requires the soap quality based on SNI 06-3532-1994. This is probably due to a reaction Sapling does not run perfect. There are some conjectures about the results of research that have not fulfilled SNI 06-3532-1994, including *semi-boiled saponification* procedure did not take place is perfect because should the heating process in this procedure should be kept constant. However, this case cannot be done due to lack of supervision on work procedures. Imperfection of saponification reaction will cause the resulting glycerol to be less than 10.33%. Although no testing of glycerol is done, it can be seen in quantities low fatty acids.

4. Conclusion

The solid soap has made from VCO oil with red betel leaf extract. By involving five quality standards specified under SNI 06-3532-1994 are only two standards of quality can be done that is water quality and amount of acidity. The percent water quality that obtained is 10% while the amount of acidity obtained is 9,32%. From the data, it can be concluded that the solid soap made not yet fulfill SNI 06-3532-1994.

References

[1] Warra A A. 2013 A report on soap making in Nigeria using indigenous technology and raw materials*African Journal of Pure and Applied Chemistry* **7**4139-145.

[2] Dileesh S,Adithya M, Sankar A, and Peter C V 2013 Determination of saponification, acid and ester values; percentage of free fatty acids and glycerol in some selected edible oils: Calculation of concentration of lye needed to prepare soap from these oils*Research Scholar ISSN* 2249–6696.

[3] Langingi R, Momuat L I, and Kumaunang M G 2012 Pembuatan Sabun Mandi Padat dari VCO yang Mengandung Karotenoid Wortel*Jurnal Mipa Unsrat Online***1**120–23.

[4] Marliyana S D andHandayani N 2013 Aktivitas Antibakteri Minyak Atsiri Daun Sirih Merah(Piper CrocatumRuiz & Pav) *ALCHEMY jurnal penelitian kimia***9**.

[5] Zubier F, Bramono K, Widaty S, Nilasari H, Louisa M, and Rosana Y 2010 Efikasi sabun ekstrak sirih merah dalam mengurangi gejala keputihan fisiologis, efikasi sabun ekstrak sirih merah dalam mengurangi gejala keputihan fisiologis*Majalah Kedokteran. Indonesia* **60**.

[6] Nisa G K, Nugroho W A, Hendrawan Y, and PertanianJKPFT 2014 Ekstraksi daun sirih merah (Piper crocatum) dengan metode microwave assisted extraction (MAE) *Skripsi Jurusan Keteknikan Pertanian Fakultas Teknologi Pertanian Universitas Brawijaya Malang.*

[7] Budimarwanti C 2014*Analisis Lipida Sederhana dan Lipida Kompleks.*

[8] Mishra D 2013*Preparation of Soap Using Different Types of Oils and Exploring its Properties.*

[9] Julia R 2011 Daya anti mikroba ekstrak dan fraksi daun sirih merah (piper betle lim) *Jurnal Ilmu Dasar***12** 1 6–12.

[10] Cavitch S M 2001*Choosing Yours Oil, Oil Propeties of Fatty Acid.*

IOP Conf. Series: Materials Science and Engineering **335** (2018) 012039 doi:10.1088/1757-899X/335/1/012039

Thermal and Surface Evaluation on The Process of Forming a Cu2O/CuO Semiconductor Photocatalyst on a Thin Copper Plate

R Zainul*, B Oktavia, I Dewata, J Efendi

Laboratory, Chemistry Dept, FMIPA Universitas Negeri Padang, Indonesia

*rahadianzmsiphd@yahoo.com

Abstract. This research aims to investigate the process of forming a multi-scale copper oxide semiconductor (CuO/Cu_2O) through a process of calcining a copper plate. The changes that occur during the formation of the oxide are thermally and surface evaluated. Evaluation using Differential Thermal Analysis (DTA) obtained by surface change of copper plate happened at temperature 380°C. Calcination of oxide formation was carried out at temperature 380°C for 1 hour. Surface evaluation process by using Scanning Electron Microscope (SEM) surface and cross-section, to determine diffusion of oxide formation on copper plate. The material composition is monitored by XRF and XRD to explain the process of structural and physical changes of the copper oxide plate formed during the heating process.The thickness of Cu plates used is 200-250 μm. SEM analysis results, the oxygen atom interruption region is in the range of 20-30 μm, and diffuses deeper during thermal oxidation process. The maximum diffusion depth of oxygen atoms reaches 129 μm.

1. Introduction

Copper oxide is a semiconductor material that is widely used as a photocatalyst(1). In its role as a photocatalyst, the electrons in the valence band will be excited to the conduction band. Semiconductor resistivity is about 5 x 10^7 μΩ cm^{-1}(2). The ability of electron excitation will be easier when the distance between PV and PK is getting smaller. The energy needed to excite the electron is called an energy gap.

The yellow copper metal surface will be oxidized by air containing oxygen to form a thin, Cu_2O, reddish layer. This color is due to the electron being filling the d orbitals with energy of 220 kJ/mol, lower than the s and p orbitals. As a result, electrons can be excited to higher energy levels by photons, the red and blue spectrum. Meanwhile, the copper atoms will reflect the red and yellow spectrum(3).

The abundance of copper metal in nature is 0.005%(4). However, the copper oxide is not present in nature independently, but rather in the form of a copper (I) iron (III) sulfide, $CuFeS_2$, a metallic-looking solid that has the two mineralogical names of chalcopyrites and copper pyrites. A rarer mineral, $CuAl_6(PO_4)_4(OH)_8.4H_2O$ is valued blue gemstone turquoise.

Cu_2O extraction process can be done with pyrometallurgy technique (thermal process) and an aqueous process (hydrometallurgy). For the pyrometallurgical process, the concentrated ore is heated (a process called roasting) in a limited supply of air. This reaction decomposes the mixed sulfide to give iron (III) oxide and copper (I) sulfide :

$$4CuFeS_2(s) + 9O_{2(g)} \rightarrow 2\ Cu_2S_{(l)} + 6\ SO_{2(g)} + 2\ Fe_2O_{3(s)}$$

Sand is added to the molten mixture, converting the iron (III) oxide into a slag of the iron (III) silicate :

$$Fe_2O_{3(s)} + 3\ SiO_{2(s)} \rightarrow Fe_2(SiO_3)_{3(l)}$$

This liquid floats on the surface and can be poured off. Air is again added, causing sulfide to be oxidized, and simultaneously copper (I) sulfide is converted to copper (I) oxide :

$$2\ CuS_{(l)} + 3\ O_{2(g)} \rightarrow 2\ Cu_2O_{(s)} + 2\ SO_{2(g)}$$

In this study, the formation of copper oxide is carried out by thermal oxidation method with a direct copper thin plate calcination technique. The process of oxide formation is evaluated by optical and gravimetric methods. The internal surface and morphological evaluation of copper plate to copper oxide was analyzed with temperature thermodynamic variables using SEM, DTA, XRF and XRD. It aims to find out how the thermodynamic condition of copper oxide thin plate synthesis process is optimum, to obtain Cu_2O photocatalyst material which can be used as light room electrode panel.

2. Experimental Section

2.1. Tools and Materials

In this study, molecular and spherical analysis of crystals using software Balls & Sticks ver. 1.81a (created by Sung J.Kang et al) using a computer brand hp pavilion processor intel core i5 with memory (ram) 12 GB. The others tools used in this research is a SEM (Hitachi S-3400N), XRD (PANalytical pw30/40), X-Ray Fluorescence, Differential Thermal Analysis (DTA), Furnace, Analytic Tools, and Scales tools glasses. The materials used in this research is the glass, Plate of Cu (PT Metalindo), NaOH (Merck), acetic acid (CH_3COOH) (Merck) and aquadest.

2.2. Methods

2.2.1. Preparation of Cu Plate. The copper thin plate is manufactured by PT Metalindo. The plate used has a thickness of 200 μm-250 μm. Plates prior to use should be cleaned using sodium hydroxide (alkaline) and acetic acid (acidic). After rinsing at acidic and alkaline conditions, followed by washing with water and drying for 15 minutes at 45°C. After drying, the next process is characterized by a thin copper plate with SEM, XRF and DTA(5-7).

2.2.2. Synthesis Cu2O Thin-Plate. The preparation of Cu_2O plates is carried out by oxidation techniques at high temperatures under oxygen conditions. The cleansed Cu plates are then calcined with the temperature of the optimization time of evaluation with DTA during optimum calcination time. The length of calcination in accordance with work procedures in previous studies reported Rahadian et al (2015), the duration of Cu calcination plate is 1 hour (6). Result of a plate of Cu_2O characterized using XRD, XRF and SEM.

2.2.3. Measurement of Mass Addition of Copper Oxide Plate. After the calcination process is done, the thin copper oxide plate is weighed using an analytic scale. The used plate is 4 cm wide and 38.5 cm high.

3. Results and Discussion

In thermodynamic conditions 25°C and 1 atm, the distribution of atoms of Cu at the surface area of $1 nm^2$ reaches 20 atoms. In volumetric $1\ nm^3$ there are 84 Cu atoms with radius size 1.278 A (127.8 pm). In this thermodynamic condition, there is no gap between atoms of Cu (Cu-Cu) as shown in Figure 1. According to Lange, the size of Cu atoms in thermodynamic conditions 1 atm and 25°C are 1.278 A (atomic radius), 1.17 A (covalent radius) and when ionized in the crystallized system will be reduced to 0.96 A (Cu^{+1}) and 0.72 A (Cu^{+2}). Analysis with Ball & Sticks ver. 1.81a, there is a spherical potential that causes oxygen interruption on the surface of the copper thin plate as shown in Figure 1b.

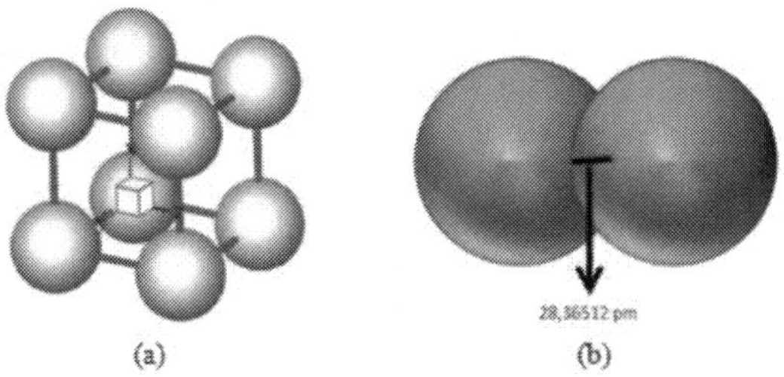

Figure 1. Atomic Cu illustration by using software Balls & Sticks ver. 1.81a (created by Sung J.Kang et al). (a) Simple Cubic for Cu Crystalline and (b) No gap between Cu atoms, but overlap of 28.36512 pm

The crystallographic analysis to Cu crystal found that Cu crystal lattice has side (a) of 3.6147 A (at 20°C), and cubic crystal (face-centered cubic, fcc) the oxygen atom and Cu bonds reach 1.86 A (Cu^{+1}) and 1.97 - 2.66 A in Cu(II) and the 1083°C liquid point as shown in Figure 2. From the Lange data, Cu metal has a density of 8920 kg/m^3 and a thermal expansion coefficient of 16.5 x 10^{-6} K^{-1}. However, if the calculation is done with Lange data, it is known that there are 4.38 Cu atoms at 25°C, in 1000 nm side length. Based on this analysis, the position of Cu atoms on the plates will be united or attached at 28.36512 pm because of the compressibility factor of the preparation of metal atoms.

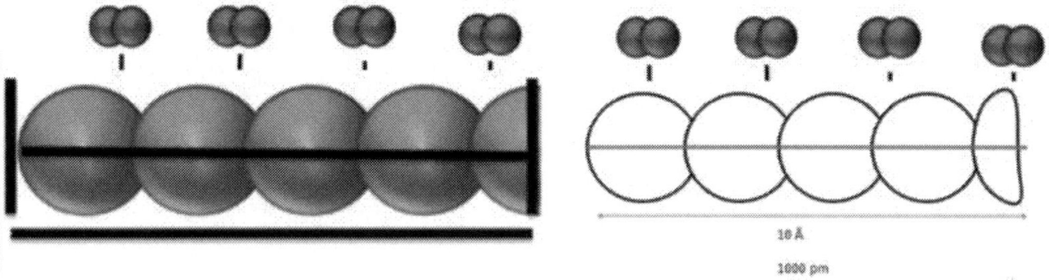

Figure 2. Position of Cu atoms according Lange

Based on spherical analysis and atomic size, there is no gap between Cu atoms at 25°C and 1 atm pressure. In this condition, the oxygen atom can only be located on the surface layer (external interface) as shown in Figure 2. This is because the oxygen molecule (O_2) has a larger size that is 60.3 pm in a straight position and 140 pm (position transverse O^{-2}) and 176 pm (O^{-1} cross-sectional position)(8). Another factor that prevents O_2 interruption from entering or migrating into Cu crystals is the Van der Wall bonding force which has 140 pm grating. DTA evaluation of the thin plate of Cu, known at 370.44°C becomes an unstable state in the surface structure of copper.

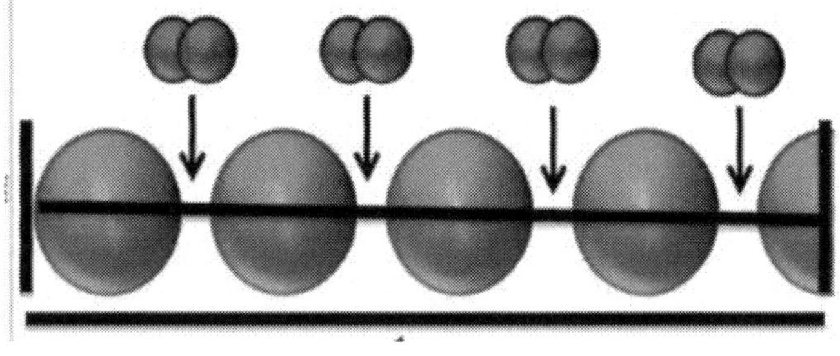

Figure 3. The gap between atoms is formed when heating at 370.44oC for 1 hour.

IOP Conf. Series: Materials Science and Engineering **335** (2018) 012039 doi:10.1088/1757-899X/335/1/012039

The DTA results inform the unstable instability of the atomic atoms of Cu due to the increase in the distance between the atoms and the electron velocity distribution in the conduction band increases so that the Cu atoms becomes smaller by the temperature increase. The decrease of the digit finger Cu with electron movement to the conduction band reaches 31.8 - 55.8 pm. Expansion of the distance between Cu atoms reaches 0.1650 pm. Although thermal expansion is insignificant, electron velocity distribution under thermodynamic conditions 370.44°C has caused a significant decrease in Cu radius radius. With this decrease, as shown in Figure 3, the gap between the Cu-Cu atoms grew to the range of 73.63 pm. in this condition, the invasion (O_2 molecular oxygen spatial migration) into the internal Cu thin plate causes the opening of the copper atomic gap deeper and lasts for a calcination process for 1 hour.

The result of SEM analysis as shown in Figure 4, the surface of Cu covered by oxygen molecule reaches 20-30 μm thickness. In this area, the molecular oxygen density reaches 1,0002713 g/cm^3. In thermodynamic conditions 25°C and 1 atm pressure, there are 25 oxygen molecules in 1000 nm^3. According to J. O. M. Bockris et al (in Gordon M. Barrow, 1988: 771), the thickness of the first metal layer and its environment ranges from 10-100 μm(9). If this assumption is used in the atomic interaction analysis between the atoms of Cu and the oxygen gas molecule, then in volumetric from the area of 1 A^2 with the thickness there are about 8,400,000 to 84,000,000 O_2 molecules. The potential of this invention will increase with the opening of a slit of copper atoms on the surface of the thin plate and the momentum of oxygen atoms increasing as a result of kinetic energy by the increase of heat during the calcination process.

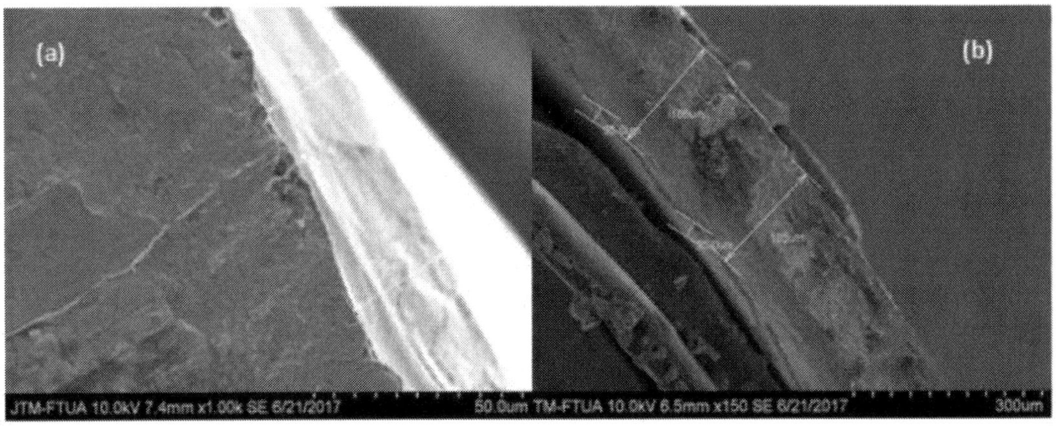

Figure 4. SEM Images of the diffusion layer of the oxygen molecule (a) and the mobility area of the oxygen atom into the thin copper plate during the formation of the copper oxide.

The calcination process lasts for 1 hour and the cooling process is reached after 12 hours. The copper oxide analysis was performed using XRD, which yielded 92.6% Cu_2O and the remaining CuO. This result is confirmed because after the oxygen migration into the copper plate internals, there is a high frequency collision resulting in the formation of ionic bonds between Cu(I) and O^{-2} atoms. In this condition, the oxide crystal structure is faced center cubic (fcc). The population of oxygen atoms that migrate during the calcination time (t=1 hour) causes an internal attack at a maximum depth of 129 μm. As the temperature begins to drop, after an hour, the internal migration of O_2 molecules begins to subside and reverse .

In this condition, molecules of Oxygen which are not experiencing ineffective momentum when collided with Cu atoms, will exit and leave space which increases the pressure on the gap and the intercellular space in the copper oxide crystals. This is what causes the transformation of Cu crystal from FCC into Monoclinic. However, the chance of transformation is still small (4%) because the energy that effects Cu atoms (Cu) becomes Cu atoms (II) is not capable of causing changes in larger quantities. The increase in temperature has led to an increase in the pressure in the intercellular space in the copper plate and causes a decrease in the size of Cu ions from a radius of 96 μm to 72 μm. This results in a decrease in the density of the oxide and increase the band gap return.

Increased Band Gap occurs, because the size of Cu(II) ions is smaller than Cu(I) ions, and the interacting Oxygen ions are 1.40 A (O^{-2} radius) and 1.76 A (O^{-1} radius). In Cu crystal (cuprite) Cu_2O (cuprite) with red density 6.0 g/cm^3, has a refractive index of 2,705 and a melting point of 1235°C(8). Meanwhile, CuO (tenorite) crystals are monoclinic with density 6.3-6.5 g/cm3, black, refractive index 2.63 and melting point 1325°C. Although the density of CuO is greater than Cu_2O, but because the crystallographic system is different so that the distance between Cu-O atoms is more separated (greater than Cu_2O), causing greater energy bandwidth. In Cu_2O the density is lower, but the lattice distance between the atoms is smaller so the energy band width is also smaller. This may lead to a change in the copper oxide photocatalytic capability, as the temperature rise causes the widening of the energy band, and the greater the gap energy to excite the electron from the valence band to the conduction band(10; 11).

The oxygen changes that migrate and bind to the surface of the copper thin plate are evaluated by XRF. The increase of the oxide compound is evaluated on the surface by calculating the mass gain. The copper oxide formed signifies the oxygen and Cu interactions on the thin plate surface has taken place at an optimum temperature of 370.44°C. The presence of a 4% shift could be due to the calcination experiment carried out at a constant temperature of 380°C.

4. Conclusion

Synthetic copper thin oxide plate successfully carried out at optimum temperature of 380°C. XRD analysis results confirmed that 92% Cu_2O and 4% CuO were formed during 1 hour of calcining and 5 hours of cooling. The thermal evaluation process through SEM and computational calculations with Ball & Sticks successfully reveals that Cu changes and the formation of Cu_2O and CuO occur as a function of the energy distribution and position of electrons and atomic atoms. Population and migration of atomic atoms can be described so that the synthesis process can be done effectively and efficiently. The level of efficiency of thin plate formation of Cu_2O reached 92%.

Acknowledgment

Thanks to the Minister of Research of Higher Education and Technology, on the Grand Research of Penelitian Terapan Unggulan Perguruan Tinggi in 2017 the first year (1), Superior Research Team of Chemistry Department State University of Padang, thank you for your suggestion and discussion with Dr. Dedy Wicaksono (Swiss Germany University), Prof. Dr. Syukri Arief, M.Eng. (Inorganic-Nanomaterial Chemistry of Andalas University), and a member of the Research Team incorporated in Zainul for Advanced Material Processing (ZAMP).

References

[1] Amy L. Linsebigler; Guangquan Lu and John T. Yates J. 1995. *Chemical Reviews* 95:735-58
[2] Tipler PA. 1982. *World Publisher Michigan*:316, 7, 686, 7
[3] Overton GR-CaT. 2010. *Descriptive Inorganic Chemistry* Fifth Edition:95-109;563-9
[4] Fahlman BD. 2007. *Springer*:3
[5] Zainul R. 2016. *Der Pharma Chemica* 8:388-95
[6] Zainul R, Alif A, Aziz H, Arief S, Dradjad S, Munaf E. 2015. *Research Journal of Pharmaceutical, Biological and Chemical Sciences* 6(4):353-61
[7] Zainul R, Alif A, Aziz H, Yasthopi A, Arief S, Syukri. 2015. *Journal of Chemical and Pharmaceutical Research* 7(11):57-67
[8] Lange NA. 1973. *Lange's Handbook of Chemistry* 11:4-90;3-119;3-21;3-8;2-45
[9] Barrow GM. 1988. *McGraw-Hill* ISBN 0-07-003905-4:825-9
[10] Butler AWaKT. 2014. *Account of Chemical Research* 47:364-72
[11]Cook WHSaEL. 1973. *J. Phys. Chem* 2:163-200

IOP Conf. Series: Materials Science and Engineering **335** (2018) 012040 doi:10.1088/1757-899X/335/1/012040

Implementation of Biplot Analysis for Mapping Elementary and Junior High Schools in West Sumatra Based on National Examination Results 2016

N Amalita*, D Fitria, V Distian

Mathematics Department, Faculty of Mathematics and Natural Science, Universitas Negeri Padang, Jl. Prof. Hamka, Padang, Sumatera Barat, Indonesia 25131

*nongaditya@gmail.com

Abstract. National examination is an assessment of learning outcomes that aims to assess the achievement of graduate competence nationally. The result of the national examination is used as a mapping of educational issues in order to arrange the national education policy. Therefore the results of National Examination are used, also, as a reference for the admission of new students to continue their education to a higher level. The results of National Examination in West Sumatra in 2016 decreased from the previous year, both elementary schools (SD) and Junior High School level (SMP). This paper aims to determine the characteristics of the National Examination results in each regency/ city in West Sumatra for elementary and junior levels by using Bi-plot analysis. The result of Bi-plot Analysis provides the information that the results of the National Examination of Regency/ City in West Sumatra Province are quite diverse. At Junior High School level there are 9 of Regencies/ Cities which have similar characteristics. English subjects are the greatest diversity among all of subjects. The calculation results of the correlation of each variable in junior high school level are positively correlated. The variables with positive correlation are mathematics that correlates with English. Based on the mark of National Examination for elementary school level in West Sumatra, there are 8 Regencies/ Cities have similar characteristics. The correlations of each variable at the elementary level are positively correlated. The variables that have positive correlation are Sciences (IPA) with Language.

1. Introduction

One of the objectives of the National Exam (UN) is to assess the achievement of national graduate competence on certain subjects in science and technology groups. The National Exam is a test conducted in the last year of each educational unit. Until now the UN is still being implemented although there are still pros and cons and society. The meaning of the National Examination is the evaluation system of basic and intermediate education standards nationally and the quality equation of education between regions conducted by the Center for Educational Assessment, Depdiknas in Indonesia based on the Law of the Republic of Indonesia number 20 of 2003.

Benefit UN in BSNP 2016, is

- Mapping the quality of educational programs and / or educational units
- Selection consideration of the next level of education; and

- The basis of guidance and provision of assistance to educational units for equity and improvement of education quality.

Although the value of the UN is not a graduation of a student, the UN is still a guideline for continuing education to a higher level, so for the students the value of the UN is very important and should strive to improve the test results in order to get a favorite school. While for the school is a pride and as one indicator as a quality school and excel. Not all subjects conducted National Examination, for Elementary School level there are three subjects namely Mathematics, Bahasa and Science. For Junior High School the subjects are Indonesian, Science, Mathematics, and English
Furthermore, the average value of junior high school and elementary level in the regency / city in West Sumatra can be seen in the following table:

Table 1: Average value of junior high school and elementary level in the regency / city in West Sumatra

No	Name of District / City	Average National Exam	
		SD	SMP
1	Padang City	75,81	66,96
2	Bukittinggi City	83,67	64,51
3	Padang Panjang City	74,38	60,12
4	Sawahlunto City	77,55	55,98
5	Solok City	55,35	54,31
6	Payakumbuh City	75,65	59,48
7	Pariaman City	75,04	55,67
8	Agam District	79,55	52,44
9	Pasaman District	72,34	49,42
10	Lima Puluh Kota District	77,92	52,91
11	Solok District	71,82	50,03
12	Padang Pariaman District	68,87	48,51
13	Pesisir Selatan District	73,69	44,03
14	Tanah Datar District	75,58	56,01
15	Sijunjung District	71,64	49,99
16	Kepulauan Mentawai District	60,08	53,76
17	Pasaman Barat District	67,63	47,58
18	Solok Selatan District	78,89	49,18
19	Dharmasraya District	75,54	50,31

In the table above shows that the average value of the UN for the highest junior high school is the city of Padang. While for the lowest value in the District of West Pasaman and Pesisir Selatan district. The highest average grade of primary school in Bukittinggi and the lowest in Solok city. It means that the value of UN both for primary and junior high school level is in big city. For that reason, it is necessary to mapping based on the result of National Examination. These results can be used to evaluate and improve the results of the National Exam the following year. Mapping used Biplot Analysis. Biplot analysis is one multivariate analysis technique that can be used to solve the problem. According to Mattjik (2011: 247), as for information that can be obtained from the analysis of biplot is to see the proximity between objects, diversity variables, relationships between variables, and the value of the variable on an object. The object of this research is 19 districts / cities in West Sumatera Province. Meanwhile, the variables in the study are subjects tested in the National Exam.

2. Method

Biplot Analysis will be employed in this work. Biplot is a graphical technique that presents together information about the object of observation and variables in a plot (Mattjik & Sumertajaya, 2011: 246). If the translation of the X data matrix is sized (n × p), then the **X** matrix can be decomposed into:

$$\mathbf{X}_{(\mathbf{n}\times\mathbf{p})} = \mathbf{G}_{(\mathbf{n}\times\mathbf{r})}\mathbf{H^t}_{(\mathbf{r}\times\mathbf{p})}$$

Biplot analysis, data analysis will be performed on the correlated data matrix against the mean. The corrected matrix of the mean is denoted by $\widetilde{\mathbf{X}}$ and can be written The corrected matrix of the mean is denoted by $\widetilde{\mathbf{X}}$ and can be written is $\widetilde{\mathbf{X}} = \begin{bmatrix} \tilde{x}_{11} & \tilde{x}_{12} & \cdots & \tilde{x}_{1p} \\ \tilde{x}_{21} & \tilde{x}_{22} & \cdots & \tilde{x}_{2p} \\ \vdots & \vdots & \ddots & \vdots \\ \tilde{x}_{n1} & \tilde{x}_{n2} & \cdots & \tilde{x}_{np} \end{bmatrix}$

If the matrix $\widetilde{\mathbf{X}}$ is a standardized data matrix, then based on the decomposition of the singular value of X matrix can be described as:

$$\widetilde{\mathbf{X}}_{(\mathrm{n}\times\mathrm{p})} = \mathbf{U}_{(\mathrm{n}\times\mathrm{r})}\mathbf{L}_{(\mathrm{r}\times\mathrm{r})}\mathbf{A^t}_{(\mathrm{r}\times\mathrm{p})} = \mathbf{G}_{(\mathbf{n}\times\mathbf{r})}\mathbf{H^t}_{(\mathbf{r}\times\mathbf{p})}$$

$$\mathbf{G} = \mathbf{U} \text{ dan } \mathbf{H} = \mathbf{AL'}$$

With the decomponment of the singular value of X data matrix, we get the following matrix U:

$\mathbf{U} = \widetilde{\mathbf{X}}(\mathbf{A^t})^{-1}\mathbf{L^{-1}}$, matriks **A** is ortogonal ortogonal with $\mathbf{A^{-1}} = \mathbf{A^t}$ so $\mathbf{U} = \widetilde{\mathbf{X}}\mathbf{AL^{-1}}$ so that G and H is

$$\boldsymbol{G} = \begin{bmatrix} \lambda_1^{-1/2}(x_1 - \overline{\mathrm{X}})^t\boldsymbol{a_1} & \lambda_2^{-1/2}(x_1 - \overline{\mathrm{X}})^t\boldsymbol{a_2} & \dots & \lambda_r^{-1/2}(x_1 - \overline{\mathrm{X}})^t\boldsymbol{a_r} \\ \lambda_1^{-1/2}(x_2 - \overline{\mathrm{X}})^t\boldsymbol{a_1} & \lambda_2^{-1/2}(x_2 - \overline{\mathrm{X}})^t\boldsymbol{a_2} & \dots & \lambda_r^{-1/2}(x_2 - \overline{\mathrm{X}})^t\boldsymbol{a_r} \\ \vdots & \vdots & \ddots & \vdots \\ \lambda_1^{-1/2}(x_n - \overline{\mathrm{X}})^t\boldsymbol{a_1} & \lambda_2^{-1/2}(x_n - \overline{\mathrm{X}})^t\boldsymbol{a_2} & \dots & \lambda_r^{-1/2}(x_n - \overline{\mathrm{X}})^t\boldsymbol{a_r} \end{bmatrix}$$

$$\boldsymbol{H} = \begin{bmatrix} a_{11}\lambda_1^{-1/2} & a_{12}\lambda_2^{-1/2} & \cdots & a_{1r}\lambda_p^{-1/2} \\ a_{21}\lambda_1^{-1/2} & a_{22}\lambda_2^{-1/2} & \cdots & a_{2r}\lambda_p^{-1/2} \\ \vdots & \vdots & \ddots & \vdots \\ a_{p1}\lambda_1^{-1/2} & a_{p2}\lambda_2^{-1/2} & \cdots & a_{pr}\lambda_p^{-1/2} \end{bmatrix}$$

Since the matrix $\widetilde{\mathbf{X}}$ is represented in the field, the matrix to be used as the best approach is the matrix $G^{(2)}$ dan $H^{(2)}$

$$\mathbf{G}^{(2)} = \begin{bmatrix} g_{11} & g_{12} \\ \vdots & \vdots \\ g_{i1} & g_{i2} \\ \vdots & \vdots \\ g_{n1} & g_{n2} \end{bmatrix} \text{ dan } \mathbf{H}^{(2)} = \begin{bmatrix} h_{11} & h_{12} \\ \vdots & \vdots \\ h_{i1} & h_{i2} \\ \vdots & \vdots \\ h_{p1} & h_{p2} \end{bmatrix}$$

The matrices $G^{(2)}$ dan $H^{(2)}$ and are the first two elements of the vector g_i and the vector h_j. The vectors g_i and h_j will be plotted in one plot, this is called the biplot, where g_i will be depicted as a point and h_i will be depicted as the line from the origin (0,0) to the point $\left(a_{j1}\lambda_1^{-1/2}, a_{j2}\lambda_2^{-1/2}\right)$ for $j = 1,2,\dots,p$.

Matjik (2011) gives information and interpretation of biplot analysis are as follows.

- Correlation between each variables described as a directed arrow. Two variables having positive correlation if angle between both variables are acute, and if its angle is obtuse, it means that they have negative correlation and also if they differ in right angle explain that no correlation between both variables.

IOP Conf. Series: Materials Science and Engineering **335** (2018) 012040 doi:10.1088/1757-899X/335/1/012040

- Variance of variables sketch as length of vector. Longer the vector, bigger the variances among variables and vice versa shorter the vector, smaller the variances.
- Neighborhood among object is sketched as objects having the same characteristics. Nearest the object, having alike characteristic.
- Variable's value on object detected from the relative position amongst variable and object. Object in a side with variables having the value upper average. If it have opposite direction, the value under average.

3. Result and Discusion

Here are the results of the analysis at junior and primary school levels:

3.1 Junior High School Education Level

Average National Examination of Regency / City level of junior high school education in West Sumatera Province in 2016 is formed into a matrix of data that is **X** matrix that size 19 × 4. Where, each row in **X** matrix district./city and each column describe the average National Exam per subject

$$\mathbf{X} = \begin{bmatrix} 75{,}03 & 69{,}56 & 57{,}24 & 65{,}99 \\ 79{,}81 & 60{,}97 & 55{,}05 & 62{,}21 \\ 78{,}36 & 56{,}21 & 48{,}80 & 57{,}10 \\ 76{,}42 & 52{,}40 & 42{,}26 & 52{,}83 \\ 75{,}64 & 50{,}40 & 41{,}40 & 49{,}78 \\ 78{,}40 & 55{,}12 & 47{,}82 & 56{,}57 \\ 72{,}88 & 56{,}32 & 45{,}02 & 48{,}44 \\ 72{,}47 & 47{,}26 & 40{,}16 & 49{,}87 \\ 70{,}72 & 43{,}68 & 36{,}54 & 46{,}75 \\ 74{,}52 & 45{,}45 & 39{,}66 & 51{,}99 \\ 71{,}66 & 43{,}48 & 37{,}20 & 47{,}38 \\ 68{,}40 & 44{,}53 & 35{,}23 & 45{,}89 \\ 63{,}97 & 39{,}53 & 32{,}11 & 40{,}5 \\ 75{,}61 & 50{,}42 & 44{,}48 & 53{,}51 \\ 72{,}38 & 43{,}10 & 37{,}03 & 47{,}43 \\ 64{,}46 & 55{,}13 & 46{,}38 & 49{,}07 \\ 68{,}80 & 41{,}42 & 35{,}01 & 45{,}09 \\ 70{,}38 & 43{,}16 & 35{,}96 & 47{,}23 \\ 71{,}53 & 44{,}43 & 36{,}95 & 48{,}31 \end{bmatrix}$$

So matriks **X** is processed so that matriks G^2 dan H^2 is :

$$G^2 = \begin{bmatrix} -0,56 & -0,29 \\ -0,42 & 0,15 \\ -0,24 & 0,17 \\ -0,08 & 0,16 \\ -0,01 & 0,12 \\ -0,21 & 0,21 \\ -0,09 & -0,26 \\ 0,06 & 0,05 \\ 0,18 & 0,03 \\ 0,05 & 0,26 \\ 0,16 & 0,08 \\ 0,20 & -0,12 \\ 0,37 & -0,29 \\ -0,08 & 0,16 \\ 0,16 & 0,15 \\ -0,06 & -0,69 \\ 0,25 & -0,02 \\ 0,19 & 0,05 \\ 0,14 & 0,08 \end{bmatrix} \text{ and } H^2 = \begin{bmatrix} -13,06 & 12,79 \\ -32,04 & -6,63 \\ -29,01 & -2,46 \\ -25,10 & 4,64 \end{bmatrix}$$

Based on G^2 dan H^2 obtain figure biplot is

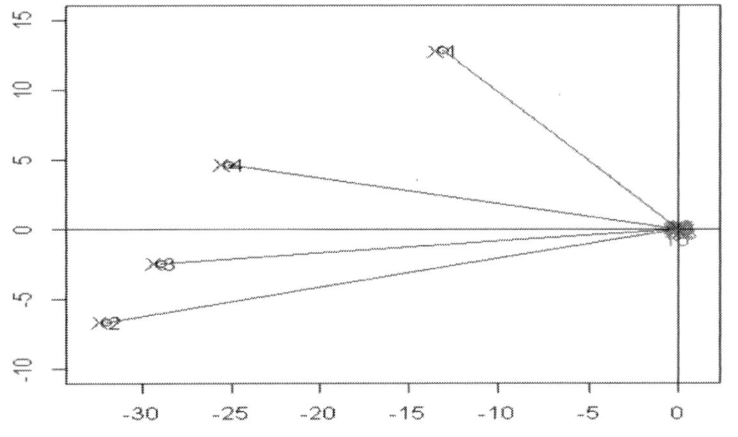

Figure 1: Biplot Regency / City based on the results of the National Examination in 2016 for Junior High schools level

Based on the above biplot can be taken information about the variance of variables. The English subject variable vector (X2) has a variable vector that is longer than the other variable vectors. The longer the vector variable, the greater the level of variability of the variable. This shows the results of the National Examination of students on English subjects in the Regency / City in West Sumatra Province varied. This means that there are some areas that have a relatively higher average National Examination and some areas have a relatively lower National Examination average.

Based on the biplot above, it can be seen that the vector of Mathematics variable (X3) forms a small angle with the vector of English variable (X2) and also the variable vector of IPA (X4). That is, students who have relatively high Mathematics National Exam score also tend to have relatively high value in English and Science subjects. Meanwhile, the English variable vector (X2) forms a considerable angle with the Bahasa variable vector (X1). This shows that students who have relatively high Bahasa National Examination score, do not always get a relatively high score on English subjects. If the enlarged Biplot graph shows that there are 9 groups :

IOP Conf. Series: Materials Science and Engineering **335** (2018) 012040 doi:10.1088/1757-899X/335/1/012040

Table 2: District / City Group based on the similarity of National Examination Results of 2016 Junior High Schools

Grup	District/City
1	Padang City
2	Bukittinggi City
3	Padang Panjang City
	Payakumbuh City
4	Pariaman City
5	Kepulauan Mentawai District
	Sawahlunto City
6	Tanah Datar District
	Solok City
	Lima Puluh Kota District
	Pasaman District
	Solok District
7	Agam District
	Sijunjung District
	Solok Selatan District
	Dharmasraya District
8	Pasaman Barat District
	Padang Pariaman District
9	Pesisir Selatan District

3.2 Elementary school Level

Average National Examination of Regency / City level of elemetary school education in West Sumatera Province in 2016 is formed into a matrix of data that is **Y** matrix that size 19 × 3. Where, each row in **Y** matrix district./city and each column describe the average National Exam per subject.

$$\mathbf{Y} = \begin{bmatrix} 78{,}96 & 70{,}95 & 77{,}53 \\ 84{,}43 & 81{,}16 & 85{,}42 \\ 78{,}62 & 68{,}55 & 75{,}97 \\ 79{,}65 & 72{,}76 & 80{,}25 \\ 79{,}03 & 69{,}76 & 77{,}75 \\ 79{,}31 & 68{,}62 & 79{,}01 \\ 78{,}38 & 69{,}44 & 77{,}30 \\ 80{,}45 & 76{,}16 & 82{,}03 \\ 75{,}56 & 67{,}72 & 73{,}75 \\ 80{,}30 & 72{,}80 & 80{,}66 \\ 75{,}93 & 66{,}12 & 73{,}41 \\ 73{,}74 & 63{,}38 & 69{,}50 \\ 76{,}69 & 70{,}19 & 74{,}18 \\ 79{,}87 & 69{,}71 & 77{,}17 \\ 76{,}61 & 64{,}34 & 73{,}98 \\ 63{,}10 & 57{,}46 & 59{,}68 \\ 73{,}05 & 60{,}17 & 69{,}68 \\ 79{,}40 & 76{,}98 & 80{,}29 \\ 80{,}51 & 66{,}78 & 79{,}33 \end{bmatrix}$$

So matriks **Y** is processed so that matriks G^2 dan H^2 is

$$G^2 = \begin{bmatrix} -0{,}07 & -0{,}02 \\ -0{,}44 & -0{,}26 \\ -0{,}001 & 0{,}10 \\ -0{,}16 & -0{,}03 \\ -0{,}06 & 0{,}08 \\ -0{,}06 & 0{,}24 \\ -0{,}04 & 0{,}06 \\ -0{,}25 & -0{,}20 \\ 0{,}08 & -0{,}08 \\ -0{,}17 & 0{,}01 \\ 0{,}12 & 0{,}05 \\ 0{,}25 & 0{,}01 \\ 0{,}03 & -0{,}21 \\ -0{,}06 & 0{,}11 \\ 0{,}13 & 0{,}26 \\ 0{,}65 & -0{,}47 \\ 0{,}31 & 0{,}24 \\ -0{,}22 & -0{,}39 \\ -0{,}05 & 0{,}48 \end{bmatrix} \qquad H^2 = \begin{bmatrix} -17{,}84 & 4{,}94 \\ -22{,}92 & -7{,}20 \\ -23{,}78 & 3{,}24 \end{bmatrix}$$

Based on G^2 dan H^2 obtain figure biplot is

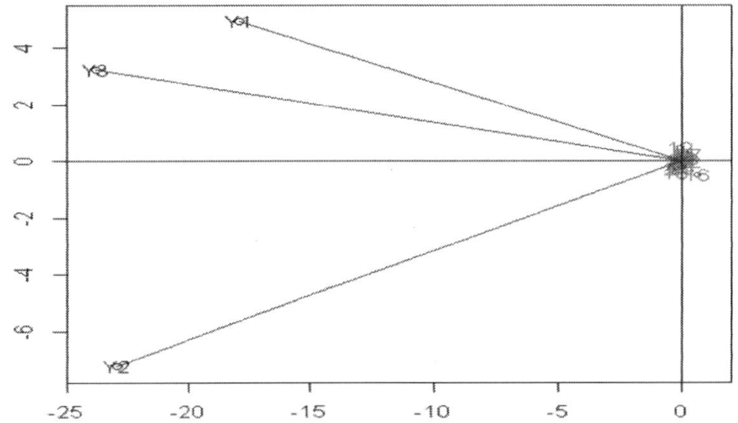

Figure 2: Biplot Regency / City based on the results of the National Examination in 2016 for elementary education level

Based on the above biplot can be taken information about the diversity of variables. The Mathematics subject variable vector (Y2) has a variable vector that is longer than the other vector variables. The longer the vector variable, the greater the level of variability of the variable. This shows the results of the National Examination of students in elementary school Mathematics level in the Regency / City in West Sumatra Province is diverse. This means that there are some areas that have a relatively higher average National Examination and some areas have a relatively lower National Examination.

The IPA variable vector (Y3) forms a small angle with the Indonesian variable vector (Y1). That is, students who have a relatively high IPA National Examination Scores, tend to also have a relatively

high value on the subjects of Bahasa . However, overall all subjects tested in the National Examination of elementary level have a positive relationship with each other.
If the biplot image is enlarged then seen grouping the city / district based on similarity

Table 3: District / City Group based on the similarity of National Examination Results of 2016 Elementary Schools

Grup	District/City
1	Bukittinggi City
2	Agam District
	Solok Selatan District
	Padang City
	Payakumbuh City
	Padang Panjang City
	Sawahlunto City
	Pariaman City
	Solok City
3	Lima Puluh Kota District
	Tanah Datar District
4	Dharmasraya District
	Sijunjung District
5	Solok District
	Pesisir Selatan District
6	Pasaman District
7	Pasaman Barat District
	Padang Pariaman District
8	Kepulauan Mentawai District

4. Conclusion

There are 9 regency / municipal groups formed based on similarities of National Examination results of junior high school in West Sumatera Province. For Basic School Level consists of 8 groups having the same characteristics. At the SMP level the English variable vector is longer than the other vector variables. This shows that English variables have the greatest diversity. While at the primary school level the greatest diversity is Mathematics. At the junior level the most powerful or positive relationship is the relationship of Mathematics and English subjects. For elementary school level IPA subjects have a strong relationship with Bahasa.

Reference

[1] BSNP dan Kalitbang Kemdikbud. 2016. *Buku Saku Ujian Nasional*. Jakarta: Departemen Pendidikan dan Kebudayaan.
[2] Gabriel, R. 1971. The Biplot Graphic Display of Matrices with Application to Principal Component Analysis. Journal of Biometrika 58: 453-467.
[3] Howard, A., dan Rorres, C. 2004. Aljabar Linier Elementer Versi Aplikasi. Jakarta: Erlangga.
[4] Mattjik, A. A., & Sumertajaya, I. M. (2011). Sidik Peubah Ganda dengan Menggunakan Aplikasi SAS. Bogor: IPB Press.
[5] Sartono, dkk. 2003. Analisis Peubah Ganda. Jurusan Statistika FMIPA IPB. Bogor
[6] Suprapto, J. 2004. *Analisis Multivariat Arti dan Interpretasi. Jakarta*: Rineka Cipta

IOP Conf. Series: Materials Science and Engineering **335** (2018) 012041 doi:10.1088/1757-899X/335/1/012041

Pricing Employee Stock Options (ESOs) with Random Lattice

E Chendra[1], L Chin[2], A Sukmana[3]

Department of Mathematics, Parahyangan Catholic University, Bandung

[1]erwinna@unpar.ac.id, [2]chin@unpar.ac.id,[3]asukmana@unpar.ac.id

Abstract. Employee Stock Options (ESOs) are stock options granted by companies to their employees. Unlike standard options that can be traded by typical institutional or individual investors, employees cannot sell or transfer their ESOs to other investors. The sale restrictions may induce the ESO's holder to exercise them earlier. In much cited paper, Hull and White propose a binomial lattice in valuing ESOs which assumes that employees will exercise voluntarily their ESOs if the stock price reaches a horizontal psychological barrier. Due to nonlinearity errors, the numerical pricing results oscillate significantly so they may lead to large pricing errors. In this paper, we use the random lattice method to price the Hull-White ESOs model. This method can reduce the nonlinearity error by aligning a layer of nodes of the random lattice with a psychological barrier.

1. Introduction

Employee Stock Option (ESO) is a call option on company's stock granted by the company to its employees. ESO has become very popular in the last twenty years, especially for start-up companies. These companies do not have the resources to pay key employees and they solve this problem by supplementing the salaries with ESOs. If the company does well and shares are sold to the public in an IPO, the ESO is likely become very valuable. Many companies also attract and keep the best employees by offering them very attractive stock option packages.

ESO has unique characteristics that distinguish it with standard options. ESO often last as long as 10 to 15 years. It cannot be exercised immediately because there is a vesting period. When employees leave their jobs during the vesting period, the ESO becomes worthless. But if they leave after the vesting period, employees forfeit ESOs that are out-of-the money and they have to exercise options that are in-the-money. Employees also are not permitted to sell their ESOs. The only way employees can realize a cash benefit from the ESO is by exercising its and selling the underlying shares. This leads to a tendency for ESOs to be exercised earlier.

In practice the early exercise behavior of employees varies widely from company to company. In a much citied paper, Hull and White [1] suggest a simple model to capture the earlier exercise behavior. They assume that employees will exercise voluntarily their ESO as soon as the option has vested and the stock price reaches a certain multiple of a strike price. In other words, there is a single psychological barrier to induce an earlier exercise in the model. Then they use a binomial tree method to value the ESO. However, this method has a disadvantage to price options that has a barrier feature. Figlewski and Gao [2] argue that the pricing results oscillate significantly due mainly to the nonlinearity error.The nonlinearity error occurs at certain critical locations such as a certain point, a price level, or a time point. For ESO, the critical price level occurs along the psychological barrier price. To overcome this, we use a random lattice method proposed by Das [3] to reduce the nonlinearity error by restructuring the tree to make the critical price level coincide with the tree's nodes.

2. The Hull-White ESO Model

The Hull-White ESO model assumes that employees will exercise their options if the ratio of the stock price to the strike price is above a certain level (M). Suppose there are N time steps of length Δt in the tree, $S_{i,j}$ is the stock price at the jth-node of the CRR Binomial tree at time $i\Delta t$, and $f_{i,j}$ is the value of the ESO at this node. Define Kas the strike price of the option, v as the time when vesting period ends, λ as the employee exit rate, and r as the riskless interest rate.Assume that the probability of employees leaving the company for each period of time Δt as $1 - e^{-\lambda\Delta t}$ [4]. At the maturity date T, the payoff of the ESO is given by

$$f_{N,j} = \max(S_{N,j} - K, 0)$$

and for $0 \leq i \leq N - 1$:

- If $i\Delta t < v$ then

$$f_{i,j} = (e^{-\lambda\Delta t}).e^{-r\Delta t}[pf_{i+1,j+1} + (1-p)f_{i+1,j}]$$

- If $i\Delta t \geq v$ and $S_{i,j} \geq MK$ then

$$f_{i,j} = S_{i,j} - K$$

- If $i\Delta t \geq v$ and $S_{i,j} < MK$ then

$$f_{i,j} = (1 - e^{-\lambda\Delta t}).\max(S_{i,j} - K, 0) + (e^{-\lambda\Delta t}).e^{-r\Delta t}[pf_{i+1,j+1} + (1-p)f_{i+1,j}]$$

Then the value of the ESO is $f_{0,0}$.

3. Random Lattice

The random lattice, also known in the literature as the stochastic-mesh method, is first proposed by Das [3] to price a European option and an Asian option. This method combines the computational benefits of Monte Carlo simulation with the ability to implement dynamic programming on a lattice. The basic idea for the lattice is to restrict the stochastic process to a finite number of discrete values, which defines a set of "buckets" into which every stock price is assigned.

Denote the number of buckets as m, which is the width of the tree, and the number of level in the tree as d, its depth. Then the random lattice takes on a matrix structureof size $m \times d$. To generate the lattice, we conduct n root-leaf random walks through it, starting with the same value at the root for each path. Assume that the stock price follows a lognormal diffusion process, then at each node, the next stock price is randomly generated by

$$S_{\Delta t} = S_0\exp[(r - 0.5\sigma^2)\Delta t + \sigma\epsilon\sqrt{\Delta t}]$$

where $\epsilon \sim N(0,1)$, σ is the volatility, and S_0 is the initial stock price. Once the next stock price is obtained, then we round it up or down to the nearest bucket and proceed to generate the next stock price and so on. For every edge on the lattice from node i at time t to node j at time $t+1$, we maintain a count of number of traversals made via the edge, denote as $c_t(i,j)$, and keep track of the total number of out-traversals from node i at time t, denote as $c_t(i)$. Therefore

$$c_t(i) = \sum_{j=1}^{m} c_t(i,j)$$

From this, we derive the transition probabilities $p_{ij}(t)$ on the lattice, i.e.

$$p_t(i,j) = \frac{c_t(i,j)}{c_t(i)}$$

Once we have the probabilities for the lattice, the stochastic process is in place. Then we can compute the final payoffs at the last level of the lattice and the price of the ESO at the root node is determined by dynamic programming on the lattice.

4. Numerical Results

Assume that$S_0 = K = 50$, $r = 7{,}5\%$, $\sigma = 30\%$, $T = 10$ years, $m = 51$, $M = 2$, $v = 3$ years, $\lambda = 6\%$,$d = 300$, and $n = 10^5$ then the ESO price with random lattice method is 19.071 while the ESO price with binomial method is 18.718. Figure 1 shows the convergence of ESO priceswith random lattice method for various values of d.

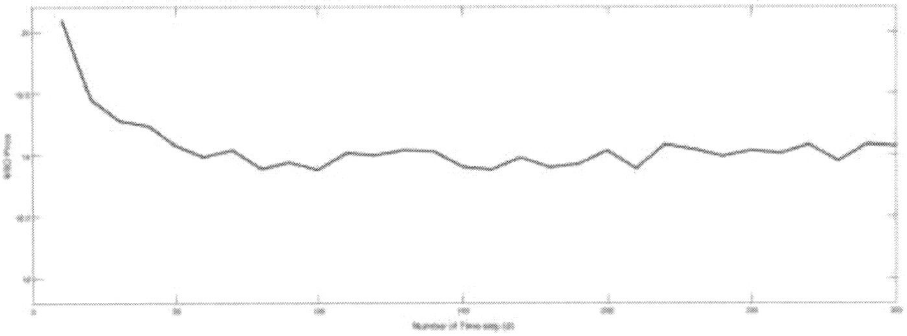

Figure 1. ESO Price vs Number of Time Steps

5. Conclusions

We have presented the random lattice method for valuing employeestock options. The method is easy to implement and flexible. For further research, we will compare the ESO price with other lattice methods, like the Bino-Trinomial Tree (BTT) method proposed by Dai and Lyuu [5].

References

[1] Hull J and White A 2004 How to value employee stock options *Financial Analysts Journal* **60** 114-9.

[2] Figlewski S and Gao B 1999 The adaptive mesh model: a new approach to efficient option pricing*Journal of Financial Economics***53(3)** 313-51.

[3] Das SR 2011 Random lattices for option pricing problems in finance *Journal of Investment Management* **9(2)** 134-52.

[4] Ammann M and Seiz R 2004 Valuing employee stock options: does the model matter? *Financial Analysts Journal***60(5)** 21-37.

[5] Dai TS and Lyuu YD 2010 The Bino-Trinomial Tree: a simple model for efficient and accurate option pricing *Journal of Derivatives***17(4)**7-24.

IOP Conf. Series: Materials Science and Engineering **335** (2018) 012042 doi:10.1088/1757-899X/335/1/012042

Ethnomathematics Exploration of the Toba Community: Elements of Geometry Transformation Contained in Gorga (Ornament on Bataks House)

C Ditasona

Mathematics Education Department, Faculty of Teacher Training and Education, Christian University of Indonesia,

*candra.ditasona@uki.ac.id

Abstract. Gorga is an ornament known to the Batak community. As one of the works of art that poured in the form of carvings, gorga become icons for batak society. Long before the Batak people knew formal education, they had made gorga. This is evidenced through several historical sources. Gorga not only contains the value of art but also has a mathematical element. There are many mathematical principles used in the process of making gorga. The principle of geometry transformation is very prominent in the gorga motifs. This article is an ethnomathematics research that will discuss how the thinking process in making gorga. Observations and interviews with the gorga craftsmen (pande) are conducted to find out how the principles of rotation, translation, dilation and reflection are used in making gorga motifs

1. Introduction

The word ornament comes from the Latin ornare, which has the meaning of adorning. According to [1] ornaments "are components of art products that are added or deliberately made for the purpose of decoration. So, based on that understanding, ornament is the application of decoration on a product. The ornamentalmain function is to beautify the objects of products or items decorated".The presence of an ornament is not merely as a filler of the empty and meaningless, like the works of the past ornaments. The various forms of ornaments actually have functions, namely (1) purely aesthetic functions, (2) symbolic functions, (3) constructive engineering functions.

A purely aesthetic function is an ornamental function to embellish the appearance of a decorated product form to become a work of art. The function of such ornaments is evident in the products of ceramics, batik, weaving, woven, jewelry, traditional weapons, household appliances, and leather and wooden crafts that emphasize the aesthetic value of the ornaments it implements.

The symbolic function of ornaments is commonly found in ceremonial or biblical objects of religion and belief, accompanying the aesthetic value. For example ornaments that use motif kala, monitor lizards, dragons, birds or *garuda*, the works of the past function symbolically. In its later development, the symbolic aspect of an ornament increasingly loses its meaning.Toba or Gorga Batak traditional ornaments are all kinds of ornaments made to beautify the traditional house (exterior of the house), which is passed down from generation to generation based on the mindset of Batak Toba society. Gorga scattered throughout the Toba region and not always evenly sub-sub Toba region. Toba Batak society in particular at this time, less or even not understand with things about the culture. One of them is the understanding of *Gorga*.

Gorga Batak is one of Batak art and culture that is old enough. A traditional sculpture made naturally. In ancient times, *gorga* was only made for a house that is considered honorable, because Batak Ancestors consider that gorga is not just an ornament, but has a meaning that reflects the life of the Batak. Usually *gorga* found on the walls of traditional houses Batak tribe. The only traditional Batak traditional house in Samosir is located in Parmonangan Village, Simanindo District. Batak house, built 350 years ago. The house belongs to *Opung Souaon Situmorang*. The house still looks complete with *gorga* or original Batak carvings.

Other mathematical concepts that we can observe one of them is the Batak ornament (gorga). As we know from the above explanation that gorga far already found before the Batak people learn geometry. The motives found in gorga use the concept of transformation geometry such as reflection, shifting, rotation, and rescaling. All science including mathematics is a product produced from a culture or human civilization. The process of creating meaning through a culture-based learning process has several components, meaningful tasks, active interaction, explanation and application of the science contextually, and the use of diverse learning resources [2][3]. In cultural-based learning, culture becomes a method for students to transform their observations into creative forms and principles about the field of science. One form of cultural-based learning is ethnomatematics (Ethnomathematics).

D'Ambrosio [4] says Ethnomatematics is the study of mathematics which takes into account cultural considerations in which mathematics arises by understanding the reasoning and mathematical systems they use. The study of ethnomatics in mathematics learning encompasses all areas: architecture, weaving, sewing, agriculture, kinship, ornament, and spiritual and religious practices often in harmony with patterns occurring in nature or ordering systems of abstract ideas. According to [5], ethnomatics includes mathematical ideas, thoughts and practices developed by all cultures. Etnomathematiics can also be considered as a program that aims to study how students to understand, articulate, process, and ultimately use math ideas, concepts, and practices that can solve problems related to their daily activities.

Ethnomatics recognizes that there are different ways of doing mathematics by considering the acquisition of mathematical academic knowledge developed by different sectors of society and taking into account the different modes in which different cultures negotiate about their mathematical practice [6-8]. The purpose of studying etnomatology is to understand the interrelation between mathematics and culture, so that students and society's perception of mathematics becomes more appropriate, and mathematics learning can be better adapted to the cultural context of students and society, and mathematics can be more easily understood because it is no longer perceived as foreign by students and society. Another goal is to optimize the application and the benefits of mathematics for the life of students and the wider community more, so that students and society obtain the optimal benefits of mathematics learning activities.

2. Method

The type of this research is qualitative research. Place of study is Toba Samosir District. Methods of data collection using interviews recorded and direct observation. Subjects to be interviewed are customary figures and some *gorga* craftsmen. Interview data will be made in the form of transcripts of the conversation, while the observed data is documented in the form of photographs. Interview transcripts along with photographs will be analyzed and elaborated to explain the concept of transformation found in gorga.

3. Result and Discussion

The transformation of geometry is to change each point coordinate (the points of a wake) into other coordinates on a plane with one particular rule. For example, the transformation (T) to point P (x, y) produces a shadow P '(x', y '), the operation can be expressed as follows:

$$P(x,y) \rightarrow P'(x',y')$$

Some gorga motifs show that there are several transformation principles used such as:

3.1. Translations

Translation is a transformation that moves each point on a plane by a certain distance and direction. The distance and direction of a translation can be represented by a directed line segment, eg $\overrightarrow{AB}$or a pair of numbers$\binom{a}{b}$, where "a" represents the distance and direction of movement horizontally (horizontally) and "b" represents the distance and direction of displacement vertically (upright). At one translation each wake does not change. The translation operation can be expressed as follows:

$$A(x,y) \xrightarrow{T=\binom{a}{b}} A'(x',y') = A'(x+a, y+b)$$

Gorga motif similar to parang batik motif is also formed from the translation of one object. Figure 1 shows the translational process along the horizontal line to form new objects such as the repetition of the previous object. The direction and shape is also exactly the same as the previous form.

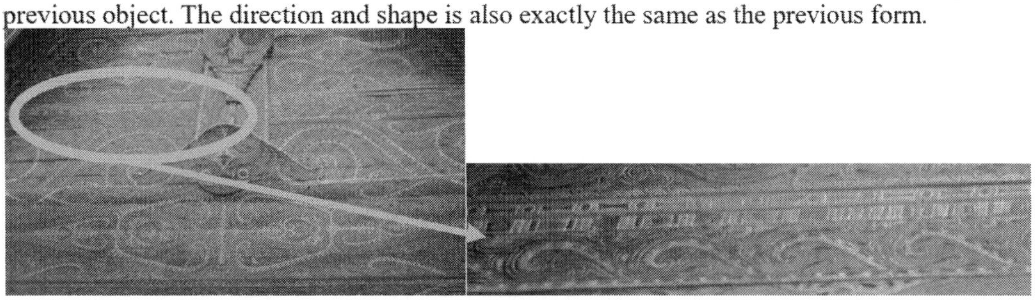

Figure 1. Translation in Gorga Motifs

3.2. Reflection

Reflection is a transformation that moves each point on a plane by using the mirror image of the points to be moved. For reflection of the y-axis (vertical), if the point shadow P (x, y) is P '(x', y ') then P' (x ', y') = P '(- x, y), the matrix form can be written as follows:

$$x' = -x$$

$$y' = y$$

$$\begin{pmatrix} x' \\ y' \end{pmatrix} = \begin{pmatrix} -1 & 0 \\ 0 & 1 \end{pmatrix} \begin{pmatrix} x \\ y \end{pmatrix}$$

The following figure is an example of a reflection on the vertical axis of the gorga motif:

(i)

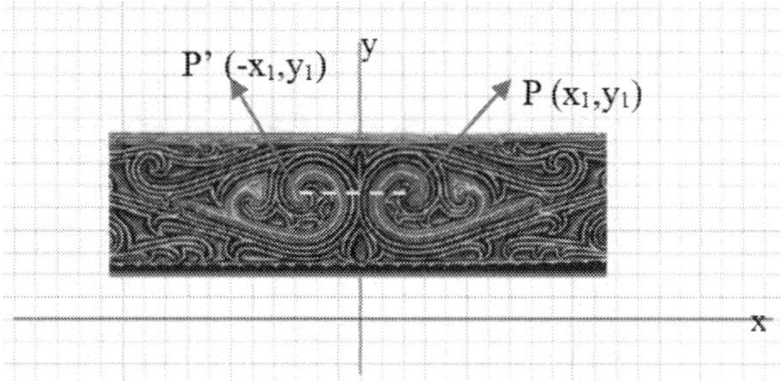

(ii)

Figure 2. Gorga motif is reflected on the axis "y" (vertical)

Figure 2 (i) shows one of the gorga motifs reflected on the vertical axis. In Figure 2 (ii) it is seen that the coordinates of point P (x, y) after being reflected then the shadow coordinates of point P or so called P 'are (-x, y). From the figure we can observe that the distance of point P to the y-axis is equal to the distance of point P 'to the y-axis. It satisfies the properties of reflection: (1) In the reflection of an object and its image will be congruent, but there is an inverted orientation; (2) Any point located on the axis of the reflection is not invariant; (3) The distance of each point on the object to the reflection axis is equal to the distance of each shadow point to the reflection axis; (4) The angle formed by the reflection axis with the line connecting each point to its shadow is right-angled.

For reflection of the x axis (horizontally), if the point shadow P (x, y) is P '(x', y ') then P' (x ', y') = P '(x, -y) the matrix can be written as follows:

$$x' = x$$

$$y' = -y$$

$$\begin{pmatrix} x' \\ y' \end{pmatrix} = \begin{pmatrix} 1 & 0 \\ 0 & -1 \end{pmatrix} \begin{pmatrix} x \\ y \end{pmatrix}$$

So $\begin{pmatrix} 1 & 0 \\ 0 & -1 \end{pmatrix}$, is the reflection matrix of the y axis.

3.3. *Rotation*

Rotation is a transformation that moves a point on a plane of another point by rotating to a particular center point. In rotation or rotation on a plane is determined by the center of rotation. Rotation center point is a fixed point or a center point used as a reference to determine the direction and magnitude of the rotation angle. On a rotation, each wake does not change its shape. The direction of rotation is agreed with the following rules: (1) If the rotation is opposite to the clockwise direction, then the rotation is positive (+). (2) If rotation is clockwise, then this rotation is negative (-). (3) The magnitude of the rotational angle determines the rotation. Deep rotation is expressed in the fractional plane against a full rotation (360°) or large angle in degrees or radians.

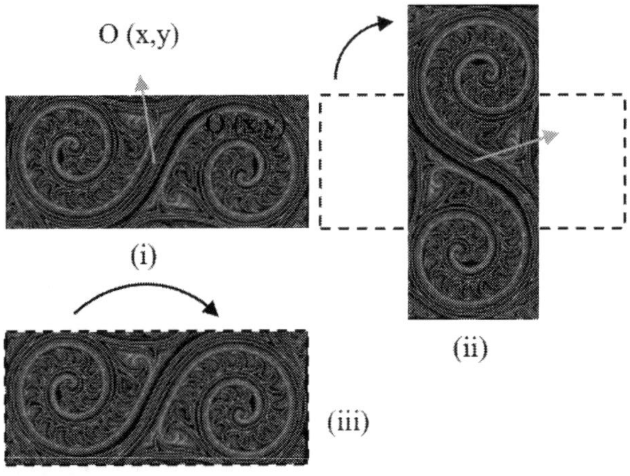

Figure 3. Rotation of *Gorga* Motif With Center Point O (x, y)

In Figure 3 one of the *gorga* motifs is rotated clockwise at 180 ° with the center point O (x, y). It turns out that the rotation does not change the shape of the motive. The interesting thing about the gorga motif is that if rotated as far as 180 ° with the position of the center point O (x, y) as in the picture of 6 parts (ii), then the motive seems to have no rotation, because the motive is the same as before rotation .

3.4. *Dilatation*

Dilatation is a transformation that changes the size or scale of a geometry build whether it magnifies or wakes up but does not change the shape of the wake. A dilatation on the plane is determined by the center of dilation and dilation factor (scale factor). Notice the dilation of some objects in the Figure 4

(i) (ii)

Figure 4. Dilatation on *Gorga* motif "Ulu Paung"

Figure 4 shows that there is a positive dilation process or magnification on the *Gorga* motif. In the process of dilatation, the result object is dilated similar to the original object. In the 6-part (ii) if it is enlarged, it appears that the shape of the dilated object is not very precise. This may happen because some factors like the process of shooting / photos are not right in the middle of the object so that change the point of view, other factors are the process of *ulu paung* is not like other *gorga* motif generally made on the plane board, while the field surface which is found in the *ulu paung* motif is curved.

4. Conclusion

Batak society has known the geometry of transformation for a long time ago. It can be seen from the principles of the transformation geometry that they use in the making of a work ornament called Gorga. From some *gorga* motifs can be concluded that the motive has been using the principle of reflection, rotation, translation and dilatation. Gorga made using traditional tools so that in some motives found not too precise. It is interesting to see that the accuracy of their calculations by using simple tools of their day requires a mathematical ability that is good enough for the craftsman in making a motif *gorga*. For further research it is possible to observe Gorga in terms of mathematical abilities used.

References

[1] Gustami. (1980). NukilanSeni Ornamen Indonesia. Yogyakarta:ASRI.

[2] D'Ambrosio, U. (1985). Ethnomathematics and its place in the history and pedagogy of mathematics. For the Learning of Mathematics, 5(1), 44-48.

[3] D'Ambrosio, U. (1991). 'Ethnomathematics and its place in the history and pedagogy of mathematics', in M. Harris (ed.). Schools,Mathematics and Work. The Falmer Press. London. pp. 15–25.

[4] Bishop, A. (1988). Mathematics Educations in its Cultural Context. Educational Studies in Mathematics, 19, 1979-1991.

[5] Edy Tandililing. (2013). Pengembangan Pembelajaran Matematika Sekolah dengan Pendekatan Etnomatematika Berbasis Budaya Lokal Sebagai Upaya Untuk Meningkatkan Kualitas Pembelajaran Matematika di Sekolah. Prosiding Seminar Nasional Matematika danPendidikan Matematika. Yogyakarta: FMIPAUNY.

[6] Izmirli, I. M. (2011). Pedagogy on the Ethnomathematics - Epistemology Nexus: A Manifesto.

Journal of Humanistic Mathematics, Vol. 1 Issue 2 (July 2011), pp. 27-50.

[7] Presmeg, Norma C. (1998). Ethnomathematics in Teacher Education. Journal of Mathematics Teacher Education. pp317-339

[8] *Zhang, W., Zhang, Q. (2010). Ethnomathematics and Its Integration within the Mathematics Curriculum. Journal of Mathematics Education. June 2010, Vol. 3, No. 1, pp.151-157.*

IOP Conf. Series: Materials Science and Engineering **335** (2018) 012043 doi:10.1088/1757-899X/335/1/012043

Factors that Affect Poverty Areas in North Sumatera Using Discriminant Analysis

D H Nasution*, P Bangun, H R Sitepu

Department of Mathematics, Faculty of Mathematics and Natural Sciences, University of Sumatera Utara, Medan 20155, Indonesia

*dewiharninasution@gmail.com, pengarapen@usu.ac.id, henry1@usu.ac.id

Abstract. In Indonesia, especially North Sumatera, the problem of poverty is one of the fundamental problems that become the focus of government both central and local government. Although the poverty rate decreased but the fact is there are many people who are poor. Poverty happens covers several aspects such as education, health, demographics, and also structural and cultural. This research will discuss about several factors such as population density, Unemployment Rate, GDP per capita ADHK, ADHB GDP per capita, economic growth and life expectancy that affect poverty in Indonesia. To determine the factors that most influence and differentiate the level of poverty of the Regency/City North Sumatra used discriminant analysis method. Discriminant analysis is one multivariate analysis technique are used to classify the data into a group based on the dependent variable and independent variable. Using discriminant analysis, it is evident that the factor affecting poverty is Unemployment Rate.

1. Introduction

Poverty is a problem that concerns economic growth and its relation to income distribution [4]. The problem of poverty is one of the fundamental problems that it becomes central government attention both central government and local government. The size of the poor is strongly influenced by the lines poverty, because the poor are people who have an average per capita spending per month below the poverty line.

The phenomenon experienced in North Sumatera is the number and percentage the many poor people. Even official data of poverty shows poverty at a higher level. Countermeasures program poverty of local government is seen at the village level, but these programs often do not meet the expectations of government officials regions and citizens due to weak implementation and supervision mode.

Discriminant analysis is one of the methods used in a multivariate analysis with dependency method (where the relationship between variables can already be distinguished where the dependent variable and which variable free). Basically discriminant analysis can be used for know the characteristic variables that distinguish population groups that exists, but it can also be used as a grouping criterion which is first known clearly grouping. Discriminant analysis aims to recognize which factors can distinguish two or more groups. These distinguishing factors will form a distinctive function (called a discriminant function).

2. Research Method

In his research suggests that essentially discriminant analysis (discriminant analysis) can be used to know the characteristic variables that differentiate the existing population group, can also be used as grouping criteria [1]. Technically, discriminant analysis is similar to regression analysis, because both have the dependent variable and the independent variable in the model. It's just in simple regression analysis and multiple, variable bound in the form of data metrics and the independent variables can be either metric or nonmetrik, while the discriminant analysis, the type of data for the dependent variable must be a category (using ordinal or nominal scales) and the independent variables are metric data (interval and ratio) [2].

2.1. Purpose of Discriminant Analysis

Since the multivariate form of discriminant analysis is bound, the dependent variable is the variable underlying the discriminant analysis, with the aim of:

a. Want to know if there is a clear distinction between groups on the dependent variable, or it can be said whether there is a difference between member of group 1 with member of group 2.
b. If there is a difference, which independent variables are on the discriminant function which makes such a difference.
c. Create a function or discriminant model, which is basically similar with regression equation.
d. Classification of objects (in SPSS terminology is called line), whether an object (can be a person's name, the name of the plant, objects or other) belonging to group 1 or group 2, or others.

2.2. Assumptions in Discriminant Analysis

In addition to the basic and discriminant goals, it turns out there are assumptions that must be fulfilled prior to discriminant analysis, ie distributed free variables normal multivariate (multivariates normal distribution) and variance in each group are equal (equal variances).

2.3. Discriminant Analysis Process

In general the basic process of discriminant analysis is:

a. Splitting the variables into dependent variables and independent variables.
b. Determine the method to create a discriminant function. Where on the principle there are two basic methods for it:
   1) Simultaneous Estimation, method by entering all variables together then process discriminant.
   2) Stepwise Estimation, method by entering one by one variable into the discriminant model. In this method, of course there are variables that remain in the model, and there are possibilities one or more independent variables are removed from model.
c. Conducting testing the significance of the discriminant function that has been formed, using Wilks Lambda, Pilai, F test and others.
d. Understand the discriminant function that has been formed.
e. Conducting the classification accuracy of the discriminant function, including knowing the accuracy of individual classification with casewise diagnostics.

2.4. Discriminant Analysis Model

The discriminant analysis model is an equation that shows a linear combination of various independent variables, namely [3]:

$$D_i = b_0 + b_1X_{i1} + b_2X_{i2} + b_3X_{i3} + \cdots + b_pX_{ip}$$

Were D is an independent variable and D_i is value of discriminant score from observation / i-responden ($i = 1,2,\dots,n$), b_j is discriminant coefficient of free variable or j-attribute ($j = 0,1,2,\dots,p$) and X_{ij} is j-independent variable (attribute) of the i-respondent.

3. Result and Discussion

This research uses quantitative research method, that is with collecting secondary data in accordance with the variables research that has been determined.

3.1. Types and Data Sources

The type of data used in this study is secondary data and data sources used are demographic, social and economic data area of North Sumatera in 2012 which has been published by the Agency Statistics Center.

3.2. Types and Data Sources

The sample in this study is all districts / cities in North Sumatera which amounted to 33 districts / cities. The variable is bound to the study this is the Percentage of Poverty (%), the dependent variable is divided into2 (two) groups, namely:

a. Group I, low poverty rates are areas with percentages poor population $< 15\%$.
b. Group II, high poverty rates are areas with percentages the poor $\geq 15\%$.

while the independent variables are:

1. X_1: Population Density (Soul / Km^2)
2. X_2: Open unemployment rate (%)
3. X_3: Per Capita PDRB Based on Constant Price (Thousand Rupiah)
4. X_4: Per Capita PDRB Based on Current Prices (Thousand Rupiah)
5. X_5: Economic Growth (%)
6. X_6: Life Expectancy (AHH) (%)

Before performing discriminant analysis, it must be met first some assumptions, ie data must be multivariate normal distribution and the variance in each group is the same.

3.3. Test Data Normality

Calculations using SPSS program, can be seen from the test *Kolmogorv Smirnov* that the value of *Asymp.Sig.* (*2-tailed*) is 0.972 more large from 0.05 so it can be concluded that the data being tested is distributed normal.

3.4. Average Equality Test

Calculations using the SPSS program, judging from the results of the value significant that the figure is significant for the Open Unemployment variable (TPT) of 0.034 and PDRB Based on Constant Price (PDRB ADHK) 0.036 below 0.05 which means that there are inter-group differences, or a regions are said to have low and high dependent poverty rates on the Unemployment and Open Unemployment Rate variables Constant.

3.5. Matrix Covariance Similarity Test

Calculations using the SPSS program, viewed from the output of *log determinant* table that the number of *log determinant* for the level category low poverty (2,900) and high poverty (2,123) are no different many, so the *covariance matrices group* would be relatively similar to both groups. This means that the data has met the assumption of discriminant analysis, so the discriminant analysis process can proceed.

Table 1. Log Determinant

Y	Rank	Log Determinant
0	1	2.900
1	1	2.123
Pooled within-groups	1	2.790

3.6. *Discriminant Analysis Process*

Calculations using the SPSS program, based on Table *Variables in the Analysis* variables of the selected into the discriminant function and significant is TPT. Or it could be said to affect high or high the low level of poverty in the District / City of Suamatera Utara. Next will be analyzed based on the selected variables using the SPSS program:

3.6.1. Significant Test of Discriminant. Functions Formed

Table 2. Wilks' Lambda

Test of Function(s)	Wilks' Lambda	Chi-square	df	Sig.
1	.862	4.517	1	.034

Viewed table *Wilks' Lambda Chi-Square figure* of 4.517 with the 0.034 significance level well below 0.05 then sufficient evidence for reject H_0, with a high significance level indicating the difference which is evident between the two regional groups of the Regency / City low poverty rates and high poverty rates.

3.6.2. Test the Power of Discriminant. Function Relationships

Table 3. Eigenvalues

Function	Eigenvalue	% of Variance	Cumulative %	Canonical Correlation
1	.160[a]	100.0	100.0	.371

Based on the Eigen Value table above, *Canonical Correlation* measures closeness relationship between *discriminant score* and group. The *Canonical Correlation* number of 0.371 indicates the closeness which is enough, with the size of the association scale between 0 to 1.

3.6.3. Discriminant Function

Table 4. The Canonical Coefficient Of Discriminant Function

	Function
	1
TPT	.248
(Constant)	-1.448

Based on the canonical coefficient table above the discriminant function, then obtained discriminant function:

$$D = -1{,}448 + 0{,}248\,TPT$$
$$D = -1{,}448 + 0{,}248\,X_2$$

3.6.4. Feasibility Assessment of Discriminant Functions

From the *Classification Results*, obtained the prediction accuracy of the model is 72.7%, so it is evident that the discriminant function has high precision prediction, because in general the accuracy in over 50% is considered valid. So the discriminant function can be used to predict factors affecting poverty district of North Sumatera. Based on the discriminant function that has been formed and tested its feasibility it can be determined the composition of group members (region low poverty rates and high poverty rates).

Table 5. Main Opportunities for Groups

		Cases Used in Analysis	
P_Kemiskinan	Prior	Unweighted	Weighted
0	0.500	26	26.000
1	0.500	7	7.000
Total	1.000	33	33.000

The main Opportunity Table for the above Group shows the composition to 33 objects, which by discriminant model yields 26 objects in groups of low poverty levels and 7 objects in groups high poverty rates, as well as *Casewise Statistics* tables the exact area predictions there are as many as 24 regions and the wrong classification there are 9 regions.

4. Conclusion

Based on the results of research on "Influencing Factors Poverty of North Sumatera Region Using Discriminant Analysis ", obtained the factors that most affect the level of poverty in The North Sumatera is open unemployment and its discriminant functions are: $D = -1{,}448 + 0{,}248\, X_2$. The discriminant (function) model obtained has precision classifies case of 72.7%. Because above 50%, the precision of the model is considered high so that the above model (function) can be used for classify a case of factors affecting poverty district of North Sumatera.

References

[1] Andriani, Yuniarti. 2010. Analisis Diskriminan untuk Mengetahui Faktor yang Memepengaruhi Pilihan Program Studi Matematika di FMIPA dan FKIP Universitas Sriwijaya. Penelitian Sains. 14403:1-2.
[2] Santoso, Singgih. 2010. Statistik Multivariat: Konsep dan Aplikasi dengan SPSS. PT Elex Media Komputindo. Jakarta.
[3] Supranto, J. 2010. Analisis Multivariat Arti dan Interpretasi. PT Rineka Cipta. Jakarta.
[4] Todaro Michael P. 1995. Ekonomi untuk Negara Berkembang. Edisi Cetakan Pertama. Bumi Aksara. Jakarta.

Logistic Regression and Path Analysis Method to Analyze Factors influencing Students' Achievement

N Noeryanti*, K Suryowati, Y Setyawan, R R Aulia

Department of Statistics, Institut Sains & Teknologi AKPRIND Yogyakarta, 55222, Indonesia

*noeryanti@akprind.ac.id

Abstract. Students' academic achievement cannot be separated from the influence of two factors namely internal and external factors. The first factors of the student (internal factors) consist of intelligence (X1), health (X2), interest (X3), and motivation of students (X4). The external factors consist of family environment (X5), school environment (X6), and society environment (X7). The objects of this research are eighth grade students of the school year 2016/2017 at SMPN 1 Jiwan Madiun sampled by using simple random sampling. Primary data are obtained by distributing questionnaires. The method used in this study is binary logistic regression analysis that aims to identify internal and external factors that affect student's achievement and how the trends of them. Path Analysis was used to determine the factors that influence directly, indirectly or totally on student's achievement. Based on the results of binary logistic regression, variables that affect student's achievement are interest and motivation. And based on the results obtained by path analysis, factors that have a direct impact on student's achievement are students' interest (59%) and students' motivation (27%). While the factors that have indirect influences on students' achievement, are family environment (97%) and school environment (37).

1. Introduction

Factors that influence learning achievement can be classified into two groups, namely internal factors that originate in students and external factors that originate from outside students themselves [1]. Internal factors consist of intelligence, attention, talent, interest, motivation, maturity, preparedness and fatigue. While external factors consist of family environment, school environment, and community environment. Both factors are very influential on student's achievement, because internal and external factors come from their selves or outside. Students achievement at school will show how far they learn and generally will be represented by the grades of their reports.

Based on data from the Educational Office of Madiun Regency in East Java Province, there are 93 Junior High Schools, one of them is SMP Negeri 1 Jiwan Madiun. SMP Negeri 1 JiwanMadiun is one of the favorite public school accredited A in Madiun Regency. The school is able to score achievements in academic and nonacademic fields. In the academic field, SMP Negeri 1 Jiwan Madiun always pay attention to academic achievement of each student, in this case is the grade of student's report. Increasing the grade of a student's report can be influenced by factors related to the student's achievement.

In order to know the influence of internal and external factors on the results of students' achievement it is necessary to analyze. In this research Binary Logistic Regression Analysis and Path

IOP Conf. Series: Materials Science and Engineering **335** (2018) 012044 doi:10.1088/1757-899X/335/1/012044

Analysis are used. Binary Logistic Regression aims to determine what factors affecting student achievement and know the tendency toward student's achievement. Path Analysis aims to determine the magnitude of direct influence, indirect influence and total influence on student's achievement. To support this research, there are previous studies that are used as references [2] and [3].

2. Binary Logistic Regression and Path Analysis

Binary logistic regression model is used to analyse the relationship between one response variable and one or more factor variables, with the response variable in the form of dichotomous data that is worth one (1) to state the occurrence of an event and zero(0) to indicate no occurrence of an event. The response variable is denoted by Y, while the factor variables can be a category, continuous or mixed of both variable. The logistic regression model is formed by expressing the value of P(Y = 1|x) as $\pi(x)$ [4], which is denoted as follows:

$$p_i = \pi(x_i) = \frac{\exp(\beta_0 + \beta_1 x_{1i} + ... + \beta_p x_{pi})}{1 + \exp(\beta_0 + \beta_1 x_{1i} + ... + \beta_p x_{pi})} \quad (1)$$

Where $x_i = (x_{1i}, x_{2i}, ... x_{pi})$: The independent variable, with $(i = 1, 2, \ldots, n)$, $\pi(x)$: Opportunity for the occurrence of success, β_j : The value of the jth parameter $(j = 0, 1, 2, \ldots, p)$.

Logit Function:

$$g(x) = \beta_0 + \beta_1 x_1 + ... + \beta_p x_p \quad (2)$$

$g(x)$is a function of the relationship of a logistic regression model called a logit relationship function.

Hypothesis Test steps

a. Significant test of the model (Test G)

Hypothesis:

$H_0 : \beta_1 = \beta_2 = ... = \beta_p = 0$

H_1 :There is at least one $\beta_j \neq 0$, j = 1, 2, ..., p

Significance levelα = 5%

Test Statistics:

$$G = -2 \ln \frac{Likelihood\ without\ independent\ variables}{Likelihood\ of\ all\ models} \quad (3)$$

Critical area : Reject H_0 if $G > \chi^2_{(\alpha,p)}$ or significant value (p-value) <α with α = 5%. [5]

Conclusion: H_0 processed or H_0 not rejected

b. Partial test of partial significance (Wald test) for $X_j$variable (j = 0, 1, ..., p) [5]

Hypothesis:

$H_0 : \beta_j = 0$

$H_1 : \beta_j \neq 0$

Significance level: 5%

$$\text{Test Statistics: } W = \frac{\hat{\beta}_j}{S_e(\hat{\beta}_j)} \quad (4)$$

Critical area : Reject H_0 if $W < -Z_{\alpha/2}$ or$W > Z_{\alpha/2}$orp-value <α with α = 5%. [5]

Conclusion: H_0 processed or H_0 not rejected

IOP Conf. Series: Materials Science and Engineering **335** (2018) 012044 doi:10.1088/1757-899X/335/1/012044

c. Odds ratio

The value of odds ratio forthe category of$X_k$interpreted as the risk or inclination of occurrence or event y = 1, is the amount of $exp(\beta_k)$times the risk of occurrence or event y = 1 in the comparison category [5].

$$odds = \frac{\pi_j}{1-\pi_j} = \frac{\exp(\beta_0 + \beta_1 x_1 + ... + \beta_p x_p)}{1+\exp(\beta_0 + \beta_1 x_1 + ... + \beta_p x_p)} \qquad (5)$$

Path Analysis Technique is a means to analyze the causal relationship between variables in order to know both the direct influence and indirect influence among the independent variables on the dependent variable.Basic principles [6]:

a. The existence of aditivitas. No interaction effects and normal distribution.
b. Data interval scale. All observed variables have scaled data intervals. If the data has not been in the form of interval scale, the data should be modified using the successive interval (MSI) method first.
c. All residual variables (not measured) are not correlated with any of the variables in the model.
d. There is no multicollinearity and Autocorrelation.
e. Recursivity. All arrows have one direction, no looping occurs.

Form the structural equation as follows:

$$Z_0 = \rho_{01} Z_1 + \rho_{02} Z_2 + ... + \rho_{0k} Z_k + \varepsilon_k \qquad (6)$$

With: Z_0 = free standardized variable, Z_1 = free standardized variable, ρ_{01} = path coefficient and ε_i = error, i=1, 2, 3, … ,k

Steps to calculate the path coefficient as follows [7]:

a. Create a path diagram
b. Overall Path Coefficient Testing

Hypothesis:

$H_0: \rho_{YX_1} = \rho_{YX_2} = ... = \rho_{YX_k} = 0$

There is no effect of exogenous variables on Endogenous variables

H_1 :is incorrect (Exogenous variable influence exogenous variable)

Significance level: α = 5% then obtained Ftable= $F_{(\alpha,\, k;\, n-k-1)}$

Test Statistics: $F = \frac{(n-k-1)(R^2_{x_u(x_1,x_1-x_k)})}{k(1-R^2_{x_u(x_1,x_1-x_k)})}$ [8] (7)

With: i = 1, 2, ..., k; k = Many exogenous variables in substructures being tested

Critical area: Reject H_0 if F arithmetic > $F_{(\alpha,\, k;\, n-k-1)}$ or significance value <α =5%.

Conclusion: H_0 processed or H_0 not rejected

c. Path Coefficient Testing

Hypothesis:

$H_0 : \rho_{YX} = 0$ (There is no effect of exogenous variables on endogenous variables)

$H_1 : H_0$ is incorrect

There exogenously exogenous variables influence on endogenous variables

Significance level: α then obtained $t_{tabel} = t_{(\alpha/2, n-k-1)}$

Test Statistic: $t = \dfrac{\rho_{x_u x_i}}{\sqrt{\dfrac{(1 - R^2_{x_u(x_1, x_1 - x_k)})c_{ii}}{n-k-1}}}$ [8] (8)

With: i = 1, 2, …, k ; k = Many exogenous variables in the substance under test, R^2 = Coefficient of determination

Critical are: Reject H_0 if t count > $t_{(\alpha/2, n-k-1)}$ or significance value <α=5%.

Conclusion: H_0 processed or H_0 not rejected

d. The Trimming Theory and Test of Goodness of Fit

Trimming theory is a method that works by removing path coefficients that are not significant and do not meet the criteria [6].

Hypothesis:

H_0 : The model corresponds to the data

H_1 : Model does not match the data

Significance level: α

Test Statistics: $W = -\left(N - d\right) \ln Q;\ Q = \dfrac{1 - R_m^2}{1 - M}$ (9)

$R_m^2 = 1 - [(1 - R_1^2)(1 - R_2^2)\ \dots\ (1 - R_k^2)]$ (10)

With: M = coefficient of multiple determinations(R_m^2)after the non-significant path coefficient isomitted

N = sample size,

d = difference of number of path coefficients,

R_m^2= coefficient of determination of proposed model

Critical area : Reject if Jika $W > \chi^2_{(\alpha, d)}$ or significance value <α=5%.

Conclusion: H_0 processed or H_0 not rejected

3. Research Methodology

The data in this study is primary data, by spreading questionnaires to students of grade eight for academic year 2016/2017 at SMP Negeri 1 Jiwan Madiun. The variables in this study consist of dependent (Y) and independent variables(X). The dependent variable (Y) is student achievement, and independent variable (X) are students' intelligence, health, interest, and motivation as internal factors, and family environment, school environment and community environment as external factors.
The procedures in this study are as follows:

3.1 Perform logistic regression analysis with the following stages:

a. Create a binary logistic regression equation model and get the parameter estimation.
b. Perform the test for parameters, there are two stages in doing this test, namely:
   1) Test of significance simultaneously
      This test is conducted to check the significance of at least one of the coefficients βj, j = 1,2,p against the dependent variable together using equation (4).
   2) Test the significance of one by one (partially)
      This test is conducted to check the significance of a coefficient βj, for each j = 1,2, ..., p against the dependent variable individually (partially) using equation (3).
c. Remove independent variables that have no effect on the dependent variable based on simultaneous test results and partial test with enter method procedure.
d. Interpret the model.
e. Predict the chance of success of the significant variables.

3.2 Conduct Path Analysis method with the following stages:

a. Create a frame of mind
b. Create causal relationships model between variables and structural equations
c. Data analysis
At this stage the steps are as follows:
1) Test Path analysis Assumptions.
2) Estimate Path Coefficient
Based on the path diagram we have determined the structural equation resulting the correlation equation to estimate.
a) Test the estimated coefficients
The path coefficients that have been generated need to be tested partially and simultaneously.
b) Test the Model matching
Model matching tests are performed to determine a feasible model or not.
c) Trim the model when needed
Model trimming is conducted if the path model has not fitted.
d) Calculate the effect
Find direct, indirect or total effects.
3) Conclusion
Draw the conclusion in order to answer the initial objectives of the study, and then make the necessary recommendations according to the analysis and discussion.

4. Results and Discussion

4.1 Student's Characteristics

After conducting an assessment of 153 respondents of grade eight students for academic year 2016/2017 at SMP Negeri 1 Jiwan Madiun, obtained:

a. Gender: male (56%) and female (44%)
b. Parents' Employment: self-employed (30%), private employee (26%), civil servant (13%), farmer (10%), army/policemen (7%), and others (14%).

4.2 Results of Logistic Regression

4.2.1. Model Significance Test (Test G). Statistics to test model significance were performed using G test, SPSS 20 output results are presented in Table 1 below:

Table 1. G test

Chi-square	df	Sig
66,443	7	0,000

Taking a significant level of $\alpha = 0.05$, from Table 1 due to the p-value (0.00) <0.05 and the G value of 66.443> $\chi^2_{(0.05,\ 7)} = 14.067$ it can be concluded that $H_o$is rejected. That is, there is at least one independent variables simultaneously (intelligence, health, interest, motivation, family environment, school environment and community environment) affect the dependent variable (student achievement).

4.2.2. Wald Test. Wald test aims to test the influence of independent variables on the response variable partially. The wald test results obtained from the SPSS *20 output are expressed in Table 2.*

Table 2. Wald Test

Variable	Coefficient	Wald	P value	Exp(B)	Conclusion
Constant	-11,694	17,502	0,000	0,000	
Intelligence (X_1)	-0,004	0,002	0,963	0,996	Not significant
Health (X_2)	-0,013	0,024	0,876	0,987	Not significant
Interests (X_3)	0,389	17,818	0,000	1,475	significant
Motivation (X_4)	0,178	4,313	0,038	1,195	significant
Family Environment (X_5)	-0,054	0,452	0,501	0,948	Not significant
School Environment (X_6)	0,018	0,050	0,823	1,018	Not significant
Community Environment (X_7)	0,077	0,909	0,340	1,080	Not significant

From Table 2 above, it can be seen there are only 2 significant variables i.e. interest and motivation variables. Therefore it is necessary to re-test by using only significant variables. Based on Wald test above only Interest and Motivation variables that significantly affect students' achievement. Therefore it is necessary to re-test by using only significant variables. In Table 3 below the Wald test results are shown on the significant variables

Table 3. Significant Wald Variable Test

Variable	Coefficient	Wald	P value	Exp(B)	Conclusion
Constant	-11,306	28,710	0,000	0,000	
Interests (X_3)	0,395	19,759	0,000	1,484	Signifikan
Motivation (X_4)	0,189	5,972	0,015	1,208	Signifikan

From Table 3 obtained logistic regression model with β coefficient value of each significant variables can be written as follows:

$$\pi(x) = \frac{\exp(-11,306 + 0,395_{(interest)} + 0,189_{(motivation)})}{1 + \exp(-11,306 + 0,395_{(interest)} + 0,189_{(motivation)})}$$

,

With logit function $g(x) = -11,306 + 0,395_{(interest)} + 0,189_{(motivation)}$

4.2.3. *Odds Ratio.* The value of the odds ratio is interpreted as the risk or inclination of student achievement. Because in the wald test only Interests and Motivation variables that affect student achievement, so that only two variables are seen tendency. The output results are presented in table 4 below:

Table 4. Odds Ratio Value

Variable	*odds ratio* value
Interests	1,484
Motivation	1,208

From Table 4, we get the odds ratio for the interest variable of 1.484 so that it can be explained that every increase of one unit on the number of interest score and motivation score remain, then the comparison between student achievement opportunities with good student achievement is not good increase 1,484 times. For the value of odds ratio of Motivation variable equal to 1.208 so it can be explained that every increase of one unit on the number of score of motivation and score of interest remain, hence comparison of good student achievement opportunity with student achievement not good will increase 1,208 times.

4.2.4. *Test the Goodness of Fit.* The goodness of fit test used to check whether the logistic regression model is well estimated. The calculation of goodness of fit by using Hosmer and Lemeshow Test is presented in Table 5.

Table 5. *Hosmer and Lemeshow Test*

Step	Chi-Square	df	*P-value*
1.	11,059	8	0,198

From Table 5 the value of χ 2 is 11,059 <χ 2 (α, db) = 5%, db = 8 is 15,507 and p-value is 0,198> 0,05 so H_0 is not rejected. The conclusion that the model has been quite able to explain the data (accordingly).

4.3 Results of Path Analysis

Path Analysis aims to investigate which variables have an effect on learning achievement and how much influence. It also can know the direct influence, indirect influence and total influence.

4.3.1. Model Conceptualization. Based on factors influencing student achievement then as exogenous variable is intelligence as X1, health as X2, interest as X3, motivation as X4, family environment as X5, school environment as X6, and society environment as X7. While the endogenous variable is student achievement as X8, then the conceptual concept of the path diagram is expressed in Figure 1 as follows:

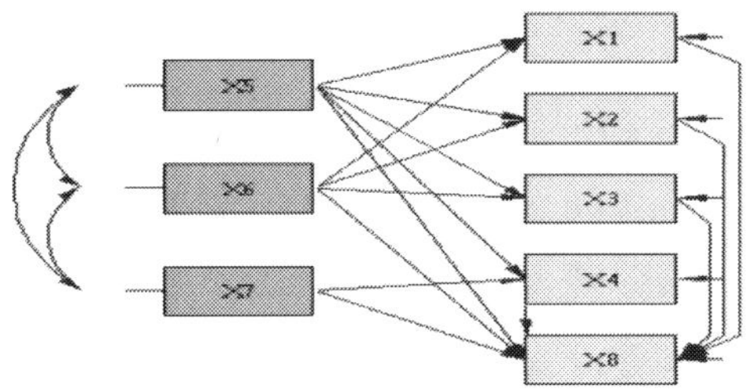

Figure 1. Proposed Line Diagram

So obtained the structural model as follows:

a. $X_1 = \rho_{15}X_5 + \rho_{16}X_6 + \varepsilon_1$

b. $X_2 = \rho_{25}X_5 + \rho_{26}X_6 + \varepsilon_2$

c. $X_3 = \rho_{35}X_5 + \rho_{36}X_6 + \varepsilon_3$

d. $X_4 = \rho_{45}X_5 + \rho_{46}X_6 + \varepsilon_4$

e. $X_8 = \rho_{81}X_1 + \rho_{82}X_2 + \rho_{83}X_3 + \rho_{84}X_4 + \rho_{85}X_5 + \rho_{86}X_6 + \rho_{87}X_7 + \varepsilon_8$

4.3.2. Significance of Path Coefficient Test. Significance Test of Path Coefficient is done in two ways, simultaneous test and partial test.

a. Simultaneous Test

The result of the SPSS 20 output used in the simultaneous test obtained F value presented in Table 6 below

Table 6. Simultaneous Test

No	Equation	Value F	*P-value*	Conclusion
1.	Structural Equation 1	16,871	0,000	Significant
2.	Structural Equation 2	20,159	0,000	Significant
3.	Structural Equation 3	27,326	0,000	Significant
4.	Structural Equation 4	33,301	0,000	Significant
5.	Structural Equation 5	26,545	0,000	Significant

From Table 6 it appears that all equations have sig (0,000) <α (5% = 0,05) so H_0 is rejected. This means that simultaneously exogenous variables have an effect simultaneously on endogenous variables.

b. Partial Test
The result of LISREL output obtained by t-value value in partial test stated in Table 7 below:

Table 7. Partial Test

Variable	t	t_{table}	Conclusion
Structural Equation 1			
Family environment	3,05	1,96	Significant
School environment	2,08	1,96	Significant
Structural Equation 2			
Family environment	4,20	1,96	Significant
School environment	1,23	1,96	Not significant
Structural Equation 3			
Family environment	3,19	1,96	Significant
School environment	3,36	1,96	Significant
Structural Equation 4			
Family environment	3,40	1,96	Significant
Community Environment	4,72	1,96	Significant
Structural Equation 5			
Intelligence	-0.40	1,96	Not significant
Health	-1.07	1,96	Not significant
Interest	9.02	1,96	Significant
Motivation	4.53	1,96	Significant
Family environment	0.53	1,96	Not significant
School environment	-1.06	1,96	Not significant
Community Environment	0.075	1,96	Not significant

From Table 7, there are no significant variables, it is necessary to re-test for significant variables only. So we get the latest path diagram presented in Figure 2 below:

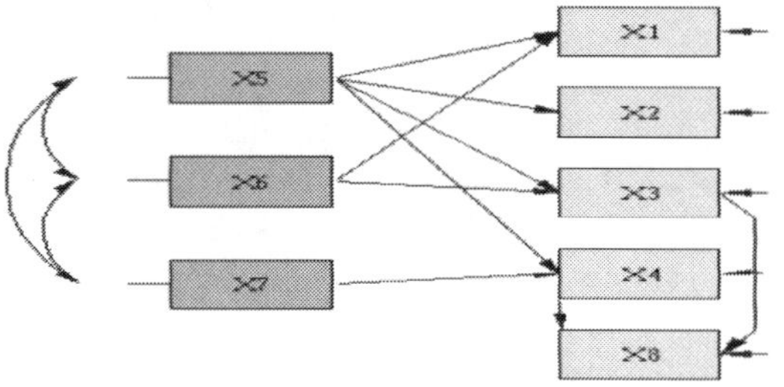

Figure 2. Significant Variable Path Chart

4.3.3. Test Goodness of Fit. Goodness of fit test results using coefficient of determination and after coefficient path is not significant removed, stated in table 8 below:

Table 8. Coefficient of Determination

$R^2$the proposed model	(1-R^2)	R^2variable model is significant	(1-R^2)
0,18	0,82	0,18	0,82
0,21	0,79	0,2	0,8
0,27	0,73	0,27	0,73
0,31	0,69	0,31	0,69
0,5	0,5	0,5	0,5

From the calculation coefficient of determination presented in Table 8 above, then using equation (10) is obtained

$$R_m^2 = 1 - [(0,82)(0,79)(0,73)(0,69)(0,5)] = 0,8369$$

$$M = 1 - [(0,82)(0,8)(0,73)(0,69)(0,5)] = 0,8348$$

$$Q = \frac{1 - R2_m}{1 - M} = 0,9875$$

$$W_{Count} = -(153 - 6)\ln(0,9875) = 1,8491$$

From the calculation obtained value of 1.8491 <by 12.592 then H0 is not rejected so it can be concluded that the model in accordance with the data.

4.3.4. Direct Effect, Indirect Effect and Total Influence. Calculation Direct influence, indirect effect and total influence are obtained from LISREL output which is standardized.

Table 9. Direct Effect, Indirect Effect and Total Influence

Exogenous Variables	Direct Effect	Indirect Effect		Total Influence
		Through	Calculation	
Interests (X_3)	(X_3=> X_8) = 0,59	-	0	0,59
Motivation (X_4)	(X_4=> X_8) = 0,27	-	0	0,27
Family Environment (X_5)	(X_5=> X_1) = 0,28 (X_5=> X_2) = 0,45	(X_5 => X_3 =>X_8) (X_5 => X_4 =>X_8)	(0,28*0,59) + (0,27*0,27) = 0,24	0,97
School Environment (X_6)	(X_6=> X_1) = 0,19	(X_6 => X_3 =>X_8)	(0,30*0,59) = 0,18	0,37
Community Environment (X_7)	-	(X_7 => X_4 =>X_8)	(0,37*0,27) = 0,10	0,10

Based on the calculation of Table 9 obtained a direct influence on student achievement of Interest of 59% and 27% Motivation. While that has an influence on student achievement indirectly among other family environment for 24%, the school environment by 18% and the community environment by 10%.

5. Conclussion

Characteristics of respondents in this study are as follows: a) Male gender (56%) and women (44%); b) Occupation of Parent entrepreneurs (30%), private employees (26%), civil servants (13%), farmers (10%), army/policemen (7%) and others (14%). Based on Binary Logistic Regression analysis, it is known that factors influencing student achievement are Interest and Motivation with tendency that is interest variable has value of odds ratio equal to 1.484 so it can be explained that every one-unit increase in the amount of interest score and motivation score remain, hence comparison of achievement opportunity good students with poor student achievement will increase 1,484 times. And Motivation variables have an odds ratio of 1.208 so it can be explained that each increase of one unit on the number of scores Motivation and interest score fixed, then the comparison of student achievement opportunities with good student achievement is not good will increase 1,208 times. With Path Analysis known to have a direct influence on student achievement of Interest of 59% and Motivation by 27%. While the influence on student achievement indirectly among other family environment by 24%, the school environment by 18% and the community environment by 10%.

Acknowledgments
Thanks to the Institut Sains & Teknologi AKPRIND Yogyakarta that support funding and research facilities.

References

[1] Purwanto N, 1990, Educational Psychology, PT RemajaRosdakarya, Jakarta.
[2] Malik H, Sudarno, Abdul H, 2015, Path Analysis of Factors Affecting the Student Statistics GPA (Statistics) UNDIP, Gaussian Journal Volume 4 Number 1, Department of FSM Statistics, Diponegoro University, http: // ejournal -s1.undip.ac.id/index.php/gaussian.
[3] Ratih B, Yuciana W, Suparti, 2015, Comparison of Binary Logistic Regression Methods and Multivariate Regression Spline (MARS) On High School (Case Study of SMA Negeri 2 Semarang), Gaussian Journal Volume 4 No. 4 Page 987-996, Statistics FSM, Diponegoro University, http://ejournal-s1.undip.ac.id/index.php/gaussian
[4] Hosmer, David W, Stanley Lemeshow, and Rodney X Sturdiven, 2013, Applied logistic Regression 3rdEd, Wiley, New Jersey.
[5] Agresti A, 2007, Analysis of Ordinal Categorical Data, Wiley and Sons, New York.
[6] Riduwan and Engkos, 2012, The Easy Way to Use and Use Path Analysis (Path Analysis), AlfaBeta, Bandung.
[7] Sarwono J, 2005, Path Analysis for Business Research with SPSS, C.V. Andi Offset, Yogyakarta
[8] Somantri A and Sambas, 2007, Statistics Application In Research, Loyal Pustaka, Bandung.

IOP Conf. Series: Materials Science and Engineering **335** (2018) 012046 doi:10.1088/1757-899X/335/1/012046

Convergence of Transition Probability Matrix in CLV-Markov Models

D Permana[1], U S Pasaribu[2], S W Indratno[2], S Suprayogi[3]

[1]Study Programme in Statistics, Faculty of Mathematics and Natural Sciences, UniversitasNegeri Padang, Indonesia
[2]Statistics Research Division, Faculty of Mathematics and Natural Sciences, InstitutTeknologi Bandung, Indonesia
[3]Industrial System and Techno-Economics Research Group, Faculty of Industrial Engineering, InstitutTeknologi Bandung, Indonesia

E-mail: donypermana@fmipa.unp.ac.id

Abstract. A transition probability matrix is an arrangement of transition probability from one states to another in a Markov chain model (MCM). One of interesting study on the MCM is its behavior for a long time in the future. The behavior is derived from one property of transition probabilty matrix for n steps. This term is called the convergence of the n-step transition matrix for n move to infinity. Mathematically, the convergence of the transition probability matrix is finding the limit of the transition matrix which is powered by n where n moves to infinity. The convergence form of the transition probability matrix is very interesting as it will bring the matrix to its stationary form. This form is useful for predicting the probability of transitions between states in the future. The method usually used to find the convergence of transition probability matrix is through the process of limiting the distribution. In this paper, the convergence of the transition probability matrix is searched using a simple concept of linear algebra that is by diagonalizing the matrix.This method has a higher level of complexity because it has to perform the process of diagonalization in its matrix. But this way has the advantage of obtaining a common form of power n of the transition probability matrix. This form is useful to see transition matrix before stationary. For example cases are taken from CLV model using MCM called Model of CLV-Markov. There are several models taken by its transition probability matrix to find its convergence form. The result is that the convergence of the matrix of transition probability through diagonalization has similarity with convergence with commonly used distribution of probability limiting method.

1. Introduction

In Markov Chain Model (MCM), matrix of transition probability is a main model or a main result describing characteristics of relationship inter states. The matrix contain probabilities of transition from one state to another states. One of interesting study on the MCM is its behavior for a long time in the future. This term is called convergency of n-step transition matrix for $n$moves to infinity. Mathematically, the convergence of the transition probability matrix is finding the limit of the transition matrix that is powered by n where n moves to infinity. In [1],[2] the long run behavior of MCM is explained by a theorem below.

Theorem 1. Let **P** be a regular transition probability matrix on the states 1,2,…,N. Then the limiting distribution $\pi = (\pi_1, \pi_{2,} \cdots, \pi_N)$ is the unique nonnegative solution of the equations

$$\pi_j = \sum_{k=0}^{N} \pi_i P_{kj}, \;_{j=1,2,\cdots,N} \text{ and } \sum_{k=0}^{N} \pi_k = 1$$

The theorem has implication,

$$\lim_{n\to\infty} \mathbf{P}^n = \begin{pmatrix} \pi_1 & \pi_2 & \cdots & \pi_N \\ \pi_1 & \pi_2 & \cdots & \pi_N \\ \vdots & \vdots & \ddots & \vdots \\ \pi_1 & \pi_2 & \cdots & \pi_N \end{pmatrix}$$

The long run behaviour of MCM is stationary form of Markov chain. The two equation in the theorem are an ordinary linear system that has simple method to get solution vector of π. This paper will explain another method to get vector of π. We use one of concept in linear algebra. That is a diagonalization of the transition matrix to find $\mathbf{P}^n$ and then limiting it for n to infinity to get the solution. This way has more advantage than the theorem. One, we can get generalization form of matrix transition to the power of *n*as *n*-steps transition probability matrix with *n* arbitrary. This result is useful to see transition matrix before stationary.

For cases, we take models of determination of Customer Lifetime Value (CLV) using Markov Chain, we called models of CLV-Markov. In the model, there are some characteristic of relationship interstate of customer. Every model has different size of transition matrix. It is based on the number of states. This paper will take three matrix from the models of CLV-Markov.

CLV is a value of a customer as expectation value of a firm revenue from him or her, start from present time to long time in the future[3]. CLV is a familiar term in marketing system. CLV is one concepts to support marketing program [4]especially to manage of cost of customer, to acquire a new customer and to maintain an old customer[5]. Many methods can be used to determine CLV[6]. One of the methods using MCM [7].

2. Diagonalization

Let P is matrix size *mxm*which has *m* different eigen values. Thus, P diagonalizable[8], write

$$\mathrm{P} = UDU^{-1} = \lambda_1 \mathrm{C}_1 + \lambda_2 \mathrm{C}_2 + \cdots + \lambda_m \mathrm{C_m}$$

Matrix C_j is come from multiplication of jth column vector of U with j^{th} row vector of U^{-1}. Thereofore, p to the power of n can rewrite,

$$\mathrm{P}^n = UD^nU^{-1} = \lambda_1{}^n \mathrm{C}_1 + \lambda_2{}^n \mathrm{C}_2 + \cdots + \lambda_m{}^n \mathrm{C_m}$$

For the simple, let m is equal two. Matrix $\mathbf{C_1}$ and $\mathbf{C_2}$ can be write,

$$\begin{aligned} \mathrm{P^n} &= \lambda_1{}^n \mathrm{C}_1 + \lambda_2{}^n \mathrm{C}_2 \\ &= \lambda_1{}^n \begin{pmatrix} k_1 & k_3 \\ k_5 & k_7 \end{pmatrix} + \lambda_2{}^n \begin{pmatrix} k_2 & k_4 \\ k_6 & k_8 \end{pmatrix} \\ &= \begin{pmatrix} k_1\lambda_1{}^n + k_2\lambda_2{}^n & k_3\lambda_1{}^n + k_4\lambda_2{}^n \\ k_5\lambda_1{}^n + k_6\lambda_2{}^n & k_7\lambda_1{}^n + k_8\lambda_2{}^n \end{pmatrix} \end{aligned}$$

Especially if P is a transition probability matrix, the last equation can be write,

$$\mathrm{P^n} = \begin{pmatrix} k_1\lambda_1{}^n + k_2\lambda_2{}^n & 1 - k_1\lambda_1{}^n - k_2\lambda_2{}^n \\ k_3\lambda_1{}^n + k_4\lambda_2{}^n & 1 - k_3\lambda_1{}^n - k_4\lambda_2{}^n \end{pmatrix}$$

Some n can be fixed to construct a linear system and solve it to get k_i, $i = 1,..4$.

3. Models of CLV-Markov

Here, we take three models of CLV-Markov. The first model contain two states, customer and former customer. The transition matrix is below,

$$\mathrm{P} = \begin{pmatrix} r & 1-r \\ q & 1-q \end{pmatrix}$$

Symbol *r*is aretention probability. It is mean a kind of retention rate or proportion customerswhich succeses to be maintained. Whereas symbol q is an aquation probability. It is mean probability of a

IOP Conf. Series: Materials Science and Engineering **335** (2018) 012046 doi:10.1088/1757-899X/335/1/012046

former customer to back to customer again. The model use this matrix is called Model CF(1,1). Index have a meaning that the number of customer state and former customer state.
The second model contain three states, prospect, Customer, and Former Customer. The matrix can be rewritebelow,

$$\mathrm{P} = \begin{pmatrix} 0 & a & 1-a \\ 0 & r & 1-r \\ 0 & q & 1-q \end{pmatrix}$$

Symbol a is aaquation probability to get a new customer. The model use this matrix is called Model PCF(1,1,1).

The last models contain three states too, one level customer state and two level former customer state. So, the model is called Model CF(1,2), representing below

$$\mathrm{P} = \begin{pmatrix} \mathrm{r} & 1-r & 0 \\ \mathrm{q} & 0 & 1-q \\ 0 & 0 & 1 \end{pmatrix}$$

4. Results

The first model has two eigen values, $\boldsymbol{\lambda_1 = 1}$ **and** $\boldsymbol{\lambda_2 = r - q}$. Thus $\mathbf{P^n}$ can write,

$$\mathrm{P}^n = \frac{1}{1-r+q}\left[\begin{pmatrix} q & 1-r \\ q & 1-r \end{pmatrix} + \begin{pmatrix} 1-r & r-1 \\ -q & q \end{pmatrix}(r-q)^n\right]$$

$$= \frac{1}{1-r+q}\begin{pmatrix} q & 1-r \\ q & 1-r \end{pmatrix} + A{\lambda_2}^n$$

Thus,

$$\lim_{n\to\infty} \mathrm{P}^n = \frac{1}{1-r+q}\begin{pmatrix} q & 1-r \\ q & 1-r \end{pmatrix}$$

The second model has three eigen values, $\boldsymbol{\lambda_1 = 1, \lambda_2 = r - q}$**, and** $\boldsymbol{\lambda_3 = 0}$. Thus $\mathbf{P^n}$ can write,

$$\mathbf{P}^n = \frac{1}{1-r+q}\begin{pmatrix} 0 & q & 1-r \\ 0 & q & 1-r \\ 0 & q & 1-r \end{pmatrix} + \left[\begin{pmatrix} 0 & a & -a \\ 0 & r & -r \\ 0 & -q & q \end{pmatrix} + \frac{q}{1-r+q}\begin{pmatrix} 0 & -1 & 1 \\ 0 & -1 & 1 \\ 0 & 0 & 0 \end{pmatrix}\right](r-q)^n$$

$$= \frac{1}{1-r+q}\begin{pmatrix} 0 & q & 1-r \\ 0 & q & 1-r \\ 0 & q & 1-r \end{pmatrix} + B{\lambda_2}^n$$

Thus,

$$\lim_{n\to\infty} \mathrm{P}^n = \frac{1}{1-r+q}\begin{pmatrix} 0 & q & 1-r \\ 0 & q & 1-r \\ 0 & q & 1-r \end{pmatrix}$$

Whereas the third model has three eigen values, $\lambda_1 = \frac{r+\sqrt{r^2+4q(1-r)}}{2}$, $\lambda_2 = \frac{r-\sqrt{r^2+4q(1-r)}}{2}$, and $\lambda_3 = 1$.
Thus $\mathbf{P^n}$ can write,

$$P^n = \begin{pmatrix} \frac{(1-r)(\lambda_2-r+q)}{(1-\lambda_1)(\lambda_2-\lambda_1)}{\lambda_1}^n + \frac{(1-r)(r-q-\lambda_1)}{(1-\lambda_2)(\lambda_2-\lambda_1)}{\lambda_2}^n & \frac{(r-1)(\lambda_2-r+1)}{(1-\lambda_1)(\lambda_2-\lambda_1)}{\lambda_1}^n + \frac{(1-r)(\lambda_1-r+1)}{(1-\lambda_2)(\lambda_2-\lambda_1)}{\lambda_2}^n & 1-{P^n}_{11}-{P^n}_{12} \\ \frac{-q(1-r+\lambda_2)}{(1-\lambda_1)(\lambda_2-\lambda_1)}{\lambda_1}^n + \frac{q(1-r+\lambda_1)}{(1-\lambda_2)(\lambda_2-\lambda_1)}{\lambda_2}^n & \frac{\lambda_2+q(1-r)}{(1-\lambda_1)(\lambda_2-\lambda_1)}{\lambda_1}^n + \frac{-(\lambda_1+q(1-r))}{(1-\lambda_2)(\lambda_2-\lambda_1)}{\lambda_2}^n & 1-{P^n}_{21}-P_{22} \\ 0 & 0 & 1 \end{pmatrix}$$

$$= G{\lambda_2}^n + H{\lambda_2}^n + \begin{pmatrix} 0 & 0 & 1 \\ 0 & 0 & 1 \\ 0 & 0 & 1 \end{pmatrix}$$

Thus,

$$\lim_{n\to\infty} \mathbf{P}^n = \begin{pmatrix} 0 & 0 & 1 \\ 0 & 0 & 1 \\ 0 & 0 & 1 \end{pmatrix}$$

5. Conclusions

Convergence of transition probability matrix in Markov Chain is an interesting subject to study properties of transition object inter states in the long future time. Commonly, convergence of the matrix is found by a familiar theorem in many stochastic books. But, here we try to find convergence the matrix use approach diagonalization of the transition matrix. This method are applicated to three models of CLV-Markov. The result, we get generalization of $\mathbf{P}^n$ from every models.

References

[1] Taylor H M, & Karlin, S 1998. *An Introduction to Stochastic Modeling 3rd Ed.* London: Academic Press
[2] Ross S 2014*Stochastic Processes 3th Ed.* New York, USA: John Wiley
[3] Permana D, Pasaribu U, & Indratno S 2014 Analysis Study of Behavior and Determination of Customer Lifetime Value (CLV) using Markov Chain Model. *AIP conference Proceedings, 1589* 456-459
[4] Cermak P 2015 Customer Profitability Analysis and Customer Life Time Value Models: Portfolio Analysis *Procedia* 14-25
[5] Pfeifer P, & Caraway R 2000 Modeling Customer Relationship As Markov Chain. *Journal of Interactive Marketing 14(2)* 43-55
[6] Gupta S H 2006 Modelling Customer Lifetime Value*Journal of Service Research 9(2)* 139-155
[7] Permana D, Pasaribu U, & Indratno S 2013 Study of Behavior and Determination of Customer Lifetime Value using Markov Chain Model for segment of Prospect, Customer, and Former Customer. *iCMS Conference Proceedings* 458-467
[8] Anton H, & Rorres C 2005*Elementary Linear Algebra 9Ed.* John Wiley and Sons

IOP Conf. Series: Materials Science and Engineering **335** (2018) 012047 doi:10.1088/1757-899X/335/1/012047

Mathematical modelling of the influenced of diffusion rate on macro nutrient availability in paddy field

Renny[1], **Supriyanto**[1]

[1]Department of Mathematics, Jenderal Soedirman University, Indonesia

E-mail: `renny.unsoed@gmail.com`

Abstract. Nutrition is the chemical compounds that needed by the organism for the growth process. In plants, nutrients are organic or inorganic compounds that are absorbed from the roots of the soil. It consist of macro and micro nutrient. Macro nutrients are nutrition that needed by plants in large quantities, such as, nitrogen, calcium, pottacium, magnesium, and sulfur. The total soil nutrient is the difference between the input nutrient and the output nutrients. Input nutrients are nutrient that derived from the decomposition of organic substances. Meanwhile, the output nutrient consists of the nutrients that absorbed by plant roots (uptake), the evaporated nutrients (volatilized) and leached nutrients. The nutrient transport can be done through diffusion process. The diffusion process is essential in removing the nutrient from one place to the root surface. It will cause the rate of absorption of nutrient by the roots will be greater. Nutrient concept in paddy filed can be represented into a mathematical modelling, by making compartment models. The rate of concentration change in the compartment model forms a system of homogeneous linear differential equations. In this research, we will use Laplaces transformation to solve the compartment model and determined the dynamics of macro nutrition due to diffusion process.

1. Introduction

Nutrition is the chemical compounds needed by the organism for growing process and normal function in the body system. In plants, nutrients are organic or inorganic compounds that are absorbed from the roots of the soil. Research has shown that plants need 17 nutrients, called "essential elements" [4]. Based on the level of needs, nutrients in plants can be classified into two parts namely macro nutrients and micro nutrients. Nutritional needs are absolute for every plant, and can not be replaced by other elements, of course with different levels according to the type of plant. Lack of nutrients will inhibit the plant life cycle itself. In rice fields, the dynamics of the quantity of macro nutrients in the soil is strongly influenced by several factors. In [3] it is mentioned that soil texture or soil composition over sand, mud and clay contents greatly affects the soil's ability to retain water and nutrients. In addition, the amount of decomposed organic matter also determines the quantity of nutrients in the soil. The more organic matter contained the more macro nutrients are produced. The process of decomposition of organic matter can not be separated from the role of microorganisms that act as decomposers. Nutrient concept in paddy filed can be represented into a mathematical modelling, by making compartment models. From concept of balance of dynamics of nutrition at paddy field, Ardiansyah, et al [1], formulate dynamic model of macro nutrition using system of differential equation. In the model, it is explained that the available macro nutrients in the soil are the difference between input nutrients (organic matter decomposition) with nutrient output (nutrients absorbed by plants, yawning nutrients, and nutrients that precipitate). The nutrient transport can be done through diffusion process. The diffusion process has an important role in spreading the macro nutrient in soil. Diffusion is very important in transferring macro nutrient from one place to another on the surface of the root, so that the nutrient can be absorbed by the plants [2].

2. Mathematics modelling

In building the model, we assume that this model consists of two compartments, which are compartment of available concentration of available nutrient (c_{ava}) and compartment concentration nutrient in the form of organic materials (c_{org}). The model compartment diagram can be seen below:

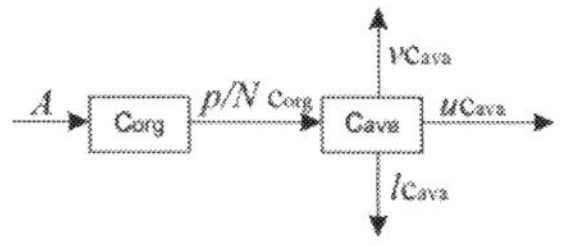

Figure 1. Compartment model of nutrient availability.

Based on Figure (1), the rate of change of nutrient concentration in the form of organic materials can be increased because of the organic input at the constant rate of A. Meanwhile, the rate of change of nutrient concentration in the form of organic materials can be decreased because of decomposition process by the microorganism, that will be spread in ready-to-absorb form, at a rate of $\frac{p}{N}c_{org}$. We claimed that N is the diffusion rate of macro nutrient.

Then, the rate of change of nutrient concentration in ready-to-absorb form can be increased because of the spreading of decomposition process of the microorganism at a rate of $\frac{p}{N}c_{org}$. It also can be decreased because there is nutrient absorption in the form of ready to be absorbed (uptake) by the roots at a rate of uc_{ava}, leaching process at a rate of lc_{ava} and evaporation process at a rate of vc_{ava}. Thus, we have the following differential equation

$$\frac{dc_{org}(t)}{dt} = A - \frac{p}{N}c_{org}(t) \quad (1)$$

$$\frac{dc_{ava}(t)}{dt} = \frac{p}{N}c_{org}(t) - uc_{ava}(t) - vcava(t) - lc_{ava}(t) \quad (2)$$

The parameters can be seen in the following table

Table 1. Parameter lists.

Parameter	Definition
A	Input rate of organic materials
p	Decomposition rate of organic materials
N	diffusion rate of macro nutrient
u	The rate of nutrient uptake by the roots
v	The rate of nutrient evaporation into the air (volatilized)
l	The rate of nutrient leaching by the water

3. Main results

Suppose $u+v+l=m$, then equation system (1) and (2), can be rewritten as

$$c_{org}' = A - \frac{p}{N}c_{org}(t) \quad (3)$$

$$c_{ava}' = \frac{p}{N}c_{org}(t) - mc_{ava}(t) \quad (4)$$

Suppose that we have intial condition $c_{org}(t_0)$ and $c_{ava}(t_0)$. Using Laplace transformation in the both side of equation (3) and (4), we obtain equation

$$\begin{aligned} \mathscr{L}\left\{c_{org}'\right\} &= \mathscr{L}\left\{A - \frac{p}{N}c_{org}\right\} \\ \mathscr{L}\left\{c_{ava}'\right\} &= \mathscr{L}\left\{\frac{p}{N}c_{org} - mc_{ava}\right\} \end{aligned} \quad (5)$$

System (5), can be write in the following equation

$$\begin{aligned} s\mathscr{L}\{c_{org}\} - c_{org}(t_0) &= A\mathscr{L}\{1\} - \frac{p}{N}\mathscr{L}\{c_{org}\} \\ s\mathscr{L}\{c_{ava}\} - c_{ava}(t_0) &= \frac{p}{N}\mathscr{L}\{c_{org}\} - m\mathscr{L}\{c_{ava}\} \end{aligned}$$

Then we have,

$$\left(s + \frac{p}{N}\right)C_{org}(s) = \frac{A}{s} + c_{org}(t_0) \quad (6)$$

$$(s+m)C_{ava}(s) - \frac{p}{N}C_{org}(s) = c_{ava}(t_0) \quad (7)$$

From equation (6), we obtained the following equation

$$C_{org}(s) = \frac{A}{s\left(s+\frac{p}{N}\right)} + \frac{c_{org}(t_0)}{\left(s+\frac{p}{N}\right)} \quad (8)$$

If we substituted equation (8) into (7), we get the following equation

$$\begin{aligned} (s+m)C_{ava}(s) - \frac{p}{N}\left(\frac{A}{s\left(s+\frac{p}{N}\right)} + \frac{c_{org}(t_0)}{\left(s+\frac{p}{N}\right)}\right) &= c_{ava}(t_0) \\ \Leftrightarrow (s+m)C_{ava}(s) &= \frac{p}{N}\left(\frac{A}{s\left(s+\frac{p}{N}\right)} + \frac{c_{org}(t_0)}{\left(s+\frac{p}{N}\right)}\right) + c_{ava}(t_0) \\ \Leftrightarrow C_{ava}(s) &= \frac{pA}{Ns\left(s+\frac{p}{N}\right)(s+m)} + \frac{pc_{org}(t_0}{N(s+m)\left(s+\frac{p}{N}\right)} + \frac{c_{ava}(t_0)}{(s+m)} \end{aligned} \quad (9)$$

The solution of $c_{org}(t)$ can be obtained from equation (8) by using the invers of Laplace transform,

$$\begin{aligned} c_{org}(t) &= \mathscr{L}^{-1}\{C_{org}(s)\} \\ &= \mathscr{L}^{-1}\left\{\frac{A}{s\left(s+\frac{p}{N}\right)} + \frac{c_{org}(t_0)}{\left(s+\frac{p}{N}\right)}\right\} \\ &= \frac{A}{p/N}\mathscr{L}^{-1}\left\{\frac{1}{s} - \frac{1}{\left(s+\frac{p}{N}\right)}\right\} + c_{org}(t_0)\mathscr{L}^{-1}\left\{\frac{1}{\left(s+\frac{p}{N}\right)}\right\} \end{aligned}$$

Then, the general solution of $c_{org}(t)$ is given by

$$c_{org}(t) = \frac{AN}{p}\left(1 - e^{-\frac{p}{N}t}\right) + c_{org}(t_0)e^{-\frac{p}{N}t} \tag{10}$$

The invers of Laplace transform, from equation (9),

$$\begin{aligned} c_{ava}(t) &= \mathscr{L}^{-1}\{C_{ava}(s)\} \\ &= \mathscr{L}^{-1}\left\{\frac{pA}{Ns\left(s+\frac{p}{N}\right)(s+m)} + \frac{pc_{org}(t_0)}{N(s+m)\left(s+\frac{p}{N}\right)} + \frac{c_{ava}(t_0)}{(s+m)}\right\} \\ &= \frac{pA}{N}\mathscr{L}^{-1}\left\{\frac{1}{s\left(s+\frac{p}{N}\right)(s+m)}\right\} + \frac{p}{N}c_{org}(t_0)\mathscr{L}^{-1}\left\{\frac{1}{(s+m)\left(s+\frac{p}{N}\right)}\right\} \\ &+ c_{ava}(t_0)\mathscr{L}^{-1}\left\{\frac{1}{(s+m)}\right\} \end{aligned}$$

We then have the general solution of $c_{ava}(t)$, namely

$$\begin{aligned} c_{ava}(t) &= \frac{A}{m} + \left(c_{ava}(t_0) - \frac{A}{m} - \frac{A}{m-\frac{p}{N}} - \frac{p}{N\left(m-\frac{p}{N}\right)}c_{org}(t_0)\right)e^{-mt} \\ &+ \left(\frac{A}{m-\frac{p}{N}} + \frac{p}{N(m-p)}c_{org}(t_0)\right)e^{-\frac{p}{N}t} \end{aligned} \tag{11}$$

The solution of the systems for a long periods of time is

$$\begin{aligned} \lim_{t\to\infty} c_{org}(t) &= \lim_{t\to\infty}\left(\frac{AN}{p}\left(1-e^{-\frac{p}{N}t}\right) + c_{org}(t_0)e^{-\frac{p}{N}t}\right) \\ &= \frac{AN}{p} \end{aligned}$$

and

$$\begin{aligned} \lim_{t\to\infty} c_{org}(t) &= \lim_{t\to\infty}\left(\frac{A}{m} + \left(c_{ava}(t_0) - \frac{A}{m} - \frac{A}{m-\frac{p}{N}} - \frac{p}{N\left(m-\frac{p}{N}\right)}c_{org}(t_0)\right)e^{-mt}\right) \\ &+ \lim_{t\to\infty}\left(\left(\frac{A}{m-\frac{p}{N}} + \frac{p}{N(m-p)}c_{org}(t_0)\right)e^{-\frac{p}{N}t}\right) \\ &= \frac{A}{m} \end{aligned}$$

4. Simulation

In this simulation, we used the following parameter values $c_{org}(0) = 1000, c_{ava}(0) = 50, p = 0.02, u = 0.3, l = 0.02, v = 0.05, A = 10$. We used different values of diffusion rate, N, namely $N = 0.02, 0.1, 0.2, 0.3, 0.8$. Then the illustration of macro nutrient in organic form can be presented in the Figure (2), From the figure, we could see that the higher diffusion rate, the macro nutrient concentration in organic form is higher. By using the same parameter values, we will see the graphics of the influence of different diffusion rate into the availability of macro nutrient in 120 days, as shown in Figure (3). We can see that, as the diffusion rate increase, then the availability of macro nutrient in soil will decrease. It is reasonable, that the macro nutrient will be uptake by the plant roots or volatilized into the air.

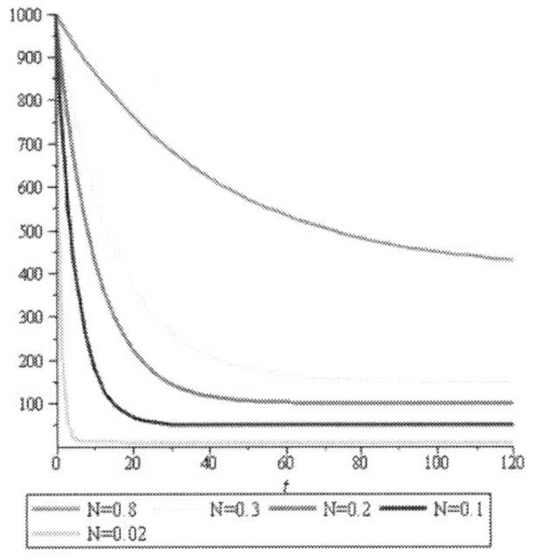

Figure 2. The change of macro nutrient concentration in the form of organic materials by using different values of diffusion rate.

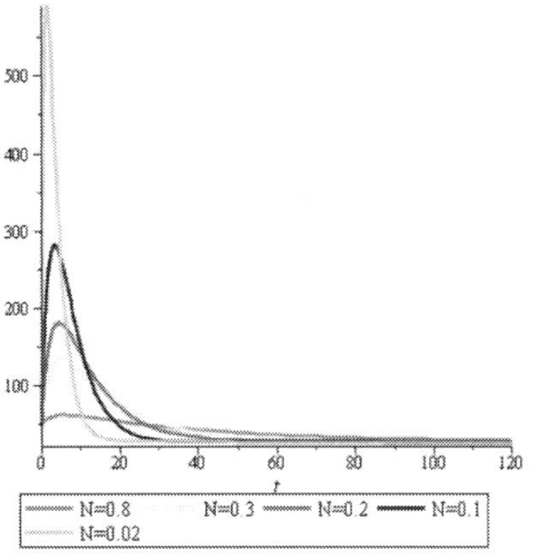

Figure 3. The influence of different diffusion rate into the availability of macro nutrient in 120 days.

5. Conclusion

The concentration of macro nutrient availability in soil will be decreased as the diffusion rate value increased. In long time periods, the concentration of macro nutrient availability will decrease into $\frac{A}{m}$.

6. Acknowledgment

Author would like to express gratitude to LPPM Unsoed for research support.

References

[1] Ardiansyah S B I, Arief C, Saptomo S K, and Masaru M 2012 *Soil Macro Nutrient (N, P, K) during Growth Stages under Conventional and SRI (System of Rice Intensification) Practices in Tropical Soil*
[2] Hillel D 1982 *Introduction into Soil Physic* (New York: Academic Press, Inc)
[3] Jones C and Jacobsen J 2001 *Plant Nutrition and Soil Fertility* (London: Montana State University)
[4] Marschner H 1995 *Mineral Nutrition of Higher Plants* (London: Academic Press)

IOP Conf. Series: Materials Science and Engineering **335** (2018) 012048 doi:10.1088/1757-899X/335/1/012048

Internal Cluster Validation on Earthquake Data in the Province of Bengkulu

D S Rini*, P Novianti, H Fransiska

Statistics Department, Faculty of Mathematics and Science, University of Bengkulu, Indonesia

*dyah.setyorini@unib.ac.id

Abstract. K-means method is an algorithm for cluster n object based on attribute to k partition, where $k < n$. There is a deficiency of algorithms that is before the algorithm is executed, k points are initialized randomly so that the resulting data clustering can be different. If the random value for initialization is not good, the clustering becomes less optimum. Cluster validation is a technique to determine the optimum cluster without knowing prior information from data. There are two types of cluster validation, which are internal cluster validation and external cluster validation. This study aims to examine and apply some internal cluster validation, including the Calinski-Harabasz (CH) Index, Sillhouette (S) Index, Davies-Bouldin (DB) Index, Dunn Index (D), and S-Dbw Index on earthquake data in the Bengkulu Province. The calculation result of optimum cluster based on internal cluster validation is CH index, S index, and S-Dbw index yield $k = 2$, DB Index with $k = 6$ and Index D with $k = 15$. Optimum cluster ($k = 6$) based on DB Index gives good results for clustering earthquake in the Bengkulu Province.

1. Introduction

Cluster analysis is a multivariate technique that has the main purpose to classify objects based on similarity characteristics. The similarity between objects is measured using the distance. Characteristics of objects in a cluster have a high level of similarity, while the characteristics of objects in a cluster with other clusters have a low level of similarity. In other words, the variance in a cluster is minimum while the variance between clusters is maximum [1].

K-means method is an algorithm for cluster n object based on attribute to k partition, where k <n. The clustering of objects is seen from the distance of the object to the closest centroid. After knowing the nearest centroid, the object will be classified as a member of that category. The K-Means algorithm is as follows: a. Select k centroid point randomly, b. Group the data so that k clusters with the point of the centroid of each cluster is the centroid point that has been selected previously, c. Update the centroid point value, 4. Repeat steps 2 and 3 until the value from the centroid point no longer changes. This shows that there is a shortage of algorithms before the algorithm is executed, the points are initialized randomly so that the grouping of the resulting data can vary, so never know the actual cluster by using the same data. If the random value for initialization is not good, then the resulting grouping becomes less optimal.

Cluster validation is a technique for determining optimal cluster number without knowing prior information from data. In general, there are two types of cluster validation, which are external cluster validation and internal cluster validation. In situations where external information is not available, the internal validation index is the only way to validate the cluster. Internal cluster validation consists of several types including the Calinski-Harabasz (CH) Index, Sillhouette (S) Index, S-Dbw, Davies-Bouldin (DB) Index, Krzanowski-Lai (KL) Index, Dunn Index (D), and etc.

Liu, *et al* (2010) performed a comparison of 11 internal cluster validations on crisp clustering. The result of the research indicates that the S-Dbw index is a good internal cluster validation index in all aspects, while other index has certain limitations [2]. Rendon, *et al* (2011) discusses the comparison of internal versus external cluster validation on the K-means and Bissecting K-means algorithms. The study found that internal cluster validation is more accurate in determining the optimal number of clusters. The cluster validation index used in the study is BIC index, CH index, DB index, S Index, NIVA index, and Dunn index [3]. Arbelaitz, *et al* (2013) examined a comparison of 30 internal cluster validation indexes on various data characteristics [4]. Novianti, *et al* (2016) classify earthquake events and create zones based on earthquake classification characteristics using K-means cluster analysis. The cluster validation used in this research is Krzanowski-Lai (KL) index with optimal cluster number of 7 clusters [5]. Therefore, the researcher is interested to apply the index of other internal cluster validation on the earthquake data of Bengkulu province so that it is expected to obtain the best clustering for earthquake area in Bengkulu province.

2. Literature Review

2.1. K-Means cluster analysis

K-means cluster analysis seeks to partition the *n* individuals in a set of multivariate data into *K* clusters, where each individual in the dataset is allocated entirely to a particular cluster. As a hard partitioning algorithm, *K*-means cluster analysis is an iterative process. First, data are initially partitioned. Each group is calculated its mean and then the data partitioned again by allocating each datum to its nearest mean cluster position [6,7].

MacQueen suggests the use of *K*-Means to describe the algorithm in the determination of an object in a certain cluster by the size of the nearest center (mean). In its simplest form, this process consists of three stages [8]:

a. Partition objects into *K* initial cluster

b. Begin by noting the objects, determine an object into a cluster that which has the closest centroid (mean). (The distance is usually calculated using the Euclidean distance with either standardized or unstandardized observation). Recalculate the cluster centroid to get a new object and for the cluster that lost object. Centroid of the group is calculated by calculating the average value of the data is as follows:

$$C_{kj} = \frac{x_{ikj} + x_{2kj} + \cdots + x_{akj}}{a}, j = 1, 2, \cdots, p$$

C_{kj} = centroid of group-k, variable-j

a = the number of members in the group k

c. Step b is repeated until no more transfer of the object.

Euclidean distance is the distance the most common type selected. Its simplicity is the geometric distance in multiple dimension of space. Euclidean distance is usually calculated from the raw data, and not of the standard data. This method has several advantages, including the distance of any two objects are not affected by the addition of new objects to be analyzed, which may be an outlier.

2.2. Cluster validation

Cluster validation is a technique for determining optimal cluster number without knowing prior information from data. In general, there are two types of cluster validation, which are external cluster validation and internal cluster validation. External cluster validation is a cluster validation technique based on prior knowledge of data. While internal cluster validation is a cluster validation technique based on intrinsic information from the data itself [3]. In situations where external information is not available, the internal validation index is the only way to validate cluster [2].

Internal cluster validation index consists of several methods, among which are as follows:

- Calinski-Harabasz (CH) Index

The Calinski-Harabasz (CH) index evaluates the cluster validity based on the average between and within cluster sum of square [2].This Index is computed by [3]:

$$CH = \frac{trace(S_B)}{trace(S_w)} \cdot \frac{n_p - 1}{n_p - k}$$

dengan

S_B : the between-cluster scatter matrix

S_w : the internal scatter matrix

n_p : the number of clustered samples

k : the number of clusters

The optimal cluster number is determined by maximizing the value of this index [2].

- Silhouette (S) Index

Silhouette (S) Index [2] validates the clustering performance based on the pairwise difference between and within cluster distance. In addition, The optimal cluster number is determined by maximizing the value of this index. It is defined as [3]:

$$S = \frac{(b(i) - a(i))}{Max\{a(i), b(i)\}}$$

where

$a(i)$: the average distance between the i-th sample and all thesamples included in X_j

$b(i)$: the minimum average distance between the i-th and all of the samples clustered in $X_k, k = 1,\ldots,c, k \neq j$

- Davies-Bouldin (DB) Index

This index aim to identify sets of clusters that are compact and well separated. The Davies-Bouldin index is defined as:

$$DB = \frac{1}{c}\sum_{i=1}^{c} Max_{i \neq j}\left\{\frac{d(X_i) + d(X_j)}{d(c_i, c_j)}\right\}$$

where

c: the number of clusters

i, j: cluster labels

$d(X_i)$: all samples in clusters i and j to their respective cluster centroid

$d(c_i, c_j)$: the distance between these centroid

Smaller value of DB indicates a better clustering solution[3].

- Dunn (D) Index

Dunn (D) index is defined as:

$$D=\min_{1\le i\le c}\left\{\min\left\{\frac{d\left(c_i,c_j\right)}{\max\limits_{1\le k\le c}\left(d\left(x_k\right)\right)}\right\}\right\}$$

with

$d\left(c_i,c_j\right)$: the intercluster distance between X_i and X_j

$d\left(X_k\right)$: the intracluster distance of cluster (X_k)

c: the number of cluster of dataset

Large values of D correspond to good clustering solution [3].

- S-Dbw Index

[4]This is a ratio-type index that has a more complex formulation based on the Euclidean norm $\|X\|=\left(X^T X\right)^{1/2}$, the standard deviation of a set of objects $\sigma(X)=1/|X|\sum_{x_i\in X}(x_i-\bar{x})^2$, andthe standard deviation of a partition $stdev(C)=1/K\sqrt{\sum_{c_k\in C}\|\sigma(c_k)\|}$. Its definition is:

$$\text{S-Dbw(C)}=\frac{1}{K\sum\limits_{c_k\in C}\frac{\|\sigma(c_k)\|}{\|\sigma(X)\|}}+\frac{1}{K(K-1)}\sum_{c_k\in C}\sum_{c_l\in C\backslash c_k}\frac{den\left(c_k,c_l\right)}{\max\left\{den\left(c_k\right),den\left(c_l\right)\right\}}$$

where

$$den(c_k)=\sum f\left(x_i,\bar{c}_k\right)$$

$$den\left(c_k,c_l\right)=\sum_{x_i\in c_k\cup c_l} f\left(x_i,\frac{\bar{c}_k+\bar{c}_l}{2}\right)$$

$$f\left(x_i,c_k\right)=\begin{cases}0 & if\ d_e\left(x_i,\bar{c}_k\right)>stdev(C)\\ 1 & otherwise\end{cases}$$

[3] The minimum value of S-Dbw indicates the optimal cluster number.

2.3. Earthquake events in Bengkulu Province

Bengkulu Province is one of the provinces on the island of Sumatra which lies to the west of the Bukit Barisan mountains. The total area of Bengkulu Province reaches approximately 1,991,933 hectares or 19,919.33 square kilometers. Bengkulu Province extends from the border of West Sumatra Province to the border of Lampung Province and its distance is approximately 567 kilometers.

Astronomically, Bengkulu Province lies between 2 ° 16 'to 3 ° 31' S and between 101 ° 01 'to 103 ° 41'E. Meanwhile, in terms of geographical location, Bengkulu Province in the north is bordered by West Sumatera Province, in the south bordering with Indonesia Ocean and Lampung Province, in the west by the Indonesian Ocean and to the east and Jambi Province and South Sumatra Province.

Bengkulu has shaken two powerful tectonic earthquakes in a relatively short period of time, i.e. in 2000 and 2007. On June 4, 2000, Bengkulu was shaken by a tectonic earthquake with a strength of 7.3 Ms. Then a major earthquake occurred in Bengkulu on 12 September 2007 with the strength of 7.9 Ms. The main earthquake following the aftershocks has caused heavy casualties, property, and destruction of public facilities. According to the history of Bengkulu Province has been shaken several times big earthquake such as on November 24, 1833 (VIII-IX MMI), August 18, 1938 (VII MMI),

August 18, 1871 (VI-VII MMI), June 26, 1914 (VII-VIII MMI) , November 24, 1933 (VIII-IX MMI) and 15 December 1979 (VIII MMI) [9].

3. Methods

Data used in this research is secondary data obtained from Badan Meteorologi Klimatologi dan Geofisika (BMKG) Kepahiang Regency. The object of this research is Bengkulu Province which is astronomically located at5°40'-2°0 'S and 100°40'-104°0'E. The data used is the catalog of earthquake occurrences sourced from single-station BMKG Kepahiang, Bengkulu from January 1970 to December 2015.

Firstly, descriptive statistics of earthquake data in Bengkulu province will be presented. The goal is to see the variance of data and other general features. The next step is to apply the internal cluster validation method to the data to obtain the optimum number of clusters. Furthermore, the optimum number of clusters will be used as the number of initial clusters in the K-means cluster analysis. Then, compare the results of K-means cluster analysis to find out the optimum cluster of earthquake area in Bengkulu province.

4. Result and discussions

4.1. Description of earthquake data in Bengkulu Province

Bengkulu Province is one of the provinces located at the meeting of tectonic plates of IndoAustralia and Eurasia which is the main generator of high earthquake activity. Data of Badan Penanggulangan Bencaña Daerah (BPBD) of Bengkulu Province shows that the epicenter of the earthquake that occurred in Bengkulu in 1970 - 2015 vary. The difference of earthquake epicenter and earthquake occurrence is caused by geology and seismicity in Bengkulu Province. Therefore, every seismic event has interconnectedness in space and time.

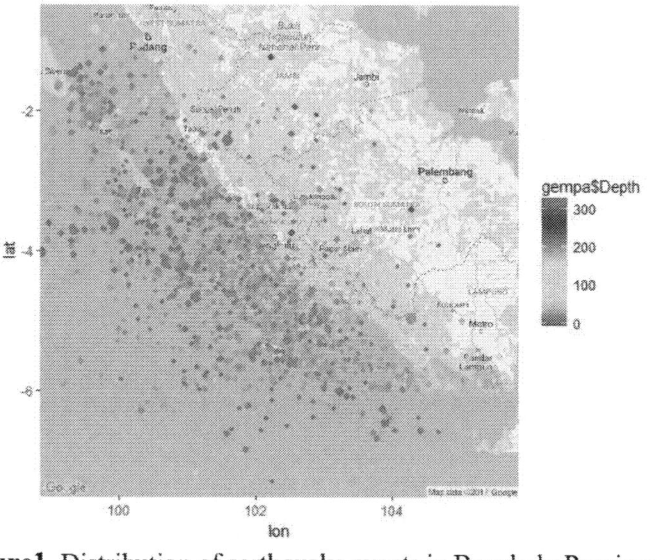

Figure1. Distribution of earthquake events in Bengkulu Province 1970-2015

From 1970 to 2015, the largest magnitude earthquake occurred in 2007, at 7.9 Ms. The position of the epicenter was at -4.67^{0}S and 101.13^{0}E and the depth of 10 KM under the sea. The occurrence of an earthquake with magnitude of 4 Ms is the occurrence of earthquakes with minimum magnitude occurring since 1970 to 2015. Figure 1 shows that the epicenter of the earthquake that occurred in the province of Bengkulu mostly located in the sea area.

4.2. Internal cluster validation

Cluster analysis is a multivariate technique that aims to classify objects based on the similarity characteristics. K-means method is an algorithm for cluster n object based on attribute to k partition, where k <n. The determination of k on this method is based on the will of the researcher, so that the number of clusters is not necessarily the optimal number of clusters.

Internal cluster validation is a technique for determining the optimal number of clusters based on the prior information. The optimal number of clusters is determined by a given index. In this study, the internal cluster validation index used is Calinski-Harabasz (CH) index, Davies-Bouldin (DB) index, Silhouette (S) index, Dunn index (D), and S-Dbw (S-Dbw) index. Each index has certain criteria in determining the optimal number of clusters. The CH, S, and D indices require that the optimal number of clusters be determined by the largest index value. While the DB index and S-Dbw index using the smallest index value to determine the optimal number of clusters. These indices are used in determining the optimal number of clusters of earthquake event data in Bengkulu province from 1970 to 2015. Furthermore, R Program will be used in calculating the value of the internal cluster validation index.

Table 1. Optimum cluster number based on CH, S, DB, S-Dbw, and D

k	CH	S	DB	S-Dbw	D
2	**1706.643**	**0.40665**	0.90756	**1.11447**	0.00806
3	1419.745	0.31115	1.12526	3.17425	0.00881
4	1263.765	0.31846	1.12314	1.90366	0.01242
5	1127.464	0.29621	1.20120	5.08853	0.00819
6	1080.015	0.28789	**0.61909**	3.59231	0.00362
7	1033.560	0.28348	1.12808	5.72832	0.00362
8	978.760	0.27560	1.12096	15.0530	0.00927
9	953.066	0.27487	1.13220	9.63439	0.01167
10	915.266	0.26860	0.98330	13.7527	0.00851
11	890.888	0.26726	1.06585	35.3362	0.01111
12	863.364	0.26456	0.92488	23.6950	0.01203
13	831.544	0.25727	1.07730	12.9756	**0.01508**
14	815.435	0.26920	0.99723	6.67181	0.00860
15	797.447	0.26563	0.97879	24.6156	0.00502

Table 1 gives information that the optimal number of clusters based on the CH, S, and S-Dbw indices are two clusters ($k = 2$) with the index values being 1706.643, 0.40665, and 1.11447, respectively. All three indices require that the largest index value represents the optimal number of clusters. The smallest DB index of 0.61909 refers to the optimal cluster number of six clusters ($k = 6$). And the optimal number of clusters obtained based on index D is 13, because the largest value of index D is 0.01508. Thus, the earthquake events data in Bengkulu province from 1970 to 2015 can be cloned in 2, 6, and 13 clusters ($k = 2, 6, 13$).

4.3. K-means method for earthquake data in the Bengkulu Province with k = 2, 6, 13

K-Means Cluster Analysis is a non-hierarchical cluster analysis method that seeks to partition existing objects into one or more clusters of objects based on their characteristics, so that objects that share the same characteristics are grouped in the same cluster and objects that have the different characteristics grouped into another cluster. The purpose of clustering is to minimize the objective function set in the clustering process, which essentially seeks to minimize variation in one cluster and maximize the variation between clusters [10]. Clustering of earthquake data in Bengkulu Province using K-means method. The attributes used in this clustering are the epicenter (longitude and latitude) and magnitude.

Based on the index CH, S, and S-Dbw, the optimal cluster number is 2 ($k = 2$). The K-means method for $k = 2$ yields:

Table 2. K-means clustering for earthquake data with k = 2

k	Means of Longitude	Means of Latitude	Means of Magnitude	Sizes	SSE
1	102.67	-4.99	4.91	1035	1874.81
2	100.62	-3.10	4.99	725	1532.56

Table 2 shows that cluster 1 is around 102.670BT and -4.990LS. The average magnitude of cluster 1 is 4.91 Ms. Cluster 1 consists of 1035 members and the sum of squares of error is 1874.81. The 725 member of cluster 2 has an average magnitude of 4.99 Ms with its centroid of 100.62^{0}E and -3.10^{0}S. The results of clustering the earthquake data in Bengkulu Province with k = 2 visualized using R program are as follows:

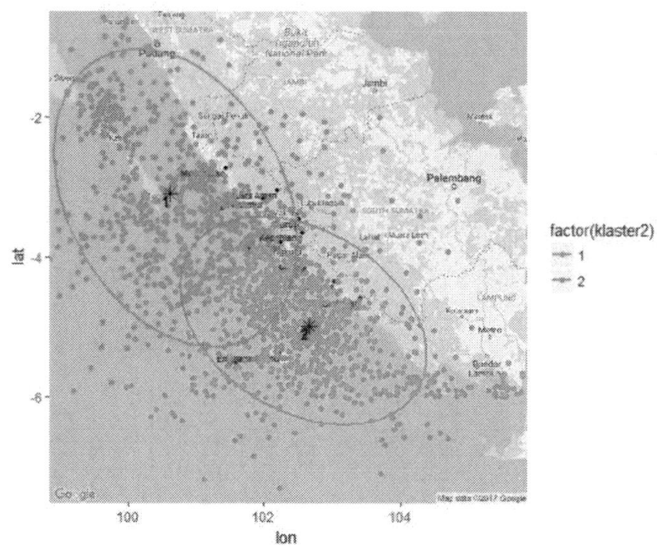

Figure 2. K-means method for earthquake data in Bengkulu Province withk = 2

The optimal cluster number for earthquake data in Bengkulu Province based on DB index is 6 (k = 6). Furthermore, the optimal number of clusters will be applied to the K-means method as the initial number of groups. The clustering results using the R program are described as follows:

Table 3. K-means clustering for earthquake data with k = 6

k	Means of Longitude	Means of Latitude	Means of Magnitude	Sizes	SSE
1	100.58	-3.45	4.98	332	380.77
2	102.76	-5.04	5.01	505	356.35
3	101.98	-3.84	4.76	397	333.49
4	104.49	-5.85	5.04	137	168.14
5	101.29	-5.28	4.64	195	207.46
6	99.85	-1.86	5.30	194	195.51

Table 3 shows that cluster 1 has 332 members, an average magnitude of 4.98 Ms, and is around 100.58^{0}E and -3.45^{0}S. Cluster 2 has 505 members, a magnitude of 5.01 Ms, and is around 102.76^{0}E and -5.04^{0}S. Cluster 3 has 397 members, an average magnitude of 4.76 Ms, and is around 101.98^{0}E and -3.84^{0}S. Cluster 4 has 137 members, an average magnitude of 5.04 Ms, and is around 104.49^{0}E and -5.85^{0}S. Cluster 5 has 195 members, an average magnitude of 4.64 Ms, and is around 101.29^{0}E and -5.28⁰S. Cluster 6 has 194 members, an average magnitude of 5.3 Ms, and is around 99.85⁰E and -1.86⁰S.The result of earthquake data clustering in Bengkulu Province for k = 6 can also be seen in the following figure:

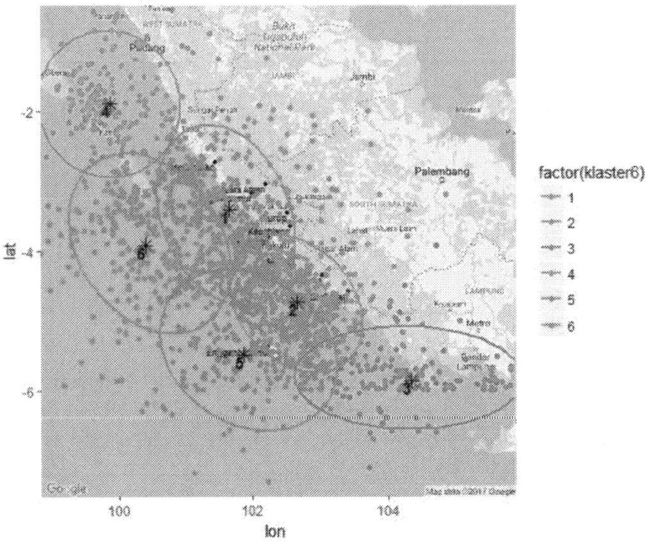

Figure 3. K-means method for earthquake data in Bengkulu Province withk = 6

Index D determines k = 13 as the optimal number of clusters in earthquake data in Bengkulu Province. The thirteen clusters are described by the R program as follows:

Table 4. K-means clustering for earthquake data with k = 13

k	Means of Longitude	Means of Latitude	Means of Magnitude	Size	SSE
1	100.53	-4.64	4.54	87	48.52
2	101.06	-3.71	5.81	109	84.60
3	99.70	-1.78	5.38	155	109.65
4	99.82	-3.49	4.95	106	77.51
5	104.62	-5.86	5.13	117	110.38
6	102.07	-4.69	5.38	198	89.80
7	102.95	-4.89	5.24	221	78.58
8	101.56	-5.96	4.62	84	86.46
9	102.75	-3.18	4.84	64	60.65
10	102.94	-5.46	4.51	165	68.01
11	101.02	-2.53	4.73	121	74.24
12	102.12	-4.39	4.36	180	55.37
13	101.40	-3.62	4.49	153	57.31

Table 4 shows that cluster 7 has the most members, i.e 221 members with an average magnitude of 5.24 Ms, and centered at 102.95⁰E and -4.89⁰S. While the number of cluster members in cluster 9 is 64 members. Cluster 9 is a cluster whose members are located around the centroid of 102.75⁰E and -3.18⁰S, and the average magnitude of this cluster is 4.84 Ms. The clustering results can also be seen in the following figure:

IOP Conf. Series: Materials Science and Engineering **335** (2018) 012048 doi:10.1088/1757-899X/335/1/012048

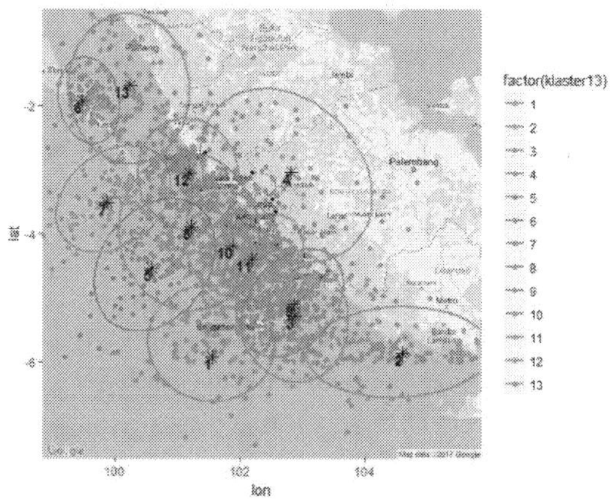

Figure 4. K-means method for earthquake data in Bengkulu Province with k = 13

Based on the three clustering results, it was found that the number of clusters k = 2 resulted in considerable variation. It can be seen from the SSE value of each cluster 1 and cluster 2 which is quite big. The number of clusters k = 13 has a fairly low SSE value for each cluster. However, when viewed in Figure 4, the clusters formed tend to overlap. In other words, there are members of a cluster who are members of other clusters. The number of clusters k = 6 is the ideal number of clusters for earthquake data in Bengkulu Province from 1970 to 2015. This is shown in Figure 3, where the clusters formed have members of each cluster that are not crossed. Although the SSE value generated by 6 clusters is greater than the value of SSE generated by 13 clusters.

5. Conclusion

The conclusions obtained in this study are as follows: The Calinski-Harabasz index, the Silhouette index, and the S-Dbw index determine the optimal number of clusters for earthquake data in Bengkulu Province 1970 - 2015 for 2 clusters. Based on the Davies-Bouldin index, the optimal number of clusters for earthquake data in Bengkulu Province in 1970 - 2015 was 6 clusters. Based on the Dunn index, the optimal number of clusters for earthquake data in the Bengkulu Province from 1970 to 2015 were 13 clusters. The optimal number of optimal clusters for earthquake data in Bengkulu Province from 1970 to 2015 is 6 clusters. If the number of clusters is 2, then there will be considerable variation. Whereas if the number of clusters is 13, then members of a cluster will become members of other clusters.

References

[1] MattjikAA and IM Sumertajaya 2011 IPBPress Bogor
[2] Liu Y, Li Z, Xiong H, Gao X, Wu J 2010*IEEE International Conference on Data Mining* DOI 10.1109/ICDM.2010.35
[3] Rendon E, Abundez I, Arizmendi A, Quiroz EM 2011 *International Journal of Computers and Communications***5**27
[4] Arbelaitz O, Ibai G, Javier M, Jesus MP, Inigo P 2013*Pattern Recognition* **46**243
[5] Novianti P, Ulfasari R, and Dyah S 2016*Laporan Hasil Penelitian Hibah Penelitian Pembinaan Universitas Bengkulu*Bengkulu
[6] WeatherillG and Burton PW 2009*Geophysical Journal International***176**565
[7] EverittB and HothornT 2011*An Introduction to Applied Multivariate Analysis with R*Springer Science + Business MediaUSA
[8] JohnsonR A and Wichern D W 2002 *Applied Multivariate Statistical Analysis*Fifthedition Pearson Education Inc USA
[9] BMG 2007 *Database Gempa BMG Kepahiang* BMG Kepahiang Bengkulu
[10] Ediyanto, Muhlasah NM, Neva S 2013 *Bimaster***2** 133

IOP Conf. Series: Materials Science and Engineering **335** (2018) 012049 doi:10.1088/1757-899X/335/1/012049

A Perishable Inventory Model with Return

S W Setiawan, D Lesmono*, T Limansyah

Department of Mathematics, Universitas Katolik Parahyangan, Jalan Ciumbuleuit 94 Bandung 40141

*jdharma@unpar.ac.id

Abstract. In this paper, we develop a mathematical model for a perishable inventory with return by assuming deterministic demand and inventory dependent demand. By inventory dependent demand, it means that demand at certain time depends on the available inventory at that time with certain rate. In dealing with perishable items, we should consider deteriorating rate factor that corresponds to the decreasing quality of goods. There are also costs involved in this model such as purchasing, ordering, holding, shortage (backordering) and returning costs. These costs compose the total costs in the model that we want to minimize. In the model we seek for the optimal return time and order quantity. We assume that after some period of time, called return time, perishable items can be returned to the supplier at some returning costs. The supplier will then replace them in the next delivery. Some numerical experiments are given to illustrate our model and sensitivity analysis is performed as well. We found that as the deteriorating rate increases, returning time becomes shorter, the optimal order quantity and total cost increases. When considering the inventory-dependent demand factor, we found that as this factor increases, assuming a certain deteriorating rate, returning time becomes shorter, optimal order quantity becomes larger and the total cost increases.

1. Introduction

In managing goods, retailer should consider several costs such as purchasing, ordering, holding and possibly shortage costs. For some special goods, especially when dealing with perishable goods such as vegetables, milk, fruit or drugs, deteriorating rate or the product shelf life should be taken into account as well by considering return costs where goods can be returned to supplier at some costs. Considering shortage cost, there are two conditions faced by the retailer when goods are not available when demand comes. Customers can wait until goods are available in the next period (backorder) or they can try to find another retailer to fulfil their demand (lost sales).

In this paper, we develop a mathematical model for perishable goods with return cost. We assume that demand is deterministic and the amount of inventory depends on the demand (inventory-dependent demand). Customers tend to buy more products when there are more goods displayed in the shelf, therefore inventory-dependent demand is realistic to be included in the model. The developed model is based on the model developed by Nafisah, et.al (2016) andNagareandDutta (2012). Numerical examples are provided to illustrate our model and some sensitivity analysis are also performed by studying the effect of changes in the deterioration rate and inventory-dependent demand factor on the optimal return time, order quantity and total cost.

2. Literature Review

Perishable inventory models have become interesting topic of research for many years. Many papers deal with perishable inventory models with many variants that related to problems faced by industries. This paper contributes in the developing mathematical model for perishable inventory with deterministic demand by considering return cost, where all shortages are handled with backorder. In the area of perishable inventory models, a model for perishable goods where the purchasing cost is modelled as a function of the credit policy of the retailer has been developed by Dye (2012). Kumar et.al developed a model by considering demand as quadratic function and uncertain holding cost, while a model with Weibull distribution for the deteriorating rate and holding cost as a linear function of time has been developed by Mishra (2013). Nagare and Dutta (2012) have developed a continuous review model by considering deteriorating rate and inventory-dependent demand. Nafisah et.al (2016) proposed a multi-item model for goods that close to their expiration date by applying discount and return. They assumed that demand rate can be divided in several intervals while their values are constant, but different in each interval. Rikardo et al (2017) proposed a continuous review model for perishable goods with all-unit discount. In this paper we consider the demand function used by Nagare and Dutta (2012) from the beginning period until the return time and then constant after the return time.

The organization of this paper is as follows. In Chapter 3, we introduce and formulate our model and give the condition for the optimal return time and order quantity. Numerical examples and sensitivity analysis are provided in Chapter 4 while conclusions and further research are relegated in the last chapter.

3. The Model

3.1 Notations

The notations we use in this paper are as follows:

TC : totalinventory cost $I(t)$: inventory level at time t.
PC : purchasing cost m : number of inventory during interval$[0, T_r]$
OC : ordering cost S : number of shortageduring interval $[T_r, T]$
HC : holding cost P : purchasing cost per unit
SC : shortage cost θ : deteriorating rate
T : time between replenishment (weeks) h : holding cost per unit per year
$D(t) = \alpha + \beta I(t)$: demandat time t A : ordering cost per order
β : inventory-dependent demand factorA_r : return cost per return
Q : order quantity R : return cost per unit
T_r : return time (weeks) π: backorder cost per unit

3.2. Model Development

In developing our model, basic assumption we use is inventory-dependent demand as given by the following equation.

$$D(t) = \begin{cases} \alpha + \beta I(t) & , 0 \le t \le T_r \\ B_1 & , T_r \le t \le T \end{cases}$$

The demand representation is given by Figure 1. When$t = 0$, the replenishment of amount$Q$arrives so the inventory reaches its maximum level. As time goes by, inventory is depleted because of the demand as depicted in Figure 1.At $t = T_r$, all unsold goods are returned to the supplier. When return takes place, they are handled by *full backorder*because all returned goods will be available in the next replenishment. Demand and deteriorating rate in the interval $[0, T]$ will decrease the inventory level according to the following first order differential equation:

$$\frac{dI(t)}{t} = -\theta I(t) - D(t) \ ,0 \leq t \leq T$$

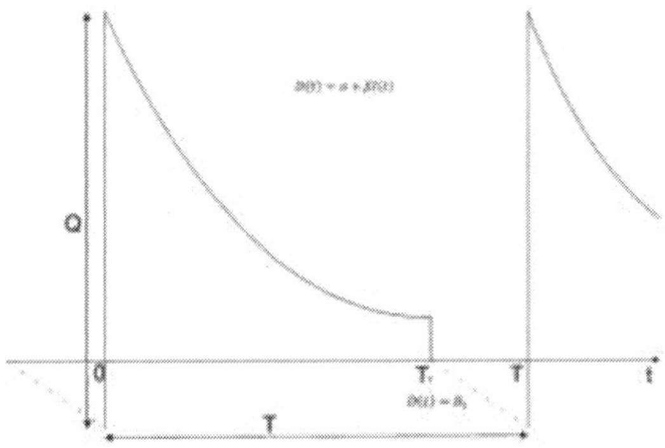

Figure1. Demand representation for one cycle T

Using the inventory-dependent demand, $D(t) = \alpha + \beta I(t)$, we have

$$\frac{dI(t)}{dt} = -\theta I(t) - (\alpha + \beta I(t)) \quad ,0 \leq t \leq T_r$$

$$\frac{dI(t)}{dt} = -\theta I(t) - B_1 \qquad ,T_r \leq t \leq T$$

Using the boundary condition $I(T_r) = 0$, we have

$$I(t) = B_1(T_r - t)$$

In the interval $0 \leq t \leq T_r$we have

$$\frac{dI(t)}{dt} = -\alpha - (\theta + \beta)I(t)$$

Let $\lambda = \theta + \beta$, and using the boundary condition $I(T_r) = 0$,we finally have

$$I(t) = \frac{\alpha}{\lambda}\left(e^{\lambda(T_r - t)} - 1\right)$$

Then

$$I(t) = \begin{cases} \frac{\alpha}{\lambda}\left(e^{\lambda(T_r - t)} - 1\right) & ,0 \leq t \leq T_r \\ B_1(T_r - t) & ,T_r \leq t \leq T \end{cases}$$

When$t = 0, I(t) = Q - B_1(T_r - t)$then we have

$$Q = \frac{\alpha}{\lambda}\left(e^{\lambda(T_r - t)} - 1\right) + B_1(T_r - t)$$

The number of inventory during $[0, T]$ is given by

$$m = \int_0^{T_r} I(t)dt = \int_0^{T_r} \frac{\alpha}{\lambda}\left(e^{\lambda(T_r - t)} - 1\right)dt = \frac{\alpha}{\lambda^2}\left(e^{\lambda T_r} - 1\right) - \frac{\alpha}{l}T_r$$

The number of shortages during $[T_r, T]$ is

$$S = -I(t)(t - T_r) = B_1(T_r - t)^2$$

3.3. Model Formulation

The objective of the developed model is to find the optimal order quantity that minimizes the total annual cost. The total cost (TC) consists of the purchasing cost (PC), the ordering cost (OC), the annual holding cost (HC), the backorder/shortage cost (SC) and the return cost (RC). All those costs are given as follows.

$$PC = P.D;\ OC = \frac{A}{T};\ HC = \frac{hm}{T2} = \frac{h}{2T}\left[\frac{\alpha}{\lambda^2}\left(e^{\lambda T_r} - 1\right) - \frac{\alpha}{l}T_r\right]$$

$$SC = \frac{\pi S}{T2} = \frac{\pi}{2T}[B_1(T_r - t)^2];\ RC = \frac{1}{T}\left[A_r + \frac{RS}{2}\right] = \frac{1}{T}\left[A_r + \frac{RB_1}{2}(T_r - t)^2\right]$$

$$TC = P.D + \frac{A}{T} + \frac{h}{2T}\left[\frac{\alpha}{\lambda^2}\left(e^{\lambda T_r} - 1\right) - \frac{\alpha}{l}T_r\right] + \frac{\pi}{2T}[B_1(T_r - t)^2] + \frac{\pi}{2T}[B_1(T_r - t)^2] + \frac{1}{T}\left[A_r + \frac{RB_1}{2}(T_r - t)^2\right]$$

Taking the first derivative of TC with respect to $T_r$and equates to zero we have

$$\frac{dTC}{dT_r} = \frac{h\alpha}{\lambda}\left(e^{\lambda T_r} - 1\right) + 2\pi B_1(T_r - t) + 2RB_1(T_r - t) = 0$$

The second derivative of TC with respect to T_r is

$$\frac{d^2TC}{dT_r^2} = h\alpha e^{\lambda T_r} + 2\pi B_1 + 2RB_1$$

Since the second derivative is always positive, then the value of T_r that satisfies $\frac{dTC}{dT_r} = 0$, will yield the optimal Qthat minimizesTC

4. Numerical Experiments and Sensitivity Analysis

In this chapter, we give numerical experiments according to the following input parameters (T = 24 weeks):
$\alpha = 1000$ unit, $\beta = 0.1$, $\theta = 0.1$, $A = \$10$, $h = \$1$, $\pi = \$1.2$, $P = \$14.5, A_r = \$10, R = \$1, B = 250$ unit.

Using Maple we have the optimal return time (weeks) and order quantity (units) that minimizes the annual total cost as follows:

$$T_r = 7.59; Q = 19{,}692; TC = \$4{,}903.80$$

Results on the sensitivity analysis are given in Table 1. In Table 1, effectsofβ and θ on T_r, Q and TC are analyzed. Smaller value of θ represent lower deteriorating rate, meaning that it takes longer for the products to deteriorate. Smaller value of β indicates that the effect of inventory on demand is smaller or the inventory-dependent demand factor is smaller. If we look at Table 1 for certain value ofβ, we can see that as θ gets larger, the optimal return time gets shorter but the order quantity becomes larger and the total cost becomes larger. This is due to the fact that goods are deteriorating faster, so the return time becomes shorter and the order quantity becomes larger (to fulfill demand), so the total cost gets larger. For certain value of θ as β gets larger, the same situation occurs, due to the increase of demand. Whenβ and θ get larger, the optimal return time becomes shorter, the optimal order quantity becomes larger and eventually the total cost becomes larger.

Table 1. Effect ofβ and θ on T_r, Q and TC

	$\beta\backslash\theta$	0.1	0.2	0.3	0.4	0.5
Tr		7.59	6.39	5.53	4.89	4.39
Q	0.1	19,692	20,934	21,698	22,246	22,668
TC		4,903.80	5,422.29	5,768.55	6,019.14	6,211.29
Tr		6.39	5.53	4.89	4.39	4.00
Q	0.2	20,902	21,694	22,246	22,668	23,004
TC		4,941.41	5,406.28	5,729.16	5,969.63	6,157.49
Tr	0.3	5.53	4.89	4.39	4.00	3.67

Q		21,690	22,245	22,668	23,004	23,278
TC		5,044.00	5,439.18	5,727.96	5,950.35	6,128.08
Tr		4.89	4.39	4.00	3.67	3.40
Q	0.4	22,245	22,668	23,004	23,278	23,508
TC		5,149.20	5,486.30	5,743.21	5,946.83	6,112.97
Tr		4.39	4.00	3.68	3.40	3.17
Q	0.5	22,667	23,004	23,278	23,508	23,702
TC		5,244.63	5,536.06	5,765.58	5,951.86	6,106.59

5. Conclusions and Further Research

We have developed a perishable inventory model for inventory-dependent demand with return. All shortages are handled using backorder. Analysis of sensitivity of the model with respect to changes on the deterioration rate and inventory-dependent demand factor is given. We found that as the deterioration rate and inventory-dependent demand factor are both increasing, the optimal return time is shortened; the optimal order size and total cost are both increasing. In developing this model, we assume that demand is deterministic which is a bit unrealistic in most cases. Considering a probabilistic demand model is an avenue for further research. Also, incorporating discount factor from the supplier whether it is all-unit discount or incremental discount is a challenging development to pursue,

Acknowledgments

The authors would like to thank The Institute of Research and Community Service Universitas Katolik Parahyangan for funding this research.

References

[1]. Dye C Y, 2012 A finite horizon deteriorating inventory model with two-phase pricing and time-varying demand and cost under trade credit financing using particle swarm optimization, *Swarm EvolComput***5**37-53.

[2]. Kumar S, Kumar P and Saini M 2012 An order level inventory model for deteriorating items with quadratic demand rate and variable holding cost,*Int. J.Sci. Res. Eng. Technol.***1**212-24

[3]. Mishra V K, Singh LS and Kumar, R 2013 An inventory model for deteriorating items with time-dependent demand and time-varying holding cost under partial back logging *J IndEngInt***9** 4.

[4]. Nafisah L, Sally Wand Puryani 2016 Model persediaan pada produk yang mendekati masa kadaluwarsa: mempertimbangkan diskon penjualandanretur *JurnalTeknikIndustri* **18** 63-72.

[5]. Nagare M and Dutta P 2012 Continuous review model for perishable products with inventory dependent demand. *Proc. of the International Multi Conference of Engineers and Computer Scientists March 24-16 2012*Hong Kong1513-17.

[6]. Rikardo C, Lesmono D and Limansyah T 2017Pengembanganmodel continuous review dengan all-unit discount danfaktorkadaluwarsa*JurnalTeknikIndustri***19**29-38.

IOP Conf. Series: Materials Science and Engineering **335** (2018) 012050 doi:10.1088/1757-899X/335/1/012050

Mathematical Model of Cytomegalovirus (CMV) Disease

R Sriningsih*, M Subhan, M L Nasution

Mathematics Department, Faculty of Mathematics and Natural Science, UniversitasNegeri Padang, Jl. Prof. Hamka, Padang, Sumatera Barat, Indonesia 25131

*srirysriningsih@yahoo.com

Abstract. The article formed the mathematical model of cytomegalovirus (CMV) disease. Cytomegalovirus (CMV) is a type of herpes virus. This virus is actually not dangerous, but if the body's immune weakens the virus can cause serious problems for health and even can cause death. This virus is also susceptible to infect pregnant women. In addition, the baby may also be infected through the placenta. If this is experienced early in pregnancy, it will increase the risk of miscarriage. If the baby is born, it can cause disability in the baby. The model is formed by determining its variables and parameters based on assumptions. The goal is to analyze the dynamics of cytomegalovirus (CMV) disease spread.

1. Introduction

Cytomegalovirus (CMV), a common virus, is one of the herpes viruses. The virus can infect people of all ages. Most people infected with CMV show no signs or symptoms. They are exposed to CMV at birth or as children. That's because a healthy person's immune system usually keeps the virus from causing illness. The virus stays in a person's body for life and can reactive. A person can also be re-infected with a different strain of the virus.

CMV is spread by direct contact of body fluids, such as saliva, blood, urine, semen, vaginal fluids and breast milk from an infected person. CMV can also be transmitted through organs transplants, stem cell transplants, blood transfusion, breastfeeding, kissing, sexual contact, infection drug use (sharing needles). Not only that, CMV transmission can also occur during pregnancy through the placenta from the mother's blood or from vaginal secretions at delivery to child (congenital CMV).

In some cases, infection in healthy people can cause mild illness that may include fever, sore throat, fatigue and swollen glands. Occasionally CMV can cause liver problem. However, CMV infection can cause serious health problems for people with weakened immune systems, as well as babies infected with the virus before they are born (congenital CMV). In people with weakened immune systems, CMV infection can attack different organs of the body and may cause blurred vision and blindness (CMV retinitis), lung infection (pneumonia), painful swallowing (esophagitis), diarrhea (colitis), inflammation of the liver (hepatitis), or inflammation of the brain (encephalitis), which may cause behavioral changes, seizures or coma. Babies born with CMV can loss hearing and vision, mental disability, seizures and growth problems.

Based on the problems then formed a mathematical model of the spread of CMV disease. The purpose is to look at the dynamics of disease spread and what to do in preventing or reducing the spread of the disease in the future.

2. Method and Design

The research is a basic (theoretical) research. The method used is the analysis of theories relevant to the issues discussed. The steps taken are to study the phenomenon of the problem, collect and link the theories related to the problem, determine the appropriate assumptions with the problems, define the variables and parameters used, to form and analyze the model, and interpret the results of model analysis obtained to answer the problem.

3. Findings and Discussion

3.1 Formation of Mathematical Model

The population consists of 3 groups. They are susceptible group (S), exposed group (E), and infected group (I). Assumption of the model: disease has latent period, every individual born of exposed and infected group is assumed to have CMV and move to exposed group, individual born in susceptible move to susceptible group, CMV transmission occur horizontally (direct contact between susceptible group and infected group (primary CMV)) and vertically (from mother to child during pregnancy (congenital CMV)), individual in exposed group moves to infected group due to the weakened immune system or reactivation of latent CMV infection, constant population and individual recovery moves to exposed group (doesn't heal permanently). Parameter of the model: transmission rate (a), birth rate (b), death rate (d), recovery rate (f), and individual transfer rate from exposed group to infected group (c).

The general overview of model showed in flow diagram.

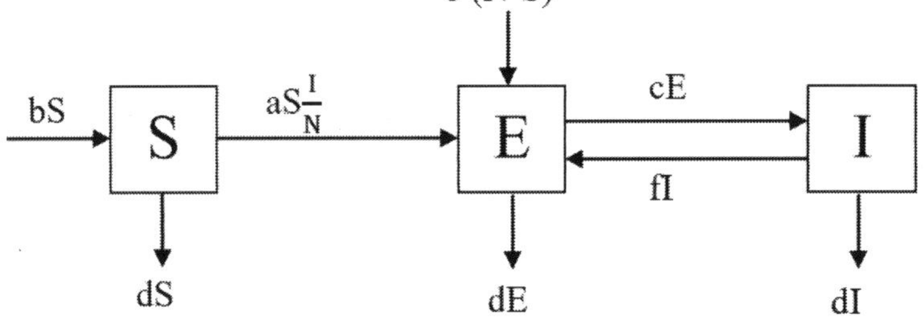

Figure 1. Flow diagram of model

Here, the mathematical model formula. That is system (1)

$$\frac{dS}{dt} = (b-d)S - aS\frac{I}{N} \quad (1)$$

$$\frac{dE}{dt} = aS\frac{I}{N} + b(N-S) + fI - (c+d)E$$

$$\frac{dI}{dt} = cE - (f+d)I$$

Where $S + E + I = N$

To simplify the system (1), let $s = \frac{S}{N}$, $e = \frac{E}{N}$, and $i = \frac{I}{N}$ and obtained system (2)

$$\frac{ds}{dt} = (b-d)s - asi \quad (2)$$

$$\frac{de}{dt} = asi + b(1-s) + fi - (c+d)e$$

$$\frac{di}{dt} = ce - (f + d)i$$

Where $s + e + i = 1$

3.2 Equilibrium Point of Model

To find the solution of system (2), we can change system (2) become to be system (3)

$$\frac{ds}{dt} = (b - d)s - asi$$

$$\frac{di}{dt} = c\big(1 - (s + i)\big) - (f + d)i \tag{3}$$

According Perko [1996], point $\hat{x} \in \mathbb{R}$ is called equilibrium point of system (3) if $f(\hat{x}) = 0$. Furthermore, we get

$$(b - d)s - asi = 0 \tag{4}$$
$$c\big(1 - (s + i)\big) - (f + d)i = 0 \tag{5}$$

Based on equations (4) and (5), we get 2 kinds equilibrium point: free and endemic equilibrium points. Free equilibrium point is $E_0 = (\hat{s}, \hat{\imath}) = (1,0)$ and endemic equilibrium point is $E_1 = (\hat{s}, \hat{\imath}) = \left(1 - \left(\frac{c+f+d}{c}\right)\left(\frac{b-d}{a}\right), \frac{b-d}{a}\right)$

3.3 Stability Equilibrium Point of Model

The stability of equilibrium point is investigated to study system dynamics. To facilitate an analytical calculation, then System (3) is linearized using the Jacobian matrix and obtained:

$$J(\hat{s}, \hat{\imath}) = \begin{bmatrix} b - d - ai & -as \\ -c & -(f + c + d) \end{bmatrix}$$

The stability of equilibrium point of the model is as follows:

3.3.1. Stability of free equilibrium point $\mathbf{E_0} = (\hat{\mathbf{s}}, \hat{\boldsymbol{\imath}}) = (\mathbf{1}, \mathbf{0})$. The stability of this point is seen from the Eigen value of Jacobian matrix around the $\mathbf{E_0}$ point, as follows:

$$\big|\lambda I - J\big(f(E_0)\big)\big| = 0$$

$$\begin{vmatrix} \lambda - (b - d) & a \\ c & \lambda + (f + c + d) \end{vmatrix} = 0$$

$$\lambda^2 + a_1\lambda + a_2 = 0$$

where, $a_1 = (f + c + d) - (b - d)$ and $a_2 = -\big(ac + (b - d)(f + c + d)\big)$

Based on the Routh-Hurwitz criteria, $\mathbf{E_0}$ point is stable when $a_1 > 0$ and $a_2 > 0$. This happens if and only if $R_0 < 1$ and $R_0 < -\frac{ac}{(f+c+d)^2}$

3.3.2. Stability of endemic equilibrium point $\boldsymbol{E_1} = (\hat{\mathbf{s}}, \hat{\boldsymbol{\imath}}) = \left(\mathbf{1} - \left(\frac{c+f+d}{c}\right)\left(\frac{b-d}{a}\right), \frac{b-d}{a}\right)$. The stability of this point is seen from the Eigen value of Jacobian matrix around the $\boldsymbol{E_1}$ point, as follows:

$$\big|\lambda I - J\big(f(E_1)\big)\big| = 0$$

$$\begin{vmatrix} \lambda & a - \dfrac{(f+c+d)(b-d)}{c} \\ c & \lambda + (f+c+d) \end{vmatrix} = 0$$

$$\begin{vmatrix} \lambda & a - \dfrac{(f+c+d)^2}{c}\left(\dfrac{b-d}{f+c+d}\right) \\ c & \lambda + (f+c+d) \end{vmatrix} = 0$$

$$\begin{vmatrix} \lambda & a - \dfrac{(f+c+d)^2 R_0}{c} \\ c & \lambda + (f+c+d) \end{vmatrix} = 0$$

$$\lambda^2 + (f+c+d)\lambda + ((f+c+d)^2 R_0 - ac) = 0$$

Based on the Routh-Hurwitz criteria, $\mathbf{E_1}$ point is stable when $(f+c+d) > 0$ and $(f+c+d)^2 R_0 - ac$. This happens if and only if $R_0 > \frac{ac}{(f+c+d)^2}$

4. Conclusion

The research found 2 kinds equilibrium points of model. It is free equilibrium point ($E_0 = (1,0)$ and endemic equilibrium point ($\boldsymbol{E_1} = \left(\mathbf{1} - \left(\frac{c+f+d}{c}\right)\left(\frac{b-d}{a}\right), \frac{b-d}{a}\right)$). E_0 point stable if eligible $R_0 < 1$ and $R_0 < -\frac{ac}{(f+c+d)^2}$. Furthermore, E_1 point stable if $R_0 > \frac{ac}{(f+c+d)^2}$, where $R_0 = \frac{b-d}{f+c+d}$

Acknowledgments

The research that we discuss in this article is a part of a research entitle *The mathematical model of TORCH disease spread* which funded by Universitas Negeri Padang, Indonesia.

References

[1] A. Lahrouz, L. Omari, D. Kionach. 2011. *Global Analysis of Deterministic and Stochastic Nonlinear SIRS Epidemic Model. J.Math.Biol 16* (1): 59-76.
[2] Brauer, Fred. Dkk. 2008. *Mathematical Epidemiology*. Math Biosci, Springer.
[3] Edelstein, Leah. Keshet. 1988. *Mathematical Model in Biology*. SIAM.
[4] Perko, Lawrence. 1996. *Differential Equation and Dynamical System*. Springer
[5] V. Capasso, G. Serio. 1978. *A Generalization of the Kermack-Mckendrick Deterministic Epidemic Model*. Math Biosci. 42: 43-61
[6] Verhulst, Ferdinant. 1990. *Nonlinear Differential Equation and Dynamical System. Heldelberg*: Springer-Verlag.
[7] Golalipur, B. Khodabakhshi and E. Ghaemi. 2009. *Possible Role of TORCH Gents Congenital Malformations in Gorgan*, Northern Islamic Republic of Iran La Revue de Sante de la Mediterance, Vol 15, No. 2.
[8] Jusuf S Effendi. 2014. *Makalah OBGYN Terkini TORCH*. Program Studi Magister KebidananFakultasKedokteran UNDIP
[9] MS Sadik, Fatima, K Jamil, C Patil. 2012. *Study of TORCH profile in Patients with Bad Obstetic History*, Research Article Biology and Medicine, Vol. 2, Issue 2, page 95-101, ISSN: 09748369
[10] Septra Julitar Syamsul Huda SKM. 2014. *Infeksi TORCH (Toxoplasma, Other Disease, Rubella, Cytomegalovirus, Herpes Simplex Virus)*. My Life

IOP Conf. Series: Materials Science and Engineering **335** (2018) 012051 doi:10.1088/1757-899X/335/1/012051

A Mathematical Model of Economic Population Dynamics in a Country That Has Optimal Zakat Management

M Subhan*

Mathematics Department, Universitas Negeri Padang, Indonesia

*13subhan@fmipa.unp.ac.id

Abstract. Zakat is the main tools against two issues in Islamic economy: economic justice and helping the poor. However, no government of Islamic countries can solve the economic disparity today. A mathematical model could give some understanding about this phenomenon. The goal of this research is to obtain a mathematical model that can describe the dynamic of economic group population. The research is theoretical based on relevance references. From the analytical and numerical simulation, we conclude that well-manage zakat and full comitment of the wealthy can achieve wealth equilibrium that represents minimum poverty.

1. Introduction

Zakat, the third Islamic pillar, may be the first pillar inIslamic economy system. Zakat is the main tools for two main sensitive issues in Islam: economic justice and providing aid for the poor. Zakat is also a base for wealth distribution and has the power to reduce poverty and other social problems [1]. The obligation for zakat is written in holy Quran more than 30 times, where 28 of them is related to obligation of pray. Some hadiths also mention the obligation. However, most of Muslim majority countries are among the poor countries in the world and for other nations the gap between the wealthy and the poor is still very deep [2].

The government's role in collecting and redistributing zakat is very important. Prophet Muhammad as Islam leader is given order to collect zakat, while the *amil* manage the distribution [3]. It indicates that zakat is not managed by individual, but by a government institution since era of Osmania caliphates. Today, we can see zakat management by government in Saudi Arabia, Malaysia, Libya, Pakistan and Sudan. In other countries like Kuwait, Jordan, Iraq, Oman, Qatar, Bahrain, Bangladesh, and Indonesia, zakat is managed by some non-governmental institution and mostly by individuals.

The potential of zakat, by some researchers, may achieve 3-3.6% of Gross Domestic Product (GDP) every year in countries that have Muslim majority. Indonesia that has largest muslim population in the world have potential of xakat up to 2% of its GDP [4]. This number is large enough to overcome issues that we mention before. If zakat is distributed among various categories of recipients, the cumulative result of this annual redistribution of wealth will substantially reduce unemployment, expand investment base, eliminate poverty and extreme disparity of wealth between rich and poor in a few years [5]. But, the goals only can be achieved if the system of collection and distribution of zakat is credible, cost-effective, loophole-free and supported by strong accountability mechanisms to assure its integrity. Today, we observe that the management of zakat is not fully integrated and also the awareness of the zakat-obligatory is low in the Islam majority countries. That makes poverty and rich-poor gap still major problems in such countries.

Increasing the role of government in collecting and distributing zakat will be effective and efficient if the problem is comprehensively understood. A mathematical model can provide some better view of the problem. In this research, we build a deterministic model about the dynamic of economic population. The main concern is population of zakat-obligatory (wealthy) and zakat-receiver (poor).

2. Method

This research is theoretical research. Using relevance theories from corresponded references, we build a concept that mathematically valid in form of a mathematical model. We began the research by studying population dynamic, economic aspect and zakat. In next step, we build a deterministic mathematical model, then analyse the model, and interpret ting the result. We also include a numerical simulation of the model.

3. Results

3.1. The initial model

The population will divide into four compartments, which is:

X:the number of individuals that have zakat obligation and pay it.

Y:the number of individuals that have zakat obligation and neglect it.

Z:the number of individuals that receive zakat

W: the number of individuals that neither zakat-obligatory nor zakat-receiver; the 'neutral' population.

Total population is N where $N = X + Y + Z + W$

Some assumptions that are used in this model are:

a. The population is in the region that has integrated zakat management (government institution) with orientation to weak economy empowerment.
b. Every wealthy individual cannot be poor immediately, but may become 'neutral population.
c. The change from individuals obeying to neglecting zakat happen when they interacted one to another, and also the contrary.
d. Zakat-receiver will have possibility of positive economic increasing. Even though some of the portion of the zakat goes to consumption, some portion are used in productive investment [6].
e. The change from zakat-obligatory to 'neutral or 'neutral' to zakat-receiver are caused by economic recession. On the contrary, from zakat-receiver to 'neutral or 'neutral' to zakat-obligatory is affected by economic growth.
f. Total population is constant because birth rate equals to death rate and effect of migration is ignored.
g. The rate and probability of change from zakat-obligatory that paying into neglecting zakat is constant, and also the contrary.

Based on the assumptions, we have some parameters in this model:

Parameter:

a. Rate of change from zakat-obligatory who neglect zakat to pay zakat affected by interaction among them is β. The value is positive if the zakat payers have better influence, and negative if the contrary.
b. Proportion of zakat payer to non-payer for new wealth is ρ.
c. Rate of change from zakat-obligatory who neglect zakat to 'neutral' compartment affected by penalty from the government when they caught breaking the rule is δ.
d. Rate of change from zakat-receiver to 'neutral' compartment affected by optimal use of zakat is α.
e. Birth rate and also death rate is represented by μ.

IOP Conf. Series: Materials Science and Engineering **335** (2018) 012051 doi:10.1088/1757-899X/335/1/012051

f. Rate of change of each population affected by economic growth,equal for all populations, is ε. Some research shows that growth can reduce poverty [7,8], but other shows that growth also increasing poverty [9].
g. Rate of business failures is γ.

From the assumptions, we build the first initial model as follows:

$$\frac{dX}{dt} = \varepsilon\rho W + \beta\left(\frac{X}{N}\right)Y - (\gamma + \mu)X$$
$$\frac{dY}{dt} = \varepsilon(1-\rho)W - \beta\left(\frac{X}{N}\right)Y - (\delta + \gamma + \mu)Y$$
$$\frac{dZ}{dt} = \gamma W - (\varepsilon + \alpha + \mu)Z$$
$$\frac{dW}{dt} = \mu N + \gamma(X + Y - W) + \delta Y + (\varepsilon + \alpha)Z - (\varepsilon + \mu)W \quad (1)$$

Since the population is constant, we can simplify the model into just three equations. Using these new scaling variables

$$P = \frac{X}{N}, U = \frac{Y}{N}, R = \frac{Z}{N}, S = \frac{W}{N}$$

we have$S = 1 - P - U - R$. Our concern is three varables P (proportion of zakat-payer population), U (proportion of not paying zakat population), and R (proportion of zakat-receiver). So, the new model is:

$$\frac{dP}{dt} = \varepsilon\rho(1 - P - U - R) + \beta PU - (\gamma + \mu)P$$
$$\frac{dU}{dt} = \varepsilon(1-\rho)(1 - P - U - R) - \beta PU - (\delta + \gamma + \mu)U$$
$$\frac{dR}{dt} = \gamma(1 - P - U - R) - (\varepsilon + \alpha + \mu)R$$

In this model, all variables are feasible on$[0,1]$. We validate the model on boundaries, and we have all $\frac{dP}{dt}, \frac{dU}{dt}, \frac{dR}{dt}$ equals zero when the value of the variable is zero, and all $\frac{dP}{dt}, \frac{dU}{dt}, \frac{dR}{dt}$ less than zero when the value of the variable is one. We can conclude that the model is well-defined and we can use the model.

3.2. Analyse of the model in case of full obedience

Since zakat has been mandatory, in sharia law countries, the population of non-paying zakat is almost zero. We have $U = 0$, $\rho = 1$ and $\delta = 0$ for this situation and our model will be as follows

$$\frac{dP}{dt} = \varepsilon(1 - P - R) - (\gamma + \mu)P$$
$$\frac{dR}{dt} = \gamma(1 - P - R) - (\varepsilon + \alpha + \mu)R$$

This system of ordinary differential equation has one fixed point

$$P = \frac{\varepsilon(\varepsilon + \alpha + \mu)}{(\alpha + \varepsilon)(\gamma + \varepsilon + \mu) + \varepsilon\mu + (\gamma + \mu)^2}$$
$$R = \frac{\gamma(\gamma + \mu)}{(\alpha + \varepsilon)(\gamma + \varepsilon + \mu) + \varepsilon\mu + (\gamma + \mu)^2}$$

Linearizing the system around the fixed point we have the Jacobian matrix:

$$\begin{bmatrix} -(\gamma + \varepsilon + \mu) & -\varepsilon \\ -\gamma & -(\gamma + \alpha + \varepsilon + \mu) \end{bmatrix}$$

The eigenvalues of the matrix are

$$-\frac{1}{2}(2\gamma + \alpha + 2\varepsilon + 2\mu) \pm \frac{1}{2}\sqrt{\alpha^2 + 4\varepsilon\gamma}$$

IOP Conf. Series: Materials Science and Engineering **335** (2018) 012051 doi:10.1088/1757-899X/335/1/012051

When $\varepsilon > 0$, both eigenvalues are negative in their real parts. So, the fixed point is stable asymptotic.

If we consider to the reducing poverty effort, then the threshold value of dR is when

$$\frac{S}{R} = \frac{\alpha + \varepsilon + \mu}{\gamma}$$

It means that proportion of zakat-receiver (poor) will decrease when ratio between S and $R$is below $\frac{\alpha+\varepsilon+\mu}{\gamma}$ and will increase when the ratio is above $\frac{\alpha+\varepsilon+\mu}{\gamma}$.

3.3. Numerical simulation

Using the following set of parameter values

α=0.0000273973/day, γ=0.000342466/day, ε= 0.001369863/day, μ = 0.000547945/day,

we have the fixed point (0.5709187875,0.05798397965) which represent proportion of zakat-payer (wealthy) and zakat-receiver (poor) are respectively around57% and 5,8%. Figures 1 and 2 show the trajectory of the dynamical system confirm our analysis before.

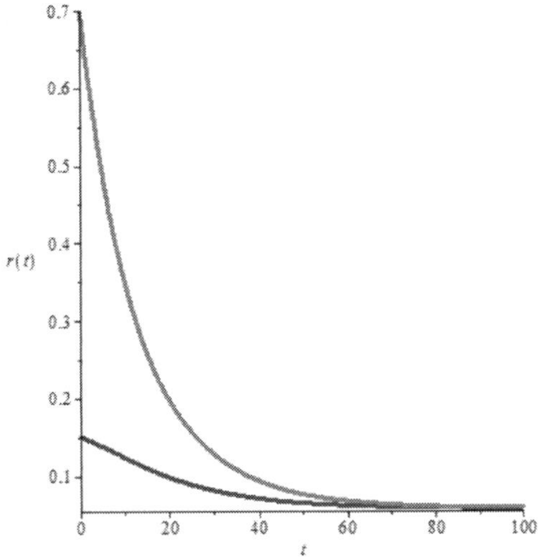

Figure 1. Trajectory of R over 100 years when annual economic growth is 5% for initial value $P(0) = 0.05, R(0) = 0.15$ and $P(0) = 0.02, R(0) = 0.7$.

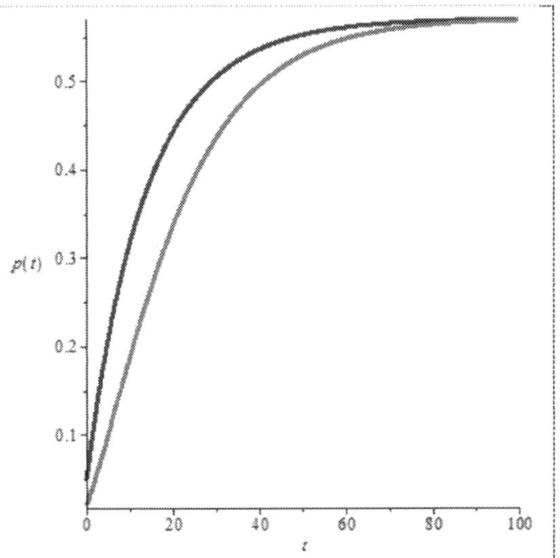

Figure 2. Trajectory of P over 100 years when annual economic growth is 5% for initial value $P(0) = 0.05, R(0) = 0.15$ and $P(0) = 0.02, R(0) = 0.7$

When the economic growth rather low, halving the ε, the dynamic's behaviour quite the same, but the proportion of the wealth and poor become about 38% and 11% respectively. The trajectories in Figures 3 and 4 shows that the fixed point also stable.

IOP Conf. Series: Materials Science and Engineering **335** (2018) 012051 doi:10.1088/1757-899X/335/1/012051

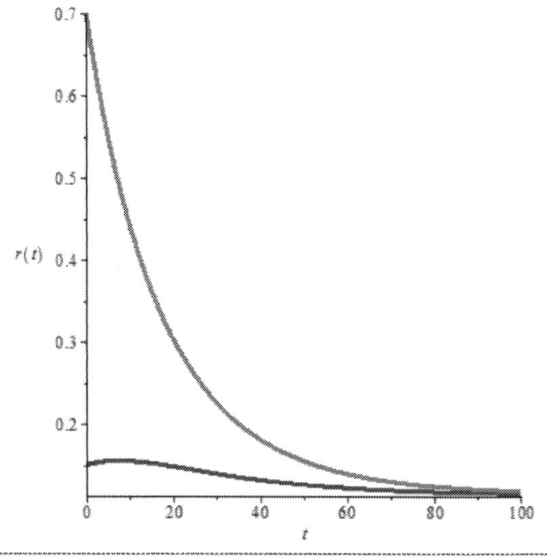

Figure 3. Trajectory of R over 100 years when annual economic growth is 2.5% for initial value $P(0) = 0.05, R(0) = 0.15$ and $P(0) = 0.02, R(0) = 0.7$.

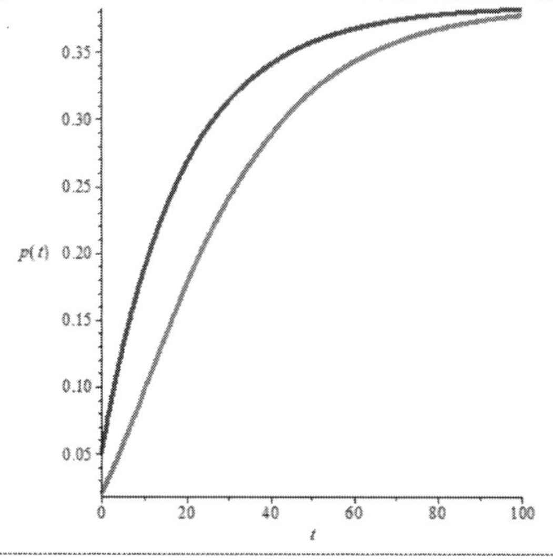

Figure 4. Trajectory of P over 100 years when annual economic growth is 2.5% for initial value $P(0) = 0.05, R(0) = 0.15$ and $P(0) = 0.02, R(0) = 0.7$

4. Conclusion

The model that we formed shows that if zakat is collected well and managed optimally, then the country will achieve its wealth equilibrium that represents the minimum numbers of poor population. Even though the poverty cannot be diminished, the numbers can be keep low. Model analysis and numerical simulation, for case fully obedience of paying zakat, confirm that zakat can contribute positively in poverty alleviation. For other case, we still need to analyze the model to have additional conclusion.

References

[1] Kusuma D B W and Sukmana R 2010 The power of zakah in poverty alleviation. *Proc.7th International Conference The Tawhidi Epistemology*
[2] Kuzudisli A 2017Fight against poverty from the Islamic point of view: the wealth distribution and share*Economics World*Vol. **5**19-15
[3] Dogarawa, AB 2008 Islamic social welfare and the role of zakah, *MPRA Paper No. 23192A*
[4] Aisyah M 2014 The role of zakah and binary economics in poverty reduction. *J. Bisnis dan Management* Vol **42**.
[5] Quraishi M A 1999 The Institution of Zakat and Its Economic Impact on Society*Proc. of the Second Harvard University Forum on Islamic Finance*pp.77-81
[6] Abdelbaki H H 2013 The Impact of Zakat on Poverty and Income Inequality in Bahrain *Review of Integrative Bussiness and Economy Research* Vol **21** pp 133-154
[7] Mahat N I and Warokka A 2013 Investigation on zakat as an indicator for moslem countries' economic growth. *J. Global Business Advancement* Vol. **61**
[8] Sboui F2012Effects of growth and inequality on poverty in Tunisia. *Region et Developpement* **35**57-80

IOP Conf. Series: Materials Science and Engineering **335** (2018) 012052 doi:10.1088/1757-899X/335/1/012052

A Comparison of Weights Matrices on Computation of Dengue Spatial Autocorrelation

K Suryowati[1*], R D Bekti[1], A Faradila[1]

[1]Department of Statistics, Institut Sains&Teknologi AKPRIND Yogyakarta, 55222, Indonesia

*suryowati@akprind.ac.id

Abstract. Spatial autocorrelation is one of spatial analysis to identify patterns of relationship or correlation between locations. This method is very important to get information on the dispersal patterns characteristic of a region and linkages between locations. In this study, it applied on the incidence of Dengue Hemorrhagic Fever (DHF) in 17 sub districts in Sleman, Daerah Istimewa Yogyakarta Province. The link among location indicated by a spatial weight matrix. It describe the structure of neighbouring and reflects the spatial influence. According to the spatial data, type of weighting matrix can be divided into two types: point type (distance) and the neighbourhood area (contiguity). Selection weighting function is one determinant of the results of the spatial analysis. This study use queen contiguity based on first order neighbour weights, queen contiguity based on second order neighbour weights, and inverse distance weights. Queen contiguity first order and inverse distance weights shows that there is the significance spatial autocorrelation in DHF, but not by queen contiguity second order. Queen contiguity first and second order compute 68 and 86 neighbour list

1. Introduction

Spatial analysis is a method that describes how geographically distributed phenomenon and how it compares with other phenomena. The spatial pattern can be shown with spatial autocorrelation. Spatial autocorrelation is the assessment of the correlation between observations on a variable. If the observations $X_1, X_2, ..., X_n$ shows the dependencies of space, then the data is said to be spatially auto correlated. So that spatial autocorrelation is used to analyze the spatial pattern of the spread of point locations to distinguish the location and attributes, or specific variables

In spatial analysis, it need weighting to show the neighboring among locations. This weighting is expressed in a spatial weighting matrix, namely**W**. This weighting is expressed in a spatial weighting matrix, namely W. This matrix gives an important role in any analysis of spatial data, because it shows the location, neighborhoods, and the relationship about distance among locations nearby. Selection of the type of matrix is also a major concern in spatial analysis. In the assessment of the spatial weighting, [1] states that the election of a weighted matrix to be done specifically and arranged convenience or convention. Specifications of the weighting matrix is represent the information and intensity of the spatial effect of a unit space locations in the geographic system.

Reference [2] state that adjacency relationships, distance relationships and comprehensive factors relationships are summarized to characterize the definitions of spatial weight matrix.According to the spatial data, type of weighting matrix can be divided into two types: point type (distance) and the neighborhood area (contiguity). Reference [3] indicate that the selection weighting function is one determinant of the results of the spatial analysis. The distance weighting matrix involve an element of distance between the locations whose value is continuous in building the weightingmatrix, so that each location will receive weighted according to that distance. Meanwhile, contiguity weighting involve an element of area or neighborhoods. The location that neighboring directly shows the higher spatial dependency than the locations far apart.

The diversity of spatial data and methods are causing the importance of selecting the appropriate weighting matrix. In most applications, the weight matrix is more likely to be based on the distance between units, or simply contiguity ([4],[5]). Research [6] use this type of distance weighting to the analysis of ICT indicators using Geographically Weighted Regression. Research[7]use four types of contiguity weighting function to calculate Moran's I. Research[2] states that the selection of weighting matrix will effect on a parameter estimation in spatial modelling. Furthermore,[8] states that the ways of weight matrix will effect on spatial autocorrelation analysis, especially the value of Moran's I greatly.

The study compared the type of contiguity and distance weighting in the analysis of spatial autocorrelation. The case study used is the incidence of dengue fever in Sleman, Yogyakarta, Indonesia.Selection of cases is based on the goal to identifyingexposures of public health concern based on spatial epidemiology. As research [9] which use spatial autocorrelation for analysis the dengue vector populations. Reference [10] which analysis the examination of dengue fever in Guayaquil, Ecuador. The results of this research was Moran's I value shows the high spatial autocorrelation and suggest a higher possibility of infection in places with already infected people.

Along with the development of spatial autocorrelation method for spatial analysis in various fields, on the other hand, the computer as a computing tool is also experiencing rapid development software. This research also make application development in R which easy to use and speed up the time to complete a job mainly related to the spatial analysis. This application is in plug-in form which use R language that available as free software and provides a wide variety of statistical. This plug-in provides three method of spatial autocorrelation: Moran's I, Geary's, and Local Indicator of Spatial Autocorrelation (LISA), with three types of weights matrices: Queen Contiguity by first order neighbor, Queen Contiguity by second order neighbor, and distance weighted.

2. ResearchMethodology

This research compute spatial weight matrix in three types of weighted: 1) queen contiguity based on first order neighbour weights, 2) queen contiguity based on second order neighbour weights, and 3) inverse distance weights. The next steps is use those weight to compute spatial autocorrelation by Moran's I, Geary's C, and LISA. This analysis will demonstrate the spatial autocorrelation in each weight. The results of the study is about: 1) the neighbour list in each weight matrices, 2) the significance of the results of autocorrelation in any type of spatial autocorrelation method, and 3) a comparison method.

The calculations are performed using the application program that has been developed, in the form ofplug-in package R. The development of computer applications aimed at assisting the spatial autocorrelation in each weight. It was done in some steps: the design of a computer program with the waterfall model, designing systems and interface, implementation and unit testing, and test of the system operational. The designing interface was done in Deducer R which compatible with RJava. Deducer is a package R which provide an intuitive graphical user interface (GUI) for R [11]. It works when used with the Java based R GUI JGR. This application also was formed into package R. It was built by RStudio and Rtools programs.

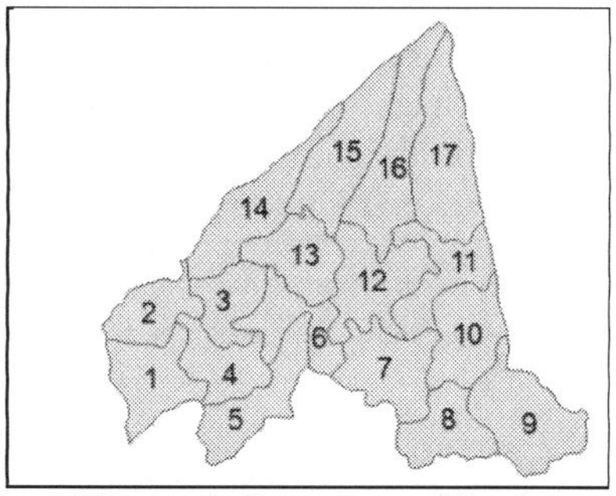

Figure 1. Location of Study

The data used in this research is the secondary data about the number of occurrences of Dengue Hemorrhagic Fever (DHF) in 17 sub districts in Sleman (see Figure 1). The data was obtained from Sleman District Health Office in 2010 - 2015. The names of 17 sub-district respectively : 1. Moyudan, 2. Minggir, 3. Seyegan, 4. Godean, 5.Gamping, 6. Mlati, 7. Depok, 8. Berbah, 9. Prambanan, 10. Kalasan, 11. Ngemplak, 12. Ngaglik, 13. Sleman, 14. Tempel, 15.Turi, 16.Pakem, 17. Cangkringan.

2.1. Spatial Autocorrelation

Spatial pattern analysis is an analysis that explains how geographic phenomena distributed and how it compares with other phenomena. It also explains how the relationship between geographic objects and comparing the object pattern in a location with object pattern found in other locations. Some form of spatial patterns are clustered, dispersed, and random. The forms of these patterns can also be tested by the method of spatial autocorrelation.

If there is a systematic pattern in the deployment of a variable, then there is spatial autocorrelation. The existence of spatial autocorrelation indicates that the attribute values in specific areas related to the attribute values in other areas that are neighboring. Reference [12] also explained that the spatial autocorrelation refers to a pattern in which the observations are derived from the locations adjacent sites tend to have the same value.

This study uses spatial autocorrelation test Morans' I, Geary's C, and LISA. Moran's I (Index) is used to measure global spatial autocorrelation among location. The formula as in equation (1).

$$I = \frac{\sum_{i=1}^{n}\sum_{j=1}^{n} w_{ij}(x_i - \bar{x})(x_j - \bar{x})}{\sum_{i=1}^{n}(x_i - \bar{x})^2} \quad (1)$$

with n is the number of sample size, x_i and $x_j$are the value of observed variable at the location i and j. $\bar{x}$ is the average of the n observed variable. w_{ij} is the element of spatial weight matrix **W**. Moran's Index value is between -1 and 1. The positive value (near +1) indicates clustering, while negative value (near -1) indicates dispersion ([12], [13]). To evaluate the statistical significance test, it use the null hypothesis that there is no spatial autocorrelation among locations. Reject the null hypothesis if $|Z|>Z_{\alpha/2}$, which $Z=(I-E(I))/\sqrt{Var(I)}$ and $Z_{\alpha/2}$ derived form normal standard distribution.

Geary's coefficient (Geary's C) is the second method which commonly used in spatial autocorrelation. The formula as in equation(2).

$$C = \frac{(n-1)}{2W}\frac{\sum_{i=1}^{n}\sum_{j=1}^{n} w_{ij}(x_i - x_j)^2}{\sum_{i=1}^{n}(x_i - \bar{x})^2} \quad (2)$$

The value of Geary's C lies between 0 and 2. It is negatively related to Moran's I, which the values less than one indicate increasing positive spatial autocorrelation and the values greater than one indicate increasing negative spatial autocorrelation [14]. As in Moran's I test, the null hypothesis that there is no spatial autocorrelation among locations. Reject the null hypothesis if $|Z|>Z_{\alpha/2}$, which $Z=(C-E(C))/\sqrt{Var(C)}$.

Moran's I and Geary's C used to identify the global autocorrelation, whereas Local Indicator of Spatial Autocorrelation (LISA)is Moran local. The motivation of LISA was to decompose global autocorrelation into the contribution of individual observation ([14], [15]). It evaluate the spatial autocorrelation or clustering in those individual units by calculating local Moran's I for each location and evaluating the statistical significance for each I_i. The formula as in equation (3).

$$I_i = z_i \sum_{i=1}^{n} w_{ij} z_j \quad (3)$$

wherez_j is the standardize of observed variable.

2.2. Spatial Weights Matrix

A spatial weights matrix is an $n \times n$ positive symmetric matrix **W**, as in equation(4). The element of these matrix is w_{ij} at location i, j for n locations. It play an important role in describing the structure of neighbouring or spatial structure among location. It typically reflects the spatial influence of location j on location i or conversely. In matrix **W**, every unit area is described as rows and columns. As example, the value of w_{12} or w_{21} shows the weighting between the first and second locations. The location which is close to the certain locations tend to have large w_{ij}, while the location which far from each other tend to have small w_{ij}.

$$\mathbf{W} = \begin{bmatrix} w_{11} & w_{12} & w_{13} & \cdots & w_{1n} \\ w_{21} & w_{13} & w_{23} & \cdots & w_{2n} \\ \vdots & \vdots & \vdots & w_{ij} & \vdots \\ w_{n1} & w_{n2} & w_{n3} & \cdots & w_{nn} \end{bmatrix} \tag{4}$$

Matrix **W** have some characteristic: 1) all diagonal elements of w_{ij} is 0, because it is assumed that locations is not adjacent to the unit itself; 2) It is a symmetric matrix where $w_{ij} = w_{ji}$, which basically describes the interrelationships of spatial relation; 3) row matrix indicates how a location spatially associated with other regions. Therefore, the total values in a row i is the total neighbors that are owned by the location i. There are many ways to define the weights, including contiguity weights, distance weights, and other weights (Zhou and Lin, 2008). The simplest form of weights is binary weight where $w_{ij} = 1$ for neighbouring locations i and j, otherwise, $w_{ij} = 0$.

Contiguity weights illustrates the relative location of a unit of spatial data with other locations based on boundaries area or neighborhood list. Research [4] states the type of contiguity weights, such as linier contiguity, rook contiguity, bishop contiguity, and queen contiguity. This research use queen contiguity matrix, which create the neighborhood list based on edges corner. The element of spatial weight w_{ij}=1 shows that the location i and j are share an edge and/or a corner. Otherwise, w_{ij}=0 shows that the location i and j don't share edge and/or a corner each other.

The spatial weights have been discussed is the type of first order neighbour weights. This study also use higher order neighbour weight, which constructs a spatial weight matrix of order n based on the spatial neighbour relationship between locations. It assumed that the value of a unit is not only affected by the immediately contiguous units, but also the second until n order contiguous units. The element $w_{i,j}$=1 if location j is adjacent to the neighbours of order n-1 of location i, and is 0 otherwise. As example for second order neighbours, as locations that are neighbours to the first order neighbour[5], the element $w_{i,j}$=1 if location j is adjacent to the first order neighbours of location i, and is 0 otherwise.

The weight matrices also can approach by Row Standardization as inequation (5).

$$w_{ij}^{*} = \frac{w_{ij}}{\sum_{j=1}^{n} w_{ij}} \tag{5}$$

There are many variants or modification in the build spatial weight. The choice of the method is also based on the assumptions about the nature phenomena, the characteristics of regions, socio-economic condition, the availability of public facilities, or others. Reference [16] was study about spatial weights matrix in diffusion of housing demand. In economic study, the choice of appropriate spatial weights also a central component of spatial models, which may or may not correspond to reality. In some cases, the contiguity weight would fail to capture a potentially significant relationship. As example of areas consisting of islands. According to the administrative boundaries, those islands are neighboring. However, based on polygon in map, they don't share edge and/or a corner each other.

In endemic disease study, especially in dengue cases, the spatial relationship among locations not only can describe by neighbouring based on share edge and/or a corner, but also how the extent to which the disease is transmitted or spread. According to this reasons, this study also use distance weight. The distance is obtained from the calculation of the distance between the point latitude and

IOP Conf. Series: Materials Science and Engineering **335** (2018) 012052 doi:10.1088/1757-899X/335/1/012052

longitude [3]. So, each location will have weighted according to the distance to other locations. The distance location i and j were calculated by Euclidean distance as inequation (6).

$$d_{ij} = \sqrt{(u_i - u_j)^2 + (v_i - v_j)^2} \quad (6)$$

whereu_i, v_i are the geographic coordinates longitude-latitude in location i and u_j, v_j are the geographic coordinates longitude-latitude in location j.

Many kinds of distance weight are k-Nearest Neighbor, radial distance, inverse distance, exponential distance, and double distance weights. This study use inverse distance weights, which the formula as in equation(7) [17].

$$w_{ij} = \frac{1}{d_{ij}^{\alpha}} \quad (7)$$

where α represent the distance friction coefficient, generally $\alpha=1$ or $\alpha=2$.

3. Results and Discussion

The results presented in this paper is about the characteristics of the spread of the incidence of Dengue Hemorrhagic Fever (DHF) in Sleman and an overview of the distribution pattern of DHF in 2014-2015. Furthermore, calculating weighted, spatial autocorrelation test, and compare the results in each weighting. All of these measures is done using the application program that has been built.

3.1. Spatial Pattern of DHF

Spatial pattern of incidence of dengue in the Sleman District used to make the priority areas need to be considered to suppress the spread of dengue incidence. According to the Data Center and Epidemiological Surveillance by Ministry of Health, incidence of dengue region can be categorized in the high (more than 55 in 100.000 population), medium (21-55 in 100.000 population) and low incidence (less than 20 in 100.000 population).

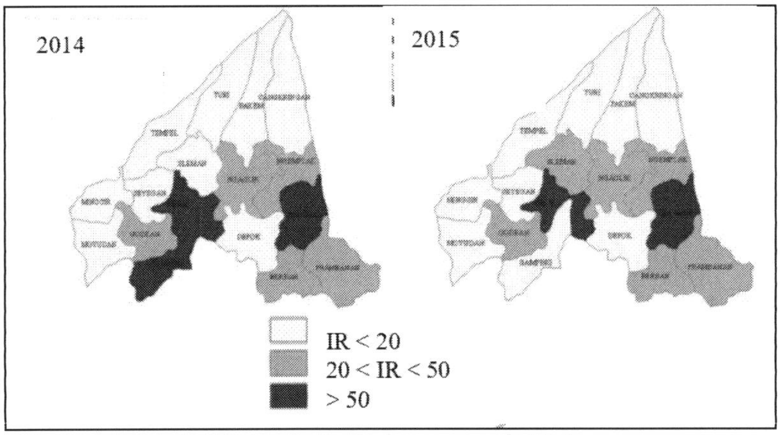

Figure 2. Spatial pattern of DHF

Based on Figure 2 can be seen that the incidence rate (IR) of DHF in 2014-2015 included in the category of low, medium and high. Area with a high incidence is Gamping, the area with the incidence being is Godean, Mlati, and Kalasan Sub Districts. The area with a medium incidence is the Godean, Ngaglik, Ngemplak, Berbah, and Prambanan Sub Districts, and the others is in category low incidence. These characteristics indicate that the incidence of relatively high DBF located in the southern part of Sleman. In addition, the sub-districts are also close to each other. This is shows the spatial pattern. It also occurs in 2015.

3.2. Weighting Calculation

Creation of spatial weights matrix was done by 'Autocorrelation Test' package which build in Deducer and RStudio. Run the syntax library(Autocorrelation Test and menu menu Autocorrelation

Test, so spatial analysis menu will appear as in Figure3. By this application programs, users can analyze the autocorrelation test in three methods (Moran's I, Geary's C, and LISA) and three spatial weight (queen contiguity based on first order neighbor weights, queen contiguity based on second order neighbour weights, and inverse distance weights). The data used is in form data frame and shape file (.shp). The data frame contains the main variable in row and column to be tested. The shape file contains the geospatial data formats which used to get the neighboring and coordinate. This application programs is development from previous research [18].

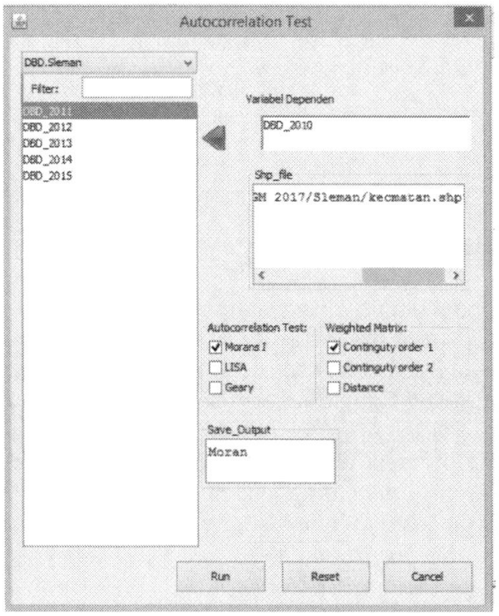

Figure 3. Autocorrelation Test in Deducer

```
Characteristics of weights list object:
Neighbour list object:
Number of regions: 17
Number of nonzero links: 68
Percentage nonzero weights: 23.52941
Average number of links: 4

Weights style: W
Weights constants summary:
   n  nn S0       S1       S2
W 17 289 17 9.170556 70.12333
```

Figure 4. Characteristic of weight list

Figure 4 ilustrate the queen contiguity based on first order neighbor weights. There were 17 regions (sub distrcicts) and 68 neighbor list. This weighting matrix is formed into a weight matrix as shown in Figure 5. Queen contiguity based on second order neighbor weights and inverse distance is presented in Figure 6 and Figure7.

$$W=\begin{bmatrix}
0 & \tfrac{1}{2} & 0 & \tfrac{1}{2} & 0 & 0 & 0 & 0 & 0 & 0 & 0 & 0 & 0 & 0 & 0 & 0 & 0\\
\tfrac{1}{4} & 0 & \tfrac{1}{4} & \tfrac{1}{4} & 0 & 0 & 0 & 0 & 0 & 0 & 0 & 0 & 0 & \tfrac{1}{4} & 0 & 0 & 0\\
0 & \tfrac{1}{5} & 0 & \tfrac{1}{5} & 0 & \tfrac{1}{5} & 0 & 0 & 0 & 0 & 0 & 0 & \tfrac{1}{5} & \tfrac{1}{5} & 0 & 0 & 0\\
\tfrac{1}{5} & \tfrac{1}{5} & \tfrac{1}{5} & 0 & \tfrac{1}{5} & \tfrac{1}{5} & 0 & 0 & 0 & 0 & 0 & 0 & 0 & 0 & 0 & 0 & 0\\
0 & 0 & 0 & \tfrac{1}{2} & 0 & \tfrac{1}{2} & 0 & 0 & 0 & 0 & 0 & 0 & 0 & 0 & 0 & 0 & 0\\
0 & 0 & \tfrac{1}{6} & \tfrac{1}{6} & \tfrac{1}{6} & 0 & \tfrac{1}{6} & 0 & 0 & 0 & 0 & \tfrac{1}{6} & \tfrac{1}{6} & 0 & 0 & 0 & 0\\
0 & 0 & 0 & 0 & 0 & \tfrac{1}{5} & 0 & \tfrac{1}{5} & 0 & \tfrac{1}{5} & \tfrac{1}{5} & \tfrac{1}{5} & 0 & 0 & 0 & 0 & 0\\
0 & 0 & 0 & 0 & 0 & 0 & \tfrac{1}{3} & 0 & \tfrac{1}{3} & \tfrac{1}{3} & 0 & 0 & 0 & 0 & 0 & 0 & 0\\
0 & 0 & 0 & 0 & 0 & 0 & 0 & \tfrac{1}{2} & 0 & \tfrac{1}{2} & 0 & 0 & 0 & 0 & 0 & 0 & 0\\
0 & 0 & 0 & 0 & 0 & 0 & \tfrac{1}{4} & \tfrac{1}{4} & \tfrac{1}{4} & 0 & \tfrac{1}{4} & 0 & 0 & 0 & 0 & 0 & 0\\
0 & 0 & 0 & 0 & 0 & 0 & \tfrac{1}{5} & 0 & 0 & \tfrac{1}{5} & 0 & \tfrac{1}{5} & 0 & 0 & 0 & \tfrac{1}{5} & \tfrac{1}{5}\\
0 & 0 & 0 & 0 & 0 & \tfrac{1}{6} & \tfrac{1}{6} & 0 & 0 & 0 & \tfrac{1}{6} & 0 & \tfrac{1}{6} & 0 & \tfrac{1}{6} & \tfrac{1}{6} & 0\\
0 & 0 & \tfrac{1}{5} & 0 & 0 & \tfrac{1}{5} & 0 & 0 & 0 & 0 & 0 & \tfrac{1}{5} & 0 & \tfrac{1}{5} & \tfrac{1}{5} & 0 & 0\\
0 & \tfrac{1}{4} & \tfrac{1}{4} & 0 & 0 & 0 & 0 & 0 & 0 & 0 & 0 & 0 & \tfrac{1}{4} & 0 & \tfrac{1}{4} & 0 & 0\\
0 & 0 & 0 & 0 & 0 & 0 & 0 & 0 & 0 & 0 & 0 & \tfrac{1}{4} & \tfrac{1}{4} & \tfrac{1}{4} & 0 & \tfrac{1}{4} & 0\\
0 & 0 & 0 & 0 & 0 & 0 & 0 & 0 & 0 & 0 & \tfrac{1}{4} & \tfrac{1}{4} & 0 & 0 & \tfrac{1}{4} & 0 & \tfrac{1}{4}\\
0 & 0 & 0 & 0 & 0 & 0 & 0 & 0 & 0 & 0 & \tfrac{1}{2} & 0 & 0 & 0 & 0 & \tfrac{1}{2} & 0
\end{bmatrix}$$

Figure 5. Queen contiguity based on first order neighbour weights

$$W=\begin{bmatrix}
0 & 0 & \tfrac{1}{4} & 0 & \tfrac{1}{4} & \tfrac{1}{4} & 0 & 0 & 0 & 0 & 0 & 0 & 0 & \tfrac{1}{4} & 0 & 0 & 0\\
0 & 0 & 0 & 0 & \tfrac{1}{4} & \tfrac{1}{4} & 0 & 0 & 0 & 0 & 0 & 0 & \tfrac{1}{4} & 0 & \tfrac{1}{4} & 0 & 0\\
\tfrac{1}{5} & 0 & 0 & 0 & \tfrac{1}{5} & 0 & \tfrac{1}{5} & 0 & 0 & 0 & 0 & \tfrac{1}{5} & 0 & 0 & \tfrac{1}{5} & 0 & 0\\
0 & 0 & 0 & 0 & 0 & 0 & \tfrac{1}{4} & 0 & 0 & 0 & 0 & \tfrac{1}{4} & \tfrac{1}{4} & \tfrac{1}{4} & 0 & 0 & 0\\
\tfrac{1}{6} & \tfrac{1}{6} & \tfrac{1}{6} & 0 & 0 & 0 & \tfrac{1}{6} & 0 & 0 & 0 & 0 & \tfrac{1}{6} & \tfrac{1}{6} & 0 & 0 & 0 & 0\\
\tfrac{1}{8} & \tfrac{1}{8} & 0 & 0 & 0 & 0 & 0 & \tfrac{1}{8} & 0 & \tfrac{1}{8} & \tfrac{1}{8} & 0 & 0 & \tfrac{1}{8} & \tfrac{1}{8} & \tfrac{1}{8} & 0\\
0 & 0 & \tfrac{1}{8} & \tfrac{1}{8} & \tfrac{1}{8} & 0 & 0 & 0 & \tfrac{1}{8} & 0 & 0 & 0 & \tfrac{1}{8} & 0 & \tfrac{1}{8} & \tfrac{1}{8} & \tfrac{1}{8}\\
0 & 0 & 0 & 0 & 0 & \tfrac{1}{3} & 0 & 0 & 0 & 0 & \tfrac{1}{3} & \tfrac{1}{3} & 0 & 0 & 0 & 0 & 0\\
0 & 0 & 0 & 0 & 0 & 0 & \tfrac{1}{2} & 0 & 0 & 0 & \tfrac{1}{2} & 0 & 0 & 0 & 0 & 0 & 0\\
0 & 0 & 0 & 0 & 0 & \tfrac{1}{4} & 0 & 0 & 0 & 0 & 0 & \tfrac{1}{4} & 0 & 0 & 0 & \tfrac{1}{4} & \tfrac{1}{4}\\
0 & 0 & 0 & 0 & 0 & \tfrac{1}{5} & 0 & \tfrac{1}{5} & \tfrac{1}{5} & 0 & 0 & 0 & \tfrac{1}{5} & 0 & \tfrac{1}{5} & 0 & 0\\
0 & 0 & \tfrac{1}{7} & \tfrac{1}{7} & \tfrac{1}{7} & 0 & 0 & \tfrac{1}{7} & 0 & \tfrac{1}{7} & 0 & 0 & 0 & \tfrac{1}{7} & 0 & 0 & \tfrac{1}{7}\\
0 & \tfrac{1}{6} & 0 & \tfrac{1}{6} & \tfrac{1}{6} & 0 & \tfrac{1}{6} & 0 & 0 & 0 & \tfrac{1}{6} & 0 & 0 & 0 & 0 & \tfrac{1}{6} & 0\\
\tfrac{1}{5} & 0 & 0 & \tfrac{1}{5} & 0 & \tfrac{1}{5} & 0 & 0 & 0 & 0 & 0 & \tfrac{1}{5} & 0 & 0 & 0 & \tfrac{1}{5} & 0\\
0 & \tfrac{1}{6} & \tfrac{1}{6} & 0 & 0 & \tfrac{1}{6} & \tfrac{1}{6} & 0 & 0 & 0 & \tfrac{1}{6} & 0 & 0 & 0 & 0 & 0 & \tfrac{1}{6}\\
0 & 0 & 0 & 0 & 0 & \tfrac{1}{5} & \tfrac{1}{5} & 0 & 0 & \tfrac{1}{5} & 0 & 0 & \tfrac{1}{5} & \tfrac{1}{5} & 0 & 0 & 0\\
0 & 0 & 0 & 0 & 0 & 0 & \tfrac{1}{4} & 0 & 0 & \tfrac{1}{4} & 0 & \tfrac{1}{4} & 0 & 0 & \tfrac{1}{4} & 0 & 0
\end{bmatrix}$$

Figure 6. Queen contiguity based on second order neighbour weights

$$W = \begin{bmatrix}
0 & 23.09 & 19.92 & 102.98 & 78.01 & 28.56 & 143.38 & 30.53 & 41.45 & 34.16 & 14.57 & 15.77 & 12.63 & 9.05 & 6.76 & 6.80 & 6.92 \\
23.09 & 0 & 144.53 & 29.76 & 17.82 & 119.73 & 27.37 & 13.15 & 14.85 & 70.18 & 39.41 & 49.62 & 27.85 & 14.88 & 9.55 & 9.64 & 9.87 \\
19.92 & 144.53 & 0 & 24.69 & 15.87 & 65.76 & 23.03 & 12.06 & 13.47 & 47.49 & 54.19 & 75.56 & 34.48 & 16.58 & 10.22 & 10.33 & 10.59 \\
102.98 & 29.76 & 24.69 & 0 & 44.44 & 39.52 & 322.78 & 23.57 & 29.62 & 51.11 & 16.97 & 18.62 & 14.39 & 9.92 & 7.23 & 7.28 & 7.41 \\
78.01 & 17.82 & 15.87 & 44.44 & 0 & 20.92 & 50.97 & 50.15 & 88.19 & 23.80 & 12.28 & 13.12 & 10.87 & 8.11 & 6.22 & 6.26 & 6.35 \\
28.56 & 119.73 & 65.76 & 39.52 & 20.92 & 0 & 35.45 & 14.77 & 16.94 & 169.23 & 29.75 & 35.21 & 22.62 & 13.24 & 8.85 & 8.93 & 9.12 \\
143.38 & 27.37 & 23.03 & 322.78 & 50.97 & 35.45 & 0 & 25.32 & 32.45 & 44.63 & 16.18 & 17.67 & 13.81 & 9.64 & 7.08 & 7.13 & 7.26 \\
30.53 & 13.15 & 12.06 & 23.57 & 50.15 & 14.77 & 25.32 & 0 & 114.73 & 16.16 & 9.87 & 10.41 & 8.94 & 6.98 & 5.53 & 5.56 & 5.64 \\
41.45 & 14.85 & 13.47 & 29.62 & 88.19 & 16.94 & 32.45 & 114.73 & 0 & 18.80 & 10.80 & 11.44 & 9.69 & 7.43 & 5.81 & 5.85 & 5.93 \\
34.16 & 70.18 & 47.49 & 51.11 & 23.80 & 169.23 & 44.63 & 16.16 & 18.80 & 0 & 25.38 & 29.24 & 20.00 & 12.30 & 8.42 & 8.49 & 8.67 \\
14.57 & 39.41 & 54.19 & 54.19 & 12.28 & 29.75 & 16.18 & 9.87 & 10.80 & 25.38 & 0 & 191.62 & 93.97 & 23.86 & 12.59 & 12.76 & 13.16 \\
15.77 & 49.62 & 75.56 & 18.62 & 13.12 & 35.21 & 17.67 & 10.41 & 11.44 & 29.24 & 191.62 & 0 & 63.26 & 21.23 & 11.82 & 11.96 & 12.32 \\
12.63 & 27.85 & 34.48 & 14.39 & 10.87 & 22.62 & 13.81 & 8.94 & 9.69 & 20.00 & 93.97 & 63.26 & 0 & 31.95 & 14.53 & 14.75 & 15.29 \\
9.05 & 14.88 & 16.58 & 9.92 & 8.11 & 13.24 & 9.64 & 6.98 & 7.43 & 12.30 & 23.86 & 21.23 & 31.95 & 0 & 26.66 & 27.39 & 29.31 \\
6.76 & 9.55 & 10.22 & 7.23 & 6.22 & 8.85 & 7.08 & 5.53 & 5.81 & 8.42 & 12.59 & 11.82 & 14.53 & 26.66 & 0 & 952.42 & 288.89 \\
6.80 & 9.64 & 10.33 & 7.28 & 6.26 & 8.93 & 7.13 & 5.56 & 5.85 & 8.49 & 12.76 & 11.96 & 14.75 & 27.39 & 952.42 & 0 & 412.78 \\
6.92 & 9.87 & 10.59 & 7.41 & 6.35 & 9.12 & 7.26 & 5.64 & 5.93 & 8.67 & 13.16 & 12.32 & 15.29 & 29.31 & 288.89 & 412.78 & 0
\end{bmatrix}$$

Figure 7. Inverse Distance weights

Based on the Figure5, can be seen that the sub-district with the highest number of neighbors is the Mlati and Ngaglik. Mlati has six neighbors which means that the incidence of dengue in this location affects and/or was affected by 6 that neighboring. Ngaglik also has six neighbors. Location that has smallest neighbors is Moyudan, Limestone, Prambanan and Cangkringan. The value in the matrix is w_{ij} is calculated by equation (4).

Queen contiguity second order in Figure 6 produce different regions of neighborhood. There were 17 regions (sub districts) and 86 neighbor list. Through weighting queen contiguity first order, Moyudon has only two neighboring list (Minggir and Godean). While, using queen contiguity second order, Moyudon has 4 neighboring list (Seyegan, Gamping, Mlati, and Tempel). This is because Sayegan and Tempel are share edge and/or a corner with Minggir, also Gamping, Mlati are share edge and/or a corner with Godean.

Inverse distance weighting matrix in Figure7 is calculated based on the Euclidean distance which defined as the shortest straight line between two points. It describes the value of the inverse of the distance from each of the sub districts. If two locations have great distance, they will produce small value of weighting. If two locations have small distance, they will produce high value of weighting. The value in the matrix is w_{ij} is calculated by Equation (6).

3.3. Spatial Autocorrelation

Table 1 shows the results of spatial autocorrelation test. It provides information that using weighted queen contiguity first order and inverse distance indicated that a significant spatial autocorrelation, both on the method of Moran's I and Geary's C. This is shown by the value $|Z| > 1.96$ or $P_{value} < 0.05$. The existence of spatial autocorrelation shows that there were dependence in incidence of dengue among sub among locations. While, the test using a weighted queen contiguity second order shows that there is no spatial autocorrelation.

Pattern of spatial autocorrelation (use Moran's I) in the queen contiguity first order is the incidence DHF which spatially clustered. As example DHF in 2014, there are 6 sub districts located in the High-High group (Gamping, Depok, Berbah, Prambanan, Kalasan, and Ngemplak). The locations in this group has a high DHF which surrounded by locations that have high DHF. The second group is Low-Low group which contains 8 sub districts (Moyudon, Minggir, Seyegan, Sleman, Tempel, Turi, Pakem, and Cangkiran). The definition of this group is the location that has a lower DHF surrounded by location that have lower DHF. There are three sub districts in High-Low group (Godean, Mlati, and Ngaglik). The definition of this group is that the location that have high DHF surrounded by location that have lower HF. Figure 8 shows a comparison of LISA test with 3 types of weighting. It can be seen that by weighting queen contiguity first order and inverse distance, there are several locations that have significant spatial autocorrelation at $\alpha = 5\%$ and $\alpha = 10\%$.

Table 1. Spatial Autocorrelation Test

Year	Index, Expectation, and Z	Queen Contiguity first order	Queen Contiguity second order	Inverse Distance
		Moran's I		
2014	I	0.4882	-0.1348	0.4217
	E(I)	-0,0625	-0,0625	-0,0625
	|Z|	3.29*	0.59	P value = 0.01*
2015	I	0.4262	-0.195	0.3879
	E(I)	-0,0625	-0,0625	-0,0625
	|Z|	2.92*	0.59	P value = 0.01*
		Geary's C		
2014	C	0.4518	1.0615	0.5995
	E(C)	1	1	1
	|Z|	3.47*	0.42	P value = 0.03*
2015	C	0.4728	1.1285	0.61
	E(C)	1	1	1
	|Z|	3.34*	0.89	P value = 0.05*

Note: *) Significance at α=5%

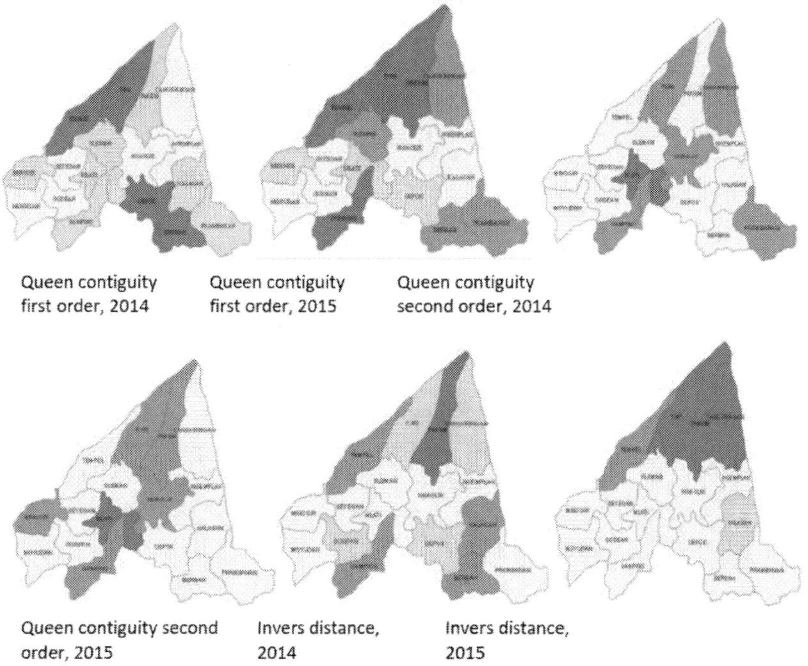

Figure 8. The Results of LISA test

4. Concussion

A spatial weights matrix, **W**, have element w_{ij} at location i, j for n locations. It describe the structure of neighbouring or spatial structure among location and reflects the spatial influence of location j on location i or conversely. There are many types of weighting based on the shape of the area and its spatial structure. Different types of weighting will create the different spatial autocorrelation. In this research, queen contiguity based on first order neighbour weights compute 68 neighbour list and shows the significance spatial autocorrelation on DHF of Sleman District. Queen contiguity based on second order neighbour weights give the different results. It compute 86 neighbour list and shows the non-significance spatial autocorrelation. Inverse distance weights show that all locations adjacent each other based on Euclidean distance. It shows that there is a significance spatial autocorrelation.

Acknowledgments
Thanks to the Institut Sains&Teknologi AKPRIND Yogyakarta which support funding and research facilities.

References

[1] Anselin L and Griffith D A 1988. Do spatial effects really matter in regression analysis?. Pap. Reg. Sci. 65(1) 11–34.
[2] Wang Z, Liang H Y, and Zhou Q 2013. Model for Effect of Spatial Weighted Matrix on Spatial Autocorrelation. International Journal of Applied Mathematics and Statistics. 44(14).431–445.
[3] Fotheringham A S, Brunsdon C, and Charlton M 2003 Geographically Weighted Regression: The Analysis of Spatially Varying Relationships (New York: John Wiley & Sons).
[4] Anselin L 2001 Spatial Econometrics (Dallas : University of Texas)
[5] LeSage J P and Pace R 2009 An introduction to spatial econometrics, 1st ed (Boca Raton: Taylor and Francis)
[6] Bekti R D, Andiyono, and Irwansyah E 2014. Mapping of Illiteracy and ICT Indicators using Geographically Weighted Regression. J. Math. Stat. 10 (2).130–138
[7] Chen Y 2012. On the Four Types of Weight Functions for Spatial Contiguity Matrix. Lett. Spat. Resour. Sci. 5 (2). 65–72.
[8] Chen J and Jiang J 2010. Analysis for spatial autocorrelation for point objects based on line buffer. Proceeding Accuracy 2010 Symp. Leic. UK. 20.
[9] Azil A H, Bruce D, and Williams C R 2014 Determining the spatial autocorrelation of dengue vector populations: influences of mosquito sampling method, covariables, and vector control. J. Vector Ecol. 39(1). 153–163
[10] Castillo K C, Körbl B, Stewart A, Gonzalez J F, and Ponce F 2011 Application of spatial analysis to the examination of dengue fever in Guayaquil. Procedia Environ. Sci. 7, 188–193
[11] Fellows I 2012 Deducer: a data analysis GUI for R. Journal of statistical Software. J. Stat. Softw. 49(8).1–15
[12] Lee J and Wong D W 2001 Statistical analysis with ArcView GIS (New York : John Wiley & Sons).
[13] Cliff A D and Ord J K 1988 Spatial autocorrelation, vol. 5. (London: Pion)
[14] Fischer M. M.andGetisA 2009 Handbook of applied spatial analysis: software tools, methods and applications. (New York: Springer Science & Business Media)
[15] Anselin L 1995. Local indicators of spatial association—LISA. Geogr. Anal. 27(2). 93–115.
[16] Bhattacharjee A and Jensen-Butler C 2006. Estimation of spatial weights matrix, with an application to diffusion in housing demand. Cent. Res. Ind. Enterp. Finance Firm Discuss. 519
[17] Zhou X and Lin H 2008 Spatial weights matrix. Encyclopedia of GIS. (US: Springer)
[18] Tanty H, Bekti R D, and Irwansyah E 2014. Package Plug-In R untukPemetaanAutokorelasiSpasialpadaKualitas Air," presented at the Seminar NasionalAplikasiSains&Teknologi (SNAST) 2014, IST AKPRIND Yogyakarta, C269–C274.

IOP Conf. Series: Materials Science and Engineering **335** (2018) 012053 doi:10.1088/1757-899X/335/1/012053

Mapping the Factors Affecting Household Food Security of Tuberculosis Patients in Coastal Region of Surabaya

D Susilaningrum, B S S Ulama, R Lathifah

Business Statistics Department, Faculty of Vocational Institut Teknologi Sepuluh Nopember

destri_s@statistika.its.ac.id

Abstract. Food security is a condition of a person who has access to adequate, safe, and nutritious food to meet the needs of a healthy life. Affordability of food is determined by the aggregate purchasing power. It is also can be measured by the amount of poverty in Indonesia which reached 28,07 million in 2014. According to data from the health department, the largest TB cases in Indonesia were in East Java Province and Surabaya accounted for the largest number, there are 48379 cases in 2015. This study mapping the factors that affect household food security of TB patients in the coastal areas of Surabaya. This study used secondary data sources from 11 clinics i.e. TB patient's addres and primary data source to survey patients about food security. Variables used in this study are variables related to socioeconomic factors and sanitary factors. Those variables will be analyzed descriptively and mapping using biplot analysis. Biplot generated based on socio-economic factors, sanitary factors, and status of households in 11 districts Surabaya that near the beach geographically. The result shows that 64% of TB patient households are food insecure than the left are food secure. More than 50% of TB patient households have a good house physically. From the result of mapping between sub-districts and factors, formed five groups consisting of: 1) Benowo and Mulyorejo; 2) Asemrowo, Gunung Anyar, Sukolilo and Pabean Cantikan; 3) Semampir, Krembangan and Bulak; 4) Kenjeran; and 5) Rungkut.

1. Introduction

Food Security is the condition when all people, at all times, have physical, social, and economic access to sufficient, safe, and nutritious food that meets their dietary needs and food preferences for an active and healthy life [1]. A good food security of the households gives a better way to ward off the disease, on the other hand the more fragile food security of a family the more vulnerable to diseases, like Tuberculosis (TB).

Tuberculosis is the type of the disease that spreads directly. The disease caused by bacteria that spreads through the air and has become the third highest cause of death in the world. According to data from the Health Department, the largest TB cases in Indonesia were in East Java Province and Surabaya accounted for the largest number, there are 48379 cases in 2015.

The earliest research had done by the researcher gave a conclusion that in the household with food secure for TB patient depend on the history of the disease, the mass of the house, and smoking habit. Furthermore, food insecure households also depend on the ventilation condition, nutritional status, and population status. This research aims to mapping the sub-districts between the percentage number of household in food security condition and the factors affecting the household of TB patients.

IOP Conf. Series: Materials Science and Engineering **335** (2018) 012053 doi:10.1088/1757-899X/335/1/012053

2. Material and Methods

2.1. *Food Security*

Food Security is the condition when all people, at all times, have physical, social, and economic access to sufficient, safe and nutritious food that meets their dietary needs and food preferences for an active and healthy life [1]. Also, there are four food security indicators used to classify each household [2], those are:

- Food Availability
- Food Stability
- Food Access
- Utilization Food

where, the categories of household food security can be assigned by combining two aspects, between continuity of food availability and quality of food [2], which is presented in Table 1.

Table 1. Classified of Food Security for Each Households

Continuity of Food Availability	Quality of Food	
	Good	Not Good (Bad)
Continu	Food-secure	Food-insecure
Not continous	Food-insecure	Food-insecure

2.2. *Tuberculosis*

Tuberculosis is an infectious disease caused by *Mycobacterium tuberculosis*. This rod-shaped bacterium is firstly discovered by Robert Koch in Berlin, in Germany on 24 March 1882.This kind of bacteria is also an Acid-Fast Bacilli (AFB) smear. The TB bacteria can infect almost all human organs, such as the lungs, brain, kidney, digestive system, bones, lymph nodes, spleen, joints, skin, intestine, urinary tract, and other organs [4].

Based on the organ which is affected, tuberculosis is divided into two groups, those are Pulmonary tuberculosis and Extra-pulmonary tuberculosis. Moreover, in regard to the Acid-Fast Bacilli (AFB), tuberculosis is classified into two [5], those are:

- *Tuberculosis* with (+) AFB:
- *Tuberculosis* with (-) AFB:

Tuberculosis is one of serious diseases that can cause the high rate mortality in the world. There is around 8.6 million cases of tuberculosis in the world where 1.3 million people (about 15%) of the total cases are died. Factors which affecting the occurrence of tuberculosis [6], such as demography factors, environment and sanitation system, the behavior of patient, economy factors, nutritional status, hospital sheet of each patient, and the treatment period of patient.

2.3. *Source of Data and Research Variables*

In this research, researchers use primary data which is collected by conducting survey using a random sampling technique [7] to 162 respondents taken from 1.338 TB patients in eleven sub-districts – Mulyorejo, Pabean Cantikan, Semampir, Krembangan, Bulak, Kenjeran, Sukolilo, Gunung Anyar, Rungkut, Benowo and Asemworo – in Surabaya.

The variables used are :

IOP Conf. Series: Materials Science and Engineering **335** (2018) 012053 doi:10.1088/1757-899X/335/1/012053

Table 2. The Variables of Food Security

X_1: Respondent status	X_7: Source of medical expense
X_2: Education background	X_8: Kind of primary food
X_3: Long suffered from TB/month	X_9: Eating frequency per day
X_4: Gender	X_{10}:The ownership status of paddy/fishing grounds/ship
X_5: Kind of TB suffered	X_{11}: Access of getting food
X_6: Number of people suffering TBC in family	X_{12}: Protein consumed

Table 3. Continuous

X_{13}: House ownership status	X_{20}: Place to defecate (frequently)
X_{14}: Type of roof	X_{21}: Septic tank distance to the toilet (m)
X_{15}: Type of wall	X_{22}: Source of water consumed
X_{16}: Type of floor	X_{23}: Electrical source
X_{17}: Building area	X_{24}: Landfills
X_{18}: Ventilation width	X_{25}: Sewerage system
X_{19}: Toilet ownership status	

The factors that affecting household's food security of Tuberculosis patient shown in the following table.

Table 4. The Factors that Affecting Households Food Security of Tuberculosis Patient

X_{26}: Percent number of household educated	X_{32}: Percent number of households have a toilet
X_{27}: Percent number of household With the head of the household have regular work	X_{33}: Percent number of household with an Electrical source from PLN
X_{28}: Percent number of household with school age children	X_{34}: Percent number of the households have Landfill
X_{29}: Percent number of household with toddlers	
X_{30} : Percent number of households with quite good ventilation	X_{35}: Percent number of households with food secure
X_{31} : Percent number Households with The density of household members sufficient	X_{36}: Percent number of households with food insecure

2.4. Biplot analysis

Biplot analysis is a method that describing the vector rows and colums simultaneously, manifested in the graphic formats two dimension. The variables /atributes and the objects are shown in the same graph. This analysis was first introduced by Gabriel in 1971.

Suppose a matrix ${}_n\mathbf{Y}_p$ is a matrix data and ${}_n\mathbf{X}_p$ is A matrix of data that has been corrected with their average scores, $\mathbf{X} = \mathbf{Y} - (\mathbf{JY})/n$, Wich ${}_n\mathbf{J}_n$ as a matrix with the elements of one. With Singular Value the composition [8] :

$${}_n\mathbf{Y}_p = {}_n\mathbf{U}_r\ {}_r\mathbf{D}_r\ {}_p\mathbf{V}_r^T \quad (1)$$

where **U** dan **V** are matrix with orthonormal colom ($\mathbf{U}^T\mathbf{U}=\mathbf{V}^T\mathbf{V}={}_r\mathbf{I}_r$) and **D** is a diagonal matrix with the diagonal element are eigen values. $\mathbf{U}=\mathbf{D}_w^{-1/2}\mathbf{U}$ dan $\mathbf{V}=\mathbf{D}_q^{-1/2}$

the equation (1) can be written as $\hat{Y} = \tilde{U}D_\beta\tilde{V}^T$ (2)

then $\mathbf{G} = I^{1/2}\mathbf{U}\mathbf{D}_\beta$ and $\mathbf{H} = J^{1/2}\mathbf{V}$.

The information that can be reach by Biplot analysis are:

- Vector long shown the variations of a variable.
- The angle between vectors show a correlation between both, The smaller angles give more highly correlation of both

3. Result and Discussion

3.1. *The characteristic of the Households TB Patient*

The characteristic of the households covering the status of food security, the percentage of the households food insecurity every sub-districts, the physical condition of buildings ownership of the toilet.

The status of food Security shown in Figure 1, Indicate that there has been 64 percent of the households are insecure and only 36% of the households in food secure.

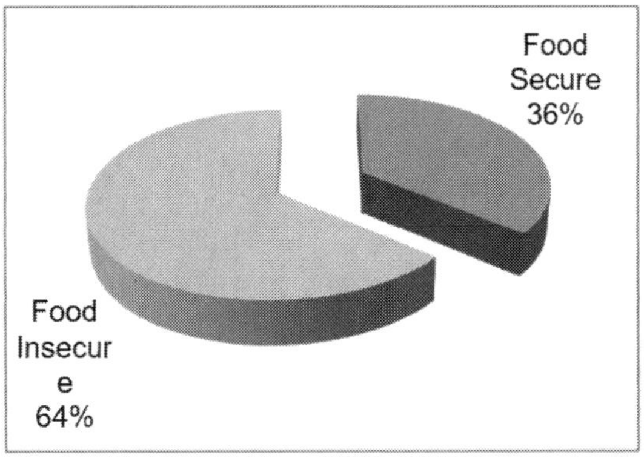

Figure 1. The Status of Food Security

The percentage of total households food insecure in each subdistricts reveals in Figure 2. Subdistrict with the highest of total number households food insecurity is Gunung Anyar (12 %) compared to others, while the lowest is Benowo (3%) and also Mulyorejo about 6 percent of households food insecure.

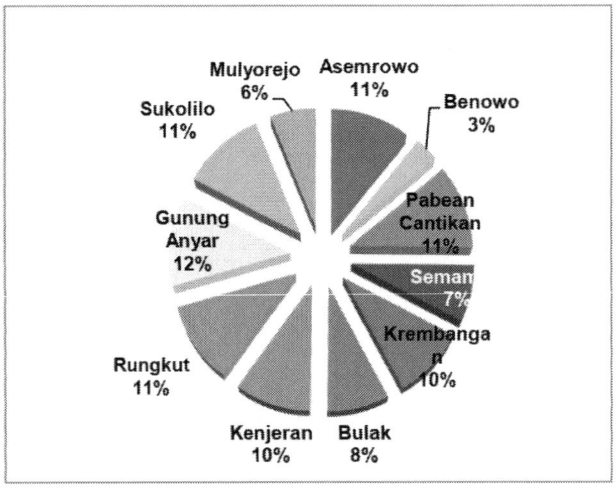

Figure 2. The Percentage of Total Households Food Insecure in Each Subdistricts

The Physical Condition of the Houses shown through the type of floor, a roof and walls. As shown in Figure 3. There are still had an asbestos roof, dirt floor and woods as a wall, although the percentage were the small ones.

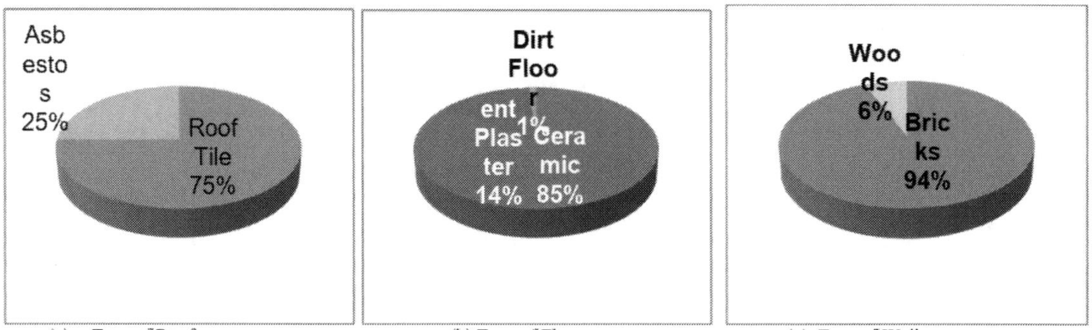

(a) Type of Roof (b) Type of Floor (c) Type of Wall

Figure 3. The Physical Condition of the Houses

Figure 4. shown of the ownership of the toilet, Eighty seven percents of the house have the privy and only thirteen percents of them use a public property as their toilet.

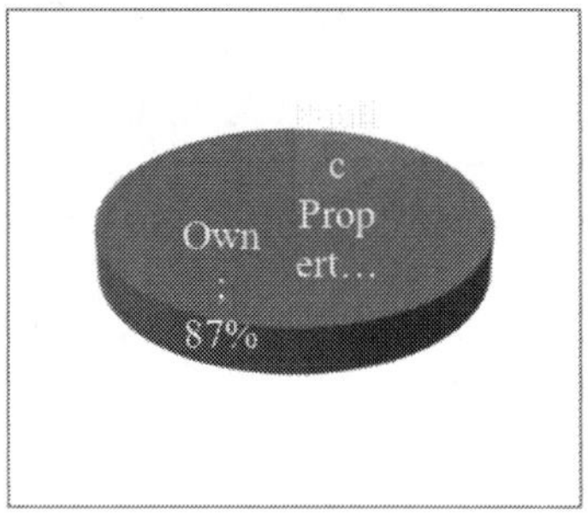

Figure 4. Ownership of the Toilet

The condition of household food security and the factors that affect describe along with subdistricts in a single picture as shown in Figure 5. The subdistricts mapped near by the factor of, means they correlated by each others.

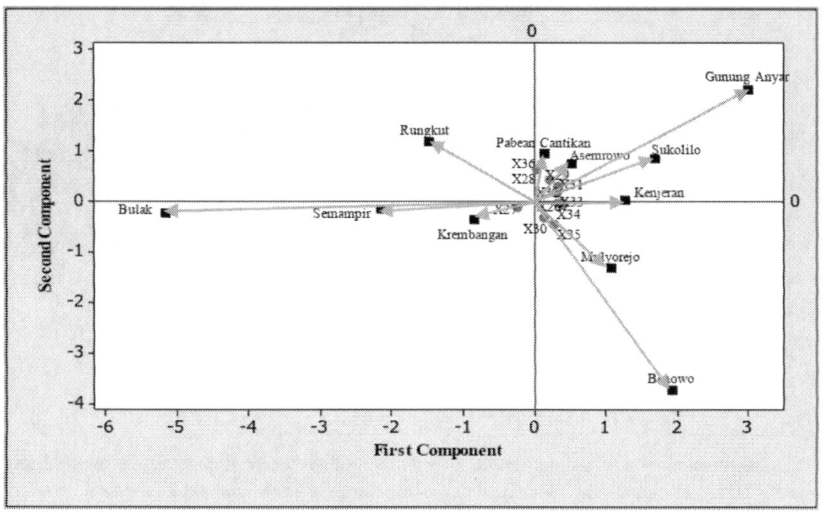

Figure 5. The Mapping of The Factors That Affecting Tuberculosis Patient Households food security Among Subdistrics

Comparing Figure 5 and Table 5., there are several groups can be made depend on the factors that close to subdistrict.

Table 5. The Angle Between Factor And Location

Factors / Location	x_{26}	x_{27}	x_{28}	x_{29}	x_{30}	x_{31}	x_{32}	X_{33}	x_{34}	x_{35}	x_{36}
ASEMROWO	57,79°	164,02°	2,89°	11,31°	99,67°	22,52°	53,77°	57,52°	58,33°	95,81°	10,75°
BENOWO	39,54°	66,70°	100,22°	86,02°	2,34°	74,80°	43,55°	39,81°	38,99°	1,52°	108,08°
PABEAN CANT	68,05°	150,96°	7,37°	21,58°	109,94°	32,78°	64,04°	67,79°	68,60°	106,08°	0,49°
SEMAMPIR	164,10°	23,11°	135,22°	144,91°	122,22°	160,64°	168,11°	168,88°	168,07°	126,07°	127,36°
KREMBANGAN	145,07°	4,08°	154,25°	144,91°	103,19°	179,66°	149,09°	145,34°	144,53°	107,06°	146,39°
BULAK	166,04°	25,05°	133,28°	144,91°	124,16°	158,70°	149,09°	145,34°	144,53°	128,01°	125,42°
RUNGKUT	136,75°	117,01°	76,07°	90,28°	178,63°	66,12°	97,37°	101,12°	101,93°	139,41°	68,21°
KENJERAN	11,97°	207,04°	48,71°	34,51°	53,85°	23,29°	7,96°	11,71°	12,52°	49,99°	56,57°
GUNUNG ANYAR	45,24°	186,23°	15,44°	1,24°	87,12°	9,97°	41,22°	44,97°	45,78°	83,26°	23,30°
SUKOLILO	36,69°	177,68°	23,99°	9,78°	78,57°	1,43°	32,67°	36,43°	87,23°	74,71°	31,85°
MULYOREJO	33,06°	107,93°	93,74°	79,54°	8,82°	68,32°	37,08°	33,33°	32,52°	4,96°	101,60°

The subdistricts : Asemrowo, Pabean Cantikan, Gunung Anyar and Sukolilo become a group that close to the household food insecure and the factors household with shcool age children and toddlers, the density of household members sufficient, especially for Sukolilo also close to the factors of the household have a toilet, and electricity source from PLN.

Benowo and Mulyorejo are the subdistricts close to the food secure household and the factor household with quite good ventilation, Mulyorejo also closed to the factors that the households have a toilet, electricity source from PLN and Landfill. Semampir, Krembangan and Bulak as a group close to the factor that the head of household have a regular work. Kenjeran is close to the factors household educated, an enough population density, have toilet, an electrical source from PLN, and have Landfill, and the last is Rungkut as a subdistrict with no factors close to. The results above can be described on a map throughout Surabaya's map as given in Figure 6.

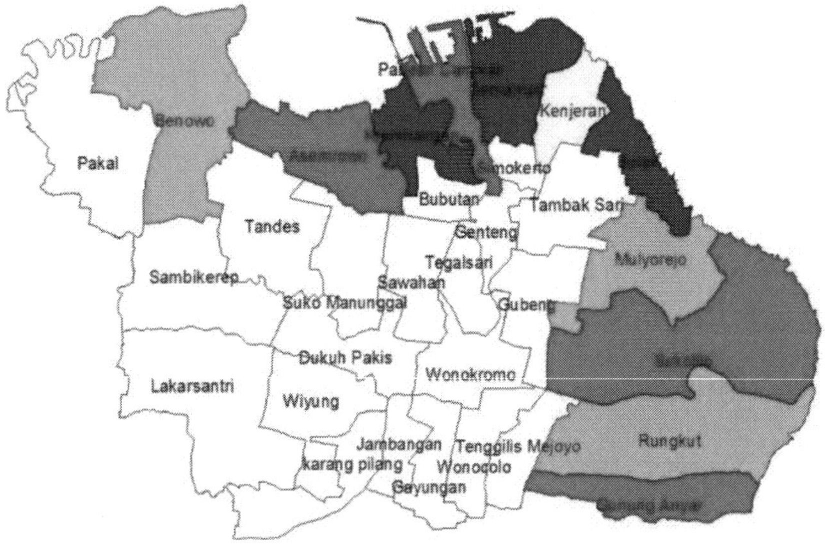

Figure 6. Map of Coastal Subdistricts of Surabaya with Percent of the total number of Households Food Secure is high, Percent of the total number of Household Food Insecure is high, Percent of the total number of Householder Have Regular Jobs, Percent of total number of Householder Educated, The Density Of Household Members Sufficient, Electrical Source from PLN and Have Landfill are high, There are no factors influential dominant

IOP Conf. Series: Materials Science and Engineering **335** (2018) 012053 doi:10.1088/1757-899X/335/1/012053

4. Conclusion

There are still households Food Insecure, with Asbestos roof, cement plaster floor and wood wall in the coastal region of Surabaya. The coastal subdistricts divide into 5 (five) groups those are :

- Asemrowo, Gunung Anyar, Sukolilo and Pabean Cantikan
- Benowo and Mulyorejo
- Semampir, Krembangan and Bulak
- Kenjeran and Rungkut each stands alone

References

[1] (FAO), F. a. (2008). *Food Security Information for Action Practical Guides.* New York: The EC-FAO Food Secury Programme.

[2] Kependudukan-LIPI, P. (2009). *Konsep dan Ukuran Ketahanan Pangan Rumah Tangga di Pedesaan.* Diambil kembali dari http://www.ppk.lipi.go.id/file/publikasi/

[3] Departemen Kesehatan RI. (2009). *Buku Saku Kader Program Penanggulangan TB.* Tangerang: Direktorat Jenderal Pengendalian Penyakit dan Penyehatan Lingkungan Departemen Kesehatan RI.

[4] Anonim_1. (2015). *Penyakit TBC.* download Januari 24, 2015, fromi http://mediacastore.com/tbc/penyakit_tbc.htm

[5] Buntuan Velma, (2014), Gambaran Basil Tahan Asam (Bta) Positif Pada Penderita Diagnosa Klinis Tuberkulosis Paru Di Rumah Sakit Islam Sitti Maryam Manado Periode Januari 2014 S/D Juni 2014, *Jurnal e-Biomedik (eBM),* Volume 2, Nomor 2, Juli

[6] Departemen Kesehatan RI. (2009). *Buku Saku Kader Program Penanggulangan TB.* Tangerang: Direktorat Jenderal Pengendalian Penyakit dan Penyehatan Lingkungan Departemen Kesehatan RI.

[7] Scheaffer R.L., M. W. (2012). *Elementary Survey Sampling, 7th ed.* Boston: Brooks/Cole, Cengage Learning.

[8] Greenacre, M. (2010). *Biplots in Practice.* Madrid: BBVA Fondation.

IOP Conf. Series: Materials Science and Engineering **335** (2018) 012054 doi:10.1088/1757-899X/335/1/012054

Modelling Status Food Security Households Disease Sufferers Pulmonary Tuberculosis Uses the Method Regression Logistics Binary

S P Wulandari*, M Salamah, A F D Rositawati

Institut Teknologi Sepuluh Nopember, Jl. Arief Rahman Hakim, Surabaya 60111 Indonesia

*sripingitwulandari@gmail.com

Abstract. Food security is the condition where the food fulfilment is managed well for the country till the individual. Indonesia is one of the country which has the commitment to create the food security becomes main priority. However, the food necessity becomes common thing means that it doesn't care about nutrient standard and the health condition of family member, so in the fulfilment of food necessity also has to consider the disease suffered by the family member, one of them is pulmonary tuberculosa. From that reasons, this research is conducted to know the factors which influence on household food security status which suffered from pulmonary tuberculosis in the coastal area of Surabaya by using binary logistic regression method. The analysis result by using binary logistic regression shows that the variables wife latest education, house density and spacious house ventilation significantly affect on household food security status which suffered from pulmonary tuberculosis in the coastal area of Surabaya, where the wife education level is University/equivalent, the house density is eligible or 8 m2/person and spacious house ventilation 10% of the floor area has the opportunity to become food secure households amounted to 0.911089. While the chance of becoming food insecure households amounted to 0.088911. The model household food security status which suffered from pulmonary tuberculosis in the coastal area of Surabaya has been conformable, and the overall percentages of those classifications are at 71.8%.

1. Introduction

Compliance toward the need of food is one of the basic components in the development of human resources. Considering the importance of food fulfillment, so every country will prioritize the development of food security as a foundation for the development of other sectors (Arumsari, 2007). Indonesia is one country that is committed to achieve food security as a main priority, as stated in article 34 of the 1945 Constitution (UUD 1945) which states that the State is responsible to fulfill their basic needs, including food. Indonesia also supports the initiation of world Food and Agriculture Organization (FAO) in an effort to eradicate hunger through zero hunger (WFP, 2012).

Adverse effects of hunger and food insecurity is the creation of a country's economic instability. Therefore, in an effort to prevent hunger and food insecurity, then the government must pay attention to the problem of food security in Indonesia. The food needs should also be oriented towards achieving the Millennium Development Goals (MDGs). Agenda to achieve the MDGs consist of eight points, five of them are in health aspect. One of them is the health sector to achieve the MDGs agenda is the fight against tuberculosis (TB) (Dinkes, 2014).

TB disease is included in a chronic infectious disease. WHO stated that Indonesia is the country with a high burden of TB. Number of patients who suffered from TB in Indonesia ranks fourth after India, China and South Africa are almost 700 thousand cases with a mortality rate 25 per 100 thousand inhabitants. TB is the leading cause of death among infectious diseases and is the third of the 10 highest killer disease in Indonesia that caused 100,000 deaths annually (Sarwani, Nurlaela, & Zahrotul, 2012).

East Java Province is one of the provinces in Indonesia with the incidence of TB has increased from year to year. In East Java, Surabaya is a city that ranks first in the spread of TB. Surabaya citizens vulnerable to the bacteria Mycobacterium tuberculosis. At least 4493 citizens of Surabaya infected with the bacteria Mycobacterium Tuberculosis (Dinkes, 2015).

Research on the factors that affect household food security have been carried out. But no studies that examine the factors that affect food security at the household which suffered from pulmonary tuberculosis in the coastal area of Surabaya. Therefore, this study was conducted to analyze the factors that affect the food security status of households with regard disease pulmonary tuberculosis suffered by members of the household.

2. Literature Review

2.1. Binary Logistic Regression

Binary logistic regression is a method of data analysis used to find the relationship between the response variables are binary or dichotomous with the predictor variables that are polichotomous (Hosmer, Lemeshow, & Sturdivant, 2013). The response variable (y) consists of two categories: "success" and "failure" are denoted by y = 1 for success and y = 0 to fail.

2.1.1. Parameters Significance Test. Parameter significance test conducted to determine whether the predictor variables in the model significantly affect the response variable. Significance parameters test consists of simultaneous test and partial test.

Simultaneous Test. Simultaneous test also called model test Chi-square, made an effort to examine the role of the predictor variables in the model together. Hypothesis significance testing coefficient parameters simultaneously are as follows.

Hypothesis:

H_0: $\beta_1 = \beta_2 = ... = \beta_p = 0$

H_1: At least one $\beta_j \neq 0$; for j = 1, 2, ..., p

Statistical test:

$$G = -2\ln\frac{\left(\frac{n_1}{n}\right)^{n_1}\left(\frac{n_0}{n}\right)^{n_0}}{\prod_{i=1}^{n}\hat{\pi}_i^{\,y_i}\left(1-\hat{\pi}_i\right)^{1-y_i}} \qquad (1)$$

The statistical test G follows the Chi-square distribution with degrees of freedom (df) = p (number of predictor variables), so that it will be obtained a decision reject H_0 if the value of G greater than the value $\chi^2_{(\alpha,p)}$ or a p-value less than α (Hosmer, Lemeshow, & Sturdivant, 2013).

Partial Test. After test parameter coefficient β simultaneously on the response variable, then doing the significance test β partially on the response variable. Partial testing hypothesis are as follows.

Hypothesis:

H_0: $\beta_j = 0$

H_1: $\beta_j \neq 0$, For j = 1,2, ..., p

Statistical test:

$$W = \frac{\hat{\beta}_j}{SE\left(\hat{\beta}_j\right)} \tag{2}$$

The statistical test *W* follows a normal distribution, so that it will be obtained a decision reject H_0 if the value of | W | greater than the value $Z_{\alpha/2}$ or a p-value less than α (Hosmer, Lemeshow, & Sturdivant, 2013).

2.1.2 The Odds Ratio Value. The odds ratio is defined as the ratio of the value of logistic regression to x = 1 toward the value of logistic regression to x = 0 (Hosmer & Lemeshow, 2000), which can be formulated in the following equation.

$$OR = \frac{\frac{\pi(1)}{[1-\pi(1)]}}{\frac{\pi(0)}{[1-\pi(0)]}} = \frac{\pi(1)[1-\pi(0)]}{\pi(0)[1-\pi(1)]} = \frac{e^{\beta_0+\beta_1}}{e^{\beta_0}} = e^{\beta_1} \tag{3}$$

2.1.3 Conformance Model Test. Conformance model test is used to determine whether there is a significant difference between the results of observations with the possible results of model predictions. Hypothesis testing the conformance of the model are as follows.

Hypothesis:

H_0: Model appropriate

H_1: The model does not appropriate

Statistical test:

$$\hat{C} = \sum_{k=1}^{g} \frac{(O_k - n'_k \bar{\pi}_k)^2}{n'_k \bar{\pi}_k (1-\bar{\pi}_k)} \tag{4}$$

The statistical test $\hat{C}$ follows the Chi-square distribution with degrees of freedom (df) = g-2, so that it will be obtained a decision reject H_0 if the value of $\hat{C}$ greater than the value $\chi^2_{(\alpha,g-2)}$ or a p-value less than α (Hosmer & Lemeshow, 2000).

2.2. Classification Accuracy Evaluation

Classification accuracy evaluation is an evaluation to see opportunities for misclassification based on the criteria or measures used. The measure used is Accuracy Total (Johnson & Winchern, 2007). Here is a classification table to get the value Accuracy Total.

Table 1. Table Classification.

		Predicted Membership	
		Positive Prediction	Negative Prediction
Actual Membership	Positive Class	True Positive (TP)	False Negative (FN)
	Negative Class	False Positive (FP)	True Negative (TN)

$$\textit{Apparent Error Rate}\ (\text{APER}) = \frac{FP+FN}{TP+TN+FP+FN} \times 100\% \tag{5}$$

$$\textit{Accuracy Total} = 1 - \text{APER} \tag{6}$$

3. Research Methodology

3.1. Data Source
The data used in this research is secondary data and primary data. Secondary data is data on the number and address of pulmonary tuberculosis patients who conduct examination in the clinic coastal region of Surabaya, in January till December 2015. The primary data is the information obtained directly through a survey of pulmonary tuberculosis patients who are in the coastal city of Surabaya.

3.2. Research Variables
The variables used in this study consist of the response variable (Y) and the predictor variables (X). The explanation for each of those variables is described as follows.

3.2.1. Response Variable. The response variable (Y) in this study is the household food security status which suffered from pulmonary tuberculosis in the coastal area of Surabaya.

Table 2. Response Variable.

Variable	Explanation	Category
Y	Food Security Status	0 = Insecure Food
		1 = Secure Food

Determination of household food security status is measured based on the following four indicators.

Table 3. Indicators Determining Food Security Status

Indicators	Explanation	Category
Sufficiency of Food Availability	Rice Stock for a Month	0 = < 20 days
		1 = ≥ 20 days
Food Stability	Mealtime Frequency Per Day	0 = < 3 times
		1 = ≥ 3 times
Food Accessibility	Market Location	0 = > 2 km
		1 = ≤ 2 km
	Number of Household Members	0 = ≥ 7 people
		1 = < 7 people
	Head of Household Education Level	0 = No School
		1 = Minimum Elementary School
	How To Obtain Food Staple	0 = Debt
		1 = Not Debt
The Quality or Food Safety	Protein Consumed	0 = Vegetable only or not at all
		1 = Animal and vegetable or animal only

3.2.2. Predictor Variables. The predictor variables (X) used in this study is as follows.

Table 4. Predictor Variables.

Variables	Explanation	Variables	Explanation
X_1	Head of Household Age	X_{12}	Home Ownership
X_2	Wife Age	X_{13}	House Density
X_3	The Head of Household Latest Education	X_{14}	The Widest Roof Type
X_4	The Wife Latest Education	X_{15}	The Widest Wall Type
X_5	The Head of Household Jobs	X_{16}	The Widest Floor Type
X_6	Employment Status of Wife	X_{17}	Spacious House Ventilation

Variables	Explanation	Variables	Explanation
X_7	Number of Household Members	X_{18}	Toilet Ownership
X_8	Number of School Children	X_{19}	Clean Water Source
X_9	Number of Children Under Five	X_{20}	Trash Place
X_{10}	Earning per Month	X_{21}	Waste Water Disposal
X_{11}	Spending per Month	X_{22}	Electricity Source

4. Analysis Result

4.1. Binary Logistic Regression Analysis
Binary logistic regression analysis aims to model and find the variables that significantly influence household food security status which suffered from pulmonary tuberculosis in the coastal area of Surabaya. Here are the steps binary logistic regression analysis in this study.

4.1.1. Parameter Significance of the Suspected Factors Affecting Household Food Security Status. The significance of the parameters is performed to determine which variables are significant effect through simultaneous test and partial test.

Test Simultaneously against Factors Suspected of Affecting the Status of Food Security Household. Simultaneous test conducted to determine the effect of predictor variables on household food security status simultaneously by using Backward method of inserting all the variables to be obtained only from the best model variables that effect. The best model that was stopped at the last iteration is step 20. Here are the simultaneous test results obtained.

Hypothesis:

H_0: $\beta_1 = \beta_2 = ... = \beta_{22} = 0$

H_1: At least one $\beta_j \neq 0$ j = 1,2, ..., 22

Significant level: $\alpha = 0.1$

Statistical test: χ^2

Rejection zone: Reject H_0 if $\chi^2 > \chi^2_{(\alpha,df)}$ or P-value < α

Here are the simultaneous test results to include all predictor variables.

Table 5. Simultaneous Test Results by Inserting Entire Predictors Variables

χ^2	df	$\chi^2_{(0.1,7)}$	P-value
59.317	7	12.017	0,000

Table 5 shows that at least one of the predictor variables is significantly effect on household food security status which suffered from pulmonary tuberculosis in the coastal area of Surabaya.

Partial Test toward Suspected Factors Affecting Household Food Security Status. Simultaneous test has been done to the conclusion that at least one of predictor variables that significantly effect on the food security status of households. Therefore, the partial test performed to determine which variables were significant with hypothesis used are as follows.

Hypothesis:

H_0: $\beta_j = 0$

H_1: $\beta_j \neq 0$, for j = 1,2, ..., 22

Significant level: $\alpha = 0.1$

Statistical test: χ^2

Rejection zone: Reject H_0 if $\chi^2 > \chi^2_{(\alpha,df)}$ or P-value < α

Table 6. Partial Test Results by Inserting Entire Predictors Variables.

Variables	χ^2	df	$\chi^2_{(\alpha,df)}$	P-value
Wife Education (X4)	9.259	4	7.779	0.055 *
Wife Education (1) (X4)	5.984	1	2.706	0,014 *
Wife Education (2) (X4)	7.097	1	2.706	0,008 *
Wife Education (3) (X4)	4.209	1	2.706	0,040 *
Wife Education (4) (X4)	7.278	1	2.706	0,007 *
Number of Household Members (1) (X7)	0,000	1	2.706	0.998
House Density (1) (X13)	6.783	1	2.706	0,009 *
Spacious House Ventilation (1) (X17)	4,350	1	2.706	0,037 *

*) Significant at α = 10%

Table 6 shows the partial test results on the last iteration (step 20). Based on these results known that the variable wife latest education, house density and spacious house ventilation significantly affect on household food security status which suffered from pulmonary tuberculosis in the coastal area of Surabaya.

4.1.2. Formation of Logistic Regression Binary Model toward Factors Suspected of Affecting the Status of Food Security Household. The method used in this modelling is a method wald by only including three predictor variables which significantly affect on household food security status in accordance with the results obtained in the partial test. The results obtained are as follows.

Hypothesis:

H_0: $\beta_j = 0$

H_1: $\beta_j \neq 0$, for j = 1,2, and 3

Significant level: $\alpha = 0.1$

Statistical test: χ^2

Rejection zone: Reject H_0 if $\chi^2 > \chi^2_{(\alpha,df)}$ or P-value < α

Partial test results of the variables used in the modelling shown by Table 7.

Table 7. Partial Test Results toward Variables Used in Model Formation

Variables	B	χ^2	df	$\chi^2_{(\alpha,df)}$	P-value
Wife Education (X4)		10.656	4	7.779	0,031 *
Wife Education (1) (X4)	3.703	7.143	1	2.706	0,008 *
Wife Education (2) (X4)	2.296	7,545	1	2.706	0,006 *
Wife Education (3) (X4)	1,453	2,434	1	2.706	0.119
Wife Education (4) (X4)	1,924	5.758	1	2.706	0,016 *
House Density (1) (X13)	0,833	3.984	1	2.706	0,046 *
Spacious House Ventilation (1) (X17)	0.761	3,505	1	2.706	0.061 *
Constant	-2.970	14.253	1	2.706	0,000

*) Significant at α = 10%

Table 7 shows the partial test results to the variables used in the formation of a binary logistic regression model. Based on these results known that the variable wife latest education, house density and spacious house ventilation significantly affect on household food security status which suffered from pulmonary tuberculosis in the coastal area of Surabaya.

4.1.3. Logit Models on Household Food Security Status which Suffered from Pulmonary Tuberculosis in the Coastal Area of Surabaya. Logit model formed based on the partial test results are shown in Table 8. Here is a logit model in binary logistic regression analysis on household food security status which suffered from pulmonary tuberculosis in the coastal area of Surabaya.

$$\hat{g}(x) = -2{,}970 + 3{,}703X4_{(1)} + 2{,}296X4_{(2)} + 1{,}453X4_{(3)} + 1{,}924X4_{(4)} + 0{,}833X13_{(1)} + 0{,}761X17_{(1)}$$

Based on the logit model, then the calculation of the value opportunity for the two categories of the response variable. Here is the value opportunities that come in specific categories, such as households with wife education level is University/equivalent, the house density is eligible or $\geq$ 8 m^2/person and spacious house ventilation $\geq$ 10% of the floor area. Households with these categories has the opportunity to become food secure households amounted to 0.911089. While the chance of becoming food insecure households amounted to 0.088911.

4.1.4. Odds Ratio Value of Variables Affecting Household Food Security Status which Suffered from Pulmonary Tuberculosis in the Coastal Area of Surabaya. Odds ratio values obtained from Exp (β). Here is the interpretation of the odds ratio value of each predictor variable that goes into a binary logistic regression model.

Table 8. Odds Ratio Value of Variable Logged in Model

Variables	*Odds Ratio*	Variables	*Odds Ratio*
Wife Education (1) (X4)	40.569	Wife Education (4) (X4)	6.846
Wife Education (2) (X4)	9.933	House Density (1) (X13)	2,299
Wife Education (3) (X4)	4,275	Spacious House Ventilation (1) (X17)	2,141

Table 8 shows that a household with the latest wife education is University/equivalent will tend to be food secure household amounted to 40.569 times compared to household with wife's education level is not school. A household with the latest wife education is Senior High School/equivalent will tend to be food secure household amounted to 9.933 times compared to household with wife's education level is not school. A household with the latest wife education is Junior High School/equivalent will tend to be food secure household amounted to 4.275 times compared to household with wife's education level is not school. While a household with the latest wife education is Elementary School/equivalent will tend to be food secure household amounted to 6.846 times compared to household with wife's education level is not school.

A household with a house density is eligible or $\geq$ 8 m^2/person will tend to be food secure household amounted 2.299 times compared to household with house density is not eligible or <8 m^2/person. A household that has spacious house ventilation $\geq$10% of the floor area will tend to be food secure household amounted 2.141 times compared to household with spacious house ventilation <10% of the floor area.

4.1.5. Model Conformance Test of Household Food Security Status which Suffered from Pulmonary Tuberculosis in the Coastal Area of Surabaya. Here is the result of the model conformance test of

IOP Conf. Series: Materials Science and Engineering **335** (2018) 012054 doi:10.1088/1757-899X/335/1/012054

household food security status which suffered from pulmonary tuberculosis in the coastal area of Surabaya.

Hypothesis:
H_0: Model has been conformable (There is not a significant difference between the results of observations with the possible results of model predictions)
H_1: The model has not been conformable (There is significant difference between the results of observations with the possible results of model predictions)
Significant level: $\alpha = 0.1$
Statistical test: χ^2
Rejection zone: Reject H_0 if $\chi^2 > \chi^2_{(\alpha,df)}$ or P-value < α
The following is a model of conformance test results.

Table 9. Model Conformance Test Results

χ^2	df	$\chi^2_{(0.1,8)}$	P-value
8.360	8	13.362	0.399

Table 9 shows that the model household food security status which suffered from pulmonary tuberculosis in the coastal area of Surabaya has been conformable, or there is not a significant difference between the results of observations with the possible results of model predictions.

4.1.6. Classification Appropriateness of Household Food Security Status which Suffered from Pulmonary Tuberculosis in the Coastal Area of Surabaya. Here is the result of the classification accuracy evaluation of household food security status which suffered from pulmonary tuberculosis in the coastal area of Surabaya.

Table 10. Classification of The Household Food Security Status which Suffered from Pulmonary Tuberculosis in the Coastal Area of Surabaya

Observation		Prediction: Food Security Status		Percentage (%)
		Insecure Food	Secure Food	
Food Security Status	Insecure Food	67	24	73.6
	Secure Food	16	35	68.6
Percentage of Overall				71.8

Table 10 shows that the percentage of classification of insecure food household is at 73.6%, from 91 households there are 67 insecure food households classified correctly as insecure food households, while 24 other households wrongly classified as secure food households. Then the percentage of classification of secure food households is 68.6%, from 51 households there are 35 secure food household classified correctly as secure food household, while 16 other households wrongly classified as insecure food household. The overall percentage of those classifications are at 71.8%, which means that of the 142 households there are 102 households were classified correctly and there are 40 households that are classified incorrectly.

5. Conclusions and Suggestions

The conclusions and suggestions from the analysis results are as follows.

5.1. Conclusions

Based on the analysis result, it can be concluded that variables wife latest education, house density and spacious house ventilation significantly affect on household food security status which suffered from pulmonary tuberculosis in the coastal area of Surabaya. The model household food security status which suffered from pulmonary tuberculosis in the coastal area of Surabaya has been conformable, or there is not a significant difference between the results of observations with the possible results of model predictions. The overall percentage of those classifications are at 71.8%, which means that of the 142 households there are 102 households were classified correctly and there are 40 households that are classified incorrectly.

5.2. Suggestions

Suggestions for key government decision makers are to improve household food security in the coastal area of Surabaya, especially in households which suffered from pulmonary tuberculosis. The efforts to improve household food security can be through the renewal of the compulsory education program in Indonesia, as well as to educate the society, especially for women to pursue education levels as high as possible, as the analysis shows that wife's education level significant effect on household food security status. In addition, households are advised to occupy a dwelling house with a density that qualify are$\geq$ 8 m^2/person and are advised to make the spacious house ventilation $\geq$10% of the floor area, because the two variables are also variables that have a significant impact to household food security.

Reference

[1] Arumsari, W. (2007). Peran Wanita Tani dalam Mewujudkan Ketahanan Pangan pada Tingkat Rumah Tangga di Kabupaten Sleman, Daerah Istimewa Yogyakarta. *Jurnal Ekonomi Pembangunan, 13.*

[2] Dinkes. (2014). Profil Kesehatan Provinsi Jawa Timur Tahun 2013. Surabaya: Dinas Kesehatan Provinsi Jawa Timur.

[3] Dinkes. (2015). Profil Kesehatan Surabaya Tahun 2014. Surabaya: Dinas Kesehatan Kota Surabaya.

[4] Hosmer, D., & Lemeshow, S. (2000). Applied Logistic Regression Second Edition. New Jersey: John Willey & Sons.

[5] Hosmer, D., Lemeshow, S., & Sturdivant, X. (2013). Applied Regression (3rd Ed.). New Jersey: John Willey & Sons.

[6] Johnson, R., & Winchern, D. (2007). Applied Multivariate Statistical Analysis (6th Ed.). New Jersey: Pearson Prentice Hall.

[7] Sarwani, D., Nurlaela, S., & Zahrotul, I. (2012). Faktor Risiko Multidrug Resistant Tuberculosis (MDR-TB). Jurnal Kesehatan Masyarakat, 60-66, 1858-1196.

[8] WFP. (2012). *WFP Indonesia 2012 in Review.* Jakarta: WFP Indonesia.

The Influence of ENSO to the Rainfall Variability in North Sumatra Province

H Irwandi[1*], **N Pusparini**[1], **J Y Ariantono**[1], **R Kurniawan**[2], **C A Tari**[1], **A Sudrajat**[1]

[1]Climatological Station of Deli Serdang, 17 Meteorologi Raya Road, Sampali Medan, North Sumatra,
[2]Research and Development Centre of Indonesian Agency for Meteorology, Climatology, and Geophysics (BMKG)

*hendrisampali@gmail.com

Abstract. The *El Niño* Southern Oscillation (ENSO) is a global phenomenon that affects the variability of rainfall in North Sumatra. The influence of ENSO will be different for each region. This review will analyse the influence of ENSO activity on seasonal and annual rainfall variability. In this research, North Sumatra Province will be divided into 4 (four) regions based on topographical conditions, such as: East Coast (EC), East Slope (ES), Mountains (MT), and West Coast (WC). The method used was statistical and descriptive analysis. Data used in this research were rainfall data from 15 stations / climate observation posts which spread in North Sumatera region and also anomaly data of Nino 3.4 region from period 1981-2016. The results showed that the active *El Niño* had an effect on the decreasing the rainfall during the period of DJF, JJA and SON in East Coast, East Slope, and Mountains with the decreasing of average percentage of annual rainfall up to 7%. On the contrary, the active La Niña had an effect on the addition of rainfall during the period DJF and JJA in the East Coast and Mountains with the increasing of average percentage of annual rainfall up to 6%.

1. Introduction

North Sumatera Province is located in the western part of Indonesia, located on 1° -4° N and 98° -100° E. The province is adjacent to the waters and land areas, which are: Aceh Province in the north, Malaysia State and the Malacca Strait in the, Riau and West Sumatra Province in the south, and Indian Ocean in the west. Based on geographical location, topography, and natural condition, North Sumatera Province is divided into 3 groups including West Coast (WC), Mountains (MT), and East Coast (EC)[1].

Climate that occurs in a region will influence the other areas as well as *El Niño*-Southern Oscillation (ENSO) or well known as *El Niño* and La Niña. ENSO is a combined phenomenon of atmospheric and ocean interactions that causes annual climate variations worldwide [2, 3, 4]. The maximum influence of ENSO occurs in almost all parts of Indonesia except the northern part of Sumatera and some parts of Kalimantan5. Several previous studies have stated that ENSO activity has impact on rainfall variability in Indonesian region [4, 6, 7, 8, 9]. Furthermore, in another study there was a decrease in rainfall in North Sumatra region during *El Niño* event [10]. This was the reason why we conducted the research of ENSO's influence on seasonal and annual rainfall in North Sumatra region during 1981-2016.

During *El Niño* event, sea surface temperature in the Eastern Equatorial Pacific becomes warmer than its normal conditions. Meanwhile, during La Niña event, sea surface temperature in the Eastern

IOP Conf. Series: Materials Science and Engineering **335** (2018) 012055 doi:10.1088/1757-899X/335/1/012055

Equatorial Pacific is colder than its normal condition (Figure 1). This both phenomena occur as a result of the wind circulation system, the movement of the atmosphere above the ocean, and the surface pressure over the ocean along the ocean in the equatorial zone. *El Niño* event make the rainy season come late and reduce the total rainfall. On the contrary, La Niña is one of the causes of increasing rainfall in Indonesia[12].

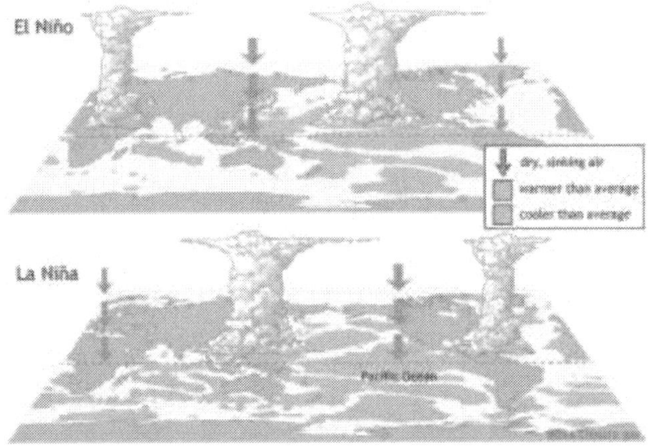

Figure 1. *El Niño* and La Niña Conditions 13

The ENSO index was obtained by calculating the mean sea surface anomaly level in the Niño 3.4 region (190° -240° E and 5° S - 5°N) based on the *El Niño* definition[14].

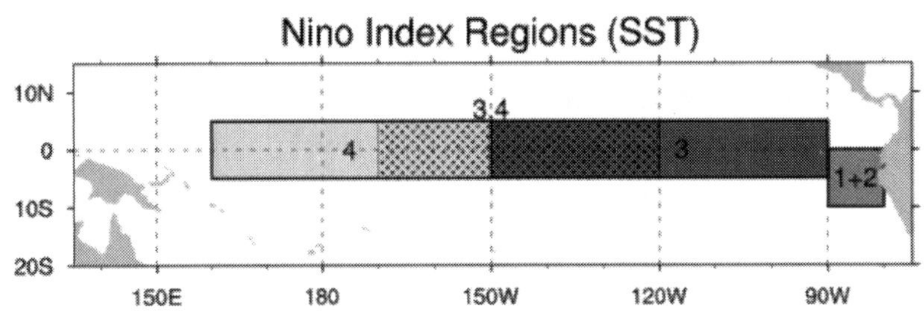

Figure 2. Nino location in Pasific Ocean[15]

The results of ocean surface analysis of the Pacific Ocean region, especially in Nino 3.4 region, indicated a positive anomaly occurred during the *El Niño* years (1982, 1983, 1986, 1987, 1991, 1992, 1997 and 1998) with the highest anomalies occurred in 1983, 1977 and 2015 with sea level anomaly > 2° C (Figure 3), whereas during La Niña (1988, 1989, 1999, 2000, 2010, and 2011), the lowest negative anomalies occurred in 1988 and 1999 (Figure 4).

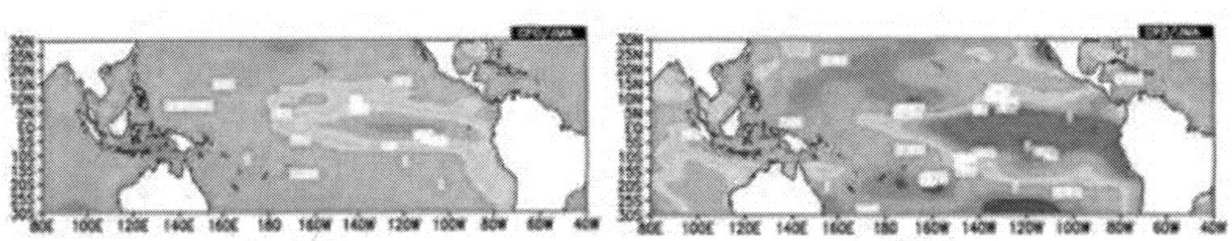

a. 1982 b. 1983

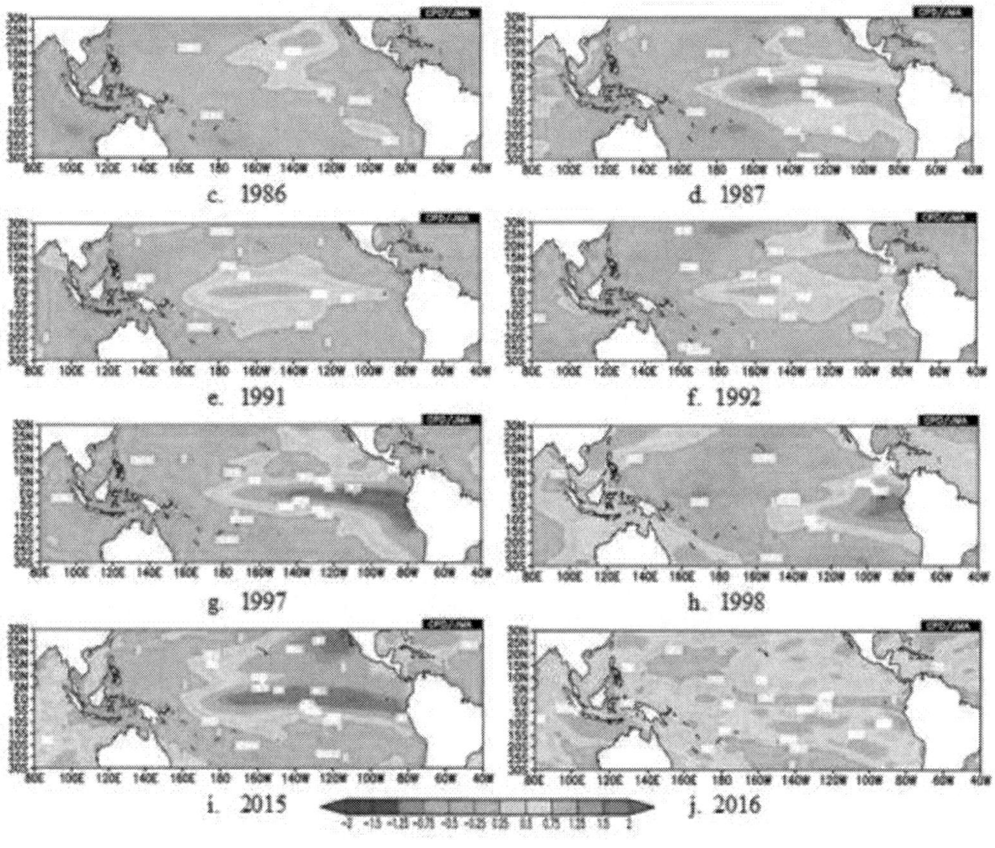

Figure 3. Sea surface temperature anomaly in Pasific Ocean and Indonesian waters during *El Niño*

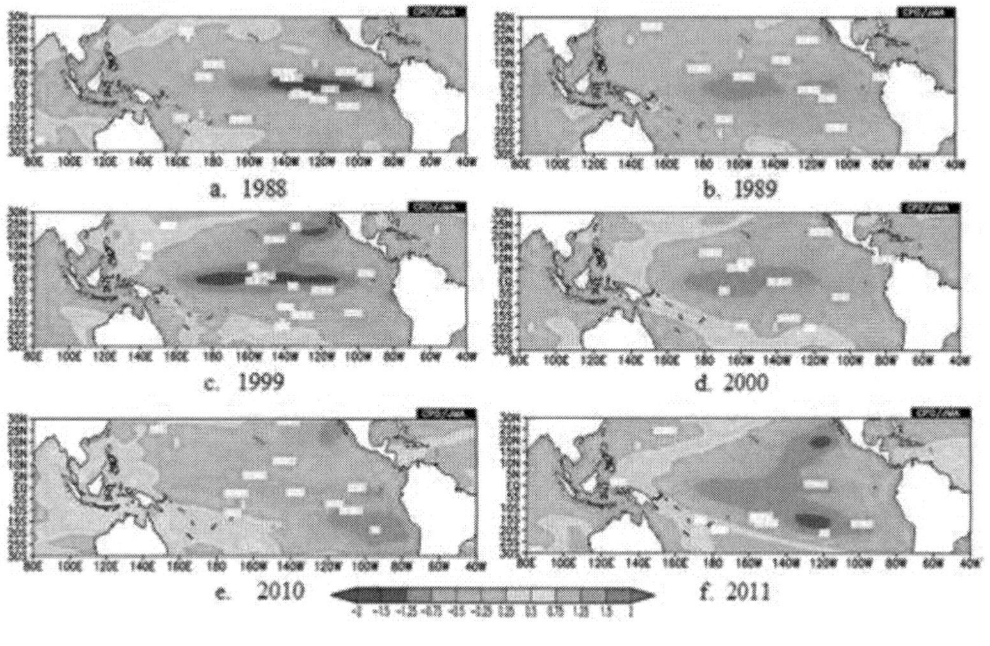

Figure 4. Sea surface temperature anomaly in Pasific Ocean and Indonesian waters during *La Niña*

IOP Conf. Series: Materials Science and Engineering **335** (2018) 012055 doi:10.1088/1757-899X/335/1/012055

2. Data and Methods

2.1. Data

The data used were observation data of monthly and annual rainfall data from 15 stations / climate observation posts with period of data from 1981-2016. Meanwhile ENSO used in this research was sea surface temperature anomaly data in Nino region 3.4. North Sumatra Province has distinctive climatic characteristics, flanked by two waters which are Indian Ocean and the Malacca Strait, as well as the Bukit Barisan Mountains that stretch from North to South of the province. Therefore, we divided North Sumatra region into 4 (four) topographic areas: East Coast (EC), East Slope (ES), Mountains (MT), and West Coast (WC) (Figure 5 and Table 1).

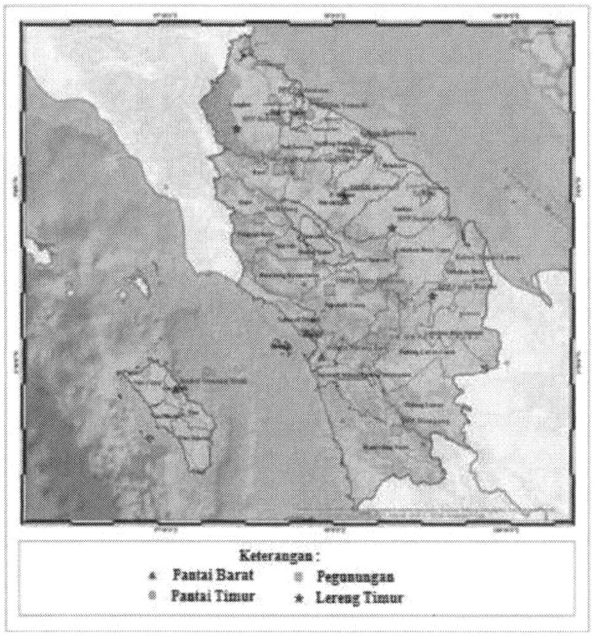

Figure 5. Research Location in North Sumatera Region

Table 1. Stations / Climate Observation Post as the research locations

No	Station	Latitude	Longitude	Period of time	Region	Information
1	Pardamean	3.75° N	98.44° E	1983-2016	East Coast	EC
2	Sampali	3.62° N	98.78° E	1981-2016	East Coast	EC
3	Rambutan	3.38° N	99.16° E	1981-2016	East Coast	EC
4	Negeri Lama	2.32° N	100.0° E	1985-2016	East Coast	EC
5	Bahorok	3.50° N	98.13° E	1982-2016	East Slope	ES
6	Marihat	2.91° N	99.08° E	1981-2016	East Slope	ES
7	Bandar Pulau	2.65° N	99.50° E	1981-2016	East Slope	ES
8	Ujung Bandar	2.06° N	99.85° E	1981-2016	East Slope	ES
9	Sidikalang	2.73° N	98.33° E	1984-2016	Mountains	MT
10	Kuta Gadung	3.15° N	98.50° E	1981-2016	Mountains	MT
11	Gabehutaraja	2.11° N	98.95° E	1981-2016	Mountains	MT
12	Arse	1.70° N	99.32° E	1985-2016	Mountains	MT
13	Mompang	0.91° N	99.53° E	1981-2016	Mountains	MT
14	Pinangsori	1.55° N	98.88° E	1981-2016	West Coast	WC
15	Gunung Sitoli	1.27° N	97.58° E	1982-2016	West Coast	WC

2.2. Methods

The method used in this research was statistical and descriptive analysis. Initial stage of the research was collecting rainfall data from climate stations / posts that exist in North Sumatra region. The rainfall data for each study area will be analyzed and linked to data of ENSO incident years, both *El Niño* (1982, 1983, 1986, 1987, 1991, 1992, 1997 and 1998) and La Niña (1988, 1989, 1999, 2000, 2010, and 2011).

This was conducted to determine the impact and influence of ENSO in North Sumatra region. The normal data used for comparison was during 1981 - 2010. ENSO conditions will be linked with the season forward or backward as well as its impact on annual and seasonal rainfall during: December-January-February (DJF), March-April-May (MAM), June-July-August (JJA), and September-October-November (SON) which were averaged by year of ENSO events and then spatially described.

3. Results and Discussion

3.1. The Influences of ENSO on Season in North Sumatera

Rainfall analysis in North Sumatra showed that the West Coast region (yellow line) and East Slope region (brown line) had a wet characteristic season along the year. This was characterized by rainfall per decade reached more than 50 mm, meanwhile the East Coast region (dark blue line) and Mountains region (light green line) had a distinct seasonal characteristics between the Dry Season (DS) and Rainy Season (RS) with 2 (two) Dry Season periods and 2 (two) Rainy Season periods (Figure 6).

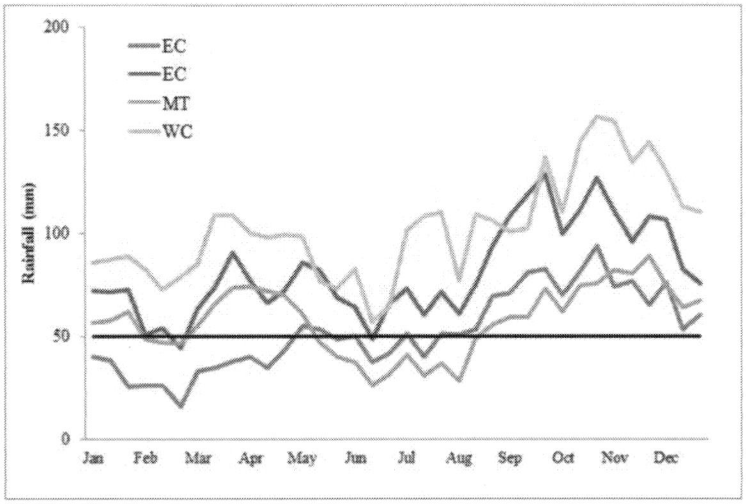

Figure 6. Normal Rainfall in North Sumatera Region

Rainfall data during El Nino and *La Niña* events was compared with the normal rainfall data of 1981 - 2010. Based on the analysis (Table 2), it showed that ENSO had no significant effect in the East Slope and West Coast region. This was due to the area of East Slope and West Coast in North Sumatra throughout the year has a rainy season criteria which was indicated by the absence of rainfall less than 50 mm throughout the year[15].

During *El Niño*, the East Coast region experienced a long Dry Season (DS) up to 23 decades, during Jan 1-Aug 2, and the Rainy Season II (RS II) retreated for 3 decades. During La Niña, the period of DS I was shorter 6 decades, DS II was earlier 2 decades and shorter 1 decade, RS I was earlier 6 decades and longer 2 decades, and RS II was advanced 3 decades. During *El Niño* condition, Mountain region (MT) experienced DS I earlier 3 decades, DS II was longer 5 decades (until Oct 1), RS I was retreated 1 decade, and RS II was retreated 5 decades. Meanwhile *La Niña* had no significant effect on RS and DS in Mountain region.

Table 2. Normal condition and the influence of ENSO in North Sumatra Region

No	Region	Condition	DS I	DS II	WS I	WS II
1	EC	**Normal**	**Jan 1 - Apr 3**	**Jun 2-Jul 2**	**May 1-Jun 1**	**Jul 3-Dec 3**
		El Niño	Jan 1-Aug 2	-	-	Aug 3-Dec 3
		La Niña	Jan 3- Feb 3	May 3-Jun 2	Mar 1- Apr 3	Jun 3-Des 3
2	ES	**Normal**	-	-	-	-
		El Niño	Jan 2 - Mar1	-	-	-
		La Niña	-	-	-	-
3	MT	**Normal**	**Feb 1-Feb 3**	**May 2-Aug 2**	**Mar 1-May 1**	**Aug 3-Dec 3**
		El Niño	Jan 1-Mar 1	May 3-Oct 1	Mar 2-May 2	Oct 2- Dec 3
		La Niña	Feb 1-Feb 3	May 1-Aug 2	Mar 1-Apr 3	Aug 3-Dec 3
4	WC	**Normal**	-	-	-	-
		El Niño	-	-	-	-
		La Niña	-	-	-	-

3.2. The Influences of El Niño on the Rainfall in North Sumatera Region

The influence of *El Niño* on rainfall in North Sumatera during the DJF period (Figure 7.a) showed that in general, *El Niño* affected on rainfall reduction between 51-100 mm in some parts of the Central and Southern part of Mountain region, the Northern part of East Slope, and the southern part of East Coast. During MAM period (Figure 7.b), *El Niño* affected on rainfall reduction ranged from 51-100 mm in the East Coast and rainfall reduction ranged from 200-300 mm in the Northern part of East Slope. On the contrary, in the Center and Southern part of Mountain region and the West Coast, rainfall increasing ranged from 100-200 mm. This condition was due to the increasing of sea surface temperature around the Indian Ocean in the western part of Sumatera.

During JJA period (Figure 7.c), in general, *El Niño* gave impact on rainfall reduction ranged from 51-100 mm in the Central part of Mountain region, the Northern part of East Slope, the Northern part of East Coast, and some parts of West Coast. During SON period (Figure 7.d), *El Niño* had an effect on the reduction of rainfall ranged from 51-100 mm in all regions of North Sumatra except the Northern part of East Coast and also the Northern and Southern part of Mountain region.

Based on the results of *El Niño* events analysis that occurred in 1982/1983, 1986/1987, 1991/1992, 1997/1998 and 2015/2016 comparing with annual rainfall data in the same year, there was a decrease in rainfall with average decrease up to 7%. The East Coast region had a decreasing rainfall ranged from 3-13% with an average decrease up to 8%. The East Slope region had a decreasing rainfall ranged from 4-21% with an average decrease up to 11%. The Mountain region had a decreasing rainfall ranged from 3-9% with an average decrease up to 5%. The West Coast region had a decreasing rainfall ranged from 1-2% with an average decrease up to 1.5% (Figure 8).

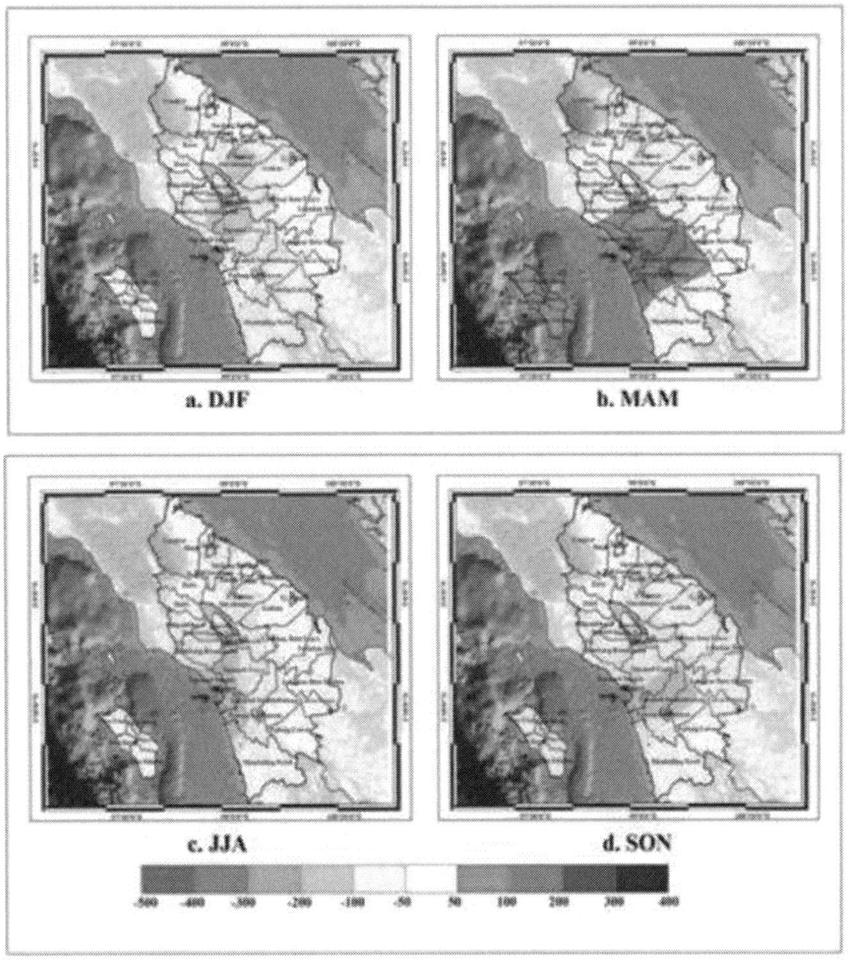

Figure 7. The influence of *El Niño* on the Rainfall

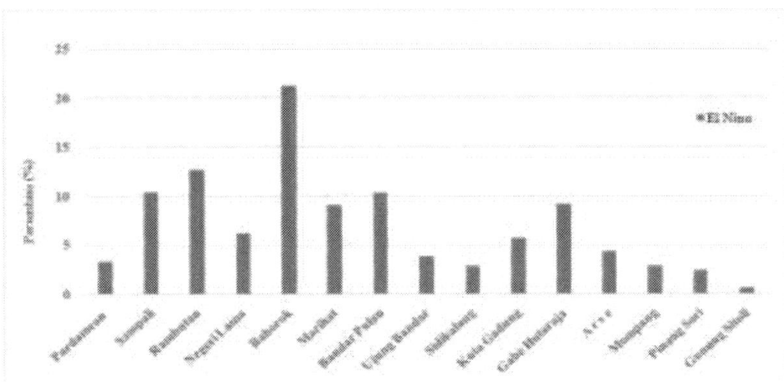

Figure 8. The percentage of rainfall decrease during *El Niño* years

3.3. The Influences of La Niña on the Rainfall in the North Sumatera Region

The influence of *La Niña* on rainfall in North Sumatra during the DJF period (Figure 9.a) showed that La Niña generally affected the addition of rainfall ranged from 51-100 mm in the Northern and Southern part of Mountain region and the Northern part of the East Coast region. The insignificant reduction of rainfall occurred in the Central part of East Slope region. During MAM period (Figure 9.b), La Niña only influenced the addition of rainfall ranged from 101-200 mm in the Northern part of East Slope and the Northern part of East Coast region, whereas in the Northern and Southern part of Mountain region and the Central part of East Slope region had a decreasing rainfall ranged from 51-100 mm.

During JJA period (Figure 9.c), it was showed that generally La Niña affected on the addition of rainfall ranged from 51-100 mm in the Central and Northern part of the East Coast region, the Southern part of Mountain region, and also the Northern and Southern part of East Slope region. During SON period (Figure 9.d), it was showed that generally La Niña influenced the addition of rainfall ranged from 51-100 mm in the Southern part of Mountain region and the East Coast region, meanwhile the Northern part of the East Slope and the Southern part of East Coast had a decreasing rainfall ranged from 51 -100 mm.

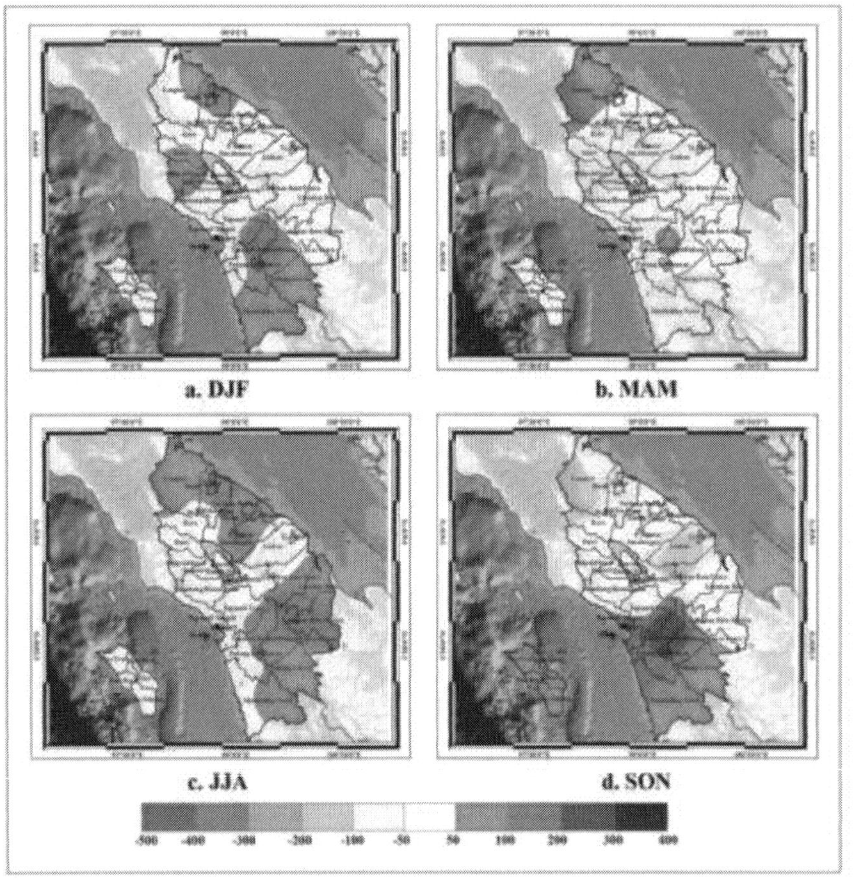

Figure 9. The influence of *La Niña* on the Rainfall

Based on the results of La Niña event analysis that occurred in 1988/1989, 1999/2000 and 2010/2011 comparing with annual rainfall data in the same year, there was an increase in rainfall with an average increase up to 6%. The East Coast region had an increasing rainfall ranged from 5-13% with an average increase up to 9%. The East Slope region had an increasing rainfall ranged from 0-4% with an average increase up to 2%. The Mountain region had an increasing rainfall ranged from 0-22%

with average increase up to 5%. The West Coast region had an increasing rainfall ranged from 5-6% with an average increase up to 5.5% (Figure 10).

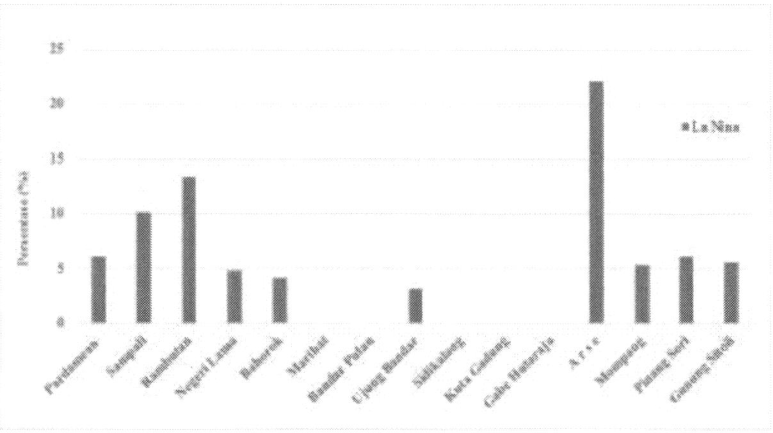

Figure 10. The percentage of rainfall decrease during *La Niña* years

It was generally seen that *El Niño* and *La Niña* were not significantly affected the rainfall in North Sumatera region. This was because besides ENSO, there were several other factors that affect North Sumatera region. The interaction between the atmosphere and the Indian Ocean in the West Coast of Sumatera and East Africa causes Dipole Mode phenomenon[13] or better known as IOD (Indian Ocean Dipole). Dipole Mode value is positive if the western coastal sea surface temperature is cooler and negative if it is warmer than the East Coast African sea surface temperature. Positive Dipole Mode causes reduced rainfall and negative value causes more rainfall in Indonesia. Variations of the IOD index have implications on climate change, especially in the western part to the southern part of Indonesia, from Sumatera, Java, to Nusa Tenggara[13].

In addition, sea surface temperature can also affect rainfall in Indonesia, which one of them is in North Sumatra. Theoretically, when sea level temperatures warm up, it will have an impact on increasing the chance of rain due to an increase in convective clouds growth. Furthermore, if the IOD event is associated with ENSO events, it will further affect the *El Niño* and La Niña events. Another factor that can affect rainfall in North Sumatra is the tropical disturbance in the form of a cyclonic pattern around the South China Sea and also in the Indian Ocean. This cyclonic pattern can increase the rainfall in North Sumatera region by establishing a convergence region.

4. Conclusions

Based on the discussion above, it can be concluded several things as follows:

1. The East Coast (EC) and Mountain (MT) region were regions in North Sumatra with a clear difference of Dry Season (DS) and Rainy Season (RS), meanwhile the West Coast (WC) and the East Slope (ES) tended to be wetter with Rainy Season criteria throughout the year.
2. *El Niño* caused longer Dry Season (DS) and shorter Rainy Season (RS). This condition was significant in the East Coast (EC), East Slope (ES), and Mountain (MT). In general, there was a reduction of rainfall ranged from 51-200 mm during DJF, JJA and SON. On the contrary, during MAM in the West Coast (WC) and the Central-Southern part of the Mountain region (MT), there was an increase of rainfall ranged from 51-200 mm.
3. La Niña caused shorter Dry Season (DS) and longer Rainy Season (RS), which was significant in the East Coast and Mountain region. In general, there was additional rainfall ranged from 51-200 mm during DJF and JJA.

4. There was a decrease in annual rainfall during the active *El Niño* in North Sumatra with an average percentage decrease up to 7%. The significant percentage decrease occurred in the East Coast, East Slopes, and Mountain region with an average decrease up to 8%, 11% and 5 %.
5. There was an increase in annual rainfall during the active La Niña in North Sumatra with an average percentage increase up to 6%. The significant percentage increase occurred in the East Coast and Mountain region with an average of increase up to 9% and 5%.

Acknowledgments
We would like to thank to the Chief of Climatology Station of Deli Serdang, Mr. Klaus J.A. Damanik, ST and all colleagues who gave so much contributions that this research can be conducted.

References

[1] BPS 2017 North Sumatra in Numbers Indonesian Agency for Statistics of North Sumatera Province. https://sumut.bps.go.id/frontend/index.php/publikasi/319 Accessed on 28 Agustus 2017
[2] Ropelewski, C.F. and Halpert, M.S 1987 Global and Regional Scale Precipitation Patterns Associated with the El Niño/ Southern Oscillation *Monthly Weather Review (115):1606 – 1626.*
[3] Mc Bride, J. Haylock, M.R. and Nicholls, N 2003 Relationships between the Maritime Continent Heat Source and the El Niño-Southern Oscillation Phenomenon *Journal of Climate (16): 2905 – 2914*
[4] Aldrian, E 2008 Meteorology of Indonesian Waters. *Indonesian Agency for Meteorology, Climatology, and Geophysics* ISBN: 978-979-1241-19-9
[5] Aldrian, E 2003 Spatial Patterns of Enso Impact on Indonesian Rainfall *Journal of Science and Technology of Weather Modification Vol. 3, No. 1: 5-15*
[6] Gutman, G., Csiszar, I. & Romanov, P. 2000 Using NOAA/AVHRR products to monitor El Ninoimpacts: focus on Indonesia in 1997-98. *Bulletin American Meteorological Society, 81,* 1189–1205.
[7] Mulyana, E 2002 The Relation between ENSO and Rainfall Variation *Journal of Science and Technology of Weather Modification, 3(*1), 1-4.
[8] Hamada, J. I., Yamanaka, M. D., Matsumoto, J., Fukao, S., Winarso. P. A., & Sribimawati, T. 2002 Spatial and Temporal Variations of The Rainy Season Over Indonesia and Their Link to ENSO.*Journal of the Meteorological Society of Japan Ser. II, 80(2), 285-310.*
[9] Hendon, H. H 2003 Indonesian rainfall variability: Impacts of ENSO and Local Air–Sea Interaction.*Journal of Climate 16*, 1775-1790.
[10] Darsiman, Sri Mulyani dan Zailand Harisda 1999 The Impact of *El Niño* 1997/1998 in North Sumatera. *Journal of Meteorology and Geophysics Vol 3, September.*
[11] Trenberth, K.E 1997 The Definition El Niño *Bulletin of the American Meteorological Society Volume 78. No 12. 2771-2777.*
[12] Qian, J.H., A.W.Robertson, V. Moron 2010 Interaction Among ENSO, The Monsoon, dan Diurnal Cycle in Rainfall Variability Over Java, Indonesia. *Journal of the Atmospheric Sciences 67: 3509 – 3524*
[13] https://www.climate.gov/enso, Diakses 25-08-2017
[14] https://climatedataguide.ucar.edu/climate-data/nino-sst-indices-nino-12-3-34-4-oni-and-tni Accessed on 28-8-2017
[15] Indonesian Agency for Meteorology, Climatology, and Geophysics 2009 The Rainy Season Forecast of 2009/2010 in Indonesia.

Seismic Travel Time Tomography in Modeling Low Velocity Anomalies between the Boreholes

A Octova[1*], R Sule[2]

[1]Mining Engineering Department, Faculty of Engineering, Universitas Negeri Padang, Indonesia
[2]Geophysical Engineering Department, Faculty of Mining and Petroleum Engineering, Institut Teknologi Bandung, Indonesia

*adree.octova@gmail.com

Abstract. Travel time cross-hole seismic tomography is applied to describing the structure of the subsurface. The sources are placed at one borehole and some receivers are placed in the others. First arrival travel time data that received by each receiver is used as the input data in seismic tomography method. This research is devided into three steps. The first step is reconstructing the synthetic model based on field parameters. Field parameters are divided into 24 receivers and 45 receivers. The second step is applying inversion process for the field data that consists of five pairs bore holes. The last step is testing quality of tomogram with resolution test. Data processing using FAST software produces an explicit shape and resemble the initial model reconstruction of synthetic model with 45 receivers. The tomography processing in field data indicates cavities in several place between the bore holes. Cavities are identified on BH2A-BH1, BH4A-BH2A and BH4A-BH5 with elongated and rounded structure. In resolution tests using a checker-board, anomalies still can be identified up to 2 meter x 2 meter size. Travel time cross-hole seismic tomography analysis proves this mothod is very good to describing subsurface structure and boundary layer. Size and anomalies position can be recognized and interpreted easily.

1. Introduction

Geophysics is one of exploration method that can be used to knowing subsurface imaging with a relatively short time and low cost. One of the geophysical methods that provides high resolution picture is seismic tomography between the boreholes. Crosshole Seismic tomography uses characteristic of artificial wave propagation to imaging the anomalies through it. The source of these artificial waves can be dynamite or modified sources for various purposes. The waves propagating from the source placed on one of the borehole will be received by the receivers at the other borehole. Travel time seismic tomography has several advantages. The results of the analyzed data are only the first wave travel time received by the receivers (first break). The other waves do not need to be analyzed further.

Geophysical modeling consists of two types that are forward modeling and inverse modeling. In seismic tomography, forward modeling is a way of seeing the movement of waves in a geological model. This modeling begins with initial modeling based on the parameter model. By doing mathematical analysis of physical phenomena, we will be known the movement of the wave of each stage of time. The Output from forward modeling is the travel time calculation. The result of this forward modeling is using to testing the algorithm for a geological model. This process is known as inverse modeling.

The medium that through the ray of wave is divided over grids that have a certain speed. Eikonal equation can be used to calculate travel time to get positions, directions and time. Various ways can be applied from the Eikonal equation. Simplicity of Eikonal equation formulation is related to speed of calculation.

In this research will be presented a suitability method of crosshole seismic tomography in providing an overview between the boreholes. The first step is reconstructing the synthetic model. The second step is applying inversion process for the field data. The last step is testing quality of tomogram.

2. Seismic Tomography Modeling

Projection and reconstruction play important role in seismic tomography [1,2]. The Projection is a observation of object responses that given disturbance. The object is subsurface structures, while the disturbance is seismic waves through to the object. The Projection process is usually called as forward modeling. The opposite of the projection is the reconstruction process. This process is object development based on projected results. By doing it, we will know the object structure. The reconstruction process is usually called as inverse modeling.

Forward modeling can describe seismic patterns from geological model. This patterns include ray path which produces travel time of waves emitted from source to receivers. The medium which through out of the ray is divided into grids that have certain velocities. Travel time calculation is performed by resolve the Eikonal equation to obtain the position, direction and time propagation. Various ways can be applied to obtain the approach of Eikonal equation. Simplicity Eikonal equation formulation is associated with computing speed. Eikonal equation is expressed by :

$$|\nabla t|^2 = \frac{1}{v^2} \text{ or } |\nabla t|^2 = s^2, \tag{1}$$

$$\frac{\partial T}{\partial X_i}\frac{\partial T}{\partial X_i} = \frac{1}{c^2} \tag{2}$$

$$\left(\frac{\partial T}{\partial x}\right)^2 + \left(\frac{\partial T}{\partial z}\right)^2 = \frac{1}{c^2} \tag{3}$$

s is slowness parameter

Eikonal equation is a solution of high frequency wave equation that can calculate the shortest travel time between source and receiver. This equation helps in the reconstruction of wave propagation in the medium in its path. Eikonal equations can be solved in several ways, one of them is finite difference methods [3-5] whose have developed FAST program.

The solution of the eikonal equation using finite difference method was proposed by [6]. The main advantage of this method is that data directly calculated to the point of the grid, no interpolation is needed next. Travel time functions are restricted to single-valued, thus this technique is suitable for first arrival time modeling.

Figure 1 shows a grid element, counter-clockwise numbering nodes, starting from the left bottom node. 0th node is a known travel time (t0). This travel time is used to determine time propagation in 1st, 2nd, and 3rd nodes (t_1, t_2, t_3).

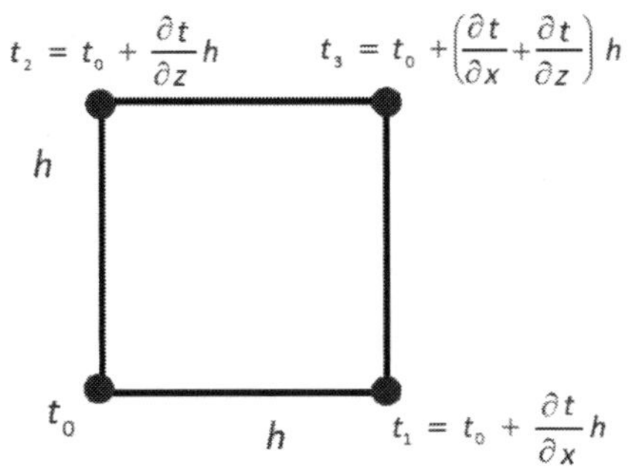

Figure 1. Diagram of a grid element using the Vidale [6] method

Inverse modeling is a porocess to get model from data. The input is delay time of wave propagating from source to receivers. This delay time is difference between observed and calculated wave travel time at specific velocity model of the earth. Travel time is determined from the observed wave arrival time reading on the receiver. The output of inverse modeling is subsurface geological model.

Travel time tomography is a nonlinear inverse problem that can be solved by linearization and iteration. Travel time of seismic wave propagation through the slowness is obtained from:

$$T_i = \int_{L_i} s(\vec{x}(l))\, dl \tag{4}$$

where dl is the segment of the ray length along the ray path Li from source to receiver. Linearization of the Eikonal equation from the slowness is resulting from the changes of travel time to parts of slowness perturbation (Δs). It can be written be:

$$T_i = \int_{L_i} \left[s_0(\vec{x}(l)) + \Delta s(\vec{x}(l))\right] dl = \int_{L_i} s_0(\vec{x}(l))\, dl + \int_{L_i} \Delta s(\vec{x}(l))\, dl \tag{5}$$

Based on Fermat's principle which states that no changes of ray path length, while the rays to be divided ($L_i \approx L_i^0$), so that the time delay can be written as:

$$\delta t_i = T_i - T_i^0 \approx T_i^0 = \int_{L_i^0} \Delta s(\vec{x}(l))\, dl \tag{6}$$

$$\text{or} \quad \delta t_i = \sum_{j=1}^{M} \delta s_j \partial L_i^0 \tag{7}$$

where M is the number of cells, ∂L_i^0 is the length of the ray path segment of ith ray in j[th] cell of the reference model, as shown in Figure 2.

Inversion problem is usually solved by least-squares minimization method. This method estimate solution by finding model parameters that minimize of length measurement from the modeled data. Solution of linear least-squares inversion is calculating the error (E) that obtained on the model parameters (m) that produces a zero value. So it can be written:

$$\underline{\mathbf{G}}^T\underline{\mathbf{G}}\mathbf{m} - \underline{\mathbf{G}}^T\mathbf{d} = 0 \tag{8}$$

This equation is a quadratic matrix equation for unknown model parameters. Meanwhile, for known model parameters, the solution of least-squares method is:

$$\mathbf{m} = [\underline{\mathbf{G}}^T\underline{\mathbf{G}}]^{-1}\, \underline{\mathbf{G}}^T\mathbf{d} \tag{9}$$

The damped least-squares inversion will cause problems if there are a lot of "a priori" information that applied in the model parameters. This information will control the model parameters which produces misinterpretation final result. These results will only explain the influence of "a priori" dominant rather than observational data. Regularization is an approach by addition of the inversion data to omit the under-determined part. The regularization scheme of [5] minimizes an objective function that includes norms that measure model roughness and data misfit. The advantage of this scheme is its flexibility to combine the smallest, flattest, and smoothest perturbation constraints in the inversion scheme [7]. The objective function Φ minimized at each iteration is:

$$\Phi(\mathbf{m}) = \delta\mathbf{t}^T\underline{\mathbf{C}}_d^{-1}\delta\mathbf{t} + \lambda[\mathbf{m}^T\underline{\mathbf{C}}_h^{-1}\mathbf{m}+\omega\mathbf{m}^T\underline{\mathbf{C}}_v^{-1}\mathbf{m}] \quad (10)$$

where $\delta\mathbf{t} = \mathbf{t} - \underline{\mathbf{G}}(\mathbf{m})$ is the data residual vector; **m** is the model vector; $\underline{C}d$ is the data covariance matrix which is a diagonal matrix and consists of picking uncertainty information; $\underline{C}h$ and $\underline{C}v$ are the roughness matrices in horizontal and vertical direction, respectively; and λ is the trade-off parameter. For 2D tomographic, $\underline{C}h$ and $\underline{C}v$ are band-diagonal matrices, which non-zero elements represent the second-order Finite Difference scheme of the Laplace operator. The parameter ω is the parameter that controls the inversion scheme which can be written by:

$$\omega = \alpha s_z + (1 - \alpha)\, s_p \quad (11)$$

where α is the parameter that controls the relative weighting of fitting the smallest perturbation (sp) constraint equations versus smoothnest/flattest constraint equation. Meanwhile sz is smoothness parameter between vertical and horizontal models.

The minimization of the objective function Φ is related to the results of the model vector **m** that contain in equation:

$$\begin{bmatrix} \underline{C}_d^{-\frac{1}{2}}\underline{L} \\ \lambda\underline{C}_h \\ s_z\lambda\underline{C}_v \end{bmatrix} \delta m = \begin{bmatrix} \underline{C}_d^{-\frac{1}{2}}\delta t \\ -\lambda\underline{C}_h m_0 \\ -s_z\lambda\underline{C}_v m_0 \end{bmatrix} \quad (12)$$

where **L** is the kernel data matrix containing the ray segment in each cell; $\mathbf{m}_0$ is the current model; $\delta\mathbf{m}$ is the model perturbation using the updated model which is given by $\mathbf{m} = \mathbf{m}_0 + \delta\mathbf{m}$

Generally, seismic tomography process can be shown in the flow chart in Figure 3 [8]. The most crusial processes in tomography are picking travel time observation and determinating initial model. Tomography gives updated model of the earth at each iteration in order to produce very small difference travel time observation data. The smallest difference can be used as the final model.

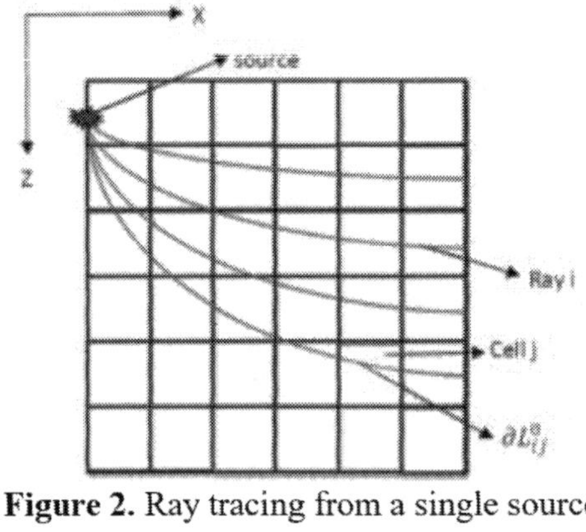

Figure 2. Ray tracing from a single source

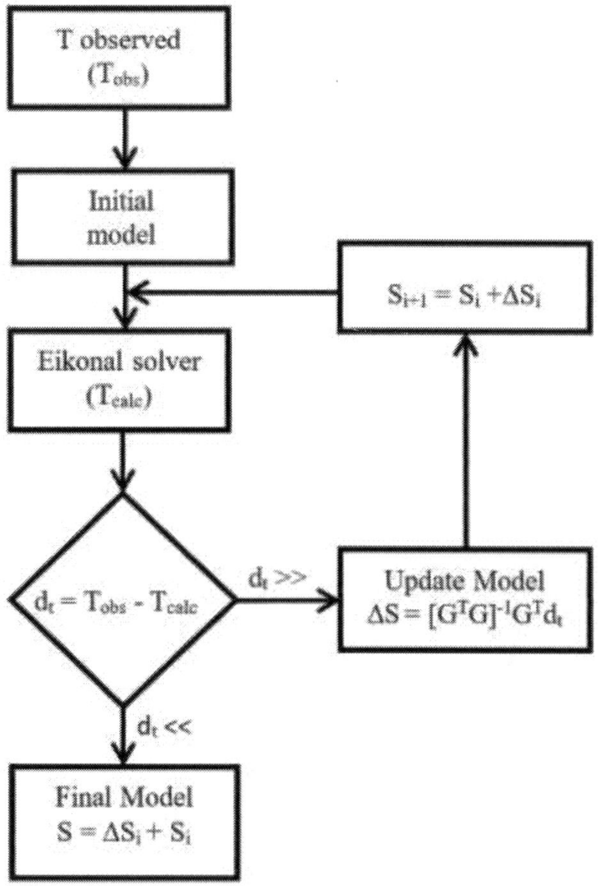

Figure 3. Seismic travel time tomography flow chart

3. Synthetic Model

Simple models are made in order to travel time tomography test. Initial velocity model (background velocity) is set to 2000 m/s whole the medium. Positive and negative anomalies are positioned in the center of the medium with 4 m x 4 m dimensions. Positive anomaly has 2500 m/s velocity (Figure 4 top left) whereas negative anomaly has 1500 m/s velocity (Figure 4 bottom left).

The inversion results can be shown in perturbation velocity cross section, i.e. absolute velocity changes of absolute average overall value of the tomogram. In Figure 4 right, top and bottom can be seen clearly that the structure of the positive and negative anomalies are corresponding to the initial model. The structure and dimension of the anomaly are similarly in size and position. It has been proven the reconstruction of synthetic data in positive and negative anomalies with this method gives good result.

IOP Conf. Series: Materials Science and Engineering **335** (2018) 012056 doi:10.1088/1757-899X/335/1/012056

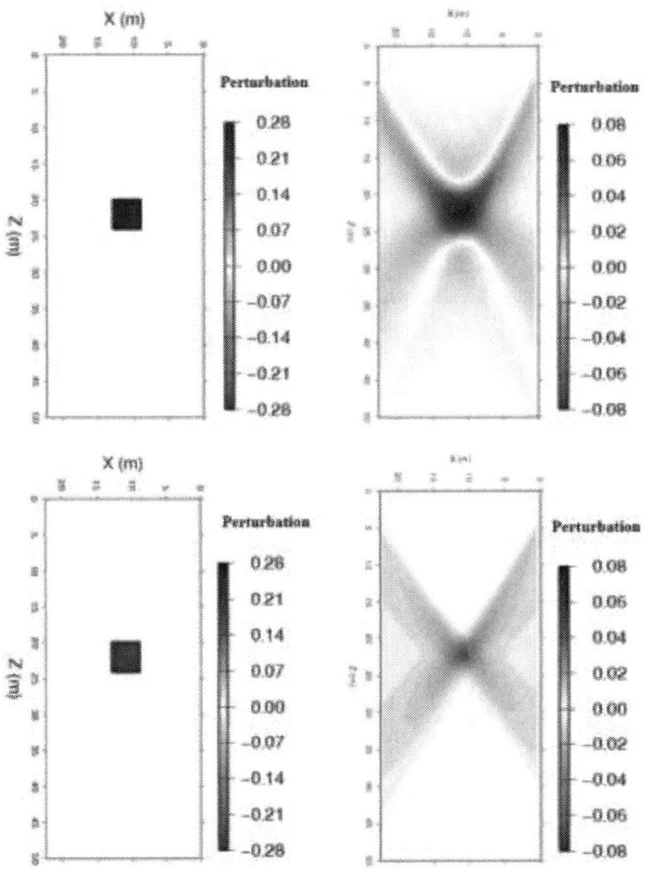

Figure 4. Top (left to right) : Initial model with positive anomaly, inversion result. Bottom (left to right) : Initial model with negative anomaly, inversion result.

4. Application In Real Data

Seismic travel time tomography has been applied in synthetic data with good reconstruction of positive and negative anomalies. Then it is used in real data to see structures between boreholes in Kuala Lumpur Malaysia. There are five boreholes were applied in this area. Position of these boreholes are shown in Figure 5.

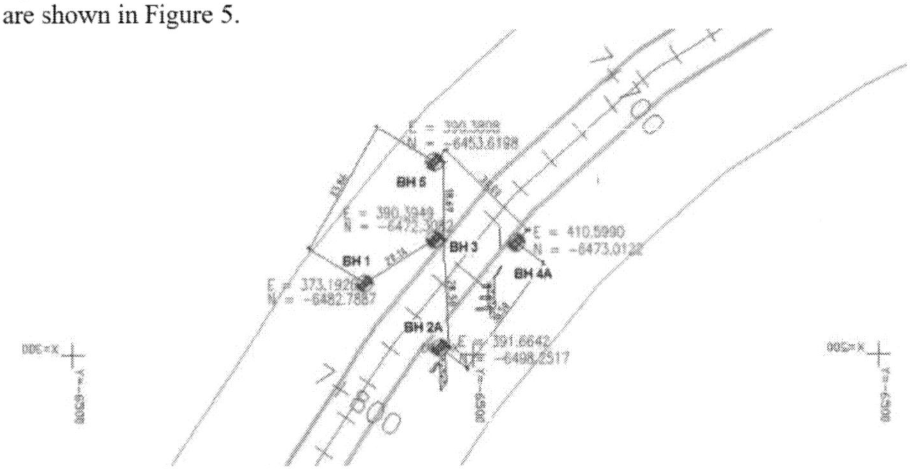

Figure 5. Boreholes position in Istana Island

IOP Conf. Series: Materials Science and Engineering **335** (2018) 012056 doi:10.1088/1757-899X/335/1/012056

Initial velocity model is the first step to choose velocity background as insitu velocity. It is obtained from travel time average against the distance of source to receiver. In Kuala Lumpur Malaysia the initial velocity model that choosen is 4521 m/s whole the model.

The model is divided into small cells in x, y, and z direction. The size distribution of this model is affects for the quality of tomogram result and computing time. The smaller grid gives better tomogram result, but with a longer computing time. In Kuala Lumpur Malaysia, the grid that choosen is 0.1 m x 0.1 m.

After inversion process with FAST program, there are five sections tomogram that produced in Istana Island Malaysia. The rays move from sources to receivers not in straight line. They are moved away and assembled in several sections. These rays can be seen in Figure 6 left, Figure 7 left, Figure 8 left, Figure 9 left and Figure 10 left. There are only three sesctions indicate cavities presence in Figure 6 left, Figure 8 left, and Figure 10 left. The gathering area become a place with high velocity anomalies and the moved away area are low velocity anomalies as predicted as cavities. The perturbation sections as shown in Figure 6 middle, Figure 7 middle, Figure 8 middle, Figure 9 middle and Figure 10 middle give clear structures and clear cavities anomalies (Figure 6 middle, Figure 8 middle, and Figure 10 middle). Cavities are predicted in rounded line. The dashed line are the limit section with no data.

The absolute velocity sections are generally composed loose soil and limestone. Weaker (fractured and more porous) limestone or perhaps related to water infilled cavities are identified between the solid limestone. These cavities give drmatically velocity changes inside the limestone layer as shown in Figure 6 right, Figure 8 right, and Figure 10 right. The absolte of the cavities have range from 3800 m/s to 4800 m/s.

The combination of five pairs boreholes produce 3D section as shown in Figure 11. This section gives excellent layer boundaris between layers of soil and marble. There is interconnected layers boundary between one drill hole to another. This proves that seismic tomographic methods between the boreholes can provide a good overview of the quality results that can be interpreted easily.

5. Checker-Board Test

To make sure quality of the tomographic result, it needs to be done of resolution test [9]. The models are made in such a way using the synthetic data. FAST program is used to reconstruct a model of the inversion process. This resolution test is using the checker-board with the different size of anomalies.

The resolution test is done by making several positive and negative anomalies inside the models. There are three types of schemes. These are 4 m x 4 m, 3 m x 3 m, and 2 m x 2 m anomalies. This process is carried out to see the accuracy of the tomographic inversion process to reconstruct the size of the anomalies. Background velocity is set to the value of 2000 m/s. Positive anomalies are set to the value of 2500 m/s and the negative anomalies are set to value of 1500 m/s. These initial model can be seen in Figure 12 top.

The shots position are placed in the left side of the boreholes from 6 m to 50 m depth. Meanwhie the receivers are positioned in the right borehole from 5 m to 50 depth. Genrally after doing the numbers of iterations, anomalies are still reconstructed with 2 m x 2m anomalies with same shape and position. But then, in the top of section, anomalies couldn't give same shape as well as initial model. It caused of no data in 0 m to 5 m depth (Figure 12 bottom).

IOP Conf. Series: Materials Science and Engineering **335** (2018) 012056 doi:10.1088/1757-899X/335/1/012056

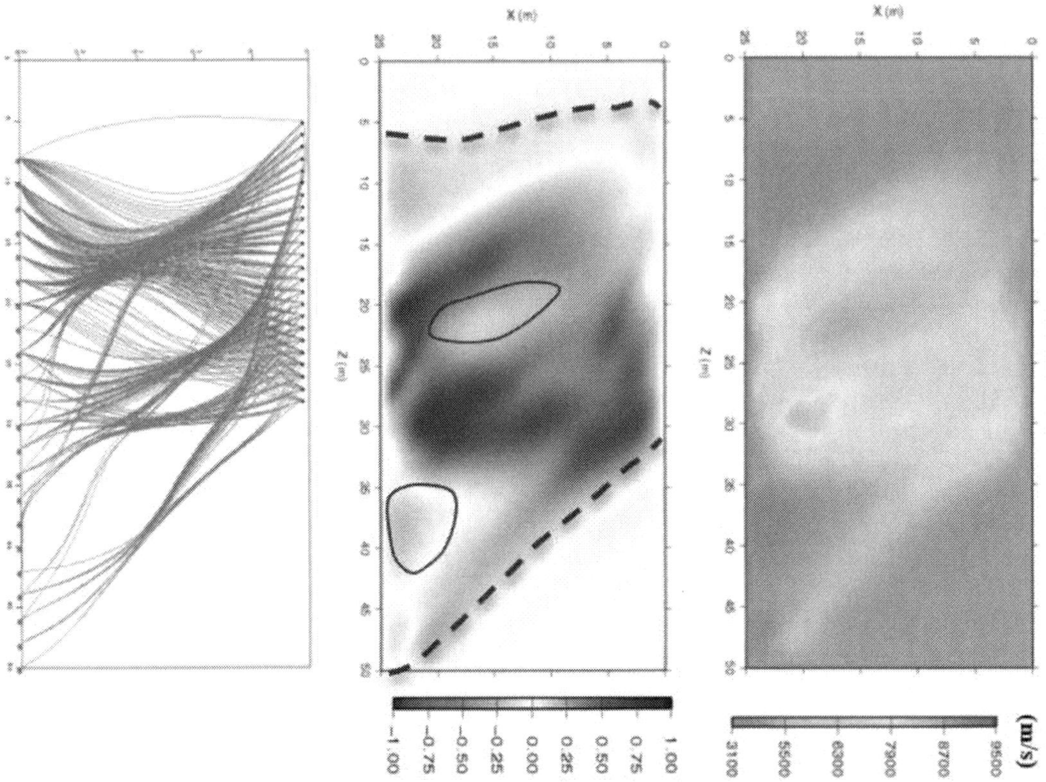

Figure 6. Inversion result of BH2A-BH1. left to right: ray tracing, perturbation section, absolute velocity section.

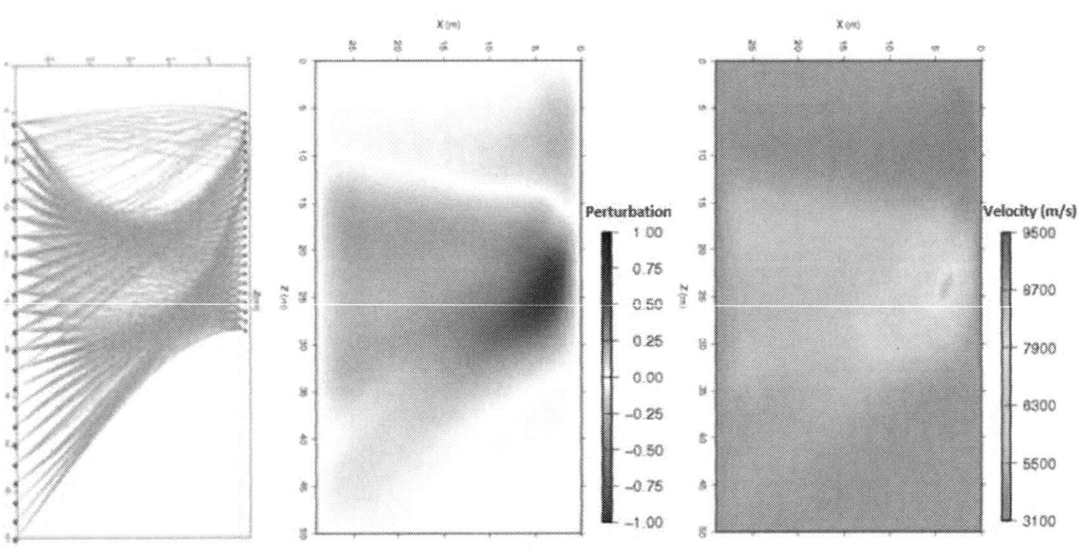

Figure 7. Inversion result of BH2A-BH3. left to right: ray tracing, perturbation section, absolute velocity section.

IOP Conf. Series: Materials Science and Engineering **335** (2018) 012056 doi:10.1088/1757-899X/335/1/012056

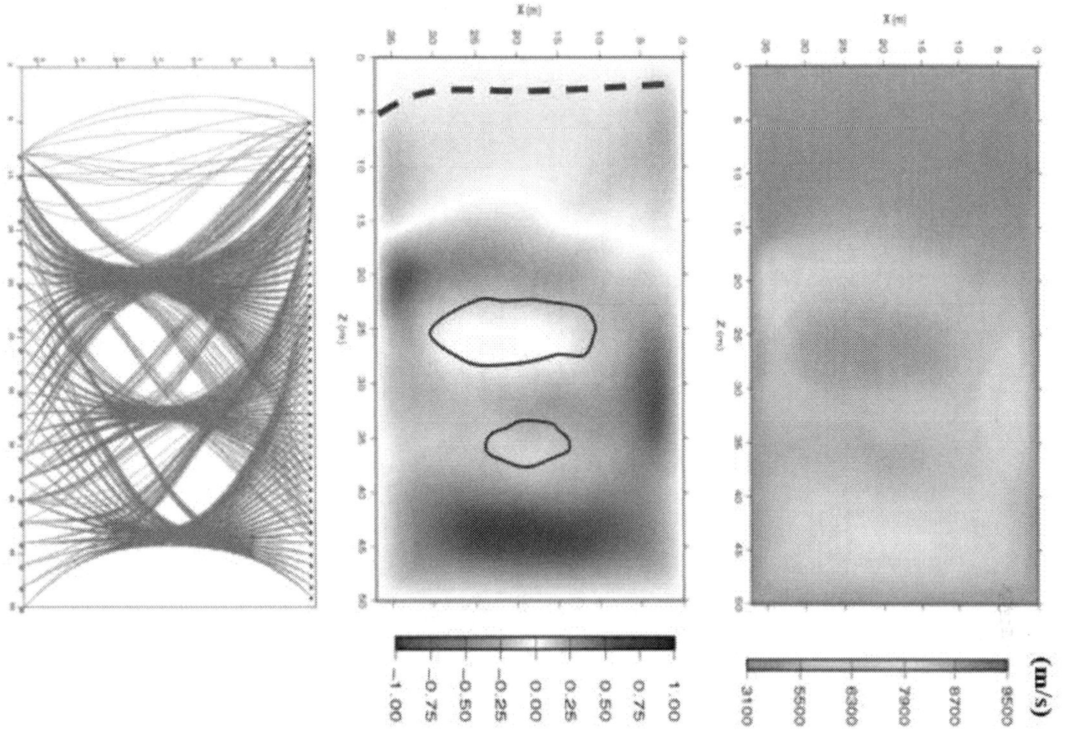

Figure 8. Inversion result of BH4A-BH2A. left to right: ray tracing, perturbation section, absolute velocity section.

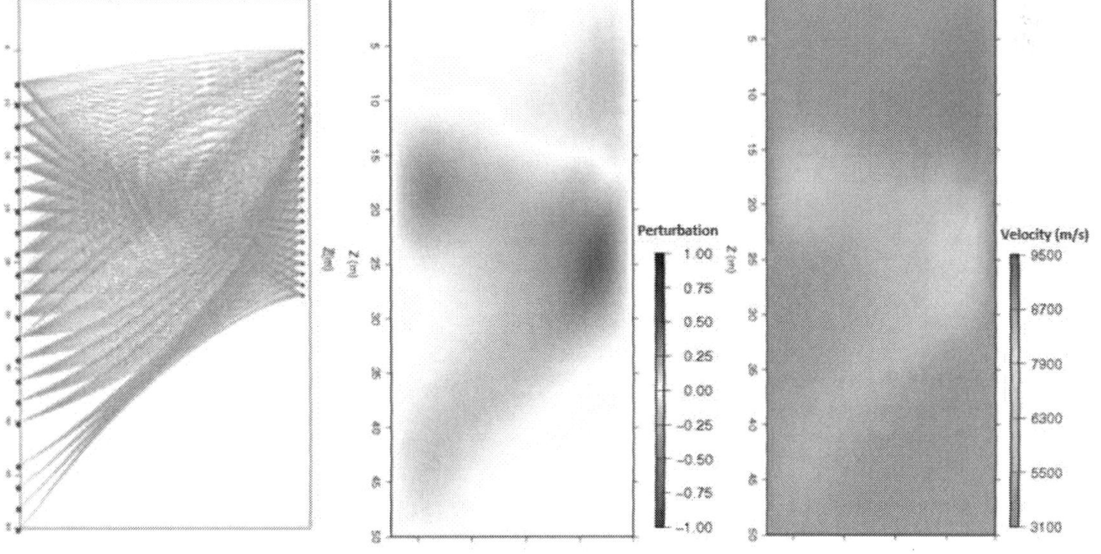

Figure 9. Inversion result of BH4A-BH3. left to right: ray tracing, perturbation section, absolute velocity section.

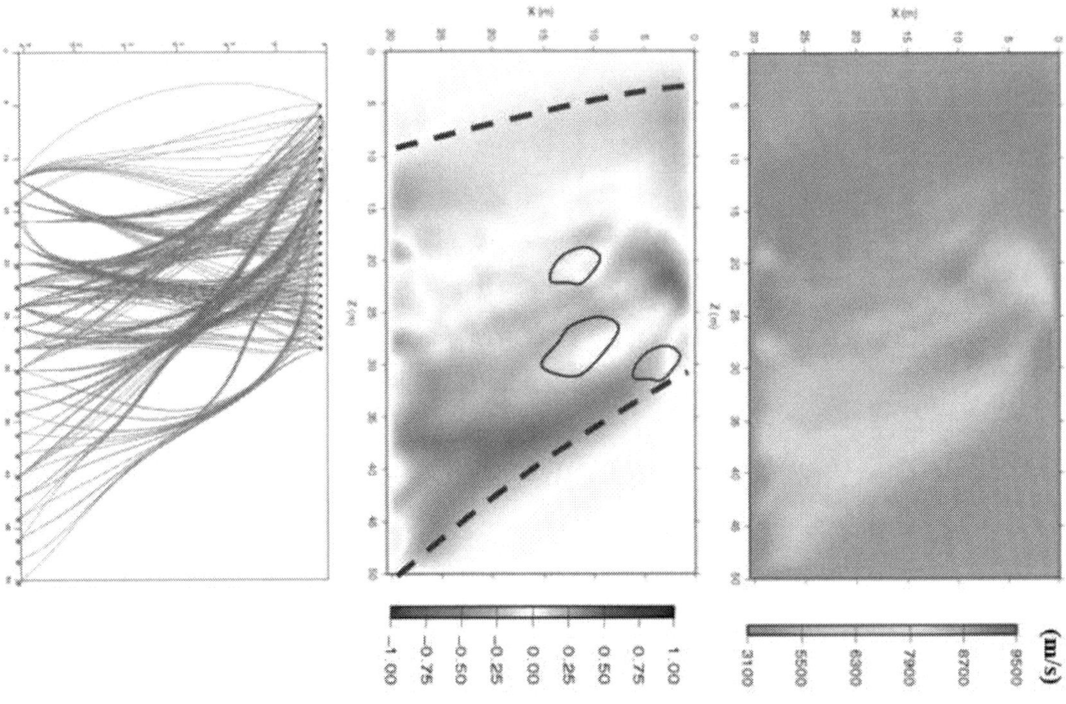

Figure 10. Inversion result of BH4A-BH5. left to right: ray tracing, perturbation section, absolute velocity section.

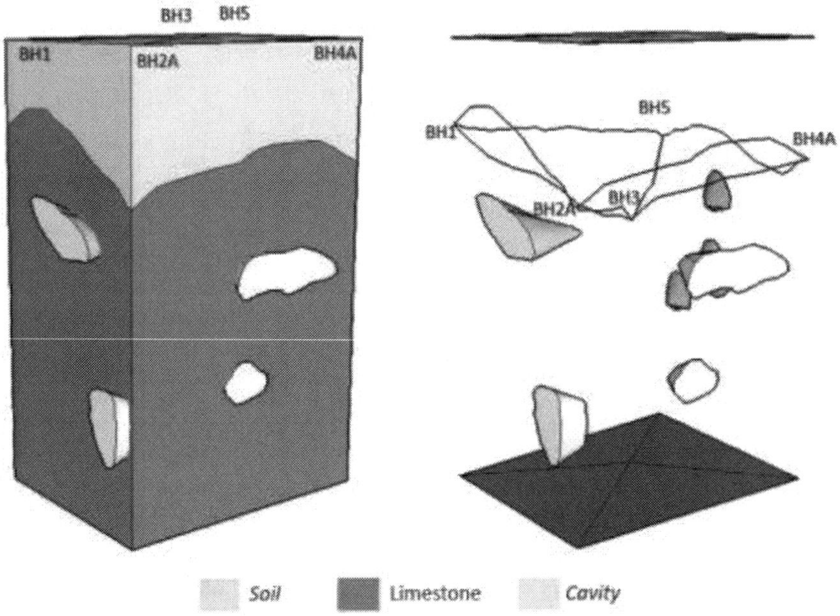

Figure 11. 3D section

IOP Conf. Series: Materials Science and Engineering **335** (2018) 012056 doi:10.1088/1757-899X/335/1/012056

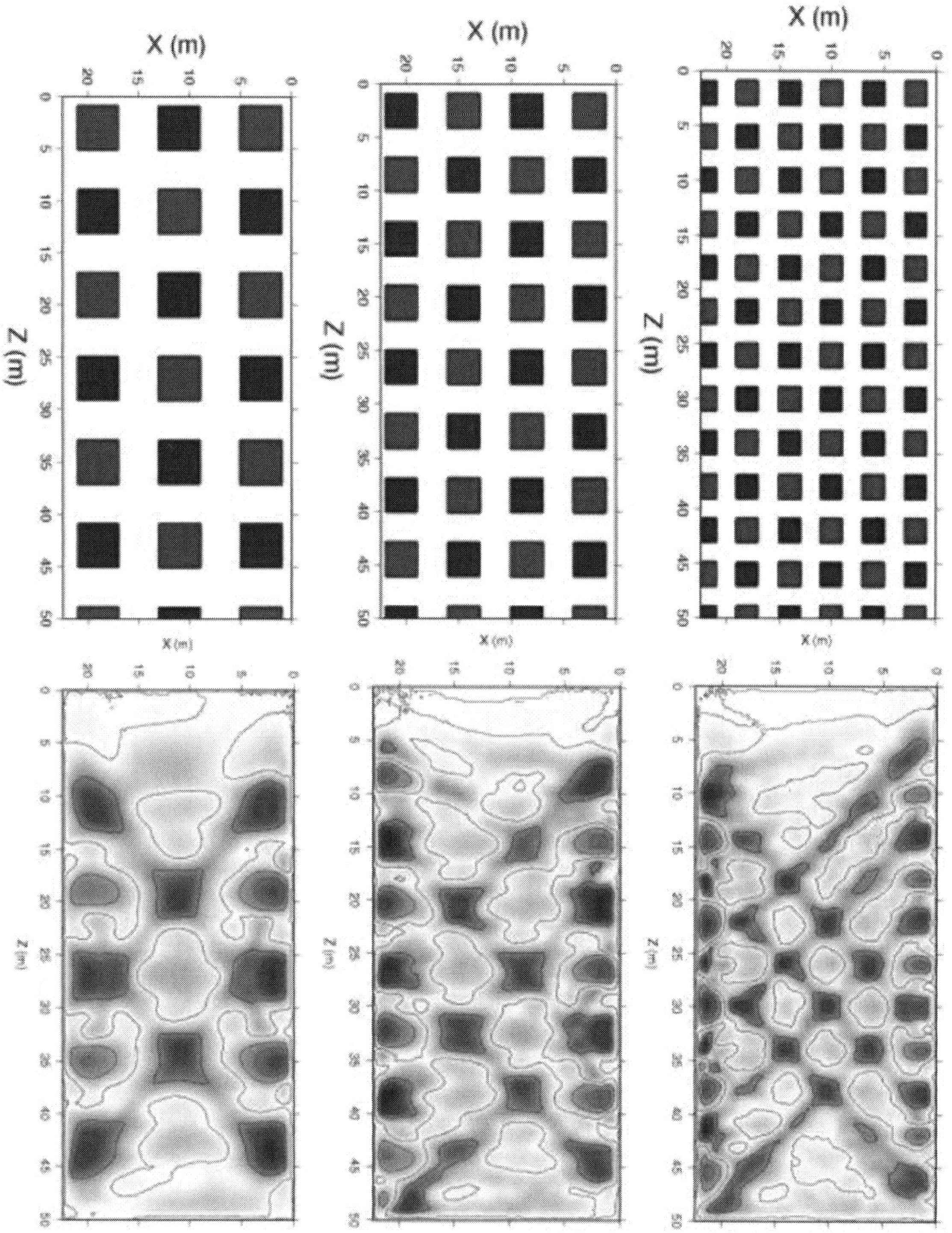

Figure 12. Checker-board test. Top : initial models. Bottom : reconstruction results. Left to right : 4 m x 4 m anomalies, 3 m x 3 m anomalies, and 2 m x 2 m anomalies

6. Conclusions

Based on the tomogram obtained, crosshole seismic travel time tomography method can give good and clear subsurface structure. It proves from anomalies detected between the boreholes. Syntetic model with square model in the middle area could reconstructed similarly. Meanwhile for real data in Malaysia, could detected the cavities in certain of depth. The accuracy of tomogram resolution, is effected on the number of rays. More ray that passes through the medium will give better and more detail tomogram result. Checker-board test gives that anomalies still can be identified up to 2 m x 2 m size

Acknowledgments

We thank the Department of Irrigation and Drainage - Malaysia for the permission in using the data.

Reference

[1]. Stewart, R.R 1991 *Exploration Seismic Tomography : Fundamentals* (Course Notes Series, Volume 3. Society of Exploration Geophysicists)

[2]. Widiyantoro, S 2000 *Diktat Tomografi Geofisika Edisi ke-1* (Program Studi Geofisika Jurusan Geofisika dan Meteorologi Fakultas Ilmu kebumian dan Teknologi Mineral Institut Teknologi Bandung)

[3]. Aldridge, D.F., D.W.Oldenburg 1993 *Two-Dimensional Tomographic Inversion with Finite-Difference Traveltimes* (Journal of Seismic Exploration 2) p 257-274

[4]. Eaton, D.W.S 1990 *Finite-difference Solution of The Eikonal Equation for Transversely Isotropic Media*

[5]. Zelt, C.A., P.J.Barton 1998 *Program Package for First Arrival Seismic Tomography* (Rice Univercity, Houston)

[6]. Vidale, J 1988 *Finite-difference Calculation of Traveltimes* (Bull. Seism.Soc.Am) p 78:2062-2076

[7]. Sule, M.R 2004 *Seismic Travel Time Tomography and Elastic Waveform Modeling Aplication to Ore-Dyke Characterizations* (Dissertation, Logos Verlag Berlin)

[8]. Grandis, H 2009 *Pengantar Pemodelan Inversi Geofisika* (Himpunan Ahli Geofisika Indonesia)

[9]. Octova, A 2014 *Rekonstruksi Model Sintetis 2D Seismik Tomografi Waktu Tempuh Antar Lubang Bor dengan Metode Beda Hingga* (Jurnal Invotek ISSN 1411-3414 Vol. XIV, No. 1 - Februari 2014 pp 2875-2886

IOP Conf. Series: Materials Science and Engineering **335** (2018) 012057 doi:10.1088/1757-899X/335/1/012057

The Impact of 200 Meter Breast Stroke Swimming Activity on Blood Glucose Level of The Student

S Syahrastani[1], H Badri[1], A Argantos[1], E Yuniarti[2]

[1]Physical Education Dept, Faculty of Sports Science, State University of Padang, Padang, Indonesia
[2]Biology Department, Faculty of Mathematics and Science, State University of Padang, Padang, Indonesia

*syahrastani@yahoo.com

Abstract. Blood sugar in the human body is very important, as it is a source of energy for humans. Fasting blood sugar needs to be recognized, because it is an indicator of a person's health. The research that has been done aims to determine the effect of physical activity on fasting blood sugar. This research is a quasi-experimental research. The research was conducted on 15 students of FIK UNP Padang, who have passed the swimming course. Blood was taken before and after doing physical exercise activity of 200 meter breaststroke swimming. Data collection was conducted with tests and measurements. The data analysis technique used inference statistics with t test formula, with $\alpha = 0.05$. From the results it is obtained that there is a very significant impact on blood sugar levels after 200 meters breaststroke swimming activity with $p < 0.05$

1. Introduction

People living in the globalization era tend to have lack of motion, because almost all human activities have been helped by technology. Lack of motion can give a bad effect on the human body, such as decrease in endurance, muscle loss, brittle bone, and susceptible to disease that are often caused by lack of motion, such as cardiovascular ability reduction, hypertension, diabetes, and others.

Diabetes mellitus (DM) is still a serious health problem of the world, including Indonesia. Indonesia is at the fourth rank with the highest prevalence of diabetes in the world after India, China and the United States. Even the number of diabetics continues to increase from year to year, especially for type 2 diabetes. WHO data estimates the number of patients with type 2 diabetes in Indonesia will increase significantly to 21.3 million in 2030 [1]. Diabetes can be resolved with regular exercise in addition to taking medication and regular control to the doctor.

Swimming is one of the sports recommended by many doctors and government agencies working in the health field, swimming becomes so important because of its benefits. Some of the benefits are: by only 30 minutes of this sport for three times per week, people can burn 900 calories which reduces the risk of developing type 2 diabetes by more than 10 percent. And if s person already has type 1 diabetes, the benefits of swimming can be very helpful, as this type of exercise can improve insulin sensitivity.

According to the American Diabetes Association, people with diabetes should get 150 minutes per week, spread over at least three days per week, with moderate intensity physical activity such as swimming to help control glycemic / blood glucose in the body. In addition to swimming, exercise with the same effect to insulin sensitivity is walking [2]. This is why swimming is also called a health sport. Health sport is an essential aim to maintain or improve health.

Swimming is one of the aerobic sport. Swimming activity uses the entire body, such as: the length of the leg, the width of the palm, foot width and leg muscle strength. This physical state will affect the speed of one's swimming. Syahrastani [3] reported, there is a significant relationship between leg length and leg muscle strength and foot width with breast stroke swimming ability. Furthermore, the motor ability also affects the learning result of swimming [4].

The role of swimming: growth, physical development, health (hypertension, asthma, pinched nerve), cardiovascular (circulatory), heart. Kushartanti [5] suggests that exercise can cause changes in the heart, blood vessels, lungs, and muscles by the type, duration, and intensity of the exercise. By doing regular exercise will improve lung elasticity and increase the number of active alveoli. Both of these can increase the storage capacity and the distribution of oxygen to the bloodstream. Muscles will also experience changes by doing regular exercise. The size of the muscle fibers will increase and the energy supply system in the muscle will increase. This will lead to increased strength, flexibility, and muscular endurance.

There is physical movement in swimming. Physical movement needs an energy source. One source of energy is glucose. Glucose produces ATP. Glucose in blood is measured with blood glucose levels. Blood glucose level is the level of glucose concentration in blood expressed in mg / dl. Blood glucose levels are divided into blood fasting and non-fasting. In this study the non-fasting is used, in which blood glucose was taken at a time without any fasting. Blood glucose levels are considered as normal when blood glucose is between 100-200 mg / dl, otherwise it is not normal if bigger than 200 mg / dl. Blood glucose measurements were performed using glucometer. Glucometer is a medical device to determine the approximate concentration of glucose in the blood [6]. High blood glucose levels in a person, signifying insulin is not working. This results in diabetes mellitus. According to research results I Gusti Bagus Aginda Dwi Pawana, I Wayan Sudhana, I Wayan Losen Adnyana [6]. from 55 samples of diabetics mellitus, as many as 9 people (16.4%) have medium activity, 25 people (45,4%) has mild activity, 21 people (38,2%) are having diabetes mellitus who are not active in the physical activity. Therefore, it takes exercise activity to decrease blood glucose levels, because with the exercise the glucose can be convert into energy. Swimming is one of the exercise that requires great energy. According to Scott, Powers Edward, Howley [7] the estimates of energy expenditure on breaststroke swimming is 10 METS which is equivalent to 7.0 kcal per minute.

Diabetes is a fairly high disease encountered in the community. To reduce the risk and overcome diabetes can be done with exercise. Exercise can cause changes in the heart, blood vessels, lungs, and muscles by the type, duration, and intensity of exercise performed. By doing regular exercise will improve lung elasticity and increase the number of active alveoli. Both of these can increase the storage capacity and the distribution of oxygen to the bloodstream.

Many forms of aerobic exercise can affect health fitness such as swimming. Swimming is an excellent exercise for fitness and health because this horizontal body position will make the blood flow even throughout the body with almost the same pressure, besides almost all parts of the body can move in balance. Thus swimming can reduce the risk of diabetes because of this massive carbohydrate burning.

Nowadays the diabetes mellitus disease increased sharply including in Indonesia. This disease needs to be aware of since it can lead to complications of various diseases. Diabetes mellitus is a metabolic disease with characteristics of hyperglycemia that occurs due to abnormalities of insulin secretion and insulin work. This is because the body can not release or use insulin adequately. Insulin is a hormone released by the pancreas and is the main substance responsible for maintaining blood glucose levels in the body to keep it in balance. Insulin serves as a tool that helps the glucose to move into the cells that can produce energy or stored as energy reserves [8]. WHO estimates the global prevalence of Diabetes Mellitus will increase from 171 million people in 2000 to 366 million in 2030 [9]. About 60% of these patients are in Asia [10]. Indonesia is ranked the 4th most cases of Diabetes Mellitus in the world [11].

Diabetes Mellitus if not treated properly will result in complications in various organs of the body such as the eyes, heart, kidneys, leg veins, nerves and others. DM patients compared with non-DM patients have a tendency to 25 times occur blind, 2 times coronary heart disease, 7 times chronic renal failure, and 5 times suffering from diabetic ulcers. The chronic complications of DM in Indonesia consist of 60% neuropathy, coronary heart disease 20.5%, diabetic ulcers 15%, 10% retinopathy, and nephropathy 7.1%.

In addition to regulating the diet, people with diabetes should also exercising diligently. Therefore, exercise helps the body to maximally absorb all the glucose circulating in the body. Swimming can reduce the risk of diabetes because it is huge enough to burn carbohydrate with just 30 minutes of this exercise for three times per week, this swimming can burn 900 calories - reduce the risk of type 2 diabetes. Proper physical exercise, regular, individually and fun can improve and inhibit the decline of body organ function, nourish the body and increase body immune against disease. In order of finding the right physical activity for diabetics has been done research on the impact of swimming activity on blood glucose level of the student of faculty of sports science UNP. But the impact of 200 meters breaststroke swimming activity and the influence of swimming 3 times a week and 4 times a week against blood glucose levels is still unknown.

2. Methodology

2.1 Type of research

This research is an experimental research with pre and post-test one-group design. The research conducted for 6 months with several research phases. The studemt chosen as sample were asked to live in the dormitory. They have been fasting for eight hours, and then being fed and tested to measure their blood glucos level. And right after, they have been asked to do 200 meters swimming. Then their blood glucose level were measured again.

2.2 Research location

This research is conducted on the swimming pool and lab of Sport Science Faculty, Padang State University. The research is guided by a doctor and paramedic.

2.3 Research object

The object of this research is the students who have been passed basic swimming course, male and healthy. They were willing to be sample on the research.

2.4 Research procedure

In the measurement of fasting blood glucose all samples were not allowed to eat (in other words no carbohydrate intakes were allowed) during eight hours. After 8 hours passed, their blood glucose is measured, and then they were fed. After the sample ate, two hours after that their blood glucose is measured again.

Then sample were given a warm up followed by 200 meters swimming. After doing swimming, the sample blood glucose level is measured. This blood glucose called blood glucose after doing activity.

2.5 Data analysis technique

All data is recorded on special sheet, and data processing is computerized and presented in a form of table or graphic. The average difference of blood glucose level between before and after breast stroke swimming and 200 meters swimming test is statistically tested using Paired sampel t-test with significant value of $p<0,05$.

3. Results and Discussion

This experimental study was conducted in August 2017 with pre and post-test group design on 15 students of Faculty of Sport Science, Padang State who have passed basic swimming course that meet inclusion and exclusion criteria. Inclusion criteria are male, with ideal BMI and willing to contribute on the research. Exclusion criteria are being sick, have a history of metabolic disease, taking drugs within the past week, active smokers, carrying out heavy activity 24 hours before intervention. The characteristic of the research subject can be seen on the following Table 1.

Table 1. The characteristic of the research subject

Name	Blood glucose		
	Fasting	Non-fasting	After Activity
Mean	90.333	109.067	93.333
Median	87	108	94
Mode	85	108	99
Standard deviation	9.340	11.367	10.356

In the Table 1, there were found 90,333 mg / dL mean of fasting blood glucose, and mean of non-fasting blood glucose was 109,067 and mean of blood glucose after activity was found 93,333 mg / dL. A median of fasting blood glucose was 87 mg / dL, a median of blood non-fasting glucose was 108 mg / dL and median blood glucose after activity was found to be 94 mg / dL.

Thereafter it was found a mode of fasting blood sugar level was 85 mg / dL, a non-fasting blood glucose mode was 108 mg / dL and a blood glucose mode after activity was found to be 99mg / dL. In this study also found a standard deviation of fasting blood glucose for 9,340, the standard deviation of non-fasting blood glucose was 11.367 and the standard deviation of blood glucose after the activity of 200 meters breaststroke swimming was found at 10,356.

Levels of blood glucose is something important and need to be controlled on someone. Blood glucose is a source of energy for all organs of the body. Meanwhile, according to Rudianto, F, B [12] Blood Glucose is a term that refers to the level or concentration of glucose in the blood, or serum glucose level, strictly regulated in the body. Blood-borne glucose is the main source of energy for the body's cells. Generally, blood glucose levels persist at the narrow limits throughout the day: 4-8 mmol / l (70-150 mg / dl) where 1 mmol / l = 18 mg / dl. This level increases after eating and is usually at the lowest level in the morning, before people eat. Here's the range of normal blood sugar levels in the body:

- Before eating: 70 – 130 mg / dL
- Two hours after eating: less than 180 mg / dL.
- After not eating (fasting) for at least eight hours: less than 100 mg / dL.

3.1. The impact of 200 meter breaststroke swimming against blood glucose of the student

The impact of 200 meter breaststroke swimming against blood glucose of the student of Faculty of Sport Science Padang State University can be seen on Table 2.

Table 2. The impact of 200 meter breaststroke swimming against blood glucose of the student

Variable	n	200 meter breaststroke swimming activity		p
		N on-Fasting blood glucose	After Activity blood glucose	
		Mean ± SD	Mean ± SD	
Levels of blood glucose	15	109,067± 11.367	93.333± 10.356	<0.001

Prior to the provision of 200 m swimming activity the non-fasting blood glucose level was 109,067 mg / dL with standard deviation (SD) 11.367 and after given activity of 200 meter breaststroke swimming was found sugar level blood of 93.33 mg / dL with the standard deviation (SD) 10.356. Statistical test results obtained p value of less than 0.001 it can be concluded there is a very significant influence between non-fasting blood glucose and blood glucose levels after given the activity of 200 meter breaststroke swimming. From the results of statistical data analysis, Hypothesis that was raised in this research is accepted. "There was a decrease in blood glucose levels of FIK UNP students after doing a 200m breaststroke swimming activity"

In this study non-fasting blood glucose levels is used, in which blood glucose was taken at a time without any fasting. Blood glucose levels are considered as controlled when blood glucose is between 100-200 mg / dl, otherwise it is considered as uncontrolled if> 200 mg / dl. Blood glucose measurements were performed using glucometer. Glucometer is a medical device to determine the approximate concentration of glucose in the blood [6].

Swimming is one of the exercise suggested by many doctors and government agencies working in the health field, swimming becomes so important because of its benefits, one of the benefit was by swimming for only 30 minutes three times a week can burn 900 calories can certainly reduce the risk of developing type 2 diabetes by more than 10 percent. If someone already has type 1 diabetes, the benefits of swimming are helpful, as this type of exercise can improve insulin sensitivity. According to the American Diabetes Association, people with diabetes should get 150 minutes per week, spread over at least three days per week, with moderate intensity physical activity such as swimming to help control glycemic / blood glucose in the body.

4. Conclusion and Recommendation

4.1. Conclusion

Based on this research, it can be concluded that the activity of 200 m breaststroke swimming can decrease blood glucose level. A decrease in blood sugar level in the 200-meter breaststroke swimming activity was statistically significant at $p < 0.05$. Thus giving a 200-meter breaststroke swimming activity is very good to be implemented by diabetics.

4.2. Recommendation

- Further research is needed on the impact of giving a 200 m breaststroke swimming activity in type 1 diabetes mellitus
- Further research is needed on the impact of freestyle swimming activity in type 1 diabetes mellitus
- Diabetics need to be given appropriate feeding or low food intake of carbohydrate content.

References

[1]. Pramono, Bowo 2017. "60 Percent of Indonesian Society Unconsciously Diabetes".http://www.dikti.go.id/60-persen-masyarakat-indonesia-tidak-sadar-mengidap-diabetes/ (accessed on April 14, 2017).

[2]. Kompasiana 2013. .kompasiana.com/lelihesti/8-manfaat-renang-bagi-kesehatan. (accessed on April 14, 2017).

[3]. Syahrastani 2015. Effect of Technical Ability and Length of Legs and Legs of Footprint to The Ability of FIK UNP Students on 200-meter Breaststroke Swimming. Padang: Padang State University.

[4]. Syahrastani 2012. Influence Learning Media and Motor Ability Againts Learning Outcomes Technique of Breaststroke Swimming of FIK Student Padang State University (Dissertation). Jakarta: Jakarta State University.

[5]. Kushartanti W 2007. Physiology and Health of Sports. Yogyakarta: FIK UNY.

[6]. I Gusti Bagus Aginda Dwi Pawana, Sudhana, I Wayan, and I Wayan Losen Adnyana 2013. Physical Activity Overview of Blood Sugar Levels in Diabetes Mellitus Patients in Karangasem I Puskesmas Working Area In September-October 2013. Bali: Udayana University.

[7]. Powers, Scott K and Howley, Edwart T. Exercise Physiology Theory and Application to Fitness and Performance. Seventh Edition; McGraw Hill, New York.
[8]. Mahdiana R 2010. Preventing Chronic Illness Early. Yogyakarta: Tora Book.
[9]. Riskesdes 2007. Ringkasan Profil kesehatan di Indonesia. http://Selasi.net.Index.Php?optim=com
[10]. Mahendra B, Krisnatuti D, Tobing A, & Alting AZB 2008. Care Yourself, Diabetes Mellitus. Jakarta: Penebar Plus.
[11]. Purnomo H 2009. Pencegahan dan Pengobatan Penyakit yang Paling Mematikan. Yogyakarta: Buana Pustaka.
[12]. Rudianto F B 2012. Conquer Hypertension and Diabetes (pp. 88; 92) Yogyakarta: Sakkhasukma.

IOP Conf. Series: Materials Science and Engineering **335** (2018) 012058 doi:10.1088/1757-899X/335/1/012058

Analysis of Calorific Value of *Tibarau* Cane Briquette

H Nurdin[1]*, H Hasanuddin[1], D Darmawi[1], F Prasetya[1]

[1]Mechanical Engineering Department, Faculty of Engineering, Universitas Negeri Padang, Indonesia

*hens2tm@yahoo.com

Abstract. The development of product diversification through tibarau cane briquettes as an effort in obtaining alternative fuels. Tibarau cane is one of the potential materials of renewable energy sources that can be processed into briquette. So as to reduce dependence on energy fuel oil, which for the middle to lower class is the main requirement. Efforts and innovations tibarau cane briquettes in producing fuel that has quality and performance can be measured with calorific value. Prior to development of this potential required the existence of test and evaluation stages according to the order of the flow of new material product development. Through process technology of briquette product making with compaction and optimization of composition content on tapioca adhesive and mesh particles suitable to get optimum calorific value. The results obtained in this research are the development of tibarau cane briquette model which is recommended as replacement fuel. Where the calorific value of tibarau cane briquette is 11.221,72 kJ / kg at composition percentage 80: 20 and its density is 0,565 gr/cm^3. The comparison of mass tibarau with tapioca, particle size, pressure force (compaction), can affect the calorific value and density of tibarau cane briquette.

1. Introduction

World energy needs continue to increase, especially in Indonesia as one of the producers of petroleum. According to reports of various print and electronic media has reached about 211 trillion in 2012 in which world energy demand will reach twice that of current needs, especially petroleum. It is estimated that the domestic oil reserves, which are the main source of Indonesia's energy, will be exhausted by 2025. By 2015, domestic fuel consumption is already above 1.5 million barrels per day, while production is below 800,000 barrels per day. This condition indicates an energy and / or energy emergency crisis. So it must import 700,000 barrels per day. The Center for Energy Studies of Asia Pacific (APERC) estimates that fuel energy needs will jump from 1,087 million tons of oil equivalent in 2002 to 1,991 million tons of oil equivalent by 2030. If there is no remarkable effort to reduce oil-based energy consumption or encourage energy use renewable, APERC estimates that Indonesia will rely on imports to meet domestic energy needs by 2030 (http://edukasi.kompas.com/). The delay in anticipation in developing alternative energy and energy sources, according to experts will cause Indonesia to experience serious energy problems entering the year 2020.

This increase in energy demand has an impact amongst the community and burdens the government with fuel subsidies that affect the joints of people's lives. With the increasing demand and fuel prices unpredictable in the last decade has encouraged the development of bioenergy and the use of renewable energy as an alternative energy source. The Government has issued Presidential Regulation RI No. 5 - 2006 on the national energy policy which contains the provision that a certain alternative energy source is a certain energy source substitute for fuel. This is an opportunity for all components of the nation to actively participate in developing energy resources, including the potential of biomass energy from industrial waste and agricultural products, such as the utilization of bagasse, durian skin, coconut shell charcoal, and so on. In the Presidential Regulation RI No. 79 - 2014 on National Energy Policy, it is formulated that it is necessary to increase the utilization of new energy sources in the form of solid and renewable energy sources of biomass and waste types. Presidential Regulation No. 41 - 2016 on the Procedures for the Determination and Crisis of Energy and / or Emergency of Energy implemented by the National Energy Council (NEC) in order to ensure national energy security. The acceleration of provision in various options and the adoption of alternative energy is one of the steps that must be taken in the future. Among these things can be achieved through the production of solid fuel types, such as tibarau cane-made briquettes. According to Grover and Mishra [1], fuel briquette technology can be done with three methods.

The development of bioenergy as an alternative energy source, outside the fossil energy source is increasingly scarce. One of the renewable energy that has great potential in Indonesia is biomass. As in the policy of renewable energy development and energy conversion of the Department of Energy and Mineral Resources (2016) [2] mentioned that the potential of biomass energy in Indonesia is quite large reaches 49.8 GWe. This biomass energy includes, wood fuel, agricultural & plantation / forest waste, organic material components from industrial and household waste. Some alternative energy that can be developed as a substitution of petroleum is geothermal, coal, and biomass (waste). For geothermal and coal is still a fossil energy although not yet optimal use and processing. In this regard, it is necessary to optimize the development of materials and potential biomass energy sources into a form of innovative fuel products that can meet the demands of energy needs and make it as a trading commodity.

Tibarau cane plant is an alternative source of potential energy because tibarau cane produces biomass of bagasse (bagasse). Biomass can be burned in the form of powder, briquette, or bar. Briquettes are an effective method for converting solid raw materials into a compact form that is easier to use.

The utilization of tibarau cane that has not been touched so far can be used as solid fuel (briquettes) processed with a combination of potential materials of other biomass energy sources. This is one form of solid fuel product innovation so it becomes the embryo of alternative energy. By diversifying sources of potential energy derived from biomass can improve the investment climate and national capabilities in implementing government programs.

The research that has been done on tibarau cane plant that get the result of low heat bio-ethanol power and also has the value of heat energy / calor to the juice by fermentation process after distillation [3]. As a naturally occurring fibrous plant containing lignocellulose, tibarau sugarcane also contains sugar with brix content of origin (niranya) ranging from 3% - 6% and slightly lower than sugarcane cultivation containing 8% - 12% brix [4]. Lalit K Singh, et al [5] from Pelgia Research Library-Indian Institute of Technology reported that tibarau cane plant has chemical content. The potential of this tibarau cane plant can be recommended for raw materials for the development of biofuels.

Before tibarau cane briquette product developed as alternative fuel, it is necessary to evaluate the quality and performance of calorific value analysis so that this innovation model can be applied as a contribution to achieve national energy security.

2. Research Methods

In this research, an experimental study on the characteristics of tibarau cane briquette test sample. The implementation orientation is focused on the manufacture and development of briquette fuel products made from raw tibarau cane (Figure 1, and 2). Sugarcane was previously chopped and made into granular particles (mesh) as a requirement.

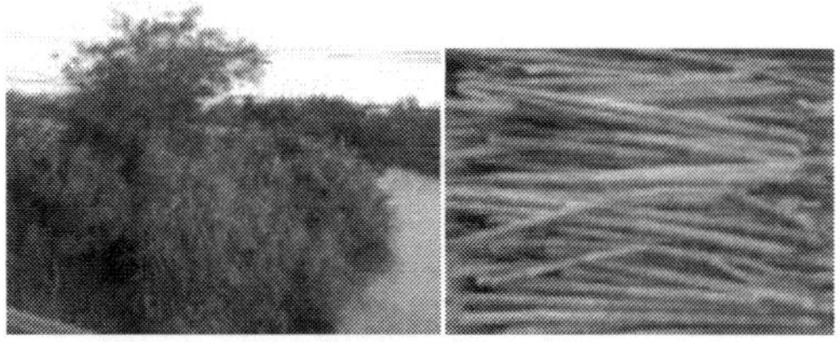

Figure 1. Tibarau Cane Plantation

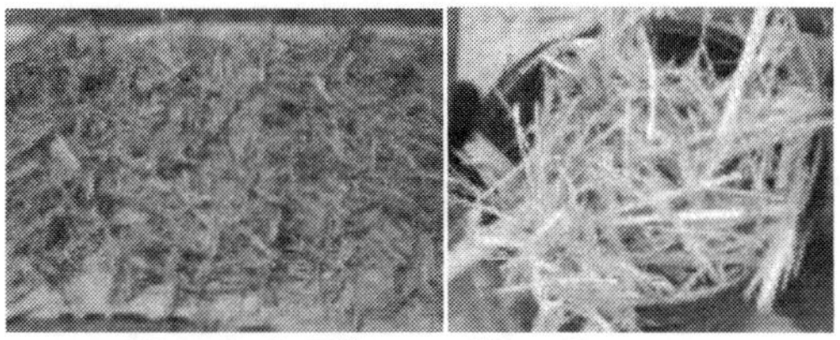

Figure 2. Tibarau Cane Bagasse

The adhesive material used in the manufacture of this briquette is tapioca (Figure 3). The method of making this briquette with optimal percentage of mix as variance, modification of addition of raw material having potency as main mixture. Percentage of main raw material mixture to adhesive is 90%: 10%, 80%: 20%, 70%: 30%, 60%: 40%, 50%: 50%. Prepared mixed variance is followed by made on compaction is 100 kgf/cm^2. The madeed briquette model is then dried in a few days heat radiation of the sun.

Figure 3. Tapioca

The next perform testing in the laboratory for each test sample produced. The objective of this research is by applying appropriate methods to inventory, measure, test or analyze the parameters and research variables that are established or arise due to treatment and conditional.

In the laboratory experiment, a device called "Bomb Calorimeter" (Figure 4) was tested with stable room temperature conditions. The test obtained is related to heat energy of combustion or calorific value.

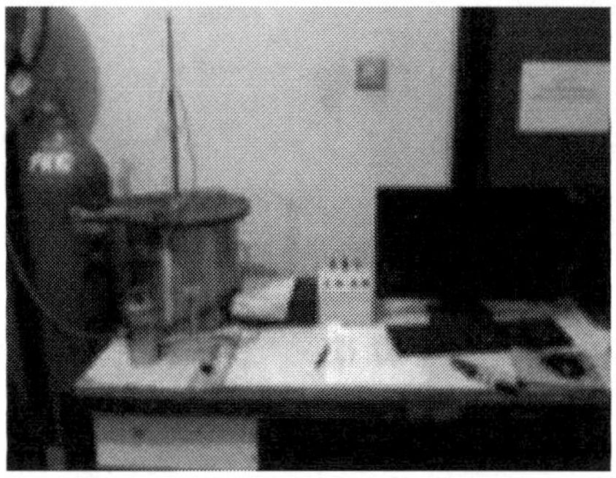

Figure 4. Bomb Calorimeter

In the analysis of each type and composition of the percentage of mixed materials and treatment parameters will be recorded and tabulated in the table data collection is designed as needed. Data analysis and calculation using the existing raw formula and computed aided processing. The highest heating value (N_{bb}) can be calculated by the equation:

$$N_{bb} = \frac{(t_2 - t_1)\, c_v}{m_{bb}} \qquad (kJ/kg) \tag{1}$$

3. Results and Discussion

In this study prepared as much raw materials as possible for the manufacture and printing of tibarau cane briquettes. The need of research materials in the form of bagasse tibarau by predicting the number of variations of the mixture between bagasse with tibarau adhesive. This is done based on the line of thought according to technical briquette making then obtained some prototype of the physical result of development (Figure 5).

Figure 5. Tibarau Cane Briquette.

Table 1. Physical Characteristic of Tibarau Cane Briquette

Composition Percentage Raw Materials and Adhesives		Briquette Mass dry	Density (ρ)	Calorific Value N_{bb}
Tibarau Cane (%)	Tapioca (%)	(gr)	(gr/cm^3)	(kJ/kg)
90	10	45,4	0,482	9479,82
80	20	53,2	0,565	11221,72
70	30	52,6	0,558	8195,95
60	40	53,2	0,565	7127,22
50	50	52,8	0,561	5059,14

Based on the order of procedure of using the Bomb Calorimeter tool, the values of combustion for tibarau cane briquettes are generated as in Table 1. The values obtained by the quantities of physics in Table 1 are derived based on the standard formula by inputting the data from the measurement results on the test equipment Bomb Calorimeter.

From the testing result of tibarau cane briquette calorific value to all variations, then from the analysis data can be stated that the composition that has the greatest value is recommended for the production more as the next test sample. Tibarau cane briquettes that have an optimum value of 80 : 20 composition variance, where calorific value is 11.221,72 kJ/kg and density 0,565 gr/cm^3.

4. Conclusion

From this research can be concluded as follows:

a. Successfully utilize tibarau cane as a candidate of raw material Briquette in the development of renewable energy to achieve national energy security.
b. The resulting tibarau cane briquettes have physical characteristics of optimum calorific value of 11221,72 kJ/kg at 80: 20 composition variance with density 0,565 gr/cm^3
c. The composition of the mixture between the raw material and the adhesive, the grain size, the pressure force (compaction) greatly affects the mass of the briquette value of the resulting calorific value.

IOP Conf. Series: Materials Science and Engineering **335** (2018) 012058 doi:10.1088/1757-899X/335/1/012058

5. References

[1]. Grover,P.D & Mishra,S.K (1996), Biomass Briquetting: Technology and Practices, Field Document No 46, FAO - Regional Wood Energy Development Program in Asia, Bangkok

[2]. Departemen Energi dan Sumberdaya Mineral (2016); Statistik Energi Indonesia

[3]. Hasanuddin & Hendri Nurdin (2016); Pengembangan Bahan Bakar *Bioethanol* Dari Tumbuhan Tebu *Tibarau* Dengan Penambahan Zat Kapur Kulit Kerang Untuk Peningkatan Energi Panas, Laporan Penelitian Desentralisasi

[4]. Hasanuddin & Hendri Nurdin (2015); Design and Contructions of Simple Distilations Unit With Reflux Column Model For Cane Tibarau (*Saccarum Spontaneous Linn*) Bioethnol Productions, Proceeding International Conference ICOMSET, pp. 441 - 447, Oktober 2015

[5]. Lalit K Singh, et al (2011), Utilization of hemicellulosic fraction of lignocellulosic biomaterial for bioethanol production, Journal Pelagia Research Library Adv. Appl. Sci. Res., 2011, 2 (5):508-521

IOP Conf. Series: Materials Science and Engineering **335** (2018) 012059 doi:10.1088/1757-899X/335/1/012059

The Combination of Coastal Resources Potential: Development of Windmill Techno Park in the Context of Edutourism and Hinterland Analysis

H Hasanuddin*, H Nurdin, W Waskito, R Refdinal

Mechanical Engineering Department, Faculty of Technology, Universitas Negeri Padang, Indonesia

*sanquansing55@gmail.com

Abstract. There are many coastal cities in Indonesia, among them the city of Padang and Pariaman in West Sumatra with the ownership of the grace of coastal resources, such as stunning coastal scenery and wind energy that can be converted to its potential. These two types of resources need to be combined utilization especially in relation to the development of tourism so as to have an impact on the regional development and the cities growth. This paper discusses of information related to the potential of coastal resources used as a parameter development of windmill techno park in the context of educational tourism. The discussion focuses on locational aspects and consideration of the design and modification for the windmill model to the enhancement of tourist attraction. In addition, it is also studying the presence of the windmill techno park as a development center in the context of hinterland in the regional analysis. Through the measurement results obtained average wind speed (2.5 - 6.5) m/s and meet the standards & criteria as a windmill design variables. While the geomorphology of the region, its location has a breathtaking coastal landscape as a tourist destination. Then through gravity analysis, the location of the windmill techno park development provides optimum strength to the growth of the city of Padang, with the value of interaction strength is higher than other regions. This means as a sub-urban area of the city the existence of windshield techno park later can support economic development and growth of cities around the coast.

1. Introduction

The tourism sector in Indonesia in general, especially in West Sumatera area is still experiencing many problems and challenges, so that the number of potential and variety of existing resources gradually does not continue to stretch and can contribute to become the prime mover role of regional economic growth. Places of visits and attractions are still stagnant and slow moving so not yet a major choice for tourists both nationally and internationally.

The area of natural tourism & coastal areas as part of sub urban can not function optimally as hinterland for development and growth of urban area. As it is known that one of the external factors that will affect the development of a city is its relation with other cities, as well as its relation with its hinterland, including the surrounding rural areas and its linkage to resource potential [1].

In relation to the problems, there should be breakthroughs and solutions to solve the problems faced by introducing an educational tourism model through the concept of windmill techno park in coastal areas and/or on scenic areas, especially on objects that are potential to be developed as nature tourism.

The purpose of tourism development will be successful with optimal when supported by the potential of the area either in the form of natural attractions and tourist attractions man-made. Travel activities in tourism in general, not least educational tours will have widespread impact for economic development in the region, because it can support the economic movement of the people as well as open pockets of art and culture, including the activities of utilization and preservation of the potential of natural resources to be known by every tourist [2].

The windmill techno park concept is based on the terms wind fields and offshore wind parks associated with an area or area for wind power generation, where many conversion installation units consist of various windmills/wind turbines built on vast land, coastal / coastal areas, and windmill rafts floating above sea level or ocean breeze fields [3].

Another aspect of the concept of wind farms (a collection of windmill installations) is to make it a tourist visiting park by modifying the constructs and their functional features, such as introducing a variety of physical models along with the essential functionality of a windmill on potential coastal areas location placement so that it becomes a special tourist attraction, and especially worth the educational content/nuance. Therefore, data and information about the potential of coastal resources in support of windmill techno park development program, as well as how to blend its existence to the development of tourism, especially education tourism and its contribution as hinterland area of urban growth.

2. Research Methods

In general, the approach used in this study is multidisciplinary or a combination of quantitative and qualitative, in accordance with the goals and target achievements in each year of implementation. Particularly in this article is oriented on methods and techniques, exploration, observation, and survey as well as quantitative (field measurement) and documentation studies.

3. Results and Discussion

3.1.Overview of Current Tourist Distination at the Research Location

So far, coastal areas within the study area are as sub urban coastal cities, especially for Padang and Pariaman areas where natural coastal tourist visits, with little physical development touch. There are two locations of visits and forms of tourism located in the area, namely the concept of religious tourism in the form of a tomb of Muslim religious figure "Syech Burhanuddin" located in the area Ulakan-Tapakis and sightseeing tours of the Pantai Tiram. Although both forms and areas of tourist visits have long existed, but has not shown its function optimally for the support of urban growth.

(a) The Tomb of Syech Burhanuddin (b) Traditional Food Sales Area

Figure 1. Ritual Tour Model "Syech Burhanuddin" Pariaman

Figure1 shows the tomb garden of Syech Burhanuddin visited on a seasonal basis by his followers of understandings from various regions of Indonesia. They make tourist visits, are seasonal and limited by performing religious/ritual activities called "basafa" traditions. In the villages around the tomb area stood and found traders area of traditional food of the local area.

In addition, in some coastal areas close to tourist destinations are also found some attractions such as beach tourism in the village of Ketaping in Batang Anai District. Specifically this area lies adjacent to the city of Padang and the location where there is an international airport of Minangkabau. Based on community information and observations in the field this area with a number of beach scenery is relatively crowded and visited by visitors and relatively potential to be addressed and developed as a tourist destination.

3.2. The combination of Coastal Resources Potential and Hinterland Analysis

The wind turbine technological park that is introduced is the engineering and the creation (work / artificial) of human being used as a tourist attraction or the attraction of the visit and also serves as a medium of learning/education for tourists, especially for students/students in the context of educational tourism. To need a number of consideration factors or development parameters in support of technical implementation.

3.2.1 Coastal Resources for the Development of Windmill Techno Park

By conducting a number of exploratory studies on coastal / coastal areas close to or as sub-urban areas of urban areas, especially in the surrounding areas of Padang & Pariaman, information about 2 (two) main supporting potentials related to resouces endowment, i.e wind power/energy and natural landscape. These two types of coastal resource potential are used as parameter measures in the windmill techno park development plan. Integrating these two types of coastal resources will make it different from other areas / tourist visits.

***a.** Information on the Speed of the Wind*

The characteristics and potential of wind energy resources in the research area in this case are taken into consideration in designing and developing windmill physical models. For that purpose, the livelihood and selection of the area that can be used as the placement area (site) of the development of windmill tekno park by setting on the nearest coastline/ lips around the city (Padang and Pariaman Cities).

Table 1. shows the potential of wind resources in coastal areas in villages around Minangkabau International Airport in Batang Anai and Ulakan-Tapakis sub-districts in Padang Pariaman and Purus beaches in Padang.

Table 1. Average Potential of Wind Speed in Coast of Padang City and Padang Pariaman

No	Name of Location / Village	Measurement Time & Wind Speed (m/s)			Average Speed (m/s)
		Morning (8 - 10a.m)	Daytime (10 a.m -2 p.m)	Afternoon (14 - 18 p.m)	
1.	Manggopoh Ulakan	0,0 - 1,5	2,0 - 5,0	5,2 - 6,9	2,4 - 4,46
2.	Tiram Ulakan	0,5 - 2,0	2,2 - 5,5	5,5 - 6,3	2,7 - 4,60
3.	Ketaping Bt Anai	1,0 - 2,2	2,5 - 5,8	6,0 - 6,5	3,2 - 4,83
4.	Purus Padang	0,0 - 1,8	2,0 - 4,5	4,7 - 6,0	2,2 - 4,10

The measured periods of wind velocity are taken / measured in the time lap for 1 week, adjusted to the way / measurement model performed by [4]. The average measurement of wind speed is not much different from the Meteorological Agency report for the surrounding city of Padang is 2.5 - 6.5 m/s. Technically the average wind speed for a windmill engineering technique is normally to move between 2.0 m/s - 6.0 m/s.

Thus, the mean locational selection and placement of wind parks planned to meet the requirements & options for wind turbine construction engineering. That is, the functional variation of wind speed is relatively safe and feasible to be used as a design parameter for the development of a wind power conversion system.

***b.** Information on Regional Landscape Characteristics*

The landscape of nature (natural lanscape) is meant a geomorphological area of various forms of the earth (valleys, rivers, lakes, mountains, sea views, etc.). These landscapes are potential or natural tourist attractions, which can be integrated into their use for tourism development by providing an attractively appealing attraction such as a set of windmill installation units. The Netherlands became famous as the country of windmills, through the attractions with wind fields and windmill hills so that set every May 13 as the day of the windmill. To support the placement of windmills, giving more impression of the appearance of the background of natural scenery, especially the scenery of the sea and coastal areas, the following are provided snapshots of photo shoots of coastal areas used for the choice of construction sites.

In Figure 2 one of the coastal area with the background of the sea wave scene when the sun with a blazing sun in broad daylight at around 12:00 pm. Coinciding with the beach breeze, not only can reduce the heat of daylight but it is a potential for energy resources that can drive windmill blades.

From the aspect of the beautiful natural scenery, although in hot weather but the impression to the memories & capture the moment like this by sebahagian visitors are highly sought and eagerly awaited his presence. Moreover, the waves whilst enjoying the natural scenery followed by the artificial attraction of the presence of windmills along the coast so that it becomes the impression as a windmill beach tour (the windmill beach).

Figure 2. Landscape Beach Lipscape on the Village of Ketaping Ujung- Pd Pariaman

Furthermore, in Figure 3 shown a natural landscape of the beach when the sun enters the afternoon threshold at around 16:00 pm which is also as one of the location of the location of the park of the windmill. Air condition began to feel a bit cool because the sun's rays have started to fall in intensity. Most visitors usually start a lot of downhill walks along the beach lips set footprints on the sand.

The tourists will be eagerly waiting for the sunset event, so they will come & be in the location while enjoying the natural scenery in the area around the tour long before the event took place. With the construction of the windmill park along with other supporting tourism facilities will give the impression and attraction for visitors, so that will give impact to the growth of the city & economy of society along with the increasing number of tourist visits.

(a) Sand overlays on the beach

(b) The sun sets in western touk

Figure 3. One of the Lanscape Coast Area in Ketaping- Batang Anai

3.2.2. The Tourist Visit Area as a Function of City Hinterland

From the aspect of growth pole theory, tourist destination (TD) or tourist visit area can act as growth centers including its function as a back-area or hinterland for the development and growth of a city. Are cities with one or several tourist areas, such as the coastal city with its natural scenery will have a special attraction for tourists to make it more memorable and fascinating or a choice of visit and may be developed faster than the cities without the ownership of tourist resources.

To see how far the interaction that occurs due to the construction of windmill techno park to the city area around the development, in this case the use of interaction theory between regions based on the method of gravity. The basic concept of this analytical tool is to discuss the size and spacing between two places, the growth center with the surrounding area, to the extent to which a region that is the center of growth influences and interacts with the surrounding area.

The theory of gravity is the most widely used model to see the magnitude of the attraction of a potential located at a location. The model is often used to look at the potential linkage of a location and the magnitude of the area of influence of that potential. That is why the model of gravity doubles as a location theory and as a tool in planning [5]. The greater the number of interaction between sub-districts as the center of economic growth with the surrounding area shows the closer interaction relationship between the growth center and the surrounding area (hinterland). The formula of determining attractiveness among sub-regions (Tij) is:

$$T_{ij} = k\frac{P_i \times P_j}{{d_{ij}}^b}$$

where:

Pi = Population sub-region i,
Pj = Population sub-region j,
d_{ij} = distance between sub-region i and sub-region j,
b = rank is taken 2.0
k = regional interaction constant, taken 1.0

For the sake of development of educational tourism, the population in this case will be approached with the number of potential trip trips conducted by the number of students and teachers who live in the area around the development of windmill techno park. Based on field information data and secondary data from district documents at sub-district and city level, the result of the analysis is obtained using the formula as in Table 2.

Table 2. Location Data of Techno Park and Interaction Analyisis

No.	District / City	Population	Amount of Student + Teacher	Distance Between Regions (km)		Level of city interaction (T_{ij})x10^6	
				Padang	Pariaman	Padang	Pariaman
1.	Batang Anai	46.424	2,346	20	36	5,86	0,56
2.	Ulakan Tapakis	19.431	3,326	35	21	0,08	0,02
3.	Padang	902.413	658,802	0	56	-	(64,67)
4.	Pariaman	86.724	30,817	56	0	(64,67)	-

Source: Survey Result and Document, 2017

Gravitational analysis in this study is used to assess the strength of the relationship (proximity) between two regions, where one region can be regarded as a mass that has an attractive appeal so that will appear the relationship of mutual influence between the two regions. In terms of regional economics, inter-regional relations can be identified as economic interactions between the growth centers and the surrounding areas. The result of this analysis shows that the highest interaction value is between Batang Anai subdistrict to Padang city, with an interaction rate of 5.86 x 10^6, compared with Ulakan Tapakis subdistrict to Padang city or Pariaman city as its sub-urban area.

High interaction rates indicate a close relationship between growth centers and surrounding areas. Interactions can be characterized through the movement of people, goods and money. In addition, it can be realized in the form of economic and social service relations of the community within the region. With this analysis it can be seen that locational placement & construction of the Windmill Park in the Batang Anai subdistrict area shows strong interaction as hinterland for Padang city, which shows the highest interaction rate compared to other cities.

IOP Conf. Series: Materials Science and Engineering **335** (2018) 012059 doi:10.1088/1757-899X/335/1/012059

4. Conclusion

From this research can be concluded as follows:

a. The combination of two types of coastal resource potential, ie landscape / geomorphology of the region and the motion / kinetic energy of coastal winds can be used as parameters of development planning of windmill techno park within the context of education tourism development.

b. The selection and location of the optimal windmills for the coastal zone, in the results of this research lies in the coastal area of Batang Anai District with the strongest interaction power to the growth of the city of Padang for 3.86×10^6 units.

c. With the presence of a windmill techno park at the selected location, it can also serve as a development center & tourist destination that will have an impact on the growth of surrounding cities.

References

[1]. Soegijoko [2005], Keterkaitan Pusat –Pusat Pertumbuhan dengan Daerah Hinterlandnya (*The Article in Indonesian Language*), http, google, Acses on Jun,2017.

[2]. Ajake, A.O. & Amalu, T.E. (2012). The Relevance of Tourism on the Economic Development of Cross River State, Nigeria. Journal of Geography and Regional Planning. 5(1), 14-20.

[3]. Scalavounos, P.D (2007), Floating Wind Turbine, Laboratory for ship and Platform flow – Dept.- of Mech.Engineering, MIT-USA

[4]. Gregory Ford & Victoria Hewey & Nicholas Lima (2013), An Analysis of Small-Scale Wind Pump Design for Use in Developing Countries, Institute in Partial Fulfillment of the Requirements for the Degree of Bachelor of Science BJS – WI13, Faculty of Worcester Polytechnic.

[5]. Haynes, K and A.Stewart Forteringham (1984), Gravity and Spatial Interaction Models, Gravity Model Overview, Beverly Hills, Sage.

IOP Conf. Series: Materials Science and Engineering **335** (2018) 012060 doi:10.1088/1757-899X/335/1/012060

MMSE Estimator for Children's Speech with Car and Weather Noise

V Sayuthi*

Department of Information System, STMIK Indonesia Padang, Padang, Indonesia.

*vetho@stmikindonesia.ac.id

Abstract. Previous research mentioned that most people need and use vehicles for various purposes, in this recent time and future, as a means of traveling. Many ways can be done in a vehicle, such as for enjoying entertainment, and doing work, so vehicles not just only as a means of traveling. In this study, we will examine the children's speech from a girl in the vehicle that affected by noise disturbances from the sound source of car noise and the weather sound noise around it, in this case, the rainy weather noise. Vehicle sounds may be from car engine or car air conditioner. The minimum mean square error (MMSE) estimator is used as an attempt to obtain or detect the children's clear speech by representing simulation research as random process signal that factored by the autocorrelation of both the child's voice and the disturbance noise signal. This MMSE estimator can be considered as wiener filter as the clear sound are reconstructed again. We expected that the results of this study can help as the basis for development of entertainment or communication technology for passengers of vehicles in the future, particularly using MMSE estimators.

1. Introduction

Previous research mentioned that most people need and use vehicles for various purposes, in this recent time and future, as a means of traveling. Many ways can be done in a vehicle, such as for enjoying entertainment, and doing work, so vehicles not just only as a means of traveling [1].

This research focus on voice and speech activities which are done by human passengers in the vehicles, for example, speech activities of daily human doing or voice activities at work, where are done in the vehicles. To have speech activity services, the vehicles must need communication system to connect within and with outside environment.

Recent architecture of vehicle communication and mobile communication technology has served those activities. Femtocell mobile in the 5G communication cellular technology can make passenger to communicate within vehicle or to outside environment [2].

Besides that 5G networks, there are other technologies to realized it, some functions in these vehicles connect persons within it and with the outside of vehicles by "*connectivities*" known as v-2-x connectivities, include vehicle-to-sensor on-board (V2S), vehicle-to-vehicle (V2V), vehicle-to-road infrastructure (V2R), and vehicle-to-internet (V2I), all can be implemented by using wireless technology [3].

Some aspects of those technologies need better performances of wireless solution, frequency band, data rate, TX power, MAC protocol, and modulation. The application purposes of the technology must be considered to work properly in v-2-x connectivities[3] and also in 5G cellular communication technology too.

The electronic computers had advanced in size and the performance, this makes the performance of communication technology had increased better, so it caused and helped the complexity and performance of speech technology increase and advance, and this trend is predicted will be continuous in the future. This system of communication technology needs the robustness of speech that will be significant to apply. Concern to have better performances of speech communication activities in vehicle by previous researches, we can conclude that the speech communication research area in vehicle consists of two main categories that must be considered, the speech processing and the speech communication technology, both are designed as speech communication system in the vehicles with vehicle based-centred, that can make human such as passenger or driver communicates each other within vehicle or outside environment, especially with voice interactive activities.

Many factors that must be concern in speech vehicle communication are robustness in environmental and internal vehicle sources noises, voice-interactive format dialogue and solutions, and the convenience and safety in mobile speech applications [4].

2. Human voice and Noise Sources

Many devices and speech processing for example in speech therapy for human voice or in communication, is intended for children user application, due to less complexity, recommended to above of six years old of age [6], and statistics said that for example in the Europe, percentages of number of passenger cars by age 5 to 10 years is around 25 % from 29 countries in Europe [7].In this study , we will examine the childern's speech from a girl of eight years old that satisfied [6] and [7] in the vehicle that affected by noise disturbances from the sound source of car noise and the weather sound noise around it, in this case, the rainy weather noise. Vehicle sounds may be from car engine or car air conditioner.

3. Car Noise

Noise in vehicle can happened in many version depends on many sources that appeared in the vehicle. Figure 2 explain some example that arise when noise within vehicle appears. [4].

We can divided the internal car noise by divided the categories of the sources of noise with specific and detail dividing, we can defined it, as follows [4];

1. First we define the type of vehicle, wether it is sedan, truck, or SUV.
2. Then if we choose sedan, which model of sedan vehicle that focused. The option can be Nissan Altima or Honda Civic
3. Then look out the condition of car we observed, wether it is stationary or in motion
4. Then we get into the car , it can be especially air conditioner , wether it is off or on
5. Any other significant real condition can be window status, or in other component in internal vehicle can be horn status.

4. Outside Environment and Weather Noise

Our research focus on vehicle that located in rainy weather condition. The geographical West Sumatra region, according to data has extreme daily rainfall values which ranges from 42.4 to 114.7 mm. These rainfall values are higher and more varied than the eastern region of Sumatra. That can be seen in figure 1 below, threshold of extreme rainfall in Sumatra island (1998-2012) [5].

This natural phenomena of rainy weather can make noise for vehicle, especially for human passengers those are doing speech activities in car or vehicles. It can be happened in the night or at noon time in West Sumatra region.

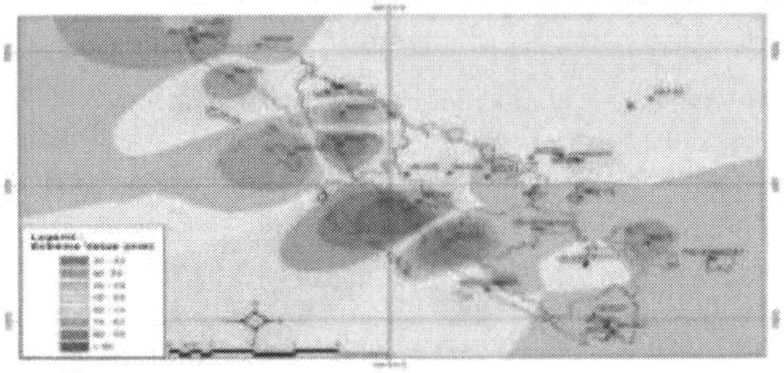

Figure 1. Threshold of extreme rainfall in Sumatra island (1998-2012) [5].

5. Speech Enhancement

Speech enhancement can be doing by three step, those are: 1. Data acquisition, 2. Analysis, computing, and method using of original signal detection, and 3. Removing, eliminating or reducing noise signal. [4], [11].

5.1. Audio Noise Reduction

Signal of speech and noise can be mixed in any area or environment for examples are in building, in outdoor street and within cars those are sources of noise which are located relatively near of person who speaks and hears the speeches.

If the signal of original speech, is wide-sense stationary, ergodic and uncorrelated with noise, it can be computed with wiener filter approach[10] and if the noise signal is white noise which has the constant magnitude spectrum along of frequency so it can be reduced with minimal error by using technique of Minimum Mean Square Error (MMSE) estimator which can be considered as a function of wiener filter [11] .

Let us denote original data speech signal as q_n, which is distorted by white noise p_n of noise sources, is received in our devices that hear it as o_n. So the equations

$$o_n = q_n + p_n \tag{1}$$

Wiener filter in this problem is Linear Time Invariant (LTI) System and Finite-duration Impulse Response (FIR) N-length Filter. The impulse response of this filter in not this interval $0 \leq n \leq N-1$ is $h(n) = 0$. [11]

By using Wiener filter approach of Simon Godsill works, until we have $J_{\min}$ and J_{actual} (Simon Godsill, 2015) [10], it can be followed as below equations, first is that we have autocorrelation function of noise p in discrete time k with σ_p^2 as,

$$r_{pp}[k] = \sigma_p^2\, \delta[k] \tag{2}$$

Autocorrelation of noisy signal denote as $\mathrm{r}_{oo}[\mathrm{k}]$, while autocorrelation of desired signal is $\mathrm{r}_{qq}[\mathrm{k}]$, this value follows the equation

$$r_{oq}[k] = r_{qq}[k] \tag{3}$$

$\mathrm{r}_{\mathrm{pp}}[\mathrm{k}]$ can be measured by

$$r_{pp}[k] \approx \frac{1}{N}\sum_{n=0}^{N-1} o_n o_{n+k} \tag{4}$$

$$r_{qq}[k] = r_{oo}[k] - r_{pp}[k]\begin{cases} r_{oo}[k] & k \neq 0 \\ r_{oo}[0] - \sigma_p^2, & k = 0 \end{cases} \quad (5)$$

The wiener filter equation of h becomes

$$\boldsymbol{h} = \boldsymbol{R}_o^{\ -1}\boldsymbol{r}_{oq} \quad (6)$$

The theoretical minimum mean square error $J_{\min}$ of wiener filter equation of h is

$$J_{min} = r_{qq}[0] - \boldsymbol{r}_{oq}^T\boldsymbol{h} \quad (7)$$

and the actual mean square error J_{actual} of the true q_n from the wiener filter approach systems of h is

$$J_{actual} = \frac{1}{N}\sum_{n=0}^{N-1}(q_n - \hat{q}_n)^2 \quad (8)$$

6. Method and preparing data measurement of noised speech in vehicle

We can process data measurement by collected children voice in Indonesian language for the word of 'pohon' that equals to 'tree' in English. Then we collected data sounds of environment as shown in figure 2.

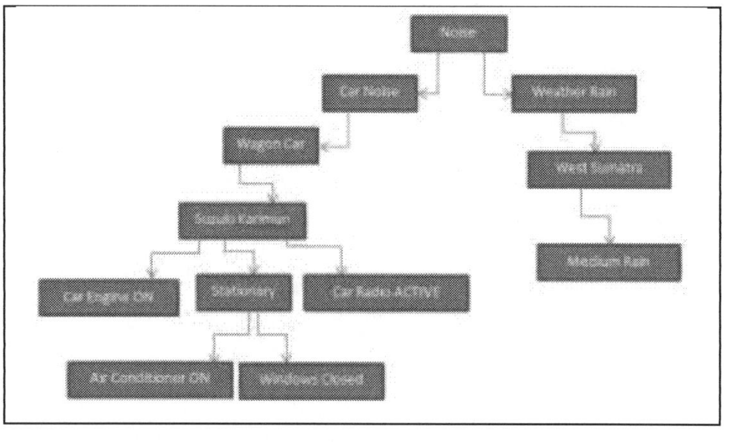

Figure 2. Car and weather noise tree structure. The leaves have most specific information [4].

7. Experimental Data Result of MMSE Enhancement Estimator

The minimum mean square error (MMSE) estimator is used as an attempt to obtain or detect the childern's clear speech by representing simulation research as random process signal that factored by the autocorrelation of both the child's voice and the disturbance noise signal. This MMSE estimator can be considered as wiener filter as the clear sound are reconstructed again.

With wiener filter of 14 taps we have the original speech signal, the noisy speech signal and the output as shown in figure 3,4,and 5. The quality result of our experiment is average in hearing of the reconstruction of children speech signal, because of the function of taps of wiener filter that values 14, and the method that we have done by conventional and basics or fundamentals of the wiener filter methods that mentioned in above explanation of our basics wiener filter methode.Result of experimentations are,

IOP Conf. Series: Materials Science and Engineering **335** (2018) 012060 doi:10.1088/1757-899X/335/1/012060

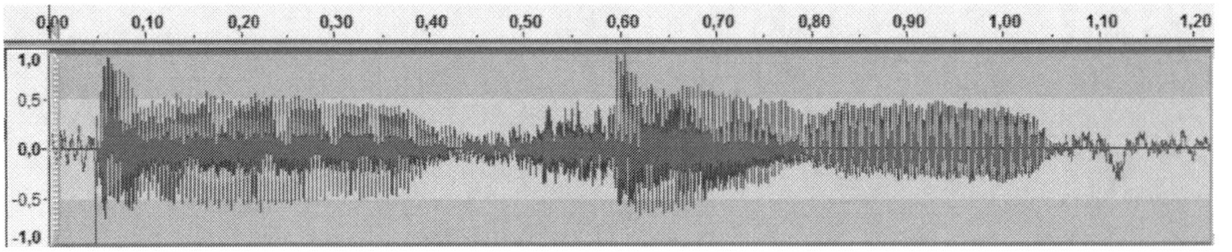

Figure 3. The original speech signal said 'pohon' in Indonesian language equals to 'tree' in English language.

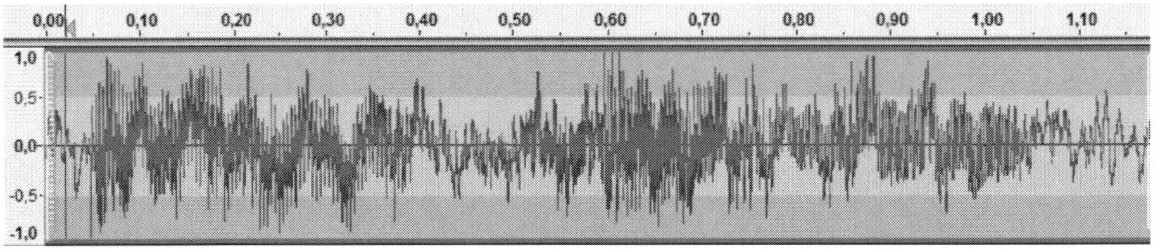

Figure 4. The noisy signal with impact of car and weather noises.

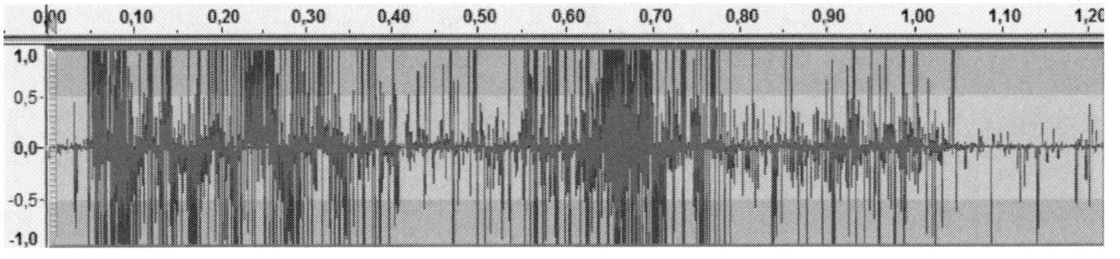

Figure 5. The output of wiener filter signal with taps 14 and autocorrelation scaling value of 0.0078.

8. Conclusion

As we expected that the results of this study can help as the basis for development of entertainment or communication technology for passengers of vehicles in the future, particularly using MMSE estimators. We have shown that with the conventional or basics or fundamentals of wiener filter we can contruct the original signal that affected by car and weather noise.With the limitation of the number of taps in wiener filter we must study in the next future of the number of other taps that may be improved our this final result of the conventional wiener filter to reconstruct or to bring out the original speech signal especially in a car.

With the used of conventional method of wiener filter it have some of difficulties for example in constructed or make formulas and algorithm in this wiener filter method, so hopefully in next future there are some ways to compute this wiener filter algorithms in the better form.

The estimator MMSE can detect speech voice especially children in destructed ways by vehicle sounds or various noises that will be happen may be from the car engine or car air conditioner noises. The techniques to improve the way for minimizing noises in these environments is needed in the future.

Acknowledgements
Our thanks to STMIK INDONESIA PADANG institute and AMAL BAKTI MUKMIN PADANG, foundation who help to succeed the research in contract number of 108/A.12/STMIK-I/2017.

References
[1] Anjan C, Joseph G, Matthew J, Haans-Werner K, and Philip W 2001 Telematics: Decision Time for Detroit *Business Strategy Review*. Vol 12 issue 2 p 21-38
[2] Cheng-Xiang Wang, Fourat Haider, Xiqi Gao and Xiao-Hu You, Yang Yang, Dongfeng Yuan, Hadi M. Aggoune, Harald Haas, Simon Fletcher, Erol Hepsaydir, 2014 5G WIRELESS COMMUNICATION SYSTEMS: PROSPECTS AND CHALLENGES Cellular Architecture and Key Technologies for 5G Wireless Communication Networks *IEEE Communications Magazine*. p 124
[3] Ning Lu, Nan Cheng, Ning Zhang, Xuemin (Sherman) Shen, Jon W. Mark, Connected Vehicles: Solutions and Challenges *IEEE Internet of Things Journal*. Vol 1 issue 4 p 1-4
[4] Nitish Krishnamurthy and John H.L. Hansen 2014 Car noise verification and applications *Interational Journal of Speech Technology*. Vol 17 issue 2 p 167
[5] S Supriyadi, R Hidayati, R Hidayat, and A Sopaheluwakan 2017 Mapping Extreme Rain Conditions in Sumatra by Influence Global Conditions *IOP Conference Series: Earth and Environmental Science*.
[6] Natalia Zharkova, Fiona E. Gibbon and Alice Lee 2016 Using Ultrasound Tongue Imaging to Identify Covert Contrasts in Children's Speech *Clinical Linguistics & Phonetics*. Vol 31 issue 1 p 32
[7] Eurostat 2017 Passenger Cars in the EU *Eurostat*.
[8] Sharmistha S. Gray, Daniel Willett, Jianhua Lu, Joel Pinto, Paul Maergner, and Nathan Bodenstab 2014 Child Automatic Speech Recognition for US English: Child Interaction with Living-Room-Electronic-Devices *Proceedings of Workshop on Child Computer Interaction*.
[9] Pardeep Sharma, Rajesh Mehra, Naveen Dubey 2015 *4th International Conference on Reliability, Infocom Technologies and Optimization (ICRITO) (Trends and Future Directions)*
[10] Simon Godsill, 2015, Random Processes - Random Processes, Optimal Filtering and Model-based Signal Processing. *Lecture Note, University of Cambridge, Signal Processing and Communications Laboratory, Department of Engineering*.
[11] Alan V. Oppenheim and George C. Verghese, 2010, Signals, Systems, and Inference. *Class Notes for 6.011: Introduction to Communication, Control and Signal Processing*.
[12] Saeed V. Vaseghi 2000, Advanced Digital Signal Processing and Noise Reduction , Second Edition. *John Wiley & Sons Ltd*.

IOP Conf. Series: Materials Science and Engineering **335** (2018) 012061 doi:10.1088/1757-899X/335/1/012061

The Influence of Protein Supplementation on Muscle Hypertrophy

A Fardi[1*], **W Welis**[1]

[1]Sport Science Faculty, Universitas Negeri Padang, Indonesia

*adnan_fardi@yahoo.com

Abstract. The problem of this study was the lack of knowledge about nutrition, so the use of protein supplements to support the occurrence of muscle hypertrophy is not optimal. The use of natural supplements is a substitute of the manufacturer's supplements. The purpose of this study was to determine the effect of natural protein supplementation to muscle hypertrophy.The method of the research was a quasi experiment. There are 26 subject and were divided two group. Instrument of this research is to use tape measure and skinfold to measure muscle rim and thickness of fat in arm and thigh muscle. Then to calculate the circumference of the arm and thigh muscles used the formula MTC - (3.14 x TSF). MTC is the arm muscle or thigh muscle and TSF is the thickness of the muscles of the arm or thigh muscles. Data analysis technique used was t test at 5% significant level. The result of the research showed that average score of arm muscle hypertrophy at pretest control group was 255.61 + 17.69 mm and posttest average score was 263.48.58 + 17.21 mm and average score of thigh muscle hypertrophy at pretest control group was 458.32 + 8.72 mm and posttest average score was 468.78 + 11.54 mm. Average score of arm muscle hypertrophy at pretest experiment group was 252.67 + 16.05 mm and posttest average score was 274.58 $\pm$ 16.89 mm and average score of thigh muscle hypertrophy at pretest experiment group was 459.49 $\pm$ 6.99 mm and posttest average score was 478.70 + 9.05 mm. It can be concluded that there was a significant effect of natural protein supplementation on muscle hypertrophy.

1. Introduction

Exercise is a human effort to obtain health and fitness, characterized by a series of organized and planned gestures or exercise activities with rhythmic and composed movements. To achieve this, it is necessary to exercise in the sport systematically and sustainably. In accordance with the opinion of Bo that exercise is a systematic exercise activity in a long time, progressively enhanced and individual that leads to the characteristics of the function and psychological man to achieve goals has been determined [1]. In addition to getting health and fitness, the goal of people doing sports is to get the ideal body and proportional, has a proportionate body muscles is the dream of everyone, because this is enough to affect the performance or appearance in the eyes of others. The best way to achieve that goal is by weight training. Weight training ground can be done in fitness centers. But there are still many fitness members who exercise has not achieved the expected goal of muscle enlargement.

Structured and programmed weight training will result in better muscle growth and formation or increased muscle mass (hypertrophy). In weight training, nutritional intake is a factor to consider. Adequate nutrition intake or in accordance with the needs of the body, will enable the occurrence of muscle hypertrophy. Formation of muscle mass (hypertrophy) and strength is determined by well-programmed exercises supported by a healthy and balanced diet [2]. The role of protein in exercise is the basic ingredient of muscle formation. On the other hand, supplements are also needed to help muscle growth. Protein is also to repair the damaged muscles resulting from exercise.

Excessive consumption of supplements will cause disorders of the body, especially in the liver and kidneys. The findings, published in the British Journal of Cancer, show that those taking muscle-building supplements for three years or more and under age 25 (especially pills and powders containing either creatine or androstenedione) is associated with an increased risk of testicular germ cell cancer (3). Furthermore, The British Dietetic Association (BDA) in Compass (2017) says high doses of additional proteins can cause various health problems such as kidney and liver damage. The UK Department of Health recommends that adults avoid protein consumption higher than recommended (55.5 grams for men and 45 grams for women) [4].

In weight training, to support the occurrence of muscle hypertrophy supplements become a mainstay for fitness members. The use of manufacturer supplements is an option to get instant and satisfying results, but many fitness members do not know the content of supplements or the composition of the supplements they consume. Lack of knowledge about nutrition, resulting in the use of supplements are not in accordance with the dose or not adjusted to physical activity, especially for beginners. Given the side effects and lack of fitness member knowledge about the use of manufacturer supplements, the alternative or substitute to support the occurrence of muscle hypertrophy in weight training is by consuming supplements in natural form. Natural protein supplements are the consumption of protein-containing ingredients such as pure cow's milk or soy milk, eggs and tofu. Compared to the manufacturer's supplements sold in the market or in fitness places, the price of natural protein supplements is much more affordable. Based on the background of the problems already raised, the researcher is interested in conducting research on the effects of weight training with natural supplementation on increasing muscle mass (hypertrophy). The purpose of this study was to analyze the effect of natural protein supplementation in weight training on increasing muscle mass (hypertrophy).

2. Research Methods

Research type is experiment with two-group pretest-posttest design. The study was conducted on One Gym Fitness Padang City. The study was conducted during June to October 2017. Subjects were recruited from registered male fitness members for less than 1 month. Subjects consisted of 13 treatment groups and 13 controls. Subjects do weight training with pyramid method with intensity 75% - 95%, reps 4 - 12, series / set 1 - 5, and interval 30 - 40 seconds. The treatment group received natural foods containing high protein of 1.5 g / kg BW given 4 times a week for 4 weeks. Types of food given are soy milk, egg whites and tofu. Measurement of hypertrophy in the muscles of the arms and thighs using a tape measure and skinfold caliper which aims to measure the muscle circumference and measure the thickness of fat in muscle. After the data obtained then converted with the formula:

$$\text{Muscle Hypertrophy} = \text{MMC} - (3.14 \times \text{TSF}) \quad (1)$$

where:

MMC : Mid Muscle Circumference

TSF : Skinfold thickness

The effect of high protein supplementation on muscle hypertrophy was analyzed using independent test t test.

3. Results and Discussion

The results showed that the mean body weight of experimental group subjects was 69.3 + 10.8 kg and control group 60.4 + 7.8 kg. The average height of the experimental group was 168.1 + 5.6 cm and the mean height of the control group was 173.6 + 6.6 cm. Statistically, there was no difference in body weight and height of study subjects (p value > 0.05).Measurement data of hypertrophy muscle of arm and thigh muscle before and after intervention can be seen Table 1.

IOP Conf. Series: Materials Science and Engineering **335** (2018) 012061 doi:10.1088/1757-899X/335/1/012061

Table 1. Data of Hypertrophy of Arm and Tigh Muscles Experimental Group

	Hypertrophy of Arm Muscle (mm)		Difference (mm)	Hypertrophy of Tigh Muscle (mm)		Difference (mm)
	Pretest	Postest		Pretest	Postest	
Maximum score	278.02	300.72	26.28	466.76	491.32	31.28
Minimum score	227.30	250.58	16.42	440.88	461.16	12.14
Mean	252.67	274.58	21.92	459.49	478.70	19.21
Deviation Standard	16.05	16.89	3.05	6.99	9.05	5.28

From the Table 1 shows that average score of arm muscle hypertrophy in the experimental group was 252.67 + 16.05 mm. After high-protein diet treatment, the average score of arm muscle hypertrophy was 274.58 $\pm$ 16.89 mm with an average change of 21.92 + 3.05 mm. The average score of thigh muscle hypertrophy was 459.49 $\pm$ 6.99 mm. After a high protein diet treatment, the average score of thigh muscle hypertrophy was 478.70 + 9.05 mm with a change of 19.21 + 5.28 mm.

Table 2. Data of Hypertrophy of Arm and Tigh Muscles Control Group

	Hypertrophy of Arm Muscle (mm)		Difference (mm)	Hypertrophy of Tigh Muscle (mm)		Difference (mm)
	Pretest	Posttest		Pretest	Posttest	
Maximum score	290.02	299.16	16.28	479.76	492.90	21.28
Minimum score	230.30	231.30	4.64	441.74	449.02	3.14
Mean	255.61	263.48	16.28	458.32	468.78	10.46
Deviation Standard	17.69	17.21	1.00	8.72	11.54	5.51

In the control group (Table 2), the average score of arm muscle hypertrophy was 255.61 + 17.69 mm. After a high-protein diet treatment, the average score of arm muscle hypertrophy was 263.48.58 + 17.21 mm with an average change of 16.28 + 1.00 mm. The average score of thigh muscle hypertrophy before treatment was 458.32 + 8.72 mm. After a high protein diet treatment, the average score of thigh muscle hypertrophy was 468.78 + 11.54 mm with a change of 10.46 + 5.51 mm. The results of statistical tests showed that there were also differences in hypertrophy of arm muscle and hypertrophy of thigh muscles in the control group. The result of statistical test shows that there was difference of hypertrophy of arm muscle and thigh muscles ($p = 0.00$). From the statistical analysis it can be concluded that there was influence of high protein feeding to hypertrophy of arm muscle and thigh muscle.

The results of this study indicate that there was the effect of high-protein feeding on subjects who are weight training against the hypertrophy of arm and thigh muscles. Hypertrophy is the enlargement or increase of the total mass of a muscle. All hypertrophy is the result of an increase in the number of actin filaments and myosin in each muscle fiber, thus causing the enlargement of each muscle fiber, which is simply called fiber hypertrophy. This event usually occurs in response to a muscle contraction that takes place at maximum or almost maximum strength. To increase muscle hypertrophy requires a good exercise program and optimal protein intake. The formation of muscle mass (hypertrophy) and strength is determined by a well programmed exercise and supported by a healthy and balanced diet [2].

Muscle protein synthesis is stimulated in the recovery period after resistance exercise [5, 6]. However, the rate of muscle protein breakdown is also increased, thereby blunting the change in the net balance between synthesis and breakdown. Although net muscle protein balance is generally improved after resistance exercise, it remains negative. Therefore, nutrient intake is necessary to achieve positive net muscle protein balance.

Hypertrophy following resistance training requires net protein synthesis of the myofibrillar proteins, and hence, a maximal stimulation of protein synthesis is favourable for the development of muscle hypertrophy. Mixed muscle protein synthesis rate is increased in humans after bouts of resistance training provided the stimulus is of a sufficient magnitude [5].

Consuming soy milk after exercising can facilitate the delivery of amino acid to muscle and protein synthesis which will all support perfect muscle formation [7].
Athletes and exercisers consume high-protein diets for a number of reasons. Probably, the most prevalent reason for high protein intake is to enhance gains of strength and mass with resistance training. Muscle mass and strength can be gained on a wide range of protein intakes, from as little as the RDA up to very large amounts. There is little support for the necessity of very high amounts, e.g. > 2g protein/kg body mass per d, for optimal muscle hypertrophy during energy balance or excess. There are more recent studies showing that protein intakes as low as 1.2 g protein/kg body mass per d are sufficient for increased muscle mass [8].

Borsheim *et a.l* proposed the existence of a dose-response relation between muscle protein synthesis and amino acid consumption after resistance exercise on the basis of a comparison of data from 2 studies [10]. They observed a postexercise stimulation of muscle protein synthesis almost twice as great after ingestion of 6 g compared with only 3 g Essential Amino Acids (EAA's) [6,9].

4. Conclusion

The research results can be concluded that the muscle hypertrophy in the experimental group is better than the control group. There was an effect of high protein supplementation against muscle hypertrophy ($p<0.05$).

References

[1]. Bompa 1990. Bompa, Tudor O.,1990. Theory and Methodology of Training; the Key to Athletic Performance. Dubuque, Iowa: Kendall / Hunt Publishing Company.

[2]. Irianto, D. P. 2006. Panduan Gizi Lengkap, Keluarga dan Olahragawan. Yogyakarta: C.V Andi Offset.

[3]. JPNN. 2015. Bahaya Mengkonsumsi Suplemen Pembentuk Otot. dari http://www.jpnn.com / news/bahaya-mengonsumsi-suplemen-pembangun-otot. Diunduh tanggal 10 Februari 2017

[4]. Kompas 2017 Kompas. (2017). *Dibalik Suplemen Protein Pembangun Otot*. Februari 2017 dari http://health.kompas.com/read/2012/09/04/15555275/ di.balik. suplemen protein .pembangun .otot. Downloaded on 15 July 2017.

[5]. Biolo G, Maggi SP, Williams BD, Tipton KD, Wolfe RR.1995. Increased rates of muscle protein turnover and amino acid transport after resistance exercise in humans. *Am J Physiol Endocrinol Metab* 268:E514–E 520

[6]. Chesley A., MacDougall, J. D., Tarnopolsky, M. A., Atkinson, S. A. & Smith, K. 1992. Changes in human muscle protein synthesis after resistance exercise. *Journal of Applied Physiology* 73, 1383–1388.

[7]. Lieby. 2013. Susu Kedelai Untuk Pembentukan Massa Otot. http://liebynews.blogspot.co.id/ 2013/04/susu-kedelai-untuk-pempentukan-massa.html. Downloaded on 10 July 2017.

[8]. Tipton, K.D. 2011. Efficacy and consequences of very-high-protein diets for athletes and exercisers. *Proceedings of the Nutrition Society*, 70, 205–214. doi:10.1017/ S002966 5111000024.

[9]. Esmarck B, Andersen JL, Olsen S, Richter EA, Mizuno M, Kjær M. 2001.Timing of post exercise protein intake is important for muscle hypertrophy with resistance training in elderly humans. *J Physiol 535:301–311.*

[10]. Borsheim E, Tipton KD, Wolf SE, Wolfe RR. 2002. Essential amino acids and muscle protein recovery from resistance exercise. *Am J Physiol Endocrinol Metab*; 283: E648–57.

IOP Conf. Series: Materials Science and Engineering **335** (2018) 012062 doi:10.1088/1757-899X/335/1/012062

Design of a C- Band Circular Polarization Microstrip Antenna

Yohandri[1], Yusna Jumiah[1] and Josaphat Tetuko Sri Sumantyo[2]

[1]Physics Department, Faculty of Mathematics and Natural Science, Universitas Negeri Padang, Jl. Prof. Hamka, Padang, Sumatera Barat, Indonesia 25131
[2]Microwave Remote Sensing Laboratory, Center for Environmental Remote Sensing, Chiba University, 1-33, Yayoi Cho, Inage Ku, Chiba Shi, Chiba, Japan 263-8522

yohandri@fmipa.unp.ac.id

Abstract. The development of circularly polarized microstrip antenna is an interesting topic in current research, due to its superiority in various applications. In this work, the design of a circular polarization antenna that will be operated in the C-band range will be described. The developed antenna is intended to be used for Synthetic Aperture Radar (SAR) applications. Through this application, various targets or areas on the surface of the earth, such as buildings, soil and land can be observed. To get the ideal antenna characteristic, in this research the various parameters in antenna design will be simulated. A software CST Studio will be operated in this simulation. Based on the simulation results, the optimum parameters are obtained in term of reflection coefficient, VSWR, axial ratio, and gain. The reflection coefficient of the antenna (S11) is obtained at -19.75 dB and VSWR of 1.23. Meanwhile, the axial ratio and gain of the antenna were obtained at 2.66 dB and 2.1 dBi, respectively. Based on this simulated results, antenna design is potential to be developed and fabricated for SAR sensor applications.

1. Introduction

In remote sensing technology, a Synthetic Aperture Radar (SAR) sensor played an important role such as mapping, oceanography, disaster monitoring, land use, land classification and various other applications. Moreover, the SAR sensor can be implemented measure land surface changing due to landslide or land subsidence [1]. The SAR technology can be operated with very high accuracy measurements to millimeter order [2]. In measurement, SAR sensors collect information from the target or area observed based on the back scattering signal returned by the object or area. Therefore, an antenna has a very important role as a transmitter and receiver signal to and from the target.

In previous works, several antennas for SAR sensor have been developed [3]-[6]. All the antennas in previous work is developed on L band frequencies, as a result, the size of the antenna in the array configuration is quite large. For the antenna to be installed on unmanned aerial vehicle (UAV), a large size of the antenna will be difficult to carry and requires a large space in the UAV. The smaller antenna, antenna is developed at c band frequency so it is easy to install on UAV and lighter. One way to reduce the size of the antenna is to change the working frequency to be higher like a C-band.

Several C-band antennas have been produced for many applications. Most of the antennas produced work with linear polarization. The purpose of this work is to present a C-band microstrip antenna with circular polarization. The circular polarization is beneficial since the future SAR requires a sensor that insensitive to Faraday rotation effect and produces more information from the target. The circular microstrip antenna with hole at center is adopted in this antenna design [7]. Circular polarization is generated by setting the phase difference on the two feeds. In next section, the design of proposed antenna and research result will be discussed in section two and three, respectively. Finally, the conclusion from this work will be presented in the last section.

2. Antenna Design

The C-band antenna is designed to satisfy the specified SAR sensor parameters as listed in Table 1. In this design, a Nippon Pillar substrate (NPC-H220A) is used having a thickness 1.6 mm, a dielectric constant $\varepsilon_r = 2.17$ and a loss tangent $\delta = 0.0005$.

Table 1. Antenna parameter for SAR sensor

Parameters	Specification
Center Frequency	5.3 GHz
S11	< -10 dB
VSWR	< 2
Axial Ratio	< 3 dB
Polarization (Tx/Rx)	RHCP/LHCP
Main Lobe	4-7 dBi

As shown in Table 1, the proposed antenna is designed to operate at C-Band with center frequency 5.3 GHz. Reflection coefficients (S11) and voltage standing wave ratio (VSWR) is targeted less than -10dB and 2, respectively. To satisfy the circular polarization requirement either Left Handed Circular Polarization (LHCP) or Right Handed Circular Polarization (RHCP), the axial ratio of the antenna design is regulated less than 3 dB.

In this design, a circular microstrip antenna with hole in the centre is impelented. To produce circular polarization, two feeds with phase difference of 1/4 λ are used in the proposed antenna. Design of the proposed antenna is illustrated in Figure 1.

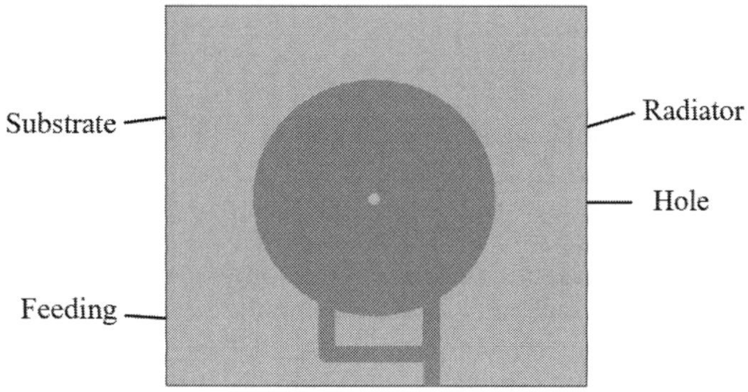

Figure 1. Geometry design of the proposed antenna.

The radiator of the proposed antenna is placed on the top with a distance of 1.6 mm from the ground. Ground of the antenna is at lower part under the substrate. A hole is positioned at center of the radiator. The feeding net work is attached to the radiator directly on the op of antenna surface.

3. Result and Discussion

Based on the simulation results, the optimum parameters of the antenna are obtained. Parameters of the antenna consist of radius of the hole, feeding width, length and width of the substrate or ground, and radiator radius. The optimum of these parameters is listed in Table 2.

Table 2. Optimum parameter of the antenna design

Parameters	Size (mm)
Radius of the hole	0.984
Feeding width	1.85
Length of the substrate	14.71
Width of the substrate	35.79
Radiator radius	19.45

Using the parameters in Table 2, the characteristics of the proposed antenna design are obtained as shown in Figures 2 to 5. In Figure 2, the reflection coefficient (S11) as a function of frequency is plotted. The reflection coefficient at centre frequency 5.3 GHz is obtained of 20 dB. The simulated result of S11 shows the working frequency of the antenna ranging from 1.26 GHz to 5.34 GHz at -10dB. Based on S11 characteristic, the antenna design can be realized and operated on C-band frequency.

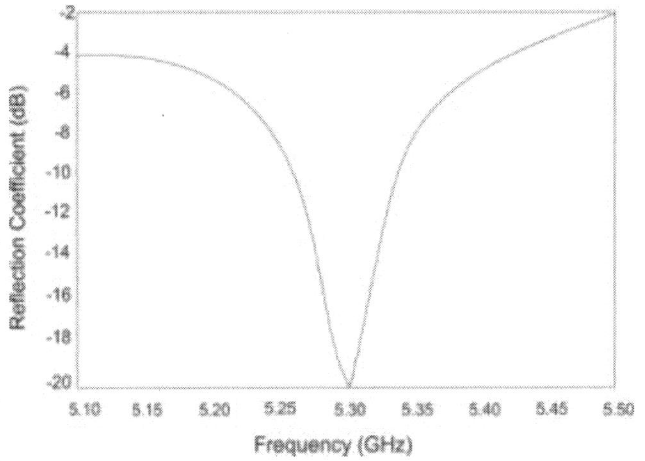

Figure 2. Reflection coefficient plotted as a function of frequency.

Figure 3 present the VSWR characteristics of the antenna. A satisfy antenna must have VSWR less than 2 in working frequency. The simulation results show the VSWR of the antenna deign obtained of 1.23. Thus, it can be concluded that the antenna works well within the designed frequency range. The polarization characteristic of the antenna is shown through the axial ratio (AR) as presented in Figure 4. The perfect circular polarization can be obtained for 0 dB axial ratio. However, the perfect circular are very difficult to produce in a circular polarization antenna. Therefore, the axial ratio of less than 3 dB can still be classified as circular polarization antennas. The simulation results show that the Axial Ratio of the antenna obtained is 1.35 dB. This result show the designed antenna can be classified as a circular polarization. A simulated axial ratio bandwidth of 2.8% is obtained from 5.22 GHz to 5.37 GHz. The direction of circular polarization can be investigated using surface current as shown in Figure 5. In this design, the surface current shows the type of polarization obtained is Left Handed Circular Polarisation (LHCP).

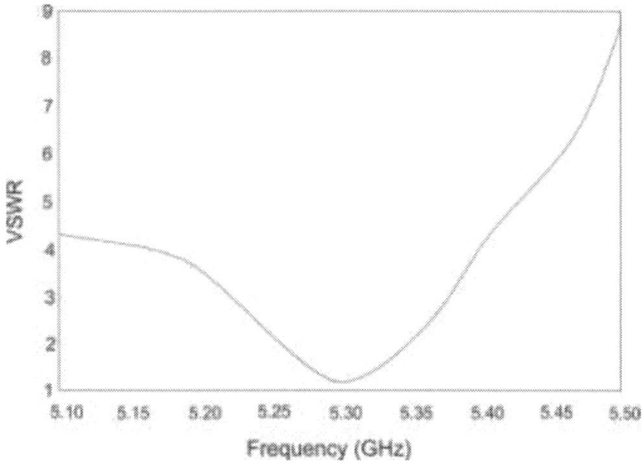

Figure 3. VSWR

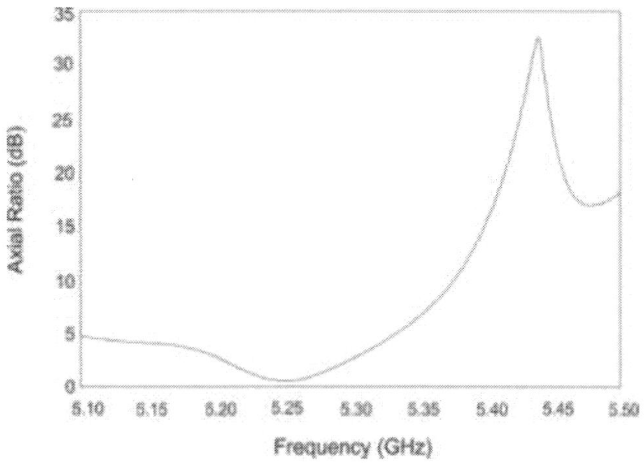

Figure 4. Axial ratio (*AR*) plotted as a function of frequency.

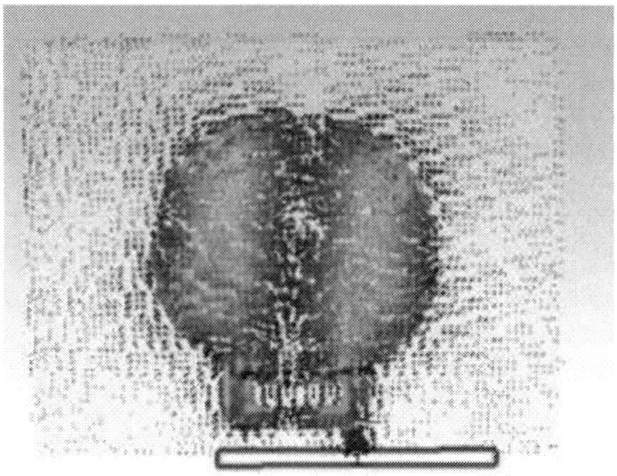

Figure 5. Surface Current.

IOP Conf. Series: Materials Science and Engineering **335** (2018) 012062 doi:10.1088/1757-899X/335/1/012062

The performance of the antenna in terms of radiation pattern is presented in Figure 6. A maximum gain of 2.1 dBic is obtained in simulation. This gain is still acceptable for the antenna at C band frequency. The main lobe of the antenna radiation is generated in 45° direction.

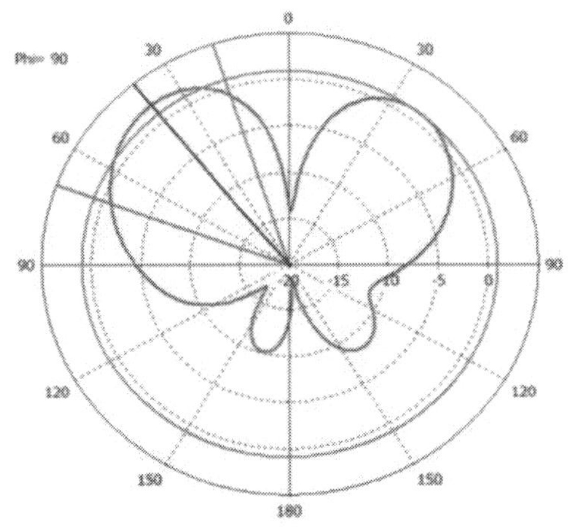

Figure 6. The radiation pattern of the antenna design in the theta plane ($x - z$ plane).

4. Conclusion

A design of C-band circular polarization microstrip antenna has been presented in this paper. In general, the designed antenna can be operated in working frequency at centre frequency 5.3 GHz. The satisfy characteristics are obtained in term of reflection coefficient, VSWR, axial ratio, surface current and radiation pattern. Based on this simulation results, the proposed antenna can be implemented as transmitter and receiver in SAR sensor system.

References

[1] V. C. Koo et al., "A new unmanned aerial vehicle synthetic aperture radar for environmental monitoring," Prog. Electromagn. Res., vol. 122, no. November 2011, pp. 245–268, 2012.

[2] J. T. Sri Sumantyo, M. Shimada, P. P. Mathieu, and H. Z. Abidin, "Long-term consecutive DInSAR for volume change estimation of Land deformation," *IEEE Trans. Geosci. Remote Sens.*, vol. 50, no. 1, pp. 259–270, 2012.

[3] M. Z. Baharuddin, Y. Izumi, J. T. S. Sumantyo, and Yohandri, "Side-Lobe Reduced, Circularly Polarized Patch Array Antenna for Synthetic Aperture Radar Imaging," IEICE Trans. Electron., vol. E99.C, no. 10, pp. 1174–1181, 2016.

[4] Yohandri, Asrizal, and J. T. S. Sumantyo, "Design of tilted beam circularly polarized antenna for CP-SAR sensor onboard UAV," in 2016 International Symposium on Antennas and Propagation (ISAP), 2016, pp. 658–659.

[5] Yohandri, J. T. Sri Sumantyo, and H. Kuze, "Circularly polarized array antennas for synthetic aperture radar," in Progress in Electromagnetics Research Symposium, 2011, pp. 1244–1247.

[6] M. Hussein, Yohandri, J.T. Sri Sumantyo and A. Yahia, 2013, A low sidelobe level of circularly polarized microstrip array antenna for CP-SAR sensor, Journal of Electromagnetic Waves and Applications, Vol. 27, No. 15, pp. 1931-1941. DOI: 10.1080/09205071.2013.828577

[7] C. E. Balanis, "Antenna Theory: Analysis and Design, 3rd Edition - Constantine A. Balanis," Book. p. 1136, 2005.

IOP Conf. Series: Materials Science and Engineering **335** (2018) 012063 doi:10.1088/1757-899X/335/1/012063

The Implementation Internet of Things(IoT) Technology in Real Time Monitoring of Electrical Quantities

D Despa[1*], G.F Nama[2], M.A Muhammad[2], K Anwar[1]

[1]Department of Electrical Engineering, University of Lampung, Indonesia
[2]Department of Informatics Engineering, University of Lampung, Indonesia

*despa@eng.unila.ac.id

Abstract. Electrical quantities such as Voltage, Current, Power, Power Factor, Energy, and Frequency in electrical power system tends to fluctuate, as a result of load changes, disturbances, or other abnormal states. The change-state in electrical quantities should be identify immediately, otherwise it can lead to serious problem for whole system. Therefore a necessity is required to determine the condition of electricity change-state quickly and appropriately in order to make effective decisions. Online monitoring of power distribution system based on Internet of Things (IoT) technology was deploy and implemented on Department of Mechanical Engineering University of Lampung (Unila), especially at three-phase main distribution panel H- building. The measurement system involve multiple sensors such current sensors and voltage sensors, while data processing conducted by Arduino, the measurement data stored in to the database server and shown in a real-time through a web-based application. This measurement system has several important features especially for real-time monitoring, robust data acquisition and logging, system reporting, so it will produce an important information that can be used for various purposes of future power analysis such estimation and planning. The result of this research shown that the condition of electrical power system at H-building performed unbalanced load, which often leads to drop-voltage condition

1. Introduction

Electricity is an important component for any organization,the electricity quality should be always monitored. University of Lampung (Unila) very concern to perform evaluation ofelectricity services, especially at Department of Mechanical Engineering Unila. Various policy have been made including the addition of power supply, installation system repairs.However, these efforts did not provided maximum results without being followed by maintaining the quality of theelectricity itself, because in reality the amount of electricitytends to change as a result of changes in the load or the powersupply itself. These changes should be identified immediatelyin order to anticipate the consequences that may occur to the system.

As a matter of fact, In general, electrical energy sources usea three-phase system, whereas the load in universities adoptone-phase system. In term of this conditions that not monitored properly as well, then it will certainly result on unbalance of electrical quantities for each phase. Manual measurement conducted was certainly not very effective, because there are so many panels that should be monitored. Besides, the value of electricity will also change every time. For that reason there should be a system that can monitor this situation, measure the changes in the value of electrical quantities, and monitored data online by utilizing the internet network facilities that are already available in University. The proposed of this research is to implementing the Internet of Things (IoT) technology in field area of electrical quantities monitoring to provide an useful information to operators, and if there any abnormal condition, then the mitigation and prevention can be done quickly.

2. Literature Review

2.1. Related works

Monitoring of electrical energy consumption, room temperature, and internet network bandwidth has been done on previous work [1], unfortunately Gigih et all in this research only provides information on the magnitude of electrical energy consumption by capturing voltage and current values, while the value of other electrical quantities did not proposed. In the other work [2], Despa et all has succeeded created a prototype system that can monitor the quantity of electricity in a 3-phase electric power panel using BCM2835 Single Board Computer. In this research, the electrical quantities that can be monitored were not only electrical energy consumption, it also included voltage, current, power, power factor. The measurement process uses current sensors and voltage sensors. Current sensors used a type of sensor that mounts in series circuit with electrical systems that are at risk of disrupting the electrical system performance in case of damage to the sensor [2].

The researcher on works [1,2] then continued their research and focus to refine the prototype described in [2]. The use of current sensors was no longer installed serially and high risk impact to the system, but uses a safer current transformer type. However, in this paper, the real-time report was not provided, in the other hand the display function and data processing was still offline, another disadvantages, theprototype described in paper was still using two embedded system (single board computer BCM2835 and Arduino) which is not efficient economically [3]. Despa et all on work [4] has created a simple prototype of a system of electrical monitoringin 3-phase electrical power panels using only Arduino. The electrical quantities that can be monitored was focused on voltage, current, power, and power factor. The measurement data was stored in to MySQL database and displayed on the web. This monitoring system is the main reference in this study which has some drawbacks, such as the absence of some important parameters of electrical power like frequency of the system and energy consumption [4], whereas those values are very important to be considered in term of electrical system reliability and estimation. Some previous research related to implementation of Internet of Things technology can be found on works [5,6,7].

2.2. Electrical Quantities

The current can be defined as the transfer of charge through a conductor. The total amount of charge passing through a random cross-section of a material per second is defined as ampere [8] and can be written as follows:

$$I = \frac{Q}{t} \; atau \; Q = I.t \qquad (2.1)$$

where Q is the symbol of charge measured in the coulomb (C), I is the current in amperes (A) and t is the time in seconds (s). An electric current can also be defined as the charge through a point in an electrical circuit, written as:

$$i = \frac{dQ}{dt} \qquad (2.2)$$

A constant electric current is known as a direct current or DC denoted by symbol I, while an electric current that changes with time is known as an alternating current or AC is represented by the symbol i or i (t). Voltage is defined as the energy required to move the electrical charge between two points [9] expressed as :

$$v = \frac{dw}{dQ} \qquad (2.3)$$

where V is the symbol of the voltage measured in volts (V), and dw is the energy in joule (J). Further, the relationship between current and voltage by George Simon Ohm known by Ohm's law [10], can be expressed as:

$$I = \frac{V}{R} \quad \text{or} \quad V = I.R \tag{2.4}$$

where R is the symbol of the resistance in ohm (Ω). Equation 2.4 can be used in analysis of electrical circuits, both in power systems with direct current and alternating current, which contain only resistive circuit elements. For circuits containing inductive (L) and capacitive (C) elements, equation 2.4 could not be applied. In a circuit containing elements L and C, the relation between current and voltage is expressed by the following equation:

$$i = \frac{v}{z} \tag{2.5}$$

where Z ; impedance (Ω) and jX ; total reactance (Ω).

$$Z = R + jX \tag{2.6}$$

The unit of power is expressed in watts, ie the amount of energy absorbed or sent within 1s. can written as [10]:

$$p = \frac{dw}{dt} \tag{2.7}$$

where p is the symbol of power measured in watts (W). Mathematically the relationship between power, current, and voltage is expressed as follows:

$$P = V.I \tag{2.8}$$

In alternating current power system, known three kinds ofpower, namely: apparent power or complex power is the result of multiplication of voltage phasor (V) and conjugate of electric current (I *). The unit of apparent power is expressed in Volt Ampere (VA). The active power or the real power of the Watt (W) unit, is the power fully absorbed by the resistive loads, while the reactive power of the Reactive Volt Ampere (VAR) unit is the power absorbed by the reactive (inductive and capacitive) loads stored in the form Electric field and electric charge [11]. Reactive power flow causes the phase of the electric current to shift preceding (leading) or lagging (lagging) from the phase voltage. This phase difference of voltage to voltage is called power factor (cos θ). The relationship between active power, reactive power, current, voltage, and power factor is indicated by the following equations:

$$P = V.I^{*}.cos\theta \tag{2.9}$$

$$Q = V.I^{*}.sin\theta \tag{2.10}$$

Where θ is the symbol of the phase shift in units of degrees (°).

2.3. Monitoring System

Monitoring system is a system that plays an important role in the management of electric power system. The development of embedded system also increased rapidly, ranging from the development of a conventional regulatory system where each sub-system (substations) requires an operator, computer-based regulatory system where the conventional system can be monitored centrally from a distance, and the last is an Integrated regulatory system where the sub system does not require any more operator which means the operator function is taken over completely by the control center device [12,13].

Monitoring system is one of the three major security system functions performed in the control center operation. The monitoring system provides up to date information to the power system operators in relation to conditions on a system. Includes the amount of supply of equipment in the load group, effective operation and measurement of critical quantities and transmits the measurement results to the control center.

2.4. Data Acquisition

Data acquisition is a process of sampling a signal by measuring the real world physical condition and converting the resulting sample into a digital numerical value that can be manipulated by a computer [14]. The basic elements of Data Acquisition (DAQ) include:

1. Sensor and transducer
2. Signal conditioning
3. Analog to Digital Converter (A / D)

Transducers can be used to detect different physical phenomena such as movement, electrical signals, radiation energy, thermal energy, magnetic, or mechanical energy. Transducer is used to convert one type of energy to another. Usage The type of input or output of the transducer depends on the type of signal being detected or the process being controlled. Devices with input functions are called sensors with the function of detecting physical events that change according to some parameters like heat or pressure. While the device with the output function is called the actuator. Transducers and sensors are used to convert physical phenomena into electrical signals (voltage or current) which are then converted into digital signals used for subsequent stages such as computers, digital systems, or storing in memory [15].

2.5. Internet of Things

Internet of things (IoT) is a network set that consist of physical devices, and other electronics embedded system, software, sensors, actuators, and network connectivity which enable these environment to collect and exchange data. IoT is also well known as a global infrastructure for some information society, enabling advanced information services by interconnecting (both physical and virtual) things based on existing and involving interoperable information and communication technologies [16].

The IoT will allows any objects to be sensed or controlled remotely wherever there any existing network infrastructure, creating opportunities for integration of the real world into computer-based systems, to intend in improved efficiency, accuracy, and economic benefit in addition to reduced human intervention [17,18]. When IoT is augmented with complex sensors and actuators, the technology will becomes an instance general class of cyber-physical systems, which also overwhelm other technologies such smart grids, virtual power plants, smart homes, intelligent transportation and smart cities.

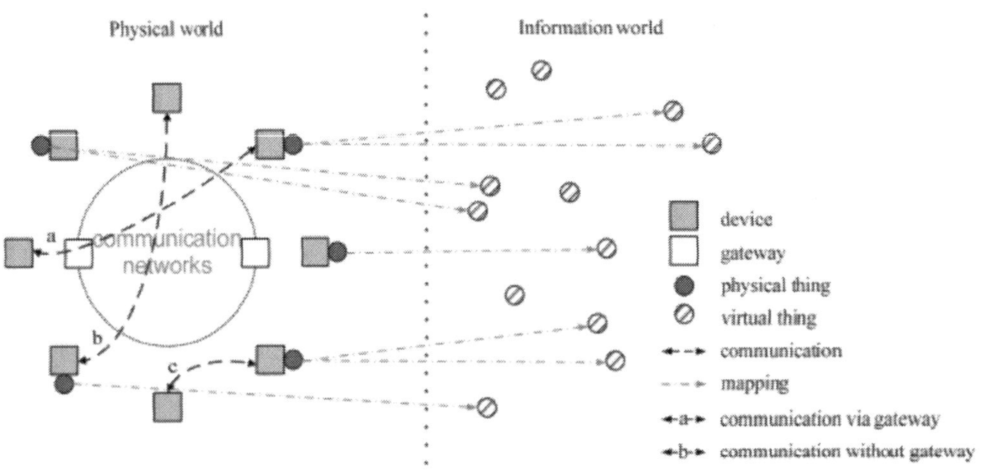

Figure 1. Technical overview of the IoT [13]

IOP Conf. Series: Materials Science and Engineering **335** (2018) 012063 doi:10.1088/1757-899X/335/1/012063

Figure 1 shown that a physical thing may represented in the information world by one or more virtual things, but a virtual thing can also exist without any associated physical thing. A device is a piece of equipment with mandatory capabilities of communication and optional capabilities for sensing, actuation, data capture, data storage, and data processing. The devices plays a role to collecting various information and provide it to the information/communication networks for further advance processing [13].

3. Research Methodology

3.1. System Design

This research proposed a system that can monitor the amount of electricity in a three-phase distribution panel by utilizing the Internet network as a means of sending data. Figure 2 shows the structure diagram of the developed system.

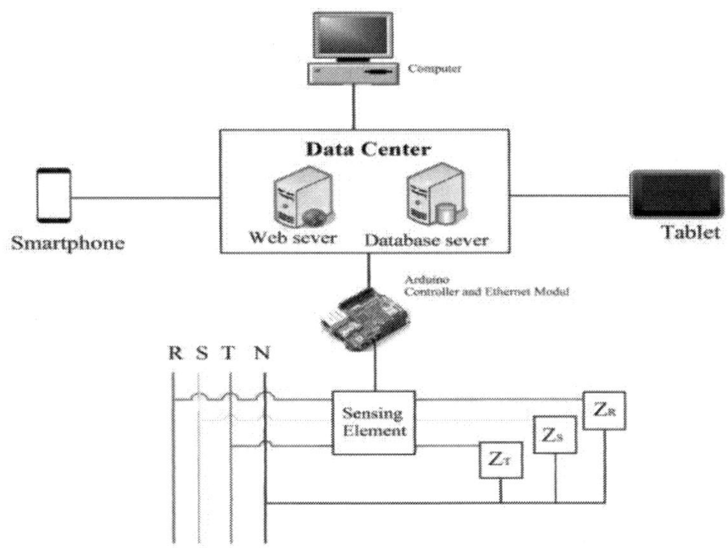

Figure 2. Structure diagram of online monitoring of distribution power system based Arduino and internet network

3.2. Hardware Design

The hardware consist of a voltage sensor which used a combination of voltage transformer and resistive voltage divider, a current sensor combined with current transformer and burden resistor, a signal conditioning circuit used an offset enhancer and a comparator where Arduino and Arduino-Ethernet-shield were used as main controller. Figure. 3, 4, 5, 6 shown the schematic wiring diagram of several sensor that be used on this research.

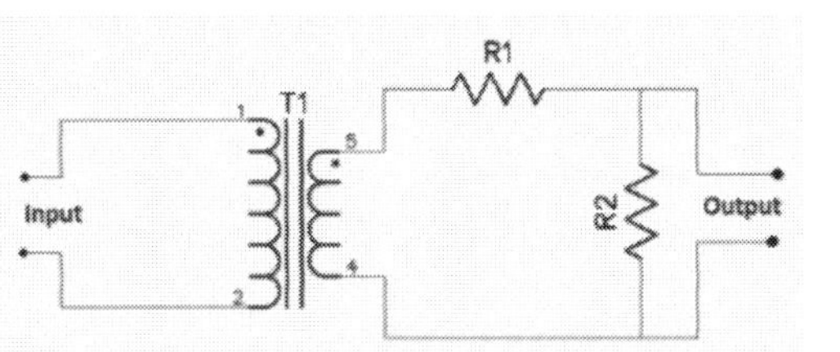

Figure 3. Structure diagram of voltage sensor

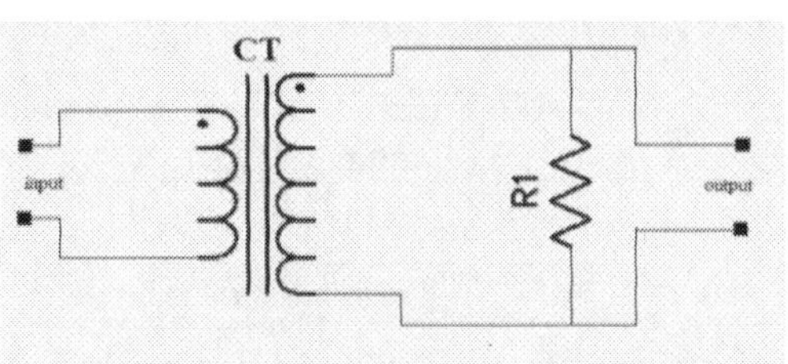

Figure 4. Structurediagram of currentsensor.

IOP Conf. Series: Materials Science and Engineering **335** (2018) 012063 doi:10.1088/1757-899X/335/1/012063

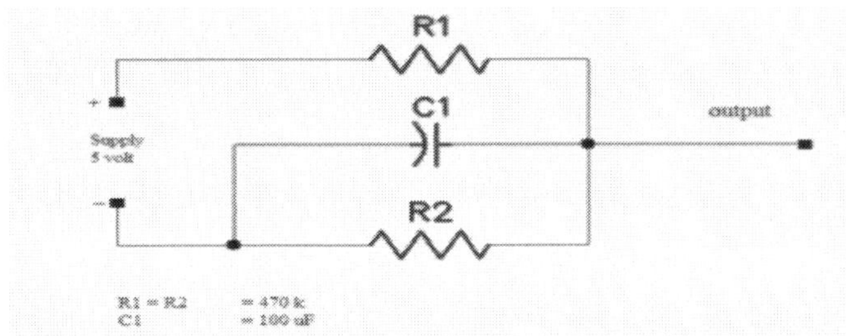

Figure 5. Structure diagram for signal conditioning of voltage and current sensor output.

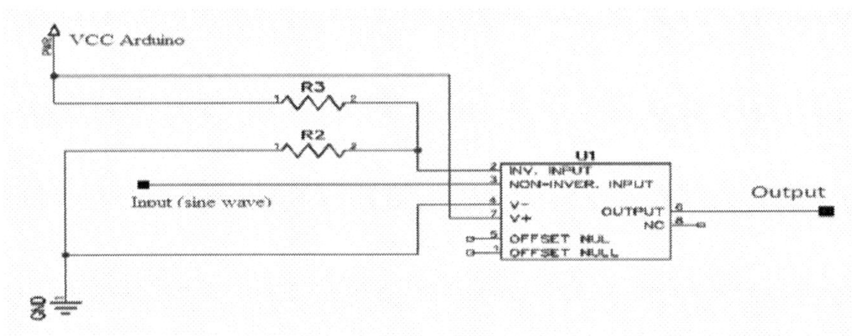

Figure 6. Structure diagram of sinus - square signal converter.

3.3. Software Design

The software in this research was developed using several programming languages, such as C programming language, PHP, HTML, and Javascript. The program on the Arduino uno controller board was designed for gathering the input from of the sensor, calculates the voltage, current, active power, apparent power, power factor and frequency of the power system, and sends the data to the server. The program was built by using the Arduino IDE (C programming language). The value of energy consumption was obtained from the calculation process on the server. This process was done by assuming the measured power is constant in the measurement range time.

4. Results and Discussion

4.1. Measurement Device Testing

To ensure that the designed system provide valid information, then before the system applied, the testing process was conducted in the Laboratory of Electrical Quantities Measurement, Electrical Engineering Department, University of Lampung (UNILA) by using standard laboratory measuring instruments.Shown on Table 1 and 2, are the data obtained based ontesting process whivch consist of hardware, software, and measurement accuracy testing.

Table 1. Results of V-T correlation coefficient test on measurement devices 1 and measurement devices 2

No	Trafo Code	Correlation Coefficient
1	Trafo RMeasurement 1	0,999990525
2	Trafo SMeasurement 1	0,99998779
3	Trafo T Measurement 1	0,9999849

IOP Conf. Series: Materials Science and Engineering **335** (2018) 012063 doi:10.1088/1757-899X/335/1/012063

Table 2. Results of CT correlation coefficient test on measurement devices1 and measurement devices 2

No	CT Code	Correlation Coefficient
	CT R Measurement 1	0,999965
	CT S Measurement 1	0,999938
	CT T Measurement1	0,999938

Based on Table 1 and Table 2 it shown that all voltage transformers and current transformers used have correlation coefficient values greater than 0.999. This indicates that this device has a good linearity and suitable for use as a sensor.

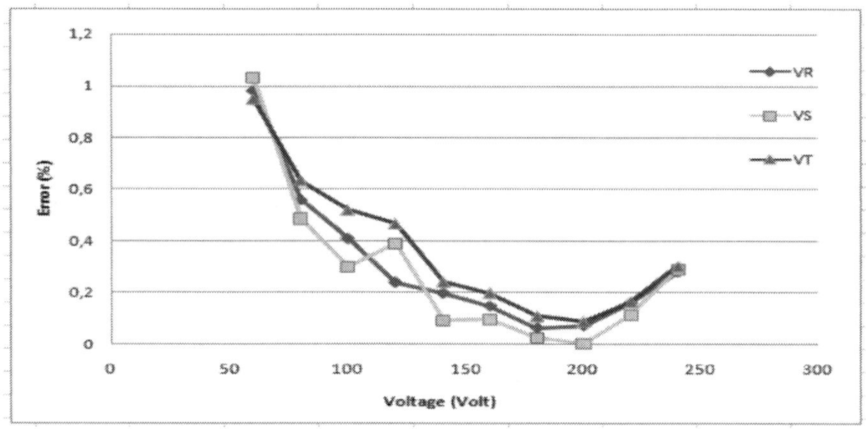

Figure 7. Percentage error of voltage measurement

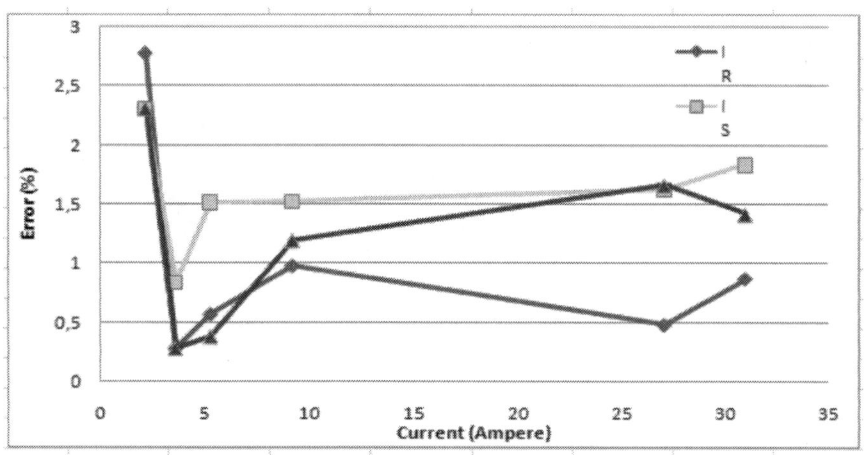

Figure 8. Percentage error of current measurement

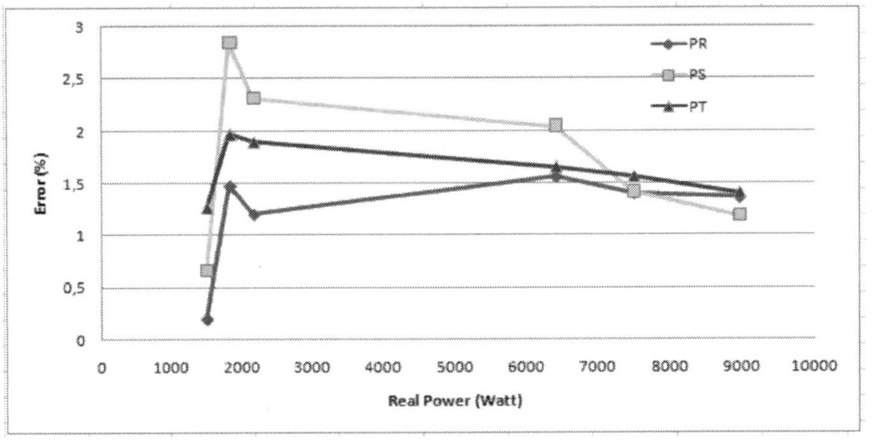

Figure 9. Percentage error of power measurement

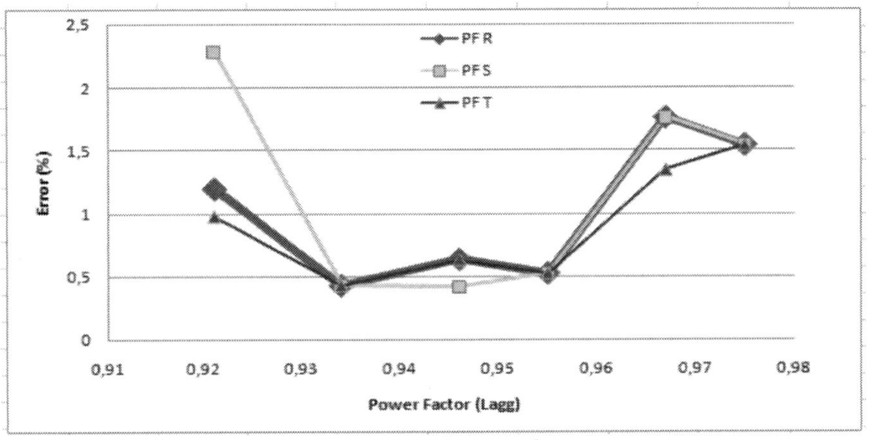

Figure 10. Percentage error of power factor measurement

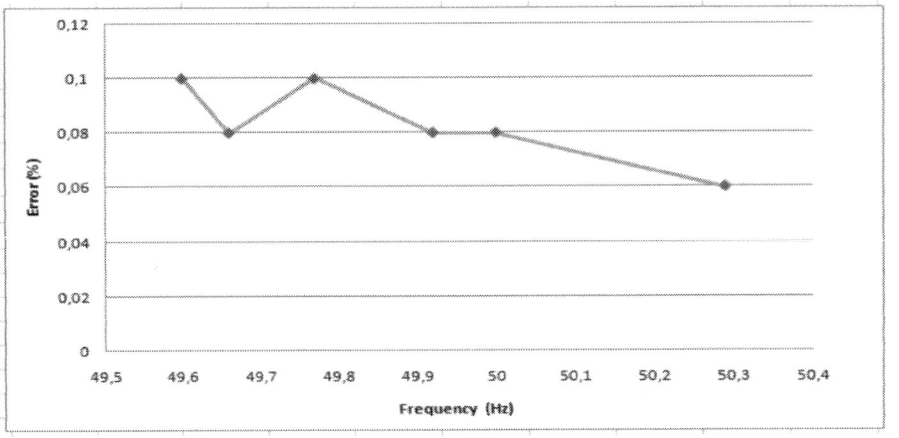

Figure 11. Percentage error of frequency measurement

Based on Figure 7, 8, 9, 10 and 11., it shown that percentage error of the measurement results of electrical quantities by using designed device in this study are quite low on all measured quantities of electricity, less than 10% . This indicates that the measuring device has medium-scale accuracy and is suitable for monitoring the electrical quantities in power distribution panels

4.2. Online Monitoring of Electrical Quantities Applied atMechanical Engineering Dept Power Panels, UNILA

Electrical quantities online monitoring has been done in power distribution panel located in Building H Faculty of Engineering UNILA. Figure 12 untill 17 are the results of monitoring within a period of one week starting on Friday, July 21, 2017 at 09.10 WIB until Friday, July 28, 2017 at 09.10 WIB

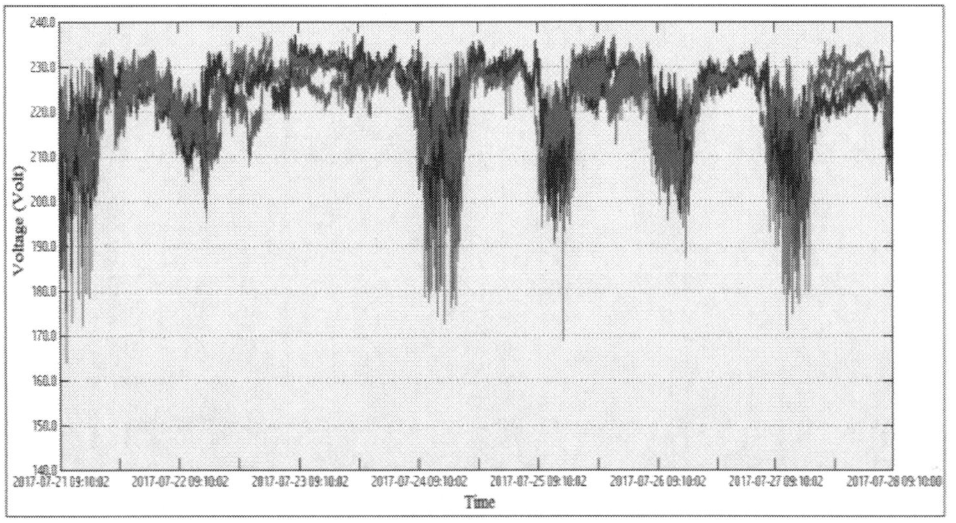

Figure 12. Voltage profile at power distribution panel of Building H Mechanical Engineering Unila 21 July 2017-28 July 2017

Figure 12 describes the fluctuates of voltage value on power panel. It can be seen that a considerable drop voltage was happened in the time span at 08.00 WIB until 17.00 WIB for every day, except on 22 and 23 July. The voltage drop occurs during working hours, due to the large number of electrical loads such as air conditioners, computers, and other equipments. While starting in the afternoon until the morning and on weekends, the use of the load is reduced, so the voltage would back to its normal condition.

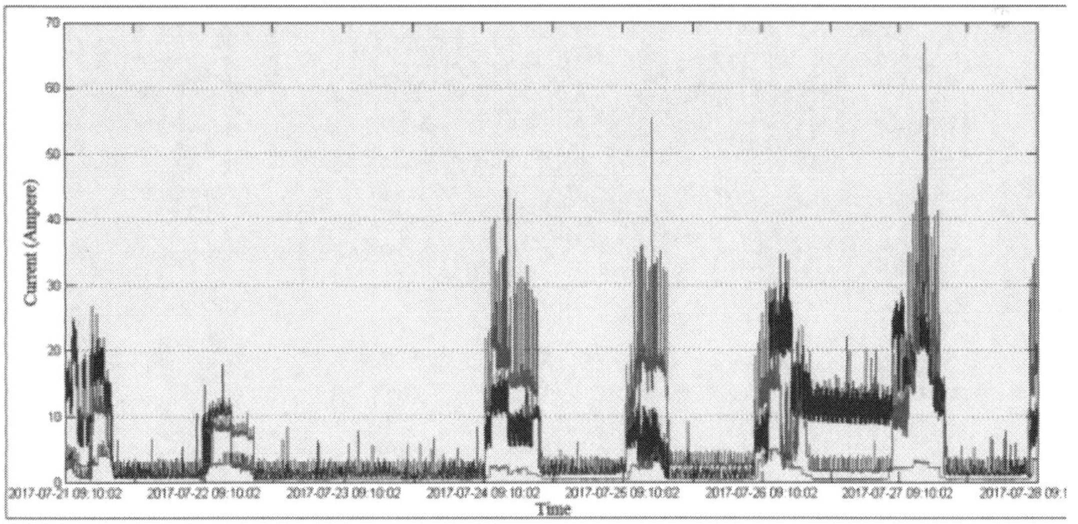

Figure 13. Current profile at power distribution panel of Building H Mechanical Engineering Unila 21 July 2017-28 July 2017

As the same case like the voltage shown in Figure 13, the current value also fluctuates. The current value will increase during working hours except holidays. This illustrates that the use of electrical energy is tends to during work hours. In addition, the important thing to note is that the current values between phases have considerable differences as a result of unbalanced loads. The load tends to accumulate on one phase which in this case is the phase R. This case means that in the future, No load is able to be connected to the phase R anymore.

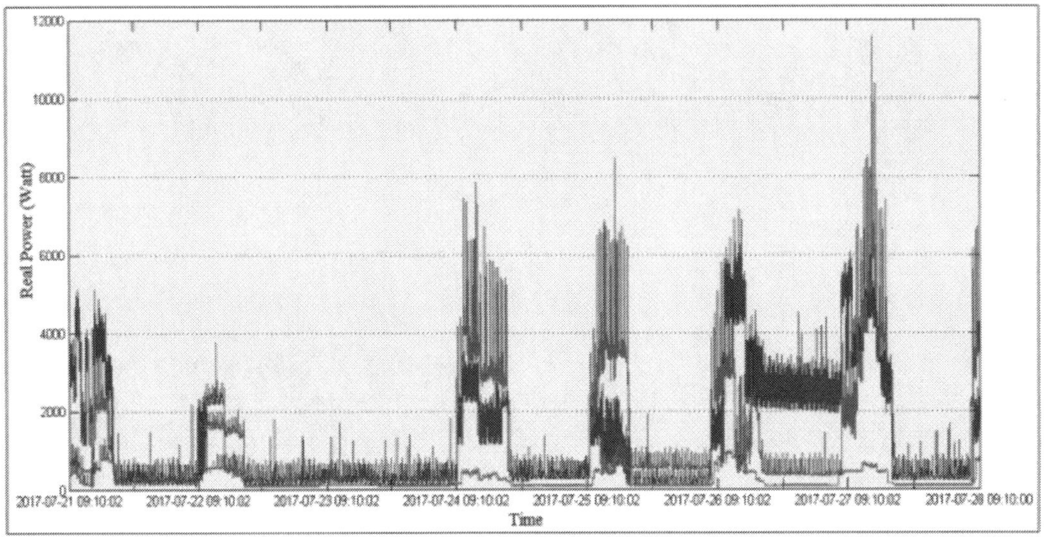

Figure 14. Power profile at power distribution panel of Building H Mechanical Engineering Unila 21 July 2017-28 July 2017

The graph in Figure 14 is an image of the power value. It appears that the pattern look ssimilar to the current profile described infigure 13.Thisshows that the power value is proportional to the current change. But the change is not always linear with current changes, because it is also influenced by the value of power factor.

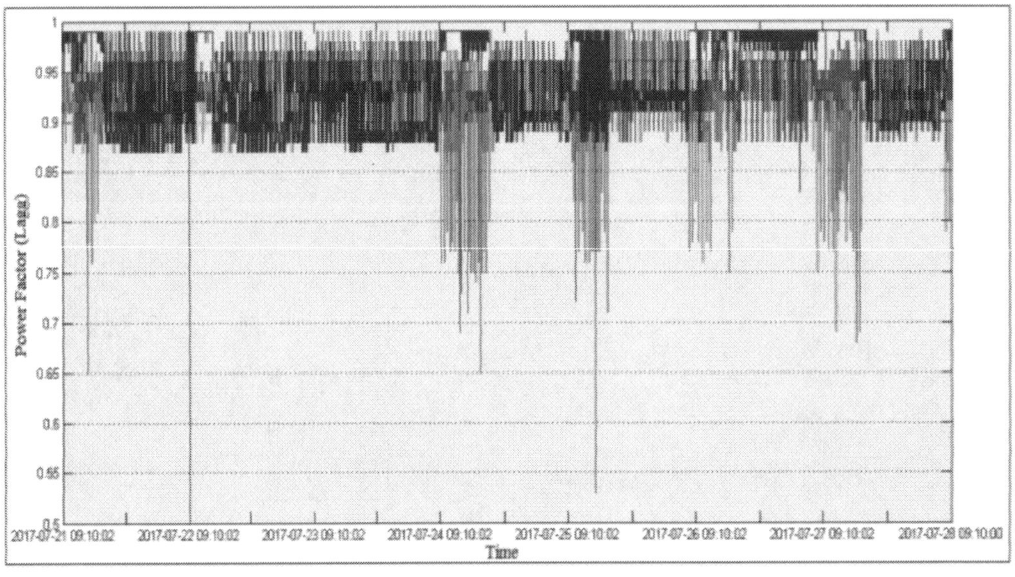

Figure 15. Power factor profile at power distribution panel of Building H Mechanical Engineering Unila 21 July 2017 – 28 July 2017

It shown on Figure 15 that the power factor also fluctuates. The observed value is still within the allowed range, even tends to be good

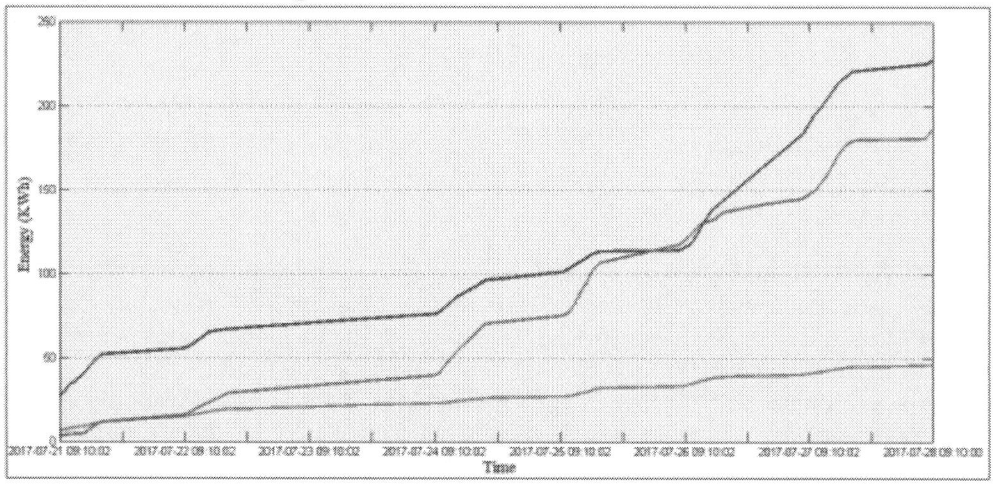

Figure16. Energy consumption profile at power distribution panel of Building H Mechanical Engineering Unila 21 July 2017- 28 July 2017

Figure 16 describes the amount of energy consumption. Based on the figure above, The magnitude of energy consumption has a considerable difference, this illustrates that there is an imbalance in the distribution of one-phase load.

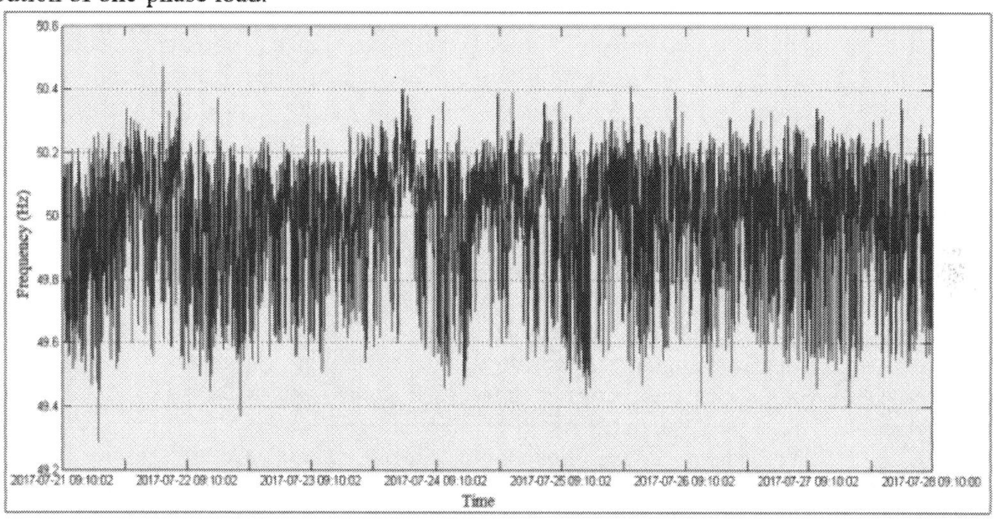

Figure 17. Frequency profile at power distribution panel of Building H Mechanical Engineering Unila 21 July 2017- 28 July 2017

Figure 17 shows the fluctuation of frequency in the H building of Mechanical Eengineering power panels. it can be understood because the load is changing at any time, besides this fluctuation is not only due to load changes in the measured panels only, but also the influenced by other loads.

5. Conclusion

Based on research, it can beconcluded that: The designed system can provide information changes of electrical quantities in building H Department of Mechanical Engineering Unila. Monitoring results showed that there is unbalance condition for almost all measured quantities. This case occurs as the result of uneven load installed for each phase, and also due to the working hour pattern in the H building Department of Mechanical Engineering which is still tends to be done during the day (working hours). Forthe future work, monitored and recorded data in server should be utilized for estimation and future improvement of electrical power system

Acknowledgement
Author would like to express gratitude to The Ministry of Research, Technology and Higher Education Indonesia and Unversity of Lampung, who have facilitated and supported this research.

References

[1]. G.F.Nama, M.Komarudin, H.Priambodo, Mardiana,H.D. Septama, 2014. Electricity, Temperature, and Network Utilization Monitoring at Lampung University Data Centre Using Low Cost Low Power Single Board Mini Computer. In The Regional Conference on Computer and Information Engineering (RCCIE), Yogyakarta Indonesia, pp 184-189.
[2]. D.Despa, A.Kurniawan, M.Komarudin, Mardiana, G.F.Nama, 2015. Smart monitoring of electrical quantities based on single board computer BCM2835, 2nd International Conference on Information Technology, Computer, and Electrical Engineering (ICITACEE)
[3]. D.Despa, Mardiana, G.F.Nama,M.Bernard, 2016. Multi-Area Smart Monitoring of Electrical Quantities Based on Mini Single Board Computer BCM 2835. 8th International Conference on Electrical Power, Electronics, Communications, Controls and Informatics System 2016 (EECCIS 2016), Malang.
[4]. D.Despa, M.A.Muhammad, H.Gusmedi, N.Amaro. 2017. Sistem Monitoring Besaran Listrik Dengan Teknologi IoT (Internet of Things). Bandar Lampung
[5]. G.F.Nama, D.Despa, Mardiana, 2016. Real-time monitoring system of electrical quantities on ICT Centre building University of Lampung based on Embedded Single Board Computer BCM2835, International Conference on Informatics and Computing (ICIC), Lombok, Indonesia.
[6]. G.F.Nama, M.Ulvan, A.Ulvan, A.M.Hanafi, 2015. Design and Implementation of Web-Based Geographic Information System for Public Services in Bandar Lampung City - Indonesia, 2015 International Conference on Science in Information Technology (ICSITech), pp 270-275, Yogyakarta, Indonesia.
[7]. G. F. Nama, M. Komarudin, H.D. Septama. 2015. Performance Analysis of Aruba TM Wireless Local Area Network Lampung University, 2015 International Conference on Science in Information Technology (ICSITech), pp 41-46, Yogyakarta, Indonesia.
[8]. N.Wasif, Concepts in Electric Circuits. 2008.Ventus Publishing ApS. Bookboon.com
[9]. S.Sudaryanto, Analisis Rangkaian Listrik Jilid1, 2012.Bandung: Darpublic.
[10]. R.J.Fowler, Electricity Principle&Applications.2008.New York: McGraw-Hill.
[11]. C.Cekmas,T.Barlian, Rangkaian Listrik, 2013. Jakarta: ANDI
[12]. W.Allen, B.F.Wollenberg. Power Generation, Operation, and Control, 1996. New York: JohnWiley&Sons,Inc.
[13]. T. Cegrell. Power System Control Technology, 1986. Prentice Hall Company.
[14]. T.H. Rosemary, Data Acquisition for Sensor System, 1997. London: Chapman &Hall.
[15]. E.M.Paolo, Data Acquisition System, 2013. Newyork: Springer.com.
[16]. V.Ovidiu, F.Peter. Internet of Things: Converging Technologies for Smart Environments and Integrated Ecosystems,2013. Aalborg, Denmark, Rivers Publishers.
[17]. S. Gérald. The Internet of Things: Between the Revolution of the Internet and the Metamorphosis of Objects. European Commission Community Research and Development Information Service. Retrieved August 2017.
[18]. Singh.S.K., Data Base Systems Concepts, Design, and Applications, 2009. NewDehli: Dorling Kindersley.

IOP Conf. Series: Materials Science and Engineering **335** (2018) 012064 doi:10.1088/1757-899X/335/1/012064

Development of Gravity Acceleration Measurement Using Simple Harmonic Motion Pendulum Method Based on Digital Technology and Photogate Sensor

Yulkifli, Zurian Afandi and Yohandri

Physics Department, Faculty of Mathematics and Natural Science, Universitas Negeri Padang, Jl. Prof. Hamka, Padang, Sumatera Barat, Indonesia 25131

yulkifliamir@gmail.com

Abstract. Development of gravitation acceleration measurement using simple harmonic motion pendulum method, digital technology and photogate sensor has been done. Digital technology is more practical and optimizes the time of experimentation. The pendulum method is a method of calculating the acceleration of gravity using a solid ball that connected to a rope attached to a stative pole. The pendulum is swung at a small angle resulted a simple harmonic motion. The measurement system consists of a power supply, Photogate sensors, Arduino pro mini and seven segments. The Arduino pro mini receives digital data from the photogate sensor and processes the digital data into the timing data of the pendulum oscillation. The calculation result of the pendulum oscillation time is displayed on seven segments. Based on measured data, the accuracy and precision of the experiment system are 98.76% and 99.81%, respectively. Based on experiment data, the system can be operated in physics experiment especially in determination of the gravity acceleration.

1. Introduction

Physics is one branch of natural science that studies the nature of the phenomenon. Natural phenomena in physics can be reviewed both theoretically and experimentally. Experiments were conducted to validate the theory while the theory was used to guide the net of an experiment. Physical experiments can be done by involving a measurement system [1][2]. The measurement system is a combination of activities, procedures, measuring instruments, software, and subjects that aim to obtain measurement data against the characteristics being measured. Measurements are utilized as a means of obtaining data to draw conclusions and means to determine the linkages between two or more variables. Measuring tools used in the measurement must be in accordance with the quantity to be measured and have a good accuracy in order to obtain the exact measurement results. The development of measurements can be observed by the transition from mechanical or manual systems to digital systems [3]. The transition from a mechanical to a digital system requires a sensor. Sensors are devices that convert physical quantities into electrical quantities. Through the sensor can be designed various systems that can work automatically and able to analyze the phenomena that occur in nature, both for measurement and for control. Physics has provided a strong basis on technological progress. Technological advances are closely related to the progress of physics. One of the areas of physics studies that are not less rapid development of today is the Electronics and

IOP Conf. Series: Materials Science and Engineering **335** (2018) 012064 doi:10.1088/1757-899X/335/1/012064

Instrumentation. Progress Electronics and Instrumentation is enough to help humans in the field of science [4]. One of his roles is the experiment of physics.

Physical experiments produce various theories. Physical theory was born and developed from Einstein era until now. Through experiments new theories develop. Many experiments have been carried out, among which there are Regular Straight Motion (GLB), Straight Circularly Changed Motion (GLBB), Circular Motion, Simple Harmonic Motion (GHS), viscosity, pendulum and many others. In general, almost all measuring devices are still used manually. The manual measurement tools make the researchers difficult because much of the data obtained is less precise than the theories studied. This incorrect data is caused by several mistakes made in the experiment, such as tool calibration, scale reading and precision in the use of tools. With the design of digital tools is expected to be used as a tool for education and facilitate students who follow the activities of practicum in recording the results of activities [5]. This can be done by students only by looking directly from the display that is provided in the form of seven segments that display the results of the experiment. In this work, a simple digital harmonic motion measurement in physical pendulum will be presented.

Motion is a physical science that we apply every day. One simple harmonic motion application is the physical pendant swing. Simple Harmonic Motion (GHS) is the motion of alternating objects through a certain equilibrium point with the number of vibrations of objects in every second always constant. Electronic components used in the form of sensors and digital signal processing using Arduino microcontroller. One of the sensors used is a photogate sensor in which the sensor is used to detect the number of oscillations pendulum. This paper will explain the development of measuring tools acceleration of earth gravity using simple experimental harmonic movement tool.

2. Method

Based on the problem proposed that the research model will be conducted is laboratory experimental research (laboratory experimentation), research that apply science into a design in order to get the performance as expected. The cause or effect relationship of an independent variable to the dependent variable can only be obtained through the experimental procedure [6]. In a pure experimental procedure, the researcher exercises full control to manipulate the independent variables. The experimental procedure intends to compare the effects of the independent variables on the dependent variable through the manipulation of the independent variables. Changes that occur in the dependent variable are returned to the cause of the different treatments given to the independent variables.

Research Variables are basically anything in the form of what is set by the researchers to be studied so that obtained information about it, then drawn conclusions. Preparation of simple harmonic motion experimental set on pendulum based photogate sensor with seven segment display consists of three variables, namely independent variable, dependent variable and control variable. The Free variable is the variable that influences or causes the change of the dependent variable (bound). The dependent variable is a variable influenced by the independent variable. Control Variables are variables that are controlled or made constant so that the relationship of independent variables to the dependent variable is not influenced by outside factors that are not examined. In this study the independent variable is the length of the rope. The dependent variable is the time of vibration and period. The control variable is the amount of vibration spring and the system builder electronics component.

2.1. Design of the Mechanic System

The design of simple harmonic motion instrument is illustrated in Figure 1. This system consists of mechanic and electronic sub system. In mechanic design, the swing ball is hanged to a stand that equipped with the photogate sensor and angle indicator.

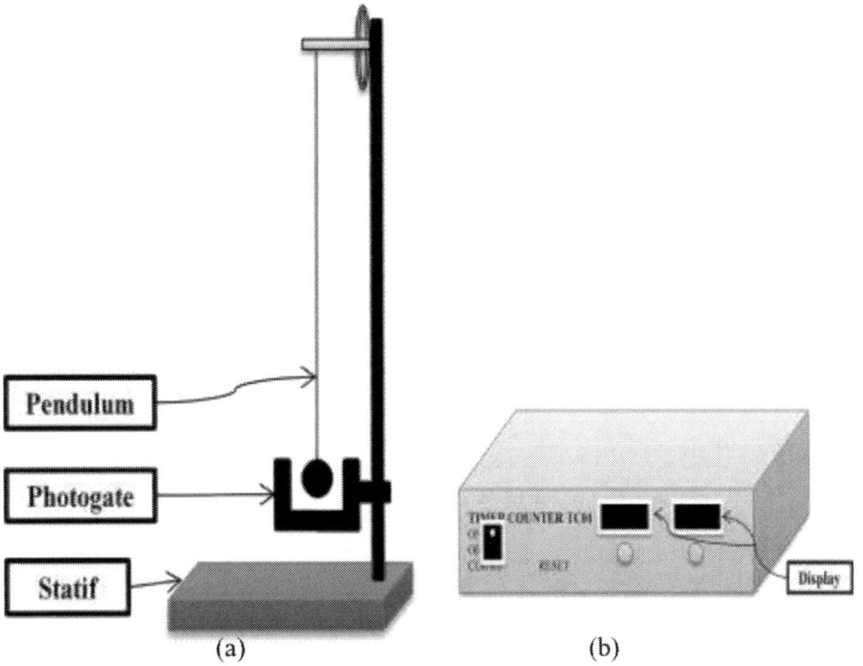

Figure 1. Design of simple harmonic motion instrument; (a) Mechanics system and (b) Electronic system.

Figure 1 shows the design of mechanics and electronic systems that have been made. Photogate sensor serves to calculate the number of oscillations. Stative serves as a hanging place pendulum. The electronic system consists of power supply, arduino pro mini and seven segment. Power Supply serves to provide power to the electronics system. Microcontroller is used to program the experimental experiment set of simple harmonic motion experiments on pendulum and seven segments used to display the measurement data [7][8].

2.2. Block diagram of the Digital Timer Counter

Digital experimental set of simple harmonic motion in pendulum consists of several series of electronics, including power supply circuit, photogate sensor circuit, Arduino Uno microcontroller circuit, and Seven Segment. The system block diagram of the simple digital harmonic motion experimental set on the pendulum can be seen in Figure 2.

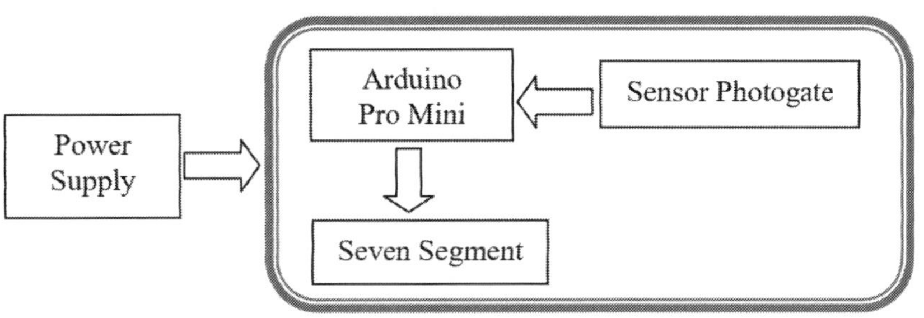

Figure 2. Block Diagram of the digital Timer Counter.

Figure 2 shows a block diagram of a set of simple harmonic motion digital experiments on the pendulum. Power supply is used as a DC voltage source, the voltage used ± 5 Volts [9]. Power supply is voltage source for photogate sensor, arduino pro mini and series of seven segments. Arduino pro mini receive digital data from photogate sensor and process digital data into timing data of oscillation pendulum. The calculation result of the pendulum oscillation time is shown on seven segments.

3. Results and Discussion

3.1. Specification of the System Performance

The set of simple harmonic digital motion experiments in the pendulum consists of mechanical systems and electronic systems. Figure 3 is a mechanical form, in which the photogate sensor is used to calculate the number of oscillations. Stative is used as a hanging place pendulum and electronic systems used as data processors and displays on seven segments. Figure 4 is an internal part of the timer counter, where (1) Represents an electronic circuit that serves as a power supply with a DC source, a voltage generated by a power supply of ± 5 Volts. Supply power circuit systems are used for voltage gators on Arduino pro mini, photogate sensors and seven segments. (2) A data processing system, Arduino pro mini receives digital data from photogate sensors and processes digital data into timing data of pendulum oscillation. (3) Represents part of digital data input; photogate sensor transmits digital data to Arduino pro mini. (4) Is display data measurement, seven segment display data in 4-digit number.

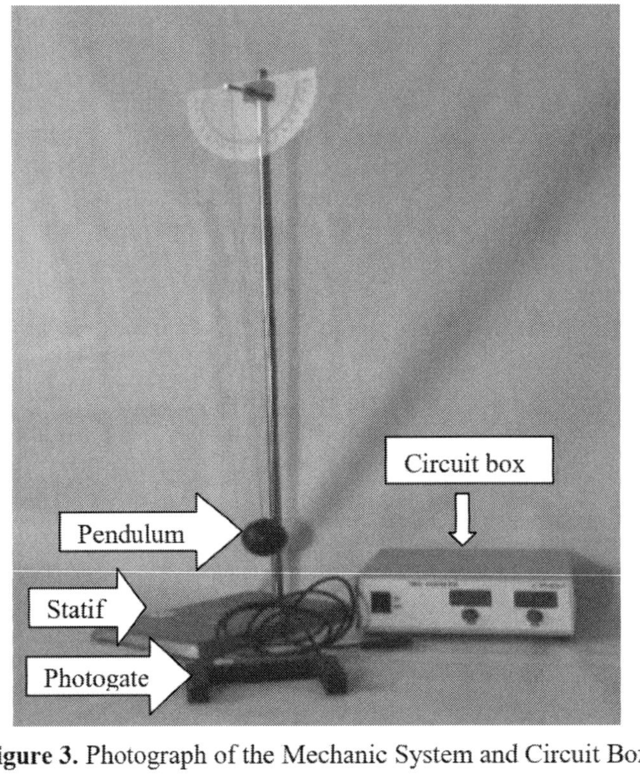

Figure 3. Photograph of the Mechanic System and Circuit Box

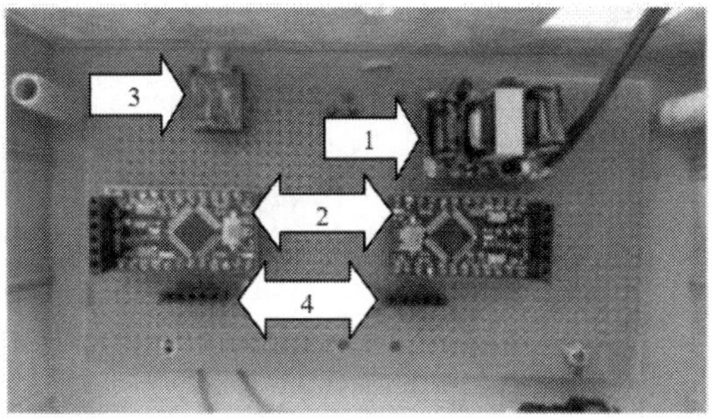

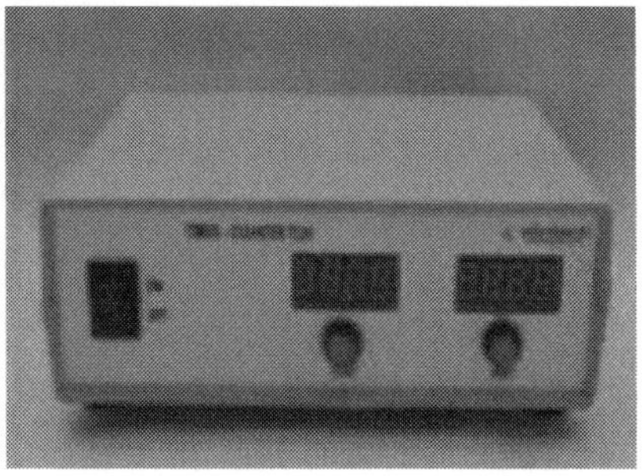

(a) (b)

Figure 4. Electronic system (a) electronic circuit and (b) front panel

3.2. Specification of the Design

3.2.1. Accuracy of the system

Accuracy or precision is obtained from the comparison between gauges made with standard gauges. A simple set of digital experimental harmonic motion on pendulum compared to measurement on oscilloscope as standard gauge. In addition to measurements using timer counters and oscilloscopes, manual measurements are also made. Figure 5 shows the relationship between the length of the rope and the value of gravity on the measurement using timer counter, standard gauge and manual measurement.

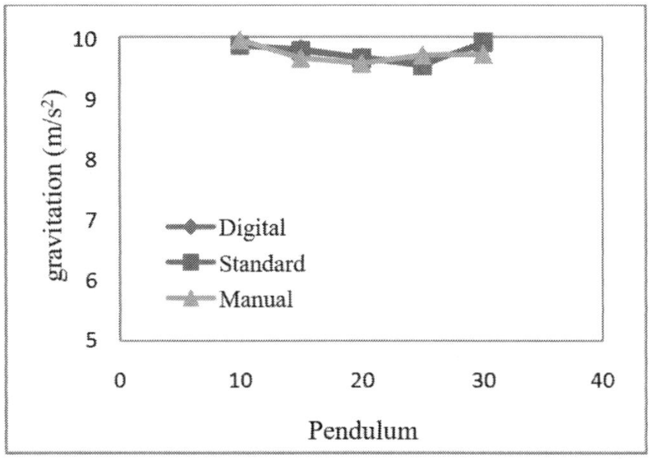

Figure 5. The relationship between the pendulum and gravitation

Figure 5 shows a comparison between digital measurements, standard measuring instruments and manual measurements. Based on the graph above, the measurement digitally produces the value of gravity closer to the standard gauge compared with the measurement manually. From result of data processing got average gravity value at timer counter is 9.797 m / s2. The average gravity score in the measurement of standard gauge is 9.79 m / s2. The measured gravity of the system has accuracy of about 98.76% and relative error 0.24%. While the average gravity value on manual measurement is 9.762 m / s2. Compare to the standard value of the gravity, accuracy of the system is 98.64% and relative error 1.36%. Figure 6 shows the relationship between the length of the rope and the oscillation time on the measurement using timer counter, standard gauge and manual measurement.

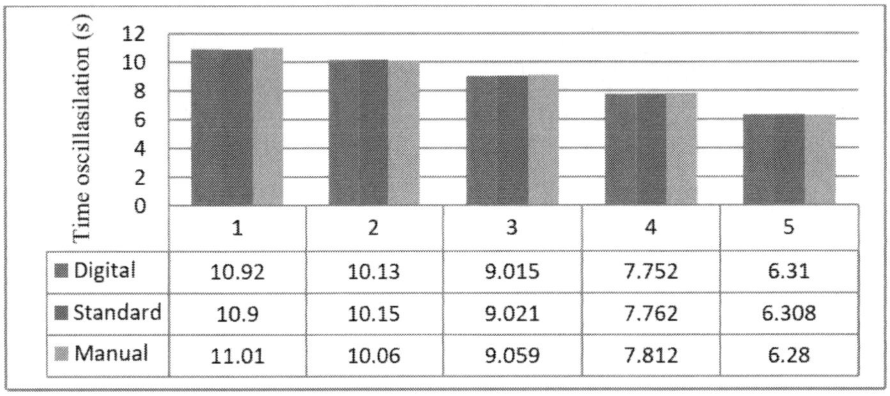

Figure 6. The relationship between the pendulum and time oscillation

3.2.2. Precision of the system

Precision or accuracy of measuring instruments obtained from the results of repeated measurements. Table 1 shows repeatable measurements in the Digital experiment set of simple harmonic motion on the pendulum.

Table 1. Precision of Digital Instrument

Pendulum (cm)	Time oscillation (s)										
	t1	t2	t3	t4	t5	t6	t7	t8	t9	t10	Average
30	10.92	10.91	10.92	10.92	10.90	10.92	10.92	10.92	10.93	10.92	10.918
25	10.13	10.14	10.14	10.13	10.12	10.13	10.13	10.13	10.12	10.13	10.130
20	9.014	9.015	9.015	9.013	9.015	9.015	9.017	9.015	9.015	9.015	9.015
15	7.753	7.752	7.754	7.752	7.751	7.752	7.752	7.752	7.751	7.752	7.752
10	6.313	6.311	6.312	6.310	6.310	6.313	6.310	6.310	6.310	6.301	6.310

Table 1 shows repeated measurements on timer counters. Based on the data processing on the value of precision Set of digital experiment simple harmonic motion on the pendulum is 99.81% and error 0.19%. Table 2, Table 3 and Table 4 show the measurement results using timer counters, standard measuring instruments and manual measurements.

Table 2. Results of manual measurement

L (cm)	t (s)	T	g (m/s^2)
30	11.01	1.101	9.760
25	10.06	1.006	9.742
20	9.059	0.905	9.611
15	7.812	0.781	9.694
10	6.280	0.628	10.00

IOP Conf. Series: Materials Science and Engineering **335** (2018) 012064 doi:10.1088/1757-899X/335/1/012064

Table 3. Results of digital measurement

L (cm)	t (s)	T	g (m/s^2)
30	10.92	1.092	9.922
25	10.13	1.013	9.608
20	9.150	0.915	9.705
15	7.752	0.775	9.844
10	6.310	0.631	9.905

Table 4. Results of standard measurement

L (cm)	t (s)	T	g (m/s^2)
30	10.90	1.090	9.958
25	10.15	1.015	9.570
20	9.021	0.902	9.693
15	7.762	0.776	9.819
10	6.301	0.630	9.911

4. Discussion

The measurement results show the performance system specification, accuracy and accuracy set of digital experiment simple harmonic motion on pendulum. The performance system specifications consist of mechanical tools and some electronic circuits, including power supply circuit, photogate sensor circuit, Arduino Uno microcontroller circuit and Seven Segment. The measurement process is initiated by power supply as voltage giver on Photogate, arduino pro mini and seven segments. The arduino pro mini receives digital data from the photogate sensor and processes the digital data into the timing data of the pendulum oscillation. The result of calculation of the pendulum oscillation time is shown on the seven segments. From the result of measurement and data processing the average gravity score on the measurement of standard gauge is 9.79 m / s2. If the value of gravity measurement digitally compared with the measurement of standard gauge then got the value of timer count accuracy is 98,76% and relative error 0,24%. While the average gravity value on manual measurement is 9.762 m / s2. If the value of gravity of manual measurement is compared with measurement of standard measurement then the value of manual measurement accuracy is 98,64% and relative error 1,36%. Precision value Set of simple digital harmonic motion experiment on pendulum is 99,81% and error 0,19%.

5. Conclusion

The development of the acceleration tool for gravity acceleration with simple harmonic motion of pendulum method based on digital technology and photogate sensor has been completed. Digital experimental set of simple harmonic motion in pendulum consists of several series of electronics, including power supply circuit, photogate sensor circuit, Arduino Uno microcontroller circuit, and Seven Segment. Based on the results and discussion, the average gravity score on the measurement of the standard gauge is 9.79 m / s2. If the value of gravity measurement digitally compared with the measurement of standard gauge then got the value of timer count accuracy is 98.76% and relative error 0.24%. While the average gravity value on manual measurement is 9.762 m / s2. If the value of gravity of manual measurement is compared with measurement of standard measurement then the value of manual measurement accuracy is 98.64% and relative error 1.36%. Precision value Set of simple digital harmonic motion experiment on pendulum is 99.81% and error 0.19%.

IOP Conf. Series: Materials Science and Engineering **335** (2018) 012064 doi:10.1088/1757-899X/335/1/012064

Acknowledgement
The authors would like to thank Directorate General of Higher Education (DIKTI), Ministry of Research, Technology and Higher Education, Indonesia, for the Research Grant (Hibah TIM Pascasarjana 2017), No. 446/UN35.2/PG/2017.

References

[1] Tipler, P. A. 1991. Fisika Untuk Sains dan Tehnik Jilid 2 (alih bahasa Dr.Bambang Soegijono). Penerbit Erlangga: Jakarta
[2] Y Dasriyani, Hufri, and Yohandri, 2016, Pembuatan set eksperimen gerak jatuh bebas berbasis mikrokontroler dengan tampilan PC. Sainstek: Jurnal Sains dan Teknologi 6 (1), 84-95.
[3] Yulkifli & Yohandri. 2016. Pengembangan Teknologi Sensor Menjadi Alat-Alat Praktikum Fisika Dalam Mendukung Implementasi Kurikulum 2013. Prosiding Semirata, 22-23 Mei 2016 Palembang. ISBN 978-60271798-1-3
[4] Ihsan, N.,Yulkifli, and Yohandri , 2017. Development of speed measurement system for Pencak Silat kick based on sensor technology , IOP Conference Series: Materials Science and Engineering, volume 180, number 1, pages 012171.
[5] Hae-Seok Park , Jun-Sik Hwang, Won-Youl Choi, Dong-Sik Shima, Kyoung-Won Naa, Sang-On Choi,Development of micro-fluxgate sensors with electroplated magnetic cores for electronic compass, Sensors and Actuators A 114. pp. 224–229.
[6] Djamal, M. dan Yulkifli, (2009): Fluxgate Sensor and Its Application, Proc. ICICI- BME, November, 23-25, 2009, Bandung
[7] Yohandri, 2013. Mikrokontroler dan Antar Muka. Padang : Universitas Negeri Padang.
[8] Yulkifli. 2013. Sistem Sensor dan Aplikasinya. Padang : Universitas Negeri Padang.
[9] Pavel Ripka & Alois Tipek. (2007). Modern Sensors Handbook. ISTE: United States.Sears, Zemansky. 1954. University Physics Mechanics, Heat, and Sound. USA: Addison-Wesley Publishing Company.

How Games are Designed to Increase Students' Motivation in Learning Physics? A Literature Review

V Tinedi[1], Y Yohandri[1], D Djamas[1]

[1]Physics Department, Universitas Negeri Padang, Jl. Prof Dr. Hamka Air Tawar, Padang, Sumatera Barat 25131, Indonesia

*vonny.tinedi@gmail.com

Abstract. Game is a promising tool to help students in understanding physics concept. It can motivate and provide the opportunities for students to become independent in learning. In order to fulfil these functions, games should be carefully designed. Thus, the objective of this paper is to present how games are designed to increase students' motivation in learning physics based on several literature reviews. The results showed that there are several ways to increase students' motivation in learning physics and to achieve that, game dimensions are needed to be considered when designing a game. This literature review may have useful to assist teachers and contribute in improving the design of games.

1. Introduction

The impact of games on life has increased significantly over the last few years. The increasing number of gamers in Indonesia is much higher compared to other countries. According to survey result conducted by Unity (Figure 1), the number of games installed in Indonesia during the year 2015-2016 has increased 192% from previous year [1]. This increase is several times more than in other countries. The number of gamers in Indonesia are about 21 million [2] and they play games on average three times a day [3]. According to these data, it can be concluded that games have been filled most of our leisure time now.

Playing games can give many positive effects. One of the positive impacts of playing games is it can enhance cognitive functions that include: problem solving, spatial skills, and creativity [4]. Another positive effect that can be obtained when playing games is it can satisfy basic psychological needs, such as: competence, autonomy, relatedness, and mastery off controls [5]. Furthermore, a more recent study indicates that when games are integrated with mathematics, it can engage and motivate students [6].

Motivation provided by games can help students when learning physics. All this time, many students still regard physics as difficult, abstract, and boring subject. With the games integrated in learning physics, students are expected to learn while playing so they can enjoy their learning process. In addition, more play will improve learning and performance [7]. Adding fun in the learning process will make it becomes enjoyable, compelling, and effective [7]. Moreover, games can also help students to be independent in learning.

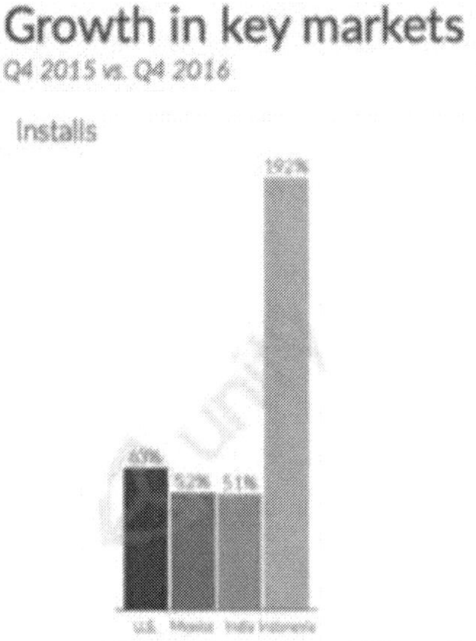

Figure 1. Increasing Number of Gamers in 2016 [1]

Although games can give positive effect in learning, games can also give negative effect. This depends on game design that is used in learning. Thus, this systematic literature review is conducted in order to find out how to design games that can increase motivation for students. Furthermore, it may also assist teachers and contribute in improving the design of games.

2. Background

2.1. Games for Education

Game is an organized play that is fun and engaging [7]. The word game is often associated with play and fun. Both of it can also make learning easier. Play can provide pleasure and increase involvement in learning [7] while fun can put students in relax and receptive mind for learning [8].

Many studies found that games can give positive impacts to student performances in school. These impacts depend on the types of games played by the students. There are 8 types of games that are action, adventure, fighting, puzzle, role playing, simulations, sports, and strategy [7].

Games have motivational power that makes learning fun and can be a powerful learning tool by engaging students to learning through doing [9]. However, game can also become a distraction for learning process. It depends on the game design and there are 6 design elements that need to be considered when designing a game. Those elements are: (a) rules, (b) goals and objectives, (c) outcomes and feedback, (d) conflict/competition/challenge/opposition, (e) interaction, (f) representation or story [7]. Besides that, there are 6 key dimensions of game, that is: (a) fantasy, (b) rules/goals, (c) sensory stimuli, (d) challenge, (e) mystery, (f) control [10]. All of these elements and dimensions are what affect students' motivation in learning physics.

2.2. Motivation in Learning Physics

There are two kinds of motivation that are intrinsic and extrinsic motivation. Intrinsic motivation is the desire to perform the task for their own sake because they find it fun and challenging and extrinsic motivation is related to rewards [11, 12]. These two motivations can give different effect to students in learning process. Intrinsic motivation results in high quality learning and creativity but teacher cannot always rely on it because not all the tasks that needed to be performed by the students are enjoyable and interesting [12]. Extrinsic motivation has limited effect and reduces students' motivation to do similar task in the future but it can also boost intrinsic motivation [11].

IOP Conf. Series: Materials Science and Engineering **335** (2018) 012065 doi:10.1088/1757-899X/335/1/012065

One way to motivate students in learning physics are making the activity interesting and make achieving the outcome becomes the important thing to do [10]. Recent study show that students are motivated to study physics when using modern information technology, computer, internet, and mobile phones [13].

3. Discussion

The effect given by games in learning is determined by its design. So, when designing a game, there are several things that are needed to be considered. One thing that needs to be considered is game dimensions, including: fantasy, rules/goals, sensory stimuli, challenge, mystery, and control. Fantasy in games allows students to do actions that have no effect in real world, but have real consequences [10]. Fantasy in games should depend on the practice of skills and vice versa (endogenous fantasy) [18]. Another way to add fantasy in games is by using meaningful stories [14, 20]. For examples, when students need to learn about vectors, they can do it by piloting a plane in a game. When they don't take the game seriously, they will get real consequences because their plane will be crashed. Thus fantasy in piloting the plane is depended on students' competence in vector and vice versa. This can give excitement to students and teach them to be responsible thus can improve their intrinsic motivation in learning. Besides that, the relevance of the situation while piloting the plane to the real life will motivate students in learning physics.

Rules/goals need to be clear, specific, challenging, and difficult goals but not impossible to do [10, 18, 19]. Giving difficult goals in games will enhance students' performance [10] and to prevent students to give up in the middle of playing, the goals should be started by an easier one, and then improved step by step. By achieving goals one by one starting from easy to hard, students will get excited and become curious about what will happen after they are making progress in game. This condition will help them learn physics concept in the game gradually and give enjoyment while learning it. Besides that, students will be more motivate if the game highlights the consequences result of a goal and emphasizes the importance of students' action within a given situation [20]. Furthermore, if the tasks in game are time-based, the number of tasks and time period to finish it should be calibrated carefully [19]. Rules use in game should encouraging reflection by embedding and rewarding reflection assignments [18], so students will learn from their mistake and elaborate their learning if they already understand the material when playing game.

Sensory stimuli in games including sound effects, dynamic graphics, and animation are needed in games to become students' attention grabbing [10, 22]. One thing to be remembered when design the sound of the game is it should not distract students' attention more than the game play itself. Other sensory stimuli that can be used are avatar, player skill level, and profile development [14, 19, 20]. These will give the sense of autonomy (ability to choose for oneself to engage in activity [24]) to students, thus increase their intrinsic motivation.

Challenge in game should be neither too easy nor too difficult, clearly specified, yet the possibility of obtaining the goal should be uncertain [10] and align the challenging level with students' competency level [18]. Performance feedback and score keeping is needed in game [10] to help students be aware of their success rate in game play. Feedback in game can give significant increase in motivation [16]. Feedback use in game can be in the form of positive and negative reinforcement. Positive reinforcement can be given after students complete a difficult task and if it is in the form of reward, it can be used to help them complete next task. This way, their extrinsic motivation will increase because of their eagerness to get the reward so they can continue their game play easier. This condition will also make achieving the outcome becomes the important thing to do by students. Another form of feedback that can be used in game are points, badges, leader boards, level, progress bars, performance graphs, quests, and team play [14, 15, 18, 20, 21, 23]. By using ranking scores among peer students, can help increase their extrinsic motivation [12].

There are several things to remember when designing rewards in game. Reward is given based on the quality (proportion of the items solved versus the items tried) not on the quantity of performance (total of items solved correctly) [18]. This condition will make students do their best in playing games, not just guessing or do trial and error until they find the right answer. Furthermore, reward should be given at the end of each playing cycle, so it will become part of the outcome processing [17] to help students evaluate their learning progress.

Mystery in game will generate students' curiosity and curiosity is one of the primary factors that drive learning [10]. For example, adventure game can be used in learning fluid concept. The game can start with someone get stranded in a deserted island and try to get back home. In this game, student can help the person by using fluid concept. The situations encountered by the person will give the sense of mystery to the students and by helping the person, students will automatically solving lots of problem which is the way to understand physics. Another condition that can help in creating mystery in games are by using randomly selected input and varies difficulties for task, thus keeping the game interesting and engaging [19].

Control is the ability to regulate, direct, or command something [10]. Students need to be given the ability to control how to proceed with the game play. This will make them enjoy the game and will give them more excitement when they can achieve the game goals with their own way. Besides that, game user interaction interface should be carefully designed in order to hinder wild guess by students. One way to achieve this condition is to avoid multiple choice that enable random clicking [18]. Furthermore, if the game is time-based, time limit and time remaining is need to be displayed through the game [19], so students will know their remaining time to complete the games' goal, thus can determine the best way for them to use the remaining time effectively to get maximal outcome. The summary of the relation between game dimensions design and motivation is concluded in Table 1.

Table 1. Relation Between Game Dimensions Design and Motivation

Game Dimensions	Design	Motivation		References
		Intrinsic	Extrinsic	
Fantasy	Design endogenous fantasy	√		[18]
	Use meaningful stories	√		[14], [20]
Rules/Goals	Set specific and challenging goal	√		[10], [18], [19]
	Rules that encouraging reflection by embedding and rewarding reflection assignments	√		[18]
	Calibrated number of task to complete and time period	√		[19]
	Clear, difficult, but achievable goal	√		[10], [20]
	Highlights the resulting consequences of a goal		√	[20]
	Emphasizes the importance of students' action within a given situation	√		[20]
Sensory Stimuli	Use player skill levels	√		[19]
	Use avatars and profile development	√		[14], [20]
	Use sound effects, dynamic graphics, and animation	√		[10], [22]
Challenge	Clearly specified, yet the possibility of obtaining the goal should be uncertain	√		[10]
	Use badges, leader boards, and performance graphs as performance feedback	√	√	[14], [15], [18],[20], [21], [23]
	Use ranking scores among peer students		√	[17]
	Game rewards received at the end of each playing cycle	√		[17]
	Align the challenging level with students' competency level	√		[18]

IOP Conf. Series: Materials Science and Engineering **335** (2018) 012065 doi:10.1088/1757-899X/335/1/012065

Table 1. Relation Between Game Dimensions Design and Motivation

Table 1. Relation Between Game Dimensions Design and Motivation (continue)

Game Dimensions	Design	Motivation Intrinsic	Motivation Extrinsic	References
				[10], [18]
	Reward is given based on the quality not on the quantity of performance		√	[18]
	Feedback should be informative elaborated, not just summative feedback	√		[18]
	Assign point for each successful output produced during the game and score summary for performance feedback	√		[19]
	Feedback in form of positive and negative reinforcement	√		[20]
	Use points, level, progress bars, quests, and team play	√	√	[20], [21], [23]
Mystery	Input are randomly selected	√		[19]
	Use varies difficulty for task	√		[19]
Control	Game user interaction interface is designed to hinder wild guess	√		[18]
	Display time limit and time remaining through the game	√		[19]

4. Conclusion

Games can increase students' motivation in learning physics because it makes learning become fun and enjoyable. It can also increase their intrinsic and extrinsic motivation in learning physics. Another motivation needed by students in learning physics are relevance to the daily life, making the activity interesting and make achieving the outcome becomes the important thing to do. In order to increase motivation, game dimensions such as fantasy, rules/goals, sensory stimuli, challenge, mystery, and control are needed to be considered when designing a game.

References

[1] SuperData and Unity 2017 *Can't Stop, Won't Stop: 2016 Mobile and VR Games Year in Review* (SuperData Research and Unity Technologies) p 8

[2] Mareza B 2016 *Ayo Cek, Kamu Termasuk Kecanduan Game Online?* https://student.cnnindonesia.com/edukasi/20160211134438-317-110344/ayo-cek-kamu-termasuk-kecanduan-game-online/

[3] Maulana R 2017 *Tingkat Perkembangan Pasar Game Mobile Indonesia Tiga Kali Lipat Amerika Serikat* https://id.techinasia.com/perkembangan-pasar-game-indonesia-salah-satu-yang-tertinggi-di-2016

[4] Isabela G, Adam L and Rudger C M E E 2014 The Benefits of Playing Video Games *American Psychological Association Vol. 69 No. 1* p 69

[5] Andrew K P, Richard M R and C Scott R 2010 A Motivational Model of Video Game Engagement *Review of General Psychology Vol. 4 No. 2* p 155-156

[6] Wijers M, Jonker V and Kerstens K 2008 *Proc. of the 2nd European Conf. on Games-Based Learning (ECGBL)* (Barcelona: Spain) p 513

[7] Prensky M 2001 *Digital Game-based Learning* (New York: McGraw-Hill)

[8] Mitchell A and Smith C S 2004 *The Use of Computer and Video Games for Learning: A Review of The Literature* (United Kingdom: Learning and Skills Development Agency) p 2

[9] Kirriemuir J and Mcfarlane A 2004 Literature Review in Games and Learning. A NESTA Futurelab Research report - report 8. 2004. p 4
[10] Garris R, Ahlers R and Driskell J E 2002 Games, Motivation, and Learning: A Research and Practice Model *Simulation and Gaming Vol. 33* p 447–451
[11] Benabou R and Tirole J 2003 Intrinsic and Extrinsic Motivation *Review of Economic Studies Vol. 70* p 490, 492
[12] Ryan R M and Deci E L 2000 Intrinsic and Extrinsic Motivation: Classic Definitions and New Directions *Contemporary Educational Psychology Vol. 25* p 55
[13] Holubova R 2015 How to Motivate Our Students to Study Physics? *Universal Journal of Education Research 3(10)* p 733
[14] Michael S, Jan U H, Sarah K M and Heinz M 2017 How Gamification Motivates: an Experimental Study of the Effects of Specific Game Design Elements on Psychological Need Satisfaction *Computers in Human Behavior Vol. 69* p 378
[15] Carloz C, Michael D H and Jesse F 2015 The Need to Achieve: Players' Perception and Uses of Extrinsic Meta-game Reward Systems for Video Game Consoles *Computers in Human Behavior* p 8
[16] S Erhel and E Jamet 2013 Digital Game-based Learning: Impact of Instructions and Feedback on Motivation and Learning Effectiveness *Computers & Education Vol. 67* p 162
[17] Wen-Hao H, Wen-Yeh H and Jill T 2010 Sustaining Iterative Game Playing Processess in DGBL: the Relationship Between Motivational Processing and Outcome Processing *Computers & Education Vol. 55* p 794-795
[18] Fengfeng K 2008 A Case Study of Computer Gaming for Math: Engaged Learning from Gameplay? *Computers & Education Vol. 51* p 1614-1619
[19] Luis V A and Laura D 2008 Designing Games with a Purpose *Communications of the ACM Vol.51* p 63-64
[20] Michael S, Jan H, Heinz M and Markus K 2014 Psychological Perspective on Motivation through Gamification p 4-5
[21] Ramine T, Markus L-R, Elena S and Wendy H 2017 An Investigation of Player Motivation in Eyewire, a Gamified Citizen Science Project *Computers in Human Behavior Vol. 73* p 538
[22] Wenhao D H, Tristan E J, and Seung-Hyun C H 2013 Impact of Online Instructional Game Features on College Students' Perceived Motivational Support and Cognitive Invesment: a Structural Equation Modeling Study *Internet and Higher Education Vol. 17* p 66
[23] Elisa D M, Florian B, Alexandre N T and Klaus O 2015 Towards Understanding the Effects of Individual Gamification Elements on Intrinsic Motivation and Performance *Computers in Human Behavior* p 9

IOP Conf. Series: Materials Science and Engineering **335** (2018) 012066 doi:10.1088/1757-899X/335/1/012066

Validity of High School Physic Module With Character Values Using Process Skill Approach In STKIP PGRI West Sumatera

M Anaperta[*1], H Helendra[1], R Zulva[1]

[1]Physics Education Program, College of Teacher Training and Education (STKIP) PGRI West Sumatra, Padang, Indonesia

*mega.syani@yahoo.com

AbstractThis study aims to describe the validity of physics module with Character Oriented Values Using Process Approach Skills at Dynamic Electrical Material in high school physics / MA and SMK. The type of research is development research. The module development model uses the development model proposed by Plomp which consists of (1) preliminary research phase, (2) the prototyping phase, and (3) assessment phase. In this research is done is initial investigation phase and designing. Data collecting technique to know validation is observation and questionnaire. In the initial investigative phase, curriculum analysis, student analysis, and concept analysis were conducted. In the design phase and the realization of module design for SMA / MA and SMK subjects in dynamic electrical materials. After that, the formative evaluation which include self evaluation, prototyping (expert reviews, one-to-one, and small group. At this stage validity is performed. This research data is obtained through the module validation sheet, which then generates a valid module.

1.Introduction

Science learning has characteristics that support student knowledge. In order for students to study physics correctly, then physics as one branch of science should be introduced intact, whether it concerns the object, the problem, or the level of organization of the objects around. Process skills become an integral part of physics learning.

Facts in the field, especially students of STKIP West Sumatra, especially Physics Education showed that students often perform activities that are less useful during the learning took place, such as when the lecturer asks students to discuss with friends sebangkunya most of the students just talk. In addition, very few students who want to convey ideas or opinions about learning materials and many students who tend to be lazy to think independently.

A process skill approach is a learning approach oriented to the IPA process. In implementing physics learning with process skill approach, it is required that the character-oriented module using the process skill approach with model innovation and learning strategy is expected to create an active learning atmosphere, facilitating the mastery of the material.

2. Research Method

The type of research conducted is research development. According [1] research and development method is a research method used to produce a particular product, and test the validity, practicality and

IOP Conf. Series: Materials Science and Engineering **335** (2018) 012066 doi:10.1088/1757-899X/335/1/012066

effectiveness of the product. It also states that development research is an effort to develop and produce a product in the form of materials, media, tools or learning strategies. This model of learning strategy development uses a development model proposed by Plomp [2] consisting of phases (1) preliminary research, (2) the prototyping phase, and (3) the assessment phase phase.

2.1. Phase Design and Realization (prototyping phase)

In this phase, the formulation or planning of the module is done. The design of this module refers to the instructional development guidelines issued by depdiknas.The design of modules is made in accordance with established indicators and based on the format that is tailored to the needs of the researcher.In this phase, there is a formative evaluation that includes self-evaluation, expert reviews, one-to-one, small group, and field test. The design flow for formative evaluation is shown in Figure 1

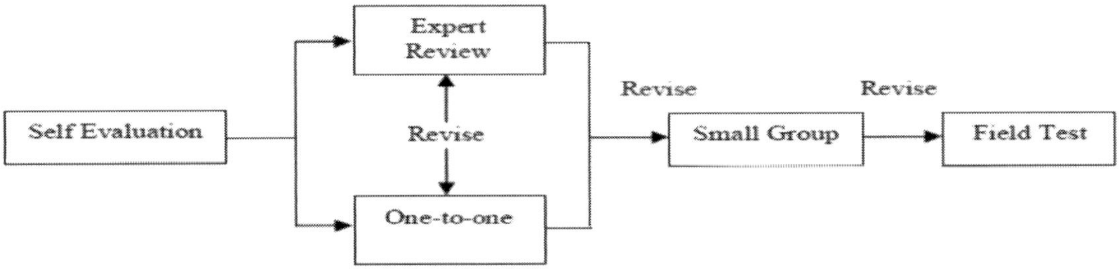

Figure 1. Formative Evaluation Design Layout

1) *Self Evaluation*

At this stage, module designing will be developed. Designing is based on the results of the initial analysis or problem identification, needs analysis and concept analysis or material content, and the literature review required in the learning strategy.

2) *Expert Review*

At the expert review stage, an expert judging and evaluating the module has been designed. The experts assess the validity of content, constructs, and language. Expert advice is used as a reference to revise developed modules. The main parts that are validated are the suitability of KD, indicators, correctness of concepts and the language used. The input of the validator is used to fix and revise the module developed so that a valid module is generated.

3) *One-to-one*

In the one-to-one stage, a design trial has been developed to students and lecturers who become testers. The results of this trial are used to revise the designs that have been made.

4) *Small group*

The revised results of the expert's study and the difficulties experienced during the pilot in the first trial were used as the basis for revising the module, then the results were tested on a small group. The results of this trial are then used for revision before testing in the field test stage. After modifying the module based on the suggestions and comments of the students in the small group, obtained the results of the analysis of the module.

5) *Field Test*

Small group revision results were tested on the subject. This test is a field . Modules that have been tested on field tests are modules that meet the already validated quality criteria.

2.2. Assessment phase
This phase is done to test the modular level of practicability.This stage is conducted to find out whether the module can be used to achieve the goal in improving the quality and achievement of student learning. In this research, the practicality of Physics Course Module of SMA / MA and SMK can be known from activity and student learning result after learning strategy.At this study, researchers only conduct preliminary research phase and design phase because at the early stages, researcher only completes the product validation stage.

3. Data Collection Techniques and Research Instruments
The data collection technique for determining Module validation is through validation and discussion with validators. Data collection techniques used for knowing the validation of the modules that characterized by the process skills approach. For this purpose the instruments used in this study are the validation sheet.

The validation sheet contains items that reveal the validity and constructs.The content of validity and construct in Modules, based onstatement from *Depdiknas* Book Center 2006 which has been modified by Salmiah (2008: 41).

The collected validation data is then tabulated. The tabulated results of each bill are searched for percentages, by formula

$$P = \frac{\sum score}{\max score} x100\% \quad (1)$$

Based on the percentage results, each bill is categorized into:

Table 1. Categories Validity of the module is oriented to character values by using a process skill approach

(%)	category
0-20	Invalid
21-40	Less valid
41-60	Quite valid
61-80	Valid
81-100	Very valid

(Refernce [4])

Descriptive technique is used to describe data of prototype validation discussion with expert.

4. Results and Discussion

Validation of high school Physics modules using the development model proposed by Plomp has the following results:

Module Validation
Module validation is performed to determine if the device created is valid to be tested or not.The validator gives an assessment by filling the validation sheet.In validating, validators are required to provide judgments and opinions on modules that have been designed and revised in advance. Validation is said to be completed if the validator has stated that the module designed is valid and feasible and ready for trial.

The validation results provided by the learning validator are developed. From the validation results are analyzed. If the result of the analysis states that the module is not valid yet, then the revision is made to obtain a valid module. Validation results have been done by the validator shown by Table 2.

Table 2: Module validation result

No	Indicator	Validator			Average	Category
		1	2	3		
1.	Feasibility of Content	92	92	94	94	Very Valid
2.	Construction Feasibility (Component of Presentation)	95	95	95	95	Very Valid
3.	Language Components	92	95	94	95	Very Valid

Based on Table 2, it shows that the contents of Modules that have been designed have a category very valid. This is seen from the feasibility of the contents of the value given 94 validator. Module Construction gets a value of 95 which means it is also categorized very valid. The use of Indonesian language scores 95. This means that the Indonesian Language used in the Module has also been very good.

5. Conclusion

Validation results from the validators show that the character-oriented module values on the process skill approach are highly valid

Acknowledgments

Acknowledgments to the research and technology ministry of the Directorate General of Higher Education for the assistance of the research fund providers of 2017 and the parties concerned.

References

[1] Sugiono, *Metode penelitian kuantitatif, kualitatif dan R&D*. Bandung : Alfabeta. (2009), p. 407-408

[2] Plomp, Tjeerd."*An Introduction to Educational Design Research*". Enschede: University of Twente.(2010), p.15

[3] BSNP. 2008. *Paduan Pengembangan Perangkat Strategi Pembelajaran* KTSP. Jakarta: Depdiknas.

[4] Riduwan.*Belajar Mudah Penelitian Untuk Guru, Karyawan Dan Peneliti Muda*.Bandung : Alfabeta. (2009), p.102

IOP Conf. Series: Materials Science and Engineering **335** (2018) 012067 doi:10.1088/1757-899X/335/1/012067

Effectiveness of Adaptive Contextual Learning Model of Integrated Science by Integrating Digital Age Literacy on Grade VIII Students

A Asrizal[1], A Amran[2], A Ananda[3] and F Festiyed[1]

[1] Physics Department, Faculty of Mathematics and Sciences, Universitas Negeri Padang, Padang 25131, Indonesia
[2] Chemistry Department, Faculty of Mathematics and Sciences, Universitas Negeri Padang, Padang 25131, Indonesia
[3] Social Science Education, Faculty of Social Sciences, Universitas Negeri Padang

asrizal@fmipa.unp.ac.id

Abstract. Educational graduates should have good competencies to compete in the 21st century. Integrated learning is a good way to develop competence of students in this century. Besides that, literacy skills are very important for students to get success in their learning and daily life. For this reason, integrated science learning and literacy skills are important in 2013 curriculum. However, integrated science learning and integration of literacy in learning can't be implemented well. Solution of this problem is to develop adaptive contextual learning model by integrating digital age literacy. The purpose of the research is to determine the effectiveness of adaptive contextual learning model to improve competence of grade VIII students in junior high school. This research is a part of the research and development or R&D. Research design which used in limited field testing was before and after treatment. The research instruments consist of three parts namely test sheet of learning outcome for assessing knowledge competence, observation sheet for assessing attitudes, and performance sheet for assessing skills of students. Data of student's competence were analyzed by three kinds of analysis, namely descriptive statistics, normality test and homogeneity test, and paired comparison test. From the data analysis result, it can be stated that the implementation of adaptive contextual learning model of integrated science by integrating digital age literacy is effective to improve the knowledge, attitude, and literacy skills competences of grade VIII students in junior high school at 95% confidence level.

1. Introduction

Educational graduates should have good competence to exist in the 21st century. For this reason, education requires preparing and creating the graduates with abilities required to compete in the 21st century. There are four skills which required by education graduates in the 21st century namely ways of thinking, ways of working, tools for working, and skills for living in the world [1]. Therefore, education needs to create and to develop fourth these skills in learning.

Literacy skills are very important for students in the 21st century because of these literacy skills are able to influence their success in learning and their daily life. In other word, literacy is one of the most important foundations for success in school and life [2]. Good literacy skills will help students in discovering and understanding various sources of learning such as oral, written, and visual text. The ability to find and to understand which relevant learning resources is very important in their learning process and their life. For this reason the development of student literacy in learning needs to be done. The development of literacy skills are necessity for preparing students in improving and widening their content knowledge and concepts which give support to other domains of life [3]. So, literacy skills of students are very important in education.

The integration of the curriculum is a way to develop skills of students in the 21st century. Curriculum integration is often seen as a way to effectively approach 21st century learning. Integrated curriculum seems like a perfect fit for developing generic 21st century competencies such as collaboration, creativity, communication, character education, literacy and critical thinking. More importantly, students are more engaged in school when they are learning in an interdisciplinary environment [4]. The existence of integration in the curriculum makes learning material is close to the real world situation. Thus, this integration is relevant to science because the essence of nature which found in daily life is integrated patter.

Science learning in the 2013 curriculum should be conducted in an integrated patter. The learning content in junior high schools which based on integrated concepts from various disciplines for educational purposes is the science subject. In essence, science is developed as a subject in integrated science. The content of science comes from the sub disciplines of biology, physics, and chemistry [5]. The concept of integration is shown in the core competence and basic competence. In one basic competence has integrated the science concepts from sub disciplines of biology, physics, chemistry, and earth and space science.

However, integrated science learning and integration of literacy in learning can't be implemented well in junior high school. These real conditions were found from the preliminary research which has been done before developing this learning model. There were three activities in the preliminary research include analyzing the implementation of science learning in schools, analyzing the learning materials in integrated science textbooks, and analyzing the result of student literacy skills. Hence, the results of this preliminary research were important for designing and developing the adaptive contextual learning model.

The first result of this preliminary research was the description of the implementation of science learning in the classroom. Instrument in observation activity was observation sheet of science learning in the classroom. The data analysis data indicates that the average value of the contextual learning strategy element is 48.3. It means that the implementation of contextual in science learning can be classified into low category. In addition, data analysis also shows that the value of integration of science material in learning process was 37.5. This average value indicates that the integration of science materials in learning was low [6]. Thus, the integration both contextual learning and learning materials in science learning process can be classified into low category.

The second result of the preliminary research relates to the description of learning material content of ten integrated science textbooks for grade VIII students in junior high school. Instrument to collect the data was a document analysis sheet. The analysis was done for each chapter of the integrated science textbooks. The result of analysis was percentage of learning materials of biology, physics and chemistry respectively 40.0%, 35.6% and 21.1%. The percentage of integrated science was only 3.3%. This integrated science materials was only found in science books in the 2013 curriculum. These results indicate that the integration of learning materials in integrated science textbooks was low [7]. Thus, learning material in integrated science textbooks was still separated in sub-discipline of science namely biology, physics and chemistry.

The last result of the preliminary research refers to the description of digital age literacy skills of students. Instrument to collect the data was literacy test sheet. Digital age literacy consists of functional literacy, scientific literacy, and visual literacy. The digital age literacy test was given to three classes of students from three schools in Padang. The average value of student's literacy from each of state junior high school respectively 39.89, 46.99 and 41.89. The average value of digital age literacy of these schools was 42.92. Therefore, this average value can be entered into low category [6].

IOP Conf. Series: Materials Science and Engineering **335** (2018) 012067 doi:10.1088/1757-899X/335/1/012067

The actual conditions which found didn't match with the expected conditions. This gap indicates a problem to be investigated. An alternative solution of this problem is to implement the adaptive contextual learning model of integrated science by integrating digital age literacy on grade VIII students. There are four theoretical review which related to the solution of this research problem those are: learning model, contextual learning, integrated science, and digital age literacy.

Learning model is a conceptual framework that can be used as a guide by teachers in learning. An instructional model as a plan or patter that can used to shape curriculum, to design to instructional materials, and to guide instruction in the classroom and other settings [8]. On the other hand, model of teaching can be defined as instructional design which describes the process of specifying and producing particular environmental situations which cause the student to interact in such a way that a specific change occurs in their behavior. It consists guidelines for designing educational activities and environment. Thus, the function of teaching model is to shape a curriculum, to design the learning materials, to guide learning in classroom, and to create learning situations [9].

Adaptive contextual learning model is a combination of contextual learning and learning model. Contextual learning is the learning which associated with the real world. Contextual learning is a concept that helps teachers relate school learning to real-world situations. It motivates learners to take charge of their own learning and to make connections between knowledge and its application [10]. Contextual learning is considered as an instructional method that engages students in a learning process by connecting the learning tasks with real-life situation issues or with contexts that guide them to seek meanings [11]. From these definitions, adaptive contextual learning model can be defined as a pattern which can be used by teachers to design instructional materials and to guide learning in the classroom and to create learning situations with connecting the learning content to real-world situations. Adaptive word in this learning model is used to describe the adaptability of learning model with contextual learning approach. For example, in the learning model doesn't use teaching and learning but it is only use learning terminology.

Integrated science is an approach which used in an integrated curriculum to describe the interrelationships between sub disciplines of science. Integrated science integrates the perspectives of sub disciplines such as biology, chemistry, physics, and earth/space science [12]. Through this integration, teachers expect students to understand the connections between the different sub disciplines and their relationship to the real world. Integrated science is defined as an approach to the learning of science in which concepts and principles are represented so as to express the fundamental unity of scientific thought and avoid premature or undue stress on the distinctions between the various scientific fields. An integrated science course eliminates the repetition of subject matter from the various sciences and does not recognize the traditional subject boundaries when presenting topics or theme [13]. In addition, integrated science provides the student with exposure and connections between the study of science and their real life.

Literacy is also important for students to get success in science learning and its application in daily life. Literacy is the ability to identify, understand, interpret, design, communicate, and learn to use printed and written materials relating to a variety of contexts [14]. Digital era is a term which used in the digital emergence, internet network, especially computer information technology. Digital age literacy is the ability in a particular field which required to find, to access, to use, and to evaluate the various information in digital technology and internet network. The digital age literacy skills consist of basic literacy, scientific literacy, economic literacy, technological literacy, visual literacy, information literacy and multicultural literacy [15].

The development of adaptive contextual learning model of integrated science needs to be done to solve this research problem. This learning model is important to be implemented to realize integrated science learning, to create the relevant context of science learning materials, and to develop literacy skills of students. For this reason, researcher is interested to conduct this research. The objective of this research is to understand the effectiveness of the implementation of adaptive contextual learning model of integrated science on knowledge, attitude and scientific processes skills competence.

2. Research methods

In general, type of this research is the research and development. This research is a research method to develop a particular product and to test its effectiveness. Product of this research and development is an adaptive contextual learning model. The research activity which reported in this article was a part of the research and development of adaptive contextual learning model of integrated science. A limited field testing was conducted to determine the effectiveness of the learning model that has been developed. Design research which used in limited field testing was before and after treatment for one group of sample. The effectiveness of learning model was determined by comparing the after and before treatment of student's competence. Thus, determining the effectiveness of the implementation of the learning model was important part of this limited field testing.

Subject of research was grade VIII students in SMP Negeri 8 Padang. Class sample of the research was one class of grade VIII students. The number of students which involved in research activity was 28 students. In research activities, adaptive contextual learning model was implemented to students for 6 meetings. Learning tools as a support system of the learning model were integrated science textbooks and integrated student worksheet. Integrated science textbook is accompanied by elements of contextual learning strategy while student worksheet is accompanied by digital age literacy. Therefore, motivation and involvement of students in the implementation of adaptive contextual learning model are essential to improve their competences.

Instrument to collect the data in this research can be grouped into three parts. First instrument was assessment sheet of student skill. This instrument was used to obtain data of scientific processes skill of students in learning model. The components of scientific processes skills include observing, asking question, formulating hypotheses, planning investigation, conducting investigation, interpreting data, summarizing investigation, and communicating the investigation result. The second instrument was observation sheet of student's attitudes in learning. Attitude aspects which observed were curiosity, self-confidence, discipline, inquiry commitment, cooperation, and communication. The third instrument was learning outcomes test sheet of knowledge competence of students. This instrument was used to obtain data of knowledge competence of students after and before implementing the adaptive contextual learning model. Therefore, these instruments were important to get the knowledge, attitude and literacy skills of students in limited field testing.

The research data were analyzed by using descriptive statistic analysis, normality test and homogeneity test, and paired comparison test. Descriptive statistics analysis was used to provide information about a data group of knowledge, attitude, and skill of student's competence. The parameters of the statistical analysis include minimum value, maximum value, range, mean value, median, mode, standard deviation and variance. The normality test was used to determine a group of data has normally distribution. Homogeneity test is used to determine the variance of two distributions have same values or different value. The paired comparison test was used to determine the difference of average value of student's competence after and before treatment. Hence, the existence of differences in student's competence between after and before of this treatment was important to indicate that the effectiveness of adaptive contextual learning model of integrated science on grade VIII students.

3. Results and discussions

In general, the result of this research consists of three parts, namely knowledge competence, attitude competence and skills competence. Knowledge competence is related to result of pretest and posttest of student's knowledge competence. Attitude is related to student attitudes which observed during the learning process of science. On the other hand, the competence of skills is related to scientific processes skills of students in conducting science investigation activity.

The first result of this research is the effectiveness of adaptive contextual teaching learning model of integrated science in knowledge competence. Pretest was given to students before implementing the

IOP Conf. Series: Materials Science and Engineering **335** (2018) 012067 doi:10.1088/1757-899X/335/1/012067

adaptive contextual learning model. The learning model was implemented for six meetings in three weeks on the use wave and optic materials in our lives. After implementing the learning model was given posttest to students. Number of students who follow the pretest and posttest activity was 28 students. The parameter values of the descriptive statistics, the normality test, the homogeneity test and paired comparison test can be seen in Table 1

Table 1. Value of Statistics Parameter of Knowledge Competence

No	Parameter of Statistics	Calculated Value	Reference Value
1	Mean of pretest	55.43	
2	Standard deviation of pretest	9.90	
3	Mean of posttest	71.29	
4	Standard deviation of posttest	10.17	
5	P-value of normality test of pretest	0.10	
6	P-value of normality test of posttest	0.15	0.05
7	P-value of F test	0.89	
8	Paired comparison test	-6.58	2.05

From the result of students pretest can be explained that the value of students varies between 28.00 to 68.00 with a range of 40.00. The average value on pretest is 55.43. This average value can be grouped into enough category. Standard deviation and variance of pretest result respectively 9.90 and 98.03. On the other hand the posttest value of students varies from 52.00 to 88.00 with range is 36.00. The average value of students on posttest activity is 71.29. The average value of students can be classified into good category. The standard deviation and variance value of the posttest respectively 10.17 and 103.47.

The P value for 28 students on pretest is 0.10, while the P value for posttest is 0.15. A significant level is established on $\alpha = 0.05$. This means that groups of data both pretest and posttest have normally distribution. F test is also done to group data of pretest and posttest to determine the homogeneity of these data. From the homogeneity test obtained the value of F is 0.89. This F value is greater than the value of $\alpha = 0.05$. This means that data groups both posttest and pretest have same variance. Therefore, data groups both pretest and posttest come from the same variance.

The effectiveness of implementation of adaptive contextual learning model of integrated science is determined by comparing posttest result and pretest result of knowledge competence. Paired comparison test is used to analyze this difference. The paired sample t test compares two mean values from the same individual of students. The two mean values present two differences of time such as pretest and posttest with a treatment between two times. The purpose of this test is to determine the difference between the results paired test which observed on a particular competency outcome. This paired t test can be entered into the parametric test.

Paired comparison test is used to analyze the difference knowledge competence of 28 students, value of mean and standard deviation of pretest respectively 5.43 and 9.90, value of mean and standard deviation of the posttest respectively 71.29 and 10.17. The value of paired t test is $t = -6.58$ and P value is 0.00. Meanwhile, the value of t in table for the number of 28 students, degree of freedom is 27 and the significant level $\alpha = 0.05$ and a half of $\alpha = 0.025$ is 2.05. The value of calculated t is outside the acceptance area of the null hypothesis. From this result can be stated that there are significant difference in students learning outcomes between after and before implementing of adaptive contextual learning model of integrated science. The existence of this difference indicates that the implementation of adaptive contextual learning model gives a significant effect on the knowledge competence of students. Therefore the implementation of adaptive contextual learning model of integrated science by integrating digital age literacy is effective to improve knowledge competence of students.

The second result of this research is the effectiveness of implementation of adaptive contextual learning model of integrated science on attitude competence of students. Before implementing the adaptive contextual learning model was observed student attitudes. Attitudes of students were observed for two meetings. After implementing the adaptive contextual learning model also observed attitude of students for six meetings. Attitude of students which observed in learning consists of six component, those are: 1). curiosity, 2). self confident, 3). discipline, 4). commitment inquiry, 5). cooperation, and 5). communication. The values of descriptive statistics parameters, P-value of normality test, P-value of homogeneity test and t-value of paired comparison test for attitude competencies of students can be shown in Table 2

Table 2. Parameters Value of Attitude Competence

No	Parameter of Statistics	Calculated Value	Reference Value
1	Mean before treatment	57.74	
2	Standard deviation before treatment	11.82	
3	Mean after treatment	77.18	
4	Standard deviation after treatment	8.51	
5	P-value of normality test before treatment	0.06	
6	P-value of normality test after treatment	0.15	0.05
7	P-value of F test	0.09	
8	Paired comparison test	-13.3	2.05

The mean values of student's attitudes and standard deviations on the initial condition were 57.74 and 11.82 respectively. Meanwhile, the mean value of student's attitudes competence and standard deviation after implementing the adaptive contextual learning model respectively 77.18 and 8.51. From normality test is obtained P value for attitude competence data before implementing adaptive contextual learning model is 0.06, while P value for attitude competence after implementing this learning model is greater than 0.15. Both p value before and after implementing the learning model is larger than α value is 0.05. From this result can be explained that the data group of attitude competence of students before and after implementing the adaptive contextual learning model are distributed normally. On the F test, the P value of two data groups of attitude competence of students was obtained 0.09. This F value is larger than α value so it can be stated that variance of the two data groups of student's attitude competency has same variance. Thus, the two data groups of student's attitude come from homogeneous variance.

For the number of students is 28 students, the mean and standard deviation of data group of attitude competence before implementing this learning model respectively 57.74 and 11.82, the mean and standard deviation after implementing this learning model were 77.18 and 8.52, respectively, the paired t test is - 13.33. The P value for this descriptive statistic parameter is 0.00. By using two-tails test for freedom degree of 27 students, significant level α = 0.05, a half of α =0.025, the t value in table is $t_{0.975}$ = 2.05. It means that the value of calculated t is outside the acceptance area of null hypothesis. With another expression, P value less than α value. From this paired comparison test, it can be stated that there is a significant difference in student's attitude competence between after implementing the adaptive contextual learning model and before implementing this learning model at 95% confidence level. Therefore, the existence of significant difference in attitudes competency of students indicates that the implementation of adaptive contextual learning model of integrated science is effective to improve attitudes competence of grade VIII students in junior high school.

The last result of this research is effectiveness of adaptive contextual teaching learning model of integrated science in skill competence on scientific process activities. Both before and after implementing the adaptive contextual learning model, skills of students were assessed by using performance assessment. Scientific processes skills of students which assessed in learning consist of eight components, namely 1). observing, 2). asking question, 3). formulating hypothesis, 4). planning investigation, 5). conducting investigation, 6). interpreting data, 7). getting conclusion, and 8). communicating. Performance of students was assessed before and after implementing the adaptive contextual learning model. These student performance data were analyzed with appropriate statistics. The values of descriptive statistics parameters, p-value of normality test, p-value of homogeneity test and t-value of paired comparison test of processes scientific skills are displayed in Table 3

Table 3. Parameters Value of Scientific Processes Skills

No	Parameters of Statistics	Calculated Value	Reference Value
1	Mean before treatment	56.25	
2	Standard deviation before treatment	6.96	
3	Mean after treatment	71.99	
4	Standard deviation after treatment	7.96	
5	P-value of normality test before treatment	0.03	0.05
6	P-value of normality test after treatment	> 0.15	
7	P-value of F test	0.49	
8	Wilcoxon paired comparison test	-4.63	1.90

Before implementing the adaptive contextual learning model, the average value of scientific processes skills of students and standard deviation are 56.25 and 6.96, respectively. On the other hand, after implementing this learning model, the average value and standard deviations of scientific processes skills of students are respectively 71.99 and 7.96. In normality test, the P value of skill competence for data group before implementing adaptive contextual learning model is 0.03, while P value of skill competence for data group after implementing this learning model is greater than 0.15. This means that the data group of scientific processes of students before implementing the learning model has normally distribution while the data group of scientific processes skills of students after implementing the learning model is also has normally distribution. On the other side, the P value on F test of the two groups of scientific processes skills of students is 0.49. This P value is greater than α value = 0.05. From result of this F test can be argued that the variance of both after group data and before group data of scientific processes skills is same. Thus, both initial data group and final data group of scientific processes skills of students have same variance.

The result of normality test indicates that the data group of scientific processes skills of students before implementing learning model hasn't normally distribution, the data group after implementing the learning model has normally distribution, and the variance of two data groups of scientific processes skills of students have same variance. For this reason, the difference test which used in data analysis is the Wilcoxon paired comparison test. The Wilcoxon test is a nonparametric test which used to analyze pairs of observations of data group with non-distributed normally. By using Wilcoxon paired comparison test it can be obtained of Z value = - 4.63 and Sig 0.00. At the significant level α= 0.05 with two-tailed test the value of a half of the significant level of 0.025, the area of the curve F (z) of 0.475, the value of Z in table is 1.95. This is mean that the value of calculated Z is outside acceptance area of null hypothesis. In other way the Sig value which obtained is smaller than the α value so that the null hypothesis is rejected. From Wilcoxon paired comparison test, it can be stated that there is a significant difference in scientific processes skills of students between after implementing the adaptive contextual learning model with before implementing the learning model at 95% confidence level. Therefore, this significant difference in scientific processes skills of students indicates that the implementation of adaptive contextual learning model of integrated science is effective to improve the scientific processes skills of grade VIII students in junior high school.

4. Conclusion

From the analysis of data which has been done it can be stated the conclusion of this research. The conclusion of this research is the implementation of adaptive contextual learning model of integrated science gives significance difference on competences of students including knowledge, attitude, and scientific processes skills of competences at 95 % confidence level. Therefore, the implementation of adaptive contextual learning model of integrated science is effective to improve knowledge, attitude and scientific processes skills competences of grade VIII students in junior high school.

References

[1] Germaine, Ron, et all. 2016. Purposeful Use of 21st Century Skills in Higher Education. *Journal of Research in Innovative Teaching*. Volume 9, Issue 1

[2] Adedokun and Olufunke, Mary. 2014. Literacy in Early Childhood: Implications for Sustainable Development. *Proceeding of 1st Annual International Interdisciplinary Conference*, AIIC, 24-26 April, Azores, Portugal

[3] Wahyuni, Dewi Sri. 2017. Promoting Literacy Skills Through Journal Writing. *The 1st International Conference on Language*, Literature and Teaching

[4] Drake, Susan M and Savage, Michael J. 2016. Negotiating Accountability and Integrated Curriculum from a Global Perspective. *International Journal of Learning, Teaching and Educational Research,* Vol. 15, No. 6, pp. 127-144

[5] Wiyanto and Widiatmoko, Arif. 2016. Preparation Model of Student Teacher Candidate in Developing Integrative Science Learning. *Journal of Education and Human Development*, Vol. 5, No. 2, pp. 169-177

[6] Asrizal, et all. 2017. Need Analysis to Develop the Adaptive Contextual Learning Model of Integrated Science by Integrating Digital Age Literacy for Grade VIII Students. *International Conference on Teacher Education* (ICTE). Faculty of Education and Teacher Training. Riau.

[7] Asrizal, ett all. 2017. Development of Integrated Science Teaching Material of Our Digital System Health by Integrating Scientific Literacy for Grade Students. *Proceeding International Conference on Global Education Education V.* Volume 1A, p 491-504.

[8] Metzler, Michael W. 2017. *Instructional Model For Physical Education*. Third Edition, Routledge Taylor & Francis Group, London and New York.

[9] Subramani, P. C. Naga. 2016. *Effective Teaching and Learning*. First Edition, Laxmi Book Publication, Solapur, Ashok Yakkaldevi

[10] Sears, Susan Jones. 2002. *Contextual Teaching and Learning: A Primer for Effective Instruction*. Phi Delta Kappa Education Foundation

[11] Sung, Han-Yu, et all. 2015. An Integrated Contextual and Web-based Issue Quest Approach to Improving Students' Learning Achievements, Attitudes and Critical Thinking. *Journal of Educational Technology & Society*, 18 (4), 299-311

[12] Mubita, Simamuna Likando and Kalimaposo, Kalisto. 2016. Factors Affecting Teaching and Learning of Home Economics in the Integrated Primary Curriculum: A Study of Selected Primary Schools in Lusaka province of Zambia. *International Journal of Multidisciplinary Research and Development*, Volume 3; Issue 11; Page No. 09-16

[13] Opara, Jacinta A. 2011. Bajah's Model and the Teaching and Learning of Integrated Science in Nigerian High School System. *International Journal of Academic Research in Business and Social Sciences*, Vol. 1, Special Issue

[14] Chew, Fong Peng. 2012. Literacy among the Secondary Schools Students in Malaysia. *International Journal of Social Science and Humanity*, Vol. 2, No. 6

[15] Turiman, Punia, et all. 2012. Fostering the 21st Century Skills through Scientific Literacy and Science Process Skills. *Elsevier, Procedia - Social and Behavioral Sciences* 59, 110 – 116

IOP Conf. Series: Materials Science and Engineering **335** (2018) 012068 doi:10.1088/1757-899X/335/1/012068

Implementation Authentic Task to Enhance Problem Solving and Self-Management for Physics College Students

Festiyed*[1], D Djamas[1] and D Pilendia[2]

[1]Physics Department, Faculty of Mathematics and Science, Universitas Negeri Padang, Jl. Prof. Hamka, Padang, Sumatera Barat, Indonesia 25131
[2]Postgraduate Student in Physics Education, Faculty of Mathematics and Science, Universitas Negeri Padang, Jl. Prof. Hamka, Padang, Sumatera Barat, Indonesia 25131

*festiyed@fmipa.unp.ac.id

Abstract. The purpose of this study is to enhance the problem solving and self-management abilities of student teachers through individual and group authentic task. Preliminary results showed that the learning outcomes in high category, nevertheless problem solving and self-management abilities are still low and average categories (scattered at interval $40 \leq N \leq 65$). Initiative to improve this condition is needed. Action research is the alternative solution for that condition through planning, acting, evaluating, and reflecting. This study is allowed in 4 cycles. The acting step result with integrated discuss method, case study, and presentation including self-assessment for individual and group. This method was effective to enhance problem solving and self-management abilities. The final learning outcomes seen from the correlation between student self-assessment and lecture-assessment (r=0.19). Its means there are unidirectional relationship between the result of self-assessment and lecture- assessment. The Conclusion of the research was effective to enhance problem solving and self-management ability.

1. Introduction

Learning in higher education preferred to offer subject that can stimulate developing critical thinking skills in problem solving, collaboration, and communication [1]. Physics evaluating development subject have been taken by graduate physical education teacher training college Padang (Now UNP). On This course there are the main points of discussion were very useful for the assessment of the future of learning physics, students who attend this course are expected to have sufficient provision to carry out the duty as educators in assessing Physics learning. The lectures learning model that had been done was by discussion and presentations method. Discussion method is an effective method for creating a conductive learning either in small groups or large group in the education involving lecture [2]. Research results on the two classes of teaching spread at intervals$_{(66 \leq N 100)}$, but for the assessment process of problem-solving and self-management ability on average is in the category of low to medium or is in the interval $45 \leq N \leq 65$ [3].

Based on these findings it is possible to find an effective learning approach so as to improve the ability of students because some study materials have learned while at the undergraduate level. The hypotheses of the causes of not been optimal of the problem-solving and self-management connected to the task and assessment system in which the lecturers have not stimulate the student's ability.

The theory of learning identifies five dimensions of learning that are essential for the success of the learning process. Implementation self-assessment at least refers to the dimensions above [4]. Self-assessment is a powerful technique for self-improvement [5]. The literature on self-assessment uses the term 'self-assessment', 'Self-reflection' and 'self-evaluation [6]. A contribution of self-assessment is the improvement of behaviour and student achievement [7]. Learning through self-assessment is developed based on critical thinking skills. The ability aspects, namely: the ability to think critically in terms of problem solving and self-management ability [8][9][10]. The Indicator of self-management consisted of: a sense of responsibility for the actions taken, making work plans systematically, implemented the work planning consistently, stay calm in the stressful situations, perform a self-evaluation and to seek improvements in order to improve performance, have confidence in the ability to complete the work, and manage various resources owned to produce the best work [11].

The Indicators of problem solving ability consisted of: identifying the problems and gather relevant information to the issues, define the problems that will be addressed, evaluated various alternative solutions as well as sort, make decisions or take action based on pre-set criteria, the consequences of the decision taken, apply the solution decided upon and examine the results, and solve problems independently. Stimulation of the problem solving and self-management required authentic learning and also authentic assessment.

2. Method

Action research can be seen as a powerful methodology for reforming education. Research is very effective measures implemented to apply theory and practice [12]. Research carried out consists of four stages, namely planning, acting, observing, and reflecting.

The subjects of this research were a student teacher from the graduate of the Physics Education program in Universitas Negeri Padang. Data was analyzed by a reflective and evaluative. The result of this action that further be used as input to determine further action. The outcomes of action level indicated in the number using a scale of 1 - 5. Score 1 showed no performance, it followed by lower levels of learning outcomes or very incompetent, while a score of 5 indicates a excellent performance indicated the outcome level is very competent.

Furthermore, the correlation test to determine the effect of learning outcomes of observations and authentic assessment, reflection on the results of self-assessment is made by students. Correlation tests conducted with product moment correlation formula. States [13] the product moment correlation formula as follows:

$$r = \frac{n\sum XY - (\sum X)(\sum Y)}{\sqrt{\{n\sum X^2 - (\sum X)^2\}\{n\sum Y^2 - (\sum Y)^2\}}} \tag{1}$$

3. Results

The results of the research can be divided into four stages: preparation, implementation, observation and reflection. The results of problem solving ability can be seen from Figure 1. Effects action to self-management ability showed in Figure 2. The Results relationship self-assessment and lecture assessment in last cycle, such as Figure 3. Based on Figure 3 showed that the correlation between self-assessment (X) and lecturer assessment (Y) is obtained coefficient value is 0.19 and a coefficient of determination ($r^2$is 0.034). Based on the results of the analysis, it is known that there is a very strong and positive values among the learning outcomes of the observations and the value of reflection and self-assessment results [14].

IOP Conf. Series: Materials Science and Engineering **335** (2018) 012068 doi:10.1088/1757-899X/335/1/012068

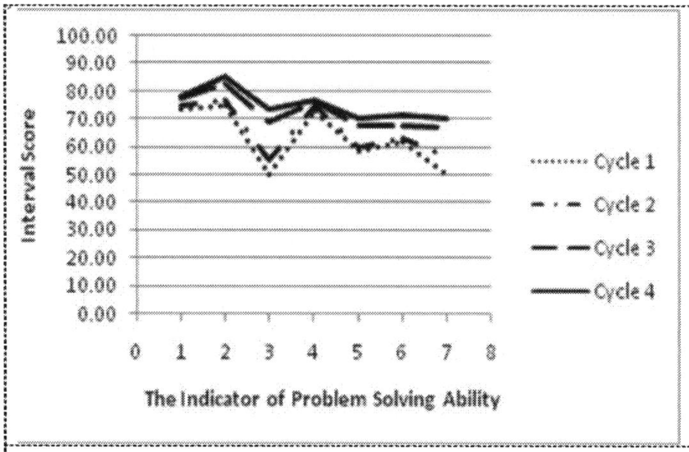

The indicators of Problem Solving Ability

1. Identify problems and gather information relevant to the problems
2. Define the problem that will be solve.
3. Evaluate various alternative solutions and sorting
4. Make decisions or take actions based on predefined criteria
5. Seeing the consequences of decisions made
6. Implement solutions that have been decided and check the result
7. Solve problems independently

Figure 1. The Average Score of Problem Solving Ability Per Cycle.

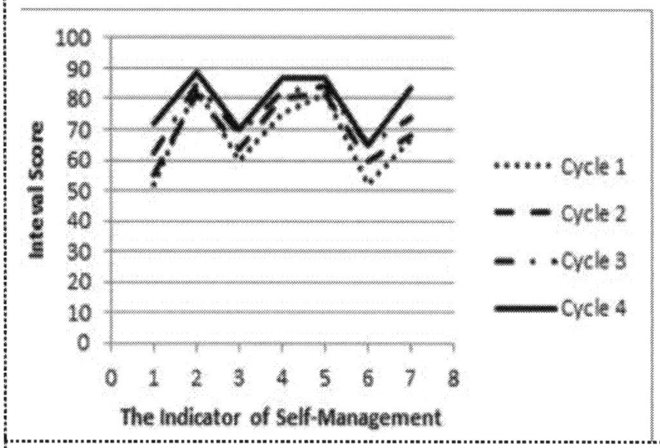

The Indicators of Self-Management

1. Responsible for actions taken
2. Creating a lesson plan systematically
3. Implementing lesson plan consistently
4. Be calm in the face of stressful situations
5. Self-evaluation and improvements effort to increase learning achievement
6. Belief in the ability to complete the work (tasks lectures).
7. Manage available resources to produce the best performance.

Figure 2. The Achievement of Self-management ability

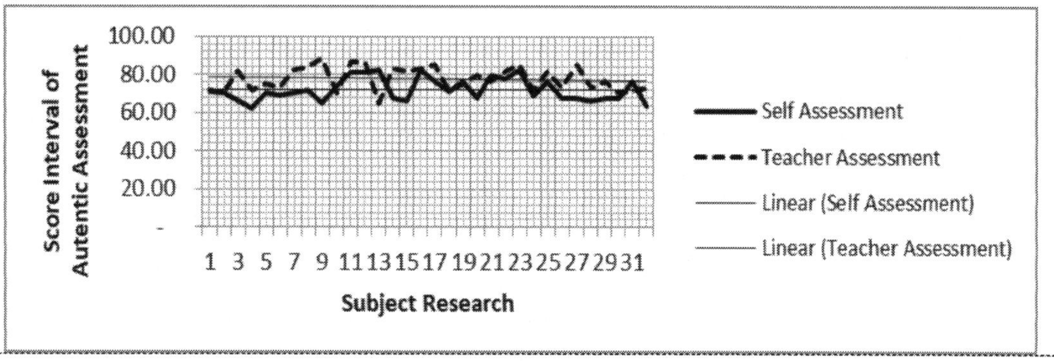

Figure 3. The relationship between learning outcomes based on self-assessment and lecture assessment

4. Discussion

The Learning approach to enhance the problem solving and self-management of student teachers through the development of authentic tasks for individuals and groups and assessment with emphasis on self-assessment, and effective for all indicators. The increase in the indicators is because students are trained to find their selves root of the problem and seek developing alternative forms of problem solving the weaknesses of the previous task to produce the best work the next.

Self-assessment requires the direct involvement of students in integrating knowledge, skills and attitudes. Because of they can control of their learning, they can decide how to use the learning resources available to them inside or outside the classroom. Parallel with the opinion of [15] which states in the self-assessment, individuals were asked to rate their skill by themselves in different domain. Authentic assessment can provide more valid data about the student teacher competence [16][17]. The analysis result showed, as many as 68.5% of students demonstrated in the problem-solving ability in the high category once, and as much as 80.0% of the students demonstrated the self-management in the high and very high category. On average, these results can be interpreted as an indication of the ability of employability skills of students in both categories. Through self-assessment, students can see the advantages and disadvantages, to further these shortcomings become improvement goals[18].

5. Conclusions

Based on the results concluded that the reflective of problem solving and self-management ability both categories. This study should be continued with the implications of implementation of authentic learning and assessment using control class, therefore the result can be differentiated. Learning outcomes seen from the end of the correlation results of student self-assessment with the assessment of the lecturers obtained $r = 0.19$ meaning that the linear relationship between self -assessment of students with the lecture assessment. It can be concluded learning approach through authentic tasks as well as individual and group self-assessment effectively to improve problem solving skills and self-management of student teachers.

6. References

[1] Alismail, H A and McGuire, P. (2015). 21st Century Standards and Curriculum: Current Research and Practice. *Journal of Education and Practice*. **26** (6) p 150
[2] Philip H, Pollock, K H and Bruce, M W. (2011). Learning Through Discussions : Comparing the Benefits of Small-Group and Large-Class Setting. *Journal of Political Science Education*. **7**(1) pp 46-64
[3] Festiyed & Djamas, D. (2016). Analisis Ketrampilan Guru Fisika Sekolah Menengah Menggunakan Model Pembelajaran Terintegrasi Pendekatan Saintifik. Makalah dalam Semirata Fisika :74
[4] Marzano et al. (1997). *Dimensions of Learning*. Alexandria : ASCD
[5] Ross, J A and Bruce, C D. (2007). Teacher self-assessment: A mechanism for facilitating professional growth. *Teaching and Teacher Education*. **23**(2) pp 146-159
[6] Andrade, H and Du, Y. (2007). Student Responses to Criteria Referenced Self-assessment. *Assessment & Evaluation in Higher Education*. **32**(2) pp 159–181
[7] Ozarslan, Y & Ozan, O. (2016). Self-Assessment Quiz Taking Behaviour Analysis in an Online Course. *European Journal of Open, Distance and e-learning*. **19**(2) : ISSN 1027-5207
[8] Bahr, N. (2010). Thinking Critically about Critical Thinking in Higher Education. International Journal for The Schools Teaching and Learning. **4**(2) art 9
[9] Styron, R A. (2014). Critical Thinking and Collaboration : A Strategy to Enhance Student Learning. *Systemic Cybernetics and Informatic*. **12**(7) : ISSN 1690-4524

[10] Muhlisin, A et al. (2016). Improving Critical Thinking Skills of College Student Through RMS Model for Learning Basic Concept in Science. Asia-Pasific Forum onScience Learning and Teaching. **17** (1) art.12
[11] Suarta, M. (2013). Hubungan Sistem Pembelajaran, Lingkungan Belajar, Konsep Diri dan Pengembangan Employability Skills Mahasiswa. *Jurnal Penelitian dan Evaluasi Pendidikan.* Edisi Dies Natalis ke 48 UNY
[12] Bobrakov, S. (2014). Student Teachers' Perceptions of Theory and Practice Integration through Action Research. *International Journal of Learning, Teaching and Educational Research.* **8**(1) pp 1-15
[13] Puth, M T et. al. (2014). Effective Use of Pearson's Product-Moment Correlation Coefficient. *Animal Behaviour.* **93** pp 183-189
[14] Grant, K A. (2014). Comparing Student and Teacher Self Assessment Practices. *Proceedings of the IDEAS : Rising to Challenge Conference.* pp 159-168
[15] Allen, J, Raamakers, G and Velden. (2005). Measuring Competencies of Higher Education Graduates. *Enhancing Alumni Research : European and American Perspectives, New Directions for Institutional Research.* (126) pp 49-59
[16] Jonsson & Enez, Omer.G. (2008). The Complexity of Planning Problems With Simple Causal Graphs. *Journal of Artificial Intelligence Research.* **31** pp 319-351
[17] McMillan, J. & Hearn, J. (2008). Student Self-Assessment: The Key to Stronger Student Motivation and Higher Achievement. *Educational Horizons.* **87** pp 40-49
[18] Keskitalo, T, Pyykkö, E, & Ruokamo, H. (2011). Exploring the Meaningful Learning of Students in Second Life. *Educational Technology & Society,* **14** (1) pp 16–26

IOP Conf. Series: Materials Science and Engineering **335** (2018) 012069 doi:10.1088/1757-899X/335/1/012069

The Development of Virtual Laboratory Using ICT for Physics in Senior High School

M Masril[1], H Hidayati[1], Y Darvina[1]

[1]Department of Physics, Faculty of Mathematics and Sciences, Universitas Negeri Padang, Padang 25131, Indonesia

masril_qch@yahoo.com

Abstract. One of the problems found in the implementation of the curriculum in 2013 is not all competency skills can be performed well. Therefore, to overcome these problems, virtual laboratory designed to improve the mastery of concepts of physics. One of the design objectives virtual laboratories is to improve the quality of education and learning in physics in high school. The method used in this study is a research method development four D model with the definition phase, design phase, development phase, and dissemination phase. Research has reached the stage of development and has been tested valid specialist. The instrument used in the research is a questionnaire consisting of: 1) the material substance; 2) The display of visual communication; 3) instructional design; 4) the use of software; and 5) Linguistic. The research results is validity in general has been a very good category (85.6), so that the design of virtual labs designed can already be used in high school.

1. Introduction

In studying physics there are two inseparable aspects, namely physics as a product (knowledge of physics that consist of facts, concepts, principles, laws, and theories) and physics as a process (scientific activity). Therefore, physics learning must consider the characteristics of physics as a process and product [1].

To study physics as a process and product, it is necessary for inviting students to learn actively, not only acquire certain knowledge or concepts but directly involved in the discovery of the concepts. One of learning approach that can be used is science process skills approach.

Science process skills can be developed by practicum method. Through practicum activities, students not only get information of facts or concepts, but also learn to find them using various aspects of the science process skills, such as formulating problems, predicting, planning experiments, observing, using tools and materials, interpreting data (interpretation), applying concepts (applications) and communicating.

Based on the characteristics of physics as a process and product, the government through the Minister of Education and Culture has made a syllabus for physics in senior high school using the latest curriculum. It can be read with Regulation of Education and Culture Minister No. 59 in 2013. The fourth of main competence describes the importance of practicum activities. Thus, schools should facilitate laboratory to do a practicum.

Based on the results of observations in senior high schools of Padang, most schools do not have complete laboratory facilities to support the implementation of the latest curriculum so many practicum projects that cannot be practiced. Based on the experience of researchers when doing research in 2015 for teaching materials, it was found some weaknesses of learning in school, such as practicum activities could not be done although the student worksheet has been designed to help physics learning.

Theoretically, the process of learning in the classroom or in the laboratory includes in developing three competencies namely cognitive, affective and psychomotor. It requires balancing those competencies between theory and practicum activities. So, the laboratory is needed. The activities in the classroom can only assess cognitive competence while the implementation of the practicum may include assessment in students' affective and psychomotor competence. Activities in the classroom and in the laboratory in the learning process are inseparable because they are related to each other to achieve the objectives in the learning.

To solve the problem, the role of teachers is needed, because the teacher is the agent of change. Therefore, it takes various strategies and learning innovation to overcome the inadequacy of laboratory facilities and infrastructure without having to abandon the achievement of cognitive, affective and psychomotor competence.

Practicum activities can be done in various ways, such as the utilization of Information and Communication Technology (ICT), which is supported by the use of computers as an ICT device for example virtual lab experiments (virtual lab). Although the virtual lab is not a real activity, students can do a practicum like a real practicum. Virtual laboratory (virtual lab) is one of ICT-based learning process that can be used as an alternative solution for learning with practice method.

With the existence of virtual laboratory, students can understand the physics concept and teachers can design of physics lab for all concepts in physics. Virtual laboratory also minimizes the cost of procuring tools and practicum materials. They are many benefits of virtual laboratory such as 1) reducing time constraints, if there is not enough time to teach all learners in the laboratory until they understand; 2) Reducing geographic barriers, if there are learners whose location is far from the school; 3) Being economical, does not require laboratory buildings, tools and materials as in conventional laboratories; 4) Improving the quality of the experiment, as it is possible to repeat it to clarify doubts in laboratory measurements; 5) Increasing the effectiveness of learning, because learners will spend more time to practice repeatedly; and 6) Improving safety and security, as it does not interact with real tools and materials [2].

Virtual laboratory (virtual lab) is one of ICT-based learning process that can be used as an alternative solution for learning with practice method. Learning-based virtual lab is one of the superior products of advances in information technology and laboratories. According to Russel et al. [3], in teaching and learning activities, teachers should help learners develop their understanding by providing: direction and organization for learning, motivation of learning, conceptual explanations, activities that can help learners recognize and fix misconceptions, and opportunities to provide direction in problem solving. Visualization of the phenomena of physics and its concepts related to animation at the microscopic level, as well as the simulations associated with the daily examples of learners can increase the learner's knowledge visually and stimulate more learners to achieve a high level of understanding of the concepts in physics [3]. Based on the results of research, learners are more motivated to learn the concept of physics when accompanied by visualization abstract concepts [4]. The ideal virtual lab is run on the internet, so participants can experiment from wherever and whenever. However, it can also be run in an intranet environment or standalone computer. With virtual labs, the building and physical lab tools are transformed into computers and virtual lab software [5]. Based on what has been described, it has become clear that virtual labs can be used as an alternative to assisting human beings in improving human productivity and well being, solving problems, predicting how options of completing and implementing solutions.

Virtual labs are widely developed, as developed by the University of Colorado, Physics Education Technology (PhET) [6]. In PhET there is a theory and experimental simulation involving the user actively. Users can manipulate activities related to experiments, so that in addition to building concepts, PhET can also be used to bring up the skills of the science process.

Therefore, in order to implement the 2013 curriculum in high school and to meet the demands of the curriculum, the authors are interested in designing virtual labs through ICT for senior high school physics.

2. Research Method

The research method is a Research and Development (R & D) with the 4-D model. According to Thiagarajan et al.[7], the stages of development research are defined, design, develop, and disseminate.

2.1 Define

At this stage, the curriculum is analyzed by analyzing the main competence in physics syllabus and listing the practicum will be done in learning activities. This analysis needs to be done in order to design the lab with a virtual lab in accordance with the curriculum demands. Teachers have done a practicum in the learning process, but it is still in limitations because the laboratory is not so optimal. Therefore, virtual laboratories developed through ICT can help students practice finding concepts in physics learning both in the classroom and outside the classroom.

2.2 Design

In this stage, the virtual lab model is designed. This model is arranged in the order of main competence in the syllabus. Each practicum activity is completed with student worksheet, all integrated with ICT. The virtual lab model is systematically designed and interesting in order to achieve the expected competencies and objectives.

2.3 Development

In this stage, the virtual lab model is validated and revised based on evaluator and practitioner (user) suggestion, and limited trial. Then a second analysis and revision was performed based on a limited trial analysis. Valid virtual labs will be tested in the form of limited tests in sample schools. In the implementation of limited tests, it uses learning tools such as lesson plans and student worksheet. To find out the effectiveness, it can be analyzed from student learning outcomes. During the first year, it was a valid, practical and effective for learning in school.

2.4 Disseminate

The second year of research is the dissemination of products in senior high school students in Padang City. The sample school is chosen by stratified sampled. The dissemination was done in the form of quasi-experimental research in sample schools. The virtual lab is re-revised until it is ready for widespread use.

The product is tested again by five people to get validity value. The instrument used for the validity test is using a questionnaire consisting of five indicators: content feasibility, language usage, presentation, graffiti, and developed virtual lab products. Technique Data analysis conducted using qualitative analysis using graph..

3. Result and Discussion

3.1 Result

3.1.1 Validity: Feasibility

The results of virtual laboratory validation for the content feasibility indicator are obtained from the value of each indicator consisting of 8 statements, namely: 1) the suitability of the material with Core Competencies and Basic Competencies; 2) Compliance with students' needs; 3) conformity with the needs of virtual laboratories; 4) the truth of material substance; 5) material conformity with daily life; 6) benefits for the addition of knowledge insight; 7) conformity of training and evaluation with material; 8) conformity with values, morality, social. The results obtained from each Feasibility indicator statement are plotted in Figure 1.

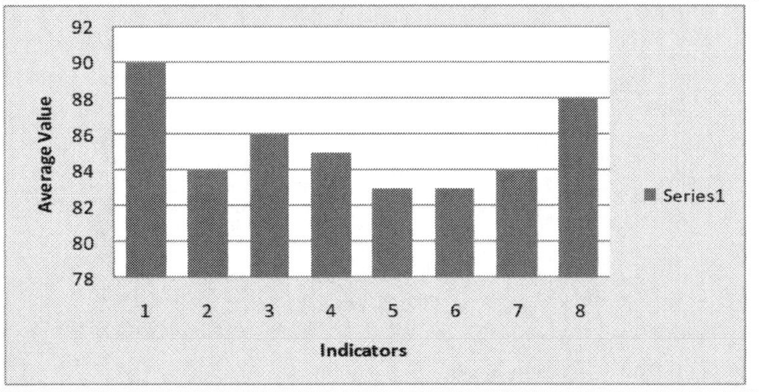

Figure 1. Validity: Feasibility.

3.1.2 Validity: Language Usage

The results of the virtual lab validation for language usage indicators are derived from the value of each indicator consisting of 12 statements namely 1) legibility, 2) idea density, 3) the beauty of the language style, 4) the use of the short length of the sentence, 5) the way of constructing a sentence, 6) how to write paragraphs, 7) the use of punctuation, 8) how to write physics terms, 9) how to write physics equations, 10) how to write table and picture titles, 11) clarity of information, and 12) the use of language effectively and efficiently . The results obtained from each indicator of language usage are plotted in Figure 2.

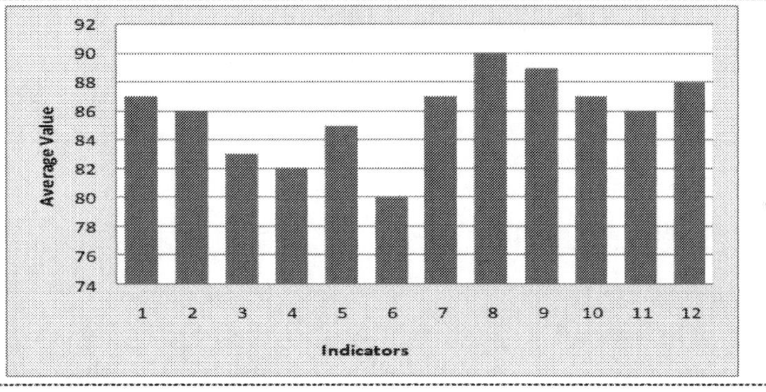

Figure 2. Validity: Language Usage.

3.1.3 Validity: Presentation

The results of the Virtual Laboratory validation for the presentation indicator are obtained from the value of each indicator consisting of 5 statements, namely: 1) Clarity of purpose; 2) Order of presentation; 3) Provision of motivation; 4) Interactivity (stimulus and response); and 5) Completeness of information. The results obtained from each statement indicator of the presentation of Virtual Laboratory are plotted in Figure 3.

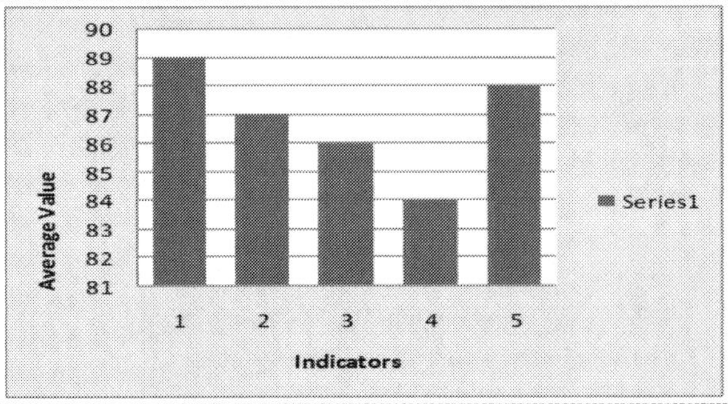

Figure 3. Validity: Presentation.

3.1.4 Validity: Graphics

The results of the Virtual Laboratory validation for the graphics indicator are obtained from the value of each indicator consisting of 5 statements: 1) the use of font (type and size); 2) layout, 3) illustrations, graphics, pictures, photographs; and 4) display design. The results obtained from each statement of virtual lab indicator are plotted in Figure 4.

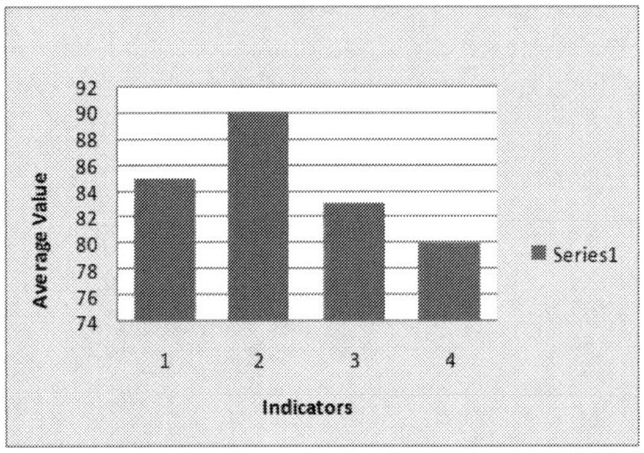

Figure 4. Validity: Graphics.

3.1.5 Validity: Development Steps

The results of validation of Virtual Laboratory for the indicator of Virtual Labor developed are obtained from the value of each indicator consisting of 8 statements namely: 1) Design; 2) Linkage to concepts; 3) Ease of access; 4) Interactive; 5) A developed guide; 6) Material depth; 7) Meet all KD; 8) Evaluation in the form of adequate quiz. The results obtained from each of the Virtual Laboratory indicator statements developed are plotted in Figure 5.

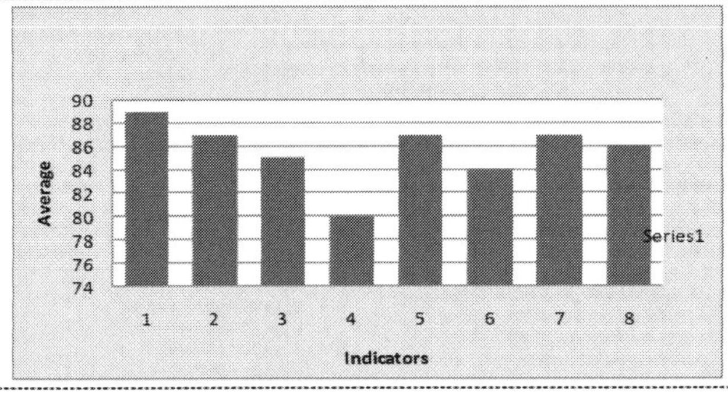

Figure 5. Validity: Development Steps.

3.1.6 Practicality: Ease of Use

The result of Virtual Laboratory practicality test for the Ease of Use indicator is obtained from the value of each indicator consisting of 5 statements, namely: 1) easy to operate; 2) can be used anytime; according to the needs of teachers; 3) easy to carry, because it can be accessed anywhere; and 4) can be used repeatedly; and 5) easy to interpret by the teacher in using interactive multimedia. Results obtained from each indicator statement Ease of Use of Student Worksheet developed with an average value of 85.94 in plotted in Figure 6.

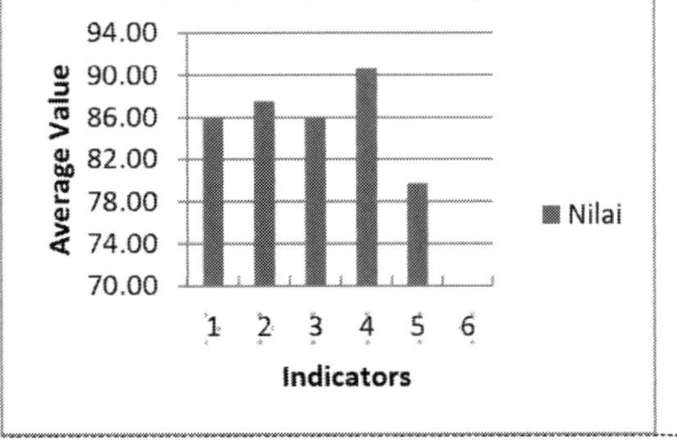

Figure 6. Validity: Ease to Use

3.1.7 Practicality: Attractiveness

The result of Virtual Laboratory practicality test for the attractiveness indicator is obtained from the value of each indicator statement consisting of 6 statements, namely: 1) Presentation of Student Worksheet based on attractive virtual laboratory; 2) The brief information in the Student Worksheet based on the virtual laboratory is equipped with an appropriate image of the material; 3) The pictures presented in the Student Worksheet are quite clear; 4) Display of Student Worksheet through interesting ICT; 5) The font type Student Worksheet read clearly; and 6) The combination of colors used is proportional. The results obtained from each indicator statement of the Student Worksheet developed with an average grade of 89.84 are plotted in Figure 7.

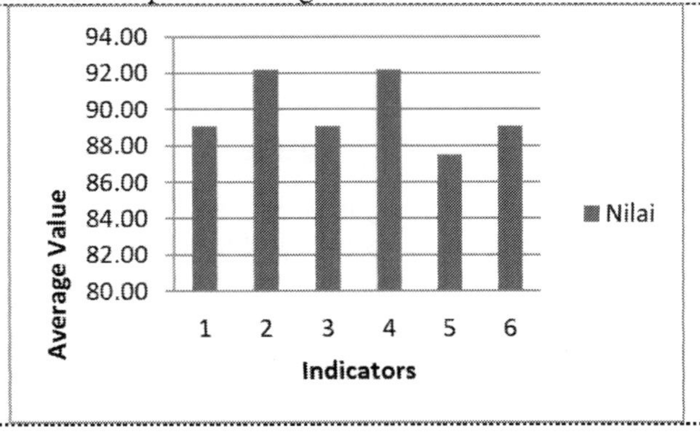

Figure 7. Validity: Attractiveness

3.1.8 Practicality: Benefits for Teacher

The result of practicality test for the utilization indicator of Student Worksheet is obtained from the value of each indicator consisting of 4 statements, namely: 1) can be a reference for the teacher to activate the student in learning; 2) can support teacher's activities in fulfilling K13's demands; 3) can be used to motivate student learning; and 4) can make learning more interesting. The results obtained from each indicator with an average value of 89.45 are plotted in Figure 8.

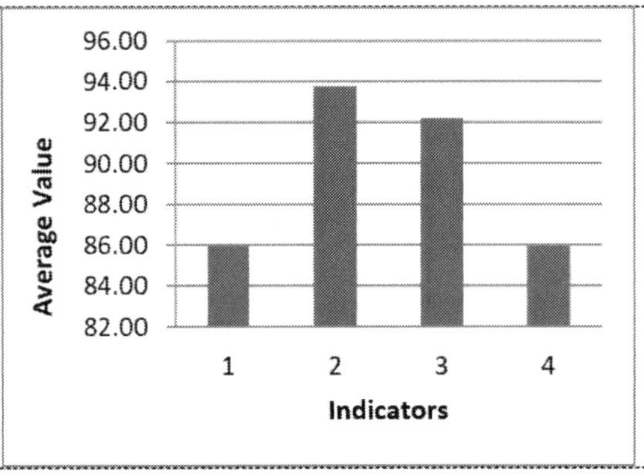

Figure 8. Practicality: Benefits for Teacher

3.1.9 Practicality: Opportunities for Implementation

The result of practicality test for LSM Implementation indicator based on virtual laboratory is obtained from the value of each indicator consisting of 5 statements, namely: 1) facilitates students to work in accordance with scientific method; 2) help students to think critically; 3) makes learning time more efficient; 4) can help students learn independently; and 5) Evaluations can be used to measure students' mastery of learning materials. The results obtained from each statement indicator of Student Worksheet implementation opportunities developed with an average value of 83.13 are plotted in Figure 9.

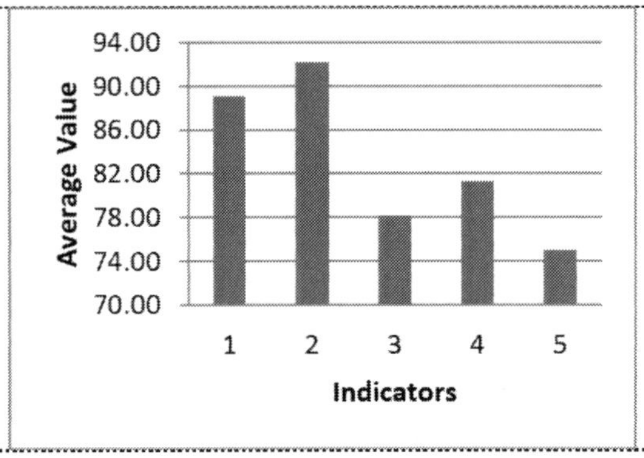

Figure 9. Practicality: Opportunities for Implementation

3.1.10Average Value for Practicality

The results of the virtual laboratory practice test are obtained from the mean values for each indicator. The average results of validity values for each indicator are: Ease of use (85.94), attractiveness (89.84), Benefit (89.45), and implementation opportunities (83.13), so the average value of the test results of practice virtual lab with four indicators is 87.09. Based on these values it can be argued that all the virtual lab indicators are in very high categories.

3.2 Discussion

From the validity test, virtual lab can be explained as follows: The virtual lab has shown the suitability of the material with the basic competencies. The virtual lab is very useful in practicum. The content is described in good explanation so students can prove the concepts. In addition, the steps to learn virtual labs are made as detailed as possible so as to make it easier for students to use the developed virtual lab.

In general, virtual lab is understandable because of the language. This is due to the density of ideas in developing sentences, how to construct sentences, how to construct paragraphs, how to write table titles and drawings, and the use of language effectively and efficiently in accordance with the enhanced spelling. Besides, the high value of validity is caused in the development of virtual lab many express new ideas that make it easier for students to learn the virtual lab developed. .

Virtual lab provides motivation and interactivity (stimulus and respond). Theoretically, giving motivation can improve the activity of the child. In addition, the presentation of virtual labs equipped with animations that appeal to students can also increase students' interest to learn so that students are interested to learn it. The presentation of virtual lab makes it easier for students to study anywhere and anytime.

Virtual lab uses varied fonts, layout and neat layouts, interesting picture illustrations and good display design. Besides, the illustration of the picture is interesting and complete with a complete description. Virtual lab is in high category because the value of each statement of the developed indicators has been high such as: design, linkage to the concept, Ease of access, Interactive, easy-to-understand guide, material depth, and Evaluation in the form of quiz interactive.

Based on the results of practical tests, virtual lab are: easy to operate, can be used anytime in accordance with the needs of teachers, easy to carry because it can be accessed anywhere, can be used repetitive, and easy to interpret by the teacher in using interactive multimedia.

The presentation of worksheet based on the virtual laboratory is interesting, the short information in the worksheet is equipped with the appropriate picture, the pictures are quite clear, font type clearly reads , the color combination used is proportional. It can be a reference for teachers in enabling students in learning; can support teacher activities in meeting the demands of Curriculum, can be used to motivate student learning; and can make learning more interesting. Virtual lab facilitate students to work in accordance with scientific method, helping students to think critically; making learning time more efficient; help students learn independently; and can be used to measure students' mastery of learning materials.

4. Conclusion

Based on the results of research and discussion that has been done can be concluded: First, The results of virtual lab validity test included high category with validation value 85.6 while practical test 87,09. Second, Virtual lab designed this can be used for physics learning of the latest curriculum in high school.

5. Acknowledgement

This research is research grants from Directorate Of Higher Education. My thanks and appreciations also go to my colleague in developing the project and people who have willingly helped me out with their abilities.

Reference

[1] Depdiknas. 2010. *Badan standar Nasional Pendidikan*, Jakarta
[2] Farreira, MJM. 2010. *Intelligent classrooms and smart software: Teaching and learning in today's university.* Springer Science and Business Media. Springer publications.
[3] Russell, J.W. et al. 1997. Use of Simultaneous-Synchronized Macroscopic, Microscopic, and Symbolic Representations to Enhance the Teaching and Learning of Chemical Concepts. *Journal of Chemical Education*. 74. (3). 330-334.
[4] Hamidah I. 2008. Developing electric field learning media using Finite Element Method Laboratory to enhance the quality of physics learning instruction. *Preceding of UPI-UPSI International Seminars*, Perak-Malaysia.
[5] I Ketut Gede Darma Putra. 2009. *Pendidikan Berbasis Teknologi Informasi*. Rakorda
[6] https://phet.colorado.edu/in/simulations/category/physics
[7] Thiagarajan, S., Semmel, D. S & Semmel, M. I. 1974. Instructional Development for Training Teachers of Expectional Children. Minneapolis, Minnesota: Leadership Training Institute/Special Education, University of Minnesota.

IOP Conf. Series: Materials Science and Engineering **335** (2018) 012070 doi:10.1088/1757-899X/335/1/012070

The Practicality of Statistical Physics Handout Based on KKNI and the Constructivist Approach

S Y Sari [1*] and R Afrizon

[1]Physics Department, Faculty of Mathematics and Science, State University of Padang, Indonesia

silviyuliasari@fmipa.unp.ac.id

Abstract. Statistical physics lecture shows that: 1) the performance of lecturers, social climate, students' competence and soft skills needed at work are in enough category, 2) students feel difficulties in following the lectures of statistical physics because it is abstract, 3) 40.72% of students needs more understanding in the form of repetition, practice questions and structured tasks, and 4) the depth of statistical physics material needs to be improved gradually and structured. This indicates that learning materials in accordance of The Indonesian National Qualification Framework or *Kerangka Kualifikasi Nasional Indonesia* (KKNI) with the appropriate learning approach are needed to help lecturers and students in lectures. The author has designed statistical physics handouts which have very valid criteria (90.89%) according to expert judgment. In addition, the practical level of handouts designed also needs to be considered in order to be easy to use, interesting and efficient in lectures. The purpose of this research is to know the practical level of statistical physics handout based on KKNI and a constructivist approach. This research is a part of research and development with 4-D model developed by Thiagarajan. This research activity has reached part of development test at Development stage. Data collection took place by using a questionnaire distributed to lecturers and students. Data analysis using descriptive data analysis techniques in the form of percentage. The analysis of the questionnaire shows that the handout of statistical physics has very practical criteria. The conclusion of this study is statistical physics handouts based on the KKNI and constructivist approach have been practically used in lectures.

1. Introduction

Abdullah (2007) [1] explains that the actual statistical physics is the mathematical study materials such permutations and combinations due to an abstract issue. The abstractness of the statistical physics material is tried to study deeper instatistical physics course on July to December, 2015 using a questionnaire. The questionnaire results provide a deeper understanding of the needs of students and lecturers in lectures. These results are recorded through a set of analysis known as the needs analysis of the development of statistical physics learning material.The results of needs analysis is preliminary research results from this study are summarized in Table 1.

Table 1. Results of Needs Analysis [2]

No	The Typeof Analysis	The Results of Needs Analysis
1.	Preliminary analysis	Performance of lecturers, social climate, student competence, soft skills and the difficulties of students in following the statistical physics lectures are in enough categories
2.	Learner analysis	• 57,14% of students have a good preliminary ability • 40,74% of students have low understanding in statistical physics • 40,72 % of students require more understanding in the form of repetition, practice questions and structured assignment • 100% of the environment affects the ability of mastery
3.	Task Analysis	• The group learning and repetition of material need to be improved • The structured assignment has been in accordance with the competence
4.	Concept Analysis	The depth of statistical physics material needs to be increased gradually and structured

The preliminary analysis showed the importance of developing lectures material based on the Indonesian National Qualification Framework (KKNI). Analysis of learners, tasks and concepts illustrates that the need for a constructivist approach applied in statistical physics lectures in order to create lectures which are structured, coherent and clear. Constructivist learning approach can improve and develop the active participation of students, student independence in learning and students knowledge construction [3].

The statistical physics lectures device had been designed in accordance with the needs analysis and learning objectives formulated aresyllabus, lesson plan, learning materials, and assessment sheets. The design of instructional materials in the form of handouts in Figure 1 is prepared based on the needs analysis that has been recorded. Handout has been developed based on KKNI as indicated on the learning outcomes and learning objectives. Orientation phase is already reflected in the introduction summary handout materials. The elicitation and restructuring ideas phase have emerged on the part of the study material handouts discussion. The designed questionsare expected to be able to reconstruct and clarify ideas on the students themselves. This discussion study material also comes with a "keep in mind!" menu. Structured assignments have embodied the steps of implementing ideas and reviews.

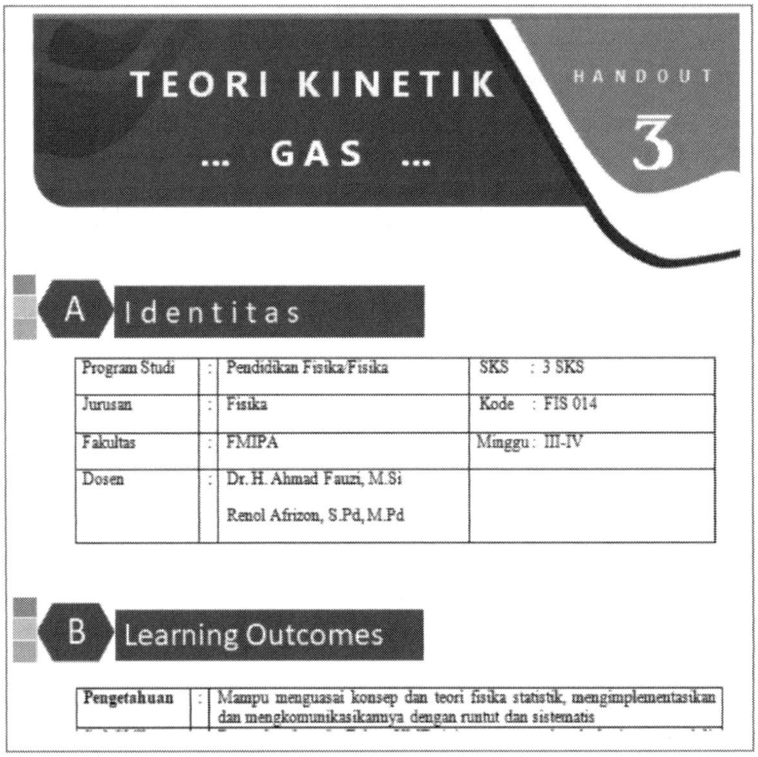

TEORI KINETIK ... GAS ...

HANDOUT 3

A Identitas

Program Studi	:	Pendidikan Fisika/Fisika	SKS : 3 SKS
Jurusan	:	Fisika	Kode : FIS 014
Fakultas	:	FMIPA	Minggu : III-IV
Dosen	:	Dr. H. Ahmad Fauzi, M.Si Renol Afrizon, S.Pd, M.Pd	

B Learning Outcomes

Pengetahuan	:	Mampu menguasai konsep dan teori fisika statistik, mengimplementasikan dan mengkomunikasikannya dengan runtut dan sistematis

Figure 1. Design Handout [2]

The design of statistical physics handout is a continuation of preliminary research.The statistical physics handout are designed based on the results of a needs analysis.The designed handouts have been validated by experts with very valid category. The validation results consisted of 89.58% of content aspects, 92.71% of construct aspects, 91.67% of language aspects, and 89.58% of presence aspects [2]. Handout construction showed the highest percentage means that the handout preparation has been based on KKNI with a constructivist approach according to expert assessment. After it validated by experts and revised, then handouts need to be tested right in the classroom. The testing of handouts in the classroom is useful to know the level of the handout practicalities.

Practical based on Indonesian Dictionary or *Kamus Besar Bahasa Indonesia* (KBBI) defined as 1) based on practice, 2) easy and happy to use, and 3) efficient. It also expressed by Jan van den Akker, et al.[4] that *'Practicality' refers to the extent that users (and other experts) consider the intervention as clear, usable and cost effective in 'normal' conditions*. Thus, the indicators are used to look at the practicalities of handout is easy to use, attractive, and efficient. The practicalities associated with use of lecture material by users such as students and lecturers and other experts. The practicality of a handout should be seen in the class with normal conditions. It means if the material kinetic theory of gases designed for the 3rd and 4th meeting in the semester plan, it should be implemented in the classroom then tested at that meetings.

Nieveen and Folmer [5] stated four quality criterias that can be applied in resolving complex issues related education world intervention to get the best solutions, including:

a. Relevance shows that intervention should be based on state-of the art (scientific). It is known as the content validity.
b. Consistency, shows that intervention should 'logically' designed. It is also called construct validity.
c. Practicality, indicating that this intervention can be used in a setting that has been designed.
d. Effectiveness, shows that interventions can produce products on the desired results.

Based on the quality criteria which have been advanced, quality practicality of the product becomes very important to be tested. Therefore, the research is aimed to see whether the handouts that are valid according to experts has been practically used in statistical physics lectures.

2. Research Methods

This study is a research and development model of the 4-D developed by Thiagarajan. This research and development should be carried out systematically and sequentially. Thus, the implementation should not be arbitrary nor abandoned the steps. This research has been conducted in the semester of July-October 2016 in the Physics Department, Faculty of Mathematics and Natural Sciences, State University of Padang. The stages that have been passed by this study are define, design and develop (expert appraisal) phase. The develop phase steps shows in Figure 2.

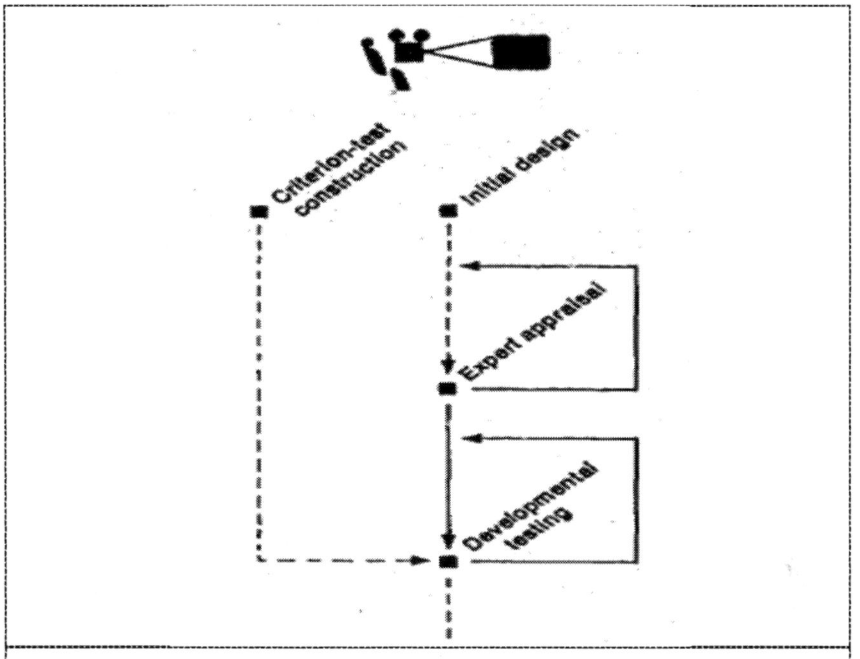

Figure 2. The Develop Stage 4-D model [6]

The research conducted has entered the final part of the develop stage is development testing. In this section, it involves the testing of learning materials for users to obtain the revised parts. Respons/reaction and comment from users, indispensable as an evaluation and revision of improving the design of handouts to get practical qualities. The cycle in development testing are test, revise and test again to be repeated so that materials used are consistent and effective [7]. Data collection was done by questionnaire. The analysis technique used to process data is the technique of qualitative data analysis with 5 criteria: 1) it is not practical (0% $<P<$20%), 2) less practical (20% $<P<$40%), 3) reasonably practicable (40% $<P<$60%), 4) practical (60% $<P<$80%), and 5) very practical (80% $<P<$100%) modified from [8].

3. Result and Discussion

The practicality of handouts in the study will describe the quality of handout development.The questionnaire responses of lecturer and students toward the use ofstatistical physics handouts in lecture can describe the level of handout practicality. The result of questionnaire analysis of lecturer and students response will be explained as follow.

IOP Conf. Series: Materials Science and Engineering **335** (2018) 012070 doi:10.1088/1757-899X/335/1/012070

3.1. The Questionnaire Response of Lecturers

The statistical physics lecture using handouts as one of the media that will be used in a lecture based on KKNI and constructivist approach will be monitored well when given questionnaires to lecturers. Two statistical physics lecturers who became a practitioner will be asked for opinions and comments related to the use of handouts in the class whether it is practical or not. The implementation of assessments took place while using handouts for students. The results of handouts practicalities by lecturers are shown in Figure 3. Figure 3 shows that the handout has been very practical for use with the practicalities value 92.15%. "Interesting" indicator is the lowest one and the indicator "easy to use" is the highest. Nevertheless, the handout quality is still in very practical criteria.

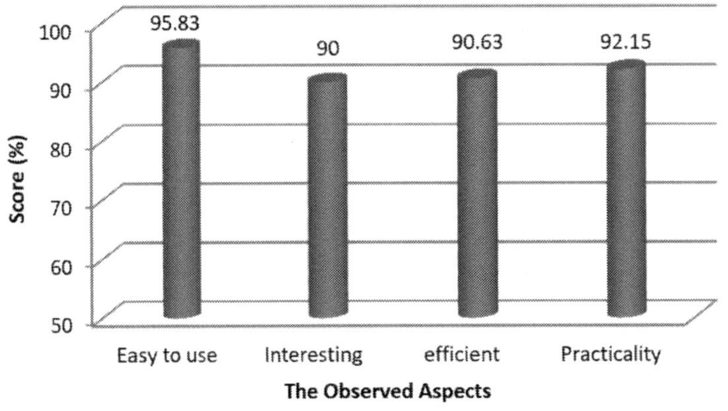

Figure 3. The Results of Handout Practicality of Lecturer Questionnaire

3.2. The Questionnaire Response of Students

Students have been involved as a user in order to obtain advice and suggestion for the improvement of statistical physics handout that has been designed. Handouts that have been designed are regularly fixed at each meeting according to students requirement and validator suggestions. The practicality of handout is obtained through a questionnaire distributed to students at the end of the 3rd meeting. The results of handout practicality is shown clearly in Figure 4.

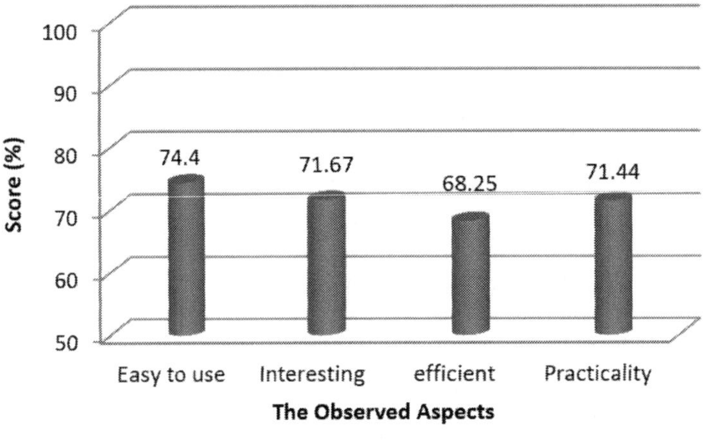

Figure 4. The Results of Handout Practicality of Students Questionnaire

Figure 4 shows that efficient component has the lowest value though overall handout is practical (71.44%). This shows that the students are still difficult in arranging time to solve discussion materials and guiding questions on the handout. "Easy to use" component gets the highest score as 74.40%.

Based on the analysis of questionnaires that have been distributed to students and lectures about practicality of handout is obtained that statistical physics based on KKNI and constructivist approach has a practical value by lecturers and students as users. This is similar with the opinion of Fauzan [9] that "the practicality is useful to see how the users (lecturer and students) and other experts consider whether the product is attractive and can be used in normal condition".

4. Conclusion

Statistical physics handouts based on KKNI and constructivist approach has been practically used in lectures. It is obtained from very practical criteria handouts based on questionnaire of lecturers and students.

Acknowledgement

This article is part of the Research for Beginner Lecturer, which is supported by DRPM Director General of Research and Development Kemenristekdikti. The author thank to practitioners who had been involved in this research. The author also did not forget to express his gratitude to Dra. Hidayati, M.Si for validating the research instrument and Weria Hendri, S.Si, M.Pd who has helped document the author's research.

References

[1] Abdullah, M. 2007. *Pengantar Fisika Statistik untuk Mahasiswa*. Bandung: ITB.

[2] Afrizon, R. Silvi Y.S. Ahmad, F. 2016. *Analisis Kebutuhan Perancangan Perangkat Perkuliahan Fisika Statistik Berbasis KKNI dengan Pendekatan konstruktivis*. Prosiding Semirata 2016 Bidang MIPA BKS-PTN Wilayah Barat. Palembang: FMIPA UNSRI.

[3] Prasojo, L.D. 2006. *Konstruktivisme dalam Pendidikan Tinggi*. Dinamika Pendidikan No. 2 Tahun III: FIP UNY.

[4] Jan van den Akker, Brenda Bannan, Anthony E. Kelly. Nienke Nieveen, Tjeerd Plomp. 2013. *Educational Design Research Part A: Introduction. Netherlands Institute for Curriculum Development (SLO)*, Enschede, the Netherlands.

[5] Nieven N and Folmer, N, 2013, Formative evaluation in educational design research in Educational Design Research, T. Plomp and N. Nieveen (Eds), Netherlands Institute for Curriculum Development (SLO), Enschede, the Netherlands, pp.152 - 169

[6] Thiagarajan, S; Semmel, D.S; & Semmel, M .I. 1974. Instructional Development for Training Teachers of Exceptional Children: A Source book.Indiana: Indiana University.

[7] Rochmad. 2012. Desain Model Pengembangan Perangkat Pembelajaran Matematika. Jurnal Kreano, ISSN : 2086-2334. Diterbitkan oleh Jurusan Matematika FMIPA UNNES Volume 3 Nomor 1, Juni 2012.

[8] Riduwan. 2009. Belajar Mudah Penelitian untuk Guru, Karyawan, dan Peneliti Pemula. Bandung: Alfabeta.

[9] Fauzan, A. 2002. Applying Realistic Mathematics Education (RME) in Teaching Geometry in Indonesian Primary Schools. Thesis. University of Twente, Enshcede.

IOP Conf. Series: Materials Science and Engineering **335** (2018) 012071 doi:10.1088/1757-899X/335/1/012071

The Study of Literacy Reinforcement of Science Teachers in Implementing 2013 Curriculum

W S Dewi, F Festiyed, H Hamdi, S Y Sari[1]

Physics Department, Faculty of Mathematics and Sciences, Universitas Negeri Padang, Padang 25131, Indonesia

wahyunisatria87@gmail.com

Abstract. This research aims to study and collect data comprehensively, new and actual about science literacy to improve the ability of educators in implementing the 2013 Curriculum at Junior High School *Padang Pariaman* District. The specific benefit of this research is to give description and to know the problem of science literacy problem in interaction among teacher, curriculum, facilities and infrastructure, evaluation, learning technology and students. This study uses explorative in deep study approach, studying and collecting data comprehensively from the interaction of education process components (curriculum, educator, learner, facilities and infrastructure, learning media technology, and evaluation) that influence the science literacy. This research was conducted in the districts of *Padang Pariaman* consisting of 17 sub-districts and 84 junior high schools managed by the government and private. The sample of this research is science teachers of *Padang Pariaman*District with sampling technique is stratified random sampling. The instrument used in this study is a questionnaire to the respondents. Research questionnaire data are processed by percentage techniques (quantitative). The results of this study explain that the understanding of science teachers in *Padang Pariaman* District towards the implementation of 2013 Curriculum is still lacking. The science teachers of *Padang Pariaman* District have not understood the scientific approach and the effectiveness of 2013 Curriculum in shaping the character of the students. To improve the understanding of the implementation of Curriculum 2013, it is necessary to strengthen the literacy toward science teachers at the Junior High School level in *Padang Pariaman* District.

1. Introduction

Efforts to improve the quality of education have been done both by the government and by the private sector through the improvement of curriculum [1,2] procurement of textbooks, improvement of teacher quality, improvement of education management and welfare improvement teacher. Our curriculum issues left behind when compared to other countries curriculum including the curriculum used in the Training Center and the Training Course. As a result, when someone entered the world of work, knowledge and labor skills have to be reformatted *(On the job training)* to fit the area to be worked on.

Efforts have been made to solve the problem, as has been done Deliberation Subject Teacher in developing and implementing a curriculum with workshops, as well as some research done with respect to the curriculum [3]. Despite many attempts, it has not solved the whole problem. This is one of the causes of unsuccessful efforts to improve the quality of process and learning outcomes so far. These studies have been successful for partial application, but have not yet addressed the problem as a whole and have not been able to find the root of the problem. Some unanswered weaknesses such as: the lack of relevance of student learning experience planning between those in the syllabus and that contained in the lesson plan; yet integrated learning experiences of students in attitude, skill and knowledge aspects; The preparation of the instrument is less refers to indicators of achievement of learning outcomes and the characteristics of each learning material. Even sometimes less consider the conditions in each school. If this continues to be allowed and there is no improvement effort, and curriculum development then the learning process will not work properly.

In connection with the review of the achievements of literacy, many studies have been published by [4,5]. The improvement of *scientific* literacy of students in the context of this course requires a teacher who has a good pedagogic competence, so as to deliver quality learning process as one of the factors that affect the learning achievement. The achievement scores of Indonesian students' science literacy are 393, 395, 395 for 2000, 2003 and 2006, respectively [6]. The average score of all participating countries is 500 with standard deviation 100. The low score indicates that Indonesian students have limited scientific knowledge. The low science literacy score reflects the general phenomenon of poor student learning achievement in Indonesian students. This raises questions for science teachers and policy makers: why such a thing can happen? What factors lead to low natural science performance? What are the implications for learning, curriculum, and school institutions? And how should the education process work to increase the literacy especially science literacy.

The educational process is an interaction of six components consisting of teachers, curriculum, facilities and infrastructure, evaluation, learning technology, and students.The curriculum is one component of education that is quite influential on the successful implementation of education. The curriculum is said to be effective when the curriculum is able to prepare graduates according to the public interest. This is important because the curriculum is the heart of education activities and processes. The curriculum has three main dimensions, namely: the intended curriculum *(intended curriculum)* related to the goals and plans; implemented curriculum *(curriculum implemented a)* with regard to the learning process and institutional arrangements; and curriculum were obtained *(the attained curriculum)* where the students gain educational experience.

In line with the implementation of regional autonomy, the curriculum development policy also embraces the decentralization principle because the centralistic approach is difficult to adapt the curriculum to the needs of the environment. Education should be relevant to the needs of learners, so that education programs should reflect the needs that they are worthy and *intelligent* enough to live in the social. With the enactment of the 2013 Curriculum in primary and secondary education, it indicates the central government's efforts to empower local governments, schools and teachers to determine their own learning experiences to be provided to students, or to emphasize on human resources (students) based on talents and interests each. This autonomy certainly provides an opportunity for the region to recognize the potential of self, potential nature, and the direction of development that will be done in the future. So there is such a freedom to creatively seek the superiority of each school.

In implementing the 2013 Curriculum well, it is necessary to increase or strengthen the literacy of teachers at the Elementary and High School. The strengthening of this literacy will affect teachers' understanding of the Curriculum 2013, an understanding of the content of textbooks, an understanding of the learning process based on the 2013 Curriculum and implementation of the evaluation according to the demands of 2013 Curriculum.

2. Research Methods

This study was designed for a three-year phases: *In the first year* with the *indeep study* approach is *exploratively,* study and collect data *comprehensively,* new and actual from the interaction process components of education (curriculum, educators, students, infrastructure, technology and media Learning, evaluation) that influenced learning science literacy.

2.1 Population and Sample

This research was conducted in pariaman regency with 17 districts consisting of 84 junior high schools run by the government and the private sector (Ministry of Education, Ministry of Religious Affairs and public/private). Decision population in Padang Pariaman regency because 30% of the number of teachers and students of West Sumatra in Padang Pariaman district, and school can also be grouped by urban and rural areas. Samples were taken 25% of each group of the population by urban and rural areas, where teachers support the theory-based subjects and experiment using stratified random sampling technique.

2.2 Data Collection and Analysis Techniques

The data used in this research is obtained from the questionnaire data to the respondents, namely the teachers of IPA. The instrument used in this study is a questionnaire to the respondents. The data taken by questionnaire is processed by percentage technique (quantitative) which is proposed by [7], that is:

$$P = \frac{F}{N} \times 100\% \quad (1)$$

where:
P = Percentage level of understanding
F = frequency of respondents
N = total number of respondents studied
The criteria used are stated in Table 1 [8].

Table 1. Criteria Level Understanding Respondents

Interval	Category
0 - 20%	Very low
21 - 40%	Low
41-60%	Enough
61-80%	Good
81-100%	Very good

3. Results and Discussion

The results of this study only for activities the first year, whereas in the first year of use approach *explorative,* study and collect data in *a comprehensive,* new and actual interaction process components of education (curriculum, educators, students, infrastructure, technology and media, As well as evaluation) that influence the learning literacy of science, as shown in Table 2.

Table 2. The relationship between the age of educators with an understanding of the 2013 curriculum.

Age	Understanding K. 13 (%)	Training K.13 (%)	MGMP (%)	KKG (%)	Mass media (%)	Following K.13 training (%)
25-30	75	33.3	0	0	66.7	33.3
31-36	25	37.5	0	0	50	50
37-42	25	25	0	0	75	25
43-48	0	100	0	0	0	100
49-54	0	80	0	0	20	80

Based on Table 2 it is known that in the range of age 25-30 years the average science teachers have understood the 2013 Curriculum well by 75%. Furthermore, it can be explained also that at this age the average respondents get more understanding about the 2013 Curriculum from the mass media information that is as much as 66.7% while who understand the 2013 Curriculum through training 2013 Curriculum as much as 33.3%. For ages 31-36 Years and 37-42 Years it is said that average respondents still lack understanding about the 2013 Curriculum with an average percentage of 25%. An understanding of the 2013 Curriculum. Most respondents gain an understanding of the mass media and the rest of the 2013 Curriculum training that has been followed. In the age range of 43-54 years it is said that there are no respondents who understand about the Curriculum 2013, although most have attended the 2013 Curriculum training. Curriculum training 2013 is mostly followed by science teachers aged 43-48 Years, then followed by science teachers Age 49-54 Years.

Table 3. The relationship between the age of educators with an understanding of the components of 2013 Curriculum

Age	Objective K.13	Scientific approach	Learning Models	Effectiveness of K.13 to Character	Integrated Science
25-30	66.7	56.3	60.4	47.9	60.4
31-36	56.3	37.5	53.1	46.9	43.8
37-42	56.3	43.8	50	37.5	50
43-48	37.5	37.5	62.5	50	37.5
49-54	50	40	45	40	45

As Table 3 above can be explained that the understanding of the science teachers aged 25-30 years in *Padang Pariaman* district on curriculum objectives in 2013, learning models are recommended and considered good *science integrated* with an average of 66.7% and 60.4 %. Understanding of science teachers to the scientific approach average 56.3% with sufficient category. The effectiveness of the 2013 Curriculum to shape the characters can be quite understood by the teachers of science that is an average of 47.9%. At the age range 31-36 Years of understanding of the objectives of the Curriculum 2013, suggested instructional models, effectiveness K.13 to Character and integrated science still enough categories, while the teachers' understanding of the scientific approach is still lacking. Likewise for the 37-54 year age understanding of the science teachers to curriculum components in 2013 (Goal Curriculum, the scientific approach, the model pembeljaran, the effectiveness of the curriculum and *integrated science)* remained at an average of 21-60% by category not understand until quite understand.

The percentage of opinion of science teachers in *Padang Pariaman* District to the Science book of SMP 2013 Curriculum published by *Kemendikbud* can be seen in the Figure 1.

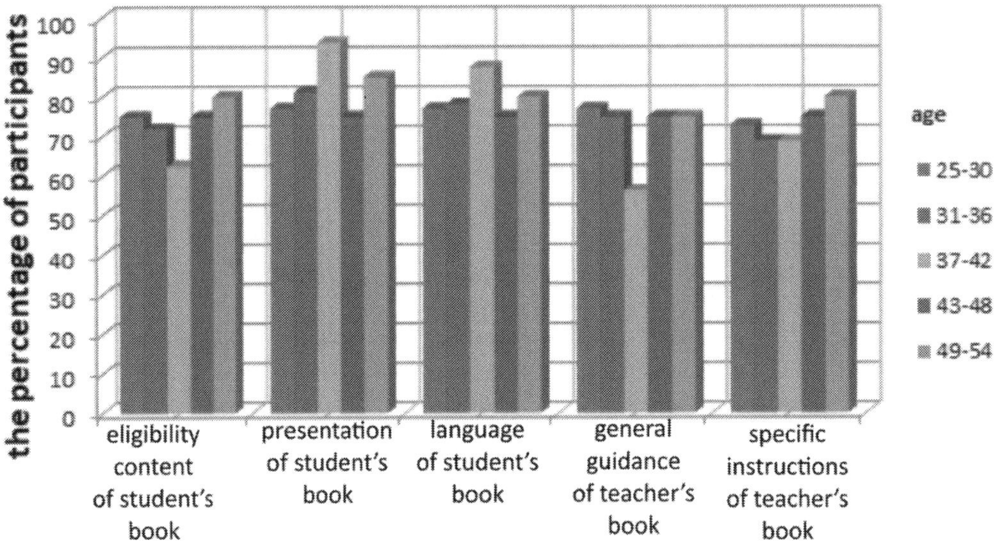

Figure 1. The percentage of opinion to science textbook in 2013 curriculum

Based on Figure 1, it can be seen that the teachers of IPA age 25-30 Years judge that the student book used is quite in accordance with the content feasibility components, components of presentation and linguistic component that is the average suitability of 77%. Science Teachers aged 31-36 Years assess that the student's book is quite in accordance with the components of content and language components contents of 71.9% and 78.9%. And for the presentation component of the teacher's book is considered appropriate where suitability equal to 81.3%. Age 37-42 Years assess that the books published students *Kemendikbud* not in accordance with content feasibility components but it is very appropriate in terms of components and linguistic components with 93.8% percentage and 87.5%. Likewise for the age of teachers IPA 43-54 Years judge that the student book used is in accordance with the components of content feasibility, presentation components and linguistic components that the average suitability of 75% and 80%.

4. Conclusion

Based on the analysis of the relationship between the age of educators with the understanding of the 2013 Curriculum can be seen that the age determines the achievement of teacher understanding of the curriculum used. The level of teacher's understanding of the curriculum is largely determined by the strengthening of literacy on the curriculum itself. So it can be concluded that the strengthening of literacy science teachers Padang Pariaman District still less in implementing Curriculum 2013. The results of in-depth study to the teachers' understanding of the curriculum in 2013 illustrates that literacy science teachers need to be improved or strengthened. The strengthening of this literacy needs to be given to teachers in order to implement the 2013 curriculum very well. Strengthening of this literacy can be done by giving workshops or seminars on the importance of literacy in implementing the 2013 Curriculum to increase literacy.

References

[1] Aljufri, B. S. 1998. *Analisis Kebutuhan Pengembangan Kurikulum FPTK-IKIP Padang dalam menghadapi era Persaingan Global*. Disampaikan pada Seminar Lokakarya Kurikulum FPTK IKIP Padang tanggal 27 Juli 1998.

[2] Ansyar, M.(1989) *Dasar-Dasar Pengembangan Kurikulum*. Jakarta: Depdiknas.
[3] Fauzi, A., Festiyed, 2011. *Pendekatan Holistik untuk Meningkatkan Penguasaan Standar kompetensi Dasar mata Pelajaran Sasaran Ujian Nasional Tingkat SMA di Kota Bukittinggi dan Kabupaten Agam*. Jakarta: Penelitian hibah Kompetensi dikti.
[4] Mahdiansyah and Rahmawati. 2014. *Literasi Matematika Siswa Pendidikan Menengah: Analisis Menggunakan Desain Tes Internasional dengan Konteks Indonesia*. Jurnal Pendidikan dan Kebudayaan. Volume 20, No 4, Desember 2014, pp. 452-469.
[5] Irianto, A., Festiyed, 2011, *Model Peningkatan Kompetensi Dasar Siswa SMA pada Mata Pelajaran Sasaran UN di Kota Solok, Kabupaten Solok dan Kabupaten Solok Selatan*. Jakarta: Penelitian hibah Kompetensi dikti.
[6] Hariadi, Eko. 2009. *Faktor-Faktor Yang Mempengaruhi Literasi Sains Siswa Indonesia Berusia 15 Tahun*. Jurnal pendidikan dasar. Volume 10 no. 1,maret 2009. pp. 29-43.
[7] Sudjana, N. 2005. *Dasar-Dasar Proses Belajar Mengajar*. Bandung: Sinar Baru Algensindo.
[8] Arikunto, S. 1989. *Dasar-Dasar* Evaluasi *Pendidikan*. Jakarta: Bumi Aksara.

IOP Conf. Series: Materials Science and Engineering **335** (2018) 012072 doi:10.1088/1757-899X/335/1/012072

Impact of Learning Model Based on Cognitive Conflict toward Student's Conceptual Understanding

F Mufit[1*], F Festiyed[1], A Fauzan[2] and L Lufri[3].

[1]Physics Department, Faculty of Mathematics and Science, Universitas Negeri Padang, Indonesia.
[2]Mathematics Department, Faculty of Mathematics and Science, Universitas Negeri Padang, Indonesia.
[3]Biology Department, Faculty of Mathematics and Science, Universitas Negeri Padang, Indonesia.

fatni.mufit@gmail.com

Abstract. The problems that often occur in the learning of physics is a matter of misconception and low understanding of the concept. Misconceptions do not only happen to students, but also happen to college students and teachers. The existing learning model has not had much impact on improving conceptual understanding and remedial efforts of student misconception. This study aims to see the impact of cognitive-based learning model in improving conceptual understanding and remediating student misconceptions. The research method used is Design / Develop Research. The product developed is a cognitive conflict-based learning model along with its components. This article reports on product design results, validity tests, and practicality test. The study resulted in the design of cognitive conflict-based learning model with 4 learning syntaxes, namely (1) preconception activation, (2) presentation of cognitive conflict, (3) discovery of concepts & equations, (4) Reflection. The results of validity tests by some experts on aspects of content, didactic, appearance or language, indicate very valid criteria. Product trial results also show a very practical product to use. Based on pretest and posttest results, cognitive conflict-based learning models have a good impact on improving conceptual understanding and remediating misconceptions, especially in high-ability students.

1. Introduction

Problems of misconception and low understanding of concepts is a problem that many occur in the learning of physics. Various studies show that misconceptions occur on almost all physics topics. Wandersee, Mintzes, and Novak, 1994 [1] suggest that misconceptions occur in all areas of physics, as many as 700 studies of the conceptual, there are 300 examining the fields of mechanics, 159 on electricity, 70 about heat, optics, and properties material, 35 about earth and space and 10 researches on modern physics. Physical misconceptions also occur in every level of education of learners, even in teachers, such as elementary teachers [2,3] and high school teachers [4,5]. Various studies also reveal misconceptions that occur in prospective students [6-11]. Misconceptions that occur in prospective teachers, potentially spreading misconceptions to students later.

Many factors become the cause of misconception and low understanding of concept in student. Suparno [1] states that among them the cause of misconception is the learner's own factor, the educator factor, the textbook, the learning context and the teaching method by the educator. Learning methods and models that do not involve learners in learning, such as lecture methods, tend to cause misconceptions. To address misconceptions, need interesting methods and learning strategies by associating topics with events. In addition, before starting a new topic, educators need to provide some sort of test to identify misconceptions on students. Some of the strategies that have been done to overcome misconceptions are teaching based on action research, teaching based on computer, and teaching based on real experiment video analysis [12]. Components of a fully integrated strategy or method will form a learning model. The development of learning model needs to be done to achieve the learning objectives, especially those that can overcome student misconception. This study aims to develop a cognitive conflict-based learning model and explore its impact on conceptual understanding and remediation of student misconceptions.

The learning model has five basic components [13], namely (1) syntax, is the sequence or operational steps of learning, (2) social system, describes the roles of educators and learners, hierarchical relationships and role-based rules (3) principles of reaction, is a picture of how teachers should see, treat, and respond to learners, (4) support system, ie all means, materials, tools, or learning environments that support learning, and (5) instructional & nurturant effects, is a learning outcome obtained directly based on the intended objectives and out-of-target learning outcomes. In this research, cognitive-based learning model is designed.

According to Lee, et al. (1) preliminary is done by presenting a cognitive conflict, (2) conflict is creation with the help of demonstration activities involving assimilation and accommodation process, and (3) resolution is discussion and concluding the discussion. While, Baser [2] states the implications of cognitive conflict learning are carried out as follows: (1) activation of alternative conceptions of students, (2) presentation of'situations that cannot be explained by existing concepts, (3) creation of cognitive conflict with anomalous situations, (4) needs of the concep(5) students construct their own knowledge actively, (6) students interact with each other to share their ideas about the anomalous situation to find a solution, and (7) to know conception is helpful to solve the problem the same that may be faced in the future.

2. Methods

This type of research is Development Research or also called Design Research, is one of the research models to develop and validate the products developed to be feasible to use. Development / Design Research by Plomp [14] is: to design and develop interventions such as teaching and learning strategies, learning materials, products and systems as a solution to complex educational problems. This design / development research uses the development of the Plomp [14] model which consists of 3 stages. The first stage of preliminary research: to analyze needs and contexts, review the literature, develop a conceptual framework or theory for research. The second phase of development or prototyping phase: is an iterative design stage consisting of iterations, each becoming a micro-cycle research with formative evaluation as the most important research activity aimed at improving and refining the intervention. The third phase of the assessment phase, ie (Semi) summative evaluation to conclude whether the solution or intervention meets the specified specifications. This stage often leads to recommendations for improvement of interventions, also called summative semifinals.

This article reports on the second stage of development or prototyping phase, which includes the prototype design process, validity test and the practical test of one to one evaluation. The process of developing / designing educational research is conducted through a systematic cycle process, from analysis, design, evaluation activities and revisions done iteratively in order to achieve the right balance between the desire and the realization achieved.

The resulting product or prototype is a cognitive-based learning model along with its components including syntactic, social systems, reaction principles, support systems and instructional impacts and counterfeit impacts contained in the model book. Model support systems in the form of Syllabus (S), Lesson Plan (LP), and Student Jobsheet (SJ) are also the products in this study. Product validity test is done to some experts on didactic, content, appearance and language aspects. Furthermore, a limited product trial was conducted through one-to-one evaluation of 9 sample students with low, medium and high ability from 3 different universities.

3. Result and Discussion

3.1. Prototype Design

The prototype was designed based on preliminary research results, that the students still experienced misconceptions and low-concept understanding despite having taken the basic physics course [12]. Students are still separating between understanding concepts and physics equations. The existing learning model has not fully addressed the problem of conceptual understanding, especially misconception. For that purpose, it is designed a learning model that can overcome the problem of concept comprehension and student misconception. The learning model is designed to reveal the initial knowledge of the student before the learning begins, especially identifying common misconceptions. The model also comes with presentation of phenomena or events that can generate conflict and dissatisfaction in students' thinking, especially phenomena that counter their misconceptions.

Based on various analysis in preliminary research stage, such as requirement analysis, context, student and study of various literatures then designed prototype of cognitive conflict based learning model with 4 syntax as follows: 1. preconception activation, 2. presentation of cognitive conflict, 3. Discovery concepts and equations , and 4. Reflection. The prototypes of model support systems in the form of S, LP, SJ are also designed and constructed based on the syntax of cognitive conflict-based learning model.

3.2. Test the validity of Prototype

Aspects of the validity assessment of the model prototype include the suitability of learning theories that support the model and the suitability and clarity of the model components, including syntax, social systems, reaction principles, support systems, instructional and nurturant effects. Assessment also includes the language and appearance aspects of the model book (MB).

The prototypes of model support systems (S, LP, SJ) are based on the syntax of the cognitive conflict learning model. Aspects of the syllabus's validity assessment include component conformity, syntax and language compatibility. Assessment of LP validity includes both syntax and language. Aspects of the assessment of the validity of the SJ include the suitability of components as well as the compatibility of syntax, content / content, appearance and language appropriateness.

Several revisions were made to the prototype according to the expert's suggestion in the areas of language, content, didactic and appearance to obtain valid product / prototype. Among the revisions were the use of punctuation and capital letters (cover), cover and lay out (appearance), allocation of learning time of each syntax (didactic), and revision of equations and images of physical phenomena (content). The validation of 8 experts on prototype can be seen in Table 1.

Table 1. Validity of prototype based on some expert review.

Expert Validator	Areas of Expertise	Percentage value of research product Validity			
		MB	S	LP	SJ
1	Appearance	80	82	85	82
2	Content	96	96	98	97
3	Content	86	78	85	83
4	Content	99	95	98	89
5	Didactic	89	86	90	87
6	Didactic	80	87	89	84
7	Didactic	85	82	82	82
8	Langaunge	90	93	93	94
Average		88	88	90	87

IOP Conf. Series: Materials Science and Engineering **335** (2018) 012072 doi:10.1088/1757-899X/335/1/012072

The validity value of the prototype cognitive conflict-based learning model and the support systems of the models (S, LP, and SJ) ranged from 88 to 90. The value is in a very valid category [15].

3.3. Practicality Test Prototype

The practicality of the product / prototype JS is limited to one-to-one evaluation of 9 students, 3 students with low, medium and high ability from 3 different universities. Practicality assessment includes aspects of ease of understanding JS, benefits and appeal of JS as well as JS working time allocation. The average value of practicality is based on these four aspects by 9 student samples, ranging from 82. - 90, which is at very practical criteria. Through interviews, in general all of the samples stated that they understood JS easily in terms of language, either the workmanship or the content of JS. Based on the level of academic ability and type of college, the practicality of JS prototype can be seen in Figures 1. Figure 1. shows that sample with low-academic ability (AA3) are more comfortable with understanding SJ and feeling the benefits. However, SJ is more interesting and the time allocation of SJ work is more effectively felt by sample with high-academic ability (AA1). Students with medium ability to feel the working time of JS less effective.

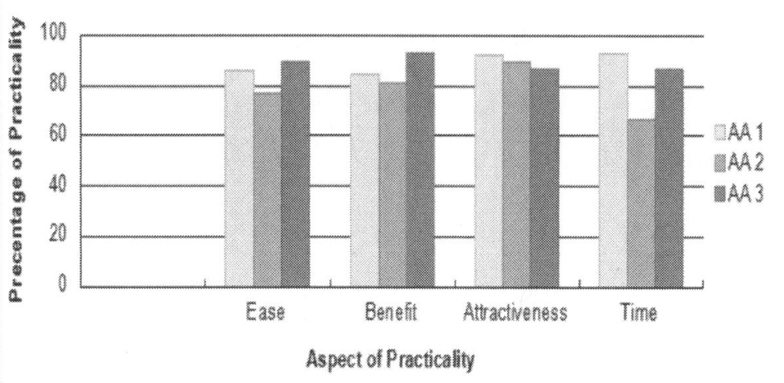

Figure 1. Practicality of SJ based on Sample Academic Ability Level

3.4. Pretest and Posttest Results

The impact of the implementation of cognitive-based learning model on student concept understanding is seen from the comparison of pretest (Pre) and posttest (Post) results given to 9 sample students. There are three categories of level of understanding of concept that is understood concept (U), misconception (M) and not understand concept (NU). In Figure 2. (a) it can be seen that the three samples with high ability have increased concept comprehension and decreased the number of misconceptions after using JS. Some concepts are not understood (UN) students on pretest, become understand (U) on posttest. Most moderately moderate samples (Fig. 2.b) also experienced increased conceptual understanding and reduced misconceptions after using JS-based cognitive conflicts. Meanwhile, low-capacity samples (Fig. 2.c) have not partially shown a good impact on increasing conceptual understanding and reduced misconceptions. In general, from 9 samples of one-to-one evaluation, it was found that the application of cognitive conflict-based learning model through the use of JS had a positive impact on increasing conceptual understanding and decreasing misconception. This is because the cognitive conflict-based learning model has a syntax that has the potential to improve conceptual understanding and remediate student misconceptions.

IOP Conf. Series: Materials Science and Engineering **335** (2018) 012072 doi:10.1088/1757-899X/335/1/012072

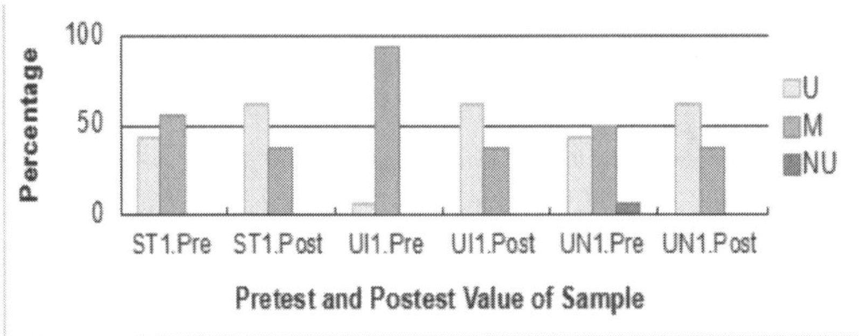

(a)

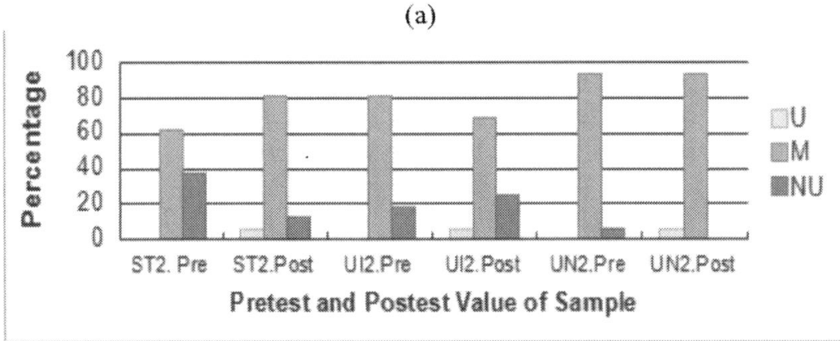

(b)

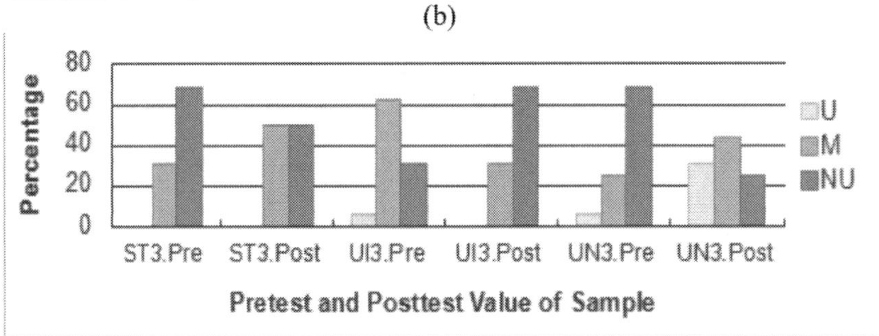

(c)

Figure 2. Level of Understanding Concept of Students based on academic ability level: (a) high, (b) medium and (c) low.

The first syntax of 'preconception activation' functions for learners to understand their misconceptions and educators can identify difficult concepts and concepts that become misconceptions. In accordance with Ormrod's [16] opinion, among the general strategies that have been shown to have an impact in overcoming misconceptions, namely: identifying existing misconceptions before learning begins and convincing learners that their beliefs need to be revised. The second syntax of 'presentation of cognitive conflict' will lead to conflict and discontent in the minds of learners and they will be motivated to find the correct concept. Trumper, 1997 (in [17]); also stated that the first step required for conceptual change is to allow learners to recognize the need to change the initial concept and to feel dissatisfaction with an explanation based on prior knowledge. The third syntax of 'discovery of concepts and equations' provides an opportunity for learners to actively engage in experiments and discussions to discover concepts and equations that support the concept. As Carin [18] points out, some strategies in guiding learners to revise their misconceptions are: helping learners learn the right information to a meaningful level rather than just rote. In addition, focusing on a deep understanding of ideas is more important than teaching all topics but superficially. The fourth syntax of 'reflection' also plays an important role in conceptual change. Carin [18] also states that providing corrective feedback to learners about responses that reflect their misconceptions including a strategy for revising misconceptions.

In addition, cognitive conflict or conceptual conflict strategies are consistent with the philosophy of constructivism theory, which allows learners to construct new concepts or knowledge. Educators do not monopolize learning by rushing to teach the content without paying attention to learners about their initial knowledge. Learners are given the opportunity to realize conceptual mistakes (misconceptions) that may occur in constructing new concepts, and presents cognitive conflicts for learners in getting a new concept right. Cognitive conflict strategies train learners to think deeply, discover and realize mistakes, discover and test ideas, and construct new concepts or knowledge.

4. Conclusion

Cognitive conflict-based learning models have been designed and produced four syntaxes: (1) preconception activation, (2) presentation of cognitive conflict, (3) discovery of concepts and equations, (4) reflection; as an attempt to overcome the concept problem, especially the problem of misconception. The prototype of cognitive conflict-based learning model resides in very valid criteria through expert review process and very practical criteria through one-to-one evaluation limited test. In general, cognitive-based learning models have a good impact on improving conceptual understanding and remediating misconceptions, especially in students with high academic ability.

Acknowledgments

Acknowledgments to the Directorate General of Higher Education who has funded this research through Doctoral Dissertation Research Grant, funding year 2017. Acknowledgments were also to the Dean of the Faculty of Mathematics and Science of Universitas Negeri Padang, Chairman of STKIP PGRI West Sumatera and the Dean of Tarbiyah and Teacher Faculty UIN Imam Bonjol Padang which has allowed its college students to participate in this research.

References

[1]. Suparno, Paul (2013). *Miskonsepsi dan Perubahan Konsep dalam Pendidikan Fisika*. PT. Gramedia Widiasarana Indonesia : Jakarta.
[2]. Baser, M. 2006. Fostering conceptual change by cognitive conflict based instruction on student understanding of heat and temperature concepts. *Eurasia Journal of Mathematics, Science and Technology Education*. 2(2). 96-114. Retrieved from http://www.ejmste.com/022006/ d6.pdf
[3]. Pujayanto (2011). *Miskonsepsi IPA (Fisika) pada Guru SD*. Jurnal Materi dan Pembelajaran Fisika (JMPF) Vol 1 No 1 (2011) 22-24. Retrieved from http://jurnal.fkip.uns.ac.id/ index.php/fisika/article / view/1856
[4]. Wahyudi, I & N Maharta (2013) *Pemahaman Konsep Dan Miskonsepsi Fisika Pada Guru Fisika SMA RSBI Di Bandar Lampung*. Jurnal Pendidikan MIPA, Volume 14, Nomor 1, April 2013. Retrieved from http://download.portalgaruda.org/
[5]. Halomoan (2008). *Analisis Konsepsi Guru Mata Pelajaran Fisika Madrasah Aliyah terhadap Konsep Gaya pada Benda Diam dan Bergerak*. Retrieved from http://sumut.kemenag.go.id/
[6]. Kabaca T, Z Karadag, M Aktumen (2011). *Misconception, cognitive conflict and conceptual changes in geometry: a case study with pre-service teachers*. Mevlana International Journal of Education (MIJE) Vol. 1(2), pp.44-55, 30 December, 2011. Retrieved from http://eujournal.org/index.php/esj/ article/download/3914/3705
[7]. Saputri D.F & C Sarwanto (2012) *Penyebab Dan Remediasi Miskonsepsi Gaya Menggunakan Multimedia Dan Modul*. Jurnal materi dan pembelajaran fisika vol 2, 2012. ISSN: 2089-6158.
[8]. Taufiq, Muhammad (2012). *Remediasi Miskonsepsi Mahasiswa Calon Guru Fisika Pada Konsep Gaya Melalui Penerapan Model Siklus Belajar (Learning Cycle) 5E*. Jurnal Pendidikan IPA Indonesia JPII no I vol 2 pp. 198-203. Retrieved from http://journal. unnes.ac.id/ index.php

[9]. Suastika, T Jhoni, T Utami (2015). *Penelusuran Miskonsepsi Mahasiswa Tentang Matriks Menggunakan Certainty Of Response Index*. Prosiding Seminar Nasional Fisika (E-Journal) SNF2015, Volume IV, Oktober 2015. Retrieved from http://snf-unj.ac.id/kumpulan-prosiding/snf2015/

[10]. Saputri, D. F & Nurussaniah. (2015) *Penyebab Miskonsepsi Pada Optika Geometris*. Prosiding Seminar Nasional Fisika (E-Journal) SNF2015, VOLUME IV, OKTOBER 2015. Retrieved from http://snf-unj.ac.id/kumpulan-prosiding/snf2015/

[11]. Mufit, Fatni. 2016. "A Study about Understanding the Concept of Force and Attitude towards Learning Physics on First-Year Students in the Course of General Physics; as Preliminary Investigation in Development Research". *Proceeding of SEA-DR (South East Asia-Design/Development Research) Conference 2016, International Conference:* April, 17th-18th 2016. Universitas Negeri Padang, West Sumatera Indonesia. ISBN: 978-602-19877-5-9. (http://seminar.fmipa.unp.ac.id/seadr16/ brosur/Prosiding% 204th %20SEADR%202016.pdf)

[12]. Mufit, Fatni. 2016. "The Study of Misconceptions on Motion's Concept and Remediate Using Real Experiment Video Analysis". *Proceeding of ACER-N (Asean Comparative Education Research Network) Conference 2016, International Conference: November*, 30th-Desember 01th 2016. Kopertis Wilayah X (West Sumatera, Riau, Jambi, Kepri Indonesia. ISBN: 978-983-2267-95-9.

[13]. Joyce, Bruce, M. Weill, E. Calhoun. 2011. *Models of Teaching*. Model-Model Pengajaran (edisi delapan). Penerjemah Achmad Fawaid dan Ateilla Mirza. Yogyakarta: Pustaka Pelajar.

[14]. Plomp T, N Nieveen (2013). *Educational Design Research, Part.A: An Introduction*. SLO. Netherlands Institute for Curricullum Development. Retrieved from http://downloads. slo.nl/ Documenten/ educational-design-research-part-a.pdf

[15]. Riduwan .2005. *Belajar Mudah Penelitian untuk Guru, Karyawan dan Peneliti Pemula*. Bandung: Alfabeta

[16]. Ormrod. J.E. 2009. *Psikologi Pendidikan;* alih bahasa Triwibowo. J.S. Jakarta: Erlangga.

[17]. Liang, S. 2016. "Teaching the concept of limit by using conceptual conflict strategy and Desmos graphing calculator". *International Journal of Research in Education and Science (IJRES)*, 2(1), 35-48.

[18]. Carin, Arthur A. (1997). *Teaching Science Through Discovery,* Eighth edition. Pearson Prentice Hall: New Jersey, Columbus Ohio.

IOP Conf. Series: Materials Science and Engineering **335** (2018) 012073 doi:10.1088/1757-899X/335/1/012073

Planning Model of Physics Learning In Senior High School To Develop Problem Solving Creativity Based On National Standard Of Education

A Putra*[1], M Masril[1] and Y Yurnetti[1]

[1]Physics Department, Faculty of Mathematics and Science, Universitas Negeri Padang, Indonesia

*amali.unp@gmail.com

Abstract. One of the causes of low achievement of student's competence in physics learning in high school is the process which they have not been able to develop student's creativity in problem solving. This is shown that the teacher's learning plan is not accordance with the National Eduction Standard. This study aims to produce a reconstruction model of physics learning that fullfil the competency standards, content standards, and assessment standards in accordance with applicable curriculum standards. The development process follows: Needs analysis, product design, product development, implementation, and product evaluation. The research process involves 2 peers judgment, 4 experts judgment and two study groups of high school students in Padang. The data obtained, in the form of qualitative and quantitative data that collected through documentation, observation, questionnaires, and tests. The result of this research up to the product development stage that obtained the physics learning plan model that meets the validity of the content and the validity of the construction in terms of the fulfillment of Basic Competence, Content Standards, Process Standards and Assessment Standards..

1. Introduction

The privileges the curriculum 2013 compared to the curriculum 2006 is the completeness of learning content with metacognitive knowledge, and the level of cognitive process that achieved to creative thinking skills. Both of these aspects are the knowledge and skills that needed in the 21st century. A preliminary study has been conducted on the implementation the curriculum 2013 in physics learning in several high schools in Padang with documentation analysis, observation, questionnaires and interviews. The results of the documentation analysis show that: a) Learning plan documents and teaching materials that used by the teachers generally was not included metacognitive knowledge and creative skills, b) Indicators of learning objective formulas, and competency assessment instruments was not included aspects of Higher Order Thinking Skills (HOTS). The results of classroom observations show that: a) the scientific approach has not done well, and b) The learning generally still focuses on the recall aspect dominated by the lecture method. The results of questionnaires and interviews to the teachers reveal that: a) they still have difficulties in presenting metacognitive knowledge, b) assessment of students learning outcomes generally is still focused on fulfillment of Lower Order Thinking Skills (LOTS), and c) they still have difficulties in prepare the planning and implementation of learning in accordance to the curriculum 2013.

The results of the preliminary study indicate that generally the teacher has not been able to arrange the learning plan in accordance with the National Education Standards through the implementation of the curriculum 2013. As a result in implementing the curriculum, the teachers experienced have some difficulties, so the purpose of the curriculum is not achieved. The objectives of the curriculum of the subject, known as "competence" According to article 1 (4) of Government Regulation of the Republic of Indonesia number 32 of 2013 [1], the competencies are: "a set of attitudes, knowledges and skills that must be possessed, and mastered by the students after learning a learning content, completing a program, or completing a particular educational unit ". Attitudes and skills gained through learning are expected as a reflection of the increased knowledge and improvement of cognitive processes of student. The point is the learning process to produce attitudes and skills in accordance with the learning objectives. It is needed to be addressed knowledge and cognitive processes that will be implemented in the learning in accordance with learning plan that arranged by the teachers.

2. Literature Review

Knowledge that learned in school is known as the subject matter or curriculum content. That is the curriculum component of the answer to the question, "what is taught?" The knowledge in question is scientific knowledge obtained as a result of the application of the scientific method [2]. More complex a person's knowledge will be the easier it will be, and many of the problems that can be solved.

What kind of knowledge is needed in learning? Based on this view, the knowledge in question is useful knowledge for life. Anderson and Krathworth (2001) [3] divide the complexity of knowledge over 4 dimensions, ie factual, conceptual, procedural, and metacognitive knowledge. These four dimensions of knowledge are contained in the Ministerial Decree of Education and Culture of the Republic of Indonesia number 20 [4] and 21 [5] in 2016, as reference of National Education Standards in formulating Graduate Competency Standards, Content Standards and learning objectives to be achieved. The weakness that has been happening in the application of the curriculum 2006 is the knowledge learned by students only in the form of conceptual, procedural, factual knowledge, knowledge of facts, concepts, theories, laws and procedures. So this knowledge is only limited to recall not to the stage of implementation.

In implementing these four dimensions of knowledge in high school physics learning, which referred to factual knowledge is knowledge related to facts, circumstances and phenomena experienced by the students. That is the knowledge gained from the activity of seeing, observing and feeling of the various circumstances and phenomena that occur in the surrounding environment. Conceptual knowledge is concepts, principles, laws, theories, meanings and rules that apply in explaining situation and natural phenomena. Procedural knowledge can be working procedures, tips, steps, or algorithms that taken to achieve or produce something in concrete and abstract sphere.

The curriculum 2013 innovation with respect to learning content is complemented by metacognitive knowledge. That is knowledge that often termed as "thinking about thinking" [6] [7], namely the ability of the students to control cognitive aspects they had when solve the problems encountered. This metacognitive knowledge is influenced by declarative knowledge, procedural knowledge, conditional knowledge, and self-knowledge. With this metacognitive knowledge that students seek to control their cognitive, guiding the knowledges they had when solve the relevant problems to their knowledge creatively. Metacognitive knowledge in learning physics will grow and develop in students through the interaction process with objects and natural phenomena which poured on the teacher learning plan in the form of student learning materials.

Each learning must ensure the acquisition of knowledge (content) and improvement of student's thinking ability. Eggen and Kauchak (2012) state that, every strategy and model applied in learning is in order to "teach content and thinking skills". Anderson and Krathwohl (2001) [8] have encountered the complexity of this cognitive (thinking) process on 6 levels include abilities: remembering, understanding, applying, analyzing, evaluating, and creating. These six levels of cognitive processes are combined with the four dimensions of knowledge as described above. The result of this combination is a refinement of Bloom's taxonomy that was conducted by Anderson and Krathwohl (2001), known as Bloom's revised taxonomy.

In relation to the levels of cognitive processes, innovations in the curriculum 2013 are to replace the knowledge aspect at the lowest level of cognitive process with the ability to remember, and place the creative ability aspect as the highest level of thought process to achieved through learning with the ability to create i.e the ability to produce something new, in the form of abstract thought or a real object as a result of the ability of creative thinking. This creative ability is influenced predominantly by the metacognitive knowledge they have.

In Permendikbud No. 22 of 2016 [9] on Standards The process of implementing learning in primary and secondary education, explained that the six levels of cognitive processes are used as students activities to obtain the four dimensions of knowledge in teacher-created learning activities. The learning process in question is: "The learning process is held interactively, inspiration, fun, challenging, motivate the sudents to participate actively, and provide enough space for initiative, creativity, and independence according to talents, interests, and physical and psychological development of students ". It is also in the mandate that the learning is done to ensure the implementation of a scientific approach that contains at least 5 activities that are observing, asking, trying, reasoning, and communicating. Strategies and learning models applied are research-based learning (discovery/inquiry learning), as well as learning that produces contestual work with problem-based learning.

In Natural Sciences, including physics, the application of a scientific approach begins with observing activities. This is activity of studying or observing the studied phenomena. If the observation is done critically and analysis will bring up the students curiosity in the form of the question that they want to know the answer (questioning activity). To find the answers, the students will compile a hypothesis as a prediction or a temporary answer. The hypothesis is the basis of the students to trying (find out) the truth. The results of investigation is data that are described in the form of tables and graphs. These data are discussed and interpreted (reasoning activity) resulting in conclusions for reporting or presenting (communicating activity) [10] in the form of student conceptions. Furthermore with teacher confirmation, the student's conception will be refined into a scientific concept that becomes student knowledge as a result of learning.

Based on the description can be concluded that the purpose of learning for the students is to demand the knowledge and improve ability of thinking. The complexity of the content (knowledge) and the level of cognitive processes of the students in learning is a combination of the four dimensions of knowledge and the six levels of cognitive processes. As the output of these two aspects is the mastery of knowledge and improvement of cognitive processes that are packaged as the achievement of knowledge competence. Reflection of the intensity of knowledge competence is dominated by students that expected to change their attitudes and improve the skills as the outcome of the learning process. This argument is recommended to develop physics learning model in high school oriented to competency standards, content standards, process standards, and assessment standards. The lesson is carried out by scientific approach as mandated by the Curriculum 2013, so that the implementation of the educational process is accordance with the National Education Standards.

The assessment standard applied to the curriculum 2013 is poured in the Regulation of the Minister of Education and Culture of the Republic of Indonesia number 23 of 2016. [11] The assessment is the process of collecting and processing information to measure the achievement of learning outcomes of students including attitude aspects (behavior), mastery of knowledge and skills of students to applying that knowledge. The purpose of the assessment is to monitoring and evaluating, process, learning progress and improvement of student's learning outcomes. Assessment of learning outcomes can be done in the form of repetition, observation, assignment, or in other forms if required by using the relevant instruments in the form of tests, observation format, fortofolio and so forth. Operationally to organize the implementation of the assessment of learning in accordance with the assessment standards, Kemdikbud has published the Guidance Assessment for Junior and Senior High Schools. According to the guidelines, the assessment is an authentic assessment. Indirect and ongoing attitude assessment and responsibility of all subjects based on positive and negative attitudes that arise during schooling both within and outside of learning. Assessment of attitudes in the context of subjects is done by teachers using observation sheets or journals. Assessment of skills in learning can also be used in observation sheets for each skill indicator. While the knowledge competency assessment can be used in the form of multiple choice test or essay test with scoring rubric.

The curriculum 2013 has adapted competency standards, content standards, process standards, and assessment standards, to the National Education Standards as set in article 3 of the RI Law No.20 of 2003 [12] on the National Education System (Sisdiknas) that aim to directing the entire education activities in Indonesia to make the students always develop their potential to become a nation that is intelligent, knowledgeable, capable, creative, independent, and become a citizen who is democratic and responsible and dignified to build theirself, society and nation. Therefore all these standards should be poured in the planning and implemented in the learning activities in the classroom to achieve the intended educational objectives.

Some previous researches on the physics learning plan include: Siti Chodijah et al (2012) developed a guided discovery learning model equipped with a portfolio assessment format to be applied in high school physics learning, especially for the concept of circular motion. The result of the development by using the instrument through expert test and field trial obtain learning media with very valid category and can be used in high school physics learning. After the effectiveness test, the overall research concludes that the result of learning media is very valid, very practical and effective in improving the competence of the students. The second research is Nuris Septa Pratama et al (2015), conducted a study of physics-based Higher Order Thinking (HOTS) learning in class X high school in Yogyakarta. This study aims to find out how the planning and implementation of physics-based higher order thinking skills (HOTS) learning in class X high school, in the form of survey research with the state of high school population in Yogyakarta. There are 10 samples that was determined by census technique. The source of information consists of 10 physics teachers of class X and 281 students. The instruments used in this study are RPP document analysis, observation guidelines, teacher questionnaires and students and documentation. Data were analyzed by qualitative descriptive technique and supported by quantitative data. The results showed that the implementation of HOTS-based physics learning planning compiled by physics teacher of class X at the state of Senior High School in Yogyakarta was in moderate category (TS). Implementation of HOTS-based physics learning conducted by physics teacher of class X at state Senior High School in Yogyakarta was in medium executed category (TS).

Based on the results obtained from the studies that have been conducted, it concludes that the existence of subjects in school must be dynamic, in the sense always up to date in accordance with policy changes, modernization and field demands, so it becomes valuable and effective in order to improve the quality of education as a whole.

3. Research Methods and Finding

To generate the model product of the planning, the selected development procedure follows the ADDIE model, an acronym of Analyze, Design, Develop, Implement, and Evaluate [13] .These five words in the acronym are the stages of model development being pursued. The reported research is up to the develop stage. The steps of each stage and the results obtained at each stage are described as follows:.

a. Needs Analysis Stage (analyze).

At this stage have been done some analysis, which are curriculum analysis, characteristic and needs of student's analysis, instructional analysis, and related theories and concepts analysis through a review of literature. The results of the curriculum 2006 analysis and the curriculum 2013 derive fundamental differences in competency standards, content standards, process standards, and assessment standards, as a basis for the reconstruction of lesson plan, can be described in Table 1. [14]

IOP Conf. Series: Materials Science and Engineering **335** (2018) 012073 doi:10.1088/1757-899X/335/1/012073

Table 1. Differences in the elaboration of the National Education Standards in the curriculum 2006 and the curriculum 2013 [15][16][17][18]

No	Distinguishing Aspect	Curriculum 2006	Curriculum 2013
1	Graduate competence standard	Described on Competency Standards and Basic Competencies	Described on Core Competencies and Basic Competencies
		Standards of competence are distinguished into 3 aspects namely the competence of cognitive, affective and psychomotor aspects	Core competence is divided into four parts: spiritual attitudes, social attitudes, knowledge and skills
2	Content Standards	Refers to Bloom's taxonomy (1956) that separates the learning objectives of the knowledge aspect with the cognitive level aspects of the process	Refers to Bloom Bloom's taxonomy (2001), by combining the learning objectives of the knowledge aspect with aspects of the cognitive process.
		Knowledge that is used as learning objectives, divided into 3 dimensions, namely: knowledge of facts, concepts and procedures.	The competence of the knowledge aspect used as the learning objectives is divided into 4 dimensions, namely: factual, conceptual, procedural, and metacognitive knowledge.
		The level of cognitive process as the learning objective refers to Bloom's taxonomy (1956) consists of six levels of adjectives: yes, knowledge, understanding, application, analysis, synthesis, and evaluation	The level of cognitive process as a learning objective refers to Bloom's taxonomy Revision (2001) consists of 6 levels of verbs: remember, understand, apply, analyze, evaluate and be creative.
3	Standard Process	Learning is oriented towards mastering the concept of science	Learning is oriented towards achieving the competence of knowledge, skills and attitudes in a balanced way.
		Learning is still centered on teacher activity	Learning-oriented student activities
		The process implanted in students in learning focuses on exploration, elaboration and confirmation activities	The process implanted in students is oriented towards the application of a scientific approach through the process of observing, questioning, trying, reasoning and

			communicating
4	Assessment Standards	More dominant judgments on mastery of knowledge and tend to ignore attitude and skill judgments	Assessment is holistic, covering the mastery of knowledge, attitude and skill changes

The results of the analysis show that the developed Physics learning model is oriented towards the fulfillment of 4 dimensions of knowledge and 6 levels of cognitive process as stated in the regulation of the Minister of Education and Culture of Republic of Indonesia number 20, 21 and 22 of 2016 as reference of the improvement of curriculum 2013

The analysis of the characteristics and needs of the students is based on the age and level of student's development as well as test results, documentation analysis and field observations, to develop the content complexity and level of cognitive processes required in the learning. For this purpose, a preliminary study involving four high schools that implemented the curriculum 2013 in Padang. Instruments for reviewing students' knowledge competencies are based on 4 dimensions of knowledge and 6 levels of cognitive processes, totally 24 items. Competence test is done to 12 students of class X which are in the ranking category 1, 2, and 3. Data collection about teacher learning process is done by questionnaire, interview and observation. The result of preliminary study concludes that the quality of student's knowledge competence achievement in physics lesson in State of Senior High School in Padang is still low, in terms of content complexity, and the level of cognitive processes shown by the majority based on evaluation results, cognitive process level is still dominated by the ability of Lof Order Thinking Skills (LOTS) level. The majority of the ability to remember is followed by the ability to understand and the ability to apply. As for the level of High Order Thinking Skills (HOTS) for the ability to analyze is still measly. The ability to evaluate and create tend not to show up. Preliminary research results concluded that from the aspect of the complexity of content need to increase the intensity. Especially on procedural knowledge and the development of metacognitive knowledge. In terms of the complexity of the level of cognitive processes achieved more developed capabilities at the HOTS level, including the ability to analyze, evaluate, and be creative.

Instructional analysis is conducted to formulate the necessary instructional design forms as part of the development of learning models to achieve learning objectives, content complexity, cognitive process complexity, learning evaluation instruments and developed syntax model. The results of the analysis conclude that all the instrumental aspects developed poured in the Syllabus, Learning Implementation Plan (RPP), self study materials and student activity sheets.

The analysis of theory, related concepts and rules, focused on the application of Minister of Education and Culture of Republic of Indonesia Regulation No. 20, 21, 22, 23 and 24 of 2016 as reference, Bloom's taxonomy revised, model of study-based learning model and information-processing learning theory, development theory ADDIE model, and supporting statistical theory for data processing and analysis. Analysis of this theory is made as the basis that supports the development of learning models

b. Designing Model Stage (design)

The aspects of design at this stage is concerning the formulation of learning objectives, the design of the complexity of learning content and its relation to the complexity of the cognitive process level trained, the design of the learning evaluation instrument, and the syntax design of the learning model so as to obtain the complete learning model design that consisting of:

The design of learning objectives formulation, conducted based on KI and KD analysis describe in the form of indicators of KD achievement, and then poured in the form of learning objectives. Learning objectives are formulated based on indicators of achievement of competencies that embrace elements ABCD + K (A = Audience/student, B = Behavior, C = learning conditions, D = Degree/ level of achievement, and K = Knowledge). For example, the formulation of learning objectives: "With LKS workmanship and teacher confirmation, students can identify 4 characteristic styles and their effect precisely" (Description: student = Audience, can identify = Behavior, 4 characteristic style and influence on the object = Knowledge, with LKS workmanship and teacher confirmation = Condition, with exact = Degree).

Design model of learning content complexity, compiled for Force and motion of objects on class X physics lessons semester 2 in high school. The designed model of learning content complexity is referred to the fulfillment of the four dimensions of knowledge as proposed by Anderson and Krathworth (2001). Learning content is arranged in the form of a declarative statement or phrase containing statements for conceptual, procedural or metacognitive factual knowledge. Learning content is often also called as essential material, built on learning objectives wrapped by theme or learning topics. Design model of cognitive process complexity to be trained to learners.

The design model of the complexity of the cognitive process level is rested on the fulfillment of 6 levels of cognitive processes. The hierarchy consists of abilities: remembering (C-1), understanding (C-2), applying (C-3), analyzing (C-4), evaluating (C-5) and creativity (C-6). The order of C-1 to C-6, is the order of ability to be reached from the lowest to the highest order level. The high ability includes the ability level below.

Design model of output evaluation and outcome of learning outcomes. Evaluation of output is done to know the achievement of students knowledge competence. The model design of this output evaluation instrument is in the form of an essay and objective test completed with an answer key and a scoring column. The contents or answers of the developed instrument can be scored 1, 2, 3, or 4. Outcome evaluation is done to know the attitude change and skill improvement as the reflection (impact) of knowledge competence achieved. The developed outcome evaluation instruments tend to be observation sheets, and student activity sheets. The instrument for these two aspects, developed by using an instrument grid that refers to the competency aspects contained in the curriculum that contained in KI and KD.

The syntax design of the developed learning model. The syntax of the designed learning model is the achievement of student competence, oriented to 3 aspects, namely: a) the complexity of the content (the fulfillment of 4 dimensions of knowledge), b) the complexity of the cognitive process level (trained 6 levels of cognitive processes including LOTS and HOTS), and c) carried out with a scientific approach (containing elements: observing, asking, trying, reasoning and communicating). The results of the orientation of these 3 aspects resulted in a syntax model consisting of 5 stages of activities implemented on the student activity sheet as follows:

Stage 1: Identify physical symptoms as reference learning and observation objects
Stage 2: Conceptualization exercise, based on factual knowledge
Stage 3: Build tips and strategies for designing activities
Stage 4: Develop creativity to solve problems
Stage 5: The exercise concludes and communicates the results of the learning activities

c. Model Development Stage (develop)

Learning lesson planning products are grouped into 2 kinds, namely: 1) Learning Implementation Plan (RPP), and 2) Teaching Materials. RPP is developed based on syllabus. The components of RPP developed consist of: a) formulation of learning objectives, b) content development, c) process development, and d) development of evaluation instruments. Development of RPP refers to the Process Standards contained in the Regulation of the Minister of Education and Culture of the Republic of Indonesia number 22 of 2016. Learning planning components developed are expressed by the flow chart in Figure 1

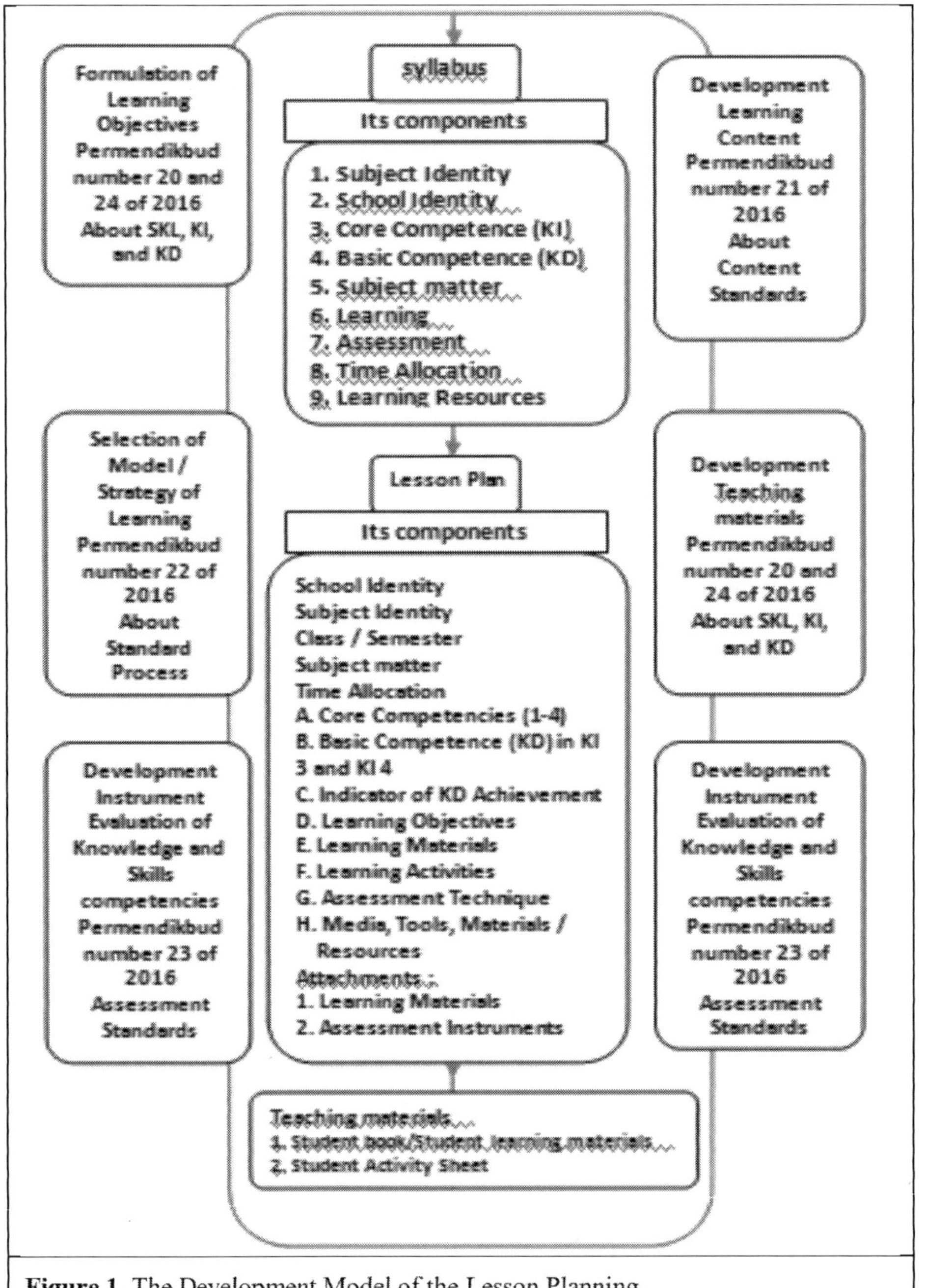

Figure 1. The Development Model of the Lesson Planning.

The developed teaching materials consist of 2 kinds, namely Student Self Study Material and Student Activity Sheet. Good teaching materials are prepared based on indicators of achievement of learning objectives. In preparing the printed materials should pay attention to some conditions that must be met, that are didactic requirements, construction requirements and technical requirements. Didactic requirements pertain to effective learning principles. Construction requirements regarding the use of language, sentences, vocabulary levels of difficulty and clarity of information. Technical requirements is relating to the writing of drawings and the appearance of teaching materials. The

IOP Conf. Series: Materials Science and Engineering **335** (2018) 012073 doi:10.1088/1757-899X/335/1/012073

requirements of good teaching materials can be evaluated by using an evaluation instrument format in the form of a Likert scale[19]

With teaching materials oriented to the complexity of content and cognitive processes, students are expected to learn physics through interaction with various natural objects that poured in teaching materials. Interaction results are described, analyzed, and applied to solve various relevant physical problems through the application of the scientific method. Thus it will be embedded and mastered students to the four dimensions of knowledge into learning content, in line with the increase of LOTS into HOTS which can be actualized in the form of the ability to solve the relevant physics problems[20]. The aspects of complexity are contained in the developed materials, presented by the diagram in Figure 2.

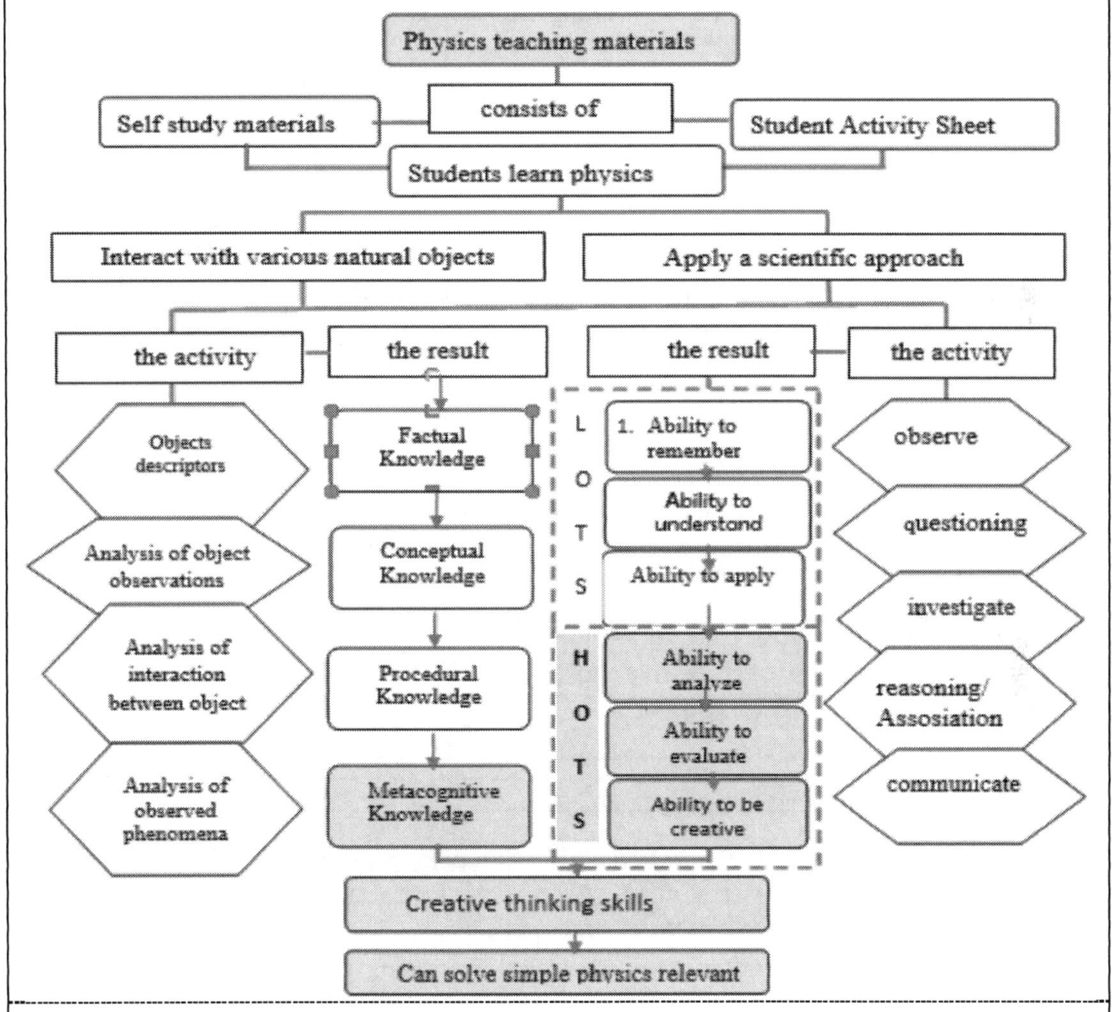

Figure 2. The aspect of complexity found in teaching materials

To find out the validity of construction, product planning and learning materials are discussed with 2 peers of judgment to get input. The product is refined based on the input obtained so that the resulting prototime is ready to be validated. Product validation activities conducted by 4 experts judgment. Recapitulation of validation results from all aspects of physics learning planning is presented in Table 2.

IOP Conf. Series: Materials Science and Engineering **335** (2018) 012073 doi:10.1088/1757-899X/335/1/012073

Table 2. Results Validation of Physics Learning Planning Components

I. VALIDITY OF LEARNING OBJECTIVES		VALIDATORS				RESULTS	
A	BASIC FORMULATION OF LEARNING OBJECTIVES	V1	V2	V3	V4	3,38	V
B	COMPONENT OF LEARNING OBJECTIVES	V1	V2	V3	V4	3,35	V
C	REQUIREMENTS OF LEARNING OBJECTIVES COMPONENTS	V1	V2	V3	V4	3,29	V
	AVERAGE	3,3	3,5	3,47	3,1	3,33	V
II.CONTENT VALIDITY OF ESSENTIAL MATERIALS		VALIDATORS				RESULTS	
A	LEARNING CONTENT REQUIREMENTS	V1	V2	V3	V4	3,4	V
B	COMPLETENESS OF LEARNING CONTENT	V1	V2	V3	V4	3,38	V
C	COMPONEN REQUIREMENTS OF LEARNING OBJECTIVES	V1	V2	V3	V4	3,38	V
	AVERAGE	3,3	3,5	3,4	3,3	3,38	V
III. VALIDITY OF STRATEGY / LEARNING MODEL		VALIDATORS				RESULTS	
		V1	V2	V3	V4	3,38	V
	AVERAGE	3,2	3,2	3,47	3,7	3,38	V
IV. VALIDITY OF LEARNING MATERIALS		VALIDATOR				RESULTS	
A	FEASIBILITY OF STUDENT ACTIVITIES	V1	V2	V3	V4	3,33	V
B	FULFILLMENT OF 4 KNOWLEDGE DIMENSIONS	V1	V2	V3	V4	3,31	V
C	TRAINING 6 LEVEL OF COGNITIVE PROCESSES	V1	V2	V3	V4	3,38	V
D	DEVELOPING ATTITUDES AND SKILLS	V1	V2	V3	V4	3,54	V
	AVERAGE	4,8	4,6	4,15	4,2	3,41	V
V. VALIDITY OF INSTRUMENTS EVALUATION		VALIDATORS				RESULTS	
A	COMPLETENESS TYPES OF EVALUATION	V1	V2	V3	V4	3,31	V
B	ASSESSMENT OF 4 KNOWLEDGE DIMENSIONS	V1	V2	V3	V4	3,31	V
C	ASSESSMENT 6 LEVELS OF COGNITIVE PROCESSES	V1	V2	V3	V4	3,50	V
D	ASSESSMENT OF ATTITUDE AND SKILLS	V1	V2	V3	V4	3,50	V
	AVERAGE	3,7	3,6	3,0	3,5	3,43	V
VI. VALIDITY OF LESSON PLAN		VALIDATORS				RESULTS	
		V1	V2	V3	V4	3,65	V
	AVERAGE	**3,6**	**3,8**	**3,47**	**3,7**	**3,65**	**SV**
VI. VALIDITY OF PHYSICS LEARNING PLANNING		VALIDATORS				RESULTS	
		V1	V2	V3	V4	3,43	V
	TOTAL AVERAGE	3,5	3,5	3,33	3,4	3,43	V

Note: V1, V2, V3, V4 = The first, second, thirth and fourth of validators

4. Discussion

The learning at school will work in the best way when it comes to achieving the objectives of the curriculum, and has been planned in advance. The experience shows that learning without a plan, tend not directed, and spend a lot of time and the purpose of learning is not achieved. A lesson planning based on experience and some benefits of good learning planning predict the results to be gained, the learning will be implemented systematically as a guideline of activities, the content of learning can be well done, can streamline the use of time and so forth.

In designing the learning plan, the aspect of objectives, content, process and assessment are arranged in such a way to achieve the learning objectives. Physical learning planning model that has been compiled refers to the National Standards of Education, especially with regard to graduate competency standards, core competencies, basic competencies, content standards, process standards and assessment standards set forth in permendikbud no 20, 21, 22, 23 and 24 of 2016. The result of the trial is limited to 4 experts judgment, concluding that the planning model is in valid category. It is expected that the actual implementation of the classroom will be effective in improving students' competence in physics learning, especially with regard to solving problem creativity

5. Conclusions

Conclusions obtained until the development of products, has been generated learning planning to meet the validity of the content and validity of the construction that ready to proceed with testing on the real class to know the effectiveness and practicality.

Acknowledgments

With the completion of the activity of developing this learning planning model to product validation stage, we would like to thank the esteemed: 1) Rector of State University of Padang and Head of LP2M who provided Fund of this Applied Product Research, 2) Dra. Yunida Herawati, M.Pd, Drs. Irwan Khalik M.Si, Dra. Mutia M.Pd, and Drs. Arnel Hendri, M.Si whose has been contained as the Validator of this research product. and 3) the students who participated in this research.

References

[1] Kemendikbud RI, "PP No. 32 Tahun 2013 Tentang Perubahan PP No 19 Tahun 2005 Tentang SNP," 2013.

[2] M. V. Christine, "Nature of science and the scientific method," *Geol. Soc. Am.*, hal. 1–11, 2006.

[3] E. G. Rozycki, M. M. Mccarthy, W. A. Carpenter, A. Ricker, R. J. Jones, dan B. W. Sorenson, "A Taxonomy for Learning , Teaching , and Assessing : A Revision of Bloom ' s Taxonomy of Educational Objectives."

[4] Kemendikbud RI, "Permendikbud No 20 Tahun 2016 Tentang SKL Pendidikan Dasar dan Menengah," hal. 1–8, 2016.

[5] Kemendikbud RI, "Permendikbud No.21 Tahun 2016 Tentang Standar Isi Dikdasmen," 2016.

[6] P. R. Pintrich, "The Role of Metacognitive Knowledge in Learning, Teaching, and Assessing," *Theory Pract.*, vol. 41, no. 4, hal. 219–225, 2002. 157–172

[7] Retrieved from : http://www.learner.org/courses/learningclassroom/support/about_guide.pdf Darling-Hammond, L., Austin, K., Cheung, M., Martin, D., Barron, B., Palincsar, A., & Shulman, L. (2003). Thinking about thinking: Metacognition. The Learning Classroom, "Thinking About Thinking :Metacognition," in *The Learning Classroom*, 2003, hal. 157–172.

[8] D. R. Krathwohl, "A Revision of Bloom ' s Taxonomy :," vol. 5841, no. October, hal. 37–41, 2017.

[9] Kemendikbud RI, "Peraturan Menteri Pendidikan dan Kebudayaan Republik Indonesia Nomor 22.Tahun 2016 Tentang Standar Proses Pendidikan Dasar dan Menengah," hal. 1–30, 2016.

[10] Kemendikbud RI, "Modul pendampingan implementasi kurikullum 2013 di SMA tahun 2014," 2014.

[11] Kemendikbud RI, "Permendikbud No.23 Tahun 2016 Tentang Standar Penilaian Dikdasmen," no. Standar Penilaian Pendidikan, 2016.

[12] Kemdiknas RI, "Undang-undang sistem pendidikan nasional," 2003.

[13] S. J. Mcgriff, "Instructional System Design (ISD): Using the ADDIE Model," *Instr. Syst. Coll. Educ. Penn State Univ.*, hal. 2, 2000.

[14] Kemendikbud RI, "Kurikulum 2013 Sekolah Menengah Atas/Madrasah Aliyah," 2014.

[15] Menteri Pendidikan Nasional Republik Indonesia, "Peraturan Menteri Pendidikan Nasional Republik Indonesia nomor 23 tahun 2006 tentang standar kompetensi lulusan untuk satuan pendidikan dasar dan menengah," *Peraturan. Menteri*, 2006.
[16] M. P. N. R. Indonesia, *Lampiran Permendiknas No. 22 Tahun 2006*. 2006, hal. 1–43.
[17] Menteri Pendidikan Nasional Republik Indonesia, "Peraturan Menteri Pendidikan Nasional Republik Indonesia Nomer 41 Tahun 2007 tentang Standar Proses untuk Satuan Pendidikan Dasar dan Menengah (Ministry of National Education of the Republic of Indonesia Regulation number 41 year 2007 on Process Standards," 2007.
[18] Menteri Pendidikan Nasional Republik Indonesia, "Permendiknas No 20 Tahun 2007," 2007.
[19] A. H. Hernawan, Permasih, dan L. Dewi, "Panduan Pengembangan Bahan Ajar," *Depdiknas Jakarta*, hal. 1–13, 2008.
[20] M. Forster, "Higher Order Thinking Skills," *Res. Dev.*, vol. 11, hal. 1–5, 2004.

IOP Conf. Series: Materials Science and Engineering **335** (2018) 012074 doi:10.1088/1757-899X/335/1/012074

Influence of Learning Strategy of Cognitive Conflict on Student Misconception in Computational Physics Course

A Akmam*[1], R Anshari[1], H Amir[1], N Jalinus[2], A Amran[3]

[1]Physics Department, Faculty of Mathematics and Sciences, Universitas Negeri Padang, Padang 25131, Indonesia
[2]Mechanical Engineering Department, Faculty of Engineering, Universitas Negeri Padang, Padang 25131, Indonesia
[3]Chemistry Department, Faculty of Mathematics and Sciences, Universitas Negeri Padang, Padang 25131, Indonesia

*akmam_db@fmipa.unp.ac.id

Abstract. Misconception is one of the factors causing students are not suitable in to choose a method for problem solving. Computational Physics course is a major subject in the Department of Physics FMIPA UNP Padang. The problem in Computational Physics learning lately is that students have difficulties in constructing knowledge. The indication of this problem was the student learning outcomes do not achieve mastery learning. The root of the problem is the ability of students to think critically weak. Student critical thinking can be improved using cognitive by conflict learning strategies. The research aims to determine the effect of cognitive conflict learning strategy to student misconception on the subject of Computational Physics Course at the Department of Physics, Faculty of Mathematics and Science, Universitas Negeri Padang. The experimental research design conducted after-before design cycles with a sample of 60 students by cluster random sampling. Data were analyzed using repeated Anova measurements. The cognitive conflict learning strategy has a significant effect on student misconception in the subject of Computational Physics Course.

1. Introduction

Physics is a unique subject because it involves many levels of abstractions in different forms, such as, conceptual, mathematical formulation, experimental, descriptive. Knowing a phenomenon's cause, effects, and how to interact with it are some of the attributes related to understanding which also fit the purposes of learning physics [1]. For this reason, students must be known and understand the concept that they studied. Consistently, students often hold conceptions that are incompatible with established knowledge that known as misconceptions. If the number of misconceptions increases, students will have difficulty in understanding the subject's material such as Computational Physics.

The learning outcome of Computational Physics course is, student can formulation of the basic technique of numerical analysis to solve physics problem algorithmically. The creation of a flow chart is independent of a specific computer language and is useful for developing and understanding the basic structure of a program [2]. Embedding computational thinking content into required science coursework [3]. In computational thinking, students construct their experience thought to think about data and ideas, and about using and combining data and ideas to solve problems. Students have skills in designing computational programs related to problem solving and physics phenomena as well as natural phenomena related to Physics using application program. Students have curiosity, logical thinking, critical, creative, and innovative, disciplined, confident, appreciative of others' work, communicate, appreciative and participative.

The learning objectives of Computational Physics Courses will achieve if the lecturer encourages students to think computationally. Computational thinking is integrating the power of human thinking with the capabilities of computers. To encourage computational thinking in the classroom, teachers must ask different questions related to problem solving and the use of technology, such as what are the power and limit of human and computer intelligence [4]. Thinking computationally is systematically with well-organized groove thinking with clear flow charts.

The problem in the Computational Physics course is the student difficult to construct their knowledge through scientific procedures. Indication of this problem was a lot of students do not get mastery learning. This problem occurs due to the generative thinking ability of the underdeveloped students. Generative thinking skills students are required to understand conceptual more. A conceptual understanding is needed student to get a quantitative solution of qualitative problems [5]. The conceptual understanding of students can grow through the ability to think critically. Critical thinking allows students to discover the truth in the midst of an abundance of events and information that surrounds them [6]. Students with critical thinking will argue in an organized manner and systematically evaluate the quality of their opinions. Critical thinking of students has been growing by cognitive conflict learning strategies [7]. When the irrelevant information is associated with a response, reliance on automatic processing might facilitate the task if this response is desirable. However, when the irrelevant, automatically processed information is associated with an inappropriate response, the resulting conflict between appropriate and inappropriate responses might be difficult to overcome [8]. In order for students to construct their knowledge conceptually in learning without misconceptions required, a model of learning that can encourage students to participate actively in learning. Based on the above reasons, the use of generative thinking model based on cognitive conflict is one of the solutions to solve the problems in Computational Physics learning. Students through learning by cognitive conflict strategy will be able to reduce their misconceptions in the course of Computational Physics.

Cognitive conflict is the discomfort one feels when his beliefs, values or behaviors contradict one another. When an individual believes very strongly in something that proves to be false, he will experience conflict [9]. Cognitive conflict strategy is a part of psychological theories of conceptual change. This strategy is effective in correcting a misconception as well as in improving performance. The conflict teaching approach as a means of helping learners reconstruct their knowledge [10].

A cognitive conflict strategy emphasises destabilising students' confidence in their existing conceptions through contradictory experiences such as discrepant events and then enabling students to replace their inaccurate preconceptions with scientifically accepted conceptions [11,12]. Among cognitive strategies, cognitive dissonance is the perception of incompatibility between two cognitions, which can be defined as any element of knowledge, including attitude, emotion, belief, or behavior. The cognitive dissonance strategy holds that contradicting cognitions serve as a driving force that compels the mind to acquire or invent new thoughts or beliefs, or to modify existing beliefs, in order to reduce conflict between cognitions [13]. Moreover, creating and resolving cognitive dissonance can have a powerful impact on students' motivation for learning. In our case-study, we use cognitive dissonance to increase motivation for learning as well as foster students' inner self-regulation [14]. Likewise, novice science teachers are unaware of student prior knowledge and its role in instruction to effectively implement constructivist teaching practices [15]. Science teacher educators have sought to raise pre-service science teachers' concern about students thinking and to assist them in learning.

Cognitive conflict strategies are effective enough to address misconceptions among learners in order to establish a higher science balance [16]. Cognitive conflict stimuli in learning helps the assimilation process become more effective and meaningful in the intellectual formation of learners. Giving cognitive conflicts helps learners reflect on conceptions, explanations of the phenomena they learn to develop the activation of phenomena obtained. Encouraging students to think critically in understanding the concept thoroughly can reduce their misconceptions. Cognitive conflict strategies are an active process of constructing the relationship between new knowledge and the old one. The essence of the cognitive conflict strategies is thinking is not a passive consumer of information [17]. Students participate in the learning process and generate knowledge by forming mental relationships between concepts. There are four elements in cognitive conflict strategies that are recalled, integration, organization, and elaboration [18]. Cognitive conflict strategies in its implementation is divided into four phases, namely preliminary phase, focus, challenge and application phase in the form of algorithm making.

Preliminary phase that is to give the initial task to the students through reading the reference according to the topic to be studied. Pre-focusing that students are asked questions are open and students answer in accordance with the knowledge they have. Phase focusing that student activity in this phase is to participate in scientific activities understanding of concepts associated with new concepts. Phase challenge that students are facilitated to make a change of views on the problem and guide students to perform numerical analysis of the given problem. Confirmation Phase that students are given consideration to all alternative answers given by students. Phase Confirmation that students are given logical consideration of all alternative answers that have been given by the group of students to lead to the achievement of learning objectives. Phase checking comprehension that students are given pretest. Phase verification that students carry out the practice of making a Matlab script in based on the algorithm. The work of the students is directly examined and given a feedback on the results of his work.

2. Research Method

In order to create cognitive conflict situations on student we selected two tasks with the potential of evoking conflict, Tasks 1 is familiar content, while Tasks 2 is an unfamiliar content. Half the students worked on Task 1 in pairs and on Task 2 in groups of four, and the other half worked on Task 1 in groups of four and on Task 2 in pairs. Each task began with an individual assignment, in which each student was asked to solve a computational physics problem on his or her own, in writing. After solving the problem alone, the members of the group were asked to discuss their solutions and reach an agreement. They were then asked to resolve the contradiction as a group. For clearer analysis, the data were divided into three categories as follows: (a) the percentage of correct responses (full understanding), (b) the percentage of students partial correct responses (misconceptions), and (c) the percentage of students' incorrect response (not understand). An open-ended question was set to ask students to explain their difficulties in each problem. Data was analyzed by Spanova statistics in order to look for the effect of cognitive conflict strategy on student misconceptions in the Computational Physics course.

3. Results and Discussion

The comparison of the increase in student learning outcomes, both the experimental and control groups are shown in Figure 1. The graph in Figure 1 shows that there is an interaction effect of the average score of the students' learning outcomes in the Computational Physics course of experimental class and control class during the three measurement periods. Figure 1 shows that learning by cognitive conflict strategy can increase the average score of student learning outcomes in Computational Physics courses. The significance of learning effect of cognitive conflict strategy can increase the average score of students' learning outcomes in the Computational Physics course.

Test of Within-Subject Contrasts analysis indicates that there is a significant linear relationship between dependent variables (score midterm-semester and end of semester). The result of pair comparison shows that after controlling with the Bonferroni method, the average score of student's learning outcomes in the Computational Physics course of midterm-second semester is lower than the end of the semester, but not significantly (differences is 0.19, $p < .05$). The average score of student's learning outcomes in Computational Physics end of the semester two is higher than mid-semester two significantly (differences is score is 3.25, $p < .05$). Table 1 shown that students achieve mastery in learning generally they experience misconception in choosing a suitable method to solve the problem given. For example, students ware calculate the electric field strength of a certain distance from a non-homogeneous charge density as shown in Figure 2. Base on electric field equations, $E = \int_{-L}^{L} \frac{1}{4\pi\varepsilon_0} \frac{e^{-x}}{r^3}\, dx$. Students generally assume r does not depend on x.

Table 1. Distribution of Student Misconception in Computational Physics Course.

No	Topic	Understanding Level			Understanding Level		
		No under-standing	Mis-concept-ion	Full Under-standing	No under-standing	Misconcept-ion	Full Under-standing
1	Interpolation with the same difference	20.00	43.33	36.67	13.33	56.67	30.00
2	Calculate the solution equations of Van der Waals equations	20.00	50.00	30.00	16.67	53.33	30.00
3	Calculates the coefficient friction of turbulent flow	23.33	53.33	23.33	13.33	60.00	26.67
4	Calculate differential based on given data	10.00	30.00	60.00	10.00	50.00	40.00
5	The Numerical Solution of the Electrical Fields on the wire	10.00	40.00	50.00	13.33	50.00	36.67
6	Determine the numerical solution of damped vibration equations	10.00	43.33	46.67	16.67	40.00	43.33
7	Using the first order Runge Kutta method to complete the LR circuit	20.00	40.00	40.00	16.67	50.00	33.33
8	Determine the numerical solution to complete the RLC circuit	23.33	26.67	50.00	23.33	26.67	50.00

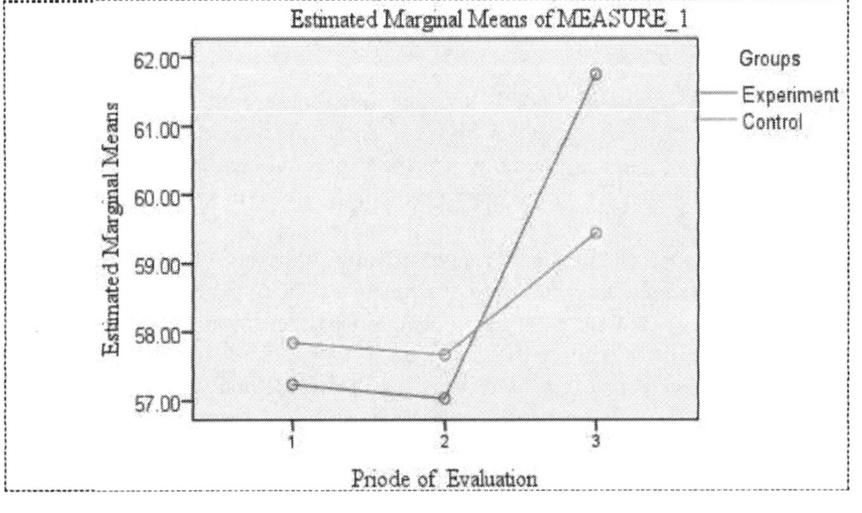

Figure 1. The mean score of student learning outcome with cognitive conflict learning strategy and without cognitive conflict learning strategy in the Computational Physics course.

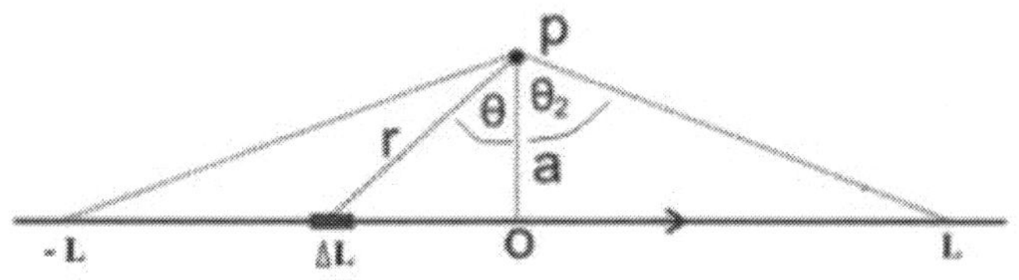

Figure 2. Electrical field in current wire

The data analysis showed that the average of the students 'learning outcomes in the group that were studied using the cognitive conflict learning strategy was lower than the average of the students' learning outcomes in the group that were given the lesson did not use cognitive conflict strategy in the Computational Physics course at the beginning of the lecture. Misconceptions can hinder student achievement in learning [19]. Misconceptions in the student self will cause cognitive conflict in a person has not been solved perfectly [11]. Cognitive conflict is not resolved properly will lead to continuous misconception, resulting in low student learning outcomes. Students fail to construct their learning experiences so that they mistakenly solve the problem [15]. This is what is expected to influence the negative learning of cognitive conflict strategy on the average of student learning outcomes at the beginning of this strategy is applied. This shows that this strategy works well. This strategy cannot work well, because students do not make a good introduction, do not read the practicum module and teaching materials before the lecture.

This problem is solved by giving students the pretest before the practicum beginning in the mid-semester and the end of the semester. This strategy has a positive impact, where the students' misconception rate in the middle of the mid-semester two and the end of the semester two may be reduced. If students are accustomed to construct their learning experiences well, their ability to build knowledge and skills will also improve [20;21]. If students who experience cognitive conflicts can be resolved properly misconceptions in the minds of students will reduce. The cognitive conflict is a mismatch in the mind (old knowledge) of the students with information gained during the learning (new information). In order to increase student learning outcomes, the old concepts that exist in the minds of students they must be consciously revised [9,16]. This research can be seen as a replication as from research [21]. Implementation is learning by cognitive conflict strategy well used in learning Computational Physics, if students have the power of reason and reference reading power well.

4. Conclusions and Recommendations

The learning strategy of cognitive conflict can increase the average score of student learning outcomes in Computational Physics at the Physics Department of FMIPA Padang. Therefore, it means that learning by cognitive conflict strategy can reduction of student misconception on Computational Physics course at the Department of Physics, Faculty of Mathematics and Science, Universitas Negeri Padang, but so far not significantly. Based on the analyzed of student's answer sheet, students that are not mastery at Computational Physics learning is caused by students have misconception in choosing a method to solve the problem. Implementation is a generative learning based on cognitive conflict strategy well used in Computational Physics learning. However, further research with more samples and other moderator variations should be undertaken. Cognitive conflict strategy needs to be proved in other courses because each subject has its own characteristics.

References

[1] Bodin, Madelen (2012), Computational problem solving in university physics education Students' beliefs, knowledge, and motivation, Electronic version available at http://umu.diva-

portal.org/Printed by: Print & Media, Umeå University, Sweden.
[2] Landau, H. Manuel, J. P., Cristian C. B., 2012, Computational Physics Problem Solving with Computers, Enlarged e-TextBook Python, 3rd Edition, Copyright Landau, Paez & Bordeianu
[3] Orton,K., Weintrop,D., Beheshti, E., Horn, M., Jona, K., and Wilensky, U., (2016), Bringing Computational Thinking Into High School Mathematics and Science Classrooms, *ICLS 2016 Proceedings*, https://www.discretesearch.com download, 26 Agust, 2017
[4] Hambrusch,S., Hoffmann, C., Korb, J.,T., Teaching Computational Thinking to Science Majors, (2010), *SIGCSE '09 Chattanooga*, Tennessee, USA, Copyright 200X , download, 16 Agust 2017.
[5] Cowan, J., (1998), *One Becoming an Innovative University Teacher. Reflection in Action.* SRHE/Open University Press, Buckingham.
[6] Johnson, Elaine B, (2012), *Contextual Teaching & Learning*, Penerbit Kaifa Bandung.
[7] Tan, W.C., Omar,M., Zuraidah, A.R., Siti, F.M.D.& Hanafi,A., (2008). Generative Learning Objects for Collaborative Learning and Critical Thinking: A Proposed Conceptual Framework, *Malaysian Journal of Distance Education* 10(1), 129–141.
[8] Vincent, van Veen, and Cameron S. Carter, (2006), Conflict and Cognitive Control in the Brain, Current Directions in Psychological Science, All content following this page was uploaded by Vincent van Veen on 09 August 2017.
[9] Toka, Y. & Askar, P. (2002). The Effect of Cognitif Conflict and Conceptual Change Text on Student's Achievement Related First Degree Equation With Unknown, *Hecettepe Universiste Egitim Fukultesi Dirgisi*. 2, 211-217.
[10] Sela, H & Zaslavsky, O, (2007), resolving cognitive conflict with peers – is there a difference between two and four, In Woo, J. H., Lew, H. C., Park, K. S. & Seo, D. Y. (Eds.).*Proceedings of the 31st Conference of the International Group for the Psychology of Mathematics Education*, Vol. 4, pp. 169-176.
[11] Kang, H., Lawrence, C. S., Sukjin K.& Taehee, N. (2010). Cognitive conflict and situational interest as factors influencing conceptual change, *International Journal of Environmental & Science Education,* Vol. 5, No. 4, 383-405.
[12] Gyoungho, L., Jaesool, K., Sang-Suk P., Jung-Whan, K., Hyeok-Gu, K. & Hac-Kyoo, P. (2003). Development of an Instrument for Measuring Cognitive Conflict in Secondary-Level Science Classes. *Journal of Research in Science Teaching*, 40 (6), 585–603.
[13] Lee, H.W., Kyu Y.L, and Barbara L. G. (2005). Generative Learning: Principles and Implications for Making Meaning, Pennsylvania State University, University Park, Pennsylvania, http://www.aect.org/edtech/edition3/ER5849x_C010.fm.pdf.
[14] Marta, A, Thanasis D,2 and Fatos X, (2016), a model for Providing Emotion Awareness and Feedback Using Fuzzy Logic in Online Learning, Soft Computing. ISSN: 1432-7643. Springer. Published online 20 October 2016. DOI 10.1007/s00500-016-2399-0. download, 26 Agust, 2017.
[15] Moussiaus, S.J. & Norman, J.T. (1997). Constructivist Teaching Practices: Perception of Teachers and Students, Wayne State University, e-journal. www. Perception.edu/ Constructivist.html.
[16] Yu-Fen Yang (2010). Cognitive Conflicts and Resolutions in Online Text Revisions: Three Profiles, *International Forum of Educational Technology & Society* (IFETS). 5(2), 61-73.
[17] Mason, G. (2006). *Generative learning*. George Mason University.
[18] Osborne, R. & M.C. Wittrock. 1985. The Generative Learning Model and Its Implications For Science Education. *Studies in Science Education*, 12: 59-87.
[19] Barke H.D, (2013), Structure of Matter – Diagnosis of Misconceptions and Challenge, *Bullettin of the Chemists and Technologists of Bosnia and Herzegovina*, 40, 21-30.
[20] Radhiah, A.R., Norah, M.N. & Norasykin, M. Z. (2015). Meta-analysis on Element of Cognitive Conflict Strategies with a Focus on Multimedia Learning Material Development, *International Education Studies*. 8(13). 913-923

[21] Trespalacios, J.H., (2008), The Effects of Two Generative Activities on Learner Comprehension of Part-Whole Meaning of Rational Numbers Using Virtual Manipulatives. *Dissertation*, submitted to the Faculty of the Virginia Polytechnic Institute and State University. Blacksburg.

Electronic Module Design with Scientifically Character-Charged Approach on Kinematics Material Learning to Improve Holistic Competence of High School Students in 10th Grade

R Anggraini[1], Y Darvina[1], H Amir[1], M Murtiani[1], Y Yulkifli*[1]

[1]Physics Department, Faculty of Mathematics and Sciences, Universitas Negeri Padang, Padang 25131, Indonesia

yulkifliamir@gmail.com

Abstract. The availability of modules in schools is currently lacking. Learners have not used the module as a source in the learning process. In accordance with the demands of the 2013 curriculum, that learning should be conducted using a scientific approach and loaded with character values as well as learning using interactive learning resources. The solution of this problem is to create an interactive module with a scientifically charged character approach. This interactive module can be used by learners outside the classroom or in the classroom. This interactive module contains straight motion material, parabolic motion and circular motion of high school physics class X semester 1. The purpose of this research is to produce an interactive module with a scientific approach charged with character and determine the validity and practicality. The research is Research and Development. This study was conducted only until the validity test and practice test. The validity test was conducted by three lecturers of Physics of FMIPA UNP as experts. The instruments used in this research are validation sheet and worksheet sheet. Data analysis technique used is product validity analysis. The object of this research is electronic module, while the subject of this research is three validator.

1. Introduction

The 2013 curriculum is a character-based and competency-based curriculum. Through education is expected to the character and competence of learners can be formed in a better direction. In learning, learners are guided by educators to participate in self-useful to establish the competence and character of learners. Educators are expected to have high creativity and innovation in creating a fun learning environment. In the formation of competence and character of learners requires a quiet and enjoyable learning. Lessons learned in the 2013 curriculum include the attitudinal competencies, knowledge and skills available in all fields of study at every level of education. According to Permendikbud Number 59 of 2014, the learning pattern of curriculum 2013 is centered on students, interactive learning that allows learners to obtain information from anywhere as a source of learning.

The module is one of the learning resources that can direct the students to master the four core competencies contained in the curriculum of 2013 as a whole. The interactive module is a module equipped with various media so that it is expected to facilitate learners' learning styles. The use of varied media can provide a quiet and enjoyable learning environment and learners can interact actively in learning as expected by the 2013 curriculum. One approach that can actively involve learners is the scientific approach.

The scientific approach is an approach that engages learners actively in the learning process. A scientific approach can involve the learner's skills in the process such as observing, questioning, trying / gathering information, reasoning and communicating. With a scientific approach is expected to form the ability of learners in solving problems.

Based on the observations that researchers do in SMA Negeri 3 Bukittinggi, it is found that overall learners stated in the learning has not been using interactive module. Learning resources used in the form of modules in print and only 10% of learners who can understand the material presented in the module. So that learners prefer to receive material explanation from teacher only, while in curriculum 2013 expected some interaction in learning process that is learners interaction with educator, learners with learners, learners with learning resources. In fact the interaction of students with the source of learning that is still in print form provided the school is still lacking so that the learners are less active in the learning process.

Seeing the importance of interesting learning resources in the learning process and can guide the character of learners, the interactive module is expected to improve the spirit of learners learn. So that the competence of learners can be improved holistically through the learning process using this interactive module. In this interactive module can load video, audio, simulation, animation and drawing so hopefully more fun learning atmosphere.

This interactive module is designed with a scientific approach consisting of the process of observing, asking, trying / gathering information, reasoning and communicating. Physics is a science that studies natural phenomena so that scientific approach suitable for use in learning physics. Application of this scientific approach will make learners will be more focused in understanding the concept of physics itself. In addition, the 2013 curriculum emphasizes on improving the knowledge, attitudes, and skills of learners. One way to improve the ability of learners holistically is to use interactive modules with scientific and character-filled approaches. The module will increase the knowledge and faith of learners to God Almighty accompanied by changes in spiritual and social attitudes towards the better.

Previous research on module making has been done by Ike Evi Yunita [7], Mukhayyarotin Niswati R. J [8], and Eka Lestari and Abdur Rahman [9]. Research conducted by Ike Evi Yunita, Eka Lestari and Mukhayyarotin Niswati indicate that the module is already in the category valid and very suitable for use in learning. The difference of this research with the previous research is the module that is made an interactive module that contains the character values. The software used in creating interactive modules is Adobeflash CS6. The kinematics material is chosen as the material in making this interactive module. In accordance with the 2013 curriculum, interactive modules are created following the steps of a scientific approach.

This paper describes the results of Research and Development (R & D) research with the title of developing character-based interactive module with a scientific approach to kinematics materials to improve the holistic competence of high school students of class X. R & D are a research method used to produce a particular product, and test to effective the product [11]. A feasibility test in the form of validity and practicality needs to be done so that the interactive module made has good quality.

2. Research Methods

The type of research used is research and development or Research and Development (R & D). According to Sugiyono (2007: 407), R & D is a process or steps to develop a new product or refine an existing product, where all its activities can be accounted for, also can be interpreted as research used to produce certain product and test the effectiveness of the product. Based on the development model used, the procedure in this research consists of five stages: potential and problems, data collection, product design, design validation, design revision, product trial, product revision, product revision, trial usage, product revision, and mass production but, in this study the stages are carried out only until the revision of the product.

Interactive module that has been validated and revised, tested on one class of students class X SMA Negeri 3 Bukittinggi. The limited trial subjects consisted of 3 Physics lecturers as validators, and 2 Physics teachers as a parent. The interactive teaching experiment was conducted in SMA Negeri 3 Bukittinggi. This research data in the form of quantitative data and qualitative data. Quantitative data is the value of the interactive module validation and the test value of the practicality of the interactive module. Qualitative data is the result of interviews with teachers and learners on the implementation of interactive modules.

Data collection is an important work in a study. Correct conclusions can only be obtained from the correct data collection. Therefore, errors in collecting data will lead to erroneous conclusions.

The data collection instruments used in this study consisted of instruments of validity and instrument of interactive module practice. The creation of interactive modules containing characters with a scientific approach must be validated first by experts to determine the accuracy of the constituent components. Instrument validity includes several aspects, namely: 1) material substance aspect, 2) display aspect, 3) grammatical aspects, 4) design aspect, 5) aspect of software utilization and 6) aspect of module of interactive.

Instruments of practicality are used to collect practicality test data of the learning process with ease of using characteristic interactive module with scientific approach. Practical tools are filled by teachers and learners. Practicality instruments filled by teachers aim to get responses and suggestions for product improvement so that the resulting nteraktif module becomes a practical product used in supporting learning. The instrument of practicality consists of five aspects: 1) completeness of interactive module, 2) interactive module coverage, 3) presentation of interactive module, 4) benefit of interactive module and 5) implementation of interactive module.

Data analysis technique is related to the calculation to answer the formulation of the problem and testing the hypothesis proposed. In this study, used data analysis techniques, namely validation analysis and practicality analysis. Interactive module validation analysis was assessed by lecturer of Physics Department of State University of Padang. The statistical test performed in validity is the descriptive analysis depicted through the table. Weighting is done on a Likert scale. Likert scale is used to measure attitudes, opinions, and perceptions of a person about social phenomena. Validity analysis using Likert scale with steps as follows: (1) give score for each item of answer, (2) summing score of each validator for all indicator, (3) giving validity value is done by using an equation (1):

$$Score = \frac{bobot\ total}{bobot\ maksimum} \times 100 \qquad (1)$$

The analysis of the practicality of the interactive module is characterized by the scientific approach according to the teacher based on the practicality instrument is done by the following steps: (1) give the score for each answer item, (2) sum up the score of each validator for all indicators, and (3) give practicality value by using the formula:

$$Final\ Score = \frac{Jumlahskor}{SkorMaksimum} \times 100 \qquad (2)$$

3. Research Results and Discussion

3.1 Research Results

In the research that has been done there are two main results. The results of this research include: interactive module design with characteristic scientific approach and quality of interactive module from the aspect of validity and practicality.

3.2 Form of Interactive Module Design

The design of an interactive module with a characteristic scientific approach to the kinematics material is made in accordance with the steps that have been prepared. This interactive module created using Adobe Flash CS6 application as main software and Photoshop, Format Factory, Audacity as supporting software, then this interactive module will be publish with exe format.

IOP Conf. Series: Materials Science and Engineering **335** (2018) 012075 doi:10.1088/1757-899X/335/1/012075

The design of this interactive module uses different types of writing. The type of writing used is adapted to its use. In writing the title using the type of writing Arial because this type of writing is not complicated and easy to read. In the material description section use the type of tuition century ghotic because this type of paper is easy to read. The colors used in this interactive module are bright and soft colors. Color selection is done based on the wishes of learners obtained from the questionnaire observation.

This interactive module contains the values of characters that can be trained in the learning process. The value of the characters contained is religious, honest, disciplined, hard work, independent, curiosity and careful guided by KI1 and KI2. The value of this character will be shown in the steps of the learning activity, the evaluation page, as well as on the part of the experimental activity using the simulation. So in every activity undertaken learners will display the values of characters that can guide learners in work. With the value of this character is expected to change the behavior of learners to a better direction along with the increase of knowledge it has.

The interactive module created has instructional instructions for educators and learners that can be read before using this interactive module. This interactive module has an evaluation question that can give the scores obtained by the learners as feedback. Each learner will score differently and depending on the answers given by the learner. This interactive module also includes simulations that can be varied variables. With this simulation learners can better understand the concept of the given material.

In the presentation of interactive module materials using scientific steps. The steps start from: (1) observe, in this step learners are given animation or video related material; (2) asks, in this step learners are given the opportunity to write questions or answer questions that have been provided based on the animation or video that has been seen earlier; (3) collect information, in this step pesrta educated given information in the form of material that can increase understanding of learners; (4) reasoning, in this step learners are given a workable problem if learners already understand the material obtained in the previous step; (5) communicate, in this step learners are required to be able to communicate the results of activities that have been done in front of the class.

This interactive module has several main components, namely: opening menu, main menu, module instruction manual, KI, KD, indicator, learning objectives, learning activities, and evaluation questions, and experimental activities using virtual lab in the form of swf. The main components each have a button to be accessible directly. This interactive module contains animations, simulations, music, videos and bright colors that can add to the charm of this interactive module.

The opening menu of the interactive module contains the module title, author name, UNP logo, and buttons to access the main menu. The background opening menu has an abstract blend of colors and edited images using Adobe Photoshop CS6. The title is created using an Arial font with a 30 letter size. The opening menu can be seen in Figure 1.

Figure 1. The opening menu.

This interactive module uses the background with bright and attractive colors. This is based on the favorite colors of learners obtained from the observations that have been done. On the opening page is shown a picture of the train and coconut fell because the events in the picture relate to the material that is straight irregular motion and straight-changing motion irregularly.

The main menu view contains slide titles, main menu buttons, exit buttons, UNP logos, main menu hints. The main menu button can make it easy to access other desired pages. The buttons consist of: (1) Usage Button, this button connects to the page about module usage instructions for learners and teachers; (2) KI buttons, KD, indicators and learning objectives, these buttons are linked to pages containing the competencies of the material to be learned as well as the objectives of the learning to be performed; (3) Button of learning activities, this button is connected to a page that presents the material in a learning activity that uses scientific steps; (4) evaluation button; the button is linked to a page containing the evaluation to test the learners' understanding; (5) Profile button, this button connects to the interactive module author's profile page.

In the main menu view of this learner can choose the desired page by clicking one of the main menu buttons. This page has a music instrument that can add to the attraction of this interactive module. The music contained on this page can be turned off by clicking the sound button above the exit button.

The KI, KD and indicator views have buttons linking with multiple pages to display core competencies, basic competencies, indicators and learning objectives that are appropriate to the material to be covered. On this page there is also a Home button to return to the main menu. On each page description of core competencies, basic competencies, indicators and learning objectives have font size 20 and type of writing Arial. Baground writing is made using colors that are more in contrast to writing. When a button is clicked or assigned to the mouse it will produce a sound that can add attraction from the interactive module. The material menu display begins with a page that leads learners to pray. This page contains the home, next and exit buttons. The exit buttons are made to make it easy to close interactive modules. The writing background is colored in contrast to the text. At the bottom there is a running banner, it contains the author's name and the title of the interactive module. The next view of the learning activity is the sub-material page to be discussed. On this page there are two sub-material buttons that connect to the presentation page of the material. This button has a sound when designated and clicked and the button color will change when designated with the mouse. Back, next, exit and home buttons are also present on this page for the purpose of making it easier to access other pages. This sub-material button will link to the page that will present the material. At the beginning of the presentation of this material there are learning reference buttons, materials and experiments using virtual lab.

In the initial view of the presentation of this material learners can trace the character values listed on the interactive module. Learners can choose the page you want to read with the navigation buttons like the next, previous and home buttons. In the view of scientific steps consists of several pages of learning activities with a scientific step. This scientific step consists of several activities that start from observing, asking, gathering information, reasoning and communicating. On the page observing the learners are asked to observe the animation or the video provided. The animations and videos contained on this page are not given an explanation, because after observing the learners are expected to generate questions related to the animation he saw earlier.

In the ask view load the back, next, blanks and exit buttons. Empty columns are reserved for where learners write questions. When learners want to go back to the page of observing it can be done by clicking a button back.Halaman also provides buttons linked to a page that displays the question that has been provided in the interactive modules. This question will lead learners to the material to be studied. Once given the questions learners can see the available explanations to help answer questions.

The initial view of the information gathering activity consists of several buttons linked to the page containing information about the material. In this information gathering display contains material that is accompanied by examples of questions that can increase the understanding of learners. On the pages of reasoning contains questions that can be done by learners. This question is given with the aim of looking at the students' reasoning abilities and using the concepts that have been obtained at the stage of collecting information. Problem given can train learners in using the concept that has been studied. Learners are asked to write answers on their respective exercise books.

At the stage of communicating the students are asked to present the results of activities that have been done. The character values are also listed on this page and can be trained by the learner. On this page there is a linking button on the page that contains material musings. After completion until communicating stage, then learners can see the reflection of the material that has been studied.

In this experimental activity view, learners perform simple experiments using virtual labs from the swf. On this page there are back, home, exit, and work step buttons linked to the page describing the trial procedure to be performed. On the work step page learners can see which part will be made during the experiment later with the picture included in the work step. In the virtual lab page learners can do a simple experiment by varying the acceleration and speed that have been provided. To retrieve the data repeatedly learners can use the reset button located on the screen. After completion of the experiment the learners are asked to create a report. Learners are required to make a report in accordance with the structure that has been given. Each group is responsible for sending group representatives to present in front of the class.

In the evaluation view the learners are first asked to write the name and class. Before starting to work on the problem learners can read the instructions and the value of characters that are loaded. This page has musical instruments that add to the charm of this interactive module. On the opening page of the evaluation there are two navigation buttons are the start and home buttons. To start answering questions learners can press the start button, but if you want to re-read the material learners can press the home button. Problems contained in this interactive module are multiple choice questions. Learners can answer the question by clicking one of the options provided answers. After choosing the correct answer, then the learners will receive feedback.

On the feedback page learners can see the scores obtained at the top right corner. When learners answer the question correctly, then the score will increase automatically, but when the answer given is wrong then the score remains. To answer the next problem learners can press the next button. After completing the problem, then the learners will get the final score. On the final display page of the evaluation the learner can see the total score obtained after answering all questions. If the score score is not satisfactory, the student can press the replay button. The replay button on this page serves to repeat the problem. When this problem is done at home, then learners can submit the printed final score of this page by using the snipping tool. When finished doing the question then learners can press the home button to return to the main menu.

3.3. Validity Test Results

The result of validity is obtained from validation instruments filled by experts. This validation result is used as a guide in improving interactive modules and to determine the eligibility of the interactive module. The validation instrument used contains six aspects of the assessment of the interactive module. Assessment aspects contained in the validation instrument are the substance of the material, the appearance, the grammar, the module design, the utilization of the software, the founder of the interactive module.

In the process of validating this interactive module the number of experts involved as many as three lecturers of physics FMIPA UNP. In each aspect the scoring has the lowest score of 1 and the highest score 5. The scores obtained will be converted in the form of numbers from range 1 to 100. Determining an assessment can be seen from the average value of indicators contained in the aspect of the assessment of the interactive module. Based on the validation instrument used in the material substance aspect there are three indicators. This aspect indicator is on the horizontal axis of the graph. Indicators are: 1) interactive module made not deviate from the truth of science, 2) interactive module presented in accordance with the depth of material, 3) interactive module presented in accordance with the development of science.

Based on the validation results it is found that the average percentage of the material substance aspect is 84.4%, the average percentage of the display aspect is 84%, the average percentage of the grammatical aspects is 76.7%, the average percentage of the design aspect is 90.4%, the average percentage of the software utilization aspect is 91.1% and the average percentage of the interactive learning faculty aspect is 85%. The overall Validation value can be seen in Figure 2.

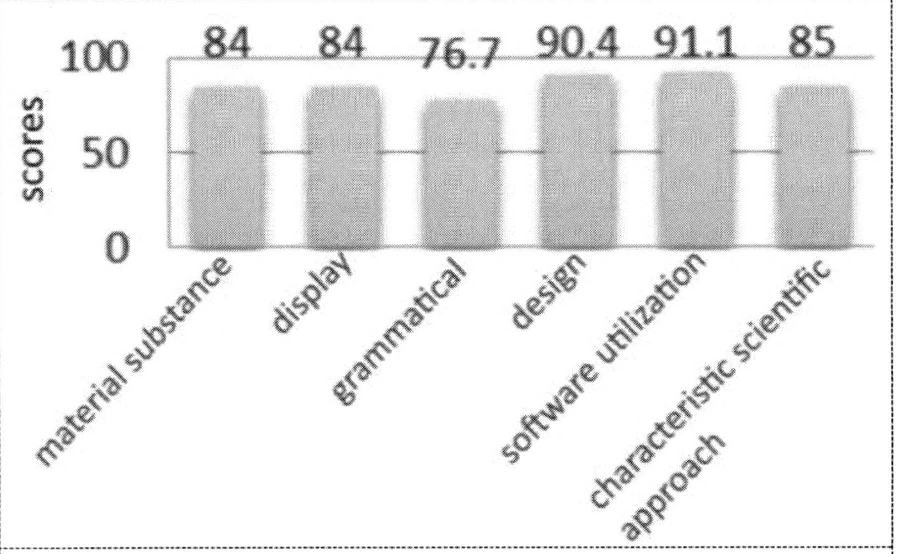

Figure 2. Overall validity value of interactive module.

From Figure 2 we can see the percentage of average validity of interactive module with characteristic scientific approach is 85.2% with very valid criteria. It shows that interactive module with characteristic of scientific approach to the kinematics material is valid, both from material susceptibility aspect, display, grammar, design, software utilization and interactive module feature so that this interactive module can be used in learning.

3.4. Practicality Test Results

The results of practicality test obtained from the analysis of instruments used. This practicality test aims to find out the practical level of interactive module with characteristic scientific approach. In assessing the practicality of this interactive module involved two physics teachers of SMA 3 Bukittinggi. The results of this practice will serve as guidelines for improving interactive modules that have been designed.

The instrument of practicality consists of five aspects of assessment. These aspects include: 1) completeness of the module, 2) module coverage, 3) module presentation, 4) module benefits, 5) module implementation opportunities. In every aspect there is an assessment indicator of the interactive module. Each indicator is scored 1 through 5. The scores obtained will be converted in percentage form. For the assessment of one instrument, it is obtained from the average of each indicator in the interactive module assessment aspect.

Based on the teacher's response to the interactive module provided it is found that the average percentage on the completeness aspect is 80%, the average percentage on the coverage aspect is 80%, the average percentage on the presentation aspect is 80%, the average percentage on the benefit aspect is 76.7%, and the average percentage on the aspect of the interactive module implementation opportunity is 83%. The overall value of practice can be seen in Figure 3.

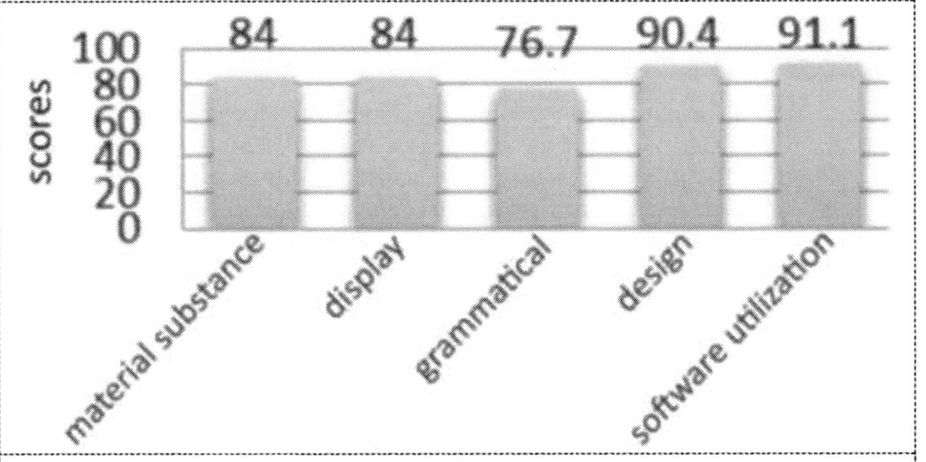

Figure 3. Overall practical value of interactive module.

From the data above, the percentage of the average practice of interactive module with characteristic scientific approach is 80%. It shows that the interactive module is characterized by a scientific approach to practical kinematics material from aspects of completeness, coverage, presentation, benefits, and implementation opportunities. Thus, this interactive module made is practical for teachers to use in learning.

4. Discussion

The main purpose of this research is to produce interactive module design and to know the quality of interactive module in terms of its validity and its practicability. This interactive module can be used in learning physics class X semester 1 on kinematics material. This interactive module can be operated with a computer / laptop so that it can be used out of the classroom by the learners. This interactive module contains the value of characters that can be trained by learners in learning. The value of the characters contained is religious, honest, disciplined, hard work, independent, curiosity and meticulous. The character values contained in this interactive module are guided by KI1 and KI2. Given the value of characters in this interactive module is expected learners can apply good character in learning.

This research is based on research and development (R&D) consisting of ten stages, that is knowing potential and problem, collecting information, designing product, design validation, design revision, product trial, product revision, trial usage, product revision, and mass production (Sugiyono, 2012). However, this study was conducted until the seventh stage. Based on research procedures that have been done can be explained the results achieved in research, constraints and limitations encountered, as well as some alternative solutions to overcome the constraints and limitations in the study. The results of this study include product design, validation results by experts and test results practicability by educators, as well as product revisions based on suggestions given validators and practitioners.

Based on the validation results by three experts it can be seen that the average validity score is 85.2% so that the interactive module made is categorized very valid for use in Physics learning process. This is in accordance with Sugiyono (2012), states that the validation is done by experts who are experienced in the field under investigation. The assessment component in the validity of this product consists of material substance, appearance, grammar, module design, software utilization and interactive module identifier. From the results of validation and suggestions given can be seen that this interactive module still needs improvement. Improvements made according to the validator's suggestions are on the objectives and reference of learning that have not been included, the value of the character of the material, and the selection of colors more interesting. As Borg and Gall (2003) argue, in development research, the initial product developed must be validated or revised based on the input of experts prior to use. After making improvements then produced a better interactive module.

Based on the results of practical test conducted at SMA 3 Bukittinggi can be concluded that the interactive module is practically used with an average value of 80%. This practicality test is performed using a practical instrument that will be filled by the educator. This interactive module is practically used judged from module completeness aspects, module coverage, module presentation, module benefits, and module implementation opportunities. After analyzing the results of practical test, it can be concluded that this interactive module can be used in physics learning as teaching material and can improve students' competence holistically. This is in line with the opinion Purwanto (2009) a product is said to have good practicality if the possibility to use the product is large. In addition to this, the product is said to be practically seen from its ease of use and the benefits of using the product.

From the results of practical tests that have been done there are suggestions given by practitioners. Improvements that have been made based on suggestions provided are, adding photos and videos that are more supportive and complete the material presented in the interactive module. After making improvements, the resulting interactive module is better than ever so it can facilitate learners in understanding the material.

In the implementation of learning using interactive module is still experiencing constraints. The constraint is a limitation on the material presented. This interactive module contains only straight uniform motion and straight uniform motion, circular motion, and parabolic motion. The solution that can be done to overcome this limitation is by developing class X material in this interactive module. Another obstacle is the limitations of laptops / computers that provide schools, so that learners less than the maximum in using interactive modules. Solutions that can be done to overcome this limitation is to form small groups in learning, so learners can learn to use interactive modules.

5. Conclusions

Based on the results of research and discussion that has been done can be put forward several conclusions as follows: An interactive module is an animation, audio, video, and simulation module created using R & D (Research and Development) steps. The steps taken are to identify potentials and problems, gather information, design products, design validation, design revisions, product trials, and product revisions. An interactive module created using a scientific approach with Adobe Flash CS6 application. Interactive modules containing characters with a scientific approach on kinematics material that is made are in the criteria very valid and practical, used to improve the holistic competence of learners. With the percentage of each of the criteria of 85.2% and 80%

Acknowledgement

The authors thank to lecturers who have provided many motivations and suggestions and to the big family of SMAN 3 Bukittinggi who has given permission and assistance in conducting the research.

References

[1] Kemendikbud. 2014. *Permendikbud Nomor 59 Tahun 2014 tentang Kurikulum 2013 Sekolah Menengah Atas/Madrasah Aliyah.* Jakarta: Kemendikbud.

[2] Kemendikbud. 2016. *Permendikbud No. 22 Tahun 2016 tentang Standar Kompetensi Lulusan Pendidikan Dasar dan Menengah.* Jakarta: Departemen Pendidikan Nasional.

[3] Kurniawan, Syamsul. 2013. *Pendidikan Karakter : Konsepsi dan Implementasinya secara Terpadu di Lingkungan Keluarga, Sekolah, Perguruan Tinggi dan Masyarakat.* Yogyakarta : Ar-Ruzz Media.

[4] Ismet Basuki dan Hariyanto.2014. *Asesmen Pembelajaran.* PT Remaja Rosdakarya Offset-Bandung.

[5] Mulyasa. 2014. *Implementasi Kurikulum 2013.* Bandung. PT Remaja Rosdakarya.

[6] Munir Abdul. 2010. *Pembuatan Bahan Ajar Multimedia Interaktif Menggunakan Authorings Tools Course Lab (http://www.vedcmalang.com*, diunduh 16 Maret 2016).
[7] Festiyed. 2008. *Disertasi Peningkatan Perhatian Belajar Siswa Berbantuan Program Komputer Interaktif.* Pasca Sarjana UNP: Padang.
[8] Busra, Festiyed, Ramli. *Analisis Kebutuhan Pengembangan Media Pembelajaran Interaktif dengan Menggunakan Power Point Berbasis Model PBL (Problem Based Learning) untuk Meningkatkan Kemampuan Berpikir Kritis Siswa*. Padang. Indonesia. 7 November 2015.
[9] Amali Putra, Asrizal. 2012. *Pengembangan Bahan Ajar Fisika Berbasis Web Interaktif dalam Bahasa Inggris dengan Software Moodle untuk Pembelajaran Siswa R-SMA-Bi Kelas X Semester 1*. Padang: UNP.
[10] Izah Nur Hidayah. 2016. *Pengemabangan Asesmen Berbasis Komputer Model Drilland Practice Materi Inti Atom untuk Siswa Sekolah Menengah Atas*. Sumatera Selatan: Universitas Sriwijaya.
[11] Sugiyono. 2012. *Metode Penelitian Kuantitatif Kualitatif, dan R&D*. Bandung: Alfabeta.
[12] Rochmad. 2012. *"Desain Model Pengembangan Perangkat Pembelajaran Matematika"*. Jurnal Kreano (Volume 3: Nomor 1).
[13] Hake, Richard. 1999. *Analyzing Change/ Gain Scores*. Dept of Physics. Indiana Univercity..
[14] Borg W.R dan Gall M.D. 2002. *Educational Research: An Introduction*. New York : Longman.
[15] Yulkifli, 2012. *Pendidikan Karakter Bangsa: Sebuah Refleksi Pendekatan dalam Ilmu Sains, Sukabina Press* ISBN 9786028124911. Padang.
[16] Yulkifli, 2014. Optimalisasi Peran Dan Fungsi Penasehat Akademik Mahasiswa Berbasis Nilai-Nilai Karakter Menuju Pembimbingan Akademik Bermutu (PAB), Prosiding Seminar Nasional MIPA 2014, Padang.

IOP Conf. Series: Materials Science and Engineering **335** (2018) 012076 doi:10.1088/1757-899X/335/1/012076

The Iimplementation of Iintegrated Natural Science Textbook of Junior High School be Charged on Character-based Shared Models to Improve the Competence of Learners' Knowledge

S Rahmiwati[*]**, Ratnawulan, Yohandri**

[1]Physics Department, Faculty of Mathematics and Science, Universitas Negeri Padang, Padang 25131, Indonesia

*kamsiah35@gmail.com

Abstract. The process of science learning can take place if there is an attempt to create an active learning atmosphere and can improve the knowledge competence of learners. One of the efforts made is to use learning resources. Textbooks are a learning resource used by learners. This study aims to describe the increase of knowledge's competence of learners with integrated Natural Science (IPA) textbook of Junior High School (SMP) be charged on character-based shared model. The method used pre-test, post-test design with one group using the class as a research subject. Pre-test was given before treatment to measure student's initial understanding of the problem, while the post-test was given to measure student's final understanding.The subject of this research is students of class VII SMP N 13 Padang. Result of gain score is 0,73. The result showed competence student's knowledge increased significantly and high categorized.

1. Introduction

The quality of education is one of the determinants of the nation's competitiveness. Educational institutions are required to produce graduates who have the skills and competencies to compete globally. This competition requires graduates who are not only skilled in their respective fields, but also must be able to communicate well to the outside world. Education is a conscious and planned effort to create an atmosphere of learning and learning process so that learners actively develop their potential, has a spiritual strength, self-control, personality, intelligence, noble character, as well as skills needed himself, society, nation and state [1].

Based on the act above can be seen that education has a goal in the learning process that will always grow. The development of the education system has been done in the educational process. One of the developmental changes and curriculum development of KTSP in curriculum 2013. Curriculum 2013 emphasizes on the competence of graduates with noble characters, possessing related skills and related knowledge in learning materials in the form of thematic integration between knowledge. The curriculum is a 'chip' that contains the wishes and wishes of a particular community to improve the situation itself, now and in the future [2]. SMP Curriculum 2013 curriculum is developed based on internal and external challenges, improvements in mindset, strengthening of curriculum and material governance.

Internal challenges relating to educational demands that refer to the National Standards of Education and the development of the Indonesian population. External challenges related to the flow of globalization and issues related to environmental issues, information technology advances, the rise of the creative and cultural industries, and the development of education at the international level. Curriculum in the educational process serves as a tool for achieving educational goals [3]. In this case, means as an educational tool the curriculum has important components and as a supporter that can support its operation well. These constituent components are related to each other. The components of curriculum development, namely component objectives, content components, component methods, and evaluation components. The components of each other are interrelated.

The component of the method or strategy is an important component because the methods and strategies used in the curriculum determine whether the material provided or the expected objectives can be achieved or not. the standard of basic and secondary education process has hinted about the need for a learning process guided by the principles of scientific in this learning process is often referred to as a characteristic and a distinct strength of the existence of the curriculum 2013, which is certainly interesting to be studied and elaborated further [4]. The approach is a learning process designed in such a way that learners are actively thinking, asking, discussing and interacting between students and teachers [5].The steps of the scientific approach, as follows:(a) Observing, presents the real object media, learners happy and challenged and easy to implement. (b) Asking : Effective teachers are able to inspire learners to improve and develop the sphere of attitude, skills, and knowledge. When the teacher asks, at that time he also guides or guides learners to learn. (c) Reasoning, The interaction pattern is done through stimulus and response (S-R). This theory was developed based on the results of Thorndike's experiment, which became known as association theory. (d) trying, To obtain real or authentic learning results, learners should try or experiment, especially for the appropriate material or substance. (e) collaborative learning is a personal philosophy, more than just a learning technique in school. In collaborative learning the authority of the teacher is more directive or the learning manager. In collaborative learning the authority of the teacher is more directive or the learning manager.

One way to apply the whole education is by integrating knowledge of Natural Science (IPA) into learning materials in the form of thematic integration between knowledge. IPA is one branch of science that discusses the phenomenon of natural phenomena systematically. . In general, IPA includes three basic sciences, including: biology, physics and chemistry. These three basic sciences are called integrated learning. Natural science is defined as systematic knowledge and constituted by linking the material nature phenomena and based on observations and inductions [6].

Integrated learning is also divided into a variety of integrated learning models as proposed [7] as many as ten integrated learning models. The ten integrated learning models are: (a) *fragmented*, (b) *connected*, (c) *nested*, (d) *sequenced*, (e) *shared*, (f) *webbed*, (g) *threaded*, (h) *integrated*, (i) *immersed*, and (j) *networked*. Of the 10 kinds of Fogarty models selected are shared models, because the shared model is a teaching-learning approach that combines two or more subjects that see the same concepts, attitudes and skills. Merging between the concepts of lessons, skills and attitudes that are interconnected with one another are shaded in one theme, so as to provide a meaningful experience for students.The characteristics of integrated learning [8] : (a) holistic: an event that became the center of attention in integrated learning is studied from several fields of study as well as to understand a phenomenon from all sides. (b) Meaning: the interrelationship between other concepts will add meaningful concepts learned and hopefully the child is able to apply the acquisition of learning to solve real problems in his life. (c) Active integrated learning is developed through the in-discovery approach. Learners are actively involved in the learning process that can indirectly motivate the child to learn. The integrated IPA learning model is recommended at SMP level, as it has several objectives, namely: improving the efficiency and effectiveness of learning; increase interest and motivation, as well as some basic competencies can be achieved at once.

Competency standards of graduates which states that every learner must have competence in 3 dimension domains, namely: attitude dimension, knowledge dimension and skill dimension [9]. In the realm of attitude dimension, the first most important indicator that becomes the most important is the faithful and devoted attitude to God which is also included from the indicator of spiritual value. In this dimension of attitudes learners are given an understanding of the fundamental aspects of the universe of science that will cultivate an attitude of admiration to God and add confidence to the greatness of God based on the beauty contained in the rules of God's created nature. The linkage between material classification materials and their amendments that are integrated into material with the theme of cohesion and adhesion to living beings can be useful for increasing the value of the character and competence of knowledge to learners as outlined in a book called textbooks.

Textbook is instructional media, education strategy, education quality and reference book obliged to deliver curriculum materials [10]. So that textbook is a mandatory component that must exist in learning. The preparation of textbooks with the 2013 curriculum must be in accordance with predetermined rules or principles. Textbook creation should pay attention to the suitability of the content of the book with SKL, KI and KD, adequacy and depth of material, scientific approach, authentic assessment [11].

Experience shows that the teaching materials used are still not integrated. In line with the above explanation, Textbooks are used to improve the efficiency and effectiveness of the numbers tailored to the needs of learners [13]. Text books are the main operational tool for the implementation of the curriculum [12]. Lesson textbooks can be a major learning resource for achieving core competencies and core competencies and are declared appropriate by the Ministry of Education and Culture for use in educational units.

Based on the literature review, the authors will develop the analysis of learners related to the knowledge of learners. The formulation of the problem in this research is how to develop the learner's knowledge analysis and its implication on the design of textbook of *SMP* science. The purpose of this research is to describe the improvement of the competence of learners' knowledge by using integrated textbook *IPA SMP* charged characters based on shared model.

2. Method

This research is a research of Pre- Experimental Design by using One Group Pre-Test and Post-Test Design. Prior to the treatment, pre-test was first given to determine the students' initial ability. After the learning by applying the textbook is given post-test to determine the ability of the end of the students so it can be known how to improve students' understanding. Subjects in this study were students of class VII N 13 Padang Year of Study 2016 / 2017. Data collection techniques used in this study is to use observation sheet used by observers in observing the implementation of learning plans, written tests used for pre-test and post-test, and questionnaire used to obtain student response data on learning.

This research is a Pre-Experimental Design study using One Group Pre-Test, Post-Test Design research design. Prior to the treatment, pre-test was first given to determine the students' initial ability. After the learning by applying the textbook is given a post-test to determine the ability of the end of the students so it can be known how to improve students' understanding.

Subjects in this study were students of class VII N 13 Padang Year of Study 2016 / 2017. Data collection techniques used in this study is to use observation sheet used by observers in observing the implementation of learning plans, written tests used for pre-test and post-test, and questionnaires used to obtain student response data on learning.

Analysis of pre-research data include problem item analysis, normality test analysis and population homogeneity test analysis. The results of the analysis of the item obtained as shown in Table 1.

Table. 1. Problem Analysis Results Item

No	Level of difficulty	Different	Validity	Conclusion
1	Easy	Medium	Enough	Used
2	Easy	Medium	Enough	Used
3	Easy	Bad	Low	Disposable
4	Average	Medium	Enough	Used

5	Difficultly	Bad	Enough	Corrected
6	Easy	Bad	Enough	Corrected
7	Easy	Medium	Low	Disposable
8	Easy	Medium	Enough	Used
9	Medium	Medium	Low	Disposable
10	Medium	Medium	Enough	Used

Based on Table 1 above, of the 10 tested questions, only 7 questions are used for the pre-test and post-test in which each consists of 5 items. While the other 3 problems are not used because they do not meet the feasibility requirements.

Data analysis of research result includes analysis of learning plan implementation, pre-test and post-test result analysis and student response questionnaire analysis. Analysis of the implementation of the lesson plan used the equation (1).

$$P(\%) = \frac{\textit{Total score of data collection}}{\textit{criteria score}} \times 100\% \quad (1)$$

The criterion of assessment score interpreted as follows:

- 25% - 43.7%: not good
- 43.8% - 62.5%: poor
- 62.6% - 81.25%: good
- 81.26% - 100%: very good

The analysis of pre-test and post-test data was performed using paired t-test as follows:
a. Determining the hypothesis:
Ho: There is no significant increase in student conceptual understanding before and after treatment.
Hi: There is a marked increase in student conceptual understanding before and after treatment
b. Determining the gain value (d) is the difference between the pre-test and post-test values of each student.

d = value of post-test - pre-test value. (2)

c. Determine the mean of the gain by formulation:

$$M_d = \frac{\varepsilon d}{N} \quad (3)$$

Where M_d is mean from gain (d), εd is amount of gain (d) and N is result of gain.

By testing criteria, reject Ho if t-count>t-table. With the level of trust and dk = (N-1). Furthermore, the calculated normalized gain score to calculate how much increase the students 'concept of understanding when t-test is obtained that the visualization of experimental video in the laboratory as a learning medium provides significance to improving students' understanding.

3. Results

To find out the improvement of science understanding of students after being given treatment, the result data of pre-test and post-test that got tested by using paired t-test, then searched how big achievement increase understanding of concept mastered by student by using gain of score normalized.

3.1 Normality Test

Before paired t-test can be used to test data of pre-test and post-test results, then test the normality of research data first. The results of normality test data pre-test and post-test are as follows;

Table 2. Results Calculation of Normality of Pre-Test and Post-Test

Test			Result
Pre-Test	4,43	11,10	Normal
Post-Test	0,48		Normal

3.2. Paired T-Test Results

After the results of the research are normally distributed, the *t*-test is performed with the following results

Table 3. Results of Paired T-Test Calculations

		Explanation
27,93	2,03	Ho is rejected

Based on the result of t-test calculation as in Table 3 above, it is concluded that Ho is rejected with 95% reliability. Thus, it can be said there is a significant increase to the understanding of the concept of students before and after giving treatment in the form learning using integrated IPA textbooks

4. Discussion

The success of the level of achievement of the students' understanding is due to several factors, namely: the use of integrated IPA textbook SMP is characterized as a learning medium. This integrated IPA textbook is tailored to the textbook structure that refers to *Kemendikbud* 2014 starting from the cover (chapter title) of the textbook. The title display of textbook can be seen in Figure 1.

Cover describes the content of the textbook in the form of cohesion and adhesion events that occur in the environment. Water is on the leaves of taro representing cohesion, wound on the fingers representing adhesion and cohesion events that make the paper not wet. The cover design is tailored to the learner's analysis, and learners are expected to be enthusiastic in science learning because the IPA is associated with natural events.

With effective learning media to convey certain concept of IPA, then understanding of the concept can be accepted by students with good and easy. The medium is suitable for junior high school students entering the formal operational stage of cognitive development where students have been able to think abstractly. With the help of textbooks, students are able to understand the concept of a well-carried IPA without having to carry out their own experiments in the laboratory. Learning using the shared model that is implemented maximally, students can obtain information well. The information can then be used by students to assimilate the new concept with the concept that students have so that there is meaningful learning that can improve students' conceptual understanding.

Assessment of learners' knowledge is done by pretesting at the beginning of learning and posttest at the end of each learning session and performs daily test. Pre-test and post-test are done in every meeting to see the learners' knowledge at the meeting. The daily test is done at the end of the material classification chapter. The average of pre-test and post-test values can be seen in Figure 2.

Figure 1. Cover of IPA Textbook.

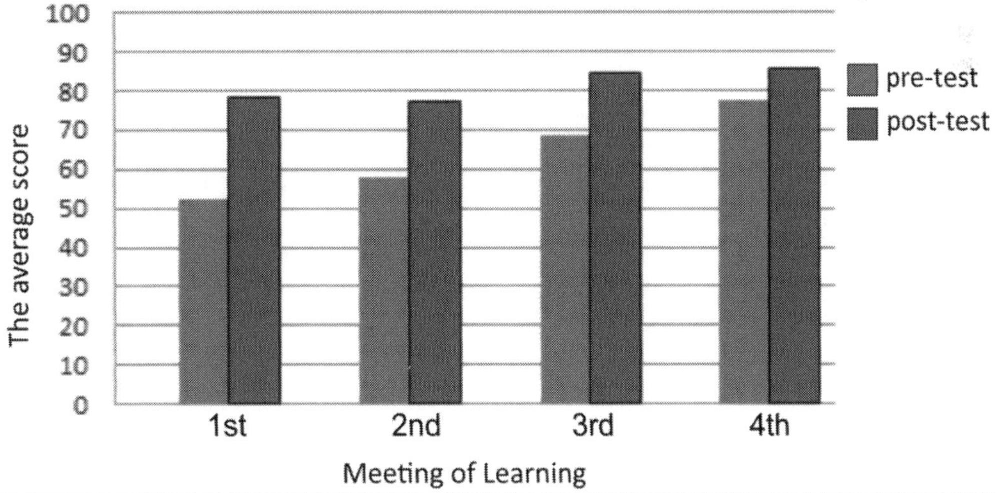

Figure 2. Average pre-test and post-test values of each meeting

Based on Figure 2, students are very interested in the concept of IPA taught by applying an integrated textbook IPA SMP charged characters. It is characterized by excellent ratings on the points of items one, two, three and four. The learning and explanation of teachers get a good response while the teacher guidance in the discussion process is considered very good. Students are enthusiastic if the learning of teaching materials in addition to teaching materials in this research is delivered with the same learning strategy as learning strategy applied in this research.

5. Conclusions

Based on the results of research and discussion that have been put forward in the above section, it can be concluded that the implementation of integrated textbooks SMP lecture-based character of the shared model as a medium of learning to understand the concept of science with better can be done well, increased understanding of science concept students on the subject matter of hydrostatic pressures by utilizing the visualization of experimental video gravity currents as a learning medium is achieved significantly, and the subject of the study responds positively to the learning being carried out.

References

[1] Undang- undang Republik Indonesia No. 20. 2003. Sistem Pendidikan Nasional, Jakarta

[2] Abduh, M. 2014. *Evaluasi Pembelajaran Tematik Dilihat Dari Hasil Belajar Siswa.* Indonesian Journal of Curriculum and Educational, 1 (1), pp. 1-9

[3] Permendikbud Nomor 68 Tahun 2013 tentang Kerangka Dasar dan Struktur Kurikulum Sekolah Menengah Pertama/Madrasah Tsanawiyah.

[4] Kusuma, M. 2015. Penerapan Pembelajaran Terpadu Tipe Nested (Tersarang) Untuk Meningkatkan Literasi Sains Siswa. *JurnalScientiae Educatia Volume 5 No.2 tahun 2015.*

[5] Permendikbud Nomor 65 Tahun 2013 tentang Standar proses Pendidikan dasar danMenengah.

[6] Chang. W. 2002. *Interactive Teaching Approach in Year One University Physics in Taiwan: Implementation and Evaluation.* Taiwan: Implementation and Evaluation (accessed: 27 maret 2016)

[7] Dallinger, S. 2016. *The effect of content and language integrated learning on students' English and history competences. Journal Learning and InstructionVol 41 (2016) 23-31.*

[8] Fogarty, R. (1991). *Ten ways to integrate curiculum.* Palatine Illinois : IRI/SkylightPublishing, Inc. (accessed 27 maret 2016).

[9] Viera, A. J. 2005. "Understanding Interobserver Agreement The Kappa Statistic. Research". *Journal Research Series.* Vol.37 No.5 page 360-363.

[10] Permendikbud Nomor 68 Tahun 2013 tentang Kerangka Dasar dan Struktur Kurikulum Sekolah Menengah Pertama/Madrasah Tsanawiyah. Permendikbud Nomor 65 Tahun 2013 tentang Standarproses Pendidikan

[11] Permendiknas Nomor 11 tahun 2005. Kementrian Pendidikan Nasional. Jakarta.

[12] Permendikbud Nomor 104 Tahun 2014. 2014. Kementrian Pendidikan dan Kebudayaan. Jakarta.

[13] Permendikbud Nomor 8 Tahun 2016 tentang Buku yang digunakan oleh satuan pendidikan. Jakarta.

IOP Conf. Series: Materials Science and Engineering **335** (2018) 012077 doi:10.1088/1757-899X/335/1/012077

The Implementation of Physics Problem Solving Strategy Combined with Concept Map in General Physics Course

H Hidayati* and R Ramli

Department of Physics, Universitas Negeri Padang , Padang 25131, Indonesia

hidayati_unp@yahoo.co.id

Abstract. This paper aims to provide a description of the implementation of Physic Problem Solving strategy combined with concept maps in General Physics learning at Department of Physics, Universitas Negeri Padang. Action research has been conducted in two cycles where each end of the cycle is reflected and improved for the next cycle. Implementation of Physics Problem Solving strategy combined with concept map can increase student activity in solving general physics problem with an average increase of 15% and can improve student learning outcomes from 42,7 in the cycle I become 62,7 in cycle II in general physics at the Universitas Negeri Padang. In the future, the implementation of Physic Problem Solving strategy combined with concept maps will need to be considered in Physics courses.

1. Introduction

Recently, research in physics education has made some major advances by adopting important findings from research on expert-beginner relationships and cognitive science [1]. These findings lead to the ability of students to solve physics problems with high-level thinking and can help students become like an expert in solving physics problems. To achieve this goal, a problem solving strategy in physics is needed to transform the student into an expert in solving physics.

There are two factors that can help a student to have better problem solving physics skills, namely: First, students must know and understand the principles of physics. Secondly, students should have a strategy to apply these physics principles to new situations where physics can help [2]. Thus, the given learning needs to apply a physics problem-solving, learning strategy to help students have better physics problem solving abilities.

The problem solving, learning strategy is part of the inquiry learning strategy. This learning strategy, emphasizes the completion of a problem by reason. Why is the importance of a problem-solving learning strategy? The reason, because in principle learning is a process of interaction between humans with the environment, which takes place gradually, ranging from receiving stimulus from the environment, to give the right response to it.

Problem solving is an individual or group effort to find answers based on knowledge, understanding, skills that have been previously held in order to meet the demands of unusual situations [3]. Meanwhile, Martinez [4] says that problem solving is "... the process of working towards a goal without a script and this is the cognitive passport to the future ...." Thus problem-solving learning is indispensable in developing student skills to solve problems in everyday life.

Research on problem solving strategies has been applied in several disciplines such as physics, medical diagnosis, engineering, design and computer programming. In Physics learning, this strategy has been done with various methods and levels of education. Heller and Hollabaugh [5] develop problem solving strategies in cooperative learning that can improve students' problem solving physics skills. Gonen and Basaran [6] use a problem solving strategy combined with a Course Management System (CMS) in distance physics learning via the internet. It effectively improves students' problem solving skills in physics. Chi and VanLehn [7] found that when high school students are trained to identify and understand the concept of the problem, the gap between good problem solvers and weak problem solvers can be minimized.

In Indonesia, the application of problem solving strategy in physics learning has also been done by several researchers. Asma and Ramli [8] developed a model of learning to improve the ability to solve high school physics problems. Sriyanti [9] apply Heuristic method in solving basic physics problems I so that can increase activity and result of student learning in subject of Basic Physics I.

Problem solving skills is one of the six competency indicators of students who have guaranteed 71.7% satisfaction from stakeholders [10]. The six indicators are: computer skills, application in major, oral comunication, problem solving skills, thinking / reasoning and writing skills.

Based on the experience of teaching the subjects of General Physics and discussions with some lecturers of General Physics subjects, the author observes that students have difficulties in understanding abstract physics concepts, and students tend to understand the physics concepts separately. In addition, there are still many students who only understand the learning materials, but are not able to solve the problems given. As a result, when the examination, the value obtained by students is still low. When given any exercise, it turns out that only a small percentage of students can do it well, most students do not know what to do. In order for students to be able to solve the problems of physics, needed mastery of physics concepts and strategies problem solving physics is good.

Based on detailed interviews with some students and a team of General Physics lecturer, author concludes that the concerns expressed in this General Physics learning, among others:

- Students have difficulty connecting physics concepts.
- Students have less problem solving skills in solving general physics problems.

Consequently, the learning process needs to be carried out with problem solving strategies combined with concept maps, to improve students' ability to solve general physics problems.

At the beginning, researchers have not used the strategy of Minnesota Physics Problem Solving, so the ability of students to solve the problem of General Physics, is still low. Further Researchers take action that is the implementation of Minnesota Physics Problem Solving strategy as much as two cycles. In the first cycle of learning using the strategy of Minnesota Physics Problem Solving, and continued in cycle II using the strategy of Minnesota Physics Problem Solving combined with concept maps so that the expected end-ability of students in solving the problem of General Physics, increased. The diagram of the framework of thinking is shown in Figure 1.

Problem-solving strategies can be divided into major steps and minor steps. Heler and Heler [11] divide the problem solving strategies in physics into five steps: Focus on the Problem, Physics Description, Plan a Solution, Execute the Plan, and Evaluate the Solution.

The five steps of the problem solving strategy above are an effective way to generate problem solutions based on a physics understanding. The quality of the solution depends on the knowledge students use in obtaining solutions. The use of strategy also makes it easier to look back through the solutions that have been made. The above strategy is called the minnesota problem solving strategy. This strategy can be used to organize abilities in order to develop students' problem solving skils, and become one of the important tools in physics learning.

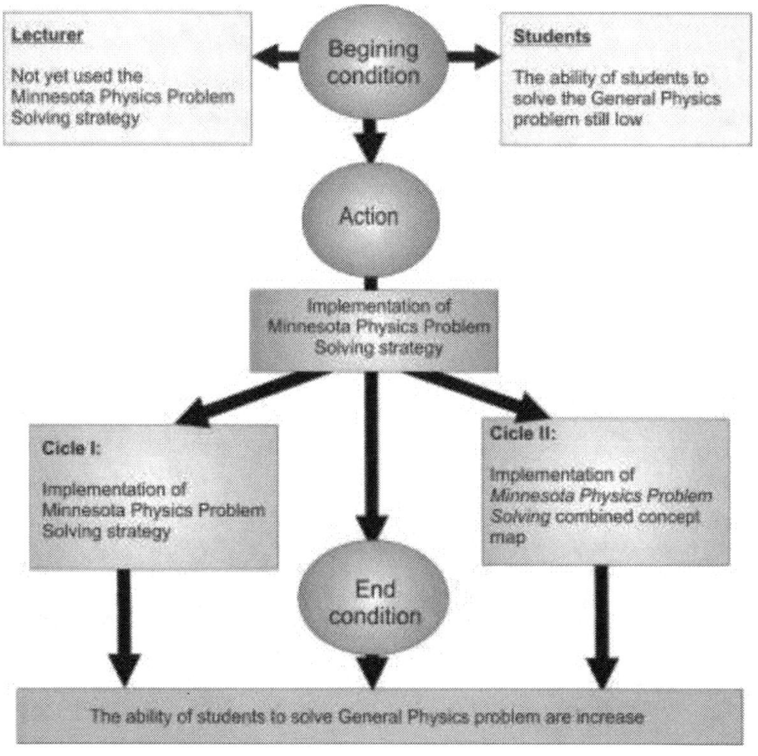

Figure 1. Framework of thinking

2. Research Methods

The type of research is Classroom Action Research. The design of each cycle consists of four stages: action planning, action implementation, observation/evaluation and reflection. The data of learning result is collected through test which its implementation is done at the end of each cycle. Meanwhile, student activeness data was collected by observation sheet. Research subjects were 46 first year students who took the General Physics course.

The steps of research activities and class actions that have been done are:

1. Preparation of Action: Preparation of action is as follows: drafting of Minnesota Physics Problem Solving strategy guidance, drafting concept maps, assessment formats, evaluation tools and questionnaires.
2. Implementation of Action: (1) The researcher prepares the lesson material, (2) The researcher explains the material and gives some examples of how to work the problem using 5 steps in Minnesota Physics Problem Solving strategy: Focus on the Problem, Physics Description, Plan a Solution , Execute the Plan, and Evaluate the Solution. At the end of the cycle / recovery researchers give tests to determine the ability of learning/learning outcomes.
3. Observation and Evaluation: During the learning activity the researcher monitors the learning process on the learning outcomes and understanding the student concept in each cycle. Observation of student learning outcomes is done at the end of each cycle by providing a written test.
4. Reflection: The results of observation / evaluation of the action is done to see if there is a stage that has not been optimal, so it needs re-diagnostic to the implementation of action in the first cycle. These diagnostic results are used to plan and implement actions in the second cycle.

3. Results and Discussion

The student activity when doing General Physics problems using Minnesota Physics Problem Solving strategy in cycle I is shown in Figure 2. It shows that focus the problem activity, has the highest value compared with other activities. This is possible because this activity is related to drawing diagrams and writing down known information from reading questions.

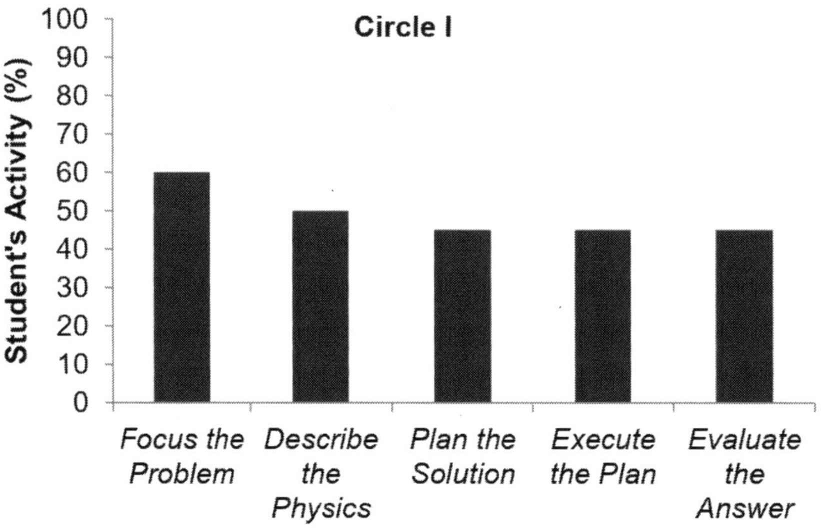

Figure 2. The average activity of students in the application of physics problem solving strategy in cycle I.

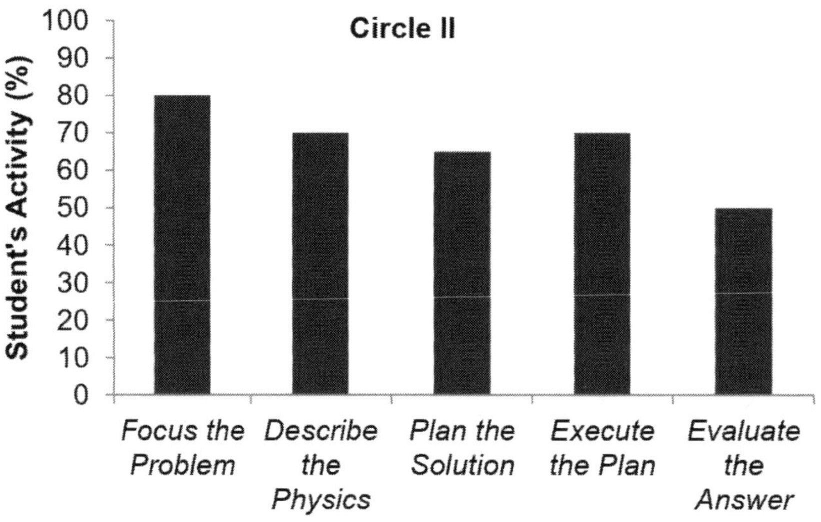

Figure 3. The average activity of students in the application of physics problem solving strategy in cycle II.

After the reflection on the results of cycle I, then in the second cycle is applied the application of physics problem solving strategy combined with concept map. The results show that the average student activity has increased, as in Figure 3. This happens because concept maps add students' understanding in solving general physics problems. In addition, the concept map is a representation of the relationship between one concept with another concept that students have [12, 13]. Comparison of student activity in applying physics problem solving strategy between cycle I and cycle II is shown in Figure 4

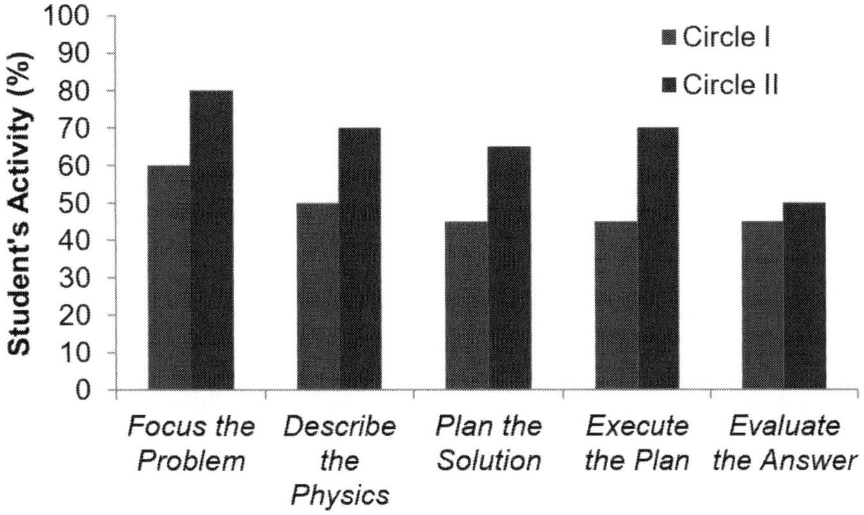

Figure 4. Comparison between student's average activity in applying physics problem solving strategy at every cycle..

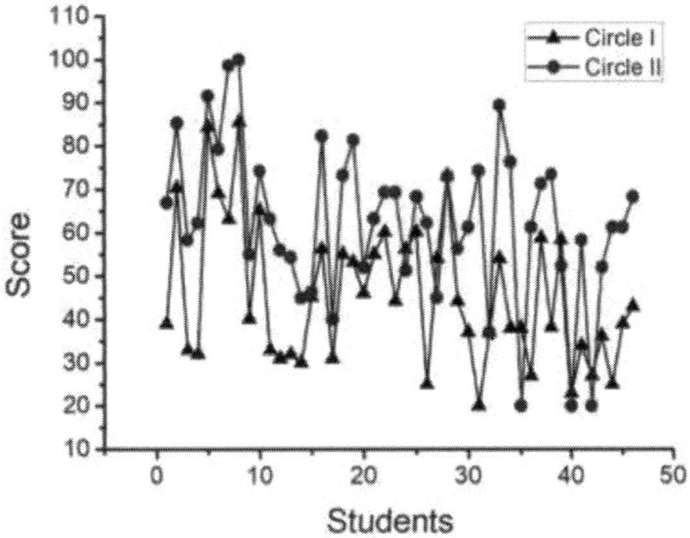

Figure 5. Distribution of student scores on cycle I and cycle II..

Meanwhile, the result of student learning as a result of application of physics problem solving strategy for cycle I and cycle II is shown in Figure 5. It can be seen that the average of student learning outcomes has increased from 42.7 in cycle I to 62,7 in cycle II. The improvement of this learning result happened because in cycle II, the students have better comprehended the strategy of solving the problems of General Physics, as can be seen from their activity in the application of physics problem solving strategy in cycle II. This results are consistent with Gok & Silai's research results [2], that learning with problem-solving strategies influences student achievement, student attitudes toward problem solving and student motivation positively.

4. Conclusion

Based on the results obtained from the actions taken during the two cycles, it can be concluded:

1. Implementation of Physics Problem Solving strategy combined with concept map can improve students' activity to solve the problem in General Physics course with an average increase of 15%.
2. Implementation of Physics Problem Solving strategy that combined concept map can improve student learning outcomes in General Physics course from 42,7 in cycle I to 62,7 in cycle II.

References

[1] Teodorescu, R.E., Bennhold, C., Feldman, G., and Medsker, L. 2013. New approach to analyzing physics problems: A Taxonomy of Introductory Physics Problems, *Physics Review Special Topics-Physics Education Reserach*, Vol. 9 No.1, 2013, 010103.1–20.
[2] Gok, T & Silay, I., 2010. The effect of Problem Solving Strategies on Student's Achievement, Attitude and Motivation, *Latin-American Journal of Physics Education*, Vol 4 No 1, 7-21.
[3] Krulik, S., & Rudnick, J. A. (1996). *The new sourcebook for teacing reasoning and problemsol ving in Junior and Senior High School*. Boston: Allyn and Bacon.
[4] Martinez, M, 1998. What is problem solving? *Phi Delta Kappan*, Vol. 79 No.8, 605-610.
[5] Heller, P dan Hollabaugh. M. 1992. Teaching Problem Solving Through Cooperative Grouping. Part 2: Designing Problems and Structuring Groups, *American Journal of Physics*, Vol. 60, No. 7, 637-644.
[6] Gonen, S., dan Basaran, B. 2008. The New Method of Problem Solving in Physics Education by Using Scorm-Compliant Content Package. *Turkish Online Journal of Distance Education* Vol. 9 Issue. 3, 112-120.
[7] Chi, M., dan VanLehn, K. 2010. Meta-Cognitive Strategy Instruction in Intelligent Tutoring Systems: How, When, and Why. *Educational Technology & Society*, Vol 13 No 1, 25–39.
[8] Asma, N dan Ramli, (2007): *Model Pembelajaran untuk Meningkatkan Kemampuan Memecahkan Masalah Fisika SMA* [Learning Model to Improve Ability to Solve High School Physics Problems], Research Report of *Hibah Bersaing XV/I*, Universitas Negeri Padang.
[9] Sriyanti, I. 2009. *Penerapan Metode Heuristik dalam Penyelesaian Soal-Soal pada Matakuliah Fisika Dasar* [Application of Heuristic Methods in Solving Problems on Basic Physics Course], *Forum Kependidikan*, Vol. 28, No.2, 114-118.
[10] Djauhari, M.A. (2001): Arah Pengembangan Kurikulum Bidang MIPA [Direction of Curriculum Development in Mathematics and Sciences], MIPANET seminar, FMIPA UGM, Yogyakarta, 30 August - 1 September 2001.
[11] Heller. P dan Heller. K. (2010): *Cooperative Problem Solving in Physics:A User's Manual, Minnesota*: University of Minnesota Press.
[12] Asan, A. 2007. Concept mapping in Science Class: A Study of fifth grade students. *Jurnal Educational Technology & Society*, Vol 10 No 1, 186-195.
[13] Vanides, J. 2005. Using Concept Maps in the Science Classroom. *Jurnal National Science Teacher Association*, Vol 28 No 8, 27-31.

Smart Aquarium as Physics Learning Media for Renewable Energy

D Desnita*[1], R Raihanati[1] and D Susanti[1]

[1]Physics Education Study Program, Faculty of Mathmatic and Natural Sciences, Universitas Negeri Jakarta, Jl. Rawamangun Muka Jakarta Timur, Indonesia

*desnita@unj.ac.id

Abstract. Smart aquarium has been developed as a learning media to visualize Micro Hydro Power Generator (MHPG). Its used aquarium water circulation system and Wind Power Generation (WPG) which generated through a wheel as a source. Its also used to teach about energy changes, circular motion and wheel connection, electromagnetic impact, and AC power circuit. The output power and system efficiency was adjusted through the adjustment of water level and wind speed. Specific targets in this research are: to achieved: (i) develop green aquarium technology that's suitable to used as a medium of physics learning, (ii) improving quality of process and learning result at a senior high school student. Research method used development research by Borg and Gall, which includes preliminary studies, design, product development, expert validation, and product feasibility test, and vinalisation. The validation test by the expert states that props feasible to use. Limited trials conducted prove that this tool can improve students science process skills.

1. Introduction

The need for renewable energy, start out in the era of fossil energy crisis in the 1970s. In this millennium, the energy crisis will be more alarming as everything need run by machines or automated which is increased needs of energy. Based on a study by Wood Mackenzie, changes in energy consumption patterns will occur in a few years to come. Its state that the demand for natural gas and low-carbon fuels will rise by 60% in 2035. It is mean, the need for coal and oil will touch its peak before 2035. The highest growth occurs in the need of new renewable energy that will rise fivefold in next 20 year. With declining oil demand, new renewable energy projects are rising in almost all countries. New renewable energy needs are expected to rise by 500% over the next 20 years [1].

Indonesia has abundant renewable energy potential that is feasible to be developed to meet domestic energy needs. Sources of renewable energy that need to be explored include plants containing carbohydrates, such as high organic compounds (corn oil [2] and palm oil [3]), or biomass (micro algae) [4], sugar cane waste [5], and castor seed [6]). Utilization of natural energy in Indonesia such as wind, solar, and water, each of it has been utilized to build wind power plants in *Bantul, Jogyakarta*; *Karangasem Bali* solar power plant; and hydroelectric power plant in *Singkarak* [7].

Research on the sources and benefits of renewable energy in the laboratory scale is also increasing. Seung et al [8], utilizes wind power as a source of electricity for water treatment by coagulation techniques. The electrical energy that flows into the electrocoagulation cell, oxidizes the anode to the hydroxide salt which will coagulate the colloid particles that can be found in the waste water. Utilization of solar energy as a source of electrical energy or better known as solar cells has also been used as a source of electrical energy for room heater [9]. Bilal [10] designed a micro hydro power plant with the help of the matlab simulink program with turbine type, flow rate, and flow height parameters. The results show that cross flow turbine with head 5 m and 1.0 m3 / second flow rate produces better electrical energy compared with using Pelton or Caplan turbine.

Micro hydro power can also be used as a source of electrical energy for the aquarium. The results of the preliminary study showed that the use of micro hydro power produces only about 20% of the electrical energy needed. Therefore, to meet the needs of electrical energy, other electrical energy sources are needed.

Smart Aquarium is the application of the concept about renewable energy in the aquarium, in the form of a combination of engineering MHPG and Wind Power Generation. It can be used as a model of renewable energy which can be applied to the fulfillment of energy needs on a small scale. Installation of energy in smart aquarium, has some connection with physics concepts, such as renewable energy (MHPG and WPG), circular motion and wheel connection, energy changes, electromagnetic impacts, and AC power circuit. Replacing the wind source with a fan, can change the smart aquarium from just a renewable energy application model into a physics learning medium for some material. To see the feasibility and effectiveness of the smart aquarium as a medium of learning, it is necessary to used expert validation and use trials in physics learning.

2. Literature Review

2.1. Instructional Media

In the education and learning field, renewable energy has started giving from high school to college level, for diploma programs, undergraduate programs as well as for master's and doctoral programs. Tara and Lars [11], reported syllabus and curriculum at several universities that run master's and doctoral programs for renewable energy. Some universities are also building renewable energy research centers such as the Institute of Technology Delhi, the Solar Energy Research Center (SERC) and the European Solar Engineering School (ESES) at Dalarna University, Borlange (Sweden).

Desnita [12] reported a syllabus for Biodiesel material applied to a physics education program at the State University of Jakarta. In another study, Desnita [13] used project based learning method on physics learning with renewable energy material. The results showed that problem-based learning can improve the skills and motivation of physics students of UNJ.

Anas et al. [14] conducted a survey of Jordan's high school students' perceptions of renewable energy. The results of his study of 617 students, indicating that students can not distinguish sources of renewable energy sources and non-renewable energy. 50% of students are not familiar with the terms biofuel, biodiesel and bioethanol. While 87% of students believe that renewable energy is a sustainable energy. In contrast, students in Germany show they prefer renewable energy to everyday activities, such as preferring to ride a bike to travel and always turn off the light when not in use [15]. Its also found for students in Finland who are very concerned about bioenergy [16]. In contrast, UK students' concern for biomass as a source of renewable energy is very low [17].

Robin and Nicholas [18] have compiled a teaching manual to introduce the concept of renewable energy to junior and senior high school students through experiments covering the manufacture of solar cells, biodiesel and assembling micro hydro. In addition to syllabus and curriculum, instructional media is also needed in teaching and learning process. Learning media is a tool used to convey the subject matter in the learning process and also to facilitate students to understand the material teaching. Furthermore, the function of learning media (i) to clarify information or learning messages, (ii) to facilitate the students to better understand certain concepts, principles and skills by using the most appropriate media in accordance with teaching materials, (iii). different and varied learning experiences that further stimulate students' interest and motivation to learn, (iv) cultivate certain attitudes and skills in that students use or operate the media directly.

According to Kemp and Dayton in Susilana [19] the use of instructional media in teaching and learning process makes learning more interesting, learning goes two ways, more quality learning, learning time becomes more effective and efficient.

2.2. Micro Hidro Power Generator (MHPG)

MHPG is a power generation system that can convert water potentials with certain heights and discharges into electricity, using water turbines and generators. MHPG works by utilizing the height difference and the amount of water discharge per second available in irrigation channel, river or waterfalls. This water flow will rotate the turbine shaft to produce mechanical energy. This energy then drives generators and generate electricity. A micro hydro scheme requires two things: water discharge and head height to generate power that can be utilized. This is an energy conversion system of altitude and flow (potential energy) energy into the form of mechanical energy and potential electric energy of water that can be converted into electrical energy The higher the water fall the greater the potential energy of water that can be converted into electrical energy (Serway and Jewett 2004).

In the smart aquarium is not only utilized potent water energy, but its also used kinetic energy or water pressure. Because water is driven using a water pump commonly used in the aquarium itself.

The most important part of MHPG is turbine. The principle of turbine work is to change the potential energy of the water, into the kinetic energy of the water that turns the mill, then the mill turns the turbine, using the gear, the turbine spins the generator so that the rotation of the magnetic coil occurs and produces an electromotive force between the two ends of the coil. When connected with the resistor will produce an electric current. GGL generated AC or DC, determined by the type of dynamo used. In this research used DC dynamo.

2.3. Wind Power Generator (WPG)

Wind power is a power plant that uses wind as a source of electrical energy, by turning wind power into electricity with the help of windmills or wind turbines. The wind will rotate the propeller on the turbine or windmill. As a result of the rotation of the propeller then, the rotor of the generator will rotate, causing flux changes in the stator cross section. This flux change produces an electromotive force. The electromotive force produced which can be AC or DC, depending on the type of dynamo used. In this research used AC dynamo.

3. Research Methods

The research objectives achieved by applying research and development method, Borg and Gall model, with required analysis, design, tool development and expert validation, testing, and evaluation stage. This research was conducted at Media Development Laboratory of Physics Education Study Program, Faculty of Mathmatic and Natural Sciences, Universitas Negeri Jakarta.

Required analysis is conducted by used a survey of the existence of learning media renewable energy and other materials that have been mentioned in school. The expert validation using a questionnaire contains four aspects, such as security and practicality, usefulness, functionality of media, and competency achievement. Four aspects are translated into 25 items of instrument. Also validated by high school physics teachers using a questionnaire containing six aspects including forty-seven instrument items. The category of media assessment is considered feasible, if the average score of all experts and teachers ≥ 80.

The effectiveness of the smart aquarium as a learning medium is seen from the skills of the science process, which is assessed using the observation format and the practicum report written by the students. Instrument consists of four aspects described into 10 points assessment. The role of smart aquarium in learning is said to be effective if the value of science skills of students ≥ 70 good category.

IOP Conf. Series: Materials Science and Engineering **335** (2018) 012078 doi:10.1088/1757-899X/335/1/012078

4. Results and Discussion
4.1 Results
4.1.1. Smart aquarium display

After several construction changes related to materials, size, layout, and installation system as a whole, final result for Smart aquarium was shown in the Figure 1 to 4.

Figure 1. Installation of smart aquarium

Figure 2. WPG

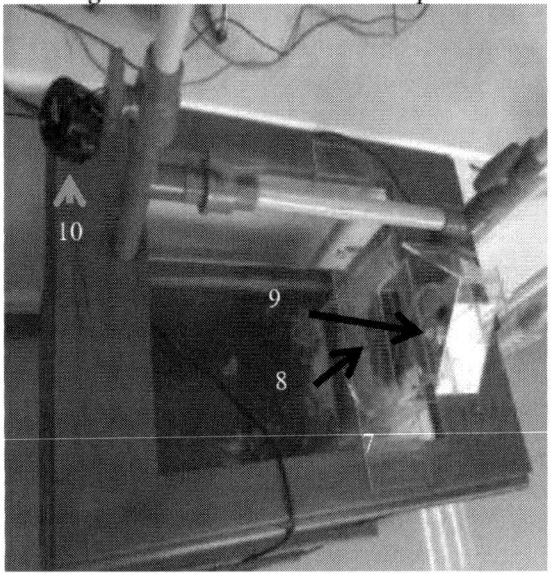

Figure 3. Instalation of MHPG

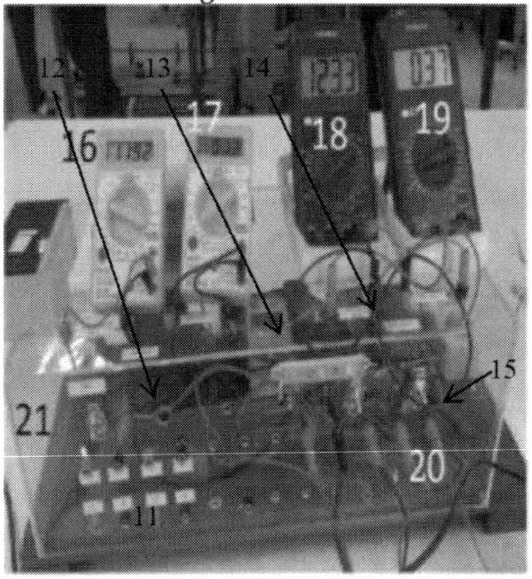

Figure 4. Electric circuit

Information:

(1) windmill arm (2) fan blades, (3) dynamo AC, (4) wind turbine stand, (5) out put cable, (6) wind turbine shaft, (7) home turbine, (8) windmill, (9) gears, (10) counter, (11) power outlet (12) batteries, (13) inverters, (14) controllers, (15) switches, (16) input voltmeter, (17) input amperage, (18) output voltmeter, and (19) output amperemeter.

4.1.2. Smart Aquarium Performance Test Results

Smart aquarium performance test aims to measure the output of electric power generated by the installation of smart energy aquarium. One experimental performance test tool for fifty measurements, presented in Table 1.

Table 1 Smart Aquarium output power

P (watt)								
WPG	MHPG	TOTAL	WPG	MHPG	TOTAL	WPG	MHPG	TOTAL
3.042	7.08	10.12	5.72	7.36	13.01	4.74	7.70	12.45
3.225	7.25	10.48	5.66	7.53	13.16	4.90	37.1	12.03
3.444	8.19	11.63	5.98	7.88	13.85	4.93	7.18	12.11
3.476	9.052	12.53	4.28	7.43	11.71	4.85	7.42	12.27
3.76	8.064	11.82	4.0	8.0	12.0	4.58	7.31	11.89
4.508	6.82	11.33	4.22	7.44	11.65	5.15	7.81	12.96
4.182	7.93	12.11	3.78	6.89	10.67	4.75	7.17	11.92
4.2	8.024	12.22	4.0	7.25	11.25	4.66	7.25	11.9
4.508	8.004	12.52	4.66	7.08	11.73	6.27	7.69	13.96
4.116	7.125	11.24	4.51	7.63	12.14	6.34	7.41	13.75
4.324	7.686	12.01	5.05	7.94	12.99	5.61	7.89	13.50
5368	7.296	12.66	5.24	74.34	12.67	4.8	7.66	12.46
4.223	6.669	10.89	5.39	7.14	12.53	5.80	7.08	12.88
5.6.5	7.192	12.84	5.26	7.38	12.65	5.19	6.9	12.09
4.14	7.192	11.33	5.2	7.75	12.95	5.2	6.96	12.16
7.375	6.832	14.21	5.38	7.95	13.31	6.08	7.19	13.80
5.985	7.788	13.77	4.87	7.56	12.44	5.72	7.32	13.04

4.1.3. Results of Correct Validation

The average score by five validators ranged from 82 to 96. The lowest score was given to item 22 on the ability of the smart aquarium to create enjoyable learning. The highest score is given for numbers 3, 4, 5, 10, 12, 17, 18, 20, 21, and 25. The mean scores for each grain awarded by expert are presented in Figure 5. Recapitulation of expert to four aspects of assessment (1) safety and practicality, (2) usefulness, (3) functionality of the media, and (4) competency attainment tested, presented in Figure 6.

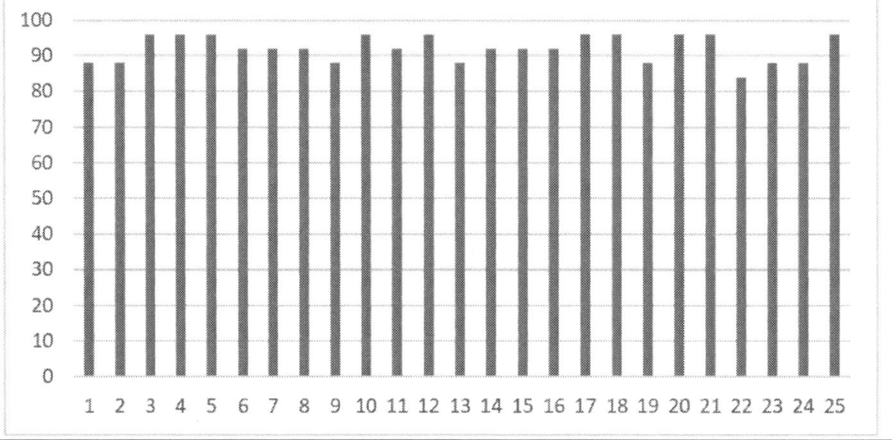

Figure 5. Results of expert validation

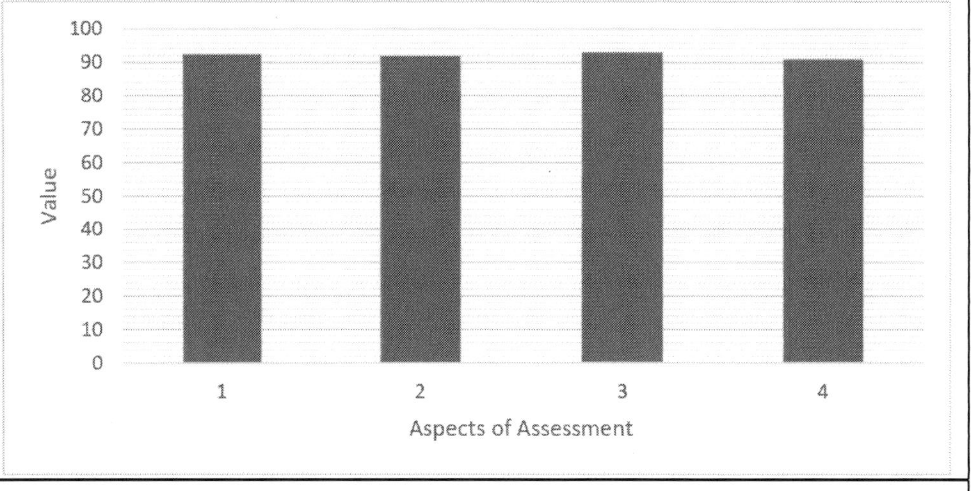

Figure 6. Expert validation

4.1.4. Teacher's Perception about the Smart Aquarium

Before students use the smart aquarium as a learning medium, the tool is assessed by the teacher. Assessment by teacher includes (1) objectives, (2) content, (3) language, (4) design, (5) conformity, (6) practice, (7) effectiveness, (8) science process skills, and (9) student worksheet. The mean values given by the teacher for each assessment aspect ranged from 75 to 88, with an average of 82.4, the distribution of scores for each aspect of the assessment by the teacher is presented in Figure 7.

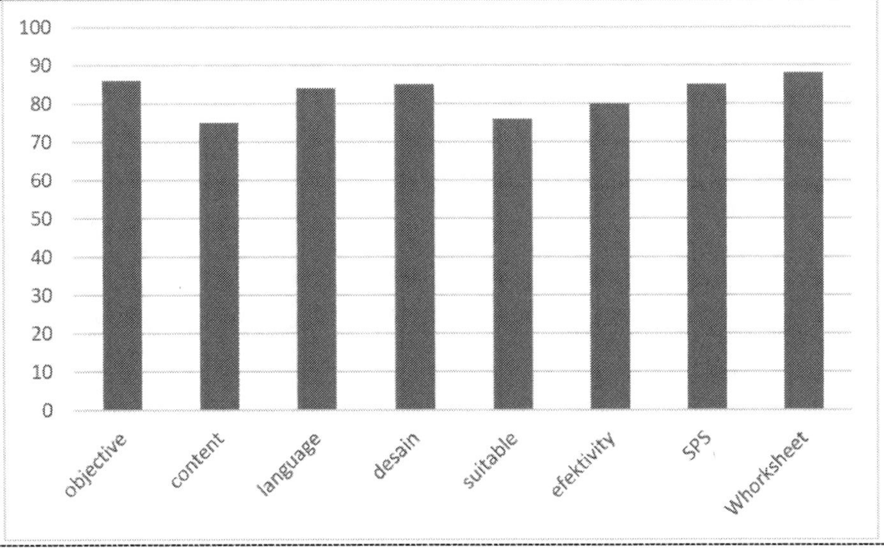

Figure 7. Teacher perception about the Smart Aquarium

4.1.5. Science Process Skills of the Student

Trial of smart aquarium use by students aims to see the effectiveness of the tool as a medium of physics learning. The effectiveness of practicum tools, among others, can be seen from the value of students' science process skills. In this research, the students science process skill consists of five aspects: planning the experiment, conducting the experiment, processing the experimental data, reporting the experimental results, and presenting, which consist of 10 skills such as (1) selecting tools and materials, (2) checking the condition of the appliance prior to use, (3) performing measuring tools, (4) assembling tools, (5) observing and measuring, (6) recording measurement results, (7) variable setting. (8) data processing, (9) drawing conclusions, and (10) presenting.

The scores of students' science process scores ranged from 73-100 with an average of 81.7. The lowest value to present and the highest value equalize the unit. The mean value distribution of the 60 students sampled is shown in Figure 8. Recapitulation of the value of science process skills to the five assessed aspects is presented in Figure 9.

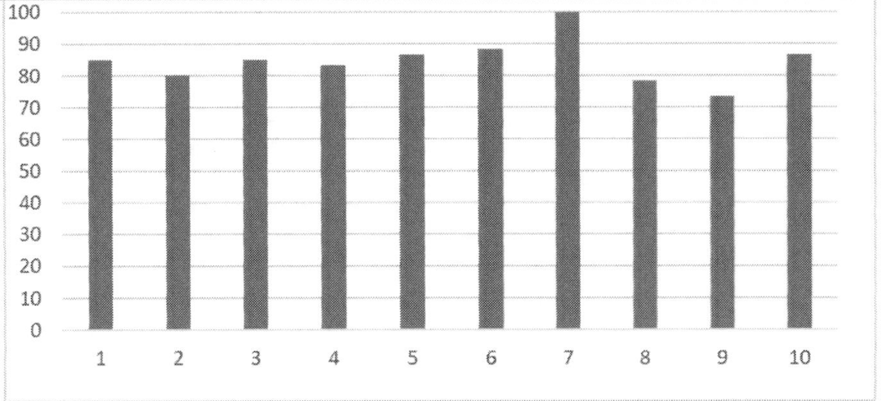

Figure 8. Score of the science process skills of the students

IOP Conf. Series: Materials Science and Engineering **335** (2018) 012078 doi:10.1088/1757-899X/335/1/012078

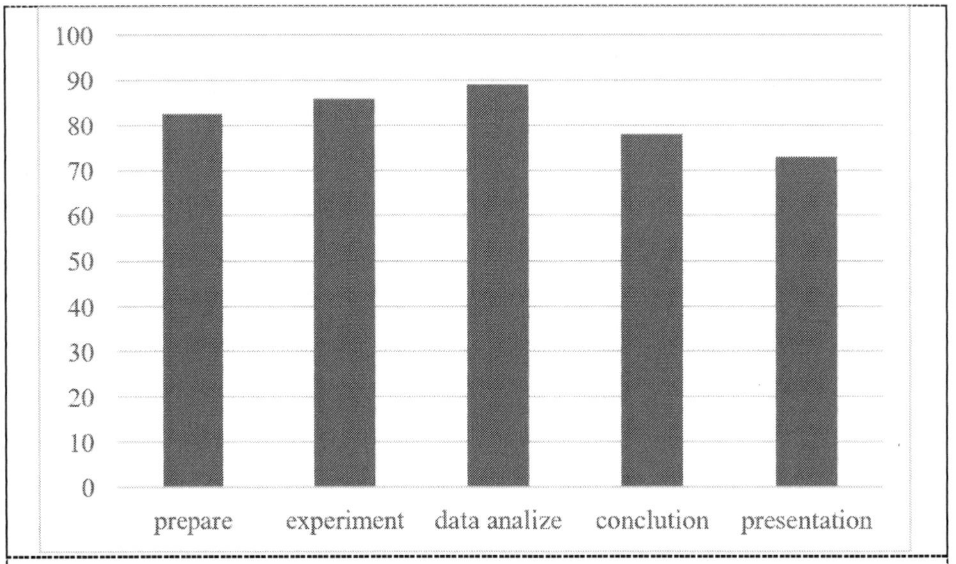

Figure 9. Score of the science process skills of the students

4.2 Discussion

Smart aquarium as a learning media is a real example of renewable energy that is utilized for aquarium pump operations. All components of energy installations are demonstrated. So that the stages of energy generation process and its utilization observed. Physics quantities associated with each stage and components are also measurable, because smart aquarium equipped with measuring devices as needed. The energy production from aquarium water circulation can be seen; as shown in figure 1 - 4. This tool gives users the opportunity to using different physical quantities, such as wind speed, number of blades, and arm length at wind power. High water fall level at a micro hydro power plant can also be varied. The energy generated by both energy sources can be used to operate the pump, turn on the fan, or for the practical purposes of the AC circuit.

The complete installation of energy from smart aquarium and various variations of physical quantities that can be arranged by students, enabling this tool can be used as a medium of learning for some subject matter physics. Practicum that can be done using this tool is renewable energy, especially wind power and micro hydro power plant, circular motion and wheel connection, electromagnetic impact, energy change, discharge, and AC power circuit.

The real application of renewable energy and multifunction are the two advantages of this tool as a learning medium. Because currently in school rarely found a renewable energy learning media that provides opportunities for students to change the amount of physics at the time of experiment. Nor are there any sources of renewable energy that can be used to study other physics in addition to renewable energy. These advantages as well as the attraction of these tools as a medium of physics learning.

Output power ranging from 10.12 to 13.04 is equivalent to 6 - 8 dry batteries commonly used in physics laboratories. Even more interesting is the voltage generated by this AC tool. The small and varied AC voltage is difficult to find in laboratorium. To meet the needs of the experiment is usually used inverter. For a micro hydro and wind power generation the micro power of this instrument is relatively large. Experiments show that the power can be used to operate the fan and small electrical appliances.

The advantages of this tool attract expert and convince them that this tool will be effective for the achievement of a number of physics competencies. So it makes sense if five colleagues as experts in physics and physics learning to give a very good appreciation of the smart aquarium as a medium of learning. Packaging tools in such a way, have considered the security for the user and also the practicality. When the tool is not used as a learning medium, it can be used as an aquarium as usual.

The multi function properties of the tool is a high value of usefulness for the tool. So it is only natural that colleagues appreciate very well for the aspect of usefulness. Many variations that might be done by the students, allowing the students to explore their own physical knowledge through the tool. Surely this will make it easier for them to achieve competence. Being directly involved in building knowledge. Teachers' duties become lighter and certainly facilitate the delivery of subject matter. Because many students work independently. The role of the teacher is only as a motivator and facilitator. Of course this is the factor that encourages expert to appreciate very well. The same is true for aspects of competency achievement. Stable power output, diverse diversity, and immediate experience that occurs with students during learning, make them very appreciative for this aspect.

Assessment with a very good category provided by expert showed that smart aquarium as a medium of learning is scientifically appropriate. Both as seen from a physics and pedagogic point of view. When this tool is used in learning, it is expected to have a positive impact on the quality of process and learning outcomes. This tool is eligible and suitable for use as a medium of physics learning.

Physics teachers are the ones that are directly related to physics learning. Teachers better understand the use of media in learning. By observing the design and display of media as well as testing the learning media, the teacher is able to predict how much the benefits of the media for the learning of physics. Though expert from higher education and learning media demonstration centers in PPIPTEK assess this media very well, based on their expertise. The teacher's opinion about the media is needed, because their experience is dealing directly with the students in the class. So before the tool used by students, first needed to validated by the teacher. Teacher's assessment of smart aquarium as a medium of learning, covering aspects of media content and content is in good category. While the other six aspects of goal, language, design, effectiveness, science process skills, and student worksheets are considered very good. The average score that the teacher gives to all aspects of the assessment is in very good category. Assessment categories provided by teachers describe, that smart aquarium worthy to be used as a medium of learning.

The main purpose of development and use of smart aquarium as a learning media is to facilitate students to achieve competence. One of the competencies that can be achieved through the use of this tool is the skill of the students' science process. The results of the average science process skills of sixty students for the five aspects considered to be in the category of good and very good. Good category for aspects of concluding and presenting. As for the aspects of preparing, implement, and analyze the data is in very good category. The average score of the aspects of the assessment of science process skills is 81.7 very good categories.

The value category of science process skill is very good showing that this tool has good ability to train the skill of the science process. Because one tool provides the opportunity to practice on a number of practical activities, it is positive to the built up their science process.

The category of expert assessment, teachers, and test results of students was in line, because it is in the same category; good and very good. This data shows that smart aquarium is suitable to be used as a medium of physics learning. The tool can be used to train students' science process skills. In other words, smart aquarium can improve the quality of the process and the results of physics learning.

5. Conclusions

Has been successfully developed smart aquarium as physics learning media for renewable energy. The smart aquarium feasible to use for instructional media for renewable energy. The smart aquarium has been able to improve student's science process skills.

References

[1] Bisnis .com Jakarta 28 November 2016
[2] El Boulifi.N., A. Bouaid, M. Martinez, and J. Aracil. 2010, Process Optimization for Biodiesel Production from Corn Oil and Its Oxidative Stability. *International Journal of Chemical Engineering*.: 1-9

[3] Mekhilef .S, S. Siga, R. Saidur .2011. A review on palm oil biodiesel as a source of renewable fuel. *Renewable and Sustainable Energy Reviews* 15:1937-1949
[4] Rawat, I., R. Ranjith Kumar, T. Mutanda, F. Bux.2013. Biodiesel from microalgae: A critical evaluation from laboratory to large scale production, *Applied Energy* 103 :444-467
[5] Magno José Alves, Ícaro Vinícius Cavalcanti, Miriam Maria de Resende, Vicelma Luiz Cardoso, Miria Hespanhol Reis.2016. Biodiesel dry purification with sugarcane bagasse. *Industrial Crops and Products* 89 : 119-127
[6] Amalia.I. Kartika, M. Yani, D. Ariono, Ph. Evon , L. Rigal.2013. Biodiesel production from jatropha seeds: Solvent extraction and in situ transesterification in a single step. *Fuel* 106 : 111-117
[7] http://ebtke.esdm.go.id/post/2016/08/21/1322/energi.terbarukan
[8] Seung-Bae Jeon, Sunmin Kim, Sang-Jae Park, Myeong-Lok Seol, Daewon Kim, Yong Keun Chang, Yang-Kyu Choi.2014 . Self-powered electro-coagulation system driven by a wind energy harvesting triboelectric nanogenerator for decentralized water treatment. *Nano Energy* 28 : 288–295
[9] Sayigh.A.A.M. 1979. Solar House Heating with Heat Pipe Collectors. Solar application in Building.12 :279-285
[10] Bilal Abdullah Nasir .2014.Design consideration of Micro hydro electric power plant. *Energy Proceedia,* 50 :19-29
[11] Tara C. Kandpal 1 , Lars Broman .2014 Renewable energy education: A global status review. *Renewable and Sustainable Energy Reviews* 34: 300-324
[12] Desnita. Program education on Biodiesel at Dept of Physic State University of Jakarta. *International Seminar of Renewable Energy*, 12-14 September 2015, Malang, Indonesia
[13] Desnita. 2017. Project-Based Learning Strategy in The Renewable Energy Education at State University of Jakarta. *International Journal of Multidisciplinary and Current Research* vol 5 : 663-668
[14] Anas Zyadin , Antero Puhakka, , Pirkkoliisa Ahponen , Tarja Cronberg , Paavo Pelkonen.2012. School students' knowledge, perceptions, and attitudes toward renewable energy in Jordan . *Renewable Energy* 45: 78-85
[15] Gossling A, Kunkel T, Schumacher K, Heck N, Birkemeyer J, Froese J, et al.2005. A target group-specific approach to "green" power retailing: students as consumers of renewable energy. *Renew Sust Energ Rev* 2;9:69–83.
[16] Devine-Wright P. *Reconsidering public attitudes and public acceptance of renewable energy technologies: a critical review.* A working paper of the research project "Beyond Nimbyism: a multidisciplinary investigation of public engagement with renewable energy technologies" funded by the ESRC under the 'Towards a Sustainable Energy Economy' Programme.Working paper 01.04.07.
[17] Halder P, Pietarinen J, Havu-Nuutinen S, Pelkonen P. 2010.Young citizens' knowledge and perceptions of bioenergy and future policy implications. *Energ Policy* 38:3058–3066.
[18] Robin Sherman and Nicholas Brooke.1994. *A Guide to Teaching Renewable Energy in Junior and Senior High School Classrooms*. Union of Concerned Scientists.

IOP Conf. Series: Materials Science and Engineering **335** (2018) 012079 doi:10.1088/1757-899X/335/1/012079

Validity of Basic Electronic 1 Module Integrated Character Value Based on Conceptual Change Teaching Model to Increase Students Physics Competency in STKIP PGRI West Sumatera

A Hidayati*[1], A Rahmi[1], Yohandri[2] and Ratnawulan[2]

[1]Department of Physics Education, STKIP PGRI Sumatera Barat, Padang 25137, Indonesia

[2]Department of Physics, Faculty of Mathematics and Science, Universitas Negeri Padang, Jl. Prof. Hamka, Padang, Sumatera Barat, Indonesia 25131

*auliyahidayati030888@gmail.com

Abstract. The importance of teaching materials in accordance with the characteristics of students became the main reason for the development of basic electronics I module integrated character values based on conceptual change teaching model. The module development in this research follows the development procedure of Plomp which includes preliminary research, prototyping phase and assessment phase. In the first year of this research, the module is validated. Content validity is seen from the conformity of the module with the development theory in accordance with the demands of learning model characteristics. The validity of the construct is seen from the linkage and consistency of each module component developed with the characteristic of the integrated learning model of character values obtained through validator assessment. The average validation value assessed by the validator belongs to a very valid category. Based on the validator assessment then revised the basic electronics I module integrated character values based on conceptual change teaching model.

1. Introduction

Education is the process of socialization towards intellectual, social, moral maturity in accordance with human dignity. In other words that education is a process of transfer and refinement that will involve and include various aspects in order to achieve the expected. Education goals plays a very important role in the survival of a nation.Good education is education that can increase the human resource character to face global competition.A nation will be faced with various developments that occur in all aspects of life both in the economic, social, cultural, science and technology (Science and Technology.One of science supporting technological development is Physics..

Physics is studied from primary and secondary education to universities.One of the compulsory courses of Physics is Basic Electronics. There are several concepts discussed in Basic Electronics, including electrical components, circuits, diodes, capacitors, transistors, operational amplifiers, and signal processing. Basic Electronics Lecture at STKIP PGRI Sumatera Barat is divided into two stages, namely Basic Electronics 1 and Basic Electronics 2

Basic Electronics 1 is the Scientific and Scientific Study Group in the Physics Education Study Program STKIP PGRI Sumatera Barat with3 credits of course. The basic materials studied in Basic Electronics 1 are the concept of electronics components, basic electrical law, voltage and current dividing circuit and equivalent circuit, charging and discharging capacitor, passive signal processing circuit, RLC circuit, and transistor as amplifier.

It is expected that by studying Basic Electronics 1 students are able to develop scientific thinking ability and apply it in various aspects of life. In the face of global challenges like today, students are required to apply theories obtained in real life, for example the manufacture of automatic alarms, light detectors, flip-flops, and many other simple tools. In addition to apply theory, the most important thing students must have is to the values of the characters so that the theory obtained can be used properly and not abused.

Character education is not a new thing but has existed since the beginning of independence. This is felt because number of graduates who are smart and skilled in solving problems and intelligent brain, but do not have a strong mentality. This is because character education requires habits to be trained to become a habit. Character education is concerned with values and norms that must be developed and practiced in everyday life.

Based on the observations that researchers do, found the lack of learning resources. The textbook used in the Basic Electronic 1 course is "Electronic Principle 3rd edition" by Albert Paul Malvino, Ph.D. the language converted by prof. Barmawi. This book is difficult to understand by students because the language of the book is still less standard, not communicative and less in line with the characteristics of students. The companion book circulating in the market is more suitable for Electrical Engineering students, so that students are less motivated to learn independently. Another cause is that the book used is more emphasized on the mission of delivering the material, but not in material understanding. As a result, lecturers have difficulties in completing the planned chapter material, as more time is spent on explaining material and practice questions. In addition, there is no Basic Electronics 1 textbook which integrates character values.

To overcome the problems, a media that can familiarize the students to learn continuously and directed in the form of a module is requiredThe module is a self-contained, self-contained unit consisting of a series of learning activities designed to help students achieve a number of clearly defined and specific objectives (Nasution: 2008). With the integrated module character values are expected to motivate students to learn as well as improve students' understanding that aims to improve the competence of Physics education program students in STKIP PGRI West Sumatra.

In the learning of physics lecture method is not fully suitable for use because in the process of learning physics is not enough just to hear and memorize. Physics learning is more emphasized on conceptual understanding. One of the learning models that emphasizes conceptual understanding is the Conceptual Change Teaching (CCT) learning model developed by Driver and Oldham (Chen Hu Lin: 2010).

Learning Model CCT is a learning model that has stages to generate student conceptual changes. CCT learning model is based on the view of constructivism paying attention to the experience and the concept of early students. Through CCT, learning is related to the context of daily life environment, so that students more easily understand the contents of the lesson, linking the content of the lesson with the environment around the students will make a more meaningful learning. CCT learning patterns with various activities leads to learning more interesting and fun for students, so as to improve student learning outcomes.

In previous research, the researcher has done the development of CCT based physics learning device on Dynamic Electrical material. From the results of these studies show that the learning devices based on CCT on dynamic electrical materials can improve student learning outcomes (Auliya: 2014). This study is limited to a single subject of dynamic electricity. Furthermore, the researcher will develop CCT based module in Basic Electronics 1 course. Based on the background, the researcher is interested to develop Basic Electronics 1 module integrated values of Character based on CCT model to improve student physics competence in STKIP PGRI West Sumatra.

2. Research Method

Development of learning module Basic Electronics 1 integrated character values based on CCT model by using Plomp model which is divided into three stages: preliminary research, prototyping phase, and assessment phase. In the preliminary research phase, the researcher performs a preliminary analysis or problem identification, needs analysis, concept analysis or material content, and a review of the literature required in the lesson.

Concept analysis aims to determine the content and learning materials needed in module development. In the concept analysis, the researcher identifies the essential concepts of Basic Electronics 1 material. Next, an analysis of student characteristics to determine the module development model will be used. In addition, researchers also conducted a study of the available literature.

The next stage is the phase of design and realization (prototyping phase) where the formulation or design of the module. The design of modules is made in accordance with established indicators and based on the format that is tailored to the needs of the researcher. In this phase, there is a formative evaluation which includes self-evaluation, expert reviews, one-to-one, and small group, and fieldtest. However, in this study, the research is only done until one to one stage.

The designed module is validated by an expert. The input of the validator is used to fix and revise the module developed so that a valid module is generated. At the one-to-one stage, a design trial has been developed to students and lecturers who become testers. The results of this trial are used to revise the designs that have been made. The result of the assessment of the validator is analyzed by using the Likert scale which is converted to the value using equation 1.

$$p = \frac{f}{n} \times 100 \qquad (1)$$

where :

p = value of validity

f= scores obtained

n = maximum score (Modified from Riduan, 2009: 89)

The module validity category based on the final value obtained can be seen in Table 1.

Table 1. Category Validity Module

Interval	Kategori
0 – 20	Very invalid
21 – 40	Invalid
41 – 60	Less valid
61 – 80	Valid
81 – 100	Very valid

3. Results and Discussion

3.1 Module Validity

Based on the initial investigation conducted on the students, the design of integrated module of conceptual values based on conceptual change teaching,researchers develop modules with some problems in advance based on the phenomenon encountered in real life with the aim to bring the idea of students to the material. Then students are guided in learning activities that will eventually reconstruct their knowledge.

Module validity is seen from the instrument of validity by experts. The results of validity by experts are used to determine the feasibility of modules and guidelines in revising the product to produce a more perfect module. Assessment by experts includes content validity, construct validity and language validity. The results of the assessment can be seen in Table 2.Based on Table 2 it can be seen that the assessment of each validator is 90.5, 74.3, and 92.6 with an average value of 85.8 with very valid category. The average value of validator for each module component can be seen in Figure 1. Based on the validation result, it can be concluded that the basic electronics 1 module integrated the values of character based on the conceptual change teaching model are in very valid category.

Table 2. Expert Validation Results

	Validator 1	Validator 2	Validator 3	Average Validator	Category
Content validity	90.4	67.3	94.2	83.9	Very valid
Construct Validity	91.7	80.6	94.4	88.9	Very valid
Language Validity	89.3	75	89.3	84.5	Very valid
Average	90.5	74.3	92.6	85.8	Very valid

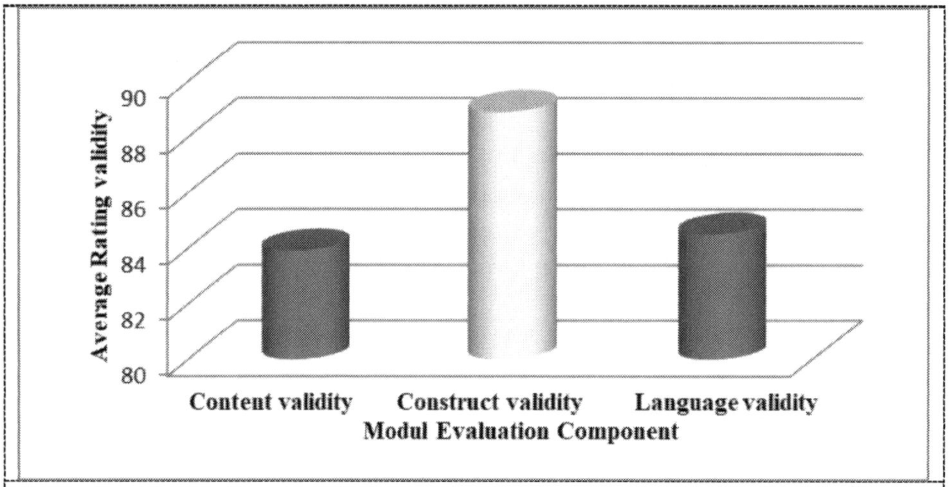

Figure 1. Average Rating Component Assessment Module

3.2. Description of Product Module

Based on the existing weakness, then the moduleis revised. Revisions include weaknesses in writing and content submitted by three experts. The weakness can be seen in Table 3.

Table 3. The Weaknesses on Modules to be Revised

	Weakness in Module
Validator 1	1. How to Write Kirchoff
	2. The use of sentences
	3. Create an equation number
Validator 2	1. The color of the module cover is less bright
	2. The formula is made oblique and numbered
	3. Review the accuracy of each phase
Validator 3	1. Index is made further down
	2. The writing is less clear

After the module is revised as per the suggestion, we get a module that can be used in learning. Modules are made in accordance with the design that has been prepared. Module design includes Cover, introduction, content, type and font size and so on. The cover, introduction and module contents are shown in Figures 2, 3 and 4.

Figure 2. Cover of the module

IOP Conf. Series: Materials Science and Engineering **335** (2018) 012079 doi:10.1088/1757-899X/335/1/012079

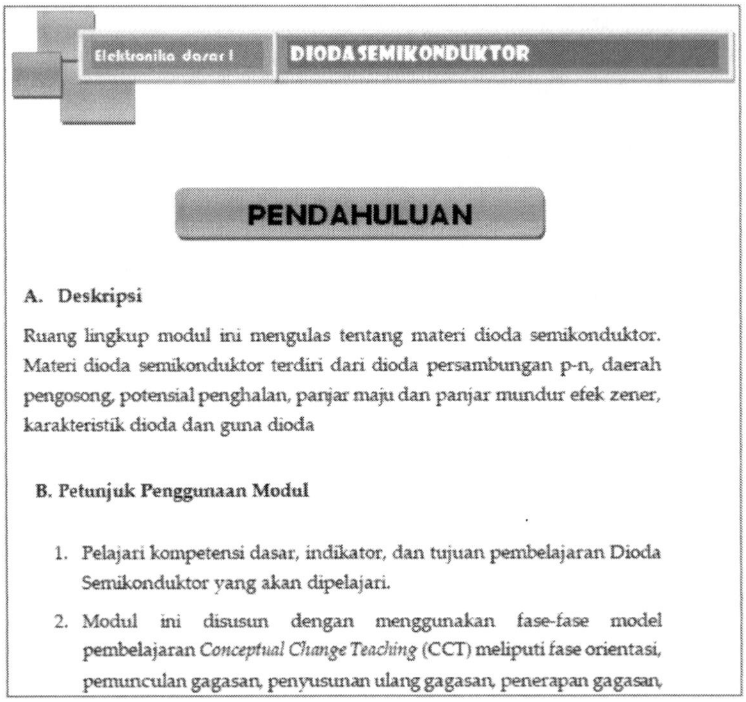

Figure 3. Print Screen of the Introduction

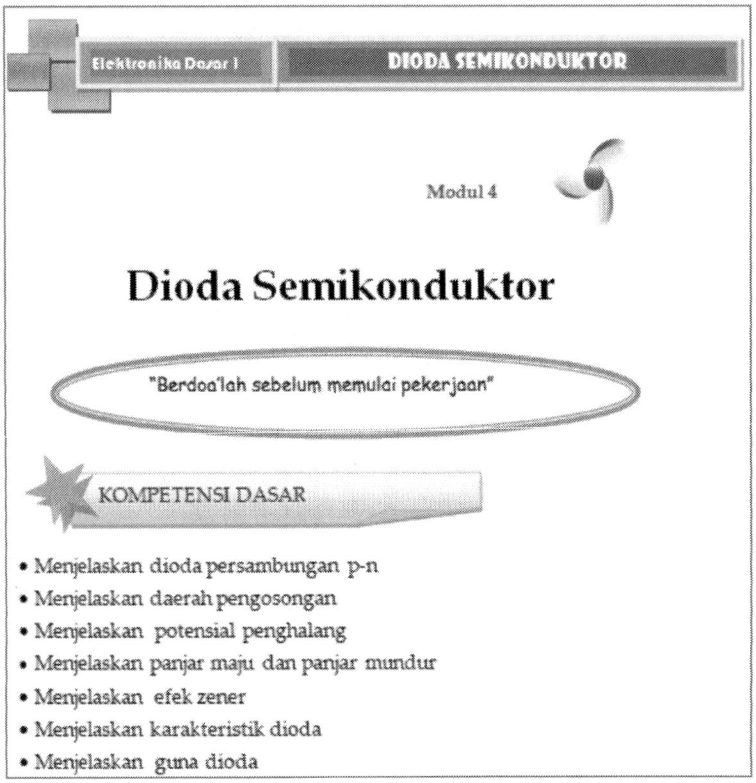

Figure 4. Content of the module

The modules developed are in accordance with the systematics of module writing. In addition, the issues presented in the module are related to the material covered and close to the student environment. The module also has a clear reference list. For the stage of rearranging ideas, the statements presented can help lead students in building their own knowledge. Problems designed to assist students in applying the knowledge they have built.

The module presentation component is also rated highly valid by the validator. This is because the module is presented systematically. The module presentation is done in sequence according to CCT model. Furthermore, the design of the display used is simple and interesting, the language used is easy to understand by students, and the font used is clearly read. The use of language in the module is already communicative and clear. Based on the assessed aspects, the validated module has been declared valid by the validators

4. Conclusions

Based on the results of the assessment of the validator obtained the result that the development of basic electronic module 1 integrated values of character-based conceptual change teaching model categorized very valid.

Acknowledgments

Thank you toMinistry of Research, Technology and Higher Education of the Republic of Indonesiathat has funded this work through Research of Inter-University Cooperation under contract No. 073/STKIP-UP3M/PGRI-SB/2017.

References

[1] Auliya Hidayati. 2013. Through our Research-Based Physics Learning Support theImplementation of the Curriculum. Proceedings of the National Seminar on Physical Learning. 2 Nov 2013 PPs UNP.
[2] Depdiknas. 2008. Guidance on Teaching Materials Development. Jakarta: Directorate of High School Development.
[3] Kemendiknas. 2011. Character Education Implementation Guide. Jakarta: Kemendiknas
[4] Lin, Chen-Hu, et al. 2010. "Utilizing A Concept Map As The Teaching Strategy Based on Conceptual Change Theory For The Course Information Technology and Society". Joint International IGIP-SEFI Annual Conference 2010, 19th-22th September 2010, Timava, Slovakia.
[5] Nasution, 2008. Berbagai Pendekatan dalam Proses Belajar Mengajar. Jakarta : Bumi Aksara.
[6] Neale, D., Smith, D. & V. Johnson, V. (1990). Implementing Conceptional Change Teaching In Primary Science. Elementary School Journal 91: 109-131.
[7] Plomp, Tjeerd. 2010. An Introduction to Educational Design Research. Enschede: University of Twente.
[8] Purwanto. 2007. Module Development. Jakarta: Depdiknas
[9] Tessmer, M. 1998. Planning and Conducting Formative Evaluations. Philadelphia: Kogan Page.

Development of Thermal Radiation Experiments Kit Based on Data Logger for Physics Learning Media

H Permana[1,*], B H Iswanto[2]

[1]Physics Education Study Program, Faculty of Mathmatic and Natural Sciences, Universitas Negeri Jakarta, Jl. Rawamangun Muka Jakarta Timur 13220, Indonesia
[2]Physics Study Program, Faculty of Mathmatic and Natural Sciences, Universitas Negeri Jakarta, Jl. Rawamangun Muka Jakarta Timur 13220, Indonesia

*handjoko@unj.ac.id

Abstract. Thermal Radiation Experiments Kit (TREK) based on data logger for physics learning media was developed. TREK will be used as a learning medium on the subject of Temperature and Heat to explain the concept of emissivity of a material in grade XI so that it can add variations of experiments which are commonly done such as thermal expansion, transfer of thermal energy (conduction, convection, and radiation), and specific heat capacity. DHT11 sensor is used to measure temperature and microcontroller Arduino-uno used as data logger. The object tested are in the form of coated glass thin films and aluminum with different colors. TREK comes with a user manual and student worksheet (LKS) to make it easier for teachers and students to use. TREK was developed using the ADDIE Development Model (Analyze, Design, Development, Implementation, and Evaluation). And validated by experts, physics teachers, and students. Validation instrument is a questionnaire with a five-item Likert response scale with reviewed aspect coverage: appropriate content and concepts, design, and user friendly. The results showed that TREK was excellent (experts 88.13%, science teachers 95.68%, and students 85.77%).

1. Introduction

Permendikbud no.65/2013 explains that the learning process in the 2013 curriculum is emphasized using a student-centered learning approach that emphasizes the scientific approach including observing, asking, trying, processing, presenting, summarizing and creating [1]. Experiments in lab are activities that make learners active and loaded with a scientific approach. The series of lab activities begins by installing equipment, observing, measuring, retrieving data to draw conclusions [2]. Experiment activities are very effective for achieving three domains, namely: (1) High cognitive skills by practicing in order to understand the theory, integrate different theoretical aspects, and apply theory to real problems. (2) Affective skills by learning to plan activities independently, working together, communicating information about the field, and appreciating the field; (3) Psychomotor skills by learning to install the equipments so that it really works [3].

However, there are some obstacles faced by students in doing experiment activities such as the limited time of learning so it makes it difficult for students to do repetition data retrieval and the accuracy of data taken / measured. Data retrieval is a boring activity in the lab. All these constraints can be overcome by using an experiment kit equipped with data logger equipment [4]. Data logger is an apparatus for data recording with sensors and probes with a certain precision. The data logger will be the main equipment in secondary science classroom technology as becomes more widespread and in common use. The use of the computer and data logger can be of enormous benefit for improvements in time efficiency, clear presentation of data to allow easier analysis and interpretation, data rapidly displayed clear visual interpretation of relationship between variables [5]. The Table 1 shows the comparison between conventional / manual experiments and using data loggers.

Table 1. Comparison between Conventional/manual Experiments and using Data Loggers.

	Conventional	Data Logger
Implemenation	• Manual • Need more time	• Automatic • More faster
Data Accuracy	Quite accurate	Very accurate
Result of experiment	• Observations should be done thoroughly • Results in the form of numbers	• Data is recorded automatically and in real time • Results in the form of numbers, graph, diagram or table

Expected competences in subject Temperature and Heat grade XI (Kurikulum 2013 Revisi) are that students are able to analyze the heat transfer, thermal characteristics of a material and able to experiment on the thermal characteristics of a material. Based on the data presented by Viajayani (2013), found that 82% of students expressed difficulty in imagining physical processes on the subject of temperature and heat, especially in understanding heating curve when the temperature of the object increased [6]. Experimental activities are needed to address this.

From the preliminary needs-analysis it is known that common topics that frequently demonstrated are thermal expansion, transfer of thermal energy (conduction, convection, and radiation), and specific heat capacity. Other topics which are not less interesting are heating curve characteristic at phase transition, characteristic of thermal radiation, and characteristic of thermal emissivity. But there is no experiment kit to show that.

Purpose

The purpose of this study was to develop a Thermal Radiation Experiments Kit (TREK) for senior high school (SMA) students. This TREK is comprised of microcontroller arduino-uno, heat sensor (DHT11), and a modified Leslie's cube. The TREK is expected to be an alternative kit that can be used in learning physics. The study investigated the level of respondent' agreement using Likert-type scale about the TREK in terms of learning content, learning concept, design, and interaction. It also investigated the perceptions of the participants towards using the TREK. Respondents are made up of experts, science teachers, and students.

2. Methods

The method used in this study was the method of research & development. The research model used is the ADDIE's Model (Analyze-Design-Develop-Implement-Evaluate) as shown in Figure 1 [7].

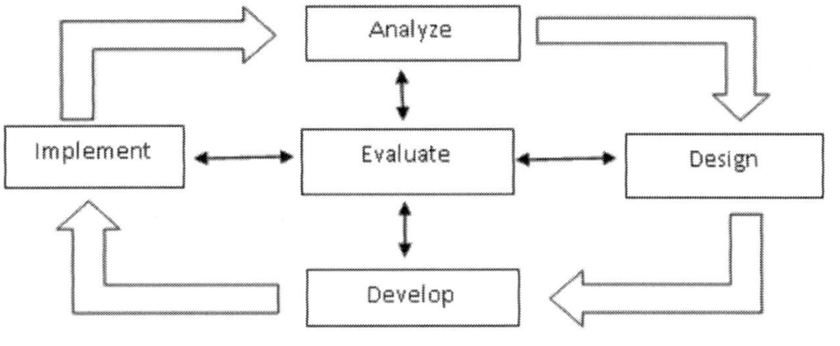

Figure 1. The development model of ADDIE

IOP Conf. Series: Materials Science and Engineering **335** (2018) 012080 doi:10.1088/1757-899X/335/1/012080

The study was conducted on March – July 2017 at laboratory of Instructional Research and Development, Faculty of Mathematics and Science, Universitas Negeri Jakarta. A small-group trial was conducted at "*SMAN-99 & SMAN-105*" senior high school, Jakarta.

Analysis phase (Analyze) begins with preliminary needs-analysis of the high school physics teachers, basic competences literature studies (KD 2013 curriculum) relating thermal radiation materials, preliminary study of tools and materials required.

- Design phase : makes product design TREK, students worksheet and manual book, collecting materials and tools to operate arduino and data logger.
- Development phase: realize the design of the previous phase,
- Implementation phase : doing test of the TREK, preparing validation instruments, collecting feedback from respondents (experts and science teachers)
- Evaluate phase: evaluation and revision of the TREK, redesign based on the feedback from respondents (experts and science teachers)

Data Analysis Technique

Likert scale is a scale used to measure attitudes, opinions, and perceptions of a person or a group of social phenomenon. The calculation is as below :

$$P_{(K)} = \frac{S}{N} \times 100\% \quad (1)$$

where,

$P_{(K)}$ = the percentage of component

S = total score of components research

N = the amount of the maximum score

Then the data were interpreted using the score table as shown in Table 2.

Table 2. Likert-scale interpretation [8]

No.	Percentage(%)	Interpretation
1	<= 20 %	Very poor
2	20% – 40%	Poor
3	40% – 60%	Average
4	60% – 80%	Good
5	80% – 100%	Very good

Instruments and Participants

There were 2 types of questionnaires used in this study. Preliminary needs-analysis questionnaires were distributed to science teachers and students. Validation instruments were questionnaires with a five-item Likert response scale with scope of reviewed aspects: appropriate content and concept, design, and user friendly. Table 3 shows number of participants.

Table 3. Number of participants

No.	Reviewed Aspects	Number of Participants		
		Expert	Science Teacher	Student
1	Appropriate Content	1	4	-
2	Appropriate Concept	1	4	-
3	Design	1	4	36
4	User friendly	1	4	36

3. Results and Discussion

The results of preliminary need-analysis of 8 physics teachers at *SMA Negeri 99* Jakarta and *SMA Negeri 105* Jakarta are as below:

- 87.5% of teachers use powerpoint slide as learning media for temperature and heat materials.
- 62,5% of teachers use students worksheet,
- 75% of teachers use laboratory activities,
- 100% of teachers said there was no thermal radiation experiment kits' available.

So it can be concluded that there was a need to develop a thermal radiation experiment kit that can perform measurements of temperature on the material in real time at a certain time interval.

The development of thermal radiation experiment kit (TREK) was developed based on the concept of radiation on the Leslie cube and some relevant research results. The TREK uses a DHT11 temperature sensor to measure the radiation absorbed by the material from the heat source (heater) over a specified time interval measured using the data logger. The measurement results are then displayed in the form of data and graphs.

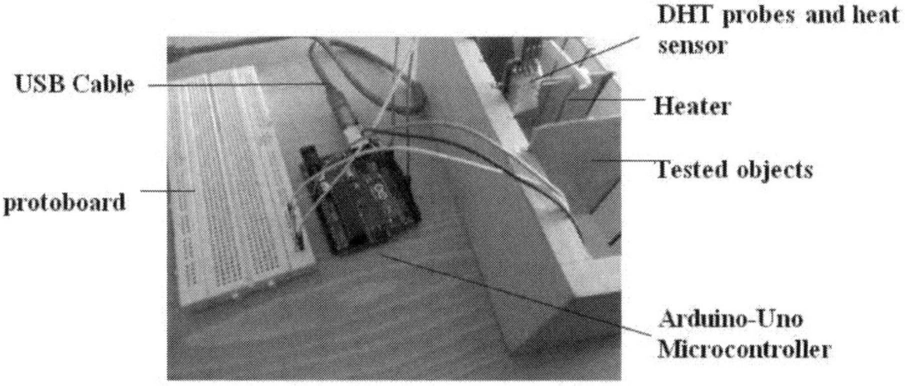

Figure 2. Thermal Radiation Experiment Kit

Figure 3. Tested Object (painted alumina and thin coated glass)

Figure 4. Students' Work-Sheet

How does the TREK work ?

- Thermometer digital + data logger (microcontroller arduino uno + 2 units thermal sensor DHT11).
  Thermal sensors (DHT11) get temperature data in real time and record in CSV format. We have prepared an interface application using python that can issue table data and temperature-time curve.
- Modified Leslie's Cube:
  Leslie's cube is a device used in the measurement or demonstration of the variations in thermal radiation emitted from different surfaces at the same temperature (BBC GCSE 2016). We used a heating element as a heater. The two sensors are used to record data temperature from inside the cube, and from the front of the two-sides of the cube.

The TREK's test results are shown in Figure 5.

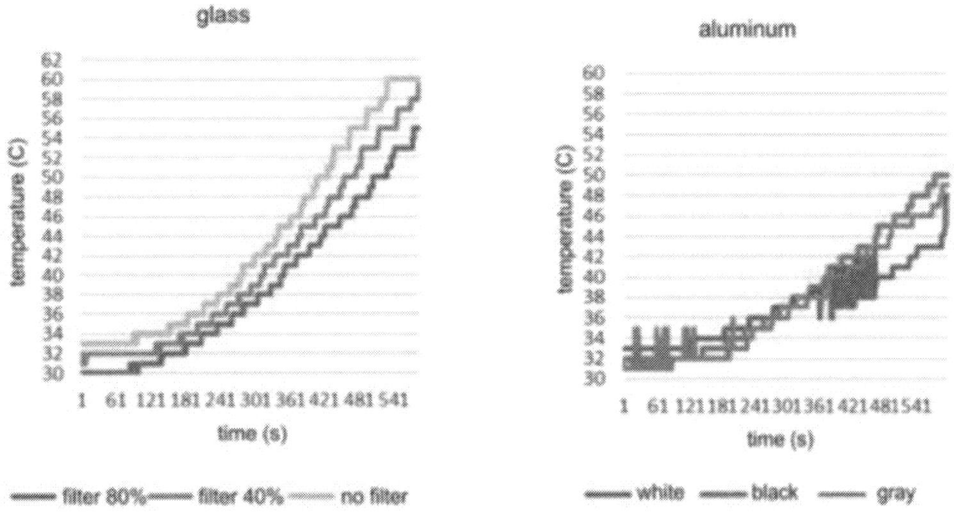

Figure 5. Test-Result of TREK

Figure 6. Small-group Trial

Validation results

The validation results by the experts, the science teachers and a small group of trial by students are shown in Table 4, 5, 6.

Table 4. Validation result by experts.

No.	Aspects	Average score	Remarks
1	Appropriate content	85.71%	Very good
2	Appropriate concept	81.54%	Very good
3	Design	97.14%	Very good
Total Average score		88.13%	Very good

Table 5. Validation results by science teachers.

No.	Aspects	Average score	Remarks
1	Appropriate content	98.58%	Very good
2	Appropriate concept	97.14%	Very good
3	Design	94.0%	Very good
4	User friendly	93.0%	Very good
Total Average score		95.68%	Very good

Table 6. Validation results by students.

No.	Aspects	Average score	Remarks
1	Design	85.24%	Very good
2	User friendly	86.30%	Very good
Total Average score		85.77%	Very good

4. Conclusion

The developed Thermal Radiation Experiment Kit is viable and valid to be used as an instructional media.Conduct a field test in schools. In the further research the study the effect of using this TREK will be investigated.

Acknowledgements

This study was supported by the Faculty of Mathematics and Natural Sciences - Universitas Negeri Jakarta under contract numbers 39 / SPK.PENELITIAN / 6.FMIPA / 2017 date: 5 Mei 2017. The authors would like to thank all the people who took part in this study.

References

[1] Permendikbud. (2013). Jurnal Lampiran Peraturan Menteri Pendidikan dan Kebudayaan Republik Indonesia Nomor 65 Tahun 2013 Tentang Standar Proses Pendidikan Dasar dan Menengah.
[2] Amien, M. (1988). *Buku Pedoman Laboratorium dan Petunjuk Praktikum Pendidikan IPA Umum (General Science) Untuk Lembaga Pendidikan Tenaga Kependidikan.* Jakarta: Departemen Pendidikan dan Kebudayaan
[3] Rahayuningsih, & Dwiyanto. (2005). *Pembelajaran di Laboratorium.* Yogyakarta: Pusat Pengembangan Pendidikan Universitas Gajah Mada.
[4] Pun-Hon, N. Y.-y. (2000). Implication of data logging on AL Physics Experiment: Preliminary study. *Asia Pacific Forum on Science Learning and Teaching*, Vol.1, Issue 2, p.1.
[5] Roberson, P. (2004). Using Data Logger. *Science Teacher Workshop*, 1-3.
[6] Viajayani, E. R., Radiyono, Y., & Rahardjo, D. T. (2013). Pengembangan Media Pembelajaran Fisika Menggunakan Macromedia Flash Pro 8 pada Pokok Bahasan Suhu dan Kalor. *Jurnal Pendidikan Fisika.*
[7] Tegeh, I. M. (2014). *Model Penelitian Pengembangan.* Yogyakarta: Graha Ilmu.
[8] Hidayat, S. (2012). *Profesi Kependidikan: Teori dan Praktik di era Otonomi.* Tangerang: Pustaka Mandiri.

Computer-assisted Biology Learning Materials:Designing and Developing an Interactive CD on Spermatogenesis

M Haviz

Institut Agama Islam Negeri (IAIN) Batusangkar, Indonesia

mhaviz@iainbatusangkar.ac.id

Abstract. The purpose of this article is to design and develop an interactive CD on spermatogenesis. This is a research and development. Procedure of development is making an outline of media program, making flowchart, making story board, gathering of materials, programming and finishing. The quantitative data obtained were analyzed by descriptive statistics. Qualitative data obtained were analyzed with Miles and Huberman techniques. The instrument used is a validation sheet. The result of CD design with a Macro flash MX program shows there are 17 slides generated. This prototype obtained a valid value after a self-review technique with many revisions, especially on sound and programming. This finding suggests that process-oriented spermatogenesis can be audio-visualized into a more comprehensive form of learning media. But this interactive CD product needs further testing to determine consistency and resistance to revisions.

1. Introduction

Research trends in technology-enhanced learning (TEL) is divided two great lines of research. The first is related to technology, including virtual learning environments, game-based learning, mobile learning and learning through communication technologies. The second is related to learning content and includes engineering, mathematics, science, reading and writing [1]. The utilization of technology for enhancing science learning has gradually become one of the major trends in science education [2]. Currently, the use of computer-assisted learning (CAL) terms such as the use of interactive CD or e-learning has been replaced by technology enhanced learning. Although the use of the term technology enhanced learning (TEL) is still debated in learning [3].

Currently, learning requires more varied and speculative ways, methods and media. The use of interactive CD for specific learning materials, including speculative media. Speculative media in digital education research are needed for researchers to be able to engage precisely in digital education [4]. Learning with interactive media can be used with teacher assignments and explanations [5]. Results of research on the effect of media multitasking on learning show that the group in the test reading with a tested video shown simultaneously condition performed significantly better than the group of the background condition [6]. A highly interactive hypermedia atlas of a rat dissection which allows a student to investigate the relationships between anatomical structures in the body and provides information on the physiological function of many of them [7]. The use of interactive multimedia also improves student achievement [8. 9].

In science education, ICT effect on in two models of a science curriculum are one based on empirical science and one on a model of scientific reasoning [10]. Computer assisted learningis still indispensable to enhance and to connect knowledge with skills in science education. Computer assisted learning materials, designed to provide experience in decision-making are being used to bridge the gap between theory and practice in the field of biotechnology [11]. Because there are several processes in learning that can only be shown with computer assist learning. The condition of this learning environment is needed in science learning.

Computer-assisted learning (CAL) still use in biological education. For example the use of computer-assisted and computer-based concept mapping on science education [12] and teaching and assessment of biological science [13]. Some research reports using interactive media in biology learning are learning on photosynthesis [14], practical work of science education [15], molecular biology [16], biology cell [17], mitosis and meiosis [18], reproduction of animals and plants [19] and traditional histology instruction [20].

Learn the material of spermatogenesis requires a media that is able to help convey material messages to students. In spermatogensis found many processes and facts in learning. The physiological function of spermatogenesis in mammals is to produce spermatozoa [21]. The mechanisms involved in establishing and maintaining a germline stem cell population in testes, as well as the factors that regulate their contribution to the pool of differentiating cells[22]. Spermatogenesis in mammals is achieved by multiple players that pursue a common goal of generating mature spermatozoa [23]. Spermatogenesis takes place in the seminiferous epithelium of the seminiferous tubule, the functional unit of the mammalian testis spermatozoa [21].The developmental processes acting on male germ cells that culminate in the production of the functional spermatozoa are regulated at both the transcription and post-transcriptional levels [23].

In this study we designed an interactive CD on spermatogenesis, because spermatogenesis became an important process in the human reproductive system [24, 25] and spermatogenesis also loads many processes in the formation of sperm that must be visualized with interactive CD media. This is a research and development [26] which will design CD media, after previous research has been designed cooperative learning device[27,28] used in the application of CD interactive. The purpose of this article is to design and develop a product in the form of CD interactive on spermatogenesis.

2. Methods

This is a research and development [26] which uses the procedure of development media [29]. The steps of the work is making the outline of media programs, making flowcharts, preparation story board, collections of materials, programming and finishing. At the outline of media programs, we identify the tittle, goal and content that write on media plan. At this stage, we also reconstruct syllabus and unit of learning. At making flowcharts and storyboard, we collect the materials and concept to make flowcharts. Flowchart is guidance of story board, because its description which contains the explanation of visual and audio of each flow in the flowchart. Each one column of storyboard is representation of one slide. Then we collect related picture and sound to programming and packagingofCD interactive. At the end we do self review of CD interactive [30].

In developing of instrument, we design of instrument of assessment the CD interactive. The indicator of the instrument of assessment are (1) theoretical: specification of objective and content, determination of strategy and time; (2) practicality: demonstration, familiarity, clarity and active learning; (3) characteristics: self instructional and contained, stand alone, adaptive, user friendly, visualization, variation, use, response and type of learning and (4) excellence: creativity, visualization of information and process, student motivation, packaging, novelty. This instrument is test for validity by expert. The result of test validity is 3.00 (valid). We use the statistic descriptive to calculate mean and standard deviation and Miles and Huberman technique to explain the qualitative data [31].

3. Results and Discussion

We have identify the aims and content of learning material on spermatogenesis. The result is write on media plan (table 1). We also designed the flowchart of CD interactive and the result is show on Figure 1.Then, we also have designeda storyboard. Storyboardcontains title, visual and audio (Table 2).The result of self review [30]is shown on Table 3.

Table 1. Media plan of spermatogenesis

Aspect	Description
Course	Developmental of biology
Topic	Spermatogenesis
Description	Spermatogenesis learns about the formation of spermatozoa cells and the hormones that govern them.
Main competencies	Students are able to explain the basic concepts and processes of development individual
Title	Cd interactiveon spermatogenesis
Supporting Competencies	Students are able to explain the process of spermatogenesis
Sub Topic	Introduction, stage of spermatogenesis and hormonal regulation

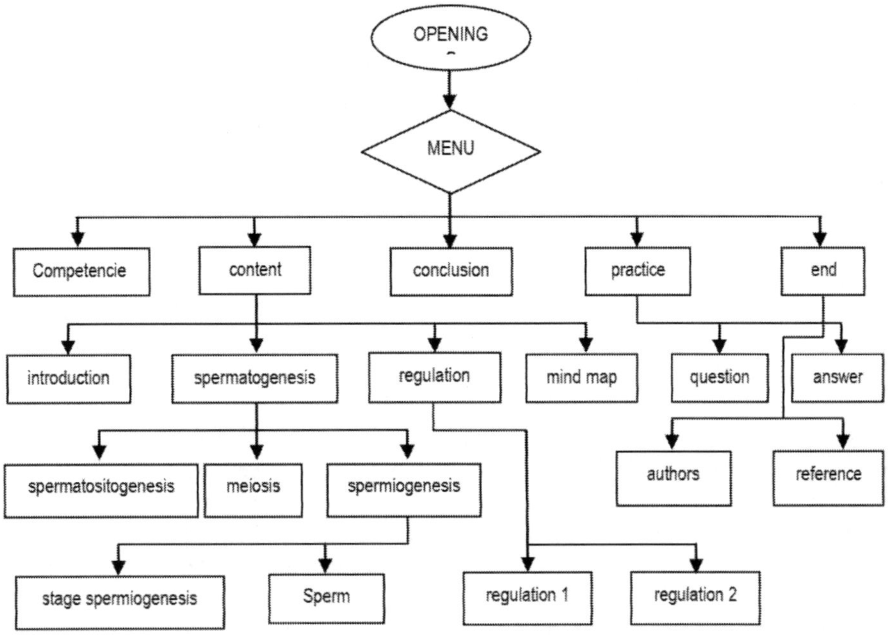

Figure 1. *Flowchart* CD interactive on spermatogenesis

Table 2. Storyboard of spermatogenesis

Part	Visual	Audio
Opening / Cover	**Animation 3D *Head*** :CD INTERACTIVE OF DEVELOPMENTAL BIOLOGY **Text**:CD INTERACTIVE OF DEVELOPMENTAL BIOLOGY WITH MACRO FLASH MX SPERMATOGENESIS BY M. HAVIZ DEPARTMENT OF BIOLOGY EDUCATION IAIN BATUSANGKAR **Image:** SPERMATOZOA **Panel:** STOP, NEXT	sound effect

Table 3. Result of self review of CD interactive on spermatogenesis

No	Aspect of review	Result of self review (Means ± SD)
1	Theoretical: specification of objective and content, determination of strategy and time	3.12 ± 0.35
2	Practicality: demonstration, familiarity, clarity and active learning	3.18 ± 0.09
3	Characteristics: self instructional and contained, stand alone, adaptive, user friendly, visualization, variation, use, response and type of learning.	3.15 ± 0.71
4	Excellence : creativity, visualization of information and process, student motivation, packaging, novelty	3.12 ± 0.35

Note: χ>3.20 is highly valid; 2.40< χ ≤3.20 is valid; 1.60< χ ≤2.40 is valid; 0.80< χ ≤1.60 is valid; χ ≤0.80 is invalid; SD = Standard deviation

In the programming stage, interactive CD programming is categorized into two forms: technical specification analysis and program work analysis. Analysis technical specifications illustrate that to be able to run this interactive CD required windows 98, ME and XP operating system with processor 128 MHz. The software needed in the design of the CD is macro flash MX 2004 as the main program. The supporting programs used for picture editing used Microsoft office power point 2007 and adobe Photoshop. Packing a CD using Ahead Nero 6.6. The hardware used to run this medium is a single computer unit. Computer equipped with CD Rom used to play and read media, S-VGA monitor to display program, keyboard and mouse standard windows for navigation purposes.

Analysis of the program work done to determine the effectiveness of navigation media that has been designed. Program work is designed with instructions and navigation buttons that will guide users to interact with media content. Interactive media that has been designed has 17 slides in the form of flash document and flash movie. Media that has been designed stored in flash movie format in CDR media with a maximum capacity of 700 Mb. The CD media consists of 3 sections: opening (2 slides), contents (13 slides) and cover (2 slides).

The result of the research is show that this research is categorized to research and development type 1, because the research focus on developing program, design process, evaluation, technique, organizing of content [32]. The finding the research also similarity with other research such developing material on science education [33], usability testing of a web site [34], interactive curriculum [35], physical education [36], financial system [37], and video teletraining [38].

The finding the research also show that the designing and developing is differ with the simple learning material. Because designing and developing is a study of systematics about design, develop, evaluation of product that meet the internal of consistency and relevancy [39]. Research and development is a study of effect of using of product, process and evaluation of product in learning [32].

The research findings show that the processes and facts present in spermatogenesis can be audio-visualized into a more comprehensive media of learning. As at spermatogenesis, sperm maturation process occurs in the epididymis [40], that its very typical cell, the core is packed so densely with acrosome found in the apical region near the nucleus and mitochondria are arranged in a spiral shape around the base of the flagellum [41]. There is a change in the topography of the spermatid cells consists of four phases, Golgi complex, cap, acrosome and maturation [42]. Spermatozoa maturation process is also accompanied by the series of changes in the composition, for example, changes in the number and presence of particular proteins in the membrane of the sperm head. This process is call polarization protein that it needed in changes in morphology and function of spermatozoa [43]. In this process, the protein is segregated from the anterior head domain or posterior head domain and found when spermatozoa momentarily left testes.

Furthermore, the result of content analysis on fact was done in the following section. The results research showed that the membrane before it mixes with ejaculated, the spermatozoa still have an important component when in the caudal epididymis. The results of the study show the structure of the sperm membrane [44] is covered by proteins, carbohydrates, lipids and other materials, that its release from the surface of the sperm membrane is an important part of capacitation [40]. The epithelial cells actively secrete fluid needed spermatozoa while in the epididymis [45] as changes in sperm surface associated with epididymis transit [46] and preventing the acrosome reaction early in the epididymis and shortly after ejaculation [47]. The others show that the material facts produced in the tubules; it will undergo a series of maturation process in the epididymis [48]. Mass, there is interaction between spermatozoa and it to the epithelial cells of the epididymis [46]. At maturation, sperm acquire the ability to move, and the ability to fertilize from epididymis protein is sperm-egg binding proteins [49] and proteins that play a role in the agglutination between spermatozoa [50, 51]. Agglutination effect decreased in the caudal epididymis before sperm are ejaculated [52, 53]. This is presumably due to the presence of other protein components that act as anti-agglutinin [54]. All of these explanations show that the many physiological processes that occur in the body during spermatogenesis.

This study has many limitations. The product of this study has not been tested for validity at next level such as expert review and small-large group test [30] and the product is also less have relevancy and consistence. So this product is not resistant to revision.

4. Conclusion

Based on the findings and discussion, it can be written that has produced a valid of interactive CD on spermatogenesis. The results show that the products that have been produced require a lot of revisions, especially on sound and programming. This finding suggests that process-oriented spermatogenesis can be audio-visualized into a more comprehensive form of learning media. But this interactive CD product needs further testing to determine consistency and resistance to revisions.

References

[1]. Badia A 2015 Research trends in technology-enhanced learning / Tendencias de la investigación en el aprendizaje favorecido por la tecnología *Infancia y Aprendizaje Journal for the Study of Education and Development* **38** 2 253-278.

[2]. Lin T C and Tsai C C 2015 *Innovative Technology-Assisted Science Learning in Taiwan in Science Education Research and Practices in Taiwan Challenges and Opportunities*, Chiu M H, Ed., (Singapore: Springer) p 189-209.

[3]. Bayne S 2015 What's the matter with technology-enhanced learning? *Learning, Media and Technology* **40** 1 5-20.

[4]. Ross J 2017 Speculative method in digital education research Learning Media and Technology 42 2 214-229.

[5]. Beemt AV D, Akkerman S and Simons P R J 2010 Pathways in interactive media practices among youths *Learning, Media and Technology* **35** 4 419-434.

[6]. Lee J, Lin L and Robertson T 2009 The impact of media multitasking on learning *Learning, Media and Technology* **37** 1 94-104.

[7]. Baxter M Q and Dewhurst D 1992 An interactive computer-based alternative to performing a rat dissection in the classroom *Journal of Biological Education* **26** 1 27-33.

[8]. Goldman J D G and Steele G T 2009 Teaching about human relationships education for primary school student-teachers using an interactive CD-ROM *Pastoral Care in Education, An International Journal of Personal, Social and Emotional Development* **27** 2 119.

[9]. Bayraktar S 2001 A Meta-analysis of the effectiveness of computer-assisted instruction in science education *Journal of Research on Technology in Education* **34** 2 173-188.

[10]. McFarlane A and Sakellariou S 2002 The role of ICT in science education *Cambridge Journal of Education* **32** 2 219-232.

[11]. Jenkins R O 1997 Computer-assisted learning materials and the practice of biotechnology *Journal of Biological Education***31** 1 65-69.

[12]. Lee S W Y and Tsai C C 2013 Technology-supported learning in secondary and undergraduate biological education: observations from literature reviewb*Journal of Science Education and Technology* **22** 2 226-233.

[13]. Sewell R D E, Stevens R G and Lewis D J A 1995 Multimedia computer technology os a tool for teaching and assessment of biological science *Journal of Biological Education* **29** 1 27-32.
[14]. Russell A W, Netherwood G M and Robinson S A 2004 Photosynthesis in silico, overcoming the challenges of photosynthesis education using a multimedia CD-ROM *Bioscience Education* **3** 1 1-14.
[15]. Kirschner P and Huisman W 1998 Dry laboratories' in science education; computer-based practical work *International Journal of Science Education* **20** 6 665-682.
[16]. Brooke N C ,Mullins R ,Antoniw J and Verrier P 1999 The Molecular Biology Notebook on CD-Rom – A step toward the virtual laboratory *Journal of Biological Education***33** 4 223-225.
[17]. Baggott L and Wright B 1995 Interactive learning in biology with Photo CD and associated software ALT-J *Research in Learning Technology* **3** 1 62-68.
[18]. Wilson J and Preece J 2005 Software review: mitosis and meiosis: an interactive approach *Journal of Biological Education* **40** 1 16-27.
[19]. Soyibo K and Hudson A 2000 Effects of computer-assisted instruction (CAI) on 11th Graders' attitudes to biology and CAI and understanding of reproduction in plants and animals *Research in Science & Technological Education* **18** 2 191-199.
[20]. Deniz H and Cakir H 2006 Design principles for computer-assisted instruction in histology education: an exploratory study *Journal of Science Education and Technology* **15** 5-6 399.
[21]. Yan C C and Mruk D D 2010 Introduction the biology of spermatogenesis: the past, present and future *Philosophical Transactions: Biological Sciences* **365** 1546 1459-463.
[22]. Fatima M D, Farias D B, Kohn A and Cohen S C 1999 Spermatogenesis and spermiogenesis in Microcotyle Sp. (Microcotylidae, Monogenea) *The Journal of Parasitology* **85** 5 832-838.
[23]. Anilkumar B and Wilkinson M F 2010 Transcription and post-transcriptional regulation of spermatogenesis *Philosophical Transactions: Biological Sciences* **365** 1546 1637-1651.
[24]. Haviz M 2014 Konsep dasar embriologi: tinjauan teoretis *Sainstek* **6** 1 96-101.
[25]. Haviz M 2013 Dua sistem tubuh: reproduksi dan endokrin *Sainstek* **5** 2 153-168.
[26]. Haviz M 2013 Research and development; penelitian di bidang kependidikan yang inovatif, produktif dan bermakna *Ta'dib* **16** 1 28-42.
[27]. Haviz M 2015 Cooperative learning model on developmental of biology *American Journal Educational Research* **3** 10 1298-1304.
[28]. Haviz M, Lufri, Helendra and Sumarmin R 2011 Pengembangan perangkat pembelajaran kooperatif pada biologi perkembangan hewan *Ta'dib* **14** 1 7-13.
[29]. Riyana C and Susilana R 2007 *Media Pembelajaran, Hakikat, Pengembangan, Pemanfaatan dan Penilaian* (Bandung: CV Wacana Prima).
[30]. Tessmer M 1993 *Planning and conducting formative evalution* (London: Kogan Page).
[31]. Gay L R, Mills G E and Airasian P W 2009 *Educational Research: Competencies for Analysis and Application 9th ed* (New Jersey: Pearson education, Inc).
[32]. Richey R C, Klein J D and Nelson W A 2002 Developmental research: studies of instructional design and development in *Handbook of research on educational communications and technology* Jonassen D (Ed). (Washington, Association for Educational Communication and Technology) p 1101.
[33]. McKenney S 2002 Computer-based support for science education materials development in Africa: exploring potentials Doctoral dissertation (The Netherlands: Universiteit Twente).
[34]. Corry M D, Frick T W and Hansen L 1997 User-centered design and usability testing of a web site: an illustrative case study *Educational Technology Research and Development* **45** 4 65–76.
[35]. Crane G and Mylonas E 1988 The perseus project: an interactive curriculum on classical Greek civilization *Educational Technology* **28** 11 25-32.

[36]. Cregger R and Metzler M 1992 PSI for a college physical education basic instructional program *Educational Technology* **32** 8 51-56.
[37]. Fischer K M, Savenye W C and Sullivan H J 2002 Formative evaluation of computer-based training for a university financial system *Performance Improvement Quarterly* **15** 1 11-24.
[38]. Martin B and Bramble W 1996 Designing effective video teletraining instruction: the florida teletraining project *Educational Technology Research and Development***44** 1 85–99.
[39]. Seels B B and Richey R C 1994 *Instructional technology: the definition and domains of the field* (Washington: Association for Educational Communications and Technology).
[40]. Visconti P E, Bailey J L and Moore G D P 1995 Capacitation of mouse spermatozoa, correlation between the capacitation state and protein tyrosine phosphorylation *Development* **121** 4 1129-1137.
[41]. Garner D L and Hafez E S 2000 Spermatozoa and seminal plasma in *Reproduction in farm animals*, 7th ed. (South Caroline: Lippincot Williem and Wilson) pp. 96-109.
[42]. Jhonson H M and Everitt B J 1995 Essential reproduction, 4th ed. (Oxford: Blackwell Scientific).
[43]. Hunnicut G R, Koppel D E and Myles D G 1997 Analysis of the process of localization of fertilin to the sperm posterior head plasma membrane domain during sperm maturation in the epididymis *Developmental Biology* **191** 1 146-159.
[44]. Kohane A I, Pinicoro L and Blaqueir J A 1987 Androgen controlled synthesis of specific protein in the rat epididymis *Endocrinology* **112** 5 1590-1596.
[45]. Setchell B P, Maddocks S and Brooks D E 1994 Anatomy, vasculature, innervation and fluids of the male reproductive tract in *The physiology of reproduction*, 2nd ed., (New York: Raven Press.Ltd.) pp. 1063-1175.
[46]. Hammerstedt R H and Parks J E 1994 Changes in sperm surface associated with epididymal transit *J. Reprod. Fertil. Suppl,* **34** 133-149.
[47]. Hunter R H, Holtz W and Hermann H 1978 Stabilizing role of epidymal plasma in relation to the capacitation time of boar spermatozoa *Anim. Reprod. Scie,* **1** 161-166.
[48]. Bredford J M 1975 Maturation, transit and fate of spermantozoa in the epidiymis in *Handbook of Physiology* (Washington D. C: Physiology Society).
[49]. Yanagimachi R 1994 Mammalian fertilization in *The physiology of reproduction*, 2nd ed. (New York: Raven Press. Ltd).
[50]. Haviz M, Agungpriyono S, Boediono A, Fahrudin M and Setiadi M A 2007 Distribution of sperm antiagglutinin in plasma collected from epididymis and ejaculate of ram *Jurnal Veteriner* **8** 1 24-31.
[51]. Dacheux J L, Paquinon M and Combarnous Y 1983 Head-to-head agglutination of ram and boar epididymal spermatozoa and evidence for an epididymal anti-aglutinin *J. Reprod Fertil* **67** 1 181-189.
[52]. Haviz M, Boediono A, Setiadi M A, Priyono S A and Fahrudin M 2008 Study on sperm agglutination with characterization of plasmacollected from epididymis and ejaculate in ram *Jurnal Veteriner* **9** 4 45-56.
[53]. Dacheux E and Dacheux J L 1998 Immunocytochemical localization of anti-aglutination in boar epididymis *Cell Tissue Res* 329-337.
[54]. Harayama H , Miyano T, Miyake M and Kusunoki 1994 Identification of anti-aglutinin in epididymal plasma of bor *Mol. Reprod. Dev* **37** 4 436-445.
[55]. Nieveen N 2010 Formative evaluation in educational design research in *The seminar conducted at the East China Normal University, Shanghai (PR China) November 23-26, 2007* (Enschede: *East China Normal University*).

IOP Conf. Series: Materials Science and Engineering 335 (2018) 012082 doi:10.1088/1757-899X/335/1/012082

Developing Guided Inquiry-Based Student Lab Worksheet for Laboratory Knowledge Course

Y L Rahmi[1*], E Novriyanti[1], A Ardi[1], R Rifandi[2]

[1]Biology Department, Faculty of Mathematics and Science,
Universitas Negeri Padang, Indonesia
[2]STKIP Adzkia, Padang, Indonesia

yosibio@fmipa.unp.ac.id

Abstract. The course of laboratory knowledge is an introductory course for biology students to follow various lectures practicing in the biology laboratory. Learning activities of laboratory knowledge course at this time in the Biology Department, Universitas Negeri Padang has not been completed by supporting learning media such as student lab worksheet. Guided inquiry learning model is one of the learning models that can be integrated into laboratory activity. The study aimed to produce student lab worksheet based on guided inquiry for laboratory knowledge course and to determine the validity of lab worksheet. The research was conducted using research and developmet (R&D) model. The instruments used in data collection in this research were questionnaire for student needed analysis and questionnaire to measure the student lab worksheet validity. The data obtained was quantitative from several validators. The validators consist of three lecturers. The percentage of a student lab worksheet validity was 94.18 which can be categorized was very good.

1. Introduction

The class of laboratory knowledge is an introductory class for biology student to conform to various lectures practicing in the biology lab. The learning outcomes of laboratory knowledge are students' own knowledge and skills, how to manage the laboratory, understanding about procedure to figure in a laboratory and having ability to keep their safety in the laboratory [1].

The learning of laboratory knowledge cannot go away from biology as a part of science that divided into two are product and operation. The attempt to realize biology characteristic as the product and operation of learning by doing practice. Inquiry is one of learning model that can be applied in practice. Learning strategies such as inquiry have long been reported to produce superior learning outcomes in teaching science education. Martin and Hansen [2] suggest that in inquiry-based learning, student conduct an investigation activity which resembles an experiment performed by the experts in the real word and state of [3] inquiry-based learning is focused on describing, explaining, predicting, and communicating scientific phenomena. The effects of inquiry-based learning implementation have investigated by [4] show that inquiry-based instruction can succesfully help students' to develop a deep comprehension of knowledge and scientific process skills which are vital to development of the students' literacy. Another research finding by [5] inquiry also encourage students science literacy.
According to [6] there are four levels of inquiry; 1) confirmation inquiry, 2) structured inquiry, 3) guided inquiry and 4) open inquiry. The kinds of inquiry that suitable for univ student is guided inquiry, because of guided inquiry is based on extensive studies of the Information Search Process

(ISP) in assigned research projects [7] and as a state of [6] guided inquiry is the third level of inquiry, the teacher provides students with only the research question, and the students design the procedure or method to test the question and the resulting explanation.

The steps of guided inquiry learning activities develops students' scientific activity by [8], [9], and [10] as follows: 1) identifiying problem, questions and issues; 2) formulating hypotheses; 3) collecting data; 4) analyzing the data; 5) drawing conclusions and generalizations. According to [11] learning by guided inquiry model can be implemented in the stages of observation, manipulation, generalization, verification and applications. However, [12] develops a guided inquiry learning cycle into five stages: orientation, exploration, concept formation or concept invention, application and closure.

According to [7] there are six characteristics of guided inquiry; 1) students learn by being actively engaged and reflecting on that experience; 2) students learn by building on what they already know; 3) students develop higher order thinking through guidance at critical points in the learning process; 4) students' development occurs in a sequence of stages; 5) students have different ways of learning; 6) students learn through social interaction with others.

A number of studies about the advantages of the use of guided inquiry based laboratory has been done by [13] inquiry-based gave students the ability to design, conduct, assemble, and troubleshoot an experiment that was not provided for them. Dwiyanti's research [14] showed that students' response to the implementation of the guided inquiry based student lab-worksheet on making pineapple flavoring were found very good. In line with these results, the study by [15] also states that development of guided inquiry based worksheet on colligative properties solution has very high validity, very high practicality and effectively used in the learning process.

The practice based on guided inquiry can be doing success if supported by student lab worksheet. According to the Ministry of National Education [15], the structure of the content of worksheet at least include: 1) the title or identity, 2) the instruction for learning; 3) the competence to be achieved; 4) learning materials; 5) the task or work step, and 6) evaluate.

Base of researcher observation, the laboratory practice class have not thoroughly established. This is caused have not a student lab worksheet to lead students' doing practical. If the practical do not accomplish it will be impacted to another course.

Established in the background, we will develop a practical guide for a laboratory practical course based on guided inquiry in the Biology Department, Faculty of Mathematics and Science Universitas Negeri Padang.

2. Method

The research was conducted using research and development (R&D) model. Development model that is used in this research is a Four-D model by [17]. The stages of Four-D models shown on Figure 1 below. The stages of research are as follows: 1) Define; 2) Design; 3) Develop and 4) Disseminate, although in this research the stages was limited to develop phase.

Sources of data in this study are two lecturers from a Biology Department and one lecturer from Language Department. The instruments used in the form of a sheet questionnaire for student needed analysis and questionnaire to measure the student lab worksheet validity. Measuring the validity of student lab worksheet contain of four aspects are didactic aspect, construct aspect, technical aspect and language aspect.

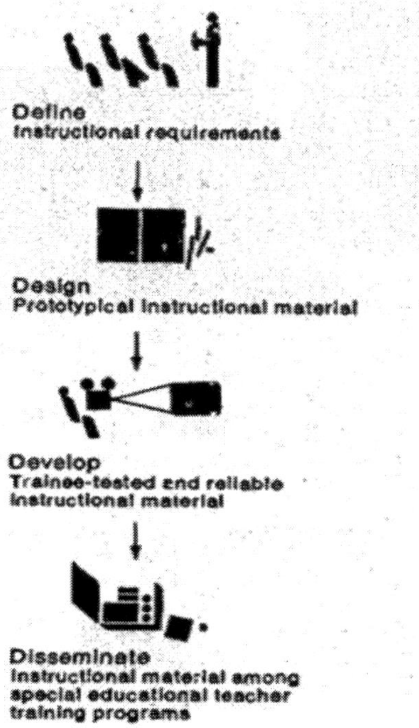

Figure 1. Four-D Models [17]

3. Results and Discussion

Development prosedure in this research is a Four-D models. The stages of research are as follows: 1) Define; 2) Design; 3) Develop and 4) Disseminate, although in this research the stages was limited to develop phase.

3.1 *Define*

The first stage of the procedure is define. The purpose of this stage is to stipulate and define instructional requirements. The steps of the stage are:

3.1.1 *Front-end analysis*

Based on interviews with leturers and students, it can be concluded that the basic problem is the laboratory practice class have not thoroughly established. Although the laboratory practise class is important to support the theory in the class. One of factor is caused have not a teaching material to lead students' doing practical.

3.1.2 *Learner analysis*

In this study students that will be served as a research subject is the first year univ students aged between 18-19 years. According to Piaget's theory of learning the way of thinking-age children in the formal operational stage of development intellectual that children are able to think logically, draw conclusions and interpret.

To get information about learner analysis by giving questionnaire for student needed. The information are students agree if lecturer will be develop a teaching material for laboratory knowledge course. They also want a teaching material is dominated by blue and green colour. For the kind of tulisan, the student choose Times New Roman and Comic S in the student lab worksheet.

3.1.3 *Concept analysis*

The way to analyze the concept is analysis the syllabus of laboratory knowledge course. Based on the syllabus, there are 12 (twelve) topics that should be learned by students for one semester: (1) Defenition, fungtion, lay out and equipments of laboratory; (2) Management of Laboratory; (3) Overview, fungtion, and procedure glasses, porselen and optical tools; (4) Overview, fungtion, and procedure measuring and fisiology tools; (5) Overview, fungtion, and procedure ecology and microbiology tools; (6) Overview chemistry materials; (7) Maintenance tools and materials; (8) Safety work in laboratory; (9) Working in laboratory; (10) Solution, consentration and dilution technique; (11) A reagent solution; (12) Planning a proposal college field.

3.2 Design

The second stage of procedure is design. The purpose of this stage is to design prototype instructional material. The steps of the stage are:

3.2.1 Media selection

Based on matching the main problem in the front-end analysis, the characterictic of the students and student needed in learner analysis and 12 (twelve) topics that sould be learned by students in concept analysis, the researcher making decicion to design a teaching material like student lab worksheet based on guided inquiry for laboratory knowledge course.

3.2.2 Initial design

In this step is the presenting of the essential instruction through appropriate media. The design of student lab worksheet containing: 1) Cover of student lab worksheet; 2) Guiding for lecturer; 3) Guiding for student; 4) The rules of practical; 5) The title or identity of the experiments performed; 6) The purpose of experiment; 7) Background problem; 8) Formulating the problem; 9) Formulating the hypotheses; 10) Collecting the data; 10) Analysis the data; 11) Referral to make conclution.

From 12 (twelve) topics that should be learned by students for one semester. The researcher contructs to be 10 (ten) practical topic. The first and second topic include in one practical, and the last topic about planning to proposal college field is not include in the practical.

3.3 Develop

The purpose of this stage is to modify the prototype instructional material. Although much has been produced since the define stage, the results must be considered an initial version of the instructional material which must be modified before it can become an effective final version. In the development stage, feedback is received throughformative evaluation and the materials are suitably revised.

3.3.1 Expert appraisal

Expert appraisal is a technique for obtaining suggestions for the improvement of the material. A number of experts are asked to evaluate the material from instructional and technical points of view. On the basis of their feedback, the material is modified to make it more appropriate, effective, usable, and of high technical quality.

In this case, there are three experts, including two lecturers from a Biology Department and one lecturer from Language Department.Validators give a check mark in the column available in the validation sheet or questionnaire. They also give feedback or suggestion to the student lab worksheet. Measuring the validity of student lab worksheet contain of four aspects are didactic aspect, construct aspect, technical aspect and language aspect. The validation results are shown in table 1 below.

Table 1. The percentage of validity student lab worksheet

No.	Validator	Validity				**Average**	**Validity Criteria**
		Didactic Aspect	Construct Aspect	Technical Aspect	Language Aspect		
1	A	87.5	91.2	93.8	100	**93.1**	**Very Valid**
2	B	96.9	83.8	83.33	93.75	**89.4**	**Very Valid**
3	C	100.0	100.0	100.0	100.0	**100.0**	**Very Valid**
		92.2	**87.5**	**92.4**	**97.9**	**94.18**	**Very Valid**

Based on the Table 1, it can be seen overall the average score on student lab worksheet validity is 94.18 % which according to [17] can be concluded by very valid criteria. Van den Akker [18] states that the validity refers to the level of a design based on knowledge (content validity) and various components related to each other (construct validity). The experts also give feedback or suggestion to make the student lab worksheet more appropriate. The feedback or suggestion from the lectures are shown in Table 2 below.

Table 2. Validators suggestion for repairing guided inquiry-based student lab worksheet.

No	Aspect	**Validators Suggestion**
1	Didaktic	-
2	Construct	Please check again the numbering
		Please repair the layout of researcher identity in the worksheet cover
		Please change the position of microscope picture with researcher identity in the worksheet cover
		Please add the space for researcher identity in the worksheet cover.
		Please give the resources of the picture and be good if personal picture.
		Proporsional the size of picture.
3	Technical	-
4	Language	Please consistent for using greeting words (Anda or Kamu).
		Please attention for using pronoun (mahasiswa or praktikan).
		Writing of a word 'dibawah ini' could be 'di bawah ini'. Because prefix, should use spaces if show place or location
		Changing a word 'matakuliah' should be 'mata kuliah'.
		Notice the use of punctuation point and coma on paragraphs contained in the preface.

Based on the Table 2, it can be seen there are four aspects that valuing by experts. There are two aspects get more suggestions from lecturers. There are construct and language aspects. Revision of student lab worksheet is made based on their feedback. Suggestions and feedback are used to improve the product, so a proper student lab worksheet can be used.

4. Conclusion

The conclusion of this study is the quality of the student lab worksheet as a product development of the Four-D model is in the very valid criteria, in terms of aspect of the didactic, construct, technical, and language. All of these aspects meet a very good criteria.

5. Acknowledgment

The authors thank to the validators Fitri Arsih, S. Si., M. Pd., Utami Dewi Pramesti, M. Pd., and Relsas Yogica, M.Pd. And also thank to the Indonesia's Ministry of Research, Technology and Higher Education of the Republic of Indonesia (Ristekdikti) for the funding of Research and Community Development through the scheme of *Penelitian Dosen Pemula* year 2017 for this work.

References

[1]. Ristiono, Syamsurizal, and Rahmi, Y.L. 2017. *A Syllabus Lecture Knowledge Laboratory*. Padang: Biology Department FMIPA Universitas Negeri Padang.

[2]. Martin and Hansen, L. 2002. Defining Inquiry, Exploring The Many Types of Inquiry in The Science Classroom. *The Science Teacher*, 69(2),34-37.

[3]. Harrison, A.G. and Treagust, D.F. 2000. A Typology of School Science Models. *International Journal of Science Education*, 9, 1011-1026.

IOP Conf. Series: Materials Science and Engineering **335** (2018) 012082 doi:10.1088/1757-899X/335/1/012082

[4]. Chabalengula, V. M., Mumba, F., Lorsbach,T., and Moore, C. 2008. Curriculum and Instructional Validity of The Scientific Literacy Themes Covered in Zambian High School Biology Curriculum. *International Journal of Enviromental and Science Education*,3 (4), 207-220.

[5]. Trna, J. and Trnova, E. 2015. The Current Paradigms of Science Education and Their Expected Impact on Curriculum. *Procedia-Social and Behaviour Sciences*, 197 (February), 271-277.

[6]. Banchi, Heather and Randy Bell. 2008. The Many Level of Inquiry. *Science and Children : 26-29*.

[7]. Kuhlthau, C. C., Maniotes, L. M., and Caspari, A. K. 2007: *Guided Inquiry: Learning in the 21st Century*. Westport, Conn.: Libraries Unlimited.

[8]. Arends, R.I. 2012. *Learning to Teach*. New York: Mc Graw-Hill Companies, Inc.

[9]. Kuhn, D., Black, J., Keselman, A., and Kaplan, D. 2000. The Development of Cognitive Skills to Support Inquiry Learning. *Cognition and Instruction*, 18 (4), 495-523.

[10]. Kipnis, M. and Hofstein, A. 2008.The Inquiry Laboratory as a Source for Development of Metacognitive Skills. *International Journal of Science and Mathematics Education*. 2008 (September), Volume 6, Issue 3, 601–627.

[11]. Wenning, C.J. 2011 The Levels of Inquiry Model of Science Teaching. *J.Phys.Tchr.Educ.* 6(2), 9-16.

[12]. Hanson , David. M. 2005. *Designing Process-Oriented Guided-Inquiry Activities. In Faculty Guidedbook: A Comprehensive Tool For Improving Faculty Performance, ed. S. W.* Beyerlein and D. K. Apple. Lisle, IL: Pacific Crest

[13]. Sundararajan, S., Faidley, L.E., and Meyer, T.R. 2012. *Developing Inquiry-based Laboratory Exercises for a Mechanical Engineering Curriculum*. American Society for Engineering Education. Iowa State University.

[14]. Dwiyanti, G. Suryatna, A. and Taibah, I. 2017. Development of Guided Inquiry-Based Student Lab Worksheet on the Making of Pineapple Flavoring. *Journal of Physics: Conf. Series 812 (2017) 012074.*

[15]. Irham, S.M., Oktavia, B. and Mawardi. 2016. The Development of Guided Inquiry-based Worksheet on Colligative Properties of Solution for Chemistry Learning. *Advances in Social Science, Education and Humanities Research (ASSEHR)*, volume 57. Atlantis Press.

[16]. Ministry of National Education. 2010. *Technical Guidelines for The Development of High School Teaching Materials*. Jakarta : The Directorate Guidance High School.

[17]. Thiagarajan, S. Semmel, D.S. and Semmel, M.I. 1974. *Instructional Development for Training Teachers of Exceptional Children: A sourcesbook*. Indiana University Bloomington, Indiana.

[18]. Riduwan, 2007. *The Scale of Measurement of The Study Variables*. Bandung: Alfabeta.

[19]. Van den Akker, J. 1999. *Design Approaches and Tools in Education and Training*. Dordrecht: Springer Science Business Media, B.V.

IOP Conf. Series: Materials Science and Engineering **335** (2018) 012083 doi:10.1088/1757-899X/335/1/012083

They do, They Get and They Know; How to Motivate Learner to Upgrade Their Learning Quality

R Yogica*, H Helendra

Biology Department, Faculty of Mathematics and Science, Universitas Negeri Padang, Indonesia

*relsasyogica.1103992@gmail.com

Abstract. A learning process that occurs in the classroom is a very important thing to note the quality, so it can be a determinant of student success in understanding the content of the lesson. The success of the learning process could be seen from the learning outcomes and the level of positive activities of students while in class. Students who are active in the classroom at the time of learning happen mean interest to the content of the lesson and will make their understanding deeper. In some learning processes in the classroom, the authors observed that in the first weeks of learning the level of student activity was very low. This is due to low student learning motivation. The author applies a method named: they do, they get, and they know. This method is very influential on the increase of learning activities because it affects the psychology of students to improve their learning motivation. After study in this method at two different courses in university, authors make a conclusion in the end that the method is effective to increase the frequency of student positive activity, so this method plays a role in improving the quality of learning.

1. Introduction

This study aims to observe how a learning method can affect students' motivation to learn. Often teachers make innovations in the implementation of learning in the form of teaching methods, but often also forgotten whether the innovation can foster students' motivation in learning or not. Students who do not have motivation in learning, will not get results in accordance with educational demands. Teachers have difficulty finding suitable methods to carry out learning because they feel that they have applied variations of learning methods, but forget about the existence of motivational components for each method used. Teachers who are able to understand how to teach and have many sources of information have a great opportunity to improve the quality of learning [1]. Briefly mentioned that any method used in learning can be adjusted to the condition of students as long as teachers are able to provide learning motivation to the students. Although teachers should not only play a role to motivate students, but also play a role to maintain the motivation that already exist in students. If students who already have the motivation but no effort to maintain it then will be at risk students will leave the classroom [1].

Researchers have observed about the students' initial motivation in 2 subjects (Animal Structure and Animal Development) in Biology Department of FMIPA Universitas Negeri Padang. The result of the observation is that the students in both courses at the beginning of the lecture have low motivation to learn. This is known by observing the frequency of student activity. The intended activity is a positive activity supporting the learning, which is asking questions, providing answers, criticizing and suggestions for the improvement of materials and conclude the concept of learning. The frequency of these activities is still low and very apprehensive if it continues to be ignored. The problem of low students' willingness to participate in the learning process resulted in a tedious and meaningless learning atmosphere. Should be meaningful learning to be a concern by educators to the realization of an interesting learning conditions and fun. Teachers are required to create a conducive and effective learning environment for students to have a pleasant learning experience [2]. A fun feeling for students while learning, will improve the effectiveness of the learning process [3]. Kreber states one of the characteristics of an excellent teacher is a teacher who can motivate students [4].

Meaningful learning can be realized by motivating students. Therefore, the researchers are interested to use innovative methods that are based on student motivation. Students need motivation to learn from the people around them to arise desire to participate actively in learning [5]. This innovative learning method "They Do, They Get and They Know". Basically, this method has long been done in the classroom during the learning process in various places, but not many realize the importance of the method to be informed to the public through a research and research reports such as in this publication. This method focuses on improving student motivation to be more active to carry out the learning with the higher the frequency of activity and the higher quality.

The assumption of this research is that there is an increase of motivation in students by applying this method. In the longer term, this method will affect the frequency of student activity in each lecture meeting towards a better direction. Students who are motivated to learn will increase participation in learning [6].

2. Methods

This research is action research. Elliot mentioned that action research is a study conducted by teachers who want to study, reflect and report on the results of their own teaching [7]. The stages of this research are:

2.1. Observation

Observe the condition of students during lectures. Focus of observation is on the number of students who participate actively.

2.2. Notify

The researcher tells the students that it will record the activity of each lecture every week. This stage is not only done once at the beginning, but periodically continues until the end.

2.3. Take notes

Writing on the handwritten sheet every positive activity that students do during the lecture occurs every week.

2.4. Recapitulation

Record collected data and create tabulations.

This research was conducted on two courses in the Department of Biology FMIPA Universitas Negeri Padang, Animal Structure and Animal Development. To understand the research stages and research assumptions, see Figure 1.

IOP Conf. Series: Materials Science and Engineering **335** (2018) 012083 doi:10.1088/1757-899X/335/1/012083

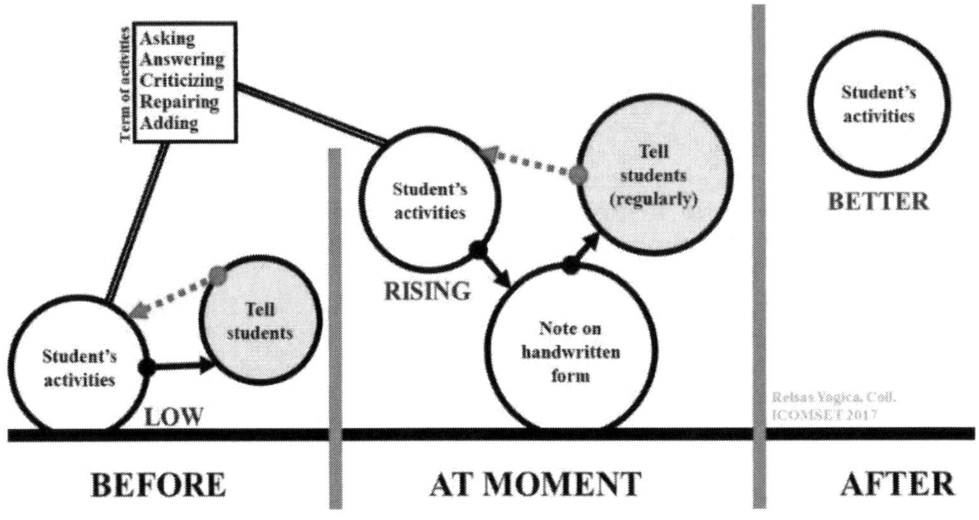

Figure 1. Research method and assumption of this research

3. Results

3.1. Implementation on the subject of Animal Structure

After application in the subject of Animal Structure, it is obtained the frequency data of student activity after observation for 14 times meeting described in Figure 2 as follows.

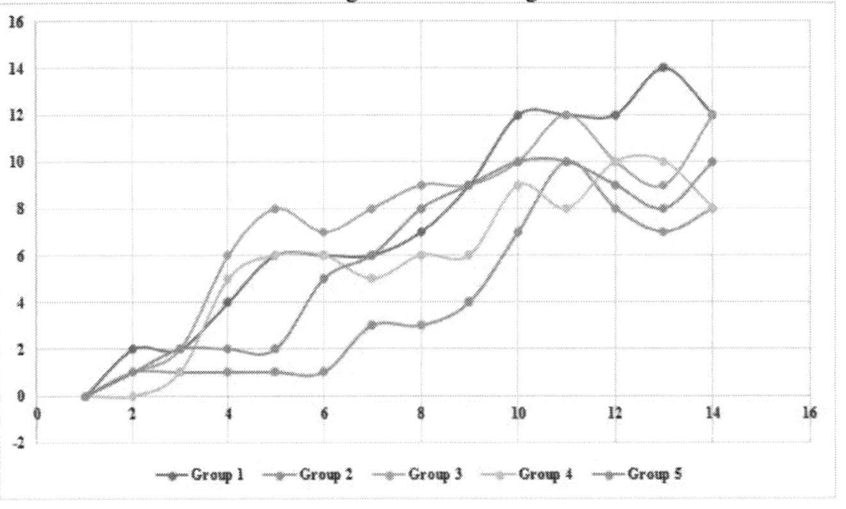

Figure 2. Frequency of student activity during 14 times meeting

3.2. Implementation on the subject of Animal Development

The frequency of student activity after 14 meetings also increased. The distribution of data in more detail can be seen in Figure 3.

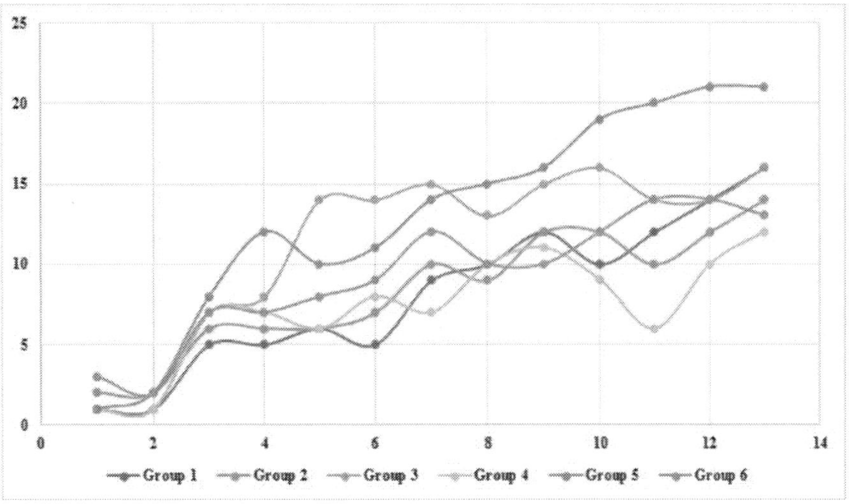

Figure 3. Frequency of student activity during 14 times meetings

4. Discussions

Based on the data obtained after the treatment in the classroom, the researcher concludes that by giving opportunity to the students to know about the development of the learning result then the students will be motivated to continuously improve their learning activity. It shows that there has been the indirect stimulation to the students to continuously improve the quality of their learning. The obligation of teachers in the learning process is to motivate students to achieve optimal learning objectives [8].

The method used in the learning process within the scope of this research has stimulated the students. The stimulus leads students to improve learning conditions that initially have little activity turned into more activity with increasing frequency. This method stimulates students' intrinsic motivation. With an intrinsic motivation, students' ability in group discussion and understanding levels will increase [9, 10].

The assumption of this research is that there is something that causes the student not to be well active in the class before the method is applied. It is that there is no internal motivation to know and understand the material. The intrinsic motivation in one's self will direct itself to a better direction. In the learning process, intrinsic motivation can improve students' understanding because it leads to an increased ability to discuss and understand more about what is being learned [9, 10]. Intrinsic motivation leads a person to act according to his inner drive and will feel satisfied after something he has done well [11].

Motivation that grows in students after learning by using this method will attract student interest to positive activity in learning. Shadrikov states that motivation is something that directs a person to perform an activity and goal-oriented [12]. Interestingly, students compete with each other to increase activity during the learning process. Zimmerman and Schunk stated that if there is one student who has good learning motivation in a class, then it will be a trigger for other students [13].

This study does not pay attention to the influence of the methods used on student learning outcomes. Further research is expected to contribute to the answer.

5. Conclusion

The applied "They Do, They Get and They Know" method stimulates students' motivation to improve their learning quality in the course "Animals Development". This method has been successful in increasing students' understanding of the subject matter as they are motivated. Students' acceptance of something depends on their inner motivation, the teacher's motivation and inspiration of their peers [14].

The researcher expressed gratitude to all parties who have cooperated in this research, because without its support then this research will not reach the level of good quality as when now. Then, the researchers also hope that this research can be continued until the experimental stage to see the effect of this method on student learning outcomes quantitatively.

References

[1] Sull, Errol Craig. 2007. *Lesson in Students Motivation*. Student Motivation, Vol. 2.
[2] Pelana, Ramdan. 2014. *Fun Learning Management for Physical Education, Sport and Health*. Asian Social Science, 10 (5), 85-90.
[3] Seung, Hoon Lee and Byeong, Min Yu. 2015. *Analysis of The Relationship among Fun, The Learning Satisfaction, and The Flow in Social Service Personal Job Training Education*. 15 (5), 674-683.
[4] Venter, Elza. 2017. *Good Teacher Education – An Exploratory Study*. The Independent Journal of Teaching and Learning, 12 (1), 63-73.
[5] Niemiec, C. P., & Ryan, R. M. 2009. *Autonomy, Competence, and Relatedness in the Classroom. Applying Self-Determination Theory to Educational Practice*. Theory and Research in Education, 7, 133-144.
[6] Ulstad, S.O.,et.all. 2016. *Motivation, Learning Strategies, and Performance in Physical Education at Secondary School*. Advances in Physical Education, 6, 27-41.
[7] Rioux, Kevin. 2014. *Teaching Social Justice in an Information Literacy Course: An Action Research Case Study*. Qualitative and Quantitative Methods in Libraries (QQML) Special Issue Social Justice, Social Inclusion, 23-30.
[8] Ryan, R. M., et.all. 2009. *Self-Determination Theory and Physical Activity: The Dynamics of Motivation in Development and Wellness*. Hellenic Journal of Psychology, 6, 107-124.
[9] Guthrie, J. T., et.all. 2007. *Contributions of Concept Oriented Reading Instruction to Knowledge about Interventions for Motivation in Reading*. Educational Psychologist, 42, 237–250.
[10] Walker, C. O., et.all. 2006. *Identification with Academics, Intrinsic/ Extrinsic Motivation, and Self-Efficacy as Predictors of Cognitive Engagement*. Learning and Individual Differences, 16, 1–12.
[11] Gagne, Marylene and Deci, Edward L. 2005. *Self Determination Theory and Work Motivation*. Journal of Organizational Behavior, 26, 331-362.
[12] Kashapov, Mergalyas M. and Leybina, Anna V. 2009. *Motivation of Professional Creative Thinking*. Psychology in Russia: State of the Art.
[13] Smit, Karin., et.all. 2017. *The Self-regulation of Motivation: Motivational Strategies as Mediator Between Motivational Beliefs and Engagement for Learning*. International Journal of Educational Research, 82, 124-134.
[14] Brackett, Virginia. 2007. Inspiring Student Self-Motivation. *Student Motivation*. Vol. 2.

IOP Conf. Series: Materials Science and Engineering **335** (2018) 012084 doi:10.1088/1757-899X/335/1/012084

Problem Solving Model for Science Learning

H Alberida[1*], Lufri[1], Festiyed[2], E Barlian[3]

[1]Biology Department, Faculty of Mathematics and Science, Universitas Negeri Padang, Indonesia
[2]Physics Department, Faculty of Mathematics and Science, Universitas Negeri Padang, Indonesia
[3]Faculty of Sport Science, Universitas Negeri Padang, Indonesia

*alberidamatua@gmail.com

Abstract. This research aims to develop problem solving model for science learning in junior high school. The learning model was developed using the ADDIE model. An analysis phase includes curriculum analysis, analysis of students of SMP Kota Padang, analysis of SMP science teachers, learning analysis, as well as the literature review. The design phase includes product planning a science-learning problem-solving model, which consists of syntax, reaction principle, social system, support system, instructional impact and support. Implementation of problem-solving model in science learning to improve students' science process skills. The development stage consists of three steps: a) designing a prototype, b) performing a formative evaluation and c) a prototype revision. Implementation stage is done through a limited trial. A limited trial was conducted on 24 and 26 August 2015 in Class VII 2 SMPN 12 Padang. The evaluation phase was conducted in the form of experiments at SMPN 1 Padang, SMPN 12 Padang and SMP National Padang. Based on the development research done, the syntax model problem solving for science learning at junior high school consists of the introduction, observation, initial problems, data collection, data organization, data analysis/generalization, and communicating.

1. Introduction

Science is the knowledge gained through the collection of experimental data, observations and deductions to produce an explanation of natural phenomena that can be trusted.According to Carin and Sund [1] there are three principles in Science,the first, science is the accumulation of knowledge organized systematically, in thesecondscience as an attitudeand lastly science as the scientific method, while Rustaman [2] statesScience contains four things: content or products, processes or methods, attitudes, and technology.Students in learning science, not just learn science product,but also processes, attitudes and technology, so that science fully understood.

Learning science should be designed to enable the activity of learning in studentsLearning activities important to establish a learning experience. The student learning experience derived from the activities it does, not from the activities of the teacher.Science learning activities undertaken by students should provide a learning experience about the discovery of a concept.Teachers should be able to design learning activities that provide a learning experience about the process of science and science products.The learning experience gained through applied learning model the teacher in the classroom.

The problem of education in Indonesia today is a weak learning process. School learning do not encourage students to develop thinking skills. The lesson is mainly directed so that students are able to remember as much fact and information as possible so the child must memorize, without understanding what is memorized [3].Weak learning process occurs almost in all educational units at every level.

The same is true with science learning. Results of research conducted TIMSS 2011, shows the average score of science grade VIII students in Indonesia is 433.This score indicates that the average science ability of Indonesian students is in the category of Low International Benchmark. Students who are in this category mean only being able to recognize some basic factsbut have not been able to communicate and associate various science topics;has not been able to apply complex concepts and abstract;has not been able to combine the information to draw a conclusion,interpret information in diagrams, graphs and charts to solve problems;has not been able to give a brief explanation that contain scientific knowledge and causality;and not able to understand the basics of scientific investigation;can not provide a written explanation to convey scientific knowledge[4].

The results of TIMSS researchers show that science learning has prioritized the mastery of procedural, factual and conceptual knowledge.Learning science more oriented to science as a product, so that students tend to memorize the concepts, principles, laws and theories[5, 6]. Science has three dimensions:first scientific attitude such as a high curiosity, critical, creative, honest, loving environment, recognizing the regularity of nature as creatures of God Almighty.The second dimension is the scientific process which deals with problem-solving procedures with scientific methods such as problem identification, hypotheses / predictions, analyze the data, draw conclusions. Third, science as a scientific product in the form of knowledge, whether it is factual, procedural or conceptual knowledge in the form of principles, law, and theory.

IPA as the process include the ability to apply the scientific method in the form of a set of skills required scientists to conduct scientific investigations known to science process skills[7]. Further Rustaman [7] says that one of the science process skills is a problem solving knowledge and skills. Essential problem-solving skills are owned by students as they relate to the ability to apply concepts- the concept of science to solve problems encountered in everyday life (life skill). Solving problems is a skill, it needs to be trained on students. Learning science in Indonesia today is less train problem solving skills to students. Learning needs to familiarize the problem solving, so that students have the process science skills.

Preliminary test results on the process science skills as a description of the problem solving ability of grade VII students of SMP *Kota Padang* showed an average below 50%.Observation skills, and designing experiments, respectively 47% and 39%.Predicting, asking questions and composing hypotheses of 18%, 40% and 27% respectively. Deducing, perform inference, applying the idea of the new situation and classify 40%, 34%, 44% and 34%,respectively.The ability of a high enough is to use simple equipment and take measurements 84% 61%.This result shows from 10 students, then 3 or 4 people who already have problem solving ability. Stenberg in [8] states the basis for implementing discovery / inquiry-based learning is problem solving. According to Carin [9] problem solving, critical thinking and creative thinking include high-order thinking skills. Carin further stated the purpose of science learning is to help students change perceptions, images, concepts to spur skills development in problem solving.Therefore it is understandable The National Curriculum recommends to use problem solving model.

The problem solving model recommended in the National Curriculum is Creative Problem Solving.*Creative Problem Solving is suitable for common problems related to everyday life.* In addition students are free to express opinions in accordance with background knowledge and interests.But Creative Problem Solving is difficult to do for problems that are not related to everyday life. Creative Problem Solving is good for students who have been accustomed to learning with problem solving model.

Carin [9]proposes the followingproblem solving steps: (1) planning, (2) collecting data, (3) organizing data, (4) data analysis, (5) generalization, (6) drawing conclusions. Planning stages from identifying problems, formulating problems, planning activities to solve problems.But the steps of problem solving is not in accordance with the scope of scientific work of the National Curriculum for junior high level. The problem solving steps that are not suited is planning the activity. Planning activities is a learning step that is suitable for high school level.Planning activities is a competency that must be achieved by high school students as stated in the scope of scientific work and work safety, consisting of: formulating problems, submitting hypotheses, determining variables, designing and conducting experiments, collect and process data systematically, draw conclusions, and communicate results oral and written experiments.While the steps problem solving for SMP as follows: formulate problems, predict, conduct experiments, collect data accurately, process data systematically, draw conclusions, and communicate the results of experiments orally and in writing.From the scope of science lesson of SMP it can be seen that the problem solving steps as presented by Carin are too high. How problem solving model which is suitable for SMP science?

2.Theory of Support

Learning models need to be applied by teachers in classroom activities.Learning model is a true learning model[10]. The application of a learning model will lead students to acquire information, ideas, skills, and values through the process of thinking and expressing themselves.Implementing a learning model means the teacher teaches how to learn. Furthermore, it is argued that the most important result of learning is the improvement of students' learning ability in the future to be more effective and easy, because they already have the knowledge and skills acquired through the learning process.

Learning model is different from strategy and learning method. The learning model has its own characteristics. Characteristic of the learning model according to Rusman [11] were: first, have the theoretical basis of certain educational experts. The foundation of this theory is logically arranged by the developer of the model.Second, have a specific mission or educational purpose. The next feature is to be used as a guide to improve learning activities in the classroom. The fourth characteristic is to have a component model: syntax, reaction principles, social systems and support systems. Meanwhile, according to Arends [12] learning model has an attribute that does not have a strategy or method of learning. Attributes of a model is the theoretical basis and teaching behavior and structure of the required classes.Jihad and Aris [13] state that the learning model is a plan or pattern used in preparing the curriculum, organizing the student material, and giving instruction to the teacher in the class in the teaching setting. Meanwhile, according to [10] learning model is built by several components; syntax, social system, reaction principles, support systems and instructional impacts. Syntax or phase, is a sequence of activities in learning.Syntax is used as a guide for teachers in implementing the learning models.Syntax is also a guide for students in learning.Syntax made the difference between models with learning methods.

The second component of the learning model is the social system. The social system is a description of the role and relationship between teachers and students, as well as the underlying rules. The role of teachers in each model is different. In one model the teacher acts as a facilitator of a group, teachers can also act as individual counselors. Joyce,et. al [10] distinguishes the social system into three levels:activity with a high structure, where teachers act as centers of activity, and sources of information.The second level is a moderate structure, the activities of teachers and students are balanced. While the third is a low structure, in this structure students as the center of activity, so that students are free to express and put forward the idea.

The third component of the learning model is the principle of reaction.The principle of reaction according to [10] is the way teachers appreciate, place and respond the student activities. For example, for one model, the teacher rewards after the student shows a particular activity, or on learning that aims to build creativity, the teacher puts himself in the same way as the student, so that the student's self-confidence appears.

The next component of the learning model is the support system. Support systems are devices that support the implementation of model. Included in the support system is a guide for teachers, student book or modules, LKS / LDS, instructional media, classroom conditions, etc.The support system is crucial to the successful implementation of a learning model.For example in science learning would be better to present the real thing than to display the image as a instructional media.

The last component of the learning model is the instructional impact and impact of the companion.Impacts are distinguished by direct and indirect impacts. The direct impact is the instructional impact,whereas the indirect impact is the impact of companions. For example, improving competence is a direct impact, but growing the competitiveness is a companion impact of a instructional model.

3. Method

The development procedure carried out follows the ADDIE development model which includes the analysis, design, development, implementation and evaluation phase.

3.1. Analysis phase

In the analysis phase, conducted the needs analysis of the problem solving model for learning science in the field today.Needs analysis includes: (a) A curriculum analysis consisting of analysis of SMP SKL, KI, and KD. The KI and KD analyzed are KI and KD in the 2013 curriculum. (b) Analysis of science teachers as users of developed products. Teacher analysis is done by observing teachers' ability and understanding of scientific approach.The ability of teachers to implement learning with a scientific approachconducted through observation of 97 science teachers of SMP Kota Padang who participated in training and socialization of Curriculum 2013 on 22-26 June 2014. Observations using the observation sheet in Appendix 1.While the teacher's understanding of the scientific approach and problem solving obtained by using a questionnaire as in Appendix 2. (c) Analysis of junior high school students in Padang City, including analysis of students' science process skills. An overview of the students' skills in the science process skills is obtained by providing a science-process exam to junior high school students who have implemented the 2013 Curriculum. Exam on process science skill given to student of SMPN 1 Padang, SMPN 12 Padang and SMP Nasional Padang. (d) Analysis of learning, the result of learning analysis is a description of the implementation of science learning in class VII, such as the model and the approach used by teachers. (e) The literature review is conducted by analyzing theories and concepts related to the research.

3.2. The design phase

The initial stage of R & D research is product planning. This stage includes: (1) the product's objectives, (2) the product's target audience, and (3) a description of the product's components and how they will be used[14]. The product developed in this research is problem solving model for science learning,and the target audience is the science teacher of SMP class VII. The resulting product component is problem solving model for science learning which consists of syntax, reaction principle, social system, support system, instructional impact and support. The use of problem solving model in science learning will improve students' science process skills.

3.3. The development phase

The development stage consists of three steps of activities: a) designing a prototype, b) perform formative evaluations andc) prototype revisions.

a. Designing a prototype

The prototype design activity begins by devising a problem solving model for science learning that is documented in book.This model comes with Student Book, and LKPD.

The preparation of students' books is based on the signs issued by Kemendikbud on the rules of student book writing. Student books are developed to support the implementation of problem solving models in the classroom.Therefore, in the book students are provided "problem solving corner" which provides an overview of the steps of problem solving for students.

In addition to student books, the model also comes with LKPD. LKPD in detail practice problem solving in students.LKPD is prepared for each meeting. LKPD contains: topic; purpose of activity, discourse, problem formulation, hypothesis / prediction / inference, tools and materials, work steps, observation data, questions, and conclusions.This step produces a draft 1, which then enters the formative evaluation phase.

b. Conduct formative evaluation

The next stage in the development phase is formative evaluation. Borg& Gall [14] call it as*justice expert or expert validity*. Formative evaluation is intended to get feedback from experts as a product revision before use. Formative evaluation in this research is validation test activity by expert. Selection of experts is based on the background of expertise, needs, and object of research.

c . Revision of prototype

The next step in the development stage is the revision of the prototype. After the validity test, a revision is made according to the input received. The revised result is a prototype 2 draft of the problem solving model for science learning that is ready to be tested on a limited basis. Revisions are made based on input from experts. Expert judgment must be in a valid category, and once the product is valid, it is new to use. If the expert recommends improper it will be revised again and the formative evaluation phase will be repeated. If the results of the expert's assessment have declared a valid prototype, then the research continues at the next stage

3.4. Implementation phase

Implementation stage is done through limited trial. A limited trial is a beginning of qualitative evaluation of the product to be produced [14].A limited trial was conducted on 24 and 26 August 2015 in Class VII 2 SMPN 12 Padang.At the stage of the trial, the researchers asked the teachers to run the model of problem solving to learning science and the use of devices that have been prepared. Researchers follow the learning process and record events during the process.After a limited trial, a discussion was held with the teacher to get feedback on the syntax of the model and learning device used. After a limited trial process, Focus Group Discussion (FGD) was conducted.FGD is product validation by practitioner that is science teacher. FGD purpose is to get feedback on products developed from practitioners in the field. The FGD was held on October 2015 at SMPN 1 Padang.

3.5. Evaluation phase

In accordance with the ADDIE model evaluation stage of which aims to obtain empirical evidence of the results of the model developed in the student, then make a diagnosis and revision according to the diagnosis. At this stage the experiments were expanded in experimental form. The aim is to determine whether the products meet its performance objectives.The expanded trial is done in experimental form.

In this study, field tests used quasi experimental design in the form of Static-Group Comparison Design. Field test model problem solving for science learning, conducted in junior high schools that have implemented the 2013 curriculum. SMP which has implemented the 2013 curriculum in Padang City is SMPN 1 and SMPN 8 Padang, for schools with good category (high level). SMPN 12 and SMPN 31 Padang for middle level. SMP SIMAK and Junior High National for school with category less (low level). For the field test each school was taken from each category [3] by lottery technique. Selected schools are SMPN 1 Padang, Padang and SMP SMP 12 National Padang.From each school 2 classes are taken for testing. In SMPN 1 Padang was chosen class VII F for experimental class and class VII E for control class. In SMPN 12 Padang was chosen VII 2 class for experimental class and class VII 1 for control class. While in the National Junior High School for class VII there is only one class that serve as experimental class.Implementation of this learning model conducted for 8 weeks (16 times of meetings).

3.6. Research instrument

The instruments used in this study can be grouped into: (1) validation instruments, (2) instruments of practicality consisting of the practicalities of the learning model; the practicability of the device according to the teacher; the practicability of the device according to the students, (3) the instrument of effectiveness includes the instrument of attitude observation; instruments of knowledge aspect assessment; and science process skill assessment instruments.

3.7. Data analysis technique

The data obtained through the instruments that have been expressed previously then analyzed by analytical techniques as follows:

1. Data analysis of validity problem solving model for science learning

Validation results are analyzed descriptively qualitatively by analyzing various inputs to obtain valid problem solving models for science learning.While the results of quantitative data are analyzed descriptively quantitative.

2. Analysis of practicality data model problem solving for science learning

Stages to determine the model's practicality were adopted from Haviz's research [15], where this practical aspect is determined from the implementation of the learning model. The model implementation data in the learning is obtained from the observer by using the observation sheet. Observers are two people. Analysis of learning implementation data is done by following procedure:

$$R = \frac{\sum_{i=1}^{n} V_i}{nm} \quad (1)$$

While R is the average result of the assessment from the observer, V_{ij} is score of assessment result of observer to-j against criterion i, n is the number of observers who judge and m is the number of criteria assessed.

Table 1. Criteria for the Implementation of Learning

Performance value	Performance Criteria
T > 3,20	Everything is done
2,40 < T ≤ 3,20	Sebagian besar terlaksana
1,61 < T ≤ 2,40	Mostly done
0,81 < T ≤ 1,60	A small part is done
T < 0,80	Nothing happened

Adopted from [16]

The criteria used to determine the problem solving model for science learning meet the practical aspect seen from the implementation,When the minimum value *T* is within the criteria largely accomplished. If the value of *T* does not reach the minimum value, then it needs to be revised, especially on the unattended aspects.

4.Results and Discussion

This research is based on the development stages of ADDIE includes: analyzing, design, develop, implement, evaluate. The results are described as follows.

a. Analyze

In the analysis phase, the need analysis of the problem solving model for science learning in the field at this time.Based on the analysis of KD in KI 2 it can be concluded that science learning in junior high school should train students' scientific attitude like curiosity, objetive, honest, meticulous, meticulous, diligent, responsible, open, critical, creative, innovative and environmental care. A scientific attitude is manifested both in classical learning and collaborative or cooperative as stated at KD 2.2. To achieve KD in KI 2, the problem solving model for science learning is applied with group learning, whereas the expected scientific attitude is achieved honestly, thoroughly and responsibly.

Analysis of KD in KI 4 then the science process skills that must be achieved in science learning of SMP is to communicate, and make observations as mentioned in KD 4.1 and 4.2.Another skill is practicum skills in KD 4.1.

Analysis on the KI KD 3 concluded Material Organization System Life has 5 sub subject that cell as a structural and functional unit of life, tissues, organs, organ systems and organisms. Material Changes in Nearby Objects have two sub-subjects: the changes of things that exist in everyday life and the separation of the mixture.

b. Teacher analysis

The analysis of teachers is done because the teacher as a user of model problem solving for science learning in the field. The results of the analysis of science teachers in Padang City are as follows. 1). 35% of science teachers in Padang City understand the steps of problem solving. 2). 41% of science teachers in Padang City understand the role of teachers in learning with problem solving. 3). 91% of IPA teachers of Padang City agree that the problem solving model is appropriate for the Curriculum 2013.4) . 91% of IPA teachers in Padang agree that problem solving skills need to be trained on students.5). . 66% of science teachers in Padang City understand that one of the goals of science learning is to practice problem solving skills.

This analysis shows that almost half of science teachers of SMP in Padang do not know that problem solving skill is one of science learning objectives in elementary education. Most teachers do not understand the steps of problem solvingand the role of the teacher in the learning with problem solvingBut almost all science's teachers of Padang City agree that problem solving needs to be trained on students

c. Student analysis

Student analysis is performed to see the characteristics of students, such as age. From the students data, it was revealed that the general student sitting in class VII have ranged between 12-13 years of age. According to Piaget's learning theory, the main characteristic of development at the formal operational stage of age 11 / 12-18 isable to think abstractly, logically, and able to develop a hypothesis.At this age, students are included into the category of individuals who are able to develop their cognitive potential that has been trained in the use of media including teaching materials LKPD.

Another characteristic is the science process skills of students of class VII.Preliminary test results on the science process skills as a description of the problem solving ability of students class VII Padang showed an average below 50%.Observation skills, and designing experiments respectively 47% and 39%.Predicting, asking questions and preparing hypotheses 18%, 40% and 27%. Compile conclusions, make inferences, apply ideas to new situations and classify 40%, 34%, 44% and 34% respectively.A high enough ability is to use simple equipment 84% and make measurements 61%.This result shows from 10 students, then 3 or 4 people who have problem solving ability.

4. Learning analysis

Analysis of the implementation of learning in class VII implemented with observation techniques. The first observation was conducted on peer-teaching of the 2013 Curriculum Implementation Training on 22-26 June 2014. At the time of application of the scientific approach in the classroom, teachers are generally not problematic in carrying out activities observing, gathering information, and communicate.But almost all teachers have difficulty when implementing the activities of the questioning.

All observed teachers (71 people) used the discovery learning model.The first phase of the discovery learning model is the problem statement. But most teachers are not able to carry out that phase.

b. Design phase

Based on literature studies related to model construction, theoretical basis, and learning activities, the syntax of problem solving model for science learning includes introduction, observation, initial problem, collecting data, organizing data, analyzing / generalizing data, communicating.This syntax is adopted from problem solving activities proposed by Carin [9].

In accordance with science characteristics consisting of science products, science processes and attitudes, then a suitable learning model is a group of information processing models, and social groups. So the problem solving learning model in this research will be developed by modifying the two groups of learning models.Based on literature study related to model construction, theoretical base, and learning activity, the syntax of problem solving model for science learning includes introduction, observation, initial problem, collecting data, organization data, analysis / generalizing data, communicating.This syntax is adopted from problem solving activities proposed by Carin [9].

In the design phase of this study formulated the goal of syntax problem solving model for science learning, description of each syntax, the role of teachers and students, as well as supporting tools. The syntax of problem solving model for science learning begins with the introduction. Introduction is important in learning, therefore it needs to be carefully planned.The thing most students remembered during the lesson is the first 10 minutes and the last ten minutes.

The observations referred to in this model are the use of one or more sense devices such as sight, listener, smell, touch, and taste for collecting information[17].Observation is useful to improve students' curiosity so that learning becomes meaningful. The science process skill contained in the observation activity is observed. Observing can be done by reading, listening, listening, watching, watching, etc. with or without tools[18]. In this research, the observation phase is done by reading the discourse that has been prepared in LKPD.

The initial problem is the next phase in the problem solving model for science learning.Formulating the problem is an important activity in problem solving.According to Piaget, problem formulation is one of the most important and creative parts of science that is often overlooked in science education[19] further states today that educators often advocate "problem solving", but we seldom hear the importance of "problem creation" and ask questions.

Problems are questions that arise from students after observation. Problems can be questions about what they do not understand, or in the form of curiosity to get additional information.Problems can be simple questions.Problems in the problem solving model for science learning, are simple questions that students will answer after learning.

In this research the initial problem phase is done by filling the "problem formulation" column in LKPD. To train the seventh grade students able to formulate the problem then in the column "problem formulation" given the signs to formulate the problem.For example, by guiding students to create questions that begin with one of the questions asked (4W + 1H), what, when, where, why and how. Next gives the question word at the beginning of the problem formulation, students proceed by filling in the prepared points. This also applies to train students to formulate hypotheses (Rezba, TT). Scientific process skills that are trained in this phase are, ask questions, formulate hypotheses, make inferences or make predictions[17].

The next phase is to collect data / information.Collecting data can be through experimental activities in the laboratory, through literature review, observing events, objects or activities and ask the experts. According to Carin [9], this activity serves to improve the students' skills in obtaining information, developing vocabulary, measuring, recording / recording data, using numbers, manipulating materials (utilizing other materials of the same function).

In order for this activity to run smoothly, then the teacher must prepare well. In this study tools, materials and procedures that must be done in student collecting data have been written in LKPD.Scientific process skills that are trained in the phase of collecting data / information is observing and measuring [17]. Activities undertaken by students in the form of exploring, trying, discussing, demonstrating, imitating, conducting experiments, reading sources other than textbooks[18].

Organization data is the activities of grouping data according to certain rules, change the data into graphs, presents data in the form of tables, charts, diagrams. This activity is to train students accustomed to record data well, grouping data, changing data in the form of tables, and graphs. Products generated in this phase are calculations of data, carta, tables, graphs, groups or groups of information, outlines, and disaggregated objects.Scientific process skills trained in this phase is to create data tables.

Phase-6 is the analysis and generalization of data included in this activity is to process the information that has been collected from the results of data collection activities on the experiments as well as the results of observation activities. This activity aims to develop skills in the identification of variables, inference, interpretation of data, develop ideas, develop models, generalize, take conclusions, and predict.To facilitate this activity teachers should prepare LKPD, which helps students develop their reasoning power through questions.

Communicating is the last stage of the problem solving model for science learning. Communicate contains activities of submitting observations and conclusions orally, in writing or other media. In this model communicate is done orally in the form of class discussion.

Communicating has several important goals, firstly improving students' thinking ability[12]. Informing students about something does not necessarily make them understand, but communicating and discussing a topic will help students strengthen and expand their knowledge.

The second goal is to increase student engagement. Communicating activities will increase the student's responsibility for his or her learning and should not be totally dependent on the teacher. Communicating also aims to help students learn important communication skills and thought processes. Developing honest, meticulous, tolerant, systematic thinking, expressing opinions briefly and clearly, and developing good and correct language skills [9, 12].

The next model component is the social system. The social system describes the role of teachers and students in learning.In accordance with the philosophy of learning that is used as the foundation of progressivism, then the learning on this model centered on students and give greater emphasis on creativity, activity, naturalistic learning, authentic assessment and peer experiences.Students perform learning activities in the form of, observing, measuring, counting, thinking, generalizing and providing assistance to peers in need.

Teachers act as a motivator, with the motivation in students when formulating the problem, when guiding the discussions etc.. The teacher also acts as a facilitator. As a facilitator, teachers prepare all the tools needed for learning, creating an atmosphere that enables the learning activities to run smoothly. Teachers also act as mentors, and reflectors.

The next model component is the reaction principle. Principle of reaction problem solving model on science learning, explaining how teachers appreciate, place and respond to actions undertaken by students. Some activities on the problem solving model, are still relatively new for students, for that teachers need patience in guiding and directing. These activities are identifying problems, formulating problems, formulating questions, formulating hypotheses.

Learning activities in this model use cooperative learning. Cooperative learning is useful for empowering peer roles to provide learning assistance to other friends. Nevertheless, sometimes cooperative learning makes students bored, therefore teachers should be creative in rewarding the most active groups. In addition to group awards, teachers should also reward individuals who exhibit positive activities, such as asking questions, responding to friends' questions, refuting, giving appropriate arguments and so on.

Teachers must be creative in determining which group should present the results of the discussion, if the performing is not all groups.Determination of the group can appear with a variety of games, so the group did not feel forced. Teachers should also reward the group performing.Awards can be stellar images of cardboard, "smile" pictures etc.This drawing card is group collected. The group that gets the most pictures means the most active group.Such awards can help motivate students to learn.

The next model component is a support system. Support system for problem solving model in science learning is model book, student book, and LKPD The support system provided helps the teacher in applying this model. In the book students have been provided "problem solving corner" which gives an idea to the students how to do problem solving steps.

In addition to student books, the model also comes with LKPD.LKPD more detail to train problem solving in students.LKPD will greatly assist teachers in applying problem solving models to science learning.

The impact of the learning model consists of instructional impact (direct impact) and impact accompaniment (indirect impact). Instructional impact of mastering the concept of science, science process skills and the impact of accompaniment in the form of values and attitudes. Mastery concepts include the concept of Science Objects and Their observations; Classification of Objects; System of Life Organization; Material Changes.

The next instructional impact is the science process skill.Scientific process skills include basic science process skills in the form of observation and classification; processing skills in the form of prediction and inference; integrated science process skills in formulating problems, asking questions, writing procedures, identifying experimental materials, inductive thinking as well as deductive in concluding, identifying variables, interpretation of data.

Impact accompaniment of scientific attitudes and social attitudes.Scientific attitude in the form of: curiosity, thorough, honest, discipline, obey rules, hard work, the ability to apply procedures. While social attitudes of being polite, respecting the opinions of others, the ability to communicate with good and polite language.Impact accompaniment is known through observation during the lesson.

*c. **Develop phase***

The development stage consists of three steps: 1 designing prototype, 2 doing formative evaluation and 3 prototype revisions.

1. Design the prototype

The activities of designing prototypes begins by devising a model problem solving for science learning which is documented in book form, as well as model supporting tools in the form of student books and LKPD.

a. Model book

The problem solving model book for science learning contains four parts: one is rational, part two supporting theory, part three component model problem solving for science learning, part four is the implementation of science learning with problem solving model.

b. Student books

Student books are developed to support the implementation of problem solving models in the classroom, so students' books are provided with the "let's experiment" section that provides an overview of problem solving steps.In this activity students are trained to develop thinking skills and solve problems that are guided by problem solving steps.

In this section there are four columns of varying colors to make it more appealing to students, each column representing a problem solving step. The giving of varied colors aims to increase students' interest and motivation in learning. In the student book there are subject matter that must be mastered by students.On some pages there is a "remember" column that contains important material that students must keep in mind.

c. LKPD

LKPD (Student Activity Sheet) is a student activity sheet that contains problem solving steps consisting of problem formulation, hypothesis, data collecting, data processing, and concluding as solution of problem solving.

Discourse columns in LKPD are useful to guide the implementation of the observation stage in the problem solving model for science learning.In this research, the observation phase is done by reading the discourse that has been prepared in LKPD. Observation is useful to improve students' curiosity so that learning becomes meaningful. The science process skill contained in the observation activity is observed.Observing can be done by reading, listening, scrutinize, take a look at, watching, etc. with or without a tool[18].

5. Conclusion

Based on the development research done, the syntax model problem solving for science learning at junior high school consists of the introduction, observation, initial problems, data collection, data organization, data analysis/ generalization, and communicating.

References

[1]. Carin, A. and Sund B. 1964. *Teaching Science through Discovery* . Ohio: Charles E Merril Books, Inc.

[2]. Rustaman N. 2005. Strategi Belajar Mengajar Biologi. Malang: Universitas Negeri Malang.

[3]. Sanjaya, W. 2011. *Strategi Pembelajaran Berorientasi Standar Proses.* Jakarta: Kencana

[4]. Tim TIMSS Indonesia. TIMSS *(Trends in International Mathematics and Science Stydy).*http://litbang.kemdik bud. go.id/detail.php?id=214 diakses tanggal 9 Nopember 2012.

[5]. Poedjiadi A. 2007. Pendidikan Sains dan Sains Terpadu. dalam *Ilmu dan Aplikasi Pendidikan* (Mohammad Ali dkk. ed). Bandung: Pedagogiana Press.

[6]. Kemendiknas. 2011. *Panduan Pengembangan Pembelajaran IPA Terpadu.* Jakarta: Dirjen Pendidikan Dasar Direktorat Pembinaan Sekolah Menengah Pertama.

[7]. Rustaman N.dkk. 2011. *Materi dan Pembelajaran IPA SD.* Buku Materi Pokok UT. Jakarta: Universitas Terbuka.

[8]. Estes, T., Mintz S., and Gunter M. 2011. *Instructions A Models Approach.* Boston: Pearson Education, Inc.

[9]. Carin, A. 1997. *Teaching Science through Discovery,* 8th ed. Ohio: Merril an imprint of Prentice Hall.

[10]. Joyce, B., Weil, M. with Shower B.1992. *Models of Teaching* 4th ed. Massachusetts: Allyn and Bacon Publisher.

[11]. Rusman. 2010. *Model- model Pembelajaran Mengembangkan Profesionalisme Guru.* Jakarta: Rajawali Press.

[12]. Arends R. I. 2007. *Learning to Teach* 7th ed. Boston: McGraw Hill.

[13]. Jihad, A., dan Haris, A. 2008. *Evaluasi Pembelajaran.* Yogyakarta: Multi Pressindo.

[14]. Borg, W.R., Gall, M.D., and Gall, J.P. 2003. *Educational Research an Introduction* 9th ed. Boston: Pearson Education Inc.

[15]. Haviz, M. 2012. Pengembangan Model Pembelajaran Integratif pada Biologi Perkembangan Hewan di Perguruan Tinggi Agama Islam. *Disertasi.* Universitas Negeri Padang.

[16]. Musdi E. 2012. Pengembangan Model Pembelajaran Geometri Berbasis Pendidikan Matematika Realistik SMPN Kota Padang. *Disertasi* (tidak diterbitkan). Universitas Negeri Padang.

[17]. Nur, M. 2009. Modul Keterampilan-Keterampilan Proses Sains disadur dari *Inquri Skills Activity Book.* Surabaya: Unesa

[18]. Kemdikbud. 2015. *Buku Guru IPA.* Jakarta: Pusat Kurikulum dan Perbukuan Kemdikbud.

[19]. Hung, W. (2016). All PBL Starts Here: The Problem. *Interdisciplinary Journal of Problem-Based Learning, 10*(2), 2.

IOP Conf. Series: Materials Science and Engineering **335** (2018) 012085 doi:10.1088/1757-899X/335/1/012085

The Effect of Using Evolution Textbook Based on ICT and Metacognitive on Cognitive Competence of Biology Students at State University of Padang

H Helendra[1*], M Fadilah[1], F Arsih[1]

Biology Department, Faculty of Mathematics and Natural Science, State University of Padang, Indonesia

*helendras@yahoo.com

Abstract. Implementation of evolution lectures at Biology Department Faculty of Mathematics and Natural Sciences State University of Padang has been considered not optimal. The reasons are the limited availability of textbooks and students' learning attitudes. Because currently the students are very familiar with the internet and even has become a necessity, it has developed textbooks of evolution based on ICT and metacognitive. Selection of ICT based is in order to optimize the utilization of multimedia, and this is very compatible with the development of learning technology. While metacognitive based is in order to train students' learning attitudes to be able to think analysis, creative and evaluative. The aim of this study is to determine the effect of the use of evolution textbooks based on ICT and metacognitive to the cognitive competence of students of Biology Department State University of Padang. The data of this research is students' cognitive competence obtained from the implementation of effectiveness test of evolution textbook in the form of student learning outcomes. The research instrument is a learning result test designed to determine students' cognitive competence. The subject of the study is a group of students in evolution course consisting of 33 students. Lectures are conducted through face-to-face and online lectures on Edmodo's platform. The result of data analysis shows that there is an increase of cognitive competence of biology students after learning using ICT and metacognitive based evolution textbook, where average achievement is 77.72 with Percentage of achievement of criteria mastery is 81.25%. Therefore, it can be concluded that the evolution textbook based on ICT and metacognitive is effective in improving cognitive competence of students of Biology Department, Universitas Negeri Padang.

1. Introduction

The nature of education, including higher education, is basically shaping and developing the patterns and ways of thinking of learners so that they can develop their potential to respond and solve problems in daily life later. The development of patterns and ways of thinking are trained intensively through the interaction between learners in the learning process. Students must build knowledge independently. For that, the source of learning, one of the teaching materials should be available and accessible to students easily and quickly. The teaching materials can be said to be good if it has the right content, interesting to use and easy to understand. Teaching materials are materials or lesson materials systematically arranged, used by teachers and students in the learning process [1]. The teaching materials are part of the learning resource. Teaching materials are a set of subject matter that is organized systematically both written and unwritten so as to create an environment/atmosphere that allows students to learn.

The teaching materials are prepared with the objective to provide materials suitable to the needs of the learner, that is in accordance with the characteristics and setting or social environment of the students, assist the learner in obtaining alternative teaching materials in addition to textbooks that are sometimes difficult to obtain, facilitate the teacher or lecturer in implement learning. Teaching materials serve as: (1) Guidelines for teachers who will direct all activities in the learning process, (2) Guidelines for students or students who will direct all activities in the learning process, (3) Evaluation tool of learning achievement result.

Textbook is one form of printed materials. Benefits that can be obtained by a teacher or lecturer if developing his own textbook, namely (1) obtained teaching materials in accordance with student learning needs, (2) students are not dependent on textbooks that are sometimes difficult to obtain, (3) teaching materials become richer because they are developed using various references (4) increase the knowledge and experience of teachers or lecturers in writing teaching materials, (5) teaching materials will be able to build effective learning communication between teachers/lecturers with students because students will feel more confident to his teacher or lecture**r. According to Depdiknas** [2], books can be used as reference material, or can be used as a quality written material.

There are three principles that are needed in the development of textbooks. These three principles are relevance, consistency, and adequacy. The principle of relevance means that instructional materials should be relevant to the achievement of competency standards and basic competencies. If the expected ability is memorized facts, then the material presented is fact. If basic competence requires the ability to do something, the subject matter is the procedure or the way to do something. The principle of consistency is grammatical in the development of teaching materials. For example, if basic competence requires students' ability to master three kinds of concepts, the material presented is also three kinds. For example, the ability expected to be mastered by the students is to construct a deductive paragraph, the material at least the meaning of deductive paragraph, how to construct a deductive paragraph, and how to revise the deductive paragraph. That is, what is being asked is what is given. Principle of adequacy, meaning that the material presented should be sufficient to achieve basic competence. The material should not be too small and should not be too much. If the material is too small, it is likely that students will not be able to achieve basic competencies by utilizing the material. If the material is too much will also take a lot of time to learn it.

Teaching materials for evolution course in Biology Department Faculty of Mathematics and Natural Sciences Universitas Negeri Padang are available. However, the implementation of evolution lectures has not been felt optimally. The author finds three conditions that become the main cause, the pattern of lecturing, limited textbooks, and attitudes of students. Prior to 2009, evolution lectures were dominated by the use of lecture methods and reading assignments. As a result, students become less creative and few are involved in understanding the concept. Starting in 2009, the authors have improved the implementation of learning in accordance with the paradigm of active learning, which involves more students through assignments, presentations, and class discussions. However, the authors still see this lecture not yet optimal because the available teaching materials have many limitations. The type of teaching materials available and used as the main teaching material so far in evolution lectures is the handout. This evolutionary teaching material (handout) has a language that is less straightforward. Some students find it difficult to find the meaning of sentences per sentence nature of the teaching materials used. In the exam, the student is only able to rewrite the contents of sentence as stated in hand out. As a result, students' thinking ability only develops at the level of recall (C1), which is the lowest level of thinking ability according to Bloom's Taxonomy.

The next weakness of this handout is the way of displaying monotonous material and incomplete material coverage. Based on interviews with several students, it was revealed that the handout structure did not interest students to read it because it was just a description of the material in the paragraphs, without any colored drawings/schemes/charts that facilitate the understanding of the material. Handout also does not have a full description of the material in accordance with the course syllabus that led. The topics described in the handout are only 5, while the topics demanded in the syllabus are 14.

The reading material in print or electronic media is actually available enough to be used by students, both written in Indonesian and English. However, the student's initiative to actively select and use various sources of learning is still low (the results of interviews with students). The reasons put forward are that the material on these available books is not very much in line with the syllabus. In addition, the material that can be found in the translation book is difficult to understand and confuses the students.

Other learning resources that are widely accessed by students are internet media, especially blogs/web that are accessible through Google search engine. Based on writer's search for a number of Indonesian blogs, the description of evolutionary material tends to be subjective and refers to the opinion of certain groups of scientists. As a result, students sometimes get caught up in the subjectivity of the opinion and find it difficult to understand the material in accordance with the objectivity of science.

This evolution handout has been used for more than 10 years. However, there has not been a good revision of the content, structure and display components. Handouts still use general and specific instructional terms. This indicates that hand out design does not support the achievement of competence, because instructional objectives describe the achievement of learning outcomes in the cognitive domain, while the competence emphasizes the achievement of learning outcomes in the cognitive, affective and psychomotor aspects. In response to the above issues, the authors have developed textbooks based on information and communication technologies (ICT) and metacognitive competence for evolution courses in Biology Department State University of Padang. ICT-based textbooks intended are teaching materials that are also used with edmodo networks.

Edmodo is a social network-based learning platform intended for teachers, students and parents. Edmodo was first developed at the end of 2008 by Nic Borg and Jeff O'hara and Edmodo itself is practically an e-learning program that implements an easy, efficient and more enjoyable learning system. With edmodo allows teachers to share content, distribute quizzes, assignments, and manage communication with students, colleagues, and parents [3].

ICT-based election is a manifestation of the demand of the 2013 curriculum, especially on the optimization of multimedia usage, and is consistent with the development of learning technology. In addition, ICT also allows textbooks to connect with several electronic learning sources such as video or process animations. ICT-based textbooks can also make it easier for students to access materials outside the lecture schedule. Development of ICT-based textbook is very compatible with e-learning courses that have been implemented in the course of evolution, but not optimal in terms of providing appropriate teaching materials. Evolutionary textbooks based on information and communication technologies (ICT) and metacognitive competencies developed have been tested for validity, and validated by validators with a validity value of 87.60. But the effectiveness of this textbook has not been tested, therefore it is necessary to conduct research to test its effectiveness. The purpose of carrying out this effectiveness test is to measure the extent to which textbooks that have been produced can improve student learning outcomes.

The development of information technology is a reason that requires the need for adjustment of learning patterns that are more flexible from face to face to learning in cyberspace. Currently many universities are responding to these trends by offering courses that can penetrate the distance or distance learning. Recent widespread availability of educational resources on the World Wide Web holds great potential for transforming education. In science education, for example, students can access real-time images from space exploration (Anonomous)[4]. Online learning resources is one component that helps the implementation of online lectures. Universities are very potential in providing online teaching materials as they consist of logical personnel who are trusted to develop content and design, although not necessarily involved in technical matters. ICT-based teaching book is a media that supports ICT-based lectures.

Since 2012, at Padang State University all lectures have been implemented through e-learning facilities that are managed directly by each lecturer. With this facility, students are expected to more easily obtain information quickly and efficiently. The source of information is no longer focused solely on the text of the book alone, but broader than that. The ability of multimedia technology that has been connected the internet increasingly adds ease in getting the information expected.

Evolution is a theoretically explicit course and an applicative example using ICT-based learning media. The main character of this course is to have a very high relevance to the various fields of science both within and outside the scope of biology, such as genetics, biochemistry, biomolecular, paleontology, archeology, and anthropology. Related material content can be easily accessed and visualized with multimedia facilities that users can download, including videos and animations that explain past events like continental separation. Therefore, lectures that were initially considered abstract and monodisciplinary will be more easily studied and understood in a multidisciplinary integrative manner. Multimedia is the channel of choice in conveying information in a more memorable way.

According to Saroso [5]' the advantages of ICT-based media in education are: (1) The learning system is more innovative and interactive, (2) Able to combine text, images, audio, music, animated picture or vidio in one mutual unity support for the achievement of learning objectives, (3) Being able to generate pleasure during the learning takes place. (4) Be able to visualize material that has been difficult to explain only with conventional explanations or props, and (4) Relatively easy and flexible storage media.

Edmodo is an electronic learning site that can be used as a supporting medium of learning process. According to Rismayanti [6], Edmodo is a social media platform often described as Facebook for schools and can function even more as needed. Edmodo is a social networking site where data sharing, events, schedules and so on for teachers/lecturers and students. In this study, the use of edmodo sites allows for a virtual classroom that can assist students in understanding the material of evolution The Edmodo website allows for a virtual classroom that will assist students in understanding the material of evolution. The Edmodo website comes with instructional materials in .doc, .pdf, .swf and .mp4 formats. Lecturers can also form discussion forums and include lecturing agenda in one week to one month. Thus learning and lectures will become more structured because students can find out a well-organized schedule. In addition students can also access Edmodo anywhere as long as there is an internet connection network.

Metacognitive is a term introduced by Flavell in 1976. According to Flavell, as quoted by Livingstone [7], metacognitive consists of metacognitive knowledge and metacognitive experiences or regulation. Metacognitive knowledge refers to the acquisition of knowledge about cognitive processes, knowledge that can be used to control cognitive processes. While metacognitive experience is the processes that can be applied to control cognitive activities and achieve cognitive goals. Livingstone (1997) defines metacognitive as thinking about thinking. Metacognitive, according to the figure is the ability to think where the object of thinking is the process of thinking that happens to yourself. Margaret W. Matlin [8] states that "Metacognition is our knowledge, awareness, and control of our cognitive process". According to Wellman (1985) quoted Usman Mulbar [9], metacognition is a form of cognition, a second or higher order thinking process which involves active control over cognitive processes. It can be simply defined as thinking about thinking or as a person's cognition about cognition.

Taccasu Project [10], describes metacognitive as follows: (1) Metacognition is the part of planning, monitoring and evaluating the learning process; (2) Metacognition is knowledge about one's own cognitive system; thinking about one's own thinking; (4) Metacognition involves both conscious awareness and consciousness of one's learning, (5) Metacognition involves both consciousness and consciousness of learning, (6) Metacognition is learning how to learn in situ learning learners encounters.

Based on several definitions that have been put forward in the above description can be identified the basic ideas about metacognitif as follows: (1) Metacognitive is the ability of the soul included in the group of cognition, (2) Metacognitif is the ability to realize, know, the process of cognition that occurs in (3) Metacognitive is the ability to direct the process of cognition that occurs in yourself, (4) Metacognitif is the ability to learn how should be learned which includes the process of planning, monitoring, and evaluation, (5) Metacognitif is a high-level thinking activity, because this activity is able to control the thinking process that is going on for yourself.

According to Preisseisen (in Paulinna Haren, et al, 2001), metacognitive consists of four skills: problems solving, decision making, critical thinking, and creative thinking. To gain tremendous learning success, teachers must train students to design what they want to learn, monitor student learning progress, and assess what has been learned.

There are 3 metacognitive strategies that can be developed to achieve student learning success, namely: (1) stage of learning conscious process, (2) learning planning stage (3) stage of monitoring and reflection learning. Metacognitive strategies are important because when students are able to design, monitor, and reflect their learning process consciously, in essence, they will become more confident and more independent in learning. Learning independence is a private possession for students to continue their long journey in meeting intellectual needs and discovering an infinite world of information. The task of the educator is to develop the metacognitive ability of all students as a learner, without exception.

Yula Mirand [11], revealed that metacognitive learning in cooperative learning strategies improves students metacognitive abilities in biological subjects. Yuni Wibowo, Asri Widowati, Titik Krisnawati [12], revealed that students' cognitive and metacognitive abilities of SMAN 1 Ngalik Sleman increased through learning using Ruondhouse diagrams. It can be reflected that the use of images, visual symbols and simple reminders can help students shape their knowledge. Maulana [13] uses a metacognitive approach as an alternative to mathematics learning to improve students' critical thinking skills.

2. Research Method

The research was conducted to test the effectiveness of evolutionary textbook based on Information and Communication Technologies and metacognitive which has been developed to the cognitive competence of the students of Biology Department, State University of Padang. So the effectiveness test in this research is the implementation stage of teaching materials that is the use of textbook in evolution lectures naturally through face-to-face and online lectures on Edmodo platform. After the textbook is used in the lecture, its influence on students' cognitive competence is tested. The test results are used for the purpose of evaluation and further improvement of teaching materials. Thus, the test design is adapted from "Single Subject Design". According Thiagarajan, this design is widely used in order to evaluate learning materials.

The subjects were 33 students of Biology Department who enrolled in the evolution course in the semester of January-June 2016. The study was conducted for one semester. The implementation of evolution lectures using learning resources in the form of textbooks based on ICT and metacognitive. The data collected is primary data in the form of students' cognitive competence. The research instrument used to collect data is a test of learning outcomes. The influence of textbook implementation is described quantitatively.

The study was conducted in two cycles. In cycle 1, students use evolutionary textbooks as learning resources in face-to-face learning. In addition, students learn also by using edmodo site with the guidance of lecturers. Different with cycle 1, in cycle 2, students in addition to learning by using textbooks and edmodo platforms, are asked to present the results of their discussion. This is done based on the reflection of learning outcomes cycle 1.

3. Results and Discussion

Cognitive competence is the result of student learning measured in two cycles of activity. From the results of this study obtained the data as follows:

Table 1. Average cognitive competence and achievement criteria mastery

Cycle	Average Cognitive Competence	Achievement Criteria Mastery (%)
1	75.46	59.37
2	77.72	81.25

The result of data analysis shows that there is an increase of cognitive competence of Biology Education students after learning using ICT and Metacognitive evolution based textbook. In cycle 1, the average achievement of students' cognitive competence is 75.46, but the percentage of achievement criteria mastery still not reached 75%. From cycle 1, as many as 40.63% of students achieve cognitive competence under the minimum mastery criteria, and students who have reached the minimum mastery criteria as much as 59.37%.

In cycle 2, the average achievement of students' cognitive competence is not too much different from cycle 1 that is 77.72, but the percentage of students who have reached the completeness criteria has reached 81.25%. So, students who have not reached the minimum completeness criteria are only 18.75%.

In construction, textbook consists of 3 parts, namely: introduction, material description, understanding reflection and competency test. In the preliminary section there is a column "explore the initial understanding", this space serves as a column to know the initial knowledge of students on the material to be studied. This column also acts as an apperception. Through apperception teachers or lecturers can focus the attention of learners on what will be studied. Metacognitive on the book is presented in the column of "critical thinking" and "inference", through this column, students trained to be able to think critically. Through inference activities students are trained to be able to analyze what has been observed. This is in line with the opinion of Trianto (2014)[14], that inclusion is the use of what has been observed to explain something that has happened.

To further improve the cognitive competence, in the second cycle, the presentation was done. The presentation was chosen based on the results of reflection, because in cycle 1, students are still not too ready with the activities in the textbook so it needs a review in the form of a presentation that aims to refresh the understanding of reading that has been done at home. Through the presentation, students are more prepared to learn and try to understand the material in textbooks in depth, so that their initial teaching schedule is sufficient. Early teaching includes the type and scope of knowledge that has been mastered and known to learners, level and stage and type of cognitive, affective and psychomotor ability that has been achieved by learners.

According to Adney (1985), Barby (1982), Griffiths and Grant (1985), and Mark (1986) as cited by Suparwoto and Yusman Wiyatmo (2005)[15], the low learning achievement in the classroom among others are: (1) erroneous opinions based on their own judgments, (2) learners develop a misconception of the experience they get in the classroom, (3) classroom learning is not always effective in reducing errors that students have. So the initial teaching of the students is necessary for more effective classroom learning. So the provision of early knowledge of students is necessary for more effective classroom learning.

4. Summary

Based on the results of the study can be concluded that the use of textbooks of evolution based on ICT and metacognitive affects the cognitive competence of Biology Department students, State University of Padang which can be seen from the improvement of the learning result of evolution.

References

[1]. Belawati, Tian, dkk. 2003. *Pengembangan Bahan Ajar*. Jakarta: Pusat Penerbitan Universitas Terbuka.

[2]. Depdiknas. 2008. *Panduan Pengembangan Bahan Ajar*. Jakarta: Direktorat Jenderal Manajemen Pendidikan Dasar dan Menengah.

[3]. Zakaria, Muhammad. 2017. *Tentang Edmodo: Pengertian, Manfaat, dan Fitur-Fiturnya yang Wajib Anda Ketahui*. http://www.nesabamedia.com/pengertian-manfaat-dan-fitur-edmodo/

[4]. Anonimous. *Introduction to Online Learning Resources*. http://ocw.usu.edu/instructional-technology-learning-sciences/producing-distance-education-resources/introduction.htm

[5]. Saroso, Siswo. 2009. *Upaya Pengembangan Pendidikan Melalui Pembelajaran Berbasis Multimedia*.

[6]. Rismayanti, A. 2012. Mengenal Lebih Dekat Edmodo (*online*). http://id.scribd.com/doc/138248827/

[7]. Livingstone, Jennifer A. 1997. *"Metacognition: An Overview"* http://www.gse.buffalo.edu/fas/shuell/CEP564/Metacog.html.

[8]. Matlin, Margaret W. 1998. *Cognition*. Philadelphia: Harcourt Brace College Publisher.

[9]. Mulbar, Usman. 2008. "Metakognitif Siswa dalam Menyelesaikan Masalah Matematika". Tersedia pada: http//www.usmanmulbar.files.wordpress.com.

[10]. Taccasu Project. 2008. "Metacognition" Tersedia pada: http://www.hku.hk/cepc/taccasu/ref/metacognition.html.

[11]. Miranda, Yula. 2010. Dampak Pembelajaran Metakognitif dengan Strategi Kooperatif terhadap Kemampuan Metakognitif Siswa dalam Mata Pelajaran Biologi di SMA Negeri Palangkaraya. *Jurnal Penelitian Kependidikan*, Th. 20, No. 2, Oktober 2010.

[12]. Wibowo, Y, Widowati, A, Krisnawati, T. 2012. Pengaruh Pembelajaran Diagram Roundhouse terhadap Kemampuan Kognitif dan Metakognitif Siswa SMAN I Ngalik Sleman Yogyakarta. *Jurnal Bioedukasi*, Vol. 5, No.2, Februari 2012.

[13]. Maulana. 2008. Pendekatan Metakognitif sebagai Alternatif Pembelajaran Matematika untuk Meningkatkan Kemampuan Berpikir Kritis Mahasiswa PGSD. *Jurnal Pendidikan Dasar*, No. 10, Oktober 2008.

[14]. Trianto. 2014. Model Pembelajaran Inovatif dalam Teori dan Praktek. Surabaya: Pustaka Ilmu.

[15]. Suparwoto dan Yusman Wiyatmo. 2005. Penjajagan Bekal Ajar Awal dan Hasil Belajar Akhir pada Mata Kuliah Fisika Dasar Program Studi Fisika FMIPA UNY. *Prosiding Seminar Nasional Penelitian, Pendidikan & Penerapan MIPA*, Yogyakarta, 8 Februari 2005.

IOP Conf. Series: Materials Science and Engineering **335** (2018) 012086 doi:10.1088/1757-899X/335/1/012086

Metacognitive Ability Relationship with Test Result of Senior High School of Biology Teacher Competence in *Sijunjung* District

A Ardi*, M. Fadilah, W Ichsani

Department of Biology, Faculty of Mathematics adn Sciences, Universitas Negeri Padang, Indonesia

ardibio@fmipa.unp.ac.id

Abstract. This research aimed to reveal how the relationship between metacognitive ability and the test result of biology teacher competence in Sijunjung District. The population of this descriptive research were all high school biology teachers in Sijunjung District, and sample is all teachers who are members of the population, which is 23 biology teachers. The instrument used in this research are a questionnaire of research on teacher's metacognitive ability and document about teacher competence test result. The questionnaire was validated first by two lecturers of biology and one lecturer of English. Data analysis using Pearson Product Moment's. Based on the results of research and discussion that have been described, it can generally be concluded that there is a low relationship between metacognitive ability with competence test results of high school biology teachers in Sijunjung District. Partially, the relationship of metacognitive ability with the test result of professional competence of biology teacher showed significant result, with correlation coefficient 0,46 and t table 1,72 while titung 2,37. The contribution of metacognitive ability to the competence test result of the teacher is 21.6%, while the other 78.4% have not been revealed in this research.

1. Introduction

Education has a strategic role in supporting the progress of the nation. In connection with this, in the Law of the National Education System of the Republic of Indonesia no. 20. Year 2003 [1], stated that education is a conscious and planned effort to create learning atmosphere and learning process so that learners actively develop their potential to have spiritual power, self-control, personality, noble character, and also skills needed himself, society, nation and state.

The educational process begins early on through the tiered stage, to cultivate the interests and talents of learners. On the other hand, parents hope that through education, their children will become useful and skilled and moral. The above expectations and desires are in accordance with the national education function contained in the Law of the National Education System of the Republic of Indonesia no. 20. Year 2003 [1]: "National Education functions to develop the ability and form the character and civilization of dignified nation in order to educate the life of the nation, aiming for the development of potential learners in order to become a man who believes and piety to God Almighty, noble character, healthy, knowledgeable, capable, creative, independent, and become a democratic and responsible citizen".

IOP Conf. Series: Materials Science and Engineering **335** (2018) 012086 doi:10.1088/1757-899X/335/1/012086

One way to realize the function of national education is to hold the process of learning in schools, which is implemented by teachers. Based on Law Number 14 Year 2005 on Teachers and Lecturers [2] it is stated that teachers are professional educators with the main task of educating, teaching, guiding, directing, training, assessing and evaluating learners on formal education, and secondary education. Professional teachers proven by their competence will encourage the realization of process and product performance that can support the improvement of education quality. Teacher competence is four, namely (1) pedagogic competence (2) professional competence, (3) social competence, and (4) personality competence. Increased teacher competence is an indicator of teacher professionalism improvement itself. In an effort to improve teacher competence, the government conducted teacher competence test.

The teacher competence test includes pedagogic competence test and professional competence. According to Ministry of Education and Culture of the Republic of Indonesia [3], the desired pedagogic competence is the consistency of pedagogic mastery between content and performance, that is not just the mastery of teachers about the introduction of learners, learning models, planning, implementation, evaluation, but tests that can predict how teachers integrate the five in the implementation of learning. While the professional competence that must be possessed is (1) the mastery of the materials, structure, concepts and mindset of science that support the subjects being taught, (2) Develop profession through reflective action, and (3) Consistency of mastering the material of the teacher between content and performance.

Based on data from the Regional Education Balance of West Sumatera Province 2015 [3], the average of teacher competence test in West Sumatra is 58.37. This value has exceeded the national average of 56.69. The results test of teacher competence test in West Sumatera Year 2016 shows that the average value of biology teacher competence test is still low. The average yield of biology teacher competence test obtained is 60.36 (Education Quality Assurance Agency of West Sumatra, 2016). From 19 districts/municipalities in West Sumatera, there are 6 districts that get low average value. This can be seen in Table 1.

Table 1. The Average of Biology Teacher Competence in West Sumatra

No	Districts	Average Pedagogic Competence	Average Professional Competence	Total Average
1	*Kepulauan Mentawai*	45,36	50,39	48,94
2	*Pesisir Selatan*	49,52	51,80	51,11
3	*Sijunjung*	53,57	55,74	55,09
4	*Pasaman*	52,71	57,72	56,22
5	*Solok Selatan*	57,74	55,96	56,49
6	*Solok*	59,02	59,30	59,21

Source: Education Quality Assurance Agency of West Sumatra (2016).

The low level of biology teacher competence results test can not be separated from the ability of teachers in planning, implementing, and evaluating the learning materials and processes undertaken. Referring to Syaiful [4], this ability is known as metacognition ability. Furthermore, it is stated that the ability of metacognition is basically a consciousness of one's thinking about the process of thinking itself. This includes two components: metacognitive knowledge and metacognitive skills. Metacognitive knowledge deals with declarative knowledge, procedural knowledge, and conditional knowledge. Metacognitive skills are related to planning skills, prediction skills, monitoring skills, and evaluation skills.

Several studies have been conducted by previous researchers about the ability of teacher metacognition. Balcikanli [5] has identified 6 factors that can be used to identify teachers' metacognition abilities. The six factors include, (1) Declarative Knowledge, (2) Procedural Knowledge, (3) Conditiona lKnowledge, (4) Planning, (5) Monitoring, and (6) Evaluating. Furthermore, Daud, Nusantari, and Dama [6] states that the ability of teacher metacognition is very important to support students' metacognitive ability. Therefore, teacher factor is very important in developing students' metacognitive ability.

Kallio, et al.[7] states that the metacognitive ability possessed by a teacher is very important. This relates to teacher professional development, to reflect on teacher work and support in the learning process.

The low average score of teacher competence test obtained by teachers in some districts in West Sumatra is allegedly inseparable from its association with the teacher's metacognition ability, but scientifically this has not been revealed. Sijunjung District is one of the regions that has an average value of Teacher Competency Exam (UKG) below the average value of the province. This area is an expansion of Sawahlunto Sijunjung district, the education sector is still low, so it has encouraged researchers to examine the metacognition skills of biology teachers related to teacher competence test results. Selection of research sites because of the low level of competence test results of biology teachers and easy to reach by researchers. The other thing that underlies is the limitation of cost and time.

Some teachers in *Sijunjung* District also stated that the low value of UKG is inseparable from the teacher factor itself. Teachers lack preparation in conducting exams and teachers' perceptions that the value gained in influencing their work. Teachers lack the mastery of all material, because the teacher who teaches class X in the school year and then still teach the class.

Based on the problems that have been presented, then has been conducted research on the relationship of metacognitive ability with the results of competence test of biology teacher in *Sijunjung* District.

2. Method

This descriptive research discloses and describes the relationship between metacognitive ability with the results of biology teacher competence test in *Sijunjung* District. The population of this descriptive research were all high school biology teachers in *Sijunjung* District, registered under the Biology UKG 2016. The sampling technique used in this study is Saturation Sampling, where all members of the population sampled are 23 biology teachers.

The author used primary and secondary data. Primary data is the ability of teacher obtained directly to the sample through a questionnaire while the secondary data is a document of teacher competence test results obtained from the Quality Assurance Agency of West Sumatra Province

The questionnaire used in this study were adopted from Metacognitive Awareness Inventory for Teachers (MAIT) [5]. The validity of questionnaire has been done in the validity of the language. Validation results serve as the basis for the revision of the questionnaire using the following formula:

$$Validity\ value = \frac{score\ obtained}{maximum\ score}\ x\ 100\% \quad (1)$$

Maximum score = number of validator x number of indicators x maximum score

Criteria for validity coefficient can be seen in Table 2.

Table 2. Criteria for Validity Coefficients

Validitas Scale	Validity Value
90% - 100%	Very valid
80% - 89%	Valid
65% - 79%	Quite valid
55% - 64%	Less valid
≤ 54%	Invalid (unusable)

Source: Purwanto [8].

Data collected through questionnaires, by spreading statement sheets to be answered by the sample. Then the questionnaire containing the statements answered by respondents with the categories of choice are Strongly Agree (SA), Agree (S), Neutral (N), Disagree (D) and Strongly Disagree (SD). The score of statement are SS 5, S 4, N 3, D 2 and SD 1.

To analyze the relationship of metacognitive ability and result of biology teacher competence test used correlational analysis technique by Product Moment Pearson correlation formula :

$$r = \frac{n\sum xy - (\sum x)(\sum y)}{\sqrt{\{(n.\sum x2 - (\sum x)^2)\}\{(n.\sum y2(\sum y)^2\}}} \quad (2)$$

Note:
r = coefficient correlation
x = metacognitive ability
y = result of biology teacher competence test
n = sample

Table 3. Correlation Criteria

r-value	Correlation Criteria
0,00 – 0,20	Very Weak or Very Low so Correlation is ignored
0,21 – 0, 040	Weak or Low
0,41 – 0,70	Medium or Fair
0,71 – 0,90	Strong or High
0,91 – 1,00	Strong or High

Source: Sudijono [9]

To determine the correlation coefficient means or not, then tested the correlation coefficient by using the following *t* -formula[10].

$$t = \frac{r\sqrt{(n-2)}}{\sqrt{(1-r)^2}} \quad (3)$$

Note:
t = count value
r = value of correlation coefficient
n = number of samples.

According to Sudjana [10] the correlation of calculated result can be said significant if the value of $t_{count} > t_{table}$ at 95% confidence level with db = 21. Then to see the percentage of the extent to which the influence of independent variable *X* to the dependent variable *Y*, then use coefficient of determination (P) can be expressed with the following formula:

$$P = r^2 \text{ x } 100\% \quad (4)$$

3. Result and Discussion

The result that the authors get after doing the data retrieval of the biology teacher's biology proficiency in *Sijunjung* Dsitrict, obtained the lowest score range that is 97 and the highest score obtained is 114. From the data analysis obtained the average value of metacognition skills of biology teachers of *SMA Negeri* in *Sijunjung* is 106.6.

Result of data of competence test of high school biology teacher of *Sijunjung* District taken from West Sumatra Educational Quality Assurance Institution 2015, then obtained the lowest value data is

35,7 until its highest value is 77,4. For the average value of *Sijunjung* District Biology Sector is 55.09. This result is below the average value of West Sumatera province that is 60,36. The results obtained from the UKG derived from two competencies namely professional competence and pedagogic competence. Each score of professional and pedagogic competence test is 54.06 and 56.20.

To know the relationship of metacognitive ability with result of UKG biology used Pearson Product Moment correlation formula. Based on the results of the analysis that has been done then in general the relationship between metacognitive ability with the test results of biology teacher competence is 0.36 which means low correlation. Partially, the result of analysis revealed that there is correlation between metacognitive ability with professional competence of teacher that is 0,46 with enough criterion, while metacognition ability with pedagogic competence of teacher there is no relation that is -0,09 with very low criterion. These results indicate that the teacher's metacognitive ability. This means that the ability of teacher metacognition in this case more shows the correlation with teacher competence test results, in the professional field, compared to pedagogic.

Based on the results of this partial analysis, then tested to determine whether there is a significant relationship between metacognitive ability with the results of teacher professional competence test. The result shows that the significance of metacognitive ability with professional ability of teacher is 2.38, with *db* = 21; t-table = 1, 72. This result is greater than t-table value, meaning that there is a significant correlation between metacognitive ability and professional ability test result of high school teacher in *Sijunjung* District.

Furthermore, to know the donation given the ability of teacher metacognition to teacher profesioanal competence, conducted determinant analysis. The result of the analysis shows that the teacher metacognitive ability contributes 21,16% to the professional ability of high school biology teacher in *Sijunjung* District.

Through this study found a picture of the ability of teacher metacognition and its relationship with the results of competence test of high school biology teachers in Sijunjung District. Based on the results of this study it was revealed that the ability of teacher metacognition, enough to show the linkage with the results of professional competence test teachers, while with the results of teacher competence test pedagogic, very low.

Referring to Law Number 14 Year 2005 regarding Teachers and Lecturers, it means that the teacher is sufficiently able to master the broad and deep learning material that includes the mastery of curriculum subject material in the school and the scientific substance that oversees the material. In other words, it can also be stated that the teacher's metacognitive ability has enough correlation with the ability of the teacher to master the source of the teaching materials or the field of their expertise study. In this connection Kallio, et al.[7], stated that the teacher's metacognitive ability in developing teacher professional competence is important. This metacognitive ability is the key for a teacher to reflect on their own work and support the learning process.

Judging from the results of teacher pedagogic competence test, it is revealed that the ability of biology teacher metacognition in *Sijunjung* District shows very low relationship with teacher competence test results pedagogically. Based on Law Number 14 Year 2005 on Teachers and Lecturers, this means that teachers 'metacognitive ability has a very low relationship with teachers' understanding of learners, the design and implementation of learning, evaluation of learning outcomes and the development of learners to actualize various the potential it has. Very low correlation between teacher metacognition competence with teacher pedagogic ability test result, allegedly will be closely related to learners learn result. In this regard, Daud [6], stated that teacher pedagogic competence is the main component to support the success of learners. Nevertheless in this case has not done further study.

Metacognitive ability that must be possessed by teacher is not just metacognition knowledge but also teacher metacognition skill in learning design which must be done by teacher well. Knowledge of metacognition of this teacher will be able to solve the problems faced by teachers in the learning process. This is supported by a statement from Kallio et al.[7], that this metacognitive ability is the key for a teacher to reflect on their own work and support the learning process.

The result of Soekisno's research [11], that metacognitive ability to explore teacher perception based on teaching experience and teacher's knowledge. By knowing the ability of teacher metakognisi hence in process of learning should teacher can dig knowledge and skill of teacher in preparing lesson. Teachers are able to optimize their capabilities. In harmony with this, Sumampouw [12] states, that in learning metacognitive knowledge possessed by a person related to his beliefs about intelligence, how conscious, his knowledge of the level of difficulty of tasks done in his own way is considered best to be able to solve them.

The result of correlation analysis with Pearson Product Moment formula did not show correlation between metacognitive ability with pedagogic competence result of biology teacher. This is because metacognition skills include planning, monitoring and evaluation. Based on the results of discussions with several teachers revealed that the plan they do only in a short time and not optimal. For supervision, teachers often neglect to do so, it is rare for teachers to evaluate what they have done.

Livingstone in Sumampouw [12] states that metacognition skills are the sequential process processes used to control cognitive activities and ensure that cognitive goals have been achieved. These processes consist of planning and monitoring of cognitive activities and evaluating the outcomes of these activities.

The result of t-test showed that there is a significant correlation between metacognition ability and biology teacher competence test result, especially professional competence. This means that the better the teacher's metacognition ability, the better the teacher professional competence test results

4. Conclusion

Based on the results of research and discussion that have been described, it can generally be concluded that there is a low relationship between metacognitive ability with competence test results of high school biology teachers in *Sijunjung* District. Partially, the relationship of metacognitive ability with the test result of professional competence of biology teacher showed significant result, with correlation coefficient 0.46 and t-table 1,72 while t-count 2,37. The contribution of metacognitive ability to the competence test result of the teacher is 21.6%, while the other 78.4% have not been revealed in this research.

Acknowledgments

We would like to thank Prof. Lufri, M. S., Mrs. Firti Arsih, S. Si., M. Pd. and Mrs. Yati Aisah Rani, M. Pd. as a validator who has provided critiques and suggestions for metacognition improvement questionnaires in this study.

References

[1]. Laws of The Republic Indonesia. 2003. *Undang-undang Sistem Pendidikan Nasional nomor 20 tahun 2003 tentang sistem pendidikan nasional*. Jakarta: Pemerintah Republik Indonesia.

[2]. Laws of The Republic Indonesia, 2005. *Undang-undang Sistem Pendidikan Nasional nomor 14 tahun 2005 tentang guru dan dosen*. Jakarta: Pemerintah Republik Indonesia.

[3]. Ministry of Education and Culture of the Republic of Indonesia. 2016. *Neraca Pendidikan Daerah 2015 Provinsi Sumatera Barat*. Jakarta: Kementerian Pendidikan dan Kebudayaan.

[4]. Syaiful. 2011. Metakognisi Siswa Dalam Pembelajaran Matematika Realistik Di Sekolah Menegah Pertama. *Jurnal Edumatica*, volume 1, no 2 p. 21.

[5]. Balcikanli, Cem. 2011. Metacognitive Awareness Inventory for Teachers (MAIT). *Electronic journal of research in educational Psychology*, volume 3, no 25, p. 1323.

[6]. Daud, Mawadda. 2013. *Deskripsi Kesadaran Metakognitif Siswa dan Guru pada Pembelajaran Biologi*. Skripsi, Jurusan Biologi, Fakultas Matematika dan IPA, Universitas Negeri Gorontalo. http://eprints.ung.ac.id/4629/

[7]. Kallio,Heli, et.al.. 2017. The Utility of the Metacognitive Awareness Inventory for Teachers among In-Service Teachers. *Journal of Education and Learning*, volume 6, no 4, p. 86.

IOP Conf. Series: Materials Science and Engineering **335** (2018) 012086 doi:10.1088/1757-899X/335/1/012086

[8]. Purwanto, Ngalim. 2012. *Prinsip-Prinsip dan Teknik Evaluasi Pengajaran*. Bandung: PT Remaja Rosda karya.
[9]. Sudijono, Anas. 2009. *Pengantar Statistik Pendidikan*. Jakarta: Raja Grafindo Persada
[10]. Sudjana. 2005. *Metode statistik*. Jakarta : Tarsito
[11]. Soekisno, R Bambang Aryan. 2016. Sikap dan Kemampuan Metakognitif Sains Matematika Guru dan Peserta Didik: Kesiapan Pencapaian Standar Kompetensi Lulusan Pda Kurikulum 2013. *Prosiding Temu Ilmiah Nasional Guru (TING) VIII*. 26 November 2016.
[12]. Sumampouw, Herry Maurits. 2011. Keterampilan Metakognitif dan Berpikir Tingkat Tinggi dalam Pembelajaran Genetika (Artikulasi Konsep dan Verifikasi Empiris). *Jurnal BIOEDUKASI*, volume 4, no 2, p. 28.

IOP Conf. Series: Materials Science and Engineering **335** (2018) 012087 doi:10.1088/1757-899X/335/1/012087

Development of Animal Physiology Practical Guidance Oriented Guided Inquiry for Student of Biology Department

Z A Z Putra*, R Sumarmin, V Violita

Biology Department, Faculty of Mathematics and Sciences, Universitas Negeri Padang, Indonesia

*zazputra1992@gmail.com

Abstract. The guides used for practicing animal physiology need to be revised and adapted to the lecture material. This is because in the subject of Animal Physiology. The guidance of animal physiology practitioners is still conventional with prescription model instructions and is so simple that it is necessary to develop a practical guide that can lead to the development of scientific work. . One of which is through practice guided inquiry guided practicum guide. This study aims to describe the process development of the practical guidance and reveal the validity, practicality, and effectiveness Guidance Physiology Animals guided inquiry inferior to the subject of Animal Physiology for students Biology Department State University of Padang. This type of research is development research. This development research uses the Plomp model. Stages performed are problem identification and analysis stage, prototype development and prototyping stage, and assessment phase. Data analysis using descriptive analysis. The instrument of data collection using validation and practical questionnaires, competence and affective field of competence observation and psychomotor and cognitive domain competence test. The result of this research shows that guidance of Inquiry Guided Initiative Guided Physiology with 3.23 valid category, practicality by lecturer with value 3.30 practical category, student with value 3.37 practical criterion. Affective effectiveness test with 93,00% criterion is very effective, psychomotor aspect 89,50% with very effective criteria and cognitive domain with value of 67, pass criterion. The conclusion of this research is Guided Inquiry Student Guided Protoxial Guidance For Students stated valid, practical and effective.

1. Introduction

The learning process that takes place at Universitas Negeri Padang, students are provided lectures by several subject groups such as: Personality Development Course, Scientific and Skills Course, Working Matter, Work Behavior Study, Joint Learning Lecture. [1]

In the Department of Biology, State University of Padang, which must be followed by biology students one of them is the subject of Animal Physiology which also holds practicum activities in the learning process. Through practicum activities, learners are directly exposed to real symptoms related to the concept of learning, both natural conditions and conditions manipulated through experiments. With the development of learning through practical work, learners are directly faced with concrete symptoms associated with the concept of learning so that the concept of learning will be more easily digested and understood [2].

Based on the results of observations that researchers do through the dissemination of student response questionnaires on September 16, 2016 at Bung Hatta University, September 21, 2016 at the

State University of Padang and questionnaire responses lecturer on the implementation of physiology of animals on September 30, 2016 at State University of Padang there are several problems in implementation of the lab. From the results of these observations found several problems such as the following:

1. The guides used for practicing animal physiology need to be revised and adapted to the lecture material
2. Guidance Animal Physiology lab used is still conventional models with manual recipe *(cookery book type)*. One of the advantages of guiding animal physiology that is still conventional and still shaped simple is in its use easy to understand. But on the other hand this is also one of the weaknesses of the guide physiology of the Animal because it is simple so not yet fully able to develop student scientific work
3. Laboratory Guides of Animal Physiology which is still conventional is less interesting for students so that the students are less interested in its use because the guide design used is still simple without any supporting pictures.

There are three approaches used in the method of inquiry: guided inquiry, free inquiry and free modification inquiry [3]. Approach according to the result of student requirement analysis to guide is guided inquiry because at that approach there are steps in presenting question / problem, making hypothesis, designing experiment, doing experiment to get information, Collecting data and Making conclusion according to student desire. This is supported by the research conducted [4] on the development of guidance for general biology guidance based on guided inkuri for students of biology education program STKIP ABDI Pendidikan Payakumbuh. The results showed that students who used guided guides oriented with guided inkuri were able to present problem formulation, hypothesis formulation, plotting experiments, conducting experiments, collecting data, analyzing data and drawing conclusions. Based on the background of the problem, it is necessary to develop a practical guide that can lead the students in designing practicum activities and developing scientific work.

2. Research Methods

This type of research used in this research is the development of research *(Research and Develompment)*. *Plomp* development model, stages of research consists of the initial investigation phase *(preliminary research phase)*, stage of development or manufacture of prototypes *(development or prototyping phase)* and a stage of assessment *(assessment phase)*. The *Plomp* development procedure has three stages. The details of the development procedure include the following steps.

1. *Preliminary Research Phase (Phase Problem Identification and Analysis)*

Activities undertaken at this stage are curriculum analysis, student analysis and practical guidance analysis.

2. *Development or Prototyping Phase (Phase Development and Prototyping)*

The results of *the prototype* design at an early stage with a *prototype* called 1 evaluated by an evaluation *(self-evaluation)* using the checklist. That is to revise itself guides that have been designed. *Prototype* 2 is a development stage by consulting with experts *(expert review)*. Revision of prototype 2 is based on expert judgment (validator). Quality criteria that want to get at this stage that is relevance, consistency and have practicality as expected. The products have been valid from the revised *prototype* 2, further evaluation of the single *(one-to-one evaluation)*. At this stage a one-to-one evaluation is performed on three students with different levels of ability (low, medium and high ability). Evaluation done at this stage using interview sheet with students. Criteria to be achieved in this activity that is product practicality. The products have been tested through the evaluation of single *(one-to-one evaluations)*, carried out the next stage of the test the practicalities of a small group *(small group)*. At this stage, the evaluation of 6 students with the ability of low, medium, high composed of three groups, each consisting of 2 students. Revision 4 *prototype* is based on a small group then continued on the stage of *the assessment phase* (assessment phase) to perform field trials.

3. *Assessment Phase (Phase Assessment)*

The development stage that has been carried out is followed by a large group test of one class to see the prototype's practicality and effectiveness. Practicality is done by filling out a questionnaire of practicality response by lecturers and students.

The subjects of this study were Biology students of FMIPA UNP who followed the Physiology of Animal Physiology Year 2016/2017 semester 4 (Four). The type of data in this study is the primary data because the data obtained directly through the provision of questionnaires practicality test, validity test, and effectiveness test. The effectiveness test data includes affective, psychomotor and student test results.

Instruments used in collecting data in this study is a questionnaire validation practice guide by experts (expert), a questionnaire questionnaire practicality practicum questionnaire filled by lecturers pengampu and guidance counselor subjects animal physiology, questionnaire effectiveness of affective observation sheet, student psychomotor and test for cognitive aspects.

The data collected from this research is the result of validation guidance of Animal Physiology based on guided inquiry approach. Validation results were analyzed using descriptive statistics. Scans for each use are provided under the following conditions [2].

Value 4 = Very Match (SS)
Value 3 = Match (S)
Value 2 = Not Match (TS)
Value 1 = Very Unsuitable (STS)

Of all items given, then tabulated and searched average by using the formula:

$$R=\frac{\sum_{i=1}^{n} Vi}{n} \quad (1)$$

Information
R = Average result of validator assessment
Vi = The score of the i validator's score
n = Many validators

3. Results and Discussion

A. Research Results

1. Preliminary Research (Phase Problem Identification and Analysis)

a. *Analysis curriculum.*

The results of the curriculum analysis are obtained from adjusting the practicum activities to the guidance developed in the order of the semester lectures. It is intended that after students study the theory of animal physiology lectures can carry out practicum in accordance with the material that has been studied previously. In conducting curriculum analysis, it is found that there is a match between learning achievement that has been developed in accordance with the demands of Indonesia National Qualification Framework (KKNI).

b. *Analysis of Students.*

The result of the student's analysis is the students want the practicum guide who has the picture in accordance with the material to be practiced in order to assist the students in understanding the practicum. Students also agree if the guides developed are equipped with guides who have guided inquiry approach that can help students to develop a scientific attitude.

c. *Practical Guidance Analysis*

The result of guidance analysis is guidance of animal physiology practicum used in practicum need to be revised to fit the syllabus in animal physiology lecture.

2. Phase Making Prototype (prototype phase)

This stage begins after the initial phase of the investigation (pleliminary research phase) is completed. This stage consists of several stages of prototyping as follows:

a. Development of Prototype I

After doing curriculum analysis, student analysis and guidance analysis, the next step is to develop guidance guide of guided inquiry in guided animal physiology according to students' need.

b. Development of Prototype 2

Prototype 2 is a development stage by consulting with experts *(expert review)*. At this stage the instrument validation and validation of the practicum guide by the expert (lecturer) in order to produce a valid lab work guide.

1. Results of Validation of Research Instruments

The instrument used in the study has a very valid category. This means that the validation practicum test instrument, practical test, and the effectiveness of the practicum guide are highly valid for use in the study.

2. Practical Guidance Validation Results

At this stage, the expert lecturers are asked to assess the guides for the practicum that has been developed. The validation of the practicum guide includes didactic, constructive, technical and linguistic requirements. Data from the validation result of practicum guide can be seen in Table 1.

Table 1. Practical Guidance Validation Results

No	Assessment criteria	Average Score	Criteria
1	Didactic	3.12	Valid
2	Construction	3.44	Very valid
3	Technique	3.36	Valid
4	Language	3.00	Valid
	Average	3.23	Very Valid

Based on the overall average of validation results of 3.23 practice labs with very valid categories.

B. Discussion

1. Initial Investigation Stage (Preliminary Investigation)

The curriculum analysis aims to look at the subject matter in the subject of animal physiology. This analysis serves as a basis for formulating indicators and objectives of practicum activities and concepts into guided inquiry guided inquiry guides developed. This is supported by [5] which suggests that the curriculum is a guideline for the conduct of teaching and learning activities.

The guidance developed was revised in accordance with the lecture semester course design of animal physiology that is to adjust the order of the practicum in accordance with the sequence of material material of animal physiology lecture. In the guides used also do not fully load the material practicum for animal physiology such as the existence of the physiological anatomy lab activities in animal physiology guides used for lab work.

Analysis is done next is student analysis, student analysis is done by observation by spreading the questionnaire needs of students to know the needs of students to guide that will be developed. This analysis also aims to recognize the characteristics of students and see the needs of students on the guide that will be used in practical activities. This is in accordance with the opinion [5] which states it is important to know and understand the students carefully so that the teacher can carefully determine the materials to be provided, use matching teaching procedures, make a diagnosis of learning difficulties. From the results of student analysis also found problems that students want a guide that has a supporting image that can help improve students understanding during the lab.

Guidance analysis The guidance of animal physiology has been found that the practicum guide has not had an approach. In the absence of approach on guidance of animal physiology practicum, students' skill in lab activities is not maximal. The approach is used to find solutions to the limitations of learning. So the animal physiology guides should be equipped with an approach that can develop student skills during the practicum process and the approach used is the inquiry approach. This is supported by [6] who argue that inquiry approach is a learning approach that directs students to find knowledge, ideas and information through their own efforts. the keyword of the inquiry approach is to find itself. the inquiry approach has a working stage, as experts use this approach in conducting experiments. So because the inquiry approach is used by experts for experiments, this approach is also suitable for practicum because in practicum activity, praktikan is led to be able to find the answer from a problem by doing an experiment. It is also supported by the [7] which states approach the *guided inquiry* learning constructivist homage that emphasizes activity into learning, the constructivist learning more students can develop its problem-solving ability. This opinion is also supported by research that has been done by [8] which states after using guided inquiry approach, student activity in the learning process is increasing active.

With the discovery of several problems at the stage of identification and initial analysis it is necessary to develop a practical guide that can develop students' scientific skills during the process lab.

2. *Stage Prototyping (prototype phase)*

Based on the analysis of the evaluation evaluation itself, it is found that the guided practicum guide has met the criteria on didactic, constructive, technical, and language requirements. From the results of this self-evaluation, the authors make improvements in the form of adding material related to the practice in the guides developed.

After a further revised product development phase conducted by consulting with experts *(exper review).* Prior to product validation, instrument validation was performed. Validated instruments are in very valid criteria. This means the instrument is valid and usable. After the instrument has been validated, the validation will be done to the practical guide to determine the feasibility of the guides that have been developed. This is supported by [9] which states that Kevalidan was obtained by seeking expert consideration. The practicum guide is stated to be very valid because it has fulfilled four aspects of didactic, construct, technical and language aspects based on the validation result done by the validator.

Overall on the didactic aspect validated by the validator declared valid by the validator lecturer. This means that the guidance that has been developed has fulfilled the requirements for the preparation of a good practicum guide, such as the suitability of the curriculum, syllabus and learning achievements and the material can provide information related to practical activities.

The construct aspect validated by the validator is declared very valid by the validator lecturer. This indicates that the guides developed for guidance developed have fulfilled the requirements for the preparation of good practicum guides, such as the guidance of practicum already having clear rules and procedures for the students. Have instructions for use for lecturers and students. The practicum guide already has a suitability between the objectives of the practicum activity with the learning achievement, the practicum activities performed in accordance with the syllabus, has clear rules and rules in the laboratory, consistency using symbols / symbols, and the guides developed are oriented inquiry and guided inquiry the inquiry stages are guided.

The next aspect is the technical aspect validated by the validator lecturer stated very valid. This means that the developed guide has been valid in terms of writing, drawing and graphics. The writing on the developed guide has used the type and size of letters that are clearly read. Images that are on the cover *(cover)* is a portrait of the contents of the guide and symbols can distinguish the procedures at each step in guided inquiry. colors on each symbol the used has been varied and interesting, and the colors on the guides can make the students interested.

Aspects of language validated by the validator lecturer, obtained valid criteria. This guideline developed has qualified the preparation of a good practice guide, such as language is already communicative, not double meaning, and is a language that is good and correct according to Indonesian rules.

4. Conclusions

Based on the development and trial of the practicum guide that has been done, obtained the following conclusions.

1. The process of developing practical guidance Physiology oriented guided inquiry through the stages of problem identification and analysis *(preliminary research),* stage of development and manufacture of prototypes *(development or prototyping phase),* and the stage of assessment *(assessment phase).*

2. Animal Physiology lab guiding oriented guided inquiry that has been developed to have the validity of the category of very valid.

References

[1]. Universitas Negeri Padang. 2014. *Academic Guide.* FMIPA: Padang

[2]. Sudjana. 2005. *Teaching and Learning Outcomes Assessment.* Bandung: PT Remaja Rosdakarya.

[3]. Anam, K. 2016. *Inquiry Based Learning.* Yokyakarta: Pustaka Pelajar

[4]. Nengsi, S. 2015. Development of Instructor-Based Biological Basic Guidance Practicum Guided For Students of Biology Education Studies Program STKIP ABDI Education Payakumbuh. *Thesis.* Padang: State University of Padang

[5]. Hamalik, Oemar. 2008. *Teaching and Learning.* Jakarta: Earth Literacy.

[6]. Sumarmin, R., Anhar, A., and Syari, E.M. 2017. Learning Motivation of Student of SMPN 1 Pariaman. Journal Sainstek 8(1): 64-72

[7]. Afandi, A. 2013. The effectiveness of *Guided Inquiry* Approach Seen and Problem Solving Ability Students of Smp. Delta-Pi: *Journal of Mathematics Education* 2 (2): 8-10

[8]. Ambarsari, W. 2010. Implementation of Guided Inquiry Learning on Basic Science Process Skills in Biology Subject of Viii Smp Negeri 7 Surakarta Class Students. Journal of biology education. 5 (1): 81-95

[9]. Hendryarto, J. 2012. Implementation of Inquiry Learning Model to Train Higher Thinking Ability of Students on Basic Material of Reaction Rate. *Journal Unesa.* (Online), (http://ejournal.unesa.ac.id/index.php/journal-of-chemical education / article / view / 2758jurnal.upi.edu / file / Hokcu.pdf), accessed 13 May 2017.

IOP Conf. Series: Materials Science and Engineering **335** (2018) 012088 doi:10.1088/1757-899X/335/1/012088

Validity of Students Worksheet Based Problem-Based Learning for 9th Grade Junior High School in living organism Inheritance and Food Biotechnology.

J Jefriadi [1*], Y Ahda[1], R Sumarmin[1]

[1]Biology Department, Universitas Negeri Padang, Padang, Indonesia

*jefriadibio@gmail.com

Abstract. Based on preliminary research of students worksheet used by teachers has several disadvantages such as students worksheet arranged directly drove learners conduct an investigation without preceded by directing learners to a problem or provide stimulation, student's worksheet not provide a concrete imageand presentation activities on the students worksheet not refer to any one learning models curicullum recommended. To address problems Reviews these students then developed a worksheet based on problem-based learning. This is a research development that using Ploom models. The phases are preliminary research, development and assessment. The instruments used in data collection that includes pieces of observation/interviews, instrument self-evaluation, instruments validity. The results of the validation expert on student worksheets get a valid result the average value 80,1%. Validity of students worksheet based problem-based learning for 9th grade junior high school in living organism inheritance and food biotechnology get valid category.

1. Introduction

Science is a very interesting subject to study, because most of the concepts of the lesson material can be viewed directly by the students in their daily lives. Therefore, the process of learning science should be conducted through a scientific approach. Learning to use this approach refers to a theory that learning is a scientific process. Learning to use this approach refers to a theory that learning is a scientific process. It is adapted to Permendikbud RI No. 103 [1] that a scientific approach based learning is more effective results compared with traditional learning. The process of learning to use a scientific approach to touch three domains, namely the attitude, knowledge and skills.

Learning science emphasis on direct experience to develop the competencies that students are able to understand the nature around through the process of "figuring out" and "do", it will help students to gain a deeper understanding. Skills in finding out or doing is called the inquiry process skills or "inquiryskills" [2].

By Permendikbud RI No. 58 [1], science learning process in junior high school from the viewpoint of the study of biology so that the presentation concepts of science in integrated science teaching materials in junior highs school begins with biological phenomena that exist around the students and discussed with a review of the three areas of study (biology, physics, and chemistry).

Biology is one branch of science teaching. It can be said biology is the science which studies the symptoms through a series of processes known to the scientific process, which is built on the basis of the scientific attitude and the results are manifested as a scientific product. Jihad and Abdul Haris [3] explains that the competence of the biology subject are grouped into two aspects: first, understanding the concept and its application, and secondly, that the scientific work. While aspects of the assessment are grouped into understanding and application of concepts and scientific performance.

The one of materials that are known to help students understand and apply the concepts of the lessons is students worksheet. Sa'idah [4], which conducts research based integrated science about development students worksheet based problem-based learning through lesson study ecosystems and environmental conservation theme. The results of the expert appraisal to get a score of 3.9 which is included in the category feasible.

Students worksheet considered to have a function in the learning process. Prastowo [5] revealed that students worksheet has four functions, namely; First, students worksheet as teaching materials that can minimize the role of educator, but more make active students. Second, students worksheet as teaching materials that facilitate learners to understand the given a lesson. Third, students worksheet as teaching materials concise and rich task to rehearse. and, fourth, students worksheet given facilitate the implementation of the teaching to the learner.

Students worksheet to be developed is based students worksheet problem based learning. Based on the opinion of Syafii and Ruhizan [6] that the application of learning models of problem based learning in a teaching materials can actually enhance your troubleshooting skills, achievements of learners and learning products of learners. Based on research Sungur [7] the effect of problem-based learning students academic achievement and performance skills in a unit on the human excretory system was investigated. Students in the experimental group appeared to be more proficient in the use and organisation of relevant information, in constructing knowledge and moving toward better conclusions. According to Ram, et al [8], Problem-based learning (PBL) is a constructivist pedagogy in which students learn in small groups by working onreal-world problems.

2. Method Development

The research is a development research using Plomp model. This development model consists of three stages of development is investigation phase (prelimenary research), phase of development and manufacture of prototypes (development or prototype), and the phases of assessment (assessment). The evaluation method used in the study as proposed by Tressmer, 1993 in Plomp [9], self-evaluation or screnning, expert review, one to one evaluation, and small group evaluation. The evaluation method to expert review.

Subject tests on validity problem based learning worksheet are 3 expert to review students worksheet be developed. The instruments used in data collection that includes pieces of observation/interviews, instrument self-evaluation, instruments validity. Data validity processed using the following equation:

$$\text{Value} = \frac{\text{obtained scores}}{\text{maximum score}} \times 100\% \quad (1)$$

Based on the values obtained, the set of criteria validity as seen in Table 1.

Table 1. Categories Validity Students Worksheet.

Value (%)	Category of Validity
0-20	Invalid
21-40	ValidLess
41-60	Enough Valid
61-80	Valid
81-100	Very Valid

Modification from [10]

IOP Conf. Series: Materials Science and Engineering **335** (2018) 012088 doi:10.1088/1757-899X/335/1/012088

3. Results and Discussion

3.1. Initial Investigation Stage (PreliminaryResearch).

According to interviews with three science teacher data showed that the main teaching materials used by teachers is a book teachers and students from Kemendikbud book, where students use the book as a handle students in the learning process. According to the teachers of all the activities and the learning process is already contained in a book teacher and student books.

According to teachers, an activity sheet on a student book contains activities that are less clear so that students feel confused to do what must be done for example on the activity sheet activity 7.1. where an activity sheet contains commands that confuse learners like to investigate the attachment of ear, ability unify the tongue and the type of thumb that family members can not be seen through photos. In addition, the presentation is not based on one model of learning that is recommended in the curriculum in 2013, it caused not optimal learning process is carried out, so that many learners who looked passive in learning.

Based on an analysis conducted by researchers student worksheets result that there are some weaknesses in an activity sheet that is used. The first, Student worksheet learners are prepared immediately drove learners conduct an investigation without preceded by directing learners to a problem or to provide stimulation to the learners. Second, students worksheet only provide field investigations without providing command perform data analysis and draw their own conclusions, Third, student worksheet not provide images of concrete, some pieces of the activities there are no pictures, it should be at the genetic material and biotechnology worksheet should display an image corresponding to the material as a stimulus for students to be able to provide motivation and real-life examples that can be seen by learners.

Analysis curriculum focused on core competencies and the basic competency that has been set in the content standards for teaching science.The results are used as a basis for formulating indicators and learning objectives and concepts necessary in the learning activities

3.2. Prototype Development Phase.

The prototype was developed based on aspects of the construct, the feasibility of the content, language and technical.

Construct aspect, student worksheet based learners problem based learning developed which contains five major components based learning problem-based learning models. 1) Orientation of Students in problems, 2) Organizing students to learn, 3) Research guiding individuals and groups, 4) Developing and presenting the works, 5) Analyze and evaluate the problem solving process.

Aspects of content feasibility, Development student worksheet refers to the curriculum tailored to core competencies and the basic competency that. Problems also equipped with an image for the learning process carried out more interesting and fun, in addition. Students worksheet also facilitate learners to be able to conduct experiments or investigations to verify and find solutions for problems that have been presented.

Aspects of language. The language used in the manufacture student worksheet is Indonesian accordance with the correct spelling. Sentences are prepared form sentences straight forward and clear so that it can be easily understood by students. The language used in students worksheet also adjusted to the level of junior high school language 9th.

Techical aspects. Cover students worksheet created by using Microsoft Office Publisher 2007. On the cover students worksheet, the combination of the colors used are blue and white. On the cover contains the title, base development, author and image products of biotechnology and form outcrossing plant species. The write colour on the identity students worksheet use white colour for more contrast with the background of the cover, the author's identity. The write used is Bernard MT Condensed with writings size is 14-22.

IOP Conf. Series: Materials Science and Engineering **335** (2018) 012088 doi:10.1088/1757-899X/335/1/012088

The formative evaluation for *Self evaluation*. The results of self evaluation are known to be found typing errors and various other errors like the images have no information sources as well as the latin name that is not italic.

3.2.1. Validity Students Worksheet. Students worksheet validation results of the experts can be seen in Table 2 below.

Table 2. Results of Validation Students Worksheet Developed

No.	Aspect Validated	Average value (%)	Category
1.	Construct	83.0	Very Valid
2.	Feasibility of contents	80.8	Valid
3.	Technical	75.0	Valid
4.	Language	81.6	Very Valid
	Average	80.1	Valid

Based on assessment from construct aspect obtained students worksheet valid. Category valid obtained from expert judgment students worksheet developed due to present indicators and clear learning objectives. The activities in students worksheet integrated using PBL learning model. Learning activities on students worksheet consists of phase to orient students on the problem by displaying a phenomenon problem that has been adapted to the topic material. According [4] the problems presented by causing curiosity of learners. This curiosity will motivate students to learn and to apply knowledge so that students are able to explain the problem. The next phase is to organize students to learn by giving directions or instructions for students to learn in groups. Problems that appear demanding learners to be able to look for solutions to problems by working together. Based on according [11] problem-based learning is an educational approach that challenges students to work together in groups to find solutions to real-world problems and to develop skills to become independent learners.

The next phase of problem based learning is to guide individual and group investigation. In this section presents the activities of learners students worksheet write temporary answer to the problem or phenomenon is presented, as well as directing students to conduct experiments and investigations to be conducted. Based on according [12] activities to formulate the problem to develop scientific thinking and develop creativity in solving problems.

The next activity there are phases develop and present work, by directing learners write the results and the command to answer some questions relating to the results of observations made learners. The next activity there are phases of analyzing and evaluating the process of solving the problem by asking the students to make conclusions from the activities carried out. According [13] drew the conclusion that the activities can develop the ability to think systematically, to express an opinion with clear and concise, and to develop good language skills and correct.

Students worksheet also has a very valid category on the feasibility aspect of the content. Category valid given by experts because separated indicators based competence base, learning objectives are also described based on indicators that have been made. Students worksheets contains a list of questions related to the concept of matter. The list of questions on the phases develop and present the works adapted to the learning objectives to be achieved learners. In comparison to traditional engineering curriculum, the PBL models appear to inspire a higher degree of involvement in study activities and, consequently, a higher level of complex comprehension [14]. A possible drawback is the risk of gaps in specific knowledge areas. Therefore, it is crucial that the students in a PBL curriculum become lifelong learners who have learned to take responsibility for their own learning process.

Category very valid given the contents of the feasibility aspect is also based on the activities of the students worksheet has direct experience of the learners. Based on research Etherington [15] obtained that, The PBL course had a positive impact on the pre-service teachers' motivation to teach science ideas within a real world context. Activity on students worksheet provide opportunities for learners to carry out activities with the activities of this invention so that learners actively in the learning process. Students worksheet activities also help develop social communication skills of learners. Students worksheets also gives learners the opportunity to ask questions. And than, in students worksheets also displayed a picture that can support material.

The next aspect is the technical aspects of obtaining a valid category of experts. Valid category of experts review because students worksheet own images and text on the cover that already reflect the content. Pictures on students worksheet also can convey the message effectively, captions are in accordance with the image presented on students worksheet. Then, the combination of colors used in students worksheet is sufficient contrast and sharp and does not disturb the focus of learners when reading students worksheet, color images can also foster the attractiveness for learners. The type and size of the letters used is also appropriate.

The next aspect that received very valid category that language. Students worksheet considered to be valid in terms of the language by experts because students worksheet designed using the rules of Indonesian true even though in some parts there are errors in the wording is less precise use of punctuation on students worksheet in some parts there are errors where there is use of punctuation less precise. The language used is adapted to the language level of ninth grade junior high school children so that students easily understand the content students worksheet.

4. Conclusion

Based on explained results and discussion able to concluded the results of the validation expert on student worksheet developed get valid result. The validation based on aspects of the construct, the feasibility of the content, language and technical.

References

[1] Permendikbud 2014 *Kerangka Dasar dan Struktur Kurikulum Sekolah Menengah Pertama/Tsanawiyah* (Indonesia: Jakarta) Kemendikbud.

[2] Trianto 2010 *Model Pembelajaran Terpadu Konsep, Strategi, dan Implementasinya dalam kurikulum tingkat satuan pendidikan (KTSP)* (Jakarta:Indonesia) Earth Literacy.

[3] Jihad A and Abdul H 2012 *Evaluasi Pembelajaran* (Indonesia: Yogyakarta) Multi Pressindo

[4] Saidah N 2014 Pengembangan LKS IPA Terpadu Berbasis Problem Based Learning Melalui Lesson Study Tema Ekosistem dan Pelestarian Lingkungan *Jurnal Unnes Science Education* Vol 3 No 2 pp 549-556.

[5] Prastowo A 2014 *Pengembangan Bahan Ajar Temati; Tinjauan Teoritis dan Praktik* (Jakarta:Indonesia) Kencana Prenadamedia Group

[6] Syafii W and Ruhizan MY 2013 Problem Solving Skills and Learning Achievements Through Problem-Based Module in Teaching and Learning Biology in High School *Journal Asian Social Science* Vol 9 No 12 pp 220-228

[7] Sungur S, C Tekkaya and O Gebaan 2006 Improving achievement through problem-based learning. Journal Research Education (Turkey: Angkara) Vol 40 No 4 pp 155-160.

[8] Ram P, A Ram, J Holzman and C Sprague 2007 A Cognitive Model of Problem-Based Learning and Its Application to Educational Software Design *Journal IADIS International Conference on e- Learning* vol (Portugal: Lisbon).

[9] Plomp T and Nieveen N 2013 *Educational Design Research: An Introduction* Enschede: SLO. Netherlands Institute for Curriculum Development.

[10] Riduwan 2009 *Belajar Mudah Penelitian untuk Guru, Karyawan, dan Peneliti Pemula.* (Indonesia: Bandung) Alfabeta.

[11] Haryadi S F E, Andreas P BP and Amin R. 2015 Designing Literacy and Problem Based Learning to Foster Critical Thinking in Biology. *Journal International Conference on Mathematics, Science, and Education* Vol 9 pp 19-23

[12] Kenengsih S 2013 Pengembangan Penuntun Praktikum Mikrobiologi Berorientasi Inkuiri Terbimbing Untuk Mahasiswa STKIP PGRI Sumbar Padang. *Tesis* tidak diterbitkan. (Padang:

Indonesia)
[13] Daryanto 2014 *Pendekatan Pembelajaran Saintifik Kurikulum 2013* (Indonesia:Yogyakarta): Gava Media.
[14] Graaff E D and Anette K 2003 Characteristics Problem-Based Learning. *Int. J. Engng.* Vol 19 No 5 pp 657-662
[15] Etheringtone M 2011 Investigative Primary Science: A Problem-based Learning Approach. *Australian Journal of Teacher Education* Vol 36 No 9 pp 53-74

IOP Conf. Series: Materials Science and Engineering **335** (2018) 012089 doi:10.1088/1757-899X/335/1/012089

Development Module Oriented Science Technology Society Indue Science Literacy Assessment for 7th-Grade Junior High School Students in 2nd -Semester

Y R Arbi [1,*], R Sumarmin [1], D H Putri [1]

[1] Postgraduate of Biology Education Program, Universitas Negeri Padang, Padang, Indonesia.

[2]Biology Department, Faculty of Mathematics and Sciences, Universitas Negeri Padang, Indonesia

[*] yulisarizkiarbi1101341@gmail.com

Abstract. The problem in the science learning process is the application of the scientific approach takes a long time in order to provide conceptual understanding to the students, there is no teaching materials that can measure students reasoning and thinking ability, and the assessment has not measured students reasoning and literacy skills.The effort can be done is to develop science technology society module indue science literacy assessment. The purpose of the research was to produce a module oriented society indue science science technology literacy assessment. The research is development research using Plomp model, consist of preliminary, prototyping, and assessment phase. Data collect by questionnaire and documantion. The result there is science technology society module indue science literacy assessment is very valid.

1. Introduction

Learning process in science using scientific approach. Implementation scientific approach is not optimal, because students take a long time to understanding of the concept. Teaching material to help students prepare for learning process is the module. Modules can make students study independently.

Module is used to give to the students understanding of the concept develop by the Science Technology Society approach. STS approach used to provide an understanding of science concept to students. STS can improve the ability to reason through problem-solving activities in the comunity.

The concept of science relevant to student's daily lives. STS according to Yager [1] a science learning in a human experience context, which emphasizes the importance of science and technology. STS approach suitable for use on materials related to the student environment. Learning material related to student environment is interaction of human organism and environment, environmental pollution and climate change.

NSTA has defined STS as the teaching and learning of science in the context of human experience. It emphasizes the importance of technology and science, noting that technology is understood better and accepted more readily as curriculum and course topics appropriate for all students even though it has not been a focus previously [2].

Reasoning ability of students can be seen by assessment. Assessment can be used to measure the ability to reasoning and high-level thinking is science literacy assessment. Science literacy assessment consists of the cognitive level of application, analyse, synthesis, and evaluation.

2. Methods

The development of the module is based on the stages of the Plomp model. This module was developed through three stages namely (1) preliminary research phase, (2) prototyping phase, and (3) assessment phase [3]. Fomative evaluation is done at every single stage in the form cycle of evaluation-revision. Implementation of this formative evaluation proposed by Tessmer (in Plomp [3]).

The trial subject is students in one class of 7th-grade of Junior High School numbers 12. Data consist of qualitative and quantitative data. The qualitative data obtained from observations and interviews in preliminary research phase and one to one evaluation in prototyping phase. The quantitative data obtained from the validity questionnaire.

Result of the research processed using the following equation.

$$\text{Value} = \frac{\text{Obtained score}}{\text{Maksimum score}} \text{ x } 100\% \quad (1)$$

This value is subsequently converted in accordance with validity criteria that can be seen in Table 1.

Table 1. Criteria level of validity

Percentase (%)	Level of validity
0 – 20	Invalid
21 – 40	Less Valid
41 – 60	Enough Valid
61 – 80	Valid
81 – 100	Very Valid

3. Results and Discussion

Module contains a description of the material that consist the science technology society component such as introduction, concept-forming, application of concepts, and consolidation of concepts. Module was developed through three stages of Plomp model. In preliminary research phase there are fourth process namely (1) problem analysis, (2) need analysis, (3) concept analysis, and (4) curiculum analysis. In prototyping phase there are process namely (1) self evaluation, (2) expert review, and (3) field tests. Validity doing by expert review. Validity of module includes aspect namely content egibility, linguistic, presentation, and graphic.

3.1. Results

Based on preliminary research analysis finds that problem in the form of non-optimal the application of scientific approach, there is no teaching materials that can measure students reasoning, and there is no scientific literacy assessment use in learning process. Student need a module that help students discover the concept itself. The module applied science technology society approach. Learning material related to environment is interaction of human organism and environment, environmental pollution and climate change.

Module contains a description of the material that consist the STS component such as introduction, concept-forming, application of concepts, and consolidation of concepts. Component of introduction contains the problem in acccordance with the indicators of learning process. Students can link the problems encountered with the learning material through discourse.Component application of concepts contains a sample application of the concepts.

Scientific literacy assessment in the module found on the competency tests. Science literacy competency in module is using scientific evidence, identification scientific issue, and explaining phenomena scientifically. The competency test contains questions that evaluate the achievement of learning indicators in one learning activity that has been studied by the students. The literacy assessment of science is an assessment with high-order thinking skills. Science literacy assessment not only assesses content but also processes in the context of everyday life of students.

Module are develop tasted for validity. Results module validation by experts are presented in Table 2.

Table 2. Result of validation module

Aspect	Persentage (%)	Criteria
Content Egibility	93.83	Very Valid
Linguistic	93.75	Very Valid
Presentation	93.33	Very Valid
Graphic	91.67	Very Valid
Average	93.14	Very Valid

Based on Table 2 can be seen that the module are develop is very valid. During the validation, experts give some advice to fix the cover, text color, and language. Cognitive competencies measured using scientific literacy assessment. The science literacy content in interaction of human organism and environment is ecosistem and population. The science literacy content in environmental pollution is earth system exchange. The science literacy content in climate change is earth system energy. The context of the literature of science on assessment is a personal, social, and global context.

Before use assessment tasted for validity. Results scientific literacy validation by experts are presented in Table 3.

Table 3. Result of validation scientific literacy assessment

Aspect	Persentage (%)	Criteria
Content Egibility	81.33	Very Valid
Linguistic	75.00	Valid
Presentation	81.25	Very Valid
Graphic	81.25	Very Valid
Average	79.70	Valid

3.2. Discussion

The result of validity test showed that the STS module indue scientific literacy assessment develop is very valid. Module have suitable with basic competences. Module also accordance with the development of the child. Student in 7th-grade is 12-13 years old that has been able to analyze problems that are gives to him. This is in accordance with the opinion of Gredler [4] states that the child has begun to thinking linked to concrete objects are developed, and comes to understand that a given operation simultaneously and necessarily implies its inverse.

Module contains a description of the material that consist the science technology society component such as introduction, concept-forming, application of concepts, and consolidation of concepts. The STS components are made intact and directed in each learning activity. Directed learning will help students to learn by themselves. This is in accordance with the opinion of Eggen [5] that concept attainment could be used to help students understand the process of Inquiry and the scientific method. This is in accordance with the opinion of Rosario [6] stated that STS approach using varied strategies to allow student exposure to issues in the context of experience which would provide opportunities for the students' construction of knowledge

The introduction component relate to issues, events that students have known, or events contained in society with the material to be discussed. It is a discourse that contains issues relating to learning indicators. Students can relate problems encountered with learning materials through discourse. This is in accordance with the opinion of Rosario [6] stated that important aspect of the approach is that activities can be derived from the local community to make the lessons more relevant.

The concept-forming component contains the materials that corresponds to the learning indicator. The materials is delivered in simple language in order to be understood by the students. This section comes with an appropriate image to support the delivery of the material. The process of the concept formation is done by discussion method. The process of forming this concept will make students find the correct concept. This is in accordance with the opinion of Dikmentepe [7] states that inclusion of the articles or books which tell about scientists' lives, documentaries about scientific and technological developments into the courses can help student to understand the nature of science and technology.

The concept application component contains examples of the application of concepts contained within the concept consolidation section. This is in accordance with the opinion of Yager [8] which states that STS focuses on the impact of science and technology to students.

Module include assessment of scientific literacy are developed with suitability indicators observed articles with material from the article in accordance with the material, derived from a newspaper article, and the article came from a magazine. Adjustments articles with incoming material on psychomotor level observed. This is in accordance with the opinion of Rosario [6] stated that familiarity with the environmental issues and problems of the community may make the experience more effective.

The concept consolidation component contains conclusions about the material. The concept of thinking is based on the material contained in the concept forming section. Stabilization of concepts is needed to fit the concept of the material that has been studied. This is in accordance with the opinion of Yager [8] states that the focus on student projects and real problems provides the way to illustrate the importance of the concepts and processes.

Students who already understand the material will be able to do reasoning on the matter and comment on the material provided. Module include scientific literacy assessment have explicit indicators and learning goals, complete materials and according to the indicators that have been made. The application of science literacy assessment can overcome the error that the assessment used is still the assessment of knowledge, understanding, and application. Science literacy assessment has difficulty ranging from understanding to synthesis. Science literacy assessment is an assessment capable of measuring high-order thinking skills. Science literacy assessment not only assesses knowledge but also assesses the ability of scientific thinking and uses it in the context of everyday life. This is in accordance with the opinion of Wright [9] states that the tests are concerned with the literacy related to problem solving in science.

4. Conclusion

The conclusion of the research that module oriented science technology society indue scientific literacy assessment are develop is very valid. Module was developed contains a description of the material that consist the science technology society component such as introduction, concept-forming, application of concepts, and consolidation of concepts. Validity of module includes aspect namely content egibility, linguistic, presentation, and graphic. Based on research that has been done, the author gives advice to test one-on-one, small group, and pitch in different grades, so that the products developed proven feasibility.

References

[1] Yager R E 1992 *The Status of Science-Technology Society reform efforts around the world* 1-2
[2] Dass P M 2005 Spring **14** 1
[3] Plomp T 2013 *Educational Design Research; An Introduction* (Netherlands: Enschende) p 19
[4] Grendler M E 2009 *Learning and Instruction : theory into practice* (New Jersey: Pearson) p 80
[5] Eggen P D 2012 *Strategies and models for teachers: teaching content and thinking skills* (Boston: Pearson) p 25.
[6] Rosario B I 2009 *Liceo Journal of Higher Education Research* **6** 3

[7] Dimentepe E, Yakar Z 2016 *International Journal of Higher Education* **5** 11
[8] Yager R E, Akcay H 2007 *Spring* **16** 7
[9] Wright R J 2008 *Educational Assessment* (California:Sage Publication) p 429

IOP Conf. Series: Materials Science and Engineering **335** (2018) 012090 doi:10.1088/1757-899X/335/1/012090

Emotional and Spiritual Quotient Approach Improve Biology Education Students' Acceptance of Evolution Theory

R Darussyamsu[1*], M Fadilah[1], D H Putri[1]

[1]Biology Department, Faculty of Mathematics and Science, State University of Padang, Indonesia

*rahmawati6786@gmail.com

Abstract. Emotional and spiritual aspect is one of main factors that influence students' acceptance of a theory. This study aim to measure university students' acceptance of evolution by learns evolution using emotional and spiritual quotient (ESQ) approach. This is a quasi-experimental research using one shot case study design with the subject 36 biology educational students at Biology Department, Faculty of Mathematics and Natural Science, Universitas Negeri Padang. Data collected using the MATE instrument by Rutledge and Warden (2000) after the students learn evolution for eight meetings since January until March 2017. The result showed that by learning evolution theory combine with ESQ aspects increase students acceptance from very low become moderate acceptance. It concluded that ESQ aspects can improve students' acceptance of evolution. Any criteria depend on it are discussed.

1. Introduction

Acceptance of evolution theory research has been conducted in some research [1], such as in United States [2], Europe [3] and Indonesia [4], because it is essential to know by educators. Some people accept the evolution theory but they do not understand it at all, and the others understand on it but do not believe this theory [5]. This condition should be a big consideration for teacher or lecturer in teaching evolution [4].

Some studies reported some factors that influence students' acceptance of evolution, for instance: students' understanding of biological evolution from school [1], the way of teacher teaching evolution [4, 6], religions [4, 7] and other factors such socio-cultural factors [8]. Religious, social, and emotional factors are the key factor for people to accept evolution, whether religious is the most dominant factor for person who rejects evolution theory [9]. Additionally, personal religious belief is the most aspect associated with acceptance of evolution [10].

In the site of religious, emotional and spiritual quotient (ESQ) is one approach to develop students literate in finding the truth [11]. ESQ can improve students' achievement in some life aspects, such as goal achievement and problem resolution, and it can be done by train [12]. Therefore, this approach predict as one solution to improve students' acceptance of evolution. Therefore this research conducted to know students' acceptance of evolution after learning evolution using ESQ approach.

Other research has reported the ambiguities of the effect of learning toward acceptance of evolution. For example, [13, 14] stated that instruction improve acceptance. On the other hand, [15, 16] reported that no progress of acceptance from learning, but it influence by other factors. This research combine instructional factor by learning evolution and other factors such emotional and spiritual aspects in learning, then measure the effect of it.

This study conducted to know the effect of using emotional and spiritual quotient approach in teaching evolution for university students toward their acceptances of evolution. It was a preliminary research to develop a product for teaching evolution in university level by lecturer when teaching evolution theory. Thus, our study assessed students' acceptances after they learn evolutionary course at Biology Department. The research question guiding this investigation was: "Do the using of emotional and spiritual quotient approach improve the level of biology students' acceptance of evolution theory?"

2. Method

This research was a quasi experimental research using one shot case study design. This design applied by focused on the changes of students' attitude toward evolution, especially their acceptance on it. This research stopped until mid semester because there was significant change for students' acceptance of evolution. The detail of this study method explains as follow.

2.1. Participants

The participants of this study were biology education students who took Evolutionary course in the January-June semester of 2017 at Biology Department, Faculty of Mathematics and Natural Science, Universitas Negeri Padang. In this semester, there were three parallel classes which amount of 36, 34 and 27 students (total 97 students). This study only used one class as the participant that was chosen by purposive sampling technique, which was the class in amount of 36 students.

2.2. Measures

Students' acceptance of evolution at the beginning (before learn evolution in university) asked using one question about their acceptance of evolution: "Do you accept the truth of evolution theory?" After students taught the evolution theory using emotional and spiritual quotient approach for eight meetings (until mid term), students' acceptance level measured with MATE instrument questionnaire [17]. Furthermore, students' perception of evolution identified by open-ended questionnaire.

The MATE instruments used because it has been tested based on the reliability and validity to assess acceptance of evolution among teachers [17], university students [18], and recommended by [1] as one of the good instrument to classify level of students' acceptance in evolution.

2.3. Analyses

There was three step of scoring the MATE instruments [18], as follow.

Step 1. Scoring of positively-phrased items (1, 3, 5, 8, 11, 12, 13, 16, 18, and 20) is as follows; strongly agree = 5; agree = 4; undecided = 3; disagree = 2; and strongly disagree =1.

Step 2. Scoring of negatively-phrased items (2, 4, 6, 7, 9, 10, 14, 15, 17, and 19) is as follows; strongly agree = 1; agree = 2, undecided = 3; disagree = 4; strongly disagree = 5.

Step 3. An individual score on the MATE is equal to the sum of the scaled responses to all 20 items.

By using those 3 steps, the MATE is scored with the lowest level of acceptance being 20 to the highest level of acceptance being 100. To determine the category level of acceptance, [17] provides the following scores and categories:

89-100 = Very High Acceptance
77-88 = High Acceptance
65-76 = Moderate Acceptance
53-64 = Low Acceptance
20-52 = Very Low Acceptance.

For scoring the questions that followed MATE instruments are: agree = 1 and disagree = 0. All the response of students calculates in percentage. The reasons that stated by students in the open-ended questionnaire is the supplementary data for MATE result.

3. Results and Discussion

At the beginning of the evolutionary course meeting, students asked to answer the questions about their perception of evolution as general, whether they accept it or not. Dominantly, students answer that they do not accept evolution as the right theory. It means that only 16.67% students accept evolution theory. This value is in very low acceptance criteria (20-52) [17].

After eight meetings of using ESQ approach in evolution learning, students were asked to tell their acceptance of evolution theory by the concepts addressed by the MATE [18], and it showed a significance change, that average students in the value of 70.17. This value is in moderate acceptance of evolution (65-76) [17]. The percentage of each acceptance criteria data presents in Table 1.

Table 1. Biology Education Students' Acceptance of Evolution Theory after Learn Evolution Using ESQ Approach

Criteria of Acceptance	Percentage (%)
High	22.22
Moderate	47.22
Low	30.56

The supplementary data from questions that answer by students after the MATE figure in Table 2.

Table 2. General Opinion of Biology Education Students about Evolution Theory after Learn Evolution Using ESQ Approach

Opinion	Percentage (%)
Evolution is linked with religion	88.89
Evolution can be learned using ESQ approach	91.67
Students accept evolution after learn it in university using ESQ approach	88.89
Students do not accept evolution theory until now	36.11
Students will use ESQ approach when teach evolution	97.22

Evolution learning that use ESQ approach has been done for eight meeting, since January until March 2017. It is a combining aspect of instruction and ESQ in teaching evolution. ESQ approach applied in classroom learning by lecturer while discussion about the material and students' weekly structural assignment. Each student asked to identify religion aspect that linked with the material every week. For instance, before learn about the origin of life, students asked to identify this aspect in Al-qur'an and hadits, because all of the students are Muslim. Additionally, emotional aspects also describe in this assignments through students analysis of religions aspect, evolution theory, and students attitude on it.

This treatment improve students' acceptance significantly from very low category become moderate acceptance. It supports by the data in Table 2 which state that most students (88.89%) accept evolution theory after learn it in university using ESQ approach, but some of them (36.11%) still do not accept evolution theory until now because of some concepts. This happens because students tend to reject evolution truth because of its conflict with religions [14, 19, 20]. Whether this aspect influence students' acceptance, but it does not make students do not want to learn about evolution [4], in the other word, students' motivation to learn well was not disturb by their acceptance. This motivation formed by their wisdom that train by emotional aspect, as suggested by [12] that emotional quotient make people well adjusted and psychological balanced.

Based on the data in Table 2, it shows that 88.89% students agree that evolution linked with religions. Concepts that discuss in evolution course also have the explanation in religion. When this explanation become linear, it will reduce the conflict between religion and evolution, as stated by [21]

that when there is no gap anymore between religion and evolution, both will be a real knowledge sources which influencing and completing each other.

Whether the students have accept evolution theory after learn it using ESQ approach, some of them (36.11%) still do not accept evolution theory until now. It makes sense because their acceptances are in moderate level of acceptance. It means that they have accept some of evolution theory, but not for some others, especially about the origin of human; as stated by some students in the open-ended questionnaire.

The last open-ended questionnaire also determine that almost students (91.22%) agree to use ESQ approach when teach evolution next time if they have been being a teacher. They state that by using ESQ approach, misconception of evolution also can resolve and it changes students' mindset of learning evolution. It shows that students feel this approach make their understanding of evolution better, and they also want to change the way of teaching evolution at school use ESQ approach. Overall, students feel their acceptances of evolution theory were better by using ESQ approach in learning.

4. Conclusion

Based on this research, concluded that the using of emotional and spiritual quotient approach in teaching evolution enhance students acceptance of evolution from very low acceptance criteria become moderate acceptance criteria. This significant change supported by the decrease of conflict between knowledge and religions, therefore students can accept evolution better. After the students get the good effect of their acceptance and understanding of evolution by learn it using ESQ approach, it motivate them to use it next time when they have been being a biology teacher. This fact also become a solution to make teaching evolution better, by using ESQ approach in learning; both in process and media.

Acknowledgements

Authors would like to thank the students of Biology Educational Program, Mathematics and Natural Science Faculty, Universitas Negeri Padang for their participation in giving information and Universitas Negeri Padang as the financial support of this research.

References

[1] Yasri, P. 2014. "A Review of Research Instruments Assesing Levels of Student Acceptance of Evolution." *Asia-Pasific Forum on Science Learning and Teaching* **15(2)** 1-16.

[2] Rutledge, M. L. & Sadler, K. C. 2007. "Reliability of the Measure of Acceptance of the Theory of Evolution (MATE) Instrument with University Students." *The American Biology Teacher* **69(6)** 332-335.

[3] Mazur, A. 2004. "Believers and Disbelievers in Evolution. *Politics and the Life Sciences* **23(2)** 55-61.

[4] Darussyamsu, R. 2015. "Biology Education Student Acceptance of Evolution Theory before Learn Evolutionary Course in Biology Department." *ICOMSET 2015 Proc.* 360-365.

[5] Nadelson, L. S. and Sinatra, G.M. 2010. "Shifting Acceptance of Evolution: Promising Evidence of the Influence of the Understanding Evolution Website." *The Researcher* **23(1)** 13-29.

[6] Moore, R. 2004. "How Well Do Biology Teachers Understand the Legal Issues Associated with Teaching of Evolution?" *Bioscience* **54(9)** 860-865.

[7] Walter, E. M., Halverson, K. M. and Boyce, C. J. 2013. "Investigating the Relationship Between College Students' Acceptance of Evolution and Tree Thinking Understanding." *Evolution: Education and Outreach* **6(26)** 1-6.

[8] Bahar, M. 2003. "Misconceptions in Biology Education and Conceptual Change Strategies." *Educational Sciences: Theory & Practice* **3(1)** 55-64.

[9] Wiles, J. R. 2014. "Gifted Students' Perceptions of Their Acceptance of Evolution, Changes in Acceptance, and Factors Involved Therein." *Evolution: Education and Outreach* **7(4)** 1-19.

[10] Sickel, AJ. & Friedrichsen, P. 2013. Examining the evolution education literature with a focus on teachers: major findings, goals for teacher preparation, and directions for future research. *Evolution: Education and Outreach* **6(23)** 1-15.

[11] Agustian, A. G. 2005. *Emotional Spiritual Quotient; Rahasia Sukses Membangun Kecerdasan Emosi dan Spiritual Berdasarkan 6 Rukun Iman dan 5 Rukun Islam.* Jakarta: Penerbit Arga, pp. 118-119.

[12] Mishra, P. and Vashist, K. 2014. "A Review Study of Spiritual Intelligence, Stress and Well-Being of Adolescents in 21st Century." *IMPACT IJRANSS: International Journal of Research in Applied, Natural and Social Sciences* **2(4)** 11-24.

[13] Wilson, DS. 2005. "Evolution for Everyone: How to Increase Acceptance of Interest in, and Knowledge About Evolution." *PloS Biology* **3(12)** e364.

[14] Ingram, E. & Nelson, CE. 2006. "Relationship between achievement and students' acceptance of evolution or creation in an upper-level evolution course. *Journal of Research in Science Teaching* **43(1)** 7-24.

[15] Bishop, BA. & Anderson, CW. 1990. Student conceptions of natural selection and its role in evolution. *Journal of Research in Science Teaching* **27(5)** 415-427.

[16] Lawson, AE. & Worsnop, WA. 1992. Learning about evolution and rejecting a belief in special creation: effects of reflective reasoning skill, prior knowledge, prior belief and religious commitment. *Journal of Research in Science Teaching* **29(2)** 143-166.

[17] Rutledge, M. L., & Warden, M.A. 1999. "The Development and Validation of the Measure of Acceptance of the Theory of Evolution Instrument." *School Science and Mathematics,* **99(1)** 13-18.

[18] Rutledge, M. L. & Sadler, K. C. 2007. "Reliability of the Measure of Acceptance of the Theory of Evolution (MATE) Instrument with University Students." *The American Biology Teacher* **69(6)** 332-335.

[19] Meadows, L., Doster, E., and Jackson, D. F. 2000. "Managing the Conflict Between Evolution and Religion." *The American Biology Teacher* **62(2)** 102-107.

[20] Woods, CS. & Scharmann, LC. 2001. High school students' perceptions of evolutionary theory. *Electronic Journal of Science Education* **6(2)**

[21] Dahler, F. 2011. *Teori Evolusi: Asal dan Tujuan Manusia.* Yogyakarta: Penerbit Kanisius, p. 127

IOP Conf. Series: Materials Science and Engineering **335** (2018) 012091 doi:10.1088/1757-899X/335/1/012091

Validity of Learning Module Natural Sciences Oriented Constructivism with the Contain of Character Education for Students of Class VIII at Yunior Hight School

K Oktarina[1*], L Lufri[2], M Chatri[2]

[1]Student of Master Degree Program of Biology Education, Faculty of Mathematics and Sciences, Universitas Negeri Padang, Indonesia
[2]Biology Department, Faculty of Mathematics and Sciences, Universitas Negeri Padang, Indonesia

*kimlin_khincai94@yahoo.co.id

Abstract. Referring to primary data collected through observation and interview to natural science teachers and some students, it is found that there is no natural science teaching materials in the form of learning modules that can make learners learn independently, build their own knowledge, and construct good character in themselves. In order to address this problem, then it is developed natural science learning module oriented to constructivism with the contain of character education. The purpose of this study is to reconstruct valid module of natural science learning materials. This type of research is a development research using the Plomp model. The development phase of the Plomp model consists of 3 stages, namely 1) preliminary research phase, 2) development or prototyping phase, and 3) assessment phase. The result of the study shows that natural science learning module oriented to constructivism with the contain of character education for students class VIII of Yunior High School 11 Sungai Penuh is valid. In future work, practicality and effectiveness will be investigated.

1. Introduction

Education is all the efforts undertaken to educate people so they can grow and develop and have the potential or ability as appropriate. The Government is seeking various things to improve the quality of learning by improving the curriculum. What has recently been done is the improvement of the education unit level curriculum (KTSP) into the 2013 curriculum.

Development of curriculum 2013 focused on the formation of competence and character of learners in the form of alloy knowledge, attitudes, and skills that can be applied learners then follow the development of times. As one of the general lessons in junior high, science becomes one of the subjects that can support the formation and integration of the students' character values. Science subjects are expected to be a vehicle for learners to learn about themselves and the environment, as well as further development processes to be applied in everyday life. Science learning should encourage learners to be able to construct their knowledge independently through learning experiences (Sudarisman, 2012).

Based on the results of preliminary observations conducted schools, researchers found things that become problems in science learning. From the source of books in the library are still lacking in meeting the needs of learners for the learning process, especially science subjects, learners rely solely on the explanation of teachers as sources and centers of learning information to them, have never used teaching materials such as modules, learners do not have Book packages or other teaching materials as a handle, learners have difficulty to build their own concepts, the ability of learners is still low in constructing the initial knowledge and still need to be guided by the teacher in constructing the initial knowledge, because subject of science lessons are still less enough according to learners, Being one of the factors that resulted in not all learners can construct knowledge initially well.

The existence of early knowledge to make learners can easily follow the learning process well. This is accordance with the theory of constructivism that the success of learning depends not only on the environment or learning conditions, but also influenced by the early knowledge of learners. Student learning outcomes are also closely related to the understanding of learners in understanding the concept of science subject matter and it is very important to do in the learning process (Patrianingsih, 2015). But not all learners have the same speed to understand the concepts of the subject matter discussed in the classroom. It is necessary for learning materials that can focus the capabilities of individual learners and can be studied independently with limited assistance from teachers with the aim of complementing the knowledge initially corresponds to the speed of learning to learners.

One that can solve this problem is teaching materials in the form of learning modules. Modules are teaching materials that are arranged systematically with language that is easily understood by learners and can be studied independently with limited help from teachers or educators (Purwanto, 2014). Modules are also printed materials that are designed to be learned independently by learners in learning, which has advantages in accordance with the principle of module writing is the feedback of the module so that learners can evaluate themselves (Director of Teaching Personnel, 2008).

In addition, the module also has many advantages compared to other written teaching materials, such as the purpose of learning more clearly and the ability of accommodation to the difference in the pace of learners in understanding the material (Nasution, 2010). Furthermore, according to Mulyasa (2006) module is a teaching material that contains learning materials are arranged in a systematic, operational and directed to be used by learners because it is accompanied by guidelines for its use. Sudjana and Rivai (2003) revealed that the purpose of learning with the module is t students able to follow the learning according to their own speed and ability.

In the results of previous studies, it turns out the module can help learners in understanding the material Biology. One of them is the result of research conducted by Elpitriani (2012) about the development of constructivism oriented biology learning module with a mind map. From this research, it is known that the module is very effective used in Biology learning in SMA and it can be seen that the use of module can improve learners' success in the material of excretion system. In learning using the module, learners can learn individually in the sense that they can adjust their learning speed with their respective abilities. In addition to the module, learners can measure their level of mastery of the given material.

A module as a delivery system in the learning process will be more optimal if oriented to a particular approach. One approach used in the development of this module is the constructivism approach. Based on these problems, the development of materials or guided learning resources, systematic, in accordance with the needs of learners, because with the exposure of structured materials will also help learners in building conceptual understanding and enable learners to learn independently. Independence of learners is expected to improve the ability to think and can measure the extent to which the level of understanding of the material through evaluation. One of the teaching materials that can make learners learn independently is the module.

The success of learning depends not only on the environment and learning conditions but also the initial knowledge of learners. Such knowledge can not be transferred intact from the teacher's mind to the learners, but actively built by the learners themselves through real experience. To support the improvement of students' understanding, so that learners can develop their initial knowledge and to learn more can be more meaningful, then the learning module is equipped with one of the learning approaches is the Constructivism approach. According to the Constructivism approach, learners construct existing knowledge with their own experience to form new knowledge. Nurhadi (2003) argues in the view of constructivism, knowledge grows and develops through experience. Understanding develops deeper and stronger if always tested with new experiences.

Constructivism learning is a learning that develops learners' thinking to learn more meaningfully by working alone, finding their own, and constructing their own new knowledge and skills (Sumiati and Asra, 2007). Basically, learners do not bring empty heads to school, but actually, they already have the initial knowledge or concept of something based on experience in everyday life (Lufri, 2007).

In learning using constructivism emphasizes the development of skills, skills and thinking of learners (Isjoni in Sutisna, 2013). With the constructivism approach, the development of knowledge for learners can be done by the students themselves through research activities or direct observation, so that learners can channel new ideas in accordance with experience by finding facts that fit with the study of theory. Learners are also required to be able to link existing knowledge with new knowledge as well as experience in the learners (Tjulifa, 2013). This is the advantage and characteristic of the Constructivism approach than any other approach.

Science learning should encourage learners to be able to construct their knowledge independently through learning experiences (Sudarisman, 2012). Science learning according to Constructivism perspective contains four core activities, which are related to initial knowledge, real experience, social interaction and environmental sensitivity (Nuryanti, 2005 in Sudarsiman, 2012). By using constructivism-oriented modules it is hoped that learners can form new understandings based on their initial knowledge and learning experience. After learners understand the learning materials they are expected to become more motivated in learning.

In the development of learning-oriented constructivism-oriented science-learning module is used curriculum 2013. Curriculum 2013 aims to change the attitude of learners to be more polite through the values of character education contained in it. This means that if you have the attitude and mentality are praise-worthy learners will be able to absorb science well and certainly be a clean generation.

Science learning is one of the substances of national education that make an important contribution to the character formation of learners. This is in accordance with Junarso's opinion in Setyaningrum and Hasamah (2012). Character as a result of education brings meaning to life in society. Therefore, the importance of understanding the value of the characters implemented in the science lesson. According to the Minister of National Education (2010), character education in Biology subject is health care, religious, independent, tolerance, friendly/ communicative, social care, responsibility, and caring environment developed in accordance with the needs and conditions of learners in society.

Character education in this module is written explicitly on the subject of learning activities so that learners will indirectly learn about the characters that must be known. With this character, education is expected that learners not only get good output in the form of cognitive learning outcomes only but also form their (affective) attitude for better, so that between the cognitive and affective learning results obtained learners to be balanced. Learning using constructivism-oriented learning modules and character education-charged allegedly can help the understanding of learners so that learners can easily digest, remember and summarize the subject matter well.

The science-oriented instructional module constructivism is characterized by a character education developed to be tailored to the 2013 curriculum that provides the opportunity for learners to learn in their own way and ability and facilitate learners in understanding the material learned and improve effectiveness and efficiency in achieving learning objectives and enriches source of reading for learners.

Departing from the observations above, it is appropriate if in science learning a teacher can use teaching materials that can bring up the initial knowledge of learners, making learners can build their own knowledge, and improve understanding of the concept and motivation of learners that will lead to learning outcomes which are desired.

Based on the background that has been described, the authors are interested to conduct research entitled "Validity of Learning Module Natural Sciences Oriented Constructivism with the Contain of Character Education for Students of Class VIII at Yunior Hight School ".

2. Development Methods

This research type is research development (development research). Another term often used for research development is design research (design research), this term is used by Plomp. Research development is an attempt to develop and produce a product in the form of materials, media, or learning strategies used to solve problems in the classroom and not to test the theory.

The stages of development of learning module of science-oriented constructivism-oriented character education using the Plomp model are as follows.

1.Preliminary Research Phase

Activities are undertaken at this stage, as follows.

1.1. Needs Analysis and Characteristics of Learners.

This analysis is conducted to determine the characteristics of the modules that students want, to examine the character of learners who become the target users of developed products and determine the module that the teacher wants as a tool in the learning process. Activities undertaken at this stage are, interviews with teachers and learners using an interview guide sheet. The results of the needs analysis are taken into consideration in the design and development of modules.

1.2. Curriculum Analysis.

At this stage, a review of the curriculum is conducted. The analysis is carried out on the competence demands set forth in core competence (KI) and basic competence (KD). This analysis is conducted with the aim of determining the order and scope of the required material in accordance with the existing basic competencies, viewing the sequence of materials, identifying the material, presenting the problems in the material relating to daily life in accordance with the indicators and learning objectives. The results of this analysis are used as guidance in developing modules.

1.3. Textbook Analysis.

The textbook analysis aims to find out the problems in the textbook such as the suitability of the content of the book with the Core Competence (KI), Basic Competence (KD) based on the 2013 curriculum so that it can be the basis for formulating the study material on developed modules.

2.Development and Prototipyng Phase

This phase begins after the preliminary research phase is completed. This stage aims to design the problem solving that has been identified at the initial identification stage. This phase has a microscope that helps in developing and improving the end result of the product. Activities undertaken in developing prototype include designing and developing the prototype, can be explained as follows.

2.1. Development of Prototype I.

The design of Prototype in the early stages is called prototype I evaluated by self-evaluation by revising the IPA module designed using a check list. The evaluation itself is done to check the design errors that exist in the module. Next prototype I was revised and then continued at prototype II development stage. In order for a questionnaire instrument to be used, it must be validated first.

2.2. Development of Prototype II.

Prototype II is a development stage by consulting with experts (expert review), ie IPA module discussed to the experts to obtain a valid IPA module. Validated aspects include aspects of the construct, content aspects, graphic aspects, and language aspects. Based on validation result, revision of IPA module is developed. In order for a questionnaire to be used, instrumentation validation was previously performed..

3. Results and Discussion

A.Research Results

1.Preliminary Research Data

1.1.Results of Needs Analysis and Characteristics of Learners.

Needs analysis is conducted to determine the characteristics of modules used in order to attract learners in the learning process. Based on interviews with learners obtained information that learners are more interested in the module with the complete material, clear writing, interesting to read, easy to understand or understand, and have a clear and interesting picture.

Based on the results of interviews with teachers SMP Negeri 11 Sungai Penuh Elni Fainofri, S.Pd obtained the information that in the learning process learners do not use teaching materials due to limited availability of teaching materials at school. This causes only teachers who hold and use instructional materials in learning while learners do not get the teaching materials. Teachers revealed that in the process of learning IPA learners still look passive, learners just copy what the teacher wrote on the blackboard and did not find other sources as supporting their understanding, learners receive more explanation from the teacher or listen to the teacher's lecture.

In addition, according to the teacher, the book of learners used by teachers in learning so far has not been presented in a complete and directed. For example in KD 3.6 should the presentation of the material must be complete and the order is clear, such as the material starts from the digestive system of food, the organ and digestion of food channels, the digestion of food both mechanically and chemically / enzymatic, and the linkage between digestive system with respiratory system and blood awareness system . While the presentation of the material in the book package only discusses the definition of food (energy needs and six types of nutrients) and food digestion channels while the linkage between the digestive system with the respiratory system, blood conscious system has not seen the discussion on the packaged book as a demand on KD 3.6.

Therefore, teachers expect the teaching materials with the delivery of complete and directed material in accordance with the KD demands that have been established, where the first submitted is a concept that must be understood besides that followed by its application in life. Then the teacher expects teaching materials that facilitate learners in understanding the lesson, can make learners can build their own concepts, and able to make all learners participate actively in learning activities.

Based on the results of the needs analysis of the characteristics of students then designed learning module science oriented constructivism charged character education. The learning-oriented science-oriented constructivism module is expected to be able to answer the expectations of teachers and learners. This is because the presentation of a systematic module and clear images will facilitate learners in understanding the subject matter. Constructivism approach will enable learners in the learning process because constructivism approach will guide learners in finding the concept of learning in accordance with predetermined objectives so that later learners will be able to build a true concept by linking the new knowledge they can with their initial knowledge have.

1.2.Results of Curriculum Analysis.

Curriculum analysis aims to determine the Core Competence (KI) and Basic Competence (KD) that has been established in the standard content of science subjects. The results of KI and KD analysis contained in the content standard are translated into indicators of learning achievement.

The character-oriented constructivism-oriented IPA module developed from KD 3.6, 4.6 and KD 3.7, 4.7 with multiple submissions of competency achievement indicators. The curriculum analysis that produces indicators of achievement is different from the presentations in the teacher manual and the learners. In teacher handbooks and students the presentation of materials in KD 3.6 and 4.6 begins with the digestion system of food, food (energy needs and six types of nutrients) and food digestion channels, whereas in the module developed the presentation of the material begins with the digestive system of food (understanding the digestive system, food and its functions, energy needs, six types of nutrients), food digestive organs, food digestion channels, food digestion process, as well as linkages of the digestive system with respiratory system and blood-conscious system. This is done because based on the KD that has been established in the content standard, that learners must understand the concept of the digestive system as well as the linkage between the digestive system with the respiratory system and circulatory system.

Whereas in KD 3.7 and 4.7 in teacher handbooks and learners the presentation of the material begins with natural and artificial additives and addictive substances-psychotropic, while in the module developed the presentation of the material begins by explaining the definition of additive and mention the type of additive substances in the food consisting of the natural and artificial coloring, sweetening,

IOP Conf. Series: Materials Science and Engineering **335** (2018) 012091 doi:10.1088/1757-899X/335/1/012091

preservatives and flavorings of food, explains the notion of addictive substances consisting of non-narcotic addictive substances and psychotropic substances, addictive substances of narcotics and psychotropic addictive substances and explains the negative effects of addictive and psychotropic substances for health and Proposed the idea of solving the circulation problem of psychotropic addictive substances in Indonesia. This is done because based on the KD that has been specified in the content standard, that learners should understand the concept of additives naturally and artificially in foods and beverages and addictive substances-psychotropic substances and their effects on health and how to solve the problem about substance abuse- psychotropic addictive.

1.3.Results of Textbook Analysis

Analysis of the book is done in the handbook of learners, this analysis aims to know the contents and presentation of content as a comparison for module development. Where the presentation of the material on the handbook of learners there that is correct based on the demand of KD is in the handbook of learners have been explained understanding the digestive system of food and organs in the digestive system of food and the handbook of learners already using a learning approach that is the scientific approach and the picture presented in the handbook of learners is quite complete only there are still some pictures that are less clear because the picture is blurred and the description is not complete. Analysis of the handbook of learners conducted by researchers with three components namely systematic, content and assessment. From the results of this analysis note that there are still some weaknesses in the handbook of learners. First, there are some material points that should be included in the picture to be clearer, but not included. Second, the image in the material presentation looks less clear because the picture is blurred and the description is not complete, such as the image of the oral cavity and anatomical structure of the small intestine, should be presented images must be clear and representative for learners more interested in reading and learners easily know the meaning of the picture presented. Third, the presentation of the material on the packaged book is still incomplete and directed so that learners need the help of others such as teachers to explain the contents of the book. Consequently, communication takes place in one direction and learners have a tendency to passive in the learning process.

2.Data of Development and Prototipyng Phase.

The results obtained at the initial investigative stage were used as guidelines in developing a constructivism-oriented IPA oriented character education module. The results of the development carried out at this stage are as follows.

2.1.Prototype I (Design).

Prototype I start from the design and manufacture of constructivism-oriented IPA oriented character education module. The module contains the cover, introduction, module instruction manual, core competency, basic competence, competency achievement indicator, learning objectives, material description, drawing, summary, competency test, answer key, feedback, glossary, and bibliography. The IPA learning module is designed with Microsoft Office Word 2007 and Microsoft Office Publisher 2007.

2.1.1. Cover.

In Cover, there are images related to the material contained in the module. The goal is that learners are interested to learn the modules developed.

2.1.2. Foreword.

The preface contains a thank-you note for the completion of the developed module, the module user's goals, the purpose of the module development, and the components contained in the module.

2.1.3. Module Use Instructions.

The instructions for using this module are designed to facilitate teachers and learners in using the module.

2.1.4. Indicators of Competence Achievement.

This indicator contains the learning objectives to be achieved in the creation of a constructivism-oriented IPA oriented module of character education.

2.1.5. Material Description.

Contains material developed in a constructivism-oriented IPA oriented character education module.

2.1.6. Picture.

Images displayed in the material-related module will improve the learner's understanding.

2.1.7. Summary.

The summary contains conclusions about the materials.

2.1.8. Competence test.

The competency test contains questions to evaluate the achievement of indicators on previously learned material.

2.1.9. Key Answer Competency Test.

The key to the competency test answers contains the correct answers to the questions given. The key answer is for students to check which answers are correct and which answers are wrong from the answers they provide after they have done the competency test.

2.1.10. Feedback.

Feedback contains instructions to calculate the learning mastery of learners. This learning feedback aims to learners can learn to learn mastery.

2.1.11. Glossary.

Specific terms that are difficult to interpret in the material discussed and the new terms that learners need to understand are explained at the end of the module in the glossary.

2.1.12. References

The references contains the literature used in the constructivism-oriented IPA oriented module of character education.

b. Prototype II (Self Evaluation)

The results of the evaluation itself indicate that the module components already exist and accordingly. However, there are some errors that occur that there are some posts whose letters are not complete so that its meaning becomes unclear, in the picture, there is not yet have the source. After the evaluation, it was revised by improving the incomplete writing and complete the source on the drawing so that the resulting learning module of science-oriented constructivism is characterized by character education.

c. Prototype III (Expert Validation)

In prototype III development stage, the formative evaluation activity is done that is seeing the validity of learning module of science-oriented constructivism based on character education based on expert's judgment which includes four aspects of construct, content, graphics, and language and legibility. Evaluation tool used is a validation sheet of learning module science oriented constructivism charged character education. Validation is done by six expert validators. In addition to assessing the module, the validator is also required to provide suggestions for improvements to the module. The suggestions provided by the validator are then used as a reference to revising the developed module.

Data of validation of learning module of science learning oriented constructivism with character education can be seen in Table 1.

Table 1. Data of validation of learning module

No	Aspects of Assessment	Average (%)	Category
1.	Aspects of constructs	89,58	Very Valid
2.	Aspects of Content	90,47	Very Valid
3.	Aspects of Graph	92,59	Very Valid
4.	Aspects of language	91,67	Very Valid
	Average	91,07	Very Valid

Based on the overall average result of validation learning module natural sciences oriented constructivism with the contain of character education 91,07% with very valid category.

B.Discussion

1.Validity of Science Constructivism-oriented Learning Module Loaded Character Education

The validity of science-oriented instructional module constructivism-charged character education is rated very valid by the validator. This is obtained from the data analysis of the validity value given by each validator with four aspects of the assessment, namely the aspects of constructs, contents, graph, and language. According to Trianto (2010) valid means that the assessment has provided accurate information about the developed teaching materials, teaching materials in the form of modules developed otherwise valid after meeting the aspects of the construct, content aspects, aspects of graft, and language aspects. The results of the product validity assessment on the constructed aspect into the category is very valid. Based on the above-mentioned modules are developed in accordance with this competence, basic competence, and defined learning objectives.

In the content aspect, the module is declared very valid. Validity in the aspect of the content provided by the validator due to indicators of achievement of competencies has been prepared and sorted by KI and KD in the curriculum 2013. The module also contains materials, learning material constraints, instruction learning activities, exercises designed systematically and interesting to achieve competence Expected. In addition, in the module, there is also a way of evaluating each submateri that aims to learners can learn by themselves. This is in line with the opinion of Sudjana and Rivai (2003), the purpose of teaching with the module is that learners can follow the teaching program according to their own speed and ability, more self-study, can know their own learning result, and emphasize the optimal mastery of learning materials Mastery learning) with a mastery level above 80%. Based on the above, the module developed has been in accordance with defined learning objectives.

Further validity is assessed from the aspect of the graph, in terms of aspects of the module of graph expressed is in very valid category. A very valid value is given a validator based on some indicators. First, it states that the module design is interesting. The developed modules are designed to be as attractive as possible using Microsoft office word 2007 and Microsoft office publisher 2007 programs. Second, fonts are used in more than one type with clear font sizes being read. The reason for the use of some types of letters is to add to the attractiveness of a character-oriented constructivism-oriented module of character education, to avoid the boredom of learners while reading, and to distinguish between material and titles. Muslich (2010) argues to distinguish and get a combination of display letters, can use variations and series of letters.

Third, the suitability of the presentation of the image. This is evidenced by the image on the module can be seen clearly, the image of the quotation includes the source and description of the image in accordance with the given image. This aims to facilitate learners in understanding the images displayed so that the images displayed can clarify the concept of material that must be understood by learners. This is in accordance with Rustaman (2002) that the module should be well designed to attract readers, both in size (paper, letters, and pictures) and from the variations. The same thing is also expressed by Galitz, et al (in Festiyet, 2008) the display of interesting learning media with images and objects in accordance with the demands of the material, will increase the interest of learners to the learning materials. Furthermore, Pratowo (2011) said that the presentation of the drawings is needed to support and clarify the contents of the material because in addition will clarify the description can also increase attractiveness and reduce the sense of boredom learners to learn it.

The last validity assessed is the aspect of language and readability. Assessment of the language aspects of the modules developed is in very valid categories. This is due to the structure of the sentences used in accordance with the EYD, in accordance with the level of understanding of learners, simple, clear and unambiguous, making it easier for learners to understand the learning activities in the module, and communicative. Hamdani (2011) stated that the things that need to be considered in the preparation of the module are the accuracy in preparing the sentence so that the module is composed communicative and easy to use as a learning guide for learners. This is also reinforced by Prastowo (2011) module is basically a teaching material that is arranged systematically with language that is easily understood learners, according to their knowledge level and age.

The four aspects of the validation of learning content learning module validated by constructivism containing character education that has been described is a unified whole and mutually supportive for the perfection of learning-oriented constructive science-oriented learning module developed character education. Based on the results of the analysis of the prevalence obtained an average value of 91,07% is with very valid category. According to Sugiyono (2008), a measuring instrument is valid if it can be used to measure what should be measured. The same thing is also stated by Arikunto (2012) if a data generated from a product is valid, then it can be said that the developed product has provided a description of the development goals correctly and in accordance with reality or the real situation. A very valid assessment of the learning-oriented constructivism-oriented IPA learning module developed this character that signifies that the module can be used as teaching material in learning..

4. Conclusion

Based on the development and experiments conducted on science-oriented learning module constructivism-filled character education obtained the following conclusions.
The science-oriented learning-oriented constructivism-oriented module of character education on the material of digestive system of food and chemicals in life for SMP class VIII developed has been declared is very valid by the expert validator with very valid category (91,07%). The validation of science-oriented learning module constructivism is characterized by character education in terms of construct, content, graph, and language.

Acknowledgments
Acknowledgments to Dr. Dwi Hilda Puteri, M.Biomed, Yohandri, M.Si, Ph.D., Dr. Darmansyah, S.T, M.Pd, Dr. Abdurahman, M.Pd, Elni Fainofri, S.Pd, and Sri Wahyuni, S.Pd as validators, and all of the big family of SMP Negeri 11 Sungai Penuh as the place of research implementation

References

[1]. Arikunto, Suharsimi. 2012. Dasar–dasar Evaluasi Pendidikan Ed. 2. Jakarta: Bumi Aksara.
[2]. Chitanana, lockias. 2012. "A Constructivist Approach to the Design and Delivery of an Online Professional Development Course: A Case of the Iearn Online Course". Research Journal of Educational Sciences. 5 (1): 23-48
[3]. Depdiknas. 2008. Panduan Pengembangan Bahan Ajar. Jakarta: Direktorat Jenderal Manajemen Pendidikan Dasar dan Menengah.
[4]. Elpitriani. 2012. Pengembangan Modul Pembelajaran Biologi Berorientasi Konstruktivisme Dilengkapi Mind Map pada Materi Sistem Ekskresi Untuk Sekolah Menengah Atas. Tesis tidak diterbitkan. Padang: PPs Universitas Negeri Padang
[5]. Festiyet. 2008. Pembelajaran Fisika Berbantuan Program Komputer Interaktif untuk Meningkatkan Perhatian Belajar Siswa (Studi Eksperimen di SMPN 7 Padang. Disertasi. Padang: Universitas Negeri Padang.
[6]. Hamdani. 2011. Strategi Belajar Mengajar. Bandung: CV Pustaka Setia.
[7]. Lufri. 2007. Strategi Pembelajaran Biologi Teori, Praktik dan Penelitian. Padang: Jurusan Biologi FMIPA Universitas Negeri Padang.
[8]. Mulyasa, E. 2006. Kurikulum Tingkat Satuan Pendidikan. Bandung: Remaja Rosdakarya Nasution. 2010. Berbagai Pendekatan dalam Proses Belajar Mengajar. Jakarta Bumi Aksara.
[9]. Muslich, M. 2010. Pendidikan Karakter Menjawab Tantangan Krisis Multidimensional. Jakarta: Bumi Aksara.
[10]. Nasution. 2010. Berbagai Pendekatan dalam Proses Belajar Mengajar. Jakarta: Bumi Aksara.
[11]. Plomp, Tjeerd dan Nieveen, Nienke. 2010. An Introduction to Educational Desain Research. Netherlands: SLO
[12]. Plomp, Tjeerd dan Nieveen, Nienke. 2013. Educational Design Research. Netherlands: Netherlands Institute for Curriculum Development (SLO).

[13]. Prastowo, A. 2011. Panduan Kreatif Membuat Bahan Ajar Inovasi: Menciptakan Metode Pembelajaran yang Menarik dan Menyenangkan. Yogyakarta: Diva Press.
[14]. Purwanto dkk. 2014. Pengembangan Modul. Jakarta: Pendidikan PUSTEKKOM Depdiknas
[15]. Rustaman, N. 2002. Perencanaan dan Penilaian Praktikum Diperguruan Tinggi(online),(http://fileupi.edu/Direktori/SPS/Prodi.Pendidikan_IPA/196 2011 51987031-pdf., diakses 21 Maret 2016).
[16]. Schneider Daniel, Paraskevi Synteta, Catherine Frete. 2002. "Community, Content and Collaboration Management Systems in Education: A New Chance for Socio-constructivist Scenarios?". Research on learning. 1 (1): 175-184
[17]. Setyaningrum, Yanur dan Husamah. 2014. Optimalisasi Penerapan Pendidikan Karakter di Sekolah Menengah berbasis Keterampilan Proses: Sebuah Perspektif Guru IPA-Biologi. (Jurnal). 1(1): Malang: UMM.
[18]. Sudjana, N dan Rivai, A. 2003. Teknologi Pengajaran. Bandung: Sinar Baru Algensindo.
[19]. Sugiyono. 2008. Metode Penelitian Pendekatan Kuantitatif, Kualitatif dan dan R&D. Bandung: Alfabeta.
[20]. Trianto. 2010. Model Pembelajaran Terpadu dalam Teori dan Praktek. Surabaya: Pustaka Ilmu.

IOP Conf. Series: Materials Science and Engineering **335** (2018) 012092 doi:10.1088/1757-899X/335/1/012092

Development Biology Worksheet Oriented Accelerated Learning on Plantae and Ecosystems for 10th-Grade Senior High School Students

D A Dipuja [1,*], L Lufri [1] and Y Ahda [1]

[1] Postgraduate of Biology Education Department, Padang State University, Padang, Indonesia

*diahdipuja@yahoo.com

Abstract. The problem that found are learning outcomes student is low on the plantae and ecosystems. Students less motivated and passive learning because learning is teacher center and teaching materials not facilitate student. Therefore, it is necessary to design a worksheet oriented accelerated learning. Accelerated learning approach that can improve motivation and learning activities. The purpose of the research was to produce worksheet oriented accelerated learning on plantae and ecosystems. This research is designed as a research and development by using Plomp model, consists of the preliminary, prototyping, and assessment phase. Data was collected through questionnaires, observation sheet, test, and documentation. The results of the research was worksheet oriented accelerated learning on plantae and ecosystems is very valid.

1. Introduction

Biology Learning aims to develop the competence of students, in order to be able to understand the natural surroundings through the process of finding and doing based on direct experience. Students in Biology learning are expected to develop process skills in the form of attitude and values. Teachers as educators to achieve the goals of learning Biology plays an important role in the success of students.

Based on the results of observation found problems of low student learning outcomes in the material Plantae and Ecosystem, students are less motivated and passive learning because the learning is still a teacher center so that learning becomes monotonous and not fun for students.

The presentation of the material on the worksheet is too succinct, there are still terms that are not understood, and the existing learning activities still do not use fast learning techniques so that they have not been able to make the students actively participate in Biology lessons that affect students' psychomotor competence.

Based on these things, activities that can accelerate students' understanding of the material to understand the concepts that exist. There is a learning model that is assumed to support the material so that learning can be accelerated and easy to remember that is Accelerated Learning. One of the learning models that can be applied to the worksheet is the Accelerated Learning model that has the syntax with the abbreviation MASTER. Accelerated Learning is a learning model that aims to make the learning process students become better and faster. The outcomes of the learning process can be in the form of additional skills, knowledge or new attitudes [1].

IOP Conf. Series: Materials Science and Engineering **335** (2018) 012092 doi:10.1088/1757-899X/335/1/012092

Availability of teaching materials in the form of worksheet Accelerated Learning that can guide students, facilitate the way students learn in learning activities so that learning becomes more focused, not monotonous and easy for teachers and students.This study aims to produce a valid Accelerated Learning Biology worksheet and reveal the effectiveness of the worksheet. Expected availability of teaching materials in the form of worksheet Accelerated Learning that can guide students, facilitate the way students learn in learning activities so that learning becomes more focused, not monotonous and easy for teachers and can improve student learning outcomes.

2. Methods

The development of the worksheet is based on the stages of the Plomp model. This worksheet was developed through three stages namely (1) preliminary research phase, (2) prototyping phase, and (3) assessment phase [2]. Fomative evaluation is done at every single stage in the form cycle of evaluation-revision.

The trial subject is students in one class of 10th grade of Senior High School numbers 5. Data consist of qualitative and quantitative data. The qualitative data obtained from observations and interviews in preliminary research phase and one to one evaluation in prototyping phase. The quantitative data obtained from the validity questionnaire.

Result of the research prosessed using the following equation.

$$\text{Value} = \frac{\text{Obtained score}}{\text{Maksimum score}} \text{ x } 100\% \qquad (1)$$

This value is subsequently converted in accordance with validity criteria that can be seen in Table 1.

Table 1. Criteria level of validity

Percentase (%)	Level of validity
0 – 20	Invalid
21 – 40	Less Valid
41 – 60	Enough Valid
61 – 80	Valid
81 – 100	Very Valid

3. Results and Discussion

Worksheet contains a description of the material that consist the Accelerated Learning steps in MASTER syntax which includes motivating, obtaining information, investigating meaning, triggering memory, displaying knowledge and reflection. Worksheet was developed through three stages of Plomp model. In preliminary research phase there are fourth process namely (1) problem analysis, (2) need analysis, (3) concept analysis, and (4) curiculum analysis. In prototyping phase there are process namely (1) self evaluation, (2) expert review, and (3) field tests. Validity doing by expert review. Validity of worksheet includes aspect namely dedactic, construct, technical, linguistic.

3.1. Results

Based on preliminary research analysis finds that problem there are known problems. student questionnaire results indicate that the interest and motivation of students in biology learning is still low, and students are less fond of reading biological materials that exist and there are still shortcomings in the worksheet prepared by the teachers discussion as in the presentation aspect that is less attractive for students because the display is not colored in particular on Plantae and Ecosystem materials. So that teachers want improvements on teaching materials in the form of worksheet that can lead students in learning.

Development on the construct aspect is on the student worksheet that implements Accelerated Learning steps. Worksheet are develop tasted for validity. Results worksheet validation by experts are presented in Table 2.

Table 2. Result of validation worksheet

Aspect	Persentage (%)	Criteria
Didactic	90.62	Very Valid
Construct	88.75	Very Valid
Technical	81.25	Very Valid
Linguistic	85.42	Very Valid

Based on Table 2 it can be seen that worksheet are develop is very valid. During the validation, expert give some advice to fix the cover, writing, writing color, and image layout.

3.2. Discussion

The result of validity test shows that the Accelerated Learning worksheet developed very valid. Worksheet have suitable with basic competences. Worksheet also accordance with the development of the child. Worksheet emphasizes more on the process of finding concepts, and most importantly in the worksheet there is variation of stimulus through various media and student activities. In line with the results of research by Ozmen and Yildirim [3], worksheet is a sheet containing work or materials that make students more active in taking meaning from the learning process.

Viewed from the construct aspect, Biology worksheet Accelerated Learning is declared valid by validator. This validation results show that worksheet has constructed the Accelerated Learning steps in MASTER syntax which includes motivating, obtaining information, investigating meaning, triggering memory, displaying knowledge and reflection. Strengthening concepts and improving memory in the Trigering of memory phase is expected to go through mind mapping, keyword and acrostic activities. According to Palmer [4], accelerated learning model combined with mind mapping gives a positive effect that can improve student achievement. In accordance with Erland [5], Accelerated Learning can improve students' cognitive, memory, and learning achievement.

Learning activities that are displayed on worksheet also help the students in understanding the concept of the material and will be long remembered by the students because the thing done alone and found itself will be remembered longer. In addition, worksheet can also be used repeatedly both at school and at home. According to Sanjaya [6] worksheet is a learning resource that contains messages related to the subject matter to facilitate student learning.

4. Conclusion

The conclusion of the research that worksheet oriented accelerated learning are develop is very valid. Worksheet was developed steps in MASTER syntax which includes motivating, obtaining information, investigating meaning, triggering memory, displaying knowledge and reflection.Validity of worksheet includes aspect namely dedactic, construct, technical, linguistic.

References

[1] Russel L 2011 *The Accelerated Learning Fieldbook* (Bandung: Nusa Media) p 15
[2] Plomp T 2013 *Educational Design Research; An Introduction* (Netherlands: Enschende) p 19
[3] Ozmen H, Yildrim N 2005 *Journal Of Turkish Science Education* p 44
[4] Palmer L 2006 *Journal of Accelerated Learning and Teaching* p 39
[5] Erland, Juyper J 1998 *Journal of Accelerated Learning and Teaching* p 29
[6] Sanjaya W 2010 *Learning Strategy Oriented Standards Education Process* (Jakarta: Kencana Prenada Media Group) p 12

Identification of Conceptual Understanding in Biotechnology Learning

E Suryanti[1,2*], **A Fitriani**[1,3], **S Redjeki**[1], **R Riandi**[1,3]

[1] Sekolah Pascasarjana, Universitas Pendidikan Indonesia, Jl. Dr. Setiabudi No. 229, Bandung 40154, Indonesia
[2] Program Studi Pendidikan Biologi, Universitas Islam Riau, Jl. Kaharuddin Nasution No. 113, Pekanbaru 28284, Indonesia
[3] Departemen Pendidikan Biologi, Universitas Pendidikan Indonesia, Jl. Dr. Setiabudi No. 229, Bandung 40154, Indonesia

*evibio@edu.uir.ac.id

Abstract: Research on the identification of conceptual understanding in the learning of Biotechnology, especially on the concept of Genetic Engineering has been done. The lesson is carried out by means of discussion and presentation mediated-powerpoint media that contains learning materials with relevant images and videos. This research is a qualitative research with one-shot case study or one-group posttest-only design. Analysis of 44 students' answers show that only 22% of students understand the concept, 18% of students lack understanding of concepts, 57% of students have misconceptions, and 3% of students are error. It can be concluded that most students has misconceptions in learning the concept of Genetic Engineering.

1. Introduction

Biotechnology is a field of science that deals with techniques to manipulate the organism or its components to produce products that are useful to humans. The development of Biotechnology as a science is supported by the advancement of DNA technology. Success in making recombinant DNA provides an opportunity for the development of sophisticated techniques to analyze genes and gene expression. According to the structure of biological science, Biotechnology discusses the structure, function, and regulation at the molecular level of living things. The study of Biotechnology almost touches all groups of living things.

The rapid development of Biotechnology must be balanced with students' understanding of Biotechnology content, especially Biology teacher candidates. A good understanding of Biotechnology content along with pedagogical knowledge will be a solid provision for Biology teacher candidates to teach Biotechnology at the high school.

Genetic engineering is one of the concepts discussed in the Biotechnology curriculum. Genetic engineering can be defined as the manipulation of genetic material for practical purposes. Genetic engineering has launched a revolution in Biotechnology, thereby greatly expanding the scope of potential applications of Biotechnology [1]. The benefits of genetic engineering such as gene therapy, the creation of new vaccines, and the formation of transgenic organisms.

Learning concepts is the primary outcome of education. The concept is a mind-building stone. The concept is the internal presentation of a group of stimuli, the concept can not be observed; concepts must be inferred from behavior. The concept is formed according to the stimulus received by a person, so that no two people have the exact same concept. However, with different concepts that people can communicate by using the names given to concepts that have been accepted together. The concept is the basis for higher mental processes to formulate the principles and generalizations [2]. The correct of conceptual understanding is an important result of cognitive outcome in science. The classical approach to knowledge acquisition and conceptual change assert that science learning to take place over a very short period of time [3]. However, according to [4] a newer approach suggests that the correct of conceptual understanding involves a lengthy process. Through this long period of time, learners will interact with new information in a socio-cultural context that enriches the conceptual understanding or restructuring existing conceptual understanding.

According to Flavell in [2] a concept has been learned when the learner can display certain behaviors. Klausmeier in [2] state there are four levels of conceptual achievement, namely concrete level, identity level, classification level, and formal level. Information about concepts that must be taught to learners of a certain age or class may be derived from a number of sources, including textbook writers, curriculum developers, educators' knowledge and experience, and learners themselves. After selecting the concepts to be taught, the educator should plan the learning strategy to teach the concepts. To that end, an educator not only master the knowledge of the field of study to be taught, but also various approaches and methods of learning and various learning theories to guide educators in applying the approach and method chosen [2].

This study aims to identify the conceptual understanding in biotechnology learning, especially on the topic of Genetic Engineering.

2. Method

This research is a qualitative or pre-experiment research with one-shot case study or one-group posttest-only design, i.e. treatment is only given to one group of subjects. Observations were made on group members to determine or assess the effect of treatment (Asher and Vockell in [5]).

Biotechnology learning is conducted by means of discussion of media-aided image and video. Data collection was conducted in December 2015 with test techniques. The test consists of six true-false-level objectives question about the concept of Genetic Engineering. The question modified from Biology book [1]. Each student was asked the reason for each question he answered and his belief in the answer given. The test was followed by 44 students of Biology Education Program, Faculty of Teacher Training and Education, Islamic University of Riau who have taken Biotechnology course. Students' answers are examined one by one and analyzed qualitatively in accordance with the provisions of [6] below:

Table 1. Categorization for respondents' answers.

Category	Answer type
1. Understanding the Concept	The answer is right, the reason is right, sure
2. Lack of understanding of the concept	The answer is right, the reason is right, not sure The answer is wrong, the reason is right, not sure The answer is right, the reason is wrong, not sure The answer is wrong, the reason is wrong, not sure
3. Error	The answer is wrong, the reason is right, sure
4. Misconception	The answer is right, the reason is wrong, sure The answer is wrong, the reason is wrong, sure

3. Results and Discussion

An examination of the test answers of 44 respondents was presented in Table 2 below:

Table 2. The analysis of respondents' answers.

Question Number	Number of answers (people)		Number of reasons (people)		Level of confidence (people)	
	Correct	Incorrect	Correct	Incorrect	Sure	Not Sure
1	19	25	20	24	40	4
2	40	4	10	34	35	9
3	14	30	10	34	30	14
4	41	3	15	29	35	9
5	39	5	9	35	38	6
6	42	2	9	35	38	6
Average					82%	18%

Based on the data in Table 2, the number of students who answered correctly is always more than the incorrect answer, except for questions number 1 and 3. With regard to question number 1, students are not expected to understand well about the types of enzymes used in genetic engineering and its function. Furthermore, students should understand the structure of plant cells and animal cells well in order to answer the number 3 correctly.

If it is related to the reasons given by the student for each question, the overall correct reason is always smaller than the number of correct answers. That is, the students have the big opportunity to guess the answer because the tested question is a matter of objectively true-false. However, this objective limitation is overcome by asking students to state the reasons for the answers they provide. This can differentiate between students who really understand the concept with students who guess answers. However, for question number 1 there is one student (respondent number 27) who is wrong in answering the question, but the reason given is true. Allegedly the student may be penalized in writing the answer, when he understood the question well.

The level of student confidence in answering the overall question and the reason given the high average is 82%.

Furthermore, the overall test results were grouped into four categories according to [6]. The results can be seen in Table 3 below:

Table 3. Categorization of respondents' answers.

Question Number	Category				Total
	Understand the concept (%)	Lack of understanding the concept (%)	*Error* (%)	Misconceptios (%)	
1	16 (36)	4 (9)	2 (5)	22 (50)	44
2	8 (18)	9 (20)	1 (2)	26 (59)	44
3	5 (11)	14 (32)	3 (7)	22 (50)	44
4	11 (25)	9 (20)	1 (2)	23 (52)	44
5	9 (20)	6 (14)	0	29 (66)	44
6	8 (18)	6 (14)	1 (2)	29 (66)	44
Total	57 (22)	48 (18)	8 (3)	151 (57)	264

Table 3 illustrates the number of students who 'understand the concept' more than any other category (lack of understanding of concepts, errors, and misconceptions) only on question number 1. In addition, for question number 2-6 the number of students 'misconception' is always more than other categories (understanding concepts, lack of understanding of concepts, and errors). Overall, the error category always exists in small numbers, except for question number 5, there is no error.

The focus of this research is to identify the conceptual understanding in learning the concept of Genetic Engineering. Based on the analysis of the data obtained can be stated that most students have misconceptions in the concept of Genetic Engineering. [7] states that misconceptions represent an interpretation of concepts that are not accepted by the learner. According to [8] misconceptions are also known as 'alternative conceptions', 'naive beliefs', 'children's ideas', 'conceptual difficulties', 'phenomenological primitives', 'mental models' , etc. According to [9], misconception is a perception of phenomena occurring in the real world that is inconsistent with the scientific explanation of the phenomenon. Misconceptions may vary from minor misconceptions to rejection of complex theories that can be fundamental barriers to students' understanding of scientific explanations.

To overcome misconceptions, [7] suggests that learners should be able to generate a logical connection between evidence and alternative conceptions. In this case, the method of discussion conducted in learning is expected to explore students' insights about the facts of genetic engineering that can be obtained from various sources of learning. In addition, in the implementation of learning is also equipped with a powerpoint and video presentation to clarify things complicated. According to Suryosubroto in [10], the method of discussion is a way of presenting teaching materials with educators providing opportunities for learners or groups of learners to hold scientific conversations to collect opinions, draw conclusions, or arrange alternative solutions to problems.

Discussion methods used in Genetic Engineering learning can not be said to be ineffective and as a cause of misconceptions in most students, as there are no other student groups as a comparison. Although discussion methods have limitations such as discussion of issues often out of context, and information obtained by students is limited [10, 11] in his study suggests that learners better assimilate ideas about genomes and genetic engineering in the field of medicine during discussion, and association of information in the mass media with discussion activities can help to disseminate scientific concepts. Meanwhile, [12] in their study stated that Genetic Engineering learning in laboratory outreach can increase substantial knowledge both in short and long periods, and motivated learners achieve higher scores.

The average level of confidence in answering the whole question is high i.e 82%, while the respondent students who understand the concept is only 22%. This is different from the opinions of [13] that students who master the concept well will have a high confidence. That is, only students who answer the question correctly, and provide the correct reasons alone who have great confidence because of his high confidence. But according to the categorization by [6], students who have confidence include students who 'understand the concept', students who experience 'misconception', and even students who 'error'. Only students who 'lack of understanding the concept' are not sure of the answers and reasons given.

4. Conclusions

Based on the results of research and discussion, it can be concluded that most students indicating misconceptions in the Genetic Engineering concept. It is advisable for educators to seek alternative learning methods that connecting the concept of Genetic Engineering to real life situations. Interviews or surveys should also be conducted to determine the causes of students' misconceptions, and need to design valid and reliable diagnostic tests to analyze student conceptions in genetic engineering learning.

References

[1]. Campbell, N.A., Reece, J.B., Urry, L.A., Cain, M.L., Wasserman, S.A., Minorsky, P.V., and Jackson, R.B. (2010). *Biologi* (terjemahan). Jakarta: Erlangga

[2]. Dahar, R.W. (2011). *Teori-teori Belajar & Pembelajaran*. Jakarta: Erlangga

[3]. Renken, M. D., dan Nunez, N. (2013). Computer simulation and clear observations do not guarantee conceptual understanding. *Learning and Instruction, 23*: 10-23

[4]. Vosniadou, S. (2007). The cognitive-situative divide and the problem of conceptual change. *Educational Psychologist, 42*(1): 55-66

[5]. Setyosari, P. (2013). *Metode Penelitian Pendidikan & Pengembangan.* Edisi Ketiga. Jakarta: Kencana Prenada Media Group.
[6]. Kaltakci, D. & Didis, D. (2007). Identification of pre-service physics teachers' misconceptions on gravity concept: A study with a 3-tier misconception test. Dalam S. A. Cetin, dan I. Hikmet (Eds), *Proceedings of the American Institute of Physics*, USA, 899, 499-500
[7]. Svandova, K. (2014). Secondary school students' misconceptions about photosynthesis and plant respiration: preliminary results. *Eurasia Journal of Mathematics, Science & Technology Education, 10*(1): 59-67
[8]. Gurel, D. K., Eryilmaz, A., dan McDermott, L. C. (2015). A review and comparison of diagnostic instruments to identify students' misconceptions in science. *Eurasia Journal of Mathematics, Science & Technology Education, 11*(5), 989-1008
[9]. Yates, T. B. dan Marek, E. A. (2014). Teachers teaching misconceptions: A study of factors contributing to high school biology students' acquisition of biological evolution-related misconceptions. *Education and Outreach, 7*:7.
[10]. Putra, S. R. (2013). *Desain Belajar Mengajar Kreatif Berbasis Sains*. Jogjakarta: DIVA Press.
[11]. Freire, A. S., Xavier, M. C. F., dan Moraes, M. O. (2013). High school students' attitudes associated with biotechnology and molecular genetics concepts in Brazil. *Creative Education, 4*(2): 149-153
[12]. Goldschmidt, M., dan Bogner, F. X. (2015). Learning about genetic engineering in an outreach laboratory: Influence of motivation and gender on students' cognitive achievement. *International Journal of Science Education, Part B: Communication and Public Engagement*. http://dx.doi.org/10.1080/21548455.2015.1031293
[13]. Kurbanoglu, N. I., dan Akin, A. (2010). The relationship between university students' chemistry laboratory anxiety, attitudes and self-efficiency beliefs. *Australian Journal of Teacher Education*, *35*(18), 48-59

IOP Conf. Series: Materials Science and Engineering 335 (2018) 012094 doi:10.1088/1757-899X/335/1/012094

Development of Learning Models Based on Problem Solving and Meaningful Learning Standards by Expert Validity for Animal Development Course

L Lufri*, R Fitri, R Yogica

Biology Department, Faculty of Mathematics and Sciences, Universitas Negeri Padang, Indonesia

*lufri_unp@yahoo.com

Abstract. The purpose of this study is to produce a learning model based on problem solving and meaningful learning standards by expert assessment or validation for the course of Animal Development. This research is a development research that produce the product in the form of learning model, which consist of sub product, namely: the syntax of learning model and student worksheets. All of these products are standardized through expert validation. The research data is the level of validity of all sub products obtained using questionnaire, filled by validators from various field of expertise (field of study, learning strategy, Bahasa). Data were analysed using descriptive statistics. The result of the research shows that the problem solving and meaningful learning model has been produced. Sub products declared appropriate by expert include the syntax of learning model and student worksheet.

1. Introduction

The demands of the government on improving the quality of higher education in Indonesia have surfaced in various mass media and Education meetings. They have taken serious action to make this happen. A similar thing is being done also by the higher education. Level of education quality can be known by looking at the quality of the learning process, which is expected to be the key to the formation of individuals who are able to apply science properly and correctly. However, based on a preliminary study on one of the subjects in college, especially in the Department of Biology FMIPA Universitas Negeri Padang (Animals Development) found unsatisfactory results. The data found is the difficulties of students understanding the concept, the students are unable to re-explain the related material, the students cannot describe the pictures and they cannot think at a higher level.

This problem is related to the student's low-mindedness to understand the concept of animal development. The ability is often known when students take the final exam of the semester. Students tend to answer wrongly about a concept, while they should not be wrong. The conclusion is that students do not understand the concept well.

Based on the facts that have been stated previously, confirmed by the results of discussions with the lecturers of the field course of Animals Development arises an idea to solve problems or find solutions. The solution offered is to develop a model that requires students to have the ability to think high level and interpret the learning materials. Grabe and Grabe suggest that with meaningful learning the students will understand the content of the lesson more deeply [1]. Meaningful learning is much better when compared with rote-based learning [2].

This model was chosen to be developed on the grounds that students' ability after learning will improve through meaningful learning. Implementation of meaningful learning can be done by lecturers because they are professional in the field of learning so that role to prepare learners to be ready to join in society or level education [3].

In addition to the model, this research also develops teaching materials that are Problem Solving based worksheets and Concept Map based worksheets. Von Glaserveld states that with problem solving learning will lead to meaningful learning, because students will feel satisfied, happy and motivated to learn [4]. Concerning the concept map, Clinton has written that meaningful learning will be beneficial to the learning strategy of "finding relationships" between concepts [5].

This research is limited to validity test of model syntax and student worksheet. The validity of the model syntax and student worksheet is very important because it indicates the level of trust in something. Guba and Lincoln stated that the level of trust has 4 aspects, namely credibility, transferability, dependability and confirmability [6]. Validity describes the precision of the data that has been found [7]. Even the truth level of a data can be indicated from the validity of the data [8]. A good research is that the researcher can validate the research instrument as well as the data obtained [9].

2. Method

The method used in the research is development research. This research development method was adopted from the instructional development cycle model developed by Fenrich in 1997 as shown in Figure 1.

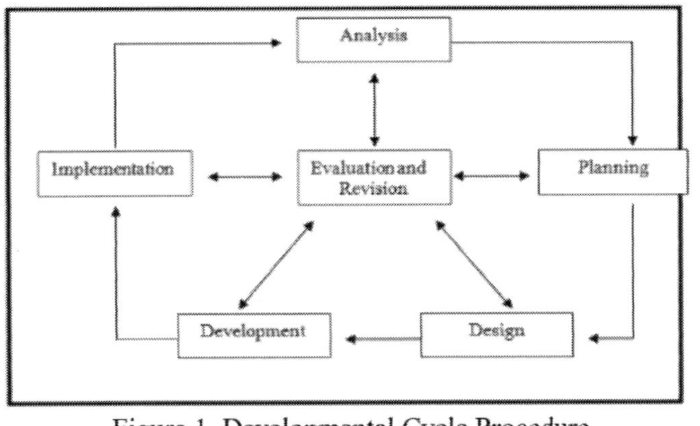

Figure 1. Developmental Cycle Procedure

The data collecting instrument in the study used a validity questionnaire that was first validated by the experts. Questionnaire has 5 scales with criteria strongly agree, agree, average, quite agree and disagree. Validator in this research is Dr. Yuni Ahda, M.Si., Dr. Ramadan Sumarmin, M.Si., Dr. Hardeli, M.Si., and Dr. Yerizon, M.Sc. The validation data obtained has been processed using the average analysis for the final result to interpret the entire product. These products include syntax module, student workbook based problem solving and student worksheet based on concept map. The interpretation values of validity data are (1) 1-1.79 is highly invalid, 1.8-2.59 is invalid, 2.6-3.39 is valid enough, 3,4-4,19 is valid and 4, 2- 5 is very valid.

3. Results and Discussion

Analysis of validation data by four validators to the syntax of the model obtained an average value of 4.3 with very valid criteria. The average value of the didactic indicator is 4.2, the average value on the construction indicator is 4.4, and the mean value for the technical indicator and linguistic is 4.3. Distribution of data more clearly can be seen in Figure 2.

IOP Conf. Series: Materials Science and Engineering **335** (2018) 012094 doi:10.1088/1757-899X/335/1/012094

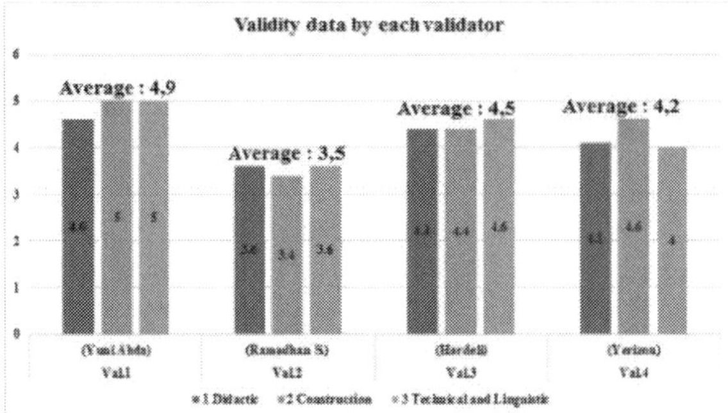

Figure 2. Validity Value of Model Syntax

Result of data analysis to student work sheet based problem solving got value 4.2 with criterion very valid. The average value on the indicator of eligibility of content got the number 4.3, the average value on the indicator language got the value 4.3, the average value for indicator presentation is 4.2, and the average value on the indicator graph is 4.2. Distribution of data more clearly can be seen in Figure 3.

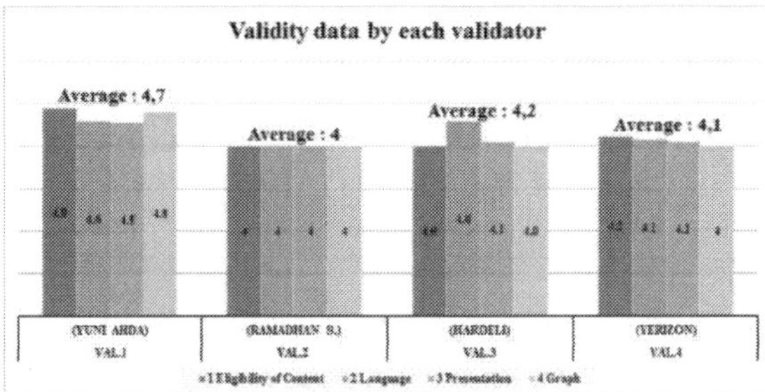

Figure 3. Validity Value of Student Worksheet Based on Problem Solving

The result of validation data analysis toward concept map based worksheet got value 4.3 with criterion very valid. The average value on all indicators (indicator eligibility of content, language, presentation, graph) is 4.3. Distribution of data more clearly can be seen in Figure 4.

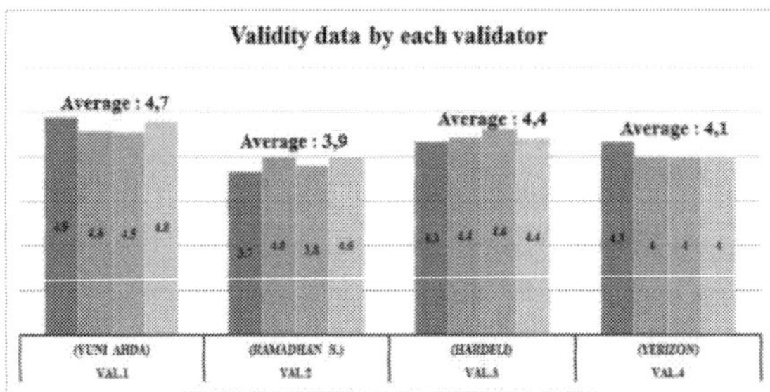

Figure 4. Validity Value of Student Worksheet Based on Concept Map

All products that have been developed and validated, are expected to be useful for lecturers as a consideration for use in learning, especially in the subject of Animal Development. This research can also be used as an example to develop the ability to produce learning model and student worksheet, so it is expected to create a better learning quality and different from usual. Different learning atmosphere than usual will be interesting for students to pay attention and understand the learning materials. Government demands that are in line with the background of this research so that teachers can create a conducive learning atmosphere in order to encourage the realization of meaningful learning will be achieved.

Meaningful learning will have an impact on satisfactory learning outcomes in accordance with the learning objectives. There are three criteria to create meaningful learning that is (1) learning materials relevant to the daily life of the students, (2) the teacher's learning materials should also be meaningful for the students 'life, and (3) the students' curiosity to connect the learning materials with the activities which is fun [10]. Novak explains that enjoyable learning is an integration between thoughts, feelings, and actions, which encourages students to be committed and responsible [10].

The syntax of the developed learning model leads the student toward meaningful learning. In essence, meaningful learning is much greater in meaning than just transmitting material from teacher to student. Meaningful learning builds opportunities for students to build their own knowledge, and the ability to share with other students [11]. Learning model that has been developed is the right choice in a learning process in general, and subjects of animal development in particular to give a positive influence on the process and student learning outcomes in the Department of Biology FMIPA UNP in the course of Animal Development.

The accuracy of choosing teaching materials that will be used in the learning process will accelerate the acceptance of the contents of the given material. If students easily accept the subject matter and they are able to understand it, then learning it will become more active and advanced. Judging from the aspect of the truth of the contents, the worksheet based on problem solving and student worksheet based on the concept map is considered very valid so that it can be stated that the student worksheet contains the material in accordance with the curriculum in the college, according to the stage of student development and their needs, the content of the material is correct, it is assumed to be able to increase the knowledge and understanding of the students and can create meaningful learning process, so it is good to be used as learning resource in learning in the subject of Animal Development.

From linguistic aspect, the work sheet of problem solving student and student worksheet of concept map given value is very valid because the description of the material in both worksheet of this student can be read clearly, clear explanation of information, spelling according to Indonesian rule and already using effective sentence and efficient. Judging from the presentation aspect, student work sheet based on problem solving and student work based on the concept map is considered very valid because the learning objectives formulated are clear, the design of student worksheets in accordance with the rules of development, stimulate learning motivation with color and picture, cover, build learning meaning and completeness of information. Then from the aspect of the graph, then the student worksheet is considered very valid because the selection of the right letter, font size, image layout and image selection.

Based on the assessment of the validators it can be drawn a conclusion that the model and worksheet of students can help students to improve their understanding in the course of Animal Development. Students are interested to learn by using this learning model because it can improve their understanding and ability to think high level. The validity test can provide a conclusion that the student models and worksheets that have been developed can be used in the learning process at different times with different specifications [13].

4. Conclusion

Based on data analysis and discussion, it can be made a generalization about the learning model and student worksheet (problem solving and concept map) for students in the course of animal development which has been developed has validity with very valid category. This development research has resulted in learning model and student worksheet for animal development students and can give description and input to related parties to continuously improve the quality of education, especially in the subject of Animal Development. The model and worksheet of students that have been developed will create a fun learning nuance and can realize satisfactory learning outcomes.

IOP Conf. Series: Materials Science and Engineering **335** (2018) 012094 doi:10.1088/1757-899X/335/1/012094

The research team suggested that it continues to develop the ability to be able to test the level of validity of an object, because each test of validity cannot be ascertained with absolute correctness. One reason is the ability of researchers to test and analyze the data validity. The calculation of the validity of an object will be influenced by the ability of researchers to collect and calculate data. Validity values cannot be separated from the possibility of error calculation techniques conducted by the researchers themselves. Therefore, the researcher receives all criticism and suggestion from the reader to improve the quality of this research.

References

[1] Huang, Yueh-Min and Chiu, Po-Sheng 2015 *The Effectiveness of A Meaningful Learning-Based Evaluation Model for Context-Aware Mobile Learning* (*British Journal of Educational Technology* vol 46) ed - (London: British Educational Research Education) chapter 2 pp 437-447

[2] Gunstone, Richard 2015 *Encyclopedia of Science Education* (Australia: Monash University)

[3] Darling, Linda and Hammond 2015 *Meaningful Learning in A New Paradigm for Educational Accountability: An Introduction* (*Education Policy Analysis Archives* vol 23) ed - (Arizona: Arizona State University) chapter 7 pp 1-9

[4] Somyurek, Sibel 2014 *An Effective Education Tool: Construction Kits for Fun and Meaningful Learning* (*International Journal Technology and Design Education* vol 25) ed - (Turkey: Gazi University) chapter 1 pp 25-41

[5] Lie, Silvia Wen-Yu et.all 2016 *Do Sophisticated Epistemics Beliefs Predict Meaningful Learning? Findings from A Structural Equation Model of Undergraduate Biology Learning* (*International Journal of Science Education* vol 38) ed - (London: Taylor&Francis Online) chapter 15 pp 1-19

[6] Morse, Janica M 2002 *Verification Strategies for Establishing Reliability and Validity in Qualitative Research* (*International Journal of Qualitative Methods* vol 1) ed - (Alberta: International Institute for Qualitative Methodology) chapter 2 pp 13-22

[7] Noble, Helen and Smith, Joanna 2015 *Issue of Validity and Reliability in Qualitative Research* (*Evid Based Nurse* vol 18) ed - (London: Research Made Simple) chapter 2 pp 34-35

[8] Golafshani, Nahid 2003 *Understanding Reliability and Validity in Qualitative Research* (*The Qualitative Report* vol 8) ed - (Florida: Nova Southeastern University) chapter 4 pp 597-607

[9] Henseler, Jorg et.all 2015 *A New Criterion for Assessing Discriminant Validity in Variance-based Structural Equation Modelling* (*Journal of The Academic Marketing Science* vol 43) ed - (US: Academy Marketing Science) chapter 1 pp 115-135

[10] Bretz, Stacey Lowery et.all 2013 *What Faculty Interviews Reveal About Meaningful Learning in The Undergraduate Chemistry Laboratory* (*Journal of Chemical Education* vol -) ed - (Washington: American Chemical Society and Division of Chemical Education) chapter 90 pp 281-288

[11] Sharan, Yael 2015 *Meaningful Learning in The Cooperative Classroom* (*Education* vol 43) ed - (London: Taylor&Francis Online) chapter 1 pp 88-94

[12] Savage, Jason W et.all 2015 *The Realibility and Validity of The Thoracolumbar Injury Classification System in Pediatric Spine Trauma* (*Spine Diagnostics* vol 40) ed - (Wolters Kluwer Health) chapter 18 pp 1014-1018

[13] Reyes, Andreas De Los et.all 2015 *The Validity of The Multi-informant Approach to Assessing Child and Adolescent Mental Health* (*Psychologycal Bulletin* vol 141) ed - (Washington: American Psychological Association) chapter 4 pp 858-900.

IOP Conf. Series: Materials Science and Engineering **335** (2018) 012095 doi:10.1088/1757-899X/335/1/012095

The Implementation of Research-based Learning on Biology Seminar Course in Biology Education Study Program of FKIP UMRAH

T Amelia*

Biology Education Study Program, Faculty of Teacher Training and Education
University of Maritim Raja Ali Haji, Tanjung Pinang, Indonesia

*trisna.amelia@umrah.ac.id

Abstract. Biology Seminar is a course in Biology Education Study Program of Faculty of Teacher Training and Education University of Maritim Raja Ali Haji (FKIP UMRAH) that requires students to have the ability to apply scientific attitudes, perform scientific writing and undertake scientific publications on a small scale. One of the learning strategies that can drive the achievement of learning outcomes in this course is Research-Based Learning. Research-Based Learning principles are considered in accordance with learning outcomes in Biology Seminar courses and generally in accordance with the purpose of higher education. On this basis, this article which is derived from a qualitative research aims at describing Research-based Learning on Biology Seminar course. Based on a case study research, it was known that Research-Based Learning on Biology Seminar courses is applied through: designing learning activities around contemporary research issues; teaching research methods, techniques and skills explicitly within program; drawing on personal research in designing and teaching courses; building small-scale research activities into undergraduate assignment; and infusing teaching with the values of researchers.

1. Introduction

According to Constitution on Higher Education in Indonesia, the functions of higher education are to develop the ability and form the character and civilization of a dignified nation; to develop innovative, responsive, creative, skilled, competitive, and cooperative academicians through the implementation of *Tridharma*; and to develop science and technology. University (college) as a provision of higher education has three duties (*Tridharma*) namely education and teaching, research and community service. Driving these functions, the synergy between lecturers and students as academicians in the implementation of *Tridharma* is needed. The lecturer as the main organizers should involve students in *Tridharma*, not just in teaching, but also in research and community service.

Based on my observation, the higher education tasks implemented in Biology Education Program, Faculty of Teacher Training and Education, University of Maritim Raja Ali Haji still partially emphasize on the implementation of education/ teaching. On the contrary, research is at a low level on quantity and quality. Research conducted by lecturers is still less nor the research involving students. Research grants won by lecturers are in the lowest category (1-3 grants in 1 year), the lecturers' motivation in conducting independent research is low, and most lecturers have not set a research roadmap. The constraints faced by lecturers to carry out research and involve students are the absence of research roadmap, and also the duties for teaching are overloaded (lecturers teaching with the maximum number of course credits).

Biology Seminar is a course in Biology Education Study Program of FKIP UMRAH that requires students to have the ability to apply scientific attitudes, perform scientific writing and undertake scientific publications on a small scale. By considering the characteristics of Biology Seminar course, it is very possible to drive a lecture while optimizes the function of higher education academicians on research. Lecture and research can be integrated in the course by applying certain learning strategy.

Research-based Learning (RBL) is considered capable to provide opportunities for lecturers to integrate teaching and research simultaneously. RBL is one of the methods in student centered learning which integrates research into the learning process. RBL is multifaceted and refers to several methods of learning. RBL gives the students an opportunity to look for information, construct a hypothesis, collect data, analyze the data, and make conclusions from the data structure. In this activity the students are learning by doing and it is clear that RBL offers the chance for the development of methods of learning. Related to optimizing lectures and lecturers' research, [1] explain that the application of RBL will provide the following benefits: 1) Encourage lecturers to conduct specific research, then update their knowledge by reading and utilizing other research results as learning materials; and 2) Encourage the role of students more actively in the learning process, and become an active partner of lecturers in research. In addition, RBL also shapes scientific attitudes of students [2].

Research is an important means to improve the quality of learning. Research consists of: background, implementation procedures, research results and analysis. These components of research have a meaningful activities, they are: formulation of the research problem, the solution of the problem, and dissemination of the research results. These activities are closely related to learning and will improve the quality of learning. RBL is a method that uses authentic learning, problem-solving, cooperative learning, contextual (hands on and minds on) and inquiry-discovery approach guided by the philosophy of constructivism [3].

Research based learning is based on the philosophy of constructivism which has four aspects, namely learning to develop the student's understanding, learning to develop prior knowledge, learning as a process of social interaction and meaningful learning obtained through actual experience [4]. This is in line with the Lockwood opinion; "Research-based learning is a system of instruction which used an authentic-learning, problem-solving, cooperative learning, hands on, and inquiry discovery approach, guided by a constructivist philosophy. Its usefulness had been recognized for many decades, but "research in classroom" had not been adopted as a teaching method by many" [5].

Considering that Research-based Learning is very much inherent in *Tridharma* duties as a lecturer, it is necessary to apply Research-based Learning in a course in Higher Education, in this case in Biology Education Study Program. In universities that have orientation to be a research-based university like University of Gajah Mada, RBL has been applied widely in teaching aspect in all study programs, even this university has compiled guidelines on applying RBL as teaching innovation. UMRAH, although not yet specifically oriented to be a research-based university, has initiated the implementation of RBL even though it has not been holistic in all courses in the Study Program.

Considering the importance of integrating research into learning, the author applied RBL in Biology Seminar courses and conducted a study to describe the RBL implementation. The research question that would be answered was: *How to implement RBL in Seminar Biology course?*

2. Method

Research-based Learning was applied to Biology Seminar course in the even semester of 2016/2017 (March to June 2017), in Biology Education Study Program, Faculty of Teacher Training and Education, University of Maritim Raja Ali Haji. The samples of this research were 19 students from 71 students in the Biology Seminar course. The Course schedule was designed for 14 classroom meeting. Observations on the implementation of RBL were conducted at each meeting, with researchers as the key instrument.

Researchers have collected data from various data sources to gain visualization of the RBL implementation process. These data sources were photographs of learning activities, questionnaires, student portfolios, and interviews. The data were analyzed in descriptive-explanative to obtain a comprehensive description about the application of RBL. Analysis of RBL implementation is done by referring to Good Practice Guide of Research-based Learning [6].

3. Results and Discussion

Based on study, it was known that RBL can be implemented in the overall Biology Seminar course consists of 13 sessions (one session was not included because it was a lecture contract session). From the research, obtained the descriptions of the RBL implementation approach as seen in Table 1.

Table 1. RBL strategies applied in Biology Seminar course

Weeks	Learning Outcomes	Students' activities	RBL Strategies Applied
1	▪ To apply the rules set out on the lecture contract ▪ To understand the orientation of the lectures	▪ Discussing the lecture contract ▪ Discussing the orientation of the lectures	-
2	▪ To understand the basic rules of scientific writing	▪ Reviewing the rules of scientific writing ▪ Discussing the basic rules of scientific writing	Designing learning activities around contemporary research issues
3, 4	▪ To understand the research methodology of biology education	▪ Discussing research biology education methodology ▪ Reviewing the research results focused on the method used	Teaching research methods, techniques and skills explicitly within program
5	▪ To obtain, understand, and analyze scientific publications of biology education	▪ Finding examples on scientific publications ▪ Discussing the research methodology in scientific publications ▪ Analyzing the article content in scientific journals	Drawing on personal research in designing and teaching courses
6-10	▪ To formulate research proposal ▪ To apply scientific work steps ▪ To report the results of the study ▪ To apply scientific attitudes and values	▪ Conducting preliminary observation ▪ Asking research questions ▪ Designing research proposals ▪ Preparing research instruments ▪ Taking the data ▪ Analyzing research data ▪ Drawing up conclusion ▪ Writing research reports	Building small skill-research activities into undergraduate assignment Infusing teaching with the values of researchers
11	▪ To create scientific writing by applying the rules of writing	▪ Writing scientific writing based on research reports	Infusing teaching with the values of researchers
12,13, and 14	▪ To conduct an oral presentation in the seminar	▪ Conducting seminar	Infusing teaching with the values of researchers

Based on the lecture conducted, there were 5 of 8 RBL strategies applied in the Course, namely; designing learning activities around contemporary research issues; teaching research methods, techniques and skills explicitly within the program; drawing on personal research in designing and teaching courses; build small-scale research activities into an undergraduate assignment; and Infuse teaching with the values of researchers. Each strategy was applied to the lectures as follows.

1. *Designing learning activities around contemporary research issues*
   This strategy applied by comparing the research results with the news report obtained by society. Students prepared a scientific research result and a popular writing that could be obtained in the campus library, or through the internet. Furthermore, the two writings were compared to gain an understanding of the basic rules of scientific writing.
2. *Teaching research methods, techniques and skills explicitly within program*
   One of the subjects that must be completed by the students before taking the Seminar Course was Research Methodology. However, lecturers had to provide reinforcement and affirmation of educational research methods that would be the main provision for students to achieve the course outcome. Lecturers facilitated students to master the concept of research methodology through group and class discussions and reviewed research methods from existing research reports.
3. *Drawing on personal research in designing and teaching courses*
   To facilitate the students to understand and analyze the existing scientific publications, the lecturer provided an example of her own scientific publications, namely scientific publications on ICMD seminar proceeding, the journal of Pedagogi Hayati, and other research reports.
4. *Building small-scale research activities into undergraduate assignment*
   In the process of learning, students were assigned to conduct mini research which aims to strengthen the mastery of scientific work, next to achieve the outcome of the course. Lecturer acted as mentor and validator who validate student research instruments. Students conducted research in the field of education in junior and senior high school. There were 13 students conducted descriptive research, 5 students conducted research and development, and 1 student conducted experimental research.
5. *Infusing teaching with the values of researchers*
   Integrating scientific values had actually been done from the 2nd meeting, but more emphasized when conducting mini research and seminars. In these assignments, students were directly required to apply scientific values and scientific attitudes namely; critical thinking skills, curiosity, objective, honest, diligent, systematic, open to criticism, and responsible. The guidance process by lecturer during the process of researching and preparing scientific publications, was aimed to guide students who were constrained, more importantly to ensure students continued to implement scientific attitudes and maintain scientific values.

Based on the results of the lectures, the implementation of RBL facilitated students to achieve learning outcomes, namely; understanding the basic rules of scientific writing; understanding the methodology of educational research; acquiring, understanding, and analysing scientific publications of biology education; formulating research proposal; applying scientific work steps; reporting on research results; applying scientific attitudes and values; creating papers by applying the rules of writing; and conducting oral presentation in the seminar. Based on the product of lectures (scientific publications: seminars and/ or articles), 17 students were able to complete scientific publications, while 2 students only completed the research proposal. Based on the final result of the lecture, it could be concluded that the application of RBL facilitated the students to complete the lecture product and achieve learning outcomes.

Based on the questionnaires, it was known that the students could follow lectures well. Students stated that they got several benefits of RBL implementation, that were; RBL improved the quality of lectures; RBL drive students more active; RBL provided real experience for students in conducting research; RBL enforced students' insights more developed; and mini research task on RBL was able to provide experience in doing thesis. Obstacle experienced by students in conducting lectures related to the time to complete the product of the Course.

4. Conclusion

RBL can be implemented in the overall Biology Seminar course by applying certain strategies, namely: designing learning activities around contemporary research issues; teaching research methods, techniques and skills explicitly within program; drawing on personal research in designing and teaching courses; building small-scale research activities into undergraduate assignment; and Infusing teaching with the values of researchers. The application of RBL facilitates the students to complete the lecture product and achieve learning outcomes. The students can follow lectures well and can feel the benefits of RBL implementation.

Acknowledgments

The author is thankful to Prof. Firdaus LN, M.Si at the University of Riau, who has reviewed this writing. Many thanks are also to all students of Biology Seminar Class, even semester of 2016/2017 who are the second subject of this study. Special thanks to Mrs. Dewi Nopita, S.Pd., M.Pd., at English Department, University of Maritim Raja Ali Haji who has helped to examine the English text of this writing.

References

[1]. Dafik 2016 Pengembangan PBR (Pembelajaran Berbasis Riset) dalam Mata Kuliah. Lembaga Pembinaan dan Pengembangan Pendidikan University of Jember Retrieve on Sept 2nd 2017 from www.dafik-fkip-unej.org/downlot.php?file=Hand%20Out%20PBR%20UNEJ.pdf

[2]. Firdaus and Darmadi 2017 Shaping Scientific Attitude of Biology Education Students Through Research-based Teaching. AIP Conf Proc 1868 100004-1–100004-5 doi 10.1063/1.4995214.

[3]. Roach M Blackmore P Dempster J 2000 Supporting High-Level Learning Through Research-based Methods: Interim Guideline for Course Design. TELRI Project-University of Warwick Retrieve on Sept 1st 2017 from http://citeseerx.ist.psu.edu/viewdoc/download

[4]. Widayati et al 2010 Guideline Research-based Learning University of Gajah Mada Retrieve on Aug 31st 2017 from http://ppp.ugm.ac.id/p3/wp-content/uploads/pupbringgris.pdf

[5]. Poonpan Suchada & Siriphan S 2001 Indicators of Research-Based Learning Instructional Process: A Case Study of Best Practice in a Primary School. Faculty of Education. Chulalongkorn University Phaya Thai Bangkok Thailand Retrieve on Sept 1st 2017 from https://www.aare.edu.au/data/publications/2005/poo05581.pdf

[6]. Griffith Institute for Higher Education 2008 Research-based Learning: Strategies for Successfully Linking Teaching and Research University of Griffith Retrieve on Sept 1st 2017 from https://www.griffith.edu.au/learning-futures/pdf/gihe_tipsheet_web_rbl.pdf

IOP Conf. Series: Materials Science and Engineering **335** (2018) 012096 doi:10.1088/1757-899X/335/1/012096

Guiding Development Based Approach Practicum Vertebrates Taxonomy Scientific Study Program for Students of Biology Education

M Arieska[1], S Syamsurizal[2], R Sumarmin [2*]

[1] Student Master of Education Sciences, Universitas Negeri Padang, Indonesia
[2] Biology Department, Faculty of Mathematics and Sciences, Universitas Negeri Padang, Indonesia

*ramadhan_unp@yahoo.com

Abstract. Students having difficulty in identifying and describing the vertebrate animals as well as less skilled in science process as practical. Increased expertise in scientific skills, one of which is through practical activities using practical guidance based on scientific approach. This study aims to produce practical guidance vertebrate taxonomy for biology education students PGRI STKIP West Sumatra valid. This study uses a model of Plomp development consisting of three phases: the initial investigation, floating or prototype stage, and the stage of assessment. Data collection instruments used in this study is a validation sheet guiding practicum. Data were analyzed descriptively based on data obtained from the field. The result of the development of practical guidance vertebrate taxonomic validity value of 3.22 is obtained with very valid category. Research and development has produced a practical guide based vertebrate taxonomic scientific approach very valid.

1. Introduction

The law on the national education system number 20 of 2003 states that education is a conscious and deliberate effort to create an atmosphere of learning and the learning process so that learners are actively developing the potential for him to have the spiritual power of religion, self-control, personality, intelligence, morality noble, as well as the skills needed him, society, nation and state. Many efforts were made to improve the quality of education continues to be done, one of which is the improvement of teaching and learning processes in all levels of education, including college. College study various fields of science, one of which is biology. Biology is the branch of science that can be studied in the laboratory than in the classroom, by doing practical work. Hovstein and Vincent (2012: 15) states that through laboratory activities will guide students to learn through understanding and construct their own knowledge.

In order to realize professional graduates, the development of the establishment of the graduate profile, the formulation of learning outcomes, competency elements, the weight of credits, learning activities, assessment and teaching materials. One of the materials used in the learning process is a guiding biology lab. Guidance practicum is a student activity sheet with instructions and questions that should be completed by the student to find a concept and presented in the form of activity in laboratory experiments (Wayan, 2010: 782). One practical guidance in learning biology is a practical guide Taxonomy of vertebrates. Morris, (2001) states that Vertebrates Taxonomy learn about animals with a backbone in a system of classification and discuss the kinship or phylogeny of vertebrate animals. The material contained in Vertebrates Taxonomy lab is as follows: 1) super class Pisces (classes Chondrichthyes and Osteichthyes 2) amphibian class 3) class Reptilia, 4) the class Aves, and 5) the class Mammalia.

Based on observations through the questionnaire responses of students STKIP PGRI West Sumatra on December 19, 2016 there are some problems in the lab Taxonomy of vertebrates ie, guiding lab using is not yet using the steps of scientific work that is clear, students difficulty in finding and formulating a problem, and student hardship formulate hypotheses. Researchers also conducted interviews with faculty on December 22, 2016, which found problems in the practical implementation difficulties identifying which students first vertebrate animals, students also have difficulty proving practical results in accordance with the theory learned during lectures and practical difficulties concludes in accordance with the purpose of the lab. This is due to the lack of supporting theories and images in a practical guide to help students identify vertebrate animals. Both students have difficulty understanding the methods and measures making it less terasahnya practicum science process skills of students in the lab. The third number is inactive when praktikan practical activities take place. Based on these problems shows that the students are still difficulties in understanding the steps in scientific learning, where students difficulty in finding and formulating a problem, students are still difficulties in formulating hypotheses and less terasahnya science process skills of students in practical activities. So it is necessary to guide the lab that is designed to help overcome the difficulties of students in the lab.

The results of the analysis of the needs of students to guidance practicum students expect that, of the students stated that they agreed that if the pilot lab Taxonomy of vertebrates to be developed have measures that exist in the approach to scientific, where students agreed that if the pilot lab to be developed to steer students on the problem will be solved when the lab work, facilitate determine hypotheses, make it easier to conduct experiments, facilitate the collection of data, making it easier to process the information that has been collected from the results of the experiments that have been carried out, allowing submit observations and make their conclusions, as well as practicum consists of several components that contain scientific approach.

Based on the expected needs of students to practical guidance Vertebrate Taxonomy, the approach to be used is a scientific approach. Wizard-based practicum scientific approach is practical guide to hone the ability of students in accordance with the level of thinking ability. In the scientific approach are the steps in the form of observations and formulate problems, formulate hypotheses, testing hypotheses, analyzing data and drawing conclusions. Of these phases can develop a variety of skills such as critical thinking skills, communication skills, and character behavior (Machin, 2014: 31).

Selection of scientific approach as the approach used in the development of this practical guide also refers to research performed by Fauziah, (2016), which has conducted research on the development of practical guidance based General Biology scientific approach to students. This study revealed that the lab by using this practical guide to improve the competence of learning (cognitive, affective and psychomotor). Rahmi, (2016) Plant Physiology develop practical guidance with scientific approach in college. The results showed this practical guide to improve the competence of learning (cognitive, affective and psychomotor). Sophia, (2017) to develop practical guidance Taxonomy of Invertebrates by scientific approach in college. The results showed this practical guide to improve the competence of learning (cognitive, affective and psychomotor). Based on the needs and problems have developed practical guidance Vertebrate Taxonomy-based scientific approaches for students of biology education.

2. Methods

This study is design and development *(design and development research)* with the aim to produce a valid practical guidance. The products developed are practical guides Vertebrate Taxonomy-based scientific approaches to students of Biology Education STKIP PGRI. This study uses a model of Plomp development, research stages consist of, the initial investigation phase *(preliminary research*

IOP Conf. Series: Materials Science and Engineering **335** (2018) 012096 doi:10.1088/1757-899X/335/1/012096

phase), stage of development or manufacture of prototypes*(development or prototyping phase*)and a stage of assessment *(assessment phase)* (Plomp, 2013: 15).

3. Results

3.1. Data Description Development Results

3.1.1. Data Preliminary Investigation

3.1.1. *Curriculum Analysis*

Of Vertebrate Taxonomy Wizard-based lab designed a scientific approach based on analysis of high diperguruan curriculum. Curriculum components that are directly related to the products produced are learning achievements, material and indicators. The translation of the indicator obtained five themes: Pisces (Chondrichthyes and Osteichthyes), Amphibia, Reptilia, Aves, and Mammalia.

3.1.2. *Interview with Lecturer Course Vertebrate Taxonomy*

The interview was conducted by a lecturer in Taxonomy of vertebrates in STKIP PGRI West Sumatra. Problems obtained is difficulty in identifying students vertebrate animals. This is due to lack of support and discrepancy theory diagrams in the practical guide to the observed object. Students difficulty in understanding the methods and measures that are less terasahnya lab science process skills of students in the lab.

3.1.3. *Problem Analysis Students in the Practical Activity*

Based on observations through questionnaire responses STKIP students PGRI West Sumatra on December 19, 2016 there are some practical problems in Vertebrates Taxonomy. A total of 40% of the students answered that the guiding lab used Vertebrate Taxonomy using the steps yet clear scientific work, where students stated that their difficulties in understanding the sequence of work contained in the practical guide. Students are also difficulties in finding and formulating the problem to be solved in practical activities, where only 53% of the students who responded were able to discover and formulate the problem to be solved when the lab. Students difficulty formulating a hypothesis in practical activities, where only 40% of students who answered able to formulate hypotheses. The number of inactive time praktikan practical activities take place, where only 40% of students who take an active role when the practicum takes place.

3.1.4. *Needs Analysis Students Against Guidance Practical*

Results of the analysis of the needs of students to guide practical to expect that, as much as 100% of the students stated that they agreed that if the pilot lab Taxonomy of vertebrates to be developed have measures that exist in the approach to scientific, where students agreed that if the guiding practicum will be developed to steer students on the problem will be solved when the lab, allows the practitioner to determine a hypothesis, enabling the practitioner to experiment, facilitate students in data collection, allows the practitioner to process the information that has been collected from the results of the experiments that have been done, and facilitate students in making In conclusion, as well as practicum consists of several components that contain scientific approach.

3.2. Development stage or Prototyping

3.2.1. Prototype I

After the design and manufacture practical guidance Vertebrate Taxonomy-based scientific approaches do, do a self-evaluation(self-evaluation). The results of *self evaluation* showed that these steps are complete scientific, but the sequence number in the sequence and picture captions yet there are some errors that occur are in writing and no pictures that do not have the source. After the evaluation, the revision to repair all the errors found.

3.2.2. Prototype 2

3.2.2.1` Results Validation Instrument

Instrument overall research beforehand validated by Dr. Darmansyah, ST, M Ed, briefly whole instrument validation results can be seen in the following table.

Table 1 Results Validation Guidance Practice-Based Research Instruments Scientific Approach

No.	Aspect Assessed	Value	Category
1.	The linkage indicators are in accordance with item questionnaire on the questionnaire statement.	4.00	Very Valid
2.	Expressions used already includes aspects of a guide containing practical validity based scientific approach.	4.00	Very Valid
3.	The phrase used by the correct Indonesian rule.	4.00	Very Valid
	Amount Overall	12.00	
	average	4.00	Very Valid

Based on the table, the general assessment validator (lecturers) to the overall research instruments obtain an average value of 4.00 with very valid criteria so that the whole instrument can be used for research.

3.2.2.2 Results Validation Guidance Practicum

Guidance Practicum that has been designed, followed by validation activities by 3 experts. At this stage, the guiding validation lab by experts (professors) in order to produce a practical guide valid. The tests can be seen in the Table 2.

Table 2. Results Validation Wizard Lab expert

No.	Assessment Criteria	Number	average validity	Category
1.	Terms Didactic	38	3.17	Valid
2.	Construction Conditions	60	3.33	Very Valid
3.	Technical Terms	80	3.33	Very Valid
4.	Terms Languages	38	3 , 17	valid
	Amount Overall		**216**	
	average		**3.22**	**very valid**

Validation lab guiding judged by three people validator is very valid. This means practical guidance that has been developed already have a very good quality and can be used for a test run to the next stage, which can be used as a guide practicum Vertebrate Taxonomy.

4. Discussion

4.1. Guidance Development Process Practical Scientific Approach Based Vertebrate Taxonomy

Wizard lab Vertebrate Taxonomy developed using scientific approach, which is designed to be used by lecturers and tutors and student practicum. Development of practical guidance tailored to the stages of the scientific approach, namely: observation, formulate problems, formulate hypotheses, testing hypotheses, analyze data and draw conclusions.

The development of practical guidance starting from the curriculum analysis, interviews with the lecturer and student analysis. The results of the analysis that has been done obtained curriculum learning outcomes that students are able to understand the characteristics, examples of species and the spread of any class of vertebrate animals. Further analysis of indicators and get five themes: Pisces (Chondrichthyes and Osteichthyes), Amphibia, Reptilia, Aves and Mammalia. According Prastowo

IOP Conf. Series: Materials Science and Engineering **335** (2018) 012096 doi:10.1088/1757-899X/335/1/012096

(2011: 119) curriculum analysis done by looking at what is taught and the core competencies and learning outcomes that should be owned by learners.

Development of practical guidance is also based on interviews with the lecturer. The results of the interview obtained obtained problem is still a lack of existing literature theory in guiding and diagrams in the guide does not match the object to be observed when the lab so students difficulty in identifying and describing the objects they observe. The next consideration in formulating practical guide is the analysis of student characteristics. Subjects in this study were students of biology education courses force of 2015 with an average age between 20-21 years old who are at the stage of formal operational period. According to Piaget's theory of learning (Budiningsih, 2008: 39) at the formal operational stage, a fundamental feature of child development have been able to think abstractly, logically, draw conclusions, interpret and develop hypotheses in using media including teaching materials self instruction. such as guiding practical

The use of scientific approach in the practicum is very appropriate, as it aims to develop the ability to think in a systematic, logical and critical student can hone his skills and science. According to McPherson (2001) approach to guide students to discover scientific knowledge on their own.

4.2. Practical Guidance validity of Vertebrate Based Taxonomy Scientific approach

Taxonomic validity vertebrates practical guidance based scientific approach validator validated by three experts / specialists. The validity of this practical guide in terms of the requirements didaktik, construction, technical, and language. Guidance practicum expressed very valid because it has met the four aspects: didactic, constructs, technical and language.

Based on the validity of the test results, the three validators declared Vertebrate Taxonomy practical guide scientific research based approach to design has been included in the category of very valid. This shows, guiding lab practicum guides developed in the course of Vertebrate Taxonomy has to be used in practical activities. According Trianto (2010: 55) valid means that the vote has been providing accurate information about the teaching materials developed. Vertebrates Taxonomy Wizard lab-based scientific approach has met with the didactic aspects valid criteria. The validation results show that the subject is presented in accordance with the applicable curriculum high diperguruan. Terms didactic closely associated with the process of finding the correct concepts and in accordance with the applicable curriculum. This is in accordance with the opinion Thoha (2001: 109-112) didactic requirements with regard to the process of determining the correct concepts in accordance with the applicable curriculum.

On the aspects of the construct also expressed very valid by the validator, because the products developed have met the terms of the preparation of practical guides. This is in accordance with the opinion of Wayan (2010: 782) there are several components of the practical guide are: introduction, objectives, tools and materials, procedures, observations, analysis and conclusions, a summary of the answer problems that have been analyzed. Penilalain on the technical aspects of Vertebrate Taxonomy practical guide based scientific approach is said to be valid. This means that all components of the technical requirements that include text, images and kegrafikaan been fulfilled properly. This is in accordance with the opinion Thoha (2001: 109-112) the technical aspects relating to the use of text, images and appearance.

Aspects of the guiding language lab Vertebrate Taxonomy-based scientific approach has a valid category in accordance with the assessment of the validator. The language requirement relating to the use of the word, spelling and punctuation, so easily understood by users. According to Hamdani (2011: 222) aspects of language is one of the aspects to be considered in the preparation of teaching materials, the language used should be simple language and easily understood. The fourth aspect of the validation votes practical guidance above is a unified whole and support each other to perfection developed practical guidance. Validation of a practical guide is very important. According to the Ministry of Education (2008: 15) validation a practical guide and other teaching materials is very important to gain recognition of conformity of teaching materials produced decent and suitable for use in lectures). In line with that Nieveen in Plomp (2013: 28) states the criteria for getting a good quality product that is valid, practical and effective

4. Conclusion
Vertebrate Taxonomy Wizard lab-based scientific approaches that have been developed have validity with the very valid category.

References
[1] Education Ministry. 2008. *Guidelines for Development of Learning Tool SBC.* Jakarta: BSNP
[2] Fauziah, N. 2016. "Development of Practical Guidance General Biology Scientific-Based Approach for Students". *thesis.* Unpublished Padang: Graduate UNP.
[3] Hamdani. 2011. *Teaching and Learning Strategies.*Bandung: Pustaka Setia CV.
[4] Hovstein, A and Vincent LL 2012. The Laboratory Science Education: Foundation of the Twenty-First Century. *International Journal of Science Education, June 2012, Vol. 52 (2), 10-51.*
[5] Machin, A. 2014. Implementation of Scientific Approach, Character and Conservation Investment In Learning Materials Eyes Growth. *Journal of Science Education Indonesia.* JPII 3 (1): 31.
[6] McPherson, Guy R. 2001. Teaching and Learning Scientific Method. *International Journal of The American Biology Teacher, Volume 63, no. 4, April 2001.*
[7] Plomp, T and Nieveen, N. 2013. *Educational Design Research in Educational Design Research: An Introduction.* Enschede, University of Twente.
[8] Prastowo, Andi. 2011. Guide to Creating Innovative Teaching Material. Yogyakarta: Diva Press.
[9] Sophia, A. 2017. Invertebrate Taxonomy Development Practical Guidance For Students. *thesis.* Unpublished UNP Graduate Program.
[10] Thoha, C. 2001. *Evaluation of Engineering Education.*Jakarta: King Grafindo Persada.
[11] Trianto. 2010. *Integrated Learning Model (Concepts, Strategies, and Implementation in Curriculum Education Unit)*.Jakarta: PT Bumi Literacy.
[12] Wayan, I. 2010. The *National Education Standards.*Jakarta: Az-Zahra Book's.

IOP Conf. Series: Materials Science and Engineering **335** (2018) 012097 doi:10.1088/1757-899X/335/1/012097

Validity and Practitality of Acid-Base Module Based on Guided Discovery Learning for Senior High School

Yerimadesi*[1], Bayharti[1], S M Jannah[1], Lufri[2], Festiyed[3], Y Kiram[4]

[1]Department of Chemistry, Faculty of Mathematics and Science, Universitas Negeri Padang, Indonesia
[2]Department of Biology, Faculty of Mathematics and Science, Universitas Negeri Padang, Indonesia
[3]Department of Physics, Faculty of Mathematics and Science, Universitas Negeri Padang, Indonesia
[4]Department of Sport Coaching Education, Faculty of Sport Science, Universitas Negeri Padang, Indonesia

*yeri@fmipa.unp.ac.id

Abstract. This Research and Development(R&D) aims to produce guided discovery learning based module on topic of acid-base and determine its validity and practicality in learning. Module development used Four D (4-D) model (define, design, develop and disseminate).This research was performed until development stage. Research's instruments were validity and practicality questionnaires. Module was validated by five experts (three chemistry lecturers of Universitas Negeri Padang and two chemistry teachers of SMAN 9 Padang). Practicality test was done by two chemistry teachers and 30 students of SMAN 9 Padang. Kappa Cohen's was used to analyze validity and practicality. The average moment kappa was 0.86 for validity and those for practicality were 0.85 by teachers and 0.76 by students revealing high category. It can be concluded that validity and practicality was proven for high school chemistry learning.

1. Introduction

Acid-base is the topic studied in semester 2 of 11th grade. Based on curriculum 2013, students are expected to understand, implement, and analyze the factual, conceptual, procedural, and metacognitive knowledge by studying this topic [1]. To achieve these competencies, students are required to learn the independent, active, and scientific learning process. One of learning models that can make students active and independent is guided discovery learning.

Guided discovery learningis a model which guides and trains students to achieve active learning, acquire knowledge, and build scientific concepts that they find for themselves[2]. Bruner considered that discovery learning is in accordance to the active invention knowledge and automatically gives the best results. Bruner also suggested learning would be more meaningful for students if they were focused on understanding the structure of information. Students should active to obtain information structure, and had to identify the key principles rather than merely accepted the teacher's explanation[3].

The application of guided discovery learning could improve student learning outcomes in science learning, such as physics [4], biology [5], chemistry [6], and geometry [7] learning in high school.

Research results revealed that guided discovery learning instruction would affect learning achievement, would influence retention, and would achieve presented significantly positive effects on learning retention[8]. Application of learning guided discovery learning also increases eight learning activities, namely oral, physical, mental, visual, and emotional, writing, listening, drawing activity [9]. In discovery learning process, students would be able to develop a more positive attitude towards learning and will improve student learning outcomes.

In applying guided discoverylearning, the teacher acts as a facilitator and motivator [10] [11]. Teachers must provide a complete learning material for students [10]. One of the learning materials that can lead students to learn independently is module.

Chemistry learning module can increase student motivation better in learning than other learning materials [12].In the other hand, it canimprove students' learning outcome, such as in topic of electrochemical analysis [13],hydrocarbons and their derivatives[14], mole concept [15], chemical equilibrium [16], and a buffer solution [17].

Based on this background, it was necessary to develop module for the other topics in chemistry. This study aims to develop a guided discovery learning module on acid-base topic for senior high school, then tested its validity and practicality.

2. Method

This research was research and development (R&D) methods that was used to produce a particular product, and tested the affectivity of the product [18]. This research produced module based on guided discovery learning on the topic of acid-base for senior high school. The development model used four D model (4D model) that consists of four phases: define, design, develop and disseminate[19]. This new research carried out until the stage of development that was the validity and the practicality of the module.

2.1 Define

Define stageconducted five analysis that describes in the following explanation. (1) Front-end analysis aimed to raise and set the basic problems faced by teachers and students in the learning process. This analysis was done by giving questionnaires to chemistry teachers and high school students; (2) Learner analysis wasa study of student characteristics. This analysis was conducted through study literature and giving questionnaires to students. (3) Task analysis aims to reidentify and analyze the master of students' ability through the determination of contents and learning units in accordance with the curriculum. The task analysis was done by analyzing the basic competencies of acid-base material, so it could be formulated in the form of learning indicators. (4) Concept analysis wascarried out by identifying and arranging them in the form of hierarchy, and detailing the main concepts and supporting material of acids and bases, through several books related to chemistry concepts which were stated in a tabular analysis of the concept. This was the basis of the preparation of concept maps. (5) Specifying instructionalobjectives, was carried out by formulating the learning objectives of the acid-base materials, that baseon indicators that have been made previously. Analysis of learning objectives and task analysis were the result of the analysis of concepts into learning objectives.

2.2 Design

This stage included:(1) the product's objectivies, (2) the product's target audience, and (3) a description of the product's components and how they will be used[19]. The products designed was learning materials in the form of acid and base module based on guided discovery learning for senior high school student. The module was designed in accordance with the stages of guided discovery learning, which included five phases: (1) motivation and problem presentation, (2) selection of learning activities, (3) data collection, (4) data processing, and (5) closure[20]. Thepreparation module format was modified from guide books of learning materials development that stated by the Education Ministry [21] and creative guides in designing innovative learning materialsby Prastowo [22].

2.3 Develop

The develop stage aimed to generate acid and base module based on guided discovery learning and tested validity, practicality and effectiveness of this module. The validity of module was appraised by experts, and practicality was tested by practitioners, that were teachers and limited or a small group students.

The research instrument used was a questionnaire. The validity of the questionnaire addressed to three chemistry lectures of Universitas Negeri Padang and three chemistry teachers of SMAN 9 Padang. Practicalities questionnaire addressed to three people a chemistry teachers and 30 students of SMAN 9 Padang academic year 2016/2017. The validity of the questionnaire was useful to assess the preliminary design of modules developed based on aspects of content, presentation, linguistic and format. Practicality questionnaire used to assess the practicalities of useful products, in the form of ease of use, efficiency and benefits of learning time aspects [23]. Validity data and practicality data obtained were analyzed by using Cohen kappaformula (equation 1), so that the moment kappa values obtained [24].

$$moment\ kappa\ (k)\ = \frac{\rho_o - \rho_e}{1 - \rho_e} \qquad (1)$$

Information:

k = momen kappa describing validity of the product.

ρ_o = Realized proportion; counted by summing the score given by validators and then divided it by maximum total score.

ρ_e = Unrealized proportion; counted by substracting the maximum total score with the sum of total score given by validator, which then divided by the maximum total score.

Kappa moment value *(k)* ranges from 0 to 1 [24]. Interpretation kappa moment value can be seen in Table 1.

Table 1. The category of decision based on Kappa moment (k) [24].

Interval	**Category**
0.81 - 1.00	Very high
0.61 - 0.80	High
0.41 - 0.60	Medium
0.21 - 0.40	Low
0.01 - 0.20	Very low
≤ 0,00	Invalid

3. Result and Discussion

Based on the purpose of research and development (R&D) by using a 4D model development, it was obtained results of the study for each stage of development. The results obtained at each stage that has been done were defined, design, and develop.

3.1 Define

At this define stage, it was found five data, those were front-end analysis, analysis of student, task analysis, concept analysis, and the analysis of learning objectives. The results were described in the following explanation.

3.1.1 Front-end Analysis Results. Based on the front end analysis, this information could be obtained. (1) Analysis of questionnaires that was given to 24 senior high school chemistry teachers in Padang, and the data showed that the learning materials used was that 63.6% of teachers used books, 66.7% of teachers used student worksheets, 6.1% of teachers used module, and 3% of them used handout. (2) The learning materials used in schools was presented verbally and concepts provided directly, even though there were pictures but had not been able to fully help the students to discover concepts independently in accordance with the demands of the 2013 curriculum. (3) 48.5% of the teachers had been still teaching in the conventional *(teacher center,* whereas in 2013 curriculum demanded learning process that was student center. Based on the above issues, it should be developed a learning materials that can make students active, creative and innovative that demanded to 2013 curriculum. This study developed a learning material, namely acid and base module based on *guided* discovery *learning* that was expected to lead students in concept invention, to enhance the activity of students in the learning process, and enhance the students' understanding of the material being studied.

3.1.2 Analysis of Students Results. According to Piaget, based on the age of high school students, theywere categorized into the formal operational stage. At this stage the child already has the ability to think abstractly, reason logically and draw conclusions from the available information. However, not all students in adolescence were be able to think abstractly as described by Piaget, but they were a lot of teenagers who still thought in concrete operations[3]. Based on the analysis of questionnaires filled out by the students, it was obtained the data that the materials used had not fully make active students in learning. Students were more like-colored materials, display, and attractive which motivated students to learn. This wasone of consideration in developing and preparing materials in the form of module.Developed and designed module had a color display which was expected to make students attractive in learning process,increasestudents activity andmotivate themin learning.

3.1.3 Task Analysis Results. Based on analysis of core competencies and the Basic Competency that stated in chemical-interest syllabus of 2013 curriculum for senior high school, some indicators of learning were formulated. Acid and base material has 2 Basic Competencies, namely (1) Basic Competency 3:10 Analyzing the nature of a solution based on the acid-base concept and pH of the solution; and (2) Basic Competency 4:10 Ascore competencies the idea of using the appropriate indicators to determine the acidity of acid / base or titration of the acid / base[25]. Indicators were formulated based on both Basic Competency 3:10 and Basic Competency 4:10.There were five indicators gained. (1) Describing the nature of the solution using Arrhenius, Brønsted-Lowry, and Lewis theory; (2) Determining the nature of the acid-base solution using litmus paper and natural indicators; (3) Analyzing the nature of the solution by pH of the solution; (4) Determining the nature of the acid-base solution using natural and chemical indicators; (5) Connecting the degree of acidity (pH) with the degree of ionization, and acid-base equilibrium constant.

3.1.4 Concept Analysis Results. Based on analysis of the main concept of acid base, the concept obtained in the acid-base materials were acid-base (acid-base theory), the strength of acid and base (strong acid, weak acid, strong base and weak base), acid-base properties, and the pH of acid and base. The concepts obtained were created from a concept analysis table comprising label concept, the definition of the concept, the type of concept, the attribute of concept, the position of concept, example and non-example[26]. This concept analysis table was useful to create concept map and guide preparing the chemistry content of the module. Concept map in the module would help students easily remember the core concepts of the acid base.

3.1.5 Specifying instructional objectives. Based on the task analysis and analysis of the concept, the learning objectives were formulated.There were 12 learning objectives. (1) Students can describe the nature of the solution based on the concept of acid-base according to Arrhenius theory through learning materialscorrectly. (2) Students can describe the nature of the solution based on the concept of acid-base according toBrønsted –Lowry theory through learning materials correctly. (3) Students can explain conjugate pairs based on reaction equation according to Brønsted-Lowry theory through learning materials clearly. (4) Students can describe the nature of the solution based on the Lewis acid-base theory through a given learning materials properly. (5) Students can determine the nature of the acid, alkaline, and neutral solutionby using litmus paper and natural indicators through experiment appropriately. (6) Students can determine the natural product that can be used as indicators of the acid base through the trial properly. (7) Students can distinguish between a weak acid and a strong acid and weak base and a strong base through a given instructional materials appropriately. (8) Students can analyze the nature of the solution by pH of the solution through trial. (9) Students can distinguish a weak acid with a strong acid and weak base with a strong base the same concentration using indicators or pH meters through a trial. (10) Students can determine the pH of the solution using several chemical indicators through experiments conducted properly. (11) Students can calculate the pH of acids and bases through the examples given correctly. (12) Students can connect a weak acid with a strong acid and weak base with a strong base to obtain the degree of ionization(α) or ionization constants (K_{α})through a given instructional materials appropriately.

The learning objectives that have been formulated were a basis in preparing the acid-base modulebased on guided discovery learning, especially in developing activity sheet, student worksheet, and evaluation sheets.

3.2 Design

This stage was the stage to producea preliminary draft of acid and base modulebased on guided discovery learning for teaching senior high school chemistry. The parts of the modules that were designed through modified from the components of the modules presented in guidebook for development of learning materials by the Education Ministry[21] and creative guides make innovative learning materials developedby Prastowo[22].There were nine parts of module,such as: (1) cover,(2) core competencies, basic competencies, indicators and learning objectives, (3) a concept map (4) learning instructions or instructions for using the module, (5) activity sheets (6) student worksheet, (7) evaluation sheets, (8) the evaluation key and (9) bibliography.

Acid and base module was designed and prepared based onmodel of guided discovery learning for teaching senios high school chemistry. Syntaxmodel of guided discoverylearning was reflected in the student activity sheet. Some students experience activities were designed to find concepts through activities in the laboratory and in the classroom. Itwas also designed how to use chemical learning approaches that emphasized at all three levels of representation. Three level of representation should be achieved by students in understanding the chemistry, and is known as the *chemistry*triangle such as macroscopic level, sub-microscopic and symbolic level[27].Example display on the module can be seen in Figure 1.

Instructions in module were a guideline for students to use the module in the learning process. Activity sheet contained activities conducted during the students' learning process and some of the question must be filled out by the students. Student worksheet contained practice questions that would be done after completing activity sheets students. The evaluation sheet contained questions related to the overall learning materials in modules.

3.3 Develop

On*develop phase,*two forms of data were obtained, the validity data were given by the expert and practicality data were resulted from test of practicality of module by teachers and students.

3.3.1 Validity of Module. Data validity of modulebased on *guided discovery learning* can be seen in Table 2. Table 2 showed that module developed had very high validity for all components of the assessment that incuded the feasibility in terms of content, presentation, linguistic, and grafics.

Table 2. Results of the module validity by 6 validators

No.	Aspects assessed	k value	Category
1	Eligibility contents	0.85	Very High
2	Presentation	0.90	Very High
3	Linguistic	0.81	Very High
4	Grafics	0.90	Very High
Average		0.86	Very High

This module had very high validity category of the contents and the feasibility aspects. It meant that: (1) the module was in accordance to the demands of core competencies and basic competencies material on the syllabus of acid and basedemanded by senior high school curriculum in 2013 as outlined by Ministry of Education Rule No. 59 in 2014; (2) the preparation of module was accordancewith the child's development; (3) the preparation of module was in accordance with the needs of learning materials; (4) module had the correct material substance, (5) the content of the module could add insight; and (6) the characteristics of the module were in accordance to moral and social values [23][28].

The module had a very high validity in aspects of presentation, and it meant that the module constructed based on learning materials development standards according to the National Education Standards Agency (BNSP) [28].This module had clarity of learning objectives, the order of presentation was coherent, structured activity sheet syntax based on guided discoverylearning model and information submitted was complete. The module also gaveapperseption and motivation. Syntaxmodel of guided discoverylearning lead students to discovering and understood the concept through self – learning, so that objectives were achieved. The module was based on the characteristics of high school students and the material characteristics of acid-base. Learning with this module was equipped with a lab activity supporting students' understanding of concepts to be learned. In addition, the topic contained in the worksheet and evaluation sheets on the module had been prepared in accordance with the learning objective, so it can be used as a measurement of achievement of the learning objectives. Evaluation questions could be used to deepen and test their understanding of the material being studied, because the modules werecompleted with an answer key.

The module had a very high validity in aspects of language which means that the module was communicative, clarity of information, in accordance with Indonesian Spelling and used the language effectively and efficiently. The module has a validity of a very high in aspect of grafics.The module developed used a type and font size in accordance with the assessment standards of learning materials, lay out or correct layout, illustration, drawings and photographs as well as the design looked clear and attractive as writing guidelines learning materials[21] [22][28].

3.3.2 Revision. Phaserevision aimed to fix the acid-base modulebased *guided discvery learning* that was considered and evaluated by the validators prior to the product being tested. Module that had been repaired and then given back to the validator to be discussedfurther before being tested. Revision was completed if the validators had been declared that module developed was valid. Some aspects were revised based on suggestions of validators including: (1) create a title of sheet activities to the contents of the module, fix the concept maps to be clearer and derivatives of each of these concepts must have properties that are equivalent, as a follow-up then register an activity sheet on the table of contents modules and concept maps revised as the validator advice; (2) adjusted ions in the drawings, the same charge ion will repel that its position is far apart, and vice versa different ions will charge mutual attraction so that the position of the ion adjacent[29];(3) given a description of the type of acid/ base on the sample table of acid / base. This module was prepared for Indonesian students, so that the module designed was in Indonesian language. Figure 1 to 4 were examples of a revised image of the module.

IOP Conf. Series: Materials Science and Engineering **335** (2018) 012097 doi:10.1088/1757-899X/335/1/012097

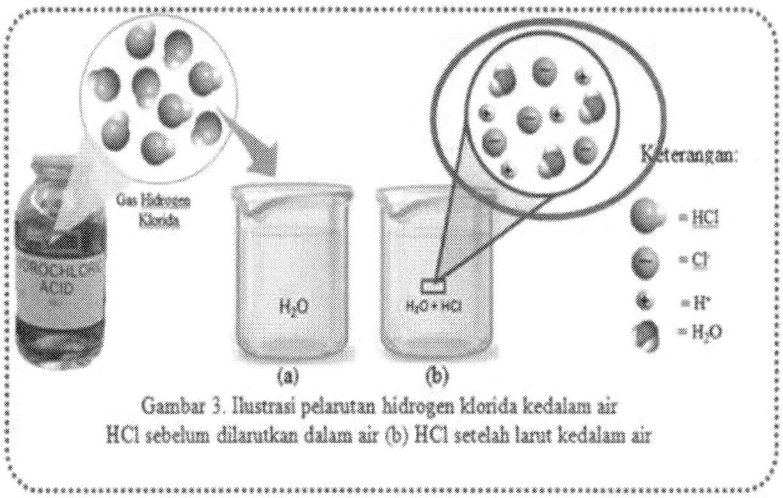

Figure 1. Location of ions in Figure Before Revision

Figure 1 needs to be repaired, the picture is a picture when the hydrogen chloride which is not an acid, but when dissolved in water will have acidic properties. With regard Pictures students are expected to explain the meaning according to Arrhenius acid. Define Arrhenius acid is a compound that releases H^+ in water and alkalis releasing OH^-[29].

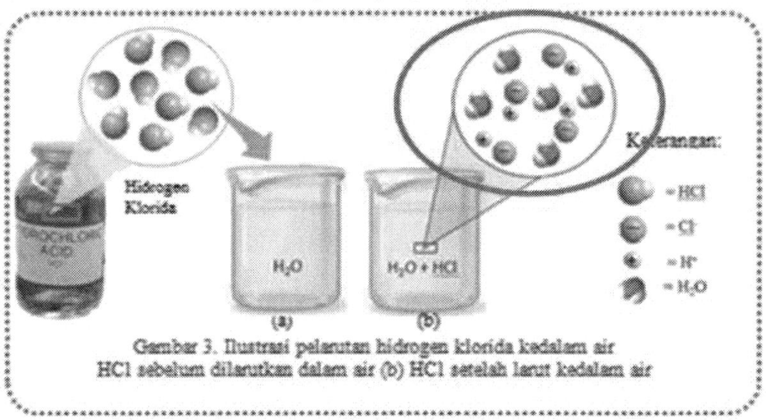

Figure 2. Location of Ions on the Image After the Revision

Rumus Basa	Nama Basa	Ionisasi Basa
NaOH	Natrium hidroksida	$NaOH(aq) \rightarrow Na^+(aq) + OH^-(aq)$
KOH	Kalium hidroksida	$KOH(aq) \rightarrow K^+(aq) + OH^-(aq)$
$Ca(OH)_2$	Kalsium hidroksida	$Ca(OH)_2(aq) \rightarrow Ca^{2+}(aq) + 2OH^-(aq)$
$Ba(OH)_2$	Barium hidroksida	$Ba(OH)_2(aq) \rightarrow Ba^{2+}(aq0 + 2OH^-(aq)$
NH_3	Amonia	$NH_3(aq) \rightarrow NH_4^+(aq) + OH^-(aq)$

Figure 3. Table Example Acid Alkali before Revised

IOP Conf. Series: Materials Science and Engineering **335** (2018) 012097 doi:10.1088/1757-899X/335/1/012097

Rumus Asam	Nama Asam	Reaksi Ionisasinya	
HF	Asam flourida	$HF(aq) \rightarrow H^+(aq) + F^-(aq)$	Asam monoprotik
HBr	Asam bromida	$HBr(aq) \rightarrow H^+(aq) + Br^-(aq)$	
HNO_3	Asam nitrat	$HNO_3(aq) \rightarrow H^+(aq) + NO_3^-(aq)$	
CH_3COOH (aq)	Asam asetat (cuka)	$CH_3COOH(aq) \rightleftharpoons H^+(aq) + CH_3COO^-$	
H_2S	Asam sulfida	$H_2S(aq) \rightleftharpoons 2H^+(aq) + S^{2-}(aq)$	Asam diprotik
H_2SO_4	Asam sulfat	$H_2SO_4(aq) \rightarrow 2H^+(aq) + SO_4^{2-}(aq)$	
H_3PO_4	Asam fosfat	$H_3PO_4(aq) \rightarrow 3H^+(aq) + PO_4^{3-}(aq)$	Asam triprotik

Figure 4. Table Example Acid Bases After Revision

3.3.3 Practicality of Module. Practicality of module was determined through testing of module by three chemistry teachers and 30 students of class XI IPA 4 SMAN 9 Padang. The results obtained were presented in table 4. It can be seen that the already practical module developed by teachers and students, in the form of the aspect ease of use, efficiency of learning, as well as benefits. The following table described the practicality of module based on the analysis of questionnaire responses of teachers and students.

Table 3. k value of the practicalities Test Module by teachers and students

No.	Aspect Assessed	k value	
		Teacher	Student
1	Ease of Use	0.84	0.77
2	Time Learning Efficiency	0.80	0.74
3	Benefits	0.90	0.79
Average of k value		0.85	0.76
category		Very high	High

3.3.3.1 Practicality of Teachers

Teacher questionnaire aimed to determine the extent of teachers' understanding of and response module that developed. Three aspects of the assessment provided by the teacher, and all aspects were considered to have a very high practicality category. These data indicated that the acid-base module guided discovery-based learning developed was easy to use and was practical in pilot schools. The module is easy to use and practical nature seen from the material presented that was very clear and simple. The module content was generally easy to understand. These data indicated that the modules developed in conformity with the practicalities of a resource criterion. Easy of using meant that product was practically used any time, matter and the language used in module was easily understood, the learning time become short, quick and precise, and learning steps provided on the module clear, and easily interpreted by the teacher[30].

3.3.3.2 Practicality of Students

Student questionnaire responses aimed to determine the extent of students' understanding of the material presented in the module. Based on the analysis of student questionnaire responses, obtained an average value of 0.76 kappa moment with high practicality category. These data indicated that students have started to exploit modules for independent learning, both individually and in groups without the presence of a teacher. The use of colors and images contained in the module made students interested in learning. The high interest or attention of students could occur because students were

IOP Conf. Series: Materials Science and Engineering **335** (2018) 012097 doi:10.1088/1757-899X/335/1/012097

delighted with their communicative pictures as in Figure 2. The materials on module were also presented in accordance with the principles of chemical learning, so it can make students more systematic mindset and help in constructing understanding, and able to increase reading interest, motivation and curiosity of students. From Figure 2 showed that the material on the module were presented using the principles of chemistry triangle, namely (1) the level of macroscopic shows the real of hydrogen chloride acid in the bottle, (2) the level of sub-microscopic, seen from figure of the model of the HCl molecule, Cl^- ion H^+ ion and H_2O molecule; and (3) the symbolic level indicated by the chemical formula. The formula of water is H_2O, the formula of hydrochloric acid is HCl, the formula of the chloride ion is Cl^-,and the formula of hydrogen ion is H^+.

Practicality module can also be seen based on the answers to questions in the activities and student worksheet by students that can be shown in Figure 5. This figure contained the score of activity sheet and student worksheet. It was obtained that the score of the student was above the minimum completeness criterion which was 80.

Data in Figure 5 showed that the learning module could improve student learning outcomes. The results obtained in accordance with the research done about modules on the learning other chemical materials, including: (1) the use of modules in chemical equilibrium based on scientific was effective the learning outcomes of students of class XI MIA SMAN 4 Padang, with the percentage of completeness of students 84%[16]; (2) the use of a buffer solution modules based on discovery learning was effective, and student learning outcomes in chemistry learning in class XI SMAN 7 Padang with an average value above minimum completeness criterion, that was 92[17].

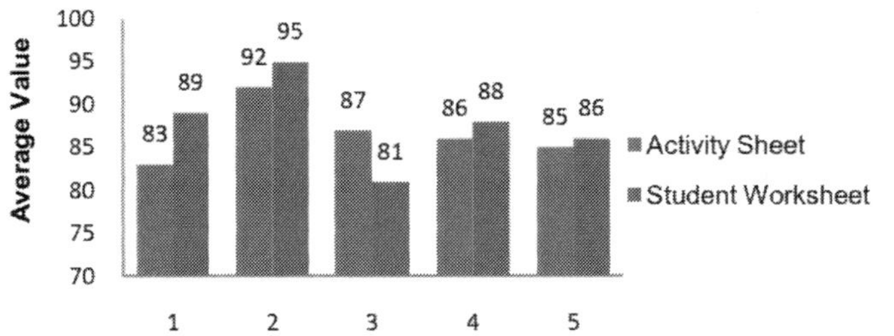

Figure 5. The Average Value Activity Sheet and Student Worksheet

Model guided discovery learning was used as a basis in preparing the proven module that can lead students to learn to find the concept and create a centralized learning to students. Student-centered learning methods by reinforcing the sense of group participation among the students motivate them to further study and enhance learning in higher levels of cognition [27]. Teaching methods promoted in guided discovery instruction are to cultivate learners' abilities of discovery, exploration, problem-solving and independent thin core competencies, and creation and invention or discovery through creative learning. Activity of students could encourage students in learning participation. It also made students construct knowledge by themselves. In other words, all knowledge was individually operated and explained, rather than passively acquired [8].

4. Conclusion

Based on the results of research and data analysis, it was concluded that the acid-base module based on guided discovery learning developed was valid and practical for high school chemistry teaching. Module developed had very high validity categories (k = 0.86) according to expert judgment, practicality was very high according to the teacher (k = 0.85) and high according to the student (k = 0.76). Therefore, this module can be recommended to use in a real learning in senior high school after this module tested in order to test the effectiveness. It is recommended that the continued study is aimed to test the effectiveness and disseminate the developed module .

Acknowledgments
The authors thank the Institute for Research and Community Services of Universitas Negeri Padang, which has funded this research through DIPA funds, Universitas Negeri Padang in accordance with the Rector Decree No. SP UNP DIPA-042.01.2.400929 / 2017 Date May 29, 2017. We also thank to Dr. Latisma Dj, M. Si.; Drs. Iswendi, M.S.; and Dra. Suryelita, M.Si., who had provided input and advice in this study; Drs. Bahrizal, M. Si.; Budhi Oktavia, M. Si., Ph.D.; Drs. Suheri; Dra. Murti and Mrs. Krisnayeni, S. Pd., as validators; chemistry teachers and students of SMAN 9 Padang 2016/2017 school year, as well as all those who have helped in the completion of the research and preparation of this article.

References

[1] Permendikbud Tahun 2014 Nomor 59. 2014. *Tentang Kurikulum 2013 Sekolah Menengah Atas/Madrasah Aliyah.*
[2] Carin, A. A. 1997. *Teaching Science Through Discovery.* 8th. Ed. Upper Saddle River, New Jersey Columbus,Ohio: Pearson Prentice Hall, Inc.
[3] Santrock. 2007. *Psikologi Pendidikan (edisi kedua).* Jakarta: Prenada Media Group.
[4] Rambe, FA and Ridwan Abd. S. 2014. The Effect of Guided Discovery Learning Model on the Student's Achievement in Physics of VII Grade in SMPN 1 Tebing Tinggi Academic Year 2013/2014. *Jurnal Inpafi*. Vol. 2, No. 3: 89-94
[5] Akanbi, A.A. dan Kolawole, C.B. 2014. Effects Of Guided-Discovery And Self-Learning Strategies On Senior Secondary School Students' Achievement In Biology. *Journal of Education and Leadership Development.* Vol 6 No. 1: 19-42.
[6] Fatokun K.V.F and Eniayeju P. A. 2014. The effect of concept mapping- guided discovery integrated teaching approach on Chemistry students' achievement and retention. *Academic Journal: Educational Research and Reviews*. Vol. 9(22): 1218-1223
[7] Luzviminda J. Achera. 2015. The Effect Of Group Guided Discovery Approach On The performance of Students In Geometry. *International Journal of Multidisciplinary Research and Modern Education (IJMRME)*
[8] *C.-J. Shieh & L. Yu*. 2016. A Study on Information Technology Integrated Guided Discovery Instruction towards Students' Learning Achievement and Learning Retention. *Eurasia Journal of Mathematics, Science & Technology Education*, 12(4), 833-842.
[9] *Utami* FAW, Sajidan, Sri D. 2015. Implementation of Guided Discovery Learning Model to Improve Student's Biology Learning Activity in Class X-2 SMA Muhammadiyah 1 Karanganyar in the Academic Year 2013/2014. BIO-PEDAGOGI. Vol.4, No. 1, 25-29.
[10] Akinbobola, A.O. dan Afolabib. F. 2010. Constructivist practices through guided discovery approach: The effect on students' cognitive achievement in Nigerian senior secondary school physics. *Eurasian J. Phys. Chem. Educ.* 2(1):16-25.
[11] Udo, M. E. 2010. Effect of Guided-Discovery, Student- Centred Demonstration and the Expository Instructional Strategies on Students' Performance in Chemistry. *Jurnal Multi-Disiplin Internasional*, Ethiopia; Vol. 4 (4): 389-398.
[12] Vaino K, Jack H and Miia R. 2012 Stimulating students 'intrinsic motivation for learning chemistry Through the use of context–based learning modules. *Chemistry Education Research and Practice.* Vol. 13,410–419.
[13] Novianty, I., Oktavia S, Neena Z. Efektivitas Penerapan Modul Materi Analisis Elektrokimia Berbasis Inkuiri Terbimbing Terhadap Hasil Belajar dan Persepsi Siswa Kelas XI Semester 1 Kompetensi Keahlian Kimia Analisis SMKN 7 Malang. UNM: *jurnal-online.um.ac.id.*
[14] Febriana, BW., Ashadi, dan M. Masykuri. 2014. Pengembangan Modul Kimia Berbasis Problem BasedLearning (PBL) Pada Materi Senyawa Hidrokarbon dan Turunannya Kelas XI SMK Kesehatan Ngawi. *jurnal.fkip.uns.ac.id*

[15] Sunaringtyas K, Sulistyo S, Mohammad M. 2015. Pengembangan Modul Kimia Berbasis Masalah pada Materi Konsep Mol Kelas X SMA/MA Sesuai Kurikulum 2013. Jurnal Inkuiri. Vol. 4. No.2: 36-46.
[16] Yerimadesi, Bayharti, Fitri H, dan Wiwit FL. 2016. Pengembangan Modul Kesetimbangan Kimia BerbasisPendekatan Saintifik Untuk KelasXI SMA/MA. *Journal of Saintek*. Vol. 8. No. 1: 85-97.
[17] Yerimadesi, Ananda P, Ririanti. 2017. Efektivitas Penggunaan Modul Larutan Penyangga Berbasis *Discovery Learning* Terhadap Hasil Belajar Siswa Kelas XI MIA SMAN 7 Padang. Jurnal Eksakta Pendidikan (JEP). Vol. 1, No. 1: 17-23
[18] Sugiyono. 2012. *Metode Penelitian Kuantitatif kualitatif dan R&D*. Bandung: Penerbit Alfabeta.
[19] Thiagarajan, Sivasailan. Dorothy S. Semmel. dan Melvin I. 1974. *Intruvtional Development for TrainingTeachers of Exceptional Children A sourcebook*. Indiana: Indiana University Bloomington.
[20] Smitha, VP. 2012. *Inquiry Training Model and Guided Discovery Learning For Fostering Critical Thinking And Scientific Attitude*. First Edition. Vilavath Publications, Kozhikode.
[21] Depdiknas. 2008. *Panduan Pengembangan Bahan Ajar*. Jakarta: Depdiknas.
[22] Prastowo, A. 2011. *Panduan Kreatif Membuat Bahan Ajar Inovatif*. Yogyakarta: DIVA press.
[23] Depdiknas. 2008. *Pengembangan Bahan Ajar*. Jakarta: Departemen Pendidikan Nasional, DirJen Manajemen Pendidikan Dasar dan Menengah, Direktorat Pembinaan SMA.
[24] Boslaugh, S dan Paul AW. 2008. *Statistics in a Nutshell, a desktop quick reference*. Beijing, Cambridge, Famham, Köln, Sebastopol, Taipei,Tokyo: O'reilly.
[25] Kemendikbud. 2016. Silabus Mata PelajaranSekolah Menengah Atas/Madrasah Aliyah(SMA/MA)Mata Pelajaran Kimia. Jakarta: Kementrian Pendidikan dan Kebudayaan.
[26] Herron, J. D., L. L. Cantu, R. Ward, & V. Srinivasan. 1977. Problems Associated with Concept Analysis. *Science Education.*Vol 61 No 2:185-199
[27] Talanguer, Vicente. 2010. "Macro, Sub micro, and Symbolic: the Many Faces of the Chemistry "Triplet". *International Journal of Science Educatoin*. Vol 33. No 2: 179-195.
[28] Badan Standar Nasional Pendidikan (BNSP). 2006. Standar Penilaian Bahan Ajar.
[29] Brady, James, E., Sense, Fred dan Jespersen, Neil, D. 2009. *Chemistry Matter and Its Changes*. John Wiley and Sons, Inc.
[30] Nieveen, N. 2010. *Formative Evaluation in Education Design Research*. Dalam Tjeer Plomp and Nienke Nieveen (Ed). An Introduction to Educational Design Research. Nederland in www.slo.nl/organisatie/international/publication.

IOP Conf. Series: Materials Science and Engineering **335** (2018) 012098 doi:10.1088/1757-899X/335/1/012098

Student Task Analysis for the Development of E-Learning Lectural System in Basic Chemistry Courses in FKIP UMMY Solok

A Afrahamiryano[1*], **D Ariani**[2]

[1]BiologyEducation Department, University of Mahaputra Muhammad Yamin, Solok
[2] Economy Education Department, University of Mahaputra Muhammad Yamin, Solok

*afrahamiryano@gmail.com

Abstract. The student task analysis is one part of the define stage in development research using the 4-D development model. Analysis of this task is useful to determine the level of understanding of students on lecture materials that have been given. The results of this task analysis serve as a measuring tool to determine the level of success of learning and as a basis in the development of lecture system. Analysis of this task is done by the method of observation and documentation study of the tasks undertaken by students. The results of this analysis are then described and after that triangulation are done to draw conclusions. The results of the analysis indicate that the students' level of understanding is high for theoretical and low material for counting material. Based on the results of this task analysis, it can be concluded that e-learning lecture system developed should be able to increase students' understanding on basic chemicals that are calculated.

1. Introduction

Basic Chemicals is one of the science courses for first year students who take the Department of Mathematics and Natural Sciences Education, Study Program of Biology Education, at Mahaputra Muhammad Yamin University. This course is taught because chemistry is closely related to biology, science and other science. Graduates of Biology Education course are expected to have the skills to become biology and science teachers. Therefore, the basic courses related to IPA are included in the curriculum of the Biology Education course.

The presence of Basic Chemicals in Biology Education course curriculum is one of the steps to fulfill the learning achievement of Biology Education program namely:

a. Apply basic concepts of educational science and design appropriate biological learning in appropriate elementary and secondary education units;
b. Develop and utilize a system of assessment and evaluation of biology learning in elementary and secondary education units appropriately;
c. Identify the characteristics of learners from the academic, physical, social, emotional aspects appropriately and apply the example to become a superior person, in the scope of elementary and secondary education units and society;
d. Analyze the basic principles of science in life, chemistry of life, cell and genetics, evolution and biodiversity, ecology and conservation, structures and functions of animals and plants and applied biology correctly and responsibly;

e. Implement strategic management and management of biology laboratories in primary and secondary education units and apply entrepreneurial skills in biology and biology education with full responsibility;
f. Publish results of research in the field of Biology and Biology Education in a quality elementary and secondary education units.

In addition NSTA [1] recommends a standard of content that must be mastered by Biology teachers namely core competencies, advanced competencies, and supporting competencies. For supportive competencies, biology teachers are also prepared to effectively apply the concepts of math and science to biology learning that include the basic concepts:

- Chemistry, including basic chemistry and biochemistry with basic laboratory techniques;
- Physics, including light, sound, optics, electricity, energy, magnetism, and thermodynamics;
- Mathematics, including probability and statistics.

The above NSTA recommendation implies that the graduates of Biology Education courses in addition to mastering the material in the field of science, but also must master the chemistry and biochemistry. As part of science, the field of chemistry studies is so wide that chemistry is dubbed as central of science. Chemistry is generally abstract and requires an understanding of concepts and the ability to calculate. So basically in studying chemistry, students need a good understanding of concepts and able to see the relationship between concepts and apply it.

Afrahamiryano in The International Conference on Mathematics, Science, Education and Technology [2] states that the students' perception of the Basic Chemicals at FKIP UMMY Solok is good. The good perception of biology education program students to Basic Chemicals is a great potential for lecturers to develop learning. Development of Basic Chemicals learning system is expected to increase student interest in this course. However, prior to development, needs to be done first needs analysis. Needs analysis can be done through student task analysis. Analysis of this task aims to determine the level of student understanding of the material that has been studied.

Based on the background of the above problems, the authors are interested to know the level of understanding of students about Basic Chemicals through research entitled "Analysis of Student Duties for the Development of e-learning lecture System in Basic Chemicals at FKIP UMMY Solok". This study aims to determine the level of students' understanding of theoretical anc calculation of Basic Chemicals. To achieve this goal, this research is focused on the analysis of student tasks in Basic Chemicals. The formulation of the problem in this research is "How is the level of understanding of the students on theoretical and calculation in Basic Chemicals?"

2. Method

This research is part of a development research entitled "Development of e-Learning Lecture System in Basic Chemicals at FKIP UMMY Solok", with the 4-D development model. This student task analysis is part of the Define stage. The analysis of student task is done by using descriptive research. The descriptive research method is a method that tries to describe and interpret the object as it is [3].

The population in this research is all students of the Biology Education Study Program for academic year 2016/2017 which takes Basic Chemicals that is eight people. Because the population is small, the sampling technique used is Total Sampling. The data collection technique is done by analyzing the student's answer sheet and then input it in the task analysis sheet, and continued by calculating the percentage of students' understanding of the material that becomes the task, using the formula:

$$\text{Percentage of Understanding} = \frac{\text{(Number of Students doing the job correctly)}}{\text{(Total Student Total)}} \times 100\% \quad (1)$$

The results obtained are interpreted using the following criteria:

Table 1. Criteria Interpretation Score Student Understanding Level [4]

Criteria Range	Percentage
0 - 20	Very Low
21 - 40	Low
41 - 60	Medium
61 - 80	Good
81 - 100	Very Good

Because this research uses the descriptive research method, the data analysis technique used is triangulation that is to draw a conclusion based on the findings in the field.

3. Results and Discussion

Basically, Basic Chemicals teaching materials consist of theoretical and calculating material. The theoretical material has five topics, namely equalizing the reaction equation, atomic structure, periodic system, chemical bond, and organic chemistry. Counting material there are six topics, namely basic concepts of chemistry, basic law of chemistry, stoichiometry, solution, thermodynamics, and thermochemistry. The data of student task analysis on Basic Chemicals is shown in Table 2 below:

Table 2. Percentage of Student Understanding Level on Basic Chemistry Material

No	Matter	Percentage of Understand	Criteria
1	Equalizing reaction equation	62,5%	Good
2	Basic Concepts of Chemistry	50%	Medium
3	Basic Chemical Law	50%	Medium
4	Atomic Structure	87,5%	Very Good
5	Periodic System	87,5%	Very Good
6	Chemical Bond	75%	Good
7	Stoichiometry	25%	Low
8	Solution	25%	Low
9	Thermodynamics	25%	Low
10	Thermochemicals	37,5%	Low
11	Organic Chemicals	25%	Low

The data in Table 1 above shows students' understanding of the "equity of equations" is 62.5%. This data shows that students are able to equate the equation of the reaction well. However, some students are not able to equate the equation of the reaction, the cause is that the student is incapable of writing a complete and equitable reaction equation when the known name of the reacting compound and the name of the reaction product compound.

The level of students' understanding of the "Basic Concepts of Chemistry" is 50%. This data indicates that the level of students' understanding of this material is included in the medium category. This material is discussed about the understanding of chemistry, scientific methods applied in chemistry, elements, compounds, mixtures, material properties, and mixed separation techniques. In general, the material discussed is theoretical, but there are some concepts that are calculated to determine the level of substances.

The level of students' understanding of the "Basic Chemistry Law" material is included in the medium category (50%). This material is related to the matter of reaction equations and is counted. This material is discussed on the Law of Conservation of Mass, Fixed Comparison Law, Comparative Multiplier Law, Mutual Reciprocal Law, Avogadro's Law, Unifying Volume Law, calculating relative atomic mass, and calculating the relative molecular mass. In general, students have difficulties in working on questions relating to this material, because this material demands an understanding of the concept of reaction equations and good math skills.

For the material "Atomic Structure and Periodic System", the level of student understanding is very good (87.5%). These two materials are theoretical and require a good conceptual understanding because between the Atomic Structure and the Periodic System are interrelated to one another. In this material, students generally have difficulty in determining the four major quantum numbers of the last electron in an electron configuration, or otherwise determining the electron configuration if the four major quantum numbers of the last electron are known.

The students' understanding of "Chemical Bonding" materials is good. The weakness found in this material is that students do not memorize members of the main group elements, so in general the students are difficult in determining the valence electron of an element, and it is difficult to determine the number of electrons involved in forming a chemical bond.

Furthermore, for stoichiometric material, solution, thermodynamics, and thermochemistry, the level of student understanding is low. These materials are quite complicated for students because in addition to demanding concept comprehension also requires the ability in the analysis of problems and calculations. In addition, these materials are also related to the reaction equation. So if students do not understand the equation of the reaction, then these materials will be very difficult.

Last is the material "Organic Chemistry". This material discusses the uniqueness of carbon atoms, hydrocarbons, functional groups and their derivatives, organic reactions, and polymers. This material is very difficult for students to understand because in this material is demanded by students understanding in nomenclature of chemical compound and reaction mechanism.

Based on the results of this analysis can be seen that the students of Biology Education Studies Program is rather weak on counting material. Besides that also added with the limited factor of reference book that can be used by students to support the learning process. In addition, student's motivation to enrich information through low online media.

4. Conclusions

Based on the results of the analysis can be concluded that the level of student understanding is high for the material that is theoretical and low for counting material. Therefore, it takes e-learning lecture system that can improve student's understanding on basic chemicals. Based on the conclusion of the results of this study, it is advisable to do Academic Potential Tests and Modality Tests to determine students' academic ability and student learning type. Next, the results of this study suggest that the development of e-learning can be applied to the lecture process conducted.

Acknowledgments

Our thanks go to the Ministry of Technology and Higher Education which has facilitated us in research, through the Beginner Lecturer Research Grant.

References

[1] NSTA. (2003). *Standars for Science Teacher Preparation.*

[2] Faculty of Mathematics and Science, S. U. of P. (2015). *The International Conference on Mathematics , Science , Education and Technology* (ICOMSET 2015) (p. 424).

[3] Sukardi. (2003). *Metodologi Penelitian Pendidikan, Kompetensi dan Praktiknya*. Yogyakarta: Bumi Aksara.

[4] Riduwan. (2005). *Belajar Mudah Penelitian untuk Guru, Karyawan, dan Peneliti Pemula.* Bandung: Alfabeta.

IOP Conf. Series: Materials Science and Engineering **335** (2018) 012099 doi:10.1088/1757-899X/335/1/012099

Validity And Practicality of Experiment Integrated Guided Inquiry-Based Module on Topic of Colloidal Chemistry for Senior High School Learning

A Andromeda[1], Lufri[2], Festiyed[3], E Ellizar[1], I Iryani[1], G Guspatni[1], L Fitri[4]

[1]Lecturer of Chemistry Department, Faculty of Mathematics and Science,UniversitasNegeri Padang, Indonesia
[2]Lecturer of Biology Department, Faculty of Mathematics and Science,UniversitasNegeri Padang, Indonesia
[3]Lecturer of Physics Department, Faculty of Mathematics and Science,UniversitasNegeri Padang, Indonesia
[4]Alumnus of Chemistry Department, Faculty of Mathematics and Science,UniversitasNegeri Padang, Indonesia

*andromedasaidir@yahoo.com

***Abstract**. This Research & Development study aims to produce a valid and practical experiment integrated guided inquiry based module on topic of colloidal chemistry. 4D instructional design model was selected in this study. Limited trial of the product was conducted at SMAN 7 Padang. Instruments used were validity and practicality questionnaires. Validity and practicality data were analyzed using Kappa moment. Analysis of the data shows that Kappa moment for validity was 0.88 indicating a very high degree of validity. Kappa moments for the practicality from students and teachers were 0.89 and 0.95 respectively iindicating high degree of practicality. Analysis on the module filled in by students shows that 91.37% students could correctly answer critical thinking, exercise, prelab,postlabandworksheet questions asked in the module. These findings indicate that the integrated guided inquiry based module on topic of colloidal chemistrywas valid and practical for chemistry learning in senior high school*

1. Introduction

Chemistry is developed through processes done in the laboratory to produce what is called scientific knowledge. As a science, chemistry is formed through interrelation among attitude and scientific processes, investigations of natural phenomena, and scientific products [1]. At school, chemistry is more frequently taught in class than in laboratory. As students rarely do experiment and discover concepts in the laboratory, they often consider chemistry as a knowledge of memorization and thus is difficult and unpopular. Students are not either actively involved in learning or given chances to develop thinking process, and thus do not experience the discovery of concepts [2]. Teachers on the other hand do not always engage affective and psychomotoric aspects in learning, so what is taught is easily forgotten by students [3]. To involve the affective, psychomotoric and cognitive aspects, chemistry is best taught through laboratory or experiement-based approach. The experiment activities can provide direct experience for meaningful learning [4].

Colloidal system is one of chemistry matters taught in grade XI senior high school (SMA) [5]. The matter consists of factual, conceptual and procedural knowledge dimensions that must be accomplished by students. Fortunately colloidal system is applicative in everyday lives. To help students learn the concepts, direct experiential learning with scientific process skill integrated can be implemented. This is in line with the command of ministry of education (permendikbud) that emphasizes science learning on the employment of scientific process skill [6]. Scientific process skill involves the advance of intellectual, social, physical skills derived from basic skills that already exist in students [7]. In addition to learning science concepts and principles through memorization and being familiar with formulas and terms through verbal exercises, students are asked to have direct experience through the processes of experiment.

Experiment method gives chances to students to be involved in discovering the concepts and developing cognitive, affective and psychomotor abilities. Experiment in the laboratory can be done by emphasizing concepts' discovery and scientific process skill [8]. Experiment activities that emphasize on discovery can be integrated in learning process (experiment integrated). Students can actively learn, investigate, collect data, analyze data, self discover the concepts, and solve problems under teacher's supervision. Learning can then be meaningful and thus is not easily forgotten by students. Studies done by Hofstein& Lunetta (2003) and Hofstein&Mamlok-Naaman (2007) show that laboratory experiences (experiment) entirely support the achievement of science education's objectives [9][10].

Observation and questionnaire data obtained from teachers and students in several schools in West Sumatra show that (1) Learning material that integrates experiment in classroom learning was not available, (2) Learning materials used could not encourage students to find the concepts by themselves, so they inclined to only memorize and do mathematics, (3) Learning materials used could not lead students to have scientific process skill, (4) Experiment activities were done after matters are taught just to confirm concepts, (5) Learning materials used could not attract students' attention and motivation to learn.

To deal with these problems, it is suggested that teachers integrate experiment in learning activities (experiment integrated) by emphasizing concept inquiry process, so students can actively learn, investigate, collect data, analyze data, find concepts by themselves, and solve problems under teacher's supervision (guided inquiry) [11]. Guided-Inquiry learning is students-centred one where students carrying their own role study in a small group while teacher's role is to make sure that all of the students are involved in learning process [12]. This type of learning can improve students' understanding and scientific process skill. To implement learning process that integrates experiment in classroom learning, an appropriate learning material is needed.

Based on the explanation above, learning material which is based on guided inquiry learning is needed to help teacher solve the problems in learning. A learning material called module is one of the appropriate one for this situation. Module consists of series of complete and various materials that allows students to learn independently, self-evaluate their achievement, determine their phase of learning to achieve learning goals with minimum help from teacher [13]-[15]. Module is a complete learning material consisting of matters, exercises, and evaluation designed systematically in certain format for students to learn and understand the concepts.

In this research, a learning material called module was developed. The module was designed by following the syntax of guided inquiry learning proposed by Lawson & Abraham (1979) and Hanson (2005). According to Lawson & Abraham (1979), the cycle of guided inquiry learning consists of exploration, concept formation, and application. According to Hanson (2005) guided inquiry learning's syntax comprises of orientation, exploration, concept formation, application, and conclusion [16][17]. To discover the concept, students are guided with a series of critical thinking questions starting from simple question to complex ones designed with indicators of scientific process skill. Students answer these questions after they observe and analyze model or do experiment on exploration stage. Experiment activities are designed by denoting to components of the laboratory investigation by The Collage Boardcontaining title, information, learning goals, tools and materials, procedure, safety precaution, prelab questions, microscopic activities, and postlab questions[16].

Researches on scientific process skill were done previously. Hesbon (2014) concluded that learning with scientific process skill integrated can significantly improve students' learning achievement[18]. Nworgu dan Otum(2013) concluded that guided inquiry has effect on the achievement of scientific process skill on students and give chances for students to use various source of information and ideas in understanding and solving various problems [19]. Research related to experiment integrated activities was done by Andromeda et. al (2015) who concluded that learning with experiment activities integrated can effectively increase students' achievement in cognitive aspect [20]. Sodikun(2015) concluded that learning that integrates experiment activities increases students' motivation and trains their scientific process skill [21]. Researches in developing valid and practical guided inquiry based worksheet for SMA learning were also done by Iryani et. al (2014), Andromeda et. al (2015,a), Andromeda et. al (2015, b)[22]-[24]. However, the worksheet did not integrate experiment in learning nor aimed to make students get scientific process skill.

In regard to the necessity of learning materials that have orientation on concepts formation and to fulfill curriculum demand and execute higher education's contribution in building and developing the Information, Technology, Science- Social and culture (IPTEKS-SOSBUD) aspect, research on developing learning material was taken. This research aims to developed guided inquiry based module that integrates experiment in learning process in order to increase scientific process skill of senior high school students.

2. Research Method

The aim of this research is to produce a valid and practical module that is based experiment integrated guided inquiry-based module on colloidal topic. The module is intended to improve scientific process skill of SMA students. The method is Research and Development (R & D) using 4D instructional design model developed by Thiagarajan, Semmel & Semmel (1974). The model consists of 4 phases called Define, Design, Develop, dan Disseminate [25]. Research in the first year was limited to validity and practicality tests. Limited trial was done at SMAN 7 Padang. The result became the basis for revising the product developed.

Define is a phase to set and define learning prerequisites. It begins with analysis of core competencies, basic competencies, and learning materials based on the content standard of curriculum 2013. This phase consists of five main steps named (1) front-end analysis, (2) students analysis, (3) task analysis, (4) concept analysis, and (5) learning goals formulation.

Design is a phase to define skills, formulate the goals, and determine the sequence of learning and small scale test that can be implemented. The focus of this phase is to design the module that is based on guided inquiry model and integrated with experiment activities. The steps of this phase includes: a) understanding basic competencies and specifying the indicators and learning goals that students must achieve after learning colloidal topic, b) determining concepts on colloidal topic, c) designing module with guided inquiry syntax developed by Hanson (2005) and experiment activities by The Collage Board and modifying it according to the needs of students, d) choosing appropriate format as described by module design guideline [26].

Develop is a phaseto develop draft of the module. This phase consists of two stages named validity and practicality testings. Validitytesting in this research was done by giving validity questionnaire to experts (four chemistry lecturers, two chemistry teachers at SMAN 7 Padang, and two chemistry teachers at SMAN 1 2 X 11 Kayu Tanam). The evaluation on validity testing were on the content, presentation, linguistics, and graphics aspects. The result was used as a basis for product revision. Furthermore, limited trial to see the practicality of the module was done. This trial was done at SMAN 7 Padang. The result was used to revise the product for further effectiveness testing of the module developed.

Disseminate is the final phase which aims to distribute the product in scholarly seminars that involve teachers and experts of chemistry, professional association and scientific journal. The quality assurance of the product must be done by standardizing it to the accepted criteria.

The instruments used to collect data in this research were validity and practicality questionnaires and the module itself. To determine validity and practicality levels, data were analysed with Kappa Cohenformula (equation 1). The interpretation of Kappa moment can be seen in Table 1[27]. For the module, it was analyzed with percentage technique.

$$Kappamoment\ (k) = \frac{\rho_0 - \rho_e}{1 - \rho_e} \quad (1)$$

Table 1. Decision category based on *Kappa moment (k)*

Interval	Category
0,81 – 1,00	Very high
0,61 – 0,80	High
0,41 – 0,60	Middle
0,21 – 0,40	Low
0,01 – 0,20	Very low
<0,00	Invalid

3. Result and discussion

3.1. Result

Experiment integrated guided inquiry-based module to improve scientific process skill was developed using 4D instructional design model. The result is as followSome text.

3.1.1. Define phase. In this phase, front-end, students, task, and concept analyses and learning goals formulation were done. In front-end analysis, problems faced by teachers and students in chemistry learning were studied. Data obtained from interview and questionnaire given to 30 teachers in West Sumatra show that (1) Learning materials that integrate experiment activities in classroom learning were not available, (2) Learning materials used could not make students find the concepts by themselves and students were only asked to calculate and memorize, (3) Learning materials used did not aim to get students have scientific process skill, (4) Experiments were done after classroom learning and had role to confirm concepts, (5) Tools and materials were available, but the experiments were not done due to insufficient time. Data obtained from questionnaire given to 50 students show that (1) 73 % students perceived chemistry as a difficult subject, 2) 88 % students explained that colloidal topic is full of memorization, 3) 70 % students explained that learning materials used were not interesting and motivating students to learn. These facts show that learning process has not fulfilled the demand of curriculum 2013. Based on the analysis, it was determined to design learning material that motivates students to learn, guides students to find concepts through direct experience using scientific process skill, and provides meaningful learning for students [6]. It also became researchers' consideration to develop experiment integrated guided inquiry-based module on topic of colloidal system to enhance scientific process skill of SMA students.

Students analysis was done through interview and questionnaire given to teachers and students. The analysis showed that guided inquiry is suitable for senior high school learning as it can improve students' motivation to learn independently and find the concept by themselves. Participant students were grade XI students whose ages range from 15 to17years old. According to Piaget developmental theory, teenagers (in this case age 15-17) have formal operational thinking. In this stage, they can think logically, can interpret and draw conclusion. The analysis of students' characteristics became the basis for developing experiment integrated guided inquiry-based module.

Task analysis was done by referring to current curriculum 2013. In this curriculum, colloidal topic is included in basic competency 3.15(categorize various types of colloidal system, explain colloidal properties and applications in daily lives) and basic competency 4.15 (make colloidal food and other products using colloidal principles). From these basic competencies, the derived indicators in this research were 1) classify solution, colloid and suspension, 2) categorize colloid based on dispersed phase and dispersing medium, 3) explain collidal properties (optical, kinetics, adsorption, coagulation, and dialysis), 4) differentiate liofil and liofob colloids, 5) explain the syntesis of colloid in laboratorium, 6) analyze the application of colloid in daily lives and industry, 7) prepare colloid in daily lives. Furthermore, concept analysis was done to determine the main concepts included in the topic. They include solution, colloid, and suspension; types of colloid; colloidal properties; liofil and liofob colloids; colloid synthesis; and the application of colloid in daily lives.

Learning goal analysis was done by specifying the indicators. Learning goals in this topic were 1) Students can explain the definition of solution, colloid, and suspension based on experimental data correctly, 2) Students can categorize types of colloid based on dispersing medium and dispersed phase correctly, 3) Students can explain the properties of colloid; the optical, kinetics, adsorption, electric, coagulation, and dialysis properties based on experimental data correctly, 4) Students can differentiate liofil and liofob colloids based on the interaction of dispersing medium and dispersed phase using model correctly, 6) Students can explain the synthesis of colloid based on dispersion and condensation ways through experimental data correctly, 7) Students can analyze the applications of colloid in daily lives and industry correctly, 8) Students can make product of colloid using its principles in daily lives through experiment activities correctly.

3.1.2. Design phase. The define phase was followed by the design of experiment integrated guided inquiry-based module on topic of colloidal system to enhacescientif process skill of SMA students. Learning material was designed with Microsoft Word with font Times New Roman, Comic Sans MS, Calibri (Body), Chitter, Ravie, Tempus Sans ITC font-styles. Background of the cover was blue, black, and white in color, while the content was white in color.

Module was designed with the syntax of guided inquiry which consists of orientation, exploration, concept formation, application, and conclusion. In exploration stage, laboratory activities ia done by denoting to thecomponents of the laboratory investigation by The Collage Board which consists of title, information, learning goals, tools and materials, procedures, safety precaution, prelab questions, microscopic activities, and postlab questions. In concept foramtion stage, students were guided to correlate facts obtained in the experiment (macroscopic) with the submicroscopic illustrations shown in the module. They were also guided with critical thinking questions to find the concepts. The module's format was following the instrcutional design guidelines by Suryosubroto. The format of the module concluded: 1) guidelines for teachers, 2) students' activities sheet, 3) worksheet, 4) the key of the activities sheet, 5) test, 6) key of the test[26].

3.1.3. Develop phase. Develop phase aimed to produce a valid and practical experiment integrated guided inquiry-based module on topic of colloidal system. Based on the analysis of validity data from eight validators, the value of Kappa moment was 0,88 indicating very high degree of validity. The result of analysis per aspect from eight validators is seen in Figure 1.

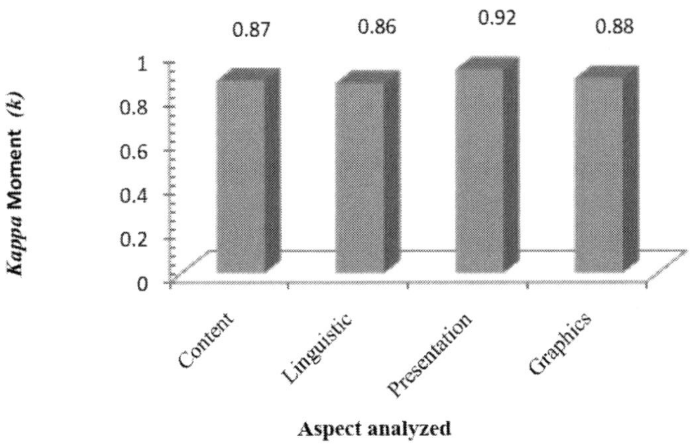

Figure 1. Analysis per aspect on validity level from validators

Although validity of the module was very high, there were several things to revise for the perfection of the module developed. Analysis on practicality data done by four teachers resulted that the average Kappa moment was 0.95 and analysis on practicality data obtained from 27 grade XI MIPA 5 students at SMAN 7 Padang revealed that the value was 0.89 indicating very high degree of practicality. The result of analysis per aspect of practicality done by students and teachers is seen in Figure 2.

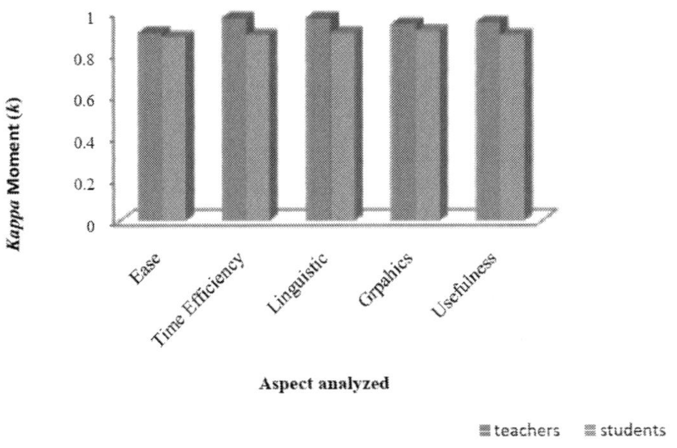

Figure 2. Analysis per aspect on practicality level from teachers and students

The result of analysis on the module filled in by students in answering critical thinking, prelab, and postlab, exercise, and worksheet questions can be seen in Figure 3.

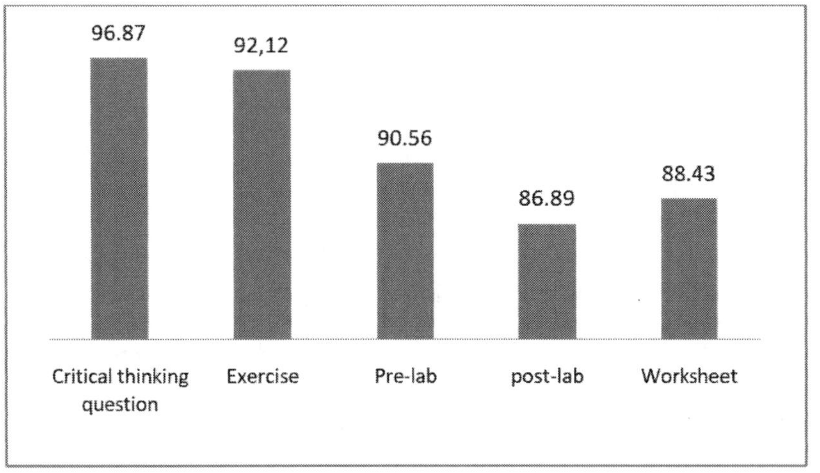

Figure 3. The percentage of student in answering the question in module

3.2. Discussion

3.2.1. Validity of the module. Analysis of validity data from eight validators on content component of the module revealed that the average of Kappa moment was 0.87 having a very high degree of validity. In other words, module developed was fulfilling the demand of basic competency 3.15 and 4.15 [6]. The module's content could make students improve their skill of scientific process using of scientific process skill indicators. This is consistent with researches done by Andromeda (2015) and Sodikun (2015) which concluded that learning which is integrated with experiment activities can improve learning outcomes, motivation to learn and practice scientific process skill of high school students [20][21]. Module provides the opportunity for students to learn continuously so their insight will be constantly increasing. Module can be said to be adaptive if it is flexible to be used and can be adapted to the development of science and technology [14].

Analysis of validity data on presentation component revealed the average Kappa moment of 0.92 indicating a very high degree of validity. In other words, the module developed was valid from presentation aspect. Colloidal system module was systematically developed and arranged accordingly to guided inquiry learning syntax consisting of orientation, exploration, concept formation, applications, and conclusion [17]. Evaluation on the writing's systematic of the module proved that the experiment integrated guided inquiry-based module on topics of colloidal system was presented in accordance with the instructional design guideline by Suryosubroto[26].

Analysis of validity data on the linguistic component revealed an average Kappa momentof 0.86 with very high degree of validity in category. This shows that the experiment integrated guided inquiry-based module on colloidal topic developed was communicative and unambiguous. The critical questions posed were clear and it provided symbols presented in consistent manner. A good module is one that is user friendly. The use of language that is simple, clear, familiar, and easy to understand is one kind of user friendliness[14].

Analysis of validity data on graphics component revealed an average Kappa momentof 0.88 indicating a very high degree of validity. Images, font-style and font-size provided in the module were clearly observed. The layout and design of the module were interesting. Therefore, module can be said user-friendly. In line with previous research, instruction and information provided should be able to assist students in understanding the content of the modules [14].

As a whole, experiment integrated guided inquiry-based module on colloidal topic had an average Kappa moment of 0.88 indicating very high degree of validity. A product is said to be valid if it can create a condition that fulfills the content and construct of the determined one [28]. Content validity shows that a model / prototype developed is based on a strong theoretical rationale while construct validity shows the fitness of components contained in the module with the defined indicators [29]. In general, module developed had a very high degree of validity, but there were some parts of the module's components needed to be fixed as suggested the validators.

3.2.2 Practicality of the module

Teachers practicality.
The average Kappa moment of ease of use of the module was 0.90 belonging to very high practicality category. This indicates that the instructions of use, the materials, the critical thinking questions, and the sequence of learning provided in the module could help teachers in learning process and enhance teacher's role as learning facilitator. The module helped students to learn independently without being fully dependent on others. That is why a module must be self-instructional [14].

The developed module had very high degree of practicality on efficiency and linguistic components (the average Kappa moment of both components = 0.97). With module, teachers could teach efficient manner. Module provided communicative, unambiguous, grammatically correct language, and clear questions. Indeed, a good module must use clear and communicative language [13].

The developed module also had very high degree of practicality on graphics and usefulness components. Analysis on graphics and usefulness components data obtained from four chemistry teachers showed Kappa moments of 0.94 and 0.95 indicating very high degree of practicality. This indicates that letters, drawings, and sub microscopic illustrations in the modules were clearly readable and understandable for students. The layout was organized; the design and appearance of module were attractive; and module was easy to carry because its size was practical. In the usefulness component, module could help students find the concept by themselves independently; it allowed teachers to increase students' interest, activities, and scientific process skills in learning.

Students Prakticality
Analysis of students' practicality data on ease of use component revealed an average Kappa moment of 0.88. This value belongs to a very high degree of practicality. This indicates that the instructions of use, the materials, the critical thinking questions, and the sequence of learning activities were straightforward and understandable for students. Analysis of students' practicality data on efficiency component revealed an average kappa moment of 0.89 showing a very high degree of practicality. It means that students could learn accordingly to their learning pace. Module could be used anytime and anywhere they needed because it was not restricted to particular time or places. Flexible learning can cultivate students' interest in learning. The use of the module can surmount the limitations of time, space, and senses on both students and teachers. Module can increase students' motivation, develop their prior capabilities, and thus encourage them to learn independently [14].

The average Kappa moments of the module on linguistic and graphics components were 0.90 and 0.91 respectively. It means that the module had very high degree of practicality on both of these components. The experiment integrated guided inquiry-based module on colloidal system topic used communicative language; it was unambiguous; and the questions asked were understandable. Letters and images used were clear and easy to read. The layout of the module was organized and neat, and it had attractive appearance. Therefore, module could attract students' interest to learn.

In terms of usefulness, the module gained an average Kappa moment of 0.89 with very high degree of practicality category. This indicates that the module developed could help students understand the concepts appropriately. Students could independently find the concept by themselves, and their scientific process skills were enhanced. This is proven by the results of analysis on sheets answered by students. In overall, student could correctly answer 96.87% of critical thinking questions, 92.12% of exercises questions, 90, 56% of prelab questions, 86.89% of post-lab questions, 88.43% of activity sheets' questions asked.

To sum up, as a whole experiment integrated guided inquiry-based module on colloidal topic increased student interest to independently learn and find the concepts by themselves. It was practical

IOP Conf. Series: Materials Science and Engineering **335** (2018) 012099 doi:10.1088/1757-899X/335/1/012099

to be used by teachers and students in learning activities. The module is a printed learning material that can be used by students with or without teachers' guidance [15].

4. Conclusion

Based on the results of research and data analysis done, it can be concluded that 1) The experiment integrated guided inquiry-based module on colloidal topic was successfully developed using 4D developmental model, 2) The experiment integrated guided inquiry-based module on colloidal topic had very high degree of validity and practicality.

Acknowledgements

Researchers would like to say thank the Institute for Research and Comunity Services of Universitas Negeri Padang, with has funded this research through DIPA funds, Universitas Negeri Padang in accordance with the Rector decree No. SP UNP DIPA-042.01.2.400929/2017 Date May 29, 2017. We also thank to Alizar Ph.D.,Yerimadesi S.Pd., M.Si, Dra. Bayharti, M.Sc., Dra. Lasmiati, S.Pd., M.Si, Fitrianis, S.Pd., Drs. Ali Munir, and Yanti Oktavia, S.Pd. for being our product validators. Lastly, we thank to participating students and teachers and all of the people who had contributed to the success of this research.

References

[1] Carin,ArthurA.1997.*TeachingScienceThroughDiscovery,8thedition*.Ohio:MerrillPubl.co

[2] Sri Rahayu, 2012. Pengembangan Model pemmbelajaran Advance Organizer untuk Meningkatkan Aktifitas dan Hasil Belajar Siswa Pokok Bahasan Koloid. Journal of Innovative Science Education, JISE 1(1) 2012, http://journal.unnes.ac.id/sju/index.php/jise

[3] Puskur. 2007. Naskah akademik : Kajian Kebijakan Kurikulum MatPel IPA. Jakarta :Balitbang Depdiknas

[4] Anderson, W.L dan Krathwohl. R. D. 2001. *Pembelajaran, Pengajaran, dan Asesmen*. Terjemahan Agung Prihantoro 2010. Yogyakarta: Pustaka Belajar.

[5] Kemendikbud. 2016. *Silabus Mata Pelajaran Sekolah Menengah Atas/ Madrasah Aliyah (SMA/MA) Mata Pelajaran Kimia.* Jakarta: Kementerian Pendidikan dan Kebudayaan

[6] Permendikbud Nomor 59 Tahun 2014 *tentang Kurikulum 2013 SMA/MA*

[7] Tawil, M,Liliasari. 2014. *Keterampilan-Keterampilan Sains dan Implementasinya dalam Pembelajaran IPA*. Makasar: UNM

[8] Maradona. 2013. *Analisis Keterampilan Proses Sains Siswa SMA kelas XI IPA pada pokok bahasan Hidrolisis Melalui Metoda Eksperimen. Prosiding Seminar Kimia* : ISBN 978-602-19421-0-9.

[9] Hofstein, Avi dan RachelMamlok-Naaman. 2007. *The Laboratory in ScienceEducation*. The State of the Art.Chemistry Education Research and Practice e journal.2007, 8(2), 105-107

[10] Hofstein, A and Lunetta V.N., 2003 *The Laboratory in ScienceEducation* : for the 21st century, Science Education Research 22, 28-54

[11] Abidin, Yunus. 2014. *Desain Sistem Pembelajaran dalam Konteks Kurikulum 2013*. Bandung: Refika Aditama.

[12] Straumanis, Andrei. 2010. *Process Oriented Guided Inquiry Learning*

[13] Daryanto. 2014. *Pengembangan Perangkat Pembelajaran (Silabus, RPP, PHB, Bahan Ajar)* . Yogyakarta: Gava Media.

[14] Departemen Pendidikan Nasional.2008. *Pengembangan Bahan Ajar*. Jakarta: Departemen Pendidikan Nasional, Direktorat Jenderal Manajemen Pendidikan Dasar dan Menengah, Direktorat Pembinaan Sekolah Menengah Atas

[15] Wena, Made. 2011. *Strategi Pembelajaran Inovatif Kontemporer*. Jakarta: Bumi Aksara

[16] The Collage Board. 2013. *AP Chemistry Guided Inquiry Experiment: Applying the Science Practices Student Manual.* College Board.

[17] Hanson, D. M. 2005. *Designing Process-Oriented Guided-Inquiry Activities. In Faculty Guidedbook: A Comprehensive Tool For Improving Faculty Performance, ed. S. W.* Beyerlein and D. K. Apple. Lisle, IL: Pacific Crest.

[18] Hesbon, E.A. 2014. *The Effect of Science Process Skills Teaching Approach on Secondary School Students Achievement in Chemistry in Nyando District Kenya*. Jurnal of Education and Social Research MCSER Publising, Rome – Italy, Vol. 4 N0.6 September 2014

[19] Nworgu L.N., dan Otum V.V. 2013 *Effect of Guided Inquiry with analogy Instructional Strategy on Students Acquisition of Science Process Skill.* Journal Education and Practic. ISSN 2222-1735(paper). ISSN 2222-288X (online). Vol 4 No.27.

[20] Andromeda, Bahrizal, Zahara A, 2015,a. *Efektifitas kegiatan praktikum terintegrasi dalam pembelajaran pada materi kesetimbangan kimia kelas XI SMA/MA*. Jurnal Eksata, Volume 1, Tahun XVII, Februari 2016. ISSN 1411-3724

[21] Sodikun. 2015. *Pengembangan Modul Berbasis Inkuiri Terbimbing Pada Materi Sistem Pencernaan Makanan untuk Meningkatkan Keterampilan Proses Sains.* Surakarta: Universitas Sebelas Maret

[22] Iryani, Andromeda, Mawardi, 2014, *Pembuatan Bahan Ajar Kimia SMA berbasis Guided Inquiry dengan Representasi Chemistry Triangle* (Laporan Penelitian HB)

[23] Andromeda, dkk.,2015,b.*Pengembangan Bahan AjarHidrolisis Garam berbasis Guided-Inquiry dengan Representasi Chemistry-triangle Untuk Siswa SMA/MA.* Prosiding Seminar SEMIRATA 2015 bidang MIPA BKS-PTN Barat

[24] Andromeda, Bayharti, MentariDalliputri. 2015. *The Development of Guided Inquiry-based Worksheet for Laboratory work on Topic of Colloidal System for Senior High School Instruction.* Prosiding Seminar International ICOMSET FMIPA UNP 2015

[25] Thiagarajan, Sivasailan, Dorothy S. S, dan Melvin 1., 1974, Instructional *Development for TraininTeachers of Exceptional Children A Sourcebook*. Indiana: Indiana

[26] Suryosubroto, B. 1983. *Sistem Pengajaran dengan Modul.* Yogyakarta: Bina Aksara

[27] Bouslaugh S. dan Watters P., A,. 2008. "*Statistics in a Nutshell, a Desktop Quick Referance*". United State of America: O'Reilley Media, Inc.

[28] Arikunto, S. 2012. *Dasar-Dasar Evaluasi Pendidikan.* Jakarta: Bumi Aksara

[29] Rochmad. 2012. *Desain Model Pengembangan Perangkat Pembelajaran Matematika*, Jurnal Kreano, Vol. 3 (1), ISSN: 2086-2334, hal. 59-71

IOP Conf. Series: Materials Science and Engineering **335** (2018) 012100 doi:10.1088/1757-899X/335/1/012100

Development of Chemistry Game Card as an Instructional Media in the Subject of Naming Chemical Compound in Grade X

Bayharti, Iswendi, and M N Arifin
Department of Chemistry, Faculty of Mathematics and Sciences, Universitas Negeri Padang, Indonesia

chembayharti@gmail.com

Abstract. The purpose of this research was to produce a chemistry game card as an instructional media in the subject of naming chemical compounds and determine the degree of validity and practicality of instructional media produced. Type of this research was Research and Development (R&D) that produced a product. The development model used was4-D model which comprises four stages incuding: (1) define, (2) design, (3) develop, and (4) disseminate. This research was restricted at the development stage. Chemistry game card developed was validated by seven validators and practicality was tested to class X_6 students of SMAN 5 Padang. Instrument of this research is questionnair that consist of validity sheet and practicality sheet. Technique in collection data was done by distributing questionnaire to the validators, chemistry teachers, and students. The data were analyzed by using formula Cohen's Kappa. Based on data analysis, validity of chemistry game card was0.87 with category highly valid and practicality of chemistry game card was 0.91 with category highly practice.

1. Introduction

Naming chemical compound is the subject that learned by grade X student in the second semester. This topic consists of fact, concept, principle, and procedure. Some facts in this subject namely chemical formula of acetic acid is CH_3COOH, chemical formula of sodium chloride is NaCl, and chemical formula of calcium carbonate is $CaCO_3$. Examples of concept in this subjectare binary compound which is a compound that consists of two different substances, and polyatomic compound which is the compound that consist of three or more different substances.

The principle in this subject is IUPAC rule for naming simple organic and inorganic compound. Procedure in this subject is the step in naming organic and inorganic compound. To understand the fact, concept, principle, and procedure and to make this subject long stay in student mind, the students have to read this subject more than two or three times and answer the exercise questions. One way to make lesson stay in the student mind is repeating the lesson that was learned [1].

Based on the results of author interviews with chemistry teacher at SMAN 5 Padang, SMAN 7 Padang, SMAN 10 Padang, and SMAN 12 Padang, it was found that the learning process of naming chemical compounds used lecture and discussion by using worksheets, textbooks, and power point media. Based on the curriculum of 2013, learning approach mandated is scientific learning that has five stages of learning activities: observation, asking question, data collection, association, and communication. However, based on interviews, on the stage of association, the students discussed and worked on the problems worksheet and textbooks exercise. This influenced student activity which was still low. The low activity of students would have an impact on the ability of academic and student learning outcomes. Therefore, in the learning of naming chemical compounds needed a media that can enhance the activity of students, easy to recall and solidify their understanding of that subject.

A game is one of the alternatives that can be used as an instructional media to increase student activity, and make students easier to remember the learning material. The game is something fun to do and allow the active participation of students in learning [2]. 80 % of teens who had an age range 12-21 years is still likely to enjoy the game, ranging from the traditional to the modern game such as online games [3] .

Game used as a instructional media was a game card that had been chemically modified. Chemistry game card contained questions that were used as exercise at the stage of association. This game would enhance students to understand the subject of naming chemical compound. Research about this game card had already done by Salmi (2014) with the title of the Periodic System Card for Game In Chemistry Learning In Class X High School. The game card of the periodic system was used effectively to improve student learning outcomes [4]. Other studies related to the use of the card was the research conducted by Sudjana (2015) with title Cation-Anion Card as Learning Media Innovation Subjects in Chemistry in High School. The results showed that the cation-anion game card could improve student learning outcomes in the subject of naming chemical compounds [5].

Based on these cases, authors were interested in designing and creating alternative instructional media in the form of a game card for chemistry subject in senior high school under the title "Development of Chemistry Game Card as an Instructional Media in the subject of Naming Chemical Compound in Grade X".

2. Research Methods

This research was Research and Development (R&D). Research and development is the research that used to produce a specific product and test the effectiveness of the product. This research was done in class X_6 of SMAN 5 Padang. The object of this research was a game cardof chemistry as an instructional mediain the subject of naming chemical compound in grade X.

The development model used was Four D model (4-D model). 4-D model consisted of four stages, namely: (1)define, (2) design, (3) develop, and (4) disseminate [6]. This research was limited to the stage of develop.

Define stage consisted of the front-end analysis, students analysis, task analysis, concept analysis, and analysis of learning objective. At the design stage, it was made a prototype of chemistry game card in the subject of naming chemical compound that consisted of a game card, the rules of the game, answer key and assessment sheet.

The third stage of this research was the develop stage. Chemicals game card that had been designed was validated by some validators. Furthermore chemistry game cards revised based on suggestions that given by validator. Chemical game card which had been revised subsequently tested to determine the level of practicality.

Instruments that used in determining the validity of chemistry game card was a validation sheet filled out by some validators, they are four Padang State University chemistry lectures and three high school chemistry teachers. The instrument used to determine the practicalities of chemistry game card is a questionnaire that completed by the students. Validity and practicality of chemistry game cards was analyzed by using formula Kappa Cohen.

$$momen\ kappa\ (k) = \frac{\rho_o - \rho_e}{1 - \rho_e} \quad (1)$$

Here k is momen kappa describing validity of the product, ρ_o is Realized proportion; counted by summing the score given by validators and then divided it by maximum total score and ρ_e Unrealized proportion; counted by substracting the maximum total score with the sum of total score given by validator, which then divided by the maximum total score. Kappa moment value *(k)* ranges from 0 to 1 [2]. Interpretation kappa moment value can be seen in Table 1

IOP Conf. Series: Materials Science and Engineering **335** (2018) 012100 doi:10.1088/1757-899X/335/1/012100

Table 1. The category of decision based on Kappa moment (k).

Interval	Category
0.81 - 1.00	Very high
0.61 - 0.80	High
0.41 - 0.60	Medium
0.21 - 0.40	Low
0.01 - 0.20	Very low
$\leq$ 0,00	Invalid

3. Results and Discussion

3.1 Result

The results of this research was the chemical game card as an instructional mediain the subject of naming chemical compounds. A set of chemical game card consisted of game card, answer key, rules of the game, and the assessment sheet.

3.1.1 A game card. This game was designed for Indonesian students, so that set of chemical game card used Indonesian language. The game cards were the cards that can be played by students.The real bridge cards had four logos; those were spade, heart, diamond and clover. Chemical game card logos were modified. These were five kinds of card with different logos, they were volumetric glass, erlemeyer, beaker glass, test tube and pipettes.

Each kind of card had fourteen cards, that consisted of number 2 to 10, A (As), K (King), Q (Queen), J (jack) and Jocker. The total number of car was 70 pieces. In the following picture, it can be seen an example of game cards.

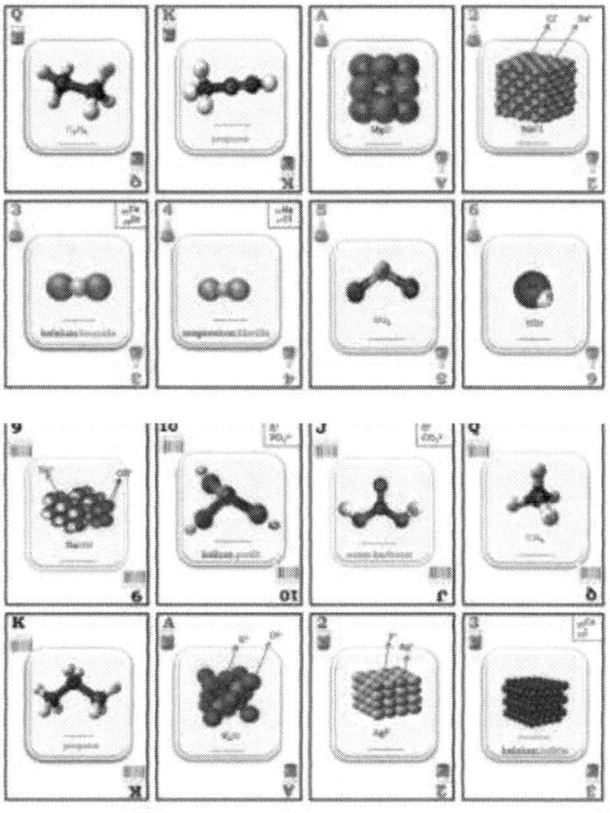

Figure 1. Example of game cards

3.1.2. Answer Key. Answer key was designed on the piesce of paper that contained the answers of of the questions in the cards. This answer key can be seen in Figure 2.

3.1.3. The rule of the game. The rule of the game was written on the piece of paper. There are five students in one group, one students as the coordinator and five students as players. Each player has six cards, and the remaining cards are put on the table as the pulling cards. The game is started when the coordinator takes a piece of pulling cards, read the question on that card, then answer that question. After that, the players continue the game. The player who has the same logo with the prior card that coordinator played picks his card, reads the question on the card and anwers it. The correct answer is scored by coordinator. If the answer is wrong, the second player will continue the game. If the player does not have a some logo's card, he has to take a card from the pulling card. The playing is continued by the next player. After playing the first logo, the gam will be continued with the other logos. The first player is the player who has the biggest card number on the prior logo playing. The winner on this game is the player who has the highes grade. The rule of the game can be seen in Figure 3.

KUNCI JAWABAN

A	aluminium oksida	kalium oksida	magnesium oksida	natrium fluorida	barium fluorida
2	kalium iodida	perak fluorida	natrium klorida	aluminium fluorida	kalsium sulfida
3	NaBr	CaI_2	$CaBr_2$	CaF_2	K_2S
4	Al_2S_3	$AlCl_3$	$MgCl_2$	Mg_3N_2	CaO
5	karbon monoksida	sulfur heksafluorida	sulfur dioksida	posfor pentaklorida	diiod pentaoksida
6	asam sulfida	asam klorida	asam bromida	asam fluorida	asam iodida
7	SF_4	SO_3	CCl_4	CO_2	CS_2
8	H_2O	NH_3	H_2O_2	PCl_5	N_2O_5
9	natrium hidroksida	kalium sianida	kalsium karbonat	amonium klorida	barium hidroksida
10	K_3PO_4	$Na_2S_2O_3$	$NaNO_3$	$NaClO_3$	Na_2CrO_4
J	H_2CO_3	H_3PO_4	K_2SO_3	H_2SO_4	CH_3COOH
Q	metana	etana	butuna	glukosa	urea
K	C_3H_8	C_2H_6	C_2H_2	C_4H_8	$CHCl_3$

Figure 2. Answer Key

3.1.4. Evaluation Sheet. This sheet was designed to record the score of each player. The right answer was scored by 4, and the wrong one was – 1. The following figure4 is the figure of evaluation sheet.

Aturan Permainan

a. Permainan ini berkelompok, tiap kelompok terdiri dari lima orang. Satu orang sebagai koordinator dan empat orang sebagai pemain.
b. Kartu bermain dipegang oleh setiap pemain, sementara kunci jawaban dipegang oleh koordinator.
c. Permainan dimulai dengan pengocokan kartu bermain oleh koordinator dan membagikan enam lembar kepada setiap pemain dan sisanya untuk kartu tarikan.

Figure 3. Rule of the game

LEMBAR PENILAIAN

Petunjuk Penilaian

Berilah poin 4 pada kolom benar apabila pemain menjawab benar, dan berilah poin -1 apabila pemain tidak dapat menjawab atau jawaban yang diberikan salah pada kolom salah.

Pemain 1		Pemain 2		Pemain 3		Pemain 4	
Benar	Salah	Benar	Salah	Benar	Salah	Benar	Salah

Figure 4. Evaluation Sheet

In the development stage, there had been done the validity and practicality test of this chemistry game card.

3.1.5 *Validity Test.* The aspects were assessed in validity test of chemistry game card was cognitive function, attention function, and compensatory function. Validity data of chemistry game card can be seen in Table 2.

Table 2. The validity of chemistry game card

No	Validity Aspect	*k*	Validity category
Cognitive Function			
1	The facts contained in chemistry game card are in accordance with the Basic Competence 3.10	0.88	Very high
2	The concepts contained in chemistry game cardare in accordance with the Basic Competence 3.10	0.78	High
3	The principles contained in chemistry game cardare in accordance with the Basic Competence 3.10	0.84	Very high
4	The procedures contained in chemistry game cardare in accordance with the Basic Competence 3.10	0.88	Very high
5	The questions contained in chemistry game cardare in accordance with the Basic Competence 3.10	0.78	High
	Kappa moment for cognitive function	0,84	Very high
Attention Function			
6	The language used in the chemistry game card use Indonesian spelling properly	0,88	Very high
7	The language used in the chemistry game cardis easy to understand	0.84	Very high
8	Images and symbols in the chemistry game card is clear and interesting	0.88	Very high
9	Design of chemistry game cardis interesting	0.92	Very high
10	The font used in chemistry game card is clear and legible	0.96	Very high
11	The size of the font used in chemistry game card is clear and legible	0.96	Very high
12	Color display in chemistry game card is interesting	0.78	High
	Kappa moment for attention function	0,89	Very high
Compensatory Function			
13	Chemistry game card as instructional mediacan strengthen student understanding	0,84	Very high
	Kappa moment for compensatory function	0,84	Very high
	Kappa moment for validity of chemistry game card	0,87	Very high

Based on the above table, it can be seen that the validity of chemistry game card as an instructional media had a moment kappa 0.87 with very high category. Nevertheless, there were some cases that need to be revised based on advice from the validator. The suggestions were:

- Change the shape of the molecule in game card, the original filling space change into ball and stick. This ball and stick make the player can see the bond that exist in a compound.
- Simplify the answer key to one sheet. Answer key that originally numbered 70 sheets made into one sheet. The aim of this is to facilitate the player coordinator in correcting the answer.
- Clarify the intent highest value at point G in the rules of the game.
- Before making a question for the game card, try to make the grating for all the learning objectives achieved.
- Using paper on both sides slippery for the game card in order not to get caught in the hand.
- Using symbols associated with chemicals to make it looks more attractive.

3.1.6 *Practicality Test.* Practicalities of chemistry game cards as an instructional media in subject of naming chemical compound obtained from the use of the product based on limited testing in the school. Practicality data were obtained from the responses of student questionnaire. This assessment data can be seen in Table 3.

Table 3. The practicality of chemistry game card

No	Practicality Aspect	*k*	Validity category
Attention Function			
1	Chemistry game cards as an instructional media has an attractive color	0,96	Very High
2	The inscription in chemistrygame card as an instructional media clear and legible	0,90	Very High
3	Language that used in chemistry game cardis easy to understand	0,90	Very High
4	Images and symbols in chemistry game card isclear and interesting	0,90	Very High
	kappa moment for attention function	0,91	Very High
Affective Function			
5	Chemistry game card as an instructional media is interesting to play	0,92	Very High
6	Chemistry game card as an instructional media is easy to play	0,81	Very High
7	Chemistry game card as an instructional media makes learning chemistry on the subject of naming chemical compound to be enjoyable	0,96	Very High
8	Chemistry game card as an instructional media makes the learning process more effective	0,96	Very High
	Kappa moment for affective function	0,92	Very High
Compensatory Function			
9	Chemistry game card as an instructional media makeseasier to understand the subject of naming chemical compounds	0,90	Very High
10	Question in chemistry game card help to understand the the subject of naming chemical compounds	0,94	Very High
11	Question in chemistry game card in accordance with the subject of naming chemical compounds	0,92	Very High
12	Chemistry game card as an instructional media is portable	0,87	Very High
13	Chemistry game card as an instructional media can be played anytime and anywhere	0,85	Very High
14	Chemistry game card as an instructional media can be played over and over again	0,90	Very High
	Kappa moment for compensatory function	0,90	Very High
	kappa moment for practicality of chemistry game card	0,91	Very High

3.2. Discussion

3.2.1. Validity of Chemistry Game Card. The validity of chemistry game card as instructional media derived from the assessment of seven validators through the validation sheet. Validity test of the instrument can be used on expert opinion (judgment experts) that the amount of at least three people [8]. Validators provided an assessment of the chemistry game cards as a instructional media in subject of naming chemical compound based on the functions of media, namely cognitive function, attention function, and compensatory function. Validity data of instructional media developed were analyzed using formula Kappa Cohen. This data can be seen on Table 2.

Cognitive Function

Based on assessment data from seven validator on cognitive function of chemistry the game card, it was obtained momen kappa 0.84 with category very high. Facts, concepts, principles and procedures contained in the chemistry game cardaccordance with the Basic Competence 3.10, which is applying the IUPAC rules for naming simple inorganic and organic compounds. In addition, the question that obtained in chemistry game card for subject of naming chemical compound correspond with learning objective that will achieved, so the instructional media in the form of chemistry game card has meet cognitive function. If the medium that used is conformity with the learning objective, the media can be used to make student learn[9].

Attention Function

The average of kappa moment for validityon attention function of chemistry game cardwas 0.89 with very high category. The language used in the media developed is easy to understand and are in compliance with Indonesian languagespelling. Media display is attracted to student attention. The increase of students attention will make the process of learning become better [10]. In addition, the type and size of font on the media that developed can be clearly legible by the user. Therefore, it can be said that the instructional media in the form of chemistry game card has fulfilled the attention function. With the fulfillment of attention function, students can concentrate and be able to direct students' attention to the content of learning [11].

Compensatory Function

Based on assessment data from seven validators on compensatory function of chemistry game cardderived momen kappa 0.84 with very high category. It shows instructional media that developed can help students who are slow in learning to be easier to understand and enhance their understanding in subject naming chemical compound.With the fulfillment of compensatoty function, the media can help students remember information and enhance understanding of the subject[11] .

Chemistry game card that developed has fulfilled the three functions of an instructional media, ie cognitive function, attention function, and compensatory function. Kappa moment for the validity of the media that developed can be seen in Table 2. The validity of chemistry game cardas a instructional media visits of cognitive function, attention function, and compensatory function, and has a gain of kappa moment consecutive 0.84; 0.89; 0.84. Thus, instructional media has been developed in accordance with the assessment instrument (validation sheet). A product is said to be valid if the instrument can measure what should be measured [12].

Although the level validity of instructional media that developed was very high, but there were some cases that need to be fixed according to the suggestion of validators to make instructional media better. Revision of this chemistry game card was being developed, and then it was be tested to the students.

3.2.2 Practicality of Chemistry Game Card. Practicality of chemistry game card as a instructional media in subject naming chemical compound was viewed from the uses of product from limited trial involving practicality and enforceability of the products. Practicality assessment of chemistry game card as an instructional media in subject of naming chemical compound was responded by the students. Practicality of media that developed assessed by students based on the function of instructional media, namely compensatory function , attention function, and affective functions.

Compensatory Function

Based on data from student questionnaire, the results of practicality oncompensatory function of chemistry game cardobtained an average kappa moment 0.90 with very high category. Chemistry game card that developed was able to increase the interest and motivate students in order to achieve the learning objectives in subject naming chemical compound. Motivation also plays a role in the achievement of learning objectives to be achieved by connecting the meaningfulness of learning objectives [13]. Chemistry game cards greatly assisted students in recalling the the subject of naming chemical compounds that have been studied previously. In addition, the media developed was nice to used in the learning process in naming chemical compound, so the media that developed in compliance with the compensatory function as in [11].

Attention Function

The results of practicality on the attentional function of chemistry game cardobtained the average of kappa moment 0.91 with very high category. Rules of the game and the language used in the media as well as the type and size of font was obvious. In addition the design or color display on chemistry game card attracted students attention. Design or color image more attracted students than black and white design, so the media that developed compliance with the attentional functions [14].

Affective Functions

Based on data from student questionnaire, the results of practicality on affective function of chemistry game card obtained kappa moment 0.92 with very high category. Chemistry game card that developed was capable to increase student interest in learning subject naming chemical compound and fun to play. Chemistry game cards greatly assisted students in establishing understanding subject naming chemical compound. In addition, the media that developed could increase the activity of students in the learning process. If the media is able to increase the pleasure of students in learning, it can be said that it has meet the affective function [15]

Based on questionnaire data processing practicalities of students obtained moment kappa 0.91. It was revealed that the chemistry game cardas a instructional media has very high category for use in the learning process ofnaming chemical compound. In addition, the game cardpacked in a box and the right size, so it was easy to carry and use anywhere. The media can be used whenever desired, so the students can play the card so many times to repeat the subject of naming chemical compound that have been studied .

Based on this, it can be concluded that the chemistry game cardas a instructional media in subject of naming chemical compound can help students in repeat and solidify the understanding in learning of naming chemical compounds in a fun way.

4. Conclusion

Based on research result and data analyzed, it can be concluded that chemistry game card as an instructional media in subject of naming chemical compound that developed has category highly valid and highly practice.

Acknowledgments

The authors thank's to Dr. Latisma Dj, Dra. Syamsi Aini, M.Sc., Ph.D., and Yerimadesi, S.Pd. M.Si, chemistry teacher and students of tenth grade of SMA 5 Padang , as well as all those who have assisted in the completion of this research and preparation of this article .

References

[1] Silberman, Melvin L. 2006. *Active Learning 101 Cara Belajar Siswa Aktif.* Bandung: Nusamedia dan Nuansa
[2] Sadiman, Arief, dkk. 2012. *Media Pendidikan Pengertian, Pengembangan, dan Pemanfaatannya.* Jakarta: Raja Grafindo Persada.
[3] Sari, Kurnia Wening. 2014. *Pengembangan Game Edukasi Kimia Berbasis Role Playing Game (RPG) pada Materi Struktur Atom sebagai Media Pembelajaran Mandiri untuk Siswa Kelas X SMA di Kabupaten Purwerejo*, Jurnal *Pendidikan Kimia (JPK)*, vol 3 No.2
[4] Salmi, Hayatus. 2014. *Pembuatan Kartu Sistem Periodik Sebagai Media Permainan Pada Pembelajaran Kimia Di Kelas X SMA.* Skripsi, Jurusan Kimia, Universitas Negeri Padang
[5] Sudjana, Jamilah. 2015. *Kartu Kation-Anion sebagai Inovasi Media Pembelajaran pada Mata Pelajaran Kimia di Sekolah Menengah Atas.* Jurnal Lingkar Widyaiswara. Edisi 2 No 1. ISSN: 2355-4118
[6] Trianto. 2012. *Model Pembelajaran Terpadu: Konsep, Strategi dan Implementasinya dalam Kurikulum Tingkat Satuan Pendidikan (KTSP).* Jakarta: Bumi Aksara
[7] Boslaugh, Sarah & Watters, Paul A. 2008. *Statistics in a Nutshell, a desktopquick reference.* Beijing, Cambridge, Famham, Köln, Sebastopol, Taipei, Tokyo: O'reilly
[8] Sugiyono. 2008. *Metode Penelitian Pendidikan (Pendekatan Kuantitatif, Kualitatif dan R & D.* Bandung: Alfabeta
[9] Sanjaya, Wina. 2006. *Strategi Pembelajaran.* Jakarta: Kencana Prenada Media Group.
[10] Haryono. 2013. *Pembelajaran IPA yang Menarik dan Mengasyikkan: Teori dan Aplikasi PAIKEM.* Yogyakarta: Kepel
[11] Kustandi, Cecep dan Bambang Sudjipto. 2011. *Media Pembelajaran Manual dan Digital.* Bogor : Ghalia Indonesia
[12] Sukardi. 2011. *Evaluasi Pendidikan, Prinsip, Dan Operasionalnya.* Yogyakarta: Bumi Aksara
[13] Jalius, Ellizar. 2009. *Pengembangan Program Pembelajaran.* Padang: UNP Press
[14] Sudjana, N. dan Rivai, A. 2011. *Media Pengajaran.* Bandung: Sinar Baru Algesindo
[15] Arsyad, Azhar. 2010. *Media Pembelajaran.* Jakarta: Raja Grafindo Persada

IOP Conf. Series: Materials Science and Engineering **335** (2018) 012101 doi:10.1088/1757-899X/335/1/012101

Development of Scientific Approach Based on Discovery Learning Module

E Ellizar*, H Hardeli, S Beltris, R Suharni

Graduate Chemistry Department, Faculty of Mathematics and Science. Universitas Negeri Padang. Indonesia

*non_jalius@yahoo.com

Abstract: Scientific Approach is a learning process, designed to make the students actively construct their own knowledge through stages of scientific method. The scientific approach in learning process can be done by using learning modules. One of the learning model is discovery based learning. Discovery learning is a learning model for the valuable things in learning through various activities, such as observation, experience, and reasoning. In fact, the students' activity to construct their own knowledge were not optimal. It's because the available learning modules were not in line with the scientific approach. The purpose of this study was to develop a scientific approach discovery based learning module on Acid Based , also on electrolyte and non-electrolyte solution. The developing process of this chemistry modules use the Plomp Model with three main stages. The stages are preliminary research, prototyping stage, and the assessment stage. The subject of this research was the 10th and 11th Grade of Senior High School students (SMAN 2 Padang). Validation were tested by the experts of Chemistry lecturers and teachers. Practicality of these modules had been tested through questionnaire. The effectiveness had been tested through experimental procedure by comparing student achievement between experiment and control groups. Based on the findings, it can be concluded that the developed scientific approach discovery based learning module significantly improve the students' learning in Acid-based and Electrolyte solution. The result of the data analysis indicated that the chemistry module was valid in content, construct, and presentation. Chemistry module also has a good practicality level and also accordance with the available time. This chemistry module was also effective, because it can help the students to understand the content of the learning material. That's proved by the result of learning student. Based on the result can conclude that chemistry module based on discovery learning and scientific approach in electrolyte and non-electrolyte solution and Acid Based for the 10th and 11th grade of senior high school students were valid, practice, and effective.

1. Introduction

Learning is a condition deliberately designed to create an atmosphere of learning that provides an opportunity for students to build their own knowledge. Currently the Curriculum is the 2013 curriculum, which has a basic view where students are subjects who have the ability to actively seek, process, construct, and use their knowledge. The implementation of the 2013 Curriculum in the learning process is using a scientific approach. Scientific approach is more highlight the dimensions of observation, reasoning, invention, validation, and explanation of a truth, which means students are required to be actively involved in the learning process.

Chemistry is the study of the structure of matter, and the changes experienced by matter [1]. Chemistry is closely related to the scientific approach, because chemistry is inseparable from the scientific process that underlies the emergence of a scientific approach. This is in accordance with the opinion of The College Board which states that scientific shortcuts will support the direct involvement of students with chemical phenomena, which means that in studying chemistry is needed a scientific approach that includes the process of solving problems and critical thinking so that students really understand the chemical phenomena [2].

The process of applying the scientific approach in chemistry learning can be realized by using the discovery learning model. This model emphasizes the importance of understanding structures or important ideas to a discipline, through active student involvement in learning. Several international studies show that discovery learning is worthy of use in the chemistry learning process. Among others, by Ali GunayBalim and Olufunmilayo I [3].

The information summaries from questionnaires which was distributed to some high school students show that learning in school has started smoothly. Students learn by using the learning package and students' activity sheet (LKS) which contains learning materials that are verbal, so that in the learning process students tend to memorize. This resulted in a lack of student interest in the learning process, and it can be said that the direct involvement of students in the concept discovery has not been maximized. This is certainly a bit contrary to the demands of learning 2013, which essentially, requires students to be able to find their own concepts through the direction and guidance of the teacher.

Discovery as a learning model has certain syntax , namely (1) stimulation to stimulate students to explore relevant materials, (2) identify problems to find out what problems are faced through observation, questioning, etc., (3) data collection to interpret the data obtained, (4) verify data to verify data, and (5) summarize concepts, and principles on the material being studied . The learning steps of this discovery learning model require students to be actively involved in discovering the concepts of the material being studied. It is expected to help the students to improve their ability in understanding learning materials [4].

Results of research conducted by Mahmoud [5] show that the effect of learning using Discovery Learning Strategy can improve the metacognitive ability of children. The scientific approach in the learning process also consists of certain steps, namely (1) Observing , (2) Questioning, (3) Collecting information (Experimenting), (4) Associating, and (5) Communicating .

The combination of these steps is actualized in the developed module. The modules as teaching materials have components such as Teachers and Students Manuals, Worksheets, Work answer-sheet, Test Sheets and Test answer- Sheet.

Based on the above descriptions, it has been carried out development research to produce a product in the form of chemical module based on discovery learning with scientific approach on the material of electrolyte and non-electrolyte solution, also for acid and base solution. The quality of the product of development is determined by three categories namely the validity, practicality, and effectiveness of the product.

2. Methods

This type of research is a developmental study using the Plomp model, ranging from the preliminary research phase, the prototyping stage development phase, and the assessment stage [6]. To obtain discovery learning based learning module with valid scientific approach done preliminary research. Based on the results of the analysis in the preliminary research phase, the learning module is designed based on discovery learning with scientific approach.

The preliminary research phase consists of needs analysis, curriculum analysis, student analysis and concept analysis. In the development or prototype phase the module based discovery learning was designed with scientific approach. In the prototyping stage, a formative evaluation was performed. The prototyping stage consists of prototype 1: self -evaluation; prototype 2: expert validation and one-to-one evaluation; prototype 3: namely small group evaluation, and prototype 4 is field test. Discovery based learning module with scientific approach that has been designed followed by self-evaluation, and in revision validated by experts (expert judgement). Module validation was carried out by three chemistry lecturers and high school chemistry teachers. To find-out the practicality and effectiveness, the module performed at SMA N 2 Padang in X class MIA $_5$ and X MIA $_6$ for electrolyte and non-electrolyte solution modules and XI class MIA_1 and XI MIA_2 for Acid-Base modules. The research data was collected through validation sheet, teacher and student response questionnaire, interview sheet, and student test result in the experimental class and control class which then tested the hypothesis by using t test.

3. Results and Discussion

3.1. Initial Investigation Phase

Activities in the initial investigative phase begin with needs analysis, curriculum analysis, student analysis and concept analysis.

3.1.1. Needs Analysis. At this stage the collection of information about the problems contained in the learning chemistry. Needs analysis is done by interviewing high school teachers, and obtained information that the school has implemented the 2013 curriculum. The learning process is done as much as possible so that students can understand the concept through discussion, question and answer and lecture method by using textbooks from publishers and LKS. This resulted in the lack of involvement of students to find and build their own knowledge, so the tendency of students in learning is memorizing, which resulted in low understanding of the material learned. Student activity during the learning process has not been fully able to make students to be active and perform the learning steps it is recommended by the 2013 curriculum that is a scientific approach. This is because the commonly used teaching materials have not yet presented the learning steps required by the 2013 curriculum clearly.

3.1.2. Curriculum Analysis. The curriculum analysis is Core Competence and Basic Competence. This analysis aims to formulate the concepts of learning on the material of Electrolyte and Non Electrolyte Solutions and Acid Bases class X and XI SMA. After analysis followed by state of indicators, then proceed to determine the concepts to be studied. Based on the results of the analysis and then identified problems that are found to then need to develop teaching materials in the form of discovery learning-based module with a scientific approach.

3.1.3. Student Analysis. Student analysis is conducted to examine the character of students who become the target users of developed products. Students who were targeted in this research were X and XI grade high school students, who were on average aged 15-16 years. According to Piaget, at the age of more than 11 years, a person is at a formal stage of operation, where the student has the ability to think abstract, which is characterized by the ability to think deductively.

3.1.4. Concept Analysis. Concept analysis is conducted to identify, detail and systematically arrange the necessary concepts and be used as a reference in developing discovery learning module with scientific approach. Based on the indicators that have been formulated from KD, the next stage of concept analysis. The results of the analysis at this early stage of investigation is the basis for the authors to develop a discovery learning-based chemical module with a scientific approach. The modules developed are tailored to the demands of the curriculum and needs in the field.

3.2. Development Phase (Prototyping)

Phase 1 prototype is the stage of design or compilation of discovery based learning module with a scientific approach on the material of electrolyte and non electrolyte solution and acid base material. The preparation of this module is based on the results of the initial investigations that have been analyzed. There are three aspects that need to be considered in the design of the module is the interface, content and support (language and legibility). The contents design, display, and language that has been designed in the module is followed by self evaluation before discussed with the experts. In addition, the researchers also checked the suitability of the content of learning material modules that must exist in accordance with KD, indicators, and learning objectives. After the self-evaluation is done then the improvement. The results of these improvements resulted in prototype 2.

The result of self evaluation which has been done is continued with module validity activity by expert on module which has been designed at prototype stage 1. The module validated by chemistry lecturer and chemistry teacher of SMA. Aspects observed in the validation are components of content, constructs, language and graphics. During the validation process there are several revisions suggested by the validator. This recommendation becomes the basis for further product design revisions once revised, the validator gives an assessment of the draft. The values given by the validator, processed using the moment kappa. The result of data analysis of validity can be seen in Table 1. Table 1 shows that the developed chemical module is valid, it is seen from the total kappa moment value is 0.89 and 0.907.

Table 1. Results of validity of the electrolyte non-electrolyte solution and acid-base solution.

No	Measurement aspect	electrolite	Level of validity	Acid & Base	Level of validity
1	Content component	0,89	Very high	0.87	Very high
2	Construct component	0,91	Very high	0.80	High
3	Language component	0,80	high	0.96	Very high
4	Graffiti component	0.93	Very high	1.00	Very high
Total Validity		0,89	Very high	0.907	Very high

The validity of the content component is to see the compatibility of chemical modules with the curriculum, the suitability of the learning steps with the discovery learning model, and the accuracy of the material content presented in the module. Validation data shows that from the aspect of content component is in very high and high category. This validation result shows that the chemistry module developed in accordance with the predefined curriculum, and the steps of learning activity on the module is very appropriate with the discovery learning step. In addition, the material presented in the module is appropriate and appropriate.

The aspects of the construct component assessed in the chemical module validation process are the module systematics, and the interrelationship between the concepts and the problems presented. Validation data for both electrolyte and non-solvent modules are in very high category, while for acid-base modules in the high category. Thus it can be said that the developed chemical module is very systematic, and has interrelationships between concepts with the problems presented. This is because the module developed consists of discovery learning steps that guide the students to find the concept of the material they are learning.

Validation of language components aims to reveal the suitability of the language used in the module with the Indonesian rule. Validation also reveals whether the language used is easy to understand and can develop students' thinking skills. The linguistic component of the developed chemical module is in the category of high validity level. This is because the language used in the discovery learning-based chemical module in accordance with the demands of Indonesian rules. In addition, the language used is a simple language that is easy to understand and can improve students' thinking ability.

Validation to see the clarity of the drawings and illustrations ,in addition, this validation also aims to see the variations in the use of the size and typeface, lay out, and colors used in the module. Based on the data of the research, the level of component validity is very high. Based on this, it can be said that the pictures or models contained in the chemical module are very clear.. The layout of each component in the module is good and the colors used make the module more interesting.

The design of the valid module then goes to the one-to-one evaluation stage. Individual evaluation aims to see the clarity of the instructions and sentences contained in the module, whether understood or not by students. Individual evaluations were conducted on three student on X th and XI thgrade from the cluster category, moderate and low group. In this activity the researchers met directly with the students and conducted informal interviews to request their response to the developed chemical module. The interview result shows that there are some sentences that are still unclear and questioned by the students. This is the subject of further product revisions. The revised product design is called prototype 3.

Activity after prototype 3 is done by doing small group evaluation. This small group evaluation aims to see the practical level of modules when used on a small scale and the suitability of time allocations with time available. This evaluation was conducted on each of 6 students of class X and XI who have high ability level, medium and low. The process of conducting this evaluation is done like a regular learning process, students are led to find the concept of the material by filling out the activity sheet on the module. During the learning process, students look enthusiastic about the use of developed chemical modules, but there are some dubious parts for some students. This is then revised to become clearer and easier to understand. The product after this revision is called prototype 4.

The activity that is done after the learning process is done is to fill the questionnaire of practicality. This small group study was conducted 5 times meeting for acid base material, while the material of electrolyte and non electrolyte solution as much as two meetings. The researcher explains how to learning by using the module. The second, third, and fourth meetings of the learning process were smoother than previous meetings because students already had modular learning experiences at the first meeting. The questionnaire results show that the developed chemical modules are practical and can be used in accordance with the time available. This can be proved by the kappa moment value (K) of student data questionnaire that is 0.88 for electrolyte and non electrolyte solution. As for the acid-base solution, K = 0.86. This value is in the range of very high level of practice, meaning that the module produced is very practical. With the completion of prototype 4 then it can be tested more widely through experiment.

The next stage of this research is the assessment phase. The activities undertaken in this evaluation were to test the use of modules developed in the actual learning process.Evaluation of large groups (field test) in this study is still a limited test conducted in SMA Negeri 2 Padang for the two modules. This evaluation uses experimental research principle with Randomized Group Control Post-test Design. The population in this study is the students of class X SMA Negeri 2 Padang registered in the academic year 2016/2017 consisting of six classes X .. After the analysis, then get two classes of samples, class X MIPA5 and X MIPA6, where X MIPA5 set as the experimental class and X MIPA6 as the control class. The experimental class for XI grade is the XI MIPA 1 and the control class is XI MIPA 2. Experimental class, applied learning using discovery learning-based chemical module, while in control class applied learning without using module. Overall, large group evaluation is useful to determine the level of effectiveness and effectiveness of developed modules. This can be explained as follows.

The level of practicality is determined from the results of the questionnaire of student responses and teacher responses to the application of module usage in learning. Data obtained, processed by using kappa cohen analysis. The result of questionnaire analysis of student response as a whole can be seen in Table 2.

Table 2. Practicality Results on Field Test

No	Measurement aspect	K-electrolyte	Practicality level	K-acid base	Practicality level
1	Attraction	0,91	Very high	0.84	Very high
2	The Use Process	0,88	Very high	0.83	Very high
3	Evaluation	0.85	Very high	0.81	Very high
Total practicality		0,88	Very high	0.826	Very High

Table 2 shows that the module developed has a very high degree of practicality of the three assessed components. The first component is the aspect of attraction with a kappa moment value of 0.91 and 0.84. The module attraction aspect relates to the overall appearance of the module, the colour composition, the font size and the clarity of the usage instructions. This is in accordance with Haryono's opinion if students can focus fully on learning, this is proven to improve learning outcomes [7].

The second component is the aspect of the use process with the kappa moment value of 0.88 and 0.83. Evaluation of the practicality of the aspect of the use process is in very high category, in means the module can help students to find the concept of learning materials independently. This is because the discovery learning steps are arranged systematically and can lead students in the discovery of the concept. This statement is in accordance with one of the advantages of discovery learning model that encourages active student involvement and training self-learning students [2].

The third component is the evaluation aspect with the K= 0.85 and 0.81. The result of the analysis of practicality shows that the evaluation aspect is in the category of high level of practicability. Based on this it can be concluded that discovery learning-based chemistry module can help students improve understanding of learning materials well through the exercises contained in the module.

The overall kappa moment value is 0.88 and 0.826 is in very high category. Chemical modules developed have high practicality, because the module is a unit of learning activities are arranged in the form of a particular package unit. The developed chemical module can be said to be a complete package that contains learning activities to find concepts, and worksheets to establish understanding of student concepts.

The effectiveness of modules developed is seen from two indicators, namely student learning outcomes and student activities. Determination of the effectiveness of chemical modules through student learning outcomes is by comparing the students' learning outcomes of the experimental class and the control class. In the experimental class applied the learning by using discovery learning-based chemical module, while the control class applied as usual learning without using module. The results of the learning process seen from the results of student learning in the experimental class and control classes are given test results of learning (final test) in the form of written tests in the form of multiple choices. To see whether the use of the module can affect student learning outcomes, hypothesis testing is performed.

The predefined hypothesis, i. e learning outcomes of students who are learning by using discovery learning-based chemical module significantly higher than the learning outcomes of students who learn without using the module.

Table 3. Hypothesis Test Results for electrolyte and non electrolyte solution module

Class	N	$\bar{x}$	S	S^2	t_{hitung}	t_{tabel}
Experiment	35	85.83	8.08	65.23	4.15	
Control	35	77.49	8.66	75.05		1.67

Table 4. Hypothesis Test Results for acid-base solutions modules

Kelas	N	$\bar{x}$	S	S^2	t_{cal}	t_{tab}
Experiment	36	85.11	10.3696	107.5032	3.83	
Control	36	76.11	10.8385	117.4731		1.67

The test result for the two modules shows the t-calc> t-table, meaning the hypothesis in this study is accepted, where there is a significant difference between the learning result of the students of the experimental class and the control class. Based on this it can be concluded that students' learning outcomes by using chemical modules are significantly higher than student learning outcomes without using modules. This result was in accordance with Balim's research results [8] which states that Discovery Learning can improve academic achievement, inquire scores and student retention scores.

4. Conclusion

From the process and the results of this study, it can be concluded that discovery learning-based chemical module with scientific approach on the material of electrolyte and non-electrolyte solution for X Grade and on Acid base Solution for XI grade can be categorized as valid, practical and effective. This chemical module can be used in a learning process without significant obstacles.

Acknowledgements

The authors thank's to LP2M Universitas Negeri Padang, Chemistry teacher and students of tenth and eleventh grade of SMAN 2 Padang, as well as those who have assisted in the completion of the research and preparation on this article.

References

[1] Hosnan, M. 2014. *Pendekatan Saintifik dan Kontekstual dalam Pembelajaran Abad 21.* Bogor: Ghalia Indonesia.
[2] The Colege Board. 2012. *AP Chemistry Guided-Inquiry Experiments: Applying the Science Practice (Teacher Manual).* New York: The College Board.
[3] Oloyede, Olufunmilayo I., (2010) Comparative Effect of the Guided Discovery and Concept Mapping Teaching Startegies on Sss Student's Chemistry Achievement. *Humanity & Social Science Journal*, ISSN 1818-4960, **5**(1): 01-06.
[4] Mulyasa, E. 2014. *Guru dalam Implementasi Kurikulum 2013.* Bandung: Remaja Rosda Karya.
[5] Mahmoud, Abdelrahman Kamel Abdelrahman. 2014. The Effect of Using Discovery Learning Strategy in Teaching Grammatical Rules tu first year General Secondary Student on Developing Their Achievement and Metacognitive Skill. *International Journal of Innovation and Scientific Research. ISSN 2351-8014. 5(2): 146-153*
[6] Plomp, Tjeerd dan Nienke Nieveen. 2007. *An Introduction to Educational Desaign Research.* Netherlands: SLO
[7] Haryono. 2013. *Pembelajaran IPA yang Menarik dan Mengasyikkan.* Yogyakarta: Kepel Press.
[8] Balim, A., G. (2009). The Effects of Discovery Learning on Students' Success and Inquiry Learning Skills. *EgitimArastirmalari-Eurasian Journal of Educational*

Validity and Reliability Testing of an e-learning Questionnaire for Chemistry Instruction

G Guspatni[1*], Y Kurniawati[2]

[1]Department of Chemistry, Universitas Negeri Padang, Indonesia
[2]Department of Mathematics, Universitas Negeri Padang, Indonesia

[*]guspatni.indo@gmail.com

Abstract. The aim of this paper is to examine validity and reliability of a questionnaire used to evaluate e-learning implementation in chemistry instruction. 48 questionnaires were filled in by students who had studied chemistry through e-learning system. The questionnaire consisted of 20 indicators evaluating students' perception on using e-learning. Parametric testing was done as data were assumed to follow normal distribution. Item validity of the questionnaire was examined through item-total correlation using Pearson's formula while its reliability was assessed with Cronbach's alpha formula. Moreover, convergent validity was assessed to see whether indicators building a factor had theoretically the same underlying construct. The result of validity testing revealed 19 valid indicators while the result of reliability testing revealed Cronbach's alpha value of .886. The result of factor analysis showed that questionnaire consisted of five factors, and each of them had indicators building the same construct. This article shows the importance of factor analysis to get a construct valid questionnaire before it is used as research instrument.

1. Introduction

E-learning is defined as learning with technology using Internet connection; be it web-based, web-distributed or web-capable learning [1]. E-learning can be done both synchronously where all learners are online at the same time during instruction and asynchronously where learners can choose the time of accessing e-learning [1][2]. E-learning offers several advantages in teaching and learning including scheduling flexibility, greater and just in-time access to knowledge, personalization and diversity, tracking improvement, information overload lessening, and cost effectiveness [2]-[5]. In chemistry instruction, e-learning increases students' interest to learn, improves students' mastery of chemistry problem solving, and has a significant positive impact on students' performance [6]-[10].

Students' attitude and satisfaction toward e-learning are affected by computer anxiety, instructor's attitude toward e-learning, e-learning course flexibility, e-learning course quality, and diversity in assessments [11][12]. Other study by Keller and Cernerud [13] revealed that the implementing strategy is more significant in influencing students' attitude toward e-learning than students' previous knowledge of computers, attitude toward new technologies, gender, age, and learning style. Challenges in e-learning implementation appear to be varied and depending on the context. In this study, e-learning was given to college students at Universitas Negeri Padang. Students' perceptions were studied by using questionnaire distributed after they used e-learning. What gives value to this study is the setting that will bring a new dimension to existing literatures.

2. Theoretical background

2.1. Validity and reliability

Validity and reliability are important concepts to consider when designing a research instrument. Validity refers to appropriateness, meaningfulness, correctness, and usefulness of an instrument or a procedure that it measures what it is supposed to measure. Reliability refers to the consistency of scores or answers from respondents over time. Reliability is a necessary but not sufficient condition for validity. It means that an instrument must be reliable for it to be valid, but a reliable instrument does not necessarily guarantee that it is a valid one [14]-[16].

There are many types of validity including content validity, criterion-related validity, construct validity, internal validity, external validity, concurrent validity, face validity, jury validity, predictive validity, consequential validity, systemic validity, catalytic validity, ecological validity, descriptive validity, interpretive validity, theoretical validity and evaluative validity [15]. Fraenkel & Wallen [16] categorize validity based on the evidence that supports validity into content-related evidence validity, criterion-related evidence validity, and construct-related evidence validity. Best & Khan [14] name criterion-related validity as evidence based on relations to other variables validity while construct validity as evidence based on internal structure validity.

Content validity refers to the extent an instrument fairly and comprehensively covers the domain or items that it purports to cover. Criterion-related validity refers to the extent one particular instrument is related to another external relevant criterion. Predictive validity and concurrent validity belong to this group of validity. Predictive validity is achieved if the data acquired at the first round of research correlate highly with data acquired at a future date. Concurrent validity can be demonstrated simultaneously with another instrument. Next, construct validity is sought on the 'operationalized' forms of a construct referring to the extent to which a measure adequately assesses the construct it purports to assess [15]. Construct validity has three components named convergent validity, discriminant validity and nomological validity [17]. Convergent validity implies that different methods for researching the same construct should give a relatively high inter-correlation while discriminant validity suggests that using similar methods for researching different constructs should yield relatively low inter-correlations.

Reliability can be defined into three principal types. First, reliability as stability, which is a measure of consistency over time and similar samples. Correlation coefficients can be calculated for the reliability of pre- and post-tests using the Spearman or Pearson statistic as appropriate. Second, reliability as equivalence that can be achieved using equivalent forms of instrument (also known as alternative forms) and through inter-rater reliability. Third, reliability as internal consistency that allows the tests to be run once. This type of reliability can be calculated by split-half method using Spearman—Brown formula, Kuder-Richardson approaches for coefficient of consistency, and Cronbach alpha formula for alpha coefficient. Moreover, standard error of measurement can also be used to measure reliability by showing how much we can expect the obtained score to differ from the individual's true score [14]-[16].

2.2. Questionnaire and its development

Questionnaire is a common instrument used in human studies especially in social and educational sciences. Questionnaire is defined as an instrument used to collect research data using a form containing set of questions that will be answered by respondents [14][18]. Two kinds of questionnaire are closed form and open form questionnaires. The closed form questionnaire calls for short response and dichotomous response, multiple choice and check-mark responses on the list of suggested ones, or rating scale and rank-order response. An example of widely used closed form questionnaire is Likert scale questionnaire. Open form questionnaire on the other hand allows respondents to response in their own words, explain, and qualify their response [14][15]. Closed form questionnaire is easy to code and classify while the open one is not. Nevertheless, closed form questionnaire gives bias and limitation to respondents on giving response [16].

2.3. Questionnaire validity and reliability

Unlike psychological tests and inventories, questionnaire has a very limited purpose and short life and is administered to a limited population. Yet its validity and reliability can be improved by several ways. To get a valid questionnaire, researchers must follow the protocol of questionnaire design and development. Then, get suggestions from colleagues and experts to reveal ambiguities or items that can be removed. Rate from the experts can effectively show the significant aspects of questionnaire's purpose thus providing estimates of content validity. Next, do follow-up observations of respondent behavior at the present time or at some time in the future to estimate its predictive validity.

To infer reliability of the questionnaire, a second administration of the instrument with a small subsample can be done. It is obtained by comparing the responses with those of the first. Reliability may also be estimated by comparing responses of an alternate form with the original form [14].

Ideally, questionnaire has to be composed and tried out over several times in a pilot work to ascertain that it can do the job for which it is intended. Sometimes we can borrow or adapt questionnaires from other researches. But there still remains the task of making quite sure that these will 'work' with our population and will yield the data we require [18]. For time and fund reasons, many researchers test validity and reliability with statistical analysis in a single instance.

3. Methodology

3.1. Context

Participants of this study were college students at Department of Chemistry, Universitas Negeri Padang. Students were instructed to learn chemistry through e-learning system. They were given cash to access e-learning off-campus and were allowed to use computer and Internet connection provided by university as well. After studying chemistry through e-learning, students were asked to fill in the questionnaire anonymously at home.

3.2. The questionnaire

Questionnaire used to evaluate e-learning implementation in chemistry learning was adopted from several models in technology and e-learning studies. Revision on some indicators was done to adapt the research's setting. English wordings were translated into Indonesian cautiously. Since the questionnaire was adopted from peer-reviewed articles, it was not pilot-tested beforehand. In general, development, distribution and analysis of the questionnaire were done appropriately as the rules and ethics suggest. Avoiding complex and leading questions, giving informed consent, allowing anonymously return, confirming truthful responses, and giving cash were done to get genuine response and high return rate [14][15].

The questionnaire distributed consisted of 20 indicators. It used Likert scales ranging from very agree to very disagree. The indicators were (1) e-learning website had organized appearance, (2) e-learning website had simple appearance, (3) e-learning website was easy to use, (4) e-learning website was understandable, (5) e-learning website run/functioned well, (6) e-learning eased me to study anytime I want, (7) e-learning eased me to learn anyway I want, (8) e-learning helped me find further concept, (9) I understood learning demand in e-learning, (10) I was sure I could learn well in e-learning, (11) I was sure I could understand the lesson in e-learning, (12) I was happy learning with e-learning, (13) I was happy I could learn at the time I want, (14) I was happy that with e-learning, I did not have to study in classroom, (15) learning content in e-learning website was adequate, (16) e-learning was rich in useful information, (17) e-learning made me want to know more about an information, (18) e-learning helped me order my own learning, (19) I want to learn with e-learning, (20) I want to learn chemistry concept with e-learning. These items were intentionally ordered based on their adjacent on linguistic meaning.

3.2.1. Item validity and reliability testing of the questionnaire

51 questionnaires returned by respondents were coded and analyzed. Three respondents left one item of the questionnaire blank. For statistical reason, data from those three respondents were not used for further analysis. 48 questionnaires were then analysed with SPSS to determine its construct validity and internal consistency. Construct validity of a questionnaire can be demonstrated by using content analysis, factor analysis, multi-trait/multi method studies, and correlation coefficient [19]. In this

IOP Conf. Series: Materials Science and Engineering **335** (2018) 012102 doi:10.1088/1757-899X/335/1/012102

study, correlation was calculated with Pearson's correlation for data were assumed to be normally distributed (parametric testing was done). Validity of each indicator was determined from the significance of its correlation with the sum score (Item-total correlation) while reliability of the questionnaire was determined as an internal consistency and assessed with Cronbach's alpha formula.

Table 1. Item-total correlation

	Indicator	1	2	3	4	5	6	7	8	9	10	11	12	13	14	15	16	17	18	19	20
SUM	r	.29*	.02	.46**	.42**	.33*	.67**	.70**	.55**	.54**	.69**	.73**	.59**	.61**	.53**	.61**	.55**	.49**	.78**	.61**	.60**
	Sig. (2-tailed)	.05	.91	.00	.00	.02	.00	.00	.00	.00	.00	.00	.00	.00	.00	.00	.00	.00	.00	.00	.00
	N	48	48	48	48	48	48	48	48	48	48	48	48	48	48	48	48	48	48	48	48

**. Correlation significant at the .01 level. *. Correlation significant at the .05 level (2-tailed).

Item-total correlation of each indicator is shown in Table 1. All indicators but one had moderate to high correlation with the sum score. Indicator 2, the exception, had correlation coefficient of .02 and significance of .91 with the sum score. It can be concluded that indicator 2 had insignificant correlation with the sum score and thus can be eliminated (called invalid indicator).

Table 2. Reliability statistic

Cronbach's Alpha	Cronbach's Alpha Based on Standardized Items	N of Items
.874	.872	20
.886	.884	19

To make sure whether or not indicator 2 should be eliminated, reliability statistic and item-total statistics were then analyzed. The value of Cronbach's alpha for 20 indicators was .874 (see Table 2). If indicator 2 was deleted, the Cronbach's alpha value increased to .886. The increasing is quite high (.012) as compared to the change of Cronbach's alpha value if other indicators such as indicator 1 and 5 were deleted (increasing of .002 and .001). The big increasing of Cronbach's alpha value if an item is deleted denotes that the indicator is appropriate for elimination. As a comparison, if Cronbach's alpha was directly calculated from 19 indicators, the value was .886.

Table 3. Item total statistics

	Scale Mean if Item Deleted	Scale Variance if Item Deleted	Corrected Item-Total Correlation	Squared Multiple Correlation	Cronbach's Alpha if Item Deleted
1	65.2917	84.637	.221	.364	.876
2	=65.6667	88.057	-.072	.552	.886
3	65.1667	81.333	.392	.685	.871
4	65.25	81.894	.346	.774	.873
5	65.4792	83.361	.256	.472	.875
6	65.9792	76.276	.611	.824	.863
7	66.125	76.707	.649	.793	.862
8	65.6667	79.631	.488	.711	.868
9	65.9375	79.805	.473	.589	.869
10	66.2708	76.585	.632	.816	.863
11	66.375	77.601	.682	.74	.862
12	66.1875	79.517	.532	.624	.867
13	65.6458	77.255	.535	.483	.866
14	66.25	77.553	.438	.652	.871
15	65.9375	77.847	.538	.717	.866
16	65.6667	80.355	.493	.689	.868
17	65.7292	81.563	.434	.421	.87
18	66.0208	74.404	.737	.798	.858
19	66.25	78.83	.546	.78	.866
20	66.0417	78.849	.538	.62	.867

3.2.2. Construct validity testing with factor analysis

In this study construct validity was assessed through convergent validity criteria. In order to demonstrate convergent validity, items that measure the same trait should correlate highly with one another. Therefore, the traits/components/factors that the questionnaire had should be determined through factor analysis. Existing relationship between indicators and big sample size are main conditions for factor analysis. Several literatures cite that sample size should be big enough; at least 5 to 15 participants per item. However, sample size is less crucial for factor analysis to the extent that the communalities of items with the other items are high, or at least relatively high [20]. Although sample size in this study was 48, the Kaiser-Meyer-Olkin measure of sampling adequacy (KMO) and the significance of Bartlett test signaled that factors could be extracted. In this study the KMO index was .742 (greater than .50) indicating sufficient items for each factor. The significance of Bartlett test was .000 (less than .05) indicating that the correlation matrix is significantly different from an identity matrix in which correlations between variables are all zero. Both test indicated factorability of the data (see Table 4).

Table 4. KMO and Bartlett's Test

Kaiser-Meyer-Olkin Measure of Sampling Adequacy.		.742
Bartlett's Test of Sphericity	Approx. Chi-Square	486.522
	df	171
	Sig.	.000

In order to extract factor, the eigenvalue greater than 1 criterion was used. As can be seen in Table 5, there are five components/factors having eigenvalue bigger than one. More than half of the variance was accounted for by the five factors (% cumulative = 69.286).

Table 5. Total Variance Explained

Component	Initial Eigenvalues			Extraction Sums of Squared Loadings			Rotation Sums of Squared Loadings		
	Total	% of Variance	Cumulative %	Total	% of Variance	Cumulative %	Total	% of Variance	Cumulative %
1	6.555	34.499	34.499	6.555	34.499	34.499	3.075	16.185	16.185
2	2.445	12.866	47.364	2.445	12.866	47.364	2.767	14.563	30.748
3	1.717	9.035	56.399	1.717	9.035	56.399	2.577	13.564	44.312
4	1.337	7.039	63.439	1.337	7.039	63.439	2.404	12.650	56.962
5	1.111	5.847	69.286	1.111	5.847	69.286	2.342	12.324	69.286
6	.921	4.848	74.134						

Extraction Method: Principal Component Analysis.

If the five components were to be retained, communalities values should be taken into account. Table 6 shows initial and extraction communalities of all indicators. The communalities in the Extracted column explain how much variance each indicator has in common with the five factors kept. Almost all of communalities after extraction had values of .50 or higher (except indicator 1,9,17), while the average communality was .69. It means that almost all of the indicators had common variance with the five factors retained. The retained factors could account big part of the variance, and adding more factors was not an obligation.

Table 6. Communalities

	Initial	Extraction
I1	1.000	.369

I3	1.000	.745
I4	1.000	.818
I5	1.000	.640
I6	1.000	.776
I7	1.000	.798
I8	1.000	.809
I9	1.000	.464
I10	1.000	.806
I11	1.000	.784
I12	1.000	.628
I13	1.000	.556
I14	1.000	.694
I15	1.000	.731
I16	1.000	.776
I17	1.000	.467
I18	1.000	.732
I19	1.000	.881
I20	1.000	.691

Extraction Method: Principal. Component Analysis

The next step was determining which item belonged to which factor. It was determined by checking the relationship between an indicator and a factor with factor loading values. A factor loading greater than .30 [21] was considered sufficiently high to assume a strong relationship between an indicator and a factor. Table 7 shows factor loading of each indicator on all of the five factors/components. The decision of which factor an indicator belongs to is by choosing the factor/component on which the factor loading is the greatest. For example, indicator 1, 3, 4, and 5 had the greatest factor loading on component/ factor 3, thus belonged to this factor. Indicator 6,7, 12, 13, and 14 had the greatest factor loading on component/ factor 1 thus belonged to factor 1. Indicator 8,9,10, and 11 belonged to factor 4; indicator 15,16,17 and 18 belonged to factor 2; and indicator 19 and 20 belonged to factor 5.

Table 7. Rotated Component Matrix[a]

	Factor				
	1	2	3	4	5
I1	-.033	.227	.543	.059	-.134
I3	.109	-.001	.820	.246	.029
I4	.181	-.159	.852	.086	.165
I5	.030	.069	.773	-.113	.155
I6	.821	.146	.038	.250	.127
I7	.781	.043	.098	.404	.116
I8	.089	.362	.179	.781	-.170
I9	.241	.093	.177	.571	.200
I10	.231	.393	-.153	.676	.342
I11	.319	.129	.044	.650	.491

I12	.534	.225	-.203	.161	.475
I13	.667	.037	.256	.150	.147
I14	.636	.398	.004	-.276	.233
I15	.250	.805	-.006	.137	.039
I16	-.035	.836	.127	.227	.094
I17	.110	.632	.012	.136	.193
I18	.479	.572	.109	.195	.356
I19	.164	.249	.036	.098	.884
I20	.239	.079	.228	.131	.747

Extraction Method: Principal Component Analysis. Rotation Method: Varimax with Kaiser Normalization. a. Rotation converged in 7 iterations.

To cross-validate the questionnaire, convergent validity was examined based on the factors obtained. A convergent valid questionnaire contains indicators that have high correlation with each other in the same trait. As can be seen in Table 8, all indicators in the same trait had significant and higher correlation with other indicators in the same trait than they did with indicators in different traits. It obviously occurred in factor 3, factor 4, and factor 5. Three exceptions occurred in factor 1. Indicator 6 and indicator 12, 13, 14 had correlation coefficient less than that of indicator 6 with indicator 18. Indicator 7 and 14 had correlation coefficient less than that of indicator 7 and indicator 8, 9, 10, 11, 18, and 20. Indicator 12 and 13 had correlation coefficient less than that of indicator 12 and indicator 10,11, 17,18, and 19. Indicator 13 and 14 had correlation coefficient less than that of indicator 13 and indicator 11 and 18. One exception occurred in factor 2 as indicator 17 and 18 had correlation coefficient less than that of indicator 17 and indicator 6, 7, 12, 14, 9, 10, and 11. Nevertheless, three of five factors that satisfied the convergent validity had implied construct validity of the questionnaire.

Table 8. Inter-item correlation

	1	3	4	5	6	7	12	13	14	8	9	10	11	15	16	17	18	19	20
1	1.00																		
3	.33	1.00																	
4	.29	.74	1.00																
5	.31	.48	0.59	1.00															
6	-.01	.18	0.19	0.08	1.00														
7	.14	.23	0.28	0.05	0.81	1.00													
12	-.05	-.03	-0.01	0.04	0.53	0.48	1.00												
13	.08	.30	0.26	0.23	0.53	0.48	0.33	1.00											
14	.03	.06	0.11	0.08	0.37	0.37	0.49	0.38	1.00										
8	.17	.30	0.19	0.14	0.30	0.38	0.20	0.19	0.00	1.00									
9	.03	0.30	0.25	0.07	0.31	0.39	0.22	0.36	0.13	0.34	1.00								
10	.07	0.08	-0.03	-0.05	0.39	0.48	0.55	0.27	0.28	0.63	0.39	1.00							
11	.09	0.25	0.17	0.09	0.45	0.54	0.53	0.44	0.29	0.49	0.49	0.74	1.00						
15	.09	0.10	-0.06	0.03	0.33	0.27	0.25	0.23	0.52	0.42	0.25	0.50	0.32	1.00					
16	.15	0.16	0.00	0.11	0.22	0.13	0.18	0.18	0.17	0.44	0.28	0.39	0.29	0.64	1.00				
17	.11	0.11	-0.03	0.03	0.33	0.26	0.36	0.13	0.19	0.28	0.20	0.35	0.22	0.37	0.51	1.00			
18	.09	0.14	0.22	0.18	0.60	0.50	0.54	0.39	0.51	0.31	0.49	0.56	0.47	0.54	0.58	0.46	1.00		

IOP Conf. Series: Materials Science and Engineering **335** (2018) 012102 doi:10.1088/1757-899X/335/1/012102

19	.00	0.11	0.18	0.15	0.34	0.30	0.54	0.28	0.38	0.06	0.21	0.52	0.59	0.32	0.31	0.30	0.53	1.00	
20	.07	0.22	0.32	0.20	0.41	0.44	0.33	0.36	0.25	0.08	0.36	0.31	0.45	0.22	0.22	0.30	0.42	0.67	1.00

4. Discussion

There have been many studies on e-learning. Most employed questionnaire using various indicators to assess factors affecting its implementation. In this study, questionnaire consisting of 20 indicators were used to assess students' perceptions on e-learning. Result showed that indicator 2 stating "e-learning website had simple appearance" was invalid because it did not have significant correlation (and thus contribution) with the sum score. The exclusion of indicator 2 was supported by higher Cronbach's alpha value than it was if the indicator was included for reliability statistic.

Inter-item correlation of 19 valid indicators was used to cross validate construct validity of the questionnaire. It was done after factor analysis revealed the number of factors and indicators that belonged to each of them. There were five factors with number of indicators ranged from two to five.

The first factor contained five indicators; (I6) e-learning eased me to study anytime I want, (I7) e-learning eased me to learn anyway I want, (I12) I was happy learning with e-learning, (I13) I was happy I could learn at the time I want, (I14) I was happy that with e-learning, I did not have to study in classroom. Statistically, half indicators had high correlation coefficient among them and higher correlation than that with most other indicators in different factors. Theoretically, those indicators shared the same trait that relates to the ease or enjoyment on using e-learning.

The second factor consisted of four indicators; (I15) learning content in e-learning website was adequate, (I16) e-learning was rich in usefull information, (I17) e-learning made me want to know more about information, (I18) e-learning helped me order my own learning. Statistically, most indicators had high correlation coefficient among them and higher correlation than that with almost all other indicators in different factors. Theoretically, those indicators shared the same trait that relates to the usefulness of e-learning.

The third factor consisted of four indicators; (I1) e-learning website had organized appearance, (I3) e-learning website was easy to use, (I4) e-learning website was understandable, (I5) e-learning website run/functioned well. Statistically, all indicators had high correlation coefficient among them and higher correlation than that with all other indicators in different factors. Theoretically, those indicators shared the same trait that relates to appearance or design of e-learning".

The fourth indicator contained four indicators; (I8) e-learning helped me find further concept, (I9) I understood learning demand in e-learning, (I10) I was sure I could learn well in e-learning, (I11) I was sure I could understand the lesson in e-learning. Statistically, all indicators had high correlation coefficient among them and higher correlation than that with all other indicators in different factors. Theoretically, those indicators shared the same trait that relates to self-efficacy of students in studying with e-learning.

The last factor (factor 5) consisted of two indicators; (I19) I want to learn with e-learning and (I20) I want to learn chemistry concept with e-learning. Statistically, the two indicators had high correlation coefficient between them and higher correlation than that with all other indicators in different factors. Theoretically, those indicators shared the same trait that relates to the intention to use e-learning".

Analysis that had been done shows that several items that were initially grouped into certain variable or factor should be grouped into other factor. Researchers cannot necessarily group items having similar meaning in the same factor. Analysis of the obtained data should be statistically done.

5. Conclusion

Item validity testing of the questionnaire revealed 19 valid indicators out of 20 indicators initially distributed. The exclusion of invalid indicators showed significant higher Cronbach's alpha value. Cronbach's alpha value was .886 revealing a high reliability (Cronbach's alpha > .70) of the questionnaire. There were five factors extracted through factor analysis that relate to enjoyment, usefulness, design, self-efficacy, and intention to use e-learning. In general, convergent validity was also accomplished through out the indicators.

References

[1] Nichols, M. (2003). A theory for eLearning. *Educational Technology & Society, 6*(2), 1-10.
[2] Welsh, E. T., Wanberg, C. R., Brown, K. G., & Simmering, M. J. (2003). E-learning: emerging uses, empirical results and future directions. International Journal of Training and Development, 7(4), 245-258.
[3] Agariya, A. K., & Singh, D. (2012). e-Learning quality: Scale development and validation in Indian context. Knowledge Management & E-Learning: An International Journal (KM&EL), 4(4), 500-517.
[4] Gunasekaran, A., McNeil, R. D., & Shaul, D. (2002). E-learning: research and applications. Industrial and Commercial Training, 34(2), 44-53.
[5] Waight, C. L., Willging, P. A., & Wentling, T. L. (2002). Recurrent Themes in e-Learning: a meta-analysis of major e-Learning reports. Unpublished work.
[6] Awad, B. (2014). Empowerment of teaching and learning chemistry through information and communication technologies. African Journal of Chemical Education, 4(3), 34-47.
[7] He, Y., Swenson, S., & Lents, N. (2012). Online video tutorials increase learning of difficult concepts in an undergraduate analytical chemistry course. Journal of Chemical Education, 89(9), 1128-1132.
[8] O'Sullivan, T. P., & Hargaden, G. C. (2014). Using structure-based organic chemistry online tutorials with automated correction for student practice and review. Journal of Chemical Education, 91(11), 1851-1854.
[9] Eichler, J. F., & Peeples, J. (2013). Online homework put to the test: a report on the impact of two online learning systems on student performance in general chemistry. Journal of Chemical Education, 90(9), 1137-1143.
[10] Dori, Y.D., Barak, M., & Adir, N. (2003). A Web-based chemistry course as a means to foster freshmen learning. Journal of Chemical Education, 80(9), 1084.
[11] Bertea, P. (2009). Measuring students'attitude towards e-learning: a case study. In Conference proceedings of" eLearning and Software for Education"(eLSE) (No. 01, pp. 417-424).
[12] Sun, P. -C., Tsai, R. J., Finger, G., Chen, Y.–Y., & Yeh, D. (2007). What drives a successful e-Learning? An empirical investigation of the critical factors influencing learner satisfaction. Computers & Education xxx (2007) xxx-xxx
[13] Keller, C., & Cernerud, L. (2002). Students' perceptions of e-learning in university education. Journal of Educational Media, 27(1-2), 55-67.
[14] Best, J.W., & Kahn, J.V. (2006). Research in education. (10th ed.). USA: Pearson Education Inc.
[15] Cohen, L., Manion, L., & Morrison, K. (2000). Research methods in education. (5th ed.). London: RoutledgeFalmer.
[16] Fraenkel, J.R., & Wallen, N.E. (2003). How to design and evaluate research in education. (6th ed.). New York: McGraw-Hill
[17] Zait, A., & Bertea, E. (2011). Methods for testing discriminant validity. Management & Marketing, IX(2), 217-224.
[18] Oppenheim, A.N. (1992) Questionnaire Design, Interviewing and Attitude Measurement. London: Pinter Publishers Ltd.
[19] Brown, J. D. (2000). What is construct validity?. Shiken: JALT Testing & Evaluation SIG Newsletter, 4 (2), 8 – 12.
[20] Leech, N.L., Barret, K. C., & Morgan, G.A. (2005). SPsS for Intermediate Statistics: Use and Interpretation. (2nd ed.). New Jersey: Lawrence Erlbaum Associates, Inc
[21] Kline, P. 1994. *An easy guide to factor analysis*. London: Routledge.

IOP Conf. Series: Materials Science and Engineering **335** (2018) 012103 doi:10.1088/1757-899X/335/1/012103

Effect of Chemistry Triangle Oriented Learning Media on Cooperative, Individual and Conventional Method on Chemistry Learning Result

L Latisma D, W Kurniawan, S Seprima, E S Nirbayani, E Ellizar, H Hardeli
Chemistry Department, Faculty of Mathematics and Science, Universitas Negeri Padang

*latismadj@gmail.com

Abstract. The purpose of this study was to see which method are well used with the Chemistry Triangle-oriented learning media. This quasi experimental research involves first grade of senior high school students in six schools namely each two SMA N in Solok city, in Pasaman and two SMKN in Pariaman. The sampling technique was done by Cluster Random Sampling. Data were collected by test and analyzed by one-way anova and Kruskall Wallish test. The results showed that the high school students in Solok learning taught by cooperative method is better than the results of student learning taught by conventional and Individual methods, both for students who have high initial ability and low-ability. Research in SMK showed that the overall student learning outcomes taught by conventional method is better than the student learning outcomes taught by cooperative and individual methods. Student learning outcomes that have high initial ability taught by individual method is better than student learning outcomes that are taught by cooperative method and for students who have low initial ability, there is no difference in student learning outcomes taught by cooperative, individual and conventional methods. Learning in high school in Pasaman showed no significant difference in learning outcomes of the three methods undertaken.

1. Introduction

Learning is the process of communication between teachers, learners, and learning resources ". Learning media is one that can be used to convey a message, stimulate the mind, feelings, attention, and willingness of students so as to encourage the learning process. Levie and Lentz in Sanaky (2009: 6-7) said that "four functions of instructional media especially on visual media that is attention function, affective function, cognitive function and compensatory function.

Learning can be effective when all components that play a role in the learning process support each other in order to achieve the goal. One of the most influential components in today's learning is the use and selection of appropriate instructional media in order to improve students' skills. According to Sadiman et al (2006: 17) the appropriate use of media can overcome the passive attitude of students, thus causing excitement of students in learning.

Each individual has different levels of ability and background of experience. Different levels of ability possessed by each individual sometimes make it difficult to cooperate with others in the group; because that is the case that individual activities are effective. Each individual has a different type or learning style. Individual learning activities can be intensified according to individual learning styles. In individual learning students are required to be able to learn independent. The positive side of the use of individual

IOP Conf. Series: Materials Science and Engineering **335** (2018) 012103 doi:10.1088/1757-899X/335/1/012103

learning is the awakening of student self-confidence, Students become independent in carrying out the learning, and students do not have dependence on others dently, without any cooperation with others.

In addition, individual learning, cooperative / group learning is a learning that can and often is used by educators. This learning comes from the concept that students will more easily find and understand difficult concepts if they are discussing with their friends. Cooperative learning can cause psychological elements of students to become aroused and become more active. Theoretically this is caused by a sense of togetherness in a group, so that students can easily communicate with a more simple language. During discussion the memory function of the students will become more active, more enthusiastic and courageous to express an opinion.

The conventional approach is a learning approach that is very often done in schools. The conventional approach that is meant here was a learning approach that is more teacher centered, done by combining learning methods of lectures and question and answer. The source of learning in conventional learning more in the form of verbal information obtained from the book / explanation of teachers (2007: 294) says that the concepts in chemistry are studied in three levels of representation, namely macroscopic, submicroscopic (molecular) and symbolic. These three levels of representation are also known as the Chemistry Triangle. Macroscopic representation is a representation that describes most of the nature of real phenomena and is seen in the experience of everyday students. A submicroscopic representation (molecular) is an explanation of phenomena at the particle level (atom, molecule or ion). Symbolic representation (symbol) is a representation using chemical symbols, formulas, equations, models and computer animations to symbolize substances. These three levels of representation are interrelated and an important character in chemistry learning. To develop students' understanding of chemistry, learning should guide students using and linking these three levels of representation (Wu, Krajcik and Soloway, 2001: 822). For that purpose, the learning media has been oriented by Chemistry Triangle in the form of multimedia that integrates animation, voice and text demonstration video to convey macroscopic, submicroscopic and symbolic chemical concepts for high school students. In the making, this learning media has been through several stages of the test validity, practicality and effectiveness. This medium has also been trialed in a limited number of schools, then the media is perfected in accordance with the validator's suggestions. Learning media that have been made is then experimented on several schools in West Sumatra.

Based on the above description has been conducted research on the use of media-oriented Chemistry Triangle on cooperative, individual and conventional learning. This study aims to determine what method is suitable to be used with learning media oriented Chemistry Triangle.

2. Method

To determine the class and the school to be sampled, cluster random sampling technique is used. The results obtained as a sample in this study are 2 SMA in Pasaman District, 2 SMAN in Solok city and 2 SMK in Pariaman city. Each school is drawn 3 classes each of which is taught by cooperative, individual and conventional learning methods using the Chemistry Triangle-oriented learning media. The level of student ability is used as a moderator variable. To obtain student learning result data used objective test given at the end of learning. Before the test, the test is tested first and analyzed to get a good question. The hypotheses to be tested in this study were:

- There are differences in student learning outcomes that learn by cooperative, individual and conventional methods using Chemistry Triangle oriented learning media
- There are differences in student learning outcomes with high ability to learn by cooperative, individual and conventional methods using Chemistry Triangle oriented learning media.
- There are differences in student learning outcomes with low ability to learn by cooperative, individual and conventional methods using Chemistry Triangle oriented learning media

Data analysis for hypothesis test was done by Anava or Kruskal Wallish technique. This experimental study used the following research design.

Table 1. Research Design

students ability (A)	learning methods (B)		
	cooperative learning (B1)	individual learning (B2)	conventional learning (B3)
Higher ability (A1)	A1B1	A1B2	A1B3
Low ability (A2)	A2B1	A2B2	A3B3

3. Results and Discussion

3.1. Result

3.1.1. *The results of high school students in Pasaman district.* Prior to hypothesis testing, the first test needs analysis. Based on the results of the analysis requirements test, hypotheses 1,2 and 3, are tested by Kruskal-Wallish Test with SPSS processing because the data was not normal but homogeneous.

Table 2. First Hypothesis Test for SMA Pasaman district

	cooperative	Individual	conventional
number of samples	74	74	74
mean	80,824	78,081	79,621
Total rank	8076	8310	8367
Quadrate rank	65221776	69056100	70006689
H count		16,1887	
Chi table		27,59	

Based on Table 2 it can be seen that H arithmetic is smaller than chi table, meaning H1 is rejected and H0.diterima. It can be concluded that there is no significant difference in student learning outcomes studied by cooperative, individual and conventional methods using Chemistry Triangle-oriented learning media. However, it can be seen that the learning outcomes of cooperative group students is highest compared to the other two groups

Table 3. Second and Third Hypothesis Test for SMA Pasaman

	High ability			Low ability		
	cooperative	individual	conventional	cooperative	Individual	conventional
number of samples	18	18	18	18	18	18
Mean	83,187	83,5	79,93	77,625	73,715	78,222
SD	11,4789	8,3485	14,0939	12,98	21,86	12,94
Varian	131,7625	70,3	198,638	168,55	478,22	167,59
H count		13,38			3,69	
chi		28,3			28,3	

From Table 3 it can be seen that for students with high or low ability, H count is also smaller than chi table, so it is concluded that there is no difference in student learning outcomes with high ability in all three groups, as well as low-ability students. In low-ability students, conventional group learning outcomes were better than the other two groups

3.1.2. *Student 's learning outcomes in Kota Solok.* Test requirements for learning outcomes in Solok city show that the data is homogeneous and normal. Hypothesis test is done with two-ways Anava

Table 4. First Hypothesis Test for SMA Kota Solok

	cooperative	individual	conventional

Number of samples	72	74	74
Mean	71,37	62,36	67,09
SD	14,42	14,72	15,26
Varian	207,82	216,59	232,88
F count		5,67	
F table		3,04	

In testing the first hypothesis, F count is greater than F table, then the null hypothesis is rejected. In other words there are differences in student learning outcomes taught with media oriented chemistry triangle on learning with cooperative, individual and conventional methods. To determine which method gives a difference in student learning outcomes, it is necessary to calculate Tukey's HSD. Based on the conclusions of the average difference table between groups in the Tukey "s HSD test, it can be seen that the cooperative group did not differ significantly with the conventional group. The individual group learning outcomes did not differ significantly with the conventional group and the cooperative group learning outcomes differed significantly individual groups. It turns out that the average of cooperative group learning outcomes is highest compared to the other two groups, and the individual group learning outcomes are lowest.

Test of Second and third Hypotheses

The second and third hypotheses test aims to see the difference in the average of high learning achievement of the first high-ability students (hypothesis 2) and low initial students. The result of test requirement shows that the data is not normal then used non parametric statistical test that is Kruskall-Wallish test.

Table 5. Test of second and third Hypotheses for SMA in Solok City

	High ability			Low ability		
	Kooperatif	individual	konvensional	Kooperatif	Individual	Konvensional
number of samples	16	17	17	16	17	17
Mean	88,4375	80,2941	85,29412	50,88235	42,05882	45,58824
SD	4,36606	5,72019	5,144958	9,552588	9,363964	9,165553
Varian	19,0625	32,7205	26,47059	91,36029	87,68382	84,00735
H count		12,5192			6,05	
chi		5,9915			5,9915	

Based on Table 5 it can be seen that the price of H> chi, either for high or low initial students. If the value of H and chi with the acceptance criteria H_0, then H_0 is accepted if the price of H is smaller than chi while for the other value H_0 is rejected. The value of chi obtained outside the acceptance criteria H_0 means that H_0 is rejected and H_1 accepted means that there is a significant difference in the average of student learning outcomes to the three groups. Based on the conclusions of the mean differences between groups in the Tukey "s HSD test, it is known that for students with high initial ability the learning outcomes of the cooperative group differed from the individual group but did not differ significantly with the conventional group. The conventional group is significantly different from the individual group.

For students with low initial ability, cooperative group learning outcomes differ significantly with conventional and individual groups. Individual group learning outcomes did not differ significantly with conventional groups.

3.1.3. *Students Learning Outcomes in Pariaman City.* Test requirements for learning outcomes in Pariaman City also show that the data is homogeneous and normal. Hypothesis test is done with Anava 2 direction.

Table 6. First Hypothesis Test

	cooperative	individual	conventional
Number of sample	58	54	59
Mean	61,983	65,741	67,542
SD	11,620	12,863	11,904
Varian	135,034	165,478	141,701
F count		6,38	
F table		3,06	

Based on the conclusion of Tukey "s HSD test, it can be seen that student learning outcomes in groups of students who learn by cooperative method is not different from student learning outcomes in groups of students taught by individual methods. Students' learning outcomes in groups that were studied by individual methods did not differ from student learning outcomes in groups of students studying by conventional methods. Students' learning outcomes in the group studying with cooperative method differed significantly with student learning outcomes in the group of students studying by conventional methods. The learning outcomes of the students who studied with the conventional method were highest compared to the other two groups who studied with cooperative and individual methods.

Table 7. Second and Third Hypotheses test

	Higherer ability			Low ability		
	cooperative	individual	conventional	cooperative	individual	conventional
Number of sample	16	14	15	16	14	15
Mean	73,75	80,357	79,33	50,938	52,857	56,333
SD	7,853	6,064	7,761	5,836	8,707	7,432
Varian	61,677	44,093	60,238	34,059	75,812	55,235
F count		3,484			2,128	
F table		3,22			3,22	

Based on the calculation in Table 7 by using one way anova test at significance level $\alpha = 0.05$, for students with high ability obtained F count 3.484 while F table = 3.22. This result shows that F calculated> F table so that H0 is rejected and H1 accepted. Means there are differences in student learning outcomes that have high ability by using cooperative, individual and conventional learning methods. Determination of which method provides different learning outcomes for groups of students with high initial ability, was tested further using Tukey's HSD test

Based on the conclusion of Tukey "s HSD calculation, it can be seen in groups of students with high initial ability there are differences in student learning outcomes between groups of students taught by cooperative methods with groups of students taught by individual methods. In groups of students with high initial ability there are differences in student learning outcomes between groups of students taught by cooperative and conventional methods. Learning by individual and conventional methods in high ability group students no significant differences in learning outcomes.

Anova test for low-ability students, obtained the price of F arithmetic 2,128 while F table = 3.22. This result shows that F calculates <F table so that H0 is accepted and H1 is rejected. This means there is no difference in student learning outcomes that learn by cooperative, individual and conventional methods using learning media Chemistry triangle

3.2. *Discussion*

The results of the study in Pariaman city suggest that overall learning outcomes, the best conventional methods, differ significantly with cooperative methods, but do not differ significantly with individual methods. Student learning outcomes with higher capabilities of individual methods are better, significantly different from cooperative methods, but not significantly different from conventional methods.Conventional method was best for students who have low ability. Individual methods are better than cooperative methods, but do not differ significantly.

In contrast to previous studies, in this study, student learning outcomes in the cooperative class did not give better results than the conventional class. This is because students are not used to study cooperatively, the discussion does not go smoothly. Individuality is more visible during group discussion, meaning that there is no good cooperation, interaction between group members has not been optimal.In the conventional method, teachers themselves who explained the media, so it make students easier to understand. Learning becomes directed and controlled and the allocation of time has been determined sufficient for the learning process.

The results of the study in Solok City showed that individual classroom learning outcomes did not differ significantly with the conventional group and the cooperative group learning outcomes differed significantly with the individual group, but did not differ significantly with the conventional class.
It turns out that the average of cooperative group learning outcomes is highest compared to the other two groups, and the individual group learning outcomes are lowest. Based on the conclusions of the mean differences between groups in the Tukey "s HSD test, it is known that for students with high initial ability the learning outcomes of the cooperative group differed from the individual group but did not differ significantly with the conventional group. The conventional group is significantly different from the individual group. For students with low initial ability, cooperative group learning outcomes were higher, significantly different from the conventional and individual groups. Individual group learning outcomes did not differ significantly with conventional groups. The high learning outcomes of cooperative classes can be attributed to the fact that they are accustomed to learning by cooperative methods. They are able to cooperate well, the discussion went smoothly

When viewed from the learning outcomes in Pasaman district, there is no significant difference in the learning outcomes of students who are learning with the three learning methods, either overall, as well as in students with high and low ability. In Pasaman district, students are accustomed to learning by conventional method, but do not show better learning outcomes than cooperative and individual methods.

The purpose of this study is mainly to find out which method is better done for chemistry-oriented media triangle. In this study conducted learning with cooperative, individual and conventional methods. The results of previous studies relating to cooperative learning generally suggest that cooperative learning is better than conventional learning. Apparently not always learning with cooperative method gives better learning outcomes than conventional methods. The results of the study in Pariaman showed that conventional methods gave higher learning outcomes (significantly different) with the other two groups. The learning outcomes at Pasaman showed no significant differences between the three groups. Only in the city of Solok gives better results to classes that learn by cooperative methods.

When viewed from the three areas where the research, it turns out the learning outcomes of students who learn by conventional methods do not differ significantly with students who learn with cooperative methods. Even in Pariaman, conventional methods show the best learning outcomes compared to the other two methods. This shows that none of the methods are good for all students. Each method has its own advantages and disadvantages. In addition, each student also has different learning styles and intelligence. There are students who like to study groups, some who like to study independently. The results also show that students with low ability, learning with conventional methods provide better learning outcomes. Different levels of ability possessed by each individual sometimes make it difficult to work with others in groups, so cooperative learning methods do not provide better learning outcomes.

In learning with individual methods, students are required to be able to learn independently. It is expected that with this individual method students with high ability will get higher learning outcomes. It turns out that only in Pariaman the result of learning by individual method gives better learning result which is significantly different with the other two methods. Teacher skills in managing learning methods will also provide different student learning outcomes. Not all teachers are able to manage cooperative

IOP Conf. Series: Materials Science and Engineering **335** (2018) 012103 doi:10.1088/1757-899X/335/1/012103

learning well. Each teacher has different professional competencies, some are more skilled at managing cooperative learning, some are happy with conventional methods and others.

4. Conclusion

The three methods were used, none of the methods are good for all students. In the sample schools in Pasaman district, there was no significant difference in the mean of student learning outcomes that were taught by cooperative, individual and conventional methods. In Solok city, cooperative learning provides better student learning outcomes than the other two methods and in Pariaman the conventional method provides the best learning outcomes compared to cooperative and individual methods. In Pasaman and Pariaman districts, students with high initial ability, individual methods provide the best learning outcomes. Students with low initial ability, conventional methods provide better results. It can be concluded that many things affect learning outcomes, not just learning methods. Student learning styles, characteristics, learning motivation will affect learning outcomes. Teaching style of teachers and the ability of teachers in managing the learning process also greatly affect the learning outcomes.

References

[1] Chandraseragan, A.L., David F. Treagust., And Mauro Mocerino. 2007. "The development of a two-tier multiple-choice diagnostic instrument for evaluating the secondary school students" ability to describe and explain chemical reactions using multiple levels of representation ". *The Royal Society of Chemistry, 8 (3): 293-307*

[2] Dimyati, Mudjiono. 2006. *Belajar dan Pembelajaran.*Jakarta: PT. Rineka Cipta.

[3] Gulo, W. 2008.*Srategi Belajar Mengajar*. Jakarta: PT Grasindo.

[4] Irianto, Agus. 2009. *Statistik konsep dasar aplikasi dan pengembangannya*. Jakarta. Kencana.

[5] Jansoon, N., Cool, RK., And Somsook, E. 2009. "Understanding Mental Models of Dilution in Thai Students". *International Journal of Environmental Science & Education. 4 (2): 147-168*

[6] Johnstone, Alex. H. 1993. The Development of Chemistry Teaching "The Forum: *Symposium on Revolution and Evaluation in Chemical Education. 70. (9) .701-704*

[7] Johnstone, Alex. H, 2001. Teaching of Chemistry- Logical or Psychological?.*Chemistry Education: Research and Practise in Eeurope*.1.(1).9-15.

[8] Olatoye, R.A., Aderogba A.A.. and Aanu E.M.. 2011. "Effect of Co-operative and Individualized Teaching Methods on Senior Secondary School Students" Achievement in Organic Chemistry. *The Pacific Journal of Science and Technology* 12.(5)310-31

[9] Rusman.2011. *Model Model Pembelajaran untukMengembangkan Profesional Guru.* Jakarta: PT Raja Grafindo Persada

[10] Russell, J.W. et al. 1997. "Use of Simultaneous-Synchronized Macroscopic, Microscopic, and Symbolic Representations to Enhance the Teaching and Learning of Chemical Concepts". *Journal of Chemical Education. 74. (3). 330-334.*

[11] Sagala, Syaiful. 2010. *Konsep dan Makna Pembelajaran*. Bandung: Alfabeta

[12] Sanaky. Hujair, AH, 2011. *Media Pembelajaran.*. Yogyakarta: Kaukaba

[13] Slavin, Robert.2009.Cooperative Learning.A Simon & Sculer company

[14] Treagust D and Gilbert J, 2009. *Multiple Representation In Chemical Education*. Australia: Springer.

[15] Trianto. 2009. *Mendesain Model Pembelajaran Inovatif-Progresif.* Jakarta: Kencana Predana Media Group

IOP Conf. Series: Materials Science and Engineering **335** (2018) 012104 doi:10.1088/1757-899X/335/1/012104

Development of Mole Concept Module Based on Structured Inquiry with Interconection of Macro, Submicro, and Symbolic Representation for Grade X of Senior High School

R Sagita, F Azra, M Azhar*
Chemistry Department, Faculty of Mathematics and Natural Sciences, Universitas Negeri Padang, Jl. Prof.Hamka, Air Tawar, Padang 25131, Indonesia

sagitaranda@gmail.com, *minda@fmipa.unp.ac.id

Abstract. The research has created the module of mole concept based on structured inquiry with interconection of macro, submicro, and symbolic representation and determined the validity and practicality of the module. The research type was Research and Development (R&D). The development model was 4-D models that consist of four steps: define, design, develop, and disseminate. The research was limited on develop step. The instrument of the research was questionnaire form that consist of validity and practicality sheets. The module was validated by 5 validators. Practicality module was tested by 2 chemistry teachers and 28 students of grade XI MIA 5 at SMAN 4 of Padang. Validity and practicality data were analysed by using the kappa Cohen formula. The moment kappa average of 5 validators was 0,95 with highest validity category. The moment kappa average of teachers and students were 0,89 and 0,91 praticality with high category. The result of the research showed that the module of mole concept based on structured inquiry with interconection of macro, submicro, and symbolic representation was valid and practice to be used on the learning chemistry.

1. Introduction

Basically, chemistry describes the structure, composition, properties and changes of matter and energy that accompany it (Brady, 2012). One of the chemistry lessons learned in class X is Mole Concept. The mole concept examines the number of particles. Mol is used as the quantity that connects the mass unit of matter, the number of particles, and the volume of gases involved in chemical reactions. Chemical calculations will be easier to apply with mole concept.

According to curriculum of 2013, the right approach in the learning process is the scientific approach. The scientific approach is believed to be the golden tool for the development of the attitudes, skills and knowledge of learners (Kurniasih, 2014). One of the learning models that suits with application of a scientific approach is inquiry.

Based on the level of teacher involvement in the learning process, the inquiry learning model is divided into inquiry confirmation, structured inquiry, guided inquiry and opened inquiry. In structured inquiry, students investigate questions provided by the teacher through the steps or procedure and receive guidance from each step to achieve the desired result. Students conduct direct investigation and develop basic skills in the investigation, such as making observations, hypotheses, collecting and processing data, and creating conclusions (Zion, 2012).

Several previous studies on structured inquiry had reported that LKS experiments and non-experiments on chemical equilibrium materials help students understand the material well, do the experiment, and improve student learning outcomes (Maryati, 2012). Chemistry laboratory guide based on structured inquiry in grade XII of SMAN 1 Indralaya Utara can improve students' understanding (Ariningsih, 2014). Learning vibration and wave materials through structured inquiry learning can increase learning interest and active student participation (Sugiarto, 2015).

In addition, learning chemistry should be emphasized on three levels of chemical representation (macroscopic, submicroscopic, and symbolic). A person's understanding of the concept of chemistry is determined by his ability to transfer and to connect the three levels of representation. In this time, learning chemistry just represents in two levels (macroscopic and symbolic), while submicroscopic level is seldom touched. Previous research had shown that inability of the students to represent chemistry at submicroscopic level can inhibit their ability to solve chemical problems related to macroscopic or symbolic levels (Sunyono, 2013). When studying a complex concept, using the multiple representations such as diagrams, graphs, and equations can be applied to facilitate the learning process (Ainsworth, 2008). The interconnection of macroscopic, submicroscopic, and symbolic is one of the keys to teaching chemistry (gilbert and treagust, 2009). This research aims to develop of mole concept module based on structured inquiry with interconection of macro, submicro, and symbolic representations for class X SMA and reveal the validity and practicality of module.

2. Methode

The type of this research was Research and Development (R & D). This research was conducted to develop of mole concept module based on structured inquiry with interconection of macro, submicro, and symbolic representations for class X senior high school and reveal the validity and practicality of the module. The object of this research is mole concept module which was developed. The instrument of this research was a questionnaire consisting of validity sheet and practicality sheet. The data obtained were analyzed using kappa moment (Boslaugh, 2008).

$$\text{Kappa moment } (k) = (\rho - \rho_e)/(1 - \rho_e) \quad (1)$$

Information:

k = Moment kappa shows the validity of the product.

ρ = The realized proportion, calculated by the sum of the values given by the validator divided by the sum of the maximum value.

ρ_e =Unrealized proportions, calculated by the sum of the maximum value minus the total number of validated entries divided by the maximum number of values.

Table 1. The decision category based kappa moment (k)

Interval	Category
0,81 – 1,00	Very High
0,61 – 0,80	High
0,41 – 0,60	Medium
0,21 – 0,40	Low
0,01 – 0,20	Very Low
$\leq$ 0,00	Unvalid

The module was designed using 4-D development model which consists of four development stages such as define, design, develop and disseminate (Trianto, 2014). This study was limited to the development stage. In the Define stage, There are 5 analyzes. They are front-end analysis, student analysis, task analysis, concept analysis, and analysis of learning objectives. Front-end analysis aims to elicit and define the basic problems faced by teachers and students in chemistry learning, especially on the mole concept material. This analysis obtained a description of facts, expectations and alternative solutions problem. Front-end analysis was done by interviewing some senior high school chemistry teachers. Student analysis was conducted to know the characteristics of students that include academic ability and student learning motivation.

Task analysis included the tasks that students do in learning and rellated with the selected material, mole concept. The steps were analized the basic competence (KD) and divided into several indicators then become the learning objectives. Concept analysis was identification the concepts to be taught and to construct them in the concept hierarchy form. The concept analysis in this research was conducted by identifying the concepts analyzed based on college chemistry books and relevant high school chemistry books.

Analysis of learning objectives was done by analyzing basic competence (KD) which was translated into several indicators then the indicator was translated into several learning objectives. The results of the define stage was carried out to the design stage. The Design stage aims to design of mole concept module based on structured inquiry with interconection of macro, submicro, and symbolic representation for class X SMA. The final stage of this research is the Development stage. This stage aims to produce a mole concept module based on structured inquiry with interconection of macro, submicro, and symbolic representation that revised based on validator suggestions.

3. Results and Discussion

3.1. Define stage

In the defining stage obtained 5 data. They were front-end analysis, student analysis, task analysis, concept analysis, and analysis of learning objectives.

3.1.1. Front-end analisys. In front-end analysis, it was found that some students still have difficulties in understanding the mole concept material. If the questions or exercises given to the students are varied or slightly different from the example discussed, students were confused about the questions.Usually, students were fixated on the formula that available on the teaching materials so that there has been no meaningful learning yet. Multiple representations provide facilities for the occurrence of meaningful learning or deep learning. Teaching materials used in schools are books, LKS, and powerpoint. Learning material in module form is not yet available. Moduleis one of the teaching materials that can help students to find concepts and enable students to learn independently.

3.1.2. Student analisys. Based on interviews conducted, senior high school students generally aged 16-18 years old. According to Piaget, the level of thinking of children aged 12-18 years is at formal operational development. At this stage the child is able to think abstractly and logically and has the ability to interpret, develop hypotheses and draw conclusions (Budiningsih, 2012). The results of interviews with chemistry teacher SMAN 4 of Padang are ability of academic and learning motivation of the student classified as "moderate". Motivation of students in learning is influenced by several factors, such as time. Students are more eager to learn in the morning than in the noon. In addition, the results of interviews with some students indicate that they prefer the learning process by using interesting teaching materials that are pictorial, colorful, and arranged in easy-to-understand language.

3.1.3. Task analisys. The task analysis was done by analyzing the basic competence(KD) which was then translated into several indicators. Basic competences (KD) of mole concept materials are:

- 3.11 Applying the concept of relative molecular mass, reaction equations, fundamental laws of chemistry and the concept of mol to complete chemical calculations.
- 4.11 Processing and analyzing relative molecular mass related data, reaction equations, basic chemical laws and the concept of moles to complete chemical calculations.

Basic competence was spelled out into several indicatorssuch as:

- Explaining the mole.
- Determining the molar mass and molar volume of the gas of a compound.
- Applying the mole concept in determining the chemical formula.
- Determining the formula of hydrate compounds.
- Calculating substance levels in the mix such as (percent, ppm, molarity, molality and mole fraction).

Based on the basic competencies and learning indicators, a mole concept module is designed to fit the task analysis so that the expected competencies can be achieved by the students at the end of the lesson.

3.1.4. Concept analisys. The concepts in this material are arranged in hierarchy concepts form. The concepts are mole concepts, molar mass, molar volume, empirical formulas, molecular formulas, hydrate compounds, and substance levels. The concepts were analyzed based on college chemistry books and relevant high school chemistry books.

3.1.5. Analysis of learning objectives. The learning objectives on mole concept materials are achieved through observation, hypothesis, collection and organization of data, and draw conclusions. The learning objectives in mole concept materials are:

- Explain the mole with the given data.
- Determine the molar mass of an element and compound known the A_r / M_r.
- Calculate the molar volume of the gas that known the mole.
- Determine the empirical formula of a compound if known mass and percentage of elements.
- Determine the molecular formula of a compound if known to the empirical formula.
- Determine the formula of a hydrate compound if the mass of the hydrate compound is known.
- Calculates substance levels (percent, ppm, molarity, molality and mole fraction) appropriately.

3.2. Design stage

The module of mole concept had been designed consists of: cover, module usage manual, core competences, basic competences, indicators, learning objectives, concept maps, activity sheets, student worksheets, and evaluation sheets. Module cover, learning indicators and objectives, activity sheets, worksheets, worksheet keys are respectively observed in Figures 1 through 5.

Figure 1. Cover

IOP Conf. Series: Materials Science and Engineering **335** (2018) 012104 doi:10.1088/1757-899X/335/1/012104

Indikator :

1. Menjelaskan pengertian Mol
2. Menentukan massa molar dan volume molar gas dari suatu senyawa.
3. Mengaplikasikan konsep mol dalam menentukan rumus kimia.
4. Menentukan rumus senyawa hidrat.
5. Menghitung kadar zat dalam campuran seperti (persen, ppm, molaritas, molalitas dan fraksi mol).

Figure 2. Indicator and Learning Objectives

Lembar Kegiatan 1: Konsep Mol

Waktu: 3 x 45 menit

Tujuan Pembelajaran:

Berdasarkan tahapan inkuiri terstruktur siswa diharapkan dapat:

1. Menjelaskan pengertian mol dari data yang diberikan.
2. Menentukan massa molar suatu unsur dan senyawa yang diketahui A_r/M_r nya.
3. Menghitung volume molar suatu gas yang diketahui molnya.

Figure 3. Activity Sheets

Lembar kerja 1

1. Jika diketahui 40 gram NH_3, hitunglah mol NH_3 tersebut?
   Jawab:
   ....
   ....

2. Berdasarkan soal nomor 1, berapakah jumlah molekul NH_3 yang terdapat dalam jumlah mol tersebut?
   Jawab:
   ....
   ....

Figure 4. Worksheets

IOP Conf. Series: Materials Science and Engineering **335** (2018) 012104 doi:10.1088/1757-899X/335/1/012104

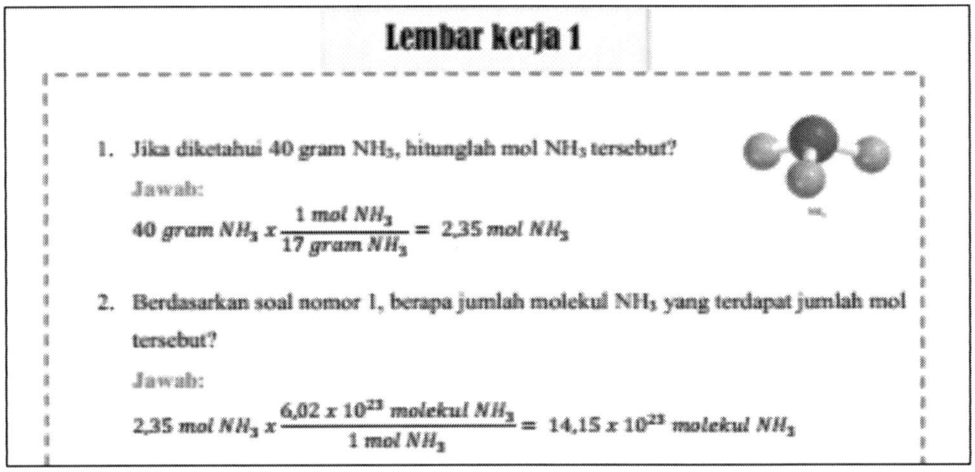

Lembar kerja 1

1. Jika diketahui 40 gram NH_3, hitunglah mol NH_3 tersebut?

Jawab:

$$40\ gram\ NH_3\ x \frac{1\ mol\ NH_3}{17\ gram\ NH_3} = 2{,}35\ mol\ NH_3$$

2. Berdasarkan soal nomor 1, berapa jumlah molekul NH_3 yang terdapat jumlah mol tersebut?

Jawab:

$$2{,}35\ mol\ NH_3\ x \frac{6{,}02\ x\ 10^{23}\ molekul\ NH_3}{1\ mol\ NH_3} = 14{,}15\ x\ 10^{23}\ molekul\ NH_3$$

Figure 5. Worksheets Keys

3.3. Develop stage

This stage aims to produce the mole concept module based on structured inquiry with interconection of macro, submicro, and symbolic representations for class X senior high shcool based on the advice of the validators. This stage consists of three steps, such as the test of validity, revision, and practicality. The validity test aims to reveal the validity of the module. Validation was done by three lecturers and two chemistry teachers. The revision stage aims to improve the module that was considered still less precise by the validator before the product is tested. The revised module is given back to the validator for further discussion before it is tested. The revision is terminated when the validator has declared the module created is valid.

Product practicality test was done to teachers and students of SMAN 4 of Padang class XI MIA 5. This practicality test was aimed to find out the practicalities of mole concept module that developed include benefit, ease of use, and efficiency of learning time by using the module. The practicality of the module was done by using a questionnaire of practicality.

*3.3.1. Validity of the mole concept module based on structured inquiry with interconection of macro, submicro, and symbolic representation for class X senior high school.*The validity of mole concept module had a value 0.95 with very high category. This validity consists of four components, such as the components of content,component of module serving, language, and graphics.

Components of the contents of the module had a kappa moment 0.94 with very high category. This indicates that the modules are in accordance with the demands of core competence and basic competence. The questions contained in the module have led the students to find the concept. In addition, the exercise is direct application of the concept learned.

The component of module serving had a kappa moment 0.95 with very high category. This indicates that the module has been systematically arranged from the title, competency standard, basic competence, indicator, and in accordance with the structured inquiry stages of observation, hypothesis, collection and organization of data, and conclusions. Structured inquiry learning guides students to discover concepts by developing their basic abilities in investigation (Zion, 2012).

The linguistic component of the module had a kappa moment 0.96 with very high category. This indicates that the module use good and correct language according to Indonesian language rules and easy to understand. In addition, the questions asked in the module are clear and consistent in using symbols.

IOP Conf. Series: Materials Science and Engineering **335** (2018) 012104 doi:10.1088/1757-899X/335/1/012104

The graphic component had a kappa moment 0.97 with very high category. This indicates that the module has used a clearly observable image, the corresponding layout, which corresponds to the size of the letters, fonts, spaces, cover appearance, module content layout, and image illustration placement. The overall kappa moment value can be seen in Figure 6.

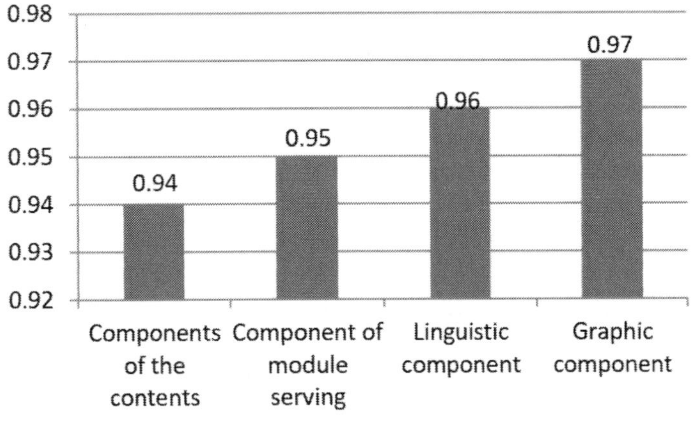

Figure 6. The curve of kappa moment value by validator

In Figure 6, it can be seen that from all aspects indicate that the value of the kappa moment is more than 0.81. The value of kappa moment 0.81-1.00 had a very high category so that the module developed otherwise valid (Sugiyono, 2013). A product is valid if product can indicate a condition that is in accordance with the contents and constructs (Arikunto, 2008).

3.3.2. Practicalityof the mole concept module based on structured inquiry with interconection of macro, submicro, and symbolic representatiosn for class X senior high school. Practicality of the module was assessed by chemistry teachers and senior high school students class XI. Assessment of practicality by teachers obtained kappa moment 0.89 with very high practicality and by students 0.91 with very high practicality. The Practicality consist of three components: ease of use, efficiency of learning time and benefits.

The ease of use component of the module had a kappa moment 0.90 from teachers and 0.91 from students with very high practicality categories. This shows the Module using easy language to understand, the questions are clear, the material is delivered in simple and portable size.

The learning time efficiency of the module had a kappa moment 0.86 from teachers and 0.88 from students with very high practicality. Learning by using modules can make learning time more efficient and students can learn at their own pace (Daryanto, 2014).

The benefit component of the module had a kappa moment 0.90 from teachers and 0.93 from students with very high practicality categories. This demonstrates the mole concept module help the student in finding the concept. The exercises allow students to have the opportunity to train themselve to learn independently. Module is also designed to attract students' interest in learning. Overall, value of kappa moments given by teachers and students can be seen in Figure 7.

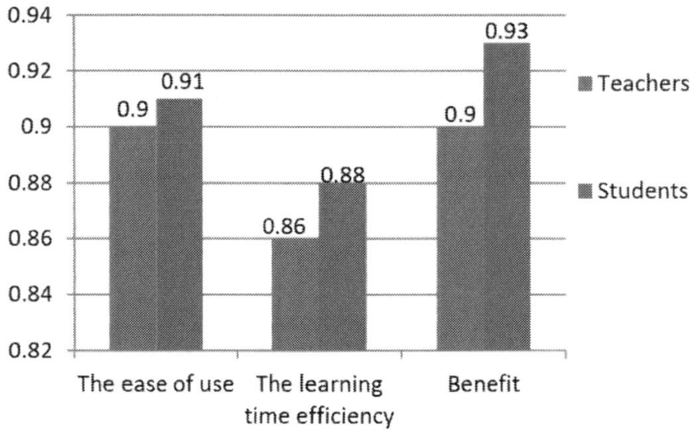

Figure 7. The kappa moment curve ofpracticality by the teachers and students.

Based on the above, it can be concluded that a mole concept module based on structured inquiry with interconection of macro, submicro, and symbolic representationhas beenpractice and can be used in the school.
follows:

- 11 point Timesor Times New Roman.
- The text should be set to single line spacing.
- Paragraphs should be justified.
- The first paragraph after a section or subsection heading should not be indented; subsequent paragraphs should be indented by 5 mm.

4. Conclusion

A mole concept module based on structured inquiry with interconection of macro, submicro, and symbolic representation has been produced using the 4-D development model. The module has a high validity and practicability both by teachers and students.

References

[1] Ainsworth, Shaaron. (2008).The Educational Value of Multiple-representations when Learning Complex Scientific Concept. *Springer.*
[2] Arikunto, S.(2008). Dasar-Dasar Evaluasi Pendidikan. Jakarta: Bumi Aksara.
[3] Arinigsih, Ismi, dkk. (2014). "Pengembangan Panduan Praktikum Kimia Berbasis Inkuiri Terstruktur di Kelas XII SMAN 1 Indralaya Utara". Program Studi Pendidikan Kimia, FKIP, Universitas Sriwijaya.
[4] Boslaugh, Sarah dan Paul A. W. (2008).Statistics in a Nutshell, a desktop quick reference.Beijing, Cambridge, Famham, Köln, Sebastopol, Taipei,Tokyo: O'reilly.
[5] Budiningsih, Asri.(2012). Belajar & Pembelajaran. Jakarta: PT Rineka Cipta.
[6] Brady, James E. (2012). Chemistry The Molecular Nature of Matter. New York: John Wiley & Sons, Inc.
[7] Daryanto. (2014).Pengembangan Perangkat Pembelajaran. Yogyakarta: PT GavaMedia
[8] Gilbert, John K dan David F.Treagust. (2009). Introduction: Macro, Submicro, and Symbolic Representations and the Relationship Between Them: Key Model in Chemical Education. Springer Science+Business Media.
[9] Kurniasih, Imas dan Berlin Sani. (2014). Sukses Mengimplementasikan Kurikulum 2013. Yogyakarta: Kata Pena.

[10] Maryati, Anita Mariana. (2015). "Lembar Kerja (LKS) Eksperimen dan Non-Eksperimen Berbasis Inkuiri Terstruktur yang dikembangkan pada Subpokok Materi Pergeseran Kesetimbangan Kimia". Prosiding Simposium Inovasi dan Pembelajaran Sains 2015.
[11] Sugiarto. (2015). Peningkatan Hasil Belajar Materi Getaran dan Gelombang Melalui Pembelajaran Inkuiri Terstruktur. Didaktikum: Jurnal Penelitian Tindakan Kelas.Vol.16. No. 4.
[12] Sugiyono. (2013). Metode Penelitian Kuantitatif Kualitatif dan R & D. Bandung: ALFABETA
[13] Sunyono.(2013). Efektifitas Model Pembelajaran Berbasis Multipel Representasi Dalam Membangun Model Mental Mahasiswa Topik Stoikiometri Reaksi.Jurnal Pendidikan Progresif. Vol.3. No. 1. Hlm. 65-79.
[14] Trianto. (2014). Model Pembelajaran Terpadu. Jakarta: Bumi Aksara
[15] Zion, Michal dan Ruthy Mendelovici. (2012). Moving from Structured to Open Inquiry: Challenges and Limits. *Science Education Internasional.* Vol.23, No.4. Hlm. 383-399.

IOP Conf. Series: Materials Science and Engineering **335** (2018) 012105 doi:10.1088/1757-899X/335/1/012105

Development of Chemistry Triangle Oriented Module on Topic of Reaction Rate for Senior High School Level Grade XI Chemistry Learning.

D R Sari, Hardeli, and Bayharti,
Chemistry Department, Faculty of Mathematics and Science, Universitas Negeri Padang, Indonesia

desiratnasari539@gmail.com, hardeli1@yahoo.com, chembayharti@gmail.com

Abstract—This study aims to produce chemistry triangle oriented module on topic of reaction rate, and to reveal the validity and practicality level of the generated module. The type of research used is EducationalDesign Research (EDR) with development model is Plompmodel. This model consists of three phases, which are preliminary research, prototyping phase, and assessment phase. The instrument used in this research is questionnaire validity and practicality. The data of the research were analyzed by using Kappa Cohen formula. The chemistry triangle oriented module validation sheet was given to 5 validators consisting of 3 chemistry lecturers and 2 high school chemistry teachers, while the practicality sheet was given to 2 chemistry teachers, 6 students of SMAN 10 Padang grade XII MIA 5 on the small groupevaluation and 25 students of SMAN 10 Padang grade XII MIA 6 on the field test. Based on the questionnaire validity analysis, the validity level of the module is very high with the value of kappa moment 0.87. The level of practicality based on teacher questionnaire response is very high category with a kappa moment value 0.96. Based on the questionnaire of student responses on small group evaluation, the level of practicality is very high category with a kappa moment 0.81, and the practicality is very high category with kappa moment value 0.83 based on questionnaire of student response on field test.

1. Introduction

Education is an effort which has a vital role to improve the quality of human resources and the individual potential. Education is a conscious and planned effort to create teaching and learning process atmosphere of students actively in developing their potential for having spiritual religion power, self control, intelligence, good manners, and skills needed by themselves, society, nation and state.

Innovations to the education as system are needed to improve the quality of education and produce competent graduates. Based on the Regulation of the Minister of National Education (Permendiknas) number 41 of 2007 on process standards, that framework of innovating educational system has been established vision, mission and strategy of national educational development[19]. Based on that regulation, a set of principles of educational implementation has been shifted paradigm of teaching to the learning paradigm. Learning is the process of interaction of learners with teachers and learning resources. Learning resources can be in the form of teaching materials. Teachers are required to develop teaching materials because by using teacher learning materials will be easier in implementing learning and students will be more helpful and easy in learning, one of the categories of teaching materials is module. Modules are a set of teaching materials that are presented systematically that can be used by students to study with or without a teacher, thus modules can serve as teacher substitutes. If a teacher has a function to explain something, then a module should be able to explain something with a language that can be easily received by learners[1].

The author has made observations about learning materials used by students in the process of chemistry learning as well as interviewing with chemistry teachers in several high schools in Padang. Based on the results of observations and interviews obtained information that the teaching materials used by teachers in the learning is a printed book that has not fully meet the demands of the curriculum 2013. In addition the learning materials used are not yet able to guide students to learn independently because of language used in the book is difficult to be understood by students. The learning materials used also have not used an attractive color combination, so it has not been fully able to motivate students to understand the concepts that exist. In addition, the learning materials used involve only two of the three levels of chemistry representation where the macroscopic and symbolic levels have not been accompanied by submicroscopic representations that can help students understand the chemistry concepts.

One of the problem in chemistry learning is that learning takes place is generally limited to two levels of representation namely macroscopic and symbolic. This is due to the learning at the sub-microscopic level is only explained through lectures and discussions, so that students consider the subject matter or learn and resulted in the quality of learning outcomes only visible from the level of good memorization and lack of deep understanding of the substance of the subject matter which he learned [2][3] .

In order to make students understand the chemistry in depth, the chemistry learning process needs to involve the three levels of chemistry representation which are the macroscopic level, sub-microscopic, and symbolic. Macroscopic level refers to phenomena that can be seen in real (concrete), such as chemistry phenomena occurring in everyday life or in a laboratory that can be observed directly. Sub-microscopic levels provide information about elementary particles such as atomic, molecular, and kinetic materials, using the theory to explain what happens at the molecular level and using representations of theoretical models, such as particles that cannot be seen directly. Symbolic levels are representations of a reality, such as representations of symbols of atoms, molecules, and compounds, both in the form of images, and algebra[4].

The linkage between the three levels representations (chemistry triangle) has been extensively researched with the conclusions of student learning outcomes by involving multi representation in chemistry learning are higher than conventional learning[5]. Another study was also conducted by Sunyono entitled "The effectiveness of multiple representation learning model in building the mental model of students on stoichiometric topics of reaction" in this research resulted in the level of practicality (cultivation and attractiveness) and high effectiveness in building mental models and increasing mastery of students' chemistry concepts[6]. Baruri reported that interactive CD learning media through chemistry triangle approach on topic of reaction rate in SMA is very suitable to be used as an alternative for teachers and students in the process of reaction rate learning[7]. In addition, Santi also reported that the electrolyte and non electrolyte based chemistry triangle learning media using android applications for the X class of SMA has a very high validity and practicality[8].

Based on the above problems, the authors conducted the development of chemistry triangle oriented module for reaction rate material which aims to facilitate learners in the learning process and can help learners to learn independently and understand abstract concepts on the material reaction rate. This research entitled "Development of Chemistry Triangleoriented Module on Topic of Reaction Rate for Senior High School Level Grade XI Chemistry Learning ".

2. Methodology

The type of research used in this study was Educational Design Research. Educational design Research is systematic study about planning, developing, and evaluating educational intervention (such as program, learning strategy and material, product, and system) as solution from several issues existing[9]

Development model used wasPlomp Model which is developed by TjeerdPlomp. This model consists of three phases, which are preliminary research, prototyping phase, and assessment phase.

The subject in this research is 3 chemistry lecturers of FMIPA UNP, 1 chemistry teacher of SMAN 10 Padang, 1 chemistry teacher of SMA Bunda Padang, and 34 students grade XII of SMAN 10 Padang.

The preliminary investigation was done to identify and analyze the requirements needed in development of chemistry triangle oriented module on topic of reaction rate. On the preliminary phase were done needs analysis, curriculum analysis, studentanalysis, and concept analysis. Based on the analysis result at preliminary investigation was done prototyping stage.

Prototyping stage produces four types of prototypes, which are prototype I, prototype II, prototype III, and prototype IV. On each of prototypes was done formative evaluation. Formative evaluation was done based on tessmer formative evaluation. Tessmer formative evaluation includes four stages, which are self-evaluation, expert review and one-to-one evaluation, small group evaluation, and field test.

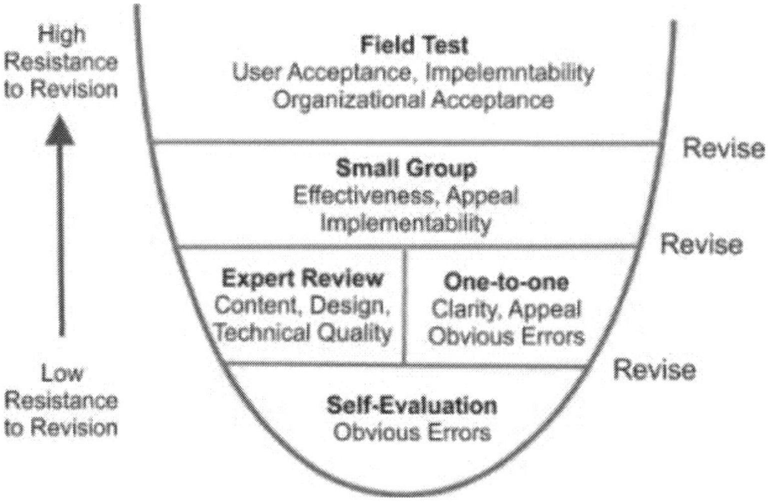

Figure 1. Layers of formative evaluation[9]

The instruments used in this research was; 1) the instruments on the preliminary stage include requirements questioner, and interview guidelines. 2) Validity sheet, 3) practicality questioners. The instrument at the preliminary investigation stage was a questionnaire requirement given to high school students of class XI, and interview guidance given to high school chemistry teacher in Padang functions to identify problems in chemistry learning process especially reaction rate. The validation sheet functions to assess the prototype generated in terms of component content, constructs, language, and graffiti. Practicality questionnaires function to evaluate the prototypes that have been generated in terms of usability, efficiency of learning time, and benefits.

Data obtained from the validation sheet and the practicality sheet were analyzed using the Cohen kappa formula to obtain the kappa moment.

$$\text{Kappa moment (k)} = \frac{P_o - P_e}{1 - P_e}$$

Were*k*is Kappa moment describing validity of theproduct, P_o isrealized proportion; counted by summing the score given by validators and then divided it by maximum total score, and $P_e$is Unrealized proportion; counted by substracing the maximum total score with the sum of total score given by validator, which then devided by maximum total score.

Table 1. The category of decision based on Kappa moment (k)[10]

Interval	Category
0,81-1,00	Very high
0,61-0,80	High
0,41-0,60	Medium
0,21-0,40	Low
0,01-0,20	Very Low
> 0,00	Invalid

3. Result and Discussion

3.1 Results

3.1.1 Preliminary Stage

Needs analysis
Based on the results of interviews conducted on 3 high school chemistry teachers in Padang, and questionnaires to 20 high school students of XI class known that the teaching materials used in the learning process of chemistry on the topic of reaction rate using printed books that have not fully meet the demands of curriculum 2013 which demands students to be more active in learning processes, learning materials cannot lead students to learn independently because the language used of books are difficult to understand by students, learning materials used only involve 2 levels of macroscopic and symbolic chemistry representations level, has not included sub-microscopic level that can help students understand the matter of reaction rate in depth. In addition the learning materials used also have not used an interesting color combination, so it has not been fully able to motivate students to understand the concepts on topic of reaction rate. Therefore, a learning material is an impportant that can enhance the students' active role in the learning process and is equipped with macroscopic, submicroscopic, and symbolic levels that are presented with attractive color combinations and easy language to understand by students so as to help students learn independently to understand concepts on topic reaction rate. One of the teaching materials that can be used is chemistry triangle oriented module using scientific step of learning.

Curriculum Analysis
The curriculum analysis was done by analyzing the curriculum and syllabus used. The curriculum used in the study of chemistry especially on the material of the reaction rate is the curriculum and syllabus 2013. Based on the analysis of curriculum 2013 it was known that the practice of learning in the curriculum 2013 using scientific learning approach system. The system of scientific learning approach guides students to learn independently and play an active role during the learning process. Learning by using scientific learning approach is a learning process that adopts scientific steps through observing, asking, collecting data, associating, and communicating activities (5M activities). Learning practice by using scientific learning approach will be done well when facilitated with appropriate learning materials. One of the learning materials that can be used is a module compiled by using scientific learning approach.

Based on the syllabus analysis of the topic reaction rate, it was known that there are four basic competence (KD) that must be mastered by students after studying the topic of reaction rate that is understanding the collision theory in chemical reaction based on the effect of temperature on the average rate of particle matter and the influence of concentration on the collision frequency (KD 3.6). Determining the reaction order and reaction rate constant based on experimental data (KD 3.7). Present ways of arranging material storage to prevent uncontrollable changes (4.6), and Designing, conducting, concluding and presenting experimental results of factors affecting reaction rate and reaction order (4.7) Based on these basic competencies can be formulated indicators of learning and objectives learning.

Students Analysis
Student analysis was done to know the characteristics of students, which include; Age, student preferences, student needs in terms of the desired teaching materials. So also with student activities in learning, and difficulties encountered students in understanding the material.

High school students, grade XI have an average age of 15-17 years and are in the formal operation phase, the main characteristic at this stage was that students begin to solve abstract problems that can be symbolically represented. Based on the results of interviews with several chemistry teachers in Padang, it was found that students who was in grade XI IPA had different academic ability, there was students who had high academic ability, moderate, and low, as well as student motivation and psychomotor depend on how students' attitude in response to the lessons given by the teacher.

Based on the results of interviews with some high school students in Padang, it was found that students prefer to study with the illustrations or images presented in learning materials related to the material and the images can be observed both in Daily life as well as observations in the lab. Students was more likely to like soft colors. In learning activities students often feel bored when the teacher explains the lesson for too long.

Based on these characteristics, the module was designed to facilitate students to be active in learning that can guide students actively construct their knowledge independently, presented with attractive colors and language that is easily understood by students, and equipped with pictures that can be observed significantly (Macroscopic images), macroscopic images are described at the submicroscopic level and supplemented by symbolic levels in order for students' understanding on topic of reaction rates.

Concept Analysis
The concept analysis aims to determine the content of the subject matter to be studied by the students and arrange it systematically according to the order of presentation. In the concept analysis is also done the identification of facts, concept, and procedural. Factual reaction rate material is the factors that influence the rate of reaction, the symbol of reaction rate, the sample fast reaction the low reaction. Conceptual rate material is the concept of reaction rate, effective collision, activation energy, reaction rate equation, reaction order and reaction rate constant. While the material of the reaction rate which is procedural is the steps to determine the equation of reaction rate, reaction order and reaction rate constant.

3.1.2 Prototyping Stage

Prototype I
Prototype I is a prototype generated from the design and realization of the preliminary research stage. Prototype I was designed using the learning stage of the scientific approach containing chemistry triangle.

At this stage a draft of teaching materials was produced in the form of chemistry triangle oriented module. The draft meant was; module's cover, module usage instructions, competencies to be achieved, student activity sheets, students' work sheet, test sheets, and test sheet locks. The cover of chemistry triangle oriented module contains the identity of module title, module name, author name, author's supervisor's name, author's agency, and target. The title and name of the module serve to inform the module user about the material discussed in the module and aspects used in the module development. The author's name, the author's supervisor's name, and the author's agency function to provide information to the module user about the module's author, the lecturer and the agency of the author.

The module cover page was designed in soft colors, with a blue background, and images related to the reaction rate material. Soft colors like blue are the most popular colors, in addition the blue color depicts trust, comfort, and also pleases the eye[11]. Images of rotted apples, rusted bridges, and fireworks show the slow, rapidly observing events of chemical reactions observable in everyday life, they are related to the rate material of the reaction.

The module usage instructions consist of instructions for teachers and instructions for students. Instructions for teachers are provided to provide teachers with information on the various activities that teachers should undertake during the lesson, while guidance for students is provided to guide students in order to learn independently. Competence to be achieved which consists of Core Competence (KI), Basic Competence (KD), Indicator and learning objectives presented aims to enable module users to know and understand the competencies that must be achieved.

The student activity sheet contains the subject matter that must be mastered by the students. The student activity sheet was designed using the learning stage of the scientific approach. Stages of the scientific approach have five stages of learning that was; stage observing, questioning, associating, gathering information, and communicating.

The worksheet was used to answer the questions or problems to be solved based on the material on the student activity sheet. The key worksheet was given to enable students to self-evaluate their work on the worksheet. A test sheet was an evaluation tool used as a measure of success or whether a goal has been achieved in the defined prototype. The test sheet key was given as a self-correcting tool to the assessment performed.

Prototype II

The prototype I that has been produced was evaluated through self evaluation. Self evaluation done with check list system aims to check the completeness of module components. Based on the results of self evaluation was know that the module did not have a list of images and a list of literature that was the components of module, in addition to the prototype I was still found a lot of deficiencies so it needs a revision. Revisions made included; 1) changing the sub-microscopic level model at the observing stage with a macroscopic model or observable phenomena in real-life daily, the macroscopic picture was related to the subject matter that the student will learn so that the student will be more motivated to learn. This is in line with that disclosed by Daryanto observing stage is the stage of cultivating students' interest to lear, because students will feel challenged, and increase students' curiosity[12], 2) add the H2O molecules and HCl molecules due to the HCl 6M solution and change the sub-microscopic level of HCl molecule into H+ and Cl- ions. The addition of H2O molecules due to the HCl 6M solution contains water molecules andHCl molecules, water molecules serve as solvents, while changing the sub-microscopic level of HCl molecules into H+ and Cl- ions based on arrhenius acid theory, HCl if dissolved in water will produce H+ ions, 3) adding the list of imges and literature on prototype I. Based on the revision made to produce prototype II.

Prototype III

Prototype III was a prototype resulted from revisions made in prototype II, at this stage formative evaluation was done in the form of one-to-one evaluation and expert review. Based on the results of interviews obtained the picture that, the existence of models in the form of images presented by involving chemistrymultirepresentation as aspects of macroscopic, sub-microscopic, and symbolic, can guide students in finding and understanding the concept on topic of reaction rate. In terms of language use in modules, the language used in the module was easy to understand. In addition, in terms of design cover and color selection, prototype II can make students interested and passion to learn it.

Assessment of this expert aims to obtain a valid prototype scientifically. Validation of chemistry triangle oriented module on topic of reaction rate was conducted by 5 validators consisting of 3 lecturers of Chemistry Department UNP and 2 high school chemistry teachers in Padang. Validation data from prototype II can be seen in Table 2.

Table 2. Assessment Data of Module Validation

No	Components	Kappa moment (k)	Category
1	Content	0,8	High
2	Construct	0,89	Very High
3	Language	0,91	Very High
4	Graphics	0,89	Very High
	Average kappa moment	0,87	Very High

Validation results from the four components assessed, showed that the generated prototype II was valid with a very high category of validity. However, there was still a need for revisions on several parts of the prototype II. Revisions made included; 1) constantly writing the phase of the substance in the equation of the reaction, before the revision phase of the substance in the equation there was a reaction that was parallel to the chemical formula and there was also written as the bottom index. The writing og the phase of the substance in the equation of the reaction according to the IUPAC rules is written as an index below or parallel to the chemical formula but in its writting tilted, 2) increase the number of solvent molecules on the sub-microscopic representations of the reaction between the Mg band and the HCl solution, that was because before the revision the number of solvent molecules was less than the number of solute molecules, based on the theory in a solution the amount of solvent is more than the solute. The revision made a valid prototype III.

Prototype IV

Prototype IV was a prototype generated from revisions made on prototype III. At this stage a formative evaluation was conducted in the form of small group trials. The prototype III practicality data on small group test can be seen in Table 3.

Table 3. Data on the results of students' practicality in small group evaluation

No	Components	Kappa moment (k)	Category
1	Ease of use	0,8	High
2	Efficiency of learning time	0,85	Very High
3	Benefits	0,78	High
	Average kappa moment	0,81	Very High

Based on the results of data analysis of practicality on small group evaluation to the prototype III, it was gotten the average kappa moment of all components of practicality is 0.81 with very high practicality category. However, although prototype III had shown very high practicality, there was still parts of the module that need to be fixed. Revisions made include; changing the sub-microscopic level of HCl 6M, before the revised HCl 6M sub-microscopic solution consisted of H^+ and Cl^- ions, and H_2O molecule, after revision, sub-microscopic level of HCl 6M solution consisted of H_3O^+ and Cl^- ions. Revisions made to produce prototypee IV. The prototype IV that has been valid and practically carried out field test to obtain the level of practicality of prototype IV.

3.1.3 Assessment Phase

Field tests were conducted for one chemistry teacher at SMA N 10 Padang, one chemistry teacher at SMA Bunda Padang, and 25 students of class XII MIA 6 SMAN 10 Padang. Practicality data of prototype IV on field test can be seen in Table 4.

Table 4. Prototype practicality data on field test

	Kappa moment (k)	Category
Teacher reponse	0,96	Very High
Student responce	0,83	Very High

Questionnaire teacher response showed that prototype IV has a very high practicality category with value of kappa moment 0,96. Questionnaire student response indicated that prototype IV has a very high practicality category with value of kappa moment 0.83. Based on practicality data given by practitioner, it can be concluded that chemistry triangle oriented module was very practical and can be used in chemistry learning process on topic of reaction rate for Senior High School level grade XI.

3.2. Discussion

3.2.1 Validity of Chemistry Triangle Oriented Module on Reaction Rate Material

Validation of the chemistry triangle oriented module on topic of reaction rate material was determined by 5 experts consisting of three chemistry lecturers FMIPA UNP and two high school chemistry teachers in Padang. This is in line with the opinion put forward by Sugiyono that to test the validity of the instrument, can be used expert opinion (judgment experts) the minimum number of three people[13].

Assessment given by the validator was analyzed by using kappa cohen formula to obtain kappa moment. The kappa moment indicates the validity of a product. Assessment done includes the assessment of component content, construction, language and graffiti.

The validity of the content of chemistry triangle oriented module on topic ofreaction rate has a kappa moment(k) was 0.80 with a high validity category which means that the material contained in the module has been in accordance with the curriculum applicable which was Curriculum 2013 and has been in accordance with the demands of Core Competence (KI) And Basic Competence (KD) which are elaborated into indicators of learning and learning objectives. As submitted by Depdiknas that to produce a meaningful module and can be easily used by learners, the module should describe the basic competencies, indicators, and learning objectives to be achieved by the students[1].In addition, the components contained in modules such as models or drawings are in accordance with the learning indicators to be achieved and can help students understand the concept and answer the questions presented in the module. This is in line with the Depdiknas opinion that the presentation of images / models in a teaching material is needed so that students can master the Basic Competence to be achieved, see the picture / model higher than the meaning of reading or listening. In addition, the questions have led the students to the achievement of predetermined indicators and learning objectives[1].

Construct validity the chemistry triangle oriented module on topic of reaction rate has a kappa moment (k) 0.89 with a very high validity category. This means that the module has been presented in a specific, systematic and clear manner ranging from titles, Core Competencies, Basic Competencies, indicators, learning objectives, models, questions, and exercises. Modules are arranged systematically, and the clarity of indicators and learning objectives to be achieved will facilitate students in learning, and students will learn in a direction. This is in accordance with opinion of Nasution that one of the advantages of learning presented clearly and specifically is the students' learning becomes directed[14]. In addition, the concepts contained in the module do not conflict with each other. This is in line with that proposed by Rochmad that the construct validation is used to check whether the components of one model are not in conflict with the other model components[15].

The validitylanguagecomponent of the chemistry triangleoriented module on topic of of reaction rate has a kappa moment(k) 0.91 with a very high validity category. This indicates that the chemistry triangle oriented module uses a clearly readable form and letter size, the instructions and information presented in the module are clear, and the language used in the module meets the Indonesian rules of good and correct, and easy to understand. In accordance with the submitted Depdiknas that to produce a good teaching material, then it is necessary to evaluate the linguistic components that include; legibility, clarity of information, conformity with good and correct Indonesian rules, and clear language utilization[1].

The validitycomponent graffiti of the chemistry triangle oriented module on topic of reaction rate has a kappa moment(k) 0.89 with a very high validity category. This means that the model presented can be clearly observed, the type of letters used can be read, regular layouts, and the colors used can attract students' attention.

Overall the average value of the validity test of the chemistry triangle oriented module on topic of reaction rate is 0.87 with the category of very high validity. This proves that the module has fulfilled the four aspects of the validity test based on the validator's assessment so that this module can be used as a teaching material in the chemistry learning process at the reaction rate topic for Senior High School students Grade XI.

3.2.2 Practicality of chemistry-oriented triangle module on reaction rate material

Practicality refers to the level of ease in use, time in the implementation, as well as the attractiveness of teaching materials to student interests in the learning process[16].The practical data of this module was analyzed by using kappa cohen formula to obtain the kappa moment value. The kappa moment indicates the product's practicability.

The result of practicability analysis toward the module produced based on questionnaire of teacher response has very high practicality category with kappa moment value 0.96 and based on questionnaire of student response on small group test result of practical analysis has very high category with kappa moment value 0,81, and result of questionnaire analysis of student response on field test has very high category with kappa moment value equal to 0,83. Therefore, assessments include assessment of easy using, efficiency of learning time, and benefits.

The results of data analysis module practicality based on teacher response questionnaire provide a very high value of practicality. The modules have been prepared based on basic competence (KD) in accordance with those listed in the curriculum subject syllabus of 2013.

In terms of ease of use, chemistry triangleoriented module is considered practical with a value 0.99. This indicates that the module already has clear and easy-to-understand usage instructions, so the teacher knows the steps to be taken in the learning process. Sudjana and Rivai states that the instruction for teachers aims to make the teacher perform the lesson efficiently[17]. In addition, modules are made practically on the ease of use aspects of the language. It is evident from the teacher's reply that the language used in the module is easy to understand. In addition, the macroscopic, submicroscopic, and symbolic representations presented in the module are easy to understand so as to help students understand the concept of reaction rate. This is in line with Chittleborough, so that students better understand the chemistry in depth then in the learning process chemistry needs to involve multi representation chemical namely; Macroscopic aspects, submicroscopic aspects, and symbolic aspects[4].

In terms of efficiency of learning time, chemistry triangle oriented module on reaction rate material has been proved to be efficient when used in learning. This is seen from the results of practicality test by teachers who claim practical with a value 0.93.

In terms of the benefits of chemistry triangle oriented module on the reaction rate material is considered practical with a value 0.98. This is indicated by the teacher's reply that the chemistry triangle oriented module on topic of reaction rate can assist or replace the teacher's role in explaining the reaction rate because the activities contained in the module are presented very clearly, systematically, and easily understood by the students, so teachers have plenty of time to monitor student learning activities.

Student module practicality test was done in three stages: one-to-one, small group, and field test. The one-to-one evaluation aims to see the real errors of the developed module. Students engaged in a one-to-one trial consisting of three look at the whole section of the module and try to answer using a small portion of the module, then assess the module that had been developed. The data collection instrument used in the one-to-one test wasan interview sheet, based on the answers given by the students, it can be concluded that the module that has been developed was interesting and can foster the spirit of the students in reading and studying the module.

Small group evaluationaim to see the practicality of modules developed. Students involved in small group evaluation are six students. Based on the analysis of the practicality results in small group evaluation obtain kappa moment 0.81 with the category of very high practicality.

Field test involves 25 students of class XII. Based on the analysis of the results of field testing practice, the kappa moment was 0.83 with very high practicality category. This practical value was the average of three aspects of the practicality test, namely ease of use, efficiency of learning time, and module benefits.

In the terms of the ease of module use, chemistry triangle was considered practical. This shows that module has a clear and easy to understand instruction, so the students know the steps that have to be done in the learning process. In addition, in the term of the ease of language use was considered practical. This was proven from the answers and the comments given by the students on the practicality questioners at the field test which states that the module was easy to understand. Thus this has addressed the obstacles often faced by students who often have difficulty in understanding the teaching materials that are presented with an elusive language.

In terms of efficiency of learning time oriented module, chemistry triangle was considered practical, this was reflected from the answers given by students. Students can learn at their own learning speed. In connection with thisNasution states that the purpose of learning with the module is to open opportunities for students to learn according to the speed of each[14]. In addition, the learning time becomes more efficient with the use of chemistry triangle oriented modules in the learning process. This is demonstrated by the answers given by the students. This is in line with the opinion of Suryosubroto states that the purpose of the use of modules in the learning process one of them is that indicators and learning objectives can be achieved efficiently and effectively[18].

In terms of the benefits, chemistry triangle oriented modules considered practical, this was illustrated by the answers given by students, which states that the module was very helpful for students to understand the concepts on the material reaction rate well, and the use of modules in the learning process can improve the spirit of students for learning, also students can learn independently and in accordance with the speed of each learning. This is in line with Suryosubroto who states that the purpose of the use of modules in the learning process one of which is that learners can understand as much as they can and conduct learning activities independently[18].

To prove the chemistry triangle oriented module on the reaction rate material has a very high practicality, it has been done an analysis of the answers of students in answering questions in the module. Based on the results of the analysis module that has been done, the result that the average percentage of student value on Student Activity Sheet is 83.44%, Student Worksheet is 94%, and on the Evaluation sheet is 85.60%. This data indicates that the module prepared can be used and understood by the students.

Overall, the results of practicality test of the triangle chemistry-oriented module are practical and can be used easily for teachers and students because the modules have been designed with scientific approach stages in accordance with the demands of the 2013 curriculum, as well as the presentation of models presented by involving all three aspects of chemical representation, that was; Macroscopic representations, sub microscopic representations, and symbolic representations that help students to understand concepts and assist teachers in teaching and learning.

4. Conclusion

Based on the result of research and development of chemistry triangle oriented module on reaction rate material for learning chemistry at Senior High School level students grade XI can be concluded that chemistry triangle oriented module on reaction rate material can be developed with Plomp development model. The final product of the developed module has a very high category of validity and a high degree of practicality.

Acknowledgements

Researchers would like to say thank to Dr.Mawardi,M.Si., Dr. Minda Azhar, M.Si., Guspatni, S.Pd, MA, Dr.Bahrizal, M.Si., Drs.Emrizal, M.Si, and Ratna Juwita, S.Pd for being our product validators. Also we thank to Rima Septiani for language translation of this article. Lastly we thank to participating students and teachers and all of the people who had contributed to the success of this research.

References

[1] Departemen Pendidikan Nasional. 2008. *Pengembangan Bahan Ajar*. Jakarta : Departemen Pendidikan Nasional, Direktorat JendralMenejemen Pendidikan Dasar dan Menengah, Direktorat Pembinaan Sekolah Menengah Atas.

[2] Sunyono. 2012. Kajian Teoritik Model Pembelajaran Kimia Berbasis Multipel Representasi (Simayang) dalam Membangun Model Mental Pebelajar. Prosiding Seminar Nasional Sains, Universitas Negeri Surabaya, 14 Januari 2012 Halaman 486-495.

[3] Mawardi, 2011, "Inovasi dalam Pembelajaran Kimia, Suatu Tinjauan Teoritis. Proseding Seminar Nasional Kimia dan Pendidikan Kimia". ISBN: 978-602-8821-28-5, pp 244-252.

[4] Chittleborough, Gail Diane. 2007. The modelling ability of non-major chemistryy student and their understanding of the sub-microscopic level. 8(3), 274-292.

[5] Herawati, R.F., Mulyani, Sri., dan Redjeki, T. 2013. "Pembelajaran Kimia Berbasis Multiple Representasi Ditinjau Dari Kemampuan Awal Terhadap Prestasi Belajar Laju Reaksi Siswa Sma Negeri 1 Karanganyar Tahun Pelajaran 2011/2012." *Jurnal Pendidikan Kimia (JPK),* Vol. 2 No. 2.

[6] Sunyono., Leny Y., dan Muslimin. 2013. "Efektivitas Model Pembelajaran Berbasis Multiple Representasi dalam Pembangunan Model Mental Mahasiswa Topik Stoikiometri Reaksi". *Journal Pendidikan Progresif (JPP),* Vol. 3 No. 1 Tahun 2013. ISSN 2087-9849. Hlm. 77-78.

[7] Baruri, Andhika., Hardeli., Oktavia, Budhi, 2013, "Pembuatan Media Pembelajaran CD Interaktif Melalui Pendekatan *Triangle Chemistry* Pokok Bahasan Laju Reaksi Di SMA". *EKSAKTA* Vol. 2 Tahun XIV Juli 2013.

[8] Santi, A, S. 2015. "Pengembangan Media Pembelajaran Kimia Berorientasi *Chemistry Triangle* Menggunakan Aplikasi *Android* Untuk Materi Larutan Elektrolit dan Non Elektrolit Kelas X SMA/MA", *Skripsi*. Padang : Universitas Negeri Padang.

[9] Akker J. V. D., Bannan B., Kelly A. E., Nieveen N., dan Plomp T. 2010. *An Introduction to Educational Design Research*. Netherlands: Netzodruk, Enschede.

[10] Boslaugh, Sarah dan Paul A. W. 2008. Statistics in a Nutshell, a desktop quick referencce. Beijing, Cambridge, Famham, Köln, Sebastopol, Taipei, Tokyo: O'reilly.

[11] Purbasari, M. " Warna Dingin Si Pemberi Nyaman", *Humaniora. Vol.5,No. 1, p.357-366.*

[12] Daryanto, 2014. *Pendekatan Pembelajaran Saintifik Kurikulum 2013*. Yogyakarta: Gava Media.

[13] Sugiyono. 2009. *Statistika untuk Penelitian*. Bandung: Alfabeta.

[14] Nasution, S. 2011. *Berbagai Pendekatan Dalam Proses Belajar dan Mengajar*. Jakarta: Bumi Aksara.

[15] Rochmad. 2012. "Desain Pengembangan Perangkat Pembelajaran Matematika". *Journal Kreano. Vol.3, No. 1, p.59-72.*

[16] Sukardi H. M. 2011. *Evaluasi Pendidikan: Prinsip dan Operasionalnya*. Jakarta: Bumi Aksara.

[17] Sudjana, N dan Rivai, A. 2009. *Teknologi Pengajaran*. Bandung: Sinar Baru Algensindo.

[18] Suryosubroto, 1983. *Sistem Pengajaran dengan Modul*. Yogyakarta : PT. Bina Aksara.

[19] Peraturan Menteri Pendidikan Republik Indonesia Nomor 41 tahun 2007 tentang Standar Proses Untuk Satuan Pendidikan Dasar dan Menegah.

IOP Conf. Series: Materials Science and Engineering **335** (2018) 012106 doi:10.1088/1757-899X/335/1/012106

Design of Chemical Literacy Assessment by Using Model of Educational Reconstruction (MER) on Solubility Topic

EYusmaita* and Edi Nasra

Departement of Chemistry education, Universitas Negeri Padang

*ekayusmaita@fmipa.unp.ac.id

Abstract. This research aims to produce instrument for measuring chemical literacy assessment in basic chemistry courses with solubility topic. The construction of this measuring instrument is adapted to the PISA (Programme for International Student Assessment) problem's characteristics and the Syllaby of Basic Chemistry in KKNI-IndonesianNational Qualification Framework. The PISA is a cross-country study conducted periodically to monitor the outcomes of learners' achievement in each participating country. So far, studies conducted by PISA include reading literacy, mathematic literacy and scientific literacy. Refered to the scientific competence of the PISA study on science literacy, an assessment designed to measure the chemical literacy of the chemistry department's students in UNP. The research model used is MER (Model of Educational Reconstruction). The validity and reliability values of discourse questions is measured using the software ANATES. Based on the acquisition of these values is obtained a valid and reliable chemical literacy questions.There are seven question items limited response on the topic of solubility with valid category, the acquisition value of test reliability is 0,86, and has a difficulty index and distinguishing good

1. Introduction

Assessment is an important component of learning process and national standards have been set demands to educators to have the ability for doing an assessment to students. Although an educator has the knowledge and teaching skills, knowledge about evaluate is requisite in indicated for effective learning. The educators' ability in assessing students learning outcomes will have a huge impact in how successful achievement of students.

In education, there are several International assessment are often performed among others TIMSS, PIRLS and PISA. PISA looked at science education have function to prepare citizens in the future, that citizens are able to participate in society increasingly affected by the progress of science and technology. Therefore, science education should develop students' ability to understand the essence of scientific, scientific procedures, as well as the strengths and limitations of scientific. Related with the ability of scientific literacy indonesian students, assessment study performed by the PISA*(Program for International Student Assessment)* revealed that science learning is less successful to improve students scientific literacy. In 2000 Indonesia was ranked 38 out of 41 participating countries in PISA with mean value tests 393; in 2003 Indonesia was ranked 38 out of 40 participating countries with a mean value tests 395; In 2006 Indonesia was ranked 50 out of 57 participating countries with mean value tests 393; in 2009, Indonesia was ranked 60 out of 65 participating countries with score 383 (OECD, 2009) and in 2012 Indonesia was ranked 64 out of 65 participating countries. The results of PISA 2015 literacy released 6th December 2016, Indonesia was ranked 62 out of 70 participating countries, tests

IOP Conf. Series: Materials Science and Engineering **335** (2018) 012106 doi:10.1088/1757-899X/335/1/012106

and PISA survey subsequent is in 2018 with test results and survey will be released in the end 2019. (Kemendikbud.go.id. 2016).

This research is important to do, the monitoring results of PISA to Indonesian students is very disappointing, therefore need a fundamental and relevant ideas to the design of chemical literacy assessment that refers to PISA is an innovation in the field of education are very possible to be realized. It is based in scientific literacy as part of PISA, making chemicals as part of the PISA too. With created chemical literacy assessment that valid and reliable, expected be able to give an overview of how the profile undergraduate difficulty in answering basic questions test chemical literacy in the basic chemistry course. In nowadays, no research yet that study about design of chemical literacy assessment in basic chemistry courses. However, there are relevant research supporting this research, among others conducted by Firman, H (2007) about analysis of students difficulties in answering basic scientific literacy test based on the results of the National PISA 2006; research conducted by Cigdemoglu Ceyhan, et al (2012) about the efforts to increase levels of chemical literacy in thermochemical and thermodynamic concepts using context-based approach; research conducted by Yusmaita,E (2013) about Construction of teaching materials based on *green chemistry* to build students' scientific literacy; and research conducted by Suat Celik (2014) about how levels of the chemical literacy in a prospective educator MIPA.

The importance of literacy assessment emphasized by Heritage (2007: 141), as follows: *"The educators learn how to teach without learning a lot about how to* assess".Furthermore, Stiggins (2004: 16) asserts that a quarter to one-third the time teachers should be used for the assessment related the learning process. Therefore, educators must know and understand the principles of valuation. Assessment is an important process because the results can be used to plan instruction, guide student learning, determining the level/order, making a difference, decisive for further education, development of educational theory, formulating policies, allocate resources and evaluate the curriculum (NRC, 1996: 76). Therefore, the assessment needs to be planned, conducted, and analyzed properly so as to function properly. Here, the plan targets and indicators of achievement outcomes expected in this study.

2. Research Methods

2.1. Study Design

Design of this study using the Model Reconstruction of education, which is a cycle of form *the cyclic* which can be alternating *(recursive).* The third component of the MER related to each other and forming a systematic groove (Duit, *et al.*,2012: 26). For more details see the Figure 1

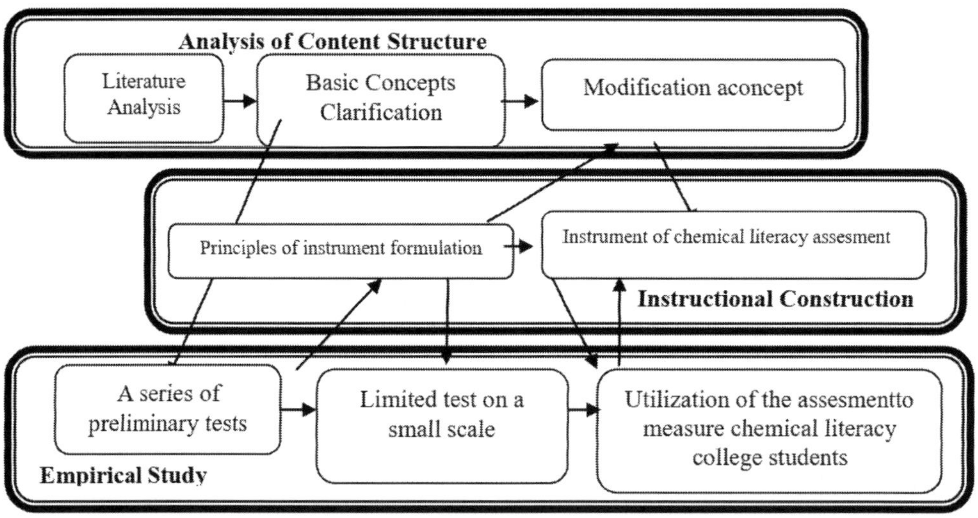

Figure 1. Modification Model of Educational Reconstruction

Figure1 Presents a repetitive process that occurs between the three components of education reconstruction model. Overall contribution to the MER can be see in the preparing the science component framework that relevant to science education research and development and thereby forming *a trilateral* relationship.

In practice, all the stages of the first component of this model does not absolutely have to be resolved first. Research process can be alternating *(recursive)* so that in perfecting the first component is affected also by the second and third components. Components of the first, second, and third performed with interleaving depend with the purpose and importance of the research will be achieved (Yusmaita, 2014: 45).

The method used in this study is a mixed methods. Another term to mention mixed methods very diverse, such as multi-method, convergence method, integrated method, and combination method (Creswell & Clark, 2007; Creswell, 2012: 22). Sources of data in this study was obtained by data triangulation, such as seeking the convergence between of qualitative methods and quantitative methods (Creswell, 2012: 22).

2.2. Research Phase

2.2.1.Analysis of Content Structure

Literature analysis

This study consists are a analysis of Basic Chemistry Course Syllabus KKNI and PISA study. Learning outcome that expected in this course are: Students can describe the condition of particle in solution and colloid and specific trait that arise as well as calculation and its usefulness; describes the determinant of the reaction rate and as well as calculation and its usefulness; redox reactions in an electrochemical cell as well as the calculation and its usefulness; describes the trait similarity elements of a class and the important compounds these elements; to explain the circumstances of nucleons in the nucleus and the changes that happened as well as the usefulness of such changes; describes organic compounds and their usefulness and reactions; and explains the important compounds in the organism and its role. In this study, basic chemical material is restricted to the concept of "Solubility".

Clarified content structures through chemistry textbook

One of component Model of Educational Reconstruction (MER) is content structure clarification. Content structure clarification on the solubility material refers to books Brady, James E and Fred Senese with title matter and itschanges Chemistry,6th edition and basic chemical book University composition sixth and the second books, Chemistry for University by Keenan, Charles W; Kleinfelter Donald C. Kleinfelter and Wood Jeese H.

Modification a concept

The process of concept modification is a process of revising the instrument after testing the test on a small scale.

2.2.2. Empirical Study. Empirical investigation conducted by conducting laboratory tests andlimited test on a small scale.

2.2.3. Assessment Construction. Construction questions chemicals literacy designed by three types of repesentation. It consist of aspek macroscopic, microscopic, dan simbolic levels.

2.3. Research sites

The research was conducted in the first year students who take Basic chemistry courses in the Department of chemistry, Padang State University. The participants consisted of 62undergraduate chemistry students.Implementation of this research was conducted in semester January-June 2017.

2.4. Instruments and Data Analysis

Analysis of the test is one of the activities in order to construct a test to get an image about the quality of the test, both the overall quality of the test and the quality of each item. Analyzes were performed

IOP Conf. Series: Materials Science and Engineering **335** (2018) 012106 doi:10.1088/1757-899X/335/1/012106

after the tests are prepared and tested to all subjects and the results fed back to the relevant test quality improvement (Mahmud, 2014).

One way to test analysis and question item is anates which is a computer application program that aims to analyze thequestion item. The program is especially useful for teachers in general observers education evaluation. The program was developed by Drs. Karno To, M.Pd. a professor of psychology at UPI and Mr. Yudi Wibisono, ST a computer consultant.

Facilities at ANATES V4

a. Scoring data, include:
   - Input score data, test outcomes
   - Weighting score data according to the required
b. Processing of data, including
   - The Reliability
   - Group superior and asor
   - Power differentiator
   - Level of difficulty Problem
   - Correlation item score with total score
   - Detractors quality (Siswoyo.2013)

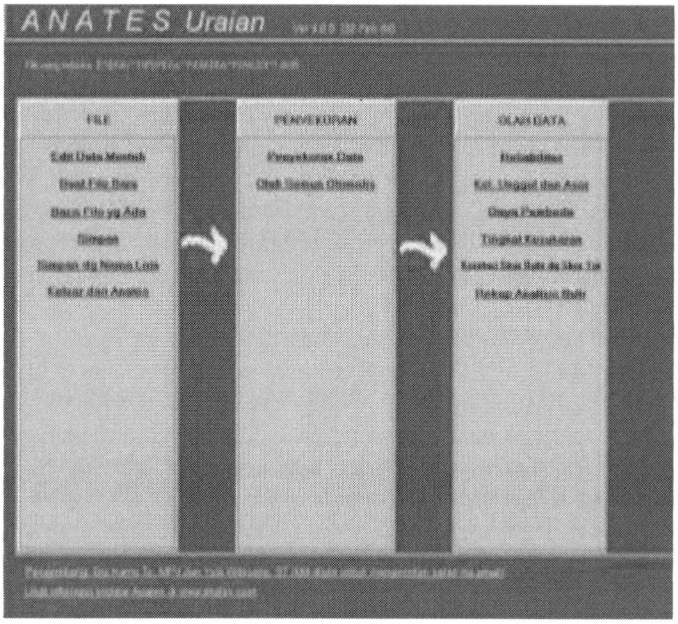

Figure 2. Interface software ANATES

3. Results

The form of the test that used in theassessment of chemical literacy undergraduate students is assessment testlimited responsethat includes two types, namely short fill type and multiple choice. In this test students are expected to be ableto analyzethe level of saturationof each solution by completing description image on the macroscopic aspects, microscopic and the symbolic. Discourse solubility text described into seven question items, where the level of thinking that used in designing this questionincludes a low level to high, with balanced proportions.

This designing assessment using the Model of Educational Reconstruction (MER), which consists of three components. Description the third cycle of these components can be seen in Figure 2.1. This cycle is back and forth, so in the implementation when the component one has not been completed in the middle of the process can be continued with two or three components. Otherwise, when the three components is being processed can return again to one or two components. The function of third components that recursive are as reinforcement between the components with one another.

In this reseach, the component is first performed analysis of content structure includes analysis of the syllabus KKNI university and analysis PISA competency that expected. Based on this analysis, obtained PISA competency about scientific phenomena which apply science knowledge on the situation-condition which it is given, describing or interpreting scientific phenomena and predicting changes, identifying the explanation and appropriate descriptions. Description from this analysis produce question indicators, wich then used as a benchmark in designing narrative discourse text. The truth of this narrative discourse text is clarified through the text book of the university and test question validation is done before the measuring instrument is tested with rational analysis through professional judgment, namely make discussion with lectures according to their scholarly expert. Measurement validity of this method is known as contentvalidity.

The next stage is tested validity of question content in the laboratory through experiments. Acquisition of data and images for this empirical study was made as complementary material of text discourse on macroscopic aspects. Then continued in the stage instrument construction by adding microscopic aspects and symbolic in the question. The trial is limited to text discourse was conducted to 31 students by using ANATES software. This test aims to obtain reability, differentiation, and level difficulty of the question.In this section, Analysis result of student's responses are presented.

3.1. Reabilitas test

As indicated in Table 1, the acquisition value of test reliability is 0,86. It can be concluded that test device classified have high reliability. It means that the test showed consistent results.

Table 1. The distribution of reliability sample

RELIABILITAS TES

Rata2= 12,97
Simpang Baku= 5,26
KorelasiXY= 0,75
Reliabilitas Tes= 0,86
Nama berkas: E:\EKA Y\VALIDITAS SOAL ANGKATAN 2014.AUR

No.Urut	No. Subyek	Kode/Nama Subyek	Skor Ganjil	Skor Genap	Skor Total
1	1	Annisa Rahim	9	7	16
2	2	Asmiyunda	4	2	6
3	3	Dilla Wahyuni	6	4	10
4	4	Dira Yulia Dasti	5	7	12
5	5	Febriani	11	5	16
6	6	Fitria Yolanda	6	6	12
7	7	Intan Selvi M...	11	5	16
8	8	Meliza Novita...	8	5	13
9	9	Mitra Susanti	5	0	5
10	10	Mutia Apriani	12	9	21
11	11	Nivelda Yani H	8	3	11
12	12	Novi A	11	6	17
13	13	Putri Handayani	11	7	18
14	14	Reza Amelia	6	5	11
15	15	Siti Nur Wili Y	12	9	21
16	16	Wulandari Dwi...	7	5	12
17	17	Andina Febria...	9	9	18
18	18	Annisatul Qaidah	3	4	7
19	19	Arni Saragih	4	2	6
20	20	Chairun Nisa	9	9	18
21	21	Diana Safitri	5	4	9
22	22	Fitria Harni ...	5	3	8
23	23	Hanifiyati Samha	10	8	18
24	24	Jusmalara Ais...	3	1	4
25	25	Mugi Santoso	10	9	19
26	26	Mugi Widodo	10	9	19
27	27	Nadia Afi Sha...	5	4	9
28	28	Resa Afrina	8	5	13
29	29	Risca Cahya A...	10	7	17
30	30	Raissa Elvia ...	2	1	3
31	31	Riris Juniati...	8	9	17

3.2. Distinguishing Power

Result in Table 2 show that most of item test have distinguishing between The top group (students who have high ability) with a lower group (students who have low ability) in good category.

Table 2. The distribution of distinguishing question

DAYA PEMBEDA
============

Jumlah Subyek= 31
Klp atas/bawah(n)= 8
Butir Soal= 7
Un: Unggul; AS: Asor; SB: Simpang Baku
Nama berkas: E:\EKA Y\VALIDITAS SOAL ANGKATAN 2014.AUR

No	No Btr Asli	Rata2Un	Rata2As	Beda	SB Un	SB As	SB Gab	t	DP(%)
1	1	2,88	2,13	0,75	0,35	0,99	0,37	2,02	25,00
2	2	2,88	1,25	1,63	0,35	1,04	0,39	4,20	54,17
3	3	2,13	1,13	1,00	0,35	0,83	0,32	3,12	33,33
4	4	3,13	0,63	2,50	0,35	0,52	0,22	1...	62,50
5	5	2,75	0,13	2,63	0,89	0,35	0,34	7,78	65,63
6	6	2,63	0,25	2,38	0,74	0,46	0,31	7,67	59,38
7	7	2,63	0,50	2,13	0,52	0,53	0,26	8,08	53,13

3.3. Difficulty Index

Based on index percentage difficulty levels of question , question number 1 in the category of easy because as many as 83.33% of students were able to answer the question correctly. This condition is commonly done in the design question that starts with an easy question then increased to medium and hard.The lowest percentage difficulty levels of question is owned by item numbre 5 and 6, wich is 35,94% and this is still included in the interpretation of average categories.

Table 3. The Distribution of difficulty index question

TINGKAT KESUKARAN
=================

Jumlah Subyek= 31
Butir Soal= 7
Nama berkas: E:\EKA Y\VALIDITAS SOAL ANGKATAN 2014.AUR

No Butir Baru	No Butir Asli	Tkt. Kesukaran(%)	Tafsiran
1	1	83,33	Mudah
2	2	68,75	Sedang
3	3	54,17	Sedang
4	4	46,88	Sedang
5	5	35,94	Sedang
6	6	35,94	Sedang
7	7	39,06	Sedang

Based on the analysis question of chemical literacy by using ANATES software, question in the bad category will be revised and modified. For the next used on a wider scale, namely as many as 62 first-year students who had been taking basic chemistry course.

4. Conclusion

Design of chemical literacy instrument on solubility topic using model of educational reconstruction. In general, this model consist of three components, namely content structure analysis, empirical study to the field and instrument constructs. The process of implementation of these three components run recursive. There are seven question items limited response on the topic of solubility with valid category, the acquisition value of test reliability is 0,86, and has a difficulty index and distinguishing good. The instrument has been designed then will be used to determine and mapped how the profil achievement chemical literacy that obtained by the first-year students deparment of chemistry.

References

[1] Allahyari, T., Rangi, N. H., Khosravi, Y., dan Zayeri, F. 2011. Development and Evaluating of A New Questionnaire for Rating of Cognitive Failures at Work. *International Journal of*

Occupational Hygiene. (3), 6-11
[2] Arifin, Zaenal. 2009. *Evaluasi Pembelajaran*. Bandung: PT Remaja Rosdakarya.
[3] Arikunto, S. 2002. Prosedur Penelitian suatu Pendekatan Praktek. Edisi V. Jakarta:
[4] Rineka Cipta.
[5] Celik, S. 2014. Chemical Literacy Levels of Science and Mathematics Teacher Candidates. *Australian Journal of Teacher Education*. 39(1): 1-13.
[6] Cigdemoglu, C., Geban, O. 2012. Improving students's Chemical Literacy Level on Thermochemical and Thermodynamics Concepts through Context-Based Approach. *Chemistry Education Research and Practice*. 1-15.
[7] Creswell, J.W. 2012. *Research Design, Pendekatan Kualitatif, Kuantitatif, dan Mixed*. A.b. Fawaid, A. Yogyakarta: Pustaka Pelajar.
[8] Duit, R. 2007. Science Educational Research Internationally: Conception, Research Method, Domain Research. *Eurasia Jurnal of Mathematics*. 3, (1), 3-15.
[9] Duit, Reinsders,dkk. 2012. The Model Of Educational Reconstruction – A Framework For Improving Teaching And Learning Science. *Science Education Research and Practice in Europe*. Netherlands: Sense Publishers.
[10] Firman, H. 2007. *Laporan Analisis Literasi Sains Berdasarkan Hasil PISA Nasional Tahun 2006*. Jakarta: Pusat Penilaian Pendidikan Balitbang Depdiknas.
[11] Flynn, A. B., Amellal, D .G. 2015. Chemical Information Literacy: pKa Values-Where Do Students Go Wrong?. *Journal of Chemical Education*. 1-18.
[12] Gabel,D.L. 1993. Handbook of Research on Science Teaching and Learning. New York: Mac. Millan Company.
[13] Laherto, A. 2012. Nanoscience Education for Scienctific Literacy. Opportunities and Challenges in Secondary School and in out-of-school Settings. Helsinki: *Academic Dissertation*.
[14] Lawshe. 1975. A Quantitative Approach to Content Validity. *Journal Personnel Psycology*. 28, 563-575.
[15] Marzono, R.J. 1994. Assesing student Outcome: Performance Assesment Using the Dimentions of Learning Model. Alexandria: Association for supervison and Curriculum Development.
[16] NRC.1996. *National Science Education Standards*.Washington: National Academic Press.
[17] OECD. 2009. PISA 2009 Assessment Framework Key competencies in reading, mathematics and science. [online]. Tersedia:http:// *www.oecd.org/dataoecd/11/40/44455820.pdf*
[18] Resnick, D.P. 1985. Standars, Curriculum, and Performance: A Historical and Comperative Perspektive. Educational Researcher.9. 5-19.
[19] Shwartz, Y., Bhen-Zvi, R., and Hofstein, A. (2006). "The Use of Scientific Literacy Taxonomy for Assessing the Development of Chemical Literacy Among High-School Students". *Chemical Education Research and Practice*. 7,(4), 203-225.
[20] Siswoyo, D. (2013). Anates V4 Software Analisis Data
[21] Stiggins, R.J. 1994. *Student-centered Classroom Assessment*. New York: Mac.millan College Publishing Company.
[22] Tola, B. 2008. *Literasi Sains Siswa Indonesia: Asesmen dan implikasinya*. Makalah pada Seminar Nasional Kimia dan Pendidikan Kimia, Bandung: FPMIPA UPI.
[23] Toharuddin, Uus, dkk. 2011. *Membangun Literasi Sains Peserta didik*. Bandung: Humaniora.
[24] Yusmaita, Eka. 2013. Konstruksi bahan ajar berbasis *green chemistry* untuk membangun literasi sains siswa. *Tesis*. Sekolah Pascasarjana UPI, Bandung.
[25] Wilson, R., Pan, W., dan Schumsky, D.A. 2012. Recalculation of the Critical Values for Lawshe's Content Validity Ratio. *Association for Assesment in Counseling and Education*. [online].

The Effect of Using Concept Maps in Elementary Linear Algebra Course on Students' Learning

H Syarifuddin

Department of Mathematics, Universitas Negeri Padang, West Sumatra, Indonesia

hendrasy@yahoo.com

Abstract. This paper presents the results of a classroom action research that was done in Elementary Linear Algebra course at Universitas Negeri Padang. The focus of the research want to see the effect of using concept maps in the course on students' learning. Data in this study were collected through classroom observation, students' reflective journal and concept maps that were created by students. The result of the study was the using of concept maps in Elementary Linera Algebra course gave positive effect on students' learning.

1. Introduction and Theoretical Background

The purpose of this research was to improve teaching and learning mathematics in higher education in Indonesia. One of the issues that have been encountered in education in Indonesia are the low level of students' involvement in mathematics learning. As a mathematics lecturer at Universitas Negeri Padang, I also found that this issue has become major problem of my teaching practice.

Students' learning in mathematics is a problem in mathematics education. According Brijlall and Maharaj[1], many students perform poorly in mathematics because they are unable to adequately handle information given in symbolic form which represent objects, for example mathematical expressions, equations and functions. Besides that problem, according to Maharaj [2] a high percentage of students have a static view about mathematics; students with such a view will experience difficulties in learning mathematics in depth and will find it difficult to engage in teaching and learning processes. To improve my teaching and learning practice and to overcome this issue I chose to implement the concept maps as a strategy.

Concept maps were used in this study as a students' activity before class. Students were required to create concept map about the topics that would be discussed in the class. Through this concept map activity students were encouraged to be actively involved in the learning process to enable them to construct their own understandings.

Concept mapping is derived from a constructivist approach to teaching and learning. Concept mapping is a technique used for representing knowledge graphically, where the knowledge that is represented is related to concepts that are interconnected. A concept map consists of nodes and links. The nodes correspond to important terms representing concepts and the links represent the relationship between the concepts [3]. The concept map is intended to demonstrate the level of deep understanding of the students through their illustration of connections between concepts [4].

Concept maps were first introduced by Novak as a research tool. The method of concept mapping has been developed specifically to tap into a learner cognitive structure and to externalise what the learner already knows [5, 6]. Although the primary intention was to use concept mapping in research, it was also found to be a useful tool in the teaching and learning process, helping students to learn.

Traditionally, people tend to have the view that the result of students' learning in mathematics can be seen through their achievement in mathematics tests. This view is demonstrated by the extensive use of tests by educators in their practices. The use of tests to evaluate students' performance in learning has many weaknesses. According to Stake [7] mathematics test scores are not a basis for indicating how well students have become successful in learning mathematics. They do not directly measure how well educated in mathematics the student is becoming. They do not identify the cognitive structures of students' thinking.

In fact, students' learning in mathematics can be obtained through a variety of ways. One of them is through authentic assessment. This kind of assessment looks directly at students' work and their performance as a result of their mathematics learning. Such assessments include students' written tasks, students' solution of problems, students' presentations, students' experiments, and teachers' classroom observations [8].

According to Battista [9] the focus of students' learning in mathematics must be on problem solving, mathematical reasoning, justifying ideas, and independently learning new ideas. Students have to be given opportunities to develop their communication ability in mathematics or to increase their ability to read, write, and discuss mathematics. The National Council of Teachers of mathematics (NCTM) has identified five process standards in teaching and learning mathematics: problem solving, reasoning and proof, communication, connection and representation. The result of students' learning in mathematics can be achieved with active involvement in learning using these processes. These processes enable them to deepen mathematical understanding that leads to increasingly sophisticated abilities required to meet mathematical challenges in meaningful ways [10, 11]. In this paper, the students' learning will be seen through problem solving process.

Students' activities that relate to the problem solving process standard are: (1) use problem solving approaches, (2) formulating problems in everyday life and mathematical situations, (3) developing and applying strategies to solve a variety of routine and non-routine problems, (4) verifying and interpreting results with respect to the original problem, and (5) distinguishing between necessary and irrelevant information in solving problems [10].

Research question to be answered in this paper is, "How is the effect of using the concept map in Elementary Linear Algebra Course on students' learning?"

2. Method

In this study, I used an action research methodology. Action research is a systematic investigation conducted by teacher researchers to gather information about how they teach and how well their students learn [12, 13]. According to Atweh [14] through action research teachers have an opportunity to improve their practice and to develop an understanding of their practice that will enable them to develop professionally. The participants in this study consisted of 37 students enrolled in the Elementary Linear Algebra course. One cycle of action research (planning, acting, observing, reflecting and creating a revised plan) was conducted during each week.

I used 'concept mapping activity' in every weekly lesson. Students created the concept maps through the discussions in their small groups. They created the maps outside the class. The groups comprising two or three students were formed based on the ease of their meeting outside the class. The presentation and discussion of mathematics concept maps was done at the beginning of the class, when it took about 15 to 20 minutes. After the presentation, the activity continued by a discussion about the concept map.

The students' learning that were observed was students' ability in problem solving. To evaluate students' ability in problem solving, the researcher would use the qualitative data from focus group interview and classroom observations.

IOP Conf. Series: Materials Science and Engineering **335** (2018) 012107 doi:10.1088/1757-899X/335/1/012107

3. Results

The aim of this section is to investigate students' ability in mathematics problem solving as consequence of the use of the concept maps approach in Elementary Linear Algebra course. The data that was used to perform the analysis in this section came from the focus group interview and the classroom observations on the discussions of group B during Week 2 and Week 6.

In focus group interview, Budi revealed his ability and his group strategy in handling problem solving,

> *In problem solving, I always found easy problems and difficult problems to solve. When I found the difficult problems I got help from friend in the group. Our strategy to solve mathematical problems in the group; if we were given a lot of problems to solve then we shared the problems so that each of us could work individually first, and then we discussed them together. So, in the group to create concept maps there was mutual exchange of information that increased my ability to solve the problem.*

Budi's comment above reflected that the use of concept maps strategy had facilitated his ability to solve the hard mathematical problems that he faced through group discussion.

In the classroom observations on the discussions of group B during Week 2, the discussions that took place were about solving selected mathematics problems. Through the discussions, students had the opportunity to sharpen their ability in solving mathematics problems. As was observed in the group discussions, Salsa and Vera were involved in a conversation to evaluate their results in solving a linear equations system.

> *What you did on number 2 is simpler than what I did Vera, you only used 5 elementary row operations.*

Salsa revealed this conclusion after she compared her job with Vera's job.

Vera gave her response,

> *Maybe you did not follow the procedure in using the elementary row operation to change its augmented matrix to the reduced row-echelon form.*

Salsa said,

> O ya, I did not pay attention to it, I will see.

Through this comparison, Salsa had the opportunity to improve her ability to solve the problem; she learnt about the technique which was used by Vera to solve the problem.

In the discussion during Week 2, they were active in finding and sharing ideas to get better answers when solving mathematics problems. Sharing ideas was really helpful for them to understand the topics and to solve the problems. After Vera's assistance about how to use the elementary row operation in solving the linear equations system, Salsa was able help Eki who did not know how to complete his job. In solving the problem, Eki said,

> *I am stuck here. I am not sure what I have to do next.*

Salsa gave her assistance,

> *You see, all entries on this row are zero, so you can interchange the row with the last row. Its meaning will not change.*

Eki's response,

> *Thanks Salsa, I understand how to do it now.*

The assistance that was given by Salsa helped Eki to better understand how to use the elementary row operation to solve the linear equations system problem.

The time allocated for discussions was limited to only 50 minutes. During Week 2 discussions, they had used most of the time to work independently. As a result, there was not enough time for them to discuss all issues that emerged from the problems so that not all members understood the solving of the problems. At the end of the discussion, Vera said, "I still don't understand yet how to decide if a system of linear equations has exactly one, infinite, or no solution by using the augmented matrix, but the time is over." This experience indicated that they needed much more time to discuss so that all members in the group were better able to understand the topics.

During Week 6, each member of the group first solved different problems. On this occasion, there were five problems to be discussed. So, one student would attempt to solve one problem, after which they discussed the solutions together to arrive at the best answer. They had arranged how long they would work independently, how long they would work together, and who was to write the final report.

The students not only asked their friends but also criticised their own work, like what was revealed by Revi in the group discussion in Week 6. Revi was not satisfied with her answer on a determinant problem; she had used many more steps to reach the result. Revi said, "Why has my answer so many steps, while your answer is so simple? How do you do that Vera?" Vera gave her response, "Let me see. I combined the expansion of cofactor and row reduction". Revi's response, "I see, I only used row reduction."

This experience encouraged them to be involved in discussions to evaluate and compare the differences of their methods. They realized that there were several ways to solve the problem. This teaching and learning strategy had enabled them to compare and discuss alternative strategies to solve the problems.

So, the use of the concept maps strategy in the Elementary Linear Algebra course had enabled students to discuss the solving of the mathematics problems. Through the discussions to create concept maps they had the opportunity to share knowledge and strategies on how to solve the problems so that their ability to solve the mathematics problems improved. In the conventional classroom that I taught before, students did not have much opportunity to share knowledge and strategies in their effort to solve the mathematics problems in the classroom. So, the use of the using of concept maps in Elementary Linera Algebra course gave positive effect on students' learning.

References

[1] Brijlall, D. and A. Maharaj *Applying APOS theory as a theoritical framework for collaborative learning in teacher education*. 2009.
[2] Maharaj, A., *An APOS analysis of students' understanding of the concept of a limit of a function.* Pythagoras: Journal of the Association for Mathematics Education of South Africa, 2011. **71**(2010): p. 41-52.
[3] Wange, Y. *Improving advanced mathematics teaching by adding some modern methods and ideas*. The China Papers, 2005.
[4] Roberts, L., *Using concept maps to measure statistical understanding.* International Journal of mathematics Education in Science and Technology, 1999. **30**(5): p. 707-717.
[5] Novak, J.D., *Concept mapping: a useful tool for science education.* Journal of Research in Science Teaching, 1990. **27**(10): p. 12.
[6] Brinkmann, A. *Knowledge maps tools for building structure in mathematics*. 2005. 1-10.
[7] Stake, R.E. *Pedagogic and Psychometric perception of mathematics achievement.* 1999.
[8] Darling-Hammond, L., J. Ancess, and B. Falk, *Authentic assessment in action*. 1995, New York: Teachers College Press.
[9] Battista, M., *The mathematics education of America's youth.* Phi Delta Kappan, 1999. **80**(6).
[10] The Oklahoma State *Priority academic students skills*. 2009.
[11] NCTM *Principles and standards for school mathematics*. 2000.
[12] Mills, G.E., *Action research: a guide for the teacher researcher*. Fourth ed. 2007, Boston: Pearson Education, Inc.
[13] Creswell, J.W., *Educational research: planning, conducting, and evaluating quantitative and qualitative research*. 3th ed. 2008, Columbus, Ohio: Pearson Merrill Prentice Hall.
[14] Atweh, B., *Understanding for changing and changing for understanding*, in *Researching the socio-political dimentions of mathematics education: Issues of power in theory and methodology*, P.V.a.R. Zevenbergen, Editor. 2004, Kluwer Academic: Netherlands. p. 1-19.

IOP Conf. Series: Materials Science and Engineering **335** (2018) 012108 doi:10.1088/1757-899X/335/1/012108

Developing Worksheet (LKS) Base on Process Skills in Curriculum 2013 at Elementary School Grade IV,V,VI

M Subhan[1], N Oktolita,M.Kn[2]

[1]Elementary Teacher Education, Education and Teacher Training Faculty, Dharmas Indonesia University, Dharmasraya, West Sumatera, Indonesia
[2]Elementary Teacher Education, Education and Teacher Training Faculty, Dharmas Indonesia University, Dharmasraya, West Sumatera, Indonesia

subhanundhari07@gmail.com

Abstract. The Lacks of students' skills in the learning process is due to lacks of exercises in the form of LKS. In the curriculum of 2013, there is no LKS as a companion to improve the students' skills. In order to solve those problem, it is necessary to develop LKS based on process skills as a teaching material to improve students' process skills. The purpose of this study is to develop LKS Process Skills based elementary school grade IV, V, VI which is integrated by process skill. The development of LKS can be used to develop the thematic process skills of elementary school students grade IV, V, VI based on curriculum 2013. The expected long-term goal is to produce teaching materials LKS Process Skill based of Thematic learning that is able to develop the process skill of elementary school students grade IV, V, VI. This development research refers to the steps developed by Borg & Gall (1983). The development process is carried out through 10 stages: preliminary research and gathering information, planning, draft development, initial test (limited trial), first product revision, final trial (field trial), product operational revision, Desemination and implementation. The limited subject of the this research is the students of SDN in Dharmasraya grade IV, V, VI. The field trial subjects in the experimental class are the students of SDN Dharmasraya grade IV, V, VI who have implemented the curriculum 2013. The data are collected by using LKS validation sheets, process skill observation sheets, and Thematic learning test (pre-test And post-test). The result of LKS development on the validity score is 81.70 (very valid), on practical score is 83.94 (very practical), and on effectiveness score is 86.67 (very effective). In the trial step the use of LKS using One Group Pretest-Posttest Design research design. The purpose of this trial is to know the effectiveness level of LKS result of development for improving the process skill of students in grade IV, V, and VI of elementary school. The data collection in this research uses the test result sheet of the process skill through pre-test and post-test. Observation results were analyzed with SPSS 16.0 software. The Result of analysis learning process of student skill of Sig value. (2-tailed) (0,000) $<\alpha$ (0.005) then H0 is rejected. There is a significant difference to the development of process skills between students using LKS with students who do not use LKS. It can be concluded that LKS have accuracy, ease and can improve result learn on aspect of skill process of student of grade IV, V and VI elementary school.

IOP Conf. Series: Materials Science and Engineering **335** (2018) 012108 doi:10.1088/1757-899X/335/1/012108

1. Introduction

Education is the key of all the progress and development of quality, because with human education can realize his potential both as a person and as a citizen. According to UU SISDIKNAS No. 20 2003. Objectives Curriculum 2013 seeks to simplify the subject matter and application of thematic-integrative learning model that starts from grade I until VI. The curriculum 2013 is prepared a generation ready in the future. Therefore, the curriculum is structured to anticipate the attitude and intellectual development. The emphasis is on encouraging students to better observe, ask, reason, and communicate (present) what they gain or know after receiving the learning materials. The objects that become learning in structuring and perfecting the curriculum 2013 emphasize on the phenomenon of nature, social, art, and culture for younger generation more creative, innovative, and productive.
Some aspects contained in the curriculum 2013 include:

1. Aspects of knowledge is the emphasis on the level of understanding of students in terms of lessons. The value of the knowledge aspect can also be obtained from Daily Deuteronomy, Middle Exam/End Semester, and Classroom Increase Test. In the curriculum of 2013, knowledge is a major aspect of curriculum in the past.
2. Aspect of skill is an effort of emphasis on student's motion or motor. In this aspect the students are guided to be able to do and be creative.
3. Aspects of attitude is attitude is the temperament of manners, attitude in learning, social, and religion. The difficulty of assessment in this aspect is mostly due to the fact that teachers are not able to supervise their students at all times.

LKS are sheets containing assignments done by students. This worksheet contains instructions and steps to complete a task given by the teacher to his students. The development objective to be achieved in this research is to produce LKS Skill Process for grade IV, V, VI Elementary School.

Carin [1] states that " Science Process Skills are often referred to as lifelong learning skills, because they can be used in daily life on learning in school in every subject. Science Process Skills is very important in the learning. Thematic learning will be meaningful for students if in practice involves motor skills to perform discovery or hands-on activities. Ostlund [2] states "

The intellectual value of mastering the science process skills is far greater than value of the ability to repeat scientific facts or principles. The process of scientific inquiry, learned not as set of rigid rules but as ways of finding answers, can be applied without limit. To be effective, methods for testing students with hands-on materials and the opportunity to demonstrate their use of science process skills.

Ostlund [2] describes process skills at level 4 until 6 as follows:

Skill
Observing: Using one or more of the five sense to gather information. May include the use equipment.
Communicating: Giving or exchanging information verbally, orally, and/or in writing.
Estimating: Approximately calculating a quantity or value based on judgment.
Measuring: Comparing objects to arbitrary units that may or may not be standardized.
Collecting data: Gathering information about observations and measurements in a systematic way.
Classifying: Grouping or ordering objects or events according to an established scheme. Based on observations.
Inferring: Develop ideas based on observations. Requires evaluation and judgment based on past experiences.
Predicting: Forming an idea of an expected result. Based on inferences.
Making models: Developing a physical or mental representation to explain an idea, object, or

event.
Interpreting Data: reading tables, graphs, and diagram: explaining the information presented in a table, a graph, or a diagram (including maps), and/or using it to answer questions.
Making Graphs: converting numerical quantities into a diagram that shoes the relationships among the quantities.
Hypothesizing: Stating a problem to be solved as a question that can be tested by an experiment.
Controlling Variables: Manipulating one factor that may affect the outcome of an event while other factors are held constant.
Defining Operationally: stating specific information about an object or phenomena based on experiances with it.
Investigating: Using observations to collect and analyze data to draw conclusions in order to solve a problem.

The application of these 15 skills is integrated into worksheet *(LKS)*. Steps to integrate the skills of the Thematic science process are integrated through the development of themes and sub themes that are adapted to teacher books and student books. The steps of developing LKS Process Skills shown in Figure 1 as follows.

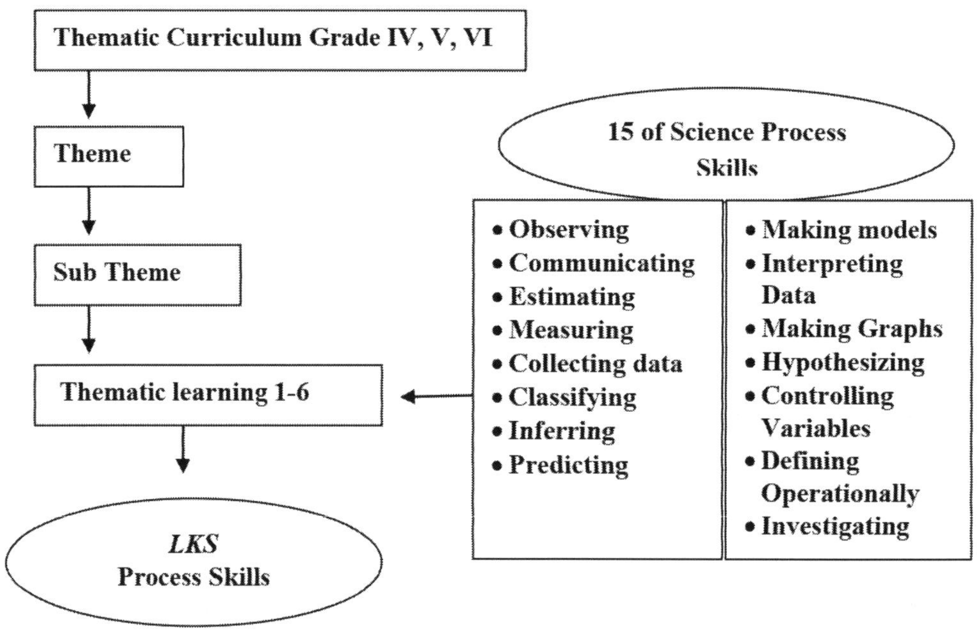

Figure 1. Steps Of developing LKS Process Skills

Osman and Kamisah [3] stated that "The characteristics of scientific process skills are discussed in the previous section with the implications for the role of primary educators. Active learning environment through observing, hypothesizing, predicting, investigating, interpreting findings and drawing conclusion and communicating presupposes teaching that stresses facilitation of learning rather than direct instruction. Teacher's role changes from one just presenting information to one of helping children to develop and use their own ideas ". Characteristics of process skills can be applied in an active learning environment through observation, hypothesis, predicting, investigating, interpreting findings and drawing conclusions and communicating presupposes teaching emphasizing the facilitation of learning from direct Thematic instructional. The teacher's role is changing from just presenting information to helping children to develop and use their own ideas.

R Russell Wilke; Straits, William J. [4] states that "There are many potential benefits for teaching process skills independently. However, the greatest benefit is that the routine teaching of these skills increases the likelihood that students will learn the skills. Students with stronger foundations in science process skills will be able to use them in other more intensive scientific inquiries, and are more likely to be successful in those inquiries. For this reason, instructors should emphasize the teaching and reinforcing of science process skill ". There are many potential benefits to teaching process skills independently. However, the greatest benefits in this teaching can improve students' process skills. Students with a strong foundation in the science process skills will be able to use them in scientific questions to be able to be used more intensely for other scientific, and more likely to be successful in the invention. For this reason, instructors should emphasize teaching and strengthen the skills of the science process.

Based on the results of the research that the development of process skills can develop aspects of student process skills as a motor of students to perform activities of discovery (inquiry). This research tries to develop LKS which is focused on developing the students process skill in Thematic learning.

2. Methods

This study uses a research and development research method developed by Borg and Gall [5]. The steps consist of preliminary research and gathering information, planning, development of the draft, initial tests (limited trial), revised first product, the end of trial (trial courts). The location of field trials was conducted at SDN 19 Sitiung, SDN 03 Tiumang, SDN 03 Koto Baru, SDN 03 Sungai Rumbai, SDN 08 Sitiung, SDN 189 / II Sari Mulya, SDN 14 Koto Baru, SDN 10 Koto Baru. In this field trial, the Design using the research design used in this research is *One Group Pretest-Posttest Design* (Purwanto&Sulistyastuti, 2007) as follows:

$$\mathbf{t\text{-}1(O_1)\,(X)t\text{+}1(O_2)}$$

- t-1 or O1 = before being treated (pretest)
- t + 1 or O2 = after being treated (postest)

The development of process skills in the data obtained by looking at the results of pretest and posttest observations. To test the results of the development of process skills by testing the hypothesis through the t-test. The first hypothesis to be tested is to compare the development of process skills, between learning before using LKS tools and after using LKS tools. The null hypothesis (Ho) and alternative hypothesis (Ha) in this study are as follows.

Ho: There is no significant difference to the development of process skills between students before using the LKS device after using the LKS device.

Ha: There is a significant difference to the development of process skills between students before using the LKS device after using the LKS device.retest-Posttest Design [6]

3. Results and Discussion

The results of the validity, practicality and effectiveness of LKS obtained are shown in Figure 2.

Table 3.1 Validity, Practicality, and Effectiveness of LKS

Validity	**Practicality**	**Effectiveness**
81.70	83.94	86.67
Very Valid	Very Practically	Very Effectively

The result of the students' skill test by applying the LKS of the development result in grade IV, V and VI classes is shown in Table 3.2 as follows.

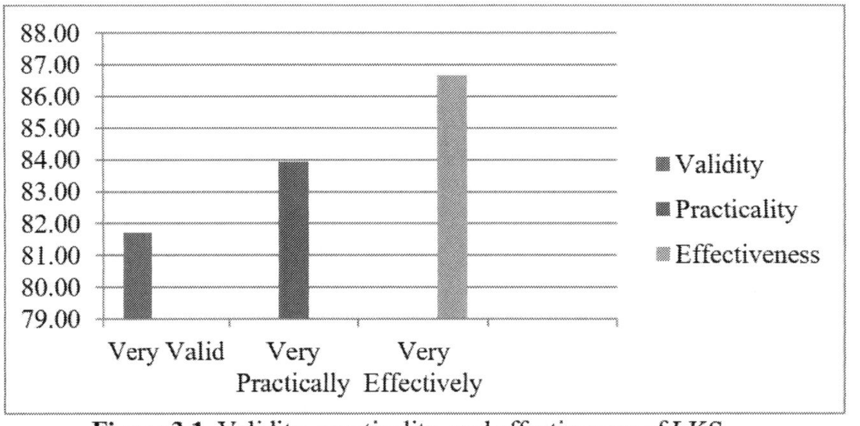

Figure 3.1. Validity, practicality, and effectiveness of LKS

Table 3.2 Pretest and Postest of process skills grade IV results

Descriptive Statistics

	N	Minimum	Maximum	Sum	Mean	Std. Deviation	Variance	Skewness		Kurtosis	
	Statistic	Statistic	Statistic	Statistic	Statistic	Statistic	Statistic	Statistic	Std. Error	Statistic	Std. Error
Pretest	72	55.00	95.00	5410.00	75.1389	6.81611	46.459	-.463	.283	1.646	.559
Postest	72	65.00	100.00	6455.00	89.6528	8.15028	66.427	-.507	.283	-.037	.559
Valid N (listwise)	72										

Table 3.3 Process Skills grade IV results

Score Process Skill	Pretest	Postest
Highest score	95.00	100.00
Lowest score	55.00	65.00
Average	75.1389	89.6528
Standard Deviation	6.81611	8.15028

Table 3.3 shows that there are differences in achievement of learning outcomes in pretest and posttest. An increase of 14.5% in postest results.

IOP Conf. Series: Materials Science and Engineering **335** (2018) 012108 doi:10.1088/1757-899X/335/1/012108

Table 3.4 Hypothesis Analysis grade IV results

Paired Samples Test

		Paired Differences					t	df	Sig. (2-tailed)
					95% Confidence Interval of the Difference				
		Mean	Std. Deviation	Std. Error Mean	Lower	Upper			
Pair 1	Pretest - Postest	-1.45139E1	7.07902	.83427	-16.17738	-12.85040	-17.397	71	.000

Table 3.5 Pretest and Postest of process skills grade V results

Descriptive Statistics

	N	Minimum	Maximum	Mean	Std. Deviation	Variance	Skewness		Kurtosis	
	Statistic	Statistic	Statistic	Statistic	Statistic	Statistic	Statistic	Std. Error	Statistic	Std. Error
Pretest	62	50.00	100.00	71.7903	10.52263	110.726	-.089	.304	-.091	.599
Postest	62	65.00	100.00	85.9194	9.73972	94.862	.235	.304	-1.205	.599
Valid N (listwise)	62									

Table 3.6 Process Skills grade V results

Score Process Skill	Pretest	Postest
Highest score	100.00	100.00
Lowest score	50.00	65.00
Average	71.7903	85.9194
Standard Deviation	10.52263	9.73972

Table 3.6 shows that there are differences in achievement of learning outcomes in pretest and posttest. An increase of 14.13% in postest results.

Table 3.7 Hypothesis Analysis grade V results

		Paired Differences					t	df	Sig. (2-tailed)
		Mean	Std. Deviation	Std. Error Mean	95% Confidence Interval of the Difference				
					Lower	Upper			
Pair 1	Pretest - Postest	-1.41290E1	10.00735	1.27093	-16.67042	-11.58764	-11.117	61	.000

Table 3.8 Pretest and Postest of process skills grade VI results

Descriptive Statistics

	N	Minimum	Maximum	Mean	Std. Deviation	Variance	Skewness		Kurtosis	
	Statistic	Statistic	Statistic	Statistic	Statistic	Statistic	Statistic	Std. Error	Statistic	Std. Error
Pretest	95	60.00	100.00	77.4737	11.00951	121.209	.123	.247	-.617	.490
Postest	95	60.00	100.00	90.6316	9.28973	86.299	-1.490	.247	2.104	.490
Valid N (listwise)	95									

Table 3.9 Hypothesis Analysis grade VI results

	Paired Differences							
				95% Confidence Interval of the Difference				
	Mean	Std. Deviation	Std. Error Mean	Lower	Upper	t	df	Sig. (2-tailed)
Pair 1 Pretest Postest	- -1.31579E1	8.22078	.84343	-14.83255	-11.48324	-15.600	94	.000

Table 3.10 Process Skills grade VI results

Score Process Skill	Pretest	Postest
Highest score	100.00	100.00
Lowest score	60.00	60.00
Average	77.4737	90.6316
Standard Deviation	11.00951	9.28973

Table 3.10 shows that there are differences in achievement of learning outcomes in pretest and posttest. An increase of 13,16% in postest results.

4. Conclusions

Table 4.1 Hypothesis Test Results Results

Grade	Paired Samples Test				
	Pretest/Postest	N	Mean	Std. Deviation	Sig. (2-tailed)
IV	Pretest	72	75.1389	6.81611	0.000
	Postest	72	89.6528	8.15028	
V	Pretest	62	71.7903	10.52263	0.000
	Postest	62	85.9194	9.73972	
VI	Pretest	95	77.4737	11.00951	0.000
	Postest	95	90.6316	9.28973	

Analysis of decision-making is a probability value. If the probability is> 0.05 then H0 is Accepted and if the probability is <0.005 then H0 is rejected. Based on Table 4.1 the results of the students' skill analysis process in class IV, V, and VI Sig values. (2-tailed) <0.005. So Ho is rejected, so there is a significant difference to the development of process skills between students before using the LKS device after using the LKS device in class IV, V, and VI. It can be concluded that LKS result of development can improve the process skill of elementary school students class IV, V, and VI.

IOP Conf. Series: Materials Science and Engineering **335** (2018) 012108 doi:10.1088/1757-899X/335/1/012108

References

[1] Carin, A. A. (1993). Teaching science through discovery. United States of America. Macmillan Publishing Company.
[2] Ostlund, K. L. (1992). Science Process Skills Assessing Hands-on Student Performance. United States of America. Addison-WesleyPublishing Company.
[3] Osman, Kamisah. (2012). Primary Science: Knowing about the World through Science Process Skills. Journal of Asian Social Science, 1-7.
[4] R Russell Wilke; Straits, William J. (2005). Practical Advice for Teaching Inquiry-Based Science Process Skills in the Biological Sciences. Journal of American Biology Teacher, 2005, 534-540.
[5] Borg, W. G., & Gall, M., D. (1983). *Educational research*. New York: Longman
[6] Erwan Agus Purwanto & Dyah Ratih Sulistyastuti. (2007). Metode penelitian kuantitatif untuk administrasi publik dan masalah-masalah sosial. Yogyakarta: Gava Media.

IOP Conf. Series: Materials Science and Engineering **335** (2018) 012109 doi:10.1088/1757-899X/335/1/012109

Teacher Mathematical Literacy: Case Study of Junior High School Teachers in Pasaman

D Ahmad, S Suherman, H Maulana

Mathematics Department, Universitas Negeri Padang, Indonesia

defri_math@fmipa.unp.ac.id, Suhermanspd_msi@yahoo.co.id, herumaulana@fmipa.unp.ac.id

Abstract. The aim of this paper was to examine the ability of junior high school mathematics teachers to solve mathematical literacy base Problems (PISA and PISA-like problems) for the case Pasaman regency. The data was collected by interviews and test. As the results of this study, teacher ability in solving mathematical literacy base problems for level 1 until 3 has been good, but for level 4 or above is still low. It is caused by teacher knowledge about mathematical literacy still few.

1. Introduction

To compare cross-nationally evidence on student performance, the Organizations for Economic Co-operation and Development (OECD) launched Program for International Student Assessment (PISA) in 1997 [5]. PISA is addressed to 15-year old pupils' scholastic performance in three domains: mathematics, science, and reading. In each domain, the average score is 500 and the standard deviations is 100 [5].

Since 2000, Indonesia has active in PISA, and the average of Indonesian student performance is always under the total average and outside of the standard deviations, especially for mathematics, for detail see the table below [6].

Table 1. Indonesian Student Performances in PISA-Mathematics.

Year	Average Score	Indonesian Rank	Total Participants
2000	367	39	41
2003	360	38	40
2006	391	50	57
2009	371	61	66
2012	375	64	65
2015	386	63	69

Based on the table 1, Indonesian score always in bottom down, even if its slowly increased, from 2009 to 2015. By this result, investigations are needed to find the causing factor or to increase the result. Is the main problem is students, teachers, curricula, or educational systems?

The investigations have been done by several researchers in various approaches. Some Indonesian researchers developed mathematics problems that is potentially increase students' mathematical literacy [2,7]. And other researchers developed teaching method for some mathematics materials due to mathematics literacy ability [1,3,9]. Differently, in this paper, we investigated teacher ability in solving PISA-like problems. As the main aspect that determine student performance, how if teacher mathematical literacy is not good enough? How can they increase student mathematical literacy?

2. Theoretical Framework

Traditionally, literacy is the ability to read, write, and use arithmetic. By expanding this definition in OECD countries, the meaning of literacy is the ability to access knowledge through technology and the ability to assess complex contexts [8]. To sum up we can say literacy as ability to read any information, to write any idea, and use arithmetic to solve problems in complex contexts.

One part of literacy is mathematical literacy, that define by PISA as the capacity to formulate, employ and interpret mathematics in variety of contexts [5]. it not only includes the ability how to use mathematics, but also reasoning mathematically and use mathematics as a tool to describe, explain, and predict phenomena [5]. Other definition also described by NCTM, i.e. The ability to read, listen, think creatively, and communicate about problem situations, mathematical representations, and the validation of solutions will help students to develop and deepen their understanding of mathematics [11]. These two definitions guide us to say that mathematical literacy is the ability to use the deepen understanding of mathematics in various contexts. In the simple ways to understand how to practice mathematical literacy, see figure below.

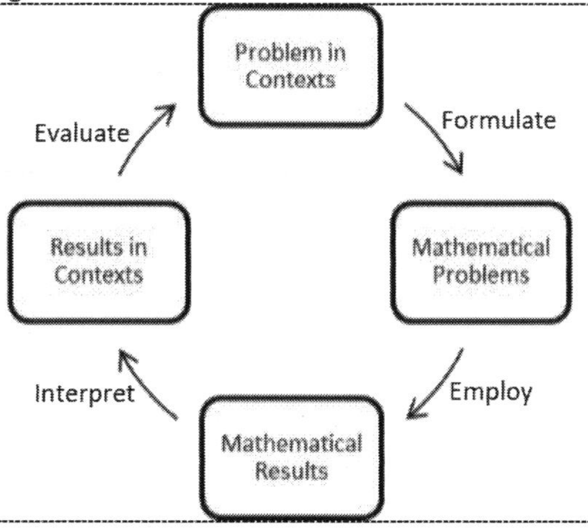

Figure 1. Model for Practicing Mathematical Literacy

PISA test is one of standardized test about mathematical literacy that hold every three years, in PISA student performance in mathematics is assessed through questions related to processes, contents, and contexts. Firstly, Processes in PISA is defined into formulating situations mathematically, using mathematics knowledges, and interpreting the mathematical results. In the other word, it includes the ability of students to change a problem in real world into mathematics world, and solve it in mathematics, and send it again to real world. Secondly, contents are divided into four categories, these are quantity, shape and space, change and relationships, and uncertainty. Quantity is about number sense and it forces the students to understand the measurements. Shape and space is not only about geometry, but also physical and visual aspects for instance pattern, positions, and orientations. And change and relationships, and uncertainty are about using variable and statistics and probability, respectively. Lastly, contexts i.e. the setting of student worlds in which the problems are placed. Is it in personal, societal, educational, or scientific?

Indonesia is one of OECD countries, so Indonesia should accommodate mathematical literacy in it curricula. Indonesian government has launched "Gerakan Literasi Sekolah (GLS)" (School Literacy Movement) as all of efforts that involve all of school society (teachers, students, and parents) and general society as part of educational ecosystem [10]. As a part of literacy, mathematical literacy should be involved in GLS programme at least in classroom activity such that mathematical literacy being the standard of classroom activities, exercises, and tests. So, we need to know teacher basic knowledges about mathematical literacy. This paper discusses about junior high school mathematics teachers in Pasaman ability in solving problem base on mathematical literacy, especially PISA or PISA-like problems.

3. Method

In this paper, case study was used to junior high school mathematics teachers. The data were collected through test and semi-structural interview about mathematical literacy. To make us easier in collecting data, we chose teachers from the same regency, in this case we chose Pasaman regency. We used 24 junior high school mathematics teachers in Pasaman as our sample.

The mathematical literacy test contained 10 problems which were adopted from PISA problems, Indonesian Mathematical Literacy Contest and some PISA-like problems from several researches [1,2,7,9]. These ten problems were chosen such that all of PISA contents include in the test. These are 3 quantity problems (in three different level those are level 1, 2, and 4), 2 uncertainty problems (level 4 and level 5), 3 space and shape problems (in three different level those are level 1, 3, and 4), and 2 change and relationship problems (both level 4). We provided around 10 minutes for each problem, so the teachers were asked to solve the problems in 100 minutes.

4. Results and Discussion

In this section, analysis of data that was collected is discussed. Generally, the score that has been achieved by teacher around 40 until 80 (scale 100) and from information that has resulted through semi-structural interview, the teachers did not really know about mathematical literacy, they only guessed its sense.

First of all, descriptive statistic of the result is presented as shown in the table below.

Table 2. Descriptive statistic of Teacher Performance in Solving PISA and PISA-like Problems

Min	Max	Average	Median	St. Deviation
40	80	60.8	59	12.3

Based on table 2, there were still some teachers that were scored lower than 50 and no one were scored more than 80. Most of them has score around 50 and 70 as the deviation standard is 12.3.

Now, the deeper discussion about teacher response is presented in four category of PISA content i.e. quantity, shape and space, change and relationship, and uncertainty. It is shown the average score and how the teachers answered the problems.

Firstly, quantity is about number sense and how to use it in various context. Three problems were provided for this category, two of them were low level problems i.e. level 1, and 2. For the low levels problem, almost all of teachers answered it perfectly, but for one other problem that was the level 4, average score of teachers is only 2.4 (scale 10) that was divided into two sub problems in which the scores are 1.9 (scale 5) and only 0.5 (scale 5), since this item separate into two problems. The example of wrong response of the teacher is shown on the figure below.

In Figure 2, it is the answer from respondent for the problems change second to minute. The respondent divided the given 949 seconds with 3600, in fact it should be divided by 60. Others response, the respondent assumed that it is 9 minutes and 49 second this information is not given in the problems (literacy problems). And most of respondent gave the blank response. Meanwhile, in figure 3, it is the answer for problem about if one part of 4-cross road has 949 second for red light, how long green light in others part of the 4-cross road? For this problem, there is no single respondent that can answer it. The selected answer in Figure 3 is the best response from respondent. Unfortunately, for this response the respondent did not realize that the way with red light, so it should be only three other ways that has chance to have green light. To sum up, for this red-light problem, teachers are not able to comprehend the number sense in the context.

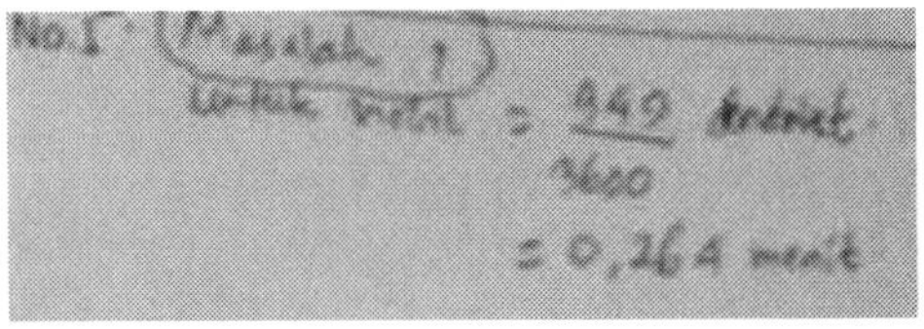

Translation:

For cars $= \frac{949}{3600}\ minutes = 0{,}246\ minutes$

Figure 2. Selected Answer 1

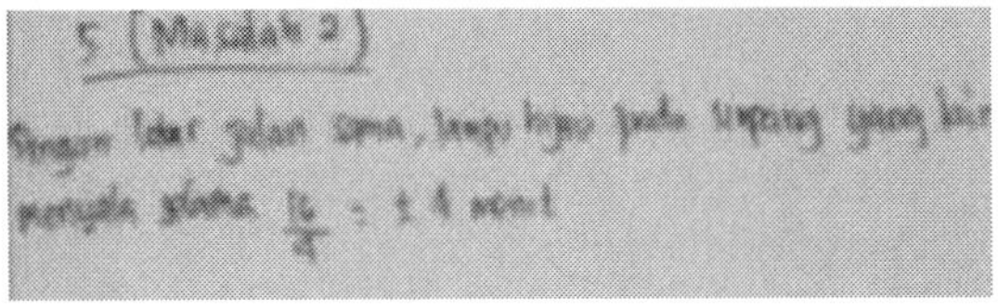

Translation:

By the same width of the roads, the green-light on others directions turn on in $\frac{16}{4} = \pm 4$ minutes

Figure 3. Selected Answer 2

Secondly, for uncertainty problems, the teachers were given two problems in level 4 and level 5. For the first item is about "Faulty Players" that is a PISA 2012 item. This problem is divided into three items, for the first item is true or false questions in three sub-items, unfortunately no correct response for this first question because no correct response for second sub-items for others sub-items more than 80% respondents gave the correct responses. The second sub-item problem is about "if in 2000 audio players with 5% of them is faulty, then In each batch of 100 video players made, exactly 5 will be faulty". All of respondents answered this sub-item by true. It means that all respondents have misunderstanding that 5% is an expected value, and is not an exact value. For other items in "Faulty Players", more than 60% respondents gave correct answers, while others did not answer it.

For the second problems in uncertainty is about "Voting and Destinations", this problem is divided into two items. The first problem is about count the numbers of voters for each destination, but the respondents assumed it as the set operations problems. After an interview, respondents gave such response because the problem just like set problem, so the respondent failed in comprehend the content of the problems. The problem and the chosen sample response are given in the figure below.

Problems: Trip Destinations

In 30 students, 4 students only choose Lawang Sewu, 8 students only choose Kebun Raya Penggoran, and 4 students only choose Kota Lama Semarang. There are 8 students that choose Lawang Sewun and Kebun Raya Penggoran, and remaining of students choose Lawang Sewu and Kota Lama Semarang. From this information, which are two destinations that will be chosen?

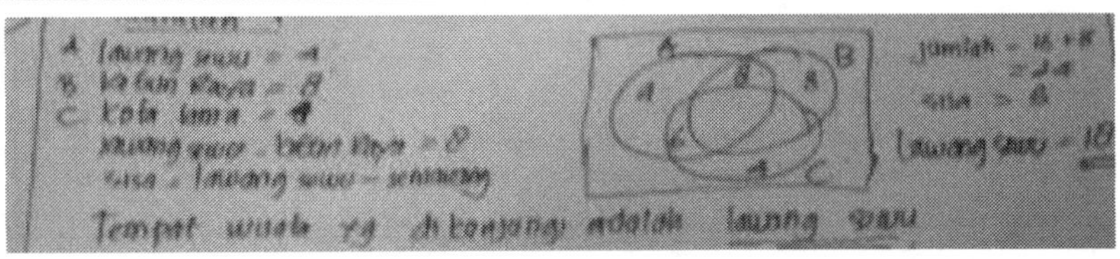

Translation:

A: Lawang Sewu = 4
B: Kebun Raya = 8
C: Kota lama = 4
Lawang Sewu-Kebun Raya = 8
Remaining =Lawang Sewu – Semarang
Sum = 16 + 8 =24
Remaining = 6
Lawang Sewu = 18
Tour destination that will be visited is Lawang Sewu.

Figure 4. Problem and Selected Answer 3

For this item, most of response is correct, but there is unneeded Venn diagram in the answer sheet as given in the sample answer above. The second problem is about choosing the cheapest transportations with time as constraint. Only 17% of respondent gave the correct response, while others only focus on the time they forgot to choose the cheapest way. They failed to see all possibility that is one of mathematical literacy needed.

Thirdly, for shape and space problems, from three problems, two of them are easy for the teachers i.e. for level 1 all of respondents gave correct answers and for level 3 problems more than 65% teachers has correct responses. For the level 3 problems teachers lose in reasoning and communicating their answered, they just assumed some that is not stated in the problems. When they are asked that "is the statement is correct?", they just process the questions with the logical "if the statement is correct, is the requirement is satisfied?". It is obviously a missing concept, because "$p \Rightarrow q$" does not mean "$q \Rightarrow p$", to see the mistaken loot at Figure 5(a) below.

Problems: Inheritance Division

Pak Bondan bequeath his inheritance field that is in a hill side with total area 216 m^2 to three of his son, as shown in the figure.

Problem: in your opinion, is the inheritance division equitable? State your reason!

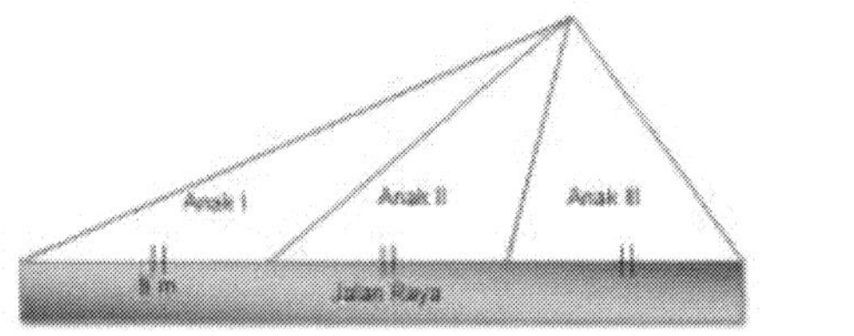

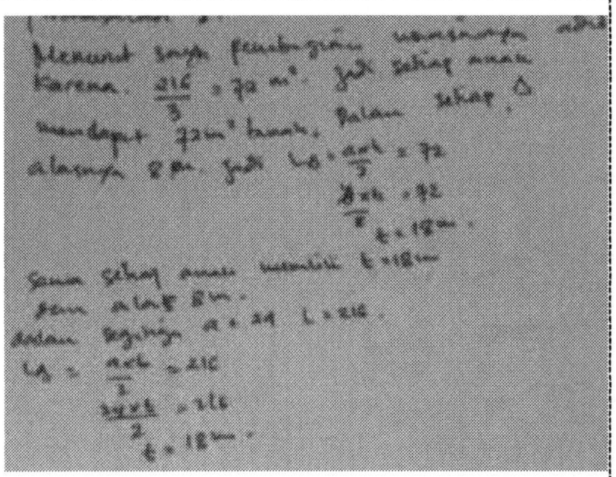

Translation:

In my opinion the inheritance division is fair.

Because $\frac{216}{3} = 72\ m^2$. So, each of son get $72\ m^2$.

In each Δ, its base is 8 m. So

$$L_\Delta = \frac{a \times t}{2} = 72$$

$$\frac{8 \times t}{2} = 72$$

$$t = 18\ m$$

The same, ech son have $t = 18m$ and its base 8m

In triangle $a = 24, L = 216$

$$L_\Delta = \frac{a \times t}{2} = 216$$

$$\frac{24 \times t}{2} = 216$$

$$t = 18\ m$$

(a)

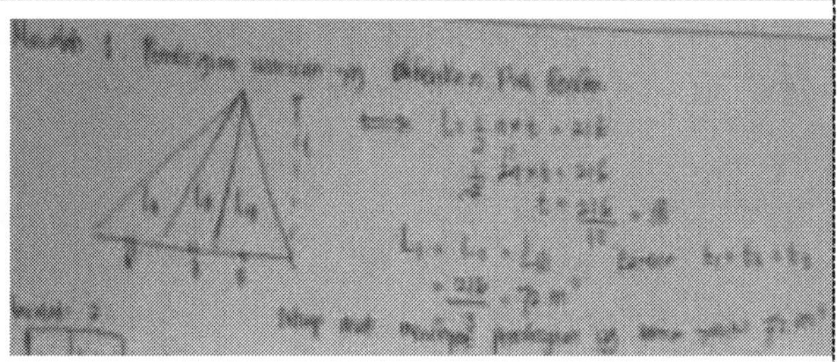

Translation:

The inheritance division that is given by Pak Bondan

$$L_\Delta = \frac{a \times t}{2} = 216$$

$$\frac{24 \times t}{2} = 216$$

$$t = \frac{216}{12} = 18\ m$$

Because of $t_1 = t_2 = t_3$

$L_1 = L_2 = L_3$

Each of son get the same area that is $72\ m^2$

(b)

Figure 5. Problem and Selected Answer 4, (a) is an incorrect response, (b) is a correct response.

Different with the answer in Figure 5(a), in Figure 5(b), the respondents show by picture that three of those triangle, each has the same height and then he counts it.

For level 4 problem in shape and space is about "Container Stack", the respondents are given a picture of container pile and they must count the number of container in the pile. For these problem most of respondent only count it one by one and do not show their method, 35% of respondent success in count it while others failed. The example of best response by respondents is given in the figure below.

Problem: Container Pile

In the figure, how many containers are in the container pile? Explain the method that you use to count it!

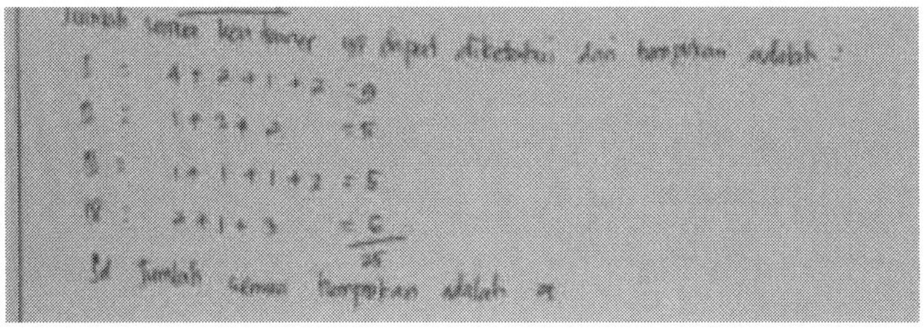

Translation:

The number of containers that is known in the pile is;
$I: 4+2+1+2=9$
$II: 1+2+2=5$
$III: 1+1+1+2=5$
$IV: 2+1+3=6$
So, the number of container in the container pile is 25.

Figure 6. Problem and Selected Answer 6

In Figure 6, the respondent state clearly how to count the number of container. So the answer in the figure has the perfect mark.

Lastly is for change and relationships problems that contains two level 4 problem. For the first problem, it is about pattern as we know it is one of material in junior high school mathematics. There are so many patterns that is appears in the book for instance triangle numbers, square numbers, e.t.c. and the problem present in the same ways with the problems in the books. In this problem, there are two pattern that should be found. Those are linear just like an arithmetic series and quadratic pattern just like quadratic numbers. For the last comment to this problem is this problem is answered correctly by 80% of respondent.

Differently, the other problem is provided in the real context but also about pattern. For this problem less than 40% respondents that answer it correctly. This problem should be easier than the first one because the pattern only in linear and repeated form, but the result says "no". May be, it is caused by the problem is presented in different ways with the fashion of problems that has been solve by the respondents before. So, they do not realize that it only linear and repeated series. Now we need to see how the respondents answered this problem. In the figure below the selected correct and incorrect answer is given.

IOP Conf. Series: Materials Science and Engineering **335** (2018) 012109 doi:10.1088/1757-899X/335/1/012109

Problem: Monas

Each Saturday night at 19.00 to 20.00, the water fountain in Monas hall is simulated. Each water fountain has their own pattern as shown in the figure bellow

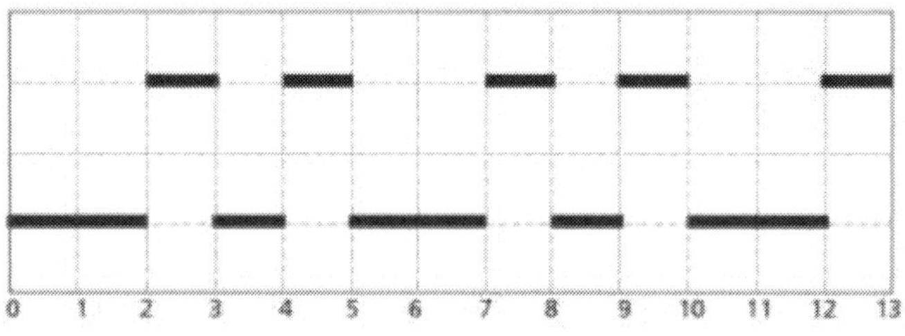

Spraying the water

Stop Spraying

Times (minute)

In total, how long the water fountain sprays the water during the simulation times? Explain your answer!

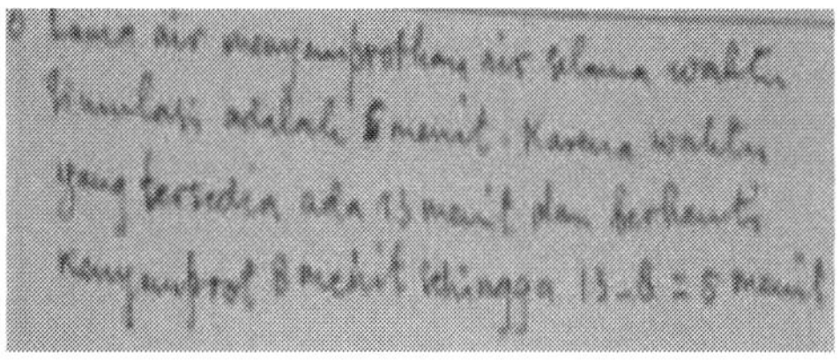

Translation:

The time for water fountain to spray the water during simulation time is 5 minutes. Because of the times that is provided is 13 minutes and stop to spray 8 minutes, then 13 – 8 = 5 minutes.

(a)

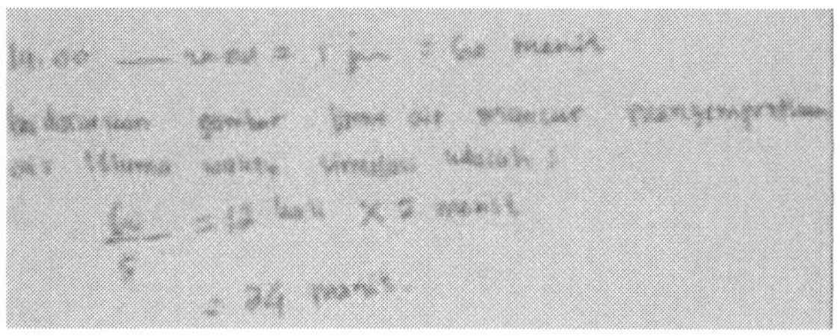

Translation:

19.00 – 20.00 = 1 hour = 60 minutes.
Base on the picture, the time for water fountain to spray the water during the simulation time is;

$$\frac{60}{5} = 12\ times \times 2\ minutes = 24\ minutes$$

(b)

Figure 7. Problem and Selected Answer 7 (a) is incorrect answer, and (b) is correct answer.

The Figure 7(a) is the selected incorrect response for the problem in the figure. The respondent by this answer cannot catch that the simulation time is one hour, he only counts the spraying time in the figure such that they cannot give the correct response. And, in Figure 7(b), the respondents state the answer correctly, but he did not give his reason and there is a mistaken in communicate his idea. Because he just wrote $\frac{60}{5} = 12\ times \times 2\ minutes = 24\ minutes$, all part of this equations is incorrect, but we just try to understand his understanding.

We have discussed the detail of teacher performance in mathematical literacy base problems. By this test, some teachers failed to solve the level 4 or above problems. Sometimes they cannot catch the context, that cause a misunderstanding in answer the problems. To get the main problems of the teachers to answers we did a semi-structural interview to the teachers.

By the interview, we got some information from the teacher about mathematical literacy. Firstly, around 25% of respondents do not have enough information to understand what is mathematical literacy. Secondly, most of teacher do not know about mathematics literacy base problems. When we asked teacher to create the mathematical literacy base problems, they only give the usual application problems. For instance, "Pak Budi has a field in form of square with length side 12 m …". By this the respondents just think that mathematical literacy is like applications problems. And when we ask the teacher about PISA, most of them has hear about it, but they never know about kind of PISA problems, and also they do not know about detail of PISA.

By this interview, we predict that why some teachers failed in solving PISA and PISA-like problems, especially for level 4 or above problems, it only because they do not accustomed this kind of problems.

5. Conclusion and Further Remarks

This study reports the ability of junior high school mathematics teachers in solving mathematical literacy base problems (PISA and PISA-like problems). As the case of Pasaman teachers, Some of them do not really know about mathematical literacy or PISA, some of them only hear about it. By the test results, in average, the teachers get grade 3 (scale 4) to the test. Around 25% of the teachers that is a respondent still hard in communicate their idea as in mathematical literacy, an individual has to reasoning their idea. By this result, we need to introduce and make mathematical literacy as the teacher habits through seminar, publications, books, e.t.c.. Because of teacher is the main factor that determine the student performance, we need more research in searching factor that is causing teacher abilities or how to improve teachers ability, the design of learning, and others.

Acknowledgments

We say thanks to all of our students that has help us in collected the data, and we also say thank to the teachers that has been our respondents.

References

[1] Ahmad F 2002 *Applying Realistic Mathematics Educations (RME) in Teaching Geometry in Indonesian Primary Schools (Published Dissertation*) (Enschede: University of Twente)

[2] Ahyan S, Zulkardi, and Darmawijoyo 2014 *Developing Mathematics Problems Based on PISA Level of Change and Relationships Content* (*Journal on Mathematics Education vol* 5 (1)) pp 47-56

[3] Ayvaz U, Gunduz N, and Bozkus F 2017 *Understanding of Prospective Mathematics Teachers of The Concept of Diagonal (IndoMS Journal on Mathematics Educations vol 8 no 2)* pp 165-184

[4] Friedland Ellen S, McMillen Susan E, and Hill Pixita del Prado 2011 *Collaborating to Cross the Mathematics–Literacy Divide: An Annotated Bibliography of Literacy Strategies for Mathematics Classrooms (Journal of Adelescent and Adult Literacy vol 55(1))* pp 57-66

[5] OECD 2016 *PISA 2015 Assessment and Analytical Framework* (Paris: OECD Publishing)

[6] OECD 2016 *Country Note – Result from PISA 2015* (Paris: OECD Publishing)

[7] Oktiningrum W, Zulkardi, and Hartono Y 2016 *Developing PISA-like mathematics task with Indonesia natural and cultural heritage as context to assess students' mathematical literacy* (*Journal on Mathematics Education vol* 7 (1)) pp 1-8

[8] Stacey K 2011 *The PISA view of mathematical literacy in indonesia, Journal on Mathematics Education (IndoMS-JME vol 2(2))* pp 95-126

[9] Zulkardi 2002 *Developing a learning environment on realistic mathematics education for Indonesian student teachers (Published Dissertation*) (Enschede: University of Twente)

[10] Kemendikbud RI 2016 *Buku Saku Gerakan Literasi Sekolah* (Jakarta: Kemendikbud

IOP Conf. Series: Materials Science and Engineering **335** (2018) 012110 doi:10.1088/1757-899X/335/1/012110

Developing Teaching Materials Using Comic Media to Enhance Students' Mathematical Communication

V N Yulian[1,2]

[1]Study Program of Mathematics Education, Universitas Subang, RA Kartini Street Km 3 Subang, Indonesia.
[2]Study Program of Mathematics Education, School of Postgraduate Studies, UPI, Jl. Dr. Setiabudhi 229, Bandung, Indonesia.

varanina15@student.upi.edu

Abstract. Teaching materials are a set of materials that are arranged systematically written or not, to create an environment or atmosphere that allows students to learn. The purpose of this study is to provide an overview of how the development of teaching materials using comic media that enhance mathematical communication, as well as feasible and effective teaching materials developed. Research method used in this research is Research and Development. In the sense of research contains about how to develop teaching materials through several stages such as validation by experts, as well as revisions. Sources of data used in this study were students and teachers SMK Bandung Barat. The results showed that the teaching materials developed feasible and effective use for students of class X SMK Bandung Barat. Teaching materials received a proper assessment of the experts after going through several stages of revision, in addition to the effective teaching materials used by students seen from the liveliness and the value of classical completeness that reaches more 85% of students. Based on the result of the research, it can be concluded that the developed teaching material gets the proper judgment from the expert, and effectively used in the learning by the students of X-1 Pharmacy class with the classical completeness reach 86% and the student activity is 91,4%.

1. Introduction

Communication is an important component in the learning process, including learning mathematics. Characteristics of mathematic is abstract, contain with terms and symbols, bring many students who accept all the knowledge without trying to understand what information was contained in there. Most students use the method of memorizing the formula to learn mathematics, whereas the essence of mathematics learning is not memorizing but as stated in Minister Rule [1].

The purpose of the fourth point of math instruction listed in Minister Rule [1] is students to be able to communicate ideas with symbols, tables, diagrams, or other media to clarify the situation or problem. It is clear that mathematical communication is one of the important capabilities that students must developed. But some research show that the ability of mathematical communication in Indonesia is still not good. According to Shadiq [8] results of the Center for the Development of Mathematics Teacher also revealed that in different parts of Indonesia, most students have difficulties in problem-solving problems and translating daily life into a mathematical model. This indicate that the communication skills and problem solving in mathematic of Indonesian students' are still poor.

This happens as a result from very rarely the students are required to provide explanations in mathematics lessons, so it is very strange for them to talk about mathematics. To reduce the occurrence of such things, students need to be accustomed communicate orally and writing ideas to others according to their own interpretation.

Communication is a very important part of mathematics and mathematics education. Communication is a way of sharing ideas and verify the reasoning. Through communication ideas can be rebound, improved, discussed, and developed. The process of communication also can helps build meaning, provide ideas and publish ideas. When students are challenged by their thinking and their ability to think about mathematics and communicate their thoughts orally or in writing, they are learning to explain and convince. Listening to other students' explanations, giving students the opportunity to develop their understanding (NCTM)[6].

We overcame this problem by introducing comic media as the development of teaching materials. Todays, teachers demands to use the right teaching materials as responsible for the learning process in the classroom, developing teaching materials in the form of media is needed to help the learning process. By applying scaffolding, strategies and a good teaching materials is expected to generate interest and motivation of students either in the form of methods or approaches through teaching materials with basis of the learning activities phase. Scaffolding is the support given to learners to learn and solve problems. Scaffolding can be a guidance, encouragement, warning, outlined problems into solving steps, giving examples, and other actions that enable learners to learn independently.

Several studies have shown that comic media have a positive impact on learning, action research shows that learning by using comics through cooperative learning model Student Team Achievement Division (STAD) can increase students' learning interest in mathematics (Wahyuni)[10]. Maulana [5] says matematikomik have significant effect to improve student's motivation and achievement; there is a significant and positive relationship between the increase of motivation with student learning achievement of student learning as the influence of matematikomik usage; and in general the students give a positive response to the learning of mathematics by using comics.

Teaching materials that will be the authors discussed in this research is the comic media, because the uniqueness of its function is as a teaching materials and can entertaining. Comics as a visual media is assumed to have an effect on the acquisition of knowledge as a result of learning, because it is able to attract interest and attention in conveying information. This is suitable with his role to visualize ideas or concept. Especially by seeing the tendency that the main consumers of comics are children from primary school to high school, although some of college students are still become consumers of comics. (Yakti)[11]

The use of comics as a teaching materials has an important role to increase students' interest in learning, because the presentation of comics takes students into a joyful atmosphere, thus creating joy in learning (DePorter, Reardon, and Nourie)[3]. The excitement in learning is a overflow of emotion that activates the nerves of the brain to be able to record lessons more easily. This is accord with Goleman's [4], "Research tells us that without emotional involvement, the brain's neural activity is less than it takes to attach a lesson to memory." Especially at school age most students still have visual learning styles that are more likely to activate their memories through images captured by the eye (DePorter and Hernacki)[2].

To overcome theabove-mentioned weaknesses, a research has been carried outby developing teaching materials using comic media as method gives students opportunities to identify problems, collect required information, investigate and analyze the result of investigation and exploration to be applied in a new situation. By developing teaching materials using comic media of learning process, they will be able to construct their own knowledge and put it in their long-term memory permanently.

2. Research Methods

This research was the Research and Development procedure. This development research, referring to the model of Sugiyono [9] which has been modified at the stage. The subjects of this study were students of X-1 Pharmacy class at vocational schools. Instruments that used in this study are student worksheets and observation sheets of learning activities. Methods of data collection conducted in this study is a validation questionnaire method, observation sheet and student learning outcomes. Data analysis method used is descriptive percentage. The first stages of research by analysing the potential and problems that exist about learning in vocational school in Bandung Barat, then make data collection to create teaching materials. The third stage start designing the initial teaching materials which are then validated by the experts, the validation process is revised according to the advice given by the expert afterwards. The revised teaching materials were tested on the use of small-scale teaching materials with 6 students, after a small-scale trial and then revised again before large-scale trials. Implementation of large-scale trials conducted learning by using teaching materials on 35 students, students give an advice to perfect the final product. Experts come from lecturers of FMIPA UPI and teachers of vocational schools, they are validating the teaching materials. Then the validation results in the form of suggestions and comments are used to improve the developed teaching materials. Subsequently validated or reassessed by experts.

3. Result and Discussions

In this case teaching materials using comic media are validated by material experts and media experts. Validation is done until the teaching materials are said to be feasible by experts. Expert validation results can be presented in Table 1.

Table 1. Data Validation Results by Experts on Teaching Materials

No.	Validator	Percentage Score
1.	Material Experts	91,5%
2.	Media Experts	90,5%

In table 1 it is clear that the feasibility percentage for validator is at very good criteria with the percentage of each is 91.5%, 90.5%, is in the range of 81% to 100%. This means teaching materials using comic media can be used in the learning process.

The effectiveness of teaching materials is measured based on student learning outcomes after learning using teaching materials that are prepared. Recapitulation of student learning outcomes on the use of developed learning materials is presented in Table 2.

Table 2. Recapitulation of Student Learning Results

No.	Category	Achievements
1.	Mean	80,5
2.	Sum of students	35
3.	Highest score	91
4.	Lowest score	55
5.	Students complete (tuntas)	30
6.	Students not complete (tidak tuntas)	5
7.	Classical completeness	86%

From Table 2 can be seen from the total number of students who completed are 30 students with students' learning completeness 86% classically with 80,5 average score. The successful of teaching materials because students can understand the teaching materials. This can be proven from the results of student responses states that 86% of students more easily understand the teaching materials using comic media. These results prove that the teaching materials using comic media effectively used for students of vocational schools. This is because the teaching materials by researchers different from the

IOP Conf. Series: Materials Science and Engineering **335** (2018) 012110 doi:10.1088/1757-899X/335/1/012110

existing teaching materials in schools. A research conducted by Purnomo [7], the application of teaching materials of research results positively affect the results of biology of students of SMP Al Irsyad, the positive influence as the impact of the application of teaching materials research results on the value of student learning outcomes. There are differences in achievement when using teaching materials due to differences in student activity in learning. Completed students are mostly students who have highly active learning and include the active student's categories. Students of high activity categorized but not complete KKM or students with low activity category but thoroughly KKM mainly due to internal factors that exist within the student. Aspects of student activity are assessed by observers using observation sheets. Recapitulation of student activity observation is presented in Table 3.

Tabel 3. Recapitulation of student activity observation

No.	Score Category	Criteria of Activity	Sum of Students	Percentage
1.	87 – 100	Very active	6	17%
2.	73 – 86	Active	26	74%
3.	59 – 72	Adequate	3	9%
4.	45 – 58	Less active	0	0
5.	≤ 44	Not active	0	0

Overall, it can be seen that the student activity in learning of the large-scale trial class is 91% taken from the percentage of very active and active student criteria, so it can be seen that student activeness level is in very active category. The results of this study indicate that teaching materials using comic media effectively used in learning, seen from the results of student activeness.

Developing of teaching materials requires various stages, starts from preliminary studies, model formulation, validation, small-scale test, large-scale test, revision and final product. The teaching materials using comic media in this research have gone through various stages. Based on the data obtained can be concluded that the compiled teaching materials has met the aspects of feasibility in terms of both theoretical, characters and language.

In addition to those aspects of teacher and student responses that show how feasible the teaching materials, an important aspect to consider is the use of teaching materials in the classroom.

Based on small and large scale trials, evidenced seen by the results of analysis of student responses and student learning outcomes that students respond positively. Teachers also respond positively the teaching materials developed by the researcher. It can be seen by the good results of teacher response analysis. From the results of research and discussion of teaching materials using comic media are feasible to be used by students of vocational schools. From the results of research and discussion of teaching materials effectively used in learning in vocational schools. It was proven by positive responses given by teachers and students as well as students' classical completeness 86% has reached the criteria that is $\geq$ 85% of students have complete learning (KKM $\geq$ 75) with average student learning outcomes of 80,5.

4. Conclusion

Based on the result of the research, it can be concluded that the teaching materials using comic media feasible based on the standard assessment of BSNP 2006 textbook that has been modified. This can be seen from the percentage of validation of experts who judge feasible results. In addition, teaching materials using comic media effectively used in learning class at vocational schools. It is seen from the positive responses given by teachers and students and obtained learning completeness by 86%. Suggestions that can be submitted by the researcher is required further research at other schools with a wider scale to determine the level of effectiveness of teaching materials products using comic media.

References

[1] Badan Standar Nasional Pendidikan 2006 *Peraturan Menteri Pendidikan Nasional Republik* Indonesia *No 22 Tahun 2006 tentang Standar Isi untuk Satuan Pendidikan Dasar dan Menengah.* Jakarta Selatan: BSNP. [online]. Tersedia: http://bsnp-indonesia.org/id/?page_id=103/
[2] DePorter B and Hernacki M 1999 *Quantum Learning: Membiasakan Belajar Nyaman dan Menyenangkan* (Indonesia:Bandung - Kaifa) p 120
[3] DePorter B, Reardon M and Nourie S 2000 *Quantum Teaching: Mempraktikkan Quantum Learning di Ruang-ruang Kelas* (Indonesia:Bandung - Kaifa) p 14
[4] Goleman D 1995 *Emotional Intelligences* (New York: Bantam Books) p 28
[5] Maulana 2010 *Pembelajaran Matematika sebagai Aktivitas yang Banyak Permainan dan Penuh Kesenangan* Jurnal Pendidikan Dasar volume 14 chapter 10 pp 32-35
[6] National Council of Teachers of Mathematics 2000 *Principles and Standards for School Mathematics* (USA)
[7] Purnomo D Indrowati M and Karyanto P 2013 *Pengaruh Penggunaan Modul Hasil Penelitian Pencemaran di Sungai Pepe Surakarta sebagai Sumber Belajar Biologi Pokok Bahasan Pencemaran Lingkungan Terhadap Hasil Belajar Siswa* Jurnal Pendidikan Biologi volume 5 pp 59-6
[8] Shadiq F 2004 *Pemecahan Masalah Penalaran dan Komunikasi* Makalah Pengembangan Matematika SMA Yogyakarta
[9] Sugiyono 2009 *Metode Penelitian Kuantitatif Kualitatif dan R&D* (Indonesia:Bandung-Alfabeta)
[10] Wahyuni F 2010 *Upaya Meningkatkan Minat Belajar Matematika Siswa Kelas VII SMP Negeri 6 Yogyakarta dengan Menggunakan Media Komik melalui Model Pembelajaran Kooperatif Tipe Student Teams Achievement Divisions (STAD)* Skripsi Universitas Negeri Yogyakarta: Tidak Diterbitkan
[11] Yakti D 2001 *Komik sebagai Motivator Siswa Melek Sains dan Teknologi* Makalah pada Seminar Nasional FPMIPA UPI Bandung

IOP Conf. Series: Materials Science and Engineering **335** (2018) 012111 doi:10.1088/1757-899X/335/1/012111

Development of Contextual Mathematics teaching Material integrated related sciences and realistic for students grade xi senior high school

H Helma[1*], M Mirna[1], E Edizon[2]

[1] Mathematics Department, FMIPA, Universitas Negeri Padang, Indonesia
[2] SMAN 1 Payakumbuh, Sumatera Barat, Indonesia

*helma667@yahoo.co.id

Abstract. Mathematics is often applied in physics, chemistry, economics, engineering, and others. Besides that, mathematics is also used in everyday life. Learning mathematics in school should be associated with other sciences and everyday life. In this way, the learning of mathematics is more realstic, interesting, and meaningful. Needs analysis shows that required contextual mathematics teaching materials integrated related sciences and realistic on learning mathematics. The purpose of research is to produce a valid and practical contextual mathematics teaching material integrated related sciences and realistic. This research is development research. The result of this research is a valid and practical contextual mathematics teaching material integrated related sciences and realistic produced

1. Introduction

Graduate competencies in the curriculum 2013 that need to be developed are three, namely good character, related knowledge, and relevant skills. It means that educational graduates should have full competence so useful in their lives.

Mathematics one of the basic science is flexible that always evolves in accordance with the demands of the times. At this time mathematics has evolved and applied to various fields and disciplines. Mathematics is not only a tool for mathematics itself, but many of its concepts are indispensable to other sciences, such as chemistry, physics, biology, engineering and pharmacy. Given the importance of mathematics it is not surprising that mathematics has been studied extensively and fundamentally since primary school.

Mathematics and science have a close relationship. Science provides problems that need to be investigated and analyzed with mathematics, while mathematics provides a useful tool for analyzing data. Often the abstract patterns learned in mathematics are very useful in science. Science and mathematics both try to find patterns and general relationships. The significance of mathematical concepts is evident when used in solving problems of science, technology and everyday life. Given this, in mathematics learning in schools, teachers should link mathematics lessons with other subjects, technology, and everyday life.

The reality in the field shows that the learning of mathematics has been an independent lesson that is separate from other knowledge. Mathematics learning in schools is highly theoretical and mechanistic. Mathematics learning only emphasizes mathematical theories and concepts without its application to other fields such as economics, science, technology, and everyday life. Such learning causes students not to know for what they are learning math. In other words math subjects felt less meaningful to his life. As a result, the learning of mathematics felt less meaningful for the students so that they could cause them less interest in mathematics.

Mathematics learning in practice usually begins with an explanation of concepts accompanied by examples, followed by practice questions. This approach of learning is dominated by the presentation of mathematical problems in a closed form of mathematical problems formulated in such a way that it has only one correct answer with one solution. In addition, these closed issues are usually presented in a structured and explicit manner, beginning with what is known, what is being asked, and what concepts are used to solve the problem. Ideas, concepts and patterns of mathematical relationships as well as strategies, techniques and problem solving algorithms are given explicitly, so students can easily guess the solution. This kind of learning approach tends to only train basic mathematics skills in a limited and isolated way.

Another fact that is found is that the problems presented in most books do not relate mathematics to the context of everyday life so that the learning of mathematics becomes away from the lives of learners. In other words, mathematical learning becomes less meaningful. Less significant learning for learners in mathematics lessons is thought to be the cause of the low interest and achievement of learning mathematics learners.

One alternative solution to solve the problems that have been expressed is to develop contextual mathematics teaching materials integrating related sciences and realistic. With this solution, learning takes advantage of real-world situations, engages students actively, and links mathematics with other relevant knowledge. Learning enables students to build and develop ideas and understanding of mathematical concepts widely and deeply, understand the interrelation of mathematics with other fields of science, and be able to apply to various problems of life and life

Mathematics education should be linked to real-world contexts. Mathematics learning [1] that offers realistic learning characteristics is Realistic Mathematics Education (RME). RME [2] is a mathematical learning approach that uses real situations or contextual problems in accordance with the reality or environment encountered and has been conceived or imagined by students. This means that mathematics is relevant to [3] everyday life.

Realistic mathematics learning has five characteristics, using context, using models, using student contributions, interactive, and integrated with other learning topics [4]. In implementing the RME, learning begins with something tangible so that students can engage in a meaningful learning process. Students can re-construct the findings in the field of mathematics through the activities and exploration of various problems, both problems in everyday life and problems in mathematics itself [5]. Through RME [6] it is expected that students are not only active individually, but there is joint activity among them.

There are three key principles in realistic mathematics learning that can be used as a basis in designing learning. First, the guided discovery and the increasingly mathematical process. Based on the principle of discovery, students are given the opportunity to experience a process similar to the process when mathematics was discovered. The history of mathematics can be used as a source of inspiration in learning materials. Second, phenomena containing didactic charge. Based on this principle the presentation of mathematical topics contained in the learning of realistic mathematics presented on two considerations that raises the variety of applications that must be anticipated in the learning process and its suitability as being influential in the progressive mathematizing process. Mathematical topics presented or contextual issues to be raised in learning must consider two things, namely the application. Third, the formation of the model by the students themselves. Based on this principle of learning while working on contextual problems [7] students are given the opportunity to develop their own models that serve to bridge the gulf between their informal and formal mathematical knowledge.

Based on the above, the development of contextual mathematics teaching materials that integrate related sciences and realistic. The purpose of this study is to analyze the validity, practicality and effectiveness of developed teaching materials.

2. Research methods

The type of research is Research and Development (R & D). R & D is a research method used to produce a particular product, and test the validity, practicality, and effectiveness of the product [8]. The development model used is Borg and Gall.

Objects in the research there are two, that are contextual mathematics teaching materials to integrate related sciences and realistic and students. Research activities are prioritized to produce contextual mathematics teaching materials to integrate related sciences and realistic that is valid, practical, and effective. Trial use of teaching materials done to students of class XI SMA Negeri 1 Payakumbuh.

Instrument of collecting data in this research there are four kinds, that are validity test sheet, test of practicality sheet, attitude observation sheet, and test result sheet. There are four kinds of data analysis techniques used, namely descriptive statistical analysis, graph method, linear regression analysis, and comparison analysis.

3. Results and discussions

Prior to contextual mathematics teaching materials integrating related sciences and realistic was developed, conducted a preliminary study of the characteristics of students in mathematics learning, Lesson Plan (LP) used by teachers and their implementation in learning, mathematics books were used, and supporting factors for the development of teaching materials. The purpose of this activity is to analyze the development needs of contextual mathematical teaching materials to integrate related sciences and realistic. The preliminary study results concluded that it is necessary to develop contextual mathematics teaching materials to integrate related sciences and realistic for the students' learning in grade XI SMA.

Characteristics of students is something to know in order that contextual mathematics teaching materials integrate related sciences and realistic on learning mathematics can be used. There are five indicators reviewed, that are a) student's background, b) student's interest in learning mathematics, c) student's attitude in learning mathematics, d) student's learning motivation, and e) student's learning style. In the indicator of attitude and learning style, although more than 85%, but based on the questionnaire obtained information that they do not have the confidence in solving math problems. Only 59% of the students have it. In addition, 35% of the students do not like the discussion in learning mathematics and explain it.

Students have no difficulty in learning mathematics in terms of characteristics,. Thus, the characteristics of students support the development of contextual mathematical teaching materials integrating related sciences and realistic. However, for the use of teaching materials the characteristics of students that determine the learning outcomes is the learning style of students. The linear regression model with the regressor background variable (L), interest (MI), attitude (S), motivation (MO), learning style (G) and learning response response variable (H),

$$H = 9.3 - 0.016\,L - 0.240\,MI + 0.198\,S + 0.035\,MO + 0.673\,G\,H$$

The regression model obtained has not shown any linear relationship between learning outcomes with at least one of the background variables, interests, attitudes, motivations, and learning styles. This can be seen from F obtained, ie 0.89. When seen t_0 values obtained for each variable, that are t_{0L} = -0.03, t_{0MI} = -0.47, t_{0S} = 0.28, t_{0MO} = 0.07, and t_{0G} = 1.53, then together none of these variables that give effect to the learning outcomes. None of these variables seem to affect the learning outcomes, if other variables also participate in the model. If the variables are made a combination of relationships and tested its significance, then obtained some meaningful linear relationship as follows.

IOP Conf. Series: Materials Science and Engineering 335 (2018) 012111 doi:10.1088/1757-899X/335/1/012111

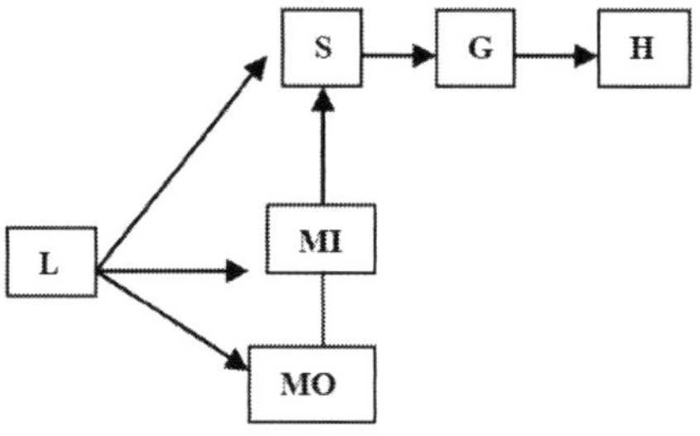

Figure 1. Relationship Among Indicators Characteristics of Students

Indicators that influence student learning outcomes directly for the development needs of contextual mathematics teaching materials integrating related scences and realistic are learning styles. While the background of students is a basic factor that will generate interest factors, attitudes, and motivation in learning mathematics.

Based on need analysis at the beginning of research, it is designed of contextual mathematics teaching material to integrate related sceinces and realistic for high school student of class XI. The teaching materials that have been made are validated. Feedback and input from experts are used as the basis for improvement of teaching materials. Validation results from experts are used to determine the feasibility and guidance in revising the teaching materials. Components of the assessment of teaching materials used are the feasibility of content, language, presentation, and graphics. According to experts, the contextual mathematics teaching materials integrating the related sciences and realistic produced have been valid.

Tested the practicality of teaching materials after repaired. The teaching materials were trial used in mathematics learning in class XI MIPA 6 and XI MIPA 8 SMA Negeri 1 Payakumbuh. The assessment component used is the ease, effectiveness, efficiency, and benefits of using teaching materials. Based on the questionnaire results, there are several indicators that need to be considered for improvement of the teaching materials. In this case less than 65% of students who assessed the teaching materials can be used. The indicator is that students feel more efficient learning with contextual mathematics teaching materials (55.6%), evaluation in contextual teaching materials more efficiently improve students' mastery of mathematics (63.9%), with contextual learning target learning material that is planned to be achieved well (63.9%), and contextual teaching materials can motivate students in learning mathematics (63.9%). Thus, it can be concluded that the contextual mathematical teaching materials integrating the related sciences and realistic produced can be used with attention to some improvements in the efficiency of material presentation, evaluation, timing, and some concrete examples related to the student environment.

After teaching materials are tested to be used in mathematics learning in class XI MIPA 6 and XI MIPA 8 SMA Negeri 1 Payakumbuh, then at the end of learning seen student learning outcomes in the class. Learning results are used to see the effectiveness of the use of teaching materials. In class XI MIPA 6 is given contextual mathematics teaching material to integrate related scences and realistic in mathematics learning, while in class XI MIPA 8 uses teaching materials commonly used by teachers in learning mathematics. The learning outcomes of the two classes before the use of teaching materials on learning mathematics is based on the value of the final exam of the odd semester of the academic year 2016/2017. The following is given the results of learning both classes before and after the use of teaching materials, as in Figure 2.

The test statistic used to determine the difference in learning outcomes of the two classes prior to the use of teaching materials is the t-test. The result of the t-test is 0.52 with a P-value of 0.604. This means that the average score of of the final exam of odd semester of the year 2016/2017 of grade XI MIPA 6 students is not different from the average of XI grade MIPA 8 exam. Thus, it can be concluded that both sample classes have the same initial ability.

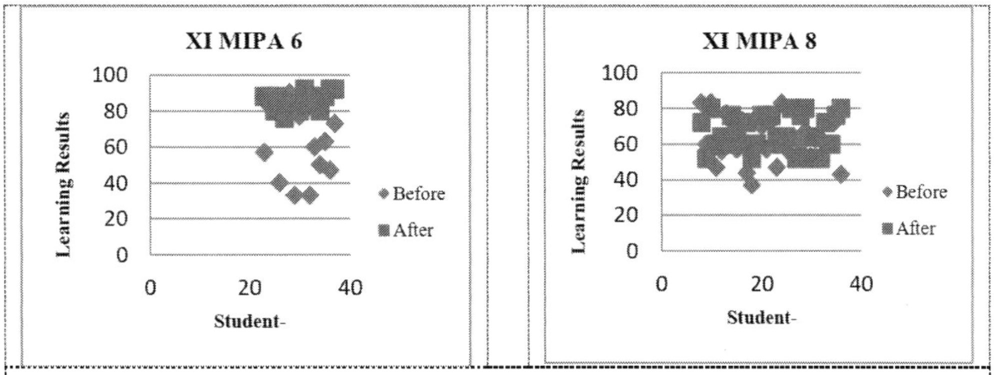

Figure 2. Results of Student Learning Before and After Use of Teaching Materials

The students' learning outcomes of both classes are seen after contextual mathematics teaching materials integrate related sciences and realistic given in learning in class XI MIPA 6,. The test statistic used is Mann-Whitney test. Mann-Whitney test result is 1812.0 with P-value obtained 0.0000. This means that the median grade of XI MIPA 6 students' exam score is different from the median of XI grade MIPA 8 exam. Thus, it can be concluded that the two sample classes have unequal end capabilities. Thus, contextual mathematics teaching materials to integrate related sciences and realistic produced can be said to be effective, that is, can change student learning outcomes for the better

4. Conclusion

Based on the results of research that has been done can be argued that the contextual mathematics teaching materials integrate related sciences and realistic has been produced is valid, practical, and effective to use. As for the use of teaching materials for learning will be more meaningful if adapted to the learning style of students, in order to achieve the expected results.

References

[1] Anita Ika Wahyu 2015 *Analisis Penerapan Pembelajaran Matematika Berbasis PMRI Pada Sekolah Dasar Kelas II di Kota Bandung* Jurnal Inovasi Pendidikan Dasar Volume I; Nomor 1

[2] Simanulang Jonny 2013 *Pengembangan Bahan Ajar Materi Himpunan Konteks Laskar Pelangi Dengan Pendekatan Pendidikan Matematika Realistik Indonesia (PMRI) Kelas VII Sekolah Menengah Pertama* Jurnal Pendidikan Matematika Volume 7, No. 2

[3] Supardi U S 2012 *Pengaruh Pembelajaran Matematika Realistik Terhadap Hasil Belajar Matematika Ditinjau Dari Motivasi Belajar* Cakrawala Pendidikan, Juni 2012, Th. XXXI, No. 2

[4] Murdani dkk 2013 *Pengembangan Perangkat Pembelajaran Matematika Dengan Pendekatan Realistik Untuk Meningkatkan Penalaran Geometri Spatial Siswa* Jurnal Peluang, Vol.1Nomor 2

[5] Muchlis Effie Efrida 2012. *Pengaruh Pendekatan Pendidikan Matematika Realistik Indonesia (PMRI) Terhadap Perkembangan Kemampuan Pemecahan Masalah Siswa Kelas II SD Kartika 1.10 Padang* Jurnal Exacta, Vol. X. No. 2

[6] Diba Farah dkk 2009 *Pengembangan Materi Pembelajaran Bilangan Berdasarkan Pendidikan Matematika Realistik Untuk Siswa Kelas V Sekolah Dasar* Jurnal Pendidikan matematika, Volume 3, No. 1

[7] Harini Susi 2015 *Pengembangan Realistik Sebagai Pilihan Dalam Peningkatan Pemahaman Konsep Pengujian Hipotesis* Jurnal Iqtishoduna Vol. 5 No. 1

[8] Sugiyono 2006 Metode Penelitian Pendidikan: Pendekatan Kuantitatif, Kualitatif, dan R&D. Alfabeta, Bandung

IOP Conf. Series: Materials Science and Engineering **335** (2018) 012112 doi:10.1088/1757-899X/335/1/012112

Visualizing Volume to Help Students Understand the Disk Method on Calculus Integral Course

F Tasman, and D Ahmad

Mathematics Department of Universitas Negeri Padang, Jl. Prof. Hamka, Padang, Sumatera Barat, Indonesia 25131

fridgo_tasman@fmipa.unp.ac.id, defriahmad88@gmail.com

Abstract. Many research shown that students have difficulty in understanding the concepts of integral calculus. Therefore this research is interested in designing a classroom activity integrated with design research method to assist students in understanding the integrals concept especially in calculating the volume of rotary objects using disc method. In order to support student development in understanding integral concepts, this research tries to use realistic mathematical approach by integrating geogebra software. First year university student who takes a calculus course (approximately 30 people) was chosen to implement the classroom activity that has been designed. The results of retrospective analysis show that visualizing volume of rotary objects using geogebra software can assist the student in understanding the disc method as one way of calculating the volume of a rotary object.

1. Introduction

Calculus is the heart of mathematics curriculum and one of the most important subjects because it is the foundation for the next course especially in the natural science, engineering, and mathematics classes[1]. This means that if students does not understand the calculus well, they will get difficulties for the next course. Because of the importance of this course, calculus becomes a compulsory course for university students in Indonesia. In Universitas Negeri Padang (UNP), this course is given to first year mathematics and science students.

One of the two important topics of calculus is integral[2]. Where on the definite integral of a function topic becomes important part. However, many studies suggest that students have difficulty in understanding the concept of integral calculus[3]. This also happened to the first year students in mathematics and science faculty of Universitas Negeri Padang. Based on the data about 30 percent of the students who took the calculus courses did not pass the course and some of them pass with unsatisfactory grade that impacted their success for next course. One of the causes that make they fail is lack of understanding of integrals concept. This happen because most of students tend to focus on integrating procedures without strengthening their understanding on integral concept itself. Tasman argues that educators should not rush to explain the procedure without a good understanding of the concept [4]. Therefore this topic needs a suitable approach to teach to the students in order to strengthen their understanding.

When students are learning about Integral concepts, they are expected to understand the rieman sum, limit, area and other concepts [5]. To understand the integral concept students must be able to connect those concepts well by building their knowledge. Integral is one part that cannot be separated from calculus. In the calculus textbook[6], integral is introduced as an inverse of a derivative and continues to application of integral. One of its application is counting solids volume where one of the methods to count solids volume is disk method.

Students answers sheet on final test on the course of calculus tell us that most of students know the formula to count solids volume. However, most of them are wrong in determining the bound of the integral of the problem. it is happens because lack of their understanding on the concept. Based on this situation, the researcher is interested to design a series of integrated classroom activities in a design research method by using geogebra software to assist students understanding the integrals concept. In this article one of the activities discussed is counting the solids volume of rotary object by using disk methods. In order to support students development in understanding the disk method researcher try to apply realistic mathematical approach.

Tasman stated five tenets of realistic mathematics approach that adapted into this research [7]. These tenets are (1) Constructions stimulated by concreteness. The research does not start with formal level but starts with a situation that is experimentally real for students by seeing the volume of solid of rotary objects using geogebra software. (2) Developing mathematical tools to move from concreteness to abstraction. This tenets bridging from concreteness to a more formal level by using models and symbols. (3) Stimulating free production and reflection. Students own construction or production assumed will be meaningful for them. (4) Stimulating the social activity of learning by interaction. the social university classroom made the students have interaction between each other that provoke them to share idea and have a good discussion to solve the problems. (5) Intertwining learning strands in order to get mathematical idea structured. The topic, application of integral, have a strong relation to the topic of integral and the derivatives.

2. Method and Design

The research was conducted under design research methodology. Edelson stated the reason to chose this methodology [8]. First, the designresearch provide a productive perspective for theory development. Second, the design research has typical usefulness of its results and third, design research directly involved the researcher in the improvement in mathematics education.

Bakker identified five tenets that apply to different types of design research [9]. First is that its purpose to develop theories about learning and the means that are designed to support learning. Instructional theory for university students are designed which means to support students in learning the application of integral especially in counting volume of an object using the disk methods. The second feature is the interventionist nature. This methodology makes the researcher not constrained to improve the design after an experiment phase that has been carry out. The third, cross-cutting feature is that design research has a prospective and reflective component that cannot be separated by an experiment. The researcher confronts conjecture in a prospective with actual learning that he observe in a reflective part. The fourth feature is the cyclic character of design research where the invention and the revision occur as iterative process. Conjecture of learning are sometime refuted and alternative conjectures can be generated and tested. The fifth crosscutting feature of design researchis that theory is relatively humble in the sense that it is developed for a specific domain that is the application integral especially counting the volume of solids using disk method therefore it must be general enough to be applicable in different contexts such as different classroom in other countries.

The main objectives of design research is to develop theories together with instructional material. This design research consist of cycle of three phases. They are:

2.1. A preparation and design phase

The first phase start with clarifications of mathematical learning goals, combined with anticipatory thought how to reach the learning goal in classroom application. The result is our conjectures of instructional activities which consist of three components, they are (1) Learning goal for students, (2) Planned instructional activities and the tools that will be used, (3) Conjectured of learning processes in which one anticipates how students' thinking and understanding could evolve when instructional activities used in classroom [10]

2.2. A teaching experiment

Instructional activities are tried in this phase to collect data. The collected data will explain how a certain set of instructional activities could work. After that the instructional activities are revised on daily basis with the purpose to develop a well considered and empirically grounded local instructional theory.

2.3. A retrospective analysis

All collected data are analyzed in this phase. Out hypothetical learning trajectory (HLT) is compared with students actual learning. The exploration is refined to form a new cycle in the emergence of local instructional theory.

3. Findings and Discussion

A learning trajectory is defined as a description of the path of learning activities that the students can follow to construct their understanding of application of integral especially counting volume of solid by disk method, where in that path considers the learning goal, the learning activities and the conjecture of learning process. The learning trajectory is hypothetical because Students' interpretations, ideas and strategy never be sure until the students really work on that problems. The general overview of the Hypothetical Learning Trajectory (HLT) of application of integral in counting solid volume using disk method showed in Table 1.

Table 1. General Overview of HLT

Name of Activity	Students Activity	Learning Goal	Mathematical Idea	Strategy
Lets count the volume	Counting the volume of rotary object using disk method	Students are able to count the volume of rotary object using disk method.	▪ Partition the surface of the object which give us the area of circle as a result of rotation. ▪ add up the area of the circle which means the integration from a (lower bound) to b (upper bound)	Giving the illustration of the volume by using geogebra software.

In, the design phase, students worksheet is designed. In the worksheet, students were asked to draw area that represented the $\int_0^4 \sqrt{x}\, dx$, where the cartesius diagram and the function already on the answer box. After that the students were asked to draw the result of rotation of the picture on x axis 360 degrees. After that the students were asked the steps to count the volume.

Our conjecture in this activity are (1) Students will have difficulties in drawing the volume. If this happens the volume will be showed by using geogebra software. (2) Some students will have idea to count the volume, if this happen they will be asked to share their idea to others. If all students do not have idea, the lecture will ask the students to see the figure on geogebra and guide them to found the volume.

In the teaching experiment, students were asked to answer problems on their answers sheet by group. Most of students can draw the area of $\int_0^4 \sqrt{x}\, dx$. However, most of them cannot draw the volume. Therefore visualization of the volume showed to the students using geogebra software. The visualization showed in figure 1.

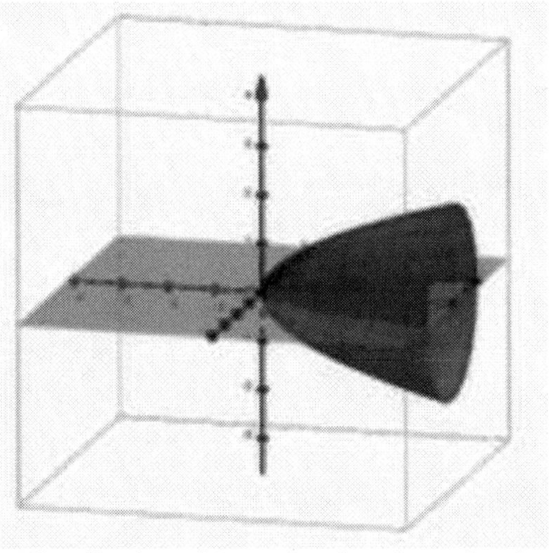

Figure 1. Visualization of Geogebra that showed to the students

After seeing the animation on the geogebra, the students are able to draw the volume of the object. Figure 2 shows some students' representation on the object. When they complete to draw the representation of the object, counting the volume became another issue to discuss in classroom discussion.

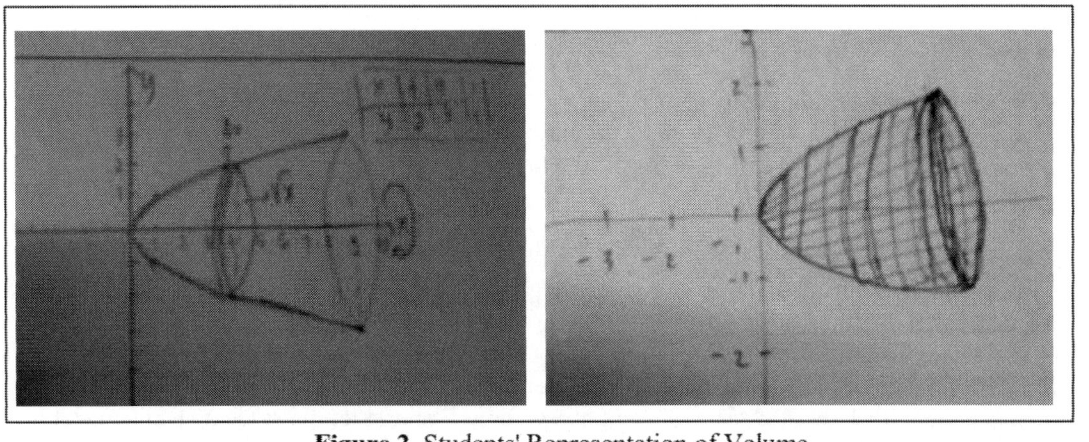

Figure 2. Students' Representation of Volume

In counting the volume of the object, students have to understand that it can be done by slicing the volume vertically (in this case), so they get many circlewith different radius. They have to sum up all of those circle. However, the radius of the circle is not a constant. In order to know the radius of the circle, the prior knowledge of the function plays an important rule. Students have to know that the radius was implied in the function. After that shifting it into integral and doing some calculation on it is the last mathematical idea that the students have to be mastered.

The classroom discussion reveals that most of students do not have idea how to count the volume of the object. However, giving the idea of the volume of rectangular prism that the base times the height can provoke the students to have idea to slice the object and get the circle. Their knowledge about the area of the circle lead and their understanding of integral lead them to know the volume of the object.

Most of the students are able to count the volume after see the visualization of the object by geogebra software. However, some of them still have mistake for its results. Figure 3 showed students error in counting volume of the object. In that figure, students already connect his knowledge about circle and their understanding about integral. However, he forgot to square the radius that implied in the function which lead them to the error. Therefore we suggest the lecture and the teacher have to ask the students to check their answers.

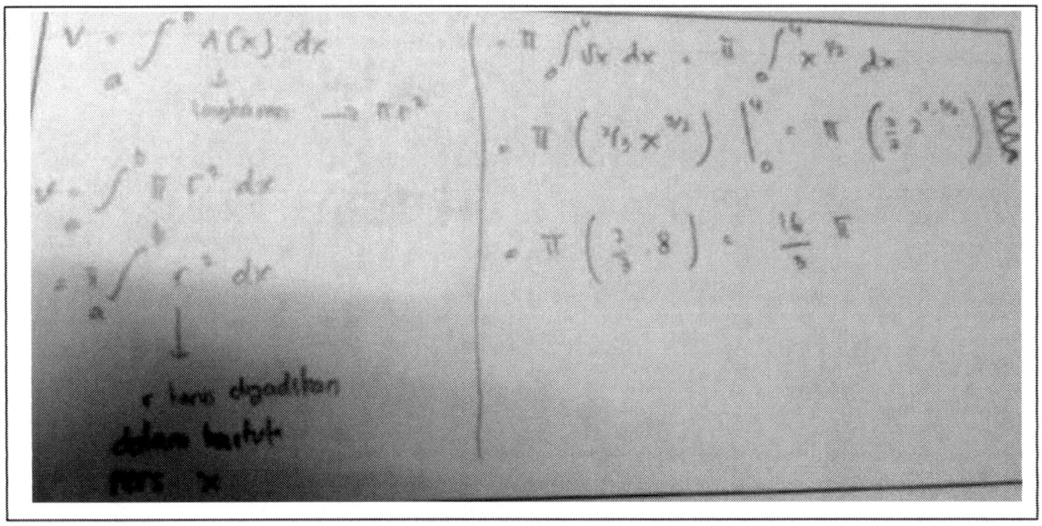

Figure 3. Student' Steps to Count the Volume

The research confirm the finding of Milanovic that the importance of using multimedia in the math classes [11]. By seeing the volume of object using geogebra software, students can see the rotation of the curve whichproduce a circle when it rotated in 360 degrees. This can make the students sure that they can use the formula πr^2 to count the volume and see that the radius implied in the function. The understanding of the radius implied in the function can build by seeing the visualization in the geogebra software. Students will see that the radius always change over the x axis.

4. Conclusion

The research found that the use of geogebra in visualizing the object in building students understanding on disk method in order to count volume of solid is important. The students need to see or imagine the object before they have an idea to count the volume of objects in disk method and build their understanding on the concept. Therefore we suggest the lecture/the teacher, in teaching the integral especially disk method topic, showed the volume first and let the students mathematize with it.

Acknowledgment

The research that we discusses in this article is a part of a research entitle *Design Research in Calculus Integral* which funded by KEMENRISTEK DIKTI Indonesia.

References

[1]. Bressoud, D. M. Why Do We Teach Calculus ? Author (s): David M . Bressoud Source : The American Mathematical Monthly , Vol . 99 , No . 7 (Aug . - Sep ., 1992), pp . 615-617 Published by : Mathematical Association of America Stable URL : http://www.jstor.org/stable/23. **99,** 615–617 (2016).

[2]. Zakaria, Effandi & Salleh, T. S. Using Technology in Learning Integral Calculus . *Mediterr. J. Soc. Sci.* **6,** 144–148 (2016).

[3]. Mahir, N. Conceptual and procedural performance of undergraduate students in integration. *Int. J. Math. Educ. Sci. Technol.***40,** 201–211 (2009).

[4]. Tasman, F., den Hertog, J. & Hartono, Y. Helping students acquainted with multiplication in rectangular model. *IndoMS J.M.E***2,** 185–198 (2011).
[5]. Serhan, D. Students ' Understanding of the Definite Integral Concept. *Int. J. Res. Educ. Sci.***1,** 84–88 (2015).
[6]. Dale Varberg, Edwin J. Purcell, S. E. R. *Calculus (9rd Edition) - Dale Varberg, Edwin Purcell and Steve Rigdon.pdf.* (Erlangga, 2010).
[7]. Tasman, F. *Supporting Second Graders' on Learning Multiplication.* (Lambert Academic Publishing, 2017).
[8]. Edelson, D. C. Design research: What we learn when we engage in design. *J. Learn. Sci.***11,** 105–121 (2002).
[9]. Bakker, A. Design research in statistics education: On symbolizing and computer tools. (2004).
[10]. Gravemeijer, K. Local instruction theories as means of support for teachers in reform mathematics education. *Math. Think. Learn.***6,** 105–128 (2004).
[11]. Milovanović, M., Takači, . & Milajić, A. Multimedia approach in teaching mathematics - example of lesson about the definite integral application for determining an area. *Int. J. Math. Educ. Sci. Technol.***42,** 175–187 (2011).

IOP Conf. Series: Materials Science and Engineering **335** (2018) 012113 doi:10.1088/1757-899X/335/1/012113

Student Responses Toward Student Worksheets Based on Discovery Learning for Students with Intrapersonal and Interpersonal Intelligence

Y Yerizon[1*], A A Putra[2], M Subhan[3]

Mathematics Department, UniversitasNegeri Padang, Padang, Indonesia

[*]yerizon@yahoo.com

Abstract. Students have a low mathematical ability because they are used to learning to hear the teacher's explanation. For that students are given activities to sharpen his ability in math. One way to do that is to create discovery learning based work sheet. The development of this worksheet took into account specific student learning styles including in schools that have classified students based on multiple intelligences. The dominant learning styles in the classroom were intrapersonal and interpersonal. The purpose of this study was to discover students' responses to the mathematics work sheets of the junior high school with a discovery learning approach suitable for students with Intrapersonal and Interpersonal Intelligence. This tool was developed using a development model adapted from the Plomp model. The development process of this tools consists of 3 phases: front-end analysis/preliminary research, development/prototype phase and assessment phase. From the results of the research, it is found that students have good response to the resulting work sheet. The worksheet was understood well by students and its helps student in understanding the concept learned.

1. Introduction

Indonesian students have a low mathematical ability compared to other countries. This is evident from the results of the 2011 Trends in International Mathematics and Science Study (TIMSS) 2011 study that shows the average score of students in class VIII is only 386 from international scale 500. This result is ranked 39th among 43 countries. This result is decreased compared to the 2007 TIMSS result, ie an average score of 397 (ranked 36th out of 48 countries). In the meantime, from the results of the 2009 Program for International Student Assessment (PISA) 2009, 76.6% found that Indonesian students were below the second level and none reached level 5 and 6. Students below the second level, according to the definition of the Organization for Economic Co- operation and Development (OECD), the quality of Indonesian mathematics learning is ranked 61st out of 65 countries with a score of 371.

This shows the learning of mathematics in the class has not been maximized. Teachers tend to develop learning tools regardless of student learning style. This makes learning less than optimal. Teachers still have difficulty in explaining the material to the students because it looks the students have different learning styles. Students with intrapersonal intelligence prefer to learn alone, while students with interpersonal intelligence prefer to study together. Students who have intelligence learning style of understanding do activities through inquiry process and do not have problem with abstract concept. Students who have musical intelligence learning style of Mastery perform activities through concrete objects, while the learning style of understanding do activities through the process of inquiry and have no problem with the concept of the abstract.

The theory of multiple intelligences suggests a pluralistic view of the human mind that everyone has different cognitive potentials. There are at least eight intelligences of the same level: linguistic intelligence, logic-mathematical, musical, physical-kinesthetic, spatial, intrapersonal, interpersonal and naturalist. For that it takes a variety of efforts to improve students' learning motivation based on the intelligence it has. Students with intrapersonal intelligence have the ability to recognize themselves [1].

Pishghadam [2] explains that interpersonal and intrapersonal intelligence and argue that the main difference between the two is that intrapersonal intelligence considers one's feelings and emotions, while interpersonal intelligence allows one to know the desires, feelings, and intentions of others. For that we need to create an atmosphere of learning that allows students to develop skills based on the intelligence they have [3].

For that need to be designed learning tools that integrate the consideration of multiple intelligences. This is supported by the research of Rabia Zonash and Irum Naqvi [4] which states that learning styles have a significant impact on math, architecture and arts students. Students will be more successful if they recognize the dominant type of intelligence within them. Although one type of intelligence can be dominant in an individual, it does not mean that another type does not exist within the individual. According to Gardner, any kind of intelligence can be developed [2], [4].

The learning tool of mathematics developed is Discovery Learning for Junior High School students suitable for students with Intrapersonal and Interpersonal Intelligence. Oemar Hamalik [5] states that discovery learning is a learning process that emphasizes the students' intellectual mental problem-solving, thus finding a concept or generalization that can be applied in the field. According to Marudin Siregar [5] discovery learning is a learning process to find something new in teaching and learning activities. In discovery learning the learning process can find something if the teacher prepares in advance a variety of material to be delivered. Students are encouraged to primarily study themselves through active involvement by finding principles in understanding concepts. Students are encouraged to identify what they want to know to continue by seeking their own information and then organizing or constructing what they know and understand in a final form.

In implementing discovery learning in the classroom, there are several procedures that must be implemented in general teaching and learning activities as follows: 1) Stimulation, the students are faced with something that causes confusion, then proceeded to not give generalization, in order to arise desire to investigate itself. Teachers can start learning activities by asking questions, suggestions for reading books, and other learning activities that lead to problem solving, 2) Problem Statement, students identify as many as possible issues relevant to the subject matter. Then one of them is chosen and formulated in the form of hypothesis (temporary answer to the question of problem), 3) Data Collection, students are given the opportunity to collect relevant information, reading literature, observing objects, interviewing responden, etc. Data Processing, students process the data obtained from reading activity, interviews, observations, etc., all data are processed, randomized, classified, tabulated, even if necessary to be calculated in a certain way and interpreted at a certain level of confidence, 4) Verification, students perform examination (5) Generalization is the process of drawing a conclusion that can be used as a general principle and applies to all occurrences or similar problems, taking into account the verification results.

2. Research Method

This research is a development research using Plomp model which is aimed to develop valid and practical mathematics learning instruction with Discovery Learning approach for junior high school students with intrapersonal and interpersonal intelligence. The development process of this instructional tools consists of three phases: front-end analysis/preliminary research, development / prototype phase and assessment phase. The front-end analysis stage consists of (a) Content Structure Analysis, (b) Concept Analysis, (c) Student Needs Analysis.

Development or Prototyping Phase (Prototyping Development Stage) produces Prototype 1, Prototype 2, and Prototype 3. Prototype 1 is designing the product and doing Self Evaluation and Expert Reviews. Revisions continue until the instructional is considered as valid. If the instructional is valid then proceed to prototype 2. After a revision of prototype 1, Prototype 2 was obtained. Prototype 2 was performed to test the device's practicality. In prototype 2 one-to-one evaluation is performed. Evaluation per person is done by asking three students to comment on the device. Based on these evaluation results, revisions are made to the device.

After the instructional is revised based on the input on one-to-one evaluation (field-by-person evaluations) a field test is conducted. The field tests were conducted under conditions similar to the actual conditions. Field tests are carried out to see the level of device practicability that has been designed. Device practice is the user's disposal rate. Practices are viewed through observation during the execution of learning, interviews and questionnaires.

At the assessment stage was tested the effectiveness of the resulting device. The effectiveness of a product means a measure that states whether or not the effect or effect of the product is developed on the user. The effectiveness of the instructional produces mathematics learning instructional with Discovery Learning approach for junior high school students can be seen from the test results of understanding the concept of mathematics students after learning by using these devices.

3. Result and Discussion

Student responses were reviewed from several aspects: ease of use, attractiveness, ease of understanding, the benefits of the worksheet, and time efficiency. Data were obtained from questionnaires filled with students.

From the aspect of ease of use, students were asked to see the instruction of manual usage and give an opinion whether the instruction is clearly understood or not. The size of the letters is not too small, not too large and the language used can be easily understood. From the aspect of ease of use obtained that 88% of students stated the worksheet was easy to use.

From the attractiveness aspect, students were asked to see the colors used, the size of the letters, and the presentation of drawings on the worksheet. From the research result, it was found that 68% of students stated that the worksheet was interesting to use.

From the aspect of ease of comprehension, the student were asked assess the given commands, illustrations of images, and sentence questions. From the research result, it was found that 74% of students stated that the worksheet was easy to understand.

From the aspect of the worksheet benefits student were asked about the benefits of the worksheet in terms of discovery activities, providing opportunities for problem solving, reasoning, and drawing conclusions. From the results of the study found that 83% of students stated that the worksheet gives benefit in learning. From the aspect of time efficiency it was found that 100% stated that the time required to work on the worksheet was sufficient

4. Conclusion

From the result of the research, it is found that the designed worksheets get positive response from the students that is from the aspect of ease of use, attractiveness, ease of understanding, the benefit of worksheet, and time efficiency.

References

[1] Perez M P and Ruz, N R 2014 Intrapersonal intelligence and motivation in foreign language learning *European Scientific* vol 10 **17**

[2] Pishghadam R 2009 Emotional and verbal intelligences in language learning. *Iranian Journal of Language Studies***3** 43–64

[3] Tamilselvi B and Geetha D 2015 Efficacy in teaching through "Multiple Intelligence" instructional strategies i-manager's *J. on School Educational Technology* Vol. 11 **2**

[4] Behjat F2012 Interpersonal and intrapersonal intelligences: Do they really work in foreign-language learning? *Proceia - Social and Behavioral Sciences***32**351 – 355
[5] Illahi MT 2012 Pembelajaran Discovery Strategy & Mental Vocational Skill. Yogyakarta: DIVA Press
[6] Whardani Sri et al 2015 Materi pelatihan guru implementasi kurikulum 2013. Yogyakarta: Kemendikbud

Working Memory in Students with Mathematical Difficulties

I R D Nur[1*], T Herman[2], S Ningsih[3]

[1]Mathematic Education, Universitas Singaperbangsa Karawang, Indonesia
[2]Mathematics Education, Universitas Pendidikan Indonesia, Bandung, Indonesia
[3]Mathematics education, UIN Antasari Banjarmasin, Indonesia

[*]iyan.rosita@fkip.unsika.ac.id, [2]tatangherman@upi.edu,

Abstract. Learning process is the activities that has important role because this process is one of the all factors that establish students success in learning. oftentimes we find so many students get the difficulties when they study mathematics. This condition is not only because of the outside factor but also it comes from the inside. The purpose of this research is to analyze and give the representation how students working memory happened in physical education students for basic statistics subjects which have mathematical difficulties. The subjects are 4 students which have a mathematical difficulties. The research method is case study and when the describe about students working memory are explanated deeply with naturalistic observation. Based on this research, it was founded that 4 students have a working memory deficit in three components. The components are *phonological loop, visuospatial sketchpad,* dan *episodic buffer*.

1. Introduction

Learning process is the activities that has important role because this process is one of the all factors that establish students success in learning. One of the most important things as an outcome from learning process is an observable change of individuals in terms of knowledge, attitude, skill, and behavior as a result from their interaction with surrounding environment. Thinking process occur in learning activities. Someone can be called as a thinker when he does mental activity and someone who learn mathematics always do the mental activity. So the result that in their thinking activity, someone can configure the relationships between a part of information as an interpretation and then arrange the conclusion. In learning process, the information will be processed and saved in their memory. Atkinson dan Shiffrin (1968) are submitted a theory or a modelling about memorys' processing in human memory which explain that information is processed and saved in 3 stages. They are sensory memory, Short-term Memory, and Long-Term Memory (Woolfolk, 2004).

Based on the preliminary research and the result of observation to the 2nd grade physical education students in singaperbangsa karawang university, there are 4 students that indicates have a mathematical difficulties. They are normal students who have a mathematical difficulties because the researcher find that the students still confuse make certain about the contents of the question, they still not understand yet the concept about the topic, they don't give the attention and the students don't have motivation to study mathematical content especially basic statistic. The students who don't have motivation for study in mathematical content will feel so hard to study mathematics. That students don't show disorder or peculiarity. They still can be invited to discuss about another things be sides mathematics and the relationship from each other look normally and general.

As we know, when we study mathematics there will involve cognitive aspects especially in working memory. It was studied in previous researchs. Ashcraft (2001) researched about the relationship between working memory, mathematic anxiety, and performance that produced the relationship between mathematic anxiety with performance specificly in mathematic tasks as a working memory temporary disturbance. De Smedt (2009) researched about working memory and individual differences in mathematics achievement that produce the individual differences related with students mathematic achievement that it is shown with uniqely in working memory component that is in visuospatial and verbal.

Be sides that, there is John Munros' article, University of Melbourne examined "*The role of working memory in mathematics learning and numeracy*" and the content from that article to verify the role of working memory in spesific mathematic assignment, the procedure to diagnose working memorys' role in mathematical difficulties and the intervention strategic to increase working memorys, process in mathematic learning. *Working memory* is an active, easy to access, and can be used to cognitive tasks in variation aptness (Cowan, 2000; Hassin 2004; Pickering 2006). This cognitive tasks are happened when the student face to face with mathematic. Cognitive stage begin from the entry of new information through study and then it will be processed in working memory and be saved in long-term memory. this collaboration between working memory and long term memory are used when finish mathematic problems in order to produce the output like a mathematics performance (Ormrod, 2008).

The research result of Ashcraft and Kirk (2001), Kane and Engle (2000) showed the capacity of someones' working memory give an influence to someones' ability to do some task in the same time. If it is linked with the case about mathematical difficulties that was described before, working memory is the cognitive area that has the role and influence in processing the knowledge that was gotten from mathematics learning. working memory maintain information in active condition with concise so that we can do something toward that information. Working memory is a mental work step when the information is manipulated and collaborated to help us undertanding discourse and verbal language, make a decision and problem solving.

2. Methodology

This research employed qualitative approach with case study method and when the describe about students working memory are explanated deeply with naturalistic observation. The population in this research are the 2nd grade physical education students in singaperbangsa karawang university and the subjects are 4 students which have a mathematical difficulties. Data collections were done through observation sheets. Observation was done in learning process in the class, when the teacher extend the material about basic statistic that are not learned yet before. The indicators that are measured as a subjects turn up respond based on verbal related something, notes result and the picture that is made, subjects ability to relate a topic and his knowledge, present his knowledge according to verbal or picture, and also his ability to solve the problem. Researcher conducted interview to some students to know the opinion of the student about their difficulty in mathematic learning.

3. Result and Findings

Students learning difficulties can be caused by some factors, not only because of the outside factor but also it comes from the inside like physiology, social factor, phaedagogic factor. We find that there is students mathematic difficulties as :

3.1. Difficulty in use the mathematics concept.

The students got a concept instruction, but not dominate yet. It is caused missing some or all memory of the mathematic concept. This missing mathematic concept can be caused by the students forget the objects name, the students inadequate explain the terminology in mathematic concept.

3.2.Difficulty in learning and use the principle

When we analize mathematic learning difficulties in use the principle, we find that generally it is caused by the students don't have concept that can be used to develop the principle as an important knowledge, poor of basic concept is the cause of mathematical difficulties, and the students don't understand with the principle of mathematic

3.3.Difficulty solve the verbal mathematics problem

- solve the verbal mathematical problem is mean that the students have to apply their konwledge in accordance with theoretical to solve the real problem. The successfull in solve the verbal mathematics problem suspended to verbal understanding ability, that is the ability to understanding the narrative question and the ability to change verbal question be mathematic modelling, generally in equation modelling appropriate with the situation in the question. There is the causes that students get difficult to solve verbal mathematic problem : They don't understand what they read, it is caused less knowledge about concept or some unknown mathematical term.
- Student not change the verbal question to mathematical modelling and their relation
- The three memory stages: first when students change information be a code or representation that can be received by memory and placed that. The second stage when the information is saved to memory. The third stage when we recall the information, and that stage called retrieval. Remembering and forgetting continous with learning and memory process. Processing information:

Information kind

Brain will choose positive memory or negative memory. The positive memory will be saved and then will be processed as an intermediate memory meanwhile a negative memory when the brain refuse to accept that information. Three factors labeling that the memory is positive or negative :

- The information for safety life. This information will be saved in long-term memory so that can make high our memory
- The information that can make raise emotion. The information that consist more emotion possibly will be recorded in memory
- Real and meaningfull information. This information has a relevance with personal experience or previous experience.

3.4.The use of memorys' technique

Memorys' technique is a skill to memorize information in according with brains' working. Something brain like are ekstrim, full color, sensori motor, humor, emotion involve, music, active learning, three dimension picture, use association, imagination, symbols, number.

3.5.Attention, fokus, and concentration

Many research and textbook said that attention, fokus and concentration cannot separated from memory. In order to some information can be accepted by the human so they have to aware/attention about the informatioan that they get.

There are 3 components in deficit working memory:

3.6. Phonological loop
Phonological loop related with the students understanding about sense of number, basic operation in mathematic, and the basic knowledge in descriptive statistic. For example how students can understand what is the differences about mean, median, and modus. In *phonological loop* process save the limited number from the voice, and the trail of memory get the damage for about 2 seconds except you repet that topic again. Because of the material in phonological loop is code according to acoustics, the same heard of sound make confuse with the others and easier to forget.

3.7. Visuospatial Sketchpad
Visuospatial Sketchpad related to how to write present the descriptive statistic data on the table, graph, or picture. As similar as phonological loop, the capacity of visuopatial sketch pad is limit. The students try to solve about graph of descriptive datas statistic in a small paper and just any heady, so that they did some mistake. As similar as when there is so many items entered into sketch pad, they can't repeat it accurately as a good repeating. However, the capacity of phonological loop dan visuospatial sketch pad is exist by itself, it means that they work separate and independent. berdiri sendiri.

3.8.Episodic Buffer
Episodic Buffer related with students ability to interpret that mean, median and modus in the graph are a part that data in descriptive statistic can be described by verbal, numeric, or picture so that the students can connect statistic concept with anything they know. From the result of observation that they cannot explain another something that related with basic statistic, they cannot descibe what is the meaning of the statistic graph.

4. Conclusion
Working memory involve personal experience. The memory about event in human life oftentimes contains much emotion. Memory can hold out for a long time. Working memory refers to how we hold on to and work with information stored in short-term memory. Kids use working memory to learn and follow directions. Working-memory boosters can be built into your child's daily life. Someone skill to make a decision is influenced by his working memory capacity. Working memory capacity give the influence to how many item can still online in one focus and influence the kind of strategies that will be used when someone do the task.

Refferences
[1] Adam, K. C., & Vogel, E. K. (2017). Confident failures: lapses of working memory reveal a metacognitive blind spot. *Attention, Perception, & Psychophysics*, 1-18.
[2] Adam, K. C., Vogel, E. K., & Awh, E. (2017). Clear evidence for item limits in visual working memory. *Cognitive Psychology*, *97*, 79-97.
[3] Alloway, T. P., Gathercole, S. E., & Pickering, S. J. (2006). Verbal and visuospatial short-term and working memory in children: Are they separable?. *Child development*, *77*(6), 1698-1716.
[4] Atkinson, R. C., & Shiffrin, R. M. (1968). Human memory: A proposed system and its control processes. *Psychology of learning and motivation*, *2*, 89-195.
[5] Ashcraft, M. H., & Kirk, E. P. (2001). The relationships among working memory, math anxiety, and performance. Journal of Experimental Psychology: General, 130(2), 224-237.
[6] Baddeley, A.D. (2001). Is working memory still working? American Psychologists, 569-864.
[7] Baddeley, A. D., & Hitch, G. J. (1974). Working memory. In G. A. Bower (Ed.). The psychology of learning and motivation, 8(00), 47-90.
[8] Baddeley, A. D., Thomson, N., & Buchanan, M. (1975). Word length and the structure of short-

term memory. Journal of Verbal Learning and Verbal Behavior, 14, 575-589.
[9] Baddeley, A. (2003). Working memory: Looking back and looking forward. Nature Reviews Neuroscience, 4, 829-839
[10] Cester, I., Mioni, G., & Cornoldi, C. (2017). Time processing in children with mathematical difficulties. *Learning and Individual Differences*, *58*, 22-30.
[11] Cowan, N. (2000). The magical number 4 in short term memory : a reconsideration of mental storage capacity. Behavioral and Brain science, 24, 87-185.
[12] De Smedt, B., Verschaffel, L., & Ghesquière, P. (2009). The predictive value of numerical magnitude comparison for individual differences in mathematics achievement. *Journal of experimental child psychology*, *103*(4), 469-479.
[13] Goddard, R. D., Hoy, W. K., & Hoy, A. W. (2004). Collective efficacy beliefs: Theoretical developments, empirical evidence, and future directions. *Educational researcher*, *33*(3), 3-13.
[14] Gruszka, A., & Nęcka, E. (2017). Limitations of working memory capacity: The cognitive and social consequences. *European Management Journal*.
[15] Hassin, R. R., Uleman, J. S., & Bargh, J. A. (Eds.). (2004). *The new unconscious*. Oxford University Press.
[16] Jones, G., Gobet, F., & Pine, J. M. (2008). Computer simulations of developmental change: the contributions of working memory capacity and long-term knowledge. *Cognitive Science*, *32*(7), 1148-1176.
[17] Kane, M. J., & Engle, R. W. (2000). Workingmemory capacity, proactive interference, and divided attention: Limits on long-term memory retrieval. Journal of Experimental Psychology: Learning, Memory, and Cognition, 26(2), 336-358.

IOP Conf. Series: Materials Science and Engineering **335** (2018) 012115 doi:10.1088/1757-899X/335/1/012115

Developing a Hypothetical Learning Trajectory for the Sampling Distribution of the Sample Means

Syafriandi

Mathematics Department, Universitas Negeri Padang, Jl. Prof. Hamka, Padang, Sumatera Barat, Indonesia 25131

syafriandi_math@fmipa.unp.ac.id

Abstract. Special types of probability distribution are sampling distributions that are important in hypothesis testing. The concept of a sampling distribution may well be the key concept in understanding how inferential procedures work. In this paper, we will design a hypothetical learning trajectory (HLT) for the sampling distribution of the sample mean, and we will discuss how the sampling distribution is used in hypothesis testing.

1. Introduction

Sampling distribution is an important topic in studying inferential statistics, which is one of topics in the elementary statistics course. Sampling distributions are studied after students learn the foundational concepts of measure of central tendency, measure of variability, probability, and normal distribution. A sampling distribution of a statistic is the probability distribution of the statistic computed from all possible random samples of the same size from the same population [1]. Especially, a sampling distribution of sample means is a distribution using the means computed from all possible random samples of a specific size taken from a population [2]. The ultimate goal of studying sampling distributions is that students are able to develop reasonable ideas of the Central Limit Theorem (CLT).

In studying the sampling distribution, the students is not only able to use the sampling distribution in inferential statistics, but the students must be able to examine how the sampling distribution is obtained. The students are expected to be able to rediscover the concept of distribution sampling as the experts first found it. However, many studies suggest that students have difficulty in understanding the concept of sampling distribution in hypothesis testing. The research undertaken by [3, 4] reveals that students often see hypothesis testing only as a set of calculation procedures, without such procedures hypothesis testing does not have much meaning for them. Furthermore, the results of research [5,6,7] reveal that students have difficulty identifying null hypotheses and alternative hypotheses, and fail to understand the role of sampling distribution in hypothesis testing.

This also happened to the students of the third semester in Mathematics Education program of Universitas Negeri Padang. Based on the data about 50 percent of the students fail is lack of understanding of a sampling distribution of sample means. This happen because most of students tend to focus on procedures without strengthening their understanding on sampling distribution concept itself. There are some misconceptions found when students study the sampling distribution, that are: the sampling distribution should look like the population, sampling distribution for large samples have more variability, and a sampling distribution is not a distribution of sample statistics. Lecturers should not rush to explain the procedure without a good understanding of the concept. If the students to make conjectures that they test with data, than they become more engaged in reasoning about the sampling distribution concepts than if the concepts were only presented to them. For that, the lecturers should use appropriate approaches to develop conceptual thinking before students are introduced to the formal ideas of sampling distribution.

Based on this situation, the researcher is interested to design a Hypothetical Learning Trajectory (HLT) which is series of integrated classroom activities in a design research method. The hypothetical learning trajectory is made up of three components: the learning goal that defines the direction, the learning activities, and the hypothetical learning process-a prediction of how the students' thinking and understanding will evolve in the context of the learning activities [8]. In order to support students development in understanding the sampling distribution concepts researcher to apply Realistic Mathematics Education (RME). The choice of RME as an underlying theory for this study relates to its potential to address issues of instructional design [9]. Realistic Mathematics Education emphasizes that of mathematics as a human activity connected to reality. Mathematical learning is doing mathematics [10]. Furthermore, [11] reveals that there are five characteristics of RME, that are (1) phenomenological exploration: constructions stimulated by concreteness; (2) bridging by vertical instrument: developing mathematical tools to move from concreteness to abstraction; (3) student contribution: students own construction or production assumed will be meaningful for them; (4) interactivity: stimulating the social activity of learning by interaction; and (5) intertwining learning strands in order to get mathematical idea structured.

2. Method and Design

This research use the educational design research as a research approach. According to [12] educational design research is the systematic study of designing, developing, and evaluating educational interventions as solutions to such problems. Educational design research aims at advancing our knowledge about the characteristics of these interventions and the processes to design and develop them. The reasons to choose this methodology are: (1) the design research provide a productive perspective for theory development; (2) the design research has typical usefulness of its results, and (3) design research directly involved the researcher in the improvement in mathematics education.

Design research consist of three phases that are done cyclically [13], (1) preparing for the experiment: drafting a preliminary HLT; (2) a teaching experiment: trying, elaborating and improving the instructional design or local instructional theory and developing an understanding of how it works, and (3) retrospective analysis: studying the entire data set to contribute to the development of a local instructional theory and more encompassing theories.

3. Findings and Discussion

The first phase in design research is preparing for the experiment. In this phase, researchers deigned preliminary HLT consisting of: learning goal, learning activities: planned instructional activities and the tools that will be used, and the hypothetical learning process or conjectured of learning processes. The general overview of the hypothetical learning trajectory of the sampling distribution of the sample means showed in Table 1.

Table 1. General overview of HLT of the sampling distribution of the sample means.

Name of activity	Learning goal	Learning activities	Hypothetical learning process	Mathematical Idea
Activity 1: let's find the mean of the sampling distribution	The students can find that the mean of the sampling distribution of the sample mean is exactly equal to the mean of the	Doing to simulation by taking all possible random samples of the same size from the same population and	Able to distinguish the mean of the sampling distribution and the mean of the population.	$\mu_{\bar{X}} = \mu$

	population from which the sample were drawn.	computed the mean of the sampling distribution and than compare with the mean of the population.		
Activity 2: let's find the standard deviation of the sampling distribution	The students can find that the standard deviation of the sampling distribution of the sample mean is exactly equal to the standard deviation of the population devided by the square root of the sample size.	Doing to simulation by taking all possible random samples of the same size from the same population and computed the standard deviation of the sampling distribution and than compare with the standar deviation of the population.	Able to distinguish the standar deviation of the sampling distribution and the standar deviation of the population.	$\sigma_{\overline{X}} = \frac{\sigma}{\sqrt{n}}$
Activity 3: Constructing a sampling distribution of the sample mean	The students are able to constructing the sampling distribution of the sample mean using the central limit theorem	Construct the frequency distribution for the sample mean, and using the normal distribution as a model.	Utilizing the frequency histogram of the sample mean for the various sample sizes	$z = \frac{\overline{X} - \mu_{\overline{X}}}{\sigma_{\overline{X}}}$

In the teaching experiment phase, the lecturer introduced the sampling distribution, and he said that sampling distribution is an important topic in studying inferential statistics. Furthermore, the lecturer invites the students to understand that the sampling distribution of a statistic is the probability distribution of the statistic computed from all possible random samples of the same size from the same population. The students worked in groups of four and there were five groups in elementary statistics class. Students worked in groups to do the simulation takes all possible random samples with the specific size of the same population. In pairs in one group, the first pair takes a sample of two and the second pair takes a sample of three with replecement from the same population. Suppose the population consists of 1, 2, 3, and 4, the sampling results as shown in Figure 1.

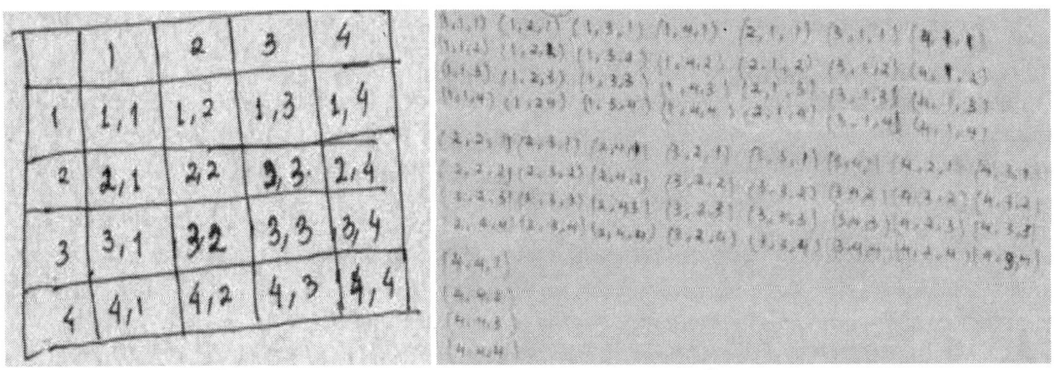

n = 2 n = 3

Figure 1. Random sample of population

IOP Conf. Series: Materials Science and Engineering **335** (2018) 012115 doi:10.1088/1757-899X/335/1/012115

The second step is to compute the statistic (the sample mean) for each random sample, the results as shown in Figure 2.

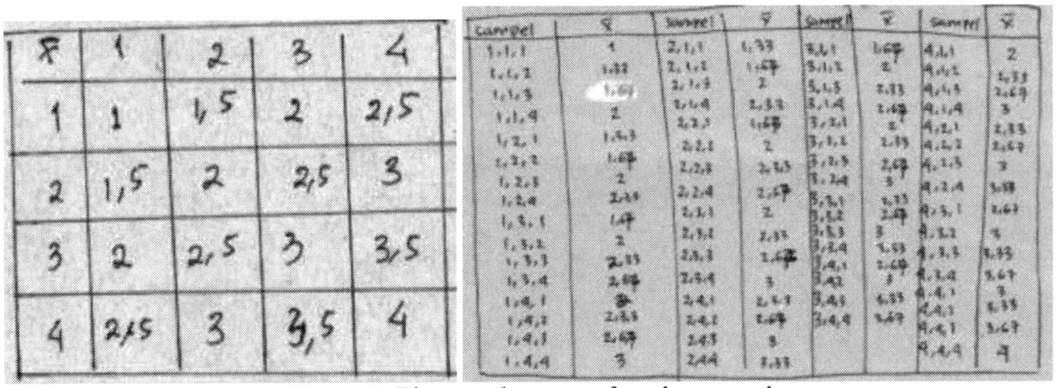

Figure 2. The sample mean of random sample

In calculating of the sample mean, the students pair having a sample size of four has a shortage of time, as there are 64 samples that are calculated of mean. For that, the lecturer gives a conjecture so that students use the software to calculate of the sample mean.

The next step is to computed the mean and the standard deviation of the sampling distribution of the sample mean and then compare with the mean and the standard deviation of the population. All students are able to calculate the mean and standard deviation of the population, but the students found a problem in calculating the standard deviation of the sampling distribution of the sample mean. The Students are confused in using the standard deviation concept of the sampling distribution of the sample mean, as shown in Figure 3.

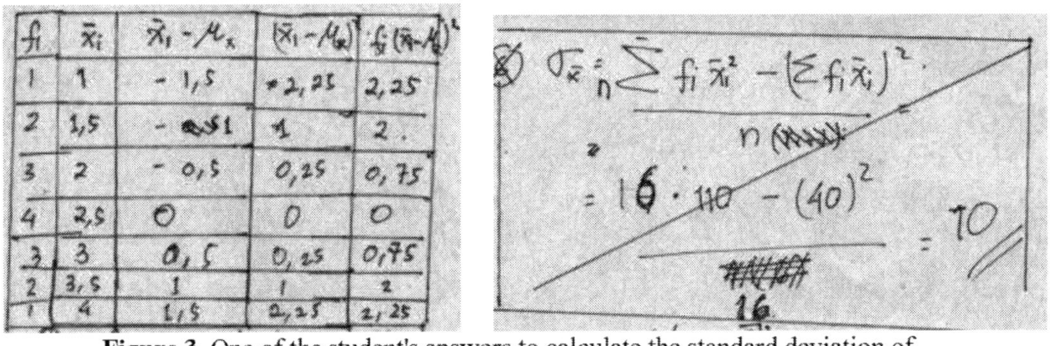

Figure 3. One of the student's answers to calculate the standard deviation of the sampling distribution.

Figure 3 shows that the student uses statistic in calculating the standard deviation of the sampling distribution of the sample mean, in which case the standard deviation of the sampling distribution of sample mean is a parameter. After the student group presents in front of the class, another group of students responds that the standard deviation of the sampling distribution is a parameter. The argument used is the sampling distribution is the probability distribution of the statistic computed from all possible random samples, so this group believes that the concept used is $\sigma_{\bar{X}} = \sqrt{\frac{\sum(\bar{X} - \mu_{\bar{X}})^2}{N_{\bar{X}}}}$. Finally, the groups of four discussed the simulation results that are summarized in tabular form.

IOP Conf. Series: Materials Science and Engineering **335** (2018) 012115 doi:10.1088/1757-899X/335/1/012115

Table 2. Simulation results

Sample size	Mean			Standard deviation		
	Population (μ_X)	Sampling Distribution $(\mu_{\bar{X}})$	Compare	Populatin (σ_X)	Sampling Distribution $(\sigma_{\bar{X}})$	Compare
n = 2	2.5	2.5	$\mu_X = \mu_{\bar{X}}$	$\sqrt{5/4}$	$\sqrt{5/8}$	$\sigma_{\bar{X}} = \sigma_X / \sqrt{2}$
n = 3	2.5	2.5	$\mu_X = \mu_{\bar{X}}$	$\sqrt{5/4}$	$\sqrt{5/12}$	$\sigma_{\bar{X}} = \sigma_X / \sqrt{3}$
Conclution : $\mu_X = \mu_{\bar{X}}$ and $\sigma_{\bar{X}} = \sigma_X / \sqrt{n}$						

In activity 3: the students constructing a sampling distribution of the sample mean using the frequency histogram of the sample mean for the various sample sizes (n = 2 and n = 3), the results are shown in Figure 4.

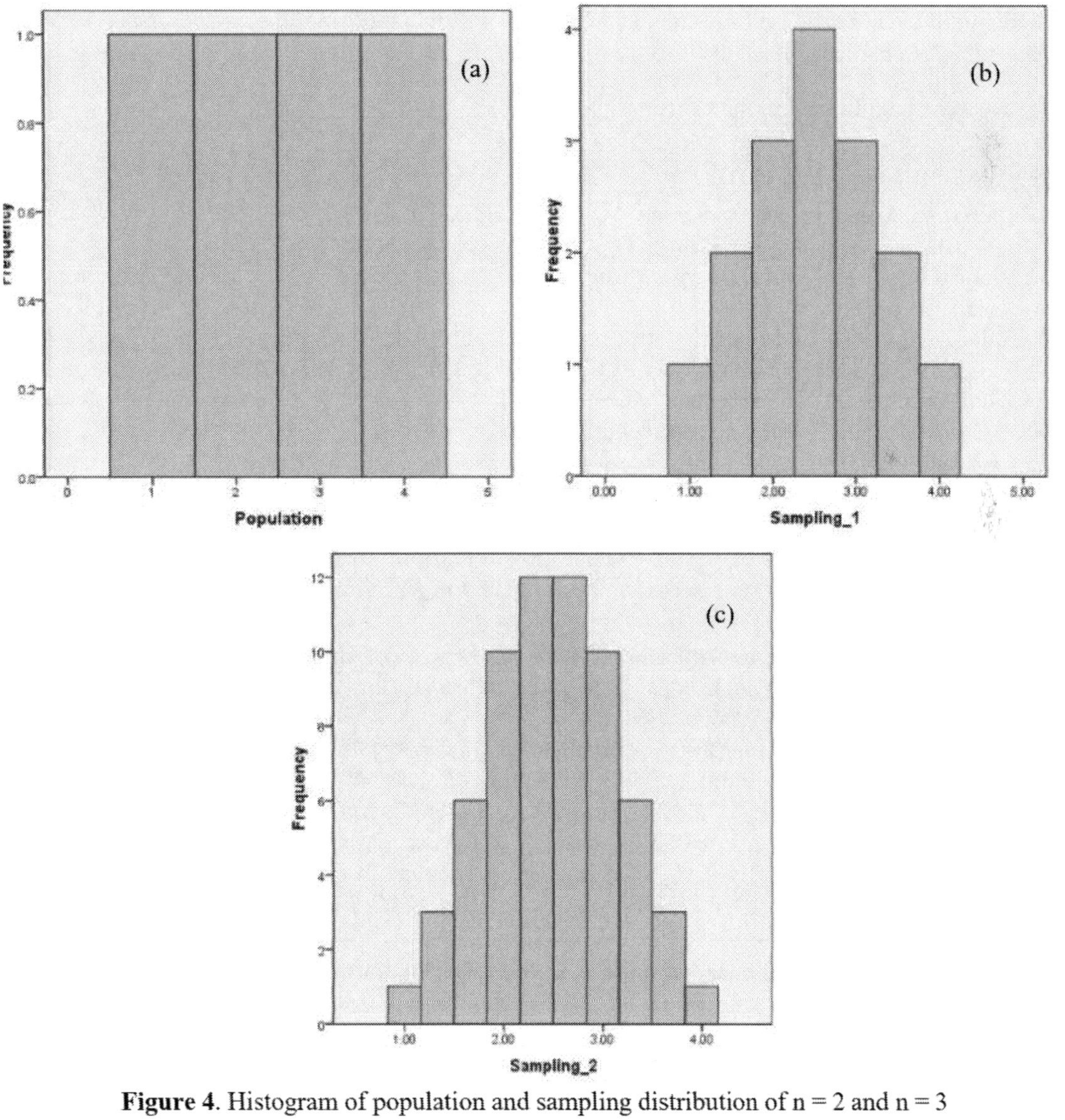

Figure 4. Histogram of population and sampling distribution of n = 2 and n = 3

The students in his group discussed the three histograms obtained. Some groups cannot find the concept differences contained in Figure (a) and (b). The students in his group discussed the three histograms obtained. Some groups cannot find the concept differences contained in Figure (b) and (c). They said that the two histograms is an approximate normal distribution, then the lecturer asks the students to observe the number of boxes of each histogram if the sample size increases. Another group finds that the larger the sample size the histogram gets closer, so the resulting frequency polygon is an approximate the normal curve. Finally, the class discussion concludes that although initially the population is not normally distributed, the sampling distribution of the sample mean approximates the normal distribution, so that the central limit theorem applies, that is $z = \frac{\bar{x}-\mu}{\sigma/\sqrt{n}}$

4. Conclusion

The research found that the three activities designed in a Hypothetical Learning Trajectory (HLT) building students understanding on the concept of the sampling distribution of the sample mean. The students need a real condition (the simulation takes all possible random samples with the specific size of the same population) and they find themselves three amazing things, that is (1) $\mu_X = \mu_{\bar{X}}$, (2) $\sigma_{\bar{X}} = \sigma_X / \sqrt{n}$, and the central limit theorem: $z = \frac{\bar{x}-\mu}{\sigma/\sqrt{n}}$.

References

[1] Glenberg, Arthur M. & Andrzejewski, M. E. 2008. *Learning from data: an introduction to statistical reasoning.* New York: Taylor & Francis Group, LLC.

[2] Bluman, A. G. 2012. *Elementary statistics: A step by step approach.* 8th ed. New York: McGraw Hill.

[3] Mendez, H. 1991. Understanding the central limit theorem. (Doctoral dissertation, University of California, Santa Barbara, 1991). *Dissertation Abstracts International,* **53**, 2722

[4] Batanero, C. 2000. Controversies around the Role of Statistical Tests in Experimental Research. *Mathematical Thinking and Learning*, **2** (1-2): 75–98.

[5] Castro Sotos, A. E., Vanhoof, S., Noortgate, W. V. den, & Onghena, P. 2007. Students' misconceptions of statistical inference: A review of the empirical evidence from research on statistics education. *Educational Research Review*, **2**, 98 – 113.

[6] Garfield, J., & Ahlgren, A. 1988. Difficulties in learning basic concepts in probability and statistics: implications for research. *Journal for Research in Mathematics Education,* **19**, 44-63.

[7] Thompson, P., Liu, Y., & Saldanha, L. 2007. Intricacies of statistical inference and teachers' understandings of them. In M. C. Lovett & P. Shah (Eds.), *Thinking with Data* (1st ed.,: 207 – 231). Mahwah, NJ: Psychology Press.

[8] Simon, M. A. 1995. Reconstructing mathematics pedagogy from a constructivist perspective. *Journal for Research in Mathematics Education, 26,* 114-145.

[9] Simpson, J. A., & Weiner, E. S. C. (Eds.). (1989). *Average. In: The Oxford English dictionary*

[10] Bakker, A. 2004. Design research in statistics education: On symbolizing and computer tools. *Dissertation Utrecht University*. Nederlands: Freudenthal Institute.

[11] Gravemeijer, K.P.E. 1994. *Developing Realistic Mathematics Education.* Utrecht: Freudenthal Institute.

[12] Plomp, Tjeerd. 2013. Educational Design Reseach: An Introduction", in Tjeerd Plomp dan Nienke Nieven (Eds.), *Educational Design Research*. Netherlands: Institute for Curriculum Development (SLO).

[13] Gravemeijer, K., dan Paul Cobb. 2006. "Design Research from A Learning Design Perspective", in Jan van den Akker, Koeno Gravermeijer, Susan McKenney, dan Nienke Nieven (Eds.), *Educational Design Research*. New York: Routledge

IOP Conf. Series: Materials Science and Engineering **335** (2018) 012116 doi:10.1088/1757-899X/335/1/012116

Errors Analysis of Students in Mathematics Department to Learn Plane Geometry

M Mirna*

Mathematics Department, Universitas Negeri Padang, Padang, Indonesia

*mirna_ujang@yahoo.com

Abstract. This article describes the results of qualitative descriptive research that reveal the locations, types and causes of student error in answering the problem of plane geometry at the problem-solving level. Answers from 59 students on three test items informed that students showed errors ranging from understanding the concepts and principles of geometry itself to the error in applying it to problem solving. Their type of error consists of concept errors, principle errors and operational errors. The results of reflection with four subjects reveal the causes of the error are: 1) student learning motivation is very low, 2) in high school learning experience, geometry has been seen as unimportant, 3) the students' experience using their reasoning in solving the problem is very less, and 4) students' reasoning ability is still very low.

1. Introduction

The reality is not everything is easy to understand and easy to do. Through logical thinking, analytical, systematic, critical, innovative and creative many difficult problems can be solved and simple solutions can be found. Such capabilities can be grown and developed in geometry learning. This ability can be grown in geometry learning. According to NCTM, geometry is a basic and important subject area of school mathematics and conceptually, the basis. In geometry class, students learn characteristic features and the relations among them with geometric shapes and structures. Spatial visualization, thinking of two or three dimensions of a geometric shape in space and looking at various aspects is the most important part of geometric thinking [1]. On this basis geometry is given at every level of education from elementary school to college. However, in geometry learning there are many problems so that the great benefits that exist in geometry cannot be possessed by students. The real facts have been said by Telima Adolphus: Some of the findings that emerged are: (1) the foundation of most mathematics teachers in geometry is poor. (2) The students have poor foundation in mathematics. (3) The teaching and learning environment is not conducive. Based on the findings, it was recommended that (a) The State government should as a matter of urgency send mathematics teachers for training and seminars for effective teaching and learning. (b) The government should endeavor to provide the necessary infrastructures and facilities that will motivate teaching and learning of mathematics [4].

Some researches in Indonesia, especially in West Sumatra have revealed that geometry is often avoided, either by students or by the math teacher itself. This is because the foundation of most mathematics teachers in geometry is poor. On the findings of Zuya, H. E. and Kwalat S.K. seen the problem of plane geometry learning, especially related to the angles on parallel lines [7]. The study revealed that most of the teachers were unable to identify the student's missing knowledge with respect to angles in parallel lines. The teachers were also unable to help the students, as they could not suggest specific ways that would help remove the student's difficulties. This problem also exists in students of Mathematics Education Study Program of Universitas Negeri Padang. If this is not resolved, then surely the problem of geometry learning will continue to be a problem in learning mathematics, both in college and in school because these students will teach in school later. As a result, the purpose of mathematics learning will not be achieved.

According Soedjadi, difficulties of students can be seen from the error of them [3]. Therefore, to overcome this problem, must be started from the lecturer as the spearhead of learning. The lecturer should identify the errors that occur and diagnose the cause as a guide in setting the solution. According to NCTM, all students should have the opportunity and the support necessary to learn significant mathematics with depth and understanding [1]. This is necessary to give them the ability to think logically, analytically, systematically, critically, innovatively and creatively, as well as the ability to solve relevant problems. A learned concept or principle will be meaningful if both can be applied in problem solving. This fact makes problem-solving skills very important in mathematics learning. In accordance with Wilson's statement, Problem solving has a special importance in study of mathematics. A primary goal of mathematics teaching and learning is develop the ability to solve a wide variety of complex mathematics problems [6]. The famous problem-solving stage proposed by Polya, that is 1) understanding the problem, 2) devising a plan, 3) carrying out the plan, and 4) looking back [2].

Errors of students in solving problems can be a clue to know the extent of student learning problems. Experience when students solve a problem they will show errors in each of problem solving stages. To find a solution to this problem, the lecturer identifies the types and causes of the errors. Hidayat in Widodo said that the errors made by students can be classified into four types, those are (1) factual errors, ie errors related to the material in question; (2) concept errors, ie misunderstanding of concepts related to matter; (3) operation errors, ie errors in calculations; and (4) the error of principle, that is, the error of misunderstanding the principle or applying the principle in the matter [5].

2. Method and Design

This study is a qualitative descriptive research to reveal the types and causes of the errors made by students of Mathematics Education Study Program in solving problems on the topic of plane geometry. For this, the data collection and analysis techniques are as follows.

First, provide written test to 59 subjects to diagnose the location and types of students errors. The tests provided in this study include the understanding of mathematical concept level (question 1) and mathematical problem solving level (questions 2 and 3). This is because all of the geometry tests that student follow, the errors that occur include all levels of mathematical ability. The questions are as follows. *Question 1*: Write definitions for the following concepts: a median of a triangle, a perpendicular bisector of a segment, a parallelogram, a parallel lines, and a concurrent lines. *Question 2:* If a segment connecting the midpoints of two sides of a triangle, then write the two properties that the segment has, and then prove your answer! *Question 3:* Given $\overline{DR}$ and $\overline{ES}$ are the median of Δ *DEF*. If $\overline{DR}$ and $\overline{ES}$ are intersect in point P and the lengths of $\overline{DP}$ and $\overline{SP}$ are 8 and 5, then determine the lengths of $\overline{EP}$ and $\overline{PR}$.

Student errors are expressed through an analysis of student answer sheets for the three questions. This analysis aims to determine the types of student error in understanding the concepts and errors in applying these concepts at each step of problem solving. The types of errors in this study consist of concept errors, principle errors and operational errors. The indicators of concept errors are (a) the student is wrong in interpreting the terms and concepts, (b) the student is wrong in using terms and concepts, (c) the student does not adjust the use of the definition to the conditions of the prerequisites, and (d) the student does not write the definition to answer the problem. The indicators of the principle errors are (a) the student is wrong in interpreting the principle or theorem, (b) the student is wrong in using the principle or theorem, (c) the student does not adjust the use of the principle or theorem to the condition of the prerequisites, and (d) the student does not write principle or theorem to answer the problem. The indicators of operational errors are (a) the steps of problem solving of students are not hierarchical and (b) the student is unable to manipulate the steps to answer a problem.

Second, interview the subject to find out the cause of the error on the student answer sheet. Subjects interviewed were defined by the rules: two persons called subject 1 (S1) and subject 2 (S2) with characteristics still experiencing errors in expressing the definition of concepts and two persons called subject 3 (S3) and subject 4 (S4) with characteristic of having errors only at the stage of applying the concepts in problem solving.

3. Findings and Discussion

3.1. Description of Student Errors

In this study the test was answered by 59 students. The result, an average 67.80% of students have not been able to reveal the definition of the concepts given in question 1. This shows the student has an error in understanding the concept of plane geometry. These conditions have a real impact on their ability to apply the concept in solving problems. At the stage of understanding the problem, on average 32.20% of students do not write what is given by the problem, 37.29% of students do not write what is asked, and 64.41% of students have not been able to declare the problem into a more operational form. Furthermore, in the devising a plan, on average 69.49% of students students have not been able to perform any activities to look for possibilities that can occur on the problem. This means, there are 30.51% of students can express the possibilities that can occur on the problem. In general, this student understands the concept and principles well. From this situation, on average only 23.73% of students are successful in solving problems on questions 2 and 3. This means, 76.27% of students fail to solve the problem.

3.2. Types of Student Errors

The first type of student error is a concept error. These errors range from their errors in interpreting terms and concepts to errors in using concepts. For example, a student interpreting the term concurrent line is the same as a congruent line and there are also students interpreting transversal line as a special line on a triangle. These unreasonable student interpretations resulted in their failure to write the definition of a given concept.

The second type of error is a principle error. These errors include errors in interpreting principles or theorems, to errors in the use of principles or theorems. For example, in answer to question 2, some students say that the diagonals of a parallelogram are bisectors from their angles. In the student's answer to question 3, on ΔDEF with three segments $\overline{DR}$, $\overline{ES}$, and $\overline{FT}$ intersecting at a point P, some students say that two isosceles triangles will be formed, ie ΔDPF and ΔEPF.

Furthermore, many students also do not adapt the use of principles or theorems to the conditions of their prerequisites. For example, in question 3, there are students who say that if two sides one triangle are equal to two sidesof another triangle, then the triangles are congruent. In ΔDEF, there is any ΔDPS. Some students apply Pythagoras's theorem to this triangle. Likewise on two lines that are not necessarily parallel, which is cut by a transversal, some students say that the number of two same-sides interior angles is 180°. So there are students who say that if two lines are cut by a transversal, then same-sides interior angles are supplemetary. In applying the principle of corresponding angle are equal, the student also does not notice the prerequisites are two parallel lines cut by a transversal. Furthermore there are also some students do not write the principle or theorem to answer the problem.

The third type of error is an operation error. Some students who understand the principle or theorem well fail in solving the problem because the student is unable to manipulate the steps to answer a problem. This is shown by some student answers that do not arrive at the expected solution.

3.3. Causes of Student Errors

Reflections with subjects S1 and S2 reveal that the main cause of student problems in studying plane geometry is a poor past experience. In the junior and senior high schools, teachers often delay the learning of this geometry even many of these geometric sections are not studied. This fact makes students see geometry as less important so they ignore it. Recognized by S1 and S2, they still ignore geometry today. This means that students do not have good motivation in learning of geometry. Furthermore, reflections with S3 and S4 reveal that past learning experiences in secondary school are the same as those delivered by S1 and S2. However, after becoming a student at a college, they realize that geometry must be mastered as a provision to become a teacher later. The problem is that they always have difficulty in solving geometry problems. According to them, learning geometry demands a high reasoning ability, unlike other courses they have undergone such as algebra and trigonometry courses.

3.4. Discussion

This research has revealed that most (67.80%) students experience problems in learning plane geometry. The problems start from understanding the basic concepts of plane geometry to apply them to problem solving. These students will become teachers later. Therefore, this is seen as a serious problem that must be resolved soon. The lecturer must increase the motivation of the students. Students must have a high awareness to understand the geometry well. Furthermore, learning must be re-planned to be more complete, details ranging from the introduction and understanding of facts and concepts, the invention of principles or theorems, and its application in problem solving. Because geometry is a science that demands a high reasoning, then the lecturer must develop the learning materials with a focus on the arrangement of students' reasoning. On reflection with upper level students (S3 and S4), it is recognized that geometry training is needed to improve students' reasoning skills so they have adequate skills in solving geometry problems. For this the students can be challenged to be more careful in formulating the statement and the correct reason for every step of problem solving. Every statement must be followed by the correct reason, which is proved by theory. There is no justification for a statement without a clear reason. This learning is known as the Statement and Reason technique.

References

[1] NCTM (2000). Principles and Standards for School Mathematics. Reston: Virginia.
[2] Polya, G. 1973. How to Solve It: a New Aspect of Mathematics Method 2nd Edition. New Jearsey: Princeton University Press.
[3] Soedjadi, R. 1996. Diagnosis Kesulitan Siswa Sekolah Dasar dalam Belajar Matematika. Team
[4] Telima Adolphus. 2011. Problems of teaching and learning of geometry in secondary schools in Rivers State, Nigeria
[5] Widodo, Sri A. 2013. Analisis Kesalahan dalam Pemecahan Masalah Divergensi Tipe Membuktikan. Jurnal Pendidikan dan Pengajaran, Jilid 46, Nomor 2, Juli 108 2013, hlm.106-113
[6] Wilson. 1993. *Mathematical Problem Solving*. New York: Macmilan Publishing Company.
[7] Zuya, H. E. dan Kwalat S.K. (2015). *Teacher Knowledge of Students about Geometry*. International Journal of Learning, Teaching and Educational Research. Vol 13, No 3. , *pp. 100-114*

Students' Mathematical Problem-Solving Abilities Through The Application of Learning Models Problem Based Learning

M L Nasution[1*], Y Yerizon[2], R Gusmiyanti[3]

Mathematics Department, Universitas Negeri Padang, Padang, Indonesia

*minora_nst@yahoo.com

Abstract. One of the purpose mathematic learning is to develop problem solving abilities. Problem solving is obtained through experience in questioning non-routine. Improving students' mathematical problem-solving abilities required an appropriate strategy in learning activities one of them is models problem based learning (PBL). Thus, the purpose of this research is to determine whether the problem solving abilities of mathematical students' who learn to use PBL better than on the ability of students' mathematical problem solving by applying conventional learning. This research included quasi experiment with static group design and population is students class XI MIA SMAN 1 Lubuk Alung. Class experiment in the class XI MIA 5 and class control in the class XI MIA 6. The instrument of final test students' mathematical problem solving used essay form. The result of data final test in analyzed with t- test. The result is students' mathematical problem solving abilities with PBL better then on the ability of students' mathematical problem solving by applying conventional learning. It's seen from the high percentage achieved by the group of students who learn to use PBL for each indicator of students' mathematical problem solving.

1. Introduction

One of the objectives of mathematics learning is developing mathematical problem solving skills [2]. The ability to solve mathematical problems is the ability to use mathematics as a tool in solving various problems of everyday life. The importance of mathematical problem-solving skills is mastered by students because most components of the standard competency and basic mathematical competence have problem-solving abilities. This indicates that the students are not only learning about mathematical concepts or theories but must be able to solve the mathematical problems that require a skill. This is in line with the opinion that the essence of learning is problems solving [1]

Based on the purposes of learning mathematics, problems in mathematics there are routine or non routine problems. The non routine problem is a new problem for students, in the sense of having a different type of problems that have been known by students. According to Cooney et al that a question will be a problem only if the question indicates a challenge that can not be solved by a routine procedure known to the perpetrator [1]. This means that with the challenges and unknown routine procedures on a given question will indicate whether the question becomes a problem or not.

To solve the non routine problem, it is not enough for students to imitate how to solve problems that have been known, but the must make additional efforts. Good mathematical problem solving skills will allow the student to be able to solve the problems that exist in everyday life gained through their experiences [4]. Various research show that children who are given a lot of practice in problem solving have high test scores compared to children who were given little practice. It was also mentioned by Capper that previous student experiences, cognitive development, and interest in mathematics were factors that greatly influenced success in problem solving [3].

Gagne also said that "high-level of intellectual skills can be developed through problem solving" [3]. This is because that problem solving is an attempt to find a way out of a difficulty in order to achieve goals that are not so immediately achieved. When student solving math problem it is possible to use the knowledge and skills already possessed to be applied in solving the problems. Therefore, by familiarizing to practice the students their problem solving ability in mathematics can improve the ability to think carefully, logically, critically and creatively to develop students' intellectual ability. In order for students to have good problem solving skills, it is necessary to designed learning that can increase the students' interest to try and to solve the given problems. One of them with Problem Based Learning (PBL) learning model. PBL learning model can be applied to learning mathematics to improve the ability of creative thinking and problem solving skills in students.

PBL is a learning that exposes students to practical problems as a foothold in learning or in other words students learn through problems. The syntax of PBL learning model consists of 5 phases: (1) student orientation to the problem, (2) organizing students to learn, (3) guiding individual and group investigation, (4) developing and presenting the work, (5) analyze and evaluate the problem-solving process. In the application of PBL learning model is used scientific approach.

The scientific approach is a science-based approach and requires students to be more active in the learning process. The steps of learning with the scientific approach are basically based on the facts of the observed object, namely (1) observing facts or phenomena; in this stage students seek information, see, hear, read, and listen; (2) asking to build knowledge, this stage is done through discussion and group work; (3) attempts to strengthen students' conceptual understanding, this is done through planning, designing, and executing experiments, and obtaining, presenting, and processing (4) reasoning data that enables students to think critically. PBL is a learning model designed in which students work on authentic issues with the intent to structure their own knowledge, develop inquiry and high-level skills, develop independence and trust. The learning model is based on problems that require real investigation and resolution so that students are encouraged to solve problems by expressing their ideas or ideas in classroom discussions.

This is in line with a scientific approach that requires students to actively construct concepts, laws or principles through observing stages (for identifying or formulating problems), proposing or formulating hypotheses, collecting data with various techniques, drawing conclusions and communicating concepts, laws or principles the "found". The application of a scientific approach to learning involves the process of observing, questioning, reasoning, trying and communicating.

Stages of PBL learning with a scientifik approach in phase (1) stage of the problem is to orient students to the problem, the teacher raises real questions in the student environment and investigated by students. Presenting certain phenomena or demonstrating an event arouses students' curiosity by giving students the opportunity to observe through viewing, listening, listening and reading activities. Teachers guide students to be able to ask questions about the problems presented, (2) the stage of organizing students to learn. Learning based on this problem students work together with each other in small groups so that teachers can guide students to be active in the group in solving the given problem. Working together to engage and to exchange ideas can train students' ability to ask questions and to communicate their ideas. Phase (3) stage guiding individual and group investigation. In this phase teachers encourage students to collect data and carry out experiments until they understand the issues and concepts or approaches and strategies used to solve the problem. The goal is for students to collect enough information to create and build their own ideas to solve problems. Questioning and associating activities strongly support the success of this phase, (4) the stages of developing and presenting the work. At this stage students write down the problem solving of a series of information that has been obtained. A series of concepts are collected in group activities and selected appropriate strategies and approaches in problem solving. In this phase the activities of trying, reasoning and communicating are highly demanded in the problem-solving process, (5) the stage of analyzing and evaluating the problem-solving process. This phase is intended to give students an opportunity to analyze and evaluate their processes and inquiry skills with the knowledge they use. The teacher's job in this phase is to guide and assist students in the investigation in the processes they use. The purpose of the study is to determine whether the mathematics problem solving skills of students who learn to use PBL better than the students who learn to use conventional learning in class XI MIA SMAN 1 Lubuk Alung.

2. Method and Design

This is a quasi-experimental study using two sample classes of experimental class and control class. The experimental class, applied the problem based learning model and the control class applied conventional learning model.To obtain the sample class the assumption test has been tested the test is the average equality with one way anova statistical test. This average equality test is performed after the assumption test of normal and homogeneous population distribution data. The experimental class is represented by the students of class XI MIA 5 and the control class is represented by class XI MIA 6 of six MIA population class SMAN 1 Lubuk Alung.

To obtain the data of research results used tests based on mathematical troubleshooting indicators is tested. Mathematical troubleshooting indicators include: organizing data and selecting relevant information; presents a problematic formula mtematically; choose and use the right approach or strategy; solve the problem; interpret the results of the answers obtained. Assessment used scoring rubrics with a score of 0 to 4. Five problem are made based on validation by some people who are experts in their field. The problems that have been used was tested in another schools that have characteristics similar to the research school. Analysis of test results is done by examining the differentiation and difficulty index questions. The results of the analysis show that all questions can be used. Data analysis one way t test with significancy level 0,05.

3.Result and Discussion

Based on the result of math problem solving test, the average score of the experimental class is 82.29 and the control class is 70.27. This indicates that the average of the experimental class is higher than the average of the control class. The standard deviations of each experimental class and control class are 10.45 and 12.25. Based on standard deviation results can be stated that the ability of solving experimental class math problems more uniform than the control class. The data analysts were performed using the t-test after the normality and homogeneity tests of both sample classes were met. The result of t-test data analysis on the real level of 0.05 shows that the average problem solving ability of experimental mathematics class is higher than the control class. This shows that the experimental class is better than the control class.

The PBL learning model begins by presenting real problems whose solutions require cooperation among students. This learning helps students to process the received information to be processed in their thinking and then develop their own knowledge to solve the problem. The facilitator guides every investigation undertaken by students from analyzing and defining problems, collecting and analyzing information, conducting experiments and formulating conclusions.

The application of the five phases of PBL is done in the experimental class, where the phases can support to improve students' mathematical problem solving skills. Phase to (1) orient the students to the problem so that students can observe problems related to the material to be studied together. In this phase students are stimulated by their curiosity questions or raises allegations so that students can investigate directly the information needed to answer these allegations.

Phase (2) organizes students to learn. In this phase students are organized to study in the form of small group discussions. In its implementation within the group, the student sets out a chairman and a notary who will record any ideas deemed appropriate in solving the problems on the worksheet. The chairperson also mediates in the event of a disagreement between the members of the group. In this phase students are given the opportunity to ask about work sheets. Then through question and answer activities students are directed to recall the material that has been studied to be able to solve the problem. Phase (3) guides individual and group investigations. Students are given guidance to investigate the problems given to the worksheet. The guidance includes the collection of information relating to the material to be discussed. Students in his group conducted experiments in solving the problem. Phase (4) develops and presents the work. In this phase after the students collect the information needed to solve the problem then the students develop the information, and choose the right solution to solve the problem. Then the selected group presented the results of the settlement that had been discussed. The last phase is the phase to (5) analyze and evaluate the problem-solving process. The facilitator with the students analyzes and evaluates the problem-solving process presented by the group as well as the entire learning activities undertaken. In this phase the facilitator provides strengthening related to the mastery of knowledge.

The five phases in the PBL can support and develop students' mathematical problem solving ability, it is because learning using the PBL model makes it a real problem as a trigger for student learning before they know the formal concept. Students will critically identify relevant information and strategies and conduct investigations to resolve the issue. By solving the problem students gain or build on certain knowledge and at the same time develop critical thinking skills and skills in solving problems. Stages of PBL are done systematically and integrated with scientific activities can develop the ability of learners in solving problems and simultaneously can master the knowledge in accordance with the basic competencies to be achieved.

Based on the results of this study can be stated that the ability of problem solving mathematics students learning applied PBL learning model is better than the problem solving skills of students who learn mathematics with conventional models. This is shown from the results of student final tests based on mathematical problem-solving indicators.

Percentage of students' test on problem solving test based on indicators for experimental class were 92%, 90%, 83%, 74%, 65%, respectively. The control classes are 87%, 80%, 72%, 57%, 52%, respectively. The percentage of acquisition of the two sample classes based on this indicator shows that the percentage of mathematical problem-solving abilities for each indicator of the experimental class is higher than the control class. The descriptions of the results of the mathematics problem-solving test are described as follows.

3.1 Organizing Data and Selecting Relevant Information

After analyzing the student answer sheets in the two sample classes, the data obtained results of students' ability in organizing data and selecting relevant information to the given problem. The percentage score of the students' experimental class ability in achieving the indicator is better than the control class. For each item the percentage of scores obtained by the experimental class in achieving the indicator is higher than the control class. The result of the percentage score of students ability for item 1 to 5 for experimental class is 91%, 89%, 93%, 92%, 94% and control class 89%, 83%, 88%, 86%, 89%.

3.2 Present a Systematic Problem Formulation

The result of the final test of the students' mathematics problem solving ability, the percentage score of students ability in presenting systematic problem formulation for the experimental class is 89%, 83%, 92%, 91%, 92% and control class is 76%, 76%, 83% , 77%, 87% Percentage score of students' ability in presenting a problem formulation in experimental class is better than control class. For each item, the score obtained by the sample class has a higher percentage than the control class.

3.3 Selecting and Using Appropriate Approaches and Strategies for Solving Problems

The results of the analysis obtained by the students for the percentage of students' ability score in choosing and using approach and strategy in solving the problems for experimental class are 85%, 73%, 85%, 83%, 83%, 87% and control classes respectively 70% , 64%, 79%, 68%, 81%. The result of the percentage of students' ability score in experimental class in choosing and using the appropriate approach and strategy is higher than the students in the control class. This suggests that the ability of the experimental class is better than the control class in mathematical problem solving.

IOP Conf. Series: Materials Science and Engineering **335** (2018) 012117 doi:10.1088/1757-899X/335/1/012117

3.4 Solving Problem

Based on the data of the final test analysis of students 'mathematical problem solving ability in the sample class in the problem solving indicator, the score of students' ability in the experiment class is better than the control class. This can be seen from the five items of the experimental class obtained a higher percentage than the control class. Obtaining the experimental class is 73%, 68%, 73%, 77%, 78% and control classes are 60%, 50%, 58%, 49%, 67%.

3.5 Interpreting the Results of Answers Obtained

Based on the results of the analysis of the student answer sheets for the experimental class were 61%, 56%, 64%, 68%, 76% and control classes were 56%, 45%, 52%, 47%, 59%. This suggests that the ability to solve mathematical problems to estimate the results of the answers obtained in the experimental class is better than the control class.

References

[1] Sadiq, Fajar, 2014. Math learning How to improve Thinking ability Student, Yogyakarta. Graha Ilmu.
[2] Permendikbud No.59 year 2014
[3] Suherman, Erman, 2003. Strategy of Contemporary Mathematics Learning, Bandung. UPI
[4] Nasution, Minora Longgom, 2014, Application of the Educational Approach Realistic Mathematics to Student Mathematical Problem-solving Abilities, Proceedings Semirata Year 2014, IPB Bogor

IOP Conf. Series: Materials Science and Engineering **335** (2018) 012118 doi:10.1088/1757-899X/335/1/012118

Efforts to Improve Mathematics Teacher Competency Through Training Program on Design Olympiad Mathematics Problems Based on Higher Order Thinking Skills in The Junior High School

A Arnellis[1], E Z Jamaan [1], N Amalita [1]

[1]Department of Mathematics, Faculty of Mathematics and Natural Sciences, Universitas Negeri Padang, Indonesia

arnellis_mat@fmipa.unp.ac.id

Abstract. The goal to analyse a improvement of teacher competence after being trained in preparing high-order math olympicad based on high order thinking skills in junior high school teachers in Pesisir Selatan Regency. The sample of these activities are teachers at the MGMP junior high school in *Pesisir Selatan* District. Evaluation of the implementation is done by giving a pre test and post test, which will measure the success rate of the implementation of this activities. The existence of the devotion activities is expected to understand the enrichment of mathematics olympiad material and training in the preparation of math olympiad questions for the teachers of South Pesisir district junior high school, motivating and raising the interest of the participants in order to follow the mathematics olympiad with the enrichment of mathematics materials and the training of problem solving about mathematics olympiad for junior high school teachers, the participants gain experience and gain insight, as well as the ins and outs of junior mathematics olympiad and implement to teachers and students in olympic competitions. The result of that the post-test is better than the result of pretest in the training of mathematics teacher competence improvement in composing the mathematics olympiad problem based on high order thinking skills of junior high school (SMP) in *Pesisir Selatan* District, West Sumatra, Indonesia.

1. Introduction

Teacher professionalism is often associated with three important factors, namely competence, certification, and professional allowance. These three factors are predicted to affect the quality of education. Given the results achieved by teachers in *Pesisir Selatan* District of junior high schools due to the poor quality of the learning process experienced by post-certified teachers and the behavior of less professional teachers. This can be seen from the low level of teacher competence, especially the absence of teachers who can represent SMP mathematics teachers olympiad from *Pesisir Selatan* district to national level. This is also supported by the achievements of students who have also never represented the mathematics olympiad to the national scale.

Based on observations and interviews with partners, some math teachers in the *Pesisir Selatan* district, have difficulties in making mathematics olympiad questions related to the material learned in school. They are used to making routine questions. Teachers' reality, it seems that most other teachers in *Pesisir Selatan* District of SMP are simply composing questions, modifying existing questions, making no questions from scratch. In terms of making the problem. It is not an easy job, at least for the mathematics olympiad. Even teachers cannot make things right, but surprisingly these teachers do not seem to have any problem with the problem-making technique. The teachers are calm, never bothered it. Indeed, there has never been a fuss about whether the teacher can make the Mathematics Olympiad question. But how can it be skillful if there is no right foundation on how to create a question, where it should be based on mathematics mathematics material.

Currently there are many books available bank of the bank that is easily obtained by the teacher or if diligently can arrange it themselves. That way the teacher with arbitrarily arranged the question, but they are constrained to make a matter of mathematics olympiad. Teachers need to get the knowledge of how to make a good and right question it is not just making a routine matter, or modify it. But it can make a problem in accordance with the demands of the Mathematics Olympiad based on high order thinking skills (HOTS) Development of teacher competence and professionalism can be done through the effort of coaching and empowerment of teachers. Thus, it is necessary to improve the quality of service learning. So that the partner of mathematics teacher conclude need to improve its special competence in making math olympiad problem based on higher order thinking skills.

To facilitate the improvement of SMP teachers 'competence in Pesisir Selatan district, the researchers conducted research that could help the problems of teachers' limitations in doing math problems for the HOTS-based Olympic level so they could provide briefings to their outstanding students. One of the programs that need to be done is the enrichment of mathematics learning materials for teachers in junior high schools, and training on making problems about mathematics based on HOTS. Thus, the improvement of teacher competence is expected to give implication to the students, namely the formation of students in the mathematics olympiad group so it is expected to carve the achievements in various mathematics events. The results of the evaluation conducted by the mathematics teacher of the SMP *Pesisir Selatan* district along with the research team of Mathematics Department of Universitas Negeri Padang, it is known that the weaknesses done by teachers concerning the main factors are the limitations of the Olympic material book, the lack of training on teacher competence improvement, especially in the preparation of Mathematics Olympiad.

Based on the above factors, then the problem formulated is "Is there a positive improvement of the competence of teachers after being given training in preparing math olympiad problems based on higher order thinking skills in junior high school teachers in Pesisir Selatan regency?". The materials that will be submitted related to the improvement of teacher competence in designing Mathematics based Mathematics based on HOTS are as follows: 1. Enrichment of mathematics learning materials that are very useful as a basis for problem solving equivalent of math mathematics olympiad problems. From the results of material enrichment produced a summary of mathematics olympiad material. This activity includes tutorials as well as discussions and explorations. 2. Training to make math mathematics olympiad problems, which emphasized more on strategies to make non-routine questions as well as on HOTS issues. From the results of training generated bank about mathematics olympiad. 3. Discussion of mathematical problem solving strategy that has been designed. 4. Implementation of the results of this training in the classroom by providing an example of the Olympic questions related to the given material.

1.1 The Olympiad of Mathematics Based High Order Thinking Skills (HOTS)

The Olympic materials are basically sourced from curricula applicable to mathematics, textbooks, supporting books and other relevant materials. For Mathematics, the materials tested are questions of exploration, reasoning, creativity and conceptual understanding through the use of props. Especially for math, the material tested is non-routine questions with a high degree of difficulty for the size of junior high school students. Non-routine questions are a) Unique issues that require creative thinking, logic, and in-depth analysis. b) Can not be solved only by using the formula alone. c) It takes longer to complete, compared to routine questions According to NCTM (National Council of Teachers of Mathematics), school mathematics standards include standard content or materials and process standards. Process standards include problem solving, reasoning and verification, interconnection, communication and representation. NCTM suggests that both material standards and process standards are collectively the basic skills and understanding that students in this 21st century need. Therefore, it is necessary to refer to international standard problems as having: 1) measuring high-order thinking ability (HOTS); 2) based on contextual issues; and 3) using various test forms. These characteristics will be used as a reference for the preparation of problems at the level of educational unit. Characteristics of international standard questions are in line with the development of assessment models in the Curriculum 2013, which lead learners to have high-level thinking skills, intelligent, creative, and able to contribute in the world civilization.

1.2 HOTS Problem Development

The development of the HOTS problem requires various criteria in terms of both the content form and the subject matter content. The technique of writing HOTS questions either in the form of multiple choice or description is generally the same as writing low level questions, but there are some distinguishing features. There are several ways that authors can be asked to write items that require high-level thinking, that the material to be questioned is measured by behavior in accordance with Bloom's cognitive domain at the level of analysis, synthesis and evaluation, in which each question is given the basic question (stimulus) and the question of measuring the ability to think critically. To develop the HOTS problem the teacher requires a prior understanding in terms of high-level thinking in Bloom's Taxonomy and how to develop the HOTS problem. Indicators of high-level thinking skills are divided into five groups (Krathwohl, 2002) namely; provide a simple explanation, build basic skills, conclude, make further explanations and organize strategies and tactics. Skills in these five high-thinking groups are detailed again as follows: a). Providing a simple explanation consists of the skills of focusing questions, analyzing arguments, asking questions and answering questions. b) Building basic skills consists of adapting to sources, observing and reporting observations. c) Summing up consists of skill in considering conclusions, generalizing and evaluating. d) Make further explanations of the examples define the terms and make definitions. e) Setting strategies and tactics for example determines an action and interacts with others and communicates. High-level thinking skills, among others, can be trained through problem-solving in the form of a variety of questions.

1.3 Teacher Competency Improvement

The purpose of improving teacher competence through teacher training is to improve the teaching and learning process in which teachers and students are involved, through a series of actions, guidance and direction. Improvement of teaching and learning process through the improvement of professional teachers is expected to contribute to improving the quality of education (Mulyasa, 2009). In relation to the training program in preparing the problem, the writing about mathematics olympiad is one of the activities of teacher profession development, not as the final destination but is actually a vehicle to report the activities that have been done by teachers to improve the quality of education, especially the learning in school. MGMP (Deliberation Teachers Subject) is a forum or a forum of professional activities of teachers of similar subjects in the studio and in each school consisting of two elements of deliberation and subject teachers. In the MGMP is expected to improve the professional competence of teachers in implementing quality learning according to the needs of learners. The container of this profession is indispensable in contributing to the increased competence of its members. The enrichment and training model for improving teachers' competence in making mathematics olympiad problems is expected to be a better value so that this enrichment and training model can be replicated more widely to teachers in other districts / cities to be formed to optimize teacher activities in MGMP Mathematics. Furthermore, it is expected to make a summary of mathematics mathematics material of SMP and bank about mathematics mathematics junior high school that can be used in MGMP Mathematics junior high school activity at provincial level

2. Methods

Implementation of research in the form of material enrichment training and workshop. Enrichment of material given about mathematics olympiad. Meanwhile, the training / workshop was used for the problem-solving activities by junior high school teachers in the MGMP Pesisir Selatan district. Implementation of the steps in conducting the research are as follows: 1. Initially, all participants are given an initial evaluation, with the quality of the matter equivalent to mathematical olympiad that has been prepared and compiled and has been tested. A preliminary evaluation of a pre-test is useful for exploring initial capabilities as a foothold to determine which materials should be given. 2. Based on the initial test, then investigated / analyzed which material material should be given intensive. 3. Based on the analytical results in step 2, the participants are given the enrichment of materials that are considered not to be mastered by the participants. 4. Furthermore, after considered sufficient enrichment of the material then the next step is to provide training in preparing math olympiad math problems. 5. To measure the success of the participants, the final evaluation will be given a useful post-test to see the level of progress after the participants have been given enrichment and training on the mathematics olympiad. The success indicators of this evaluation activity are: 1. The teacher's knowledge of mathematics mathematics material increases, with pay attention to the increase of pre test value to the test post 2. More than 75% of teachers who participated in this activity were able to compile the problem of HOTS-based mathematics. 3. The level of participants' satisfaction from the results of the activity questionnaire exceeds 3 on scale 4.

3. Results and Discussion

Description of data in mathematics teacher competence was obtained after being given pretest and postest on the sample, that is teacher of mathematics study of SMP Negeri as Pesisir Selatan Regency in Lesson Year 2017/2018. Pretest and Postest competence of math teacher as Pesisir Selatan Regency followed by 66 people. The results of pretest and posttest analysis can be seen in Table 1 below:

Table 1. The Results of Analysis *Pretest* and *Postest*

Test	Number of Teacher	Max Score	Min Score	Mean	Variance	Deviation Standar
Pretest	66	62.5	17.5	41.52	167.96	12.96
Postest	66	90	12.5	54.85	170.04	13.04

Table 1 shows that the highest value obtained from the pretest result is 62.5 while the highest value obtained from the postest result is 90. In addition, the average pretest result obtained is 41.52 whereas the average postest result obtained is 90. From the description, it shows that there is an increase of mathematics teacher competency of SMPN in Pesisir Selatan Regency after being given training. When viewed from the variance of pretest and posttest results that the variance of posttest results is greater than the variance of pretest results. This shows that posttest result of SMPN Mathematics Teacher Competence in South Pesisir Regency is more varied than the result of pretest of SMPN Mathematics Teacher Competition in South Pesisir District. Thus it can be concluded that the postest result is better than the result of pretest on the competence of mathematics teacher of SMPN in Pesisir Selatan Regency in the academic year 2017/2018. Data analysis This analysis aims to test the hypothesis. To test the hypothesis first tested the normality and homogeneity test variance data pretest and postest results using minitab software. Test of Normality of Pretest and Postest Result Data The normality test of pretest and posttest data is performed using Anderson-Darling test. The hypothesis used is as follows:

H_0: The data follows a normal distribution.

H_1: The data does not follow a normal distribution.

Based on normality test result that P value from pretest result and P value from posttest result are 0.048 and < 0.005 respectively. Since the value of P obtained is smaller than the specified significant

IOP Conf. Series: Materials Science and Engineering **335** (2018) 012118 doi:10.1088/1757-899X/335/1/012118

level, ie $\alpha = 0.05$ it can be concluded that the pretest and posttest data are not normally distributed (reject H_0).

Hypothesis testing

Based on normality test results obtained that the data pretest and postest results are not normally distributed. Therefore, hypothesis testing is done by nonparametric test, ie Mann-Whitney Test. The hypothesis used in the Mann-Whitney test is as follows:

$H_0 : \widetilde{\mu_1} = \widetilde{\mu_2}$

$H_1 : \widetilde{\mu_1} > \widetilde{\mu_2}$

Information:

$\widetilde{\mu_1}$ = Median from postest result

$\widetilde{\mu_2}$ = Median from pretest result

Based on hypothesis test results obtained that the value of P is 0.000. That is, the P value is less than the specified significance level, ie $\alpha = 0.05$. This shows that with the result of hypothesis test that median value of result of postest is higher than median value of pretest result. Thus, it can be concluded that the posttest result is better than the result of pretest on the competence of mathematics teacher SMPN Pesisir Selatan

4. Conclusion

There is a positive improvement of teacher competence after being trained in preparing high-order math olympics based on high order thinking skills in junior high school teachers in Pesisir Selatan Regency. This is indicated by the result of the hypothesis test that the median of postest result is higher than median of pretest result. So it can be concluded that the posttest result is better than the result of pretest in the training of mathematics teacher competence improvement in composing the mathematics olympics problem based on high order thinking skills of SMPN in Pesisir Selatan Regency.

5. Acknowledgments

The authors would like thank to Education Office of *Pesisir Selatan* District, principal of MGMP *Pesisir Selatan* and Teachers for supporting facilities in this research.

References

[1] Kemendikbud . 2013 .*Peraturan Menteri Pendidikan Dan Kebudayaan Nomor 64 Tahun 2013 Tentang Standar Isi Pendidikan Dasar dan Menengah Jakarta*

[2] Krathwohl, D. R. (Eds.) 2002. *A Taxonomy for Learning, Teaching, and Assessing:A revision of Bloom's Taxonomy an Overview.*. New York: Longman.

[3] Mulyasa, E. 2009. *Menjadi Guru Profesional. Bandung*: PT. Remaja Rosdakarya Offset.

IOP Conf. Series: Materials Science and Engineering **335** (2018) 012119 doi:10.1088/1757-899X/335/1/012119

Improving the Ability of Mathematic Representation Capabilities and Students Skills in Importing Square Forms to Square Using Variation Solutions

R Nirawati*

Mathematics Study Program of STKIP Singkawang, Jl. STKIP Kelurahan Naram Kecamatan Singkawang Utara, West Kalimantan, Indonesia

*resynirawaty@gmail.com

Abstract. This research was conducted to see whether the variation of the solution is acceptable and easy to understand by students with different level of ability so that it can be seen the difference of students ability in facilitating the quadratic form in the upper, middle and lower groups. This research used experimental method with factorial design. Based on the result of final test analysis, there were differences of students ability in upper group, medium group, and lower group in putting squared form based on the use certain variation of solution.

1. Introduction

Representations are expressions of mathematical ideas that students display as models or substitutes of a problem situation used to find solutions to a problem they are facing (Fadillah, 2010: 34). Some forms of mathematical representation can be diagrammed, graphs, tables or mathematical notations and write in their own language. According to NCTM (2000) "representing involves translating a problem or an a new form, representing includes the translation of diagram or physical model into the symbol or word, representing is also used in translating or analyzing a verbal problem to make its meaning clear".

In line with Hudino (2005: 19) states that the ability of representation can be support students in understanding the mathematical concepts studied and their interrelationship to communicate the mathematical ideas learned and connection between mathematical concepts or apply mathematics to mathematical problems. Therefore the mastery of mathematical concepts must be instilled in the students. It is expected that students can understand and master mathematical concepts. Reinforced by Suherman (2001:113-114) states that the learning activities of mathematics should bring students in answer problem in various ways. This is closely related to the ability of mathematical representation that students have in answering mathematical problems given, because the ability of mathematical representation support students in understanding the concepts of mathematics learned (Hudiono, 2005:19). Further stated the mathemaical ideas presented by teachers through various representations will bring enormous influence on student understanding in learning mathematics (Hudiono, 2005).

One of the subjects taught in secondary school is algebraic material. This algebraic material has sub-subjects including fact form squared. According to data compiled by Puspendik Balitbang Kemdiknas about the result of national exam junior high school the academic year 2015/2016, the material algebra for the national scope reached 52.97% while specifically for the province of west kalimantan with the number of students for the material agebra reached 47.92%. While the number operation fact squared form and its properties for the school SMP Negeri 1 Sungai Kunyit with the number of students who take the exam as many as 81 students, reaching 40.46% (Puspendik, 2016). These data indicate that the low absorption of students in junior high school Negeri 1 Sungai Kunyit on algebra material on the subject of fact formatting squared. Based on data of daily test result on the material factual quiz for the academic year 2015/2016 states that of 108 students only 48,35 students who reach KKM that has been established from the school that is 70. From the results of interviews of teachers in tne field of mathematics study obtained information that in teaching and learning activities on the material factarizing the quadratic form, the material taught by using formula only. This is in line with Hudiono's research (2005: 3) in his research on mathematics learning in Junior High School conclude that the limitations of teacher knowledge of teachers and student's habits of learning in class in the conventional way has not been possible to develop student representation power optimally.

Furthermore from the question given to the students of grade eight in junior high school on the material factarized the following quadratic form.

a. Factarizing the quadratic form$ax^2 + bx + c$ with $a = 1$ has a percentage of errors 43,58 % with the following question form.
   1) $x^2 + 5x + 6$
   2) $x^2 + 2x - 15$
b. Factarizing the quadratic form $ax^2 + bx + c$ with $a \neq 1$ has a percentage of errors 73,08 % with the following question form
   1) $3x^2 + 7x + 2$ 3). $6x^2 + 7x - 20$
   2) $4x^2 - 4x + 1$

As for the mistakes made by the student among others:

a. Students are mistaken in determining the multiplication factor of the tribes, especially the quadratic form $ax^2 + bx + c$ with $a \neq 1$.
   Sample student answers
   1) $4x^2 - 4x + 2 = 4x^2 - 2x + 2x - 1$
      $= (x-2)(4x+4)$
   2) $3x^2 + 7x + 2 = 3x^2 + 4x + 3x + 2$
      $= x(x+3) + 4(x+4)$
      $= (x+3)(x+4)$
b. Students are mistaken in determining the multiplication factor of the tribes.
   Sample student answers
   1) $4x^2 - 4x + 1 = 4x^2 - 2x + 2x - 1$
      $= 2(2x-1) + (x-1)$
      $= (2x+1)(2x-1)$
   2) $6x^2 + 7x - 20 = 6x^2 - 15x + 8x - 20$
      $= (2x-5)(3x+4)$

From the results prariset students solve problems given by working directly to the symbolic form, has not solved the problem prariset in other alternative form. This matter still not understand student in factarizing the quadratic form wich resulted weak student representation power. Based on the above problems, researchers intend to provide alternative solution varitions in fact aquaring form without neglecting the way previously taught so that the existing solutions will be more varied and students more skilled in solving mathematical problems in various ways.

2. Methods

The method used in this research is experiment using a factorial design. In this study a group is given a pretest to find students who are capable of upper, medium and lower. Then the three groups were given treatment in the form of enrichment teaching a variety of solutions. Variables contained in this research that is variation of solution and level of ability of student in factarizing quadratic form as independent variables and the students learning outcomes at a certain level of ability based on the use of a certain solution variation in factoring the quadratic form.

The data collection instrument is written essay form. The data collection instrument is written essay form. Data analysis techniques used are descriptive statistical analysis and inferential statistics.Descriptive statistical analysis is done in the form of : 1) presentation of data include frequency list aand histogram; 2) central size includes mean, median and mode; 3) the size standard deviation.As for inferential analysis using two-way anova statistical test through with SPSS.17 program.

3. Result and Discussion

3.1 Improving student's mathematical representation

Table 1. Avarage recapitulation and percentage of ability scores mathematical representation of control and experimental groups on each on indicator.

No	Indicator of the ability mathematical representation	AverageScor and percentage	
		Pre-test	Post-test
1	Visual ability	1,9 (49,08%)	2,8 (69,78%)
2	Kemampuan representasi dengan kata kata yaitu kemampuan menuliskan intrepetasi dari suatu representasi	2,2 (54,3%)	2,8 (70,93%)
3	Kemampuan eksperesi matematis Kemampuan membuat persamaan model matematika dan langkah –langkah penyelesaian masalah matematika	1,9 (47,36%)	2,6 (66,03%)

Based on he result of the research, the average and percentage score of representation ability is seen from indicator of mathematical representation ability descriptively. The mean score of posttest result is higher than the average score of pretest result.

3.2 Improving student's math skills

In this discussion the data is processed there are 28 students. From the pretest result , the students are divide according to the level of ability (upper, middle and lower) among students who have the upper ability level ($x \geq \bar{x}$ + STD),students who have middle ability level ($\bar{x}$ - STD < x <$\bar{x}$ + STD), and students who have lower ability level ($x \leq \bar{x}$ – STD). For the above group the result of the pretest scores were $x \geq 10$ (upper students were 7), the middle group of pretest scores ranges of $7 < x < 10$ (medium skilled students was 12) and the lower group score the pretest is $x \leq 7$ (under skilled students is 9). In grade VIII C students, the result of pretest scores of top students with an average score of 11.2, intermediate ability 8.5 and ability under 6.64. Then compared with the postest score upper students have an average grade of 14.4, intermediate ability 12.5 and ability down 12.27. The average overall student ability seen from the pretest and posttest score in squaring the root form of 8.25 and 12.65 (ideal score 16), then there is a difference is mean score between pretest result and postest both seen from the upper, middle and lower level ability as well as the average score of students viewed as a whole.

Furthermore, increasing the ability of students in mastering the material squaring the form of the root. In general, it was found that there was in increase in the ability to be seen from the minimum mastery criteria, the established score was 8.8 with an ideal score of 16. From 28 students on the pretest result there were 11 students categorized as complete, while in posttest learning enrichment using a variety of solution, all students are categorized as complate (capable). Based on the test of the effect of treatment obtained t arithmetic (10.81) > t table (2.05), then H_0 is rejected means there is a significant difference between pretest and posttest result. This shows there is an increase in the ability of students in mastering the material squaring the root form.

3.3 The effect of the students math ability on the ability of mathematical representation in squaring the root shape based on the solution variation.

Table 2. Average scores for use of solution variations

Group	Average score				
	Solution 1	Solution2	Solution 3	Solution 4	Solution5
Upper	1,6 40%	3,8 95%	2,8 70%	3,0 75%	2,2 55%
Middle	1,9 47,5%	3,4 85%	2,4 60%	2,5 62,5%	2,4 60%
Lower	1,4 35%	2,2 55%	1,3 32,5%	1,7 42,5%	0,7 17,5%
All student	1,63 40,75%	3,13 78,25%	2,17 54,25%	2,4 60%	1,77 (44,25%

Based on the data analysis of research result, it is known that the average score for all students in facting the form of squares by distributive way is 1.63 and the persentage is 40.75%, by directly determining the factors is 3.13 and the persentage is 78.255, by trial and error is 2.17 and the percentage is 54.25%, by means of algebraic tile modeling is 2.40 and the percentage is 60% and by determining pairs of eligible numbers with the help of the shift box is 1.77 and the ercentage 44.25%. This suggests that the mathematical representation of all students grade VIII C in the material facts the square squares better by directly determining the factors.

Of the three groups, the upper, middle and lower groups show the ability of mathematical representation in fact squaring better using a direct means of determining the factors.With the average score in the upper group was 3.8 (95%) while the mean score of the middle group was 3.4 (85%) and the mean score in the lower group was 2.2 (78.25%).

From the results of statistical tests seen because $\alpha = 0.05$, Fcount = 14.591> Ftable = 3.156. This means that the ability of the mathematical representation at different levels of ability in the matter of factacting the quadratic form with the variation of the solution, has improved meaning that there is a difference that means the ability of the mathematical representation of the students taught by using variations of solution solutions in the form of distributive nature, determining factors, Experiment, algebraic tile modeling and determine pairs of numbers that meet certain requirements with the help of a sliding box.

4. Conclusions and Suggestions

4.1. Conclusions

Based on the findings and discussions that have been described, it can be concluded several things, among others: 1) the ability of mathematical representation perindikator increased this is indicated by the average score of posttest results is higher than the average score of pretest results; 2) the ability of students in mastering the material squaring the form of increased root, which is shown by the effect of treatment test obtained tcount (10,81)> ttable (2.05), then Ho rejected means there is a significant difference between the pre-test and post-test And 3) the ability of the mathematical representation at different levels of ability in the material factored quadratic form with the increasing solution variation shown from the results of the statistical test seen Because $\alpha = 0.05$, Fcount = 14.591> Ftable = 3.156

4.2. Suggestions

Based on the conclusions obtained then there are some suggestions that researchers can provide related research that has been conducted researchers that teachers must know the initial knowledge of students and teach the prerequisite materials well; In conveying teaching materials should use concrete props and teach students in solving mathematical problems in various ways or various solutions (ways of completion) so that students are motivated and skilled in generating creative ideas in solving mathematical problems given. Then for the next researcher is suggested to carry out research ability of mathematical representation by using variation of solution at review of different aspect so that can give more information and deeply related research done

Reference

[1] Hudiono, Bambang (2005). *Representation In A Structuralist and Realistic Math View*. Paper: FKIP UNTAN

[2] National Council for Education of Teachers of Mathematics. (2000). *Principle and Standard for School Mathematics*. Reston, VA: NCTM

[3] Puspendik. (2016). *Report of National Exam Result of SMP / Mts of academic Year 2015/2016.* Jakarta: Depdikbud

[4] Syarifah Fadillah. (2010). *Improving Mathematical Mathematical Representation Multiple Mathematical, Mathematical Problem Solving and Self Esteem of Junior High School Students Through Learning with Open Ended Approach*. Bandung: UPI Dissertation

IOP Conf. Series: Materials Science and Engineering **335** (2018) 012120 doi:10.1088/1757-899X/335/1/012120

Developing Guided Worksheet for Cognitive Apprenticeship Approach in teaching Formal Definition of The Limit of A Function

R Oktaviyanthi[1,2], J A Dahlan[3]

[1]School of Postgraduate Studies, Doctorate Program of Mathematics Education, Universitas Pendidikan Indonesia, Jl. Dr. Setiabudhi No. 229, Bandung, Jawa Barat 40164, Indonesia
[2]Math Education Department, Universitas Serang Raya, Jl. Raya Serang – Cilegon Km. 5, Taman Drangong, Serang, Banten 42314, Indonesia
[3]Department of Mathematics Education, Universitas Pendidikan Indonesia, Jl. Dr. Setiabudhi No. 229, Bandung, Jawa Barat 40164, Indonesia

[1]rinaokta@student.upi.edu, [2]rinaokta@unsera.ac.id, [3]jarnawi@upi.edu

Abstract. This study aims to develop student worksheets that correspond to the Cognitive Apprenticeship learning approach. The main subject in this student worksheet is Functions and Limits with the branch of the main subject is Continuity and Limits of Functions. There are two indicators of the achievement of this learning that are intended to be developed in the student worksheet (1) the student can explain the concept of limit by using the formal definition of limit and (2) the student can evaluate the value of limit of a function using epsilon and delta. The type of research used is development research that refers to the development of Plomp products. The research flow starts from literature review, observation, interviews, work sheet design, expert validity test, and limited trial on first-year students in academic year 2016-2017 in Universitas Serang Raya, STKIP Pelita Pratama Al-Azhar Serang, and Universitas Mathla'ul Anwar Pandeglang. Based on the product development result obtained the student worksheets that correspond to the Cognitive Apprenticeship learning approach are valid and reliable.

1. Introduction

Research on the students' understanding in the concept of limit of a function has been studied by researchers for decades. Some research focuses on how the understanding acquired by students and common misconceptions are made through the teaching of limit of a function and its reflection [1]. Other studies take attention to the perception of students on the practice of teaching that students have been done and its impact on the development of students' understanding on the concept of limit of a function [2]. Another case emphasizes the design of learning a limit of a function that can be used as an alternative to teach it [3][4]. In Indonesia, the research area on the development of mathematics teaching materials about a certain material has grown almost more than ten years and still become one of the research trends. The development of instructional materials is aimed to support the implementation of learning that is adapted to the standard competencies to be achieved, the learning approach used and the indicators of learning achievement designed by the teacher.

IOP Conf. Series: Materials Science and Engineering **335** (2018) 012120 doi:10.1088/1757-899X/335/1/012120

The standard competencies to be accomplished in Calculus course with the subject matter of Function and Limit in Mathematics Education Program of Universitas Serang Raya is that students can recognize the basic concept of limit, arrange the calculation using limit and apply both in real-world problem solving by optimizing adaptive reasoning and decision making. The basic competencies to be realized that students are able to understand the limit of a function by formal definition and apply it in evaluating the limit values.

This study is part of a research on the development of mathematics teaching materials using the Cognitive Apprenticeship learning approach which is discussed separately in other papers. At a glance at the Cognitive Apprenticeship approach, Cognitive Apprenticeship instruction is more effective in teaching technical math than traditional instruction because it teaches learners to become problem solvers, emphasizes application-based issues, and encourages collaborative problem solving [5][6]. The Cognitive Apprenticeship contains interconnected aspects of the learning environment: modeling and explaining, multiple perspectives, exploration, self-study, collaboration and social negotiation, and authentic learning activities [7]. The development and implementation of Cognitive Apprenticeship by many researchers shows that Cognitive Apprenticeship has contributed to the problem-solving process.

The instructional material presented in this discussion is a student worksheet that corresponds to the Cognitive Apprenticeship learning approach that has the stages of modeling, coaching, scaffolding, articulation, reflection and exploring activities. The achievement indicator that became the focus of this worksheet design is that students are expected to (1) apply the concept of limit of a function by formal definition and (2) determine and evaluate the limit of a function using ε-δ. Based on the indicators of achievement, the objectives of this worksheet are (1) to guide the thoughts flow of the students in understanding the structure of the formal definition of limit of a function, (2) to guide the student's work steps to be directed and avoid misconceptions, and (3) to give a complete representation of the procedure of evaluating the limit using a formal definition of limit. Therefore the research question in this study is how to develop a guided worksheet in evaluating a limit of a function using a formal definition of the limit corresponding to the Cognitive Apprenticeship learning approach.

2. Methods

This study is a development research that refers to the development of Plomp products with the stages of activity are literature review, field observation, design of guided worksheets, expert validation and limited testing [8]. Stages of literature review, field observation and design of guided worksheets produce prototype worksheets which are examples presented in the table below.

Table 1. Example prototype of guided worksheet

Visualization of guided worksheet	Description
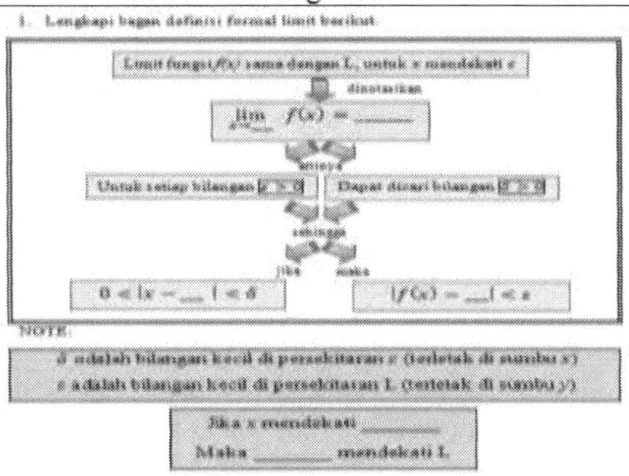	Modeling, the lecturer explains the way of the formal definition and shows how to solve the problem so that students can observe, understand and build the crucial conceptual modeling process.

2. Dengan menggunakan informasi pada poin a di atas, lengkapi dimana posisi δ dan ε.

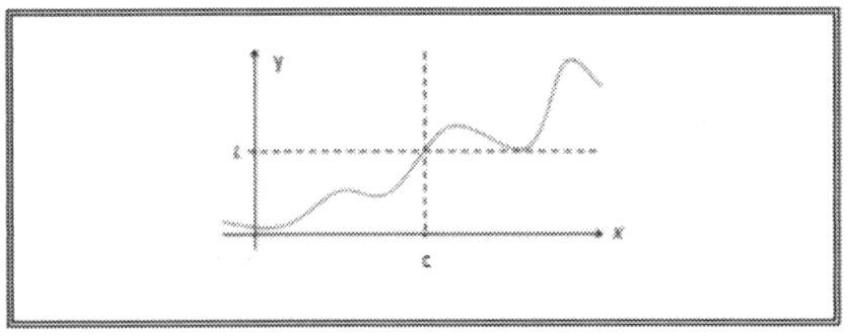

Lengkapi bagian kosong di bawah jika diketahui $\lim_{x\to 3} 2x = 6$ dengan $\varepsilon = 0{,}4$, tentukan δ.

Dari soal yang diketahui, tentukan: c = ____ f(x) = ____ L = ____	Untuk setiap $\varepsilon > 0$, ada $\delta > 0$, sehingga: Jika $0 < \lvert x - \_\_ \rvert < \delta$ Maka $\lvert \_\_ - \_\_ \rvert < \varepsilon$
Gambar di sebelah kanan adalah sketsa grafik $f(x) = 2x$	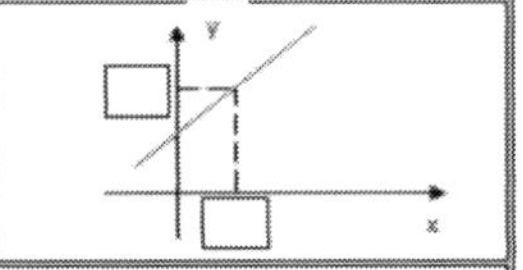

Jika diketahui $\varepsilon = 0{,}4$ tentukan letak ε dan δ pada grafik.

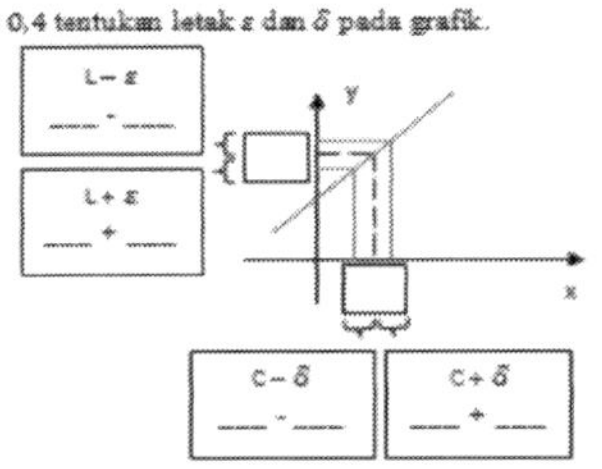

Coaching lecturers guide students through the activities of guiding directions and stages of assistance, evaluating activities and diagnosing students' difficulties, providing challenges and providing feedback.

Pembuktian:

Ikuti langkah-langkah pembuktian pada tabel berikut dengan melengkapi bagian yang kosong.

Tahap 1: Persiapan Pembuktian		
a.	Menentukan nilai δ melalui pernyataan terakhir definisi formal limit.	$\lvert f(x) - L \rvert < \varepsilon$
b.	Substitusi nilai $f(x)$ dan L yang diketahui dalam soal pada pernyataan poin a.	$\lvert (\_\_ - \_\_) - (\_\_) \rvert < \varepsilon$
c.	Sederhanakan nilai dalam tanda mutlak pada poin b menjadi seperti bentuk $\lvert x - c \rvert < \delta$.	$\lvert \_\_ - \_\_ \rvert < \varepsilon$
d.	Menyelesaikan ketaksamaan pada poin c sehingga membentuk $\lvert x - c \rvert < \delta$.	$\lvert \_\_ (\_\_ - \_\_) \rvert < \varepsilon$ $\lvert \_\_ \rvert \lvert \_\_ - \_\_ \rvert < \varepsilon$

Scaffolding, teachers support students through increasing the complexity of the problem and reducing the level of assistance so that students can complete their tasks independently.

The next procedure is validation made by experts both in terms of language and appropriateness of the problem with the indicator. In this study involved five experts consisting of a doctorate in mathematics, two math educations and two senior lecturers of Calculus. To distinguish the consistency of the validation results made the expert then used statistical test of Q-Cochran.

The last stage is limited trials to 90 first year students who take Calculus courses from three private universities in Banten namely Universitas Serang Raya, STKIP Pelita Pratama Al-Azhar Serang, and Universitas Mathla'ul Anwar Pandeglang. This limited trial aims to obtain validity and reliability of guided worksheets according to the preferred standards

3. Results

To test the result of the validity of the work sheet written by the five valuators, used statistical test of Q-Cochran with hypothesis formulation as follows:

H_0: All five valuators give uniform consideration

H_1: All five valuators give different uniform consideration

The testing criteria for hypothesis formulation is that if the Asymp.Sig value is greater than $\alpha = 0.05$ then H0 is accepted, meaning all five valuator give a uniform consideration.

Table 2. Test the uniformity of the face validity of the guided work sheet

N	6
Cochran's Q	6.857[a]
Df	5
Asymp. Sig.	0.144

a. 1 is treated as a success.

The table above shows the result that Asymp value. Sig = 0.144 is greater than $\alpha = 0.05$. Thus H0 is accepted and H1 is rejected, so it can be concluded that all five valuators have given uniform consideration to the validity of the face for this guided worksheet. This means that the use of language in the worksheet has met the effectiveness and then viewed the test of the validity and content of the guided worksheet. With the same steps and procedures, the results of the fifth valuator consideration for the contents obtained results as follows.

Table 3. Test the uniformity of the validity of the contents of the guided worksheet

N	6
Cochran's Q	5.007[a]
Df	5
Asymp. Sig.	0.287

a. 1 is treated as a success.

The result of Q-Cochran statistic test in table 3 shows that the value of Asymp. Sig = 0.287 is greater than $\alpha = 0.05$. Thus H0 is accepted and H1 is rejected so it can be concluded that all five valuators have given uniform consideration to the content validity for the guided worksheet.

Furthermore, after the uniformity test of the valuator, followed by a limited trial to 90 students who follow the Calculus course to obtain the validity and reliability of the worksheet according to the desired standards. For the validity of each item in the guided worksheet, the test criterion used is if $r_{xy} \geq r_{table}$ then it can be said the matter is valid. As for reliability, the test criteria used is if $R_{11} \geq r_{table}$, then the matter is said reliable. From the test results obtained data on the validity and reliability of guided worksheets as follows.

Table 4. Test results validity and reliability of guided worksheets

Nomor Soal	Validitas		Reliabilitas	
	r_{xy}	Keterangan	R_{11}	Kategori
1	0,535	Valid	0,807	Sangat Tinggi
2	0,432	Valid		
3	0,705	Valid		
4	0,518	Valid		
5	0,279	Valid		
6	0,324	Valid		

Table 4 shows that the questions in the guided worksheet can be used because all given questions are valid and have very high reliability.

4. Conclusion

Based on the results of the discussion it can be concluded that the guided worksheet in evaluating limit of a function using the formal definition of limit corresponding to the Cognitive Apprenticeship learning approach are valid and reliable, means that it can be used on a wider sample.

5. Acknowledgments

The researcher thanks and gives the highest appreciation to the Directorate of Research and Community Service, Ministry of Research, Technology and Higher Education (KEMENRISTEK DIKTI), Republic of Indonesia who has funded this study in Doctoral Dissertation Research scheme.

References

[1] Kula S and Guzel E B 2014 Misconceptions emerging in mathematics student teachers' limit instruction and their reflections *Qual Quant* **48** 3355-72 (doi: 10.1007/s11135-013-9961-y)

[2] Hardy N 2009 Students' perceptions of institutional practices: the case of limits of functions in college level Calculus courses *Educ Stud Math* **72** 341-358 (doi: 10.1007/s10649-009-9199-8)

[3] Oktaviyanthi R and Supriani Y 2015 Utilizing Microsoft Mathematics in Teaching and Learning Calculus *Journal on Mathematics Education* **6 (1)** 63-76. (http://jims-b.org/wp-content/uploads/2015/03/7-Full-IndoMS-JME-61-Rina-Oktaviyanthi.pdf)

[4] Oktaviyanthi R and Herman T 2016 A delivery mode study: The effect of self-paced video learning on first-year college students' achievement in calculus (doi: http://dx.doi.org/10.1063/1.4966102. (http://aip.scitation.org/doi/abs/10.1063/1.4966102)

[5] Um E R and deHaan J W 2005 *How Video Games Can be Used An Effective Learning Environment for Cognitive Apprenticeship Theory-Based Learning* (US: New York University)

[6] Oktaviyanthi R 2017 Kajian Model Pembelajaran: Pendekatan Cognitive Apprenticeship Model Case Based Reasoning dalam Pembelajaran Matematika *Preprint* osf.io/preprints/inarxiv/h7er2

[7] Cave A 2010 Learning Math in Second Grade: An Application of Cognitive Apprenticeship *National Forum of Applied Educational Research Journal* **23 (3)** 1-16

[8] Plomp T (2007) *An Introduction to Educational Design Research* Plomp J and Nieveen N The East China Normal University (Shanghai, China) pp 9-36

IOP Conf. Series: Materials Science and Engineering **335** (2018) 012121 doi:10.1088/1757-899X/335/1/012121

A Learning Trajectory for Teaching Social Arithmetic using RME Approach

A Fauzan*, A Armiati, C Ceria

Department of Mathematics, Universitas Negeri Padang, Padang, Indonesia

*ahmad.zan66@gmail.com

Abstract. This paper discusses the role of a learning trajectory (LT) in promoting students' reasoning when they learn social arithmetic using Realistic Mathematics Education (RME) approach. In our LT, we built the intertwining of the concepts such as profit, loss, percentage, discount, and interest rate, so that the students understand the relations among them. The LT was developed through a design research that consisted of a cyclic process of preparing for the experiment, conducting the experiment, and retrospective analysis. The research's subject was 32 students at grade 7 MTsN Sintoga, Pariaman, Indonesia. Data were collected through observations, interviews, checklist, videotaping, and analyzing the students' works. The results showed that the LT could help the students to reinvent the concepts in social arithmetic. The students had more confidence to use their own strategies in solving contextual problems. The most important thing, we discovered the growth in the students' mathematical reasoning.

1. Introduction

The needs to develop a learning trajectory (LT) for teaching a certain mathematics topic are increased nowadays. That is because researchers and mathematics educators realize that LTs play very important roles in building students' understanding of mathematical concepts. Gravemeijer [1] mentioned that if we want students to reinvent mathematics by doing mathematics, teachers have to adapt to how their students reason and help them build on their own thinking. To do so, teachers need to design a framework of reference (LT). LTs are very helpful for bridging the work of researchers and practitioners [2]. LTs also can help teachers evaluating and rethink teaching, which enable them to have a general vision of the class before they start teaching [3,4].

A LT is the sequences of activities and tasks that might support the development of students' understanding of a specific instructional goal [5]. Gravemeijer [6] said that neither teachers, nor researchers can rely on fixed teaching sequences, since a teacher continuously has to adapt to the actual thinking and learning of her students. Therefore, the preliminary version of a LT is in the form of a hypothetical one, and it is called *a hypothetical learning trajectory* (HLT)[5,6,7].

A HLT consists of three components: the learning goal that defines the direction, the learning activities, and the hypothetical learning processes-a prediction and anticipation of how the students' thinking and understanding will evolve in the context of the learning activities [5]. After a cyclic process of designing, testing, and re-designing, a HLT becomes a theory (LT) that can be used as a lesson learned by other mathematics educators to teach a certain mathematics topic. Gravemeijer [6] and Liljekvist [8] called the theory as a local instructional theory (LIT), while Cobb at al[9] called it as

IOP Conf. Series: Materials Science and Engineering **335** (2018) 012121 doi:10.1088/1757-899X/335/1/012121

domain specific theories. A local instruction theory consists of theories about both the process of learning a specific topic and the means to support that learning [1].

Many LTs in mathematics were developed by researchers. They were not only for teaching mathematics in primary and secondary education (see e.g. [10,11,12,13]) but also for teaching certain courses in higher education (see e.g. [14,15,16]). In general, the results of the researches revealed that the LTs were very helpful in building student's conceptual understanding.

In this research, we developed LT for teaching social arithmetic using RME approach. The LT was designed in the contrary to the way the concepts of social arithmetic were presented in mathematics textbooks which tend to be mechanistic, as can be seen on an example below.

1. ***Persentase Keuntungan***
   Persentase keuntungan digunakan untuk mengetahui persentase keuntungan dari suatu penjualan terhadap modal yang dikeluarkan.
   Misal : PU = Persentase keuntungan
   HB = Harga beli (modal)
   HJ = Harga jual (total pemasukan)
   Persentase keuntungan dapat ditentukan dengan rumus
   $$PU = \frac{HJ - HB}{HB} x100\%$$

Figure 1. A formula is presented in a mechanistic way

We can see in Figure 1 that the formula to calculate loss was given directly followed by the examples to show how the formula worked. This condition also influenced the way teachers teach mathematics which tend to be mechanistic [10,13,17].

We employed RME approach in this research because the idea of developing a LT is in line with the idea of how mathematics has to be taught in RME. The process of learning mathematics in RME can be described as a phenomenon of an iceberg below [16].

Photo: *Frans Moerlands*

Figure 2. RME as a phenomenon of an iceberg

A very strong foundation is needed to support the top of the iceberg to appear on sea surface. In relation to this phenomenon, formal and abstract mathematical concepts are situated on the top of the iceberg. Mathematics educators or researchers need to provide a strong foundation and 'a best trajectory' for students to reach the top of the iceberg. To do so, at the beginning of the lesson, students are provided with contextual problems that can be solved using their informal knowledge. The contextual problems will also facilitate students to use their own symbols or their own strategy.

This process is called *horizontal mathematization*. After experiencing similar processes and empowering by *simplification* and *formalization* (see [18]), students will use more formal language or strategies in solving contextual problems. The journey, that will bring students to re-invent a formal mathematical, is called *vertical mathematization* [6, 19, 20].

The LT designed in this research was based on three key principles of RME for instructional designed, namely *guided reinvention trough progressive mathematization, didactical phenomenology,* and *emerging models* [6, 19, 21]. Meanwhile, the implementation of the LT in the classrooms was guided by RME's characteristics [17, 18, 22, 23]. To focus the research, we formulated two research questions. Firstly, what are the characteristics of a valid and practical learning trajectory for teaching social arithmetic using RME approach? Secondly, to what extend the learning trajectory for teaching social arithmetic using RME approach could stimulate student's reasoning?

2. Method

This research used design research approach proposed by Gravemeijer and Cobb [7]. We used the approach because design research aims at understanding more of the interrelatedness between teaching and learning in order to improve teaching [8]. Design research in this study consisted of a cyclic process of preparing for the experiment, conducting the experiment, and retrospective analysis. Gravemeijer and Cobb illustrated the cyclic process as can be seen in Figure 3.

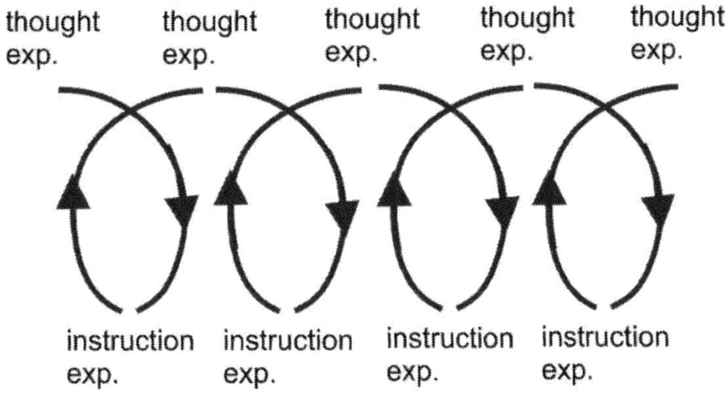

Figure 3. A cyclic process of thought and instruction experiment

In preparing for the experiment we determined the end points of the instructions. The goals of our social arithmetic lessons where the students reinvent the concepts of profit, loss, discount, interest rate, tax, and the way to determine their percentages. As the students were already familiar with the activities such as counting money, buying or selling some things, then we used such activities as the starting point of the lessons. After we set the end and the starting points, we designed the HLT that consisted of five main activities and thirteen sub-activities of solving contextual problems that would facilitate students to do horizontal and vertical mathematization as well as stimulate students' thinking and reasoning. In addition, we also formulated the predictions of students' thinking and solutions, and the anticipations.

In the experimental phase, we tried out the HLT in two cycles. The first try out was conducted in small groups that involved six students at grade seven MTsN Sintoga Padang Pariaman, Indonesia. After the retrospective analysis and re-design processes, the HLT was tried out to 32 students at grade seven in the same school. The retrospective analysis involved the research team, a teacher, and an observer. Besides focusing our attention to develop the HLT, we also observed and analyzed the impact of the HLT on the development of students' confidence in using their own strategies when solving the contextual problems and the development of students' reasoning during the tryout. Data of the research were collected through observations, interviews, checklist, videotaping, and analyzing the students' works.

3. Results and Discussion

The HLT for teaching social arithmetic was validated by three mathematic education experts in Indonesia during preparing for the experiment phase. The result showed that the HLT met the criteria of validity [24], with the characteristics; the activities of solving contextual problems in the HLT were potential to facilitate the students to reinvent the concepts in social arithmetic; the activities were well sequenced, the HLT suit the key principles and characteristics of RME; and the components in the HLT were well designed and consistent between one and another.

The HLT also satisfied the criteria of practicality [24], in which it worked as intended during the tryout. The students understood the contextual problems and they conducted 'doing math' activities without major obstacle. The probing questions that were prepared as the anticipations of students' thinking and solutions also helped the students to achieve the goals of the activities. In addition, the time provided for doing the activities of solving contextual problems was well planned.

The next example (Figure 4) is a contextual problem provided in the HLT that aimed at facilitating the students to find the concepts of profit and loss.

Mr. Andi was a fruit seller. At the end of the week he calculated his earning after making the following list.

Fruit	Quantity (kg)	Buying Price (Rp/kg)	Selling Price (Rp/kg)	Earning
Oranges	50	10,000	15,000	600,000
Apples	25	15,000	20,000	300,000
Salak	40	12,500	15,000	750,000
Mangoes	60	15,000	20,000	800,000

Mr. Andi also noted that he still had 10 kg Oranges, Apples and Salak respectively, and they could not be sold anymore, while Mangoes were sold out.

a. What do you think about the earning of Mr. Andi regarding to each fruit that he sold? Explain your answer!

b. If you want to present your findings in the form of percentages, how are you going to do that?

Figure 4. An example is a contextual problem provided in the HLT

Although the students were not formally introduced yet to the concepts of loss and profit, and the way to calculate loss, profit, and their percentages, but most of the students could reinvent the concepts using their own strategies and then mentioned the concepts using their own words. We observed that the contextual problem could stimulate the students to think about profit and loss by analyzing the context and relations among the data on the table. An example of student's answer can be seen in Figure 5.

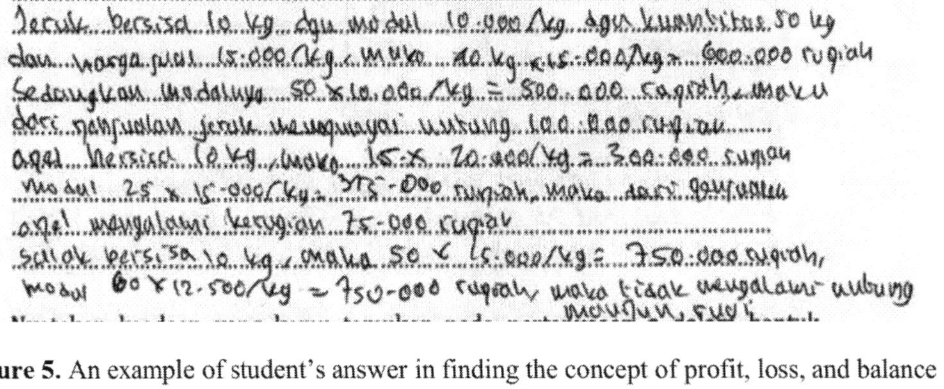
- Jeruk bersisa 10 kg dan modal 10.000/kg. Dan kuantitas 50 kg dan harga jual 15.000/kg, maka 40 kg x 15.000/kg = 600.000 rupiah sedangkan modalnya 50 x 10.000/kg = 500.000 rupiah, maka dari penjualan jeruk mendapat untung 100.000 rupiah
- apel bersisa 10 kg, maka 15 x 20.000/kg = 300.000 rupiah modal 25 x 15.000/kg = 375.000 rupiah, maka dari penjualan apel mengalami kerugian 75.000 rupiah
- salak bersisa 10 kg, maka 50 x 15.000/kg = 750.000 rupiah, modal 60 x 12.500/kg = 750.000 rupiah, maka tidak mengalami untung maupun rugi

Figure 5. An example of student's answer in finding the concept of profit, loss, and balance

In Figure 5, the student mentioned that Mr. Andi got profit =100,000 *rupiah* from selling the Oranges because he earned 40 x 15,000 = 600,000 *rupiah*, while the capital was 50 x 10,000 = 500,000 *rupiah*. For the Apples, the earning was 15 x 20,000 = 300,000 *rupiah* and the capital was 25 x 15,000 = 375,000 *rupiah*, so he got loss = 75,000 *rupiah*. The earnings for Salak was 50 x 15,000 = 750,000, the capital = 60 x 12,500 = 750,000, so he did not get profit or loss.

Most students also found the way to calculate the percentages of profit and loss after they discuss the contextual problem in the group. They found by themselves that the profit or loss was compared to the capital, in calculating the percentage, as can be seen in Figure 6.

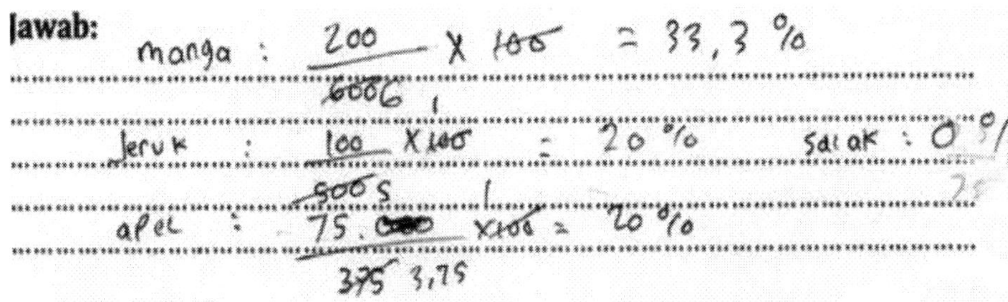

Figure 6. An example of student's answer in finding the percentage of profit and loss

The student's answers presented above indicate that the didactics phenomenology, as a key principle of RME [6], worked as intended in our experiment. The students used the context in the problem to reinvent the concepts. Besides, the contexts in the problems were facilitated the students to use their own strategies, which met one of the characteristics of RMW namely 'student's free production' [17, 18]. This situation helped the students building their confidence in learning mathematics. Finally, we observed the development in students' reasoning ability. They started giving an argument, a reason, or an explanation when solving the contextual problems. The reasoning test that we gave at the end of the experiment showed that 78,1% of the students achieved the score greater than 75. This finding confirmed that LT and RME approach are potential to improve students' reasoning [25,26].

4. Conclusion

The LT for teaching social arithmetic using the RME approach developed in this research met the criteria of validity, practicality, and effectiveness. A design research approach that we used to develop the LT was very helpful in reaching our goal. The LT for teaching social arithmetic reflected the state of the art of RME and it worked as intended in the classroom. Moreover, the LT could help the students to reinvent the concepts in social arithmetic. The students had more confidence to use their own strategies in solving contextual problems. The most important thing, we discovered the growth in the student's mathematical reasoning.

References

[1] Gravemeijer K, 2015, *Proc. MADIF9 Development of mathematics teaching: design, scale, effects* (Linköping: SMDF) p 1-3.

[2] Larson CA, WawroM and Zandieh M, 2017, A hypothetical learning trajectory for conceptualizing matrices as linear transformations, *Int.J.Math.Educ.Scie.Techno.* DOI: 10.1080/0020739X.2016.1276225

[3] Daro P, Mosher F and Corcoran T, 2011, *Learning Trajectories in Mathematics: a Foundation for standards, curriculum, assessment, and instruction* (Philadelphia (USA): Consortium for Policy Research in Education)

[4] Bahamonde ADC, Aymemi JMF and Urgelles JVG, 2016, Mathematical modelling and the learning trajectory: tools to support the teaching of linear algebra,

Int.J.Math.Educ.Scie.Techno DOI:10.1080/0020739X.2016.1241436
[5] Simon MA, 1995, Reconstructing mathematics pedagogy from a constructivist perspective, *J. Res.Math.Educ.* **26** 114-145.
[6] Gravemeijer K, 2004, *Developing Realistic Mathematics Education,* (Utrecht: Freudenthal Institute)
[7] Gravemeijer K and Cobb P, 2013, Educational Design Research Part A: an Introductioned T Plomp and N Nieveen (Enschede: SLO) p 72-113
[8] Liljekvist I, Mellroth E, Ollson J and Boesen J, 2016, *Proc. MADIF10 Development of mathematics teaching: design, scale, effects,* (Linköping: SMDF) p 119-127
[9] Cobb P, Confrey J, Lehrer R and Schauble L, 2003, Design experiments in educational research, *J.Ed. Researcher***32** 9–13.
[10] Fauzan A, Plomp T and Gravemeijer K, 2013, Educational Design Research Part B: an Introduction Ed T Plomp and N Nieveen (Enschede: SLO) p 159-178
[11] Ramirez RE and Solis AH 2016 Hypothetical learning trajectories that use digital technology to tackle an optimization problem *Int.J.Techno.Math.Educ* **24,** p 51-57
[12] Widjaja W, Dolk M and Fauzan A, 2010, The role of context and teachers' questioning to enhance students' thinking *J.Scie.Math.Educ in Southeast Asia* **33**, p.168-186
[13] Fauzan A, 2002, *Applying Realistic Mathematics Education (RME) in Teaching Geometry in Indonesian Primary Schools* (Enschede: Print Partners Ipskamp)
[14] Kwon ON, 2002, *Proc.2nd Int.Con.on the Teaching of Mathematics* (Crete: EDRS)
[15] Prahmana R C I and Kusumah Y S, 2016, The Hypothetical Learning Trajectory on Research in Mathematics Education Using Research-Based Learning, *J Pedagogika* **123**42
[16] Webb D C, Kooij H vd and Geist M R, 2011, Design Research in the Netherlands: Introducing Lograritms Using Realistic Mathematics Education, *J.Math.Edu.Technol.Coll.* **2** 47
[17] Streefland L, 1991, *Realistic Mathematics Education in Primary Schools* (Utrecht: Freudenthal Institute)
[18] de Lange J 1987 *Mathematics, Insight, and Meaning* (Utrecht: OW & OC)
[19] Gravemeijer K, 1999, How emergent models may foster the constitution of formal mathematics *J.Math. Thinking and Learning* **1** 155
[20] Gravemeijer K, Muurling GB, Kraeme JM and Stiphout I, 2016. Shortcoming of Mathematics Education Reform in The Netherlands: A Paradigm Case? *J. Math. Thinking and Learning,* **18** 25
[21] Gravemeijer K 2010 A decade of PMRI in Indonesia, Ed: R Sembiring K Hoogland M Dolk (Utrecht: Tenbrink) p 41-50
[22] Treffers A, 1991, *Three Dimensions: A Model of Goal and Theory Description in Mathematics Education,* (Dordrecht: Reidel)
[23] Freudenthal H, 1991, *Revisiting Mathematics Education* (Dordrecht: Kluwer Academic)
[24] Nieeven N and Folmer E, 2013, *Educational Design Research Part A: an Introduction* Ed: T Plomp and N Nieveen (Enschede: SLO) p 152-169
[25] Larsen S and Zandieh M, 2017, Proofs and refutations in the undergraduate mathematics classroom, *J. Educ.Studies in M*ath. **67**, p. 205
[26] Rasmussen CL and King KD, 2000, Locating starting points in differential equations: a realistic mathematics education approach. *Int.J.Math.Educ.in Scie.Techno.***31**, p.161

IOP Conf. Series: Materials Science and Engineering **335** (2018) 012122 doi:10.1088/1757-899X/335/1/012122

Practicality of Elementary Statistics Module Based on CTL Completed by Instructions on Using Software R

H Delyana[1*], S Rismen[2], and S Handayani[2]

[1] Program Studi Pendidikan Matematika, STKIP PGRI Sumatera Barat, Jl. Gunung Pangilun, Padang 25143 , Indonesia

*hafizahdelyana@gmail.com

Abstract. This research is a development research using 4-D design model (define, design, develop, and disseminate). The results of the define stage are analyzed for the needs of the following; Syllabus analysis, textbook analysis, student characteristics analysis and literature analysis. The results of textbook analysis obtained the description that of the two textbooks that must be owned by students also still difficulty in understanding it, the form of presentation also has not facilitated students to be independent in learning to find the concept, textbooks are also not equipped with data processing referrals by using software R. The developed module is considered valid by the experts. Further field trials are conducted to determine the practicality and effectiveness. The trial was conducted to the students of Mathematics Education Study Program of STKIP PGRI which was taken randomly which has not taken Basic Statistics Course that is as many as 4 people. Practical aspects of attention are easy, time efficient, easy to interpret, and equivalence. The practical value in each aspect is 3.7; 3.79, 3.7 and 3.78. Based on the results of the test students considered that the module has been very practical use in learning. This means that the module developed can be used by students in Elementary Statistics learning.

1. Introduction

Statistics is a branch of science that is needed in everyday life so that humans can not be separated from the use of statistics. Statistics became the focus of mathematics education reform as an important aspect of daily life [4]. Statistics is one of the branches of mathematics that learn about data collection, processing, analyzing data, and drawing conclusions based on data analysis [6]. Statistics have an important role in doing research, therefore statistics need to be well understood by students.

Nowadays many software or developing programs that can help us in doing data processing, presentation of data into graphical form, and data analysis. R is a free version of the S language from a similar paid software that is S-PLUS that can help us in doing data processing, presentation of data into the form of graphs, and data analysis. R has a complete and reliable feature and legal liability factor is no longer a concern in its use because it can be obtained for free. Utilization of functions and packages provided by R will make it easier to create animation programs to demonstrate a theory in statistics [12].

Through Elementary Statistics lectures students are expected to understand the basic concepts of statistics, describe data, present and analyze data and draw conclusions. In addition, students are expected to improve the ability to use the software in performing data processing, presenting data into the form of graphs, and data analysis. Students are expected to be able to understand the course as a whole. The student's understanding is derived from self-study. This is because the lectures in universities are required to learn independently from students.

Teaching material is a set of tool containing learning materials, methods, limitations, and how to evaluate the designed product systematically and interestingly in order to reach the expected purpose, i.e. reaching the competencies and sub competencies with all of the complexities [10]. Based on the definition above, it is clear that a teaching material needs to be designed well by following the rules, so that the teaching material can help the students, teachers/lecturers in the learning process. Teaching material is also defined as a systematical designed material that enables the students to learn individually and is designed based on the valid curriculum. Through the teaching material, teachers will teach the materials well-managed to the students and all the expected competencies are reached.

However, the teaching materials used by the lecturers and students in the lecture of Elementary Statistics is a translated textbook. The interview result to the lecturers show that language of the textbook is difficult to understand; though the presentation was systematically, it has not constructed the student's knowledge and been aimed to find out the concept. Generally, the materials are complete. Levels of the difficulties are generally easy to be understood. The expected teaching material is a module teaching material because it can help the students to study autonomously. Mostly the lecturers have already used the *software* in analyzing the data especially for MINITAB and SPSS, but it never uses *software R*. From the student's interview, it is found that: generally, students state that they are difficult to understand the materials. Book's presentation has been set systematically. However, they get bad grades. Most of them say that the textbook presentation is difficult to understand and lack of systematic. The materials and textbook have been complete. Generally, they inform that the questions on the textbook are difficult to understand especially in language. They expect that the teaching materials could help them to study autonomously. Mostly, they haven't used the software in analyzing the data and even they don't know about *software R*.

The other problem raising in Elementary Statistics lecture is student's low ability in analytical thinking to solve the problem. Many knowledge and information are possessed but it is difficult to relate the situation faced. The facts affect the student's learning outcome, as the following table.

Table 1.Grades of Elementary Statistics Subject of Math Education Study Program Academic Year 2013/2014 and 2014/2015

Grades	**Academic Year**			
	2013/2014		2014/2015	
	Total	%	Total	%
A	23	45,70	17	33,50
B	57		53	
C	35	54,28	64	66,51
D	35		41	
E	25		34	

Table 1 shows that students grade, which is less than 65 (C, D, and E) is 54,28 % and 66,51%. The facts are far from the expectation.

Based on the problems above, it is important to show an effort to develop a module teaching material used for students to study autonomously and can develop their own knowledge. So that they have high analytical thinking in solving the problem. One of the teaching materials developed is *Contextual Teaching and Learning Based Module.* This kind of approach can construct the student's knowledge and find out the concepts of the materials. In order the developing module can increase the ability in using the technology, the developing module is completed by directions of data analyzing using *software* R.

Previous research shows that student's response about *Contextual Teaching and Learning Based Module* is very good [11]. Learning outcome of students using the software is higher than the students who haven't used the software. Learning using *Contextual Teaching and Learning Based Module* can increase the student's skill in the scientific process and behavior [5]. So that learning by using *Contextual Teaching and Learning Based Module* can give a positive contribution towards the student's learning outcome, behavior and skill. This is also in accordance with the results of research conducted Kadir. the result of his research is inquiry approach able to improve activity, response, and mathematical creative thinking skill of student at state junior high School in Indonesia [3]. The inquiry is one part of a contextual learning approach. Besides inquiry, constructivism is also one part of contextual learning. According to Hamdunah, modules developed with a constructivist model are considered effective. This is seen from the final test in a circle and the sphere material [2].

In addition, in relation to the subject materials of Elementary Statistics connecting to the data, so that the analysis can be helped by using the software. Nowadays, there are many statistics software used to analyze the data such as SPPS, MINITAB, etc. However, that software is not free. Therefore, a free software for the student is software R. A research result is from Ulya observing the effect of using contraception tools toward the population number and density in Pemalang Regency in 2014 using *software R, AMOS, and LISREL.* That software show the similar results. Yet, software R is better to use because it can be used by notebook/computer with 32Bit and 64Bit. From the classical assumption test, software R can operate all the classical assumption test. The installation process does not need a license code. Data used in double regression consist of the data from excel, SPSS, Minitab [9]. It means that software R can be free to use it.

2. Method

This research is a Research and Development (R&D). R&D is a research method used to produce a certain product and test the product effectiveness [7]. Product developed is a CTL based module as the aids of lecturing. The research used RnD with 4-D design [8]. The development itself consists of 4 stages *define*, *design*, *develop*, and *disseminate*.The participant is validity three statisticians.

The Instrument used in this research was questionnaire practicality. It is to collect the data of module practicality and data analysis directions. Questionnaire practicality is arranged based on the module development guidance using Likert Scale. The Questionnaire contains student responses to modules that have been developed. Student response questionnaires were used to elicit student responses to the practicality of the CTL-based basic statistics module with the use of developed software R.

Practicality data are collected through questionnaires, observation sheets and interviews with students and lecturers. The questionnaire results are given scores from 0 to 4. Then the mean is determined and confirmed by the criteria: Very Practical if $3.20 < \text{average} \leq 4.00$; Practical if $2.40 < \text{average} \leq 3.20$; Practical enough if $1.6 < \text{average} \leq 2.40$; Less Practical if $0.80 < \text{mean} \leq 1.60$; and Not Practical if the $\text{mean} \leq 0.80$ [1].

3. Result and Discussion

The developed module is considered valid by the experts. Further field trials are conducted to determine the practicality and effectiveness. The trial was conducted to the students of Mathematics Education Study Program of STKIP PGRI in West Sumatera which was taken at random which had not taken Basic Statistics Course ie as many as 4 people. Practicality test results are presented in Table 2 below:

Table 2. Test Results of practicality

NO	ASPECT	AVERAGE	CRITERIA
1	Easiness	3,7	Very practical
2	Time efficiency	3,79	Very practical
3	Easy to interpret	3,7	Very practical
4	Equivalence	3,78	Very practical

Based on the results of the test students considered that the module is very practical to use in learning. This means that the module developed can be used by students in Elementary Statistics learning.

Things to consider on the ease of using modules (a) Modules using simple and easy-to-understand language, (b) Submission of materials can assist students in understanding concepts, (c) Learning to use modules capable of connecting materials with everyday life, (d) Material is easy to understand because it is associated with the real world, (e) The module is easy to use because of its usage instructions, and (f) The use of R software helps simplify the data processing. All things considered in the convenience aspect of using modules are in very practical categories except for points (a) in the practical category.

The next aspect to consider is the efficient time consisting of (a) By using this module the learning time becomes effective, (b) The module can understand the material with a relatively short time, (c) The module can make it easier to understand tasks in the available time, (e) Use of software R saves time in processing data, and (f) The approach used in the presentation of materials helps the student in understanding the material. All things considered in the time efficient aspect are in very practical categories.

Furthermore, ease in interpreting also becomes one of the aspects in module practicality. This includes the following: (a) The module contains clear words so that it is easy to understand, (b) The sentences used in the module are easy to understand, (c) Examples of questions given make it easy to complete the assigned tasks, (d) The commands given are easy to understand, (e) The module helps know the benefits of the lesson presented, and (f) The results of data processing using software R easy to understand. All the things that are concerned with the ease of interpreting are in a very practical category.

The last aspect to consider in the questionnaire is to have the same equivalence. This part of the aspect is: (a) This module may be a companion book of existing textbooks, (b) This module may serve as a means of working training provided by lecturers. All these sections are in the very practical category.

Here is an example of using the software in solving CTL-based problems. This question relates to the way the data is presented

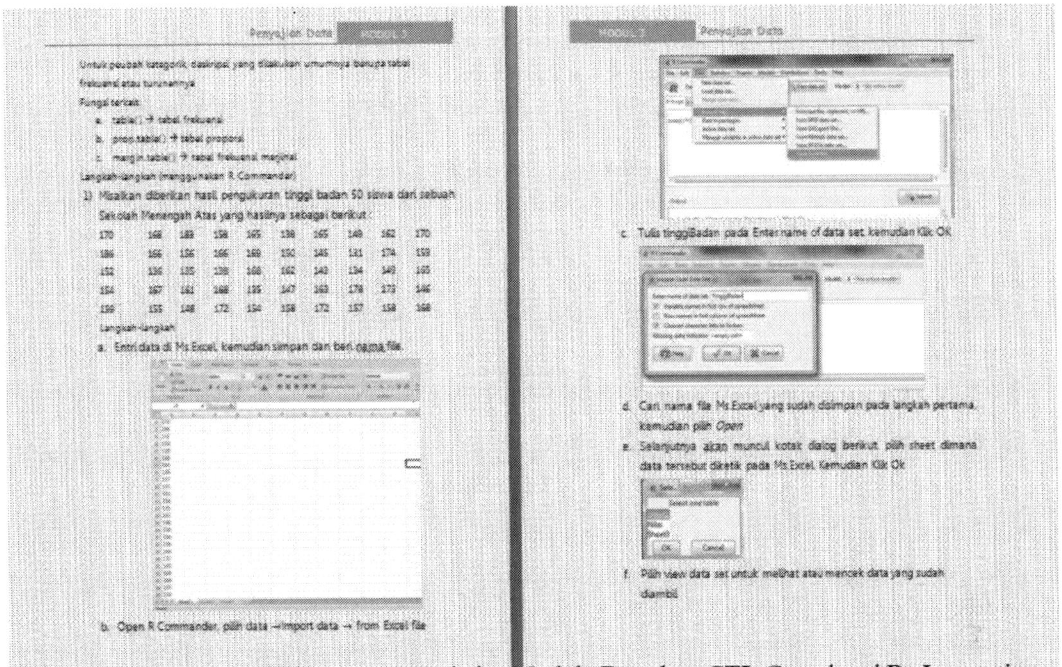

Figure 1. Part Contents of Elementary Statistics Module Based on CTL Completed By Instructions on Using Software R

4. Conclusion

The students' appraisal of the developed modules is very practical with the practical value in each aspect is 3.7; 3.79, 3.7 and 3.78. This means that the module can be used by students in learning.

The module as a companion book for students in understanding the material needs to be made with the best. Therefore the module should proceed with the next stage of the effectiveness stage. Elementary Statistics Module can be used in the lecturing. It is important a development find out the effectiveness of the module.

5. Acknowledgments

The writer expresses a great thank to *DP2M Dikti*, with the Agreement Letter of research Number: 025/E3/2017/ dated on January, 6th, 2017 as the research funder. Next, she also expresses her great gratitude to President of STKIP PGRI West Sumatra for giving the research permit to conduct the research.

References

[1] Arikunto,Suharsimi.2010. Dasar-Dasar Evaluasi Pendidikan. Jakarta: Bumi Aksara.

[2] Hamdunah. 2016. Development A Constructivist Module and Web on Circle and Sphere Material with Wingeom Software . Journal on Mathematics Education. Volume 7 No. 2: 109-116

[3] Kadir. 2017. The Implementation Of Open Inquiry Approach To Improve Students' Learning Activities, Responses, and Mathematical Creative Thinking Skills. Journal on Mathematics Education. Volume 8 No. 1: 103-114

[4] Mavrotheris, M. M. dan Mavrotheris, E. 2007. Online Communities of Practice Enhancing Statistics Instruction: The European Project Early Statistics. The Electronic Journal of e-Learning.Vol5,No.2:113-122.

[5] Rinsiyah,Iis. 2016. Pengembangan Modul Fisika Berbasis CTL untuk meningkatkan KPS dan Sikap Ilmiah Siswa Madrasah Aliyah. Jurnal Pendidikan Matematika dan Sains, 4 (2),

IOP Conf. Series: Materials Science and Engineering **335** (2018) 012122 doi:10.1088/1757-899X/335/1/012122

2016, 152-162.

[6] Sudjana. 1996. Metode Statistika. Bandung: Tarsito.

[7] Sugiyono, 2008. Metode Penelitian Kunatitatif Kualitatif dan R&D. Bandung. Alfabeta

[8] Trianto. 2007. Mendesain Model Pembelajaran Inovatif-Progresif. Surabaya. Kencana Prenada Media Group.

[9] Ulya, Inarotul, 2015. Aplikasi Software R, Amos, dan Lisrel untuk Analisis Pengaruh penggunaan jenis alat kontrasepsi terhadap jumlah penduduk dan kepadatan pendudukan di Kabupaten Pamulang.Tugas Akhir (tidak diterbitkan) Jurusan Matematika FMIPA UNNES.

[10] Widodo, Chomsin S, dan Jasmadi. 2008. Panduan Penyusunan Bahan Ajar Berbasis Kompetensi, Jakarta. Gramedia.

[11] Widarto,Suparmi, Sarwanto .2016. Pengembangan Modul Fisika Bebrbasis CTL pada Fluida Statis dan Fuida dinamis untuk meningktkan prestasi Fisika SMA kelas XI IPA. Jurnal Inkuiri Vol 5. No 1, 2016 ((hal 9-20)

[12] Yudistira, I.G.A. Anom. 2013. Desain Animasi Statistika Berbasis Bahasa Pemrograman R. E-Jurnal WIDYA Eksakta Vol 1. No 1, Juli Desember 2013 ISSN: 2338-8102

Influence of Discussion Rating in Cooperative Learning Type Numbered Head Together on Learning Results Students VII MTSN Model Padang

E Sasmita, S Edriati*), A Yunita*)

Study Program of Mathematics Education, STKIP PGRI Sumatera Barat, Padang, Indonesia

englasasmita@gmail.com

Abstract. Related to the math score of the first semester in class at seventh grade of MTSN Model Padang which much the score that low (less than KKM). It because of the students who feel less involved in learning process because the teacher don't do assessment the discussions. The solution of the problem is discussion assessment in Cooperative Learning Model type Numbered Head Together. This study aims to determine whether the discussion assessment in NHT effect on student learning outcomes of class VII MTsN Model Padang. The instrument used in this study is discussion assessment and final tests. The data analysis technique used is the simple linear regression analysis. Hypothesis test results F_{count} greater than the value of F_{table} then the hypothesis in this study received. So it concluded that the assessment of the discussion in NHT effect on student learning outcomes of class VII MTsN Model Padang.

1. Introduction

There are several reasons why mathematics is mandatory for students to learn, according to Cornelius's statement in Abdurrahman (2010: 253) there are five reasons for the need to learn mathematics because mathematics is a clear and logical means of thinking, a means to solve problems in everyday life, relationship patterns and experience generalizations, means to develop creativity, a means to raise awareness of cultural development.

Based on the observations made in class VII MTsN Model Padang that the subject matter is still explained by the teacher in detail, and in the student's mathematics learning looks monotonous, when the teacher explains the subject matter most students are busy with their own activities such as disturbing the next friend, small groups, discussing things that are not related to the subject matter. Based on the results of interviews with teachers when teachers conduct the discussion activities in the classroom, students are less active and students who come to the front of the class tend to be the same, this is because at the time of discussion took place the teacher did not do the assessment. So students do not carry out the discussion task well. As a result the value of the first semester of class VII examination is under KKM (Minimum Criteria of Completeness). The KKM applied in this school is 75.

Based on the problems that have been described above, then in the learning process required a learning strategy that can motivate and enable students in learning and discussion. Learning strategy that makes students as subject and not object in learning, as well as learning strategy that can make students understand material concepts and able to improve learning outcomes, and conduct assessment of students during the discussion. Assessment of the discussion is done so that students really pay attention to the lesson during the discussion, and make the students active in the discussion, because each discussion is done who is active will be given additional value.

According to Nikson in Muliyardi (2002: 3) "Mathematics learning is an effort to help students to construct mathematical concepts or principles with their own ability through internalization process so that the concept or principle can be rebuilt". Based on the quote can be concluded that in learning mathematics students are accustomed to gain understanding through experience about the properties possessed of an object. Mathematics learning is also a teacher's effort to help students to construct mathematical concepts or principles so that with their own ability and through the process of internalization students can build those concepts back.

One way teachers might be able to do this is by applying the discussion in the model of cooperative learning Numbered Head Together type. The discussion activities in the lesson will give the students a lot of information for their group's performances. According to Nurgiyantoro (2008: 254) "assessment of the discussion is one of the assessment process that allows for direct measurement that requires students to work in concrete and meaningful situations which also reflects the mastery of students to learning materials. Model of cooperative learning Numbered Head Together type is one form of learning with group discussion. According to Kagen in Abdul Majid (2013) states that "NHT is an approach developed to involve more students in reviewing the material covered in a lesson and checking their understanding of the content of the lesson", by applying the NHT strategy students are expected to be more independent and results student learning is increasing.

This study aims to determine whether the assessment of discussion in the model of cooperative learning Type Number Head Together have an effect on student learning outcomes class VII MTsN Model Padang.

2. Research Methods

The type of research is experimental research with the design of The One-Shot Case Study (Arikunto, 2010: 124). The population of this research is all students of class VII MTsN Padang Model Year Lesson 2014/2015 with sample class VII.7 as experiment class. The research instrument is an instrument of assessment of discussion and final test with 4 items of essay, which was tested in class VII.8 MTsN Model Padang with different class teacher with 42 students on February 25, 2015. The analysis technique used is linear regression analysis test simple because the sample data is normally distributed according to Sudjana (2005: 311) form of linear regression equation:

$$y = \beta_0 + \beta_1 x + \varepsilon, \quad (1)$$

with

$$b_0 = \bar{Y} - b_1\bar{X}, \text{ and } b_1 = \frac{\sum X_i Y_i - (\sum Y_i)\left(\frac{\sum X_i}{n}\right)}{\sum X_i^2 - \left(\frac{\sum X_i}{n}\right)}, \quad (2)$$

To prove the hypothesis used F test. For the interpretation of the test by looking at the value of Fcount and Ftable value. If Fcount obtained is less than Ftable then reject H0 and vice-verse H0.

3. Research Result and Discussion

Based on the calculation of the value of discussion each meeting obtained increase in value at each meeting as in Table 1. Table 1 shows the average value of the students' grade discussion in the experimental class experiencing an increase of each meeting. The research hypothesis is "discussion assessment in cooperative learning model type Numbered Head Together have an effect on student learning result class VII MTsN Model Padang".

Table 1. Average Rating Result of Discussion Assessment

No	Average Results of Discussion Assessment of each Meeting			
	I	**II**	**III**	**IV**
1	58,13	--	--	--
2	--	72,82	--	--
3	--	--	86,11	--
4	--	--	--	95,24

The sample class is normally distributed and tested by regression analysis to determine the contribution of the influence of the discussion assessment in the NHT type cooperative learning model to the student learning outcomes. So, hypothesis test is done by F test, obtained Fhitung = 7.20 whereas value F_ (42,0,0) = 4,07, this means Fcount> Ftable. If Fcount> Ftable, then reject H0. So it can be concluded that the assessment of discussion in the model of cooperative learning type Numbered Head Together affect student learning outcomes class VII MTsN Model Padang.

Model of learning Numbered Head Together is one model of learning that includes the type of group learning. Group activities that have been done by teachers in the sample class are regular group activities without any assessment during the activity process. So not all students feel involved in the discussion activities. Assessment of the discussion is an assessment activity undertaken during the discussion process. This assessment is expected to make the students feel involved in the activities of the Numbered Head Together, because this assessment is aimed at each individual within the group.

Based on the tests conducted, the experimental class score of student test scores above the KKM as many as 29 people while previously 21 people. The average value of the experimental class test was 78.64 while the major contribution of the discussion assessment to the final test was 15.26%. It can be seen that the students' discussion of students' influence on students' learning outcomes.

4. Conclusions and Recommendations

Based on the results of data analysis and discussion that has been raised then obtained the conclusion that the assessment of discussion in Numbered Head Together affect the results of learning mathematics students VII MTsN Model Padang. The variation of the students' learning outcomes by 15.26% is influenced by the assessment of the discussion in NHT learning, which means there is a fairly positive relationship or there is a direct relationship between the assessment of discussion in learning mathematics using NHT on student learning outcomes with the regression model $y = -3.03 + 1.05x$ and correlation coefficient of 0.39.

Based on the results of the research, it is expected that math teachers can apply Numbered Head Together (NHT) as an alternative to achieve learning objectives. For future researchers to be done on other subjects and can allocate time well for the results achieved in accordance with the expected.

References

[1] Abdurrahman, Mulyono. 2010. *Pendidikan Bagi Anak Berkesulitan Belajar*. Jakarta: Rineka Cipta.
[2] Arikunto, Suharsimi. 2010. *Prosedur Penelitian Suatu Pendekatan Praktik*. Jakarta: Rineka Cipta.
[3] Dimyati dan Mudjiono. 2002. *Belajar dan Pembelajaran*. Jakarta: PT Rineka Cipta.
[4] Majid, Abdul. 2013. *Implementasi Kurikulum 2013*. Bandung: Interes Media.
[5] Muliyardi. 2002. *Stratgi Pembelajaran Matematika* . Padang: FMIPA UNP.
[6] Nurgiyantoro, Burhan. 2008. *Penilaian Otentik Cakrawala Pendidikan*.
[7] Sudjana. 2005. *Metoda Statistika*. Bandung: PT Tarsito.
[8] Trianto. 2011. *Mendesain Model Pembelajaran Inovatif-Progresif.* Jakarta:Kencana Prenada Media Group

IOP Conf. Series: Materials Science and Engineering **335** (2018) 012124 doi:10.1088/1757-899X/335/1/012124

Influence of Problem-Based Learning Model of Learning to the Mathematical Communication Ability of Students of Grade XI IPA SMAN 14 Padang

I M Nisa[1*]

[1]Student of Mathematics Education Study Program, Universitas Negeri Padang, Indonesia

*izemeidia0594@gmail.com

Abstract. The ability of mathematical communication is one of the goals of learning mathematics expected to be mastered by students. However, reality in the field found that the ability of mathematical communication the students of grade XI IPA SMA Negeri 14 Padang have not developed optimally. This is evident from the low test results of communication skills mathematically done. One of the factors that causes this happens is learning that has not been fully able to facilitate students to develop mathematical communication skills well. By therefore, to improve students' mathematical communication skills required a model in the learning activities. One of the models learning that can be used is Problem Based learning model Learning (PBL). The purpose of this study is to see whether the ability the students' mathematical communication using the PBL model better than the students' mathematical communication skills of the learning using conventional learning in Class XI IPA SMAN 14 Padang. This research type is quasi experiment with design Randomized Group Only Design. Population in this research that is student of class XI IPA SMAN 14 Padang with sample class XI IPA 3 and class XI IPA 4. Data retrieval is done by using communication skill test mathematically shaped essay. To test the hypothesis used U-Mann test Whitney. Based on the results of data analysis, it can be concluded that the ability mathematical communication of students whose learning apply more PBL model better than the students' mathematical communication skills of their learning apply conventional learning in class XI IPA SMA 14 Padang at $\alpha = 0.05$. This indicates that the PBL learning model effect on students' mathematical communication ability.

1. Preliminary

Is a compulsory subject at every level of education and need to be given so that learners are able to think logically, analytically, systematically, critical, and creative, and cooperate. Therefore, it is fitting for math teachers preparing lessons that facilitate students to achieve goals learning mathematics with the maximum. The purpose of learning The mathematics is that learners have abilities such as understanding concepts, using patterns, using reasoning, communicate ideas, have the right attitude of respect and behavior with math, doing motor activities as well as using simple props in performing mathematical activities (Permendikbud, 2014). In the process of learning mathematics, the eight abilities are very important and expected to be achieved by students. Ability to understand concept is a basic ability that must be achieved by students. After students understand the mathematical concepts

IOP Conf. Series: Materials Science and Engineering **335** (2018) 012124 doi:10.1088/1757-899X/335/1/012124

so hopefully they can communicating math ideas or ideas well. In fact, many students who have understood the concept well but have not been able expressed his mathematical ideas correctly.

This shows that students' mathematical communication ability is still low. The issues raised are supported by the facts based on observations made in grade XI IPA SMA N 14 Padang on September 7 to 12, 2015. It is seen that within learning mathematics students have difficulty in communicating a clear mathematical idea to a friend or teacher through spoken language and writing. When the teacher asks questions about a concept or process, students have difficulty explaining the method they use. Student have difficulty in modeling a problem into the model mathematics. Many students do not read and use notation right or mathematical symbols when asked by the teacher. In addition, students too difficulties in understanding mathematical problems in the form of stories, so they also can not make a mathematical model of that problem given. Therefore, in the mathematics of teachers are required to choose a strategy, method, or model that involves a lot students in learning. One of the models that can facilitate students to developing the ability of mathematical communication is a Problem model based learning, hereinafter abbreviated as PBL. In learning PBL students are required to actively develop skills think of them. The PBL model is focused on giving the problem to the participants educate and do solve problems with concepts, principles and knowledge appropriate knowledge.

Giving a problem will trigger it good mathematical communication between teacher and student, student with source learning, or students with fellow students. By giving the problem to students, they will practice solving problems in the form of stories. Modeling mathematical problems, and can train students in the use of symbols and notations appropriately. Problem-based learning can also optimize development students' thinking ability. In line with the statement, Tan (Rusman, 2012: 229) suggests that "Problem-based learning is innovation in learning because students' thinking skills are optimized through systematic or group work ".Therefore, PBL is seen to improve students' mathematical communication skills.

Based on the formulation of the above problem, the purpose of this study is to knowing whether the mathematical communication ability of grade XI student of high school Negeri 14 Padang learning with PBL model better than siwa who learn with conventional learning. Research relevant to this research is Rahmi Research (2009) taking sample of junior high school students while research Riza (2013) took a sample of high school students. The results of the study shows that the mathematical communication ability shows developments after the implementation of different strategies with learning conventional.

2. Research Methods

Based on the problems studied and the proposed research objectives, the type of research conducted is quasi experimental (quasi experiments). according to the type of research research design used is Randomized Control Group Only Design. In this design, two random groups, the first group of experimental classes were given treatment, while the second group control class without conventional treatment or learning. in this study, samples were randomly selected as experimental class and control class. population in this study is all students of class XI IPA SMAN 14 Padang. selected class XI IPA 3 as experimental class and class XI IPA 4 as control class.

independent variables in this research is the treatment given to the sample class, which consists of learning mathematics with Problem Based Learning model in the experimental group and conventional learning in the control group. The dependent variable in this research is the mathematical communication ability of the students in the subjects of mathematics class XI IPA SMA N 14 Padang. The instrument used in this research is the ability test students' mathematical communication. This test aims to measure ability students on each indicator of mathematical communication ability.

3. Research Result

Mathematical communication ability data obtained through test with problem shaped essay. The test was conducted at the end of the study on the 21st May 2016 in both sample groups followed by 30 students of the experimental class and 29 control class students. Results of the description of the data obtained by the test which has been done can be seen in Table 1.

Table 1. Results Description Data Test Mathematical Communication Skills

Group	The number of students	The lowest score	Highest score	Average	Standard deviation
Experiment	30	4	12	9,20	1,97
Control	29	3	12	7,66	2,33

Table 1 shows that the mean of the experimental group test results is higher than the average of control group test results. the average of experimental group test result is 9,20 while control group that is 7,66. standard deviation in the experimental class is lower than the control class. these results indicate that students' mathematical communication ability of the experimental class is more uniform. the data of both classes of samples are normally distributed and have homogeneous variances. then test the hypothesis using Mann-Whitney test .

Based on Mann-Whitney test results obtained P-Value is 0.0062 which means reject H_0 and accept H_1 it means mathematical communication ability of experimental class students. better than the students' mathematical communication skills of the control class. so, it can be concluded that the mathematical communication ability of students whose learning using PBL learning model is better than students' mathematical communication ability with conventional learning. The conclusion of this hypothesis test is the PBL model gives effect to students' mathematical communication abilities, in which ability the mathematical communication of students learning with the PBL model is better than students who study with conventional learning in real level of 0.05.

4. Discussion

Students' mathematical communication ability in this research is seen through five indicators that are (1) present mathematical statement in writing or in the form of drawings; (2) Explaining ideas, situations, and mathematical relations, writing, with real objects, drawings, graphics, and algebra; (3) Declare everyday events in language or mathematical symbols; (4) Provide reason or evidence against the solution; and (5) Checking the validity of an argument.

Result of hypothesis test give result that mathematical communication ability of experiment class student is better than control class student. In the experimental class applied the PBL model, while in the control class using conventional learning. Hypothesis test results are general conclusions.

In indicator 1 students are required to present a mathematical statement into a graphical form as well as interpreting the graph. in indicator 2, the percentage of the experimental class students who achieved the maximum score is much higher than the control class students. percentage of experimental class students who are able to achieve maximum score on this indicator that is equal to 86,67% and control class student equal to 41,38%. the average score obtained by the experimental class students in indicator 2 is also higher than the control class students that is 2.87 while the average score of the students' control class score is 2.00 for the indicator 3 percentage of experimental class students who achieve maximum score is much higher than the control class student percentage of experimental class students who can achieve maximum score on this indicator that is equal to 63,33% and control class student equal to 27,59%. The average score is obtained by experimental class students on indicator 3 is also higher compared to control class students that is 2.37 while the average gain student grade score of 1.24. For this indicator 4, the percentage of control class students who achieve maximum score is higher than the experimental class students. Percentage of students control class that is able to achieve maximum score on this indicator ie equal to 65,51% and experiment class student equal to 53,33%. The average score is obtained the control class students in indicator 4 is also higher than the experimental class is 2.31 while the average score of the experimental class student score is 2.00. For this indicator 5, the percentage of the experiment class students who achieved the score maximum higher than control class students. Percentage of class students experiments that can achieve maximum score on this indicator that is equal to 63.33% and control class students 58.62%. Average score gained the experimental class students on this indicator 5 are also higher than the students control class that is equal to 1.50 while the average score of student control class score is 1.45

Based on the description of each indicator of communication ability mathematical and result of description and analysis of data that have been done can it is said that the mathematical communication ability of experiment class students more better than the students' mathematical communication skills of the control class. This matter because in the experimental classroom learning is carried out with using PBL models that engage students actively in the process learning through the five stages contained in it. Learning activities using this PBL model start with giving problem to student, where with giving of this problem student given the opportunity to conduct an investigation to solve the problem the. Through this activity students are trained to explain the idea as well strategy in solving the problem. Besides, as it has been it is known beforehand that the problems given in this PBL model is a real-world problem, so it can train students to declaring everyday events into math or language modeling the given problem. At the time of modeling the problem this, of course students are also trained to use notation or symbols mathematics well. Learning activities use this PBL model also train students to define problems and find mathematical ideas in solving the problem. Furthermore students are also trained for summarizes the solution of the problems they find using their own language. This is what makes the PBL model can improve students' mathematical communication skills in experimental class. So students able to explain ideas, situations, and mathematical relationships, in writing, with real objects, images, graphics, and algebra, able to declare the day-to-day in a language or mathematical symbol, and able to check validity an argument

5. Conclusion

The conclusion of this research is PBL model gives influence to students' mathematical communication abilities, in which ability the mathematical communication of students learning with the PBL model is better than mathematical communication skills of students who learn by using conventional learning. This is because in this PBL model learning is based on giving problems to students. As is this problem, certainly in the process of completion of trained students in modeling or declaring daily events in the language of mathematics. With these modeling activities students are also trained in using, symbols or mathematical notation. In addition, in using lessons this PBL model students are trained in explaining ideas or strategies through problem solving activities are given and students are also required to summed up the solution they found from solving that problem given in their own language. This is what makes the ability mathematical communication of experimental class students can be better than students control class.

Reference

[1] Arikunto, Suharsimi. 2008. *Dasar-Dasar Evaluasi*. Jakarta: Bumi Aksara.
[2] National Council of Teachers of Mathematics (NCTM). 1996. *Communication in Mathematics*. Reston. Virginia.
[3] Sumarmo, Utari. 2005. *Pembelajaran Matematika untuk Mendukung Pelaksanaan Kurikulum Tahun 2002 Sekolah Menengah*. Makalah Seminar Pendidikan Matematika di FMIPA Universitas Negeri Gorontalo, Gorontalo.
[4] Suprijono, Agus. 2009. *Cooperative Learning (teori & aplikasi PAIKEM)*. Surabaya: Pustaka Pelajar.
[5] Suryabrata, Sumadi. 2014. *Metodologi Penelitian*. Jakarta: PT. Raja Grafindo Persada.
[6] Warsono dan Hariyanto. 2012. *Pembelajaran Aktif Teori dan Asesmen*. Bandung: PT Remaja Rosdakarya Offset.

IOP Conf. Series: Materials Science and Engineering **335** (2018) 012125 doi:10.1088/1757-899X/335/1/012125

Understanding Mathematic Concept in Relation and Function Method through Active Learning Type Group to Group Distributed LKS

F Kudri[1], R Rahmi[2], Y Haryono[2]

[1]Student of Mathematics Education Study Program STKIP PGRI SUMBAR
[2]Lecturer of Mathematics Education Study Program STKIP PGRI SUMBAR

Abstract. This research is motivated by the lack of understanding of mathematical concepts students and teachers have not familiarize students discussed in groups. This researchaims to determine whether an understanding of mathematical concepts junior class VIII SMPN 2 in Ranah Batahan Kabupaten Pasaman Barat by applying active learning strategy group to group types with LKS better than conventional learning. The type of research is experimental the design of randomized trials on the subject. The population in the study were all students VIII SMPN 2 Ranah Batahan Kabupaten Pasaman Barat in year 2012/2013 which consists of our class room experiment to determine the grade and control class with do nerandomly, so that classes VIII1 elected as a experiment class and class VIII4 as a control class. The instruments used in the test empirically understanding mathematical concepts are shaped by the essay with rt= 0,82 greater than rt=0,468 means reliable tests used. The data analysis technique used is the test with the help of MINITAB. Based on the results of the data analisis known that both of the sample are normal and homogenity in real rate α = 0,05, so the hypothesis of this research is received. So, it can be concluded students' understanding mathematical concept applied the active Group to Group learning strategy with LKS is better than the students' understanding mathematical concept with Conventional Learning.

1. Introduction

Mathematics is a universal science that underlies the development of modern technology. The importance of the role of mathematics is seen in the implementation of education, where mathematics is a subject studied by every level of education. The importance of mathematics also shows that the mathematics lesson is one of the subjects that determine the students' passing during the national examination. Given the importance of mathematics lessons, teachers are expected to be able to educate and teach students that the objectives of mathematics learning in schools can be achieved optimally.

Mathematical learning will give good results if goals can be met. The ability that must be possessed by the students so that the learning of mathematics runs in accordance with the objectives of the ability of conceptual understanding, communication, and problem solving. One of the skills that determine the success of students in learning mathematics is the ability of conceptual understanding, where the ability of conceptual understanding requires students to be able to define concepts, identify and give examples or not examples of concepts. To that end, the government has done a lot of efforts, including improving the curriculum, improving education facilities and infrastructure, providing training for teachers, providing opportunities for teachers to continue their education, as well as improving the quality of teachers through certification.

Based on the results of observations and interviews conducted in SMPN 2 Ranah Batahan Pasaman Barat Regency on May 8, 2012 note that the process of learning mathematics students are still dominated by teachers. Students are not active during the learning, it can be seen still low motivation of students to ask, teachers have not familiarized the students discussion group so that smart students do not want to help students who are not smart in learning, students do not dare to convey their ideas when given question .

One of the alternatives to overcome these problems is to apply group-to-group active learning strategy with Student Activity Sheet (LKS) that can help students and teachers in learning process to run well and can develop students' mathematical concept understanding.

According to Silberman (2007: 166) "on the Group to Group active strategy, different tasks are assigned to different groups of learners. Each group teaches what has been learned for the rest of the class ". Student Activity Sheets are worksheets that have to be done by Prastowo students, 2011: 203).

This study aims to determine the understanding of mathematical concepts of students who applied the active learning strategy of Group to Group with LKS is better than understanding the mathematical concepts of students by applying conventional learning to students of class VIII SMPN 2 Ranah Batahan.

Relevant research with this research is Nurhasanah (2009). The result of this research is the result of student learning using active learning strategy of Group to Group Exchange with LKS better than student learning result using conventional learning.

2. Research Methods

The type of this research is experimental research with Random Against Subject design, based on Arikunto (2010). In this study, samples were randomly selected as experimental and control class. The population in this study is the students of class VIII SMPN 2 Ranah Batahan. Selected class VIII.1 as experimental class and VIII.4 as control class.

The independent variable in this research is mathematics learning by using Group to Group strategy with LKS in experiment class and conventional learning in control class. The dependent variable in this research is the understanding of the mathematical concept of the students of the sample class.

The instrument used is a test of understanding of mathematical concepts in the form of essays and rubric scoring concept understanding used is a holistic rubric, based on Iryanti (2004).

The material tested is relation and function. The instrument used is the comprehension test of mathematical concept in the form of essay with r11 = 0.82 bigger than rt = 0.468 means test question used reliabel. Data analysis techniques in hypothesis testing using t test with the help of MINITAB. Testing is done to find the hypothesis accepted or rejected.

3. Research Result

Based on the results of data analysis obtained a description of students' understanding of mathematical concepts of sample classes that can be seen in Table 1 below.

Table 1. Results of Student Concept Understanding Test

Sample Class	$\bar{X}$	S	X_{maks}	X_{min}
Experiment	66,1	13,89	97	37
Control	42,2	17,82	87	18

In Table 1 it can be seen that the average value of the students in the experimental class is greater than the average value of the control class, whereas the experimental standard deviation is lower than the control class, meaning that the control class has different values than the experimental class.

The data of both sample classes are normally distributed and have homogeneous variances. Therefore, hypothesis testing is done by using one party t test, obtained P-value 0.000 at 95%

confidence level. Since P-value is smaller than the specified α, then the hypothesis is accepted. So it can be concluded that the learning outcomes of mathematical concepts of students who applied the active learning strategy of Group to Group with LKS is better than the understanding of mathematical concepts of students applied conventional learning in grade VIII SMPN 2 Ranah Batahan students.

4. Discussion

Understanding the concept is one of the aspects assessed in the learning outcomes, in conceptual understanding there are six indicators. However, in this study only seen three indicators that is, presents the concept in various forms of mathematical representation, re-state a concept and apply the concept or algorithm solving the problem.

Based on the observation during the research, the students in the experimental class are more eager to learn. This can be seen at each meeting from the first meeting until the fourth meeting.

The first meeting of students' enthusiasm for group learning has not been seen because teachers have not used the students to learn the previous group. At the time of distributing LKS class atmosphere a bit noisy because students are afraid not get LKS. At the time of discussion there was no visible cohesiveness between groups. The appointed spokesperson presented the results of the group discussion still afraid to speak in front of the class, as well as the groups who responded were still shy to ask.

Subsequently at the second meeting, for this meeting each representative of the group distributed LKS to members of his group. At the time of responding the students have begun to dare to ask this matter seen students who responded to more than one person and there are also from other groups who obtained the material is not the same.

Student group learning spirit was also seen at the third meeting. At this meeting, learning with the Group to Group strategy runs better when compared to the first and second meetings.

While at the fourth meeting the spirit of student learning with Group to Group strategy has been better, At this meeting at the end of the learning the researcher gives the opportunity back to the students to ask about the material that is still not understood from the first meeting until the fourth meeting before the final test. in this session there are some students who ask about the problem that has not been understood. The fifth meeting held the final test for both sample classes.

Based on the final test and analysis that has been done can be concluded that the understanding of mathematical concepts of students by applying Group to Group learning strategy with LKS better than the understanding of mathematical concepts of students with conventional learning. This can be seen from the achievement of indicators of students' mathematical concepts that are visible from several pieces of final test answers given to the two sample classes.

1. Given P = Q = {0, 1, 2, 3}
   a. Draw an arrow diagram that states the relation "over" from the set P to the set Q
   b. Express the relation as a set of consecutive pairs!
   c. Also state the relation with cartesius diagram!

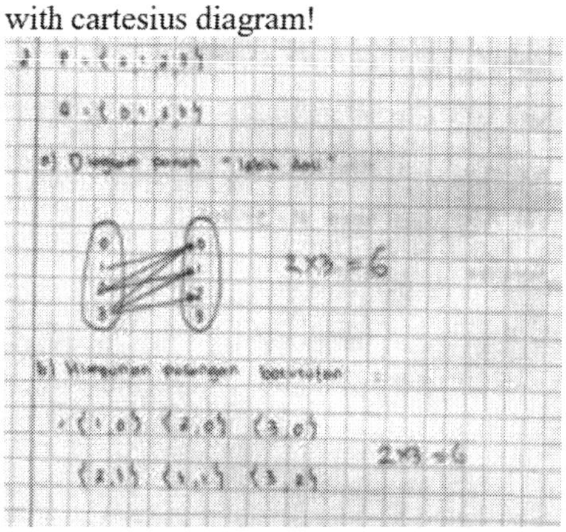

Figure 1. Answer sheets about the 2 students of the experimental class

The answer of the experimental class students in Figure 1 shows that the students have been able to understand and do the problem correctly, where the students have been able to present the concept in various forms of mathematical representation in accordance with the indicator expected in question number two. But some students still make mistakes and do not answer, examples of mistakes that students still do on the control class are as follows:

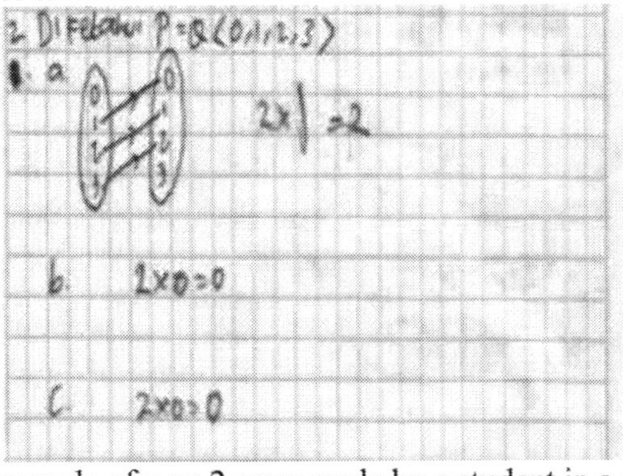

Figure 2. Example of a no.2 error made by a student in a control class

Students' answers in Figure 2 students are not able to understand the problem, it is seen that students do not answer about number 2b this shows the students are not able to present the concept in various forms of mathematical representation in accordance with the expected indicators on problem number two. when compared with student experiments there are still many control class students who make mistakes.

Analysis of the description of students 'answers to the experimental class can be concluded that the indicators of students' concept concepts of experimental class is better than the control class students.

5. Conclusion

Based on the results obtained after the analysis and discussion of the problems that have been raised in this study, it can be concluded that the understanding of mathematical concepts of students by applying Group to Group strategy with LKS is better than the understanding of mathematical concepts of students by applying conventional learning grade VIII SMPN 2 The domain of Batahan Pasaman Barat Regency.

Reference

[1] Arikunto, Suharsimi. (2010). *Prosedur Penelitian Suatu Pendekatan Praktik*. Jakarta. Rineka Cipta.
[2] Iryanti, Puji. (2004). *Penilaian Unjuk Kerja*. Yogyakarta: Depdiknas
[3] Prastowo, Andi (2011). Panduan Kreatif Membuat Bahan Ajar Inovatif. Jogjakarta: Depdiknas
[4] Nurhasanah, (2009).Penerapan *Active Learning* Tipe *Group To Group Exchange* (GGE) Dengan LKS Pada Pembelajaran Matematika Kelas VIII Mts Alwashlihah Labuhan Batu Kabupaten Labuhan Batu Tahun Pelajaran 2008/2009. Skripsi
[5] Silberman, Melvin L. (2007). *Active Learning*. Bandung: Nuas Madia.

IOP Conf. Series: Materials Science and Engineering **335** (2018) 012126 doi:10.1088/1757-899X/335/1/012126

Application of Learning Engineering Techniques Thinking Aloud Pair Problem Solving in Learning Mathematics Students Class VII SMPN 15 Padang

S Y S Widuri[1*], L Almash[2], F Zuzano[1]

[1]Department of Mathematics and Natural Sciences Education, Faculty of Teacher Training and Education Bung Hatta University
[2]Department of Mathematics Education, Faculty of Mathematics and Natural Sciences State University of Padang

*yolawiduri@yahoo.com

Abstract. The students activity and responsible in studying mathematic is still lack. It gives an effect for the bad result in studying mathematic. There is one of learning technic to increase students activity in the classroom and the result of studying mathematic with applying a learning technic. It is "Thinking Aloud Pair Problem Solving (TAPPS)". The purpose of this research is to recognize the developing of students activity in mathematic subject during applying that technic "TAPPS" in seven grade at SMPN 15 Padang and compare the students proportion in learning mathematic with TAPPS between learning process without it in seven grade at SMPN 15 Padang. Students activity for indicators 1, 2, 3, 4, 5, 6 at each meeting is likely to increase and students activity for indicator 7 at each meeting is likely to decrease. The finding of this research is $\chi^2 = 9{,}42$ and the value of p is $0{,}0005 < p < 0{,}005$. Therefore $p < 0{,}05$ has means H_0 was rejected and H_1 was accepted. Thus, it was concluded that the activities and result in studying mathematic increased after applying learning technic the TAPPS.

1. Preliminary

Education plays an important role to improve the quality of human resources. Improving the quality of human resources at a given time is determined by relevant educational innovations. If education does not follow changes occurring at a certain time, then education will be out of date. Therefore, the government made various innovations in the teaching system including in mathematics learning such as: improving the quality of mathematics teachers by doing training / training, equipping educational facilities and infrastructure as well as perfecting the curriculum.

Based on the observations that researchers do in class VII.2, VII.3, and VII.5 at SMPN 15 Padang on November 28, 2014 up to December 2, 2014, it appears that students are less active in the learning process. When the teacher asks students to ask things that have not been understood no students are asking. However, when the teacher asks only one or two students are trying to answer and people tend to be the same. While the other students much silent and not motivated in studying the material. This can be seen when the teacher explains the material of some students chatting, annoying his friends, and scribbling his notebook.

Further problems in group discussions, student participation are uneven, there are students who want to fully take over the work of the group and vice versa most students ignore their responsibilities in group work and only rely on their friends. So not all members of the working group, most of the students chatting outside the topic of discussion with a group of friends, just waiting for an answer from a group of friends and do not want to try to complete the exercise. In addition, in group presentations dominated by smart students and other students simply copying the results of group discussions that are presentations on each note.

Based on the results of interviews that researchers conducted on December 2, 2014 with one of the teachers of mathematics in class VII SMPN 15 Padang, Ifawani mother obtained information that students rarely express opinions during the learning and students difficulty in understanding the example of the problems contained in the handbook. When given the form of the problem slightly different from the example problem, only some students who can do well even most of the students difficult in working on problem solving problems. In solving problems students are less motivated to think independently, students difficulty in understanding the purpose of the problem, choosing the right strategy used in problem solving and students difficulty in developing the strategy to solve the problem. This is because students have not understood the subject matter so that students can not analyze the problems contained in the problem.

Furthermore, the teacher also stated that the result of the students' mathematics learning in the middle of odd semester examination of class VII SMPN 15 Padang in the academic year 2014/2015 is still low that can be seen in Table 1.

Table 1. Number and Percentage of Students Achieving Mathematics Learning Completion at Middle Trial Odd Examination Class VII SMPN 15 Padang Lesson Year 2014/2015

Class	Number of Student	Completed Students	
		Total	Percentages
VII.1	36	25	69,44
VII.2	36	7	19,44
VII.3	36	6	16,67
VII.4	35	4	11,43
VII.5	33	5	15,15
Total	176	47	26,70

Source: Math Teacher Class VII SMPN 15 Padang

Based on Table 1 it can be seen that there are still many students who have not fulfilled the Minimum Exhaustiveness Criteria (KKM) set at SMPN 15 Padang that is 75 except class VII.1 because the class belongs to the superior class.

In overcoming this problem, needed a learning technique that can improve activity and result of student learning that is Thinking Aloud Pair Problem Solving Technique (TAPPS). In this TAPPS learning technique, students are divided into pairs, one acts as a problem solver, and the other acts as a listener. For the next question there was a change of role. With TAPPS learning techniques each student has responsibility for the task given so that each student to take an active role in learning.

The purpose of this study is to determine the development of student activity in learning mathematics during applying TAPPS learning techniques and compare the proportion of students who achieve mastery learning mathematics by applying TAPPS learning techniques and the proportion of students who achieve mastery learning mathematics by applying regular learning in class VII SMPN 15 Padang .

Learning is the unity of the learning components that are related to each other. These components include: objectives, materials, methods and evaluation. In choosing and determining the media, methods, strategies, techniques and approaches that will be used in the teaching and learning process, the teacher should pay attention to these four components of learning.

This is in line with Hamalik's view (in Hosnan 2014: 18) which says that :"learning as a combination is composed of human elements, materials, facilities, equipment and procedures that influence each other to achieve learning goals".

In the learning of mathematics the active role of students is necessary. Gagne's learning theory cited by Suherman (2003: 33) states that:

In learning mathematics there are two objects that can be obtained by students that is direct object and indirect object. Indirect objects include the ability to investigate and solve problems, self-study, be positive about mathematics and know how to learn. While direct objects in the form of facts, skills, concepts, and rules.

Based on above learning theory, in learning mathematics students will get various facts, skills, concepts, and rules as well have the ability to investigate, solve problems, learn independently, be positive about mathematics, and know how to learn the right way.

Cooperative Learning is a learning that divides students into small groups to solve a problem. Suherman (2003: 260) states that:

Cooperative learning includes a small group of students who work as a team to solve a problem, complete a task, or do something to achieve another shared goal.

Based on the quotation, it can be concluded that cooperative learning emphasizes students to work together and be responsible to a team in solving a problem or task. So that high-ability students will try to help other members who have low ability to achieve group goals.

Cooperative learning is very concerned about heterogeneity. Students are divided into multiple pairs. The division of pairs in this study considers students' academic abilities for example, high-ability students are paired with low-ability students.

TAPPS learning techniques include one of the structures of cooperative learning methods. TAPPS is expressed by Lochead and Whimbey (in Warsono and Hariyanto 2013: 92) "as a way of developing problem-solving skills by verbally declaring, reading aloud the problems to be solved". The word loud has enough understanding to hear her partner. In line with Kotsopolus's opinion in (Kani, 2015: 20) TAPPS aims to develop students' cognitive processes related to problem solving In TAPPS learning techniques students are divided into pairs, one of which acts as a problem solver, the other acts as a listener. The problem solver reads the written problem that the teacher asks loudly. Then the problem solver also verbally resolved the problem to the listener. The listener follows all the steps done by the problem solver, listens to what the problem is, how the solution proposed problem solver, advises the problem solver, and the listener must also understand the reasoning process behind the problem solving steps posed by the problem solver. This involves asking the problem solver if the expression is not clear. For the next question there was a change of role, problem solver being listener and vice versa. And so on until the question runs out or the time provided for the learning runs out. A. Whimbely and J. Lochhead (in Ilie, 2016 : 278) described this strategy that involves thinking and learning in pairs, by rotating roles. Students are more involved and become more active.

Articulating the problem-solving process yourself and listening carefully to solving other people's problems will help students practice what they have read or heard in their learning. In addition, Kotsopoulus (in Kani, 2015 : 21) highlighted in her study on examining instances of talking aloud during peer collaborations in mathematics that it is important to teach the students both on how to express their thinking and their learning needs and how to attend to each other's thinking and learning needs in such settings.

The steps of TAPPS learning techniques (in Barkley 2012: 260) are:

1. Ask students to form partners and explain to them the roles of problem solvers and listeners. The problem solving role is to read the problem verbally and to articulate the reasoning process used in solving the problem. The role of the listener is to encourage problem solving to think verbally, and to illustrate the problem-solving steps. Listeners can also ask clarification questions and offer suggestions, but should refrain from solving problems.
2. Ask students to solve a number of problems, change roles for each new problem.

3. Activities be stopped if the student has successfully solved all the problems.

The steps of the application of TAPPS learning techniques that researchers apply are as follows: (1) The teacher explains the rules of the game and the time limit for each activity, (2) The teacher asks the students to sit in pairs based on the partner who has been determined, (3) (4) The teacher explores the initial knowledge of the students by conducting question and answer on the material that has been submitted, (5) The teacher gives the question sheet and answer sheet to each pair. (6) For the first problem solver 1 read out the problem orally and use his skills in solving the problem, while explaining each step of completion to listener 1, (7) Listener 1 observed the process solving problems, responding if there is a wrong solution, advising if problem solver finds it difficult and asking if there is anything that is not understood, (8) The teacher walking around the classroom provides guidance and helps smooth the discussion. If a partner has difficulty in solving the problem, the teacher can help by becoming a listener, by asking questions that lead to problem solving, (9) Problem solver 1 and listener 1 exchanging roles and re-discussing such as steps 6 and 7 to complete question the other, (10) The group that has solved the problem faster than the specified time will get the bonus as an additional value, (11) After all the questions have been answered, the teacher selects a random pair to convey the problem-solving steps in front of the class (12) Students and teachers conclude the lesson of the day, (14) The teacher rewards the couple who can explain the problem-solving steps well in front of the class, (15) The teacher give homework to students.

Learning is a process by which students must be active. Teachers only present lesson materials, while processing and digesting are the students themselves according to their will and their respective abilities. Montessori (in Sardiman 2010: 96) asserted that "children have the power to develop on their own, forming themselves. Educators will act as mentors and observe how the development of their students ".

Based on the description Montessori can be explained that students who do more activities in the learning, while the teacher is only tasked to provide guidance and observe the development of students during the learning process takes place. Thus, it is clear that the subject students / students must be active in learning activities. So, in learning is necessary activity. Without activity, the learning process may not take place properly.

The types of activities in learning according to Paul B. Diedrich (in Sardiman 2010: 101) are:

a. Visual activities, which include, for example, reading, watching images of demonstrations, experiments, other people's work.
b. Oral activities, such as: stating, formulating, asking, advising, issuing opinions, conducting interviews, discussions, interruptions.
c. Listening activities, for example listening: descriptions, conversations, discussions, music, speeches.
d. Writing activities, such as writing stories, essays, reports, questionnaires, copying.
e. Drawing activities, for example: drawing, graphs, maps, diagrams.
f. Motor activities, which include: experimenting, constructing, refit modeling, playing, gardening, raising.
g. Mental activities, for example: responding, remembering, solving problems, analyzing, looking at relationships, making decisions.
h. Emotional activities, like for example, are interested, bored, excited, passionate, passionate, brave, calm, nervous.

Of the eight activities proposed by Diedrich, the researchers limit the activities to be observed in this study of oral activities, listening activities, writing activities and emotional activities. Indicator used to know the student activity during the learning process take place that is:

a. Oral activities
- Ask questions related to the material being studied
- The problem solving steps to their partner.

- Responding to the settlement of the partner.

b. Listening activities
   - Listen well the explanation of the partner in solving the problem.
   - Listen well the results of the answers presented by other couples.

c. Writing activities
   - Recording or copying the material learned

d. Emotional activities
   - Doing distorted behaviors, for example: talking outside topics, interrupting friends, making a fuss, doing other work and going out of the classroom.

2. Methodology

The type of this research is experimental research. According Sukardi (2007: 16) suggests that experimental research is a study that divides the object or subject studied into two groups, the treatment group or who get treatment and the control group who did not get treatment. This research is used to see the relationship and influence between one variable with other variables.

Based on the type of research above, this research is conducted on two classes, namely experimental class and control class. The experimental class is a class whose learning uses TAPPS learning techniques and control classes is a class whose learning uses ordinary learning.

The population is the whole of the sample under study. The population in this research is the students of class VII SMPN 15 Padang. The sample is part of the population, all the sample characteristics are the same as the population. In this study, the experimental class is class VII.2 and the control class is class VII.3.

Data collection tool in this research is observation sheet and test results of mathematics learning. From the observation sheet, the improvement and decrease of the students' learning activity during the application of TAPPS learning technique in the experimental class. The test of learning result is used to find out whether the result of learning mathematics of student by applying TAPPS learning technique better than result of learning of student mathematics by applying ordinary learning.

The data analysis techniques in this study are:

2.1 Student Learning Activity

To know the development of student learning activities during applying TAPPS learning techniques used observation sheet. Data on student learning activities were analyzed using the formula put forward by Sudjana (2013: 130) namely:

$$P = \frac{F}{N} X\ 100\%$$

Information:

P = Percentage of students performing activities
F = Number of students doing the activity
N = Number of students

2.2 Learning outcomes

Find out whether there are differences in learning outcomes from two independent sample classes, the experimental class applies the lesson using TAPPS learning technique and the learning control class using ordinary learning then hypothesis test. The hypothesis to be tested in this research is as follows:

H_0: The proportion of students who achieve mathematics learning completion by applying TAPPS learning techniques is similar to the proportion of students who achieve mathematical learning by applying ordinary learning.

H_1: The proportion of students who achieve the mastery of mathematics learning taught by applying TAPPS learning techniques is higher than the proportion of students who achieve mathematical learning by applying ordinary learning.

To test this hypothesis we used the test χ^2 for two independent samples. The steps in using the test $\chi^2$for two independent samples put forward by Siegel (1985: 136-137) as follows:

a. Enter observation frequencies in a 2 × 2 contingency table, as shown in Table 2.

Table 2. Format Table Number of Students Experiment Class and Control Class According to KKM Achievement

	Experiment Class	Control Class	$\sum$
Value ≥ KKM	A	B	A+B
Value < KKM	C	D	C+D
$\sum$	A+C	B+D	N

Information:
A: Number of completed students in the experimental class
B: Number of students who complete the control class
C: Number of unfinished students in experiment class
D: Number of unfinished students in the control class
N: Total number of students in the experimental class and control class

b. Compute χ^2 with the formula:

$$\chi^2 = \frac{N\left(|AD-BC|-\frac{N}{2}\right)^2}{(A+B)(C+D)(A+C)(B+D)}$$

with db = 1

c. Determine the significance of χ^2 observation with reference table χ^2 . For a one-sided test, for two levels of designated significance. If the probability (p) given by Table χ^2 is equal to or less than α, then reject H_0 and accep H_1.

3. Results and Discussion

1. Student Learning Mathematics Activities

In learning is necessary activity. Without activity, the learning process may not take place properly. To know the development of student learning activity in experiment class is used observation sheet. At the time of the study, this observation sheet was filled at each meeting by two observers ie teachers of mathematics studies SMPN 15 Padang and colleagues.

The description of student activity obtained from the observations on each indicator observed during applying TAPPS learning technique is as follows:

1) Submit a Question Related to the Lesson being Learning

Activity is still relatively small. This is because students are still embarrassed to ask questions.

2) Stating the Problem Solving Steps to the Spouse

The percentage of students who do listening activities with good explanation of their partner in solving the problem has increased from the first meeting until the sixth meeting. This is because students have begun to understand what kind of responses will be given and students are getting used to writing their responses on the answer sheets provided specifically for students who act as listener.

3) Provide Response to the Settlement of the Spouse

The percentage of students performing activities responding to the settlement of their partner has improved from the first meeting until the sixth meeting. This is because students have begun to understand what kind of responses will be given and students are getting used to writing their responses on the answer sheets provided specifically for students who act as listener.

4) Listening Well Explanation of the Couple in Solving Problems

The percentage of students who do listening activities with both explanations from their partners in solving the problem increased from the first meeting until the sixth meeting. This is due to the increasing number of students who serve as problem solver states the steps to solve the problem to his partner who acts as a listener and the less students do other activities that are not related to learning mathematics.

5) Listen Well the Results of Answers Presented by Other Couples

The percentage of students who do a good listening activity results of answers presented by other couples increased from the first meeting until the sixth meeting. This is due to good class control and motivation by researchers to students.

6) Recording or Copying Learning Materials

The second meeting there is a decline. This is due to the lack of students who recorded at this meeting. When the researcher gives the students time to take note of the material that has been described, there is a teacher picket who enters the class to ask the students for each student's contribution and announce to the students that in the farewell event each class should have an entertainment show. As long as the teacher is in the class there are some students who are busy discussing with his friends about what appearances they will perform at the farewell show, some even just listening to the teacher speak. After the teacher came out of the class, the researcher directly gave the answer sheet and answer sheets and asked the students to do it with the TAPPS technique, this is done because the researchers remember the remaining time for only 40 minutes and this is why many students do not record at the second meeting.

7) Doing Distorted Behavior, For example: Speaking Outside Topics Discussion, Interfere Friend, Fussy, Working on Other Tasks and Entry Exit Class

The second meeting the percentage of students who do the deviant behavior increases which causes the same as the declining percentage of students who perform activities to record or copy the material that has been described previously. On third to sixth meetings, the percentage of students who behaved misbehaved continued to decline.

In the experimental class the researcher is assisted by two observers who monitor the students' activeness during the learning process takes place. The student activity is seen from the seven indicators contained in the observation sheet. In general at indicators 1, 2, 3, 4, 5, and 6 there is an increase in the percentage of students who perform activities for each meeting. Although, there are still activities that percentage up and down. In the 7th indicator of student activity performing deviant behavior, for example: talking outside topics, disturbing friends, fussing, doing other tasks and going in and out of class tends to decrease from first meeting till the sixth meeting. This happens because most students have performed activities on indicators 1, 2, 3, 4, 5, and 6 so that only a few students are still doing deviant behavior.

2. Student learning outcomes

To find out whether there are differences in learning outcomes between the experimental class and control class, data analysis is done by testing the hypothesis. To test this hypothesis test usedχ^2. Before determining the value of χ^2, first the number of experimental class and control class students according to the achievement of KKM as in Table 3.

Table 3. Number of Students of Experiment Class and Control Class According to KKM Achievement After Final Test

	Experiment Class	Control Class	Σ
Value≥ KKM	24	10	34
Value $< KKM$	12	26	38
Σ	36	36	72

Based on the above table, using the test χ^2 for two independent samples is calculated and obtained values χ^2 = 9,42 and $0{,}0005 < p < 0{,}005$. Therefore p <0.05 means reject H_0 and accept H_1. Thus, it is concluded that the proportion of students who achieve mathematics learning completion by applying TAPPS learning techniques is higher than the proportion of students who achieve mathematics learning completion by applying ordinary learning. This means that students 'mathematics learning outcomes by applying TAPPS learning techniques are better than the students' mathematics learning outcomes by applying ordinary learning.

During the application of TAPPS learning techniques in the experimental class, it appears that students are more active in learning. This is visible when the student given the opportunity to ask, asks people who are willing to ask questions. In the discussion of students who do not understand the material to ask his partner without shame because his partner is a peer. In addition, students also provide suggestions to their partners if there is a wrong solution so that a positive impact on mastery learning. Thus, by applying TAPPS learning techniques can give a good influence on the activity and mastery of mathematics learning outcomes.

The results of the TAPPS learning techniques will help students practice what they have read and hear in the learning so that students can get better learning outcomes. Percentage mastery of student learning outcomes by class in experiment class is 66,67% and control class is 27,78%. Although the percentage in the experimental class is higher but according to Depdikbud in Trianto (2010: 241) "A class is said to be complete learning (classical thoroughness) if there are $\geq$ 85% of students who have completed their study". This means the percentage of learning mastery in the experimental class has not been achieved well. Thus, students who have not completed completeness must make improvements (remedial) or be given additional tasks.

4. Conclusions

Based on the results of the discussion that has been done in the previous chapter it can be concluded:

1) Student learning activities in the classroom applying the TAPPS learning techniques from the first meeting to the sixth meeting on mathematics learning in grade VII SMPN 15 Padang tended to increase for indicators 1, 2, 3, 4, 5, and 6 and tended to decrease for indicator 7.
2) Students 'mathematics learning outcomes by applying TAPPS learning techniques are better than the students' mathematics learning outcomes that apply ordinary learning.

Reference

[1] Barkley, Elizabert. E., Cross, K.P., dan Major, C.H. 2012. *Collaborative Learning Techniques*. Jakarta: Nusa Media.

[2] Hosnan, M. 2014. *Pendekatan Scientifik dan Kontekstual dalam Pembelajaran Abad 21*. Bogor: Ghalia Indonesia.

[3] Ilie, Vali. 2016. *How to Stimulate Future Teacher-Studens to Participate in Specific Instructive-Eduacational Activities*.Romania : Journal Plus Education.

[4] Kani, Nekmahtul Hafizah Budi dan Masitah Shahrill. 2015. *Applying the Thinking Aloud Pair Problem Solving Strategy in Mathematics Lessons* . Brunei Darussalam : Asian Journal of Management Sciences & Education

[5] Sardiman, A.M. 2010. *Interaksi dan Motivasi Belajar Mengajar*. Jakarta: Grafindo.

Siegel, Sidney. 1985. *Statistika Nonparametrik Untuk Ilmu-ilmu Sosial*. Jakarta: PT Gramedia

[6] Sudjana, Nana. 2013. *Penilaian Hasil Proses Belajar Mengajar*. Bandung: PT. Remaja Rosdakarya.

[7] Suherman, Erman. 2003. *Strategi Pembelajaran Matematika Kontemporer*. Bandung : JICA

[8] Sukardi. 2007. *Metodologi Penelitian Pendidikan*. Jakarta: Bumi Aksara.

[9] Trianto. 2010. *Mendesain Model Pembelajaran Inovatif-Progresif*. Jakarta: Kencana Prenada Media Group.

[10] Warsono & Hariyanto. 2013. *Pembelajaran Aktif Teori dan Asesmen*. Bandung: PT Remaja Rosdakarya.

IOP Conf. Series: Materials Science and Engineering **335** (2018) 012127 doi:10.1088/1757-899X/335/1/012127

College Students Perceptions of Web-Based Leaning in Basic Mathematics Subject

H Husna[1*], T Septia[2], A Cesaria[3]
[1,2,3]STKIP PGRI Sumatera Barat

[*]husnarifai64@gmail.com

Abstract. With the emergence of the Internet, e-learning has increasingly become the promising solution that continues to grow day after day. Considering students' perception toward e-learning is important in successful development of e-learning in higher education, since attitude of user towards application of information technology is one of the most effective factors. This study aims to determine students' perceptions of using basic math textbook for physics with web-based. Students' perceptions are closely related to their achievement. The learning media in accordance with the desire of students will make them motivated. This research is qualitative data analysis was done descriptively. The data obtained in this study comes from researchers as the main instrument, the data questionnaire and interview data is unstructured. The results are students' perceptions of using basic math textbook for physics with web-based are in the positive category.

1. Intoduction

Introduction to mathematics is a subject which is taught to students in STKIP PGRI West of Sumatra to prepare or equip students about mathematics. Introduction to mathematics is not only learned in mathematics education department but also in physics. The introduction to mathematics for the study program physics learn about The system of real numbers and its properties, Inequality and absolute value, Coordinates Cartesian, Functions and graphics functions, Trigonometry, Systems of linear equations, Probability and statistics, Geometry, Logarithms, Limits and continuity, Derivatives and applications in physics, Integral and applications in the physical science.

The results of interviews with students who had been taken introduction to mathematics regarding the difficulties in understanding the material. They often only find numbers, formulas, charts, and die pictures, thus all things make the students feel less interested and they think the materials are boring. Students commonly use the textbook as an aid in studying for exams, but many students do not read the textbook to the extent that they should (Burchfield & Sappington, 2000; Clump, Bauer, & Bradley, 2004; Sappington, Kinsey, & Munsayac, 2002; Sikorski et al.,2002). Therefore, we need a media that is able to increase the attractiveness to students so that it can improve students' motivation and understanding. This is in accordance with the views expressed by Daryanto (2010: 5), the usage of media in the learning process can clarify the message that is not too verbal, overcoming the limitations of space and time, and excite student's learning.

IOP Conf. Series: Materials Science and Engineering 335 (2018) 012127 doi:10.1088/1757-899X/335/1/012127

The use of instructional media should provide positive impact on student achievement. Like awise the use of web as a multimedia learning for Basic Mathematics subject should have positive impact for student achievement. Use of media or multimedia learning that is not in accordance with the students expectations can make the students fell less interested in learning, therefore the student can not achieve the achievement desired learning both in process and learning outcomes(Islami, 2009).

Several studies have been published exploring student perceptions and expectations regarding e-learning (Wang, 2004). Recent studies by the National Center for Education statistics show a growing demand and acceptance of online learning (Waits & Lewis, 2003). Online access can affect how successful students will be in Web-based classes (Anawati & Craig, 2006). Previous research into online collaboration and education in international environments indicates that maintaining contact and access is essential to educational success or students' affective and cognitive development. This contact involves (a) keeping students involved in online projects and (b) building a kind of online community essential to educational success (Vogel et al., 2001).

2. Method

The research method used is qualitative method. Data collection techniques used interviews, observation, and documentation. While the technique of selecting informants using purposive sampling is in some students that take Basic Mathematics Courses. The technique of data validity using triangulation technique of source and method with data analysis start from data collecting, data reduction, data presentation, verification. Research subjects are Physics education study program students who take the basic mathematics courses amounted to 25 people. Research data were analyzed using the percentage distribution (%).

3. Result

The result of filling the questionnaire for 25 students.

Table 1 Students Questionnaire result

No	Indicator	Yes	No
1	Do you prefer learning using web-based instead of conventional learning?	80%	20%
2	Is the web-based learning application useful?	88%	12%
3	Do you interesting with web based learning?	84%	16%
4	Do you think web based learning can help you to understand the material?	88%	12%
5	Are you motivated to study material with using web based learning?	76%	24%
6	Whether with your web-based learning is not necessary face to face with lecturers every day?	20%	80%
7	Whether with web-based learning you can apply the materials in everyday life?	56%	44%
8	Are you more diligent with web based learning?	60%	40%
9	Is the web based learning increase you learning achievement?	60%	40%
10	Is the web based learning can make your time in study the material more efficient?	68%	32%
11	Do you find any difficulties in understanding the material with web based learning?	16%	84%
12	Do you like the web-based assignment method?	76%	24%
13	Do you agree if web-based learning is applied to the Basic Mathematics subject?	76%	24%

Criteria: < 50% not good, 50-70% good and >70% very good.

Based on the results of questionnaires by 25 students in table 1, then basically the perception of students on web-based learning process is good enough, useful, interesting and motivating students to study the materials. This is seen with the percentage obtained > 70% of students answered yes with very good category. Based on table 1, there are also some questions that are categorized as unfavorable percentage of student answers <50% that is related to student statement that although using web-based learning still need to face to face with lecturer, and web-based learning still can't apply learning in daily life.

The results of the questionnaire that has been analyzed also stated that there is difficulty in understanding the material by using web. This is due to several reasons such as internet connection, students' ICT skills are still low, and students are not familiar with web based learning. As explained by Surjono (2009), that the access gap in Indonesia between those residing in urban areas with rural areas and supporting facilities is an obstacle to the use of the internet. Web-based learning is a learning that can be accessed anytime and from anywhere, besides learning materials can be enriched with various learning resources including multimedia and can also be updated quickly by teachers. This learning also facilitates interaction between students and lecturers. Web-based learning can also support government programs in reducing paper use. Therefore familiarize students learn to use the web needs to be done in order to improve the learning process flexibility and interactivity can be realized.

4. Conclusion

The findings showed that the students' perceptions of the web-based learning in basic mathematics courses responded well by the students.

References

[1] Burchfield, C. M., & Sappington, J. (2000). Compliance with required reading assignments. *Teaching of Psychology, 27,* 58-60.

[2] Clump, M. A., Bauer, H., & Bradley, C. (2004). The extent to which psychology students read textbooks: A multiple class analysis of reading across the psychology curriculum. *Journal of Instructional Psychology, 31,* 227-232.

[3] Sappington, J., Kinsey, K., & Munsayac, K. (2002). Two studies of reading compliance among college students. *Teaching of Psychology, 29*(4), 272-274. doi:10.1207/S15328023TOP2904_02.

[4] Sikorski, J. F., Rich, K., Saville, B. K., Buskist, W., Drogan, O., & Davis, S. F. (2002). Student use of introductory texts: Comparative survey findings from two universities. *Teaching of Psychology, 29*(4), 312-313.doi:10.1207/S15328023TOP2904_13.

[5] Daryanto. 2010. *Media Pembelajaran.* Yogyakarta: Gava Media.

[6] Islami, Nadu Azizah. 2009. *Pengaruh Persepsi Siswa Mengenai Media Pembelajaran Guru Dan Keaktifan Siswa Dalam Proses Pembelajaran Terhadap Prestasi Belajar*. Universitas Muhammadiyah Surakarta.

[7] Wang, Y. S.(2004); "Assessment of Learner Satisfaction with Asynchronous Electronic Learning Systems"; *Information & Management*, Vol. 41, pp. 75-86.

[8] Waits, T. & Lewis, L (2003). Distance education at degree-granting postsecondary institutions: 2000-2001 (NCES 2003-017). U.S. Department of Education, National Center for Education Statistics. Retrieved January 8, 2005 from http://nces.ed.gov/pubsearch/pubsinfo.asp?pubid=2003017.

[9] Anawati, D., & Craig, A. (2006). Behavioral adaptation within cross-cultural virtual teams. IEEE Transactions on Professional Communication, 49, 44-56.

[10] Vogel, D. R., van Genuchten, M., Lou, D., Verveen, S., van Eekout, M., & Adams, A. (2001). Exploratory research on the role of national and professional cultures in a distributed learning project. IEEE Transactions on Professional Communication, 44, 114-125.

[11] Surjono, H. 2009. *Pengantar E-learning dan Penyiapan Materi Pembelajaran.* http://blog.uny.ac.id/hermansurjono/files/2009

IOP Conf. Series: Materials Science and Engineering **335** (2018) 012128 doi:10.1088/1757-899X/335/1/012128

Influence of Problem Based Learning on Critical Thinking Skills and Competence Class VIII SMPN 1 Gunuang Omeh, 2016/2017

D M Aswan[1]*, L Lufri2, R Sumarmin

[1]Master of Educational Sciences, Faculty of Mathematics and Science, State University of Padang, Indonesia
[2]Biology Department, Faculty of Mathematics and Science, State University of Padang, Indonesia

*daramutiaraaswan@gmail.com

Abstract. This research intends to determine the effect of Problem Based Learning models on students' critical thinking skills and competences. This study was a quasi-experimental research. The population of the study was the students of class VIII SMPN 1 Subdistrict Gunuang Omeh. Random sample selection is done by randomizing the class. Sample class that was chosenVIII3 as an experimental class given that treatment study based on problems and class VIII1 as control class that treatment usually given study. Instrument that used to consist of critical thinking test, cognitive tests, observation sheet of affective and psychomotor. Independent t-test and Mann Whitney U test was used for the analysis. Results showed that there was significant difference (sig <0.05) between control and experimental group. The conclusion of this study was Problem Based Learning models affected the students' critical thinking skills and competences.

1. Introduction

The development of education today directing the process of learning, the students learn to build his own knowledge. This is done to achieve the goals of education contained in the curriculum in 2013, namely that the students have the necessary competence for lives. According [1], competence meant that competence of attitude (affective) in the form of a religious attitude and high social ethics in public life, the competence of knowledge (cognitive), and competence of skills (psychomotor) in the form of skills or the ability to apply knowledge.

One of the subjects contained in junior high school subjects were biology belonging to the subjects of Natural Sciences (IPA). Science education is one aspect of science education that is expected to deliver learners to meet the skills needed in the 21st century. Some of these capabilities by [1] is a learned skill and innovation that includes critical thinking and able to solve problems, creative and innovative, and able to communicate and collaborate.

Critical thinking skills is an ability that needs to face the global challenges of the 21st century. Critical thinking skills are needed so students are able to take decisions in everyday life. Learners must be taught to develop critical thinking skills in solving problems, so the ability to problem-solving will also be developed. This is supported by [2] which showed that critical thinking skills are important, because with automatic critical thinking someone would be able to resolve simple or complex in everyday life.

Through observation and interviews conducted to teachers teaching science in class VIII SMPN 1 Subdistrict Gunuang Omeh, it is known that the learning process is focused on knowledge and understanding of the topics. As a result, students become less trained in developing thinking and problem solving skills and apply the concepts learned in school to the real world. The active role of learners in the learning process is still lacking, only a few learners are shown to be active in asking and answering questions. Questions and answers posed are still limited to questions and answers on the level of knowledge alone.

In addition to the value of critical thinking skills of students is still low, learners' cognitive competence of data obtained from a science teacher at SMPN 1 Subdistrict Gunuang Omeh also shows that the competence of learners cognitive aspect is still below the minimum completeness criteria that has been set is 75.

To improve critical thinking skills and competence of learners is started lessons with a problem or question [2]. Starting the learning with problems can be done through the application of PBL. In PBL, the problem presented in early learning and serves as a stimulus for learning activities [3]. PBL can also improve the competence of learners, both cognitive and affective competencies and psychomotor. The learning model is capable of making the students actively engaged in learning, so that learning becomes meaningful and can improve learning competencies possessed by learners [4].

2. Method

The research is a quasi-experimental. Research was conducted on two classes are experimental class and control class. The experimental class were treated in the form of a model *Problem Based Learning* (PBL) and untreated control class conventional model.

The population in this study were all students of class VIII SMPN 1 Subdistrict Gunuang Omeh academic year 2016/2017. Sampling was conducted through *random sampling* by drawing using a roll of paper which consists of 3 pieces containing the class name. Drawn first class is a class VIII_3 designated as experimental class and the second class is a class that is picked VIII_1 as the control class.

Instruments that used in this research is a matter of describing adjusted indicator to determine the value of critical thinking skills of students, the objective test to assess the cognitive competence, and observation sheets and evaluation sheets to determine the competence of learners' affective and psychomotor aspects. The data analysis technique used is the t-test for critical thinking skills and cognitive competencies and the u-test for affective and psychomotor.

3. Results

The data obtained in this study were: 1) the critical thinking skills of experimental class and control class, 2) cognitive competencies of experimental class and control class, 3) the affective competencies of experimental class and control class, and 4) psychomotor competence of experimental class and control class.

3.1 Data of Learners' Critical Thinking Skills

Data critical thinking skills of learners are presented in Table 1.

Table 1. ValuesCritical Thinking Skills of Students

Class	Average	X_{max}	X_{min}	S	S^2
Experiment	74.69	93.75	54.17	10.56	111.6
Control	66.56	85.42	47.92	9.94	98.90

Based on Table 1, it can be seen that the average of students' critical thinking skills in the experimental class is higher than the control class.

3.2 Data of learners' cognitive competence

Data of learners' cognitive competence are presented in Table 2.

Table 2. Data Competence Cognitive Students

Class	Average	X_{max}	X_{min}	S	S^2
Experiments	80.5	97.5	65	9.16	83.95
Control	70.38	92.5	50	11.45	131.10

Table 2, it can be seen that the average competence learn science students in the experimental class is higher than the control class, ie 80.5 experimental class and control class 70.38.

3.3 Data of learners' affective competence

Data affective competencies of learners are presented in Table 3.

Table 3. Data Affective Competence Students

Class	Average	Predicate
Experiment	78.82	B
Control	70.83	B

Based on Table 3, it can be seen the total value of the average affective learners in the experimental class earned an average higher than the control class.

3.4 Data of learners' psychomotor competence

Learners' psychomotor competency data are presented in Table 4.

Table 4. Data Competence Psychomotor Students

Class	Average	Predicate
Experiment	75.25	B
Control	65.25	C

Based on Table 4, it can be seen the average value of students' psychomotor competence in the experimental class is higher than the control class.

4. Discussion

The application of PBL give a positive impact on critical thinking skills of learners. PBL enhance the critical thinking skills of learners. This difference is due to the PBL, the learners are trained to develop their critical thinking skills to solve problems through activities that constitute the core of PBL. Core activities of PBL located on the second, third, and fourth phase provides an opportunity for learners to actively construct knowledge through problem-solving activities that can develop the mindset so that learners are accustomed to think critically [5].

Critical thinking skills should be developed through a process of direct experience of learners in solving problems. The critical thinking will grow if learners are accustomed to use critical thinking skills to solve problems. This is in accordance with the opinion of Masek and Yamin [6], which states that the measures contained in PBL is able to support the development of critical thinking skills of learners.

Differences in study results is due to the experimental class that uses PBL, learners are given the opportunity to build his own knowledge. The same thing was stated by Orhan and Ruhan in [7], that PBL give a positive impact on students' academic achievement. Implementation of PBL in medical school is also give a positive impact on cognitive competence. Problem-solving activities carried out on PBL is a good technique for learners to understand the content of the lessons. These activities will also make learners find his own knowledge, so that learning becomes more meaningful and can improve cognitive competence of learners. Masek and Yamin [6] states that PBL is believed to provide an environment that is conducive to learning more meaningful, so that it can affect the ability of learners to apply their knowledge. In line with Tan [7] that PBL contribute to cognitive competencies.

The observation of learners' affective competencies conducted by the observer, data showed affective competencies of students in the experimental group was significantly better than learners' affective competence control class. Concerns raised at the beginning of each study can provide a stimulus that stimulates the curiosity of learners. This curiosity arise as a result of learning that begins with the provision of the issues or issues of interest that exists around the learner. This is in line with Atikasari, et al. [8] suggest that the pattern of teaching that focuses the problem and explore science positive effect of increasing analysis capabilities and curiosity of students to provide relevant solutions to problems. PBL requires students to arrive on time and notify members if the group can not be present, so as to train the learner discipline.

Results competency assessment psychomotor learners performed, the data obtained psychomotor competence learners experimental class significantly better than the control class. The high acquisition psychomotor competence of students in the experimental class is a positive influence on learning to use PBL which in problem-solving activities, the students are required to develop and present work that requires creativity of learners. Not much different from the views expressed by [9] that PBL have special characteristics that produce products or work and show off the product. When learners have the creativity, competence psychomotor learning will increase as psychomotor competency assessment of learners is done through product ratings.

5. Conclusion

Based on the results of research and discussion that has been done it could be concluded as follows.
1. Problem Based Learning model has significant effect on the critical thinking skills of learners.
2. Problem Based Learning model has significant effect on the learners' cognitive competency.
3. Problem Based Learning model has significant effect on the lerners' affective competency.
4. Problem Based Learning model has a significant effect on the learners' psychomotor competency.

References

[1]. Ministry of Education and Culture. 2016. *Syllabus School Subjects SMP / MTs (SMP / MTs) Subjects Natural Sciences.* Jakarta: Ministry of Education and Culture.

[2]. Synder, LG and Synder, MJ 2008. Teaching Critical Thinking and Problem Solving Skills. *The Delta Pi Epilson Journal,* L (2), 90-99.

[3]. Chin, C. and Chia, LG 2005. *Problem Based Learning: Using Ill-Structured Problems in Biology Project*Work.Singapore: Wiley InterScience.

[4]. Zaduqisti. 2010. Problem Based Learning: Concept of Ideal Model for Improving Learning Achievement Learning and Achievement Motivation. *MT,* 8 (2), 181-191.

[5]. Karim and Normaya. Kemamuan 2015. Students Critical Thinking in Mathematics Learning by Using Model JUCAMA in Junior High School. *Journal of Mathematics Education,* 3 (1), 92-104.

[6]. Masek, A. and Yamin, S. 2011. The Effect of Problem Based Learning on Critical Thinking Ability: A Theoretical and Empirical Review. *International Review of Social Sciences and Humanities,* 2 (1), 215-221.

[7]. Tan, OS 2009. *Problem Based Learning Innovation, Using Problem Based Learning in the 21th Century.* Singapore: Cengage Learning Asia Pte. Ltd.
[8]. Atikasari, S., WISI, I. and Andreas, PB 2012. Effects of Problem Based Learning Approach In the Material Analysis Capability Against Environmental Pollution. *Unnes Journal of Biology Education.* 1 (3): 17-25.
[9]. Dewi, EK, and Jatiningsih, O. 2015. Influence of Learning Model *Problem Based Learning* on Critical Thinking Ability in Subjects PPKn Students of Class X at SMAN 22 Surabaya. *Moral and Citizenship Studies,* 02 (03), 936-950.

IOP Conf. Series: Materials Science and Engineering **335** (2018) 012129 doi:10.1088/1757-899X/335/1/012129

The Effect of Teacher Performance in Implementation of The 2013 Curriculum Toward Chemistry Learning Achievement

L P Dewi*, A Djohar

Curriculum Development Master Program, School of Postgraduate Studies, Indonesia University of Education, Indonesia

*4lia.puspita@gmail.com

Abstract. This research is a study about implementation of the 2013 Curriculum on Chemistry subject. This study aims to determine the effect of teacher performance toward chemistry learning achievement. The research design involves the independent variable, namely the performance of Chemistry teacher, and the dependent variable that is Chemistry learning achievement which includes the achievement in knowledge and skill domain. The subject of this research are Chemistry teachers and High School students in Bandung City. The research data is obtained from questionnaire about teacher performance assessed by student and Chemistry learning achievement from the students' report. Data were analyzed by using MANOVA test. The result of multivariate significance test shows that there is a significant effect of teacher performance toward Chemistry learning achievement in knowledge and skill domain with medium effect size.

1. Introduction

The educational curriculum in Indonesia has been progressing from time to time. The curriculum is constantly improving for a better Indonesia education until Curriculum 2013 emerges. Curriculum 2013 is raised to address the internal challenges of Indonesian education and also global challenges.

In the term of implementation of the 2013 Curriculum, the teachers have a very important role. The 2013 Curriculum can be implemented well if the teacher has the ability to translate document of the 2013 Curriculum into learning process in classroom. The teachers' understanding of 2013 Curriculum will affect teacher performance in teaching and learning activities. The learning essence of Chemistry should be understood by Chemistry teachers so that the teachers not only transfer knowledge but also demonstrate good performance in instilling values and optimizing students' competency in the domain of knowledge, attitude, and skill.

Curriculum implementation can not be separated with student achievement. This is because student achievement is the result of teaching and learning process which is conducted by teachers. The performance of Chemistry teachers in the course of teaching and learning process will have an impact or effect toward students learning achievement. Meanwhile, student achievement becomes one measure of the success of curriculum implementation.

Due to the importance of curriculum implementation by teachers as well as the students achievement, this research aims to determine the effect of teacher performance in implementation of the 2013 Curriculum toward student learning achievement in Chemistry subject.

2. Conceptual framework

2.1. *Teacher performance*
Performance or job performance is a universal and multidimensional concept. Many definition of job performance have been proposed by several researchers. Job performance can be define as scalable actions, behaviour and outcomes that employees contribute to organizational goals [1]. In the context of teacher performance, teacher is a proffesional educator which has the main task of educating, teaching, guiding, directing, training, assessing, and evaluating students. Teacher performance also can be described as the duties performed by a teacher at a particular period in the school system in achieving organizational goals [2].

2.2. *Implementation of the 2013 Curriculum*
The implementation of curriculum means the actualization of written curriculum in the form of learning. The implementation of curriculum is the application of curriculum ideas, concepts or programs into leaning practices by constantly adjusting the field situation and characteristics of students [3]. In term of the 2013 Curriculum, the teacher has to make a planning of curriculum implementation in the form of lesson plan based on syllabus that has been set in the 2013 Curriculum. Then, implementation of the 2013 Curriculum in learning process uses scientific approach and several teaching model such as discovery/inquiry, problem based learning, and project based learning. The learning process and result must be evaluated by using authentic assessment.

2.3. *Student achievement*
Student achievement can be defined as the status of subject-matter knowledge, understanding, and skills at one point in time [4]. In term of the 2013 Curriculum, student achievement should cover the domain of attitude, knowledge, and skill. However, Chemistry teacher only evaluates the student achievement in the domain of knowledge and skill; while for the student achievement in attitude domain is evaluated by civic teacher and religion teacher.

3. Methodology
The research method is descriptive with quantitative approach. The research design involves the independent variable, namely the performance of Chemistry teacher, and the dependent variable that is Chemistry learning achievement which includes the achievement in knowledge and skill domain. The subject of this research are 7 Chemistry teachers and 255 students from 7 High School in Bandung City. The research data is obtained from students ratings questionnaire about teacher performance in implementation of the 2013 Curriculum that includes aspects of learning planning, implementation of learning, evaluation of learning, and teacher personality; and Chemistry learning achievement from the odd semester report of academic year 2016/2017. Then, data were analyzed with MANOVA test by using SPSS IBM 22.0 Statistic software.

The effect of independent variable towards dependent variable is called effect size. In MANOVA calculations, the effect size can be known from eta squared (η^2). Cohen gives reference to the magnitude of effect size based on the value of eta squared, that is $\eta^2 = 0.01$ for small effect size, $\eta^2 = 0.06$ for medium effect size, and $\eta^2 = 0.14$ for large effect size [5]. However, the result of analysis data using SPSS do not include the value of eta squared, but partial eta squared(η_p^2). The value of both can be different. Yet, if there is one independent variable in the analysis of variance, the value of the eta squared equals to partial eta squared [6].

4. Results and discussion

4.1. *Results*
The result of multivariate significance test in MANOVA can be seen in Table 1. The value of efffect size in this research can be seen in the result of Pillai's Trace test.

Table 1. Multivariate significance test

Effect		Value	F	Hypothesis df	Error df	Sig.	Partial Eta Squared
Intercept	Pillai's Trace	0.995	27604.683	2.000	251.000	0.000	0.995
	Wilks' Lambda	0.005	27604.683	2.000	251.000	0.000	0.995
	Hotelling's Trace	219.958	27604.683	2.000	251.000	0.000	0.995
	Roy's Largest Root	219.958	27604.683	2.000	251.000	0.000	0.995
Kinerja	Pillai's Trace	0.122	8.205	4.000	504.000	0.000	0.061
	Wilks' Lambda	0.879	8.329	4.000	502.000	0.000	0.062
	Hotelling's Trace	0.135	8.452	4.000	500.000	0.000	0.063
	Roy's Largest Root	0.119	15.020	2.000	252.000	0.000	0.107

The result of Pillai's Trace test shows that the significancy value is less than 0.05. It means that teacher performance gives a significant effect toward Chemistry learning achievement in the domain of knowledge and skill. The value of effect size can be known from the value of partial eta squared in Pillai's Trace test that is 0.061. Based on Cohen, the value of effect size included in medium category.

4.2. Discussion
The result of MANOVA test shows that there is a significant effect of teacher performance toward Chemistry learning achievement in knowledge and skill domain with medium effect size. It shows the need for teacher performance improvement.

Curriculum implementation is influenced by teachers' belief and school characteristic [7]. The teachers' belief, knowledge, and ability in implementing the curriculum will bring a good performance. The teacher performance in implemetation of the 2013 Curriculum also emphasizes the active engangement from students in the learning process, so that the teacher should be proficient in selecting and applying strategies and learning methods. As Marzano pointed out that the major independent impact on student achievement is instructional strategies [8]. The teachers are fully responsiblefor deciding how to use the resources available to students for advancing students to some extent. The teachers need to assess the instructional needs of each student then align strategies appropriate to their needs.

To achieve maximum student learning achievement, collaborative effort between teachers and students is needed. Teacher performance in learning should be as great and as best as possible. Students should also participate actively in learning to build their own knowledge so that they get good academic achievement.

5. Conclusion
Teacher performance plays a very important role in implemetation of the 2013 Curriculum. Teacher performance in implemetation of the 2013 Curriculum gives a significant effect toward Chemistry learning achivement in knowledge and skill domain. The value of effect size is 0.061 which included in medium effect size. Collaborative effort between teacher and students is needed for the success of curriculum implementation. Some improvement in teacher performance is needed too.

References
[1] Viswesvaran C and Ones D S 2000 *Int. J. Sel. Ass* **8** 216-26
[2] Amin M, Shah R U, Ayaz M and Atta M A 2013 *Gom. Univ. J. Res* **29** 100-4
[3] Hamalik O 2008 *Dasar-Dasar Pengembangan Kurikulum* (Bandung: Remaja Rosdakarya) p 238
[4] National Board For Professional Teaching Standards 2011 *Student Learning, Student*

IOP Conf. Series: Materials Science and Engineering **335** (2018) 012129 doi:10.1088/1757-899X/335/1/012129

Achievement:How do Teachers Measure Up? ERIC_ED517573
[5] Fritz C O, Morri P E and Richler J J 2012 *J. Exp. Psy* **141** 2-18
[6] Levine T R and Hullet C R 2002 *Hum. Com. Res* **28** 612-25
[7] Roehrig G H, Kruse R A and Kern A 2007 *J. Res. Sci. Teac* **44** 883-907
[8] Thomas I A and Green R L 2015 *Nat. For. Teac. Edu. J* **25** 1-18

IOP Conf. Series: Materials Science and Engineering **335** (2018) 012130 doi:10.1088/1757-899X/335/1/012130

Designing of Holistic Mathematic Education Model Based-"System Among" at Low Grade Elementary School

R Hayati[1]*, A Fauzan[2], M Iswari[3], A Khaidir[4]

[1]Program Study of Primary Teacher Education, Faculty of Teacher Training and Education, Dharmas Indonesia University, Indonesia
[2]Department of Mathematics, Faculty of Mathematics and Science, Universitas Negeri Padang, Indonesia
[3]Department of Non Formal Education, Universitas Negeri Padang, Indonesia
[4]Department of Science of State Administration, Universitas Negeri Padang, Indonesia

*rahmatulwahyu341@gmail.com

Abstract. The purpose of this study was to develop a model of Holistic Mathematics Education (HME) among systems based on low-grade primary school students so that students have a solid foundation when entering a higher behavior. This type of research is desaign research developed by Plomp to have three stages, namely the preliminary research, development or prototyping phase, and assessement Phase. This research resulted in a model Holistic Mathematics Education (HME) -based system is among the primary school students low grade consists of 10 stages, namely 1) Recap through the neighborhood, 2) Discussion groups by exploiting the environment, 3) Demonstration Group, 4) Exercise individuals, 5) mathematical modeling, 6) Demonstration of individuals, 7) Reflections, 8) impressions and messages, and giving meaning, 9) Celebrations and 10) A thorough assessment. Furthermore, this model also produces 7 important components that should be developed teacher, namely 1) constructivism, 2) the nature of nature, 3) independence, 4) parable, 5) inquiry, 6) cooperation, and 7) strengthening. This model will produce a model in the form of books, student books and teacher's guide book as a support system that can help users in its application.

1. Introduction

This study will develop a model Holistic Mathematics Education (HME) among systems based on low-grade primary school students, because of the need to develop early mathematical concepts in class early so that students have a solid mathematical foundation when entering a higher level [1]. HME model development among systems based on low-grade primary school students is one way that is effective and can improve students' mathematical ability, motivation to learn, and foster the values of character in students. Some research says that interest and motivation, as well as character development is very influential on student learning outcomes [2-5]. Furthermore, Lovat, et al [6]; Newmann and Associates [7]; Bryk et al [8]; Bryk and Schneider [9]; Benninga et al [4]; Berkowitz et al [10], also explains that values education can provide an opportunity to deepen the relationship of teachers and students, making learning more effective, and can create the atmosphere of a good learning environment. Watson [5] also stated that given character education at the primary level provides a long-term positive influence in the lives of students. For that, the development of models of mathematics learning in a holistic manner, which is based system is among the primary school students a low grade, because through this concept of learning not only develop cognitive aspects, but also the potential emotional, social, psychomotor, aesthetic, spiritual and very useful for students for their lives later, as described by Watson [5].

This model rests on the concept of holistic education which is a concept of education that develops students' potential in a harmonious whole (integrated and balanced), include the potential for intellectual, emotional, psychomotor, social, aesthetic, and spiritual. Through the concept of holistic education, the learning of mathematics can help students develop the full potential of individuals in a learning environment that is fun, exciting, democratic and humanist through experience in interacting with the environment. This is confirmed by Bernold et al [11] and Lovat, et al [6], that the holistic teaching can help students who have the diversity to be successful, and to make learning more effective.

In the model HME-based system is among the students a low grade will be developed a concept that can make the learning of mathematics more meaningful and interesting, as well as customized developmentally elementary students a low grade because of the model HME-based system among students will be given experience in interacting with its environment through mathematics, so it becomes more realistic mathematics for students. With a system based among teachers as tutors to educate children on the basis of independence and freedom so that students can grow as nature. Firdiansah and Agus [12] also explained that the system among educates students how to: 1) provide freedom to the students, 2) keep children from learning stress and force, and 3) embed the nature of freedom is very important to make the child's personal bolder students in learning. Among system also means guiding the child with love and put the interests of children, so that children can develop in accordance with his nature. With a pleasant learning stage is expected to develop the full potential of students, whether intellectual, emotional, spiritual, psychomotor, social and aesthetic, so as to foster the values of character for students and improve understanding of mathematical concepts, as well as student motivation.

2. Literature Review

2.1 Concept of Holistic Education

Holistic Education is the concept of education that develops the potential of the child's physical, emotional, social, spiritual aesthetic, and intellectual [13,14]. Holistic paradigm can be defined as a holistic perspective in perceiving reality. Minded holistic view overall aspect means more than the parts, patterned systemic, integrated, complex, dynamic, non-mechanical, and non-linear. Thus a holistic is a point of view which states that overall as a whole is more important than its parts, while in education, holistic is an educational philosophy that comes from thinking that it is basically an individual can find the identity, purpose, meaning of life through his relationship with society, spiritual values and the natural environment. Based on the concept of holistic education, it is necessary to develop a holistic concept of mathematics education for learning mathematics not only develop cognitive aspects, but also the social, emotional, psychomotor, aesthetic, and spiritual. With a holistic mathematics education is expected to make learning math become a fun, exciting, and humanist through experience gained from the surrounding environment, and is expected to foster the values of character for students that are very useful for the rest of their lives. This holistic concept of mathematics education grounded in Gestalt theory which looks at things as a whole. On the basis of Gestalt theory, then the students will easily formulated the idea with its own language, can easily think about his experience so be creative and imaginative, and students have the opportunity to try out new ideas (Tytler in [15])

2.2 System Among

Systems Among is how education is used in the Taman Siswa education system requires teachers for the purpose of considering and importance that nature-iradatnya children without forgetting all the circumstances surrounding it. The cornerstone of the system lies in the motto among *Tutwuri Handayani* which means providing freedom, opportunity, attention and guidance that allows the child or the young generation at the initiative themselves and on their own experience to develop along the lines of the personal nature of each [16]. Tutors only an impact from the rear when there are students out of education. If it is associated with the learning of mathematics then by the application of methods among students will be guided and given the freedom to solve the problems they encounter in math as long as it does not get out of math concepts. With this kind of learning will enhance students' understanding in mathematics for students directly experience the learning process. As described by [17] that the provision of direct experience will improve and facilitate students to the learning content. This concept is grounded in humanistic theories because, according to this theory, the purpose of learning to humanize humans. Humanistic theories including the Constructivists, because this theory involves cognitive and affective behaviors [18].

2.3 The concept ofModel HolisticMathematics Education (HME) Among System Based on Low Grade Elementary School Students

Model HME among systems based education is a concept that is most appropriate to the lower grade elementary school students. Basically the age range of low-grade primary school students still see everything as a whole (holistic) so that learning is still dependent on concrete objects and experiences that happened. By looking at the nature of education low grade primary school students are then mathematics must be taught in concrete terms to students so that students are able to interpret the purpose of learning mathematics. For that we need a model that is able to develop the full potential of students, and foster the values of character for students so that learning mathematics becomes meaningful for students. One model that suits karakteristis low grade primary school students is a model-based HME Among Systems.

HME models based on this system will be awarded among the corresponding mathematical problems with the potential of students, so that the problem is given not only able to develop the cognitive aspects but also the spiritual, emotional, social, aesthetic, and psychomotor. In solving the problems of a given student is given the freedom to solve problems. From the problems students are able to work closely with members of the group and mutual appreciation and respect his fellow group members. In this case the student is guided to be honest, tolerant, disciplined, democratic, creative, communicative, curious, as well as learning to interpret their lives later on. With the character is formed, it will be able to develop the full potential of students, whether intellectual, social, emotional, psychomotor, aesthetic, and spiritual. Furthermore, with a system based among the learning of mathematics will be more meaningful and attractive to students because the foundation of the system among lies in the motto Tutwuri Handayani which means providing freedom, opportunity, attention and guidance that allows the child or the young generation on the initiative himself and on his own experiences to developing along the lines of the personal nature of each. Teachers act as

IOP Conf. Series: Materials Science and Engineering **335** (2018) 012130 doi:10.1088/1757-899X/335/1/012130

guardian. Based on these explanations, then the model HME among systems based on low-grade primary school students is grounded in behavioristic theory, cognitive, and humanistic. In behavioristic learning theory, to learn whether or not a person depends on factors conditionally granted by the environment. For it was on this model *apersepsi* through the neighborhood is the first phase of which should be applied in teaching teachers. Furthermore, in the cognitive learning theory more emphasis on learning is a process that occurs in the human mind. Learning is a process of one's efforts to acquire a new behavior changes as a whole as a result of his own experience in interaction with the environment [19]. Humanistic theory also strongly supports this model, because this model emphasizes the goal is not only cognitive aspects but also create human character. In humanistic theories explained that humans are basically different from each other, so that they can be independent and choose and organize their lives as among the system concept underlying this model.

3. Methodology

Development Sistem based HME models Among Low Grade Elementary School uses a design development Plomp [20], which has three stages or phases, namely: 1) Preliminary Research; 2) Development or Prototyping Phase; and 3) Assessment Phase.

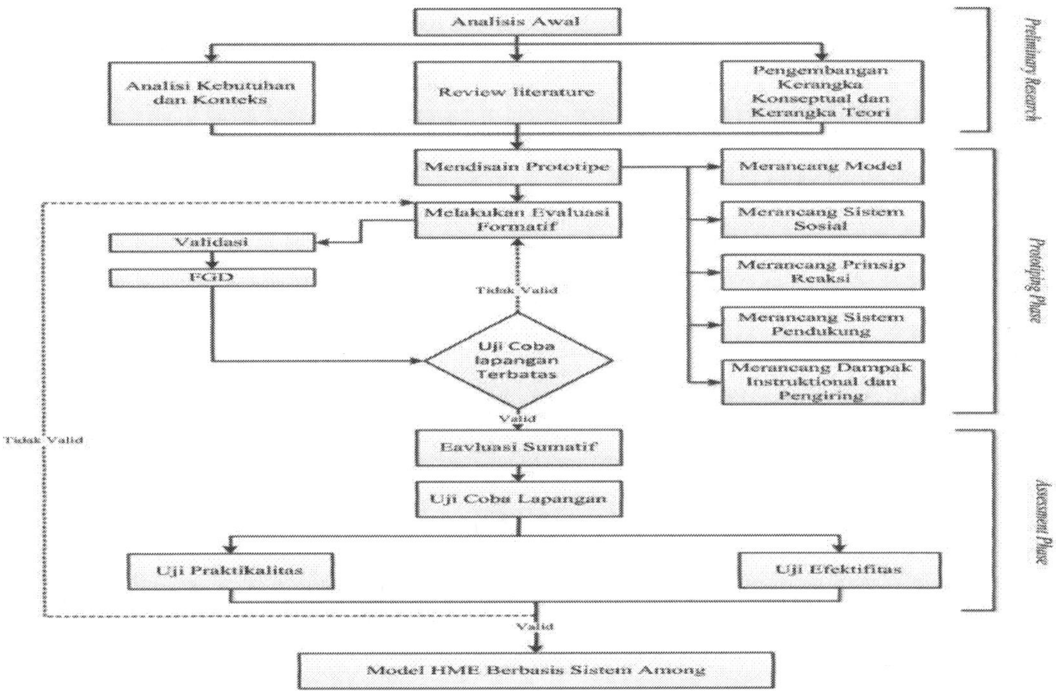

Figure 1. Procedure Development Model-based HME Among Sistems (Modification of Plomp [20])

In this study, researchers only discuss until the stage or Prototyping Development Phase, the prototype design stage. Here are the steps being taken to or Prototyping Development Phase.

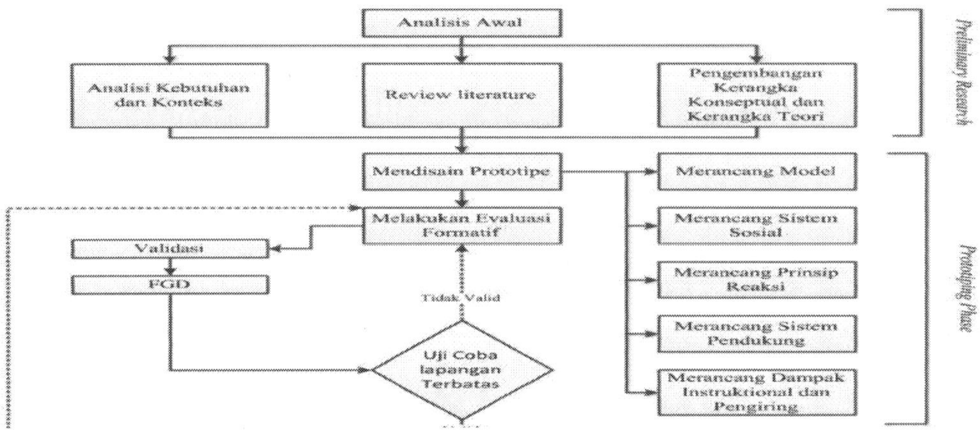

Figure 2. Procedure Development Model-based HME System Prototyping Among until stage Phase (Modification of Plomp [20])

4. Results and Discussion

Based on research carried out at the preliminary stages of research and prototyping phase which is the first and second phases of design development Plomp [20], then the results are as follows:

1) Preliminary research

a. Analysis of the needs and context

At this stage of the initial investigation the need for HME models based on the SD Class System Among Low. From preliminary studies have been conducted in several primary schools in the District of Pantai Cermin data showed that generally the teachers still use the conventional model of learning in which students are not actively involved in the learningprocess. This affects the motivation and understanding of math concepts students because they are used to receive and memorize the course material. Furthermore, in the models used have not seen the values of the desired character, such as curiosity, self-confidence, hard work, creative, democratic, and cooperation. This is because students do not participate actively in learning so that students generally only receive what is described by the teacher. In the model used by students are not taught to solve mathematical problems in groups, so the social potential students are less well developed. Furthermore, based on the objectives and course content lower grade elementary mathematics shows that the material presented in abstract still so difficult to comprehend grade students is low given the low grade primary school age are in the concrete operational stage of development.

b. Literature review

Based on the analysis of theory and concepts associated with the model-based HME among systems, this model rests on behavioristic theory, cognitive, and humanistic. This model is also based on the concept of holistic education where the concept was developed potential of children as a whole, both the potential for intellectual, emotional, social, psychomotor, aesthetic, and spiritual. This is in accordance with the characteristics of low-grade elementary school students. Furthermore, the characteristics of elementary school students a low grade, the concept among is one concept that is very suitable to be applied for this concept lies in the motto Tutwuri Handayani which means providing freedom, opportunity, attention and

IOP Conf. Series: Materials Science and Engineering **335** (2018) 012130 doi:10.1088/1757-899X/335/1/012130

guidance that allows the child or the young generation on the initiative himself and on his own experiences to develop along the lines of the personal nature of each.

c. Development of the theoretical framework

Based on a needs analysis and literature review, it is developing the theories that support the model HME-based systems among, including the theory of holistic education system among, learning theories underlying the models of HME-based systems among, and other theories that are relevant to HME models among systems based on low-grade elementary school students.

2) Prototyping phase

In this study, carried out the prototype stage to the stage of the initial product revision. From the results of the design was done until the prototyping phase (development phase Plomp), there are five aspects which are designed among other things, 1) modeling, 2) designing a social system, 3) designing principles of reaction, 4) design a support system, 5) designing impact instructional and accompanist. Based on the initial draft of which has been designed, there are several aspects that need to be fixed in accordance with the suggestion of several validators. Here are some suggestions 7 validator on a model that has been designed (3 votes mathematics education expert, 1 person linguist, one person early childhood education experts, and the first person of character education expert).

Table 1. Suggestions to the Draft Preliminary Validator HME Model Based System on Elementary Students Grades Among Low

No.	Preliminary Draft	Recommendations of Validator
1	Draft Model	a) Each syntax is designed to be explained clearly why it is very important stages in the developed model. Present theories relevant to a syntax that is designed to support the model b) Each supporting component model that is designed to be supported with relevant theory that these components really can support the model of
2	social systems	describe a concrete example, so that the social system which is expected to be seenclearly
3	principles reaction	Describe concrete examples, so that the reaction principle expected in the model seen
4	support system	Describe clearlysupport system that can support this model and adapt to the characteristics of low-grade elementary school students.
5	cover design	Covercustomized products produced with the characteristics of low-grade elementary school students
6	language used	Use simpler language so easily understood by a low-grade elementary school students.

Based on the results of the initial validation, then obtained some very useful overview for the repair of the product. From the results of these improvements, obtained 10 syntax supports this model, including: 1) Recap through the neighborhood, 2) Discussions with utilizing the environment, 3) Demonstration group, 4) Exercise individuals, 5) Mathematical modeling, 6) Demonstration of individuals, 7) Reflections, 8) impression, and messages, and giving meaning, 9) Celebrations and 10) A thorough assessment. Furthermore, as a supporter of the model was developed seven components that need to be applied to teachers so that these models can be done well, including: 1) Constructivism, 2) The nature of nature, 3) Liberation, 4) Parable, 5) Inquiry, 6) Cooperation, and 7) Strengthening.

Based on these components seventh, early learning teachers must be able to construct the student's knowledge of prior knowledge that has been held. This is very important because it is based on the theory described by Piaget and Vygotsky, increased knowledge of the construction of learning and learners, not something that is fed from the others. So with that, the need for teachers to construct knowledge so that students more familiar with the material being studied. Furthermore, the nature of nature is a second component to be developed teacher is one aspect among systems that support the concept introduced by Ki Hajar Dewantara [21]. According Dewantara, that the essence of man as a creature of God is one with the nature of Nature. Through this component, the learning process adapted to developmentally lower grade students and teachers guide and direct students with a sense of love, compassion and affection in accordance with the concept among systems. On the concept of independence, the teacher gives freedom to students to complete math problems given in accordance with the material being studied. Teachers guide and direct students to avoid misconceptions. According Dewantara in Pranarka [22], independence implies that life should be filled with happiness and peace. This is consistent with the concept of the model-based HME among systems. In the parable of components, the teacher must be able to illustrate the problems which are abstract to the concrete so easily understood by a low-grade students, for low-grade student is still at the pre-operational stage concrete. Benefits parable method are: 1) Contains elements interesting and fun, 2) Clarify meaning by associating an abstract to something concrete, 3) Encourage a positive attitude, and 4) Leaving a negative attitude [23]. Next on the component inquiry, the teacher invites students to find their own solutions to problems are given. In this component of the teacher can invite students to undertake the following activities: (1) observation, (2) ask, (3) submitting allegations (hypothesis), (4) the collection of data, and (5) inference. By that stage, it will enhance the students' understanding of the material provided for students are directly involved in learning as described by De Potter [17]. Further cooperation component also needs to be developed by teachers to support the model HME among systems based on low-grade primary school students, because of good cooperation skills, then it will also be able to develop the potential of the social, emotional, psychomotor, aesthetic, and spiritual. Strengthening which is the last component in the model is also important to be applied teacher. According to the theory Skiner (in Ruseffendi [24]), reward and reinforcement has a very important role in learning. Therefore the importance of strengthening in mathematics learning for learning mathematics is no longer a scary subject for students, but a lesson fun.

Based on the syntax and components of the model that has been described, it will further strengthen the HME models among systems based on low-grade primary school students, because the syntax and the components designed in accordance with the concept of holistic education and among system concept introduced by Ki Hajar Dewantara. Syntax and these components are also designed according to the characteristics of low-grade elementary school students, so this model is really suitable to be applied for low-grade elementary school students.

5. Conclusion

Based on the results of the initial validation of the model has been designed, then the resulting 10 stages HME models support among systems based on low-grade primary school students, namely: 1) Recap through the neighborhood, 2) Discussions with utilizing the environment, 3) Demonstration group , 4) Exercise individuals, 5) mathematical modeling, 6) Demonstration of individuals, 7) Reflections, 8) impression, and messages, and giving meaning, 9) Celebrations and 10) A thorough assessment. Further models of HME-based system is among the primary school students a low grade also resulted in seven key components that should be developed teacher, namely: 1) Constructivism, 2) The nature of nature, 3) Liberation, 4) Parable, 5) Inquiry, 6) Cooperation, and 7) Reinforcement.

Acknowledgments

The completion of the study must not be separated from the support of various parties. Therefore, the authors would like thank to Strengthening the Directorate General Research and Development of the Ministry of Research, Technology and Higher Education of the Republic of Indonesia who have given doctoral dissertation research grants to researchers

References

[1]. Stylianides, Andreas J. 2007. The Notion of Proof in the Context of Elementary School Mathematics. Educational Studies in Mathematics (2007) 65: 1–20 (DOI: 10.1007/s10649-006-9038-0)
[2]. Mayer, Richard E. 1998. Cognitive, metacognitive, and motivational aspects of problem solving. Netherlands: Kluwer Academic Publishers
[3]. Munoz, Marco A dan Judi E. Vanderhaar. 2006. Literacy-Embedded Character Education In a Large Urban District (Effects of the child development project on elementary school students and teachers). Journal of Research in Character Education, 4(1&2), pp. 47-64 (ISSN 1543-1223)
[4]. Beninga, Jacques S; dkk. 2003. The Relationship of Character Education Implementation and Academic Achievement in Elementary Schools. Journal of Research in Character Education Vol 1 No. 1 (ISSN 1543-1223)
[5]. Watson, Marilyn. 2006. Long-Term Effects of Moral/Character Education in Elementary School. Journal of Research in Character Education Vol 4 No. 1 & 2 (ISSN 1543-1223)
[6]. Lovat, Terence; dkk. 2010. Value Education as Holistic Development for all Sectors: Researching for Effective Pedagogy. Oxford Review of Education: Vol. 36, No. 6, December 2010, pp. 713-729
[7]. Newmann, F. M. & Associates. 1996. Authentic Achievement: Restructuring Schools for Intellectual Quality. San Francisco: Jossey-Bass
[8]. Bryk, A. S. & Schneider, B. 2003. Trust in schools: a core resource for school reform. Educational Leadership, 60(6), 40–44
[9]. Bryk, A. S., Lee, V. E. & Holland, P. B. 1993. Catholic Schools and the Common Good. Cambridge, MA: Harvard University Press
[10]. Berkowitz, M. W., Battistich, V. A. & Bier, M. C. 2008. What works in character education: what is known and needs to be known, in: L. P. Nucci & D. Narvaez (Eds) Handbook of moral and character education. New York: Routledge, 414–430.

[11]. Bernold, L.E; dkk. 2000. Impact of Holistic and Learning-Oriented Teaching on Academic Success. Journal of Engineering Education, April 2000
[12]. Firdiansah, Fikri dan Agus Suprijono. 2013. Pendekatan Sistem Among pada Proses Pembelajaran di SMK Taman Siswa Kediri Tahun 2006-2012/2013 dalam Dinamika Modernisasi. Avatara: e-journal Pendidikan Sejarah, Oktober 2013, 1 (3).
[13]. Latifah, Melly dan Neti Hernawati. 2009. Dampak Pendidikan Holistik pada Pembentukan Karakter dan Kecerdasan Majemuk Anak Usia Prasekolah. Jur. Ilm. Kel. Dan Kons., Januari 2009, p: 32-40.
[14]. Miller, John P. 2005. Introduction: Holistic Learning (In Miller, John P; dkk (Ed): Holistic Learning and Spirituality in education. New York: State University of New York Press
[15]. Hidayati, Titin Nur. 2011. Imlementasi Teori Belajar Gestalt pada Proses Pembelajaran. Jurnal Falasifa Vol. 2 No. 1, Maret 2011
[16]. Reksohadiprojo, Mohamad Said. 1989. Masalah Pendidikan Nasional beberapa sumbangan pemikiran. Jakarta: CV Haji Masagung
[17]. Depotter, Bobbi; dkk. 2004. Quantum Teaching. Bandung: KAIFA
[18]. Schunk, Dale H. 2012. Learning Theories: An Educational Perspectives.Boston: Pearson
[19]. Slameto. 2010. Belajar & Faktor-faktor yang mepengaruhinya. Jakarta: Rineka Cipta
[20]. Plomp, Tjeerd. 2013. Educational Design Research: An Introduction. Dalam Tjeer Plomp and Nienke Nieeveen (Ed). An Introduction to Educational Design Research. (p:9-35). Netherlands in www.slo.nl/organisatie/international/ publications.
[21]. Dewantara, Ki Hadjar. 1977. Karya Ki Hadjar Dewantara bagian pertama pendidikan. Yogyakarta: Majelis Luhur Persatuan Taman Siswa.
[22]. Pranarka, A.M.W. 1986. Relevansi Ajaran-ajaran Ki Hadjar Dewantara Dewasa ini dan di Masa yang akan Datang", dalam Wawasan Kebangsaan, Ketahanan Nasional dan Wawasan Nusantara, Lembaga Pengkajian Kebudayaan Sarjana Wiyata Tamansiswa, Yogyakarta
[23]. Ajami, Al dan Muhammad Abdussalam. 2006. Al-Tarbiyah al-Islâmiyah: Al- Ushûl wa Al-Tathbîqât (Cetakan I). Riyadh: Dâr Al-Nâsyir Al-Daulî.
[24]. Ruseffendi, E.T, dkk. 1992. Pendidikan Matematika 3. Jakarta: Depdikbud

IOP Conf. Series: Materials Science and Engineering **335** (2018) 012131 doi:10.1088/1757-899X/335/1/012131

Differences of Fundamental Motor Skills Stunting and Non Stunting Preschool Children in Kindergarten in North Padang

A Komaini*, R Mardela

Sport Science Faculty; Universitas Negeri Padang, Indonesia

*anton_chzicho@yahoo.com

Abstract. The problem that emerged is based on the result of research done by the writer in kindergarten in North Padang Sub-district which concluded that: there were kindergarten students in this sub-district who were still lack of motor ability, research data shows that 59 people (37,34%) and then 34 people (21, 52%) were in very good category, 35 people (22.15%), were in moderate category, 22 people (13.92%) were in the poor category, and 5 (5,06%) were in the very poor category. Based on this data, the authors thought that the dominant factors that affect the above situation was a nutritional factor. It could be seen from the physical appearance of kindergarten children who tend to slow growth. The purpose of this study is to explain the description and differences in stunting and non stunting Fundamental motor skills capabilities in early childhood (preschool) children. This research is comparative study with cross sectional approach. The population in this study was the students of Kindergarten of Perwari II which consisted of 60 people consisting of 37 children of stunting and 23 non stunting children in Kindergarten of North Padang Sub district, the sample was taken as a whole. The data were collected with Fundamental motor skills tests including jumping, walking, running, balance exercises, throwing and catching the ball. Technique of data analysis in this research was descriptive statistic. The result of data analysis shows that there is difference of Fundamental motor skills between stunting and non stunting children. Fundamental motor skills of non stunting or normal children are better than those who were stunting or short. While the results of Fundamental motor skills of kindergarten children in North Padang District as a whole is at a good level.

1. Introduction

There are two components of children movement ability that needed to be developed including fundamental and fine motor skills aspects. Fundamental motor skills is a movement used big muscles such as walking, running, two legs jumping, and hopping the obstacles. On the other hand, fine motor skills is a movement used small muscles such as cutting, sticking, tearing, drawing, coloring, writing and arranging the blocks. Each child has different motor ability.

Some children have a good fundamental motor ability; however, some children do not. As a matter of fact, the children who lack of fundamental motor skills (FMS) such as locomotors and non-locomotors movements can be given trainings including jumping, climbing, running, and walking. Whereas, the children who lack of fine motor ability can be trained through the trainings of folding, forming, etc. In general, the differences between fundamental and fine motor skills can be seen through its coordination and biomechanics. FMS are the building blocks to future physical activities and sport and are the movement equivalent to the ABCs in reading literacy. FMS consist of two groups of skills including locomotor and manipulation skills. Locomotor skills are skills such as running, jumping, hopping, leaping, sliding, galloping and skipping, in other words a child moves his/her body from one point to another [1].

The fundamental motor skill needed to be developed since it will affect their other developments involving physiology, cognitive, and socio-emotional. Also, socio-economic factor has an influence toward the children's gross motor ability. By having an established financial condition, the children's nutrition will be fulfilled. In developing the children's gross motor aspect in kindergarten education, it is hoped that kindergarten adjusted the curriculum to the factors which affect the children's gross motor aspect so that the children's gross motor can be easily improved. Factors affecting the development of the movement, among other things, genetic, environmental stimuli, and nutritional status [2]. To allow for optimal development, required the development of broad and balanced between the physical aspects of motor skills, intelligence, emotional and social development [3].

According to Soetjiningsih, the factors that influence the child growth are: child's learning motivation, mother's knowledge and stimulation, peer group, love and affection, number of siblings, reward or punishment, environment, household stability, mother's income, and nutrient level [4] . The results of research conducted by the author on kindergarten in North Padang Sub-district concluded that: there are still kindergarten students in sub-district lack of Fundamental motor skills, research data mentioned 59 people (37.34%), then 34 people (21.52%) are in category very good, as many as 35 people (22.15%), are in the moderate category, 22 people (13.92%) are in the poor category, and as many as 5 people (5.06%) are in the category very once[5].

Based on this data the writer sees one of dominant factors that affect the above situation was the nutritional factor. It could be seen from the physical appearance of kindergarten children who tend to slow growth, for example short body. Nutrition is one of the environmental factors and a support for the growth and development process to run satisfactorily. This means, giving qualified food in good quantity supports the growth of children, so that toddlers can grow normally and healthy. Toddlers who are not given qualified food in good quantity can experience underweight, malnutrition, short (stunting), and thin (wasting).

Stunting or height problems were caused by chronic malnutrition as the impact of lack of nutrient intake in a long time since a fetus was still in the womb. Malnutrition at an early age increases infant and child mortality, causing the sufferer to be easily sick and have no maximal posture in adulthood, as well as the cognitive ability of the sufferers is also reduced. Basic Health Research (RISKESDAS 2013) records a stunting prevalence by province and national level. The nationally short prevalence in 2013 is 37.2 percent, which means an increase compared to 2010 (35.6%) and 2007 (36.8%). The short prevalence of 37.2 percent consists of very short 18.0 percent and 19.2 percent short. In 2013 a very short prevalence showed a decline, from 18.8 percent in 2007 and 18.5 percent in 2010. The short prevalence increased from 18.0 percent in 2007 to 19.2 percent in 2013. [6].

Stunting is a problem of chronic malnutrition caused by poor nutrient intake in a long time due to feeding that is not in accordance with nutritional needs. Stunting is a sign of systemic dysfunction during the sensitive phase of a child's development. Alive and Thrive (2010) mentioned that during the first year after birth, nutrient requirements are very high to support rapid growth and development. Growth disturbances during early childhood including stunting can lessen the cognitive and motor abilities in children, as well as negatively affect the development of emotions, behavior, education and other abilities. Stunting may lead to a decrease in IQ in pre-school age children and school-aged children [7].

Stunting occurring in childhood is a risk factor which increased mortality, decrease cognitive abilities, and low motor development and unbalanced body functions [8] (Allen & Gillespie, 2001). Stunting may occur in the short and long term. According to Onis and Branca (2016) the failure of linear growth at an early age may increase morbidity and mortality from infection, loss of potential physical growth, decreased brain development and intelligence function and increase the risk of chronic diseases as adults [9]. Dewey et al (2013) argue that stunting at an early age is associated shortly after adulthood and is a fundamental cause of slow growth and causing other adverse outcomes [10].

Motor movement cannot be done perfectly if the muscle mechanism has not developed well. This occurs in children who have stunted growth disorders, striped muscle or striated muscle that controls unconscious movements which develop at a somewhat slower rate, before the child is in normal condition, there can be no coordinated voluntary action [11] (Hurlock, 2002). This study aims at obtaining data on differences in basic motion skills between stunting and non stunting children of kindergarten in North Padang Sub-district

2. Research Methodology

This study was comparative study with Cross Sectional approach which would explain the description and the difference of basic motion skills in stunting and non stunting pre-school children. The population in this study was Kindergarten Perwari II located in North Padang Sub-district amount to 60 people. The sample was taken by total sampling technique.

3. Results and Discussion

3.1.Data Distribution of Respondent Basic Motion Skills.

Based on the research that has been done on the basic motion skills of respondents, it was found that the lowest basic movement skills of respondents was 12 and the highest score was 17. After analyzing the data, it was obtained that the mean score was 14.7, standard deviation was 1.45, median was 15 and mode was 15.

Table 1. Frequency Distribution of Respondents Basic Motion Skills

No	Interval Class	Frequency Absolute	Relative (%)
1	12-13	16	26,67
2	14-15	24	40,00
3	16-17	20	33,33
Total		**60**	**100** (%)

Based on the table above, it can be seen that the basic movement skills of respondents were mostly in the interval class 14-15, that is as many as 24 people (40%), then 20 people (33.33%) were at interval class 16-17, and as many as 16 people (26 , 67%) were in interval class 12-13

3.2.Stunting and Non-Stunting Kindergarten Students

Based on data analysis conducted to 60 respondents, the following is the data of stunting and non stunting kindergarten students (Table 7).

Table 2. Frequency Distribution of Stunting and Non-Stunting respondents

No	Z-Score	Category	Number of Children	Persentage (%)
1	z-score < -3,0	Very short	4	6,67 (%)
2	z-score ≥ -3,0 s/d z-score < -2,0	Short	33	55 (%)
3	z-score ≥ -2,0	Normal	23	38, 33(%)
Total			**60**	

3.3. Overview of stunting and non-stunting children basic motion skill

An overview of basic skills of stunting and non stunting children gotten from respondents can be seen in Table 3:

Table 3. Description of Basic Motion Skill of Stunting and Non-Stunting Children

Basic motion skills	Nutritional status categories			
	Stunting		Non Stunting	
	N	%	N	%
Medium	13	35,14 %	3	13,04 %
Good	13	35,14 %	11	47,83 %
Very Good	11	29,72 %	9	39,13 %
Total	**37**	**100 %**	**23**	**100 %**

The numbers of stunting children were 37 toddlers from 60 respondents. 13 people (35.14%) of them had moderate movement skills, 13 people (35.14%) had good basic movement skills and 11 people (29.72%) had excellent motor skills. Besides, the numbers of non stunting children were 23 toddlers from 60 respondents. 3 people (13.04%) of them had moderate movement skills, 11 people (47.83%) had good basic movement skills and 9 people (39.13%) had excellent basic motion skills. So when it is viewed as a whole of the respondents, there were16 people (26.67%) had moderate basic movement skills, 24 people (40.00%) had good basic motion skills and 20 people (33.33%) had very good basic motion skills. Toddlers with stunting have a risk of decreased intellectual ability, productivity, and increased risk of future degenerative diseases. This is because stunting children also tend to be more susceptible to infectious diseases, thus risking a decrease in the quality of school learning and more frequent absences.

4. Conclusion

Based on the research findings and the results of the research discussion, it can be concluded as follows:

a. Basic movement skills of pre-school stunting children in Kecamatan Padang Utara are in good and medium category where 13 toddlers are in moderate category, and 13 toddlers in good category and 11 toddlers in very good category (37 stunting children).
b. Basic movement skills of non stunting children are 3 toddlers are in medium category, 11 toddlers were in good category and 9 toddlers were in very good category (23 non stunting children).

5. Suggestions

Based on the discussion and the results of this study, there are some suggestions as follows:

a. It is expected that kindergarten teachers in Kecamatan Padang Utara can develop and improve the basic motion skills of Kindergarten children with various methods or variations of learning.
b. Parents are expected to improve the nutritional quality that children consume in order to improve children's basic movement skills.

References

[1]. Goodway J, Famelia, R, Bakhtiar, S. Future Directions in Physical Education & Sport: Developing Fundamental Motor Competence in the Early Years Is Paramount to Lifelong Physical Activity. Asian Social Sci J [Internet] 2014 Feb [cited 2016 Sept 14]; Vol. 10, No. 5. pp 44-54
[2]. Graham G, Holt S A, Parker M. Children Moving A Reflective Approach tu Teaching Phisycal Educatin. New York: Mc Graw Hill, 2007.
[3]. Bakhtiar S, Strategi Pembelajaran, Lokasi Sekolah, dan Kemampuan Gerak Dasar Siswa Sekolah Dasar. J Ilmu Pendidikan Jilid 20 Nomor 2, Desember 2014, p. 127-252
[4]. Soetjiningsih. Tumbuh kembang anak. Jakarta: EGC, 1995.
[5]. A, Komaini, Fundamental Motor Skills of Kindergarten Students (A Survey Study of the Influence of Financial Condition, Playing Activity, and Nutritional Status). IOP Publishing. IOP Conf. Series: Materials Science and Engineering 180 (2017) 012156 doi:10.1088/1757-899X/180/1/012156

[6]. Kemenkes RI. 2013. Laporan Hasil Riset Kesehatan Dasar Sumatera Barat tahun 2013. Badan Penelitian dan Pengembangan : Jakarta.

[7]. Alive & Thrive. 2010. Infant and Young Child Feeding in Communities: A Rapid Assessment in Tigray and SNNPR, Ethiopia. Addis Ababa, Ethiopia; Alive & Thrive.

[8]. Allen, Lindsay H., and Stuart R. Gillespie. 2001. *What Works? A Review of the Efficacy and Effectiveness of Nutrition Interventions.* Geneva: United Nations Administrative Committee on Coordination Subcommittee on Nutrition, Asian Development Bank, and International Food Policy Research Institute.

[9]. Onis, M.D, Branca, F. 2016. Childhood stunting: a global perspective. Maternal and Child Nutrition.Volume 12, Issue Supplement S1May 2016 p12–26.

[10]. Dewey, KG, Begum K. 2011. Long term consequences of stunting in early life. Maternal and Child Nutrition, 7(suppl.3). pp.5-18.

[11]. Hurlock, Elizabeth B. Perkembangan Anak. Jakarta: Erlangga, 2002

IOP Conf. Series: Materials Science and Engineering **335** (2018) 012132 doi:10.1088/1757-899X/335/1/012132

Application of Student Book Based On Integrated Learning Model Of Networked Type With Heart Electrical Activity Theme For Junior High School

G Gusnedi*, R Ratnawulan, L Triana

Physics Department, Universitas Negeri Padang, Indonesia

*fisikaediunp@gmail.com

Abstract. The purpose of this study is to determine the effect of the use of Integrated Science IPA books Using Networked Learning Model of knowledge competence through improved learning outcomes obtained. The experimental design used is one group pre test post test design to know the results before and after being treated. The number of samples used is one class that is divided into two categories of initial ability to see the improvement of knowledge competence. The sample used was taken from the students of grade VIII SMPN 2 Sawahlunto, Indonesia. The results of this study indicate that most students have increased knowledge competence.

1. Introduction

In the 2013 Curriculum, learning is conducted using a scientific approach or a scientific-based approach. The scientific approach can use some strategies such as contextual learning (Kemendikbud, 2015: 19). To realize it required the facilities and infrastructure that support the learning process, one of which is the book of science students. According Kemendikbud (2015: 61) student book is a book of learning resources for learners. Each chapter is equipped with concept maps, introductory, student activity sections, both experimental and non-experimental or discussion, exercise questions, evaluation summaries, and assignments for learners.

Science students' books are developed and used as a source of information in which there are facts relating to the concept of learning needed to enhance the creativity and skills of learners in order to form learners who are characterized. Therefore, it takes a book that can stimulate the mindset of learners from aspects, namely attitude, knowledge and skills. According to the Curriculum 2013, the science students book used for junior high school level is a thematic IPA thematic book. The integrated science student book contains physics, biology, and chemistry studies that have been synergized in a student book.

Student's books that have been there are still being revised because there are still deficiencies, so that the achievement of the existing goals is not maximized, therefore it is necessary to develop the book of science students in accordance with the characteristics of learners as the target. Based on the results of observations and analyzes conducted on the books of students used in schools have not met the criteria of students' books for science lessons according to the curriculum 2013. Based on Permendikbud No.24 of 2016 states that the book students for science subjects junior level in science lesson has become a book Integrated IPA (physics, biology, and chemistry) thematic.

The importance of conducting this research is to obtain data on how the integrity of IPA in IPA textbooks in accordance with the demands of the curriculum 2013. Furthermore, the results of this study is expected to be followed up as a consideration for the Government to re-publish the second edition of the revised book that has better suitability. In addition, this research can also be a consideration for teachers to create and implement lessons that are more in line with the demands of the 2013 curriculum.

2. Methods
This experimental research used one group pre test post test design as shown in Table 1. The subjects of this research are the students of grade VIII SMP N 2 *Sawahlunto* academic year 2016/2017.

Table 1. One Group Pre Test Post Test Design

Pre Test	Treatment	*Post Test*
O_1	X	O_2

Source: Sugiyono, (2013)

The variables in this study consisted of independent variables, namely Integrated Science Student Book Using Networked Learning Model with Heart Electrical Theory, knowledge competence as independent variable, and initial ability as moderator variable. Increased competence of knowledge can be seen using N-Gain according to the customized category (Hake, 2002). While, to know the influence of independent variable to dependent variable used hypothesis test using paired test using t_{test}.

$$N - gain = \frac{S_{post} - S_{pre}}{S_{max} - S_{pre}} X\ 100\% \qquad (1)$$

The hypothesis taken in this study are as follows: H_0: No influence of Integrated Science Student Book Using Networked Learning Model with Heart Power Theme. H_1: There is an effect of the book of Integrated Science Students Using Networked Learning Model with Heart Power Themes.

3. Results and Discussion
The results showed that students who have low competence seen from low pre test. (Figure 1). Initially low knowledge competence is able to obtain higher values. This means that there is an influence of the book of Integrated Science Students Using Networked Learning Model with the Heart Electrical Theme.

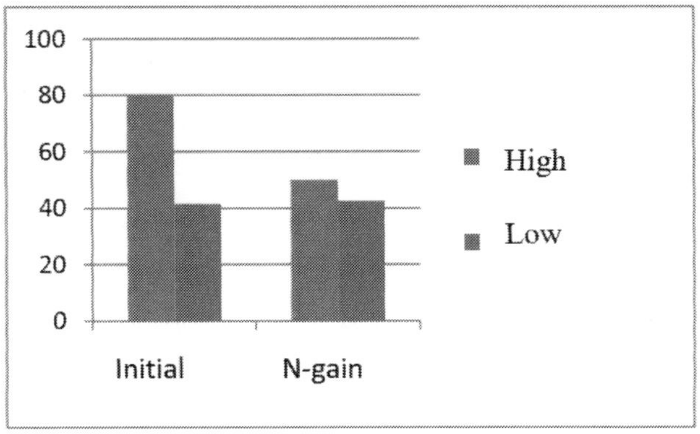

Figure 1. The Relation of the Values of Knowledge Competence with N-Gain is based on initial capability

In Permendikbud Number 21 of 2016 on Basic and Secondary Education Content Standards, it is stated explicitly that the achievement of learning outcome learning in the realm of knowledge has been revised by Lorin Anderson and David Krathwohl (2001). The domain of knowledge is a combination of dimensions of knowledge classified into factual, conceptual, procedural, and metacognitive with the dimensions of cognitive processes are arranged hierarchically ranging from remembering, understanding, applying, analyzing, evaluating), and creating (creating). Therefore, if there is an increase of the competence value of knowledge, then it can be said factual, conceptual, procedural and metacognitive students also increased.

Pursuant to result of analysis of pre test and post test value got average score of pre test which initially 45,83, after implementing learning process by using book of student of IPA Integrated Using Networked Learning Model with Heart Electrical Theme have average score 70,00. Integrated IPA learning is chosen because students in SMP / Ms have characteristics that are appropriate for this lesson. In addition, integrated learning can improve students' learning motivation. Learners feel that science lessons are very close to their lives, such as plants, animals, the environment and themselves. This will certainly have a good impact for the development of competence not only knowledge but also social and spiritual competence will also develop very well.

4. Conclusion

From the results of research conducted by the researcher can be concluded that the book of Integrated Science Student Using Networked Learning Model can influence student's knowledge competence. Increasing the value of knowledge competence gained depends on the value of initial competence before the student's book is given.

References

[1]. An Nisaa, Ranti. 2015. *Analisis Buku Biologi Kelas X Berdasarkan Muatan Literasi Sains*. Prosiding Seminar Nasional Pendidikan Biologi 2015 (Malang, 21 Maret 2015).

[2]. Arikunto, Suharsimi. 2010. *Prosedur Penelitian: Suatu Pendekatan Praktis*. Jakarta: Rhineka Cipta. hal 274.

[3]. Kemendikbud. 2015. *Materi Pelatihan Guru Implementasi Kurikulum 2013 Tahun 2015*: SMP/MTs Mata Pelajaran IPA. Jakarta : Kemendikbud.

[4]. Kemendikbud. 2015. *Materi Pelatihan Guru Implementasi Kurikulum 2013 Tahun 2015 SMP/MTs Mata Pelajaran IPA*. Jakarta : Badan Pengembangan Sumber Daya Manusia Pendidikan dan Kebudayaan dan Penjaminan Mutu Pendidikan Kementerian Pendidikan dan Kebudayaan.

[5]. Kemendikbud. 2016. Panduan Penilaian oleh Pendidik dan Satuan Pendidikan untuk Sekolah Menengah Pertama. Jakarta: Direktorat Pembinaan Sekolah Menengah Pertama.

[6]. Kementerian Pendidikan dan Kebudayaan. 2015. *Buku IPA Siswa Kelas IX SMP/MTs*. Jakarta: Kementerian Pendidikan dan Kebudayaan.

[7]. Moleong, Lexy J. 2004. Met*ode Penelitian Kualitatif.* Bandung: CV Remadja Karya.

[8]. Yusmium, Ika. 2015. Analisis Buku Teks Mata Pelajaran Ilmu Pengetahuan Alam SMP/MTS Kurikulum 2013. Semarang: Fakultas Ilmu Tarbiyah dan Keguruan Universitas Islam Negeri Walisongo.

IOP Conf. Series: Materials Science and Engineering **335** (2018) 012133 doi:10.1088/1757-899X/335/1/012133

Health Promotion and Preventive Contents Performed During Reproduction System Learning; Observation in Senior High School

E Yuniarti*, M Fadilah, R Darussyamsu, N Nurhayati

Biology Department, Faculty of Mathematics and Natural Science, Universitas Negeri Padang, Jl. Prof. Hamka, Padang, Sumatera Barat, Indonesia 25131

*dr_elsa@fmipa.unp.ac.id

Abstract. The higher numbers of cases around sexual behavioral deviance on adolescence are significantly related to their knowledge level about the health of the reproduction system. Thus, teenagers, especially school-aged, have to receive the complete information which emphasizes on recognize promotion and prevention knowledge. This article aims to describe information about health promotion and prevention, which delivered by the teacher in Senior High School learning process on topic reproduction system. The data gained through focused observation using observation sheet and camera recorder. Further, data analyzed descriptively. The result show promotion and preventive approach have been inadequately presented. There are two reasons. Firstly, the promotion and preventive value are not technically requested in the final assessment. The second, the explanation tend to refer to consequences existed in the term of the social and religious norm rather than a scientific basis. It can be concluded suggestion to promote health reproduction and prevent the risk of health reproduction need to be implemented more practice with a scientific explanation which is included in a specific program for adolescence reproductive health improvement.

1. Introduction

Teens sexual behavior is influenced by knowledge about reproductive health. Learners who have knowledge of reproductive health know about themselves, the maturity of their reproductive organs, understand their physical and psychological changes, avoid themselves from risky behavior (free sex) and be responsible for the reproductive process [1]. WHO (World Health Organization) added that healthy reproduction is not solely free from disease or disability in all matters relating to the reproductive system, as well as its functions and processes, but also has a complete physical, mental, and social well-being.

Although there is no statistical data that explains the correlation between low knowledge of reproductive health and sexual behavior distortion, [2] have proven a link between reproductive health knowledge and sexual behavior. A research involving 2000 adolescents aged 18-24 in West Java in 2004 -2006 informs about respondents' reproductive health, in general, that is "very low" (over 75% of respondents) and indicates lack of information and education related to sexual and reproductive health, so that risky adolescent sexual behavior is feared to increase [3]. These concerns are proven by the increase in free sex behavior and the spread of HIV/AIDS in Padang City, West Sumatera which gained in the eighth position in the term of HIV/AIDS cases in Indonesia [4]. In 2015, 227 HIV cases were found, consisting 181 men and 46 women, and 81 cases of AIDS, consisting of 63 men and 18 women [5].

Several studies have highlighted the important role of reproductive health promotion programs, including schools, on knowledge of the reproductive health of school-age adolescents. Teenagers who drop out, tend to experience unprotected sexual behaviors that are at risk for HIV and pregnancy. Several studies existed has developed HIV prevention school-based program along with an evaluation project related to sexual and reproductive health promotion for South African and Tanzanian adolescents [10]. In addition, a school-based HIV prevention program in South Africa which suggested that the program can be more effective if directly targeted to actual targets-adolescents who are sexually active, proposed that research on linking school quality to academic performance on aspects of sexual desire, pregnancy, and marriage is likely limited. A survey study informed that parents and schools are a source of reliable information for a number of young women in Nigeria [7].

The availability of today's reproductive health knowledge resources can be said to be varied and easily accessible. Television turned out to be the main source of information on reproductive health for some high school students in Banguntapan [6]. Not many different, Nigerian girls in Nigeria favor the internet [7] because online sources are easily accessible [8]. Both of these media may provide unlimited reproductive health information, but do not automatically guarantee the performance of healthy sexual behaviors. For example, the majority of students in Iran have an adequate level of confidence about the risks of AIDS and other sexually transmitted diseases, but they had low ability to practice healthy life behaviors [9]. Respondents were also aware of the importance of reproductive health knowledge, unfortunately services for this were not sufficient.

Refer to these phenomena, schools have vital role to participate reproductive health education because educational institutions are an integral part [11]. Sex education programs have been implemented in secondary schools [12]. This research touched on the pedagogical aspects of learning materials on high school reproduction system. Specific study in the term of learning process contribution is required because more students prefer teacher explanation around health reproduction information, besides school involvement is needed to provide a more responsible sources. Therefore, our research aims to describe how promotion and preventive of health reproduction are performed by teacher during learning process. On the substance of the national high school curriculum, adolescence health reproductive education is contained in the standard contents, namely Basic Competence 3.12, 3.13, and 4.12. The ideas contained in the principles of competence are the principles of human resource quality improvement; and to present analysis result of free affairs, disease, disorders on structure and function of reproduction organ, the technology of the reproductive system.

The preliminary survey of researchers in December 2016 towards high school students in Padang indicated that the students rarely known that regularly bathing as way to maintain health reproduction. This achievement indicated that students' knowledge were far from coverage of reproductive health component

2. Method

This study describes the promotion and preventive reproductive health information delivered in the learning of human reproductive system in Senior High School level, consisting of 2 high schools in the city of Padang. The selection of sample and sites is based on the assumption that both school could represent the average state of learning of reproduction system materials, as both school exist on higher and lower category. Data are collected from observation to learning, interviews and student responses during learning. Those, we used the instrument in the form of check list of promotion content and reproductive health preventive, interview guides and video recording. The reproductive health promotion and preventive content in the check list refer to the 2008 national curriculum content standards and promotion- preventive concept according to Kholid, 2015; Mubarak, 2007, and Irianto, 2015. Instrument validity reviewed by educational experts and medical fields. Data analysis adopted qualitative description techniques.

3. Results and Discussion

Based on observational data, the delivery of reproduction system materials has integrated promotion and preventive values. The promotion value in this study consisted of knowledge of the internal and external human organs of reproduction in men and women (including disorders and disorders), and knowledge of human reproductive functions including spermatogenesis, oogenesis, menstruation, fertilization, pregnancy, lactation, sexually transmitted diseases and contraception. The appearance of promotion value on aspects of knowledge about reproductive organs arises through verbal explanation, task discussions in Student Activity Sheets, and teacher-student questions and answers.

With these methods, teachers verify students' knowledge of the structure and function of internal and external reproductive organs in men and women. Through the study of female reproductive organs, the teacher presented the importance selecting the safe sanitary napkins because existence of threat on carcinogenic constituents. In addition, students are emphasized to maintain the reproductive organs in the sense of avoiding contact with risky sexual behaviors and behaviors that affect HIV, such as consuming narcotics and drugs.

However, promotion and preventive health knowledge of reproductive health have not been completely presented in the promotion and preventive indicator. In addition, the explanation is not technical, such as guarding the reproductive organs by wearing cotton underwear, do not wear tight underwear and thick and hot, change underwear 2-3 times a day, do not wear humid underwear, cut pubic hair regularly, wash with clean utensils every finished urination / large, do not consume alcohol, consume nutritious food, exercise regularly, do not use drugs, and do not smoke [13].

The teacher's promotion appearance on the knowledge aspect of the reproductive function has included knowledge of gametogenesis, menstruation, pregnancy, birth, contraception, and sexually transmitted diseases. An interesting finding of researchers captures the statement "Do not arouse before the time comes, because coercive things are not going to be good results" and "Do not do premature ejaculation" that arises when discussing spermatogenesis. The statement indicates that there are two different views of the teacher on spermatogenesis. First, teachers equate spermatogenesis with sperm release, which can be stimulated through sexual activity. Whereas spermatogenesis is a process of sperm formation in the testes that are routine processes that occur internally male reproductive organs. Second, the teacher's warning that students do not do premature ejaculation reflects that premature ejaculation is understood as a sexual activity. Whereas premature ejaculation is a condition of the disorder of the workings of the organ that causes sperm out early.

Overall, the promotion approach on the sub-topic of male reproductive function has included preventive content. The use of the words "do not" has a preventive effect that maintains the cycle of gametogenesis. The next promotion approach is shown when discussing menstruation. Along with the explanation of the normal menstrual cycle, the explanation of menstrual disorders successfully attracts students' attention. This is proven by the emergence of 2 (two) questions related to menstruation "Suppose I menstruate on April 4. What is my date of ovulation "? and "Does painful menstruation (dismenore) can lead to infertility" ?. Content analysis of both questions indicates that female students have an awareness of recognizing the reproductive function that occurs in themselves. The menstrual discussion is associated with pregnancy because adolescents who have experienced menstruation have a chance of having a pregnancy. But must be passed by referring to religious and social norm so as not to risk with abortion behavior in case of pregnancy out of wedlock. "... halalness in marriage. Do not try to have young age marriage." This series of explanations is actually very potential if associated with the necessity of the ideal age to marry. However, there is no record of promotion knowledge about the risk of young age marriage. In addition, teacher said, "getting married is not only physically ready but also emotionally ready". More, there is no scientific explanation.

Two subtopics that are not explicitly discussed are interference in the reproductive system and an explanation of contraception. The subtopic diseases of male and female reproductive organs are not discussed in detail, set forth in the independent reading duty instructions on Sexually Transmitted Diseases, HIV and AIDS. In fact, the exposure of disorders and disorders can serve as a warning and raises students' awareness. The contraceptive subtopic is only implicitly mentioned when discussing the female reproductive organs that continue with the call to protect the reproductive organs from HIV infection, using condoms. But the teacher stressed that it does not mean there is any suggestion of using condoms that may be misleadingly interpreted as "using a condom safe for sexual intercourse".

Observational facts suggest that the promotion and preventing approach to reproductive health tends to be oriented towards adherence to social values, not content-oriented. Generally, teachers choose a normative approach rather than a logical content approach. Teachers are worried about the overt delivery of knowledge about organs and reproductive function can inspire teens to try to show deviant sexual behavior. In contrast to Ann's et. al study results, 1995, who found that practical knowledge of reproduction such as contraception is not proven to encourage risky sexual behavior, rather purity-oriented knowledge has been shown to encourage sexual experimentation. Other similar studies based on the results of a study of empirical literature that examines the effects of HIV / AIDS and health and sexual education on adolescent sexual behavior among informants that education on reproductive health or HIV / AIDS does not increase adolescent sexual activity and begins before sexual activity desires [14].

High school students are teenagers who are being formed thinking realistically, critically and scientifically. The knowledge of reproductive health should be presented in the context of biological science and its consequences to the physiological disorders of the body. For example masturbation and masturbation problems. These two sexual activities are categorized as sexual activity that is prohibited in religious teachings because such behavior deviates from the fairness. Fulfillment of sexual desire is through marriage which is regulated by the religious norm. This statement actually has a scientific explanation that the effects of masturbation vary according to individual circumstances. Masturbation is seen as instrumental in maintaining the functioning of the circulatory system, the nervous system, and the muscular system [15]. Other experts argue masturbation is dangerous if the intensity is excessive. Another example, the logical approach to the ban on marriage at an early age. What should be said is the risk of a young pregnancy that can lead to high blood pressure, infant disability, cervical cancer, premature birth, anemia and even maternal deaths because the reproductive system is immature to experience pregnancy and childbirth. This explanation can be attributed to the statement "Do not approach adultery. It is a great sin, not only in the world, but also the till hereafter, since you have been embarrassed. " Teachers should explain that the impact of adultery is early pregnancy which finally cause some dangerous.

The application of non-specific promotion and preventive approaches relates to the topic demands being tested in the final exam. The results of the interviews indicate that promotion and preventive materials are not subject to the subject matter test. For that, teachers should prioritize the delivery of materials that become indicators of the final evaluation of learning outcomes. However, if there are questions related to reproductive health, it will be answered in accordance with the needs of students.

4. Conclusion

This study is limited in terms of the number of research sites to be observed simultaneously because the learning of the reproduction system material is relatively simultaneous in all SMAs. Nevertheless, it can be concluded that the appearance of promotion and preventive approach to reproductive health in biology teaching reproduction system material tends to be implicit and more to make religious and social norms as a preventive reference. The study of preventive content has not provided adequate provision for adolescent-aged schooling because reproductive health education has not been a legitimate mandatory program as school responsibility. Although the proportion of urban adolescents with a high risk of sexual risk is only 16.7% (out of 350 responders), [4] survey shows that West Sumatra occupies the eighth position of HIV / AIDS in Indonesia indicating an increased risk of HIV risk if there is no clear prevention program. In relation to research findings in which teachers have not explicitly implemented promotion and preventive health reproduction, the recommendation of training program improvement and teacher supervision need to be considered [16]. Further, the researcher emphasizes the need for national level programs to take advantage of the subject matter, introducing alternative teaching methods, helping to improve teacher-student relationships and community-

community counsellors. In fact, we need to think together to implement the concept of "promoting school health" for school-age teenagers who have been internationally existing and proven to be effective [17].

Acknowledgment.
The researchers would like to thank the school, teachers, and students of SMA 1 Padang, SMA 8 Padang who have been willing to provide the required information, education experts reviewing the instruments and representatives of the Indonesian General Medical Association in Padang for the contribution of reviewing health reproduction content. Equally important, this research can be done with funding support from Universitas Negeri Padang.

References

[1] I Kumalasari, 2012, *Kesehatan Reproduksi*. Jakarta: Salemba Medika.
[2] Y Endarto and P S Purnomo, 2007, Hubungan Tingkat Pengetahuan tentang Kesehatan Reproduksi dengan Perilaku Seksual Berisiko pada Remaja di SMK Negeri 4 Yogyakarta. *Jurnal kesehatan Surya Medika Yogyakarta*.
[3] A Suryoputro N J Ford and Z Shaluhiyah, 2006, Faktor-Faktor yang Mempengaruhi Perilaku Seksual Remaja di Jawa Tengah: Implikasinya Terhadap Kebijakan Dan Layanan Kesehatan Seksual Dan Reproduksi, *Journal Makara Kesehatan,* Vol. 10 No. 1 Juni 2006: 29-40
[4] Y Media, 2014, *Kajian Pengembangan Strategi Penanggulangan HIV/AIDS melalui Pendekatan Sosial Budaya Padang*: Badan Perencanaan Pembangunan Daerah.
[5] Dinas Kesehatan Kota Padang, 2016, *Profil Kesehatan Kota Padang Tahun 2015,* Padang: Kepala Dinas Kesehatan.
[6] H Adnani and C Widowati, 2009, Motivasi Belajar dan Sumber-sumber Informasi tentang Kesehatan Reproduksi dengan Perilaku Seksual Remaja di SMUN 2 Banguntapan Bantul. *Jurnal Kesehatan Surya Medika*
[7] W E Nwagwu, 2007, The Internet As The Source of Reproductive Health Information Among Adolescent Girls in Urban City in Nigeria BMC Public Media 7:354
[8] E M Selkie M Benson and M Moreno, 2013, Adolescents' Views Regarding Uses of Social Networking Websites and Text Messaging for Adolescent Sexual Health Education American *Journal of Health Education*, Volume 42 p 205-212
[9] M Simbar F R Tehrani and Z Hashemi, 2005, Reproductive Health Knowledge, Attitudes And Practices Of Iranian College Students, *La Revue de Santé de la Méditerranée orientale*, Vol. 11 N
[10] L E Aarø, 2006, Promoting sexual and reproductive health in early adolescence in outh Africa and Tanzania: Development of a theory- and evidencebased intervention programme Scandinavian *Journal of Public Health*, 2006; 34: 150–158
[11] R K Kurniawati 2011 Pendidikan Kesehatan Reproduksi Remaja Sebagai Upaya Promotif Dan Preventif Dalam Rangka Menyelamatkan Generasi Muda Indonesia, Remaja Harapan Masa Depan Bangsa, *Bhakti Kencana Medika*, Volume 1, No. 2, Juli 2011, hal 50 – 56
[12] K I N Nwalo and S T N I Anasi, 2012, Access To And Use Of Reproductive Health Information Among In-School Adolescent Girls In Lagos State, *Nigeria Health Education Journal* Vol 71, Issue 1
[13] K Irianto, 2015, *Kesehatan Reproduksi Teori dan Praktek*. IKAPI: Alfabeta.
[14] A Grunseit and S Kippax, 1997, *Impact of HIV and Sexual Health Education on the Sexual Behaviour of Young People*: A Review Update Joint United Nations Programme on HIV / AIDS UNAIDS/97.4
[15] R J Levin, 2007, Sexual Activity, Health And Well-Being – The Beneficial Roles Of Coitus And Masturbation, *Journal Sexual and Relationship Therapy,* Volume 22, 2007 - Issue 1
[16] M L Plummer D Wight J Wamovi K Nyalali T Ingall G Mshana Z S Shingongo A I N Obashi and D A Ross 2007 Are schools a good setting for adolescent sexual health promotion in rural

Africa? A qualitative assessment from Tanzania *Health Education Research*, Volume 22, Issue 4, 1 August 2007, Pages 483–499
[17] W Mũkoma and A JJ Flisher 2004 Evaluations of Health Promoting Schools: A Review of Nine Studies Health Promotion International, Volume 19, Issue 3, 1 p 357–368, http

IOP Conf. Series: Materials Science and Engineering **335** (2018) 012134 doi:10.1088/1757-899X/335/1/012134

Biodegradation Capability of Some Bacteria Isolates to Use Lubricant Oil *in Vitro*

Y Ahda[1*], M Azhar[2], L Fitri[1], A Afnida[1], G S Adha[1], W N Alifa[1], D Handayani[1], D H Putri[1], I Irdawati[1], M Chatri[1]

[1]Biology Department, Faculty of Mathematics and Sciences, Universitas Negeri Padang, Indonesia
[2]Chemistry Department, Faculty of Mathematics and Sciences, Universitas Negeri Padang, Indonesia

*ahdayuni@fmipa.unp.ac.id

Abstract. Our previous study identified three species of bacteria, i.e. Alcaligenes sp., Bacillus sp1, and Bacillus sp2 isolated from using lubricant oil-contaminated soil in a Padang's workshop. However, its ability to degrade hydrocarbon were not known yet. In this extension study, we explore a wider area to find more hydrocarbonoclastic bacteria and examined its capability to degrade hydrocarbon in vitro. Seventeen isolates were characterized its capability using NA + used lubricant oil + tween + neutral red medium. Isolates A1, B2, D1 and D4 shows the high degradation index, whereas isolates A2, A3, A5, D2, B1, B3 and isolates A4, B4, D3 have medium and low degradation index, respectively. These potential hydrocarbonoclastic bacteria need in situ characterization to know their actual activities for bioremediation.

1. Introduction

Bacteria are ubiquitous and incredible microorganisms on earth. Bacteria can live in water, soil, air, and living thing organisms. Moreover, bacteria have the incredible ability to degrade the natural or synthetic materials. This degradation process is known as biodegradation. Nowadays, bacteria or other microorganisms is utilized for clean up the polluted environment. This activity is known as bioremediation [1].

Waste lubricant oil is one of the environment pollutant [2,3,4]. It could damaged our health when it is oxidated, nitration, cracking of polymers and decomposition of organometallic compounds in diesel engine for long period and high temperature [4]. Lubricant oil is one of the petroleum derivative. Waste lubricating oil also contain other dangerous compound like polyaromatic hydrocarbons (PAHs), toxic heavy metal dan chlorinated hydrocarbons (PCBs) [2]. These compunds are carcinogenic dan neurotoxic on human life system [5,6]. Uncontrolled disposal of waste lubricant oil to environment cause human and other living organisms to be contaminated.

In plant, waste lubricant oil caused the decrease of chlorophyl and protein [7,8], while lower photosinthesis observed on phytoplankton [9]. Human as the key actor of pollution also get the negative impact of carelessly throw waste lubricant away. In human, PAHs and PCBs could induce liver or kidney disease, possible damage to the bone marrow and increased risk of cancer [10,11].

Regarding of high risk of waste lubricant oil to living organisms especially human being, it is an urgent way to do bioremediation. Many studies found some potential genus of bacterial for waste lubricant oil biodegradation like *Acinetobacter, Achromobacter, Arthrobacter, Flavobacterium, Pseudomas* and many others [12,13,14,15,16,17]. These bacteria known as hydrocarbonoclastic bacteria [18]. Our previous study in a Padang's workshop also isolated and identified three species of bacteria, i.e. *Bacillus* sp1, *Bacillus* sp2 and Alcaligenes from waste lubricant oil contaminated soil [19].

In this extended study, we explore wider area to find out more indigenous bacteria in waste lubricant oil contaminated soil of workshops and examined its capacity *in vitro*. The aim of this tudy is to identify the potential clones that can degrade waste lubricant oil effectively *in vitro*.

2. Materials and Methods

Isolation of microorganisms

Waste lubricant contaminated soil was taken from four automobile or motorcycle workshops in West Sumatera i.e. *Padang*, *Pasaman Barat*, *Pesisir Selatan* and *Bukit Tinggi*, and brought asepticaly to our laboratory for bacterial isolation. One gram soil was added to 99 ml sterilized water and diluted until 10^{-3}. The suspension was grown on 2% NA medium by a streak quadrant method for 48 hours at room temperature. Mineral salt medium (MSM) with the following composition (g/L): 1.2 g NH_4Cl; 1.6 g K_2HPO_4; 0,4 g KH_2PO_4; 0.1 g NaCl; 1 g KNO_3; 20 g $MgSO_4.7H_2O$; 10 g $CaCl_2.2H_2O$; 0.05 g $FeCl_3$, 1 mL of trace element solution [20,21] per liter containing 50 mg MnCl2.H2O, 300 mg H3BO3, 1.1 mg FeSO4.7H2O, 190 mg CoCl2.6H2O, 2 mg CuCl2.2H2O, 24 mg NiCl2.6H2O, 18 mg NaMoO4.2H2O, 42 mg ZnCl2.7H2O, 1 ml vitamin solution [21] were used as selective medium for hidrocarbon-biodegradation bacterial isolation. Single colony was inoculated to 10 mL MSM medium, incubated in a shaker incubator at 37°C at 200 rpm for 12 hours. After 12 hours of incubation, the culture was isolated as single colony and spread out on Nutrient Agar (NA) media by streak quadrant method. Single colony of each NA-containing petri dishes then were maintained in slant cultures by preserving at 4°C and subculturing at 2 weeks interval [2].

Determination of biodegradation capabilities in vitro

Single colony was inoculated to 2% NA medium enriched 10 mL waste lubricating oil, 10 mL Tween 80 and 0.5 g neutral red [22] by toothpick method. Isolates then incubated at 37°C till found colonies with clear zones. The degradation - potential of bacterial *in vitro* is revealed through degradation index.

$$Degradation\ index = \frac{HZ - CZ}{CZ} \times 100 \qquad (1)$$

3. Results and Discussion

a. Results

Exploration in four automobile or motorcycle workshops in West Sumatera found 17 isolates (Tabel 1). These isolates were characterized its potential to degradate waste lubricant oil *in vitro*. Thirteen out of seventeen isolates shows its potential to degradate waste lubricating oil while the other four isolates were not. The potential of isolate to degradate the waste lubricant oil is showed by the clear area surrounding the colony. It is known as halo area (Figure 1).

Table 1. Bacterial isolates from several workshops in West Sumatera

Location	Isolates
Bukit Tinggi	A1, A2, A3, A4, A5
Pesisir Selatan	B1, B2, B3, B4
Pasaman	C1, C2, C3, C4
Padang	D1, D2, D3, D4

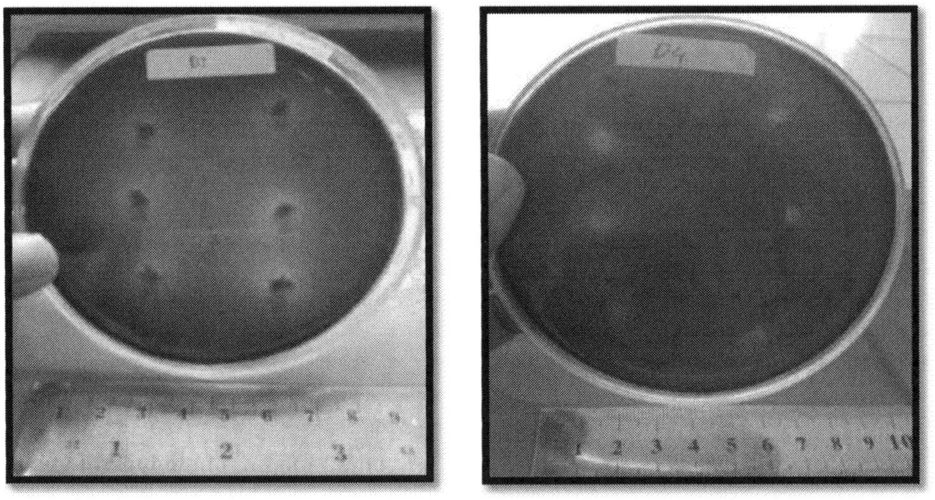

Figure 1. The representative (D1 and D4) colonies showing halo area indicated by clear circles.

Isolates that have potential to degrade waste lubricant oil are D1, D2, D3, D4 (Padang workshop), C1, C3, C4 (Pasaman workshop), A1, A2, A3, A4, A5 (Bukit Tinggi workshop) and B4 (Pesisir Selatan workshop) (Figure 2). Isolates A1, B2, D1 and D4 shows high degradation index. However, the highest potential is shown by D4. In contrast, A2, A3, A5, D2, B1, B3 and isolates A4, B4, D3 have medium and low degradation index, respectively.

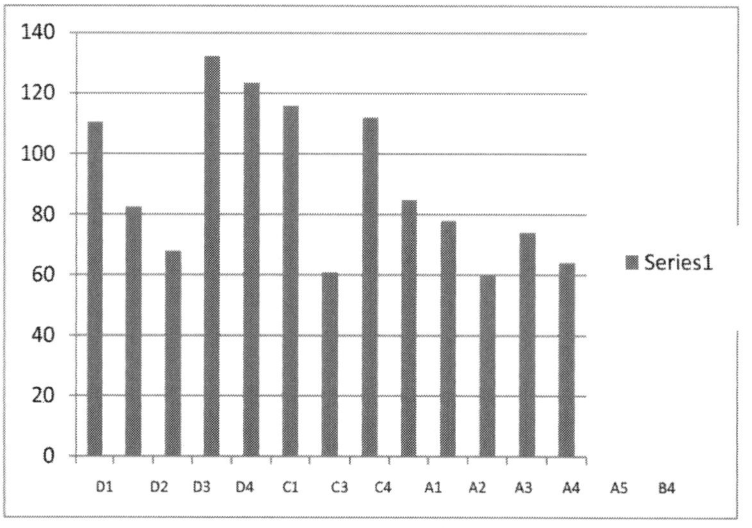

Figure 2. Mean of Degradation Index of Isolates from *Padang*, *Pasaman*, *Pesisir Selatan* and *Bukit Tinggi* workshops. D1 – D4 are isolates from *Padang*, C1 – C4 are isolates from *Pasaman*, A1 – A5 are isolates from *Bukit Tinggi*, and B4 is isolate from *Pesisir Selatan*.

b. Discussion

Bioremediation including bio-degradation in its process. Bio-degradation is a process involving microorganisms to separate the toxic macromolecules to be simpler components that are environmentally friendly. One of the microorganisms that play a major role in the bio-degradation process is bacteria [23,24,25].

There are many research that proved the ability of various species of bacteria to degrade hydrocarbons. Bacteria can degrade marine oil spills [26] and petroleum hydrocarbon in a polluted tropical stream [27]. Bacteria also have capability to degrade petroleum in soil [28,29]. Other studies revealed that bacteria can degrade hydrocarbon in using engine oil [3,4, 30,31]. Moreover, Bhattacharya [2] found a species of bacteria that have the capability of degrading hydrocarbons in different waste lubricant oil sources. In this study, we also identified some bacteria isolates that could degrade the waste engine oil. Our bacteria could be same with other published studies, but did not rule out the discovery of new species. It is therefore necessary to identify the isolates.

Our identification revealed that thirteen of seventeen isolates have the potential to degradate waste lubricant oil. This is also indicated by Ibrahim's study [3]. Two of ten isolates have the degradation capacities more than 45%. Our study have not either identify the characteristic of carbon source for optimal bacteria activies yet. However, Bhattacharya's study [2] show that Ochrobactrum sp.C1 could degrade several kind of hydrocarbon like naphtalene, xylene, diesel, and kerosene. The best result is found in naphthalene biodegradation. It is one of our focus in the next study.

Moreover, bacteria have the ability to degrade heavy metal in waste lubricant oil. *Pseudomonas aeruginosa* strain RM1 and strain SK1 could digest zinc, iron and nickel [4]. Those three metals were found in large quantities in lubricant oil. Our study identify two best isolates namely D4 and C1. These two isolates have chance to be identified with its potential to degrade heavy metals.

4. Conclusion

This study has established 13 bacteria isolates that have ability to degrade waste lubricant oil. Highest activity is shown by isolate D4. For the future, our study will be focussed on several data like the species of isolates, the optimal activity of isolates to degrade several hydrocarbon sources and the ability of isolates to degrade heavy metal contained in waste lubricant oil.

References

[1] Habe H, Chung JS, Lee JH, Kasuga K, Yoshida T, Nojiri H and Omori T 2001 Degradation of chlorinated dibenzofurans and dibenzo-p-dioxins by two types of bacteria having angular dioxygenases with different features. Appl Environ Microbiol 67:3610-17

[2] Bhattacharya M, Biswas D, Sana S and Datta S 2015 Biodegradation of waste lubricants by a newly isolated Ochrobactrum sp. C1. 3 Biotech 5:807-17 doi:10.1007/s13205-015-0282-9

[3] Ibrahim HMM 2016 Biodegradation of used engine oil by novel strains of Ochrobactrum anthropi HM-1 and Citrobacter freundii HM-2 isolated from oil-contaminated soil. 3 Biotech 6:226 doi:10.1007/s13205-016-0540-5

[4] Salam LB 2016 Metabolism of waste engine oil by *Pseudomonas* species. 3 Biotech 6:98 doi:10.1007/s13205-016-0419-5

[5] Ray S, Khillare PS, Agarwal T and Shridhar V 2008 Assessment of PAHs in soil around the International Airport in Delhi, India. J Hazard Mater 156:9-16

[6] Obayori OS, Ilori MO, Adebusoye SA, Oyetibo GO and Amund OO 2008 Pyrene degradation potentials of *Pseudomonas* species isolated from polluted tropical soils. World J Microbiol Biotechnol 24:2639-46

[7] Oluwole OS, Makinde SCO and Philips DA 2005 The impact of spent engine oil pollution on the growth of Celosia argentea. Proc. of the Fac. of Sci. Conf. Lagos States Univ. (LASU) (Nigeria)

[8] Umechuruba CI 2005 Health impact assessment of mangrove vegetation in an oil spilled site at the Bodo West field in Rivers State, Nigeria. J Appl Sci Environ Manag 9:69-73

[9] Vazquez-Duhalt R 1989 Environmental impact of used motor oil. Sci Total Environ 79:1-23

[10] Llyod CA and Cackette TA 2001 Diesel engines: environmental impact and control. Air Waste Manag Assoc 51:805-47

[11] Mishra S, Jyot J, Kuhad R and Lal B 2001 Evaluation of inoculum addition to stimulate in situ bioremediation of oily sludge-contaminated soil. Appl Environ Microbiol 67(4):1675-81

[12] Adelowo OO, Alagbe SO and Ayandele AA 2006 Time-dependent stability of used engine oil degradation by cultures of *Pseudomonas fragi* and *Achromobacter aerogenes*. Afr J Biotechnol 5:2476-9

[13] Mandri T and Lin J 2007 Isolation and characterization of engine oil degrading indigenous microorganism in Kwazulu-Natal, South Africa. Afr J Biotechnol 62:23-6

[14] Bagherzadeh-Namazi A, Shojaosadati SA and Hashemi-Najafabadi S 2008 Biodegradation of used engine oil using mixed and isolated cultures. Int J Environ Res 2:431-40.

[15] Basuki W, Syahputra K, Suryani AT, and Pradipta I 2011 Biodegradation of used engine oil. Indones J Biotechnol 16:132-8

[16] Obayori OS, Salam LB and Ogunwumi OS 2014 Biodegradation of fresh and used engine oils by Pseudomonas aeruginosa LP5. J Bioremediat Biodegrad 5:213. Doi:10.4172/2155-6199.1000213

[17] Salam LB, Obayori OS and Raji SA 2015 Biodegradation of used engine oil by a methylotrophic bacterium, *Methylobacterium mesophilicum* isolated from tropical hydrocarbon-contaminated soil. Petroleum Sci Technol 33:186-195

[18] Nursyirwani dan Amolle KC 2007. Isolasi dan Karakterisasi Bakteri Hidrokarbonoklastik dari Perairan Dumai dengan Sekuens 16S rDNA. Ilmu Kelautan; 12(1):12-17.

[19] Ahda dan Fitri 2016 Karakterisasi bakteri potensial pendegradasi oli bekas pada tanah bengkel di kota Padang Sainstek 8:2:98-103

[20] Brito EMS, Guyonraud R, Goni-Urriza M, Ranchou-Peyruse A, Verbaere A, Crapez MAC, Wasserman JCA and Duran R 2006 Characterization of hidrocarbonoclastic bacterial communities from mangrove sediments in Guanabara Bay, Brazil Res Microbiol 157:752-62

[21] M'rassi AG, Bensalah F, Gury J and Duran R 2015 Isolation and characterization of different bacterial strains for bioremediation of n-alkanes, and polycyclic aromatic hydrocarbon Environ Sci Pollut Res doi:10.1007/s11356-015-4343-8

[22] Yolantika H, Periadnadi and Nurmiati 2015 Isolation of the hydrocarbon degrading bacteria in the contaminated soil of the automotive workshop. J Bio UA 4(3):153-7

[23] Rahman KSM, Thahira-Rahman J, Lakshmanaperumalsamy P, and Banat IM 2002 Toward efficient crude oil degradation by a mixed bacterial consortium. Bio Techno 85(3):257-61

[24] Capello S, Denaro R, Genovese M, Giuliano I and Yakimov MM 2007 Predominant growth of Alcanivorax during experiments on oil spill bioremediation in mesocosms. Microbiol Res 162(2):185-90

[25] Margesin R, Moertelmaiyer C and Mair J 2013 Low-temperature biodegradation of petroleum hydrocarbons (n-alkanes, phenol, anthracene, pyrene) by four actinobacterial strains. Int Biodeterio Biodeg 84:185-91.

[26] Prince RC, Lessard RR and Clark JR 2003 Bioremediation of marine oil spills Oil & Gas Tech Rev 58(4):463-8

[27] Adebusoye SA, Ilori MO, Amund OO, Teniola OD and Olatope SO 2007 Microbial degradation of petroleum hydrocarbons in a polluted tropical stream. World J Microb Biotechnol 23:1149-59

[28] Batista SB, Mounteer AH, Amorim FR and Totola MR 2006 Isolation and characterization of biosurfactant/bioemulsifier-producing bacteria from petroleum contaminated sites. Bioresour Technol 97:868-75

[29] Chandra S, Sharma R, Singh K and Sharma A 2013 Application of bioremediation technology in the environment contaminated with petroleum hidrocarbon. Ann Microbiol 63:417-31

[30] Abioye PO, Agamuthu P and Aziz AR 2012 Biodegradation of used motor oil in soil amended with organic wastes. Biotechnol Res Int. doi:10.1155/2012/587041

[31] Dadrasnia A and Ismail S 2015 Biosurfactant production by Bacillus salmalaya for lubricating oil solubilization and biodegradation. Int J Environ Res Public Health 12:9848-63 doi:10.3390/ijerph120809848

IOP Conf. Series: Materials Science and Engineering **335** (2018) 012135 doi:10.1088/1757-899X/335/1/012135

Image Segmentation Using Minimum Spanning Tree

M P Dewi*, A Armiati, S Alvini

Mathematics Department, Faculty of Mathematics and Natural Science, Universitas Negeri Padang, Jl. Prof. Hamka, Padang, Sumatera Barat, Indonesia 25131

*meiradaud@gmail.com

Abstract. This research aim to segmented the digital image. The process of segmentation is to separate the object from the background. So the main object can be processed for the other purposes. Along with the development of technology in digital image processing application, the segmentation process becomes increasingly necessary. The segmented image which is the result of the segmentation process should accurate due to the next process need the interpretation of the information on the image. This article discussed the application of minimum spanning tree on graph in segmentation process of digital image. This method is able to separate an object from the background and the image will change to be the binary images. In this case, the object that being the focus is set in white, while the background is black or otherwise.

1. Introduction

One of the nessesity in human living is the information that can people get from everithing arround them. In this era of information technology the information can be obtained from digital device or multimedia component included digital image. Digital image is one of the multimedia components that play an important role as a form of visual information [3]. Image processing is needed for getting the information from the digital image. One of digital image processing is image segmentation.

Image segmentation is a process that separates the main picture in image from the background [6]. The aim of image segmentation are to obtain the information of the image and to find the meaning of the image which can be used to next image processing, for example to identification of finger print, barcode reader, detection of cancer cell and so on.

Nowadays, there is variety of image segmentation methods. One of them is watershed method. This method produces some objects as main image as the result of segmentation process that call as over segmentations [5]. Another method in image segmentation is global thresholding. This method is suitable for an image which uneven lighting [4]. Those weaknesses of the methods can give the non accuracy of segmentation product.

This research uses the minimum spanning tree method for image segmentation. This method is based graph and in the selection side adds the criteria of similarity. That is, the selected side is not only the minimum weighted but also meets the criteria of similarity. The neighboring pixels and having a relatively small gray level difference will be on the same object. Thus, images with a very tight gray level can be overcome if the selection of criteria is appropriate. Then, this method is not done repeatedly to get the results of segmentation, so there is no possibility of the result of segmentation of image overload (over segmentation).

IOP Conf. Series: Materials Science and Engineering **335** (2018) 012135 doi:10.1088/1757-899X/335/1/012135

2. Method and Design

This research is a basic or theoretical research. Method was conducted in this study is literature study and in the solution of the problem, the steps taken are as follows:

a. Make the stages in segmenting the image by using the concept of minimum range trees.
b. Create a process design for each stage.
c. Applying process design in the form of computer program assisted matlab application program.
d. Conducting a trial of the application program that has been created.
e. Make the conclusions from the results and discussions that have been done.

Image segmentation by using the concept of minimum spanning tree is divided into three processes, which are :

a. to convert the image to graph form
b. to process the minimum range of graph formed
c. to form the result of image segmentation.

The minimum range of tree processes performed not only takes into account the minimum weight but also considers the similarity criteria in side selection, resulting in more than one minimum range tree.

3. Findings and Discussion

3.1. Image Segmentation Using Minimum Range Tree Concepts.

Image segmentation using the minimum spanning tree concept is divided into three stages:

3.1.1 Convert Image to Graph Shape. The gray level of each pixel in the image is represented in matrix form, where the value of the matrix elements expresses the level of image, while the position of the matrix element represents the pixel coordinates of the image. Each pixel in the image is identified as the neighboring dots and pixels with the right, left, up, down, right top right, lower right corner, upper left corner of the lower left corner connected by a side. The weight of each side is obtained by using the formula :

$$w(v_i, v_j) = |I(v_i) - (v_j)| + 1 \quad (1)$$

where I (v) is the gray level at the point (pixels) v.
So we will get the weighted graph represented in the form of adjacency matrix

3.1.2 Minimum Range Tree Process. In this process, the algorithm used is the Prim algorithm by adding the criteria of similarity. That is, the selection of sides is not only based on the minimum weight but also considering the criteria of similarity. A side is said to meet the criteria of similarity if the weights are less than the value of the likeness criterion in which this value is any value. As a result, there will be some minimum range trees.

3.1.3 Shaping Image Segmentation Results. Each point (pixel) in a minimum range tree has the same color. The color of each pixel on a minimum range tree is determined by looking for the average pixel gray level in the minimum range tree. Then, the average obtained will determine the color of pixels on condition,

$$\begin{cases} black, & if\ average\ I > T \\ white, & if\ average\ I \leq T \end{cases} \quad (2)$$

where

$$T = \frac{maxv+minv}{2} \tag{3}$$

Change the gray level of the input image with 0 (white) and 1 (black) based on the above provision, to obtain the result of image segmentation in the form of binary image. For example if we have the image :

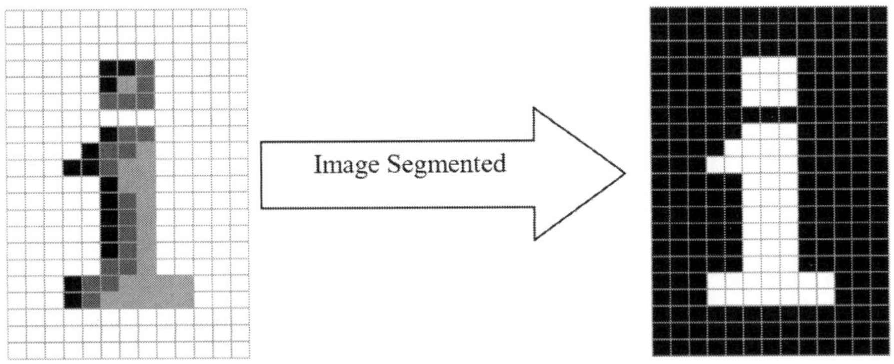

Figure 1. Image Segmentation Process

Figure 1 is a process of image segmentation where the first image is a gray scale image. This image has a lot of information, where the information locate in the structure of the letter i. It appears that in the letter i there is a bright part and there is a dark part. In order to gray scale image is only focused on the object only and do not consider the structure of the letter i, it is necessary to do image segmentation. This needs to be done so that the information from the image only focuses on the shape of the object only. The result of image segmentation can be seen in second image which is binary image.

3.2. Process Design of Image Segmentation Application Program with Minimum Range Tree Concept
In this process the modification of Prim algorithm will be translate application programming which is Matlab.

3.2.1 Algorithm Design of Matrix Establishment of Image. The color image needs to be converted to a gray scale image. Gray scale image will provide information in the form of image size and gray image level. The image size determines the many points of the graph and the gray level determines the side weights of the graph. This information will form the adjacency matrix of the image. The flow chart is shown in Figure 2.

3.2.2 Design of Minimum Range Tree Search Algorithm. In the design of this algorithm, the initial values given by intree {1} and notintree {2,3, ..., many points} where the in tree is a minimum range tree and notintree is not a minimum range tree. Then, from this initial value will be investigated the weights of the sides connecting the point on the intree with the point on notintree. The selected side weights are side weights that meet the criteria for similarity and have minimum weight. Side weights are said to meet the similarity criteria if the side weights are less than the value of the likeness criterion, whereby the value is chosen in any way. The selected side will be stored on the intree so that the point on the intree will increase by one and the point on notintree will be reduced by one. If there is no side that meets the criteria of similarity and has a minimum weight then we select the point on notintree and save in next intree, so that will form the second minimum range tree. If there is still a side that does not meet, then obtained the third minimum range tree, and so on. If each point has been investigated, then the work is completed and obtained some minimum range trees. The flow chart is shown in Figure 3.

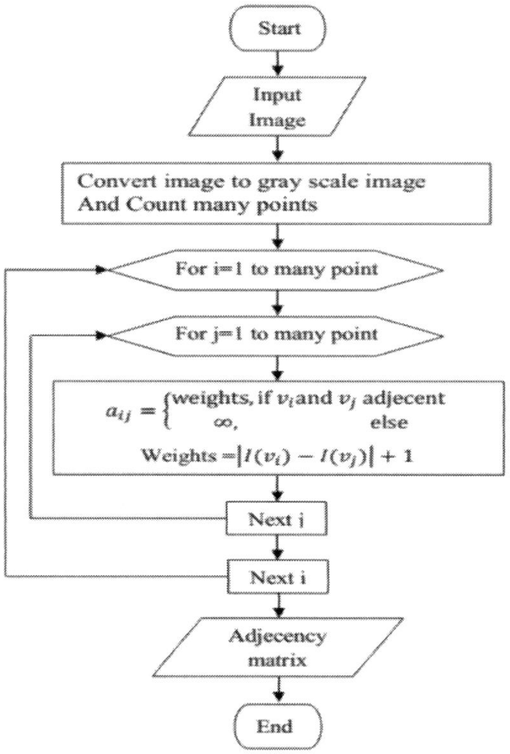

Figure 2. Flowchart of formation of adjacency matrix

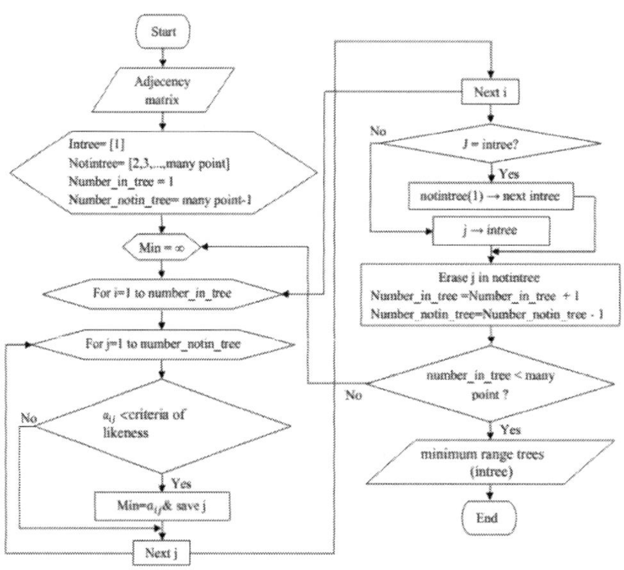

Figure 3. The Flowchart of Finding Minimum Spanning Tree

3.2.3 Design of Image Segmentation Result Generation Algorithm. In the design of this algorithm, some of the minimum range trees obtained will be searched for the average pixel gray level (dots) in each of the minimum range trees. Then find the threshold value by searching the midpoint of the highest gray level and the lowest gray level of the image. This threshold value will determine the color of the segmentation result. If the minimum range tree that has an average gray level is less or equal than the threshold then each point (pixel) in the minimum range tree will be white. Conversely, if the minimum range tree has an average gray level over threshold then each point (pixel) in the minimum range tree will be black. The flow chart is shown in Figure 4.

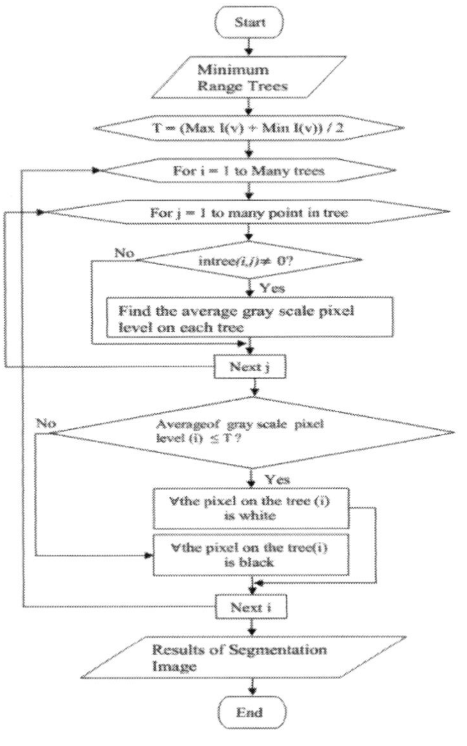

Figure 4. Flowchart of Establishment Image Segmentation Results

3.3. Determining Results of Image Segmentation in Simulation Data

Suppose we have a UNP symbol image. In order for the object in the image can be read, it is necessary to do image segmentation with the concept of the minimum range of trees, with the following segmentation results:

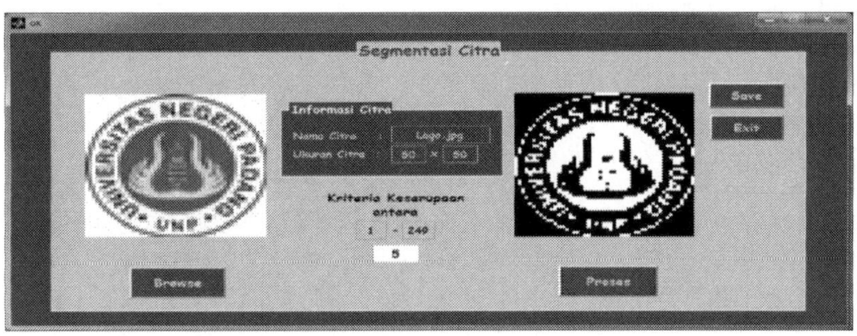

Figure 5. Result of Segmentation in Application Program

Based on the results of segmentation in Figure ,5 the left image is the image entered and the right image is the result of image segmentation. If observed from the simulation results, it is seen that the object on the UNP symbol image can be read where the circle, letters and other objects can be seen in the image segmentation results. The selection of the criterion value of similarity should be appropriate, so that the results of segmentation in accordance with the expected.

4. Conclusion

Each pixel in the image is identified as a dot. The neighboring pixels to the right, left, up, down, upper right corner, lower right corner, upper left corner of the lower left corner are connected by a side. The weight of each side of the graph grid is determined based on the gray level of each point (pixels) using the formula (1). Image segmentation using the minimum spanning tree concept is divided into several processes. The first process is to form the image into a graph form that is represented in the form of an adjacency matrix, the process of finding the minimum range of trees, and forming the result of image segmentation. The minimum range search process uses a modified algorithm of the Prim algorithm, where the algorithm produces some minimum range trees. Application program to determine the result of image segmentation with the concept of minimum assisted range of Matlab R2013a. Use of this application program on some simulation data, resulting in image segmentation in accordance with expectations. Based on the evaluation, the program can run well on the image is not too large, the object on the image is not too much, no noise (noise) on the image and the selection of appropriate criteria of similarity.

References

[1] Munir, Rinaldi. 2012. *Matematika Diskrit*. Bandung: Informatika Bandung.

[2] Budayasa, Ketut. 2007. *Teori Graph dan Aplikasinya*. Surabaya: Universitas Negeri Surabaya.

[3] Munir, Rinaldi. 2004. *Pengolahan Citra Digital dengan Pendekatan Algoritmik*. Bandung: Informatika Bandung.

[4] McAndrew, Alasdair. 2004. "*An Introduction ti Digital Image Processing with Matlab*." School of Computer Science and Mathematics Victoria University of Technology.

[5] Gonzalez, Rafael C. dan Richard E. Woods. 2002. Digital Image Processing. New Jersey: Prentice Hall, Inc., Upper Saddle river.

[6] Janakiraman,T.N.danChandra Mouli. 2010. "Image Segmentation Using Graph Euler." *Int. J. of Computers, Communications & Control*.

[7] Gonzalez, Rafael C., Richard E. Woods, & Steven L. Eddins. 2009. *Digital Image Processing Using MATLAB*. United States of Amerika: Gatesmark Publishing.

[8] Away, Gunaidi Abdia. 2006. *The Shortcut of MATLAB Programming. Bandung*: Informatika Bandung.

[9] Wijaya, Marvin Ch., & Agus Prijono. 2007. *Pengolahan Citra Digital Menggunakan Matlab*. Bandung: Informatika Bandung.

Test of Fruit Extract Pare (Momordica charantia L.) to Quality of Ejaculated Spermatozoa Mice (Mus musculus L.)

M Fifendy[1*], G Indriati[2]

[1]Departement of Biology Faculty Mathematic and Sciences, State University of Padang.Air Tawar Padang, Indonesia.
[2]Programs Study of Biology Education, STKIP PGRI Sumatera Barat, Padang Indonesia

*madesfifendy@yahoo.co.id

Abstract Pare (Momordica charantia L.) can be used in the treatment of various diseases, such as influenza, cancer, anti-inflammatory, anti-HIV, antimitotic and antifertilitas. This study aimed to determine the effect of the herbal bitter (Momordica charantia L.) to ejaculated sperm quality mice (Mus musculus L.). This research was conducted using Completely Randomized Design (CRD) with 4 treatments and 6 replications, water and fed adlibitum. First treatment is given solvent extract. Second treatments extract were given 0.2 gram, third treatment were given 0.4 gram of extracts and fourth treatment were treated exstrac 0.6 gram were orally for 30 days. After the mice decapitated, dissected and take sperm from vas deferens. Then, the sperm preparation determined using the improved Neubauer. Data were analyzed by ANOVA (Analysis of Varians). The results shoured at doses of 0,2 gram, the average sperm count was 19.89. decrease significant when compared with the control in which the average number of sperm 29.13. So with this research the effective doses to decrease sperm count and can be used as a contraception medication dosage was 0,2 gram. It can be conclude that the extract of bitter (Momordica charantia L.) can decrease the quality of the ejaculated sperm of mice (Mus musculus L.)

1. Introduction

Family Planning Program has been declared by the Indonesian government as a national program. One of the efforts that have been implemented in the program is the provision of contraception. The use of contraceptives in its principle is to prevent fertilization or the fusion of a sperm cell and an egg cell. Means of contraception is more aimed at women, while men are still limited, so the development of male contraceptives is far behind compared to female contraceptives [1].

One of the plants is expected to be an antifertility is the fruit of bitter pare (*Momordica charantia* L.) which has a bitter taste that comes from cucurbitacins substances. The fruit is included in the rate Cucurbitaceae (cucumbers).

Thrives in the lowlands and can grow wild in the yard of the house to be cultivated. This fruit has a lot of content that is good for the body. Such as high calcium, fiber, vitamin C, carotene, a chemical compound lutein, lycopene, which may help prevent cancer, as an antibiotic, antiviral, balancing blood pressure, blood sugar and helping stimulate the production of insulin.

Pare plants contain flavonoid compounds that can hamper enzyme aromatase, an enzyme that serves to catalyze the conversion of androgens to estrogen increases the hormone testosteron. The high concentration of testosterone will affect the negative feedback to the pituitary is not releasing FSH and LH, thereby inhibiting spermatogenesis.

The enzyme also catalyzes the conversion of testosterone to estradiol so mepengaruhi ovulation process. Bitter pare fruit extract also contains cytotoxic compounds such as saponins, triterpene momordikosida, and cucurbitacin which can reduce the quality and quantity of sperm cells. Seeds pare contains triterpenoids which have anti spermatozoa activity, so the use of traditional seeds pare with a view to preventing AIDS can result in infertility in men [2].

Pare consumption in the long term, either in the form of juice, salad or vegetables, can kill sperm, triggering impotension, damage the testicles and male hormones, even potentially damage the liver [3]. Bitter pare fruit extract Research has also been reported by [4], that can affect the morphology and motility of rats. The higher levels of fruit extracts Pare and the longer the administration, the motility and viability of spermatozoa is getting low, otherwise abnormal sperm morphology-tozoa increasing. This may be caused by active ingredient group triterpene glycoside contained in fruit Pare [4]. Bitter pare fruit extract 500 mg / kg / day for 14 days was found to affect the quality of spermatozoa that is, the occurrence of agglutination between head motion and circular motion in place. Circular motion can be caused due to abnormal morphology, rotational energy penghantaran no or irregular and asymmetrical tail circum stances. Furthermore, when administered for 49 days, got into abnormal morphology of spermatozoa. The abnormalities appeared in the neck bulging spermatozoa (swollen). It was allegedly caused by the swelling of mitochondria.

Sutyarso [5] concluded that the research results Pare fruit extract tends to be antifetilitas because it can inhibit spermatogenesis and the higher the dose of bitter melon fruit extract, the fewer the number of children produced mice. In addition, the effect of the fruit extracts for blocking spermatogenesis Pare is temporary (reversible). To see how it affects, then do research on Effect of Pare Fruit Extract (Momordica charantia L) on the Quality of ejaculated spermatozoa of mice (Mus musculus L.).

2. Material and Methods

Material

This study uses a test animal mice (*Mus musculus* L.) Swiss Webster aged 11-12 weeks weighing 20-30 grams were obtained from the Department of Biological Science Laboratory UNP. Bitter melon fruit extract (*Momordica charantia* L) obtained from various processes performed at the Laboratory of Chemistry UNP.

Research Methods

Mice were divided into 3 treatment groups and control. Each group consisted of six mice. The control group was given only solvent extract and the food was the same as the treatment group. Treatment B, C and D are the group treated by administration of bitter pare fruit extract orally for 30 days with a dose of 0.2 g, 0.4 g and 0.6 g.

After the test of finishes in each group, the test animals were decapitated, dissected and cut vas deferens of the left and right, place it in petridish has been charged with 10 micron PBS (*Phosphate Buffered Saline*). To calculate the number of sperm, using the method of Improved Neubauer. As for seeing the morphology of spermatozoa, using eosin staining. The data were analyzed by ANOVA (*Analysis Of Variance*).

3. Results and Discussion

From this research it can be seen the effect of bitter pare fruit extract (*Momordica charantia* L) against the number of ejaculated spermatozoa of mice (*Mus musculus* L.). After giving the fruit extract of bitter pare (*Momordica charantia* L) at a dose of 0.2 g, 0.4 g and 0.6 g in mice (*Mus musculus* L.) for 30 days there was a decline ejaculated spermatozoa (Table 1).

Table 1. Results The mean sperm, sperm Normal and Abnormal sperm ejaculated *Mus musculus* L. extract treatment with various *Momordica charantea* L.

Treatment dose	Mean Sperm Count	Mean Normal Sperm	Mean Abnormal Sperm
A (0,6 g/kgbb)	9.25 a	18.33 a	38.33 a
B (0,4 g/kgbb)	12.18 b	24.59 a	30.50 b
C (0,2 g/kgbb)	19.89 c	33.67 b	23.50 c
D (control)	29.13 d	44.78 b	19.33 c

Note: The numbers in the same column followed by the same lowercase not significant,whereas the same letter are not significantly different at α level of 5%.

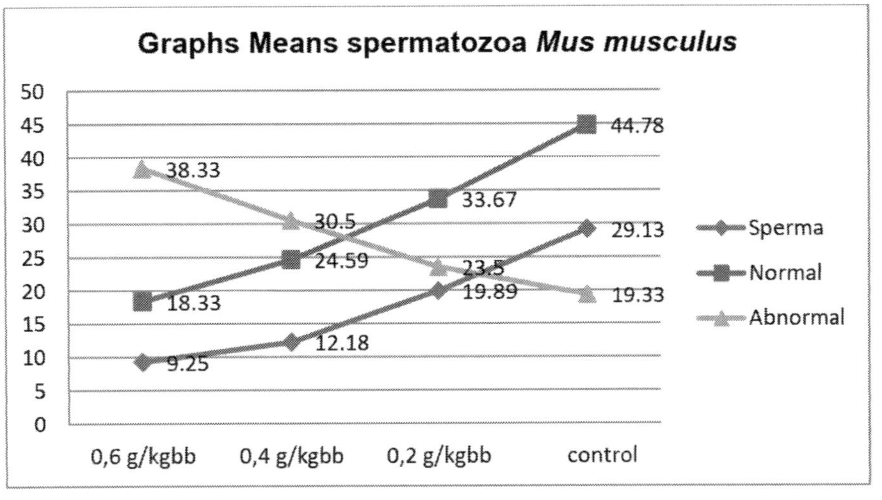

Figure 1. Graph Means Spermatozoa *Mus musculus*.

According to Hartini [1] the number of spermatozoa produced is highly dependent on the direct process that occurs during spermatogenesis in the seminiferous tubules. When spermatogenesis normal place it will produce normal sperm count as well. Conversely, if an interruption occurs during spermatogenesis process, the development of spermatogonial cells will affect the number spermatozoa formed. So with the disorder cleavage during spermatogenesis, it is expected to result in the alteration or destruction of spermatozoa produced thus might change and the damage will effect a decrease in the number of sperm ejaculated.

Based on the test results can be seen that the use of 0.2 g dose was the most effective to reduce the number of sperm ejaculated and used as an antifertility drugs. At a dose of 0.2 g is obtained by the average number of sperm 19.89 and has already seen a significant reduction in the number of spermatozoa were compared with controls who had an average of 29.13.

From examination of normal sperm morphology Table on the vas deferens mice (*Mus musculus* L.) obtained an average percentage of normal sperm morphology dose of 0.2 g , 0.4 g and 0.6 g lower when compared to the control and after a further test is statistically there are significant as a result of extract of bitter pare fruit (*Momordica charantia* L.).

In this study, the higher dose given the fewer the number of normal spermatozoa be produced. So the dose that can be used as an anti-fertility drugs are on treatment with a dose of 0.2 g B because it can inhibit or reduce the number of normal sperm.

Table it can be seen that the extract of bitter pare fruit (*Momordica charantia* L.) also cause abnormal sperm. Abnormalities in the sperm of mice (*Mus musculus* L.) allegedly because of interference maturation phase of spermiogenesis spermatozoa in this case in accordance with the theory put forward by Bardin [6] that in the seminiferous tubules, androgen function in controlling the process of spermatogenesis in meiosis and spermiogenesis process. In this study suspected antimitotic compounds contained in andrographolide has hindered produced androgen, causing disturbances in spermio genesis phases. So if the hormone androgen is a hormone that plays a role in the process of spermatogenesis when availability is a little, it will cause the process was interrupted and can lead to abnormalities in sperm during spermiogenesis.

4. Conclusion

Bitter pare fruit extract (*Momordica charantia* L.) can reduce the average amount of the ejaculate sperm of mice (*Mus musculus* L.) and reducing the number of normal sperm.

References

[1]. Hartini. 2011. Pengaruh Dekok Daun Jambu Biji Merah (Psidium guajava L.) Terhadap Jumlah Kecepatan dan Morfologi Spermatozoa Tikus Putih Jantan (Rattus norvegicus). Available at: https://www.scribd.com/ Acessed: 24 April 2013.

[2]. Girini. MM, Ahamed RN, Aladakatti RH, 2005. Effect of graded doses of *Momordica charantia* seed extract on rat sperm: scanning electron microscope study, *J Basic Clin Physiol Pharmacol.*, 16(1):53-66.

[3]. Basch E, Gabardi S, Ulbricht C, 2003, Bitter melon (Momordica charantia): a review of efficacy andsafety, *Am J Health Syst Pharm.*, 60(4): 356-9.

[4]. Wardojo BPE. 1990. *Pengaruh Fraksi Kioroform dan Air Buah Pare terhadap Spermatozoa Epididimis Tikus*. Thesis Fak. Pascasarjana UGM, h.53–102.

[5]. Sutyarso. 1992. *Pengaruh Pemberian Ekstrak Buah Pare (Momordica charantia L.) Terhadap Fertilitas Mencit Jantan Mus musculus L. Strain LMR*. Thesis Fakultas Pasca-sarjana Universitas Indonesia, Bidang Ilmu Kedokteran Dasar. Jakarta, h. 123.

[6]. Bardin,C.W. 1986 *Pituitary Testiscular Axis in Reproductive Endo-crionology* (2nd). Eds. S.S C Yen and R. B. Jaffe. Philadhepia Saunders.

AUTHOR INDEX

AUTHOR INDEX

AUTHOR INDEX